New Technological Solutions, Research Methods, Simulation and Analytical Models That Support the Development of Modern Transport Systems

New Technological Solutions, Research Methods, Simulation and Analytical Models That Support the Development of Modern Transport Systems

Topic Editors

Tomasz Nowakowski
Artur Kierzkowski
Agnieszka A. Tubis
Franciszek Restel
Tomasz Kisiel
Anna Jodejko-Pietruczuk
Mateusz Zając

MDPI

Basel • Beijing • Wuhan • Barcelona • Belgrade • Novi Sad • Cluj • Manchester

Topic Editors
Tomasz Nowakowski
Wroclaw University of Science and Technology
Wroclaw
Poland

Artur Kierzkowski
Wroclaw University of Science and Technology
Wroclaw
Poland

Agnieszka A. Tubis
Wroclaw University of Science and Technology
Wroclaw
Poland

Franciszek Restel
Wroclaw University of Science and Technology
Wroclaw
Poland

Tomasz Kisiel
Wroclaw University of Science and Technology
Wroclaw
Poland

Anna Jodejko-Pietruczuk
Wroclaw University of Science and Technology
Wroclaw
Poland

Mateusz Zając
Wroclaw University of Science and Technology
Wroclaw
Poland

Editorial Office
MDPI AG
Grosspeteranlage 5
4052 Basel, Switzerland

This is a reprint of the Topic, published open access by the journals *Applied Sciences* (ISSN 2076-3417), *Drones* (ISSN 2504-446X), *Energies* (ISSN 1996-1073), *Infrastructures* (ISSN 2412-3811), *Logistics* (ISSN 2305-6290) and *Modelling* (ISSN 2673-3951), freely accessible at: https://www.mdpi.com/topics/0JGNXR3YE7.

For citation purposes, cite each article independently as indicated on the article page online and as indicated below:

Lastname, A.A.; Lastname, B.B. Article Title. *Journal Name* **Year**, *Volume Number*, Page Range.

ISBN 978-3-7258-3101-2 (Hbk)
ISBN 978-3-7258-3102-9 (PDF)
https://doi.org/10.3390/books978-3-7258-3102-9

Contents

About the Editors

Tomasz Nowakowski

Tomasz Nowakowski is Vice-Rector for Digital Development at the Wrocław University of Science and Technology. Throughout his entire professional career, he was associated with the WUST. Research interests concern the operation and reliability of machines (including buses and trucks, working machines, inland ships, and bucket-wheel excavators), the resilience of transport and logistics systems, decision making in conditions of uncertainty, and business continuity management of university-type organizations. His achievements include over 350 works, including 280 scientific publications and 70 works in reports and statements. Number of citations according to WoS: 340. Most frequently cited work: WoS: 63 times. He supervised 15 defended doctoral dissertations and 1 initiated doctoral work. He has prepared many reviews of doctoral dissertations and reviews of the procedure for awarding the degree of habilitated doctor and the procedure for awarding an academic title and doctor honoris causa title. Since 1993, he has held various management positions at the University, including the Dean of the Faculty of Mechanical Engineering and Vice-Rector of Wrocław Tech. Member of the Transport Committee of the Polish Academy of Sciences. Since 2009, he has been the chairman of the Program Council of the Winter School on Reliability.

Artur Kierzkowski

Artur Kierzkowski is head of the Department of Technical Systems Operation and Maintenance, Faculty of Mechanical Engineering, Wroclaw University of Science and Technology. He deals mainly with assessing the resilience of systems to disruptions and risk assessment of transportation systems. The main subjects of his research are multi-criteria evaluation models and simulation models of air transport systems, reliability assessment, and data stochastic modelling. Mr. Kierzkowski is a reviewer of the National Centre for Research and Development for transport, dependability, and safety. Mr. Kierzkowski regularly publishes the results of his work at conferences and in various technical journals.

Agnieszka A. Tubis

Agnieszka A. Tubis is an associate professor at the Faculty of Mechanical Engineering, Department of Operation and Maintenance of Technical Systems of Wroclaw University of Science and Technology, Poland. Her current field of interest is related to the issues of digital transformation in logistics and transport systems, analyzing and managing risk in cyber–social–physical systems, modelling and measuring the performance of transport and logistic systems operations, and addressing supply chain resilience and vulnerability problems. She has published more than 160 scientific works. She is the President of the Board of the Polish Logistics Association and a member of the Board of the Polish Maintenance Society. She is also a member of the Management Board of the European Safety, Reliability, and Data Association and a member of the General Assembly of the European Federation of National Maintenance Societies.

Franciszek J. Restel

Franciszek J. Restel is head of the Railway Operation and Maintenance Research Group, Department of Technical Systems Operation and Maintenance, Faculty of Mechanical Engineering, Wroclaw University of Science and Technology. He deals mainly with resilience and robustness to undesirable events and their consequences influencing the operation processes of socio-technical systems. The main subjects of his research are evaluation methods of rail transport systems, reliability, and safety assessment, as well as operational research and probabilistic modelling. Mr. Restel bridges the gap between scientific theory and industrial practice and regularly publishes the results of his work at conferences and in various journals.

Tomasz Kisiel

Tomasz Kisiel, Ph.D. Eng. is an assistant professor in the Department of Technical Systems Operation and Maintenance at the Faculty of Mechanical Engineering at Wrocław University of Science and Technology, Poland, with extensive engineering management and operations expertise. His research is centred on improving the efficiency and safety of complex technical systems, where operational effectiveness and high levels of safety are paramount. Dr. Kisiel has contributed significantly to his field, authoring or co-authoring 36 scientific papers, enhancing the understanding of various engineering challenges. Throughout his career, he has been involved in nine research projects, furthering the development of practical solutions for optimizing and developing technical systems. He led one of these research projects, showcasing his leadership abilities and capacity to manage complex, multidisciplinary teams.

Anna Jodejko-Pietruczuk

Anna Jodejko-Pietruczuk works at the Department of Technical Systems Operation and Maintenance, Faculty of Mechanical Engineering, Wroclaw University of Science and Technology. Her scientific interest covers optimization theory and the simulation modelling of stochastic processes applied in machine reliability and logistic/transportation processes. She is experienced in the application of Genetic Algorithms as well as VR and AR solutions in process improvements. Mrs. A. Jodejko-Pietruczuk is a reviewer and an author of scientific research in worldwide technical journals.

Mateusz Zając

Mateusz Zając is a member of the Department of Technical Systems Operation and Maintenance, Faculty of Mechanical Engineering, Wroclaw University of Science and Technology member. He deals mainly with assessing the vulnerability and availability of transportation systems. The main subjects of his research are multi-criteria evaluation models and simulation models of the intermodal transport system, availability assessment, and data stochastic modelling. Mr. Zając is a reviewer of the Horizon RIA projects. Mr. Zając regularly publishes the results of his work at conferences and in various technical journals.

Preface

Transport systems are currently facing numerous challenges due to technological, economic, organizational, and environmental changes. As part of the critical infrastructure, transport infrastructures have to meet high requirements in terms of technical reliability and resistance to disruptions that occur. At the same time, there are ever-increasing customer demands for transport processes that link global and local supply chains and determine the level of logistics service offered to the consumer market. All this is accompanied by ever-changing legislation on providing transport services and increasing requirements to incorporate social and environmental considerations into the implemented solutions to improve current systems. All these aspects of modern transport systems mean that they must undergo numerous transformation processes to meet the challenges of the modern market.

The required change processes in transport systems are now supported by digital transformation. Solutions from the area of Industry 4.0 allow better real-time monitoring of the functioning of individual elements of the transport system, support the collection of big data, prepare multi-criteria analyses to support managers' decisions, and optimize the implementation of transport service processes. Technological solutions can be applied at various stages of the research process, from the tasks related to collecting actual data through its processing to its use in autonomous solutions.

The Topic aims to present new technological solutions, research methods, and analytical models that support the development of modern transport systems (considering the requirements of passengers, freight, and urban transport). The research results presented in our Topic show many interesting technological solutions that can improve the management of modern transport systems. Several publications also point to the possibility of using digital solutions in analytical models to support optimization processes. Ensuring the required level of safety and responding to sustainability requirements are also important issues addressed in the presented research methods.

The presented research results provide interesting source material for other researchers. The research gaps identified in the publications and the potential for further development of the issues addressed can inspire and guide further research. The issues described are a vehicle for new solutions that can be implemented by industry and government entities in real systems. The proposed research results exploit the potential provided by new technologies and may constitute innovative solutions, the implementation of which will improve the market competitiveness of organizations. We also recommend our Topic for use by students of technical and business universities and business trainers. The presented research results comprehensively complement the classic textbooks used in the education of engineers and managers for transport systems.

The Topic was edited by a team of experts representing active researchers researching various transport systems (rail, air, road, and urban), dealing with technical system reliability, digital transformation, and improving operational and decision-making processes. This allowed for a comprehensive evaluation of publications submitted by authors from all over the world. The dedicated reviewers played an important role in improving the publications, always taking on the challenge of preparing critical comments and recommendations with due diligence. The Editorial Team would like to thank them for this.

Tomasz Nowakowski, Artur Kierzkowski, Agnieszka A. Tubis, Franciszek Restel, Tomasz Kisiel, Anna Jodejko-Pietruczuk, and Mateusz Zając
Topic Editors

drones

MDPI

Article

Determining the Efficiency of Small-Scale Propellers via Slipstream Monitoring

Jaan Susi *, Karl-Eerik Unt and Siim Heering

Department of Aeronautical Engineering, Estonian Aviation Academy, Lennu 40, Reola, 61707 Kambja Parish, Tartu County, Estonia; karl-eerik.unt@eava.ee (K.-E.U.); siim.heering@eava.ee (S.H.)
* Correspondence: jaan.susi@eava.ee

Abstract: A large part of small-sized UAVs that are used for surface scanning, video- and photography, or other similar applications are of the multirotor type. These small aircraft perform mainly in hovering or nearly hovering flight mode, and the endurance of these vehicles depends greatly on the efficiency of their motors and the aerodynamic efficiency of their thrust-generating systems, including propellers, ducted fans, etc. Propellers may therefore work in different regimes: in a regime where the propeller performs work to move the vehicle through the air, and the static or hovering regime, in which standing air is accelerated. In both cases, the concept of efficiency can be used to describe the propeller's performance. There have been several previous studies on static and advancing propellers' performances. In these studies, when determining the efficiency of a static propeller, the thrust and power coefficients are most commonly compared to evaluate the propeller's performance. Sometimes, the inducted velocities are calculated via the momentum theory. As small-scale propellers work on very low Reynolds (Re) numbers below 500,000, the flow type transition and boundary layer separation make it very hard to predict the actual efficiency of the propellers in static mode. Therefore, the aim of this paper is to introduce a method to determine the static efficiency of small-scale propellers directly and empirically via a comparison between the output and input power, wherein the output power is determined via the measured thrust and mean induced velocity. The used method combines thrust, torque, and angular velocity measurements with slipstream monitoring. The performed tests showed a decrease in efficiency, with the Re number rising in spite of the rising values of the thrust coefficient. This study led to two main conclusions: thrust and power coefficients are not always the key parameters to determine the efficiency of a propeller; the role of the Re number in the propeller's efficiency is not yet clear and requires further investigation. The presence of Re number effects has been proven in numerous works, but the impact of those effects seems to not be as trivial as the claim that the lower the Re number, the weaker the propeller's performance.

Keywords: propeller efficiency; applied aerodynamics; small-scale propellers; multirotors

Citation: Susi, J.; Unt, K.-E.; Heering, S. Determining the Efficiency of Small-Scale Propellers via Slipstream Monitoring. *Drones* **2023**, *7*, 381. https://doi.org/10.3390/drones7060381

Academic Editors: Tomasz Nowakowski, Artur Kierzkowski, Agnieszka A. Tubis, Franciszek Restel, Tomasz Kisiel, Anna Jodejko-Pietruczuk and Mateusz Zając

Received: 27 April 2023
Revised: 23 May 2023
Accepted: 29 May 2023
Published: 6 June 2023

1. Introduction

The number of unmanned aerial vehicles (UAVs) used worldwide continues to increase, and this leads to problems in the efficiency and environmental friendliness of these devices. More and more UAVs are used not only as remote control devices for entertainment but also for professional use for civil and military purposes, such as surveillance, photography, and transporting small cargo loads. A large part of UAVs are of the multirotor type and perform mainly in hovering or nearly hovering flight mode and are powered by electric motors and onboard batteries. The endurance of any aerial vehicle depends greatly on the efficiency of the aerodynamic layout of the vehicle and on the aerodynamic efficiency of its thrust-generating systems (propellers, ducted fans, etc.). The efficiency of a UAV is also relevant from the point of view of environmental protection, as more efficient devices use less fuel or prolong the recycling period of onboard batteries. In multirotor-type small UAVs, thrust is generated with small-scale propellers, often working at Reynolds numbers in the region well below 500,000. The Reynolds number is considered critical for aerodynamic wings and propeller blades. The definition of the term "Small Scale Propeller" is determined by the fact of whether the main parts of the propeller (as a rule, with the blade sections occupying nearly 75% of the propeller radius) are working either at lower or higher Reynolds numbers than the critical Reynolds number. Propellers working at Reynolds numbers lower than the critical Reynolds number are often called small-scale propellers and are the propellers investigated in the present paper (with Reynolds numbers between 70,000 and 200,000). The problem is that this region lies between the two modes of air flow: streamline flow and turbulent flow. Investigations have shown that with the Reynolds number decreasing from a value of 500,000, the efficiency of a propeller also decreases due to the transition between flow types and the earlier boundary layer separation from the propeller blades in the streamline (laminar) flow mode [1–3]. It is often difficult to predict the performance of small-scale propellers [4–6]. This is understandable due to the uncertainty in determining the flow type in this region of Reynolds numbers. Propellers may work in different regimes. The two basic regimes are the propelling regime, wherein the propeller performs work to move the vehicle through the air, and the static or hovering regime, wherein the propeller performs work to induce a slipstream, i.e., to accelerate standing air. In both cases, one can use the concept of efficiency to describe the propeller's performance. Several studies have investigated both the static and advancing performance of small-scale propellers [7–12]. In studies on static performance, the concept of propeller efficiency has not been defined in the usual sense of the word. The problem is that useful power or output power has not been measured empirically, and only the thrust and power coefficients have been compared to evaluate a propeller´s performance. Sometimes the ratio of those coefficients is used to evaluate the performance of a propeller. Thrust is the only output characteristic in these cases. The induced velocities for a propeller may be calculated via the momentum theory [13], but for the reasons mentioned above, these calculations may not have enough precision at low Reynolds numbers. This leads to a need for the direct measuring of the induced velocity to define the static efficiency of a propeller via that parameter. The aim of the present paper is to introduce a method to determine the static efficiency of small-scale propellers directly and empirically via a comparison between the output and input powers, whereby the output power is determined via the measured thrust and mean induced velocity. This enables us, with no reservations, to use the term "efficiency" for static regimes of propellers. This approach is quite different from previous similar works wherein only the coefficients of thrust and power were compared to characterize the performance of propellers working in the static regime. The proposed method enables us to quickly and effectively compare the efficiencies of different propellers at different Reynolds numbers. In addition, slipstream monitoring provides the possibility to determine many aerodynamic parameters of propeller blades and acquire information about the behavior of the slipstreams.

2. Materials and Methods

When speaking about the efficiency of any kind of power-transforming system, it is usual to compare the output power and input power. This is carried out using the ratio of the output power to the input power, which is often expressed as a percentage. For propellers, the input power is defined as

$$P = M \cdot \omega. \tag{1}$$

The definition of output or useful power depends on the working mode of the propeller. In the case of an advancing propeller, the propulsive efficiency is defined via the airspeed of the vehicle or the velocity of the freestream flow:

$$\eta = \frac{T \cdot v_0}{P}. \tag{2}$$

The thrust and input power of a propeller can be presented as

$$T = C_T \cdot \rho \cdot \omega^2 \cdot D^4,\ P = C_p \cdot \rho \cdot \omega^3 \cdot D^5. \quad (3)$$

The propulsive efficiency is often defined via the advance ratio:

$$\eta = J\frac{C_T}{C_p},\ J = \frac{v_0}{\omega R}. \quad (4)$$

In both cases, it is easy to measure the parameters necessary to determine the efficiency. In the static case, i.e., when the freestream velocity equals zero, the output (induced) power of the propeller and the efficiency are defined as

$$P_i = T,\ \eta = \frac{P_i}{P}. \quad (5)$$

The mean induced velocity $\overline{v_i}$ cannot be correctly measured directly but can be derived from the results of slipstream monitoring which, in our case, means measuring all three components of the induced velocity. In theory, the induced velocity is the velocity of air in the plane of rotation. In empirical investigations, the induced velocity should be measured as closely as possible to the trailing edge of the propeller blade. Herein, when carrying out the experiment, one must consider the possibility of propeller deformation under the resulting forces of inertia and thrust. This may lead to the need to change the axial position of the measuring probe depending on the propeller's rotating speed. The axial distance of the airspeed measurement must not be far from the trailing edge of the propeller blade to minimize mass (or volumetric) rate losses in the slipstream that will grow with distance. The mean induced velocity is calculated using the measured volumetric rate and full propeller diameter. The simplest model of a slipstream (in the case of an ideal propeller) predicts the contraction of the slipstream according to the momentum theory [13]. The contraction takes place due to static pressure differences inside and outside of the slipstream. The simplest approach in slipstream theory says that the contraction of a slipstream ends and the full pressure balance is achieved when the slipstream velocity is doubled in comparison with the induced velocity. Herein, we see that the simplest model of a slipstream predicts the presence of the radial flow component of the slipstream toward the axis due to contraction. In practice, the situation is more complicated as the slipstream is rotating. This means that in any location inside a slipstream, we have three components of airflow velocities (Figure 1): the axial, the radial, and the tangential components.

In propeller efficiency, only the axial component plays a positive role in generating useful power as it is directed in the thrust line. The tangential component is caused by the rotation of the slipstream. The latter is caused by the torque of the propeller and, furthermore, by the drag of the blade sections. The drag of the blade sections depends on their airfoil parameters, angle of attack, flow separation from the blade back, etc. It also depends not only on the angle of attack but also on the blade angle itself (see Figure 2). The higher the blade angle, the less it contributes to the lift force of a blade section into the thrust and the more it contributes to the drag moment (torque). This effect is present even when the blade section has the most effective angle of attack.

The rotation of the slipstream that creates the velocity component v_t has a direct impact on the remaining two velocity components of the slipstream and on the propeller's performance in general. First, let us suppose that the whole slipstream rotates with an angular velocity $\frac{v_t}{r}$. The centrifugal acceleration on the slipstream surface is then $\frac{v_t^2}{r}$ and zero on the axis. In an air column reaching from the slipstream surface to the axis (Figure 3), the radial centrifugal pressure $\frac{\rho v_t^2}{2}$ built by the centrifugal force is determined by one half of the centrifugal acceleration on the slipstream surface.

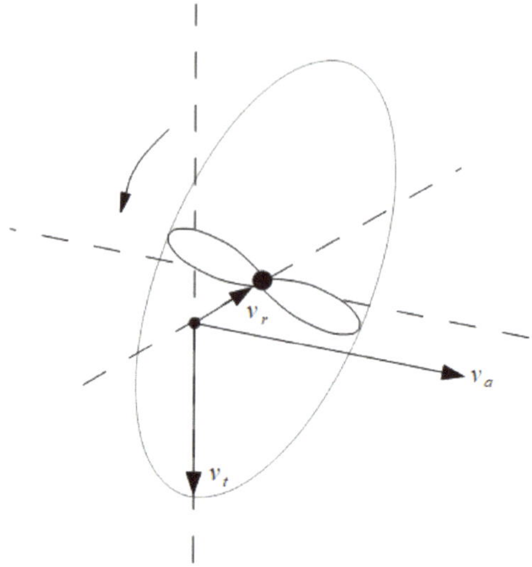

Figure 1. The three components of air velocities in a slipstream: v_a—axial component, v_t—tangential component, and v_r—radial component.

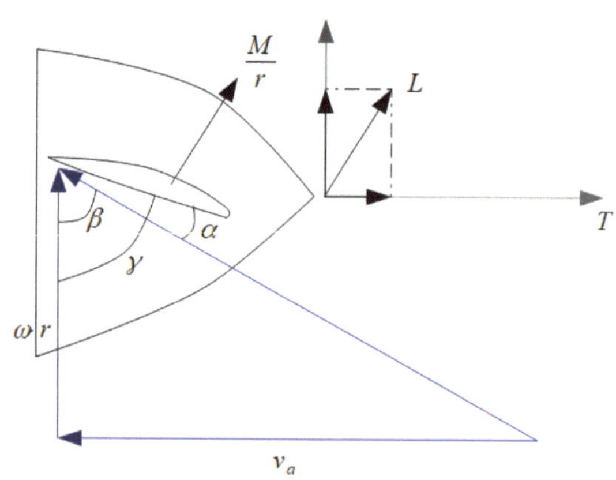

Figure 2. The role of lift force of a blade section in creating thrust and torque: L—lift force of a blade section, M—drag moment of blade section, r—distance of blade section from the propeller axis, T—thrust generated by blade section, ω—angular velocity, v_a—axial component of the airflow, γ—blade angle, β—helix angle, and α—angle of attack.

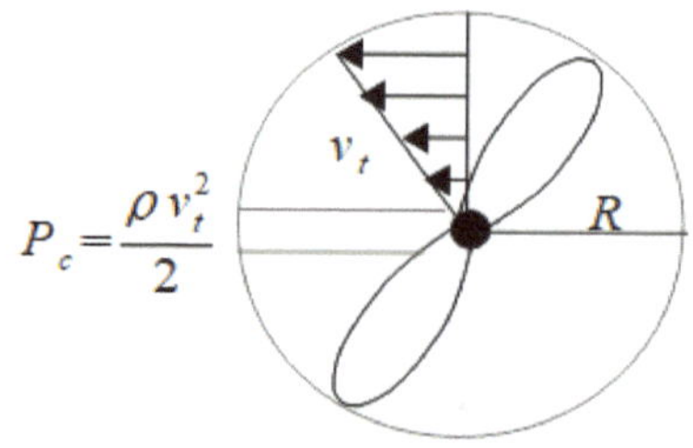

Figure 3. The buildup of the centrifugal pressure in a slipstream: P_c—centrifugal pressure, ρ—air density, v_t—tangential component of air velocity and R—propeller radius.

This centrifugal pressure counteracts the static pressure difference in and outside the slipstream and consequently counteracts the slipstream contraction. In the case of high torque (due to too high blade angles and/or low Reynolds number effects), the slipstream may even be expanding. Indeed, by increasing the blade angle to high values, the propeller begins to work more and more as a rotor of a centrifugal compressor, forcing the slipstream to expand and decreasing the role of the axial flow component. In the case of a contracting slipstream, the dynamic pressure (that is, the density of the kinetic energy in the flow) rises and provides feedback to the propeller blade faces in the form of additional pressure. That additional pressure on the blades has a direct impact on the thrust and thus on the efficiency of the propeller. The smaller the effect of the contraction is, the less thrust there is, and thus, the efficiency will decrease. However, the assumption of the angular velocity being constant for the whole slipstream results in a simplified or first-step model. In a more precise model, we must assume counteraction between the velocity components v_r and v_t. This means that if the slipstream is contracting, there will also be a direct influence of v_r on v_t along with the inverse effect caused by the centrifugal force. This influence is often called the Coriolis effect. The conservation of angular momentum in a slipstream entails that $v_t \cdot r = const$ for any radially moving air parcel. From this, the Coriolis acceleration is

$$\frac{dv_t}{dt} = \frac{const}{r^2} \cdot \frac{dr}{dt} = -\frac{v_t}{r} \cdot v_r \tag{6}$$

The Coriolis effect counteracts the slipstream contraction because it raises v_t, and thus, the centrifugal pressure increases in the central parts of the contracting slipstream in comparison with the previous model presented in Figure 3. In the case of an expanding slipstream, the Coriolis effect counteracts the expansion as well, because then v_t slows down and decreases the centrifugal pressure. Neither of the models described above considers the energy dissipation in a slipstream due to viscosity and turbulence. From this comes the reason slipstream measurements, if carried out, must be performed as closely as possible to the trailing edge of the propeller blades to minimize the effect of slipstream energy losses.

Thus, one can conclude that the behavior of a slipstream plays an important role in the efficiency of a propeller and offers several parameters to be investigated to improve a propeller's performance. Herein, the most important aspect is to gather calibrated measurements of all three velocity components of a slipstream flow. This can be performed with the help of special equipment containing a thin complex multi-hole tube as a sensor to measure the combinations of total and static Bernoulli pressures in the slipstream. All three components are then distinguished, and the radial distributions of all components can also be determined. To determine the propeller efficiency, the value of the mean induced velocity must be used. The latter is determined using the radial distribution of the measured axial velocity component (Figure 4).

First, the volumetric rate of the slipstream is determined. If the radial distribution of the axial component is presented as $v_a(r)$, the volumetric rate is calculated as

$$v_v = 2\pi \int_0^R v_a(r) \cdot r \cdot dr\,. \tag{7}$$

The mean induced velocity and the mass rate are then determined as

$$\overline{v_i} = \frac{v_v}{\pi R^2}\,,\ v_m = \rho \cdot v_v\,. \tag{8}$$

The propeller efficiency is now calculated using the measured values of thrust, torque, and angular velocity via Formula (5). In the case of advancing movement wherein the propeller efficiency is defined using the freestream air velocity or advance ratio, efficiency strongly depends on the blade angles (geometric pitch) and freestream velocity. For propellers working in the static regime, we should assume the existence of an optimal blade angle radial distribution to ensure the most effective angle of attack for the blade sections.

These blade angles, however, may depend on the propeller planform, airfoil, Reynolds number, etc., and can also be hardly predictable using simulations. In the present work, the radial distribution of the angle of attack can be determined from the experiment in which the airflow velocity components v_a and v_t are measured. For a blade section, according to Figure 2, the angle of attack is determined as the difference between the blade angle γ and the helix angle β. In the three-dimensional case, one must also consider the radial velocity component v_r. Firstly, the effective helix angle and effective chord length must be defined:

$$\beta_{eff} = arcsin\,\frac{v_a}{v_T}\,,\; d_{eff} = \mathrm{d}cos\;\varphi. \tag{9}$$

were φ is the measured azimuthal angle of the airflow, v_T is the total airspeed over a blade section, and d is the measured chord length.

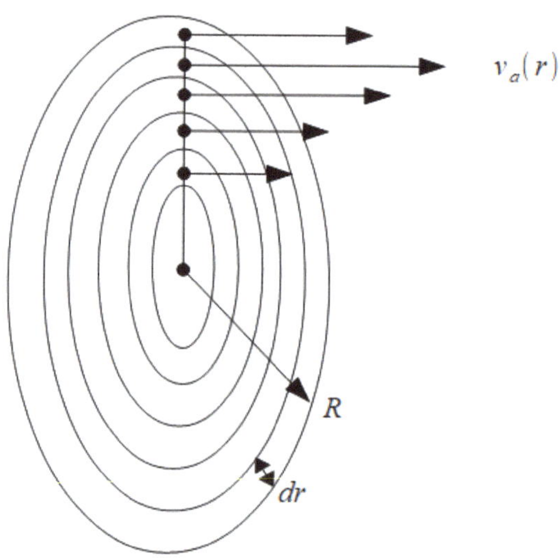

Figure 4. Determining the volumetric rate and mean induced velocity: $v_a(r)$—radial distribution of axial component, R—propeller radius, and dr—distance between measuring points.

For the effective blade angle and the angle of attack, one must consider that these parameters depend on the blade planform and twist geometry. Herein, we can use the simplified approach based on the assumption that the planform and blade angle do not change rapidly along the propeller radius. Under these circumstances, we derive formulae for those parameters as

$$\gamma_{eff} = arcsin\;(sin\;\gamma \cdot cos\;\varphi\;)\;,\; \alpha = \gamma_{eff} - \beta_{eff}. \tag{10}$$

Slipstream monitoring also provides an empirical way to determine the lift and drag coefficients for different propeller blade sections at particular Reynolds numbers and angles of attack. Hence, if a parcel of air interacts with a blade section, the air parcel will then obtain two additional velocity components after that interaction: v_L as the downwash velocity and v_D as the drag velocity (Figure 5).

Supposing the mass rate of the air interacting with the airfoil section is $\frac{dm}{dt}$, the lift and drag forces for that particular airfoil section will then be expressed as

$$L = \frac{dm}{dt}v_L,\; D = \frac{dm}{dt}v_D. \tag{11}$$

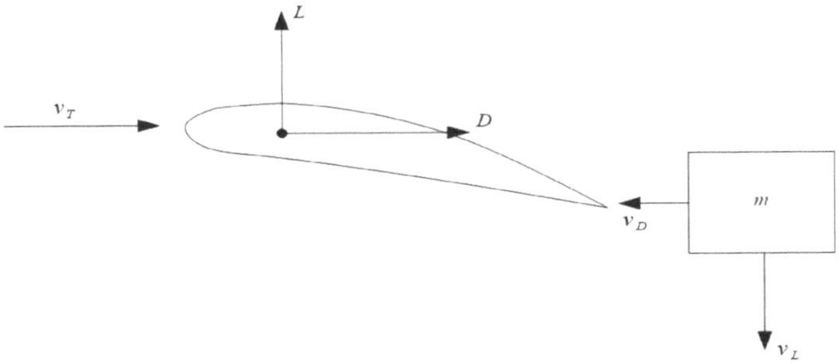

Figure 5. Downwash and drag velocities of a parcel of air with mass m after interaction with an airfoil section: v_T—total airspeed, L—lift force, D—drag force, v_L—downwash velocity, and v_D—drag velocity.

On the other hand, according to the conventional approach,

$$L = C_L \frac{\rho v_T^2}{2} S, \; D = C_D \frac{\rho v_T^2}{2} S, \tag{12}$$

where v_T is the total airspeed over the blade section, and S is the reference area. In the present context, we can express that

$$\frac{dm}{dt} = \rho v_T S \tag{13}$$

and from this,

$$C_L = \frac{2v_L}{v_T}, \; C_D = \frac{2v_D}{v_T} \tag{14}$$

The total airspeed over a blade section is calculated considering all the slipstream flow components:

$$v_T = \sqrt{(\omega r - v_t)^2 + v_a^2 + v_r^2} \tag{15}$$

The velocities induced by the blade section, v_L and v_D, are calculated from the measured values of v_a, v_t, and v_r:

$$v_L = v_a cos\ \beta_{eff} + v_t sin\ \beta_{eff}\ , \; v_D = v_t cos\ \beta_{eff} - v_a sin\ \beta_{eff}. \tag{16}$$

It must be mentioned that the lift coefficient calculated via Formula (14) is not directly valid for a rotating propeller, because it is derived for an airfoil section free from external influences. In the case of a rotating propeller, one must consider the increased pressure on the blade faces caused by the contracting slipstream. The contracting slipstream achieves some radially averaged maximum velocity value $\overline{v_{max}}$ that is not measured but can be found via measured parameters as

$$\overline{v_{max}} = \frac{T}{v_m} = \frac{T}{\rho \pi R^2 \overline{v_i}}\ . \tag{17}$$

The pressure on the blade faces is directly proportional to $\overline{v_{max}}$ and so is the lift coefficient. As in Formula (14), only the values of axial velocity v_a were used, and the calculated lift coefficient is valid only for a non-contracting slipstream. Otherwise, the effective lift coefficient for a blade section can be found by multiplying the lift coefficient found via (14) by the speed ratio:

$$C_{Leff} = C_L \frac{\overline{v_{max}}}{\overline{v_i}}. \tag{18}$$

This situation has some common features with the ground effect, whereby an airfoil section reaches higher values of lift coefficient at lower values of angle of attack due to pressure changes between the blade face and back caused by external influences on the airflow. In calculating the Re number for a blade section, the total airspeed and effective chord length must be used.

3. Test Stand and Measurement Technology

The test stand contains two main elements: the fixed propeller post and the movable airflow sensor post (Figure 6). The air data probe is driven using two stepper motors and jackscrews. The main movement of the air data probe is radial to scan the slipstream from its center up to the propeller radius.

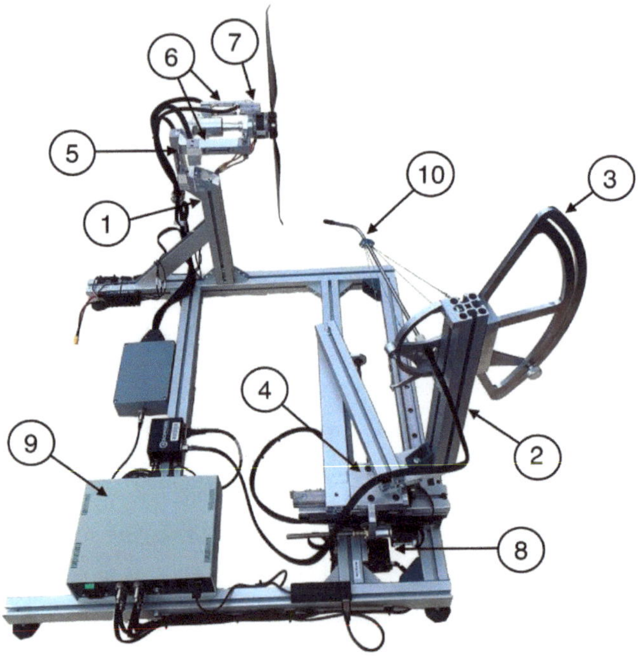

Figure 6. The test stand. 1—propeller post, 2—airflow sensor post, 3—initial angle of attack setting system, 4—initial sideslip angle setting system, 5—thrust sensor, 6—torque sensors, 7—RPM sensor, 8—two-coordinate scanning mechanism, 9—control unit, and 10—air data probe with anti-vibration cables.

The axial movement of the airflow sensor indicates the necessary scanning trajectory of the sensor tip, depending on the shape of the propeller blades and the possible flexing of the blades due to propeller thrust. The probe is secured with thin steel cables to avoid possible vibrations of the probe tube in the airflow. The torque and thrust sensors, as well as the optical tachometer, are mounted on the propeller post. As the air data probe can detect the angle of attack and sideslip angle in a region between +21 degrees and −21 degrees, the initial angle setting systems were incorporated into the airflow sensor post to widen the range of measured airflow directions. An external strobe light source can be used to visually monitor the propeller rotation and detect possible blade vibrations and the stability of the angular velocity. A high-speed camera can also be used to photograph the rotating propeller to determine the exact amount of blade bending that is necessary to determine the scanning trajectory of the sensor tip at a constant distance from the trailing edge of the blade (Figure 7).

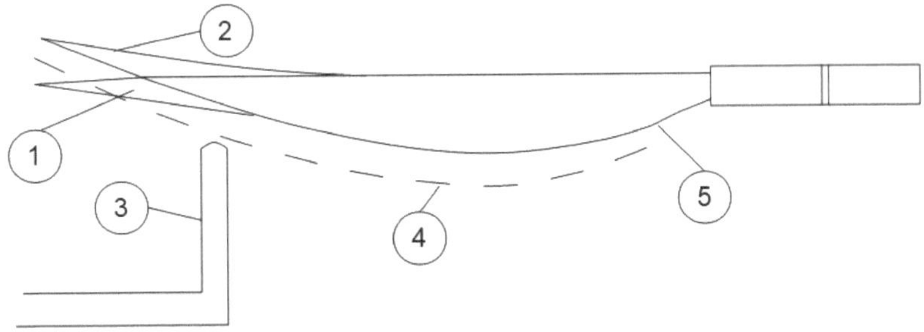

Figure 7. Determining the scanning trajectory at a constant distance from the trailing edge of a propeller blade. 1—standing blade (side projection), 2—rotating blade flexing under thrust, 3—airflow sensor tube, 4—scanning trajectory, and 5—trailing edge.

One Zemic L6D single-point load cell is used to measure torque, and another is used to measure thrust. The cells can easily be changed according to the necessary measuring range. The cells have 0.014% precision, and if calibrated according to force, torque, and thrust, these can be measured with an average relative error of 0.1%. With an optical tachometer, it is possible to stabilize the angular velocity with a precision of 0.05 rad/s, which, in the case of a propeller working at 3000 RPM, produces a relative error of 0.02% for angular velocity. This makes it possible to calculate the thrust and power coefficients with relative errors of approximately 0.5%, including an air density error that is estimated to be 0.1%. The slipstream is monitored with the Aeroprobe Micro Air Data System using a calibrated multihole air data probe that detects the full airspeed and two perpendicular angles: the angle of attack and the sideslip angle. The absolute error of the airflow velocity measurement is 1 m/s in terms of the absolute calibration of the sensor. The relative changes in the airflow velocity can be detected with a precision of 0.2 m/s without any difficulties. In the case of a typical radial distribution of the induced velocity with a peak value of 25 m/s, the relative error for the mean induced velocity is estimated to be 5% at a 95% trust level. This means that when calculating the propeller efficiency, the error of that parameter is mainly determined by the error of the induced velocity. Thus, the expected relative error of the efficiency will also be approximately 5% from the calculated efficiency, not from the input power, which is considered to be 100%. For instance, if the propeller efficiency is determined to be 70%, the full-scale relative error will be 3.5%. The changes in the efficiency can be detected with approximately 5 times higher precision, i.e., with an error of approximately 0.7% on the full (100%) scale.

4. Results

The computer-controlled measurement process includes the direct measuring of the propeller angular velocity, torque, and thrust at the maximum possible stability of those parameters. The slipstream scanning is performed by moving the five-channel sensor tube radially, starting from the center of the slipstream and following the predicted trajectory to keep the measuring distance from the propeller trailing edge constant. It was found that a distance of approximately 4 to 5 mm from the trailing edge is sufficient from the point of view of both accuracy and safety. The measuring device enables the radial scanning of the slipstream with a step of 1 mm, but taking into account the diameter of the measuring probe (0.25 of an inch), the measuring step was set to be 3 mm, corresponding to one-half of the probe's diameter. In this case, all the measurements within a 3 mm step are averaged in the output data. As the propeller slipstream is turbulent, there may be considerable temporal fluctuations in the parameters measured with the air probe. These fluctuations depend very much on the propeller's configuration and its working regime. To obtain temporally averaged results, the number of radial scanning sweeps was increased depending on the propeller type and speed of rotation. Usually, it was enough to average the results of 10 radial sweeps to obtain acceptably smooth radial distributions of the output parameters.

The dependence of the measured results on the air pressure and temperature was not investigated in this work, but as the measuring procedure is quite quick, there was no problem carrying out all the measurements during a short period of time, which means at a constant air pressure and temperature. The air probe measures the module of the total airspeed vector in the slipstream that we call the total slipstream velocity and two angles, the one that the manufacturer calls the angle of attack and the other being the sideslip angle. In terms of the spherical coordinates, the first of those is called the polar angle and the second is called the azimuthal angle. The directly measured parameters in a test are torque, thrust, angular velocity, total slipstream velocity, polar angle, and azimuthal angle. The technical parameters of the propeller such as diameter, radial distribution of blade angle, and airfoil chord are also measured directly and used as input data in further calculations of different parameters. The calculated parameters may be divided into two groups, the integral and differential ones. The integral parameters characterize the whole propeller and contain the input power P, induced power P_i, mean induced velocity $\overline{v_i}$, volumetric rate v_v, mass rate v_m, and propeller efficiency η, thrust coefficient C_T, and power coefficient C_P. Various differential parameters include those with a radial distribution and are calculated via the formulae discussed above. They can be presented in the form of table columns and/or graphs as needed and are listed as follows: axial velocity v_a, tangential velocity v_t, radial velocity v_r, effective blade angle γ, effective helix angle β_{eff}, angle of attack α, total airspeed over a blade section v_T, downwash velocity induced by a blade section v_L, drag velocity induced by a blade section v_D, lift coefficient and effective lift coefficient of a blade section C_{Leff}, drag coefficient of a blade section C_D, effective aerodynamic chord of a blade section d_{eff}, and Reynolds number for a blade section. The three components of the slipstream velocity—axial velocity v_a, tangential velocity v_t, and radial velocity v_r—are calculated from the directly measured air probe data by transferring the measured spherical coordinates into a Cartesian system.

First, a widely used multirotor-oriented Tarot 1755 propeller was tested. It is a carbon fiber propeller that is 17″ in diameter with a 5.5″ geometric pitch. It is quite thin (with a relative thickness of 0.75 R, approximately 4.3%) but has quite a high camber (with a relative camber of 0.75 R, approximately 5.3%) airfoil. It also has blade tip modifications such as propeller tips bent toward the blade back (approximately 15 degrees near the tip). The characteristic views of the propeller are presented in Figure 8.

(a) (b) (c)

Figure 8. Planform (**a**), 83-degree side projection (**b**), and 0.75 R airfoil shape (**c**) of a Tarot 1755 propeller.

The measured radial distribution of the blade angle and airfoil chord length are presented in Figure 9.

The propeller was tested at angular velocities between 220 rad/s and 600 rad/s. It generates a Reynolds number ranging from $7.0 \cdot 10^4$ to $2.0 \cdot 10^5$ for the blade section at 75% of the propeller radius. Monitoring was carried out for five different angular velocities including the minimum and maximum values. The radial distributions of the three slipstream velocity components at the minimum and maximum angular velocities are presented in Figure 10.

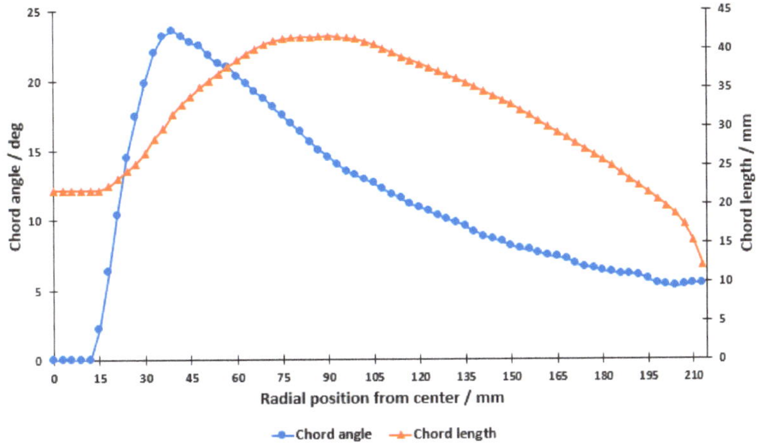

Figure 9. Radial distribution of blade angle and chord length of a Tarot 1755 propeller.

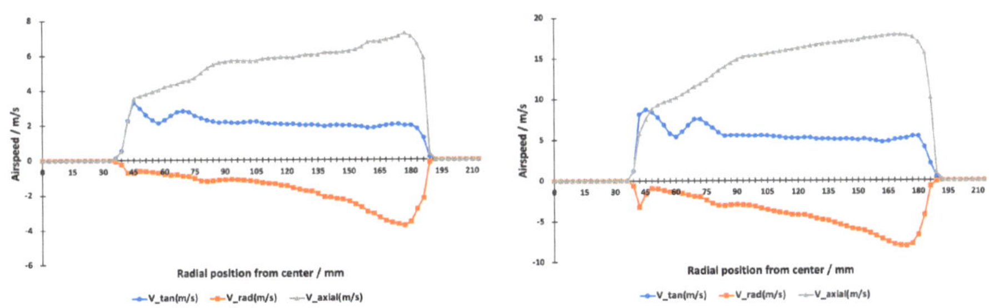

(a) (b)

Figure 10. Radial distributions of v_a, v_t, and v_r at approximately 2000 RPM (**a**) and approximately 6000 RPM (**b**). The radial velocity toward the axis is considered negative.

In Figure 11, the dependence of the thrust coefficient, power coefficient, and efficiency on the angular velocity is presented.

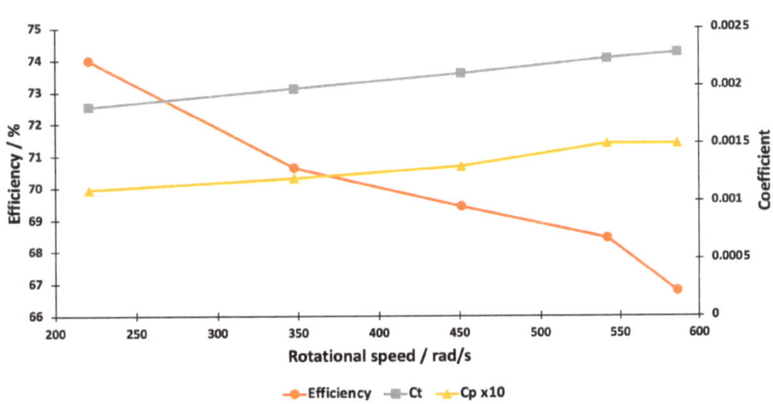

Figure 11. The dependence of C_T, C_P, and η on ω of a Tarot 1755 propeller.

In Figure 12 the radial distributions of the angle of attack for both the minimum and maximum values of the angular velocity are presented.

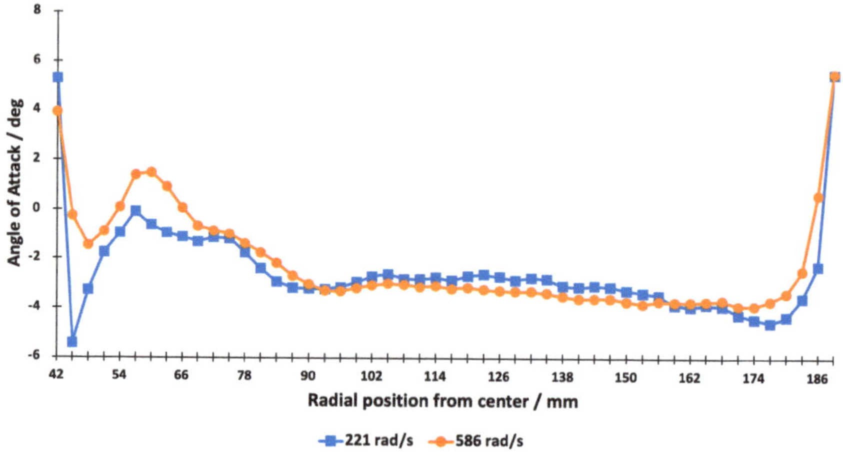

Figure 12. Radial distributions of angle of attack of 0.75 R blade section at minimum and maximum values of ω.

Figure 13 presents the dependence of the effective lift coefficient and drag coefficient for the blade section at 75% of the propeller radius on the angular velocity and the Reynolds number of that particular blade section.

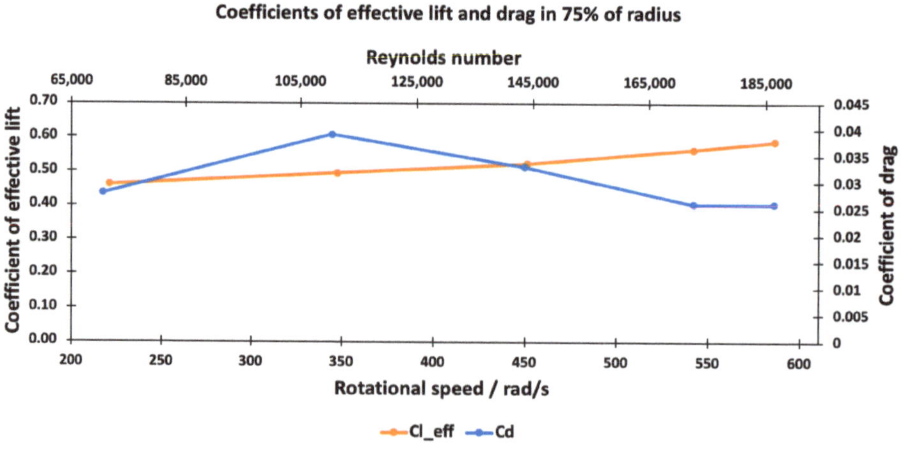

Figure 13. C_{Leff} and C_D for 0.75 R blade section against ω and Re for a Tarot 1755 propeller.

For comparison, another push-type static-use-oriented carbon-composite T-Motor FA20.2*6.6 propeller (20.2″ in diameter and 6.6″ in geometric pitch) was tested. This propeller uses a more advanced airfoil with a thin trailing edge. Its 0.75 R section has 8% of the relative thickness and 3.5% of the relative camber. The tip modifications used herein are blade tips bent toward the blade face forming approximately 20 mm long winglets with a positive sweep. The planform, side view, and 0.75 R airfoil are presented in Figure 14.

(a) (b) (c)

Figure 14. Planform (**a**), 83-degree side projection (**b**), and 0.75 R airfoil of a T-Motor FA20.2*6.6 propeller (**c**).

The radial distribution of the blade angle and cord length is presented in Figure 15.

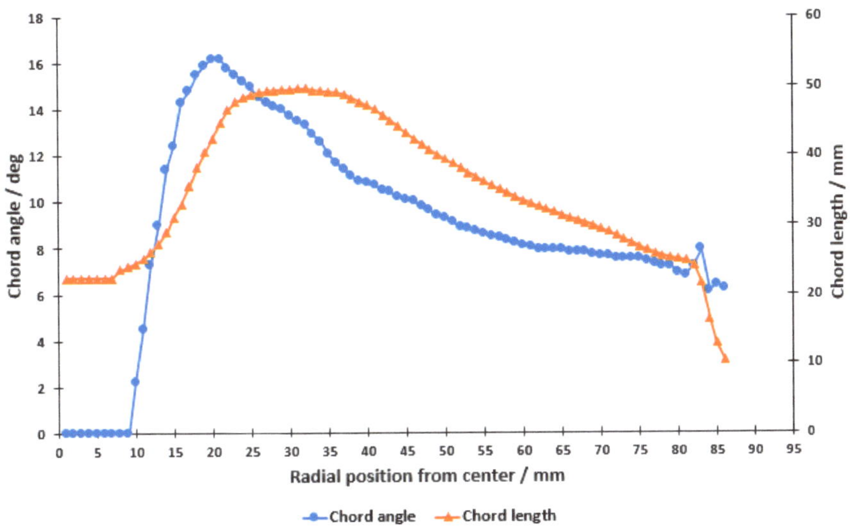

Figure 15. Radial distribution of cord length and blade angle of a T-Motor FA20.2*6.6 propeller.

The following graphs (Figure 16a,b) present the measured results for the T-Motor FA20.2*6.6 propeller in the same form as they were presented for the Tarot 1755 propeller.

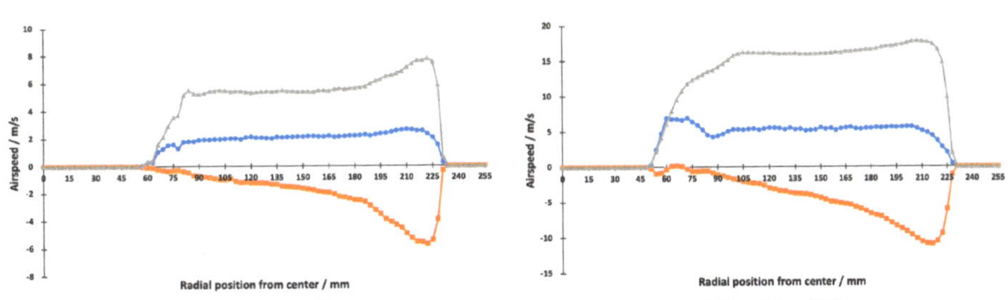

(a) (b)

Figure 16. Radial distributions of v_a, v_t, and v_r at approximately 2000 RPM (**a**) and approximately 6000 RPM (**b**) for a T-Motor FA20.2*6.6 propeller.

In Figure 17, the dependence of the thrust coefficient, power coefficient, and efficiency on the angular velocity is presented.

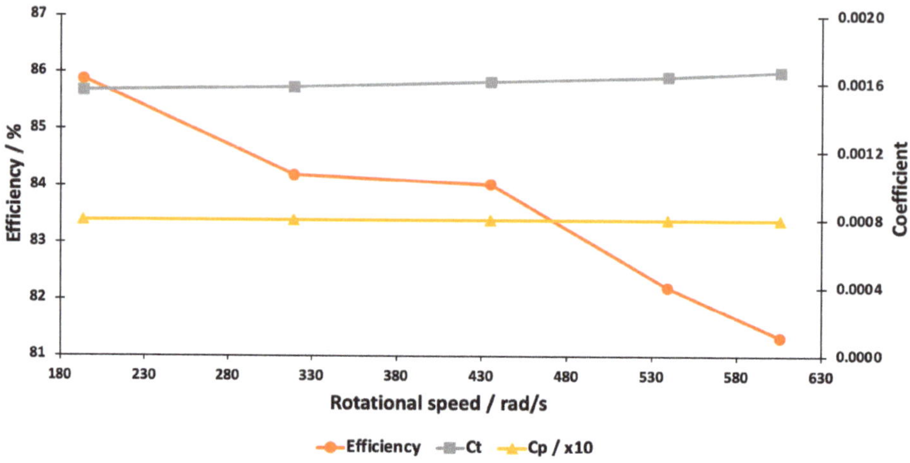

Figure 17. The dependence of C_P, C_T, and η on ω for a T-Motor FA20.2*6.6 propeller.

In Figure 18 the radial distributions of the angle of attack for both the minimum and maximum values of the angular velocity are presented.

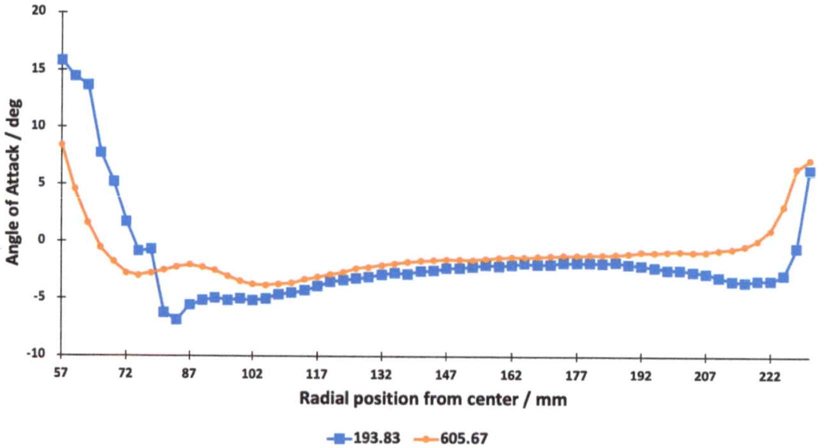

Figure 18. The dependence of angle of attack for 0.75 R blade section at minimum and maximum values of ω.

In this paper, the measured and calculated parameters presented were chosen to demonstrate some possibilities of this method of investigation. Depending on the concrete aim of the investigation, many other parameters and characteristics can be measured using slipstream monitoring together with thrust and torque measurements.

5. Discussion

The two propellers that were tested are both oriented for hovering or nearly hovering flight mode. For this reason, these propellers are of quite a low geometric pitch (approximately 30% of the diameter) and highly cambered airfoil. Despite many similarities between the propellers, there are some significant differences including in efficiency, which could be called a basic characteristic of a propeller. The basic behaviors of the components of the induced velocity along the propeller radius are quite similar for both propellers. As for differences, the Tarot 1755 propeller has a tendency for a higher $\frac{v_r}{v_a}$ ratio at the position of the axial speed maximum (66% for the Tarot 1755 propeller vs. 46% for the T-Motor FA20.2*6.6 propeller). The radial component of the Tarot 1755 propeller (Tarot) tends to rise toward the axis of rotation, while for the T-Motor FA20.2*6.6 propeller (T-Motor), it remains nearly constant. The blade characteristic points also have similar distances from the propeller axes. These points are the starting point, cut-off point, maximum speed point, center of pressure, and center of power.

The starting point is the point closest to the axis where the slipstream airflow becomes detectable (at an axial velocity of over 1 m/s). This point is located at 0.18 R for the Tarot propeller and 0.21 R for the T-Motor propeller.

The cut-off point is the point near the blade tip where the slipstream airflow drops below the detection level due to tip vortices. The position of this point is 0.88 R for Tarot and 0.89 R for T-Motor.

The maximum speed point is the blade section that generates the highest value of axial velocity. For both propellers, this section lies quite close to the break-off point, at 0.81 R for Tarot and 0.83 R for T-Motor.

The center of pressure is the thrust center of the blade. The location of this point is calculated based on the assumption that the thrust generated by a blade section is proportional to the blade width and square of the axial velocity. This point is located at 0.57 R for Tarot and 0.51 R for T-Motor.

The center of power is the blade section from which the blade half toward the axis generates the same power as the blade half toward the tip. This location is calculated assuming the generated power is proportional to the blade width and cube of the axial velocity. This point is at 0.61 R for Tarot and 0.58 R for T-Motor.

All these characteristic points were estimated at approximately 6000 RPM for both propellers. The well-known and widely recognized reference blade section to characterize the whole propeller is the section at 0.75 R. In this paper, this conventional blade reference section is also used, although none of the characteristic sections of the tested propellers are even close to that distance value.

In the literature, the thrust coefficient and power coefficient are presented as the basic parameters to characterize the efficiency of small-scale propellers. In most cases, with an increasing Re number, an increase in C_T and/or a decrease in C_P is registered. For instance, static measurements carried out for a 9″ propeller showed a 20% increase in C_T, with C_P remaining constant as the Re number increased from 2×10^4 to 7×10^4 [14]. These tendencies are generally interpreted as the impact of the Re number on the propeller's performance. The basic trend here seems to indicate that the lower the Re number of a propeller (for the 0.75 R section), the lower the propeller's performance. Measurements performed in the present work with Re numbers ranging from 7×10^4 to 2×10^5 showed an increase in C_T for both propellers, at 26% for Tarot and 7% for T-Motor. As for the power coefficient, the results were different. For T-Motor, the value of C_P remained strictly constant, but for Tarot, a 36% increase in C_P was found. This result shows that the performance of a propeller is not predictable simply via the Re number and may depend additionally on some other parameters of the propeller. The measurements of propeller efficiency ended in results one could call unexpected. For both propellers, a decrease in efficiency was found with a rise in the Re number. For Tarot, the efficiency dropped from 74% at approximately 2000 RPM to 67% at approximately 6000 RPM. The same tendency, although lower, was registered for T-Motor, with η dropping from 86% to 81%. It should be pointed out that there is a notable difference between the values of η for these two propellers, which correlates with the differences in the values of C_T and C_P. At 6000 RPM, the C_T of Tarot was approximately 37% higher than that of T-Motor, but the C_P of Tarot was nearly twice as high as that of T-Motor. The behavior of C_P with the changing Re number indicates once more that the Re number is not the only key parameter by which a propeller's performance is determined. In the case of T-Motor, we could see a decrease in η with the increasing Re number, although a significant 7% increase in C_T and constancy in C_P were detected. These results lead to the hypothesis that the efficiency of a small-scale propeller is strongly dependent on the behavior of the slipstream. The behavior of the slipstream, leaving from the plane of rotation, depends on three velocity components induced by the propeller. It is known that if a slipstream contracts, the axial velocity increases and so does thrust. However, the radial component measured in the plane of rotation does not elucidate what that component will be at greater distances downstream. In this work, no systematic monitoring of the slipstream at greater distances was carried out, but some qualitative measurements at 7 cm from the plane of rotation were conducted. These measurements showed that there were no detectable mass rate losses at that distance, but significant changes in the radial distributions of the three velocity components were observed. It was noticed that at the minimum speed of rotation (2000 RPM) the radial flow component nearly reached zero at 7 cm from the rotating propeller. This indicates a very weak contraction of the slipstream at 2000 RPM and confirms the results obtained for the mean axial speed ratio of 1.05 for Tarot and 1.04 for T-Motor. The mean axial speed ratios were calculated from the measured thrust and mean induced velocity values. The mean axial speed ratio at 6000 RPM was found to be 1.48 for Tarot and 1.52 for T-Motor. This indicates a stronger slipstream contraction at higher values of RPM, and this greater contraction was also confirmed by the presence of a considerably higher value of the radial velocity at 7 cm from the propeller. It is also clear that the rise in C_T for both propellers cannot be explained only by the slipstream contraction. The rise in the mean axial speed ratios was much greater (approximately 50%) than the rise in the thrust coefficients of the propellers. According to [15], the axial velocity reaches its maximum value at approximately 0.7 R–0.8 R downstream from the plane of rotation. The radial distribution also changes significantly; from a distance of approximately R, the slipstream starts to expand, and the region of the highest velocity moves toward the axis. It is possible to determine the mean maximum velocity from slipstream measurements when monitoring the slipstream at maximum contraction and integrating the axial component over the effective cross-section of the slipstream. This procedure is similar to that for determining the mean induced velocity in the present paper, wherein the effective cross-section was considered to be the propeller disc area. Slipstream monitoring further downstream together with induced velocity measurements provides a method to determine the energy losses in the slipstream and determine the thrust of the propeller. In [15], this method was called a slipstream method, and it showed a precision of 5–15% in comparison with direct thrust measurements, with no losses taken into consideration.

As for the angle of attack, both tested propellers worked in the region of negative angles of attack. In the main working region of the blade (0.4 R to 0.8 R), the angle of attack was constantly near −3 degrees for Tarot and rose from −3 degrees to nearly 0 degrees for T-Motor. For highly cambered airfoils such as those of our propellers, the zero-lift angle of attack is nearly −5 degrees. A slight dependence of the angle of attack distribution on the RPM value was noticed, but no interpretation was found for that effect. It must be noted that the fast rise in the angle of attack near the cut-off point was not adequate and was caused by the calculation procedure. The angle of attack is calculated as the difference between the blade angle and helix angle. As the slipstream velocity (but not $\omega \cdot r$) drops to zero, the calculated helix angle will also drop to zero and the angle of attack will show the value of the blade angle. We do not have methods to determine the angle of attack in the region where the slipstream velocity cannot be detected. In theory, the value of α near the blade tips must be somewhere near −5 degrees to produce no axial velocity. As the present method of testing enables to us to determine the effective lift coefficients and drag coefficients of all the working blade elements, this was also performed, and the results are presented only for the 0.75 R blade section for both propellers. The effective lift coefficient means that the found lift coefficient is multiplied by the speed ratio $\frac{\overline{v_{max}}}{\overline{v_i}}$. This is necessary because the lift coefficient is determined from the measured slipstream velocity components that do not take into account the impact of the pressure rise in the slipstream due to contraction and the axial velocity rise. The distributions of C_{Leff} and C_D

along the blade are quite smooth (constant) in the main working region (0.4 R to 0.8 R). A very interesting result that could also be considered unexpected was the fact that the drag coefficients of the T-Motor blade sections were approximately 75% higher than those of Tarot despite the much better efficiency of T-Motor. At the same time, the effective lift coefficients were nearly comparable but still a little higher for Tarot. A 21 % rise in C_{Leff} with the increasing Re number was detected for both propellers. The dependence of C_D on the Re number was more complicated, as seen in the graphs in Figures 13 and 19, but it showed a general tendency to decrease with the increasing Re number. The airfoil of T-Motor at 0.75 R was compared with the NACA 6404 airfoil in order to compare the efficiencies of those airfoils by comparing data from the Airfoil Database with the present experiment. NACA 6409 was found to be the most similar airfoil to the 0.75 R airfoil of T-Motor. Unfortunately, it was not possible to find a similarly described airfoil for Tarot because this airfoil seems to be unfinished and hard to compare with the described airfoils. In Table 1, the data on the T-Motor 0.75 R airfoil and NACA 6409 are presented. The corresponding data for Tarot are also included. The data in Table 1 are not of high precision and are only able to illustrate stronger tendencies and differences.

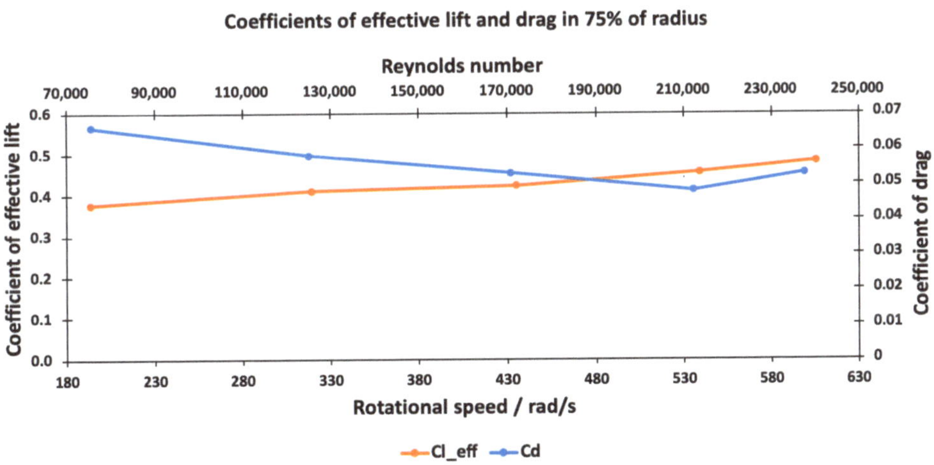

Figure 19. The dependence of effective lift coefficient and drag coefficient of 0.75 R blade section on ω and Re number.

Table 1. Comparison of airfoil efficiencies for T-Motor 0.75 R airfoil and NACA 6409 at similar angles of attack and Reynolds numbers, and data for Tarot 0.75 R airfoil.

Airfoil	Re	α (deg)	C_L	C_D	C_L/C_D
NACA 6409	70,000	−2	0.35	0.03	12
T-Motor	74,000	−2	0.36	0.073	5
NACA 6409	200,000	−1.5	0.5	0.012	42
T-Motor	230,000	−1.5	0.33	0.054	6
Tarot	70,000	−3.5	0.41	0.041	10
Tarot	190,000	−3.8	0.41	0.03	14

It is seen in Table 1 that if the lift coefficients of NACA 6409 and T-Motor airfoils are almost comparable at both values of Re numbers, the drag coefficients for the propellers are much (approximately 2.5 to 4 times) higher. This means that the airfoil working in a propeller blade does not show the same efficiency as predicted in the Airfoil Database. The data for Tarot show an even better airfoil efficiency even at lower angles of attack in spite of the fact that this particular airfoil is far from what one could call "aerodynamic", i.e., having a constant thickness and blunt trailing edge (see Figure 8c). However, the better efficiency of T-Motor is correlated with the ratio of the thrust coefficient to the power coefficient, which is 15 for Tarot and 21 for T-Motor at Re numbers near 200 000. This shows that the better efficiency of T-Motor is not explainable by the efficiency of the airfoil used near the region of 0.75 R. It should also be mentioned that this kind of comparison does not take into account the fact that the effective airfoil is not the same as that directly measured for a propeller blade because of the radial airflow component in the plane of rotation. The radial component makes the effective airfoil thinner and decreases the camber. As the aim of the present paper was to introduce a method to quickly determine the efficiency of a propeller, not much attention was paid to all the possible applications of this method.

6. Conclusions

The presented method for small-scale propeller testing that combines thrust, torque, and angular velocity measurements with slipstream monitoring is a quick and effective way to obtain necessary data on propellers and to determine static efficiency. It enables us to study the dependence of propellers' performance on geometry, pitch, tip modifications, Re number, etc. It also enables us to carry out further investigations to find answers to the questions that arose during the present investigations. The most important question is the impact of the Re number on the performance of a small-scale propeller. The tests showed a decrease in efficiency with the rising Re number in spite of the rising values of the thrust coefficient. This leads to two main conclusions.

First, the thrust and power coefficients are not always the key parameters to determine the efficiency of a propeller, and second, the role of the Re number in a propeller´s efficiency is not yet clear. The presence of Re number effects has been proven in numerous works, but the impact of those effects seems to not be as trivial as the claim that the lower the Re number, the weaker the performance. To determine the real impact of the Re number, small-scale propellers of significantly different diameters at different angular velocities must be tested with the Re number remaining constant. This would enable one to exclude the Re number effects and find out the possible role of other parameters. In this paper, the slipstream was monitored closest to the trailing edge of the propeller to detect the induced airspeed most precisely. This method would also be successful for slipstream monitoring further downstream to investigate the behavior of a slipstream. This may be necessary to find out what factors determine the rate of slipstream contraction between the plane of rotation and the point of maximum axial velocity. It is also possible to investigate not only propellers but also devices such as ducted fans and jets of any origin because this would enable us to determine the axial mass rate at any point on a slipstream or jet and the mass rate losses along the stream. As all three components of the slipstream velocity are measured, it would be possible to detect the effect of different duct constructions to decrease slipstream whirling because the tangential velocity in a slipstream is quite high, approximately one-third of the axial velocity. Decreasing the tangential velocity may produce a significant effect on the performance of ducted fans. Currently, the only serious limitation of this method is its usability only in static conditions, i.e., not in advancing flow. The test stand is not intended for use in wind tunnels because, for that purpose, it must be have a fundamentally different construction. However, many investigations will be carried out on thrust-producing small-scale devices in static conditions.

Author Contributions: Conceptualization and methodology, J.S. and K.-E.U.; software, validation, and experiments, S.H.; writing—original draft preparation, J.S.; writing—review and editing, K.-E.U. All authors have read and agreed to the published version of the manuscript.

Funding: This research received no external funding.

Data Availability Statement: The results and experimental data can be provided upon request.

Conflicts of Interest: The authors declare no conflict of interest. Other parties had no role in the design of this study; in the collection, analysis, or interpretation of the data; in the writing of the manuscript; or in the decision to publish the results.

Abbreviations

This paper uses SI units and symbols as follows:

M	Torque (Nm)
T	Thrust (N)
P	Input power (W)
P_i	Induced power (W)
C_T	Thrust coefficient
C_p	Power coefficient
L	Lift force (N)
D	Drag force (N)
C_L	Lift coefficient
C_D	Drag coefficient
v_v	Volumetric rate (m/s)
v_i	Induced velocity (m/s)
$\overline{v_i}$	Mean induced velocity (m/s)
v_a	Axial component of the induced velocity (m/s)
v_r	Radial component of the induced velocity (m/s)
v_t	Tangential component of the induced velocity (m/s)
v_0	Freestream flow velocity (m/s)
v_T	Total airspeed (m/s)
v_L	Downwash velocity (m/s)
v_D	Drag velocity (m/s)
J	Advance ratio
η	Propeller efficiency
D	Propeller diameter (m)
R	Propeller radius (m)
r	Radial parameter (m)
α	Angle of attack (deg)
β	Helix angle (deg)
γ	Blade angle (deg)
ω	Angular velocity (rad/s)

References

1. Williamson, G.A.; McGranahan, B.D.; Broughton, B.A.; Deters, R.W.; Brandt, J.B.; Selig, M.S. *Summary of Low-Speed Airfoil Data*; University of Illinois Urbana-Champaign: Champaign, IL, USA, 2012; Volume 5.
2. Selig, M.S.; McGranahan, B.D. *Wind Tunnel Aerodynamic Tests of Six Airfoils for Use on Small Wind Turbines*; NREL/SR-500-34515; National Renewable Energy Laboratory: Golden, CO, USA, 2004.
3. Lyon, C.A.; Broeren, A.P.; Giguère, P.; Gopalarathnam, A.; Selig, M.S. *Summary of Low-Speed Airfoil Data*; SoarTech Publications: Virginia Beach, VA, USA, 1998; Volume 3.
4. Bass, R.M. *Small Scale Wind Tunnel Testing of Model Propellers*; AIAA Paper 86-0392; American Institute of Aeronautics and Astronautics: Reston, VA, USA, 1986.
5. Michael, O.L.; Zeune, C.; Logan, M. *Analytical-Experimental Comparison for Small Electric Unmanned Air Vehicle Propellers*; AIAA Paper 2008-7345; American Institute of Aeronautics and Astronautics: Reston, VA, USA, 2008.
6. Moffitt, B.A.; Bradley, T.H.; Parekh, D.E.; Mavris, D. *Validation of Vortex Propeller Theory for UAV Design with Uncertainty Analysis*; AIAA Paper 2008-0406; American Institute of Aeronautics and Astronautics: Reston, VA, USA, 2008.
7. Brandt, J.B. Small-Scale Propeller Performance at Low Speeds. Master's Thesis, University of Illinois at Urbana Champaign, Urbana, IL, USA, 2005.
8. Uhlig, D.V. Post Stall Propeller Behavior at Low Reynolds Numbers. Master's Thesis, University of Illinois at Urbana Champaign, Urbana, IL, USA, 2007.
9. Merchant, M.P.; Miller, L.S. *Propeller Performance Measurement for Low Reynolds Number UAV Applications*; AIAA Paper 2006-1127; American Institute of Aeronautics and Astronautics: Reston, VA, USA, 2006.
10. Gamble, D.E.; Arena, A. *Automated Dynamic Propeller Testing at Low Reynolds Numbers*; AIAA Paper 2010-0853; American Institute of Aeronautics and Astronautics: Reston, VA, USA, 2010.

11. Baranski, J.A.; Fernelius, M.H.; Hoke, J.L.; Wilson, C.W.; Litke, P.J. *Characterization of Propeller Performance and Engine Mission Matching for Small Remotely Piloted Aircraft*; AIAA Paper 2011-6058; American Institute of Aeronautics and Astronautics: Reston, VA, USA, 2011.
12. Brandt, J.B.; Selig, M.S. *Propeller Performance Data at Low Reynolds Numbers*; AIAA Paper 2011-1255; American Institute of Aeronautics and Astronautics: Reston, VA, USA, 2011.
13. McCormick, B.W. *Aerodynamics, Aeronautics and Flight Mechanics*, 2nd ed.; Wiley: New York, NY, USA, 1995.
14. Deters, R.W.; Ananda, G.K.; Selig, M.S. *Reynolds Number Effects on the Performance of Small-Scale Propellers*; AIAA Paper 2014-2151; American Institute of Aeronautics and Astronautics: Reston, VA, USA, 2014.
15. Deters, R.W.; Ananda, G.K.; Selig, M.S. *Slipstream Measurements of Small Scale Propellers at Low Reynolds Numbers*; AIAA Paper 2015-2265; American Institute of Aeronautics and Astronautics: Reston, VA, USA, 2015.

drones

MDPI

Article

Retrieving Relevant EU Drone Legislation with Citation Analysis

Gijs van Dijck [1,2,*], Alexandru-Daniel On [3], Jasper Snel [2] and Rohan Nanda [1,4,5]

1 Maastricht Law and Tech Lab., Maastricht University, 6211 LK Maastricht, The Netherlands; r.nanda@maastrichtuniversity.nl
2 Brightlands Institute for Smart Society, Maastricht University, 6211 LK Maastricht, The Netherlands
3 Maastricht European Private Law Institute, Maastricht University, 6211 LK Maastricht, The Netherlands; daniel.on@maastrichtuniversity.nl
4 Institute of Data Science, Maastricht University, 6211 LK Maastricht, The Netherlands
5 Department of Advanced Computing Sciences, Maastricht University, 6211 LK Maastricht, The Netherlands
* Correspondence: gijs.vandijck@maastrichtuniversity.nl

Abstract: Can the retrieval of relevant unmanned aircraft systems (UAS) legislation be automated? In this article, references from and to EU legislation are used to create an overview that is subsequently compared to legislation considered relevant by subject-matter experts. The overlap between the results of the citation analysis and the expert overview is promising. Additionally, an approach was proposed and tested where, first, a relatively large number of laws were identified and, second, the laws that were considered relevant were selected. The findings reveal that this approach was successful at retrieving the majority of relevant laws. The results are relevant to researchers, policymakers, practitioners, and laypeople searching for relevant EU legislation on UAS.

Keywords: UAS; drones; legislation; automation; linked data; network analysis

Citation: van Dijck, G.; On, A.-D.; Snel, J.; Nanda, R. Retrieving Relevant EU Drone Legislation with Citation Analysis. *Drones* **2023**, *7*, 490. https://doi.org/10.3390/drones7080490

Academic Editors: Tomasz Nowakowski, Artur Kierzkowski, Agnieszka A. Tubis, Franciszek Restel, Tomasz Kisiel, Anna Jodejko-Pietruczuk and Mateusz Zając

Received: 14 June 2023
Revised: 18 July 2023
Accepted: 19 July 2023
Published: 26 July 2023

1. Introduction

1.1. Research Problem

This study focuses on the findability of relevant legislation. In recent years, several European laws have been enacted or proposed on unmanned aircraft systems (UAS). The most prevalent, UAS-specific laws are regulations on unmanned aircraft systems and on third-country operators of unmanned aircraft systems [1], on the rules and procedures for the operation of unmanned aircraft [2], and on a regulatory framework for the U-space (i.e., an unmanned traffic management system) [3–5]. Other laws that can be applicable and relevant in addition to UAS-specific legislation can be categorized as aviation-specific and general-purpose legislation.

Previous research on UAS and UAS-related legislation has focused on analyzing and assessing the regulatory framework applicable to UAS operations [6–12], identifying the potential impact of UAS regulations [13], exploring various civil [14–17] and criminal liability [18,19] issues, including the liability of AI systems (which are and will be, more and more integrated into UAS technology) [20–22], exploring issues of privacy [23], cybersecurity [24] and data breaches [25], discussing the (potential) use of UAS in Law Enforcement and Capacity Monitoring [26], identifying processes and operational preventive measures to avoid malfunctioning and user negligence [27], and flagging potential gaps and uncertainties within the regulatory and liability frameworks that apply to UAS [28].

It can be difficult, even for those with legal expertise, to identify relevant UAS legislation. At the European level, the applicable rules are scattered across different pieces of regulations and directives. Moreover, which legislation can be considered relevant depends on the legal question one aims to answer. For instance, a UAS that causes personal injury due to a crash will likely lead to questions of liability for which liability law (or even criminal law) will need to be consulted. A UAS that flies too close to buildings and captures confidential work within the facility is likely to violate privacy laws, whereas a UAS that closely flies over one's home may violate property law, more specifically, ownership rules that entail that property ownership includes not only the property itself but also a part of the airspace above the property. Considering that the applicable rules depend on the facts, a method for finding relevant UAS legislation should allow the user to select or focus on legislation that is applicable to the use case relevant to them.

This article discusses the automation of finding relevant UAS legislation. In this article, it is explored whether the finding of possible relevant drone legislation can be automated. Electronic legal information retrieval systems generally carry out the automatic identification of relevant legislation [29]. Prior research on legal information retrieval systems can be broadly classified into Boolean retrieval, Natural Language Processing (NLP)-based retrieval, and ontology-based retrieval. In Boolean information retrieval, documents are represented as sets of terms. A query is formulated in the form of a Boolean expression of terms. A document is retrieved if it satisfies the query expression. Many legal databases, such as Westlaw, Lexis, and Bloomberg Law, offer Boolean search functionality. Legal Boolean retrieval methods have been known to result in the retrieval of a large number of irrelevant documents and missing relevant documents that do not match the query expression [30]. In addition, they achieve a much lower recall (20%) compared to what manual (over 75% recall) legal researchers retrieve [31].

NLP-based retrieval methods utilize text similarity techniques for finding the most relevant legal documents for a user query. Text similarity techniques have been used in different legal information retrieval tasks, such as the identification of similar cases and judgments [32,33], identification of similar patents [34,35], identification of relevant legal statutes, and provisions for a user query [36,37]. The major text similarity techniques utilized in NLP-based retrieval include lexical, semantic, and language model-based representations. Text similarity techniques offer the advantage of ranking the retrieved legal documents on the basis of a similarity score in the order of relevance.

In ontology-based retrieval, external legal databases, thesauruses, vocabularies, or ontologies are utilized to incorporate semantic information for retrieving legal documents [38,39]. A major advantage of using an ontology-based information retrieval system is the availability of external semantic knowledge in a machine-readable form to enrich the retrieval process with the meaning of the terms. However, the user of such a system should be a domain expert or someone very familiar with the underlying ontology, semantics, and syntax of the query [40].

The present study will focus on reference-based retrieval, more specifically on whether the references in legislation to other laws can assist in identifying possibly relevant UAS legislation. It poses the question of whether the process of finding relevant UAS legislation can be automated by means of references from and to UAS legislation.

Any systematic search for relevant UAS legislation needs to grapple with several complicating factors that arise from the way the law is written, adopted, enforced, and organized in Europe. There are multiple layers of legislation that apply to UAS operations (Section 1.2), and different types of legislation can be found (Section 1.3).

1.2. Multi-Layered Legislation

The legislative and regulatory provisions applicable to UAS-related activities and operations can be organized according to their formal normative sources into four broad layers (or categories):

1. The International Layer: consisting primarily of international conventions, such as the Convention on International Civil Aviation (also known as Chicago Convention) [41] and its annexes, the Convention for the Unification of Certain Rules for International Carriage by Air (also known as the Montreal Convention) [42], and the non-binding yet highly influential global Standards and Recommended Practices (SARPs), Procedures, Model Regulations, and Guidance material for unmanned aviation developed by the International Civil Aviation Organization (ICAO) [43].

2. The European Layer: consisting of European Union legislation, together with other international conventions (such as the European Convention on Human Rights), as well as norms derived from the case law of the Court of Justice of the European Union and the case law of the European Court of Human Rights.
3. The National Layer: consisting of legislative acts and regulations adopted by individual Member States. National courts apply the legislative acts and regulations adopted by the Member states (and, in certain instances, also further develop the legal system).
4. The Local Layer: consisting of legislative acts and regulations adopted by administrative subdivisions of the state, such as regions, provinces, or municipalities.

An example of multi-layered UAS legislation concerns the rules regarding the geographical zones in which UAS operations can or cannot take place. Article 15 of EU Regulation 2019/947 prescribes the possibility for a Member State to prohibit certain or all UAS operations when defining the UAS geographical zones [44]. A subsequent inspection of Dutch national law leads to Article 9 of the Flight Operations Decree ('Besluit vluchtuitvoering'), a general administrative order that states that geographical zones, as stated in Article 15 of EU Regulation 2019/947, can be defined or designated by ministerial orders [45]. Based on this above-mentioned provision, the ministerial order Zoning Regulations for Unmanned Aerial Vehicles ('Regeling zonering onbemande luchtvaartuigen') has been introduced. This order prohibits the operation of flights in the open category in specified geographical zones, which are visualized on a website [46]. The ministerial order sets rules in the following way: it first defines the specific type of area in which the flight is prohibited, and it then determines the specific coordinates surrounding those areas concerned. For example, based on the first paragraph of Article 2 'Regeling zonering onbemande luchtvaartuigen', it is prohibited to fly a UAS in the open category higher than 30 m in the geographical areas in which low-level flight by civil or military aircraft is allowed. By way of derogation from this first paragraph prescribing 30 m as the limit, there is an additional full prohibition in the second paragraph for flights of UAS subcategory A3 [47] in the areas defined in the article [48]. Then the article provides concrete coordinates to define which areas fall under these prohibitions. One of these areas is, for example, the 'GLV I (North Drenthe)' (the area enclosed by a line connecting the following positions: 53°03′45.00″ N 006°43′30.35″ E; 52°59′14.96″ N 006°39′00.31″ E; 3°01′00.20″ N 006°38′00.54″ E; 53°03′45.00″ N 006°40′30.40″ E; 53°03′45.00″ N 006°43′30.35″ E) [49]. Because explicit references between different legal orders (e.g., from EU law to Member State Law, or vice versa) are not readily available, this study focuses on UAS legislation in the EU.

1.3. Types of EU Legislation

In the most abstract terms, EU law can be divided into two basic categories: primary EU law and secondary EU law [50] (p. 53). Primary EU law consists of the founding treaties of the European Union [50] (p. 53), the most recent of which is the Treaty on the Functioning of the European Union (TFEU) [51]. Secondary EU law consists of legislation enacted by EU Institutions, case law (especially from the Court of Justice of the European Union), general principles of EU law, and international agreements entered into by the EU [50] (p. 53).

There are three types of binding legislative acts that EU law-making institutions can adopt: Regulations, Directives, and Decisions. In addition, EU institutions can adopt non-binding Recommendations or Opinions. Regulations are acts that are directly binding across the entirety of the European Union [52]. They do not require implementation in order to apply in each individual Member State [53]. Directives are legislative acts that, in principle, aim only to formulate the goals that European Union Member states should achieve in a certain area of policy [54]. Directives require implementation by the Member states, which means that the Member States must adopt national legislation to specify further how such goals are to be achieved [52]. Decisions are acts that are addressed to a determinate or determinable legal or natural person (such as a country, a company, a group of countries, or companies in a certain area of activity) and are binding only on the parties to whom they are addressed [52]. Recommendations and Opinions allow EU institutions to formulate policy and statements in a non-binding way. They are used to suggest lines of action without imposing any obligations on the Member States or on private parties [52].

Of the four categories of legislation, the most relevant for the purpose of this article are Regulations and Directives because they are binding types of EU legislation. Both Regulations and Directives are of general applicability, but some Regulations and Directives have narrower domains than others. For instance, some regulations, such as the General Data Protection Regulation (GDPR) [55], are very general and create obligations for all data controllers and processors, whereas a Regulation such as 2019/947 [2] applies only to UAS operators and remote pilots. Whether a Directive or Regulation is adopted in any specific field depends on the type of competence the EU holds in that area, as well as the ever-evolving policy aims of the Union. Air traffic safety legislation is almost always adopted as a Regulation, whereas other domains, such as liability or the security of network and information systems, are regulated mainly through Directives. Transport is a shared competence between the EU and the Member States (see Article 4 of the TFEU), and air transport specifically can be regulated at the EU level according to Article 100(2) of the TFEU. This does not mean, however, that a search for UAS legislation can be limited to acts adopted on the basis of Article 100(2) of the TFEU. Some generally applicable legislative acts have a different legal basis, such as Article 114 of the TEFU (dealing with the harmonization of laws that have as their object the establishment and functioning of the internal market) or Article 16 of the TFEU (the right to protection of personal data).

For the purposes of this article, it is useful to further divide EU legislation into the following three categories: (1) UAS-specific; (2) aviation-specific; and (3) general-purpose legislation. UAS-specific legislation is legislation that was enacted with the explicit goal of regulating activities, products, and organizations connected with UAS technology. Aviation-specific legislation is more general, covering all types of aviation activities, products, and organizations involved in aviation. Because of its more general character, aviation-specific legislation applies to UAS as a subcategory of "aircraft", and it also compliments and influences UAS-specific legislation because of its more general character. Finally, general-purpose legislation is legislation that is not particularly designed for UAS or aircraft but can, nevertheless, be or become relevant to UASs. The application of general-purpose legislation to UASs and UAS activities can be said to be incidental because this kind of legislation was not written with UAS activities in mind. A great deal of general-purpose legislation even precedes the existence of the technologies that make UAS flights possible. Examples include the rules of property law, liability law, criminal law, or the laws that regulate cybersecurity and privacy. These rules create rights and obligations that apply to a variety of persons, goods, and services, which may also become relevant under the right circumstances to UAS activities. For instance, a liability rule that states that "a person who commits a wrongful act against another person, which can be attributed to him, is obliged to compensate the damage that the other person suffers as a result" [56] does not apply to most UAS operations because most UAS operations will go according to plan and will be perfectly safe. However, should a risk to life, bodily integrity, or property materialize (for instance, if a UAS collides with a person or a building), then general liability rules come into play, and any person who is at fault may have a duty to compensate the victim of the collision.

In the specific case of UASs, the normative and regulatory landscape is enriched but also made more complex because a large number of relevant norms are created by an agency of the European Union—the European Union Aviation Safety Agency (EASA). The EASA plays a key role in shaping EU secondary legislation by drafting and advising the European Commission on novel aviation-specific legislation and by proposing amendments during the legislative process. Additionally, the EASA adopts non-binding rules, such as the Acceptable Means of Compliance (AMC) and Guidance Material (GM), which provide much-needed guidance to all stakeholders involved in UAS operations. Finally, the EASA also assists the national competent authorities in carrying out their legislative and regulatory tasks.

In the following sections, only hard law is analyzed; thus, the scope of the present study Recommendations and Opinions from EU law-making bodies, as well as Acceptable Means of Compliance, Guidance Material, and Opinions issued by the EASA, were excluded.

The remainder of the paper is organized as follows. Section 2 presents the materials and methods. The results are discussed in Section 3. Section 4 presents the discussion. The paper concludes in Section 5.

2. Materials and Methods

Our starting point was at the European level and, more specifically, the hard law of the European Union. From a practical perspective, this was an obvious choice because of the recent regulatory activity of the EU with regard to UAS operations [1,2], as well as the regulations regarding the U-space and U-space providers [3].

It was examined whether relevant UAS legislation, as defined by subject-matter experts, can be found or approximated by applying a linked data approach. Linked data is, in essence, a collection of best practices for publishing and connecting structured data such that it is machine-readable. It should have an explicit meaning and must be linked to external datasets, which are also linked to other external datasets on the Web. From a more technical point of view, linked data uses the Resource Description Framework (RDF) format to represent information on the Web. The advantage of using RDF is that it allows different data or things to connect on the Web using typed statements [57].

In the context of this paper, the references from one law to another as linked data were used. The European Commission stores these references as metadata in a repository (called CELLAR) and publishes them on EUR-Lex. By varying the degrees of separation, it was possible to control and vary the extent to which the references were removed from the source law, with the source law being the law (or laws) that were taken as a starting point. The CELEX-id of each source law—all laws have a unique identifier called CELEX—was taken, searched for in CELLAR for incoming and/or outgoing citations related to the particular CELEX-ID(s), and returned in the form of an edges list with CELEX-IDs that consisted of sources and targets, where the sources were the laws that were citing other laws and where the targets were the laws that were cited. Based on these edges lists, citation networks could be created.

It was possible to vary both the source depth and target depth. The source depth concerns the degree of separation regarding the references from other laws. A source depth of '1' means that the CELEX-IDs were retrieved from the laws that cite the source law. A source depth of '2' means that the CELEX-IDs returned cited the source law as well as the CELEX-IDs that cited the CELEX-IDs that cited the source law. This approach can be considered a forward-looking approach, given that generally and chronologically speaking, laws only cite laws from the past, not future laws (exceptions do exist, for instance, if two or more laws are drafted or enacted around the same time and, as a result, cite one another). The target depth is the degree of separation from the source law to other laws. For instance, a target depth of '2' means that the CELEX-IDs were returned from the laws that were cited in the source law as well as the laws that were cited in the laws that cited the source law. The target depth was backward-looking. Testing different combinations of source and target depths and comparing the results with the laws identified by the experts allowed us to calculate the overlap between the laws in the expert overview and the ones found by means of our linked data approach.

The strategy deployed was as follows. One or more starting laws (e.g., Regulation 2019/947) were selected, and references from and to this law were identified. Each analysis would start with a source depth and target depth of zero (meaning that no references to or from the source law are sought). After this, the source and target depth were incrementally increased by 1, in the sense that in every iteration, either the source depth or the target depth was increased (not both at the same time). The laws that were returned at each step were compared to the laws mentioned by the experts. For this (1) the laws identified in the expert overview that were also found with our linked data approach (true positives), (2) the laws in the expert overview that were not found in our search (false negatives), and (3) the retrieved laws that were not listed in the expert overview (false positives) were selected. Based on this information, the precision (True positives/(True positives + False positives)) and recall (True positives/(True positives + False negatives)) were calculated, with precision indicating how many of the retrieved laws with the linked data approach were relevant, whereas recall was about how many of the relevant laws were retrieved. F1 scores were a tradeoff between precision and recall.

The relevant legislation was identified by consulting partners in a research project on UASs, including project partners with legal expertise in UAS legislation, comparative law, and practical expertise with UAS deployment (current or future). Representatives of nine project partners attended sessions in which an overview of relevant laws was created, of which two knowledge institutions and one UAS assessments and certification company provided the actual input. The corresponding CELEX number retrieved for each law mentioned in the overview was included in the analysis, meaning that the CELEX numbers in the expert overview were compared with the CELEX number in the searches that were conducted.

The scope of the expert overview was inherently arbitrary (see Section 1.3). Although we are confident that the vast majority of the UAS-specific legislation, such as the 2019/945 and 2019/947 Regulations [1,2], was identified by other experts, this is not necessarily the case for non-drone-specific legislation, such as laws on liability, cybersecurity, or trespassing. As explained above, what one would consider relevant for a UAS operation very much depends on the circumstances of the intended operation and on the legal questions that could or are expected to occur before, during, or after the operation is conducted.

The references were obtained through EUR-Lex, the official portal for providing access to European Union law. The information presented on EUR-Lex was retrieved from the common repository of metadata and content (CELLAR), which can be accessed through a SPARQL endpoint or through a RESTful API [58]. The available metadata includes the relationships between legislative documents. Outgoing and incoming references to or from documents can be retrieved one document at a time from either of these endpoints. This process can start from a distinct list of source documents. The resulting documents from this step can then be used as input for the next iteration of the retrieval. This process was repeated until the desired 'depth' of references was achieved.

With the references from and to the laws identified in our search, a network of laws was created where the nodes in the network consisted of the laws and where the citations form the edges that link the nodes. The application of network analysis allows us to automatically divide the laws into clusters, identify meaningful sub-topics from those clusters, and find central (relevant) laws within the clusters that can be distinguished from less central laws. The networks were analyzed and visualized in Gephi, an open-source software for the creation, manipulation, and visualization of networks [59].

The data collection was conducted on 13 June 2023. The data and code are publicly available at [anonymized].

3. Results

The laws that were identified by the experts vary in terms of the duty holder and what they aim to protect. They apply to UAS manufacturers, operators in the open, specific or certified categories of UAS operations, U-space service providers, all of the aforementioned, or to the Member States. The legislation that was considered relevant could be distinguished into six main categories: safety, more specifically, air safety, product safety, and safety investigations, cybersecurity, privacy, data governance, digital services, and liability. Table 1 provides an overview of the relevant legislation.

Table 1. Expert Overview of Relevant EU Legislation (M = UAS Manufacturers; O = UAS Operators in category open; S = UAS operators in a specific category; C = UAS Operators in a certified category; U = providers of U-space services; All = All of the aforementioned; MS = Member States). The overview builds on an initial overview provided by Filippo Tomasello (EuroUSC).

Domain	Applicable to	Reference
Safety	All	2018/1139 [60]
	MS	996/2010 [61]
	None	1025/2012 [62]
	M	765/2008 [63]
		768/2008 [64]
		2006/42 [65]
		2009/48 [66]
		748/2012 [67]
		2015/640 [68]
		2019/945 [1]
	All	300/2008 [69]
		376/2014 [70]
	O, S	2019/947 [2]
	S, C	923/2012 [71]
		2021/666 [5]
	C	1178/2011 [72]
		965/2012 [73]
		1321/2014 [74]
		1332/2011 [75]
		2018/1048 [76]
	U	2017/373 [77]
		2021/664 [3]
		2021/665 [4]
Cybersecurity	All	2016/1148 [78]
	All	2019/881 [79]
Privacy	All	2016/679 [55]
Data Governance	All	2022/868 [80]
Digital Services	All	2022/2065 [81]
Digital Identity	All	European Digital Identity (reform proposal) [82]
Liability and Insurance	M	85/374 [83]
	O, S, C	785/2004 [84]

One of the laws mentioned in the expert overview concerned, at the time of writing this paper, a proposal for reform. It was decided not to include the proposal in our analysis. This left us with 30 laws identified by the experts.

Citation Analysis

As explained above, the citation analysis consisted of the following:

1. The selection of one or more laws. These laws served as a starting point for the search (hereafter: source law(s)).
2. The collection of citations with different source and target depths, starting with zero (i.e., no references to or from the source law) and incrementally increasing the depth by 1.
3. A comparison of the results of the citation analysis (referred to as 'search results') to the laws in the expert overview. Precision, recall, and F1 scores were used to evaluate the findings.

We were first interested in the recall: do the laws in the expert overview appear in the search results? A high recall would suggest that the citation analysis was able to reproduce the results by the experts. The results reveal that a recall could be achieved that approximates 0.70 (Tables 2–5) with just a few steps (degree depth), meaning that around 70% of the laws mentioned in the expert overview could be retrieved.

Table 2. Precision, Recall, and F1 scores for different source and target depths, with Regulation 2021/664 and 2019/945 as source laws.

Target depth	0	0	0	0	1	1	1	1	1	2	2	2	2	2	3	3	3	3	3	4	4	4	4	4
Source depth	1	2	3	4	0	1	2	3	4	0	1	2	3	4	0	1	2	3	4	0	1	2	3	4
Precision	0.50	0.44	0.36	0.08	0.76	0.64	0.62	0.57	0.25	0.21	0.21	0.20	0.20	0.15	0.07	0.07	0.07	0.07	0.06	0.03	0.03	0.03	0.03	0.03
Recall	0.13	0.13	0.13	0.13	0.53	0.53	0.53	0.53	0.53	0.70	0.70	0.70	0.70	0.70	0.70	0.70	0.70	0.70	0.70	0.70	0.70	0.70	0.70	0.70
F1	0.21	0.21	0.20	0.10	0.63	0.58	0.55	0.55	0.34	0.33	0.32	0.32	0.31	0.31	0.13	0.12	0.12	0.12	0.11	0.06	0.06	0.06	0.06	0.05

Table 3. Precision, Recall, and F1 scores for different source and target depths, with Regulation 2021/664 as source law.

Target depth	0	0	0	0	1	1	1	1	1	2	2	2	2	2	3	3	3	3	3	4	4	4	4	4
Source depth	1	2	3	4	0	1	2	3	4	0	1	2	3	4	0	1	2	3	4	0	1	2	3	4
Precision	0.67	0.67	0.67	0.67	1.00	0.88	0.88	0.88	0.88	0.71	0.69	0.69	0.69	0.69	0.18	0.18	0.18	0.18	0.18	0.06	0.06	0.06	0.06	0.06
Recall	0.07	0.07	0.07	0.07	0.23	0.23	0.23	0.23	0.23	0.67	0.67	0.67	0.67	0.67	0.70	0.70	0.70	0.70	0.70	0.70	0.70	0.70	0.70	0.70
F1	0.12	0.12	0.12	0.12	0.38	0.38	0.38	0.38	0.38	0.69	0.68	0.68	0.68	0.68	0.29	0.29	0.29	0.29	0.29	0.11	0.11	0.11	0.11	0.11

Table 4. Precision, Recall, and F1 scores for different source and target depths, with Regulation 2021/665 as source law.

Target depth	0	0	0	0	1	1	1	1	1	2	2	2	2	2	3	3	3	3	3	4	4	4	4	4
Source depth	1	2	3	4	0	1	2	3	4	0	1	2	3	4	0	1	2	3	4	0	1	2	3	4
Precision	1.00	0.67	0.67	0.67	1.00	1.00	0.67	0.67	0.67	1.00	1.00	0.88	0.88	0.88	0.71	0.71	0.69	0.69	0.69	0.18	0.18	0.18	0.18	0.18
Recall	0.07	0.07	0.07	0.07	0.07	0.07	0.07	0.07	0.07	0.23	0.23	0.23	0.23	0.23	0.67	0.67	0.67	0.67	0.67	0.70	0.70	0.70	0.70	0.70
F1	0.12	0.12	0.12	0.12	0.12	0.12	0.12	0.12	0.12	0.38	0.38	0.37	0.37	0.37	0.69	0.69	0.68	0.68	0.68	0.29	0.29	0.29	0.29	0.29

Table 5. Precision, Recall, and F1 scores for different source and target depths, with Regulation 2019/947 as source law.

Target depth	0	0	0	0	1	1	1	1	1	2	2	2	2	2	3	3	3	3	3	4	4	4	4	4
Source depth	1	2	3	4	0	1	2	3	4	0	1	2	3	4	0	1	2	3	4	0	1	2	3	4
Precision	0.60	0.44	0.24	0.04	1.00	0.83	0.69	0.46	0.10	0.32	0.32	0.32	0.28	0.12	0.07	0.07	0.07	0.07	0.05	0.03	0.03	0.03	0.03	0.02
Recall	0.10	0.13	0.13	0.13	0.30	0.33	0.37	0.37	0.37	0.53	0.57	0.60	0.60	0.60	0.53	0.57	0.60	0.60	0.60	0.57	0.57	0.60	0.60	0.60
F1	0.17	0.21	0.17	0.06	0.46	0.48	0.48	0.41	0.16	0.40	0.41	0.41	0.38	0.20	0.12	0.13	0.13	0.13	0.10	0.05	0.05	0.05	0.05	0.05

The focus on recall is likely to lead to an overinclusion of laws that are not relevant or not highly relevant. This is not necessarily problematic, particularly when those laws can be clustered, as it would allow for ignoring one or more clusters of laws. Network analysis, community detection in particular, makes it possible to cluster laws. One strategy could, therefore, be first to select all possibly relevant laws (high recall) and to narrow down the selection (increase precision) subsequently. This strategy fits the notion that the legislation that can be considered relevant depends on the legal question one aims to answer, which in turn, depends on the circumstances at hand.

Increasing the depth (source and/or target depth) was the most likely strategy to result in the highest possible recall. To explore whether a visualized network could be reasonably interpreted when it includes a large number of laws that appear in the expert overview (high recall) and possibly a large number of laws that appear in the search results but not in the expert overview (low precision), the combination of source and target depth was selected that yielded the highest precision given the highest recall. The best results were obtained when selecting Regulations 2019/945 and 2021/664 as source laws, with source depth = 0 and target depth = 2. With these parameters, a recall of 0.70 and a precision of 0.21 was obtained (F1 = 0.33) (Table 2).

The network is visualized below in Figure 1.

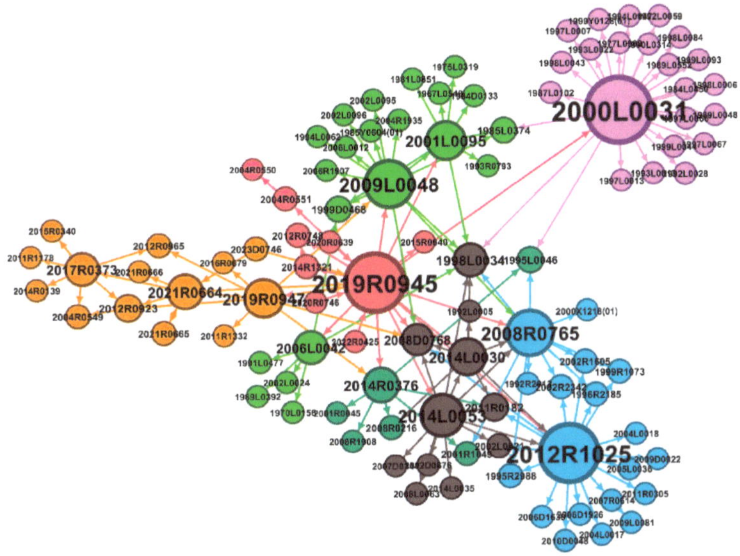

Figure 1. Network visualization (source laws = Regulation 2021/664, Regulation 2019/945, target depth = 2, source depth = 0). Notes: Communities are represented by different colors and based on the Louvain method for community detection. Node size is adjusted for ingoing and outgoing references combined. The labels are formatted as follows: year, type of legislation (R = Regulation, L = Directive, D = Decision), and legislation ID. The visualization can be inspected in more detail by downloading the Gephi file in the repository that is made publicly available at https://doi.org/10.34894/HLWZLL.

What is noticeable in this network is that the UAS-specific legislation was clustered in two communities (in red and orange), whereas the central laws in the other clusters concern subjects like electronic commerce (in purple), toy and product safety (in green), electromagnetic compatibility and radio equipment (in brown), actions following incidents in civil aviation (in dark green/turquoise), and standardization, accreditation, and market surveillance (in blue).

The network contains 21 laws out of the 30 that also appear in the expert overview. Compared to the previous results, the first network in particular, the search yielded one additional law that also appeared in the expert overview: 1985L0374 (on product liability). Because the precision is of secondary relevance, the network also includes 77 laws that are not in the expert overview that may turn out to be relevant, depending on the legal question that is at stake.

A high recall alone is not necessarily informative. It could mean that the search results include many relevant laws, but the results may also include many irrelevant laws. A high proportion of irrelevant laws can lead to relevant results that remain hidden in a pile of irrelevant ones. We, consequently, given a high recall (close to 0.70), sought source nodes that achieved higher precision while losing minimal recall. We will call this 'efficient retrieval' moving forward.

To obtain efficient retrieval, we first tried to simplify the search by minimizing the depth (source depth, target depth). The highest F1 score (0.63) was obtained when selecting a target depth of 1 and a source depth of 0 (Figure 2), but then the recall is lower (0.53) than when selecting Regulation 2021/664 as the sole starting point (recall = 0.67).

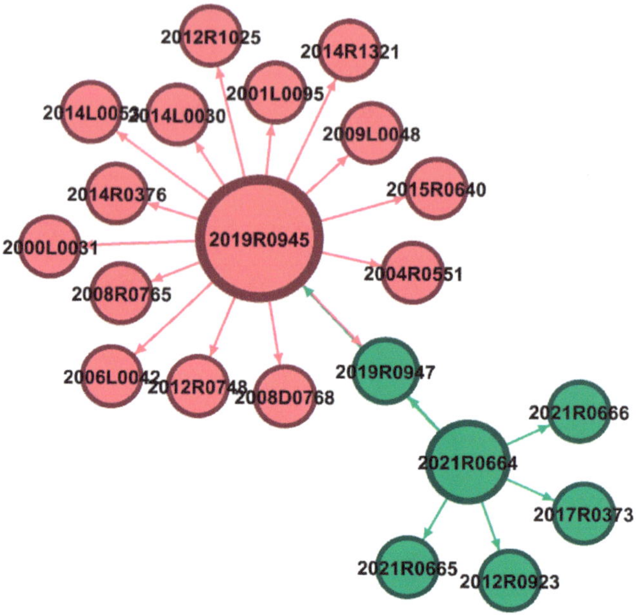

Figure 2. Network visualization (source laws = Regulation 2021/664, Regulation 2019/945, target depth = 1, source depth = 0). Notes: Communities are represented by different colors and based on the Louvain method for community detection. Node size is adjusted for ingoing and outgoing references combined. The labels are formatted as follows: year, type of legislation (R = Regulation, L = Directive, D = Decision), and legislation ID. The visualization can be inspected in more detail by downloading the Gephi file in the repository that is made publicly available at https://doi.org/10.34894/HLWZLL.

A more efficient retrieval was obtained when selecting Regulation 2021/664 [3] or Regulation 2021/665 [4] as the source law. Both will regulate the so-called U-space, a set of services and procedures related to the access to airspace for UASs. A recall of 0.67, a precision of 0.71, and an F1 score of 0.69 were obtained with target depth = 2 and source depth = 0, thereby selecting the references in Regulation 2021/664, as well as the references mentioned in the references (Table 3).The findings entail that 67% of the laws in the expert overview could be retrieved and that 71% of the laws in the search results also appear in the expert overview and appear in the search results. Increasing the target depth resulted in only a few more laws that also appear in the expert overview (recall increases from 0.67 to 0.70) and in a substantially larger number of laws not mentioned by the experts (precision dropped from 0.71 to 0.18) (Table 3). An increase in the source depth did not impact the results, which can be explained by the fact that Regulation 2021/664 is a recent piece of legislation that has simply not yet been cited often.

The laws missing in the search results concern:

- Regulation 2018/1139 (on civil aviation safety);
- Regulation 2010/996 (on accident and incident investigation and prevention);
- Regulation 2008/300 (on civil aviation security);
- Regulation 2018/1048 (on airspace usage);
- Directive 2016/1148 (on the security of network and information systems);
- Regulation 2019/881 (Cybersecurity Act);
- Directive 374/1985 (on product liability);
- Regulation 2004/785 (on insurance requirements for air carriers and aircraft operators);
- Regulation (EU) 2022/868 (Data Governance Act);
- Regulation (EU) 2022/2065 (Digital Services Act).

In turn, the legislation that was found in the search results but that did not appear in the expert overview concerned:

- Regulation 139/2014 (on aerodromes) [85];
- Regulation 549/2004 (on laying out a framework for the harmonization of European air traffic control) [86];
- Regulation 551/2004 (on the organization and use of European airspace) [87];
- Regulation 2015/340 (on air traffic controllers' licenses and certificates) [88];
- Directive 2014/53 (on the making available on the market of radio equipment) [89];
- Directive 2001/95 (on product safety) [90];
- Directive 2000/31 (on electronic commerce) [91];
- Directive 2014/30 (on electromagnetic compatibility) [92].

Various laws in the search results that did not appear in the expert overview can arguably be relevant in the context of UASs. Airspace and air traffic control legislation applies to and can affect UAS operations, whereas laws on product safety, radio equipment, and electromagnetic compatibility can be relevant for assessing the features and liability of UASs.

Here, we observe four communities when visualizing the network (Figure 3). One community (in blue) consists of legislation on U-space, one (in green) on air traffic related-matters, one (in purple) of laws mostly related to the specifications UASs are subjected to, and one community (in orange) on the laws regarding the carrying out of the operation, but also on data protection (General Data Protection Regulation—GDPR) and laws that could be brought under one of the other clusters.

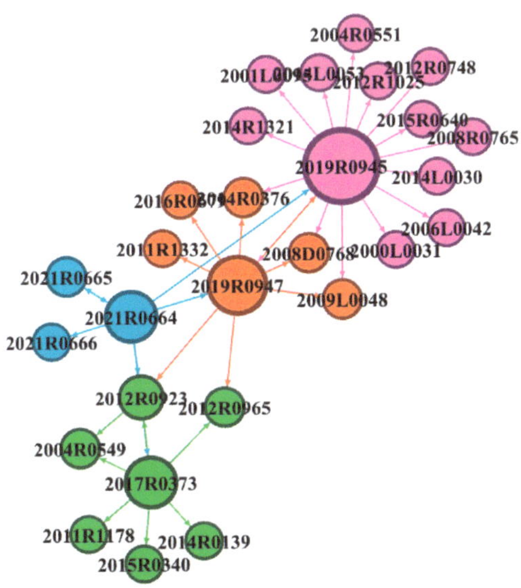

Figure 3. Network visualization (source law = Regulation 2021/664, target depth = 2, source depth = 0). Notes: Communities are represented by different colors and based on the Louvain method for community detection. Node size is adjusted for ingoing and outgoing references combined. The labels are formatted as follows: year, type of legislation (R = Regulation, L = Directive, D = Decision), and legislation ID. The visualization can be inspected in more detail by downloading the Gephi file in the repository that is made publicly available at https://doi.org/10.34894/HLWZLL.

The results are similar when selecting Regulation 2021/665 as the source law (Table 4). The only difference is that efficient retrieval was found with a target depth of 3 instead of 2 (F1 = 0.69, precision = 0.71, and recall = 0.67). The reason for this is that Regulation 2021/665 only cites Regulation 2021/664—one additional degree of separation.

The overlap between the expert overview and the search results decreased when other UAS-specific legislation was selected as a starting point (Table 5). For instance, when selecting Regulation 2019/947 [2], lower recall and precision scores were obtained. Depending on the source and target depths that were selected, either fairly high recall or precision scores could be obtained. For instance, if a source and target depth of 2 was selected, the recall was 0.60, and the precision was 0.32. A source depth and target depth of 1 increased the precision to 0.83 (more of the search results also appeared in the expert overview) but lowered the recall to 0.33 (fewer of the laws in the expert overview were retrieved).

The selection of Regulation 2019/945 [1] as the source law resulted in a maximum recall score of 0.60 and in a maximum precision score of 0.69 (Table 6). Also, here, there are no instances where both the recall and the precision were relatively high, as illustrated by the low F1 scores.

Table 6. Precision, Recall, and F1 scores for different source and target depths, with Regulation 2019/945 as source law.

Target depth	0	0	0	0	1	1	1	1	1	2	2	2	2	2	3	3	3	3	3	4	4	4	4	4
Source depth	1	2	3	4	0	1	2	3	4	0	1	2	3	4	0	1	2	3	4	0	1	2	3	4
Precision	0.43	0.44	0.36	0.08	0.69	0.57	0.57	0.52	0.21	0.18	0.18	0.19	0.18	0.13	0.05	0.06	0.06	0.06	0.05	0.02	0.02	0.03	0.03	0.03
Recall	0.10	0.13	0.13	0.13	0.37	0.40	0.43	0.43	0.43	0.53	0.53	0.60	0.60	0.60	0.53	0.57	0.60	0.60	0.60	0.53	0.57	0.60	0.60	0.60
F1	0.16	0.21	0.20	0.10	0.48	0.47	0.49	0.47	0.47	0.27	0.27	0.28	0.28	0.22	0.10	0.10	0.11	0.11	0.10	0.04	0.05	0.05	0.05	0.05

The search with the highest F1 score resulted in an ego network of the regulation, along with a cluster that groups together the rules on the operation and the airspace or U-space (Figure 4).

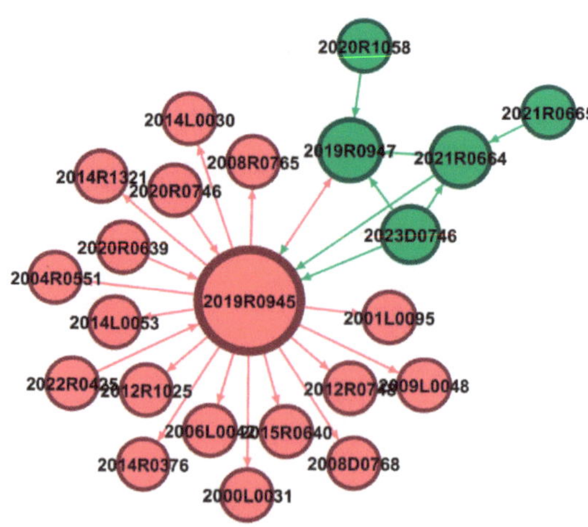

Figure 4. Network visualization (source law = Regulation 2019/945, target depth = 1, source depth = 2). Notes: Communities are represented by different colors and based on the Louvain method for community detection. Node size is adjusted for ingoing and outgoing references combined. The labels are formatted as follows: year, type of legislation (R = Regulation, L = Directive, D = Decision), and legislation ID. The visualization can be inspected in more detail by downloading the Gephi file in the repository that is made publicly available at https://doi.org/10.34894/HLWZLL.

Another strategy to increase the precision without losing substantial recall is to filter the network based on the most central laws. Here, we filtered the most central laws using the concept of degree: the combined number of incoming and outgoing references from or to other legislation (multiple citations from the same law to another (same) law counts as one citation). The idea of filtering by degree is that it is preferred to exclude laws that are rarely referenced or are referencing other laws. Ideally, the most central laws in the network coincide with the laws that appear in the expert overview.

As a first iteration, the laws with at least two incoming and/or outgoing references were selected (Degree > 2—note that all nodes in the other networks had a degree of 1). The results reveal that the overlap between the laws in the filtered network and the ones in the expert overview proved to be limited. Although the precision increased substantially, from 0.21 to 0.57, this came at a loss of recall (from 0.75 to 0.52). These results are substantially worse than the findings obtained in previous networks, the first network in particular, where the precision was 0.70 and the recall 0.68.

4. Discussion

This study focused on the automation of finding relevant unmanned aircraft systems (UAS) legislation through the use of linked data and network analysis. The overlap that was found between the expert overview and the linked data approach suggests that EU legislation relevant to UAS can, for its majority, be found by means of an automated approach that leverages the availability of linked data as metadata. At the same time, not all laws deemed relevant by the experts could be retrieved, meaning that a linked data and network approach is likely not to be complete. This particularly applies to aviation-specific and general-purpose legislation. In contrast, all of the UAS-specific legislation could be retrieved in the searches performed in this study.

The retrieval would have been more successful if the source and target depth could be determined for each individual node (e.g., source depth = 1 for node X, source depth =3 for node Y, etc.). This approach, however, has limited practical relevance, as an expert overview is required for a proper evaluation, which is often lacking in practice. The results may also have been better if the experts had different 'use cases' (circumstances) in mind. The laws listed in the expert overview suggest that the experts had drone operations in mind when selecting the laws, along with regulation on data (Data Governance Act, GDPR, and Cybersecurity) and digital services (Digital Services Act), although the applicability of the latter to UASs is debatable. Because an almost infinite number of situations can occur in reality, each of which may affect the laws that are applicable to the particular situation, and because experts may not always be readily available or affordable, a citation approach can be helpful or complimentary for identifying relevant or applicable legislation, perhaps even for experts. What makes a citation analysis promising is that it empowers users to construct networks iteratively and to zoom in on sub-networks, thereby allowing users to select the results relevant to their respective use cases.

Using citation data is not the only possibility for identifying relevant legislation. Other data-science-based or AI-based approaches exist for identifying relevant legislation in an automated manner. One option is to develop an ontology that maps the relationships between laws, provisions within laws, or relationships between concepts mentioned in legal provisions. Developing an ontology, however, requires substantial human input, whereas ontology development in an automated way is an avenue that needs further exploration and experimentation. One may also search for semantically similar laws or provisions given a selected input text. Semantic similarity may not be a promising approach since it often puts emphasis on word frequency, whereas frequently used terms are unlikely to correlate with the importance of the legal topic at hand. Previous research on court cases showed that, although there is a substantial overlap, the similarity of the full texts of court decisions does not closely mirror citation behavior [32].

Another limitation of this study is that it focuses on legislation at the EU level. Applicable and relevant UAS legislation is found on the European level as well as on the national level and in standards. One could conduct a similar exercise as performed in this article in order to find applicable legislation on the Member State level or in standards. However, this does require (1) explicit references from national legislation and/or standards to EU legislation (or vice versa) and (2) the automatic detection of the references (preferably available as metadata). For instance, references in Member State law to Regulation 2019/947 are often not recognized by the reference detection software currently in place (if such software exists), which complicates the automated detection of references, in this case, to EU legislation on UASs. Moreover, reference detection software does not exist or is not applied in several countries, making automatic detection based on citation (meta)data not realistic.

Finally, mapping the norms applicable to UASs requires us to go beyond hard law. Standards complement and further specify hard law norms. Sometimes, EU legislation or national legislation may refer to a standard or a set of standards, thus integrating the technical norms from the referred-to-standard into the system of laws, consequently turning said technical standards into something that can perhaps be called *quasi-hard-law*. Examples of references to standards in EU legislation can be found in Article 14 of Regulation 2019/947, UAS.LUC.030 para. 2(a) of Regulation 2019/947, as well as Articles 6(4), 12, 13(2)(a), 22(7)(c), 23, 30(3), 36(5)(b), 37(3), and 40(4) of Regulation 2019/945. Most of the above-mentioned references are generic and do not explicitly identify the relevant standards (a noteworthy exception is Article 40(4) of Regulation 2019/945, which refers explicitly to standard ANSI/CTA-2063-A-2019, Small Unmanned Aerial Systems Serial Numbers, 2019), thus making the European norm more flexible and adaptable to technological innovation. At the same time, this complicates the automated identification of the most relevant standards. The inclusion of standards in the automated identification of UAS rules is an avenue for future research.

5. Conclusions

This study explored whether references from and to EU laws could be leveraged to identify relevant EU legislation on UASs. To analyze this, the results of a linked data and network analysis approach were compared to the laws identified by subject-matter experts, which were used as ground truth. The overlap was calculated, that is, the laws identified by the experts compared to the laws found through the citation analysis. The comparison of findings relevant to UAS legislation on the EU level by consulting experts and based on references from and to legislation showed a substantial overlap between the two approaches.

The findings suggest that various strategies may be adopted when searching for relevant legislation by means of citation analysis. One strategy is to take recent legislation as a starting point and identify potentially relevant laws by searching for references in (and perhaps to) the source law. This approach is likely to strike a balance between precision (the proportion of the search results that is relevant) and recall (the proportion of potentially relevant laws that is in the search results). Another strategy is to maximize the recall by starting with one or more laws and incrementally increase the target and/or source depth until, for the user, an acceptable network is created. The user can subsequently select the groups of laws (communities) of interest, create a subnetwork of the laws within those communities, and further inspect the laws within their respective communities. Because the laws that are relevant depend on the circumstances or the legal question at hand, the latter approach might prove useful for both legal experts and those who have some but lack in-depth legal expertise.

The results and approaches that were explored proved promising for retrieving relevant UAS legislation. Future research could take up the challenge of creating citation networks on multiple levels, for example, the EU level and the national (Member State) level. With a linked data approach, it will then become possible to connect laws on one level (e.g., EU) to all legislation on the national level. It is also possible to include other sources, such as standards, yet this requires the legislation in standards to be published in a linked data format. Finally, a citation analysis like the one presented in this article may be connected to the work that parses legislative provisions into obligations, prohibitions, permissions, IF-THEN statements, etc. This way, it will become possible to, given a set of relevant laws, extract, for instance, all obligations that apply to an operator and extract the conditions that are related to the obligation. Such a combination of approaches has the potential to benefit automated compliance checking, whether it concerns UASs or other phenomena.

Author Contributions: Conceptualization, G.v.D.; methodology, G.v.D.; software, J.S.; validation, G.v.D.; formal analysis, G.v.D. and A.-D.O.; investigation, G.v.D.; resources, R.N., A.-D.O. and G.v.D.; data curation, G.v.D.; writing—original draft preparation, G.v.D., A.-D.O. and R.N.; writing—review and editing, G.v.D., A.-D.O. and R.N.; visualization, G.v.D.; supervision, G.v.D.; funding acquisition, G.v.D. All authors have read and agreed to the published version of the manuscript.

Funding: This research was funded by the European Union H2020 Research and Innovation Program under Grant Agreement No. 101006828—Flying Forward 2020.

Data Availability Statement: The data and code are available at https://doi.org/10.34894/HLWZLL.

Acknowledgments: We thank Filippo Tomasello of EuroUSC and other project partners for their valuable input for the expert overview.

Conflicts of Interest: The authors declare no conflict of interest. The funders had no role in the design of the study; in the collection, analyses, or interpretation of data; in the writing of the manuscript, or in the decision to publish the results.

References

1. Commission Delegated Regulation (EU) 2019/945 of 12 March 2019 on Unmanned Aircraft Systems and on Third-Country Operators of Unmanned Aircraft Systems [2019] OJ L 152/1. Available online: https://eur-lex.europa.eu/legal-content/EN/TXT/?uri=CELEX%3A32019R0945 (accessed on 18 July 2023).
2. Commission Implementing Regulation (EU) 2019/947 of 24 May 2019 on the Rules and Procedures for the Operation of Unmanned Aircraft [2019] OJ L 152/45. Available online: https://eur-lex.europa.eu/legal-content/EN/TXT/?uri=CELEX%3A32019R0947 (accessed on 18 July 2023).
3. Commission Implementing Regulation (EU) 2021/664 of 22 April 2021 on a Regulatory Framework for the U-Space [2021] OJ L 139/161. Available online: https://eur-lex.europa.eu/legal-content/EN/TXT/?uri=CELEX%3A32021R0664 (accessed on 18 July 2023).
4. Commission Implementing Regulation (EU) 2021/665 of 22 April 2021 Amending Implementing Regulation (EU) 2017/373 as Regards Requirements for Providers of Air Traffic Management/Air Navigation Services and Other Air Traffic Management Network Functions in the U-Space Airspace Designated in Controlled Airspace [2021] OJ L 139/184. Available online: https://eur-lex.europa.eu/legal-content/EN/TXT/?uri=CELEX%3A32021R0665 (accessed on 18 July 2023).
5. Commission Implementing Regulation (EU) 2021/666 of 22 April 2021 Amending Regulation (EU) No 923/2012 as Regards Requirements for Manned Aviation Operating in U-space Airspace [2021] OJ L 139/187. Available online: https://eur-lex.europa.eu/legal-content/EN/TXT/?uri=CELEX%3A32021R0666 (accessed on 18 July 2023).
6. Scott, B.I. *The Law of Unmanned Aircraft Systems*; Wolters Kluwer: Alphen aan den Rijn, The Netherlands, 2022.
7. Jones, T. *International Commercial Drone Regulation and Drone Delivery Services*; RAND: Santa Monica, CA, USA, 2017.
8. MacPherson, E. Is the World Ready for Drones? *Air Space Law* **2018**, *43*, 149–178. [CrossRef]
9. Ilker, A. Regulating Commercial Drones: Bridging the Gap between American and Europe Drone Regulations. *J. Int. Bus. Law* **2015**, *15*, 313–335.
10. Mendes de Leon, P.; Scott, B.I. An analysis of Unmanned Aircraft Systems under air law. In *Drones and Unmanned Aerial Systems: Legal and Social Implications for Security and Surveillance*; Springer: Berlin/Heidelberg, Germany, 2016; pp. 185–213. [CrossRef]
11. Gregorski, M. Analysis of international law on Unmanned Aerial Vehicles through the prism of European Union law. *Przegląd Eur.* **2018**, *4*, 73–93. [CrossRef]
12. International Civil Aviation Organization. Unmanned Aircraft Systems (UAS). Available online: https://www.icao.int/meetings/uas/documents/circular%20328_en.pdf (accessed on 17 July 2023).
13. Alamouri, A.; Lampert, A.; Gerke, M. An exploratory investigation of UAS regulations in europe and the impact on effective use and economic potential. *Drones* **2021**, *5*, 63. [CrossRef]
14. Bertolini, A. Artificial Intelligence and Civil Law; Liability Rules for Drones. Available online: https://www.europarl.europa.eu/RegData/etudes/STUD/2018/608848/IPOL_STU(2018)608848_EN.pdf (accessed on 17 July 2023).
15. Konert, A.; Kotliński, M. U-Space–Civil Liability for damages caused by Unmanned Aircraft. *Transp. Res. Procedia* **2020**, *51*, 304–312. [CrossRef]
16. Harris, K.-K. Drones: Proposed standards of liability. *Santa Clara High Technol. Law J.* **2018**, *35*, 65–109.
17. Sehrawat, V. Liability issue of domestic drones. *Santa Clara High Technol. Law J.* **2018**, *35*, 110–134.
18. Bhatti, Y. Criminal Liability. In *The Law of Unmanned Aircraft Systems*; Scott, B.I., Ed.; Wolters Kluwer: Alphen aan den Rijn, The Netherlands, 2022.

19. Balcerzak, T.; Sieradzka, M. Polish Criminal Liability of the UAV Operator in Connection with the Unauthorized Flight in Operations Area of Fire Services in the Context of the Possibility of the U-Space Using. *J. Intell. Robot. Syst.* **2022**, *106*, 25. [CrossRef]
20. Beckers, A.; Teubner, G. *Three Liability Regimes for Artificial Intelligence: Algorithmic Actants, Hybrids, Crowds*; Bloomsbury Publishing: London, UK, 2021.
21. Bertolini, A.; Episcopo, F. The expert Group's report on liability for artificial Intelligence and other emerging digital technologies: A critical assessment. *Eur. J. Risk Regul.* **2021**, *12*, 644–659. [CrossRef]
22. Giuffrida, I. Liability for AI decision-making: Some legal and ethical considerations. *Fordham Law Rev.* **2019**, *88*, 439–456.
23. Mckenna, A. Privacy and Legal Enforcement. In *Unmanned Aircraft Systems*, 2nd ed.; Scott, B.I., Ed.; Wolters Kluwer: Alphen aan den Rijn, The Netherlands, 2022.
24. Șcheau, M.C.; Achim, M.V.; Găbudeanu, L.; Brici, I.; Vîlcea, A.-L. Legal, Economic and Cyber Security Framework Considerations for Drone Usage. *Appl. Sci.* **2022**, *12*, 4663. [CrossRef]
25. Vacek, J.J. The next frontier in drone law: Liability for cybersecurity negligence and data breaches for UAS operators. *Campbell Law Rev.* **2017**, *39*, 135–164.
26. Royo, P.; Asenjo, À.; Trujillo, J.; Çetin, E.; Barrado, C. Enhancing Drones for Law Enforcement and Capacity Monitoring at Open Large Events. *Drones* **2022**, *6*, 359. [CrossRef]
27. Șcheau, M.C.; Achim, M.V.; Găbudeanu, L.; Văidean, V.L.; Vîlcea, A.L.; Apetri, L. Proposals of Processes and Organizational Preventive Measures against Malfunctioning of Drones and User Negligence. *Drones* **2023**, *7*, 64. [CrossRef]
28. Hartmann, J.; Jueptner, E.; Matalonga, S.; Riordan, J.; White, S. Artificial intelligence, autonomous drones and legal uncertainties. *Eur. J. Risk Regul.* **2023**, *14*, 31–48. [CrossRef]
29. Maxwell, K.T.; Schafer, B. Concept and context in legal information retrieval. In *Legal Knowledge and Information Systems*; IOS Press: Amsterdam, The Netherlands, 2008; pp. 63–72.
30. Gelbart, D.; Smith, J. Beyond boolean search: FLEXICON, a legal tex-based intelligent system. In Proceedings of the 3rd International Conference on Artificial Intelligence and Law, Oxford, UK, 25–28 June 1991; pp. 225–234.
31. Blair, D.C.; Maron, M.E. An evaluation of retrieval effectiveness for a full-text document-retrieval system. *Commun. ACM* **1985**, *28*, 289–299. [CrossRef]
32. Moodley, K.; Serrano, P.V.H.; van Dijck, G.; Dumontier, M. Similarity and relevance of court decisions: A computational study on CJEU cases. In *Legal Knowledge and Information Systems*; IOS Press: Amsterdam, The Netherlands, 2019; pp. 63–72.
33. Mandal, A.; Chaki, R.; Saha, S.; Ghosh, K.; Pal, A.; Ghosh, S. Measuring similarity among legal court case documents. In Proceedings of the 10th Annual ACM India Compute Conference, New York, NY, USA, 16–18 November 2017; pp. 1–9.
34. Indukuri, K.V.; Ambekar, A.A.; Sureka, A. Similarity analysis of patent claims using natural language processing techniques. In Proceedings of the International Conference on Computational Intelligence and Multimedia Applications (ICCIMA 2007), Sivakasi, Tamil Nadu, India, 13–15 December 2007; pp. 169–175.
35. Zhang, Y.; Shang, L.; Huang, L.; Porter, A.L.; Zhang, G.; Lu, J.; Zhu, D. A hybrid similarity measure method for patent portfolio analysis. *J. Informetr.* **2016**, *10*, 1108–1130. [CrossRef]
36. Liu, Y.-H.; Chen, Y.-L.; Ho, W.-L. Predicting associated statutes for legal problems. *Inf. Process. Manag.* **2015**, *51*, 194–211. [CrossRef]
37. Humphreys, L.; Santos, C.; Di Caro, L.; Boella, G.; Van Der Torre, L.; Robaldo, L. Mapping Recitals to Normative Provisions in EU Legislation to Assist Legal Interpretation. In Proceedings of the JURIX, Braga, Portugal, 10–11 December 2015; pp. 41–49.
38. Gómez-Pérez, A.; Ortiz-Rodríguez, F.; Villazón-Terrazas, B. Ontology-based legal information retrieval to improve the information access in e-government. In Proceedings of the 15th International Conference on World Wide Web, Scotland, UK, 23–26 May 2006; pp. 1007–1008.
39. Dini, L.; Peters, W.; Liebwald, D.; Schweighofer, E.; Mommers, L.; Voermans, W. Cross-lingual legal information retrieval using a WordNet architecture. In Proceedings of the 10th International Conference on Artificial Intelligence and Law, Bologna, Italy, 6–11 June 2005; pp. 163–167.
40. Stojanovic, N. *Ontology-Based Information Retrieval: Methods and Tools for Cooperative Query Answering*; Karlsruhe Institute of Technology: Karlsruhe, Germany, 2005.
41. Convention on Civil Aviation ("Chicago Convention"), 7 December 1944, (1994) 15 U.N.T.S. 295. Available online: https://www.icao.int/publications/Documents/7300_cons.pdf (accessed on 17 July 2023).
42. Convention for the Unification of Certain Rules for International Carriage by Air ("Montreal Convention"), (2004) 2242 U.N.T.S. 309. Available online: https://www.icao.int/Meetings/AirCargoDevelopmentForum-Togo/Documents/9740.pdf (accessed on 17 July 2023).
43. ICAO. About ICAO. Available online: https://www.icao.int/about-icao/Pages/default.aspx (accessed on 17 July 2023).
44. Commission Implementing Regulation (EU) 2019/947 of 24 May 2019 on the Rules and Procedures for the Operation of Unmanned Aircraft [2019] OJ L 152/45, Article 15 (1) Sub 1. Available online: https://eur-lex.europa.eu/legal-content/EN/TXT/?uri=CELEX%3A32019R0947 (accessed on 18 July 2023).
45. Article 9 Besluit Vluchtuitvoering. Available online: https://wetten.overheid.nl/BWBR0020111/2020-12-31 (accessed on 18 July 2023).
46. The Map Containing the Geographical Zones for the Netherlands. Available online: https://kaart.pdok.nl/api/api.html?mapdiv=map%5Fvialink&zoom=3&geocoder=%7B%7D&loc=162379%2E23990482%2C%20459428%2E68528338&wmsurl=

https%3A%2F%2Fgeodata%2Enationaalgeoregister%2Enl%2Fdronenoflyzones%2Fwms%3F&wmslayers=luchtvaartgebieden&markersdef=https%3A%2F%2Fkaart%2Epdok%2Enl%2Fapi%2Fjs%2Fpdok%2Dmarkers%2Ejs&layersdef=https%3A%2F%2Fkaart%2Epdok%2Enl%2Fapi%2Fjs%2Fpdok%2Dlayers%2Ejs&pdoklayers=BRTPASTEL (accessed on 18 July 2023).
47. Commission Implementing Regulation (EU) 2019/947 of 24 May 2019 on the Rules and Procedures for the Operation of Unmanned Aircraft [2019] OJ L 152/45, Annex Part A. Available online: https://eur-lex.europa.eu/legal-content/EN/TXT/?uri=CELEX%3A32019R0947 (accessed on 18 July 2023).
48. Article 2 para 2 Regeling Zonering Onbemande Luchtvaartuigen. Available online: https://wetten.overheid.nl/BWBR0044597/2022-11-10 (accessed on 14 June 2023).
49. Article 2 para 1 sub b Regeling Zonering Onbemande Luchtvaartuigen. Available online: https://wetten.overheid.nl/BWBR0044597/2022-11-10 (accessed on 14 June 2023).
50. Davies, K.; Van Munster, M.; Düsterhöft, I. *Understanding European Union Law*; Taylor & Francis: Abingdon, UK, 2022.
51. Consolidated Version of the Treaty on the Functioning of the European Union (2012) OJ C 326/47. Available online: https://eur-lex.europa.eu/legal-content/EN/TXT/?uri=celex%3A12012E%2FTXT (accessed on 17 July 2023).
52. European Union. Types of Legislation. Available online: https://europa.eu/european-union/law/legal-acts_en (accessed on 17 July 2023).
53. See Article 288 (2) TFEU: "A Regulation Shall Have General Application. It Shall Be Binding in Its Entirety and Directly Applicable in All Member States". Available online: https://lexparency.org/eu/TFEU/ART_288/#:~:text=A%20regulation%20shall%20have%20general,choice%20of%20form%20and%20methods (accessed on 18 July 2023).
54. See Article 288 (3) TFEU: "A Directive Shall Be Binding, as to the Result to Be Achieved, upon Each Member State to Which It Is Addressed, But Shall Leave to the National Authorities the Choice of Form and Methods". Available online: https://eur-lex.europa.eu/legal-content/EN/TXT/HTML/?uri=CELEX%3A12008E288 (accessed on 18 July 2023).
55. Regulation (EU) 2016/679 of the European Parliament and of the Council of 27 April 2016 on the Protection of Natural Persons with Regard to the Processing of Personal Data and on the Free Movement of Such Data, and Repealing Directive 95/46/EC (General Data Protection Regulation) [2016] OJ L 119/1. Available online: https://eur-lex.europa.eu/legal-content/EN/TXT/PDF/?uri=CELEX:32016R0679 (accessed on 18 July 2023).
56. See Article 6:162 of the Dutch Civil Code. Similar Rules Exist in almost all EU Member States (e.g., § 823I of the German Civil Code, Articles 1240-1241 of the French Civil Code, Article 1382 of the Belgian Civil Code, §1295 of the Austrian Civil Code, Article 2043 of the Italian Civil Code, Article 1902 of the Spanish Civil Code, Article 1349 of the Romanian Civil Code, Article 41 of the Swiss Code of Obligations). Available online: https://www.amsadvocaten.com/practice-areas/law-of-obligations-and-contract-law/tort/ (accessed on 14 June 2023).
57. Bizer, C.; Heath, T.; Berners-Lee, T. Linked data: The story so far. In *Semantic Services, Interoperability and Web Applications: Emerging Concepts*; IGI Global: Hershey, PA, USA, 2011; pp. 205–227.
58. European Union. Data Reuse. Available online: https://eur-lex.europa.eu/content/welcome/data-reuse.html (accessed on 17 July 2023).
59. Bastian, M.; Heymann, S.; Jacomy, M. Gephi: An open source software for exploring and manipulating networks. In Proceedings of the International AAAI Conference on Web and Social Media, San Jose, CA, USA, 17–20 May 2009; pp. 361–362.
60. Regulation (EU) 2018/1139 of the European Parliament and of the Council of 4 July 2018 on Common Rules in the Field of Civil Aviation and Establishing a European Union Aviation Safety Agency, and Amending Regulations (EC) No 2111/2005, (EC) No 1008/2008, (EU) No 996/2010, (EU) No 376/2014 and Directives 2014/30/EU and 2014/53/EU of the European Parliament and of the Council, and Repealing Regulations (EC) No 552/2004 and (EC) No 216/2008 of the European Parliament and of the Council and Council Regulation (EEC) No 3922/91 [2018] OJ L 212/1. Available online: https://eur-lex.europa.eu/legal-content/EN/TXT/?uri=celex%3A32018R1139 (accessed on 18 July 2023).
61. Regulation (EU) No 996/2010 of the European Parliament and of the Council of 20 October 2010 on the Investigation and Prevention of Accidents and Incidents in Civil Aviation and Repealing Directive 94/56/EC [2010] OJ L 295/35. Available online: https://eur-lex.europa.eu/legal-content/EN/TXT/?uri=celex%3A32010R0996 (accessed on 18 July 2023).
62. Regulation (EU) No 1025/2012 of the European Parliament and of the Council of 25 October 2012 on European Standardisation, Amending Council Directives 89/686/EEC and 93/15/EEC and Directives 94/9/EC, 94/25/EC, 95/16/EC, 97/23/EC, 98/34/EC, 2004/22/EC, 2007/23/EC, 2009/23/EC and 2009/105/EC of the European Parliament and of the Council and Repealing Council Decision 87/95/EEC and Decision No 1673/2006/EC of the European Parliament and of the Council [2012] OJ L 316/12. Available online: https://eur-lex.europa.eu/legal-content/EN/TXT/?uri=CELEX%3A32012R1025 (accessed on 18 July 2023).
63. Regulation (EC) No 765/2008 of the European Parliament and of the Council of 9 July 2008 Setting Out the Requirements for Accreditation and Market Surveillance Relating to the Marketing of Products and Repealing Regulation (EEC) No 339/93 [2008] OJ L 218/30. Available online: https://eur-lex.europa.eu/legal-content/EN/TXT/?uri=CELEX%3A32008R0765 (accessed on 18 July 2023).
64. Decision No 768/2008/EC of the European Parliament and of the Council of 9 July 2008 on a Common Framework for the Marketing of Products, and Repealing Council Decision 93/465/EEC [2008] OJ L 218/82. Available online: https://eur-lex.europa.eu/legal-content/EN/TXT/?uri=celex%3A32008D0768 (accessed on 18 July 2023).

65. Directive 2006/42/EC of the European Parliament and of the Council of 17 May 2006 on Machinery, and Amending Directive 95/16/EC (recast) [2006] OJ L 157/24. Available online: https://eur-lex.europa.eu/legal-content/EN/TXT/?uri=celex%3A32006L0042 (accessed on 18 July 2023).
66. Directive 2009/48/EC of the European Parliament and of the Council of 18 June 2009 on the Safety of Toys [2009] OJ L 170/1. Available online: https://eur-lex.europa.eu/legal-content/EN/TXT/?uri=CELEX%3A32009L0048 (accessed on 18 July 2023).
67. Commission Regulation (EU) No 748/2012 of 3 August 2012 Laying down Implementing Rules for the Airworthiness and Environmental Certification of Aircraft and Related Products, Parts and Appliances, as well as for the Certification of Design and Production Organisations [2012] OJ L 224/1. Available online: https://eur-lex.europa.eu/legal-content/EN/TXT/?uri=celex%3A32012R0748 (accessed on 18 July 2023).
68. Commission Regulation (EU) 2015/640 of 23 April 2015 on Additional Airworthiness Specifications for a Given Type of Operations and Amending Regulation (EU) No 965/2012 [2015] OJ L 106/18. Available online: https://eur-lex.europa.eu/legal-content/EN/TXT/?uri=celex%3A32015R0640 (accessed on 18 July 2023).
69. Regulation (EC) No 300/2008 of the European Parliament and of the Council of 11 March 2008 on Common Rules in the Field of Civil Aviation Security and Repealing Regulation (EC) No 2320/2002 [2008] OJ L 97/72. Available online: https://eur-lex.europa.eu/legal-content/EN/TXT/?uri=celex%3A32008R0300 (accessed on 18 July 2023).
70. Regulation (EU) No 376/2014 of the European Parliament and of the Council of 3 April 2014 on the Reporting, Analysis and Follow-Up of Occurrences in Civil Aviation, Amending Regulation (EU) No 996/2010 of the European Parliament and of the Council and Repealing Directive 2003/42/EC of the European Parliament and of the Council and Commission Regulations (EC) No 1321/2007 and (EC) No 1330/2007 [2014] OJ L 122/18. Available online: https://eur-lex.europa.eu/legal-content/EN/TXT/?uri=celex%3A32014R0376 (accessed on 18 July 2023).
71. Commission Implementing Regulation (EU) No 923/2012 of 26 September 2012 Laying down the Common Rules of the Air and Operational Provisions Regarding Services and Procedures in Air Navigation and Amending Implementing Regulation (EU) No 1035/2011 and Regulations (EC) No 1265/2007, (EC) No 1794/2006, (EC) No 730/2006, (EC) No 1033/2006 and (EU) No 255/2010 [2012] OJ L 281/1. Available online: https://eur-lex.europa.eu/legal-content/EN/TXT/?uri=celex%3A32012R0923 (accessed on 18 July 2023).
72. Commission Regulation (EU) No 1178/2011 of 3 November 2011 Laying down Technical Requirements and Administrative Procedures Related to Civil Aviation Aircrew Pursuant to Regulation (EC) No 216/2008 of the European Parliament and of the Council [2011] OJ L 311/1. Available online: https://eur-lex.europa.eu/legal-content/EN/TXT/?uri=celex%3A32011R1178 (accessed on 18 July 2023).
73. Commission Regulation (EU) No 965/2012 of 5 October 2012 Laying down Technical Requirements and Administrative Procedures Related to Air Operations Pursuant to Regulation (EC) No 216/2008 of the European Parliament and of the Council [2012] OJ L 296/1. Available online: https://eur-lex.europa.eu/legal-content/EN/TXT/?uri=celex%3A32012R0965 (accessed on 18 July 2023).
74. Commission Regulation (EU) No 1321/2014 of 26 November 2014 on the Continuing Airworthiness of Aircraft and Aeronautical Products, Parts and Appliances, and on the Approval of Organisations and Personnel Involved in These Tasks [2014] OJ L 362/1. Available online: https://eur-lex.europa.eu/legal-content/EN/TXT/?uri=celex%3A32014R1321 (accessed on 18 July 2023).
75. Commission Regulation (EU) No 1332/2011 of 16 December 2011 Laying down Common Airspace Usage Requirements and Operating Procedures for Airborne Collision Avoidance [2011] OJ L 336/20. Available online: https://eur-lex.europa.eu/legal-content/EN/TXT/?uri=CELEX%3A32011R1332 (accessed on 18 July 2023).
76. Commission Implementing Regulation (EU) 2018/1048 of 18 July 2018 Laying down Airspace Usage Requirements and Operating Procedures Concerning Performance-Based Navigation [2018] OJ L 189/3. Available online: https://eur-lex.europa.eu/legal-content/EN/TXT/?uri=CELEX%3A32018R1048 (accessed on 18 July 2023).
77. Commission Implementing Regulation (EU) 2017/373 of 1 March 2017 Laying down Common Requirements for Providers of Air Traffic Management/Air Navigation Services and Other Air Traffic Management Network Functions and Their Oversight, Repealing Regulation (EC) No 482/2008, Implementing Regulations (EU) No 1034/2011, (EU) No 1035/2011 and (EU) 2016/1377 and Amending Regulation (EU) No 677/2011 [2017] OJ L 62/1. Available online: https://eur-lex.europa.eu/legal-content/EN/TXT/?uri=CELEX%3A32017R0373 (accessed on 18 July 2023).
78. Directive (EU) 2016/1148 of the European Parliament and of the Council of 6 July 2016 Concerning Measures for a High Common Level of Security of Network and Information Systems across the Union [2016] OJ L 194/1. Available online: https://eur-lex.europa.eu/eli/dir/2016/1148/oj (accessed on 18 July 2023).
79. Regulation (EU) 2019/881 of the European Parliament and of the Council of 17 April 2019 on ENISA (the European Union Agency for Cybersecurity) and on Information and Communications Technology Cybersecurity Certification and Repealing Regulation (EU) No 526/2013 (Cybersecurity Act) [2019] OJ L 151/15. Available online: https://eur-lex.europa.eu/eli/reg/2019/881 (accessed on 18 July 2023).
80. Regulation (EU) 2022/868 of the European Parliament and of the Council of 30 May 2022 on European Data Governance and Amending Regulation (EU) 2018/1724 (Data Governance Act) [2022] OJ L 152/1. Available online: https://eur-lex.europa.eu/legal-content/EN/TXT/?uri=celex%3A32022R0868 (accessed on 18 July 2023).
81. Regulation (EU) 2022/2065 of the European Parliament and of the Council of 19 October 2022 on a Single Market For Digital Services and Amending Directive 2000/31/EC (Digital Services Act) [2022] OJ L 277/1. Available online: https://eur-lex.europa.eu/legal-content/EN/TXT/?uri=celex%3A32022R2065 (accessed on 18 July 2023).

82. Proposal for a Regulation of the European Parliament and of the Council Amending Regulation (EU) No 910/2014 as Regards Establishing a Framework for a European Digital Identity COM/2021/281 (Final). Available online: https://eur-lex.europa.eu/resource.html?uri=cellar:5d88943a-c458-11eb-a925-01aa75ed71a1.0001.02/DOC_1&format=PDF (accessed on 18 July 2023).
83. Council Directive 85/374/EEC of 25 July 1985 on the Approximation of the Laws, Regulations and Administrative Provisions of the Member States Concerning Liability for Defective Products [1985] OJ L 210/29. Available online: https://eur-lex.europa.eu/legal-content/EN/TXT/?uri=celex%3A31985L0374 (accessed on 18 July 2023).
84. Regulation (EC) No 785/2004 of the European Parliament and of the Council of 21 April 2004 on Insurance Requirements for Air Carriers and Aircraft Operators [2004] OJ L 138/1. Available online: https://eur-lex.europa.eu/legal-content/EN/ALL/?uri=CELEX%3A32004R0785 (accessed on 18 July 2023).
85. Commission Regulation (EU) No 139/2014 of 12 February 2014 Laying down Requirements and Administrative Procedures Related to Aerodromes Pursuant to Regulation (EC) No 216/2008 of the European Parliament and of the Council [2014] OJ L 44/1. Available online: https://eur-lex.europa.eu/legal-content/EN/TXT/?uri=CELEX%3A32014R0139 (accessed on 18 July 2023).
86. Regulation (EC) No 549/2004 of the European Parliament and of the Council of 10 March 2004 Laying down the Framework for the Creation of the Single European Sky (the Framework Regulation) [2004] OJ L 96/1. Available online: https://eur-lex.europa.eu/legal-content/EN/ALL/?uri=CELEX%3A32004R0549 (accessed on 18 July 2023).
87. Regulation (EC) No 551/2004 of the European Parliament and of the Council of 10 March 2004 on the Organisation and Use of the Airspace in the Single European Sky (the Airspace Regulation) [2004] OJ L 96/20. Available online: https://eur-lex.europa.eu/legal-content/en/ALL/?uri=CELEX%3A32004R0551 (accessed on 18 July 2023).
88. Commission Regulation (EU) 2015/340 of 20 February 2015 Laying down Technical Requirements and Administrative Procedures Relating to Air Traffic Controllers' Licences and Certificates Pursuant to Regulation (EC) No 216/2008 of the European Parliament and of the Council, Amending Commission Implementing Regulation (EU) No 923/2012 and Repealing Commission Regulation (EU) No 805/2011 [2015] OJ L 63/1. Available online: https://eur-lex.europa.eu/legal-content/en/ALL/?uri=CELEX%3A32015R0340 (accessed on 18 July 2023).
89. Directive 2014/53/EU of the European Parliament and of the Council of 16 April 2014 on the Harmonisation of the Laws of the Member States Relating to the Making Available on the Market of Radio Equipment and Repealing Directive 1999/5/EC [2014] OJ L 153/62. Available online: https://eur-lex.europa.eu/legal-content/EN/TXT/?uri=CELEX:32014L0053 (accessed on 18 July 2023).
90. Directive 2001/95/EC of the European Parliament and of the Council of 3 December 2001 on General Product Safety [2002] OJ L 11/4. Available online: https://eur-lex.europa.eu/legal-content/EN/ALL/?uri=celex%3A32001L0095 (accessed on 18 July 2023).
91. Directive 2000/31/EC of the European Parliament and of the Council of 8 June 2000 on Certain Legal Aspects of Information Society Services, in Particular Electronic Commerce, in the Internal Market ('Directive on Electronic Commerce') [2000] OJ L 178/1. Available online: https://eur-lex.europa.eu/legal-content/EN/ALL/?uri=celex%3A32000L0031 (accessed on 18 July 2023).
92. Directive 2014/30/EU of the European Parliament and of the Council of 26 February 2014 on the Harmonisation of the Laws of the Member States Relating to Electromagnetic Compatibility (Recast) [2014] OJ L 96/79. Available online: https://eur-lex.europa.eu/legal-content/EN/TXT/?uri=celex%3A32014L0030 (accessed on 18 July 2023).

logistics

MDPI

Article

Transport Task Models with Variable Supplier Availabilities

Julian Vasilev [1,*], Rosen Nikolaev [2] and Tanka Milkova [2,*]

1 Department of Informatics, University of Economics Varna, 9002 Varna, Bulgaria
2 Department of Statistics and Applied Mathematics, University of Economics Varna, 9002 Varna, Bulgaria; nikolaev_rosen@ue-varna.bg
* Correspondence: vasilev@ue-varna.bg (J.V.); tankamilkova@ue-varna.bg (T.M.)

Abstract: *Background*: With regard to the definition of an optimal transport plan for some material flow in the logistics system in literature and practice, the classical transport task model is developed. The minimization of total transport costs is usually considered for optimality. Some modifications to the classical transport task have also been developed. *Methods*: The article uses the methods of linear optimization. Based on these methods, two modified transport task models have been constructed, which consider the possibility of planning in advance the quantities available from the suppliers of the transported cargo. These models are applicable in SCM for pharmaceuticals with a national logistics hub. Furthermore, a solver in MS Excel is used to determine the optimal solution of optimization models. *Results*: Two new (modified, extended) models of the transport task have been constructed, in which a preliminary planning of the available quantities of the transported cargo at the suppliers is made. These quantities shall be planned in such a way as to ensure a minimum total transport cost. *Conclusions*: By applying the proposed new transport task models, lower total transport costs for carrying out imported pharmaceuticals can be ensured compared to an application of the classical transport task model.

Keywords: optimization; transportation problems; logistics system; national logistics hub

Citation: Vasilev, J.; Nikolaev, R.; Milkova, T. Transport Task Models with Variable Supplier Availabilities. *Logistics* **2023**, *7*, 45. https://doi.org/10.3390/logistics7030045

Academic Editors: Tomasz Nowakowski, Artur Kierzkowski, Agnieszka A. Tubis, Franciszek Restel, Tomasz Kisiel, Anna Jodejko-Pietruczuk and Mateusz Zając

Received: 10 May 2023
Revised: 7 July 2023
Accepted: 26 July 2023
Published: 27 July 2023

1. Introduction

One of the main tasks of any rationally functioning business organization is related to the need to make efforts to preserve its competitive advantages and increase the efficiency of its operation. In order to achieve these goals, it is necessary to seek and apply different opportunities for the optimal implementation of all activities. One major activity associated with significant financial costs is the need for the physical movement of material flows. This necessitates the application of various science-based methods and approaches to determine optimal management decisions.

A number of methods and approaches are available in the specialized literature, leading to optimal solutions regarding the transport of material flows [1], but the specific features of each specific economic system necessitate the continuous adaptation and further development of existing models and methods of optimization. One of the classic methods related to providing opportunities for optimizing transport activities is the famous model of transport tasks. Numerous modifications of this model are also known in the scientific literature.

Nowadays is the post-COVID-19 period [2,3]. Many supply chains are modified. Many air carriers transformed their passenger airplanes into cargo ones. Late supplies from Eastern and Central Asia to Europe are often practiced in many industries. Theories concerning Vendor management inventories (VMI), stock replenishment, and supply chain management (SCM) helped the logistics managers find adequate solutions to the recent changes in supply chains. These theories and good logistics practices were useful before and during COVID-19, but nowadays the search for new theories and good practices for inventory movement (and transportation) and inventory replenishment is needed. That is why the authors of this article, on the basis of their professional knowledge in mathematics (especially in the sphere of the transport task models), from one side, and their direct observations in the sphere of retail with the pharmaceutical industry, from another side, have created two new models for the transport task. The novelty of these models is that they are not only theoretical but also practical for other retail businesses. The novelty of the proposed model gives the possibility of extending existing transportation models and making them adequate to the up-to-date reality of SCM in pharmaceuticals.

The structure of the paper is according to the well-known IMRAD format. The introduction outlines the current state of supply chains and the need for new transportation models. The literature review discusses the current state of the transport model and its extensions. The section "Methods and Research Designs" gives a formal description of the created models. The "Results" section presents the results and shows some comparisons of the performance of the created models compared to the classical transport task model. The conclusion section summarizes the findings, outlines future research, and gives practical implications.

2. Literature Review

The most well-known is the classical model of the transport task [4–6], the purpose of which is to determine an optimal plan of transport at certain starting points with known quantities of stocks in them and certain reception points with known quantities of demand. As a criterion for optimality in the transport task, minimization of total transport costs is usually considered [7–9], i.e., at certain transport costs from all points of departure [10,11] to all reception points, it is necessary to define a transport plan [12] that satisfies the stated needs of the receiving points with the available quantities of stocks at the points of departure at minimum total transport costs [13–15]. This model is known as the transport problem by criterion value [16–18]. Some modifications of the transport task by criterion value are also well known [7,19], taking into account the specifics of transport activities. For example, when it is impossible to carry out direct carriage of goods between a specific point of departure and a specific receiving point, it is transferred to the so-called Freight Point—transport task with blocked transports. Another example of a modification of the transport task by criterion value is the two-stage transport task, which is passed in cases where it is necessary for transported goods to pass through intermediate warehouses before reaching the point of departure to the receiving point.

In some cases, the nature of the goods carried and the operations carried out with them could require the distribution of cargo between points of departure and reception to take place as soon as possible. In such cases, it is necessary, in addition to the points of departure with their available quantities and the points of destination with their stated needs, to allow time for moving the cargo from each point of departure to each receiving point. Furthermore, the optimality criterion "minimum total transport costs" is replaced by the optimality criterion "minimum transport execution time", and the model [20] thus obtained is called the transport problem by time criterion. The time taken for the completion of the entire plan of carriage shall be taken as the greatest time of movement from a point of departure to a point of destination where carriage is to be carried out. To find the optimal solutions to the problems thus set, well-known methods of linear optimization are used.

Solving a transport task considering only one factor—transport costs or transport execution time—is not always justified [21]. Modification of the transport task with simultaneous consideration of the time and cost of plan implementation are known.

A major limitation of the classical transport task is that a homogeneous load (one type of cargo) is carried. In view of this, modifications to a multiproduct transport task that provides minimum total transport costs subject to certain restrictive conditions in the process of delivering several different types of products from one supplier to multiple users are also proposed in the literature [22–24].

In all modifications of a transport task (by criterion, total costs) that are described, the availability of the suppliers and the needs of the users [25] have been determined in advance. On the basis of these stocks and stated needs, an optimal transport plan is determined, which ensures a minimum total cost of transportation.

The previous studies do not cover modified versions of the classical transportation task where the SCM is with a national logistics hub (serving as a delivery point for imported pharmaceuticals), several suppliers, and several corporate customers, where the quantities in the suppliers may vary, and where additional restrictions to the quantities in the suppliers are present.

In this paper, the authors aim to present a modification of a transport task in which preplanning of the suppliers' availabilities is carried out. New constraint conditions are introduced that provide inventory planning so as to obtain smaller total transport costs compared to the classical transport task model [26–29]. The constructed transport task models can be easily solved using suitable, specially developed software products [30–33]. Furthermore, corporate KPIs may be improved [34–37].

3. Methods and Research Design

3.1. Existing Transport Task Models

The classical model of a transport problem is used as a basis for creating the new models. The formal description of the classical model of a transport problem is, in general, as follows:

m, at starting points $A_1, A_2, \ldots, A_m$ (called suppliers), has some homogeneous products, respectively in quantities $a_1, a_2, \ldots, a_m$ (referred to as stocks). These items must be delivered to n points $B_1, B_2, \ldots, B_n$ (called consumers), respectively, in quantities $b_1, b_2, \ldots, b_n$ (called needs). It is assumed that the value of transport costs $c_{ij} \geq 0$ $(i = 1,2,\ldots,m;\ j = 1,2,\ldots,n)$ for the transport of a unit of cargo (goods) by the supplier A_i is known to the consumer B_j. To define the classical transport task, a number of assumptions, detailed in the specialized literature, should be considered, mainly the assumption that there is a balance between stock quantities and needs, i.e.,

$$\sum_{i=1}^{m} a_i = \sum_{j=1}^{n} b_j.$$

For the construction of the economic-mathematical model of the transport problem, variables $x_{ij}(i = 1,2,\ldots,m;\ j = 1,2,\ldots,n)$ are introduced. They represent the unknown quantity of product to be transported from the point of departure $A_i(i = 1,2,\ldots,m)$ to the final point B_j $(j = 1,2,\ldots,n)$.

The purpose of the classical transport model is to minimize total transportation costs (Z).

The transport task model can be written in the form of final sums as follows:

$$\min Z(x_{ij}) = \sum_{i=1}^{m}\sum_{j=1}^{n} c_{ij}x_{ij}$$

under restrictive conditions:

$$\sum_{j=1}^{n} x_{ij} = a_i(i = 1,2,\ldots,m)$$

$$\sum_{i=1}^{m} x_{ij} = b_j(j = 1,2,\ldots,n)$$

$$x_{ij} \geq 0\ (i = 1,2,\ldots,m;\ j = 1,2,\ldots,n).$$

We will not dwell on methods of solving a transportation problem. Appropriate software products are used to determine the optimal solutions in the article.

We will consider the following Example 1. An optimal plan for the distribution of cargo between suppliers and users should be defined, and the following are known:

Three suppliers of one cargo are available, which have respective stocks of 300, 300, and 400 units (a_1, a_2, a_3). Four users are available who have requested needs of 200, 210, 250, and 340 cargo units, respectively (b_1, b_2, b_3, b_4). Between each supplier and user there is a transport link, and the transport costs c_{ij} $(i = 1,2,3;\ j = 1,2,3,4)$ for the carriage of one unit of cargo from the *i*-th supplier to the *j*-th user are given in matrix C:

$$C = \|c_{ij}\| = \begin{Vmatrix} 2 & 6 & 4 & 2 \\ 3 & 1 & 5 & 2 \\ 5 & 4 & 2 & 3 \end{Vmatrix}$$

After entering the variables $x_{ij} \geq 0 (i = 1,2,3;\ j = 1,2,3,4)$ the task model takes the form:

$$\begin{aligned} min : Z(x_{ij}) &= 2x_{11} + 6x_{12} + 4x_{13} + 2x_{14} \\ &+3x_{21} + x_{22} + 5x_{23} + 2x_{24} \\ &+5x_{31} + 4x_{32} + 2x_{33} + 3x_{34} \end{aligned}$$

under restrictive conditions:

$$\begin{aligned} x_{11} + x_{12} + x_{13} + x_{14} &= 300 \\ x_{21} + x_{22} + x_{23} + x_{24} &= 300 \\ x_{31} + x_{32} + x_{33} + x_{34} &= 400 \\ x_{11} + x_{21} + x_{31} &= 200 \\ x_{12} + x_{22} + x_{32} &= 210 \\ x_{13} + x_{23} + x_{33} &= 250 \\ x_{14} + x_{24} + x_{34} &= 340 \\ x_{ij} \geq 0\ (i = 1,2,3;\ j &= 1,2,3,4) \end{aligned}$$

The instrument Solver in MS Excel is used to determine an optimal solution. The optimal solution for this model is the following:

$$X^1_{opt.} = \left\|x^1_{ij}\right\| = \begin{Vmatrix} 200 & 0 & 0 & 100 \\ 0 & 210 & 0 & 90 \\ 0 & 0 & 250 & 150 \end{Vmatrix},\ Z\left(X^1_{opt.}\right) = 1940.$$

According to this task plan, the first vendor should deliver 200 units to the first user and 100 units to the fourth user; the second supplier must deliver 210 units to the second user and 90 units to the fourth user; and the third supplier must deliver 250 units to the third user and 150 units to the fourth user. Under this transportation plan, the minimum total transportation costs of 1940 are obtained.

The classical model of the transport task assumes that the quantities at suppliers and the quantities at consumers are known. The real situation in business is sometimes different. Let us take, for example, the pharmaceutical business when importing medicines from abroad. All imported medicines come to a national logistics hub. Afterwards, these quantities are distributed to all suppliers. All suppliers move the stocks to consumers (usually corporate customers). This is a classical supply chain in the pharmaceutical business. The classical model of the transport task (with formal description in Example 1) serves partially the decision support process because it assumes known quantities at suppliers and known needed quantities at consumers. The real situation with SCM in pharmaceutical business with a national logistics hub opens the need for extending the classical transport model. Since we have a national logistics hub, we may have different distributions of stock quantities among suppliers. In this case, a different distribution of stock quantities may lead to lower total transportation costs (TTC). To illustrate this idea, Example 2 is created where the stock quantities at suppliers are different from Example 1, but the total quantity of the stock (at suppliers) is the same as in Example 1. The TTC in Example 1 is 1940. The TTC in Example 2 is 1840.

We will consider Example 2, which differs from Example 1 only in the quantities of cargo available at each of the suppliers. Let it be known that each of the three suppliers has 300, 400, and 300 units, respectively.

After entering the variables $x_{ij} \geq 0$ $(i = 1,2,3;\ j = 1,2,3,4)$ the task model in this case takes the type:

$$\begin{aligned} min: Z\left(x_{ij}\right) &= 2x_{11} + 6x_{12} + 4x_{13} + 2x_{14} \\ &+3x_{21} + x_{22} + 5x_{23} + 2x_{24} \\ &+5x_{31} + 4x_{32} + 2x_{33} + 3x_{34} \end{aligned}$$

under restrictive conditions:

$$\begin{aligned} x_{11} + x_{12} + x_{13} + x_{14} &= 300 \\ x_{21} + x_{22} + x_{23} + x_{24} &= 400 \\ x_{31} + x_{32} + x_{33} + x_{34} &= 300 \\ x_{11} + x_{21} + x_{31} &= 200 \\ x_{12} + x_{22} + x_{32} &= 210 \\ x_{13} + x_{23} + x_{33} &= 250 \\ x_{14} + x_{24} + x_{34} &= 340 \\ x_{ij} \geq 0\ (i = 1,2,3;\ j &= 1,2,3,4) \end{aligned}$$

The Solver instrument in MS Excel is used again to determine an optimal solution. The optimal solution for this model is the following:

$$X^2_{opt.} = \left\|x^2_{ij}\right\| = \left\|\begin{matrix} 200 & 0 & 0 & 100 \\ 0 & 210 & 0 & 190 \\ 0 & 0 & 250 & 50 \end{matrix}\right\|,\ Z\left(X^2_{opt.}\right) = 1840.$$

According to this task plan, the first vendor (supplier) should deliver 200 units to the first user (consumer) and 100 units to the fourth user; the second supplier must deliver 210 units to the second user and 190 units to the fourth user; and the third supplier must deliver 250 units to the third user and 50 units to the fourth user. This transportation plan results in a minimum total transportation cost of 1840.

In the second variant of the task (Example 2), smaller total costs are obtained than those obtained in the first option (Example 1). This shows that the distribution of stocks among suppliers' warehouses results in lowering the total transport costs.

3.2. Formal Description of the New Transport Task Model 1

In view of this, in cases where it is permissible to plan in advance the stock of the cargo at the suppliers, we propose the following modified model of the classical transport task: What is specific about this new model 1 is that the quantities available in suppliers' warehouses are unknown. This new model 1 (Figure 1) is applicable when having a national logistics hub in the pharmaceutics business (when importing medicines from abroad) with known quantities in the logistics hub but unknown quantities in suppliers and unknown quantities to be transported (from each supplier to each corporate customer).

The classical transportation task model assumes that the stock quantities at suppliers are known. However, the situation in SCM with pharmaceutical products and a supply chain with a national logistics hub and downstream partners (suppliers and customers) needs a formal description of an extended transportation model. In this work, it is called "new transportation task model 1". Its formal description is as follows:

From m the starting point $A_1, A_2, \ldots, A_m$ (referred to as suppliers), some unknown quantities of homogeneous products must be supplied to several consumers (corporate customers). However, the total sum of these unknown quantities by suppliers is known. This is available at the national logistics hub. These unknown quantities are denoted by $a_1, a_2, \ldots, a_m$ (They are called inventories, stocks, freight, products, goods, and cargo).

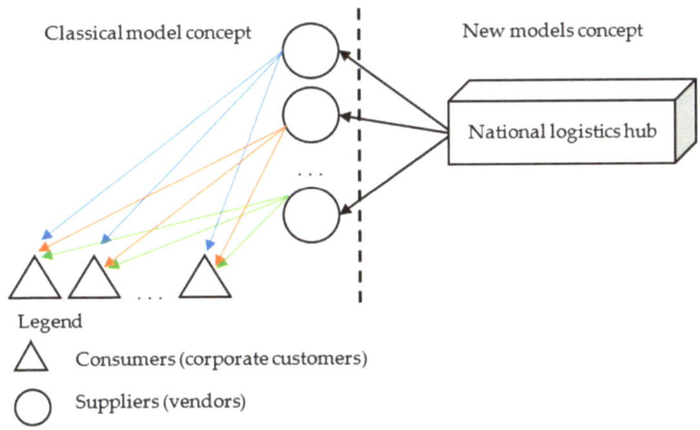

Figure 1. The concept of the new two models and the classical transportation model.

One of the assumptions of the classical transport model is that the available quantities at suppliers $(a_1, a_2, \ldots, a_m)$ are known. They are constants in the classical transport model. The new transportation task model 1 assumes that the available quantities at suppliers $(a_1, a_2, \ldots, a_m)$ are unknown. The new transportation task cannot be solved by a classic model because the classical model assumes that the available quantities at suppliers $(a_1, a_2, \ldots, a_m)$ are known, but in actual business practice with a logistics hub (the new transport task), these quantities are unknown. Each transport task has restrictions. The proposed model has all the limitations of the classical transportation task.

There are no restrictions on the quantities that can be provided in advance to each supplier. This product must be delivered to n points $B_1, B_2, \ldots, B_n$ (called consumers, users, and corporate customers), respectively, in quantities $b_1, b_2, \ldots, b_n$ (called needs). It is assumed that the value of transport costs $c_{ij} \geq 0$ $(i = 1, 2, \ldots, m;\ j = 1, 2, \ldots, n)$ for the transport of a unit of cargo (goods) by the supplier A_i is known to the user B_j. As in the classical transport problem, a number of assumptions must be considered, mainly the assumption that there is a balance between stocks and needs, i.e.,

$$\sum_{i=1}^{m} a_i = \sum_{j=1}^{n} b_j.$$

For the construction of the mathematical model of the transport problem, variables $x_{ij} (i = 1, 2, \ldots, m;\ j = 1, 2, \ldots, n)$ are introduced, representing the unknown quantity of product to be transported from the point of departure $A_i (i = 1, 2, \ldots, m)$ to the final point $B_j (j = 1, 2, \ldots, n)$. The new transport task model 1 can be written in the form of final sums as follows:

$$\min Z(x_{ij}) = \sum_{i=1}^{m} \sum_{j=1}^{n} c_{ij} x_{ij} \tag{1}$$

under restrictive conditions:

$$\sum_{j=1}^{n} x_{ij} = a_i \ (i = 1, 2, \ldots, m), \tag{2}$$

$$\sum_{i=1}^{m} x_{ij} = b_j \ (j = 1, 2, \ldots, n), \tag{3}$$

$$\sum_{i=1}^{m} a_i = \sum_{j=1}^{n} b_j, \tag{4}$$

$$a_i \geq 0,\ a_i \leq \sum_{j=1}^{n} b_j\ (i = 1,2,\ldots,m), \tag{5}$$

$$x_{ij} \geq 0\ (i = 1,2,\ldots,m;\ j = 1,2,\ldots,n). \tag{6}$$

The target function (1) expresses the total transport costs of carrying out the transport of quantities $x_{ij}(i = 1,2,\ldots,m;\ j = 1,2,\ldots,n)$ from every supplier to every consumer. The optimal solution is the minimum of the total transport costs (min Z). Restrictive conditions (2) ensure that the transport plan will be adapted to the availability of each of the suppliers, i.e., that the sum of the quantities transported by each supplier to consumers will be equal to its availability. Here we will emphasize once again that the availabilities $a_1, a_2, \ldots, a_m$ are also unknown quantities, but their sum is known—the availability in the national logistics hub. The quantities in the suppliers and the transported quantities from suppliers to consumers should be planned so as to ensure minimal overall transport costs. We assume that there are no restrictions on the capacity of each of the warehouses, i.e., that it is possible to deliver the entire quantity of the product only to a part of the suppliers. In this way, it is possible for some of the suppliers to receive zero amounts of the product. Warehouse capacity restrictions may be for minimal and/or maximum quantities.

Restrictive conditions (3) ensure that the stated needs of each of the users will be met, i.e., that each user will be delivered $b_1, b_2, \ldots, b_n$ units of the product, respectively.

According to the restrictive condition (4), stocks in suppliers will be planned in such a way that the sum of all stocks is equal to the total needs requested by consumers.

The constraint conditions (5) and (6) ensure that the variables in the model are non-negative and that no supplier receives more than the total stated product needs.

The model thus obtained is linear and can again be solved using the methods of linear optimization.

3.3. *Numeric Example of the New Transport Task Model 1*

To show the application of the new transportation model 1 (1)–(6), the following Example 3 is created: We use sample data again from Example 1, but with the condition that the quantities of the product in the warehouses of the suppliers are not known. The total quantity in the national logistics hub is known (1000 units). There are four users who have requested 200, 210, 250, and 340 units of product, respectively. These quantities of the product must be supplied by three suppliers.

Between each supplier and user there is a transport link, and the transport costs $c_{ij}(i = 1,2,3;\ j = 1,2,3,4)$ for the carriage of one unit of cargo from the i-th supplier to the j-th user are given in matrix C:

$$C = \|c_{ij}\| = \begin{Vmatrix} 2 & 6 & 4 & 2 \\ 3 & 1 & 5 & 2 \\ 5 & 4 & 2 & 3 \end{Vmatrix}.$$

It must be determined how many quantities of the product are to be delivered to the suppliers' warehouses so that, after determining a transport plan, a minimum total transport cost is obtained compared to any other option for planning availability with suppliers.

We introduce the variables $a_1 \geq 0,\ a_2 \geq 0,\ a_3 \geq 0$, which express the availabilities of each of the three suppliers. In addition, it is necessary $a_i \leq 1000\ (i = 1,2,3)$, where 1000 is the total availability in the national logistics hub.

After entering the variables $x_{ij} \geq 0\ (i = 1,2,3;\ j = 1,2,3,4)$ expressing the quantities that each supplier must transport to each user, the task model takes the form:

$$\begin{aligned} min : Z(x_{ij}) &= 2x_{11} + 6x_{12} + 4x_{13} + 2x_{14} \\ &+3x_{21} + x_{22} + 5x_{23} + 2x_{24} \\ &+5x_{31} + 4x_{32} + 2x_{33} + 3x_{34} \end{aligned}$$

under restrictive conditions:

$$\begin{gathered}x_{11}+x_{12}+x_{13}+x_{14}=a_1\\x_{21}+x_{22}+x_{23}+x_{24}=a_2\\x_{31}+x_{32}+x_{33}+x_{34}=a_3\\x_{11}+x_{21}+x_{31}=200\\x_{12}+x_{22}+x_{32}=210\\x_{13}+x_{23}+x_{33}=250\\x_{14}+x_{24}+x_{34}=340\\a_1+a_2+a_3=1000\\a_1\geq 0,\ a_2\geq 0,\ a_3\geq 0\\x_{ij}\geq 0\ (i=1,2,3;\ j=1,2,3,4)\end{gathered}$$

The Solver instrument in MS Excel is used to determine an optimal solution. The optimal solution for this new model is the following:

$$a_1=200,\ a_2=550,\ a_3=250$$

$$X^3_{opt.}=\left\|x^3_{ij}\right\|=\begin{Vmatrix}200&0&0&0\\0&210&0&340\\0&0&250&0\end{Vmatrix},\ Z\left(X^3_{opt.}\right)=1790.$$

From the optimal solution, it can be seen that with the needs of the users (consumers) as stated and the known transport costs, the possible minimum total transport costs for the implementation of the transport plan are 1790. They are obtained when 200 units of the product are provided by the national logistics hub to the first supplier, 550 units of the product are sent from the national logistics hub to the second supplier, and 250 units of the product are sent from the national logistics hub to the third supplier. Following the downstream partners of the supply chain with pharmaceuticals, the following allocations must then be made: the first supplier must deliver 200 units to the first user; the second supplier must deliver 210 units to the second user and 340 units to the fourth user; and the third supplier must deliver 250 units to a third user.

The transport costs of transporting one unit of the product from a specific supplier to each of the users depend on a number of factors, e.g., the distance between the two locations. This means that it is possible that one of the suppliers (labeled i^1) has higher transport costs for all users compared to another provider (labeled with i^2). Moreover, there will be the following dependence: $c_{i^1j}\geq c_{i^2j}$ $(j=1,2,\ldots,n)$. It can also be said that a i^2 supplier "dominates" a supplier i^1, i.e., provides less transport costs to each of the users. In such a situation, it will happen that in the optimal plan of the supplier i^1 will be set at zero amount of the product, i.e., $a_{i1}=0$. Thus, naturally, the transport task model 1 will show that a supplier must be excluded from the transport plan. It is possible, of course, that other conditions may require each of the suppliers to receive some amount of the product. This can be ensured by imposing additional constraints in transport task model 1.

3.4. Formal Description of the New Transport Task Model 2 with Additional Constraints

Depending on the particular conditions of the economic situation itself, other constraints may be added to the restrictive conditions of models (1)–(6). For example, there may be a limit on the amount of product that can be found at each of the suppliers. This condition may be imposed in terms of the supplier's warehouse capacity. Furthermore, a new group of restrictive conditions can be added to model 1. It may be assumed that in each warehouse there may be not less than $a_1^{min},a_2^{min},\ldots,a_m^{min}$ units of the product and not more than $a_1^{max},a_2^{max},\ldots,a_m^{max}$ units of the product. Furthermore, to model 1 (1)–(6), we add a new group of restrictive conditions (7), and model 2 is created:

$$a_i^{min}\leq a_i\leq a_i^{max}\ (i=1,2,\ldots,m). \tag{7}$$

These restrictive conditions ensure that the availability of each of the suppliers will lie within the specified limits between a_i^{min} and $a_i^{max} (i = 1, 2, \ldots, m)$.

3.5. Numeric Example of the New Transport Task Model 2 with Additional Constraints

To illustrate the new transport task model 2, a new numeric example is created (Example 4).

Let us look at another primer (Example 4). We will use all the sample data from Example 3 by imposing one group of additional restrictive conditions. Let the stocks in the warehouses of each of the suppliers be within the following limits:

$$200 \leq a_1 \leq 400,\ 200 \leq a_2 \leq 400,\ 200 \leq a_3 \leq 400$$

After adding these constraint conditions, the model is again solved using the Solver tool in MS Excel. The optimal solution under these conditions is:

$$a_1 = 400,\ a_2 = 350,\ a_3 = 250$$
$$X_{opt.}^4 = \left\| x_{ij}^4 \right\| = \begin{Vmatrix} 200 & 0 & 0 & 200 \\ 0 & 210 & 0 & 140 \\ 0 & 0 & 250 & 0 \end{Vmatrix},\ Z\left(X_{opt.}^4\right) = 1790.$$

From the optimal solution thus obtained, it can be seen that, again, the possible minimum total transport costs for the realization of the transport plan are 1790. They are obtained if 400 units of the product are provided as the availability of the first supplier (products sent from the national logistics hub to the first supplier), 350 units of the product as the availability of the second supplier, and 250 units of the product as the availability of the third supplier. Following the supply chain, the following distribution should then be made: the first supplier should deliver 200 units to the first user and 200 units to the fourth user; the second supplier must deliver 210 units to the second user and 140 units to the fourth user; and the third supplier must deliver 250 units to the third user.

We can notice that the minimum total costs of the plan $X_{opt.}^4$ (when there are additional restrictions on the quantities of stocks in the suppliers) coincide with those of the plan $X_{opt.}^3$ (when there are no additional restrictions).

However, this will not always be the case in every business. In general, the imposition of additional restrictions in the model may lead to higher overall costs for the implementation of the transport plan compared to those of the plan without restrictions. For example, if we consider the following additional restrictions (Example 5):

$$200 \leq a_1 \leq 250,\ 200 \leq a_2 \leq 450,\ 200 \leq a_3 \leq 450$$

The optimal solution under these conditions is:

$$a_1 = 250,\ a_2 = 450,\ a_3 = 300$$
$$X_{opt.}^5 = \left\| x_{ij}^5 \right\| = \begin{Vmatrix} 200 & 0 & 0 & 50 \\ 0 & 210 & 0 & 240 \\ 0 & 0 & 250 & 50 \end{Vmatrix},\ Z\left(X_{opt.}^5\right) = 1840.$$

It can be seen that in this case, the minimum total transport costs for the realization of the transport plan are 1840. They are obtained if 250 units of the product are provided as availability by the first supplier, 450 units of the product as availability by the second supplier, and 300 units of the product as availability by the third supplier. The following allocation must then be made: the first supplier must deliver 200 units to the first user and 50 units to the fourth user; the second supplier must deliver 210 units to the second user and 240 units to the fourth user; and the third supplier must deliver 250 units to the third user and 50 units to the fourth user.

4. Results

The paper discusses the known (classical) transport task model, in which the availability of suppliers and the needs of users for a particular product are known in advance. Transport costs from each supplier to each user (consumer) are also known, and an optimal transport plan must be determined on this basis. The classical model of the transport task (with formal description in Example 1) partially serves the decision support process in supply chains with a national logistics hub, several suppliers, and several corporate customers of pharmaceutical products. It assumes known quantities at suppliers and known quantities needed at consumers. The SCM with pharmaceutical products and a national logistics hub open up the need for extending the classical transport model. With a national logistics hub, different distributions of stock quantities among suppliers may exist. In this case, a specific distribution of stock quantities may lead to lower total transportation costs (TTC). To illustrate this idea, Example 2 is created where the stock quantities at suppliers are different from Example 1, but the total quantity of the stock (at suppliers) is the same as in Example 1. In Example 2, the TTC is lower than in Example 1.

On this basis, a new (modified) model 1 of the transport task is constructed, in which a preliminary planning of the available quantities of the transported cargo at the suppliers is made. On the basis of sample data (Example 3), it is shown that through this model it is possible to obtain lower total transport costs for carrying out the transport compared to the classical model of the transport task (Example 1).

The numeric example (Example 3) of the transport task model 1 is designed to check the validity and applicability of the created transport task model 1. The numeric example (Example 4) of the transport task model 2 is designed to check the validity and applicability of the created transport task model 2. Both models have a formal description that allows their adaptation in practice—as part of a software system or as part of a specific calculator in Excel. Both models are adequate for the new SCM reality with a national logistics hub. The obtained results show better corporate logistics in terms of inventory management at the supplier level. The new model 1 assumes that each supplier has an unlimited warehouse. However, in practice, there are limitations. That is why the new model 2 was created. In the new model 2, each supplier has warehouse limitations (minimal and/or maximum quantity). In this case, the newly created model 2 formally describes the real situation in business with importing pharmaceuticals into a national logistics hub and their further distribution in the supply chain with suppliers and customers.

The possibility of adding additional constraint conditions is created by the creation of the new transport task model 2. It describes the existence of constraints in the capacity of suppliers. Two examples (Example 4 and Example 5) are created on the basis of the formal description of the new model 2. In some business cases, additional restrictions in the model may lead to higher overall costs for the implementation of the transport plan compared to those of the plan without restrictions. The proposed two new models have formal descriptions and numerical examples.

The new models do not change the algorithm for calculating the target function (linear optimization). The direct economic effect of the two new models created was acquiring lower total transportation costs compared to the classical model by planning quantities at suppliers and adding restrictions for these quantities. The indirect effect of the two models created is their better formal description of the real business situation in SCM with pharmaceuticals through a national logistics hub and a network of suppliers and corporate consumers.

5. Conclusions

Two new (modified) transport task models are proposed in this paper by which lower total transport costs can be provided to perform the transports compared to an application of the classical (baseline) transport task model. The classical model of the transport task assumes that the quantities at suppliers and the quantities at consumers are known. The pharmaceutical business, when importing medicines from abroad into a national logistics hub, opens up the need for extending the classical transportation model. That is why two new models have been created. The newly created model 1 has known quantities in the logistics hub, unknown quantities in suppliers, and unknown quantities to be transported. The newly created model 2 has known quantities in the logistics hub but unknown quantities at suppliers, unknown quantities to be transported, and restrictions on the quantities at suppliers. The newly created formal descriptions of models 1 and 2 allow their implementation in practice and in ERP systems. The relevant numeric examples prove their efficiency compared to the classical transportation model. Moreover, the two new models better fit the real situation in the pharmaceutical business.

It should not be forgotten that the model of a transport problem, like other mathematical models in economics, is applicable subject to certain conditions and the presence of imposed environmental restrictions. Nevertheless, the application of this model can lead to better economic results in moving material flow from suppliers to consumers. The constructed transport task models can be easily solved using suitable, specially developed software products. Businessmen and practitioners may easily adapt these new models in practice.

The two new models do not consider the costs of transportation between the national logistics hub and the suppliers. Future research may address extending the two new models with a formal description of the costs of transportation between the national logistics hub and the suppliers. In this case, both models will better fit the SCM in the pharmaceutical business.

Author Contributions: All authors contributed equally to the paper. All authors have read and agreed to the published version of the manuscript.

Funding: This research received no external funding.

Data Availability Statement: Not applicable.

Conflicts of Interest: The authors declare no conflict of interest.

References

1. Tarasyev, A.M.; Turygina, V.F. Application of Material Flow Analysis in the Industry. In Proceedings of the Ecology, Economics, Education and Legislation, Albena, Bulgaria, 18–24 June 2015; Stef92 Technology Ltd.: Sofia, Bulgaria, 2015; Volume III, pp. 71–76.
2. Georgescu, M.R.; Stoica, E.A.; Bogoslov, I.A.; Lungu, A.E. Managing Efficiency in Digital Transformation-EU Member States Performance during the COVID-19 Pandemic. *Procedia Comput. Sci.* **2022**, *204*, 432–439. [CrossRef]
3. Lungu, A.E.; Bogoslov, I.A.; Stoica, E.A.; Georgescu, M.R. From Decision to Survival—Shifting the Paradigm in Entrepreneurship during the COVID-19 Pandemic. *Sustainability* **2021**, *13*, 7674. [CrossRef]
4. Habiba, U.; Quddoos, A. A New Method to Solve Interval Transportation Problems. *Pak. J. Stat. Oper. Res.* **2020**, *16*, 802–811.
5. Servos, N.; Liu, X.; Teucke, M.; Freitag, M. Travel Time Prediction in a Multimodal Freight Transport Relation Using Machine Learning Algorithms. *Logistics* **2019**, *4*, 1. [CrossRef]
6. de Brito, I., Jr.; Uneddu, S.; Maspero, E.; Gonçalves, P. Optimizing Long-Lasting Insecticidal Nets Campaign in Ivory Coast. *Logistics* **2020**, *4*, 18. [CrossRef]
7. Miryanov, R.; Chalakova, K. One Type of Problems with Infinite Number of Cube Roots. *Math. Inform.* **2019**, *62*, 284–289.
8. Miryanov, R.; Petkov, J. An Application of the Relation Between Arithmetic Mean and Geometric Mean for Rational Proving Some Inequalities. *Math. Inform.* **2017**, *60*, 363–369.
9. Miryanov, R.; Yordanova, V. Optimizing the Positioning of Serving Units in the Tourism Business. *Math. Inform.* **2017**, *60*, 515–520.
10. Milusheva, P. Aspects Of The Relationships Of The Companies With The Suppliers. *Econ. Comput. Sci.* **2016**, *2*, 6–10.
11. Milusheva, P. Some Aspects of the Decision to Buy, Not to Produce Parts and Components. *Econ. Comput. Sci.* **2019**, *5*, 64–67.
12. Adrian, R.C.; Eduard, S. A Modern 2.0 E-Learning Platform Implemented at Lucian Blaga University of Sibiu. In Proceedings of the 6th Balkan Region Conference on Engineering and Business Education & 5th International Conference on Engineering and Business Education & 4th International Conference on Innovation and Entrepreneurship, Sibiu, Romania, 18–21 October 2012; Oprean, C., Grunwald, N., Kifor, C.V., Eds.; Lucian Blaga Univ Sibiu: Sibiu, Romania, 2012; pp. 619–622.
13. Dai, B.; Li, F. Joint Inventory Replenishment Planning of an E-Commerce Distribution System with Distribution Centers at Producers' Locations. *Logistics* **2021**, *5*, 45. [CrossRef]
14. Sajjad, W.; Ullah, M.; Khan, R.; Hayat, M. Developing a Comprehensive Shipment Policy through Modified EPQ Model Considering Process Imperfections, Transportation Cost, and Backorders. *Logistics* **2022**, *6*, 41. [CrossRef]
15. Mousa, A.A.A.; Abo-Elnaga, Y. Stability of Solutions for Parametric Inverse Nonlinear Cost Transportation Problem. *Mathematics* **2020**, *8*, 2027. [CrossRef]

16. Yordanova, V. Functioning of the Construction Enterprise in Conditions of Risk and Uncertainty. *Izvestiya* **2021**, *65*, 213–227.
17. Mihaylov, D.; Yordanova, V. Some Math Quizzes for Presenting a Number by Using Equal Digits and Math Operations. *Math. Inform.* **2019**, *62*, 290–300.
18. Singh, A.; Arora, R.; Arora, S. Bilevel Transportation Problem in Neutrosophic Environment. *Comput. Appl. Math.* **2022**, *41*, 1–25. [CrossRef]
19. Sulova, S. Models for Web Applications Data Analysis. In *Proceedings of the Computer Systems and Technologies*; Vassilev, T., Smrikarov, A., Eds.; Assoc Computing Machinery: New York, NY, USA, 2019; pp. 246–250.
20. Cristescu, M.P.; Flori, M.; Nerisanu, R.A. Applying a Sustainable Vector Model to Generate Innovation. In *Education, Research and Business Technologies*; Springer: Berlin/Heidelberg, Germany, 2022; pp. 149–161.
21. Sulova, S. Association Rule Mining for Improvement of IT Project Management. *Tem J.-Technol. Educ. Manag. Inform.* **2018**, *7*, 717–722. [CrossRef]
22. Miryanov, R. About the Role of the Digital Technologies in the Education in Mathematics. *Econ. Comput. Sci.* **2022**, *8*, 29–34.
23. Lyubenov, L. The Impact of the USD/EUR Exchange Rate on Bulgaria's Foreign Trade. *Econ. Comput. Sci.* **2022**, *8*, 21–28.
24. Stoyanova, M. Good Practices and Recommendations for Success in Construction Digitalization. *Tem J.-Technol. Educ. Manag. Inform.* **2020**, *9*, 42–47. [CrossRef]
25. Stanimirov, E. Customer Relationship Management in the Field of Business Services. *Izvestiya* **2017**, *2017*, 329–344.
26. Dogaru, V.; Brandas, C.; Cristescu, M. An Urban System Optimization Model Based on CO_2 Sequestration Index: A Big Data Analytics Approach. *Sustainability* **2019**, *11*, 4821.
27. Polkowski, Z.; Prakash Mishra, J.; Sourav Prasad, S.; Kumar Mishra, S. Evaluation of Aggregated Query Plans Using Heuristic Approach. In Proceedings of the 2020 12th International Conference on Electronics, Computers and Artificial Intelligence (ecai-2020), Bucharest, Romania, 25–27 June 2020; IEEE: New York, NY, USA, 2020.
28. Prakash Mishra, J.; Polkowski, Z.; Kumar Mishra, S. Performance of Cloudlets in Task Implementation Using Ant Colony Optimization Technique. In Proceedings of the 2020 12th International Conference on Electronics, Computers and Artificial Intelligence (ecai-2020), Bucharest, Romania, 25–27 June 2020; IEEE: New York, NY, USA, 2020.
29. Polkowski, Z.; Nycz-Lukaszewska, M. Infrastructure as a Service: Cloud Computing Model for Micro Companies. In *Proceedings of the Advances in Data Science and Management*; Borah, S., Balas, V.E., Polkowski, Z., Eds.; Springer: Singapore, 2020; Volume 37, pp. 307–320.
30. Nacheva, R.V. Architecture of Web-Based System for Usability Evaluation of Mobile Applications. *Izvestiya* **2017**, *2017*, 187–201.
31. Petrov, P.; Radev, M.; Dimitrov, G.; Pasat, A.; Buevich, A. A Systematic Design Approach in Building Digitalization Services Supporting Infrastructure. *Tem J.-Technol. Educ. Manag. Inform.* **2021**, *10*, 31–37. [CrossRef]
32. Aleksandrova, Y. Comparing Performance of Machine Learning Algorithms for Default Risk Prediction in Peer to Peer Lending. *Tem J.-Technol. Educ. Manag. Inform.* **2021**, *10*, 133–143. [CrossRef]
33. Petrov, P.; Krumovich, S.; Nikolov, N.; Dimitrov, G.; Sulov, V. Web Technologies Used in the Commercial Banks in Finland. In Proceedings of the Computer Systems and Technologies (compsystech'18), Ruse, Bulgaria, 13–14 September 2018; Rachev, B., Smrikarov, A., Eds.; Assoc Computing Machinery: New York, NY, USA, 2018; Volume 1641, pp. 94–98.
34. Stoyanov, M. Measuring Key Performance Indicators in Retail Trade. *Izvestiya* **2021**, *2021*, 161–175. [CrossRef]
35. Kuyumdzhiev, I.O. Controls Mitigating the Risk of Confidential Information Disclosure by Facebook: Essential Concern in Auditing Information Security. *Tem J.-Technol. Educ. Manag. Inform.* **2014**, 3, 113–119.
36. Polkowski, Z.; Vora, J.; Tanwar, S.; Tyagi, S.; Singh, P.K.; Singh, Y. Machine Learning-Based Software Effort Estimation: An Analysis. In Proceedings of the 11th International Conference on Electronics, Computers and Artificial Intelligence (ecai-2019), Pitesti, Romania, 27–29 June 2019; IEEE: New York, NY, USA, 2019.
37. Nacheva, R.; Sulov, V.; Czaplewski, M. The Impact of M-Learning on Sustainable Information Society. In *Proceedings of the International Conference on Electronic Governance and Open Society: Challenges in Eurasia*; Springer: Berlin/Heidelberg, Germany, 2022; pp. 244–262.

applied sciences

MDPI

Article

An Indirect Identification Method for Train Basic Resistance Parameters

Fei Xia [1], Tianxiang Li [2,*] and Pengfei Sun [2]

1 State Key Laboratory for Traction and Control System of EMU and Locomotive, China Academy of Railway Sciences Corporation Limited, Beijing 100081, China; xiafei@zemt.cn
2 School of Electrical Engineering, Southwest Jiaotong University, Chengdu 610031, China; pengfeisun@swjtu.edu.cn
* Correspondence: etec_ltx@my.swjtu.edu.cn

Abstract: The train basic resistance characteristic is crucial for ATO performance. However, this characteristic is slow time-varying with wheel rail wear and fast time-varying with weather conditions. The parameters of train basic resistance are difficult to measure directly, usually only a set of empirical parameters can be obtained through repeated experiments. These factors result in an inconsistency between the model parameters in the ATO controller and the actual train basic resistance parameters (TBRP), leading to a decrease in the effectiveness of the model-based controller. Therefore, this paper proposes an indirect TBRP identification method based on speed trajectory fitting to improve the effectiveness of the TBRP for model-based ATO controllers. Firstly, the original TBRP identification problem is transformed into an optimization problem, which is to minimize the deviation between the actual speed trajectory and the model-calculated one. Then, the Newton's method is used to accelerate the search for the best set of TBRPs with minimum deviation. Finally, case studies are used to verify the effectiveness of the proposed method.

Keywords: train basic resistance; parameter identification; Newton's method

Citation: Xia, F.; Li, T.; Sun, P. An Indirect Identification Method for Train Basic Resistance Parameters. *Appl. Sci.* **2023**, *13*, 8843. https://doi.org/10.3390/app13158843

Academic Editors: Tomasz Nowakowski, Artur Kierzkowski, Agnieszka A. Tubis, Franciszek Restel, Tomasz Kisiel, Anna Jodejko-Pietruczuk and Mateusz Zając

Received: 2 July 2023
Revised: 27 July 2023
Accepted: 28 July 2023
Published: 31 July 2023

1. Introduction

Automatic Train Operation (ATO) technology plays a key role in the Intelligent Railway System, which executes the control of automatic train operation process and determines the quality of train operation [1]. The train basic resistance is an important component of train dynamics, which consists of axle bearing friction, rolling, and sliding friction between wheels and rails, impact resistance and air resistance. Its value accuracy affects the control performance of ATO system, but itself is affected by train formation, geometry, weight, wheel rail, climatic conditions, train running speed [2], etc. These factors have complex high-order coupling characteristics, making it difficult to directly calculate accurate resistance value using mechanism models. As a consequence, TBRPs are usually measured under experimental conditions to obtain a set of empirical parameters [3]. However, due to environmental conditions, such as wind speed, weather, air humidity, altitude, and air pressure, as well as the body wear and replacement of vehicle components, the TBRP change both fast and slowly over time. This leads to a deviation between the empirical parameters and the actual parameters, which may cause severe overshoot and oscillation of the model-based controller. In addition, basic resistance measurement techniques based on sensors are difficult to develop and costly. In response to this issue, researchers mainly approach it from two aspects: Basic resistance modeling and TBRPs identification.

The first basic resistance formula was determined by Davis [4], as shown in Equation (1):

$$w(v) = a + bv + cv^2 \tag{1}$$

where v is running speed of train. The parameters a, b, c are resistance characteristic parameters. a is related to wheel rolling friction and bearing friction, b is related to wheel rail sliding friction, c is related to air resistance. Parameters a, b, c can be confirmed in the official train parameter specification when vehicle category has been determined. Subsequently, Stanley [5] built models for freight train and rolling resistance, and proposed introducing suspension resistance in the Davis formula. Xue [6] analyzed the influence of air pressure and temperature change on the basic resistance of locomotive and vehicle operation by studying the data of the Sichuan-Tibet Railway. Kang [7] analyzed the composition of high-speed train basic resistance and its relationship with train weight, achieving the decoupling of air resistance and other resistances.

The above research has achieved significant results in basic resistance modeling. However, the high-order and complex basic resistance models are not suitable for the design of some linear controllers. The ever-changing operating environment and vehicle mechanical wear still cause basic resistance uncertainty. These factors led to some identification methods being proposed.

One of the identification methods is the basic resistance characteristics identification method [8–12]. This method uses speed differences to calculate acceleration, then obtain the calculated basic resistance by Newton's second law. The objective function is to minimize the deviation between the characteristic curve of a set of TBRPs and the calculated basic resistance, as shown in Equation (2):

$$\min_{\theta \in \Theta} \boldsymbol{O}_{LS} = \|\boldsymbol{Y} - \boldsymbol{\Phi}\boldsymbol{\theta}\| \tag{2}$$

where $\boldsymbol{Y}$ is the calculated basic resistance, $\boldsymbol{\theta}$ is basic resistance parameters, $\boldsymbol{\Phi} = [1, v, v^2]$ is the input-output data information vector, Θ is the definition domain of $\boldsymbol{\theta}$. The objective function is greatly affected by the accuracy and precision of data acquisition. When data shaping or high amplitude sporadic noise exists, the error of basic resistance is large. In terms of algorithm design, researchers have proposed least squares method [8,9], heuristic algorithm [10,11], and the multi-innovation theory-based identification algorithm [12] to solve the problems of this method, but issues with computational efficiency and stability still exist. On the other hand, researchers consider model inconsistency as a disturbance and use adaptive control strategies to address parameter uncertainty [13,14]. These methods can handle parameter uncertainty problems, but some control accuracy and stationarity may be lost due to the need for real-time adjustment of deviation.

Faced with these problems, this paper proposes an indirect train basic resistance parameter identification method based on speed-fitting objective (TBRPI-SF). Unlike in Equation (2), the parameter identification problem of TBRPI-SF is transformed into an optimization problem, which is to find a set of basic resistance parameters that can minimize the deviation between the fitting curve obtained from the model-based calculation and the actual speed trajectory curve. The Newton's method is used to find the optimal TBRP adjustment direction. In addition, this paper discusses the boundary constraints and local solution problems of this algorithm. Finally, cases are set to verify the effectiveness and stability of this algorithm.

2. Train Dynamics Modeling

The objective function of BRPI-SF in this paper is to minimize the deviation between the fitting curve and the actual operating curve. The fitting curve is calculated using the traction calculation method based on train operation data and the train dynamics model.

The state of the train is affected by train weight, traction or braking force, basic resistance and additional resistance during operation. Considering the uneven additional resistance caused by different positions of each vehicle, this paper establishes the multi-mass train model, and its schematic diagram is shown in Figure 1.

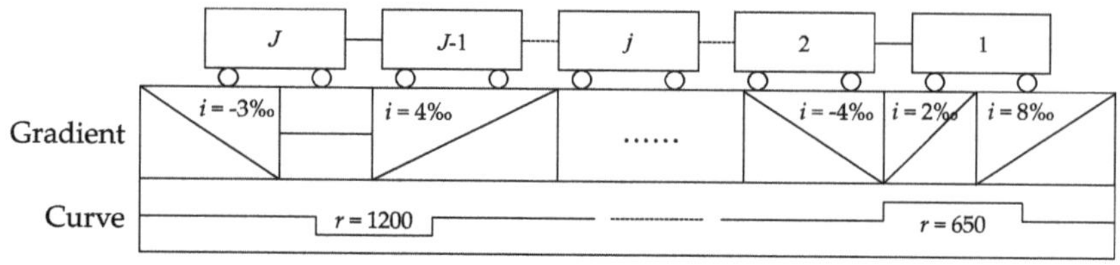

Figure 1. Schematic diagram of the multi-mass train model.

According to Newton's second law, the dynamic functions are described as:

$$\begin{cases} \frac{ds}{dt} = v \\ \frac{dv}{dt} = \frac{F - W_0(\boldsymbol{\theta},v) - G(s)}{(1+\gamma)\cdot M} \end{cases} \tag{3}$$

where v is running speed (m/s) of train, t is running time (s), s is train position (m). M is train mass (t), γ is the rotating mass coefficient. F represents the train control force (kN), with positive and negative indicating traction and braking forces, respectively. $W_0(\boldsymbol{\theta},v)$ is train basic resistance (kN). According to Davis formula, $W_0(\boldsymbol{\theta},v)$ can be expressed as a univariate quadratic equation with train running speed as a variable, as shown in Equation (4):

$$W_0(\boldsymbol{\theta}, v) = Mgw_0(\boldsymbol{\theta}, v)/1000 = Mg(a + bv + cv^2)/1000 \tag{4}$$

where $w_0(\boldsymbol{\theta},v)$ represents unit basic resistance (N/kN). g is the gravity coefficient, taken as 9.81 (N/kg or m/s^2). $\boldsymbol{\theta} = [a, b, c]$, a, b, and c are basic resistance parameters. $G(s)$ represents the additional resistance (kN), which is affected by the train operation position s, and its calculation formula is written as follows:

$$\begin{aligned} G(s) &= Mg(w_i + w_r)/1000 \\ &= \sum_{j=1}^{J} m_i g \left[i(s_j) + \frac{\alpha}{r(s_j)} \right] /1000 \end{aligned} \tag{5}$$

where w_i and w_r represent unit additional resistance (N/kN) of gradient and curve. m_j is the mass (t) of j-th car. s_j is the position of the j-th car's centroid. $i(s_j)$ means the gradient thousand fraction at position s_j. $r(s_j)$ means the curve radius (m) at position s_j. α is usually 600.

To facilitate analysis, train speed v and running time t are expressed as functions with train position s as independent variables, using Equation (3) to get:

$$\begin{cases} \frac{dv}{ds}(1+\gamma)Mv = F - W_0(\boldsymbol{\theta}, v) - G(s) \\ \frac{dt}{ds} = \frac{1}{v} \end{cases} \tag{6}$$

With the changes in speed and position, the combined force acting on the train will also change. Therefore, the train operation process is a complicated non-uniform motion, which cannot be solved directly by Equation (6). The traction calculation method divides the train operation process into some stages through discretization and assumes that combined force acting remains constant within each stage. Therefore, the train acceleration remains unchanged within a single stage and is considered constant acceleration operation.

The train operation process is divided into I stages. In the i-th stage, the train start position is $s_{i\text{-}1}$. End position is s_i. Force is $F_{i\text{-}1}$. Initial speed is $v_{i\text{-}1}$, which is obtained in (*i*-1)-th stage. We can obtain v_i by combining Equations (4)–(6), which is shown in Equation (7):

$$v_i(\boldsymbol{\theta}) = v_i(a,b,c) = \sqrt{v_{i-1}^2 + (s_i - s_{i-1})\frac{2g}{1+\gamma}\left[\frac{F_{i-1}}{Mg} - \frac{a + bv_{i-1} + cv_{i-1}^2 + G(s_{i-1})}{1000}\right]} \quad (7)$$

where $v_i(\boldsymbol{\theta})$ is the calculation speed under parameter $\boldsymbol{\theta}$. Therefore, based on train position and control force data, we can obtain the fitting curves $\boldsymbol{v}$ under a set of $\boldsymbol{\theta}$, where $\boldsymbol{v} \in \mathbf{R}^{\boldsymbol{I}+1}$.

Figure 2 illustrates the fitting curves under three sets of TBRPs. We can see that accurate parameters $\boldsymbol{\theta}_1$ can better reproduce the train operation process. Incorrect parameters, like $\boldsymbol{\theta}_2$ and $\boldsymbol{\theta}_3$, cause serious deviation from the fitting curve. It will lead to the unenforceability of the optimization curve and a decrease in controller control accuracy.

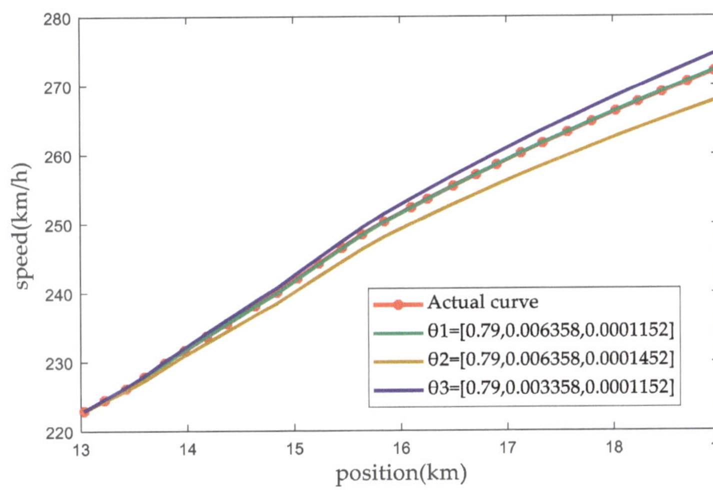

Figure 2. The fitting curves under three sets of TBRPs.

3. Algorithm

3.1. Objective Function

From a modeling perspective, if the parameters can make the train operation process calculated by the train dynamics model closer to the real value, it will be considered more accurate. Therefore, the objective of TBRPI-SF is to obtain the best set of TBRPs. After selecting them for traction calculation, the square mean of speed error between the fitting position speed curve and the actual position speed curve is the smallest.

Train operation data is obtained through two methods: Timed and event-triggered. Divide the train operation process into I stages using the acquisition interval of operation data. The operation data can be organized into: $\widetilde{\boldsymbol{F}}, \widetilde{\boldsymbol{s}}, \widetilde{\boldsymbol{v}}$, where $\widetilde{\boldsymbol{F}}, \widetilde{\boldsymbol{s}}, \widetilde{\boldsymbol{v}} \in \mathbf{R}^{\boldsymbol{I}+1}$, they are the control force, position, and speed information in the actual train operation data. Let $\boldsymbol{F} = \widetilde{\boldsymbol{F}}$, $\boldsymbol{s} = \widetilde{\boldsymbol{s}}$, the fitting curves $\boldsymbol{v}$ for this set of operational data under a set of basic resistance parameters $\boldsymbol{\theta}$ is obtained by Equation (7).

The accuracy of the parameter is assessed using the mean square error between the fitted curve and the actual curve, which is expressed as:

$$f(\boldsymbol{\theta}) = \|\boldsymbol{v}(\boldsymbol{\theta}) - \widetilde{\boldsymbol{v}}\|_2^2 = \frac{1}{I}\sum_{i=1}^{I}(v_i(\boldsymbol{\theta}) - \widetilde{v}_i)^2 \quad (8)$$

As a result, the optimization problem of BRPI-SF can be described as:

$$\min_{\theta \in \Theta} f(\boldsymbol{\theta}) \tag{9}$$

where, Θ is the definition domain of $\boldsymbol{\theta}$. Each basic resistance parameter has a corresponding physical significance. Therefore, there exists a boundary $\Theta = [\boldsymbol{\theta}_{\min}, \boldsymbol{\theta}_{\max}]$ that satisfies $\boldsymbol{\theta}_{\min} < \boldsymbol{\theta} < \boldsymbol{\theta}_{\max}$. From Equation (9), it can be seen that BRPI-SF is a constraint estimation problem.

In summarize, the main steps of the parameter identification method proposed in this paper are: (1) Configure train formation and line information, set train operation data, (2) establish the train dynamics model to obtain fitted curves for any set of basic parameters, (3) optimize to find the best set of TBRP.

For this optimization problem, heuristic algorithms are easy to apply to solve, but their solving efficiency is relatively low. In order to meet the requirement of high efficiency when this method is applied to the train controller, this paper uses the Newton's method to solve this problem.

3.2. Newton's Method

The Newton's method is an algorithm that uses a derivative to calculate the fastest descent direction of the function value. The updated formula for the Newton's method is:

$$\boldsymbol{\theta}_{n+1} = \boldsymbol{\theta}_n - \left[\nabla^2 f(\boldsymbol{\theta}_n)\right]^{-1} \nabla f(\boldsymbol{\theta}_n) = \boldsymbol{\theta}_n - \boldsymbol{H}_n{}^{-1} \nabla f(\boldsymbol{\theta}_n) \tag{10}$$

where $\boldsymbol{\theta}_n$ and $\boldsymbol{\theta}_{n+1}$ are the parameters of the n-th and (n+1)-th iteration period. $\boldsymbol{H}_n = \nabla^2 f(\boldsymbol{\theta}_n)$ is the Hessian matrix of the $f(\boldsymbol{\theta})$. $\nabla f(\boldsymbol{\theta}_n)$ is the gradient of the $f(\boldsymbol{\theta})$.

Derive the composition of $\nabla f(\boldsymbol{\theta}_n)$ and $\boldsymbol{H}_n$:

Calculate the derivative of Equation (8):

$$\nabla f(\boldsymbol{\theta}_n) = 2(v_n - \widetilde{v}) \nabla \mathrm{v}_n = \frac{2}{I} \sum_{i=1}^{I} (v_{n,i} - \widetilde{v}_i) \nabla v_{n,i} \tag{11}$$

where $\nabla v_{n,i}$ is the gradient of the i-th stage's calculated speed in n-th iteration period. The composition of $\nabla f(\boldsymbol{\theta}_n)$ and $\nabla v_{n,i}$ is:

$$\nabla f(\boldsymbol{\theta}_n) = \left[\frac{\partial f(\boldsymbol{\theta}_n)}{\partial a_n}, \frac{\partial f(\boldsymbol{\theta}_n)}{\partial b_n}, \frac{\partial f(\boldsymbol{\theta}_n)}{\partial c_n}\right]^T, \nabla v_{n,i} = \left[\frac{\partial v_{n,i}}{\partial a_n}, \frac{\partial v_{n,i}}{\partial b_n}, \frac{\partial v_{n,i}}{\partial c_n}\right]^T \tag{12}$$

According to Equation (7), we can obtain $\partial v_{n,i} / \partial a_n$:

$$\begin{aligned} \frac{\partial v_{n,i}}{\partial a_n} &= \frac{v_{n,i-1} \frac{\partial v_{n,i-1}}{\partial a_n} - (s_i - s_{i-1}) \frac{g}{1000(1+\gamma)} \left(1 + b_n \frac{\partial v_{n,i-1}}{\partial a_n} + 2 c_n v_{n,i-1} \frac{\partial v_{n,i-1}}{\partial a_n}\right)}{v_{n,i}} \\ &= \frac{v_{n,i-1} - \lambda_i (b_n + 2 c_n v_{n,i-1})}{v_{n,i}} \frac{\partial v_{n,i-1}}{\partial a_n} - \frac{\lambda_i}{v_{n,i}} = k_{n,i} \frac{\partial v_{n,i-1}}{\partial a_n} - p_{n,i} \end{aligned} \tag{13}$$

where $k_{n,i} = v_{n,i-1} - \lambda_i (b_n + 2 c_n v_{n,i-1}) / v_{n,i}, \lambda_i = g(s_i - s_{i-1})/1000/(1+\gamma), p_{n,i} = \lambda_i / v_{n,i}$. When $i = 1$, $\partial v_{n,0} / \partial a_n = 0$, so $\partial v_{n,1} / \partial a_n = -p_{n,i}$.

Similarly, we can obtain $\partial v_{n,i} / \partial b_n$ and $\partial v_{n,i} / \partial c_n$:

$$\frac{\partial v_{n,i}}{\partial b_n} = k_{n,i} \frac{\partial v_{n,i-1}}{\partial a_n} - p_{n,i} v_{n,i-1} \tag{14}$$

$$\frac{\partial v_{n,i}}{\partial c_n} = k_{n,i} \frac{\partial v_{n,i-1}}{\partial c_n} - p_{n,i} v_{n,i-1}^2 \tag{15}$$

As a result, $\nabla f(\boldsymbol{\theta}_n)$ can be calculated iteratively by combining Equations (11)–(15).

Furthermore, we derive the Hessian matrix and calculate the derivative of Equation (11):

$$\boldsymbol{H}_n = \nabla^2 f(\boldsymbol{\theta}_n) = 2\nabla v_n \nabla^T v_n - 2(v_n - \widetilde{v})\nabla^2 v_n = \frac{2}{I}\sum_{i=1}^{I}\left[\nabla v_{n,i}\nabla^T v_{n,i} + (v_{n,i} - \widetilde{v}_i)\nabla^2 v_{n,i}\right] \tag{16}$$

The composition of $\boldsymbol{H}_n$ and $\nabla^2 v_{n,i}$ is:

$$H_n = \begin{bmatrix} \frac{\partial^2 f(\boldsymbol{\theta})}{\partial a_n^2}, \frac{\partial^2 f(\boldsymbol{\theta})}{\partial a_n b_n}, \frac{\partial^2 f(\boldsymbol{\theta})}{\partial a_n c_n} \\ \frac{\partial^2 f(\boldsymbol{\theta})}{\partial a_n b_n}, \frac{\partial^2 f(\boldsymbol{\theta})}{\partial b_n^2}, \frac{\partial^2 f(\boldsymbol{\theta})}{\partial b_n c_n} \\ \frac{\partial^2 f(\boldsymbol{\theta})}{\partial a_n c_n}, \frac{\partial^2 f(\boldsymbol{\theta})}{\partial b_n c_n}, \frac{\partial^2 f(\boldsymbol{\theta})}{\partial c_n^2} \end{bmatrix}, \nabla^2 v_{n,i} = \begin{bmatrix} \frac{\partial^2 v_{n,i}}{\partial a_n^2}, \frac{\partial^2 v_{n,i}}{\partial a_n b_n}, \frac{\partial^2 v_{n,i}}{\partial a_n c_n} \\ \frac{\partial^2 v_{n,i}}{\partial a_n b_n}, \frac{\partial^2 v_{n,i}}{\partial b_n^2}, \frac{\partial^2 v_{n,i}}{\partial b_n c_n} \\ \frac{\partial^2 v_{n,i}}{\partial a_n c_n}, \frac{\partial^2 v_{n,i}}{\partial b_n c_n}, \frac{\partial^2 v_{n,i}}{\partial c_n^2} \end{bmatrix} \tag{17}$$

According to Equation (13), we can obtain $\partial^2 v_{n,i}/\partial a_n^2$:

$$\begin{aligned} \frac{\partial^2 v_{n,i}}{\partial a_n^2} &= -\frac{1}{v_{n,i}}\frac{\partial v_{n,i}}{\partial a_n}\frac{\partial v_{n,i}}{\partial a_n} + \frac{v_{n,i-1}-\lambda_i(b_n+2c_n v_{n,i-1})}{v_{n,i}}\frac{\partial^2 v_{n,i-1}}{\partial a_n^2} + \frac{1-\lambda_i 2c_n}{v_{n,i}}\frac{\partial v_{n,i-1}}{\partial a_n}\frac{\partial v_{n,i-1}}{\partial a_n} \\ &= \frac{v_{n,i-1}-\lambda_i(b_n+2c_n v_{n,i-1})}{v_{n,i}}\frac{\partial^2 v_{n,i-1}}{\partial a_n^2} - \frac{1-\lambda_i 2c_n}{v_{n,i}}\frac{\partial v_{n,i-1}}{\partial a_n}\frac{\partial v_{n,i-1}}{\partial a_n} - \frac{1}{v_{n,i}}\frac{\partial v_{n,i}}{\partial a_n}\frac{\partial v_{n,i}}{\partial a_n} \\ &= k_{n,i}\frac{\partial^2 v_{n,i-1}}{\partial a_n^2} - q_{n,i}\frac{\partial v_{n,i-1}}{\partial a_n}\frac{\partial v_{n,i-1}}{\partial a_n} - u_{n,i}\frac{\partial v_{n,i}}{\partial a_n}\frac{\partial v_{n,i}}{\partial a_n} \end{aligned} \tag{18}$$

where $q_{n,i} = (1 - \lambda_i 2c_n)/v_{n,i}, u_{n,i} = 1/v_{n,i}$. When $i = 1$, $\partial^2 v_{n,0}/\partial a_n^2 = 0$, so $\partial^2 v_{n,1}/\partial a_n^2 = -u_{n,i}(\partial v_{n,1}/\partial a_n)^2$.

Similarly, we can obtain $\partial^2 v_{n,i}/\partial b_n^2$, $\partial^2 v_{n,i}/\partial c_n^2$, $\partial^2 v_{n,i}/\partial a_n b_n$, $\partial^2 v_{n,i}/\partial a_n c_n$ and $\partial^2 v_{n,i}/\partial b_n c_n$:

$$\frac{\partial^2 v_{n,i}}{\partial b_n^2} = k_{n,i}\frac{\partial^2 v_{n,i-1}}{\partial b_n^2} - q_{n,i}\frac{\partial v_{n,i-1}}{\partial b_n}\frac{\partial v_{n,i-1}}{\partial b_n} - u_{n,i}\frac{\partial v_{n,i}}{\partial b_n}\frac{\partial v_{n,i}}{\partial b_n} - 2p_{n,i}\frac{\partial v_{n,i-1}}{\partial b_n} \tag{19}$$

$$\frac{\partial^2 v_{n,i}}{\partial c_n^2} = k_{n,i}\frac{\partial^2 v_{n,i-1}}{\partial c_n^2} - q_{n,i}\frac{\partial v_{n,i-1}}{\partial c_n}\frac{\partial v_{n,i-1}}{\partial c_n} - u_{n,i}\frac{\partial v_{n,i}}{\partial c_n}\frac{\partial v_{n,i}}{\partial c_n} - 4p_{n,i}v_{n,i-1}\frac{\partial v_{n,i-1}}{\partial c_n} \tag{20}$$

$$\frac{\partial^2 v_{n,i}}{\partial a_n b_n} = k_{n,i}\frac{\partial^2 v_{n,i-1}}{\partial a_n b_n} - q_{n,i}\frac{\partial v_{n,i-1}}{\partial a_n}\frac{\partial v_{n,i-1}}{\partial b_n} - u_{n,i}\frac{\partial v_{n,i}}{\partial a_n}\frac{\partial v_{n,i}}{\partial b_n} - p_{n,i}\frac{\partial v_{n,i-1}}{\partial a_n} \tag{21}$$

$$\frac{\partial^2 v_{n,i}}{\partial a_n c_n} = k_{n,i}\frac{\partial^2 v_{n,i-1}}{\partial a_n c_n} - q_{n,i}\frac{\partial v_{n,i-1}}{\partial a_n}\frac{\partial v_{n,i-1}}{\partial c_n} - u_{n,i}\frac{\partial v_{n,i}}{\partial a_n}\frac{\partial v_{n,i}}{\partial c_n} - 2p_{n,i}v_{n,i-1}\frac{\partial v_{n,i-1}}{\partial a_n} \tag{22}$$

$$\frac{\partial^2 v_{n,i}}{\partial b_n c_n} = k_{n,i}\frac{\partial^2 v_{n,i-1}}{\partial b_n c_n} - q_{n,i}\frac{\partial v_{n,i-1}}{\partial b_n}\frac{\partial v_{n,i-1}}{\partial c_n} - u_{n,i}\frac{\partial v_{n,i}}{\partial b_n}\frac{\partial v_{n,i}}{\partial c_n} - 2p_{n,i}v_{n,i-1}\frac{\partial v_{n,i-1}}{\partial b_n} - p_{n,i}\frac{\partial v_{n,i-1}}{\partial c_n} \tag{23}$$

As a result, $\boldsymbol{H}_n$ can be calculated iteratively by combining Equations (13)–(23).

3.3. Boundary Constraints and Local Solution Problems

The Newton's method is generally used to solve unconstrained optimization problems, but the TBRPI-SF problem has a limited range of parameters. On the other hand, the TBRPI-SF problem is a non-convex problem with multiple local optimal solutions. Therefore, the algorithm proposed in this paper needs to be improved to solve these issues.

For the boundary constraint problem of BRPI-SF, this paper adopts a step size adjustment method based on the Wolfe criterion:

When $\boldsymbol{\theta}_n - \boldsymbol{H}_n{}^{-1}\nabla f(\boldsymbol{\theta}_n) \notin \Theta$, find a step size α ($0 < \alpha < 1$), which meet:

$$\begin{cases} f\left[\boldsymbol{\theta}_n - \alpha\boldsymbol{H}_n^{-1}\nabla f(\boldsymbol{\theta}_n)\right] \le f(\boldsymbol{\theta}_n) - c_1\alpha\nabla f(\boldsymbol{\theta}_n)^T\boldsymbol{H}_n^{-1}\nabla f(\boldsymbol{\theta}_n) \\ -\nabla f\left[\boldsymbol{\theta}_n - \alpha\boldsymbol{H}_n^{-1}\nabla f(\boldsymbol{\theta}_n)\right]^T\boldsymbol{H}_n^{-1}\nabla f(\boldsymbol{\theta}_n) \ge -c_2\nabla f(\boldsymbol{\theta}_n)^T\boldsymbol{H}_n^{-1}\nabla f(\boldsymbol{\theta}_n) \end{cases} \tag{24}$$

where c_1 and c_2 are constants, and $0 < c_1 < c_2 < 1$. Starting from $\alpha = 1$, we search for α that satisfies Equation (24) and $\boldsymbol{\theta}_n - \alpha \boldsymbol{H}_n{}^{-1}\nabla f(\boldsymbol{\theta}_n) \in \Theta$. The parameters update formula in this iteration period is $\boldsymbol{\theta}_{n+1} = \boldsymbol{\theta}_n - \alpha \boldsymbol{H}_n{}^{-1}\nabla f(\boldsymbol{\theta}_n)$. Wolfe criterion preserves the descent direction of the optimal gradient and makes function value drop sufficiently under the constraint conditions.

The ending criterion of this algorithm is:

$$f(\boldsymbol{\theta}_{n+1}) < \varepsilon_1 \tag{25}$$

where ε_1 is maximum acceptable deviation, as the stop condition for this algorithm.

For non-convex problem, $\nabla f(\boldsymbol{\theta}_n) = 0$ does not mean that the optimal solution has been found. Therefore, when $\boldsymbol{\theta}_{n+1} - \boldsymbol{\theta}_n < \Delta\boldsymbol{\theta}_T$ and Equation (25) is not satisfied, it means falling into local optima. Where $\Delta\boldsymbol{\theta}_T$ means the minimum acceptable update distance in one iteration period. We expect the algorithm to jump out of the local solution and start searching again now. At this time, a parameter random updated method is used to search for a parameter that satisfies $\boldsymbol{\theta}_{n+1} \in \Theta, \boldsymbol{\theta}_{n+1} - \boldsymbol{\theta}_n > \Delta\,\hat{\boldsymbol{\theta}}$, where $\Delta\,\hat{\boldsymbol{\theta}}$ is a parameter limitation, reducing the possibility of re-entering local solutions.

In summary, the proposed basic resistance parameter identification based on the Newton's method is shown as Algorithm 1:

Algorithm 1: Basic resistance parameter identification based on the Newton's method
1. Initialize train formation information, including number of vehicles J, length l_j, and weight m_j of each vehicle.
2. Set train operation data, including control force $\widetilde{\boldsymbol{F}}$ and actual speed $\widetilde{v}$ and position $\widetilde{s}$.
3. Initialize line information, including gradient $i(s)$ and curve $r(s)$ at each position s.
4. Initialize basic resistance parameters $\boldsymbol{\theta}_1$ and calculate $f(\boldsymbol{\theta}_1)$.
5. **for** episode = 1,. . .n,. . .N do
6. Use $\boldsymbol{\theta}_n$ to calculate the calculated speed $\boldsymbol{v}_n$.
7. Calculate the coefficient $\boldsymbol{k}_n, \boldsymbol{p}_n, \boldsymbol{q}_n, \boldsymbol{u}_n$.
8. Calculate the gradient $\nabla \mathrm{v}_n$ and $\nabla f(\boldsymbol{\theta}_n)$.
9. Calculate the Hessian matrix $\nabla^2 \mathrm{v}_n$ and $\boldsymbol{H}_n$.
10. Determine if $\boldsymbol{\theta}_n - \boldsymbol{H}_n{}^{-1}\nabla f(\boldsymbol{\theta}_n) \in \Theta$. If so, update parameters as Equation (10). If not, find α based on the Wolfe criterion and update parameters as $\boldsymbol{\theta}_{n+1} = \boldsymbol{\theta}_n - \alpha \boldsymbol{H}_n{}^{-1}\nabla f(\boldsymbol{\theta}_n)$.
11. Calculate $f(\boldsymbol{\theta}_{n+1})$ and determine if $f(\boldsymbol{\theta}_{n+1}) < \varepsilon_1$. If so, **end for**. If not, randomly search for a set of parameters $\boldsymbol{\theta}_{n+1}$ that satisfies $\boldsymbol{\theta}_{n+1} \in \Theta, \boldsymbol{\theta}_{n+1} - \boldsymbol{\theta}_n > \Delta\,\hat{\boldsymbol{\theta}}$.
12. **End for**

4. Case Analysis

4.1. Simulation Setting

This paper selects the Wuhan-Guangzhou high-speed railway as the simulation line, with a main line length of 1069 kilometers. Sample train operation data form the train monitoring device. These operation data include train position, speed, traction, and braking force. The train parameters are shown in Table 1.

Table 1. Train parameters.

Train Type	**CRH380C**
Train formation	8M8T
Length (m)	400.47
Weight (t)	981.7

Some parameters setting of the Newton's method are shown in Table 2. The range of basic resistance parameters is set to match the characteristics of the actual high-speed train. The value of c_1 is chosen to easier to facilitate the identification of an appropriate step size α. The value of c_2 is chosen to achieve faster convergence. As for ε_1, its value depends on the magnitude of data disturbance, and a generally suitable value is considered to be 0.05.

Table 2. Parameters setting of the Newton's method.

N	100
$\boldsymbol{\theta}_{\max}$	[3, 0.2, 0.01]
$\boldsymbol{\theta}_{\min}$	[0.4, 0, 0]
c_1	0.1
c_2	0.9
ε_1	0.05
$\Delta\boldsymbol{\theta}_T$	$[10^{-5}, 10^{-6}, 10^{-7}]$
$\Delta\,\hat{\boldsymbol{\theta}}$	[0.1, 0.002, 0.0001]

4.2. Case Studies

4.2.1. Effectiveness Test without Disturbance

Firstly, to verify the effectiveness of this method under no disturbance, we created four sets of simulation data with different TBRPs. The actual basic resistance parameter values are shown in Table 3. This method was used to identify the TBRPs with random initial parameters. The acceptable deviation ε_1 was adjusted to 10^{-5}. The four identification results are recorded in Table 3. In the case of no data disturbance, the TBRPs obtained by this method were basically the same as the actual values. In these four identification processes, the minimum number of iterations was 4 and the maximum was 7. The mean square errors in each case were very small and met the acceptable deviation ε_1, which proves the effectiveness of the parameters and this method. The same conclusion can be drawn from Figure 3, where the fitting curve calculated by the identified parameters is consistent with the actual curve.

Table 3. Parameter identification results for four cases.

Case Number	Actual Parameters			Identified Parameters			Iterations	Mean Square Error
	a	b	c	a	b	c		
1	0.79	0.0229	0.001493	0.7901	0.0229	0.001493	5	2.8×10^{-9}
2	0.2955	0.0242	0.00159	0.2956	0.0242	0.00159	4	5.6×10^{-8}
3	1.0501	0.0253	0.00140	1.0525	0.0252	0.00140	6	1.7×10^{-6}
4	0.3488	0.0596	0.00112	0.3485	0.0596	0.00112	7	1.9×10^{-7}

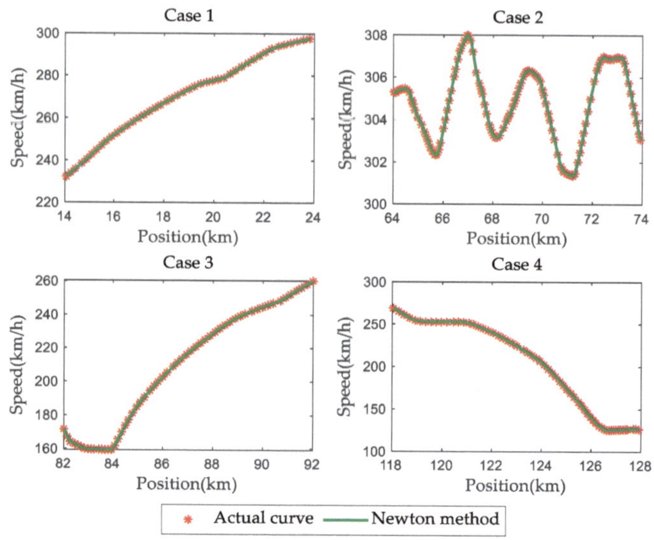

Figure 3. Comparison between fitted curves and actual curves for four cases.

4.2.2. Effectiveness Test under Random Disturbance

Secondly, we set up a set of simulation data and introduced white noise acquisition disturbances into the velocity data. The basic resistance parameter of the simulation data is set to $\boldsymbol{\theta} = [0.79, 0.0229, 0.001493]$. Using the Newton's method to identify the TBRPs of the data. The acceptable deviation ε_1 is adjusted to 0.05. Compare the curve calculated from the Newton's method with the actual curve and obtain Figure 4a. The results verify that the basic resistance parameter obtained from the Newton's method are accurate and can effectively reproduce the operation process of the train. The simulation results in Figure 4b shows that the standard deviation between the fitted curve and the actual curve was reduced from 904 to 0.033 within 5 iterations. The computational efficiency of the algorithm is pretty good. From Figure 4c, we can see the process of TBRPs changing with iterations. The final parameter is [0.483, 0.0273, 0.001492]. Compared to the actual parameters, the parameter c is the same, but error in parameter a and b. These errors are considered normal because of the existence of random disturbance. Actually, the unit basic resistance corresponding to the two sets of parameters is basically consistent in the speed domain [270, 305] (km/h), as shown in Figure 4d.

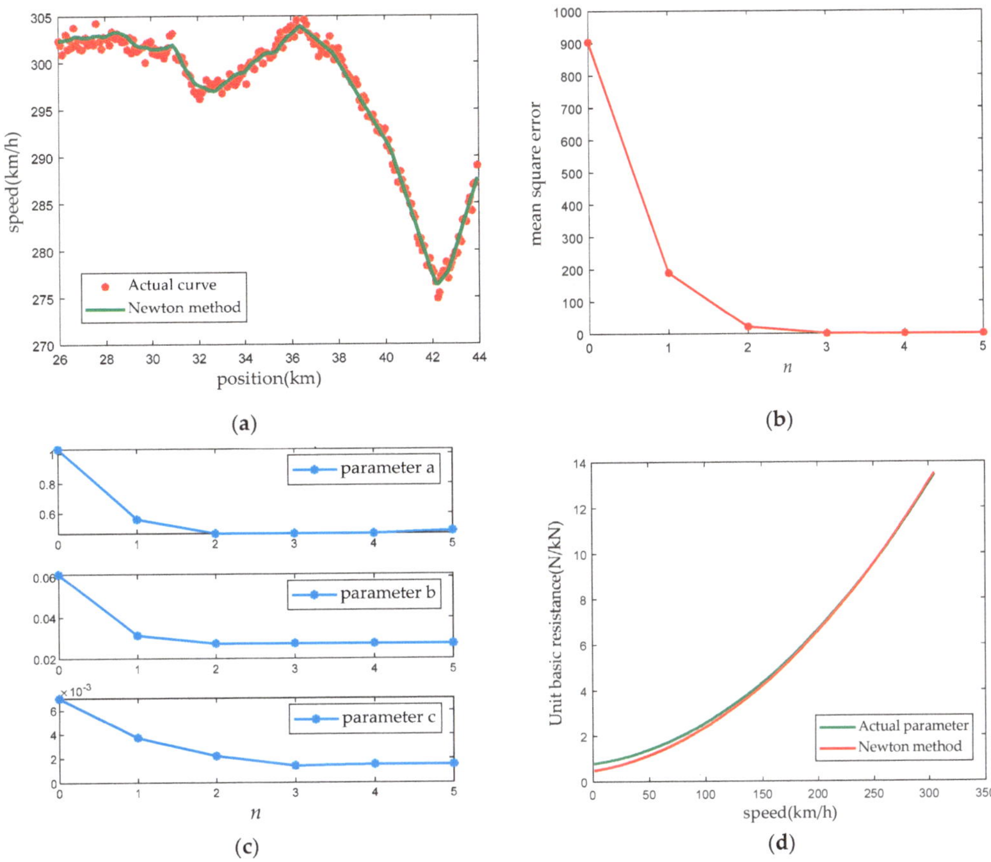

Figure 4. Parameter identification results. (**a**) Comparison between fitted curves and actual curves, (**b**) standard deviation changes with iteration, (**c**) TBRPs with iteration, (**d**) comparison between actual parameter characteristics and parameter characteristics obtained by Newton's method.

4.2.3. Online Identification Test

Finally, we set up a set of simulation experiments on the TBRPs changing with the position of the line. This simulation simulates the scenarios of reduced friction caused by smooth track after rain and increased air resistance caused by increased external wind speed. The simulation conditions of TBRPs in different sections are shown in Table 4. We introduced random disturbances in the last two sections. In this simulation process, the TBRPs of the current section were identified online based on real-time operation data generated by the train. The sampling interval was 100 m, and the data capacity I was 20.

Table 4. The simulation conditions of TBRPs in different sections.

Section	a	b	c	Disturbance
[0, 22] km	0.79	0.002288	0.001368	No
[22, 52] km	0.49	0.001208	0.001368	No
[52, 64] km	0.79	0.002288	0.001368	No
[64, 82] km	0.79	0.002288	0.001368	Yes
[82, 120] km	0.79	0.002288	0.002557	Yes

Figure 5 illustrates the results of this simulation. The identification algorithm accurately predicted the basic resistance values of most positions, but significant deviations exist in the positions of parameter switching. Similar to the previous simulation case, in the section of no data disturbance, the TBRPs obtained by this method are basically the same as the actual values, but in the section of existing data disturbance, the identification results of parameter *c* are relatively close to the actual situation, and deviations exist in the identification results parameters *b* and *c*.

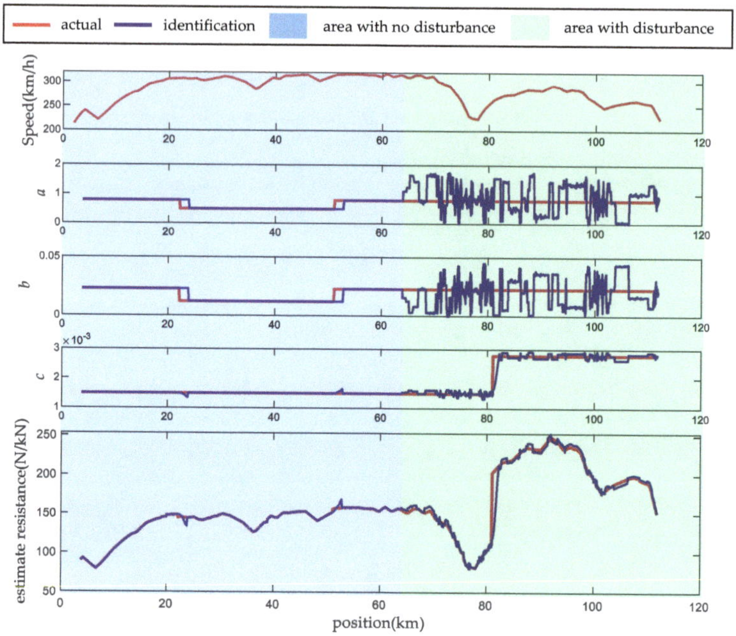

Figure 5. Results of this simulation on the TBRPs changing with the position.

4.2.4. Summary

In summary, the TBRPs identification method based on the Newton's method can find more accurate TBRP to improve the accuracy of traction calculation. The above cases prove the effectiveness of the algorithm under the condition of disturbance or not. The algorithm can find the optimal solution within several generations, which is short enough for designing periodicity real-time identification to use in the train controller.

5. Conclusions

This paper proposes an indirect TBRP identification method based on speed trajectory fitting. We transform the original TBRP identification problem into an optimization problem, which is to minimize the deviation between the actual speed trajectory and the model calculated one. By deducing the gradient and Hessian matrix of the objective function, the Newton's method is used to determine the optimal updating direction of the TBRPs. This paper discusses the boundary constraints and local solution problems of this algorithm and solves them by Wolfe criterion and random movement. Finally, simulation cases are set up to verify its effectiveness. The results show that this method can accurately identify the TBRPs under the condition without disturbance and obtain the accurate basic train resistance under the condition with random disturbance. In the online identification test, this method accurately predicted the basic resistance value of most sections. This parameter identification method can be embedded into the train control equipment to realize the online identification, so as to improve the performance of train controller.

Author Contributions: Conceptualization, F.X. and T.L.; methodology, T.L.; software, F.X.; validation, F.X., T.L. and P.S.; data curation, F.X.; writing—original draft preparation, F.X. and T.L.; writing—review and editing, F.X., T.L. and P.S.; visualization, T.L.; supervision, F.X. All authors have read and agreed to the published version of the manuscript.

Funding: This work was funded by the Open Project Funding of State Key Laboratory for Traction and Control System of EMU and Locomotive, China Academy of Railway Sciences Corporation Limited (Contract No. 2022YJ294), by project K2021J048 supported by CR.

Institutional Review Board Statement: Not applicable.

Informed Consent Statement: Not applicable.

Data Availability Statement: The simulation data can be downloaded at: https://www.researchgate.net/publication/372656434_Simulation_data_An_indirect_identification_method_for_train_basic_resistance_parameters (accessed on 20 May 2023). It involves the reported data in Figure 2 and the simulation data in Chapter 4.

Acknowledgments: The author thanks the anonymous reviewers for their constructive comments and the editor for handling the paper.

Conflicts of Interest: The author declares no conflict of interest.

References

1. Huang, W.Y.; Yang, N.Q. Ponderation on railway train basic resistance. *China Railw. Sci.* **2000**, *21*, 44–57.
2. Yu, Z.D.; Chen, D.W. Modeling and system identification of the braking system of urban rail vehicles. *J. China Railw. Soc.* **2011**, *33*, 37–40.
3. Tian, H.Q. Study on the characteristics of train air resistance under wind environment. *China Railw. Sci.* **2008**, *29*, 108–112.
4. Lukaszewicz, P. A simple method to determine train running resistance from full-scale measurements. *Proc. Inst. Mech. Eng. Part F J. Rail Rapid Transit* **2007**, *221*, 331–337. [CrossRef]
5. Stanley, A.B.; Richard, A.U.; James, P.R. The interpretation of train rolling resistance from fundamental mechanics. *IEEE Trans. Ind. Appl.* **1983**, *IA-19*, 802–817.
6. Xue, Y.H.; Ding, J.J.; Wang, L.J.; Li, Y.; Chen, K. Research on the basic resistance of locomotives and rolling stocks in Sichuan-Tibet Railway. *Electr. Drive Locomot.* **2021**, *10*, 67–72.
7. Kang, X.; Zeng, Y.; Zhang, B. Measurement method for the air resistance of highspeed train. *China Railw. Sci.* **2012**, *33*, 54–59.
8. Xie, G.; Zhang, D.; Hei, X.H.; Qian, F.C.; Cao, B.G.; Sei, T.; Hiroshi, M. Online identification method of time - varying parameters for Longitudinal dynamics model of high-speed train. *J. Traffic Transp. Eng.* **2017**, *17*, 71–81.
9. Wang, W. Parameter Identification and Its Application in Urban Train Control Model. Master's Thesis, Beijing Jiaotong University, Beijing, China, July 2018.
10. Hansen, H.S.; Nawaz, M.U.; Olsson, N. Using operational data to estimate the running resistance of trains. Estimation of the resistance in a set of Norwegian tunnels. *J. Rail Transp. Plan. Manag.* **2017**, *7*, 62–76. [CrossRef]
11. Lei, Y.; Zeng, Q.; Zhu, T.M. Train basic resistance identification and its online update algorithm. *J. Inf. Comput. Sci.* **2015**, *12*, 4161–4171. [CrossRef]
12. Liu, X.Y.; Ning, B.; Xun, J.; Wang, C.; Xiao, X.; Liu, T. Parameter identification of train basic resistance using multi-innovation theory. *IFAC-Pap.* **2018**, *51*, 637–642. [CrossRef]
13. Tan, C.; Zhang, L.L.; Yang, H.; Zhang, J.H. Adaptive error compensation control for high-speed train based on characteristic model. *J. East China Jiaotong Univ.* **2023**, *40*, 77–86.
14. Shi, W.S. Research on automatic train operation base on model-free adaptive control. *J. China Railw. Soc.* **2016**, *38*, 72–77.

logistics

MDPI

Article

Order Picking Problem: A Model for the Joint Optimisation of Order Batching, Batch Assignment Sequencing, and Picking Routing

Antonio Maria Coruzzolo [1,*], Francesco Lolli [1], Elia Balugani [1], Elisa Magnani [1] and Miguel Afonso Sellitto [2]

1 Department of Sciences and Methods for Engineering, University of Modena and Reggio Emilia, Via Amendola 2, 42122 Reggio Emilia, Italy; francesco.lolli@unimore.it (F.L.); elia.balugani@unimore.it (E.B.); magnanielli@gmail.com (E.M.)
2 Production and Systems Graduate Program, University of Vale do Rio dos Sinos, Avenida Unisinos 950, São Leopoldo 93022-180, Brazil; sellitto@unisinos.br
* Correspondence: antoniomaria.coruzzolo@unimore.it

Abstract: *Background:* Order picking is a critical activity in end-product warehouses, particularly using the picker-to-part system, entail substantial manual labor, representing approximately 60% of warehouse work. *Methods:* This study develops a new linear model to perform batching, which allows for defining, assigning, and sequencing batches and determining the best routing strategy. Its goal is to minimise the completion time and the weighted sum of tardiness and earliness of orders. We developed a second linear model without the constraints related to the picking routing to reduce complexity. This model searches for the best routing using the closest neighbour approach. As both models were too complex to test, the earliest due date constructive heuristic algorithm was developed. To improve the solution, we implemented various algorithms, from multi-start with random ordering to more complex like iterated local search. *Results:* The proposed models were tested on a real case study where the picking time was reduced by 57% compared to single-order strategy. *Conclusions:* The results showed that the iterated local search multiple perturbation algorithms could successfully identify the minimum solution and significantly improve the solution initially obtained with the heuristic earliest due date algorithm.

Keywords: order picking problem; order batching; batch assignment-sequencing; picking routing; heuristics

Citation: Coruzzolo, A.M.; Lolli, F.; Balugani, E.; Magnani, E.; Sellitto, M.A. Order Picking Problem: A Model for the Joint Optimisation of Order Batching, Batch Assignment Sequencing, and Picking Routing. *Logistics* **2023**, *7*, 61. https://doi.org/10.3390/logistics7030061

Academic Editors: Tomasz Nowakowski, Artur Kierzkowski, Agnieszka A. Tubis, Franciszek Restel, Tomasz Kisiel, Anna Jodejko-Pietruczuk, Mateusz Zając and Mladen Krstić

Received: 26 June 2023
Revised: 21 August 2023
Accepted: 29 August 2023
Published: 11 September 2023

1. Introduction

End-product warehouses play a fundamental role within the supply chain as platforms to temporarily store products before sending them to end customers. They are used in most industrial and commercial activities to maximise the service level and minimise storage costs simultaneously [1]. Although warehouses account for approximately 20% of the total cost for organisations, they are essential because they contribute to many company missions [2]. In a typical scenario, the following activities take place inside a warehouse: receiving and storing goods; preparation and order processing; shipping and exit of goods. In general, logistics companies deal with equipment of different origins that must operate according to different strategies, such as line-feed kitting (intracell) or milk-run (intercell) line feed [3]. In both cases, features such as modularity and scalability of the system can be helpful if demand changes over time [4].

Order picking is critical to manage and operate warehouses efficiently. Order picking aims to transform the load unit, with which a product is generally stored in the warehouse, into a format more suited to the customer's needs according to the service level at the minimum cost. Orders consist of order lines referring to a specific product and indicating the specific quantity requested by a customer [5]. When an order contains multiple lines, these must be accumulated and ordered before being transported to the shipping area. Therefore, the picking operation consists of planning, preparing, and releasing the customer's order by taking the products required from the various picking boxes in which they are stored [6].

As indicated in [7], the picking activity is characterised by a high intensity of manual labor. It is estimated to account for approximately 60% of warehouse work, with a consequent impact on both the overall logistic costs and the service level provided to the customer. By breaking down the warehouse costs, the picking activity is typically the most laborious and expensive, with a cost of up to 55% of the total operating costs of the warehouse [8].

The picking system must be organised and executed as effectively as possible because any malfunction of this phase can lead to unsatisfactory service, high operating costs, and, consequently, the deterioration of the entire supply chain. An efficient picking process must be designed in a robust and optimally controlled manner. The literature solves a related problem: how to determine a batch sequence and, at the same time, assign workers to batches according to pre-determined requirements [9].

This study focuses on one of the picking systems most used in warehouses, namely, the picker-to-part system. Over 80% of Western European organisations use this methodology, which is the primary system for the picking activity [7]. In this activity, each operator carries out a "picking mission" in the picking area guided by a paper or an RF/voice terminal. Pickers walk or drive along the corridors to pick up the required items and complete a single order or batch of orders, depending on the picking logic.

This study analyses the possibility of grouping customer orders in batches so that the items belonging to a batch are collected during a single picking tour. Each customer order is generally characterised by the quantity to be picked and a loading date; the latter indicates when the order must be completed and ready for shipment. A breach of the loading date entails additional costs and a lower service level. The feasibility of meeting the deadlines for a set of customer orders depends on the following factors:

1. How the customer orders are grouped in batches (order batching problem);
2. How the batches are assigned and sequenced to the operators (batch assignment sequencing problem);
3. How each order picker is routed to collect the items of each picking order (picking routing problem).

The three issues are highly interrelated and may involve trade-off considerations, necessitating an optimisation approach. If addressed independently, there is a possibility of encountering a local, suboptimal, or even unacceptable solution. Consequently, a collective, balanced resolution is required to address these challenges effectively. Therefore, the purpose of this article was to propose a joint optimisation model to reach an acceptable solution concurrently for the batching, assigning, sequencing, and routing problems. The objective function seeks to minimise the overall completion time while ensuring that total tardiness is not compromised and that orders are not excessively anticipated. The specific objectives are to theoretically formulate a model, simulate an application in a real case, and compare the application's result with other usual strategies. The primary novelty of this study lies in introducing a comprehensive model that addresses multiple correlated problems while simultaneously reducing the number of variables that need to be managed. To the best of our knowledge and supported by a search on the Scopus database relying on related keywords, this is the first study that tackles the abovementioned problems jointly.

The remaining content is organised as follows. Section 2 reviews some of the most relevant studies and highlights the novelties of this study. Section 3 describes the problem and introduces the model notation. Section 4 describes the mathematical model and explains the heuristic approaches used. Section 5 describes the experimental analysis. Section 6 outlines the conclusions along with some scope for further research.

2. Literature Review

Only recently, researchers have started to focus on topics related to the ordering, sequencing, and picking routing problems to identify a simultaneous, complete, and advantageous solution to these problems. By simultaneously considering the ordering, sequencing, and picking routing activities, the operating costs can be reduced, and the service level can be improved, e.g., by reducing the overall travel time by more than 35% [10]. Most studies dealt with these problems separately or focused on the simultaneous resolution of two or more subproblems of the joint order batching, assignment, sequencing, and routing problems (JOBASRPs). In order to position the study in front of others that are seminal and current, a brief survey in databases using the keywords "problem-solving", "warehousing", "batching orders", "assigning orders", "sequencing orders", and "routing" provided a sample of articles that, after being screened, produced the content in Table 1. Such content ensures the novelty of the proposed study.

Table 1. Comparison of this study with others in JOBASRP.

Study	Batching	Assigning	Sequencing	Routing
[11]	x		x	
[12]	x		x	
[13]	x		x	x
[14]	x			x
[15]	x		x	
[16]	x			x
[17]	x	x	x	
[18]	x		x	x
[19]	x			x
[20]	x			x
[21]	x	x		x
[22]	x			x
[23]	x		x	
[24]	x	x	x	
[25]	x			
Our study	x	x	x	x

The order-picking problems needed to be considered jointly because the selection of orders composing the batches and the sequence in which the batches are assigned and processed influences both the planning of the picking routes end the fulfilment of loading dates [13]. Ref. [18] proposed a mathematical model for solving the joint order batching, sequencing, assignment, and routing problem (JOBSPRP). In this model, which includes only linear constraints, the number of variables and constraints increases polynomially with increasing customer orders. It is significantly more advantageous than traditional models.

The first subproblem is the order batching problem, which involves grouping a set of customer orders in batches to minimise the total length of a tour. From a mathematical point of view, the order batching problem can be seen as a travelling salesman problem (TSP) with unique properties derived from the warehouse layout [26]. Several approaches have been proposed to solve this problem (e.g., [7,27]).

The second subproblem is the picking routing problem, generally solved as a Steiner TSP, where some nodes must not be visited, and others can be visited more than once [7]. The Steiner TSP can also be formulated as a classic TSP if the minimum distance between each pair of storage locations is previously calculated [14,28]. The picking routing problem involves picking a given set of items stored in known storage locations to minimise the total distance travelled.

The last subproblem is the order batch assignment and sequencing problem, which involves the assignment of batches to the minimum number of pickers and sequencing the batches for each picker so that the overall tardiness is reduced. To tackle this problem, researchers have considered different factors, such as due dates [28], and scenarios, such as multiple pickers [29], heterogeneous pickers [17], and multiple pickers in the online environment [30].

Some mathematical formulations have also been proposed for the JOBASRP considering the due dates [31], 3D warehouses [18], and multiple pickers [31,32]. Ref. [32] published a series of mathematical models examining actual distribution centres, and considering the minimisation of the total distance, tardiness, and earliness. However, their models are highly computationally complex; in fact, the number of binary variables and constraints increases exponentially with the complexity of the problem, and in particular with an increase in warehouse locations. The JOBASRP is a recent problem that has been extensively studied and represents a promising path for research. The most accredited and complete formulations of this problem are those proposed in [30,33] as indicated in [34].

The proposed model only requires the definition of some feasible batches and their execution time, which is the main difference in front of other models already mentioned. As the number of feasible batches grows exponentially with the increase in the number of variables, this activity involves a significant amount of computational time and the risk of being unable to identify an acceptable solution, even for small instances, owing to memory restrictions. To limit the number of variables, in this study we developed an integer linear model in which the number of variables and constraints grows polynomially with the increasing complexity of the problem.

Furthermore, this study changes the constraints for defining the routing strategy. The proposed mathematical model assumes that a picker always starts the tour from an initial depot and concludes it in a final depot to make the model more flexible and realistic. It is worth noting that this assumption is not critical because the input data can be easily set to make these two depots physically coincident and separated only from a logical point of view.

3. Problem Description and Notation

This study considers a warehouse with a manual, low-level picker-to-parts order-picking system from which a given set of items must be retrieved. The items are stored on pallets or in bins directly accessible to the operators [27]. More operators work simultaneously in the picking area, and each travels on a picking device, allowing him/her to collect more items during a single tour. Each tour starts from the initial depot and ends at the final depot. The initial depot is where the operator starts the tour, and the pick list and picking cart are assigned to the picker. In the final depot, the operator delivers the picking orders to be prepared for shipment. In the proposed model, these depots are assumed to be separate to increase the flexibility of the proposed model; however, they can be considered coincident if necessary. The order picker is guided by a pick list containing a set of orders along with the storage positions of the requested items and their respective quantities to retrieve them on the same tour. The pick list also indicates the operator's path to reach the required locations. The maximum number of stops a picker can make during the same tour depends on the capacity of the picking vehicle.

A batch is an aggregation of a set of customer orders which is picked in a single picking tour. The time spent by an order picker to retrieve all the items of a batch can be divided as follows [7]:

1. The setup time, i.e., the time spent preparing a tour;
2. The search time, i.e., the time spent identifying the correct item;
3. The picking time, i.e., the time spent physically retrieving the items from their storage locations;
4. The travel time, i.e., the time spent to reach the picking locations of the batch items.

The execution time of a batch starts at the batch start time, i.e., when the picker to whom the batch has been assigned starts preparing the picking tour. The batch is completed when the operator has picked all the items needed for a batch and returns them to the final depot. The start time of an order corresponds to the start time of the batch to which it is assigned; similarly, the completion time of the customer order corresponds to the completion time of the batch to which it is assigned.

Customer orders must be picked before a specific due date to guarantee an adequate service level to the end customers. Whether such due dates are met depends on how the orders are assigned to the batches, their sequencing, and the defined routing strategy. More precisely, the tardiness ta_n of an order n is defined as the non-negative difference between the completion time v_n and its due date d_n [16], expressed as $ta_n = max\{v_n - d_n; 0\}$. The sum of the tardiness of all customer orders, called total tardiness, must be null to obtain an acceptable solution at the company level. The earliness ea_n of an order n is defined as the non-negative difference between its due date d_n and the completion time v_n, which can be expressed as $ea_n = max\{d_n - v_n; 0\}$; the sum of the earliness of all customer orders is called total earliness. The excessively early preparation of orders is not desirable because it leads to the unnecessary use of warehouse space.

The JOBASRP can be stated as follows. Given a non-empty set of orders, some items with known storage locations must be removed from the warehouse. Each order is characterised by a due date by which all requested items should be retrieved and brought to the final depot. Some order pickers are available for the necessary picking operations. Then, the following questions must be answered (simultaneously) in such a way that the total tardiness, earliness, and completion time are minimised [28]:

- How should the set of customer orders be grouped into picking orders? (Order Batching Problem);
- How and in which sequence should the set of picking orders be assigned to the order pickers? (Batch Assignment-Sequencing Problem);
- In which sequence should the respective pick locations be visited for each order? (Picking Routing Problem).

4. The Problem Formulation

4.1. Sets and Parameters

Before presenting the formulation of the model in detail, it is worth introducing the sets, parameters, and variables used.

Sets:

- Set of costumer orders: $N = 1..num\ order$.
- Set of locations to visit: $V = 1..num\ location$.
- Set of missions that can be performed: $H = 1..num\ mission$.
- Set of periods, for each picker, in which a batch can be collected: $K = 1..num\ period$.
- Set of pickers: $P = 1..num\ picker$
- Set of feasible batches: $B = 1..num\ batch$

Parameters:

- $w_1 = Weight\ assigned\ to\ completion\ time$
- $w_2 = Weight\ assigned\ to\ the\ weighted\ sum\ of\ earlinesses\ and\ tardinesses$
- $\gamma = Weight\ assigned\ to\ the\ total\ earliness$
- $\partial = Weight\ assigned\ to\ the\ total\ tardiness$
- $d_{ij} = distance\ between\ picking\ locations\ i \in V\ and\ j \in V$
- $cap_n = capacity\ request\ by\ order\ n \in N$
- $Cap_{vehicle} = capacity\ of\ the\ picking\ vehicle$
- $a_{ni} = \begin{cases} 1\ If\ the\ order\ n\ \in N\ must\ visit\ the\ picking\ location\ i\ \in V \\ 0\ otherwise \end{cases}$
- $\beta^{time} = time\ required\ to\ travel\ by\ a\ unit\ of\ space$
- $\beta^{setup} = setup\ time\ required\ for\ each\ batch$
- $\beta^{picking} = time\ required\ for\ searching\ and\ picking\ items$
- $due\_carico_n = due\ date\ for\ the\ order\ n \in N$
- $M = sufficiently\ large\ positive\ number$
- $dis_i = distance\ travelled\ to\ complete\ the\ batch\ i \in B$

- $capbatch_i = capacity\ request\ by\ batch\ i \in B$
- $b_{in} = \begin{cases} 1\ If\ the\ batch\ i\ \in B\ contains\ the\ order\ i\ \in N \\ 0\ otherwise \end{cases}$

Variables:

- $ta_n = tardiness\ of\ the\ order\ n \in N$
- $ea_n = earliness\ of\ the\ order\ n \in N$
- $t\_exec_h = execution\ time\ of\ the\ mission\ h \in H$
- $u_{pk} = processing\ time\ for\ the\ batch\ assigned\ to\ the\ position\ k \in K\ of\ picker\ p \in P$
- $v_{pk} = completion\ time\ for\ the\ batch\ assigned\ to\ the\ position\ k \in K\ of\ picker\ p \in P$
- $x_{pkh} = \begin{cases} 1\ If\ the\ mission\ h \in H\ is\ assigned\ to\ the\ position\ k \in K\ of\ picker\ p \in P \\ 0\ otherwise \end{cases}$
- $y_{nh} = \begin{cases} 1\ If\ the\ order\ n\ \in N\ is\ assigned\ to\ mission\ h\ \in H \\ 0\ otherwise \end{cases}$
- $z_{ijh} = \begin{cases} 1\ if\ the\ arc\ that\ connects\ the\ picking\ locations \\ i \in V\ and\ j \in V\ are\ included\ in\ the\ mission\ h \in H \\ \\ 0\ otherwise \end{cases}$
- $f_{ijh} = Number\ of\ units\ passing\ through\ the\ arc\ that\ connects\ the\ picking\ locations\ i \in V\ and\ j \in V\ in\ the\ mission\ h \in H$
- $t\_exec_i = execution\ time\ of\ the\ batch\ i \in B$
- $x_{ikh} = \begin{cases} 1\ If\ the\ batch\ i \in B\ is\ assigned\ to\ the\ position\ k \in K\ of\ picker\ p \in P \\ 0\ otherwise \end{cases}$

The proposed model generates a solution that minimises the completion time and the weighted sum of the total tardiness and earliness. The aim is to generate a solution to increase the picking area's productivity by reducing the acceptable preparation time at the organisational level. To make the definition of the solution more flexible, each of these factors is assigned a weight, which can be modified according to organisational needs.

For the formulation of the model, the capacity of the picking vehicle is measured in several items, as in [28,29,35]. Note that this is not a critical assumption because the number of items can be easily modified to consider other capacity constraints (e.g., the maximum number of customer orders or the maximum total weight of items).

The mathematical model formulation can be divided into two parts. The first part is related to the joint order batching problem and picking routing problem; here, the customer orders are grouped into batches, and the corresponding tours are constructed. The second part is related to the batch assignment and sequencing problem. Here, the batches are assigned to an operator and arranged in a specific sequence. The picker starts processing the batch assigned to the first position, and after reaching the final depot and returning to the initial one, he/she processes the batch assigned to the next position. This process is repeated until the sequence is completed. This procedure is related to the total completion time and the completion time of each batch, which are necessary to calculate the tardiness and earliness of each customer order.

The model is defined using Equations (1)–(32).

$$min\left(w_1 \times \left(\left(\sum_{p=1}^{N}\sum_{k=1}^{K} v_{pk}\right)\right) + w_2 \times \left(\gamma \times \sum_{n=1}^{N} ea_n + \partial \times \sum_{n=1}^{N} ta_n\right)\right) \quad (1)$$

$$\sum_{h\in H} x_{pkh} \leq 1; \qquad \forall p \in P, k \in K \quad (2)$$

$$\sum_{p\in P}\sum_{k\in K} x_{pkh} = 1; \quad \forall h \in H \quad (3)$$

$$\sum_{h\in H} y_{nh} = 1; \qquad \forall n \in N \quad (4)$$

$$\sum_{n \in N} cap_n \times y_{nh} \leq Cap_{Vehicle}; \qquad \forall h \in H \tag{5}$$

$$\sum_{j \in J | j > 1} z_{1jh} = 1; \qquad \forall h \in H \tag{6}$$

$$\sum_{j \in J | j < num\_location} z_{inum\_locationh} = 1; \quad \forall h \in H \tag{7}$$

$$\sum_{i \in V | i > 1} \sum_{h \in H} z_{i1h} = 0 \tag{8}$$

$$\sum_{j \in V | j < num\_location} \sum_{h \in H} z_{num\_locationjh} = 0 \tag{9}$$

$$\sum_{i \in V | i < num\_location} z_{ijh} \geq a_{nj} \times y_{nh}; \quad \forall h \in H, n \in N, j \in V | j \geq 2 \tag{10}$$

$$\sum_{j \in V | j > 1} z_{ijh} \geq a_{ni} \times y_{nh}; \qquad \forall h \in H, n \in N, i \in V | i \leq num\ location - 1 \tag{11}$$

$$f_{ijh} \leq Cap_{Vehicle} \times z_{ijh}; \quad \forall h \in H, i \in V | i \leq num\ location - 1, j \in V | j \geq 2 \tag{12}$$

$$\sum_{g \in V | g < num\_location} z_{gih} = \sum_{j \in J | j > 2} z_{ijh}; \qquad \forall h \in H, i \in V | 2 \leq i \leq num\ location - 1 \tag{13}$$

$$\sum_{g \in V | g < num\_location} z_{gih} \leq 1; \qquad \forall h \in H, i \in V | 2 \leq i \leq num\ location - 1 \tag{14}$$

$$\sum_{j \in J | j > 2} z_{ijh} \leq 1; \qquad \forall h \in H, i \in V | 2 \leq i \leq num\ location - 1 \tag{15}$$

$$\sum_{g \in V | g < num\_location} f_{gih} = \sum_{j \in J | j > 2} f_{ijh} \geq a_{ni} \times y_{nh} \qquad \forall h \in H, n \in N, i \in V | 2 \leq i \leq num\ location - 1 \tag{16}$$

$$z_{iih} = 0; \qquad \forall h \in H, i \in V | 2 \leq num\ location - 1 \tag{17}$$

$$\beta^{setup} + \beta^{picking} \times \left(\sum_{n \in N} cap_n \times y_{nh} \right) + \beta^{time} \times \left(\sum_{i \in V | i < num\_location} \sum_{j \in J | j > 2} d_{ij} \times z_{ijh} \right) \leq t_{exech}; \qquad \forall h \in H \tag{18}$$

$$t_{exech} - M \times \left(1 - x_{pkh} \right) \leq u_{pk} \qquad \forall p \in P, k \in K, h \in H \tag{19}$$

$$u_{p1} \leq v_{p1}; \ \forall p \in P \tag{20}$$

$$u_{pk} + v_{pk-1} \leq v_{pk}; \ \forall p \in P, k \in K, | k \geq 2 \tag{21}$$

$$v_{pk} - due\_date_n - M \times \left(2 - x_{pkh} - y_{nh}\right) \le ta_n;\ \forall p \in P, k \in K, h \in H, n \in N \quad (22)$$

$$due\_date_n - v_{pk} - M \times \left(2 - x_{pkh} - y_{nh}\right) \le ea_n;\ \forall p \in P, k \in K, h \in H, n \in N \quad (23)$$

$$x_{pkh} \in \{0,1\};\forall\ p \in P, k \in K, h \in H \quad (24)$$

$$y_{nh} \in \{0,1\};\ \forall h \in H, n \in N \quad (25)$$

$$z_{ijh} \in \{0,1\};\ \forall h \in H, i \in V | i \le 1..num\ location - 1, j \in V | j \ge 2 \quad (26)$$

$$t\_exec_h \ge 0;\ \forall h \in H \quad (27)$$

$$ta_n \ge 0; \quad \forall n \in N \quad (28)$$

$$ea_n \ge 0; \quad \forall n \in N \quad (29)$$

$$u_{pk} \ge 0; \quad \forall p \in P, k \in K \quad (30)$$

$$v_{pk} \ge 0; \quad \forall p \in P, k \in K \quad (31)$$

$$f_{ijh} \ge 0 \quad \forall h \in H, i \in V | i \le 1..num\ location - 1, j \in V | j \ge 2 \quad (32)$$

Equation (1) is a mono-objective function. The first term minimises the total completion time, and the second term is the weighted sum of the total tardiness and earliness. The weights can give more relevance to either term; in this way, the solution that best fits the specific needs of each operative environment can be determined.

Equations (2)–(17) show the model's constraints. Equation (2) ensures that a maximum of one mission is assigned to each picker at each position, and Equation (3) guarantees that each mission is performed. Equations (4) and (5) ensure that each customer order is executed on a specific picking mission and that the capacity of a picking vehicle is not exceeded, respectively.

Equations (6)–(17) show the routing constraints. Equations (6) and (7) ensure that each mission starts from the initial depot and ends in the final one, respectively; Equations (8) and (9) ensure that the initial depot has no inputs and the final one has no outputs, respectively. Equations (10) and (11) ensure that each picking location has only one entrance and one exit, respectively. Equation (12) represents the capacity constraint on the arc, Equation (13) ensures that the number of incoming arcs in a vertex is equal to the number of outgoing arcs, and Equations (14) and (15) guarantee that each vertex has only one incoming arc and one outgoing arc, respectively. Equations (16) and (17) prevent the formation of cycles in the path and itineraries that end without crossing all the required locations, respectively. These constraints force each mission to pass only once from each compartment, with the consideration that the initial and final deposits cannot have entries and exits, respectively.

Equation (18) determines the execution time for each mission, composed of the setup time, the time to search and pick up the items, and the travel time. The execution time of a mission assigned to a particular position of a given operator is calculated with Equation (19). Equation (20) determines the completion time in the first period, which corresponds precisely to the execution time of the mission, and Equation (21) determines the completion time of the missions assigned to successive periods. In this case, it is necessary to consider the mission assigned to the period considered plus the completion time of the previous periods. Finally, Equations (22) and (23) determine the tardiness and earliness of each order, respectively. The third value subtracted allows the consideration that it is necessary to calculate the delay or advance of an order to evaluate the batch's completion time. Equations (24)–(25) define the domains of the variables.

4.2. Simplified Model

This study modified the mathematical model proposed in the previous section for defining and sequencing batches. Based on the model proposed in [36], this study eliminated the constraints and variables related to the picking routing problem to reduce the complexity of the model, particularly that of the picking routing problem.

However, eliminating the constraints and variables related to the picking routing problem complicates the inputs. In particular, the proposed modified model requires the input of all the executable batches as initial data. Therefore, as the number of orders increases, the computational times increase significantly because the number of batches also increases. Hence, it may not be possible to identify an improved solution even for small batches due to memory restrictions. As the model does not deal with the definition of routing strategies, this study implemented a heuristic algorithm that defines the path and distances of each possible feasible batch before the execution of the model. The closest neighbour is a standard heuristic algorithm for the TSP that optimises the route by joining the nearest nodes each time. The proposed modified model can be mathematically expressed using Equations (33) to (48).

$$min\left(w_1 \times \left(\sum_{p\in P}\sum_{k\in K} v_{pk}\right) + w_2 \times \left(\gamma \times \sum_{n\in N} ea_n + \partial \times \sum_{n\in N} ta_n\right)\right) \tag{33}$$

$$\sum_{i\in B} x_{ipk} \leq 1; \quad \forall p \in P, k \in K \tag{34}$$

$$\sum_{i\in B}\sum_{p\in P}\sum_{k\in K} x_{ipk} \times b_{in} = 1; \quad \forall n \in N \tag{35}$$

$$Cap\ batch_i \times x_{ipk} \leq Cap_{vehicle}; \quad \forall p \in P, k \in K, i \in B \tag{36}$$

$$\beta^{setup} + \beta^{picking} \times \left(Cap\ batch_i \times x_{ipk}\right) + \beta^{time} \times \left(\sum_{i\in B} dis_i \times x_{ipk}\right) \leq t\_exec_i\ i \in B \tag{37}$$

$$t\_exec_i - M \times \left(1 - x_{ipk}\right) \leq u_{pk}; \quad \forall p \in P, k \in K, i \in B \tag{38}$$

$$u_{p1} \leq v_{p1}; \quad \forall p \in P \tag{39}$$

$$u_{pk} + v_{pk-1} \leq v_{pk}; \qquad \forall p \in P, k \in K, |k \geq 2 \tag{40}$$

$$v_{pk} - due_{caricon} - M \times \left(2 - x_{ipk} - b_{in}\right) \leq ta_n; \quad \forall p \in P, k \in K, h \in H, n \in N \tag{41}$$

$$due\_carico_n - v_{pk} - M \times \left(2 - x_{ipk} - b_{in}\right) \leq ea_n; \ \forall p \in P, k \in K, h \in H, n \in N \tag{42}$$

$$x_{ipk} \in \{0,1\}; \ \forall p \in P, k \in K, i \in B \tag{43}$$

$$t\_exec_i \geq 0; \forall i \in B \tag{44}$$

$$ta_n \geq 0; \quad \forall n \in N \tag{45}$$

$$ea_n \geq 0; \quad \forall n \in N \tag{46}$$

$$u_{pk} \geq 0; \quad \forall p \in P, k \in K \tag{47}$$

$$v_{pk} \geq 0; \quad \forall p \in P, k \in K \tag{48}$$

Equation (33) is the objective function to be minimised and is the same as that used in the previous model. However, compared to the previous model, this model does not contain the picking routing constraints to reduce its complexity.

4.3. Heuristic Solving Approach

Ref. [37] showed that the problem of minimising the total tardiness for a set of independent tasks on a single machine is NP-hard. This problem can be interpreted as a special case of the order batching and sequencing problem, in which the capacity of the picking device is equal to only one customer's order. Therefore, it is possible to conclude that the order batching and sequencing problem is also NP-hard and using a heuristics approach is advantageous [28].

As both previously implemented models are complex and, therefore, were not solvable within acceptable time limits, this study developed a constructive heuristic algorithm. Constructive algorithms use common sense rules to solve the problem quickly but without the certainty of reaching the globally optimal solution. Various heuristic solution algorithms have been proposed for the order batching and sequencing problem. Such algorithms can be categorized into four groups: priority rule-based, seed, saving, and metaheuristic algorithms. The first three groups use constructive approaches, whereas metaheuristic algorithms focus on improving given solutions. Metaheuristic algorithms have been shown to significantly improve the results obtained using constructive algorithms through several numerical experiments [38].

This study used an approach based on order sequencing according to the earliest due date rule to generate the initial solution, i.e., the orders with the nearest due date are assigned first. Therefore, the proposed algorithm is priority-based, i.e., it assigns orders to the batches and the operator's position one at a time. Priority rule-based algorithms have a two-step procedure: first, priorities are assigned to customer orders, and second, these orders are assigned to different batches. This procedure ensures that the capacity constraint is not violated.

In the pseudo-code, U represents the set of orders not yet assigned to a picker; k_p is the sequence position of the operator p to whom the next batch can be assigned to; B_{pk_p} represents the set of orders included in the batch assigned to the position k_p of operator p; C_p and v_p are the numbers of items contained in the batch under consideration and the completion time of the batch, respectively. The execution time of the current batch is calculated as $\beta^{time} \times \left(TSP\left(B_{pk_p}\right)\right)$, where TSP indicates the call to the heuristic method for the calculation of the routing strategy of the batch under consideration and β^{time} is the time required to travel a unit of space. Initially, all the orders belong to the vector U because they have not yet been assigned. Then, the orders are successively assigned to specific batches, starting from the order n^* not yet assigned with the earliest due date. First, this study assigns the current order to each operator p and calculates the completion time $\widetilde{v}_p$ for each. The assignment consists of adding the order n^* to the batch B_{pk_p} if the capacity constraint is not violated; otherwise, the assignment includes opening a new batch containing n^*. Second, the algorithm decides which operator p^* the order under consideration is assigned to. This order is assigned to the operator who has generated the minimum completion time in the previous step. Third, the algorithm evaluates the completion times obtained for each picker after the assignment of n^* and selects the one with the smallest value. Fourth, the order n^* is assigned to the corresponding picker considering the capacity constraint. The algorithm ends when all the orders have been assigned to a batch. Algorithm 1 shows the procedure's pseudo-code, algorithm principle, input, and output.

Algorithm 1 Pseudo-code.

Algorithm principle: Assign the orders individually, starting with the ones with the closest due date. The current order is assigned to the operator who is least loaded.
Input: set of N customer orders sorted in ascending order of due dates; the number of items requested c_n ($n \in N$) for each order $n \in N$; set of operators P; capacity of the pickup vehicle $Cap_{vehicle}$.
Output: heuristic value z^*, which represents the objective function to be minimised, and the corresponding values of total travel time, tardiness, and earliness.

$U := set\ of\ N\ customer\ orders\ sorted\ in\ ascending\ order\ of\ due\ dates;$
for p ∈ P **do**
  $k_p = 1;\ B_{pk_p} = \varnothing; C_p := 0; v_p = 0;$
end for
while $U \neq \varnothing$ **do**
  $n^* = argmin\{due\_carico_n \mid n \in U\};$
  for $p \in P$ **do**
    if $C_p + c_{n^*} \leq Cap_{vehicle}$
      then $\tilde{v}_p = v_p + \beta^{time} \times \left(TSP\left(B_{pk_p} \cup \{n^*\}\right)\right);$
    else
$\tilde{v}_p = v_p + \beta^{time} \times (TSP(\{n^*\}));$
    end if
  end for
$p^* = argmin\ \{\tilde{v}_p \mid p \in P\};$
$U = U \backslash \{n^*\}; v_{p^*} = \tilde{v}_{p^*};$
if $C_{p^*} + c_{n^*} \leq Cap_{vehicle}$
then $B_{p^*k_{p^*}} = B_{p^*k_{p^*}} \cup \{n^*\}; C_{p^*} = C_{p^*} + c_{n^*};$
else
$k_{p^*} = k_{p^*} + 1; B_{p^*k_{p^*}} = \{n^*\}; C_{p^*} = c_{n^*};$
end if
end while

The heuristic algorithm proposed does not guarantee the identification of the optimal solution but runs the lowest possible risk of having delays. The initial solution was improved by applying different types of local search. A multi-start approach was implemented to evaluate the largest number of possible solutions. The goal was to generate a random sequence at each iteration and, based on this sequence, find the solution; for each execution, a different objective function value is obtained. By comparing the different results, the best order, i.e., the order that generates the lowest value of z, was identified. This study implemented a local search swap and insert move approach, which improved the heuristic solution produced by applying small changes that led to local improvements. Two types of local search exist, namely, the first improvement and best improvement, which define the end criterion of the algorithm. With the first improvement, when the algorithm reaches a local minimum, it stops and returns this minimum as the output value. With the best improvement, the algorithm does not stop when it reaches a local minimum but continues the search until the end criterion is satisfied.

In this study, the following approaches were developed:

1. Swap—first improvement: exchanges two orders of the input sequence and verifies the solution. Once the algorithm identifies a sequence that generates a better objective function, it stops and releases the sequence as the output.
2. Insert move—best improvement: inserts each order in each position, and once it finds an improvement, it returns the sequence as the output.

This study implemented the iterated local search algorithm to increase the objective function's value generated by the heuristic algorithm. The basic principle of this approach was to originate a sequence of possible solutions starting from a generic one. The solutions were obtained by perturbing the current solution and implementing a local search procedure. An acceptance criterion was applied to decide which solution among the candidates to consider in the next step. This study used a Markovian acceptance criterion, which allows an extreme intensification because it accepts s'^{*} if, and only if, $f(s'^{*}) < f(s^{*})$ [37]. Algorithm 2 shows the local search procedure.

Algorithm 2 The local search procedure.

procedure *IteratedLocalSearch*
  $s_0 = GenerateInitialSolution$
  $s^* = LocalSearch(s_0)$ %*Optional*
  repeat
    $s' = Perturbation\ (s^*)$
    $s^{*'} = LocalSearch\ (s')$
    $s^* = AcceptanceCriterion\ \left(s^*, s^{*'}\right)$
  until *the termination condition met*
end

The heuristic earliest due date generated the initial solution in this approach, and a multi-start algorithm performed the first local search. The first solution was perturbed through a swap move perturbation algorithm and improved through a local search insert move. As the acceptance criterion was Markovian, the solution generated was accepted only if it was better than that generated with the multi-start algorithm. If this was the case, the value of the best result found was inserted, and the number of iterations without improvement was reset. At the beginning of the procedure, it was necessary to set the maximum number of cycles without improvement. The cycle stopped when no more solutions existed for the consecutive number of cycles set. Algorithm 3 shows the earliest due date procedure.

Algorithm 3 The earliest due date procedure.

procedure *IteratedLocalSearch*
  $s_0 = Earliest\ Due\ Date\ Algorithm$
  $s^* = Multi-Start\ (s_0)$
  repeat
    $s' = Swap\ Move\ Perturbation\ (s^*)$
    $s^{*'} = Local\ Search\ Insert\ Move\ (s')$
    $s^* = AcceptanceCriterion\ \left(s^*, s^{*'}\right)$
  until $Iteration\ without\ improvement > Max\ Interaction\ without\ improvement$
end

Finally, this study implemented an iterated local search multiple perturbation. This approach had the same acceptance characteristics as the previous one, with the only difference being that it chooses whether to perform an insert move perturbation or a swap perturbation randomly.

5. Numerical Experiments

5.1. Implementation of Models

This study designed a simple problem that could be solved within acceptable time limits and allowed us to understand how the results were generated. As mentioned above, the order picking time comprises the items' travel time, search and pickup time, and setup time. This study used the same parameters of the test phase of the algorithm presented in [29] to assign realistic values to the parameters βtime, βsetup, and βpicking. It was estimated that an operator travels 20 units of length per minute, which corresponds to 3 s per unit of length. Furthermore, it was estimated that it takes 10 s to search and retrieve an item from the storage location and that each batch requires a setup time of 3 min (180 s). The capacity of the picking vehicle was assumed to be 20 items. As inputs, both models need the characteristics of each order: the due date, the number of items requested, and the picking locations that need to be visited.

It is important to note that all the time and distance values input had units of seconds and meters, respectively. However, the specific units used can be changed if they are consistent. These examples demonstrate the model's effectiveness and show how consistent solutions can be generated by assigning more or less priority to the different factors. Another emergent factor is that even slight variations can lead to significant changes to the solution in the case of more critical input data.

5.2. Heuristic Implementation

This study considered empirical data to test the proposed heuristic algorithm's correctness. For clarity, it was decided to initially test only a tiny portion of this data. The steps taken to manage and implement the data were as follows. First, to simplify the extensive shipping database, this study selected only the necessary data, i.e., the name of the order, picking locations, due date, and quantities needed in each order. It was essential to know the distances that separate all the storage locations to minimise the completion time and evaluate the distances travelled. As this information is not currently available within the company considered and the number of picking locations was too high to allow manual measurement (there are approximately 550 total spaces in the picking warehouse), this study implemented an algorithm on the resolution software IVE Xpress to calculate the distance matrix automatically. As the code assigned to each location was not random, it can be used to identify the actual position of the location in the warehouse. The basic idea of the algorithm was to exploit this knowledge to calculate distances: the first three letters of the code represent the specific aisle, and the following two numbers identify the position of the location inside the aisle. All this information allows the determination of the precise picking location.

The warehouse layout was used to calculate the distances between the picking locations. As the warehouse was developed from successive enlargements and in compliance with strong spatial constraints, its layout is complex and non-linear. The layout can be described as the union of two different blocks with a longitudinal layout. It consists of several parallel picking aisles with storage locations on both sides and a single central aisle (transverse aisle) without storage locations that enable order pickers to enter or exit a picking aisle.

Another data input not directly available is the matrix a, which is related to the picking location that each order must visit; it is defined as a $num\_order \times num\_location$ matrix with elements a_{ij} equal to 1 if order I visits compartment j and 0 otherwise. An algorithm was developed to calculate this matrix automatically, starting from orders and the corresponding locations. The algorithm used the data derived from the database as the input. However, one order was repeated several times. Each replica was associated with the value of a picking location where in the storage a different article was requested. Starting from the procedure just described, the matrix was constructed iteratively: given two vectors containing the orders and the non-duplicated spaces, the matrix was defined by setting a_{ij} to 1 if the order i required a visit to location j and 0 otherwise.

This study evaluated the bulk orders in the first week of July 2019, which represents a critical period for the organisation owing to the seasonality of the products. To evaluate the implementation of the batching strategy, this study compared the times obtained with the algorithms with those obtained with the strategy commonly implemented in the warehouse, i.e., the single order strategy. With this objective, it was decided to compare the time required to pick orders, assuming that only one operator was involved in picking individual orders. Table 2 lists the results of the case study.

Table 2. Travel time with different strategies.

Date	Travel Time (min)		Difference
	Single Order Strategy	Batching Strategy	
1 July	118.38	88.77	29.61
2 July	152.74	102.39	50.35
3 July	168.07	127.90	40.17
4 July	160.97	144.73	16.24
5 July	202.17	144.68	57.49
Average	160.47	121.69	38.77 (24%)

With the batching strategy, the average travel time was 38.77 min/d. The variability in the different days was due to the different levels of order overlap, which strongly affected the time savings obtained through the batching.

In the evaluation, this study only considered the travel time component, i.e., the time required to reach each picking location. Next, this study evaluated the setup times expected to equal 180 s. As the setup time is involved every time the picker starts a new pick list, the overall setup time in a day under the single order strategy could be calculated as the fixed setup time for the number of orders. In contrast, in the batching strategy, the time required for this activity could be calculated as the product of the fixed time and the batch number. Therefore, with the single order strategy, the average setup time was 201.00 min/d, while with batching strategy, it decreased to 31.80 min/d, which represents an average daily time saving of 169.20 min. Table 3 shows the setup time for different strategies. Since the earliest due date heuristic sequence is in ascending order according to the due date (i.e., it considers one order at a time from the one with the closest due date up to the one that represents the least criticality, and already takes into account the delays), we chose to assign the maximum weight (1) to the completion time. However, the proposed model can be applied with different weights.

Table 3. Setup time with different strategies.

Date	Total Orders	Total Batches	Setup Time (min)		Difference
			Single Order Strategy	Batching Strategy	
1 July	50	8	150	24	126
2 July	71	7	213	21	192
3 July	74	12	222	36	186
4 July	71	12	213	36	177
5 July	69	14	207	42	165
Average	67.0	10.6	201.0	31.8	169.2 (84%)

The picking time was constant on average because the number of items to be picked did not change. Therefore, the average time saved by applying the batching strategy was 207.97 min/d or approximately 3.47 h/d. Table 4 shows the total time for different strategies.

However, the batching strategy requires a final sorting activity, dividing the batch into individual customer orders. Some choices were made in the model definition phase to reduce the time and space required for this activity. First, each order was assumed to belong to only one batch, i.e., there was no fractioning; moreover, the capacity constraint guaranteed that, on average, each mission contained few orders. However, this activity was still necessary and introduces additional time not present within the single order strategy. As the time required to sort the batch is not known, this study calculated the sorting threshold time that guaranteed a gain in terms of time; if this value was exceeded, there were no time gains. Table 5 shows the sorting time with the batching strategy.

Table 4. Picking total time with different strategies.

Date	Picking Total Time (min)		Difference
	Single Order Strategy	Batching Strategy	
1 July	268.38	112.77	155.61
2 July	365.74	123.39	242.35
3 July	390.07	163.90	226.17
4 July	373.97	180.73	193.24
5 July	409.17	186.68	222.49
Average	361.47	153.49	207.97 (57%)

Table 5. Sorting time with the batching strategy.

Date	Total Difference [min]	Total Batches	Sorting Time for Batch (min)	Average Order/Batch (min)	Sorting Time for Order (min)
1 July	155.61	8	19.45	6.25	3.11
2 July	242.35	7	34.62	8.88	3.90
3 July	226.17	12	18.85	9.25	2.04
4 July	193.24	12	16.10	8.88	1.81
5 July	222.49	14	15.89	8.63	1.84
Average	207.97	10.60	20.98	8.38	2.54

5.3. Discussion on the Implementation

An alternative to the batching strategy is the sort while picking method. In this method, an operator uses a picking cart containing several pallets to place objects in the same order at the same point. The advantage of this strategy is that it does not need a sorting activity after the picking is concluded; instead, the sorting activity takes place simultaneously with the picking. This picking strategy is preferred when handling orders consisting of small items and a relatively small number of pieces.

The results showed that the traditional pick-and-sort batching strategy is advantageous only if the sorting activity requires an average time of less than 20.98 min per batch or less than 2.54 min per order. In this case, the overall time savings strongly depend on the time required for the final sorting activity. On the other hand, a sort while picking batching strategy might be more convenient because the batches contain a limited number of orders and items. Therefore, the traditional pick-and-sort batching strategy might be convenient for picking bulk orders. This way, the sorting time would be zero, with a consequent average daily saving of 207.97 min (57%). Establishing a comparison, refs. [23,24] reported savings between 10 and 25%.

Further research should also focus on a more accurate and structured analysis of the different possible sorting systems, identifying the respective characteristics, advantages, and disadvantages with particular attention to their sorting times. Furthermore, a more accurate feasibility study of sort while picking systems could be carried out, focusing on the need to use picking vehicles of greater length, with particular attention to possible consequences on traffic congestion.

All the heuristic approaches developed were tested to define the solution for each day examined. It was found that the iterated local search algorithm and iterated local search multiple perturbation generated the best solution by four and one times, respectively. Therefore, the local search effectively leads to an improvement in the initial solution. However, specific challenges might emerge during the algorithm's implementation. The primary obstacle involves the uncertainty surrounding the measurement of parameters such as order-related timings. Another complexity lies in the necessity for on-site equipment within the factory premises, accessible to operators. Lastly, should the operation's evolution surpass the algorithm's predefined limits, it would necessitate a fresh implementation.

The overall results of the study have implications.

The advances highlighted by the heuristic can be leveraged by companies developing solutions for logistics companies, either as open innovation initiatives or as software licenses. Stakeholders such as logistics companies, distribution centers, freight forwarders, and users of logistics services can benefit from such an investment by reducing their storage costs, increasing the reliability of delivery times, reducing errors in the quantity and identification of materials, or increasing the flexibility of their operations. All these effects can increase the competitiveness of logistics companies and their customers, usually supply chain managers, in their industries. In particular, due to mathematical requirements, supply chain managers should use the resource as software add-ons to their usual warehouse management systems, which are already operating.

6. Conclusions

This study considered JOBSPRP and tried adapting it to the needs of an organisation that provides logistics services. Two depots were assumed in the definition of the routing strategy that were logically separated but may be physically coincident in practice. Unlike most previous models, with more than proportional growth, the size of the proposed model grew polynomially with the number of orders. To limit the amount of input data, the study developed another model to solve the picking routing problem. This model only needed some of the picking locations of the warehouse as the input, namely, the locations that required at least one visit on a day. Both proposed models could solve minor problems with acceptable computational times. To solve real, large problems, the study developed approaches based on the local search algorithm. The results showed that, for large problems, the iterated local search and iterated local search multiple perturbation algorithms could successfully identify the minimum solution and significantly improve the solution initially obtained with the heuristic earliest due date algorithm.

Future research should focus on picker blocking in real warehouses, which generally need optimal layouts. Picker blocking represents an essential factor that can cause significant inconvenience and slowdowns during picking articles, increasing the picking times and reducing productivity. When traffic congestion is considered, the tours of different operators can no longer be independently determined, making the problem significantly more complex. Furthermore, cases where customer orders are not known in advance but are made dynamically during the day could be considered [31]. Further research should also focus on the impact of picking and batching efficiency on the other steps of the supply chain, such as road and urban transportation [39].

The findings provide room for future advances. Regarding the involved technology, the next step is to develop a software add-on to be implemented in existing warehouse management systems. Furthermore, a parallel decision support system could increase the likelihood of logistics operators meeting order requirements, such as due date, quantity, and quality of the loads, while reducing costs associated with a lack of efficiency. Regarding the implications, a further step is to lead various real-world implementations to measure the gains such an add-on could provide to customers interested in managing their orders' dispatch operations.

Author Contributions: Conceptualization, E.M., F.L. and E.B.; methodology, A.M.C., E.M., F.L. and M.A.S.; software, E.M. and E.B.; validation, F.L. and M.A.S.; formal analysis, E.M. and A.M.C.; investigation, F.L. and E.M.; data curation, A.M.C. and E.B.; writing—original draft preparation A.M.C., F.L. and M.A.S.; writing—review and editing A.M.C. and M.A.S.; supervision F.L.; project administration F.L. and E.B. All authors have read and agreed to the published version of the manuscript.

Funding: This research received no external funding.

Data Availability Statement: The data presented in this study are available on request from the corresponding author.

Conflicts of Interest: The authors declare no conflict of interest.

References

1. Song, J.S.; van Houtum, G.J.; van Mieghem, J.A. Capacity and Inventory Management: Review, Trends, and Projections. *Manuf. Serv. Oper. Manag.* **2020**, *22*, 36–46. [CrossRef]
2. Zheng, B.; Sun, C. Research on Logistics Warehousing Cost of Z Delivery Enterprise and Its Optimization Strategy. *Front. Bus. Econ. Manag.* **2023**, *8*, 12–22. [CrossRef]
3. Yilmaz, O.F.; Cevikcan, E.; Durmusoglu, M.B. Scheduling Batches in Multi Hybrid Cell Manufacturing System Considering Worker Resources: A Case Study from Pipeline Industry. *Adv. Prod. Eng. Manag.* **2016**, *11*, 192–206. [CrossRef]
4. Gauss, L.; Lacerda, D.P.; Sellitto, M.A. Module-Based Machinery Design: A Method to Support the Design of Modular Machine Families for Reconfigurable Manufacturing Systems. *Int. J. Adv. Manuf. Technol.* **2019**, *102*, 3911–3936. [CrossRef]
5. Lolli, F.; Lodi, F.; Giberti, C.; Coruzzolo, A.M.; Marinello, S. Order Picking Systems: A Queue Model for Dimensioning the Storage Capacity, the Crew of Pickers, and the AGV Fleet. *Math. Probl. Eng.* **2022**, *2022*, 6318659. [CrossRef]
6. Custodio, L.; Machado, R. Flexible Automated Warehouse: A Literature Review and an Innovative Framework. *Int. J. Adv. Manuf. Technol.* **2020**, *106*, 533–558. [CrossRef]
7. de Koster, R.; Le-Duc, T.; Roodbergen, K.J. Design and Control of Warehouse Order Picking: A Literature Review. *Eur. J. Oper. Res.* **2007**, *182*, 481–501. [CrossRef]
8. Tompkins, J.A.; White, J.A.; Bozer, Y.A.; Tanchoco, J.M.A. *Facilities Planning*; Wiley: Hoboken, NJ, USA, 2003; ISBN 9780470444047.
9. Yilmaz, O.F.; Oztaysi, B.; Durmusoglu, M.B.; Oner, S.C. Determination of Material Handling Equipment for Lean In-Plant Logistics Using Fuzzy Analytical Network Process Considering Risk Attitudes of the Experts. *Int. J. Ind. Eng. Theory Appl. Pract.* **2017**, *24*. [CrossRef]
10. Lee, M.K.; Scherer, E.; Elsayed, E.A.; Kim, S. Sequencing and Batching Procedures for Minimizing Earliness and Tardiness Penalty of Order Retrievals. *Int. J. Prod. Res.* **1993**, *31*, 727–738. [CrossRef]
11. Elsayed, E.A.; Lee, M.K. Order Processing in Automated Storage/Retrieval Systems with Due Dates. *IIE Trans.* **1996**, *28*, 567–577. [CrossRef]
12. Tsai, C.Y.; Liou, J.J.H.; Huang, T.M. Using a Multiple-GA Method to Solve the Batch Picking Problem: Considering Travel Distance and Order Due Time. *Int. J. Prod. Res.* **2008**, *46*, 6533–6555. [CrossRef]
13. Kulak, O.; Sahin, Y.; Taner, M.E. Joint Order Batching and Picker Routing in Single and Multiple-Cross-Aisle Warehouses Using Cluster-Based Tabu Search Algorithms. *Flex. Serv. Manuf. J.* **2012**, *24*, 52–80. [CrossRef]
14. Henn, S.; Schmid, V. Metaheuristics for Order Batching and Sequencing in Manual Order Picking Systems. *Comput. Ind. Eng.* **2013**, *66*, 338–351. [CrossRef]
15. Grosse, E.H.; Glock, C.H.; Ballester-Ripoll, R. A Simulated Annealing Approach for the Joint Order Batching and Order Picker Routing Problem with Weight Restrictions. *Int. J. Oper. Quant. Manag.* **2014**, *20*, 65–83.
16. Henn, S. Order Batching and Sequencing for the Minimization of the Total Tardiness in Picker-to-Part Warehouses. *Flex. Serv. Manuf. J.* **2015**, *27*, 86–114. [CrossRef]
17. Chen, T.L.; Cheng, C.Y.; Chen, Y.Y.; Chan, L.K. An Efficient Hybrid Algorithm for Integrated Order Batching, Sequencing and Routing Problem. *Int. J. Prod. Econ.* **2015**, *159*, 158–167. [CrossRef]
18. Scholz, A.; Schubert, D.; Wäscher, G. Order Picking with Multiple Pickers and Due Dates—Simultaneous Solution of Order Batching, Batch Assignment and Sequencing, and Picker Routing Problems. *Eur. J. Oper. Res.* **2017**, *263*, 461–478. [CrossRef]
19. Ardjmand, E.; Sanei Bajgiran, O.; Youssef, E. Using List-Based Simulated Annealing and Genetic Algorithm for Order Batching and Picker Routing in Put Wall Based Picking Systems. *Appl. Soft Comput.* **2019**, *75*, 106–119. [CrossRef]
20. Kübler, P.; Glock, C.H.; Bauernhansl, T. A New Iterative Method for Solving the Joint Dynamic Storage Location Assignment, Order Batching and Picker Routing Problem in Manual Picker-to-Parts Warehouses. *Comput. Ind. Eng.* **2020**, *147*, 106645. [CrossRef]
21. Attari, M.Y.N.; Torkayesh, A.E.; Malmir, B.; Jami, E.N. Robust Possibilistic Programming for Joint Order Batching and Picker Routing Problem in Warehouse Management. *Int. J. Prod. Res.* **2020**, *59*, 4434–4452. [CrossRef]
22. Cano, J.A.; Cortés, P.; Campo, E.A.; Correa-Espinal, A.A. Solving the Order Batching and Sequencing Problem with Multiple Pickers: A Grouped Genetic Algorithm. *Int. J. Electr. Comput. Eng. (IJECE)* **2021**, *11*, 2516–2524. [CrossRef]
23. Cano, J.A.; Cortés, P.; Campo, E.A.; Correa-Espinal, A.A. Multi-Objective Grouping Genetic Algorithm for the Joint Order Batching, Batch Assignment, and Sequencing Problem. *Int. J. Manag. Sci. Eng. Manag.* **2021**, *17*, 188–204. [CrossRef]
24. Cergibozan, Ç.; Tasan, A.S. Genetic Algorithm Based Approaches to Solve the Order Batching Problem and a Case Study in a Distribution Center. *J. Intell. Manuf.* **2022**, *33*, 137–149. [CrossRef]
25. Lin, C.C.; Kang, J.R.; Hou, C.C.; Cheng, C.Y. Joint Order Batching and Picker Manhattan Routing Problem. *Comput. Ind. Eng.* **2016**, *95*, 164–174. [CrossRef]
26. Henn, S.; Koch, S.; Wäscher, G. Order Batching in Order Picking Warehouses: A Survey of Solution Approaches. In *Warehousing in the Global Supply Chain*; Springer: London, UK, 2012; pp. 105–137. ISBN 978-1-4471-2273-9.
27. Lu, W.; McFarlane, D.; Giannikas, V.; Zhang, Q. An Algorithm for Dynamic Order-Picking in Warehouse Operations. *Eur. J. Oper. Res.* **2016**, *248*, 107–122. [CrossRef]
28. Valle, C.A.; Beasley, J.E.; da Cunha, A.S. Optimally Solving the Joint Order Batching and Picker Routing Problem. *Eur. J. Oper. Res.* **2017**, *262*, 817–834. [CrossRef]

29. Matusiak, M.; de Koster, R.; Saarinen, J. Utilizing Individual Picker Skills to Improve Order Batching in a Warehouse. *Eur. J. Oper. Res.* **2017**, *263*, 888–899. [CrossRef]
30. Cano, J.A.; Correa-Espinal, A.A.; Gómez-Montoya, R.A. An Evaluation of Picking Routing Policies to Improve Warehouse Efficiency. *Int. J. Ind. Eng. Manag.* **2017**, *8*, 229–238. [CrossRef]
31. Azadnia, A.H.; Taheri, S.; Ghadimi, P.; Mat Saman, M.Z.; Wong, K.Y. Order Batching in Warehouses by Minimizing Total Tardiness: A Hybrid Approach of Weighted Association Rule Mining and Genetic Algorithms. *Sci. World J.* **2013**, *2013*, 246578. [CrossRef]
32. Scholz, A.; Wäscher, G. Order Batching and Picker Routing in Manual Order Picking Systems: The Benefits of Integrated Routing. *Cent. Eur. J. Oper. Res.* **2017**, *25*, 491–520. [CrossRef]
33. Bozer, Y.A.; Kile, J.W. Order Batching in Walk-and-Pick Order Picking Systems. *Int. J. Prod. Res.* **2008**, *46*, 1887–1909. [CrossRef]
34. Cano, J.A.; Correa-Espinal, A.A.; Gómez-Montoya, R.A. Mathematical Programming Modeling for Joint Order Batching, Sequencing and Picker Routing Problems in Manual Order Picking Systems. *J. King Saud Univ. Eng. Sci.* **2020**, *32*, 219–228. [CrossRef]
35. Gademann, N.; van de Velde, S. Order Batching to Minimize Total Travel Time in a Parallel-Aisle Warehouse. *IIE Trans.* **2007**, *37*, 63–75. [CrossRef]
36. Du, J.; Leung, J.Y.-T. Minimizing Total Tardiness on One Machine Is NP-Hard. *Math. Oper. Res.* **1990**, *15*, 483–495. [CrossRef]
37. Henn, S.; Koch, S.; Doerner, K.F.; Strauss, C.; Wäscher, G. Metaheuristics for the Order Batching Problem in Manual Order Picking Systems. *Bus. Res.* **2010**, *3*, 82–105. [CrossRef]
38. Lourenço, H.R.; Martin, O.C.; Stützle, T. Iterated Local Search. In *Handbook of Metaheuristics*; Springer: Boston, MA, USA, 2003; pp. 320–353. [CrossRef]
39. Sellitto, M.A.; Borchardt, M.; Pereira, G.M.; Gomes, L.P. Environmental performance assessment of a provider of logistical services in an industrial supply chain. *Theor. Found. Chem. Eng.* **2012**, *46*, 691–703. [CrossRef]

logistics

MDPI

Systematic Review

Transformative Procurement Trends: Integrating Industry 4.0 Technologies for Enhanced Procurement Processes

Areej Althabatah [1,2], Mohammed Yaqot [1], Brenno Menezes [1] and Laoucine Kerbache [1,*]

[1] Division of Engineering Management and Decision Sciences, College of Science and Engineering, Hamad Bin Khalifa University, Qatar Foundation, Doha 34110, Qatar; aral33949@hbku.edu.qa (A.A.); myaqot@hbku.edu.qa (M.Y.); bmenezes@hbku.edu.qa (B.M.)
[2] QatarGas, Doha 22666, Qatar
[*] Correspondence: lakerbache@hbku.edu.qa

Abstract: *Background*: the advent of Industry 4.0 (I4.0) innovations has revolutionized supply chain management through technologies like the Internet of Things (IoT) and Artificial Intelligence (AI) integrated into procurement processes. *Methods*: this study addresses a critical knowledge gap by conducting a comprehensive review of *111* papers sourced from the Scopus database. These papers are classified into seven sub-themes encompassing I4.0 or procurement 4.0 (P4.0), big data, IoT, additive manufacturing, blockchain, e-procurement, and AI. *Results*: the investigation reveals that I4.0 technologies, particularly e-procurement and blockchain, have garnered substantial attention. Such technologies offer diverse value propositions, encompassing streamlined supplier evaluation, lead time reduction, cost optimization, and enhanced data security. *Conclusion*: the paper underscores pivotal trends and insights for the evolution of Procurement 4.0, illuminating a path toward more efficient supply chain management.

Keywords: procurement; industry 4.0; digitalization; additive manufacturing; blockchain; procurement 4.0; internet of things; e-procurement

Citation: Althabatah, A.; Yaqot, M.; Menezes, B.; Kerbache, L. Transformative Procurement Trends: Integrating Industry 4.0 Technologies for Enhanced Procurement Processes. *Logistics* **2023**, *7*, 63. https://doi.org/10.3390/logistics7030063

Academic Editors: Tomasz Nowakowski, Artur Kierzkowski, Agnieszka A. Tubis, Franciszek Restel, Tomasz Kisiel, Anna Jodejko-Pietruczuk and Mateusz Zając

Received: 24 June 2023
Revised: 24 August 2023
Accepted: 28 August 2023
Published: 13 September 2023

1. Introduction

The 4th Industrial Revolution, or Industry 4.0 (I4.0), represents generally the convergence of information and communication technologies (ICT), and the automation of machinery and infrastructure [1,2]. In manufacturing, it stands for the digitalization of factories and their related supply chain processes. As a result of advanced digitalization in manufacturing, a new concept known as 'smart manufacturing' has come into existence. Smart manufacturing revolutionizes the coordination of manufacturing elements by leveraging an integrated data chain system facilitated by the Internet, which made a critical improvement in strategic supply chain management [3]. Procurement is considered a critical part of the supply chain since it affects directly the overall performance of the whole supply chain.

Procurement is defined broadly in terms of any form of the buying process, whether it is in the form of purchasing, leasing, renting, or otherwise acquiring supplies, services, or construction from external suppliers [4,5]. This concept includes all management responsibilities which the managers are supposed to follow such as identifying, sourcing, accessing, and managing resources outside the company. Also, it suggests fulfilling of strategic objectives of the company to make significant business value and impact profit. From the perspective of supply chain management, the role of procurement is considered indispensable, as it plays a crucial role in highlighting the importance of ongoing enhancements within the procurement division [1].

Procurement 4.0 (P4.0), which is a term that encompasses a component of I4.0, introduces a network between all supply chain upstream partners, enabling dynamic and quick cooperation and coordination beyond organizational frontiers [6]. Companies have increasingly embraced P4.0 techniques in their supply chain (SC) operating models, offering valuable support to procurement managers in overcoming purchasing challenges. This is made possible through the utilization of cutting-edge technologies such as the Internet of Things, e-procurement, big data, additive manufacturing (AM) or 3D printing, blockchain, artificial intelligence, and modularity. Table 1 provides a brief definition for each I4.0 component. These technologies affect every aspect of the value chain and empower procurement managers to optimize their processes and enhance decision-making capabilities for more efficient and effective procurement operations, including supplier sourcing, distribution, transportation, warehousing, and most importantly, customer satisfaction. Moreover, they have the potential to contribute to effectiveness, efficiency, cost reduction, and shorter lead times in the supply chain, thereby providing significant benefits for the supply chain industry [7].

Table 1. Brief definition for each I4.0 component [8].

I4.0	Brief Definition
Modularity	Strategic division of complex systems into independent modules for efficient application management.
E-procurement	Electronic procurement processes enable streamlined sourcing, ordering, and transactions.
Big Data	Large-scale data analysis to uncover insights and patterns, enhancing decision-making.
IoT	Network of interconnected devices collecting and exchanging data for informed actions.
AM or 3D Printing	Layer-by-layer creation of objects, revolutionizing production and customization.
Blockchain	Decentralized and secure digital ledger ensuring transparency and trust in transactions.
AI	Simulation of human intelligence in machines, enabling data-driven insights and optimized decisions.

In procurement, the three primary components of I4.0 that play a pivotal role are automation, connectivity, and advanced analytics. With these evolving capabilities, procurement processes can be almost further automated, and decision-making processes can be supported through advanced analytics tools [9]. Although information technology systems have been in use to support procurement functions for many years, their emerging holistic use in each stage of the procurement process in the supply chain at present is unique and constitutes Procurement 4.0. It is believed that the scope of procurement processes will be even more expanded in the future as a result of new technological advancements. For instance, the development of modern intelligent systems, which will have the computing power, connectivity, and autonomy to be able to analyze historical and future-oriented data may bring newer mechanics to procurement processes. This will be capable of offering holistic, autonomous, and real-time task completions. Electronic procurement (e-procurement), for example, can offer electronic solutions to procurement processes. This is the first step for digitalization that gives procurement departments smart support for operational and tactical decisions [10]. However, these advantages of procurement are not well taken into the companies' operations. Bogaschwesky and Müller [11] point out, in a survey made for procurement managers in different German companies, the majority of the interviewees (56%) are still reluctant to proceed with digitalization in their procurement department and this attitude must be overcome to adopt these developments.

There is an increasing trend for procurement decision-makers to invest in digital technologies in their quest for greater access to global markets to obtain competitive advantages through the interconnection between buyers and suppliers [12]. Executives in procurement anticipate cost reductions without losing efficiency throughout the whole supply chain. However, many of them opt for digitalization in order to avoid being outdated [13]. In fact, a study by the European Economic and Social Committee states that people work more efficiently due to procurement digitalization [14]. Digitalization brings forth innovations and economic growth that create new methods and technologies. In short, one could argue that digitalization has multiple consequences as it creates and reduces jobs but its overall impact is yet to be determined [15].

The adoption of a P4.0 program requires an essential restructuring of business organizations. It calls for a new organizational strategy because several operational processes have to be modified in this new stage [16]. Through this restructuring, many processes of the companies such as procurement, production, logistics, and IT departments were connected to each other in terms of accounting and controlling their activities. Therefore, it is important to restructure interactions within supply chain management completely [17]. I4.0 is widely acknowledged as having the ability to provide new capabilities to procurement functions [9]. By using automated and internally integrated systems, a completely autonomous operational procurement system can be enabled. In p4.0, interactions between internal and external shareholders would be increased and more reliable data would be generated in which advanced analytics offer up-to-date information and new ways of analysis supporting decision-making processes.

To demonstrate the paradigm-shifting influence of I4.0 technologies on procurement procedures, the following few leading examples from diverse industries can be illustrated. In the manufacturing domain, Toyota emerges as a paradigmatic case. The seamless integration of I4.0 elements such as the IoT and AI has endowed Toyota with a streamlined supply chain marked by enhanced efficiency, minimized downtime, and amplified productivity. In the service sector, Amazon exemplifies the transformative potential of I4.0. Skillful application of advanced data analytics, automation, and AI-driven decision-making has not only expedited order fulfillment from 80% to 90% but also cultivated tailor-made customer experiences, epitomizing the burgeoning importance of these technologies in the contemporary consumer landscape [18]. Shifting the focus to the public sphere, the World Health Organization (WHO) magnifies the pivotal role of I4.0 technologies in global health management. Through data-driven insights and predictive analytics, WHO proactively addresses disease outbreaks, optimizes medical supply chains, and refines resource allocation [19]. In the private sphere, Walmart's adroit integration of I4.0 technologies has ushered in a new era of astute procurement. Capitalizing on the potential of IoT and real-time analytics, Walmart deftly manages inventory levels, augments demand forecasting accuracy, and mitigates stockouts, thereby augmenting customer satisfaction and operational seamlessness [20]. These tangible examples demonstrate the undeniable pertinence and urgency of I4.0 technologies in the realm of procurement, motivating small and medium enterprises (SMEs) across industries to expertly employ these tools for strategic advantage.

Given the transformative potential of I4.0 technologies, the correlation between I4.0 and procurement demands further clarification. I4.0 introduces intelligent automation, data-driven insights, and enhanced connectivity, all of which can significantly enhance procurement processes. The significance lies in the amplified efficiency, strategic agility, and value-driven decision-making that these technologies facilitate. By harnessing I4.0 tools, procurement can achieve enhanced transparency, sustainability, and innovation, ultimately contributing to the overarching success of organizations.

To the best of our knowledge, an up-to-date, comprehensive review of the literature on the role of I4.0 in procurement management has yet to be explored. This work introduces a novel perspective by comprehensively synthesizing the existing literature on the integration of I4.0 technologies in procurement. While previous studies have explored individual applications of these technologies, this review uniquely consolidates diverse research to present a comprehensive understanding of the enhanced efficiency in supply chain management through the P4.0 process. This study seeks to explore the profound impact of I4.0 technologies on procurement processes, identify crucial drivers and challenges for adoption, and assess how these technologies can optimize procurement efficiency, reduce costs, and revolutionize supply chain management. Additionally, we aim to unveil the current trend in smart procurement and anticipate its future direction, providing valuable insights into the evolving landscape of modern procurement practices. As such, we propose the following three research questions as primary motivators for presenting and discussing this structured review:

- Which procurement processes are predominantly impacted by Industry 4.0 technologies?
- What are the primary enablers and challenges in the adoption of Industry 4.0 technologies?
- What is the current trend in smart procurement and the future direction?

The purpose of this work is to give a thorough and well-structured review of the literature and explore the role of I4.0 technologies on P4.0 to address the aforementioned research questions. The article's primary contribution is to describe the existing state-of-the-art literature on this topic.

The article is outlined as follows. Section 2 presents a brief on the journey to Procurement 4.0. Section 3 draws the specified methodology used to conduct the literature review and the criteria utilized to gather the listed publications. The review is broken down in Section 4 where the proposed categories are offered and discussed in Section 5. Section 6 provides the reflections and a conclusion.

2. Procurement Journey

Procurement, sourcing, and purchasing are essential activities for acquiring goods and services in a company. While often used interchangeably, it is crucial to understand their differences and develop effective procedures for each process. Strohmer et al. [21] emphasize the importance of recognizing the distinctions, especially considering the demographic range of operations, where purchasing and sourcing may vary. Procurement encompasses the entire procedure from identifying the need to payment, purchasing refers to acquiring goods or services, and sourcing involves evaluating potential providers.

Procurement entails acquiring goods or services and encompasses activities such as purchasing, transportation, quality control, and inspection. Supplier selection decisions are influenced by the total cost of ownership (TCO), making procurement relevant in project environments [22,23].

Sourcing is a strategic process that involves managing supplier relationships to gain a competitive advantage. It includes decisions on supplier quantity, relationship type, contract duration, negotiation type, and local or global sourcing. Strategic sourcing aims to reduce costs, ensure long-term supply stability, and minimize supply risk [24]. It involves building relationships, allocating business, and making decisions on terms, often guided by tools like Portfolio Analysis for segmenting suppliers and products based on their strategic importance and risk profile [25].

Purchasing focuses on identifying user requirements, finding suppliers, developing agreements, ensuring prompt payment, and evaluating effectiveness. It does not include responsibilities for materials planning, scheduling, inventory management, or quality control. Procurement refers to the transactional aspect and should be streamlined [26].

Table 2 provides a concise overview of the main differences between procurement, sourcing, and purchasing [21]. It outlines key criteria such as the definition of each term, the scope of their focus, the main activities involved, decision-making considerations, the aims pursued, and the relationships established.

Before exploring the role of I4.0 in procurement processes, this section briefly discusses the previous procurement stages that laid the foundation for P4.0. Historically, the evolving methods for procurement processes started along with the industrial revolutions. While procurement or purchasing was almost a clerical process in earlier times, it became a separate but crucial part of production as a result of mass production followed by the Industrial Revolution. Traditionally, procurement has been understood as a cost-reducing mechanism by ensuring goods are at the lowest possible cost for production. But now it has grown to a proactive and strategic one as a result of advanced digital integration into procurement processes. In effect, purchasing, which has been hitherto understood as an important part of this process, has become just one part of a larger procurement operation today. This section aims to present a brief historical journey of procurement and its evolution from Procurement 1.0 to Procurement 4.0 (Smart Procurement) [2].

Table 2. Brief on main differences between Procurement, Sourcing, and Purchasing.

Criteria	Procurement	Sourcing	Purchasing
Definition	The entire process from identification to payment	Locating and evaluating potential providers	Acquiring goods or services
Scope	Industry-specific focus	Strategic advantage in supplier management	Identifying user requirements and supplier evaluation
Main Activities	Purchasing, transportation, quality control	Supplier management, relationship building	Supplier identification, agreement development
Decision-making	Total cost of ownership, project considerations	Supplier selection, contract negotiation	User needs evaluation, prompt payment
Aim	Cost efficiency, governance	Cost reduction, supply stability	User satisfaction, continuous improvement
Relationship	Related to materials planning, inventory management	Related to strategic sourcing and supplier management	Related to user satisfaction, continuous improvement

Based on the Institute for Supply Management in India [27], the traditional procurement cycle can be summarized in the following main dimensions as illustrated in Figure 1: (1) Identify the needs, a business recognizes a required product or service for its own operational purposes; (2) Purchase approvals, making a purchase request and seeking approvals required from the appropriate stakeholders; (3) Vendor evaluation and selection, supply chain professionals solicit proposals from vendors, negotiate possible working relationships, and select their final supplier; (4) Contracting and payment, the purchase order will be approved by the vendor, at which point the relationship becomes one of the vendor relationship management processes.

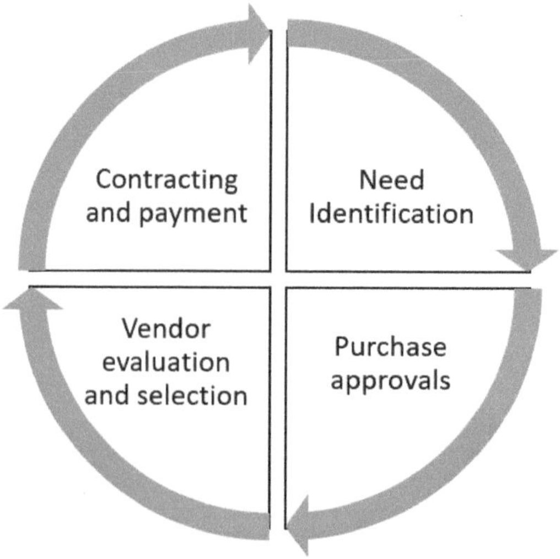

Figure 1. Main procurement dimensions.

Procurement management has developed in tandem with every successive technological advancement. There is a progression in the attention paid to and the strategy used by the procurement processes and practices at each stage as summarized in Table 3.

Table 3. Key characteristics and technological emphasis of each procurement stage.

Stage	Focus and Characteristics	Technological Emphasis
Procurement 1.0	Tactical procurement, cost reduction, operational efficiency	Transactional approach, cost containment
Procurement 2.0	Value creation, category management, cross-functional approach	Information flow control, supplier relationship management
Procurement 3.0	Supply-side optimization, cloud-based tools, automation and recording	Contextual data recording, AI-enabled predictive models

2.1. Procurement 1.0: Tactical Procurement

This stage refers to an early form of procurement processes when procurement was understood as synonymous with purchasing. This stage is characterized by an emphasis on cost reduction and operational efficiency. Procurement is predominantly regarded as a tactical function, with a transactional approach serving as the predominant mode of operation, in which obtaining products and services at the lowest possible price while keeping procurement costs to a minimum level is the main objective at this stage [28]. Although procurement management works as a supporting function with a focus on cost containment, this basic form of procurement has been adopted by the vast majority of companies today [29]. In addition, in this conventional procurement phase, the focus is on the execution of tactical tasks including dealing with contract requests, operating RFXs (requests for information, requests for qualification, and requests for proposal), and facilitating communications and relationships with suppliers [30]. Therefore, it also can be called tactical procurement where procurement professionals are assigned to take care of multiple functions in a reactive and transaction mode.

With more automation options in the interaction between both ends of the supply chain, procurement has turned away from cost containment to deliver accelerated and broader value. It is a holistic process, that consists of risk mitigation strategies, supplier innovation, demand management, global perspectives, sustainability, and more.

2.2. Procurement 2.0: From Cost Containment to Value Creation

Historically, procurement pursued a singular strategy of obtaining the best price, product, and situation available. However, with the advent of advanced internet technologies, procurement become able to create more complex and nuanced strategies [31]. This development is known as Procurement 2.0, where enterprises began to exert greater control over the flow of information, beyond basic informational tools. Procurement 2.0 is no longer solely focused on cost management [32] but instead emphasizes category management, a cross-functional approach, and supplier relationship management [33]. These enhancements mark a significant shift from the previous iteration of procurement, Procurement 1.0.

Recently, there has been a growing recognition of the potential value that procurement can bring to an organization through technological collaboration. By leveraging advanced technologies such as big data analytics and software tools, procurement can add value by addressing several key areas of concern, including working capital management, risk mitigation, and exploring new markets or business lines. The increased information flow and automation capabilities provided by these technologies can improve collaboration with suppliers and drive greater efficiency throughout the procurement process. This shift in focus from cost management to value creation is a central feature of Procurement 2.0. As a result, it has become necessary for organizations to re-evaluate the procurement function itself and consider how it can be optimized to better support broader business objectives [34]. Businesses are recognizing the important impact that procurement can have on various business models and their competitive advantages [35]. By effectively communicating the value of procurement to various departments and business functions, organizations can better align their procurement strategies with overall business objectives and drive greater value creation.

The procurement landscape has undergone a significant transformation, largely driven by the increasing share of revenue attributed to procurement. To align procurement with corporate objectives, companies have developed category strategies [36]. This transformation has led to an increased role of procurement role in cross-organizational collaboration, which has been facilitated by technological advancements connecting all departments across the business. To streamline supply chain processes and achieve greater operational efficiency, a movement to centralize purchases and reorganize vendors has emerged [37]. This has resulted in procurement playing an increasingly important role in managing working capital and identifying risks to business performance. Through enhanced information flow and the development of software to identify risks, the efficiency of these areas has been greatly improved [38].

2.3. Procurement 3.0 (Supply-Side Optimization)

Procurement 3.0 is all about supply-side optimization. It is more than just an advanced digitalization in the procurement process. It signifies a drastic change in procurement structure and the use of high-edge technologies for procurement purposes. The E-Commerce revolution of the 1990s, which laid the foundation for Procurement 2.0, had facilitated a shift from on-premise toward cloud-based tools. It also formed the basis for emerging Procurement 3.0 technologies by making advancements in categorial improvement in functional efficiency. In this phase, the application of digital technology is more focused on automation and recording of the process: a transaction executed, an invoice paid, an item purchased, and a contract signed. A system of record-keeping (in the form of software) has been used in the procurement process which helps to understand contextual reasons by probing into what happened and why. This recording is important: it makes a foundation for AI-enabled predictive models which will make progress in future decision making. A fully matured application of such a technology can be seen in the next stage of procurement, which is known as Procurement 4.0.

The key objective of Procurement 3.0 is to optimize the supply side. In this stage, business stage, business strategy has been linked to the enterprise by ensuring that a good supplier base under the right commercial agreements and delivering the right goods and services. There are clear building blocks required to accomplish this evolution.

With the emergence of Procurement 3.0, procurement and business were connected to the information existing outside of their own data ecosystem and leveraging intelligent capabilities that do more than execute transactions to guide business decisions. Most of these emerging technologies effectuated enhancing the value of legacy systems which consider small investments, have low requirements for integration, and have payback periods measured in months not years [27]. Scholars like Mavidis and Folinas [39] have vividly discussed the e-procurement journey in the public sector and its results in improving access to information and transparency in management. In this study, they evaluate comparatively the challenges and problems between procurement technologies 3.0 and 4.0 and lay down a roadmap for achieving new procurement management in Industry 4.0.

3. Methodology

The approach that is used in this research draws relevant works from an online citation index (Scopus) that is easily accessible. These works have been published in peer-reviewed journals along with other reliable sources. This section describes the search strategy used to identify relevant works and discusses how those works were categorized in relation to I4.0 in the context of procurement management.

3.1. Literature Survey Process

This work comprises a literature review of the published papers on the role of I4.0 technologies in procurement processes, aiming (1) to collect information about unveiling the current trend in smart procurement and (2) to identify crucial drivers and challenges for adoption, and assess how these technologies can optimize procurement efficiency, reduce costs, and revolutionize supply chain management. The present study adhered to the rigorous PRISMA (Preferred Reporting Items for Systematic Reviews and Meta-Analyses) guidelines [40] to ensure transparency and methodological robustness. It involves formulating research questions, conducting comprehensive database searches, applying strict inclusion/exclusion criteria, analyzing and extracting data from selected studies, synthesizing and interpreting findings, and producing a comprehensive review report analysis. Five stages are conducted in the identification, screening, and eligibility of works, as shown in Figure 2, which summarizes the stages in the review process.

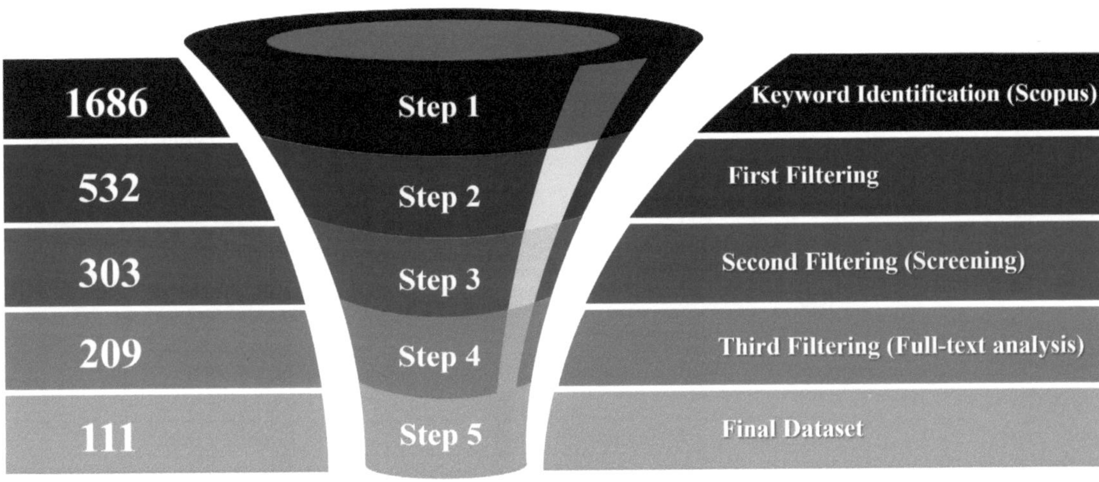

Figure 2. Literature review process–Funnel approach.

In Step 1, keywords included in the title, abstract, and keywords (TITLE-ABS-KEY) relevant to the topic are identified as, TITLE-ABS-KEY((procurement AND 4.0) OR (procurement AND ("additive manufacturing" OR "3D print*" OR print*)) OR (procurement AND blockchain) OR (procurement AND digit*) OR ("digital procurement"), (procurement AND digitalization) OR (procurement AND "digital transformation") OR ("agile procurement") OR (e-procurement) OR ("smart procurement") OR (procurement AND auto*) OR (procurement AND ("smart contract" OR "digital contract")) OR (procurement AND "big data") OR (procurement AND "artificial intelligence") OR (procurement AND "Internet of Things") OR (procurement AND (price* OR cost*)) OR (procurement AND risk*) OR (procurement AND suppl*) OR (procurement AND data)). These keywords are used for literature retrieval from the Scopus database, since this web search engine is globally recognized as the primary source of peer-reviewed literature, and its effectiveness as an index database has been demonstrated. Running the aforementioned research query resulted in a total of 1686 documents. In Step 2, the initial stage of filtering search criteria can be described as (a) the language of the article should be in English, (b) the timeframe should cover the range of years (2012–2022) as this range is considered to be the flourishing start of the fourth industrial revolution applications in manufacturing systems and supply chains, (c) document type chosen to include only articles or review articles, (d) published in journals only. The dataset after the first filtering procedure is comprised of 532 papers. The second filtering criterion, Step 3, represented in reading the titles and abstracts of the papers found and then selecting only the most relevant ones to the field of engineering management and decision sciences, e.g., engineering, business, management, decision science, economics, finance, and accounting. The pool of papers selected after this step amounted to 303 papers, which undergo a third filtering procedure, Step 4, that aims to ensure that all works to be reviewed are meaningful contributions to the topic and add significant value to the literature scope, which results in 209 papers. The third filtering criterion goes for in-depth full-text reading of the papers to ensure they are relevant to the proposed scope. Tables 4 and 5 present the inclusion and exclusion criteria and the final data set in Step 5 consists of *111* papers as a total number of articles.

Table 4. Inclusion criteria (IC).

IC	Description
IC-1	Papers should be written in English
IC-2	The papers should be published between (2012 and 2022)
IC-3	Only journal articles and review papers are considered
IC-4	Papers that examine the use of Industry 4.0 technologies in procurement processes
IC-5	Papers that address one or more of the following procurement functions: sourcing, pricing, purchasing, supplier relationship management, or supplier performance management
IC-6	Papers that focus on data sharing, data security, or risk management in procurement processes
IC-7	Papers that report on improvements or enhancements to procurement key performance indicators (KPIs) facilitated by Industry 4.0 technologies

Table 5. Exclusion criteria (ExC).

ExC	Description
ExC-1	Conference, book, editorial, and report papers are not considered.
ExC-2	Papers that are written in languages other than English.
ExC-3	Papers that focus solely on technologies and are not related to procurement processes, e.g., automation in manufacturing or artificial intelligence in marketing, etc.
ExC-4	Papers that do not clearly state how Industry 4.0 technologies are utilized in procurement processes or do not provide enough details to be relevant to the review.
ExC-5	Papers that have a very narrow focus or only cover a specific aspect of procurement processes and are not relevant to the role of technologies within its scope.

3.2. Literature Review Classifications

Figure 3 summarizes the distribution of the selected papers per technology and per publication year. As shown, the number of publications has been rapidly growing over the last three years, which confirms the high interest in the use of I4.0 technologies in the procurement area.

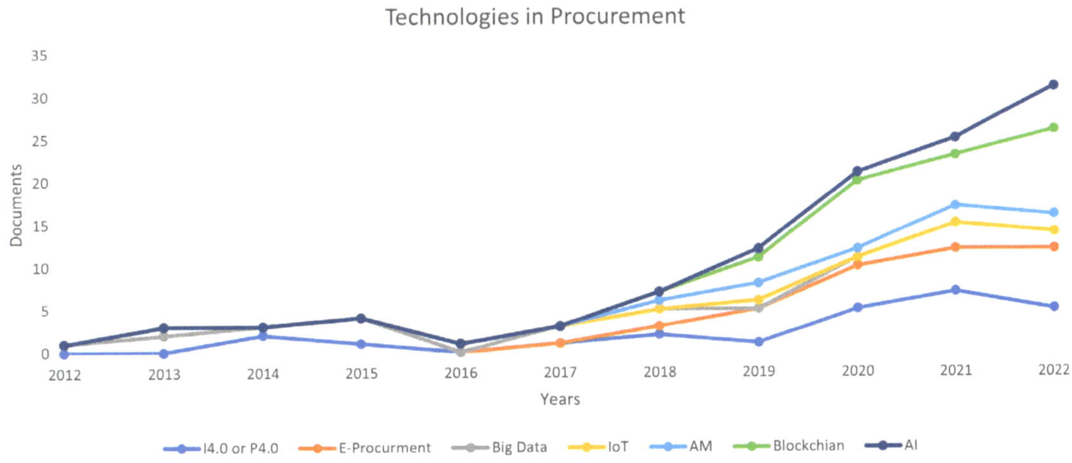

Figure 3. Distribution of the data set per technology and publication year.

Figure 4 depicts the final number of papers (n = 111) included in the review, highlighting the prominence of e-procurement, blockchain, and I4.0 or P4.0 technologies in the literature. After diagnosing the data set, it becomes evident that e-procurement, blockchain, and I4.0 or P4.0 technologies have the highest number of published papers. These technologies have attracted significant attention and research interest in the field of procurement. The bar representing e-procurement stands out prominently, indicating a substantial body of literature dedicated to this area. It signifies the importance of e-procurement in the context of I4.0, as organizations seek to automate and digitize their procurement processes for improved efficiency and effectiveness. Similarly, the bar representing blockchain technology is notable, highlighting the growing interest in leveraging blockchain for secure and transparent transactions in procurement. Lastly, the bar representing I4.0 or P4.0 technologies demonstrates the significance of I4.0 in shaping the future of procurement. The inclusion of papers related to I4.0 or P4.0 indicates the recognition of its transformative potential and the need to understand its implications for procurement practices.

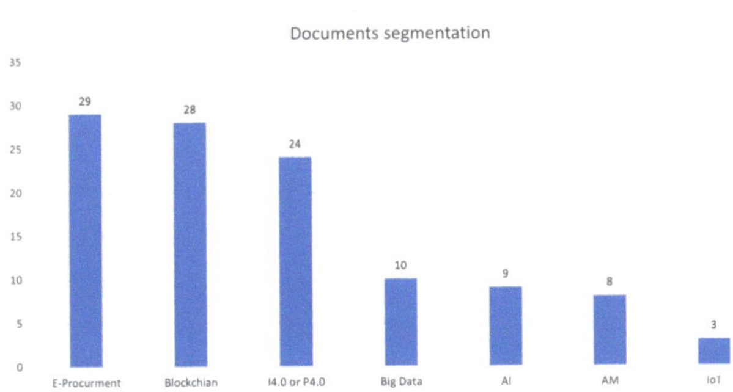

Figure 4. Distribution of Papers Included in the Review.

Figure 5 illustrates the title and the number of journals utilized in this study. The top three journals with the highest number of related selected articles are Sustainability Switzerland, IEEE Access, and the Journal of Cleaner Production. This figure provides an overview of the distribution of articles among different journals, highlighting the journals that have contributed significantly to the research in this field.

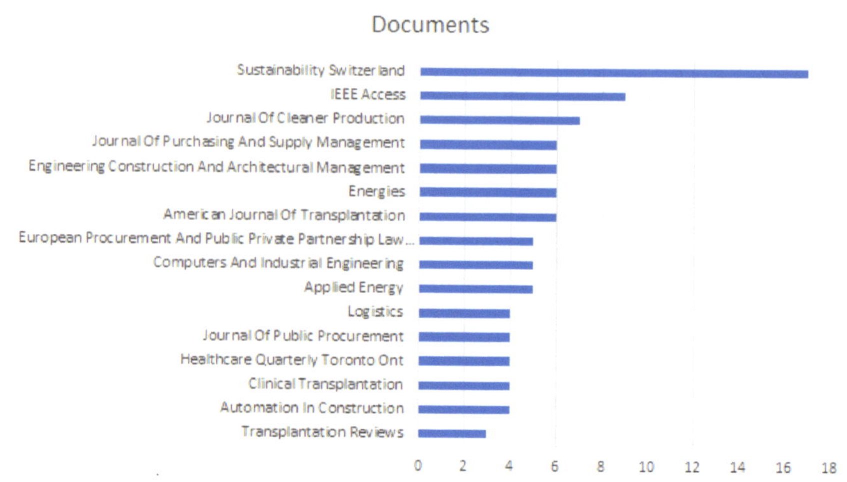

Figure 5. Titles and Number of Journals Used in the Study.

The current distribution of articles per country is visually presented in Figure 6, highlighting that the United States has the highest number of publications related to the topic, followed by the United Kingdom and India.

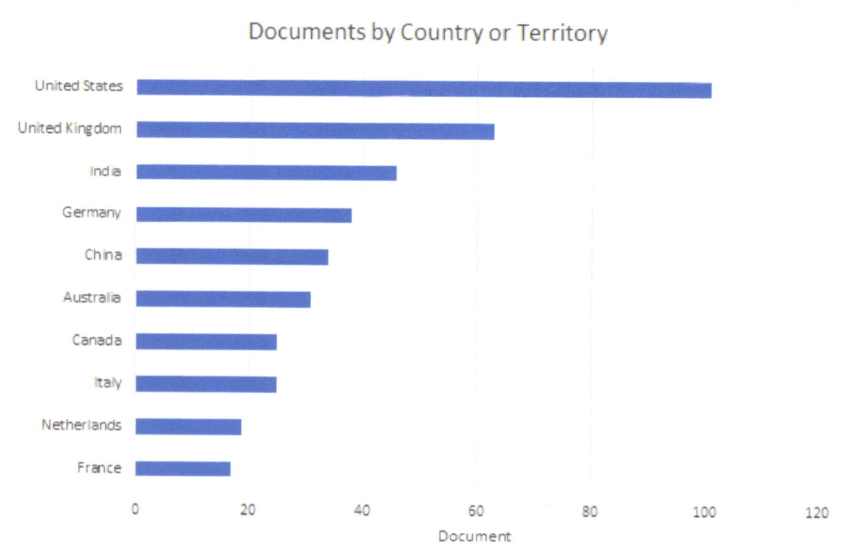

Figure 6. Distribution of Articles Per Country.

3.3. Relevant Literature

Table 6 exhibits the most relevant literature related to the procurement topic. The papers are classified into seven sub-themes, which include I4.0 or P4.0, E-procurement, big data, Internet of Things (IoT), Additive Manufacturing (AM), Blockchain, and Artificial Intelligence (AI). Most of the literature overlaps these sub-themes. However, they are classified differently for analytical purposes. Detailed reviews of these papers are discussed in the following section.

Table 6. Thematic division of the reviewed literature.

Theme	Number of publications	References
I4.0 or P4.0	24	[41–51], [6,52–59] [5,60–63]
E-Procurement	29	[64–73], [74–81], [82–92]
Big Data	10	[93–102]
Internet of Things	3	[103–105],
Additive Manufacturing	8	[106–113]
Blockchain	28	[4,92,114–121], [122–130], [131–139]
Artificial Intelligence	9	[105,140–147]

4. Literature Review

In this section, we present a comprehensive review of previous research papers related to the use of I4.0 technologies for enhancing and improving procurement processes. The literature review covers most of the I4.0 technologies and their applications in procurement. Although the implications of I4.0 concepts have been discussed individually in different industrial contexts, a limited number of studies have discussed the use of emerging technologies in procurement processes comprehensively. Therefore, this work is meant to potentially provide a comprehensive picture of the application of such technologies and offer some value propositions in order to enhance the procurement processes in a wide variety of industries.

The collected articles are divided into sub-themes, including I4.0 or P4.0, e-procurement, big data, IoT, additive manufacturing, blockchain, and AI. By analyzing separately the literature related to these themes, this study discusses the impact of different I4.0 technologies in the procurement processes and their roles within various areas of supply chain management.

Supply chain resilience has emerged as a key element in I4.0 to manage risk during out-of-stock conditions at a time of crisis. It has been used to optimize supply chains from end to end, leading to major impacts on value delivery. Researchers have discussed various ways in which supply chain resilience could be improved during crisis periods. In their insightful paper, Pereira et al. [148] discuss how procurement plays a significant role in detecting and dealing with intra- and inter-organizational problems to impact supply chain resilience. Using a systematic literature review, based on 30 published papers between 2000 and 2013, they revealed that achieving resilience in the supply chain requires efforts from both internal and external elements of the extended enterprise.

4.1. *Industry 4.0 (I4.0) and Procurement 4.0 (P4.0)*

Needless to say, digital transformation is a key component of the fourth industrial revolution, and it has increasingly grown overall in industrial sectors. According to Lasi et al. [51], I4.0 means advanced digitalization within industrial factories. As a result, it facilitates cyber-physical systems that describe the merger of digital with physical workflow [55]. It stands for a combination of internet technologies with future-oriented technologies in the field of *"smart"* objects (machines and products). Thus, it equips the industrial manufacturing system in such a way that products can control their manufacturing process. With the high-level digitalization in the industrial field, new ways of production have been created through the interactions between multiple technologies [52]. Scholars have discussed how the application of such technologies transformed the overall business management structure, and individual aspects of companies including the value chain model, e.g., [59].

To shed light on the applications of I4.0 technologies in various contexts, it is essential to provide a concise overview of the relevant literature. In this regard, Chou and Liao [63] contribute to the field by presenting a procurement decision-making model specifically designed for the food processing machinery sector. The food processing industry has witnessed a significant influx of smart manufacturing equipment with advanced and fully automated technologies, enabled by technological advancements. Within this context, I4.0 serves as an integrating force, leveraging high-equipped technological innovations to harmonize demands, sales, and manufacturing resources in the procurement process. Employing such a tool in the food processing industry offers benefits such as accuracy, reliability, real-time applications, and a reduction in uncertainties associated with human-based decision-making [63]. I4.0's emphasis on procuring subsystems for assembly into other manufactured parts is a critical feature that aligns with the evolving needs of modern manufacturing. It allows for modular production, customization, rapid prototyping, and collaborative supply chain practices. This approach empowers manufacturers to be more agile, innovative, and efficient, ultimately leading to increased competitiveness and improved customer satisfaction.

The advent of I4.0 has given rise to a paradigm shift in the realm of procurement, leading to the emergence of what is known as Procurement 4.0. This transformative development has introduced a range of significant advantages for companies, particularly in the domain of supply chain management [44]. It has been made possible through the implementation of advanced digitalization techniques within organizational structures and supply chain operations. Within an organization, the procurement division has assumed a pivotal role in driving this transformation. By transcending its traditional cost-centric approach, procurement has adopted a profit-centric perspective, thereby redefining not only what is procured but also how these acquisitions are made. This shift has prompted the procurement function to explore novel business models, enabling them to unlock untapped opportunities and enhance their strategic contributions to the organization. By embracing the principles of P4.0, companies can leverage the potential of digital technologies and data-driven insights to optimize their procurement processes, enhance supply chain efficiency, and achieve sustainable competitive advantages in the dynamic business landscape. Such a transition holds the promise of reshaping procurement practices and revolutionizing the way organizations approach their purchasing strategies and supplier relationships.

In a similar vein, Gottge et al. [5] underscore the importance of I4.0 technologies, such as big data, business intelligence, and the IoT, in the procurement process. Their study specifically focuses on the practical implications of these technologies on the purchasing activities of automotive manufacturers, employing a transaction cost economic approach to examine both theoretical and practical impacts. The authors adopt a multiple-case study methodology, utilizing qualitative content analysis and cross-case synthesis. The findings reveal that strategic purchasing processes benefit from the co-creation of specifications, parameter-based negotiations, an additional sub-process, and automated pre-qualification. Notable changes in operative purchasing processes include interactive call-offs and proactive troubleshooting. Through the application of these technologies, transaction costs decrease as uncertainty and supplier specificity are minimized, alongside reduced information search, monetary costs, and negotiations.

Ghadge and Goswami [62] delve into the impact of I4.0 implementation on the supply chain, offering a comprehensive framework to enhance its application across various industries. Through a critical literature review, the study investigates key drivers and barriers across organizational, strategic, technological, legal, and ethical dimensions. Building upon these identified factors, they develop a systematic assessment system to evaluate the impact of I4.0 on supply chain processes. Raja et al. [50] support the study and both shed light on the challenges encountered during implementation and put forward a novel framework aimed at facilitating the effective integration of I4.0 concepts into the supply chain, including the procurement function as an integral part of these processes.

Rane and Narvel [41] examine the utilization of I4.0 technologies to enhance the agility of project procurement management (PPM) processes. Given the growing market demands and shorter time-to-market requirements, the existing PPM encounters challenges in responding with agility. To address this, the authors propose strategies for implementing I4.0 tools and leveraging selective critical success factors to enhance the agility of executive PPM processes. Furthermore, they explore the use of additional tools such as IoT, mobility, business intelligence, blockchain, and robotic process automation (RPA) to further improve agility in PPM.

Digitalization has ushered in a new era for procurement, enabling the utilization of data analytics to meet evolving business requirements. Referred to as Industry 4.0, this transformative paradigm offers promising opportunities and a host of capabilities to enhance procurement processes [42]. The continuous development and refinement of various technologies and tools have significantly bolstered procurement, fostering improvements in efficiency, effectiveness, and overall performance [57]. However, it is crucial to acknowledge the substantial costs associated with the development, acquisition, and implementation of these technologies. Therefore, meticulous analyses are indispensable to determine the potential impact, scope, and applicability of investments [6]. To cultivate a competitive environment, organizations endeavor to establish digitized procurement systems that prioritize customer preferences and exhibit the necessary agility to meet growing customer satisfaction. In this context, scholars have engaged in discussions concerning the systematic application of digitalization in procurement and its consequential effects on supply chain dynamics.

Rejeb and Appolloni [54] and Bag and Sivarajah [46], delve into the role of procurement digitalization in manufacturing operations and the circular economy. Through the analysis of survey data collected from working professionals in South Africa, the latter study identifies key resources required in a P4.0 environment to enhance remanufacturing operations and circular economy performance in emerging economies. Drawing on the resource-based view theory, the researchers explore the influence of resources on P4.0 and its impact on productivity in remanufacturing operations. They argue that improving the performance of remanufacturing operations can contribute to achieving circular economic outcomes. In a similar vein, Bag et al. [46] and Bag et al. [43] investigate the effects of digital transformation on P4.0 processes in terms of productivity within a circular economy context. Findings from a survey conducted among South African manufacturers suggest that P4.0 can be leveraged to optimize the buyer's intentions and enhance circular economy performance.

Rejeb et al. [61], Pires et al. [53], and Richnak [56] investigate the utilization of emerging technologies in procurement processes to address the complexity of exchanging data and information in procurement logistics. Their study highlights the significance of digitalization and the adoption of technologies such as big data analytics, robotics, IoT, blockchain, and smart contracts in enhancing industrial procurement functions and tools. These technologies offer benefits such as improved predictability, transactional automation, and proactive supplier relationship management. The authors also discuss the potential of emerging technologies in procurement activities, as well as organizational barriers that impede their widespread implementation and suggest possible remedies.

Klünder et al. [58] examine the impact of digitalization on procurement costs, highlighting its potential to further reduce costs by ensuring supplies at a lower cost. Through empirical examination, explanatory surveys, and survey-based studies, the authors investigate the effects of digitalization on procurement costs, specifically material and commodity costs, and support the implementation of I4.0 technologies to achieve cost reduction in the procurement process. In a related study, Lorentz et al. [47] propose a procurement digitalization strategy, focusing on key drivers, technology interventions, and mechanisms. Based on data collected from interviews and workshops involving 48 digital intervention projects, the study reveals the significant influence of internal-external dynamics on procurement digitalization. The findings indicate that internal complexity facilitates digitalization by enabling communication support and process structuring interventions, while external dynamism drives interventions for information processing and decision-making, particularly related to supply market knowledge, strategic alignment, and supplier capability assessment.

Hallikas et al. [60] delve into the role of data analytics in the digitalization of procurement and supply chains, specifically examining the relationship between digital procurement processes and supply chain performance. The analysis of survey data reveals several significant findings. First, there is a direct and positive association between digital procurement processes and supply chain performance. Second, the study demonstrates that data analytics capabilities play a crucial role in enhancing a firm's competitiveness in procurement and supply chain management. However, it is important to note that the impact of data analytics differs when considering internal versus external sources. Internal data analytics have a positive influence on supply chain performance, whereas external data analytics directly enhance digital procurement capabilities, which in turn mediate their impact on supply chain performance. These findings highlight the importance of investing in data analytics capabilities to improve supply chain performance and optimize the digital procurement process.

Flechsig et al. [49] and Schiele [48] shed light on the application of Robotic Process Automation (RPA) in purchasing and supply management. While RPA has gained significant attention in digital transformation, its potential within purchasing and supply management, particularly in the public sector, remains relatively unexplored. Drawing upon multiple case studies encompassing both the public and private sectors, the authors offer valuable insights into the challenges, appropriate processes, and effective strategies for implementing RPA. The findings underscore the importance of organizations' digital procurement readiness and maturity in determining the success of RPA adoption.

To classify and describe the many digital technologies and techniques used in building procurement, Ibem and Laryea [45] conduct a thorough literature survey. Their research highlights 36 unique digital technologies and solutions, such as web-based applications that improve construction supply chain interactions among different stakeholders in real time. The study highlights the need for a consolidated digital solution that integrates and unites all main construction procurement processes, providing purchasers with a full-fledged platform from which to oversee the whole procurement lifecycle.

4.2. e-Procurement

The increasing demand for automation in the business sector has led to the emergence of e-procurement as a vital component of digitalized procurement processes [89]. Chan et al. [84] provide a comprehensive overview of the evolution and development of e-procurement, highlighting its significance in optimizing the supply chain by reducing costs and improving efficiency. E-procurement encompasses various elements, such as e-sourcing, e-coordination, and e-communities [85]. In the context of public procurement, the application of e-procurement has been instrumental in promoting good governance [88,90] and reducing corruption within organizations [91].

Within this context, Basak [69] highlights the significance of e-procurement in the aviation industry's maintenance, repair, and overhaul (MRO) sector by comparing it with traditional procurement methods. E-procurement offers advantages, such as simplified processes and increased transparency, making it a valuable tool for maximizing supply chain performance. It reduces costs throughout the value chain and enhances efficiency. In a highly agile supply chain, Chibani et al. [71] demonstrate the use of e-procurement for dynamic optimization. The buyer-supplier interactions in this context are volatile, requiring decisions on the quantity and selection of products based on price and capacity variations. Traditional optimization approaches are inadequate for addressing these dynamic challenges.

Kim et al. [70] investigate the influence of strategic sourcing and e-procurement on firm performance. Through a structured survey of 137 managers in US manufacturing firms, the study employs partial least squares-based structural equation modeling for data analysis. The findings reveal a positive impact of strategic sourcing and e-procurement on firm performance. Additionally, the study demonstrates that business characteristics and environmental factors, such as competition, market turbulence, firm size, and product life cycle stage, moderate these relationships. This research contributes to theory and practice by examining the effects of e-procurement on firm performance.

Sánchez-Rodríguez et al. [92] investigate the impact of e-procurement on small and medium-sized enterprises (SMEs), focusing on facilitators and obstacles that affect business performance. The study examines data from 199 managers in the manufacturing sector and identifies significant relationships between SMEs and factors, such as support of executive management, information technology obstacles, and strategic procuring. Furthermore, the research reveals positive associations between e-procurement, procurement efficiency, and overall business performance. These findings highlight the importance of e-procurement in enhancing the performance of SMEs.

Belisari et al. [73] investigate the positive and negative impacts of e-procurement adoption in the Italian market, encompassing both the public and private sectors. Through an examination of two Italian companies that have implemented over 150 e-procurement projects, the study highlights the effects of e-procurement solutions. The findings demonstrate that the adoption of e-procurement leads to improved purchasing performance and compliance with the procurement code. However, the manifestation of these benefits differs between the public and private sectors.

Hung [68] investigates the effects of a Web-based e-procurement system on companies' organizational performance within the supply chain. Their model incorporates various contextual factors and was tested through a survey of 105 manufacturing firms in Taiwan. Utilizing the partial least-squares regression method, the study finds that e-purchasing activities significantly improve organizational and inter-organizational efficiency. Additionally, an enhanced partnership approach positively affects supplier and buyer performance. These findings highlight the positive impact of web-based e-procurement on organizational performance.

Kamarulzaman et al. [66,67] investigate the utilization of e-procurement technologies in the Malaysian Palm Oil Industry (MPOI). Through in-depth interviews and comprehensive analysis, the authors in both articles highlight the potential of e-procurement in enhancing purchasing efficiency by streamlining the search process, providing material information, and reducing human errors. However, the study reveals that certain products, such as fresh fruit bunch, crude palm oil, and crude palm kernel oil, face limitations in adopting e-procurement tools due to established suppliers' limited utilization of such technologies in their purchasing activities.

Gupta [64] examines the impact of information technology (IT) on e-procurement in Indian organizations. Based on a survey conducted among 36 organizations, the author highlights the positive effects of IT usage, including improved business process efficiency, such as enhanced inventory control, cost and time reduction, and improved customer service. However, the study also identifies persistent security concerns that hinder the widespread adoption of IT infrastructure for e-procurement. Specifically, issues related to internet security, loss of information sharing, and legal legitimacy must be effectively addressed to establish trust in online transactions.

Adowo et al. [65] investigate the potential of e-procurement in combating corruption and unethical practices in construction project delivery in Nigeria. While e-procurement has been widely adopted to enhance supply chain management in various sectors, its role in addressing corruption has received limited attention. This is particularly significant in sub-Saharan countries where corruption levels are substantial. Through 759 cross-sectional surveys conducted across different fields in Nigeria, the researchers identify 18 anti-corruption capabilities associated with e-procurement in construction project delivery. These capabilities focus on ensuring inventory management, accountability, and reducing human involvement during the bidding process. Transparency in bidding fosters competition and aids in curbing corruption within construction procurement activities. Also, the same author, Aduwo et al. [77] investigate the presence of critical gray areas in the public procurement process and highlight the potential of e-procurement's digital solutions to combat corrupt practices. Public procurement plays a vital role in developing and maintaining city infrastructure to enhance citizens' quality of life. However, corrupt practices hinder its success. Through a survey of 759 construction stakeholders across Nigeria's six geopolitical zones, this study identifies e-procurement technologies utilized for project announcement, information exchange, tendering, proposal submission, progress monitoring, and payment. The findings emphasize the importance of digitalization in eliminating corruption within procurement processes. Thus, this paper explores digital solutions for enhancing public procurement in the Nigerian construction sector.

Ibem et al. [79] investigate the impact of e-procurement on the user experience within the Nigerian construction industry. While e-procurement adoption is widespread in developing countries like Nigeria, its influence on construction stakeholders and their experiences remains understudied. This survey-based study includes 759 stakeholders and reveals that e-procurement tools are extensively used for information exchange and expanding tender opportunities. However, their utilization for tracking construction site progress is limited. Factors such as internet infrastructure adequacy, operational environment, technology benefits, and system accessibility directly influence user experiences, emphasizing the need for their consideration to ensure sustainable e-procurement implementation in the construction sector.

Ramkumar et al. [75] emphasize the significance of the Technology Acceptance Model (TAM) and Quality-TAM (Q-TAM) in e-procurement services. Through an analysis of data from organizational buyers, The research shows how organizational purchasers' perceptions of procurement services are affected by information flow and logistical fulfillment. This research introduces a quality technology acceptance model that predicts organizational buyers' intentions to continue using e-procurement services, focusing on user satisfaction within a mandated usage setting. It distinguishes itself from existing literature by exclusively focusing on user intentions of organizational buyers, rather than firm-level adoption issues. Furthermore, the application of e-procurement in SMEs is crucial for developing competitive advantages. E-procurement systems facilitate connections between SMEs and other business activities, such as supply chain integration, resulting in rapid development in a globalized world. Madzimure et al. [76] investigate the relationship between e-procurement, supplier integration, and supply chain performance in retail SMEs through case studies in South Africa. Their quantitative research approach evaluates various e-procurement components, highlighting the impact of e-design and e-negotiation on supply chain integration. Additionally, the study demonstrates that supply chain integration positively influences supply chain performance.

Mélon et al. [80] examine the impact of large-scale e-procurement adoption on institutional quality in Denmark, the Netherlands, and Portugal, comparing it with other countries. Using data from 108 countries between 1996 and 2017, the study employs a differences-in-differences policy analysis approach. The findings indicate that the effects of e-procurement reform on institutional quality are not uniform. While Denmark and the Netherlands experience substantial improvements in the rule of law and government effectiveness post-reform, the overall impact on institutional quality is less pronounced. This paper highlights the contributions of e-procurement to institutional quality, revealing variations in its effects across different countries.

Masudin et al. [81] and Susantya et al. [86] investigate the effects of e-procurement adoption in Indonesian manufacturing companies by examining the relationships between top management support, information quality, e-procurement implementation, and company performance. Through two surveys administered to staff members, the study finds that the support of senior management and information reliability significantly influence e-procurement implementation, leading to enhanced business performance.

It is well-established that prior experience in e-procurement platforms enhances project performance. To gain deeper insights into the effectiveness of prior experience, Hong and Shao [82] explore the moderating factors that influence the impact of buyer-supplier experience in these platforms. Specifically, they examine the role of temporal distance, language difference, and routine tasks in determining the success of the prior experience. Their findings shed light on the significance of prior experience in e-procurement platforms.

Sanchez et al. [72] conducted a study to explore the impact of digital transformation (DT) on supply chain procurement (SCP) and competitive advantage (CAD). Through a survey of 221 supply chain professionals, the researchers determined that DT positively influences both SCP and CAD, with SCP mediating the relationship between DT and CAD. The findings suggest that digital procurement has the potential to revolutionize competitive markets, and the study offers preliminary recommendations for shifting from standard to smart procurement practices.

Chen et al. [83] conducted a study to explore the relationship between procurement complexity, government structure, and e-procurement adoption in local governments. Analyzing survey data from over 400 cities, the researchers found that a centralized government structure increases the likelihood of e-procurement adoption in response to procurement complexity. In contrast, governments with a coordinated structure are less likely to adopt e-procurement due to their reliance on intra-organizational collaboration and information-sharing to manage complex procurements. These findings shed light on how different government structures can effectively handle complex administrative operations and enhance the adoption of e-procurement.

Fu et al. [74] present a case study on the implementation of a green e-procurement system using a cloud model. The study focuses on private hospitals in Taiwan and highlights the challenges associated with conventional systems, including high resource requirements, regular updates of green specifications, and system crashes affecting procurement efficiency. To address these issues, the authors propose a cloud-based e-procurement system that improves efficiency and reduces information maintenance costs. The results demonstrate the successful optimization of resource allocation and timely understanding of user needs, enabling the development of tailored marketing strategies. The cloud-green e-procurement model proves beneficial for enhancing procurement activities in a sustainable manner.

E-procurement has revolutionized public procurement systems, enhancing government administration and governance [87]. Micea et al. [78] explore the impact of an integrated model of Blockchain and the Internet of Things, known as (BIoT) on achieving sustainable and smart procurement in the public sector. Additionally, Chen et al. [83] investigate the adoption of e-procurement systems in local government. They argue that the complex structure of local government procurement can be effectively addressed through the implementation of an e-procurement system. The study findings highlight the potential of different local government structures in handling complex managerial tasks and facilitating e-procurement adoption.

4.3. Big Data

Big data is an emerging and relevant topic for industrial operations. In the supply chain, the application of big data will bring benefits like higher visibility, enhanced bargaining position in negotiations, better risk management, and better decision-making. Big data helps to control spending with data analysis and reduce risks by identifying potential supply chain disruptions [93]. Therefore, it plays a significant role in enhancing the procurement process.

In a recent review essay, Lee and Mangalaraj [100] discuss the uses of big data analytics in various business functions particularly in the context of the rapid expansion of machine learning and technological infrastructure. The role of big data in supply chain management (SCM) for handling customer preferences, visibility improvements, and resilience has been emphasized in the paper but, interestingly, from the vantage point of an interdisciplinary perspective, which will offer an increased and holistic understanding of its functions. By combining existing perspectives in big data research in SCM, namely organizational and technical, and holding together a bottom-up approach, i.e., using bibliometric analysis in classifying those studies, and a top-down approach, i.e., employing a framework to classify findings, this paper analyzes prominent theories used in the study of big data and SCM. Moreover, a focus has been given to sustainability within this research domain, which has not been given due attention in previous studies.

Zeisel et al. [101] present a conceptual architecture for implementing big data applications in the procurement function. The architecture facilitates the integration of diverse data sources, skilled teams, and big data analytics. Additionally, the authors identify and discuss a comprehensive range of both existing and innovative use cases for big data in procurement.

Moretto et al. [94] analyze the value of big data technology in supporting the procurement process and the implementation of procurement practices. Using a focus group discussion among procurement professionals and experts and analyzing four case studies supported by secondary data analysis, this work illustrates how big data technology influences different procurement processes and what kind of benefits can be expected from adopting this technology for procurement performance. Big data technology is shown to have the greatest potential impact on strategic sourcing phases, mainly in strategy configuration, reverse marketing, spend analysis, supplier evaluation, negotiation, and selection. In short, the application of the technology enables better data management and improves the quality of the business environment.

Tiwari et al. [95] illustrate the implications of big data in the procurement process of various industries and businesses and argue that the application of big data technology assists companies in making informed decisions by analyzing massive amounts of data in a short time. More specifically, the author explores the implication of big data for purchasing and supply purposes that enhance companies' visibility in the supply market. Data gathered from interviews suggest that big data technologies enhance companies' ability to identify new suppliers, manage risks, and analyze suppliers. In addition, predictive analytics techniques can be used in market analyses and negotiations.

Song et al. [96] discuss the significance of a big-data-driven approach for emergency drug procurement planning. In the context of uncertainty of diseases, traditional approaches to drug procurement planning are inefficient, which often causes overstocking or understocking in hospitals. Consequently, public health care services are negatively affected. Thus, the authors use a heuristic optimization algorithm based on a deep neural network to predict morbidities of acute gastrointestinal infections based on a massive amount of environmental data. Two case studies illustrate that such methodology provides quick and efficient solutions, which are better than the methods currently used in drug procurement.

Narwane et al. [102] investigate the role of big data analytics in mediating the relationship between Supply Chain 4.0 company performance and various performance parameters. The study employs a two-stage hybrid approach combining statistical assessment and an artificial neural network (ANN) algorithm, based on data obtained from 321 responses from 40 Indian industrial organizations. The findings highlight the significant impact of organizational performance characteristics, ecologically conscious procurement and sourcing, endogenous and exogenous factors, shipping and distribution, operation-related aspects, technical expertise, and collaboration on the adoption of big data analytics.

Kuo et al. [98] discuss the implementation of the smart support system of material procurement based on big data and predictive analytics in order to reduce waste and thereby make sustainable enterprises. In this system, an intelligent event-driven feedback control method is used for controlling process plans. A decision support system is also designed to reduce material usage based on usage data. Moreover, data science and predictive analytics techniques have been applied to make better procurement decisions. The authors argue that the application of the system is highly effective in semiconductor manufacturing.

Focusing on big data's impacts on supply chain activities, Murad et al. [99] analyzes optimizing procurement as an intermediate factor in which big data's impacts on the supply chain's sustainability performance could be measured. They argue that the benefits of having big data capabilities are neither visible immediately nor achievable straightforwardly, instead, they have to be sought in intermediate factors such as procurement optimization. By analyzing existing literature, they offer a conceptual framework to examine procurement optimization with big data capabilities, specifically focusing on healthcare-sustainable supply chains.

It is an already acknowledged fact that green procurement is a necessary component in any organization to attain higher environmental performance. However, this is not an easy task because of the potential risks such absence of appropriate information, tools, and training. Alnuami et al. [97] demonstrate that big data technology is a solution to this problem. Its concept can transform the traditional e-procurement system into green procurement and ensure better environmental performance. By analyzing resource orchestration theory, the authors argue that if organizations reorganize resources and capabilities in terms of the big data e-procurement method, a higher environmental performance is achievable. Empirical samples of 216 procurement professionals indicate that big data analytic capabilities provide enhanced environmental performance and sustainability.

4.4. Internet of Things

IoT provides useful capabilities for achieving better procurement with more efficient operations. IoT refers to the interconnection of information with the help of sensing and actuating devices that facilitate a platform so that information may be exchanged across systems in an integrated manner, leading to a unified framework that can be used to power cutting-edge applications [103]. Scholars have discussed the significance of IoT as an enabler, empowering companies to gain real-time data and continuous evaluation for better decision-making by identifying consumers' consumption patterns.

Fang et al. [104] illustrate an integrated three-stage model based on IoT for the optimization of procurement, production, and product recovery. Due to increasing environmental concerns worldwide, the recovery of end-of-use products has become an important part of the manufacturing industry today. In this context, manufacturers must develop proper strategies to reduce waste, raw material usage, and production costs by recycling and refurbing used items. The authors propose the use of IoT to address the questions of how to collect end-of-use items from customers and acquire knowledge about each item. The products can be identified with codes and monitored and tracked using wireless sensor networks. A particle swarm optimization algorithm is used, and there are three recovery options for increasing product demand in each stage of the product life cycle. IoT facilitates manufacturers to collect and share data for providing products that are functional and less expensive. By doing so, manufacturers can make a global chain comprising suppliers, factories, and recyclers, which help them assemble products at a lower cost. In this context, manufacturers are forced to select appropriate suppliers for specific components, ensuring lower procurement costs of greenhouse gas emission-contained materials.

Jemmali et al. [105] present an intelligent algorithm aimed at addressing supplier selection challenges. The proposed algorithm utilizes a set of determinants and sophisticated formulas to convert predefined preferences into quantitative measurements. These measurements enable the differentiation between suppliers and their proposals. Experimental results confirm the effectiveness of the D3S model in selecting the most suitable offer from a range of proposals provided by various vendors.

4.5. Additive Manufacturing

Additive manufacturing (AM), also called 3D printing [106], is yet another technological advancement that takes a transformative approach to industrial production by facilitating the creation of lighter, stronger parts and systems. It employs a method of data-driven computer-aided design that has a huge potential for developing novel product concepts and enhancing the performance of existing ones. It also facilitates better design freedom and functional benefits such as weight reduction.

Revilla et al. [107] discuss the significance of AM technology and its potential to alter and disrupt a wide range of industrial sectors. AM technology is not only used for new product designs but also for redesigning production networks. The buyer uses AM technology to produce parts with data provided by suppliers. Using supplier-provided data, the client (buyer) could manufacture parts using additive manufacturing. This is one manner in which AM technology is influencing the future of business partnerships between buyers and suppliers. However, studies examining the effects of AM on industrial procurement are still limited. AM technologies potentially allow the decentralization of the manufacturing and restructuring of logistical aspects of supply chains that were never explored well.

Sgarbossa et al. [108] examine the logistical considerations and challenges associated with AM and decentralization. The study also investigates the potential of AM as a solution for strategic sourcing in procurement. Moreover, the authors explore the application of AM technologies in spare parts management and provide decision trees to assist industrial managers and practitioners in their decision-making processes.

The effects of using AM in spare parts procurement are investigated by Bhattacharyya et al. [112], in particular in terms of supply chain resilience and effective inventory management for spare parts. The results of the research show that using AM might lessen reliance on original equipment manufacturers (OEMs) for acquiring some replacement components. The study also found that using AM might save costs in production and lead times significantly.

The adoption of AM technologies has increased the freedom of design by expanding the possibility of producing unconventional geometrics. However, such decisions are complex since optimizing AM requires integrating data from several disciplines. For instance, to make unique and challenging products, there must be a proper assessment of the qualities of printed materials and their compatibility with a certain application, redesign forms with respect to AM's limitations, and so on. Moreover, procurement and logistics evaluation must be included in this process. In this regard, Raffaeli et al. [109] investigate the adoption of AM technology in production design. The study examines the various factors involved in this decision-making process and proposes a systematic approach that guides designers through structured algorithm procedures. This approach enables designers to evaluate the suitability of AM adoption and to develop optimal designs considering the constraints of AM technology.

Muhammad et al. [110] examine the application of AM in enhancing the resilience of the automotive supply chain network and addressing disruptions. The global supply system, particularly the automotive sector, has faced significant challenges due to the COVID-19 pandemic. To assess the potential of AM in strengthening the automotive supply chain, the authors focus on the Indian automotive industry as a case study. They identify key barriers to widespread AM adoption, such as insufficient government regulations, limited management support, restricted availability of raw materials, and workforce constraints. The study also discusses mitigation strategies that leverage AM in the upstream automotive supply chain.

Chowdhury et al. [111] present an innovative optimization framework for the design and operation of an integrated AM supply chain network. This framework considers the interdependencies among flow networks, resource limitations, and costs at the process and system levels within a unified decision framework. A two-stage stochastic programming model is proposed, where the first stage focuses on minimizing facility site selection and capacity decisions before any customer demand information is known. Once demand information is available, second-stage decisions are made, including determining the optimal layer thickness for AM products, production planning, post-processing, procurement, storage, and transportation.

Knofius et al. [113] conduct a comprehensive analysis of the total costs associated with consolidation using AM, encompassing logistics, manufacturing, and repair expenses. Their findings indicate that consolidation with AM often leads to higher overall costs, primarily due to a loss of flexibility. In contrast to conventionally constructed components that allow for partial replacements, consolidated components require complete replacement in the event of failure. Moreover, the perceived advantage of shorter lead times for resupplying consolidated spare parts proves to be insignificant. These conclusions highlight the necessity of considering a holistic cost perspective when evaluating the outcomes of design enhancements using AM. Neglecting such an approach may result in unintended consequences that cast doubt on the viability of design changes, even in the presence of significant functionality improvements.

4.6. Blockchain

Blockchain is yet another important I4.0 technology that has been predominantly used in procurement for making smart contracts. Smart Contracts play an essential role in procurement and they have been increasingly used in many industries due to many features, including automatic execution, nontempering in the blockchain environment, and transparency [4]. However, blockchain technology and its recent developments in applications are moderately well explored in the literature.

In 2017, the initial scholarly publications on the application of blockchain technology in the construction industry were introduced, comprising a total of three articles. Over the span of a few years, a body of literature has emerged with an average annual expansion rate of 184%, resulting in a cumulative count of 121 documents by early 2021. To explore the topic's growth and advancement, all 121 papers were reviewed. A literature study and scientometric analysis were used in a mixed methods approach to examine the present environment [133]. In total, 33 blockchain application categories in construction were identified and organized into seven subject areas: (1) procurement and supply chain, (2) design and construction, (3) operations and life cycle, (4) smart cities, (5) intelligent systems, (6) energy and carbon footprint, and (7) decentralized organizations.

The possibilities of blockchain technology in making a digitalized supply chain (DSC) have been explored and multiple models are experimented with in different procurement fields [114]. Dutta et al. [115] explore industrial sectors such as shipping, automotive, aviation, manufacturing energy, healthcare, etc., in which the use of blockchain technology can enhance efficiency in business process management. Similarly, another study pointed out a blockchain-enabled smart contract system has been used to mitigate unethical practices of bid shopping and peddling during the subcontractor procurement process [116].

One of the important qualities of blockchain technology is its capacity to record data in a predetermined way, in a cryptographic mode. Many enterprises employ this technology in their respective fields in order to make their procurement processes more efficient and cost-effective. In this context, Hofbauer and Sangl [117] present a comprehensive analysis of blockchain technology's application potential in digitizing transaction processes. They delve into the underlying principles and implications of blockchain in transaction activities. Through an examination of statistical data on users' expectations and requirements, the authors highlight how blockchain can meet these needs. By integrating theoretical foundations with practical applications, this study effectively showcases the transformative possibilities of blockchain technology in the digitalization of transaction processes.

A methodology for green procurement using blockchain and Internet of Things integration is proposed by Rane and Thakker [118]. By reviewing extensive literature and analyzing interviews conducted among procurement managers from different industries, they identify challenges in green procurement and then discuss how the integration of blockchain and IoT technologies provides capabilities to overcome procurement challenges, making a sustainable business model. The authors argue that the proposed architecture can create a green supply chain, and such green initiatives enhance the competitiveness of the industries at the global level. Also, a successful green procurement will contribute to saving energy consumption and efficient waste disposal, thereby realizing a sustainable business model.

Government agencies and companies widely use computing solutions because of the ease of use of procurement processes. However, the security of cloud file storage has always been a threat. Pinheiro et al. [119] designed a new method to monitor the integrity of files stored in cloud storage using smart contracts in the blockchain network, symmetric encryption, and computational trust. This consists of a protocol that provides important features such as confidentiality, decentralization, audit availability, etc. The proposed method is feasible and accurate in detecting corrupted files. Therefore, it has the potential to improve the e-procurement process in various stages of a business.

Guo et al. [127] present the significance of blockchain technology in contract management that offers better security to the traditional electronic contract system. Industries have generally been adopting the e-contract system in modern times because it reduces time and human labor and improves contract management efficiency. However, since it has a centralized database-based storage schema in the e-contract system, there is a high possibility of data hacking and information draining; therefore, it raises serious security concerns. In this context, this study proposes a blockchain technology integrated contract management system. It develops a process-oriented contract management system (BE contractor) and discusses its implementation with a focus on Hangzhou (in China) power grid enterprise.

Omar et al. [128] discuss the integration of blockchain technology in the healthcare supply chain process for automating procurement contracts. Effective healthcare supply chain (HCSC) management is crucial for a better health system, especially during pandemics such as COVID-19. Although new technological advancement has been integrated for better management, the healthcare system still suffers from poor procurement processes. Contracting through Group Purchasing Organizations (GPOs), which is an important stakeholder in HCSC, is time-consuming and lacks efficiency. In this context, the authors suggest a solution in which blockchain technology is integrated into a decentralized storage system, which provides better transparency and communication with various stakeholders. In this new method, all stakeholders such as manufacturers, GPO distributors, and providers connect through the Ethereum network. In short, this paper proposes an automatic blockchain solution model for smart contracts for GPO contract processes. Similarly, Raj et al. [129] discuss the importance of blockchain-based smart contracts for supply chain transactions that help enterprises reduce problems related to payments at various stages of the supply chain process, especially when supplying is based on third-party logistic services. Considering the inefficiency and risk even in the most modern form of electronic transactions, they introduce a new platform of transaction based on an Ethereum-based smart contract that enables a centralized payment system along with decentralized authorization and information sharing possibilities between various supply chain stakeholders.

Studying the potential applications of blockchain technology in public sector E-Procurement in Bangladesh, Khalfan et al. [130] argue that many developing countries are suffering seriously in terms of the development project and identify 38 such issues in public sector issues in Bangladesh. Most of these failures are attached to a lack of a proper development project, including poor project management, poor procurement management, and lack of transparency. One of the significant reasons for this problem is the awarding of the lowest bidder during the procurement process because of the current regulatory requirements in the country. The authors observe and analyze the Oracle platform, which is the prominent E-procurement platform in Bangladesh built on blockchain technologies and cloud computing applications, and identify that the application of blockchain technologies solves 25 out of 38 issues in the procurement process. It shows the potential of blockchain technologies in handling the majority of issues in the successful implementation of public sector projects.

Kim et al. [121] discuss the significance of blockchain technology in the construction industry. One of the main advantages of this technology is the reduction of transaction costs, which is crucial to the construction industry because it consists of a high volume of transactions among various entities. Therefore, their study aims to apply this technology in the construction industry to achieve better performance. The authors conduct a survey based on a questionnaire that collects data on the construction life cycle and project management knowledge. The data are analyzed through the method of importance-performance analysis (IPA) in order to figure out the relationship between the application of these technologies and the anticipated impacts of using this technology. In the construction project lifecycle, there are three parts, including project cost, contract bidding formation, and procurement evaluation, in which blockchain technology shows high applicability and impact.

Gunasekara et al. [124] analyze the application of digital innovations such as blockchain technology within facility management. The authors observe that the facility management industry is still lagging behind in adapting new technologies for its process, compared to other industries. A framework is proposed to enable blockchain technology in the facility management procurement process. By using interviews, templates, and content analysis for collecting data, this study suggests steps for transforming into a digitalized technology-based industry. According to the authors, applying blockchain technology in different steps of procurement through a systematic framework can solve issues related to data security, integration, and transparency.

Faccia and Petratos [125] examine the application and integration of blockchain to Accounting Information Systems (AIS) and Enterprise Resource Planning (ERP). The authors argue that it can be successfully integrated and can have significant benefits. In order to overcome security and privacy limitations, facilitate integration at multiple levels, and reduce risks in auditing and management, the authors analyze e-procurement systems and operations. In short, this study illustrates the significance of the integration of blockchain techniques, decentralized finance (DeFI), and financial technology (FinTech) into AISs and ERP systems which can benefit efficiency, security, and productivity. Similarly, Prajapati et al. [126] analyze an approach for the virtual closed-loop supply chain that utilizes blockchain and IoT technology, in order to accelerate e-commerce thereby bringing forth sustainability in the circular economy.

Xu et al. [122] analyze the benefits of smart contract applications in various industries from a procurement perspective and provide a systematic literature review with 174 publications. The authors argue that smart contract applications have been widely accepted at the global level, especially in the fields of information computing technologies (ICT), public management, supply chain, energy, finance, and healthcare.

Rane and Narvel [123] propose a data-driven decision-making technology by presenting an integrated model of Blockchain-IoT technologies for enhancing project resource management (PRM). For several reasons, including fierce competition, volatile markets, innovative new business models, and the ever-increasing complexity of technologies and developments, today's PRM methods are insufficient. In this context, the authors discuss a new formula for handling this crisis by using an integrated model of Blockchain-IoT technologies in order to provide business intelligence to advance the PRM process. Further, the authors analyze the challenges faced by the Engineering, Procurement, and Construction (EPC) industry in handling resources and argue that The integration of Blockchain and IoT technologies can provide various capabilities such as instantaneous data acquisition and self-governing coordination of resources, which can enhance the decentralization, trustworthiness, security, and transparency of transactions. These advancements can lead to an improved level of agility for EPC industries. In summary, this paper identifies a new way of utilizing Blockchain-IoT integration technologies to boost the agility of resource handling in asset-intensive industries such as EPC.

Rodríguez et al. [120] discuss the potential of emergent disruptive technologies to advance the humanitarian supply chain. While increasing the importance of humanitarian operations in the supply chain process, challenges that these operations commonly face, such as delays, congestions, poor communication, lack of accountabilities, and advantages of disruptive technologies to counter these problems, are being discussed in various literature. However, this literature on the humanitarian supply chain lacks a framework for understanding the challenges and solutions. They fill this gap by discussing the application of three such emergent disruptive technologies in detail: Artificial intelligence, Blockchain, and 3D Printing. By analyzing a case study based on the flood of Tabasco of 2007 in Mexico, they discuss the potentials of these emergent technologies in reducing congestion in the supply chain, enhancing simultaneous collaborations of different stakeholders, decreasing lead times, and increasing transparency, traceability, and accountability of material and financial resources.

In a recent study, Nodehi et al. [131] present a comprehensive blockchain solution, called the Enterprise Blockchain Design Framework (EBDF), in order to ensure a decentralized contracting and use of store of value validated through the emergence of cryptocurrencies with multiple stakeholder ecosystem. The contention posits that the implementation of blockchain technology remains in its nascent phase, and numerous corporate blockchain initiatives falter as a result of suboptimal utilization and inadequate motivation for all parties involved, stemming from a perceived lack of necessity. In this context, they propose a comprehensive framework that is not only relevant to certain cases but creates an enterprise blockchain system that is useful to multiple industries. This is certainly a novel step with regard to blockchain technology applications in private companies.

Yadav et al. [114] discuss the optimization of procurement costs through the utilization and integration of blockchain technology. The proposed model is a mathematical formulation that utilizes Mixed Integer-Linear Programming (MILP) to incorporate the expenses incurred in the various stages of block development, including procurement, placing orders, transportation, and storage or inventory procedures. An integrated MILP model is proposed to address potential leakage from the system during the four vulnerable stages, thereby enhancing the accessibility, traceability, and immutability of information for DSC. Ultimately, the LINGO 10.0 software is employed to address procurement obstacles in the digital realm, utilizing randomized datasets. The model is subsequently subjected to validation and testing in three distinct scenarios.

Prakash et al. [134] use a phenomenological method with six experienced experts to better understand the construction industry's readiness to use blockchain technology. Subsequently, the research emphasizes the necessary sequential procedural measures for the efficacious integration of blockchain technology in construction industry use cases. The study presented evidence of the potential benefits of utilizing blockchain technology in the construction sector to enhance productivity. This was achieved through the examination of various case studies, including payments, procurements, building information systems, and smart asset management. This study contributes to the advancement of efficient and feasible tactics for expediting its implementation.

A comprehensive systematic review of 153 distributed ledger technology (DLT) and smart contract publications relevant to the design, construction, and operation of built assets is conducted by Li et al. [132]. The findings are fully transparent, accountable, repeatable, and updatable since they were generated using the steps and techniques of a systematic review. In order to provide the necessary change and impact from the widespread adoption of DLT and smart contracts, a more comprehensive socio-technical approach to the solution is required. The results of this comprehensive study provide valuable information for academics, professionals, and public officials concerned with the evolution of DLT and applications of smart contracts in the construction industry.

Sanchez et al. [92] present a comprehensive assessment of the successful application of DLTs in public procurement governance, while also addressing potential legal challenges. By examining various proof-of-concept projects in the industry, DLTs demonstrate the potential to enhance transparency, integrity, autonomy, and speed in the procurement cycle. Although the nascent nature of blockchain-based public procurement architecture poses certain limitations, it has been observed that the use of private ledgers or public-permissioned ledgers holds greater suitability for tender procedures in the public procurement context.

Li et al. [135] explore the design and management of a distributed energy system encompassing renewable energy generation and diverse end-users across residential, commercial, and industrial sectors. A hierarchical structure is developed to facilitate the real-time peer-to-peer exchange of energy information, accounting for the varying demand flexibility among different user types through inter-sectorial interactions modeled as a non-cooperative game. Additionally, the inherent unpredictability of renewable generation is addressed using the receding horizon optimization technique. To establish a seamless, secure, and efficient distributed energy system, blockchain-enabled smart contracts and decentralized identifiers are employed.

Radanovic et al. [136] explore the potential applications of blockchain technology in the field of medicine. The authors pinpoint multiple domains in which blockchain technology can be utilized, including but not limited to digital healthcare records, medical insurance, clinical studies, medication supply systems, etc. However, the adoption of blockchain in medicine is not without challenges. The technology is still in its early stages and lacks widespread understanding, hindering the development of a clear strategic vision for its future potential. Scalability, smart contract security, and user adoption are among the current concerns that need to be addressed.

Amiri et al. [137] present a novel blockchain system architecture aimed at enhancing the supply chain of engineering, procurement, and construction (EPC) companies involved in oil and gas infrastructure projects. The study addresses the issue of information flow disruptions within the supply chain, which contribute to cost and time inefficiencies. The proposed blockchain system architecture demonstrates promising results, with a reduction of 12.4% in cost inefficiencies and a 36.5% reduction in operational lead times. This approach offers a viable solution for improving the efficiency of EPC supply chains in the oil and gas industry.

Mohd et al. [139] present a descriptive study that investigates challenges in Engineering, Procurement, and Construction (EPC) contracts within the Malaysian oil and gas industry. The research methodology involves a literature review on EPC contracts and cloud computing, as well as the findings from a pilot study. The analysis reveals that the characteristics of cloud computing when integrated with EPC contracts, empower stakeholders and enable highly automated operations. This integration improves the performance of the upstream oil and gas industry by enhancing speed, minimizing financial risks and schedule delays, and elevating the quality of work.

Niu et al. [138] examine the tradeoffs associated with adopting blockchain technology in the context of a common online retailer. The study focuses on the costs of blockchain adoption, the potential increase in market opportunities, and the enhanced competitiveness with local providers for overseas suppliers. The study endeavors to identify potential mutually beneficial outcomes that could result from the implementation of blockchain technology by the overseas supplier, through an assessment of the profitability of all stakeholders in the supply chain.

Hofbauer et al. [117] aim to achieve two scientific objectives in their research. First, they provide a comprehensive description of key features of blockchain technology to comprehend its advantages in transaction processes. Second, they present practical applications of blockchain in transaction activities, specifically in sales and procurement. The methodology employed in this study is analytical and theoretical, combining users' expectations and requirements with statistical data to assess the compatibility between blockchain characteristics and critical needs such as efficiency, cost reduction, security, and trust. The theoretical foundation of the research enables conceptual considerations and deductive exploration of potential application possibilities.

4.7. Artificial Intelligence

In I4.0, Artificial Intelligence (AI) tools and systems have made real impacts in all aspects of production and distribution. However, its influence on the procurement process is less examined in the literature. Scholars have acknowledged the potential benefits of AI in functions such as finance, production, marketing, and sales [143], and underscores that it could be used in procurement organizations that yield solutions to many complex problems more efficiently or effectively with the help of using smart computer algorithms [147].

The applications of AI technologies by public sector organizations have been increasingly discussed and variously debated. Sloane [141] broadly discusses the application of AI technology in the public sector and measures to tackle potential risks to society while employing it. AI technologies can harm citizens and pose a risk to social life in many ways. Therefore, the existing procurement process must be revised and innovated, considering these concerns. Noordt and Misuraca [145] analyze the implications of artificial intelligence within the public sector and its impacts in strengthing governance or administrative functions in three important ways: enhancing public service delivery, improving internal management, and policy decision-making, although in some limited ways. In short, all these studies acknowledge that an increasingly AI-powered mechanism in the public sector, especially in the field of procurement, will boost better governance and efficiency in public service delivery.

The traditional compliance management for procurement's internal auditing has been coupled with many mistakes with a lack of accuracy due to manual audit systems and faults in handling large-scale paper-based repositories. As a solution, Wang et al. [140] designed a continuous compliance awareness framework with the help of AI technology that alleviates potential risks during the earlier manual auditing, including a low level of accuracy, efficiency, accountability, high expense and laborious and time-consuming. This system automatically audits the organization's purchases by intelligently understanding compliance policies using text extraction technologies, automated processing methods, and report rating systems. In addition, it shows 95.6% accuracy with efficiency and quality and procurement's internal audit.

Jemmali et al. [105] discuss the application of a multi-criteria intelligent algorithm for efficient supply chain management. In the traditional model of the supply chain system, the process of choosing suitable suppliers of commodities is a difficult process due to the wide range of suppliers in the fields. The decision-making in this context will be time-consuming. The authors propose an intelligent algorithm model as a solution that employs a set of intelligent formulas by converting predefined preferences into quantitative measurements. The application results of this system show that there is a high level of success in choosing the most appropriate suppliers in a short time by using this technology. Similarly, in a recent study, Manimuthu et al. [144] propose a new design framework for smart contracts and decision-making by using Federated Learning-Artificial Intelligence (FAI). In such an automotive industry system, the cost function, excessive utilization of embedded apparatus, and other related tools will be moderated because of the automated execution and control enabled by FAI. This paper thus shows the significance of AI in decentralized blockchain with smart contracts in order to handle market risk assessment during socio-economic crisis periods.

Mahmoudi et al. [142] intend to create a novel decision-making system to address supplier selection challenges in unprecedented disruption situations. During the COVID-19 outbreak, for instance, essential and non-essential goods and services are disrupted unprecedentedly; therefore, a recovery from supply chain disruptions is necessary. In this context, this paper suggests a new innovative decision-making technology by integrating green and resilience aspects of the supply chain, which the author called "gresilience". In this gresilient supply chain management, a two-fold decomposition of the core algorithm of the Ordinal Priority Approach (OPA). The study employs OPA and extends it to the Fuzzy OPA (OPA-F) as a solution for supplier selection problems. The proposed model can solve supply disruptions during the time of the outbreak and sustain the supply chain.

Notwithstanding these benefits, some other scholars problematized the other side of AI-powered solutions and their potential dangers and risks in the public sector. For instance, in a recent work, Nagitta et al. [146] call for a human-centered AI (HCAI) discourse in procurement systems. By conducting research among procurement practitioners in Kenya, the author argues the importance of having a customized AI-powered procurement system in the public sector with a human-centric manual process in order to tackle potential challenges of AI technologies among humans. This approach to AI is crucial in the world and many companies are adopting such a strategy in order to keep dignity and respect for humankind.

5. Discussion

The advent of I4.0 has brought about significant transformations in the procurement landscape. From the analysis of the literature, several key themes emerged, namely I4.0 or P4.0, e-procurement, big data, the Internet of Things, additive manufacturing, blockchain and smart contracts, and artificial intelligence. Each of these themes has distinct implications for procurement, and their integration can unlock numerous benefits while also presenting challenges. The application of I4.0 technologies in procurement processes is not a novel topic because copious publications have been published on the same over the last few years. However, most of the literature discussed individual applications of I4.0 technologies in the field of procurement. The review presented herein brings this systematic literature together in order to make a comprehensive picture of the P4.0 process and enhanced digitalization that it brings forth for better efficiency in supply chain management.

Digitalization in the procurement processes enables a possibility of collaboration between various procurement processes which helps to sort out challenges related to procurement like supplier-related risks, misplacing orders, and others. This review supports that the utilization of I4.0-enabling technology is expected to yield noteworthy enhancements within procurement in particular and supply chain management in general. This is attributed to the technology's ability to enhance the efficiency and effectiveness of supply chain management, facilitate extensive supply chain integration, and improve information management. I4.0 integrates supplies, demands, sales, and manufacturing resources by using highly equipped technological innovations in the procurement process. P4.0 represents a paradigm shift in procurement practices, where digital technologies are harnessed to enable more efficient and effective processes. The studies reviewed highlight the potential of P4.0 in improving supply chain visibility, streamlining processes, enhancing collaboration, and enabling data-driven decision-making. However, the adoption of P4.0 is not without challenges, including the need for organizational change and overcoming resistance to digital transformation [39].

E-procurement emerged as a critical aspect of I4.0, enabling the automation and digitization of procurement processes. The literature demonstrates that e-procurement systems can enhance efficiency, reduce costs, improve transparency, and facilitate supplier collaboration. Nevertheless, the success of e-procurement implementation relies on factors such as organizational readiness, supplier integration [149], data security, and user acceptance [150]. There are several I4.0 technologies that help procurement managers predict prices [52]. One of them is cloud manufacturing, which takes several factors into account when estimating prices such as production conditions, variety of offerings, production capacity, customer demands, and price-supplier selection optimization.

Big data analytics has emerged as a powerful tool in procurement, enabling organizations to extract valuable insights from vast amounts of data. The reviewed studies highlight the potential of big data analytics in demand forecasting, supplier performance evaluation, risk management, and strategic decision-making. Big data can help achieve lower procurement costs through improved knowledge of the supply processes at various stages, such as supplier appraisal and marketing through which a supplier can negotiate cost reductions. Similarly, negotiations between parties could take place without the need for a professional intervention if big data analytics approaches are used. It can also cut down on negotiating times, thus lowering most costs [95]. However, challenges such as data quality, privacy concerns, and the need for analytical expertise need to be addressed for successful implementation [151]. In practice, big data has the potential to assess material prices by accessing the commodities' database of each supplier's cost structure. As a result, the best timing and price for making a purchase might be determined. Because of the important uncertainties and risks involved, selecting and assessing vendors in an e-procurement system is more difficult than in traditional procurement. In all, there are three types of hazards here: internal risks that occur within a business, supply chain risks, and external risks that occur outside the supply chain [152].

The Internet of Things has the potential to revolutionize procurement by connecting physical objects and systems to digital networks. The literature indicates that IoT-enabled procurement can enable real-time monitoring of inventory, enhance traceability, enable predictive maintenance, and support automated replenishment. Although there are different types of costs related to the supply chain, two of them are important in this context. They are the costs of purchasing raw materials and the costs of maintaining inventory in production sites [104]. Companies are utilizing the IoT by tagging their products with identification tags and linking the important data from these tags to the cloud via the Internet. As a result, these stored data can be accessed easily from anywhere on the globe, and their affiliated companies (suppliers or vendors) can be linked to them for purchases. In today's commercial context, the IoT-supported system has become a cost-effective solution by allowing faster decision-making. However, concerns related to data security, interoperability, and scalability need to be addressed to fully leverage the potential of IoT in procurement [153].

Additive manufacturing, also known as 3D printing, has emerged as a transformative technology [106]. It offers new possibilities for product design, enabling lighter and stronger parts [107]. AM provides increased design freedom, allowing unconventional geometries to be produced [109]. Furthermore, it facilitates collaboration between buyers and suppliers, leading to redesigned production networks [107]. The adoption of AM can decentralize manufacturing and reshape supply chains [108]. However, decision-making complexities arise from material properties, compatibility, and cost considerations. Despite challenges, AM has the potential to enhance supply chain resilience and reduce dependency on original equipment manufacturers [112]. It is crucial to evaluate the trade-offs between flexibility, lead times, and overall costs when incorporating AM in design enhancements [113]. In addition to its impact on product design and supply chains, AM plays a significant role in addressing disruptions and enhancing resilience in the automotive sector. The COVID-19 pandemic highlighted the vulnerabilities of global supply chains, emphasizing the need for innovative solutions [110]. However, widespread adoption of AM faces barriers such as limited government regulations, management support, raw material availability, and workforce constraints [154]. To fully leverage the benefits of AM, a systematic and interdisciplinary approach is necessary. This approach should consider the entire cost perspective, including logistics, manufacturing, repair expenses, and trade-offs between functionality improvements and design changes [113]. By addressing these challenges and embracing a comprehensive perspective, AM has the potential to optimize supply chains, enhance industrial performance, and unlock new possibilities for future advancements [109].

Blockchain technology and smart contracts have gained huge attention as enablers of secure and transparent transactions in procurement. The literature review reveals that blockchain can enhance trust, improve traceability, streamline payment processes, and facilitate supplier verification [133]. Additionally, blockchain technology gives a permanent review succession, which permits members/actors/stakeholders to know what activity was performed by whom and when it was performed [117]. This characteristic of blockchain provides straightforwardness and lessons related to the danger of data loss with outsider frameworks [119]. Huge information can assist with controlling the dangers related to providers when a presentation pointer is characterized, which prompts perceiving proper obtaining. Blockchain provides security, guarantees secrecy, and upgrades the information uprightness of exchanges without the outsider's contribution. Information security is critical for private organizations too. In any event, for privately owned business providers, for instance, save and secure the examined duplicates of procurement arranges by exchanging them straightforwardly and by utilizing electronic information exchange [122]. All things considered, there is a lot of protection from these new emerging advances. However, challenges such as scalability, regulatory considerations, and the need for industry-wide collaboration need to be addressed for widespread adoption [155]. Despite these complexities, the transformative potential of blockchain and smart contracts in revolutionizing industries and creating new business models cannot be overlooked [156].

Artificial intelligence has emerged as a powerful tool in procurement, offering capabilities such as intelligent automation, predictive analytics, and cognitive decision support. The integration of AI in procurement processes has gained attention in the literature, with studies highlighting its potential benefits in various functions and areas, including compliance management, supply chain optimization, supplier selection, and risk assessment [141]. However, it is important to consider ethical considerations, data quality, and the need for human-AI collaboration. In the public sector, AI technology has been discussed for enhancing public service delivery, improving internal management, and policy decision-making, albeit with certain limitations [145]. The existing literature also recognizes the need to address potential risks and revise procurement processes accordingly. Emphasizing a human-centric approach in AI-powered procurement systems can help tackle these challenges and ensure effective and responsible implementation [157,158].

Table 7 summarizes the drivers and challenges of Industry 4.0 technologies in procurement.

Additionally, in the context of I4.0 and its integration into procurement and supply chain management, modularization plays a pivotal role. Modularization refers to the strategic approach of breaking down complex systems or processes into modular components that can be developed, managed, and optimized independently. This approach facilitates flexibility, scalability, and customization, allowing for swift adjustments and improvements in response to changing market demands [159], and by adopting modular concepts, procurement processes can become more agile and adaptable. Components of the supply chain, such as sourcing, production, and distribution, can be structured as modules, enabling streamlined management of each element. This modularity empowers organizations to efficiently respond to dynamic market shifts, mitigate risks, and optimize resource allocation. Furthermore, modularization aligns with I4.0 technologies like IoT and AI, enabling data-driven decision-making and predictive analytics in each module. Through this integration, procurement and supply chain management can achieve enhanced efficiency, reduced operational costs, improved collaboration, and rapid responsiveness to market changes, fostering a dynamic interconnected environment and optimized for greater resilience and adaptability in an ever-evolving business landscape [160].

Comparing and contrasting the studies reviewed in these thematic areas, it becomes evident that the integration of multiple I4.0 technologies can unlock synergistic effects. For example, the combination of e-procurement, big data analytics, and AI can enable organizations to make data-driven decisions, automate processes, and optimize supply chain operations. Similarly, the integration of IoT, additive manufacturing, and blockchain can create a decentralized and transparent procurement ecosystem with enhanced traceability and customization capabilities. Moreover, comparing the adoption of I4.0 technologies in various industries and contexts, such as manufacturing, services, and the public sector, reveals both similarities and differences in their application and challenges. In manufacturing, I4.0 drives operational efficiency through automated production and supply chain processes. Similarly, the services sector benefits from improved customer experiences through data-driven insights. In the public sector, I4.0 enhances transparency and accountability in procurement processes. However, challenges vary: manufacturing deals with optimizing physical processes, services navigate intangible offerings, and the public sector focuses on regulatory compliance. While all sectors face data security concerns, manufacturing grapples with integrating IoT in complex machinery, services emphasize AI-driven personalization, and the public sector navigates policy frameworks. The paper's comparative analysis thus highlights the need for tailored strategies in each context while emphasizing the overarching transformative potential of I4.0 technologies in procurement.

Table 7. Drivers and challenges of the technologies in procurement.

Technology	Drivers	Challenges
Industry 4.0	Improved supply chain visibility, Streamlined processes, Enhanced collaboration, Data-driven decision-making	Organizational change, Resistance to digital transformation
E-Procurement	Automation and digitization, Increased efficiency, Cost reduction, Improved transparency, Supplier collaboration	Organizational readiness, Supplier integration, Data security, User acceptance
Big Data	Valuable insights, Demand forecasting, Supplier performance evaluation, Risk management, Strategic decision-making	Data quality, Privacy concerns, Need for analytical expertise
Internet of Things	Real-time inventory monitoring, Enhanced traceability, Predictive maintenance, Automated replenishment	Data security, Interoperability, Scalability
Additive Manufacturing	Decentralized production, Customization, Reduced lead times, Lower costs for low-volume production, Increased design flexibility	Limited material options, Intellectual property issues, Need for new skill sets
Blockchain	Secure and transparent transactions, Enhanced trust, Improved traceability, Streamlined payment processes, Supplier verification	Scalability Regulatory considerations Need for industry-wide collaboration
Artificial Intelligence	Intelligent automation, Predictive analytics, Cognitive decision support, Compliance management, Supply chain optimization, Risk assessment	Ethical considerations Data quality Need for human-AI collaboration

However, it is important to acknowledge the limitations and challenges associated with the adoption of I4.0 technologies in procurement. The studies reviewed often focus on specific aspects or use cases, limiting the generalizability of the findings. Additionally, there are organizational, technical, and cultural barriers that need to be overcome for successful implementation [161,162].

Organizations need to carefully consider factors such as data security, privacy, interoperability, and regulatory compliance when adopting I4.0 technologies. The potential risks associated with increased reliance on digital systems, such as cyber threats and data breaches, should not be overlooked. It is essential to develop robust risk management strategies and invest in cybersecurity measures to mitigate these risks. Moreover, the human factor should not be underestimated. While I4.0 technologies offer automation and efficiency gains, they should be seen as tools to augment human capabilities rather than replace human involvement entirely. Organizations should prioritize the development of digital skills and provide training and support to ensure the successful adoption and utilization of these technologies. From a broader perspective, the implications of I4.0 in procurement extend beyond operational improvements. The integration of digital technologies can have far-reaching effects on the procurement profession itself, including changes in job roles, skill requirements, and the nature of supplier relationships [1,163,164]. Procurement professionals need to adapt and embrace continuous learning to remain relevant in the era of I4.0.

In the realm of procurement, a range of I4.0 technologies have emerged as influential forces in different processes to enhance efficiency, visibility, predictability, and decision-making. Among them, the IoT stands out for its profound impact on supply chain management. IoT devices equipped with sensors and connectivity capabilities provide real-time data on inventory levels, product quality, and equipment performance, thereby enhancing visibility and enabling organizations to optimize their inventory management, demand forecasting, order fulfillment processes, and facilitating E-processes. Also, AI has proven to be a game-changer in procurement, empowering professionals with advanced big data analytics and decision-making capabilities. AI-powered algorithms can analyze vast amounts of data, identify patterns, and generate valuable insights on strategic sourcing, risk management, and contract management. Table 8 summarizes the potential applications in procurement functions with respect to different KPIs criteria.

Table 8. Summary of the potential applications.

Technology	Applications	Criteria
Industry 4.0	- Enhanced operational efficiency - Improved supply chain visibility - Predictive analytics - Proactive decision-making - Support modularization	Cost savings Process automation Scalability Flexibility More product variety in a shorter lead time and lower cost
E-Procurement	- Streamlined procurement processes - Enhanced supplier relationship management - Increased procurement transparency	Ease of use Integration with existing systems Security and data privacy
Big Data	- Demand forecasting - Supply chain optimization - Risk management -Performance measurement and tracking	Data accuracy and quality Real-time analytics Scalability Integration with other systems
Internet of Things	- Real-time asset tracking and monitoring - Condition-based maintenance - Inventory management optimization - Improved logistics and supply chain visibility - Facilitating E-Design, E-Sourcing, E- Negotiation, E-Evaluation	Connectivity and interoperability Data security Scalability Energy efficiency Lean processes
Additive Manufacturing	- On-demand and localized production - Customized and personalized products - Rapid prototyping and iteration - Reduced lead times and transportation costs	Design flexibility Material selection Quality control Environmental sustainability
Blockchain	- Transparent and secure transactions - Supply chain traceability and verification - Simplified procurement processes	Data immutability and integrity Smart contract functionality Interoperability with existing systems
Artificial Intelligence	- Strategic sourcing and supplier evaluations - Contract management and compliance - Risk management and predictive analytics - Process automation and optimization	Data analysis and insights Automation and efficiency Ethical considerations Scalability and adaptability

The paper highlights the transformative potential of I4.0 technologies in the procurement journey. The integration of e-procurement, big data analytics, IoT, additive manufacturing, blockchain, and AI can lead to enhanced efficiency, transparency, collaboration, and decision-making in procurement processes. However, the adoption of these technologies should be accompanied by careful consideration of risks, challenges, and the human element. Organizations need to develop comprehensive strategies, invest in cybersecurity, foster digital skills, and adapt to the evolving procurement landscape to fully harness the benefits of I4.0. Future research should focus on addressing the identified limitations, exploring new applications and use cases, and investigating the long-term impact of I4.0 on the procurement profession and supply chain ecosystems.

Moreover, it is important to recognize that the adoption of these technologies is not without challenges. Implementing I4.0 technologies requires significant investments in infrastructure, data management systems, and workforce training. Organizations need to address issues related to data security, privacy, and interoperability when leveraging technologies like big data and IoT. Moreover, the integration of multiple technologies may pose compatibility issues and necessitate changes in organizational structures and processes. To fully harness the potential of I4.0 technologies in procurement, organizations should develop a comprehensive strategy that aligns technology adoption with their business objectives. This strategy should consider factors such as organizational culture, change management, and stakeholder engagement. Collaboration with technology providers, industry partners, and research institutions can also facilitate knowledge sharing and accelerate the adoption and implementation process.

Future research in this area should focus on surveying and addressing the practical challenges and barriers to implementing I4.0 technologies in procurement. Empirical studies and case examples can provide valuable insights into the real-world impact of these technologies on procurement performance, cost-effectiveness, and sustainability. Additionally, exploring the ethical implications of technologies like AI and blockchain in procurement, including issues of bias, privacy, and fairness, is essential to ensure responsible and equitable adoption. The paper's significance can be amplified by delving into specific research questions: How do individual I4.0 technologies impact different procurement stages? What are the implications of AI-driven supplier evaluation models on decision-making? Challenges like data privacy in IoT implementation and biases in AI algorithms require scrutiny. Future research should address how AI enhances contract management and how big data accuracy impacts procurement. Contributions lie in models for technology-driven supplier selection, frameworks for IoT integration, and ethical guidelines for AI and blockchain, fostering innovative, responsible procurement practices.

6. Conclusions

In this paper, the role of I4.0 in the procurement industry has been examined, focusing on its potential benefits and implications. The procurement journey has undergone significant transformations in recent years, and a crucial role has been played by the integration of I4.0 technologies. Through the employed systematic review, various aspects of I4.0 in procurement have been explored, including the impact on supply chain management, data analytics, modularization, automation, and collaboration. The literature review conducted has provided insights into the current state of research and practice in this field, with a focus on the key trends and challenges faced by organizations.

Based on the analysis and discussion presented, it can be concluded that significant opportunities exist for procurement professionals to enhance their processes and achieve greater efficiency, cost savings, and strategic value through the adoption of I4.0. The integration of advanced technologies such as the IoT, AI, and blockchain, among others, can revolutionize traditional procurement practices and enable data-driven decision-making.

One of the major benefits of I4.0 in procurement is the ability to leverage real-time data and analytics for gaining insights into supplier performance, demand patterns, and market trends. This allows for the optimization of sourcing strategies, identification of potential risks, and enhancement of supplier relationship management. Furthermore, the automation and digitization of procurement processes streamline operations, reduce manual errors, and result in time and resource savings. Collaboration also stands as a significant aspect facilitated by I4.0 in procurement. With connected systems and platforms, seamless communication, information sharing, and effective collaboration are made possible among stakeholders across the supply chain. This leads to improved coordination, enhanced visibility, and expedited decision-making processes.

The study underscores several key implications for various stakeholders in the realm of I4.0 technologies' integration in procurement. Manufacturers can leverage the synergistic effects of I4.0 technologies to enhance operational efficiency and automate production and supply chain processes, while vendors and suppliers stand to benefit from the transparency and accountability brought about by I4.0 in procurement processes. The services sector can utilize data-driven insights for improved customer experiences, and the public sector can enhance regulatory compliance and transparency in procurement practices. Scholars and researchers gain insights into the complex interplay of I4.0 technologies across different sectors, offering valuable knowledge to inform future studies and practical implementations. As the procurement landscape continues to evolve, it is imperative for organizations to stay abreast of the latest advancements and adapt to the changing dynamics. Embracing I4.0 is not just a choice but a necessity to thrive in a digital and interconnected world. The successful adoption of I4.0 in procurement requires a holistic approach involving a combination of technological investments, organizational change management, and continuous learning.

However, it should be noted that the implementation of I4.0 in procurement is not without challenges. Concerns regarding data privacy and security, the need for upskilling and reskilling the workforce, and the integration of legacy systems with new technologies present hurdles that organizations must address. Robust strategies must be developed to fully harness the potential of I4.0 in procurement. In conclusion, a significant paradigm shift is evident with the introduction of I4.0 in the procurement industry. Unprecedented opportunities arise for organizations to transform their procurement processes and achieve strategic objectives. Through the adoption of advanced technologies, utilization of data analytics, and fostering collaboration, procurement professionals can enhance their decision-making capabilities, optimize supply chain operations, and drive sustainable growth. As the procurement landscape continues to evolve, it is imperative for organizations to stay abreast of the latest advancements and adapt to the changing dynamics. Embracing I4.0 is not just a choice but a necessity to thrive in a digital and interconnected world. The successful adoption of I4.0 in procurement requires a holistic approach involving a combination of technological investments, organizational change management, and continuous learning.

Author Contributions: Conceptualization, A.A., L.K., M.Y. and B.M.; methodology, A.A. and M.Y.; formal analysis, A.A. and M.Y.; investigation, A.A. and M.Y.; resources, A.A. and M.Y.; writing—original draft preparation, A.A. and M.Y.; writing—review and editing, A.A., M.Y., B.M. and L.K.; supervision, L.K.; project administration, A.A. All authors have read and agreed to the published version of the manuscript.

Funding: This research received no external funding.

Data Availability Statement: Not applicable.

Acknowledgments: The open-access funding provided by the Qatar National Library (QNL).

Conflicts of Interest: The authors declare no conflict of interest.

References

1. Glas, A.H.; Kleemann, F.C. The impact of industry 4.0 on procurement and supply management: A conceptual and qualitative analysis. *Int. J. Bus. Manag. Invent.* **2016**, *5*, 55–66.
2. Nicoletti, B. Procurement 4.0 and the Fourth Industrial Revolution. In *The Opportunities and Challenges of a Digital World*; Palgrave Macmillan: Cham, Switzerland, 2020.

3. Hahn, G.J. Industry 4.0: A supply chain innovation perspective. *Int. J. Prod. Res.* **2020**, *58*, 1425–1441. [CrossRef]
4. Thio-ac, A.; Serut, A.K.; Torrejos, R.L.; Rivo, K.D.; Velasco, J. Blockchain-based system evaluation: The effectiveness of blockchain on E-procurements. *Int. J. Adv. Trends Comput. Sci. Eng.* **2019**, *8*, *5*, 2673–2676. [CrossRef]
5. Gottge, S.; Menzel, T.; Forslund, H. Industry 4.0 technologies in the purchasing process. *Ind. Manag. Data Syst.* **2020**, *120*, 730–748. [CrossRef]
6. Jahani, N.; Sepehri, A.; Vandchali, H.R.; Tirkolaee, E.B. Application of industry 4.0 in the procurement processes of supply chains: A systematic literature review. *Sustainability* **2021**, *13*, 7520. [CrossRef]
7. Bueno, R.E.; Almeida dos Santos, H.; de Junior Freitas, M.; Toloi, R.C.; Gonçalves, R.F. Procurement 4.0: A Systematic Review of Its Technological Evolution. In Proceedings of the Advances in Production Management Systems. Smart Manufacturing and Logistics Systems: Turning Ideas into Action: IFIP WG 5.7 International Conference, APMS 2022, Gyeongju, South Korea, 25–29 September 2022; Proceedings, Part II; Springer: Berlin/Heidelberg, Germany, 2022; pp. 148–156.
8. Ustundag, A.; Cevikcan, E. *Industry 4.0: Managing the Digital Transformation*; Springer: Berlin/Heidelberg, Germany, 2017.
9. Nicoletti, B.; Nicoletti, B. The future: Procurement 4.0. *Agile Procurement: Volume II: Designing and Implementing a Digital Transformation*; Palgrave MacMillan: London, UK, 2018, pp. 189–230.
10. Herold, S.; Heller, J.; Rozemeijer, F.; Mahr, D. Dynamic capabilities for digital procurement transformation: A systematic literature review. *Int. J. Phys. Distrib. Logist. Manag.* **2023**, *53*, 424–447. [CrossRef]
11. Bogaschewsky, R.; Müller, H. *BME-Barometer "Elektronische Beschaffung"*; Frankfurt: BME Verband: Eschborn, Germany, 2018.
12. Radell, C.; Schannon, D. Digital procurement: The benefits go far beyond efficiency. *Supply Chain Manag. Rev.* **2019**, *6*, 14–21.
13. Sjödin, D.; Kamalaldin, A.; Parida, V.; Islam, N. Procurement 4.0: How industrial customers transform procurement processes to capitalize on digital servitization. *IEEE Trans. Eng. Manag.* **2021**, *68* 1–16. [CrossRef]
14. Economic, E.; Committee, S. Digitalisation—Challenges for Europe. 2019. Available online: https://www.eesc.europa.eu/en/our-work/publications-other-work/publications/digitalisation-challenges-europe (accessed on 29 August 2023). ISBN: 978-92-830-4488-8.
15. De Groen, W.P.; Lenaerts, K.; Bosc, R.; Paquier, F. *Impact of Digitalisation and the On-Demand Economy on Labour Markets and the Consequences for Employment and Industrial Relations*; Final Study. CEPS Special Report; European Economic and Social Committee: Brussels, Belgium, 2017; ISBN: 978-92-830-3420-9.
16. Nicoletti, B.; Nicoletti, B. Industry 4.0 and Procurement 4.0. In *Procurement 4.0 and the Fourth Industrial Revolution: The Opportunities and Challenges of a Digital World*; Springer: Berlin/Heidelberg, Germany, 2020, pp. 5–51.
17. Sánchez-Flores, R.B.; Cruz-Sotelo, S.E.; Ojeda-Benitez, S.; Ramírez-Barreto, M.E. Sustainable supply chain management—A literature review on emerging economies. *Sustainability* **2020**, *12*, 6972. [CrossRef]
18. Banerjee, S.; Kankaria, G.H.; Shivaprasad, K.T. From Forecasting Demand to Ordering—An Automated Machine Learning Approach with Amazon Forecast to Decrease Stockouts, Excess Inventory, and Costs. 2021. Available online: https://aws.amazon.com/blogs/machine-learning/from-forecasting-demand-to-ordering-an-automated-machine-learning-approach-with-amazon-forecast-to-decrease-stock-outs-excess-inventory-and-costs/ (accessed on 17 August 2023).
19. World Health Organization and the United Nations Children's Fund (UNICEF). Global Report on Assistive Technology. Geneva, Licence: CC BY-NC-SA 3.0 IGO, 2022; ISBN: 9789240049451. Available online: https://www.who.int/publications/i/item/9789240049451 (accessed on 30 August 2023).
20. Oliveira-Dias, D.; Maqueira-Marín, J.M.; Moyano-Fuentes, J. The link between information and digital technologies of industry 4.0 and agile supply chain: Mapping current research and establishing new research avenues. *Comput. Ind. Eng.* **2022**, *167*, 108000. [CrossRef]
21. Strohmer, M.F.; Easton, S.; Eisenhut, M.; Epstein, E.; Kromoser, R.; Peterson, E.R.; Rizzon, E. *Disruptive Procurement: Winning in a Digital World*; Springer: Berlin/Heidelberg, Germany, 2020.
22. Johnsen, T.E.; Howard, M.; Miemczyk, J. *Purchasing and Supply Chain Management: A Sustainability Perspective*; Routledge: London, UK, 2018.
23. Van Weele, A. *Purchasing and Supply Chain Management*; Cengage Learning EMEA: Boston, MA, USA, 2018.
24. Monczka, R.M.; Handfield, R.B.; Giunipero, L.C.; Patterson, J.L. *Purchasing and Supply Chain Management*; Cengage Learning: Boston, MA, USA, 2020.
25. Trautrims, A.; MacCarthy, B.L.; Okade, C. Building an innovation-based supplier portfolio: The use of patent analysis in strategic supplier selection in the automotive sector. *Int. J. Prod. Econ.* **2017**, *194*, 228–236. [CrossRef]
26. Mogre, R.; Lindgreen, A.; Hingley, M. Tracing the evolution of purchasing research: Future trends and directions for purchasing practices. *J. Bus. Ind. Mark.* **2017**, *32*, 251–257. [CrossRef]
27. ISM-INDIA. PROCUREMENT JOURNEY (From Procurement 1.0 to Procurement 4.0). 2020. Available online: http://ipscm.org/wp-content/uploads/2020/09/Knowledge-Paper-Procurement-Journey-2.pdf (accessed on 30 August 2023).
28. Ghadimi, P.; Azadnia, A.H.; Heavey, C.; Dolgui, A.; Can, B. A review on the buyer–supplier dyad relationships in sustainable procurement context: Past, present and future. *Int. J. Prod. Res.* **2016**, *54*, 1443–1462. [CrossRef]
29. Spina, G.; Caniato, F.; Luzzini, D.; Ronchi, S. Past, present and future trends of purchasing and supply management: An extensive literature review. *Ind. Mark. Manag.* **2013**, *42*, 1202–1212. [CrossRef]
30. Nicoletti, B.; Nicoletti, B. Systems in Agile Procurement. In *Agile Procurement: Volume II: Designing and Implementing a Digital Transformation*; Palgrave MacMillan: London, UK, **2018**, pp. 29–76.

31. Srai, J.S.; Lorentz, H. Developing design principles for the digitalisation of purchasing and supply management. *J. Purch. Supply Manag.* **2019**, *25*, 78–98. [CrossRef]
32. O'brien, J. *Category Management in Purchasing: A Strategic Approach to Maximize Business Profitability*; Kogan Page Publishers: London, UK, 2019.
33. Foerstl, K.; Hartmann, E.; Wynstra, F.; Moser, R. Cross-functional integration and functional coordination in purchasing and supply management: Antecedents and effects on purchasing and firm performance. *Int. J. Oper. Prod. Manag.* **2013**, *33*, 689–721. [CrossRef]
34. Malacina, I.; Karttunen, E.; Jääskeläinen, A.; Lintukangas, K.; Heikkilä, J.; Kähkönen, A.K. Capturing the value creation in public procurement: A practice-based view. *J. Purch. Supply Manag.* **2022**, *28*, 100745. [CrossRef]
35. Vitasek, K. Strategic sourcing business models. *Strateg. Outsourcing Int. J.* **2016**, *9*, 126–138. [CrossRef]
36. Knight, A.K.; Blessner, P.; Olson, B.A.; Blackburn, T.D. Strategic sourcing and corporate social responsibility: Aligning a healthcare organization's strategic objectives. *J. Purch. Supply Manag.* **2017**, *23*, 94–104. [CrossRef]
37. Allal-Chérif, O.; Simón-Moya, V.; Ballester, A.C.C. Intelligent purchasing: How artificial intelligence can redefine the purchasing function. *J. Bus. Res.* **2021**, *124*, 69–76. [CrossRef]
38. Suhonen, N.; Tammi, T.; Saastamoinen, J.; Pesu, J.; Turtiainen, M.; Okkonen, L. Incentives and risk-sharing in public procurement of innovations: towards contracting strategy framework. *J. Public Procure.* **2019**, *19*, 129–145. [CrossRef]
39. Mavidis, A.; Folinas, D. From Public E-Procurement 3.0 to E-Procurement 4.0; A Critical Literature Review. *Sustainability* **2022**, *14*, 11252. [CrossRef]
40. Page, M.J.; McKenzie, J.E.; Bossuyt, P.M.; Boutron, I.; Hoffmann, T.C.; Mulrow, C.D.; Shamseer, L.; Tetzlaff, J.M.; Akl, E.A.; Brennan, S.E.; et al. The PRISMA 2020 statement: An updated guideline for reporting systematic reviews. *Int. J. Surg.* **2021**, *88*, 105906. [CrossRef] [PubMed]
41. Rane, S.B.; Narvel, Y.A. Leveraging the industry 4.0 technologies for improving agility of project procurement management processes. *Int. J. Syst. Assur. Eng. Manag.* **2021**, *12*, 1146–1172. [CrossRef]
42. Prebanić, K.R.; Vukomanović, M. Realizing the need for digital transformation of stakeholder management: A systematic review in the construction industry. *Sustainability* **2021**, *13*, 12690. [CrossRef]
43. Bag, S.; Dhamija, P.; Gupta, S.; Sivarajah, U. Examining the role of procurement 4.0 towards remanufacturing operations and circular economy. *Prod. Plan. Control* **2021**, *32*, 1368–1383. [CrossRef]
44. Alabdali, M.A.; Salam, M.A. The impact of digital transformation on supply chain procurement for creating competitive advantage: An empirical study. *Sustainability* **2022**, *14*, 12269. [CrossRef]
45. Ibem, E.O.; Laryea, S. Survey of digital technologies in procurement of construction projects. *Autom. Constr.* **2014**, *46*, 11–21. [CrossRef]
46. Bag, S.; Wood, L.C.; Mangla, S.K.; Luthra, S. Procurement 4.0 and its implications on business process performance in a circular economy. *Resour. Conserv. Recycl.* **2020**, *152*, 104502. [CrossRef]
47. Lorentz, H.; Aminoff, A.; Kaipia, R.; Srai, J.S. Structuring the phenomenon of procurement digitalisation: Contexts, interventions and mechanisms. *Int. J. Oper. Prod. Manag.* **2021**, *41*, 157–192. [CrossRef]
48. Schiele, H.; Torn, R.J. Cyber-physical systems with autonomous machine-to-machine communication: Industry 4.0 and its particular potential for purchasing and supply management. *Int. J. Procure. Manag.* **2020**, *13*, 507–530. [CrossRef]
49. Flechsig, C.; Anslinger, F.; Lasch, R. Robotic Process Automation in purchasing and supply management: A multiple case study on potentials, barriers, and implementation. *J. Purch. Supply Manag.* **2022**, *28*, 100718. [CrossRef]
50. Raja Santhi, A.; Muthuswamy, P. Pandemic, War, Natural Calamities, and Sustainability: Industry 4.0 Technologies to Overcome Traditional and Contemporary Supply Chain Challenges. *Logistics* **2022**, *6*, 81. [CrossRef]
51. Lasi, H.; Fettke, P.; Kemper, H.G.; Feld, T.; Hoffmann, M. Industry 4.0. *Bus. Inf. Syst. Eng.* **2014**, *6*, 239–242. [CrossRef]
52. Wang, P.; Shen, J.; Guo, W.; Zhang, C.; Zhang, B. Cloud-based Government Procurement Information Integration Platform. *J. Digit. Inf. Manag.* **2015**, *13*. 1–9.
53. Pires, M.C.; Parreira, R.; Frazzon, E.M. Integrated operational supply chain planning in Industry 4.0. *Int. J. Integr. Supply Manag.* **2021**, *14*, 28–49. [CrossRef]
54. Rejeb, A.; Appolloni, A. The Nexus of Industry 4.0 and Circular Procurement: A Systematic Literature Review and Research Agenda. *Sustainability* **2022**, *14*, 15633. [CrossRef]
55. Choi, S.; Kang, G.; Jun, C.; Lee, J.Y.; Han, S. Cyber-physical systems: A case study of development for manufacturing industry. *Int. J. Comput. Appl. Technol.* **2017**, *55*, 289–297. [CrossRef]
56. Richnák, P. Current Trend of Industry 4.0 in Logistics and Transformation of Logistics Processes Using Digital Technologies: An Empirical Study in the Slovak Republic. *Logistics* **2022**, *6*, 79. [CrossRef]
57. Bienhaus, F.; Haddud, A. Procurement 4.0: Factors influencing the digitisation of procurement and supply chains. *Bus. Process. Manag. J.* **2018**, *24*, 965–984. [CrossRef]
58. Klünder, T.; Dörseln, J.N.; Steven, M. Procurement 4.0: How the digital disruption supports cost-reduction in Procurement. *Production* **2019**, *29*, e20180104. [CrossRef]
59. Kohnová, L.; Salajová, N. Impact of Industry 4.0 on Companies: Value Chain Model Analysis. *Adm. Sci.* **2023**, *13*, 35. [CrossRef]
60. Hallikas, J.; Immonen, M.; Brax, S. Digitalizing procurement: The impact of data analytics on supply chain performance. *Supply Chain. Manag. Int. J.* **2021**, *26*, 629–646. [CrossRef]

61. Rejeb, A.; Sűle, E.; G Keogh, J. Exploring New Technologies Procurement. *Transp. Logist. Int. J.* **2018**, *18*, 45.
62. Ghadge, A.; Er Kara, M.; Moradlou, H.; Goswami, M. The impact of Industry 4.0 implementation on supply chains. *J. Manuf. Technol. Manag.* **2020**, *31*, 669–686. [CrossRef]
63. Chou, T.Y.; Liao, J.P. Using MEC and FAHP to Establish an Evaluation Model for Food Processing Machinery Procurement. *Math. Probl. Eng.* **2020**, *2020*, 1–10. [CrossRef]
64. Gupta, M.; Narain, R. A study on usage of IT and its implications on e-procurement in Indian organisations. *Int. J. Bus. Inf. Syst.* **2012**, *10*, 222–244. [CrossRef]
65. Aduwo, E.B.; Ibem, E.O.; Afolabi, A.O.; Oluwnmi, A.O.; Tunji-Olayeni, P.F.; Ayo-Vaughan, E.A.; Uwakonye, U.O.; Oni, A.A. Exploring anti-corruption capabilities of e-procurement in construction project delivery in Nigeria. *Constr. Econ. Build.* **2020**, *20*, 56–76. [CrossRef]
66. Kamarulzaman, N.H.; Eglese, R.W. The potential use of e-procurement in the Malaysian palm oil industry supply chain. *Pertanika J. Soc. Sci. Hum* **2013**, *21*, 95–108.
67. Kamarulzaman, N.; Mohamed, Z. Application of e-procurement technologies for selecting suppliers of agro-based SMEs in Malaysia. *Int. J. Econ. Manag.* **2013** *7*, 45–66.
68. Hung, W.H.; Lin, C.P.; Tai, Y.M.; Ho, C.F.; Jou, J.J. Exploring the impact of Web-based e-procurement on performance: organisational, interorganisational, and systems perspectives. *Int. J. Logist. Res. Appl.* **2014**, *17*, 200–215. [CrossRef]
69. Basak, M. Achieving e-procurement benefits in an aviation MRO environment. *Oper. Supply Chain Manag. Int. J.* **2015**, *9*, 50–60. [CrossRef]
70. Kim, M.; Suresh, N.C.; Kocabasoglu-Hillmer, C. A contextual analysis of the impact of strategic sourcing and E-procurement on performance. *J. Bus. Ind. Mark.* **2015**, *30*, 1–16. [CrossRef]
71. Chibani, A.; Delorme, X.; Dolgui, A.; Pierreval, H. Dynamic optimisation for highly agile supply chains in e-procurement context. *Int. J. Prod. Res.* **2018**, *56*, 5904–5929. [CrossRef]
72. Sánchez-Rodríguez, C.; Martínez-Lorente, A.R.; Hemsworth, D. E-procurement in small and medium sized enterprises; facilitators, obstacles and effect on performance. *Benchmarking Int. J.* **2020**, *27*, 839–866. [CrossRef]
73. Belisari, S.; Appolloni, A.; Cerruti, C. Positive and negative impacts of the adoption of e-procurement solutions: The Italian market case. *Int. J. Procure. Manag.* **2019**, *12*, 219–241.
74. Fu, H.P.; Chang, T.S.; Yeh, H.P.; Chen, Y.X. Analysis of factors influencing hospitals' implementation of a green e-procurement system using a cloud model. *Int. J. Environ. Res. Public Health* **2019**, *16*, 5137. [CrossRef]
75. Ramkumar, M.; Schoenherr, T.; Wagner, S.M.; Jenamani, M. Q-TAM: A quality technology acceptance model for predicting organizational buyers' continuance intentions for e-procurement services. *Int. J. Prod. Econ.* **2019**, *216*, 333–348. [CrossRef]
76. Mafini, C.; Dhurup, M.; Madzimure, J. E-procurement, supplier integration and supply chain performance in small and medium enterprises in South Africa. *S. Afr. J. Bus. Manag.* **2020**, *51*, 1–12.
77. Aduwo, E.B.; Ibem, E.O.; Ayo-Vaughan, E.A.; Afolabi, A.O.; Uwakonye, U.O.; Oluwunmi, A.A. Determinants of e-procurement implementation in construction in Nigeria. *Int. J. Emerg. Technol.* **2020**, *11*, 746–755.
78. Mircea, M.; Stoica, M.; Ghilic-Micu, B. Analysis of the Impact of Blockchain and Internet of Things (BIoT) on Public Procurement. *IEEE Access* **2022**, *10*, 63353–63374. [CrossRef]
79. Ibem, E.O.; Aduwo, E.B.; Afolabi, A.O.; Oluwunmi, A.O.; Tunji-Olayeni, P.F.; Ayo-Vaughan, E.A.; Uwakonye, U.O. Electronic (e-) procurement adoption and users' experience in the Nigerian construction sector. *Int. J. Constr. Educ. Res.* **2021**, *17*, 258–276. [CrossRef]
80. Mélon, L.; Spruk, R. The impact of e-procurement on institutional quality. *J. Public Procure.* **2020**, *20*, 333–375. [CrossRef]
81. Masudin, I.; Aprilia, G.D.; Nugraha, A.; Restuputri, D.P. Impact of E-procurement adoption on company performance: Evidence from Indonesian manufacturing industry. *Logistics* **2021**, *5*, 16. [CrossRef]
82. Hong, Y.; Shao, B.B. On factors that moderate the effect of buyer-supplier experience on e-procurement platforms. *Prod. Oper. Manag.* **2021**, *30*, 1034–1051. [CrossRef]
83. Chen, Y.; Bretschneider, S.; Stritch, J.M.; Darnall, N.; Hsueh, L. E-procurement system adoption in local governments: The role of procurement complexity and organizational structure. *Public Manag. Rev.* **2022**, *24*, 903–925. [CrossRef]
84. Chan, A.P.; Owusu, E.K. Evolution of electronic procurement: Contemporary review of adoption and implementation strategies. *Buildings* **2022**, *12*, 198. [CrossRef]
85. Sitompul, A. E-Procurement System In The Mechanism Of Procurement Of Goods And Services Electronically. *Int. Asia Law Money Laund. (IAML)* **2022**, *1*, 57–63. [CrossRef]
86. Susantya, S.A.; Aswar, K.; Sumardjo, M.; Panjaitan, I. E-Procurement Adoption in Indonesian Government: A Study of Toe Model. *Innov. Mark.* **2022**, *18*, 3, 99–109. [CrossRef]
87. Sirotkina, N.; Lazarevich, S. Electronic Public Procurement: Case of Russia. *Int. J. Public Adm.* **2023**, *11*, 783–794. [CrossRef]
88. De Cazalet, B.; Zapatrina, I. E-Procurement for PPPs and Concessions: Current Trends and Opportunities. *Eur. Procure. Pub. Priv. Partnersh. L. Rev.* **2021**, *16*, 131.
89. Khahro, S.H.; Hassan, S.; Zainun, N.Y.B.; Javed, Y. Digital transformation and e-commerce in construction industry: A prospective assessment. *Acad. Strateg. Manag. J.* **2021**, *20*, 1–8.
90. Haim Faridian, P. Innovation in public management: Is public e-procurement a wave of the future? A theoretical and exploratory analysis. *Int. J. Public Adm.* **2015**, *38*, 654–662. [CrossRef]

91. Jiménez, A.; Hanoteau, J.; Barkemeyer, R. E-procurement and firm corruption to secure public contracts: The moderating role of governance institutions and supranational support. *J. Bus. Res.* **2022**, *149*, 640–650. [CrossRef]
92. Sánchez, S.N. The implementation of decentralised ledger technologies for public procurement. *Eur. Procure. Public Priv. Partnersh. Law Rev.* **2019**, *14*, 180–196. [CrossRef]
93. Lamba, K.; Singh, S.P. Big data in operations and supply chain management: Current trends and future perspectives. *Prod. Plan. Control* **2017**, *28*, 877–890. [CrossRef]
94. Moretto, A.; Ronchi, S.; Patrucco, A.S. Increasing the effectiveness of procurement decisions: The value of big data in the procurement process. *Int. J. RF Technol.* **2017**, *8*, 79–103. [CrossRef]
95. Tiwari, Sunil and Wee, Hui-Ming and Daryanto, Yosef. Big data analytics in supply chain management between 2010 and 2016: Insights to industries. *Comput. Ind. Eng.* **2018**, *115*, 319–330. [CrossRef]
96. Song, Q.; Zheng, Y.J.; Huang, Y.J.; Xu, Z.G.; Sheng, W.G.; Yang, J. Emergency drug procurement planning based on big-data driven morbidity prediction. *IEEE Trans. Ind. Inform.* **2018**, *15*, 6379–6388. [CrossRef]
97. AlNuaimi, B.K.; Khan, M.; Ajmal, M.M. The role of big data analytics capabilities in greening e-procurement: A higher order PLS-SEM analysis. *Technol. Forecast. Soc. Chang.* **2021**, *169*, 120808. [CrossRef]
98. Kuo, T.C.; Peng, C.Y.; Kuo, C.J. Smart support system of material procurement for waste reduction based on big data and predictive analytics. *Int. J. Logist. Res. Appl.* **2021**, *24*, 1–18. [CrossRef]
99. Uddin Murad, M.A.; Cetindamar, D.; Chakraborty, S. Identifying the Key Big Data Analytics Capabilities in Bangladesh's Healthcare Sector. *Sustainability* **2022**, *14*, 7077. [CrossRef]
100. Lee, I.; Mangalaraj, G. Big data analytics in supply chain management: A systematic literature review and research directions. *Big Data Cogn. Comput.* **2022**, *6*, 17. [CrossRef]
101. Zeisel, S. A framework and use cases for big data in procurement. *Int. J. Procure. Manag.* **2020**, *13*, 775–793.
102. Narwane, V.S.; Raut, R.D.; Yadav, V.S.; Cheikhrouhou, N.; Narkhede, B.E.; Priyadarshinee, P. The role of big data for Supply Chain 4.0 in manufacturing organisations of developing countries. *J. Enterp. Inf. Manag.* **2021**, *34*, 1452–1480. [CrossRef]
103. Gubbi, J.; Buyya, R.; Marusic, S.; Palaniswami, M. Internet of Things (IoT): A vision, architectural elements, and future directions. *Future Gener. Comput. Syst.* **2013**, *29*, 1645–1660. [CrossRef]
104. Fang, C.; Liu, X.; Pardalos, P.M.; Pei, J. Optimization for a three-stage production system in the Internet of Things: Procurement, production and product recovery, and acquisition. *Int. J. Adv. Manuf. Technol.* **2016**, *83*, 689–710. [CrossRef]
105. Jemmali, M.; Melhim, L.K.B.; Alharbi, M. Multi-criteria intelligent algorithm for supply chain management. *Int. J. Adv. Comput. Sci. Appl.* **2019**, *10*, 1–7. [CrossRef]
106. Ngo, T.D.; Kashani, A.; Imbalzano, G.; Nguyen, K.T.; Hui, D. Additive manufacturing (3D printing): A review of materials, methods, applications and challenges. *Compos. Part B Eng.* **2018**, *143*, 172–196. [CrossRef]
107. Revilla-León, M.; Meyer, M.J.; Zandinejad, A.; Özcan, M. Additive manufacturing technologies for processing zirconia in dental applications. *Int. J. Comput. Dent.* **2020**, *23*, 27–37.
108. Sgarbossa, F.; Peron, M.; Lolli, F.; Balugani, E. Conventional or additive manufacturing for spare parts management: An extensive comparison for Poisson demand. *Int. J. Prod. Econ.* **2021**, 233, 107993. [CrossRef]
109. Raffaeli, R.; Lettori, J.; Schmidt, J.; Peruzzini, M.; Pellicciari, M. A systematic approach for evaluating the adoption of additive manufacturing in the product design process. *Appl. Sci.* **2021**, *11*, 1210. [CrossRef]
110. Muhammad, M.S.; Kerbache, L.; Elomri, A. Potential of additive manufacturing for upstream automotive supply chains. In *Supply Chain Forum: An International Journal*; Taylor & Francis: Abingdon, UK, 2022; Volume 23, pp. 1–19.
111. Chowdhury, S.; Shahvari, O.; Marufuzzaman, M.; Francis, J.; Bian, L. Sustainable design of on-demand supply chain network for additive manufacturing. *Iise Trans.* **2019**, *51*, 744–765. [CrossRef]
112. Bhattacharyya, S.S.; Kulkarni, O.; Mishra, A. Study of emerging avenues in supply chain resilience; the case of integration of additive manufacturing with spare parts procurement. *Benchmarking Int. J.* **2022**, *29*. . [CrossRef]
113. Knofius, N.; van der Heijden, M.C.; Zijm, W.H. Consolidating spare parts for asset maintenance with additive manufacturing. *Int. J. Prod. Econ.* **2019**, *208*, 269–280. [CrossRef]
114. Yadav, S.; Singh, S.P. Modelling procurement problems in the environment of blockchain technology. *Comput. Ind. Eng.* **2022**, *172*, 108546. [CrossRef]
115. Dutta, P.; Choi, T.M.; Somani, S.; Butala, R. Blockchain technology in supply chain operations: Applications, challenges and research opportunities. *Transp. Res. Part E Logist. Transp. Rev.* **2020**, *142*, 102067. [CrossRef]
116. Pishdad-Bozorgi, P.; Yoon, J.H. Transformational approach to subcontractor selection using blockchain-enabled smart contract as trust-enhancing technology. *Autom. Constr.* **2022**, *142*, 104538. [CrossRef]
117. Hofbauer, G.; Sangl, A. Blockchain technology and application possibilities in the digital transformation of transaction processes. In *Forum Scientiae Oeconomia*; Wydawnictwo Naukowe Akademii WSB: Dabrowa Gornicza, Poland, 2019; Volume 7, pp. 25–40.
118. Rane, S.B.; Thakker, S.V. Green procurement process model based on blockchain–IoT integrated architecture for a sustainable business. *Manag. Environ. Qual. Int. J.* **2020**, *31*, 741–763. [CrossRef]
119. Pinheiro, A.; Canedo, E.D.; De Sousa, R.T.; Albuquerque, R.D.O. Monitoring file integrity using blockchain and smart contracts. *IEEE Access* **2020**, *8*, 198548–198579. [CrossRef]

120. Rodríguez-Espíndola, O.; Chowdhury, S.; Beltagui, A.; Albores, P. The potential of emergent disruptive technologies for humanitarian supply chains: The integration of blockchain, Artificial Intelligence and 3D printing. *Int. J. Prod. Res.* **2020**, *58*, 4610–4630. [CrossRef]
121. Kim, K.; Lee, G.; Kim, S. A study on the application of blockchain technology in the construction industry. *KSCE J. Civ. Eng.* **2020**, *24*, 2561–2571. [CrossRef]
122. Xu, Y.; Chong, H.Y.; Chi, M. A review of smart contracts applications in various industries: A procurement perspective. *Adv. Civ. Eng.* **2021**, *2021*, 1–25. [CrossRef]
123. Rane, S.B.; Narvel, Y.A.M. Data-driven decision making with Blockchain-IoT integrated architecture: A project resource management agility perspective of industry 4.0. *Int. J. Syst. Assur. Eng. Manag.* **2022**, *13*, 1005–1023. [CrossRef]
124. Gunasekara, H.G.; Sridarran, P.; Rajaratnam, D. Effective use of blockchain technology for facilities management procurement process. *J. Facil. Manag.* **2022**, *20*, 452–468. [CrossRef]
125. Faccia, A.; Petratos, P. Blockchain, enterprise resource planning (ERP) and accounting information systems (AIS): Research on e-procurement and system integration. *Appl. Sci.* **2021**, *11*, 6792. [CrossRef]
126. Prajapati, D.; Jauhar, S.K.; Gunasekaran, A.; Kamble, S.S.; Pratap, S. Blockchain and IoT embedded sustainable virtual closed-loop supply chain in E-commerce towards the circular economy. *Comput. Ind. Eng.* **2022**, *172*, 108530. [CrossRef]
127. Guo, L.; Liu, Q.; Shi, K.; Gao, Y.; Luo, J.; Chen, J. A blockchain-driven electronic contract management system for commodity procurement in electronic power industry. *IEEE Access* **2021**, *9*, 9473–9480. [CrossRef]
128. Omar, I.A.; Jayaraman, R.; Debe, M.S.; Salah, K.; Yaqoob, I.; Omar, M. Automating procurement contracts in the healthcare supply chain using blockchain smart contracts. *IEEE Access* **2021**, *9*, 37397–37409. [CrossRef]
129. Raj, P.V.R.P.; Jauhar, S.K.; Ramkumar, M.; Pratap, S. Procurement, traceability and advance cash credit payment transactions in supply chain using blockchain smart contracts. *Comput. Ind. Eng.* **2022**, *167*, 108038. [CrossRef]
130. Khalfan, M.; Azizi, N.; Haass, O.; Maqsood, T.; Ahmed, I. Blockchain technology: Potential applications for public sector E-procurement and project management. *Sustainability* **2022**, *14*, 5791. [CrossRef]
131. Nodehi, T.; Zutshi, A.; Grilo, A.; Rizvanovic, B. EBDF: The enterprise blockchain design framework and its application to an e-Procurement ecosystem. *Comput. Ind. Eng.* **2022**, *171*, 108360. [CrossRef]
132. Li, J.; Kassem, M. Applications of distributed ledger technology (DLT) and Blockchain-enabled smart contracts in construction. *Autom. Constr.* **2021**, *132*, 103955. [CrossRef]
133. Scott, D.J.; Broyd, T.; Ma, L. Exploratory literature review of blockchain in the construction industry. *Autom. Constr.* **2021**, *132*, 103914. [CrossRef]
134. Prakash, A.; Ambekar, S. Digital transformation using blockchain technology in the construction industry. *J. Inf. Technol. Case Appl. Res.* **2020**, *22*, 256–278. [CrossRef]
135. Li, Y.; Yang, W.; He, P.; Chen, C.; Wang, X. Design and management of a distributed hybrid energy system through smart contract and blockchain. *Appl. Energy* **2019**, *248*, 390–405. [CrossRef]
136. Radanović, I.; Likić, R. Opportunities for use of blockchain technology in medicine. *Appl. Health Econ. Health Policy* **2018**, *16*, 583–590. [CrossRef]
137. Amiri Ara, R.; Paardenkooper, K.; van Duin, R. A new blockchain system design to improve the supply chain of engineering, procurement and construction (EPC) companies–a case study in the oil and gas sector. *J. Eng. Des. Technol.* **2022**, *20*, 887–913. [CrossRef]
138. Niu, B.; Dong, J.; Dai, Z.; Jin, J.Y. Market expansion vs. intensified competition: Overseas supplier's adoption of blockchain in a cross-border agricultural supply chain. *Electron. Commer. Res. Appl.* **2022**, *51*, 101113. [CrossRef]
139. Mohd Said, N.A.; Mustaffa, N.E.; Tajul Ariffin, H.L. Integrating cloud in engineering, procurement and construction contract. *J. Comput. Theor. Nanosci.* **2020**, *17*, 893–901. [CrossRef]
140. Wang, K.; Zipperle, M.; Becherer, M.; Gottwalt, F.; Zhang, Y. An AI-based automated continuous compliance awareness framework (CoCAF) for procurement auditing. *Big Data Cogn. Comput.* **2020**, *4*, 23. [CrossRef]
141. Sloane, M.; Chowdhury, R.; Havens, J.C.; Lazovich, T.; Rincon Alba, L. AI and Procurement-A Primer. *Univ. Bologna Law Rev.* 2018, 3, 180–218.
142. Mahmoudi, A.; Javed, S.A.; Mardani, A. Gresilient supplier selection through fuzzy ordinal priority approach: Decision-making in post-COVID era. *Oper. Manag. Res.* **2022**, *15*, 208–232. [CrossRef]
143. Spreitzenbarth, J.M.; Stuckenschmidt, H.; Bode, C. The state of artificial intelligence: Procurement versus sales and marketing. In *Supply Management Research: Aktuelle Forschungsergebnisse 2022*; Springer: Berlin/Heidelberg, Germany, 2022; pp. 173–193.
144. Manimuthu, A.; Venkatesh, V.; Shi, Y.; Sreedharan, V.R.; Koh, S.L. Design and development of automobile assembly model using federated artificial intelligence with smart contract. *Int. J. Prod. Res.* **2022**, *60*, 111–135. [CrossRef]
145. van Noordt, C.; Misuraca, G. Artificial intelligence for the public sector: Results of landscaping the use of AI in government across the European Union. *Gov. Inf. Q.* **2022**, *39*, 101714. [CrossRef]
146. Nagitta, P.O.; Mugurusi, G.; Obicci, P.A.; Awuor, E. Human-centered artificial intelligence for the public sector: The gate keeping role of the public procurement professional. *Procedia Comput. Sci.* **2022**, *200*, 1084–1092. [CrossRef]
147. EL ASRI, HAYAT and BENHLIMA, LAILA Artificial Intelligence-Based Process Automation in E-procurement: A Systematic Literature Review. *J. Theor. Appl. Inf. Technol.* **2022**, *100*, 4560–4581.

148. Roberta Pereira, C.; Christopher, M.; Lago Da Silva, A. Achieving supply chain resilience: The role of procurement. *Supply Chain Manag. Int. J.* **2014**, *19*, 626–642. [CrossRef]
149. Zulkarnain, Z.; Muda, I.; Kesuma, S.A. Factors Determining The Adoption of E-Procurement in Developing Countries: A Systematic Literature Review. *Int. J. Soc. Serv. Res.* **2023**, *3*, 585–594. [CrossRef]
150. Mose, J.M.; Njihia, J.M.; Magutu, P.O. The critical success factors and challenges in e-procurement adoption among large scale manufacturing firms in Nairobi, Kenya. *Eur. Sci. J.* **2013**, *9*, 375–401.
151. Papadopoulos, T.; Gunasekaran, A.; Dubey, R.; Fosso Wamba, S. Big data and analytics in operations and supply chain management: managerial aspects and practical challenges. *Prod. Plan. Control* **2017**, *28*, 873–876. [CrossRef]
152. Mani, V.; Delgado, C.; Hazen, B.T.; Patel, P. Mitigating supply chain risk via sustainability using big data analytics: Evidence from the manufacturing supply chain. *Sustainability* **2017**, *9*, 608. [CrossRef]
153. Rejeb, A.; Keogh, J.G.; Treiblmaier, H. Leveraging the internet of things and blockchain technology in supply chain management. *Future Internet* **2019**, *11*, 161. [CrossRef]
154. Kunovjanek, M.; Knofius, N.; Reiner, G. Additive manufacturing and supply chains–a systematic review. *Prod. Plan. Control* **2022**, *33*, 1231–1251. [CrossRef]
155. Sava, N.A.; Dragoș, D. The legal regime of smart contracts in public procurement. *Transylv. Rev. Adm. Sci.* **2022**, *18*, 99–112. [CrossRef]
156. Davtyan-Davydova, D.N.; Martirosyan, M.G.; Bortenev, A.I.; Sergacheva, O.A. Implementation and realization of technologies to distributed registers (Blockchain) and smart-contracts in public purchases. In Proceedings of the Competitive Russia: Foresight Model of Economic and Legal Development in the Digital Age: Proceedings of the International Scientific Conference in Memory of Oleg Inshakov (1952–2018), Volgograd, Russia, 19–20 September 2019; Springer: Berlin/Heidelberg, Germany, 2020; pp. 569–576.
157. Guida, M.; Caniato, F.; Moretto, A.; Ronchi, S. The role of artificial intelligence in the procurement process: State of the art and research agenda. *J. Purch. Supply Manag.* **2023**, *29*, 100823. [CrossRef]
158. Burger, M.; Nitsche, A.M.; Arlinghaus, J. Hybrid intelligence in procurement: Disillusionment with AI's superiority? *Comput. Ind.* **2023**, *150*, 103946. [CrossRef]
159. Sonego, M.; Echeveste, M.E.S.; Debarba, H.G. The role of modularity in sustainable design: A systematic review. *J. Clean. Prod.* **2018**, *176*, 196–209. [CrossRef]
160. Gauss, L.; Lacerda, D.P.; Sellitto, M.A. Module-based machinery design: A method to support the design of modular machine families for reconfigurable manufacturing systems. *Int. J. Adv. Manuf. Technol.* **2019**, *102*, 3911–3936. [CrossRef]
161. Vogelsang, K.; Liere-Netheler, K.; Packmohr, S.; Hoppe, U. Barriers to digital transformation in manufacturing: Development of a research agenda. In Proceedings of the 52nd Hawaii International Conference on System Sciences (2019), Maui, HI, USA, 4937–4946 January 2019; Association for Information Systems e-Library: Atlanta, GA, USA, 2019; pp. 4937–4946.
162. Jones, M.D.; Hutcheson, S.; Camba, J.D. Past, present, and future barriers to digital transformation in manufacturing: A review. *J. Manuf. Syst.* **2021**, *60*, 936–948. [CrossRef]
163. Zangiacomi, A.; Pessot, E.; Fornasiero, R.; Bertetti, M.; Sacco, M. Moving towards digitalization: A multiple case study in manufacturing. *Prod. Plan. Control* **2020**, *31*, 143–157. [CrossRef]
164. Wehrle, M.; Birkel, H.; von der Gracht, H.A.; Hartmann, E. The impact of digitalization on the future of the PSM function managing purchasing and innovation in new product development–Evidence from a Delphi study. *J. Purch. Supply Manag.* **2022**, *28*, 100732. [CrossRef]

energies MDPI

Article

The Impact of the Level of Training of Airport Security Control Operators on the Energy Consumption of the Baggage Control Process

Jacek Ryczyński * and Artur Kierzkowski

Department of Technical Systems Operation and Maintenance, Faculty of Mechanical Engineering, Wroclaw University of Science and Technology, Wybrzeże Wyspiańskiego 27, 50-370 Wrocław, Poland; artur.kierzkowski@pwr.edu.pl
* Correspondence: jacek.ryczynski@pwr.edu.pl

Abstract: The article focuses on the study of the impact of the level of training of security control operators (SCOs) at airports on the energy consumption of the passenger baggage control process. With the constant growing emphasis on security at airports, the intensification of training processes for security personnel, especially those dealing with baggage control, has become very dynamic. An essential aspect in times of sustainable development is optimizing all kinds of processes (including training processes) to reduce energy consumption. The analysis of the demand for energy used to conduct this type of training and the impact of the operator's training level on the energy consumption of the control process are entirely ignored and have not been the subject of research by scientists so far. Therefore, this is a research gap that the authors are trying to fill in this article. The impact of safety system operator training levels on ensuring optimal energy efficiency was critically analyzed. The added value of the article is the authors' model assessing the influence of the level of training of the SCO on the energy consumption of the control process. The effects of the frequency, duration and level of operator training on energy consumption rates were investigated. The authors' activities aimed to identify the most energy-efficient approaches to training without compromising its quality and, thus, the safety of passengers. The article discusses potential strategies for minimizing energy use and draws conclusions that can help airport administrations and training providers adopt sustainable and energy-efficient training practices.

Keywords: energy consumption; fuzzy logic; training; security control operator; SCO

Citation: Ryczyński, J.; Kierzkowski, A. The Impact of the Level of Training of Airport Security Control Operators on the Energy Consumption of the Baggage Control Process. *Energies* **2023**, *16*, 6957. https://doi.org/10.3390/en16196957

Academic Editor: Marco Marengo

Received: 28 July 2023
Revised: 28 September 2023
Accepted: 3 October 2023
Published: 5 October 2023

1. Introduction

Over the past two decades, there has been a significant increase in primary energy consumption and CO_2 emissions globally by 50% [1]. Although there is no certainty about economic progress, forecasts indicate a significant escalation of global energy demand, with an expected increase of one-third from 2015 to 2040 [1,2]. This surge in energy consumption is expected to have a significant environmental impact in the coming years and could potentially lead to energy supply problems in some regions.

Typically, energy consumers are divided into three main segments: industrial, transport and other. Air transport is one of the most essential elements in the transport segment.

Airports are characterized by significant energy consumption, which is regulated by many factors. The nature of airport energy consumption is inherently unpredictable, non-linear and constantly changing. In academic circles, attention is often drawn to the energy consumption of terminal structures [1–4], even though they are only one (but a crucial) segment of the entire airport. As such, extensive exploration of the factors affecting airport energy efficiency has excellent potential for future research.

The primary energy consumption at airports can be divided into two parts: the aviation part and the land part. The aviation part includes mainly airport lighting and radio navigation equipment. In turn, the airport terminal building dominates the land, partly due to its role as a central hub for passenger and cargo handling and the number of facilities necessary for its efficient operation. Therefore, innovative research methods to reduce energy consumption in such facilities are essential.

Electricity is the primary source for meeting the basic energy needs of airports and ensuring the safe handling of air traffic. Typically, electricity is sourced from the commercial grid through utilities. However, recent trends and the scientific literature [5–9] reveal the emergence of alternative energy sources such as combined heat and power plants (CHP) and renewable energy technologies. Nevertheless, given their different attributes and potential implications for air traffic safety, it is essential to establish rules that will harmonize airport operations with these alternative energy sources for the foreseeable future.

Nowadays, the issue of reducing energy consumption is the highest priority for airport administrators. The key strategies can be divided into the following groups: improvements to management systems and energy infrastructure, improvements to HVAC (heating, ventilation, air conditioning) and lighting systems, and the introduction of modern operational management systems to improve and optimize the energy efficiency of airports. In addition, using models and simulations to analyze airport energy consumption can play a crucial role in reducing consumption [3,4]. This requires the development of precise methodologies that cater to the unique environment of airports, with an emphasis that extends beyond the issue of terminal buildings. Moreover, while energy efficiency indicators (EEIs) provide energy managers with data on energy efficiency, they lack insight into the reasons for efficient or inefficient energy use. Therefore, there is a need to develop new analytical methods adapted to the specificity of airports, which will enable a comprehensive look at the issues of sustainable energy development in airports.

2. Literature Review

Optimization of the energy consumption management process to reduce its levels in air transport is widely discussed in the scientific literature. A comprehensive review of this issue was carried out in [2]. Taking into account the analyses contained in this work and supplementing them with the most recent research results, a clear research gap can be noticed in the research on the area of energy consumption in terminal passenger service, and in particular, in all kinds of aspects related to the security control of passenger luggage. It seems necessary to place appropriate emphasis on modeling and simulating energy consumption to ensure a high level of passenger safety while maintaining the principles of sustainable development in terms of energy consumption. It should be emphasized that the analyses included the entire airport and terminal buildings, with a dominant emphasis on HVAC systems [10–13]. Some studies have taken a more focused approach to specific topics. Ma et al. [8] studied the correlation between indoor airflow and indoor space to enhance indoor comfort. Parker et al. [9] aimed to reduce the carbon footprint by expanding the glass roof at a specific airport. Meanwhile, Gowresuunker et al. [14] evaluated the effectiveness of displacement ventilation in an airport terminal.

A group of articles also focuses on forecasting energy consumption in airport terminals. In this regard, Chen [15] used the objective Markov model, Huang et al. [16] used neural networks and Fan et al. [17] constructed a model based on probability density functions.

Mambo et al. [18] observed that the dynamic regulation of the internal environment by the flight schedule could bring improvements of up to 25%. Works [19–21] similarly recognized the possibility of dynamic thermal comfort and lighting management in various terminal spaces. In the literature, only a single paper refers to the energy consumption of terminal operations, and only two [3,4] directly correlate with the analysis of energy consumption by operating systems. In the work [4], a simulation model was developed that allows for the simultaneous analysis of the efficiency of the security checkpoint at the airport and energy consumption per serviced passenger. In the work [3], a simulation model was used to analyze the sensitivity of energy consumption by the baggage handling system at the airport to a change in the resource allocation strategy.

To sum up, a critical gap in the existing literature is the lack of research documenting the energy consumption of passenger service procedures, in particular baggage security checks. So far, attention has focused on other aspects. An extensive literature review on modeling passenger service processes is presented in [22]. The review found that the main factors were the efficiency of the process and the quality of passenger service. This approach has worked, and recent research confirms that these elements remain central to research efforts, as exemplified by the stochastic passenger boarding model developed to increase process efficiency [23].

In addition, uncertainty is considered when planning operations in airport terminals [24], and optimizing service processes remains the goal [25]. Optimal allocation of resources is often used to increase process efficiency [26,27].

Airports' energy management is a crucial matter from both economic and ecological perspectives. One of the primary challenges in this field is that the data about the total energy consumption of airports are classified as sensitive information, significantly restricting their public accessibility. Therefore, researchers and analysts must rely solely on estimated data to understand which airport processes consume the most energy.

A particularly scrutinized area is baggage control. This process is time-consuming and demands a substantial amount of energy, contributing a significant portion of the airport's overall energy balance. According to certain analyses cited in references [10,11,14], this process can account for 15% to even 20% of the annual energy consumption in terminal buildings, especially in large airports. Lighting and cooling have the largest share of energy consumption in airports. They account for 46% of total energy consumption. Hence, it is not a negligibly small value compared to the total consumption, indicating that undertaking analyses concerning efficient control process management is advisable.

Nevertheless, the developed solutions omit the aspects directly affecting energy consumption. This article tries to fill this gap. It aims to develop a model focusing on aspects that have so far been overlooked in research, i.e., the impact of the level of training of baggage screening operators at airports on the amount of energy consumption in the entire screening process.

3. The Process of Training Security Control Operators

The selection of appropriate personnel is crucial at passenger security checkpoints. SCOs play various roles, with the most important being the security screening of individuals and their carry-on luggage. Other duties include access control, monitoring the environment of the checkpoint, and inspecting gates and vehicles [21]. All these tasks require professional training. Before an SCO is assigned to a control position, they must undergo a cycle of preparatory training. The details and scope of each stage are presented in Figure 1. The training cycle concludes with certification, required by national or international regulatory bodies [28–35]. Obtaining a certificate and starting work does not mean the end of training. This process is ongoing and requires continuous improvement, especially in the use of emerging technical innovations and the introduction of modern technologies.

The implementation of advanced scanning systems, image analysis software, and other technological tools requires operators to keep up with changes and effectively use these tools in their work.

Figure 1. The stages of the training system for SCOs.

4. Methodology

The main research problem presented in this study is to create a model and an expert system that will allow us to assess the impact of the SCO training level at the airport on the energy consumption of the entire control process.

The proposed SCO training level assessment system is based on integrated knowledge of the critical skills of the operator. Preliminary research indicates the subjectivity of assessments and the difficulty in combining various indicators into one coherent assessment system, which inspired the authors to build such a system. To understand the complexity of the problem, the research concept is presented in Figure 2.

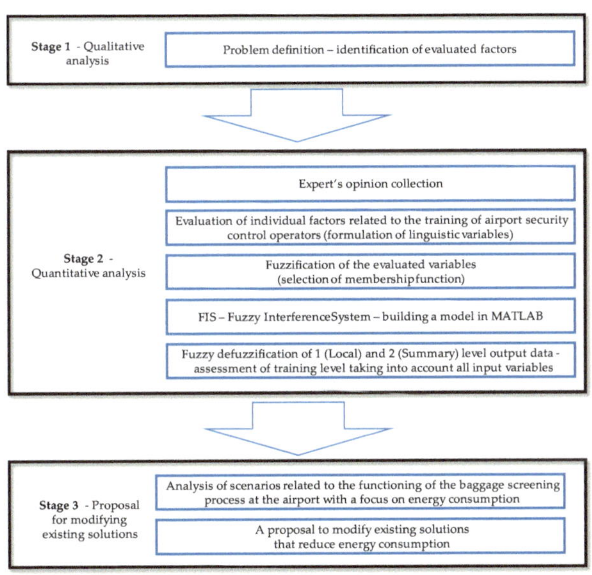

Figure 2. Research concept.

The above observations are the basis for adopting the concept in which the system for assessing the level of training and energy consumption of the inspection process will be based on an appropriate mathematical model that considers and integrates factors relevant to the assessment. In this situation, it is required to use methods that consider the imprecise and uncertain nature of the input variables [36–38]. In our study, we use the fuzzy set theory and its extension, the theory of fuzzy reasoning [39].

Fuzzy logic, being an extension of classical two-valued logic, is an indispensable tool in solving many engineering problems for various reasons. First and foremost, it allows for the modeling of uncertainty and imprecision, which often occur in real technical systems. It also enables the simulation of human reasoning, which naturally relies on fuzzy concepts rather than strict values. In the context of complex systems, fuzzy logic facilitates better management and control, which is crucial given the high level of complexity present in many systems (an example being the analyzed passenger service system at airports).

Moreover, it allows for the optimization of systems by considering various degrees of membership in fuzzy sets, which practically translates to the adaptability of systems, enabling them to learn and adapt to changing environmental conditions. Another advantage is the potential reduction in costs associated with designing and implementing new solutions in the field of system management or reducing the costs of using technical facilities. This, in turn, promotes the responsiveness of systems, which can react more effectively to changes in the environment, even when working with incomplete or imprecise data. It is also worth emphasizing the intuitiveness of designing systems using fuzzy logic, which allows for the use of linguistic variables that are closer to the human way of thinking. This compatibility with human reasoning means that fuzzy logic can be easily integrated with other computational intelligence techniques, creating hybrid intelligent systems.

Lastly, it cannot be overlooked that fuzzy logic has already found wide applications in practical technical systems, including industrial automation, robotics and risk modeling, which attests to its effectiveness in solving real technical problems.

Schematically, the fuzzy inference system is shown in Figure 3.

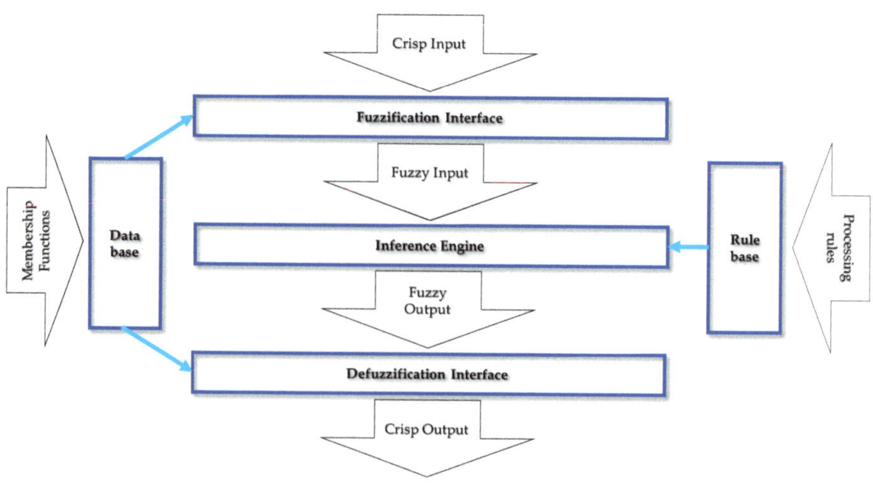

Figure 3. Fuzzy inference system.

The input of the fuzzy block is given the unfuzzy values from observations or measurements. In the blurring block, based on specific membership functions (MFs), they are associated with the values of linguistic variables, such as, for example, small, medium and large. The issue of determining the number of values of linguistic variables covering the entire space of consideration of a given variable, their form and the degree of overlapping of individual values must be determined each time when building a fuzzy reasoning system.

It is worth noting here that each unfuzzy input value may correspond to several linguistic variables with different degrees of membership. The fuzzy values are the input for the inference block. This block uses a fuzzy rule base, which, in the case of the methods used in this case, were created by experts: practitioners in the field of security control at the airport. Fuzzy inference rules are conditional sentences: IF premise THEN conclusion. Based on the fuzzy premises and all the rules met, the inference block determines the conclusion as a fuzzy linguistic variable. This conclusion is the input to the sharpening block, which, based on a defined MF, associates the fuzzy quantity with the original non-fuzzy quantity. It is the result of the fuzzy reasoning system. It is usually a number from some specific range.

Our solution was prepared in MATLAB version R2023a. The prepared fuzzy reasoning system uses local models of the Mamdani type. The fuzzy logic modules in MATLAB offer two analytical systems based on different approaches: Sugeno and Mamdani. The differences between these approaches are very significant. The Mamdani approach, which is more traditional, involves formulating fuzzy rules in a linguistic context, allowing for more intuitive system modeling. In contrast, the Sugeno approach focuses on formulating rules in a more mathematical context, facilitating precise modeling of functional relationships. In practice, Mamdani systems utilize membership functions described with words such as low, medium or high, while Sugeno systems operate on membership functions described by mathematical equations. This makes the Sugeno approach more oriented towards precision and mathematical analysis, while Mamdani focuses on linguistic interpretation. Moreover, Mamdani systems are generally easier to optimize and offer better computational performance compared to Sugeno systems. The Mamdani approach is often preferred due to its intuitiveness and ease of interpretation, making it more accessible to individuals without a deep mathematical background. It is important to note that both approaches have their place in designing fuzzy systems, and the choice between them depends on specific project requirements and the designer's personal preferences.

Mamdani-type systems allow the user to define input and output variables, determining their ranges and membership functions, which is a fundamental step in constructing a fuzzy system. Additionally, they can create various membership functions, including triangular, trapezoidal or Gaussian functions, enabling precise modeling of fuzzy relationships. A crucial element is also the ability to formulate fuzzy rules linguistically, facilitating the use of expert experience in the system design process. This tool also allows for the editing of existing rules, adapting them to changing project requirements. Users can utilize visualization options, including membership function plots and response surface charts, to better understand and analyze the system's operation. Within this tool, users also have the opportunity to simulate the functioning of the fuzzy system, observing its reactions to various inputs, which is essential for system verification and validation before implementation. A more advanced feature is the ability to optimize the built model, including adjusting membership functions and rules to achieve better results. After completing the design process, the fuzzy system can be exported to various formats or integrated with other systems in MATLAB, enhancing its utility in different application contexts. It is also worth noting the intuitive graphical user interface, which facilitates the design of fuzzy systems without the need to write code. Finally, Mamdani systems in MATLAB support various defuzzification methods, including the centroid, bisector, weighted average or maximum methods, allowing for controlled and user-expectation-aligned system output generation. Consequently, this tool represents a comprehensive solution for designing and analyzing fuzzy systems.

The Mamdani model performs its functions in the following stages:

1. Blurring: at this stage, the input values are transformed into fuzzy values using the MF. Each input variable has a specific MF that assigns to the input values the degree of membership of particular fuzzy sets.
2. Aggregation of premises (combining conditions): the conditions of individual rules are combined. If a given rule has several conditions, a fuzzy operator (e.g., T-norm) combines the degrees of membership of the conditions into one value.

3. Activation of rules: based on the aggregation value of the premises, the relevant rules are activated. Activation assigns a truth value (membership degree) to each rule. The truth value of a rule is equal to the aggregation value of the premises for the rule.
4. Activation of output sets: after activating the rules, the output sets are assigned activation values based on the conclusion of the rules. For each rule, the degree of membership for the conclusion is computed based on the rule's truth value. In practice, this means "trimming" the MF of the output sets to the truth values of the rules.
5. Aggregation of output sets: all activated output sets are combined into one output set using the aggregation operator (e.g., maximum, minimum or simultaneous minimum and maximum—MIN/MAX); details are in Figure 4.
6. Sharpening: transforming the fuzzy output set into one numerical value.

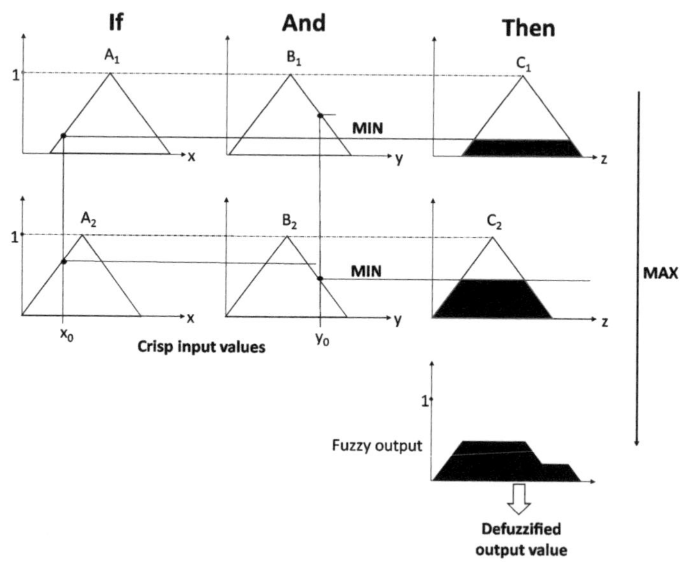

Figure 4. Functioning of FIS of Mamdani model using MIN/MAX operator for two activated output sets into one output (triangle MF).

The analysis of the problem of evaluating the level of training of SCOs drew attention to the need to include in the model and to have a thorough knowledge of the operators' skills in assessing X-ray images at security checkpoints regarding various categories of prohibited items. There are four categories: sharp objects, explosives, firearms and other prohibited items and substances. This observation was used to create a model. Thanks to this, it was possible to obtain an unambiguous assessment of the operator's level of training and to analyze partial results within four categories of hazardous objects.

A unique test stand was constructed to obtain the data necessary to build our fuzzy model. The core of the stand was the system that SCOs use daily. Such a station has not only been equipped with all the functionalities of a typical baggage control system at the control station. It was additionally equipped with the operator's eyeball tracking system, the so-called eye-tracker system. Additional equipment allowed us to obtain data to conduct analyses that are impossible to estimate in traditional systems, e.g., the time duration that the operator focuses their eyes on a dangerous object and many other parameters. The details will be discussed in Section 5. Figure 5 shows the appearance of the test stand equipped with the eye-tracker system.

Figure 5. Test stand equipped with the eye-tracker system (an oblong black element mounted directly below the monitors for image observation).

5. The Model for Assessing the Impact of the SCO Training Level on the Energy Consumption of the Baggage Control Process

The model for assessing the impact of the SCO training level on the energy consumption of the control process was implemented as a fuzzy reasoning system. The analysis of the issues allowed us to propose a general structure for the model. Due to the nature of the problem, it has a hierarchical structure in which the outputs of the first-level fuzzy local models are inputs to the second-level fuzzy local models. Thanks to this, it was possible to obtain and analyze partial results, which are crucial for the conducted analyses. The model also includes the possibility of calibrating partial grades (and thus also the final grade) by taking into account the importance factor (issue importance) determined by experts. Systems of this type are based on the knowledge created on the basis of measurements and observations, as well as the knowledge of domain experts.

Details are shown in Figure 6.

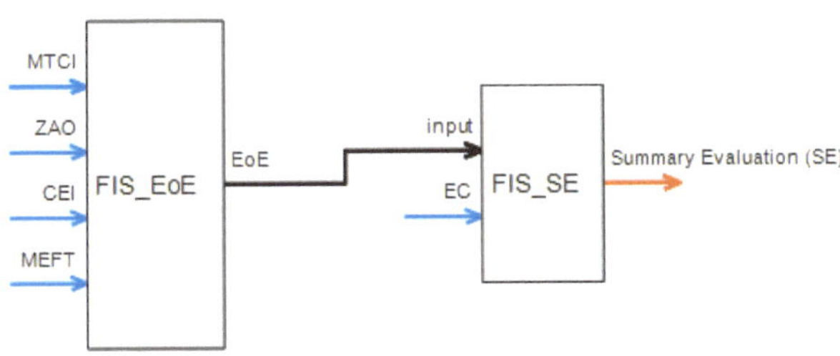

Legend:
MTCI—Mean time to correctly indicate a dangerous item.
ZAO—Zone analysis order (colors).
CEI—Correct effectiveness indication.
MEFT—Mean eye focus time on dangerous item.
EoE—Evaluation of the effectiveness of hazard identification.
EC—Energy consumption.
SE—Summary Evaluation.

Figure 6. A fuzzy model for evaluating the impact of the SCO training level on the energy consumption of the baggage control process.

5.1. First-Level Local Models

Different mapping details on the analysis images characterize the dangerous element/object categories. Research shows that security control operators' training levels may vary for the four listed categories. However, the feature of the work of SCOs, consisting of looking for patterns in images of X-rayed baggage, consistent with typical images of prohibited items, is common to all categories. Therefore, the constructed model has four first-level linguistic input variables. They include the following indicators:

1. Mean time to correctly indicate a dangerous item (MTCI): the average time expressed in seconds (s) necessary for the operator to indicate a dangerous object in the image, calculated based on the operator's eye movements tracked by the eye-tracker system;
2. Zone analysis order (colors) (ZAO): a dimensionless value, determined based on the operator's eye movements tracked by the eye-tracker system (details in the description of the variable);
3. Correct effectiveness indication (CEI): a variable describing the effectiveness of detection by the operator of objects, materials and substances considered prohibited; expressed as a percentage (%), calculated based on the quotient of correctly indicated dangerous objects among all those hidden in the images;
4. Mean eye focus time on dangerous item (MEFT): the average time expressed in seconds (s) that the operator focused on a dangerous object hidden in the image, calculated based on the operator's eye movements tracked by the eye-tracker system.

5.1.1. Mean Time to Correctly Indicate a Dangerous Item (MTCI)

The critical point affecting the overall throughput of an airport is the security checkpoint, where operators must quickly screen people and baggage effectively. The time it takes to handle a single passenger depends, among other things, on the level of training and experience of the security control operators. It is assumed that this time should not exceed 15 s. The research stage showed that, in the case of experienced operators, we rarely have a situation where the operator needs more than 11 s to analyze the image and thus indicate the prohibited element. Therefore, when describing a linguistic variable, the limiting value is 11 s (measured by tracking eye movements using the eye-tracker system). If the operator does not make a binding decision within this time, it means that they cannot assess the contents of the baggage, and proper intensive training is necessary at a certain level of difficulty. In real life, this factor is much more critical regarding potential consequences. Incorrectly conducted assessments may lead to a threat to the health or life of traveling passengers.

The linguistic variable "Mean time to correctly indicate a dangerous item" is represented by three values that describe the assessment of the level of training in the context of detecting prohibited items from this group: fast, medium and slow. Figure 7 shows the fuzzy sets describing the values for the linguistic variable "Mean time to correctly indicate a dangerous item". These values are specified in seconds, allowing values to be read directly from the stand. Based on the research conducted on a control group of experienced SCOs, the following descriptions were adopted for MFs:

- Fast: time less than 7 s;
- Medium: time in the range of 7 ÷ 11 s;
- Slow: time over 11 s.

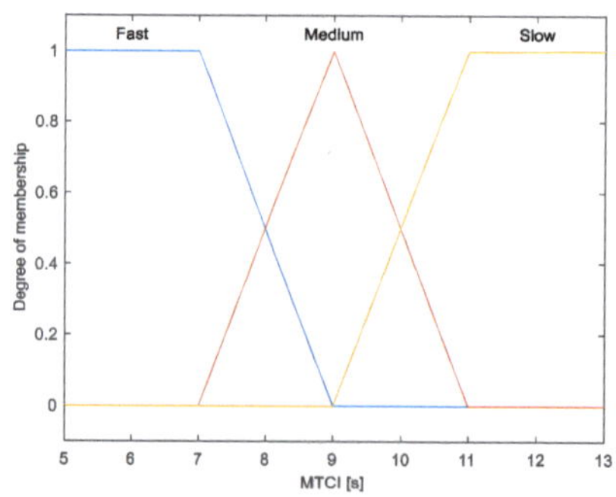

Figure 7. The form of the MF of the input linguistic variable "Mean time to correctly indicate a dangerous item".

5.1.2. Zone Analysis Order (Colors) (ZAO)

The conducted research showed that, in the case of experienced operators, there is a specific pattern in the field of image analysis.

However, to fully understand the contents described in this section, it seems necessary to discuss how the image is obtained during the X-ray luggage screening with an RTG device. As a result of the screening, we obtain a layered image: the color represents the type of material, and the color intensity indicates its thickness. The thicker the object, the greater its intensity. In the basic view, the image color depends on the atomic number of the chemical elements constituting the object and is, accordingly:

1. Orange, for elements with an atomic number Z: $1 < Z < 10$;
2. Green, for elements with an atomic number Z: $11 < Z < 18$;
3. Blue, for elements with an atomic number Z: $Z > 19$.

The first group (orange color) includes objects such as light elements, hydrogen, carbon, nitrogen, oxygen and its chemical compounds, organic materials including many explosives, plastics like acrylic, paper, textiles, food items, wood and water. The second group (green color) includes medium–heavy objects: pure aluminum, sodium, chlorine, table salt, etc. The third group includes, among others, thin elements: metals, titanium, chrome, iron, nickel, copper, zinc, tin, silver, etc.

Depending on the thickness of the screened material, the color will be more or less intense, as the force with which the radiation is absorbed is visible in the form of correspondingly higher (thicker objects) or lower (thinner objects) color intensity. Objects that completely absorb radiation will appear black in the image. Similarly, when overlaying several layers of different objects, the color intensity increases.

The essential dangerous items detected in carry-on luggage include:

1. Liquids and other organic substances: the view depends on the material density, mainly visible in the primary image in green or orange;
2. Metal knives/gun parts: mainly visible in the image in the form of a blue shade of varying intensity;
3. Imitations of safe objects: no dominant color, the view depends on the material the object is made of.

Eye-tracking software can determine the share of the operator's gaze focused on individual colors in the image and, therefore, on dangerous objects. By analyzing how the operator's eye focus is positioned during image analysis, we can also determine the order in which individual colors are tracked, the mean time to correctly indicate a dangerous item and the mean eye focus time on a dangerous item.

Each operator first analyzes the blue areas, i.e., areas where weapons or sharp objects may be hidden. If they do not notice the shapes characteristic of prohibited items, they analyze the orange areas in search of explosive materials. At the very end, without indicating prohibited items, an analysis is carried out of the other colors in the image. The eye-tracker system enables accurate tracking of the eyeball movement and thus enables accurate analysis of zones and areas of observation. The linguistic variable "Zone analysis order" is represented by three values related to the observation time where the gaze is focused on the danger zones: short, medium and long. The analysis is carried out using the eye-tracker software, and the data are obtained directly from the stand. Figure 8 shows the MF describing the values for the linguistic variable "Zone analysis order". These values are expressed as a percentage. Based on the research conducted on a control group of experienced SCOs, the following descriptions were adopted for individual MFs:

- Short: less than 20% of observation time focused on hazardous areas;
- Medium: in the range of $21 \div 80\%$ of observation time focused on hazardous areas;
- Long: over 80% of observation time focused on hazardous areas.

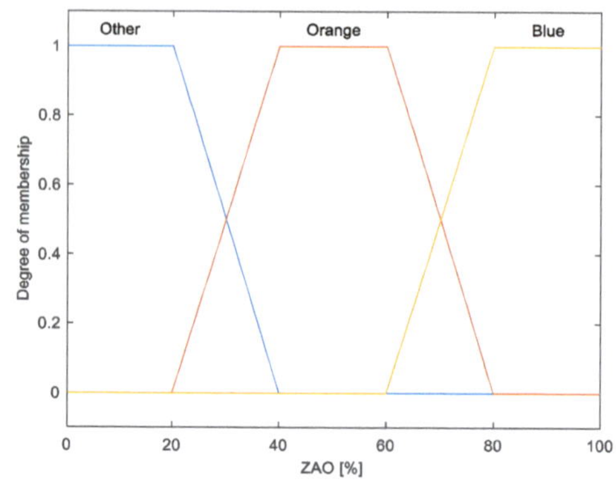

Figure 8. The form of the MF of the input linguistic variable "Zone analysis order".

5.1.3. Correct Effectiveness Indication (CEI)

SCOs are expected to perform their task of detecting prohibited items and substances in baggage with a predetermined success rate. The minimum efficiency required from operators during their monthly examinations should not be lower than 75%. Of course, in the case of experienced operators, it is required that they achieve the best possible result, confirming the appropriate preparation for work at the security checkpoint, following the principle: the higher the detection rate, the greater the level of security guaranteed by the control point. In real-life conditions, where an employee experiences time pressure or a sense of great responsibility for passengers and is subject to constant control by supervisory authorities, achieving 100% efficiency in detecting prohibited items and substances in passengers' luggage is almost impossible. Therefore, a threshold of 90% was adopted for the calculation, with the described linguistic variable as high efficiency. The linguistic variable "Correct effectiveness indication" is represented by three values that describe the assessment of the level of training in the context of detecting prohibited items from this group: unacceptable, medium and high. Figure 9 shows the MF describing the values for the linguistic variable "Correct indication effectiveness". Based on research conducted on a group of experienced SCOs, the following descriptions were adopted for individual MFs:

- Unacceptable: score less than or equal to 75%;
- Medium: score in the range of 76 ÷ 95%;
- High: score over 95%.

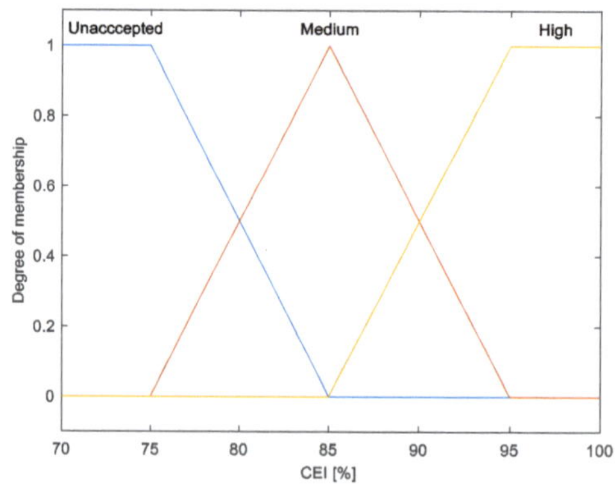

Figure 9. The form of the MF of the input linguistic variable "Correct effectiveness indication".

5.1.4. Mean Eye Focus Time on Dangerous Item (MEFT)

The eye-tracker system allows you to add up the time the operator focuses his eyes on the prohibited item when analyzing the baggage image. The times read directly from the eye-tracker system described the linguistic variable "Mean eye focus time on dangerous item". Research has shown that in the case of experienced operators, these times are similar (but consistently lower) than the average times for correctly indicating a dangerous object. This proves that an experienced operator detects a prohibited item quickly and spends the remaining time ensuring it is right before making a decision. This variable is represented by three values that describe the assessment of the level of training in the context of detecting prohibited items from this group: short, medium and long. Figure 10 shows the fuzzy sets' MFs describing the values for the linguistic variable "Mean eye focus time on dangerous item". These values are specified in seconds, allowing them to be read directly from the stand.

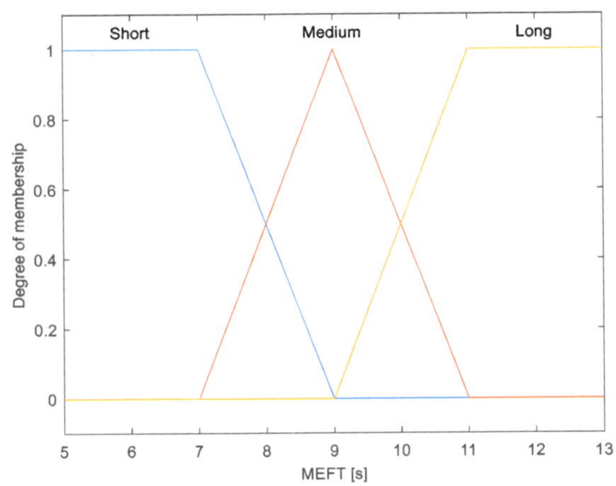

Figure 10. The form of the MF of the input linguistic variable "Mean eye focus time on dangerous item".

Based on the research conducted on a control group of experienced SCOs, the following descriptions were adopted for individual MFs:

- Short: less than 7 s;
- Average: in the range of 7 ÷ 11 s;
- Long: above 11 s.

5.1.5. The First-Level Local Output Variable: Evaluation of the Effectiveness of Hazard Identification

The first-level local output variable, which is also the second-level input variable, describes the assessment of the level of training in the context of detecting forbidden objects in images using four values: beginner, intermediate, advanced and experienced. Figure 11 shows the MF describing the values for the linguistic variables "Evaluation of the effectiveness of hazard identification". These values are expressed as percentages on a scale from 0 to 100, which allows (after the defuzzification process) to compare them and perform mathematical operations directly. Based on the research conducted on a control group of experienced SCOs, the following descriptions were adopted for individual membership functions:

- Beginner: total score less than or equal to 50%;
- Intermediate: total score in the range of 51 ÷ 65%;
- Advanced: total score in the range of 66 ÷ 80%;
- Experienced: total score above 80%.

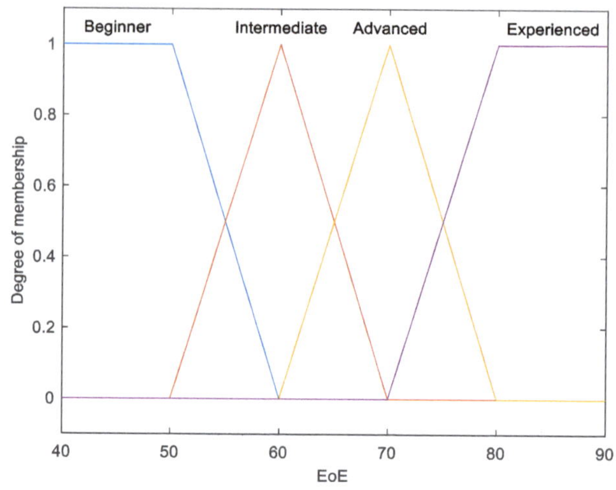

Figure 11. The form of the MF of the output(first-level)/input(second-level) linguistic variable "Evaluation of the effectiveness of hazard identification".

5.2. Second-Level Local Models

The outputs of the first-level local models determine the training level of the SCO for each category related to image analysis to detect prohibited items. On this basis, in the second-level local model, the final, aggregated assessment of the SCO training level is made, considering the energy consumption of the entire process (variable Summary Evaluation—SE). We have two input variables at the second local level: Evaluation of hazard identification (EoE) effectiveness and Energy Consumption (EC). The first one is described in Section 5.1.

5.2.1. Energy Consumption (EC)

The "Energy Consumption" variable directly relates to the X-ray machine used to scan the baggage. In all our analyses, we considered an X-ray device with a smaller inspection tunnel size (100 × 100 cm and a standard penetration of 35 mm): scanner low power (LP). The advantage of this type of device is its small size and limited demand for energy consumption. The device consumes 0.262 kWh in the standby mode and 0.725 kWh in the operating mode (scanner). Due to the specificity of the magazine, it would be necessary to precisely describe the method of estimating the MF for the "Energy Consumption" variable. Research conducted by the team revealed that during one shift (lasting 8 h), an SCO participates in five sessions consisting of analyzing 50 images in each session. The operator analyzes 250 images during one workday, so the device is in permanent working mode. An experienced operator needs about 7 s to analyze a single image; an inexperienced operator may even need 15 s. The total working time of an experienced SCO is 1750 s for the task, and that of an inexperienced SCO is 3750 s. During this time, the former consumes 0.352 kWh (due to the very fast analysis of the device, it is constantly in scanner mode), and the latter consumes 0.497 kWh of energy (0.352 kWh from device operation and 0.145 kWh from the device being in standby during image analysis by an inexperienced operator). On this basis, limited values related to energy consumption during operation were adopted: low as up to 0.370 kWh; medium as 0.371–0.480 kWh; and high as above 0.480 kWh. The details are in Figure 12.

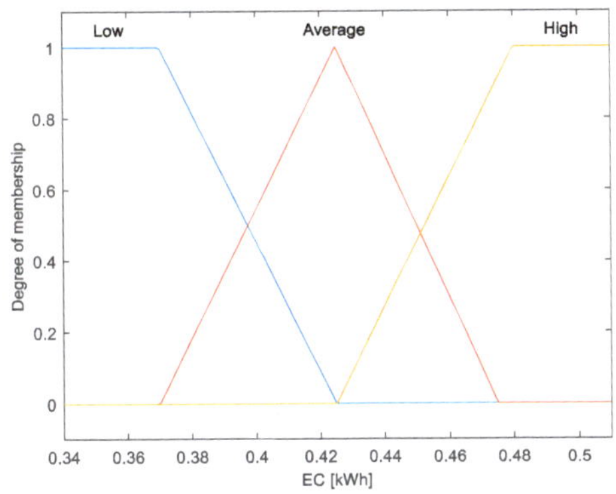

Figure 12. The form of the MF of the input linguistic variable "Energy consumption".

5.2.2. Final Output Variable: Summary Evaluation (SE)

The final assessment of the level of SCO training is made using a five-point scale with three linguistic variables. To enable an unambiguous assessment, three levels of advancement were assumed, taking into account both purely technical aspects related to the security screening process and the impact of training on the level of energy consumption during the baggage screening process itself. Adopting a five-point rating scale with three linguistic variables greatly facilitates the construction of FIS engine rules (see Section 5.3) and limits their number, facilitating the analysis of results obtained during empirical or simulation tests. Based on research conducted on a control group of experienced SCOs, the following descriptions were adopted for individual MFs (details in Figure 13):

- Weak: sum of points below 1;
- Average: total score in the range of 1 ÷ 4;
- Experienced: total score above 4.

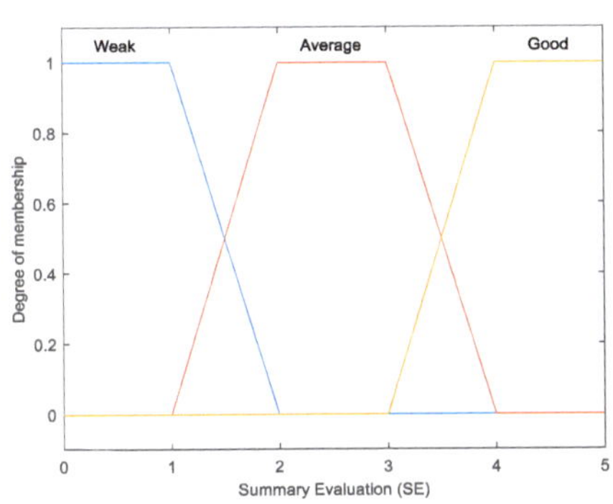

Figure 13. The form of the MF of the output linguistic variable "Summary Evaluation".

5.3. Knowledge Base: FIS (Fuzzy Inference System) Rules

In the case of the presented model of assessing the SCO training level on the energy consumption of the control process, the knowledge base, which is a set of rules of the "engine" of the blurring system (FIS), is of a hybrid nature because it was created based on two different approaches. The first was acquiring knowledge from experts: specialists in creating and operating airport security control systems. The experts defined only part of the rules presenting the most important relationships between the input variables and the assessment of the level of training for a given category of prohibited items. In particular, experts indicated the importance of individual factors (input variables) in the final assessment of the level of training. The remaining rules were generated automatically (and then verified for correctness) by the capabilities of the Fuzzy Logic Designer module of the MATLAB environment. The proposed method may involve machine learning algorithms such as unsupervised, supervised learning, optimization or genetic algorithms. Automatic rule generation can help fill in missing rules, optimize existing rules, or adapt the model to changing conditions.

A well-trained operator is an operator who quickly and effectively eliminates dangerous items from luggage, thus minimizing the energy consumption of devices supporting the screening process. In the case of creating the rules of the knowledge base, special attention was paid to the effectiveness of the correct indication of dangerous items. It is impossible to obtain the "experienced" level of training without sufficiently high efficiency. In the case of the presented model, 81 rules have been defined at the first level and 12 at the second level. An example fragment of the knowledge base is presented in Figure 14.

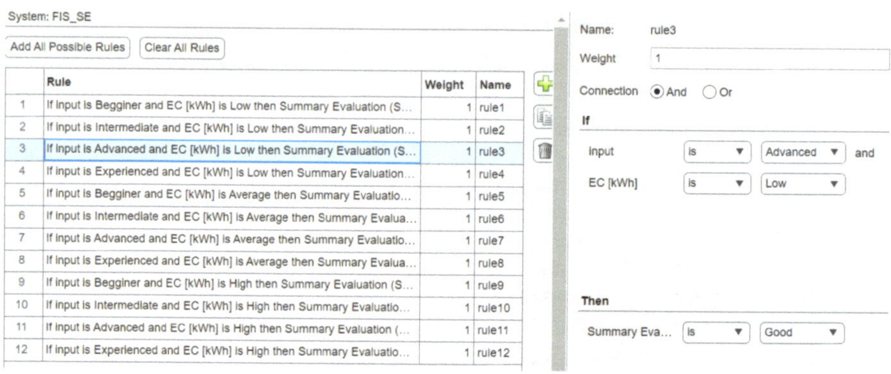

Figure 14. MATLAB window screenshot: example fragment of the knowledge base of SE.

6. Results

Experimental SCO tests were conducted to verify the built model's correct functioning. In total, 18 operators were tested. The research was carried out in three groups:

- Operators with 0 ÷ 24 months of experience in the position;
- Operators with 24 ÷ 90 months of experience in the position;
- Operators with 90 ÷ 150 months of experience in the position.

The tests were carried out on a built test stand using eye-tracking technology (see Section 4). As in real conditions, each operator was tested in threat recognition during five sessions with 50 images in each session. In total, each SCO analyzed 250 images. The images of baggage with (or without) threats that the operators analyzed were previously prepared by selecting groups of 50 photos from a database of almost 10,000 photos (the database was built as part of the project in cooperation with the airport). Each subsequent session was characterized by a greater difficulty scale (from the easiest to the most difficult). All operators were tested on the same group of images (displayed randomly within one session) to compare results. The test bench software allows you to read the total duration of an entire session (the total analysis time of 250 images). Based on this time, a script written in MATLAB calculated the energy consumption for a real X-ray station used in an airport for baggage screening. Details on how the time is calculated are described in Section 5.2, in the discussion of the Energy Consumption variable. The calculated variable was transferred automatically as input EC to the prepared fuzzy model. Based on the results of the tests, the total assessment of the operator's training level was calculated. The results of the conducted research are presented in Tables 1–3.

Table 1. Results of experimental test SCOs.

No. of SCO	Experience in Position (Months)	The Value of the Parameter (Variable) Obtained in the Test					EC * (kWh)	SE
		MTCI (s)	ZAO (%)	CEI (%)	MEFT (s)	Summary Time of Sessions (s)		
Group 0 ÷ 24 months of experience in the position								
1	7	11.22	47	77	5.11	3123	0.45	1.95
2	9	10.56	48	79	7.23	2756	0.46	**1.71**
3	6	11.65	39	81	7.56	3154	0.45	1.89
4	7	10.82	48	77	8.33	3100	0.45	2.00
5	14	10.76	56	76	5.45	2974	0.45	1.99
6	9	12.03	43	78	7.41	2843	0.45	1.91
Group 24 ÷ 90 months of experience in the position								
7	48	9.11	65	79	8.56	2343	0.39	3.22
8	34	8.03	76	86	7.65	2245	0.38	3.47
9	32	9.31	67	96	5.44	2346	0.39	3.25
10	56	9.77	72	77	5.76	2567	0.41	2.87
11	48	8.44	66	88	6.43	2134	0.39	3.76
12	35	7.56	69	83	6.66	2056	0.37	4.03
Group 90 ÷ 150 months of experience in the position								
13	120	6.11	88	97	5.12	1536	0.31	4.24
14	96	7.07	74	92	5.83	1765	0.35	4.20
15	123	6.33	71	94	5.44	1654	0.33	4.17
16	134	5.45	89	92	4.08	1438	0.28	**4.24**
17	135	6.11	87	96	5.23	1546	0.31	4.24
18	94	6.34	66	94	5.06	1747	0.35	4.15

Legend: MTCI—mean time to correctly indicate a dangerous item; ZAO—zone analysis order (colors); CEI—correct effectiveness indication; MEFT—mean eye focus time on dangerous item; EoE—evaluation of the effectiveness of hazard identification; EC *—energy consumption of one SCO per day (during one shift: 8 h); SE—Summary Evaluation.

Table 2. Results of calculation of energy consumption of SCOs.

No. of SCO	Experience in Position (Months)	EC per Day (8 h) (kWh)	Mean EC per Day in Group (kWh)	Weekly EC per One SCO * (kWh)	Yearly EC per One SCO ** (kWh)
Group 0 ÷ 24 months of experience in the position					
1	7	0.45	0.45	2.25	**135**
2	9	0.46			
3	6	0.45			
4	7	0.45			
5	14	0.45			
6	9	0.45			
Group 24 ÷ 90 months of experience in the position					
7	48	0.39	0.39	1.95	117
8	34	0.38			
9	32	0.39			
10	56	0.41			
11	48	0.39			
12	35	0.37			
Group 90 ÷ 150 months of experience in the position					
13	120	0.31	0.32	1.60	**96**
14	96	0.35			
15	123	0.33			
16	134	0.29			
17	135	0.31			
18	94	0.35			

Legend: Weekly EC per one SCO *—5 working days a week (8 h each shift); yearly EC per one SCO **—300 working days a year.

Table 3. Scenario analysis results for different SCO group compositions depending on experience.

No. of Scenario	Group Experience in Position	No. of SCO	Yearly EC per SCO (kWh)	Summary Yearly EC per 1 Group of SCOs on 1 × RTG Scanner LP * (kWh)	Summary Yearly EC per 3 Groups of SCOs on 1 × RTG Scanner LP (kWh)	Summary Yearly EC per 3 Groups of SCOs on 3 × Lines of Control with RTG Scanner LP (kWh)	Summary Yearly EC for Big Airport with 6 Lines of Control (kWh)
1	0 ÷ 24 months 24 ÷ 90 months 90 ÷ 150 months	10 0 0	135 117 96	1350	**4050**	**12,150**	**364,500**
2	0 ÷ 24 months 24 ÷ 90 months 90 ÷ 150 months	0 10 0	135 117 96	1170	3510	10,530	315,900
3	0 ÷ 24 months 24 ÷ 90 months 90 ÷ 150 months	0 0 10	135 117 96	960	**2880**	**8640**	**259,200**
4	0 ÷ 24 months 24 ÷ 90 months 90 ÷ 150 months	5 5 0	135 117 96	1260	3780	11,340	340,200
5	0 ÷ 24 months 24 ÷ 90 months 90 ÷ 150 months	5 0 5	135 117 96	1155	3465	10,395	311,850
6	0 ÷ 24 months 24 ÷ 90 months 90 ÷ 150 months	0 5 5	135 117 96	1065	3195	9585	287,550

Legend: RTG scanner LP *—low-power X-ray scanner.

Based on the analysis presented in Table 3, it can be seen that the annual difference in energy consumption during the inspection of hand luggage using a low-power X-ray scanner between experienced and inexperienced SCO is 30%.

Experimental studies have shown that the area of passenger baggage security control requires the examination of factors that have been completely overlooked so far. One of these factors is the economic factor (represented by the energy costs of the entire baggage control process), which the authors included in this publication. A control group consisting of six SCOs, with two operators with experience from each group, which were subjected to the tests presented in the article, was used to calibrate the model (determining the boundary values during the estimation of MF values representing individual linguistic variables). Based on the studies of these operators at real workstations operated at the airport, all input variables of the model were calculated. The expansion of the research group and the obtained results (Table 1) indicate that the authors did not make mistakes while building the model. The research results show that, from an economic point of view, the effectiveness of detecting dangerous objects in passenger luggage cannot be the only criterion for assessing SCOs during the verification of their capabilities. The energy difference used to assess 250 images between the best of the operators (No. 16) and the weakest (No. 2) was over 0.17 kWh, to the disadvantage of the less experienced operator. The difference of 0.17 kWh is only seemingly small. It should be emphasized that this is the difference in energy consumption of a very low-power RTG scanner during baggage control, operated by operators with different levels of experience during one full working day (one shift lasts 8 h). During one shift, the security control team usually consists of 10 people (at medium-capacity airports). In the case of larger airports, this number can be higher. Moreover, one full day consists of three shifts, each with 10 operators. So, the seemingly small difference is multiplied by 30 (see Tables 2 and 3). Even in small local airports, there are at least three parallel security control lines, so this value increases even more. To show the impact of the level of training on the value of consumption, the authors analyzed six different scenarios (Table 3). In each scenario, the team of operators on one shift lasting 8 h consisted of SCOs with a uniform (from the point of view of the obtained SE result) level of experience (scenarios 1, 2 and 3), and for comparison, a similar analysis was carried out for mixed teams (scenarios 4, 5 and 6). The authors emphasize once again that while preparing the evaluation tool, presented in the article, they took into account a very low-power RTG scanner. Scanners commonly used at airports have five times more power than the one used to build the tool. The tool presented in the article has a very universal character. To illustrate the scale of the problem to the reader, the authors presented the results of calculations for a large European airport equipped with control stations with high-power RTG scanners and six parallel security control lines for comparison in Table 3 (last column). In this case, the annual energy consumption difference in the entire process between teams of experienced and inexperienced operators amounted to over 105 000 kWh, thus showing savings reaching tens of thousands of euros.

At the same time, it was observed that with the increase in the operator's experience, the time of identification of dangerous objects (MTCI) in the luggage and the effectiveness of correct indication (CEI) were shortened. The same effect can be achieved by training the operator at appropriately prepared training stations devoid of the most energy-consuming elements, such as RTG scanners. Analyzing the data contained in Table 1, it can be seen that the overall training assessments (SE) of operators from the first group (experience at the position of less than 24 months) practically disqualify them from working as SCOs from an economic point of view (they formally meet the requirements regarding the CEI indicator value). This observation may be a hint for people managing the entire process of supervising baggage security control to train these least-experienced operators very intensively on simulators before they are allowed to work at real stations. Simulator training is stress-free and pressure-free, which additionally shapes habits and develops a certain automatism in the decisions made related to image analysis. The research showed that the most experienced operators almost automatically focused their gaze on dangerous objects for up to 82% (MFET/MTCI ratio: the average value for the entire group) of the time needed for correct indication, while beginner operators focused for only 60% of the time needed. The conclusions drawn allow people responsible for managing the entire security control process (supervising the work of operators) to create individual training strategies for each operator, which will significantly reduce the costs of the entire process.

7. Conclusions and Discussions

During the research, a large amount of experimental data was collected, which enabled the evaluation of selected parameters. These were then used to validate the assumptions made in the model.

The results obtained confirm the correctness of the assumptions adopted during the model construction stage. The analyses carried out indicate the possibility of effectively using the constructed model to make quick and highly effective diagnoses in terms of the level of SCO training (both in terms of general assessment and individual categories). Therefore, they provide the opportunity to appropriately plan the improvement training process, especially for operators showing deficiencies in individual areas.

The model presented in the article can be used to analyze the impact of various factors on the overall assessment of the level of operator training. The key conclusion from the conducted experimental research is that raising the level of SCO training greatly affects the level of electricity consumption, which has been proven in Section 6. To use its advantages even more effectively in the training process itself or at the stage of planning a training cycle, it is necessary to test a larger number of operators (especially those with the most experience) and possibly modify the model itself (e.g., in terms of the number of input variables). However, based on the experience gained so far, it is possible to propose the following scheme of work with the model and its implementation in the training process:

1. Preparation of input data: collecting information on individual factors affecting the level of SCO training, such as skills, experience, theoretical knowledge, etc. These data should be transformed into a form suitable for the inputs of fuzzy models.

2. Starting the simulation: based on the input data, a simulation is carried out, which allows the assessment of the impact of individual factors on the final assessment of the SCO training level.
3. Analysis of the results: analysis of the simulation results in terms of understanding how individual factors affect the assessment of SCO training. On this basis, training strategies can be developed to improve weaknesses and use strengths.
4. Experiments with different scenarios: it is possible to conduct various simulation experiments to study the influence of different combinations of factors on the final score. This can help determine the optimal training path for the SCO that will deliver the best results.
5. Verification and validation: after experimentation, the results can be verified and validated by comparing them with actual training results or other assessment methods.
6. Further improvement of the model: modifying the number of inputs to the model, MF, inference rules or hierarchical structure to obtain better results.

The quick and effective assessment of the SCO training level allows for the preparation of practical, personalized training strategies that will allow you to achieve the best results and thus contribute to improving passenger safety while guaranteeing optimization in electricity use and sustainable development.

In summary:

1. The research showed that differences in SCO experience significantly affect the energy consumption of the baggage control process. Operators with more experience consume less energy, which translates into significant savings, especially in large airports with multiple control lines.
2. Simulators, being less energy-consuming than actual RTG scanners, offer significant potential for training new operators, allowing them to develop skills in a stress-free and pressure-free environment, which additionally shapes habits and develops a certain automatism in the decisions made.
3. The built evaluation tool opens the way to developing individual training strategies for each operator, which can significantly reduce the costs of the entire process, with particular emphasis on training operators with less experience.
4. The authors of the study emphasize that economic analysis, represented by the energy costs of the entire baggage control process, is a significant factor that has been completely overlooked so far, suggesting that it should be included in the assessment of SCO operators' efficiency.
5. The tool presented in the article has a universal character and can be used to analyze various control systems, not only those based on low-power RTG scanners.

These conclusions suggest that both the training and professional experience of SCO operators are crucial for the efficiency and economy of the security control process, with the potential to significantly reduce operational costs and increase the efficiency of threat detection.

Author Contributions: Conceptualization, J.R. and A.K.; methodology, J.R.; software, J.R.; validation, J.R. and A.K.; formal analysis, A.K.; investigation, J.R.; resources, A.K.; data curation, J.R.; writing—original draft preparation, J.R.; writing—review and editing, A.K.; visualization, J.R.; supervision, A.K.; project administration, A.K.; funding acquisition, A.K. All authors have read and agreed to the published version of the manuscript.

Funding: This research received no external funding.

Data Availability Statement: Not applicable.

Conflicts of Interest: The authors declare no conflict of interest.

References

1. Airports. Available online: https://snopud.bizenergyadvisor.com/article/airports (accessed on 29 June 2026).
2. Ortega Alba, S.; Manana, M. Energy Research in Airports: A Review. *Energies* **2016**, *9*, 349. [CrossRef]

3. Kierzkowski, A.; Kisiel, T. Feasibility of Reducing Operator-to-Passenger Contact for Passenger Screening at the Airport with Respect to the Power Consumption of the System. *Energies* **2021**, *14*, 5943. [CrossRef]
4. Ramsey, D.; Bouscayrol, A.; Boulon, L.; Desreveaux, A.; Vaudrey, A. Flexible Simulation of an Electric Vehicle to Estimate the Impact of Thermal Comfort on the Energy Consumption. *IEEE Trans. Transp. Electrif.* **2022**, *8*, 2288–2298. [CrossRef]
5. ACI. *Annual Report 2016*; Airports Council International: Montreal, QC, Canada, 2017.
6. Correia, A.R.; Wirasinghe, S.C.; De Barros, A.G. Overall level of service measures for airport passenger terminals. *Transp. Res. Part A* **2008**, *42*, 330–346. [CrossRef]
7. De Barros, A.G.; Somasundaraswaran, A.; Wirasinghe, S.C. Evaluation of level of service for transfer passengers at airports. *J. Air Transp. Manag.* **2007**, *13*, 293–298. [CrossRef]
8. Ma, J.S.; Liu, X.T.; Zhuang, D.M.; Wang, S.G. CFD-based design of the natural ventilation system of the traffic center of T3 Beijing International Airport. *Adv. Mater. Res.* **2011**, *291–294*, 3292–3295. [CrossRef]
9. Parker, J.; Cropper, P.; Shao, L. Using building simulation to evaluate low carbon refurbishment options for airport buildings. In Proceedings of the 12th Conference of International Building Performance Simulation Association, Sydney, Australia, 14–16 November 2011.
10. Feng, Y.; Huang, Y.; Shang, H.; Lou, J.; Knefaty, A.d.; Yao, J.; Zheng, R. Prediction of Hourly Air-Conditioning Energy Consumption in Office Buildings Based on Gaussian Process Regression. *Energies* **2022**, *15*, 4626. [CrossRef]
11. Liu, J.; Yu, N.; Lei, B.; Rong, X.; Yang, L. Research on indoor environment for the terminal 1 of Chengdu Shuangliu international airport. In Proceedings of the 11th International IBPSA Conference, Glasgow, UK, 27–30 July 2009.
12. Metsä-Eerola, I.; Pulkkinen, J.; Niemitalo, O.; Koskela, O. On Hourly Forecasting Heating Energy Consumption of HVAC with Recurrent Neural Networks. *Energies* **2022**, *15*, 5084. [CrossRef]
13. Balaras, C.A.; Dascalaki, E.; Gaglia, A.; Droutsa, K. Energy conservation potential. HVAC installations and operational issues in Hellenic airports. *Energy Build.* **2003**, *35*, 1105–1120. [CrossRef]
14. Gowreesunker, B.L.; Tassou, S.A.; Kolokotroni, M. Coupled TRNSYS-CFD simulations evaluating the performance of PCM plate heat exchangers in an airport terminal building displacement conditioning system. *Build. Environ.* **2013**, *65*, 32–145. [CrossRef]
15. Chen, J.; Xie, K. A prediction model based on unbiased grey Markov for airport energy consumption prediction. In Proceedings of the 2013 Chinese Automation Congress (CAC), Changsha, China, 7–8 November 2013.
16. Huang, H.; Chen, L. A new model predictive control scheme for energy and cost saving in commercial buildings: An airport terminal building study. *Build. Environ.* **2015**, *89*, 203–216. [CrossRef]
17. Fan, B.; Jin, X.; Du, Z. Optimal control strategies for multi-chiller system based on probability density distribution of cooling load ratio. *Energy Build.* **2011**, *43*, 2813–2821. [CrossRef]
18. Mambo, A.D.; Efthekhari, M.; Thomas, S.; Steffen, T. Designing an occupancy flow-based controller for airport terminals. *Build. Serv. Eng. Res. Technol.* **2015**, *36*, 51–66. [CrossRef]
19. Mambo, A.D.; Efthekhari, M.; Thomas, S.; Steffen, T. Evaluation of Indoor Environment System Performance for Airport Buildings. *Int. J. Sustain. Green Energy* **2015**, *4*, 73–84. [CrossRef]
20. Mambo, A.D.; Efthekhari, M.; Thomas, S.; Steffen, T. Fuzzy supervisory control strategies to minimise energy use of airport terminal buildings. In Proceedings of the 18th International Conference on Automation and Computing (ICAC), Loughborough, UK, 7–8 September 2012.
21. Soukour, A.; Devendeville, L.; Lucet, C.; Moukrim, A.A. Memetic Algorithm for staff scheduling problem in airport security service. *Expert Syst. Appl.* **2013**, *40*, 7504–7512. [CrossRef]
22. Kierzkowski, A.; Kisiel, T.; Uchroński, P. Simulation Model of Airport Security Lanes with Power Consumption Estimation. *Energies* **2021**, *14*, 6725. [CrossRef]
23. Scala, P.; Mujica Mota, M.; Cheng-Lung, W.; Delahaye, D. An optimization–simulation closed-loop feedback framework for modeling the airport capacity management problem under uncertainty. *Transp. Res. Part C* **2021**, *124*, 102937. [CrossRef]
24. Mujica, M.; Scala, P.; Delahaye, D. Improving Airport Performance through a Model-based and Optimization Approach. In *Applied Simulation and Optimization 2*; Mota, M.M., De La Mota, I.F., Eds.; Springer: Cham, Switzerland, 2017. [CrossRef]
25. Skorupski, J.; Grabarek, I.; Kwasiborska, A.; Czyżo, S. Assessing the suitability of airport ground handling agents. *J. Air Transp. Manag.* **2020**, *83*, 101763. [CrossRef]
26. Schultz, M.; Evler, J.; Asadi, E.; Preis, H.; Fricke, H.; Cheng-Lung, W. Future aircraft turnaround operations considering post-pandemic requirements. *J. Air Transp. Manag.* **2020**, *89*, 101886. [CrossRef]
27. Skorupski, J.; Żarów, P. Dynamic management of aircraft stand allocation. *J. Air Transp. Manag.* **2021**, *90*, 101964. [CrossRef]
28. Act of July 3 2002 on Civil Airports (Journal of Laws of 2020. Item 993)—Poland. Available online: https://www.gov.pl/attachment/82798da7-0271-4984-a9cb-8e450445ca44 (accessed on 29 June 2023).
29. Commission Regulation (EU) 2015/1998 of 5 November 2015 Establishing Training Guidelines for Screening Operators and Accompanying Staff. Available online: https://eur-lex.europa.eu/legal-content/EN/TXT/PDF/?uri=CELEX:32015R1998&from=ET (accessed on 26 June 2023).
30. Commission Regulation (EU) 2015/340 of 20 February 2015 on the Training of Screening Operators and Accompanying Staff. Available online: https://eur-lex.europa.eu/legal-content/EN/TXT/PDF/?uri=CELEX:32015R0340 (accessed on 26 June 2026).

31. Directive 2016/681 of the European Parliament and of the Council of 27 April 2016 on the Use of Passenger Data for the Prevention of Terrorism and Serious Crime. Available online: https://eur-lex.europa.eu/legal-content/EN/TXT/PDF/?uri=CELEX:32016L0681 (accessed on 27 June 2023).
32. Council Regulation (EEC) No 3922/91 of 16 December 1991 on the Harmonization of Technical Requirements and Administrative Procedures in the Field of Civil Aviation. Available online: https://eur-lex.europa.eu/legal-content/EN/TXT/HTML/?uri=CELEX:31991R3922 (accessed on 24 June 2023).
33. Regulation of the Minister of Infrastructure of July 22. 2022. (on Security Checks at Airports)—Poland. Available online: https://isap.sejm.gov.pl/isap.nsf/DocDetails.xsp?id=WDU20220001785 (accessed on 25 June 2023).
34. Regulation of the Minister of Transport and Maritime Economy of July 5 2013. (on the Technical Conditions to Be Met by Selected Airport Equipment)—Poland. Available online: https://isap.sejm.gov.pl/isap.nsf/DocDetails.xsp?id=WDU20130000926 (accessed on 23 June 2025).
35. Regulation of the Minister of Transport. Construction and Maritime Economy of July 26 2021. on the Training of Security Control Operators at Civil Airports—Poland. Available online: https://isap.sejm.gov.pl/isap.nsf/DocDetails.xsp?id=WDU20210001526 (accessed on 25 June 2023).
36. Dubois, D.; Prade, H. Putting Rough Sets and Fuzzy Sets Together. In *Intelligent Decision Support*; Springer: Dordrecht, The Netherlands, 1992; pp. 203–232. [CrossRef]
37. Greco, S.; Matarazzo, B.; Slowinski, R. Rough sets theory for multicriteria decision analysis. *Eur. J. Oper. Res.* **2001**, *129*, 1–47. [CrossRef]
38. Zadeh, L.A. Similarity relations and fuzzy orderings. *Inf. Sci.* **1971**, *3*, 177–200. [CrossRef]
39. Siler, W.; Buckley, J. *Fuzzy Expert Systems and Fuzzy Reasoning*; John Wiley & Sons, Inc.: Hoboken, NJ, USA, 2004. [CrossRef]

applied sciences

MDPI

Article

Development of a Virtual Reality Tool for Train Crew Training

Agnieszka A. Tubis *, Franciszek Restel and Anna Jodejko-Pietruczuk

Faculty of Mechanical Engineering, Department of Technical Systems Operation and Maintenance, Wroclaw University of Science and Technology, Wyspianskiego Street 27, 50-370 Wroclaw, Poland; franciszek.restel@pwr.edu.pl (F.R.); anna.jodejko@pwr.edu.pl (A.J.-P.)
* Correspondence: agnieszka.tubis@pwr.edu.pl

Abstract: (1) Background: The article presents the results of research carried out as part of one of the project's tasks, aiming to develop training scenarios for selected workstations related to the operation of passenger rail transport. The article aims to present the results of a training experiment to verify the effectiveness of the developed training scenario for conductors regarding lift operations for people with disabilities. (2) Methods: The study used an analysis of the execution times of individual training tasks, accompanying observation, face-to-face interviews, and a comparative analysis of the effectiveness of various training strategies. The research was carried out following the developed five-stage research procedure. (3) Results: The obtained research results confirm the benefits of using virtual reality in training conductors and allow for determining the most effective training strategy. The measurements of the execution times of individual training operations made it possible to identify the tasks in the training program that were the most time-consuming or characterized by highly diversified execution times. (4) Conclusions: Based on the research, it was possible to develop recommendations regarding the changes that should be introduced in the training tool to better adapt to the real conditions of service operations. The effectiveness of using this form of training in relation to the improvement of a selected group of employees was confirmed.

Keywords: training effectiveness; immersive VR; training scenarios; conductor skills; rail transport

Citation: Tubis, A.A.; Restel, F.; Jodejko-Pietruczuk, A. Development of a Virtual Reality Tool for Train Crew Training. *Appl. Sci.* **2023**, *13*, 11415. https://doi.org/10.3390/app132011415

Academic Editor: Junfeng Wang

Received: 17 August 2023
Revised: 12 October 2023
Accepted: 16 October 2023
Published: 18 October 2023

1. Introduction

The current employee market in Poland results in a high turnover of employees in many sectors. The employee turnover rate in 2022 in Poland was 21% [1], but in the railway sector, this level is lower due to task specificity. However, it should be noted that we can also observe an increased turnover rate compared to previous years in this transport branch, which is a challenge in the current management of railway personnel. Increased rotation causes an increased risk of human errors in service processes. Every newly hired employee needs to be prepared each time in order to perform their duties. An employee starting work in a new position is usually characterized by a lack of skill and limited knowledge of the tasks performed. They also do not have a sufficiently large experience curve (number of repetitions of a given operation), which determines the speed and precision of performed actions. For this reason, such an employee is not only exposed to making mistakes but also the work he/she provides is characterized by extended execution times. It is also worth noting the research on risk assessments for rail freight transport operations, which is presented in [2]. According to them, the majority of adverse events recorded in the selected rail transport system result from errors and the negligence of crew members.

Many researchers and practitioners point to the need for changes in training systems for new employees and training for experienced employees [3]. Solutions are sought to allow employees to acquire and improve the required competencies related to their daily duties faster and more effectively. Traditional forms of training are currently rated low by both participants and employers, who point to their low effectiveness in shaping practical skills and the extended time to acquire knowledge [4]. The preferred employee training system is the active teaching form, which enables employees to achieve the required knowledge and improve their practice skills. The answer to these needs is virtual reality (VR) technology, which enables a faithful representation of the working environment of operational employees. The developed training scenarios allow trainees to practically participate in the supported processes without exposing them to damage and disruptions. Thanks to training tools developed in this way, it is possible to reduce the occurrence of adverse events in which employees are the source of risks. For this reason, VR technology is widely used in various sectors, including rail transport. It should be noted, however, that the number of publications on the use of VR technology in the training of railway personnel is very limited. Searching the database by keys "VR" AND "rail" OR "virtual reality" AND "rail" only returned 85 references. Among them, only eight items concerned the direct use of VR technology in training people to operate rail transport. These publications primarily discussed the use of VR technology to prepare personnel serving passengers in the evacuation process (among others: [5–7]), training in the use of railway signaling systems [8], and maintenance training [9]. Only one publication concerned the use of VR technology in rail transit personnel training [10], but this was directed to managers and maintenance personnel. It should be stated that there are currently no publications on the use of immersive virtual reality for training passenger-serving staff on trains, especially with respect to conductors. The application area of VR technology defined in this way currently comprises a research gap, which is filled by the research presented in this article.

The article presents the results of research carried out as part of one of the project's tasks, aiming to develop training scenarios for selected workstations related to the operation of passenger rail transport. These scenarios were developed based on accompanying observations, direct interviews with railway personnel and instructors, an analysis of the managerial staff's needs, and analyses of the risk of errors made by employees employed at the examined workstations. The article aims to present the results of a training experiment conducted on a group of 57 volunteers and a proposal for a strategy for training based on VR tools with the support of an instructor. The conducted training experiment aimed at verifying the effectiveness of the developed training scenario for conductors regarding lift operations for people with disabilities. It became the basis for the modifications introduced in the developed VR tool. The structure of the article is shown in Figure 1.

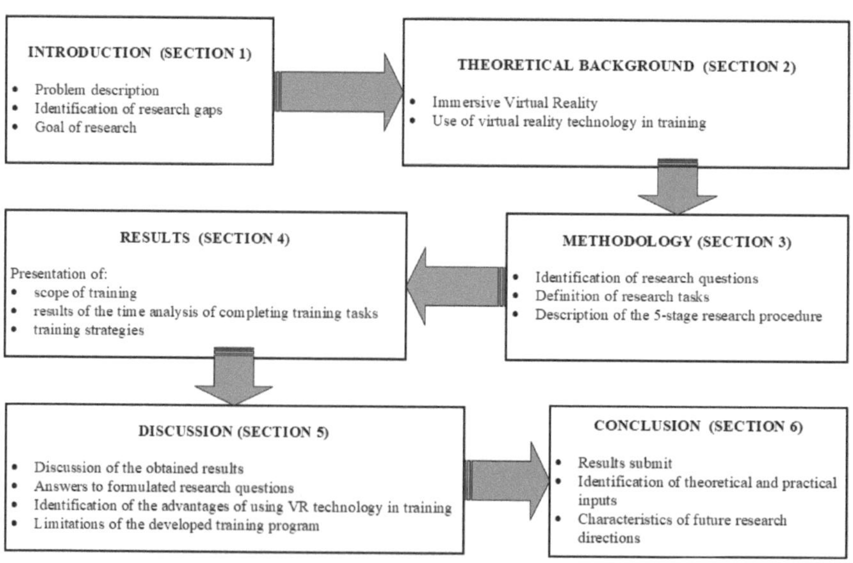

Figure 1. The structure of the article.

2. Literature Review

2.1. Immersive Virtual Reality (IVR)

Virtual reality is an artificially created digital environment, and its task is to reproduce real life. Two attributes characterize this environment:

1. "immersion"—means an objective level of sensory fidelity that a VR system provides [11];
2. "presence"—refers to the user's subjective experience resulting from being in the immersive environment [12].

However, some authors (e.g., [13,14]) indicate that one more attribute should be added to the basic features of VR technology—"interactivity", which Steuer [15] defined as the degree to which a user can modify the VR environment in real time.

The level of immersion determines the extent to which high-fidelity physical stimuli (e.g., light patterns, sound waves) are delivered to various senses (sight, hearing, and touch) to enhance the illusion of reality [16]. For this reason, in many publications, VR technology is divided into two major types based on the level of interaction and immersive environment [16–18]: (1) non-immersive virtual environment (desktop VR) and (2) immersive virtual environment (IVE). In a non-immersive virtual environment, the training simulation is presented on a conventional personal computer and is typically explored using a keyboard, mouse, wand, joystick, or touchscreen. This is the simplest type of virtual reality application. The training is presented on multiple room-size screens or a stereoscopic, head-mounted display unit in an immersive virtual environment. Audio, haptic, and sensory interfaces may enhance these systems. In his research, Mandal distinguished the third group of VR technologies, which is an intermediate between the abovementioned groups. He referred to it as semi-immersive (fish tank VR) systems [16]. These systems support head tracking and improve the feeling of "being there" thanks to the motion parallax effect. They still use a conventional monitor but generally do not support sensory output.

The "Immersive training system for railway personnel using virtual reality technology" project aims to develop training solutions using IVR technology. IVR tools provide a seamless, scalable environment capable of simulating a full range of sensory stimuli and allowing the user to experience 360° visual immersion and manipulate virtual objects in real time [18,19]. For this reason, some authors emphasize that the three main features provided by IVR are as follows [11]: the sense of immersion, real-time interaction, and imagination. IVR in the training system allows isolating the participant from external cues and uncontrolled stimuli from his/her physical word. This allows the participant to experience immersive and interactive scenery and interact with objects and tools in a desirable manner [20]. However, attention should be paid to the results of studies by Janssen et al. [21], according to which learners with certain individual traits and characteristics will benefit less from learning in VR (e.g., people with more anxious or reserved personalities). In their research, Jensen and Konradsen [12] proved that immersive HMD learners were more engaged; spent more time on learning tasks; and acquired better cognitive, psychomotor, and affective skills. However, they also noted that the graphical quality of VR and the awareness when using VR, for instance, can reduce the sense of presence.

Some authors classify the phenomenon of immersion used in VR solutions. Two classifications presented in Figure 2 deserve attention here.

Adams distinguished three dimensions of immersion in his research: tactical, strategic, and narrative immersion. A similar classification was proposed by Björk and Holopainen, who used the term sensory-motoric, cognitive, and emotional immersion for analogous dimensions. In addition, these authors supplemented the classification with three additional dimensions marked in yellow in Figure 2. Thanks to this, immersion provided in a virtual environment can be considered according to six proposed impact areas on the user.

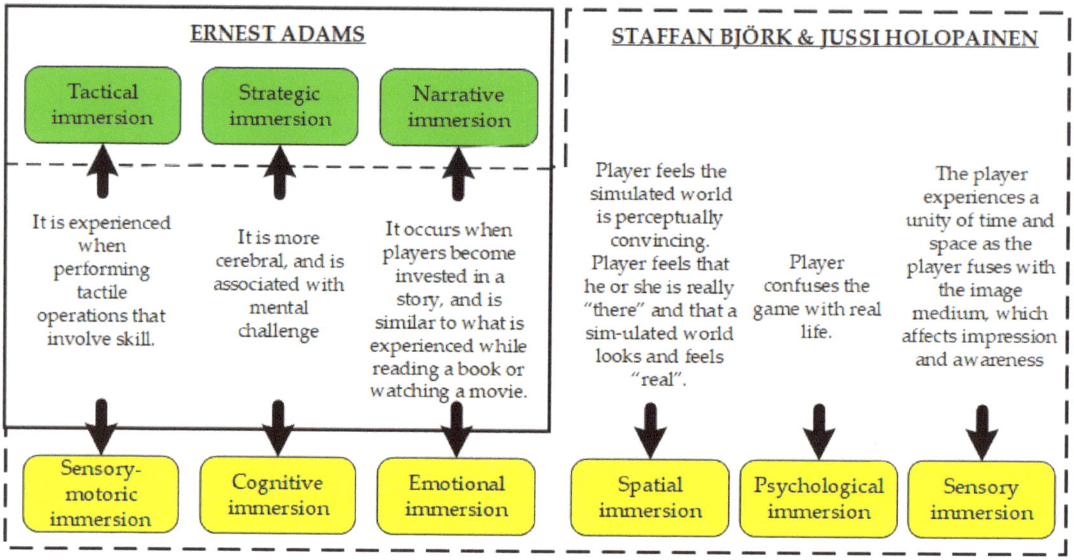

Figure 2. Classification of the phenomenon of immersion (based on [22,23]).

2.2. Use of Virtual Reality in Training

The use of VR technology in training programs is a response to the limitations of traditional training forms. Among the main limitations of real-world training, the authors usually mention the following [4]:

1. It could be time-consuming due to the efforts and time needed to set up the real-world training site and to travel to the site;
2. It could be expensive due to the cost of preparing real-world training materials and hiring human coaches;
3. It could be unappealing and unintuitive due to the lack of visual hints, such as 3D animations for illustrating skills and processes;
4. It could not be possible to train some skills in the real world, such as emergency procedures that can only be safely trained in simulators.

Although VR is not new, recent developments in immersive technologies—in terms of visualization and interactions—have made VR increasingly attractive to scholars [20]. Virtual reality is now widely used in various types of training for multiple sectors of the economy. Rogers [24], in his article in Forbes, referred to it as "the learning aid of the 21st century", and the results of research, among others, Krokos et al. [25], suggest that students retain more information and can better apply what they had learned after participating in VR exercises. For this reason, an important area of scientific and practical importance is currently the improvement of the so-called "industrial skills" via virtual reality training tools. According to Radhakrishnan [26], these skills include four groups of competencies, which are presented in Figure 3.

As noted by [27], training in industrial skills is critical to nations' economic development and competitiveness. Their effectiveness reduces the number of errors made by industrial workers, which can have catastrophic consequences for the product and total production costs [28]. Meanwhile, performance metrics used in VR tools make it possible to identify mistakes made by the trainee, and on this basis, their weaknesses are indicated [18]. Thanks to this, it is possible to adjust the learning parameters to the needs of the training participants so that they can deliberately practice critical skills and increase exposure to a wide variety of both common and low-volume cases [29]. At the same time, VR tools provide a so-called safe training environment with minimal exposure to dangerous situations (e.g., fires [30]; explosions, and natural disasters [31]). For this reason, in the initial development phase, VR technology was primarily used to improve KSAOs, for which training processes were associated with the risk of loss of the health or life of training participants or very high implementation costs [32]. Therefore, the first applications of VR refer to the training of pilots, paratroopers, firefighters, and military personnel [32].

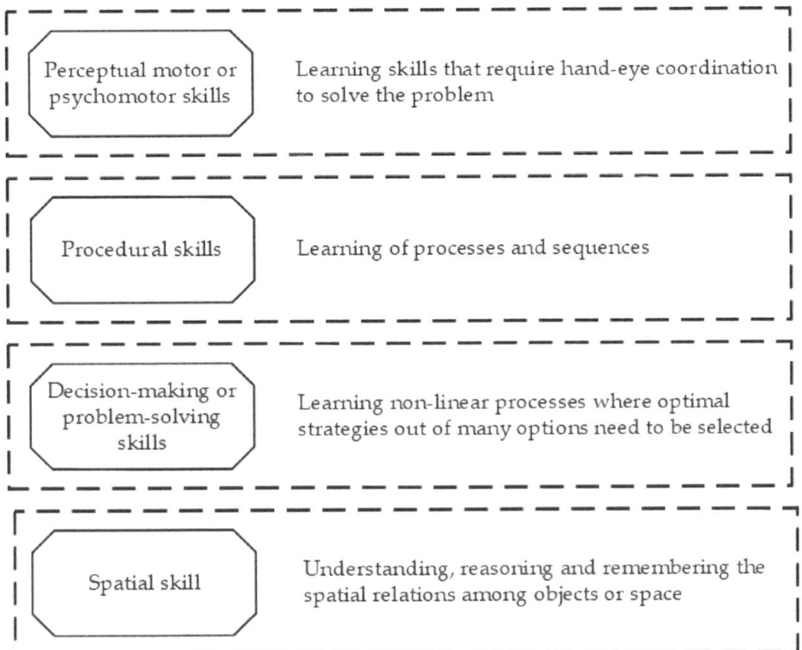

Figure 3. Industrial skills improved by VR training tools (based on [26]).

It is worth noting that training tools using VR technology meet the requirements for an effective learning process today, which include the following [33]: focus, meaningful representation of information, multiple mappings of information, and reflective learning. At the same time, currently created virtual reality environments (VREs) have unique contributions to learning scientific visualization, instructions, sensory-motor performance, and training [34]. Xie et al. point to two crucial aspects of the use of VR technology in the training process [4]:

1. The possibility of using an increased number of training scenarios while reducing training costs: This is because VR training scenarios primarily involve computer-generated 3D graphics. Developers can, therefore, create various scenarios from existing 3D resources (both proprietary and available online) that can be reused to train different people.
2. The possibility of training in the comfort of one's own space: This is especially important in cases where the trainee may feel uncomfortable about their actions in the real-world training process due to the presence of other participants (observers). At the same time, the same tool can be used in training situations that need to happen in the presence of instructors as they need to provide early feedback on the trainee's performance and alert the trainee to issues that they may be experiencing, including negative trends in their performance.

The increasing popularity of VR technology is positively influenced by the advantages of this solution, which are identified in numerous experiments described in the literature. The key benefits offered by VR technology in the field of employee training are usually indicated [35]:

1. Increased involvement in training—immersive environments and 3D representations make users more interested and committed to their training;
2. Speed—workers can obtain and retain information efficiently due to the creative and attractive method that this digital tool offers. Many studies emphasize that VR allows for training through emotional responses, the best method to not forget what has been learned;
3. Measurability—training is carried out in computer-generated environments; thus, researchers can quickly obtain any required statistics, such as the time it takes an operator to carry out a particular activity, the most common accidents, or the procedure with less difficulty;
4. Reduction in work accidents—VR offers a much safer scenario where users can practice obtaining the necessary skills without risking their integrity;
5. Personalization—scenarios can adapt to the requirements of each worker;
6. Reduction in costs—VR helps to reduce infrastructure, materials, time, and personnel expenses that are required by real simulations.

When assessing the effectiveness of VR training, it is worth considering the results of the literature review in this area, which was prepared by Strojny and Dużmańska-Misiarczyk [36]. In their research, the authors emphasize that the effectiveness of VR training can be assessed in many ways. The analysis of the publication proves the following [36]:

1. The most common approach involved using some objective method (e.g., knowledge test): such methods were applied in 82% of the reviewed studies;
2. The subjective evaluation of the training tool, learning outcomes, motivation, or other psychological constructs was also used in 72% of the studies;
3. Observation by an expert as a form of assessment of learning effectiveness was used sparingly (6%). It was primarily used where objective assessment methods could not be applied or for the qualitative evaluation of learners' reactions during learning;
4. Physiological measurements were the least frequently applied (3%).

The review's authors also emphasize that more than one method was used in more than half of the analyzed publications; in particular, combining methods from two different categories was the most popular choice. Only 15 studies combined three different methods (this accounted for about 5% of the researched publications). In the remaining analyses, only one selected effectiveness assessment method was used.

Due to VR technology's opportunities, it is currently used in training programs for many industrial and service sectors. Xie et al. [4] reviewed a total of 48 VR training applications and works. Based on their research, the dominant areas using VR in the conducted training can be distinguished. They are as follows [4]:

- First responder training (e.g., police officers, firefighters, and emergency medical services);
- Medical training;
- Military training;
- Transportation sector;
- Workforce training;
- Interpersonal skills training (in particular, openness, empathy, verbal, and nonverbal communication).

However, it should be clearly emphasized that these are not the only sectors that currently use VR tools in training.

3. Methodology

The project task concerned verifying the developed training tool that uses IVR technology and offers a dedicated elevator operation scenario for people with disabilities by railway staff (conductors). The study involved 53 volunteers who had not previously operated the device. They were also informed that the obtained results would be used for scientific purposes (including publications) and to improve the developed training program.

The research carried out as part of the presented project task aimed to answer four research questions. Therefore, six research tasks were set, the implementation of which made it possible to answer the formulated questions. The research questions and developed tasks are presented in Figure 4.

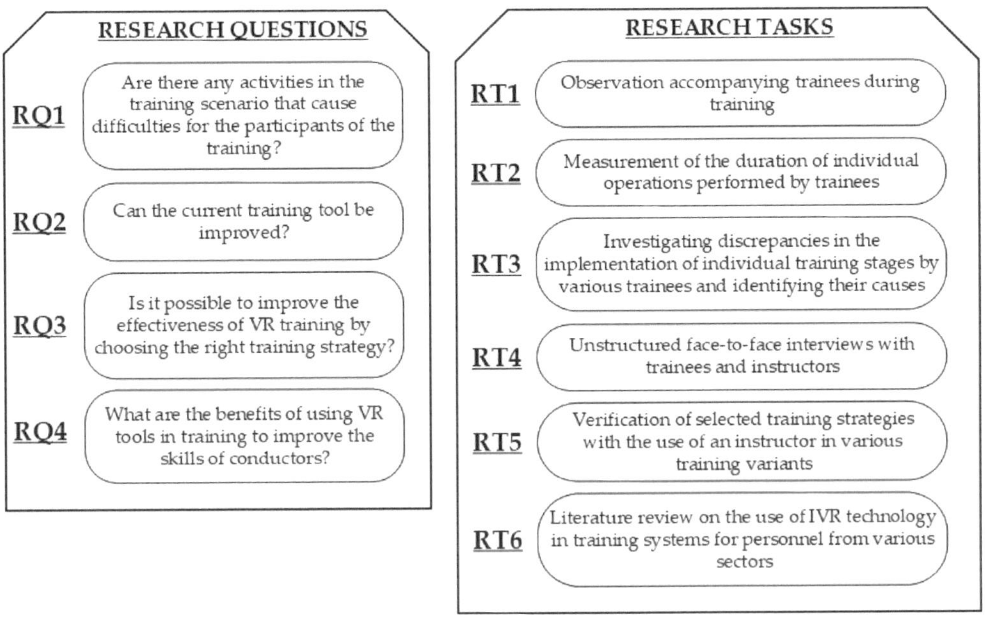

Figure 4. Research questions and tasks.

The entire experiment was carried out following the developed research methodology, which consisted of five stages of the procedure. The research procedure with the assigned tasks is presented in Figure 5.

In the conducted research, the project team was supported by four professional instructors who completed training for conductors at a selected railway carrier. The experiment was carried out in a specially prepared training room. The first stage of the investigation was to introduce the IVR tool to the course participants and discuss all the activities that should be performed with the instructor during the entire training period according to the training scenario. The instructor confirmed the understanding of all instructions by the course participants after discussing the guidelines for the implementation of the training. The training was carried out in several training groups so that it was possible to verify various training strategies.

All performed tasks were measured during the implementation of individual operations by the training participants. The measurement was carried out traditionally outside the virtual environment. Members of the research team measured each trainee individually. The analysis of the duration of the operation by the trainees was based primarily on three statistical parameters: arithmetic mean, standard deviation, and coefficient of variation. The calculated arithmetic average for each activity formed the basis for estimating the time-consuming execution of each procedural step. On this basis, activities with long execution times and that suggested a high degree of difficulty in their implementation were identified. The calculated standard deviation formed the basis for identifying discrepancies in the execution times of individual participants. The coefficient of variation is calculated based on the ratio of the standard deviation to the arithmetic mean. This measure provided the basis for identifying operations for which their performance depends on the skills and psycho-physical characteristics of the trainee.

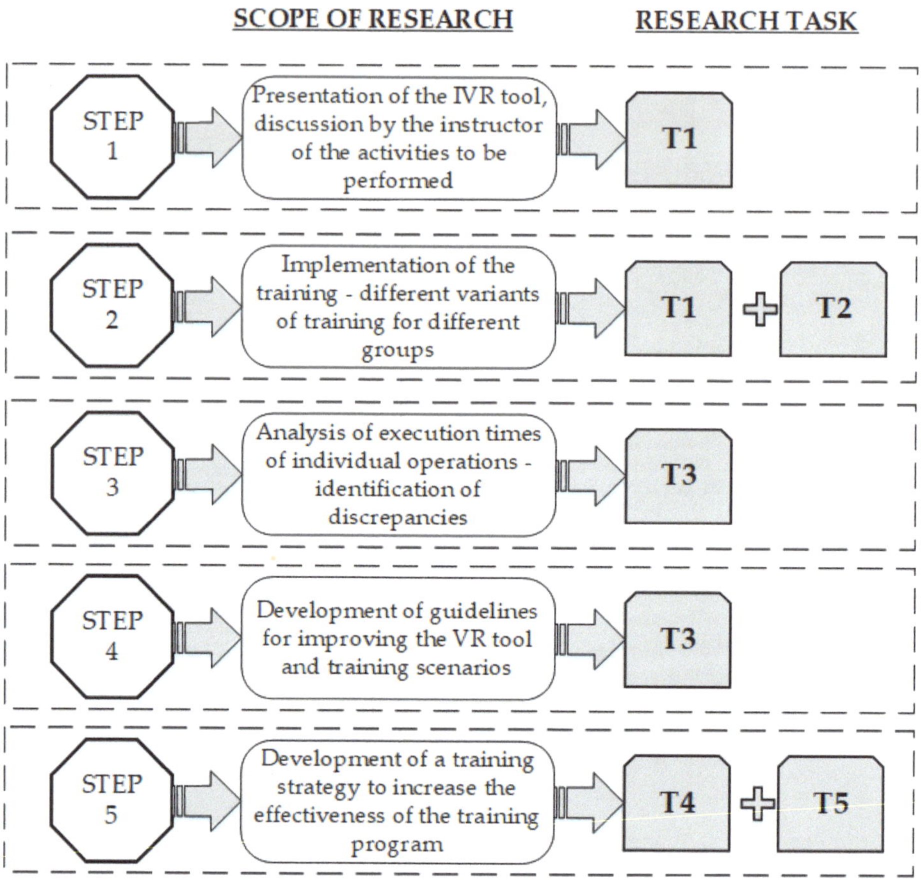

Figure 5. Research procedure.

Interviews with trainees and instructors were conducted by the members of the project team immediately after the end of the training. The survey took the form of direct, unstructured interviews. The study aimed to collect participants' opinions about their experiences (emotions) during the training, opinions about the training tool, and the identification of potential for improvement.

Training strategies were developed based on literature studies and direct observations. Two methods were used to assess the effectiveness of the developed training program—a practical test carried out on a real object and an accompanying observation carried out by two experts (a member of the project team and an experienced instructor). On this basis, a static analysis was carried out using time measurements collected during the conducted observation.

4. Results

4.1. Scope of Training Using Virtual Reality

One of the training scenarios developed within the framework of the project was the operation of an elevator for people with disabilities, which is used in Polish long-distance train vehicles made by Stadler. This training is significant for two reasons: (1) because of the level of service offered to passengers with special needs, particularly the time it takes to handle them during the exchange of passengers on the platform, and (2) traditional training requires temporarily taking the vehicle out of service, which is usually difficult to correlate with the timing of the training. Thus, this operation requires quick and error-free execution by the conductor. At the same time, it is not a standard operation and is often repeated during the course of their duties, so new personnel do not have the opportunity to develop a routine. In the training process, the problem is ensuring the availability of a vehicle equipped with such an elevator and out-of-current operation, allowing many repetitions to be performed without time constraints. Conducting this training using a virtual reality tool provides an opportunity for the trainee to perform multiple repetitions, allowing him to gain experience and proficiency in performing this operation.

The study of the fit and effectiveness of the developed training scenario was conducted using an experiment with 53 volunteers. The group of volunteers consisted of people aged 22–29 who want to pursue the conductor profession in the future, as well as people interested in rail transportation. Approx. 28% of those surveyed indicated previous experience with virtual reality, while 69% said they regularly play or have played computer games.

In the experiment, participants were tasked with the nine activities illustrated in Table 1.

Table 1. Milestone activities performed by the participants during the experiments in the VR tool.

Symbol	Activities	Visualization
A1	Taking a board with the description of the procedure and carrying it under the train	
A2	Opening a tray and retrieving the square key	
A3	Opening the lid and lever, storing the key in the tray, and opening the entire elevator	

Table 1. *Cont.*

Symbol	Activities	Visualization
A4	Unfolding the platform and wings of the elevator and securing the carriage from rolling away from the side of the carriage	
A5	Lowering the elevator to the station platform level using the remote control	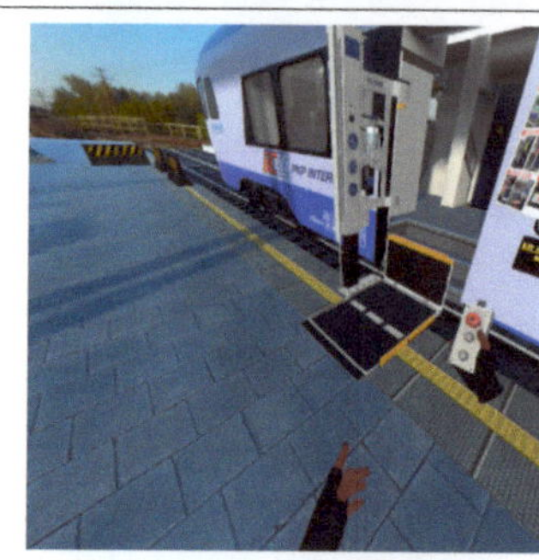
A6	Raising the elevator by remote control	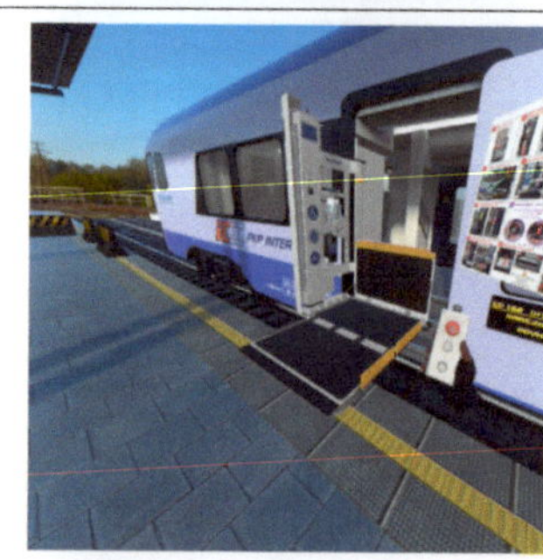
A7	Lowering the elevator hydraulically using two valves	

Table 1. *Cont.*

Symbol	Activities	Visualization
A8	Elevator raising with manual lever	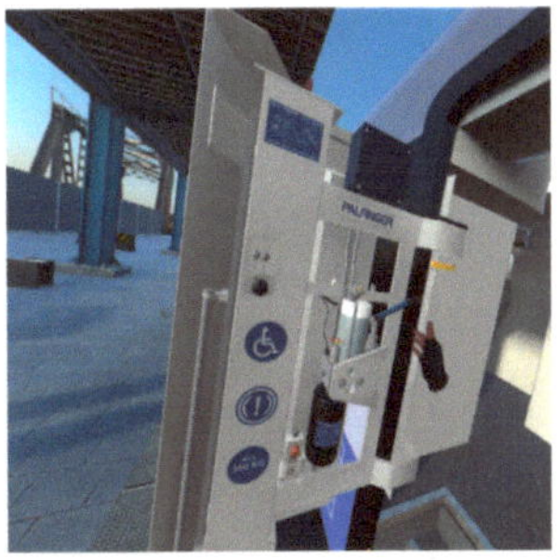
A9	Closing the elevator and exiting the train	

The first activity, i.e., moving the board, verifies the ability to observe the virtual reality tool in terms of grasping, moving, and orienting. These skills are required in further stages of training; therefore, their verification in the initial activity is essential. Subsequent activities already result from the elevator operation procedure. This procedure describes in detail the next steps of the procedure regarding the preparation of the device for operation (activities A2–A4), the ongoing operation of the device during different modes of operation (activities A5–A8), and the termination of the device (activity A9). It is also worth noting that the virtual reality training element includes procedures for securing the correct operation of a person with a disability (under activity A4).

4.2. Analysis of the Processing Times of Training Activities

In the first stage of the study, the execution time of contracted operations was measured using a virtual training tool. This measurement was to determine the following:

1. The time intensity of the execution of individual operations;
2. Identification of activities with high variation in execution times, which may indicate a strong dependence of the execution of the operation on the predisposition of the trainee;
3. Identification of activities with low variation in execution times but a high parameter of average execution time—identification of activities that are difficult in virtual reality, regardless of the predisposition of trainees.

To carry out the research task defined in this way, the obtained time measurements were subjected to quantitative analysis based on three parameters: arithmetic mean, standard deviation, and coefficient of variation. Table 2 shows the obtained results arranged in descending order according to the coefficient of variation.

Table 2. Operations performed within a training iteration.

Symbol	Activity	Average (min)	St. Dev. (min)	Variation Coef. (-)
A4	Unfolding the platform and wings of the elevator and securing the carriage from rolling away from the side of the carriage	03:22	03:56	1.17
A9	Closing the elevator and exiting the train	03:35	03:25	0.95
A7	Lowering the elevator hydraulically using two valves	02:16	01:50	0.81
A2	Opening a tray and retrieving the square key	01:47	01:23	0.77
A5	Lowering the elevator to the station platform level using the remote control	01:19	01:00	0.76
A3	Opening the lid and lever, storing the key in the tray, and opening the entire elevator	03:15	02:27	0.75
A6	Raising the elevator by remote control	01:14	00:54	0.73
A8	Elevator raising with manual lever	02:24	01:29	0.62
A1	Taking a board with the description of the procedure and carrying it under the train	00:33	00:13	0.41

The analysis of the most time-consuming execution of the indicated handling operations by individual participants in the experiment was aimed at identifying which operations cause the most difficulty in execution and whether the identified difficulties are due to the lack of skill relative to trainees, the incorrect representation of reality in the virtual environment, or the complexity of the operation itself.

Therefore, an additional analysis was carried out on the impact of a trainee's experience on the time he or she takes to perform specific tasks. For this purpose, data from a pre-training survey of participants were used, in which they specified, among other things, their experience of regularly playing computer games and virtual reality.

As can be observed in Table 3, participants with experience using VR tools obtained significantly shorter execution times per iteration of the training cycle, even if they did not play regularly on computers. In contrast, the times are significantly longer for participants using VR tools without experience. For those who additionally do not use computer games, the average time to complete a full training iteration increased by 7 min.

Table 3. Comparison of times for completing a full training iteration by people with different experiences using computer games and VR tools.

Comparing	Statistics of VR Tool Use		
	Average (min)	St. Dev. (min)	Variation Coef. (-)
VR and gaming experience	11:42	02:11	0.19
VR but no gaming experience	11:46	04:58	0.42
No VR but gaming experience	17:11	07:03	0.41
No VR and no gaming experience	18:47	05:38	0.30

The most time-consuming operations were found to be A2 (03:35), A1 (03:22), and A6 (03:15). Two of them (A1 and A2) also have the highest variance in their performance by trainees. In the case of activity A1, the value of the coefficient of variation was higher than 1, which means that in the group of volunteers, there were people who performed this activity in a relatively short period of time. At the same time, some people could not cope with the activity. This could mean that the activity was mapped incorrectly in the virtual environment and requires a certain correction.

Activities that required in-depth qualitative analyses were singled out based on the analysis of the variability index's value. These activities included the following:

- Unfolding the platform and elevator wings;
- Closing the elevator and exiting the train;
- Lowering the elevator hydraulically using two valves;
- Opening the lid and lever, stowing the key in the tray, and opening the entire elevator.

Two research methods were used for the qualitative analysis: (1) accompanying observation and (2) unstructured face-to-face interviews with training participants. Based on the collected results from both research paths, inferences were prepared regarding the possibilities for improving the developed training tool. Noteworthy among them are the following recommendations:

1. Introducing a uniform way of locating the square-type key for closed/open positions: e.g., a key in the vertical position means that the lock in question is open, and the horizontal position means it is closed;
2. Verifying the correct direction of turning the square-type key to open/close locks;
3. Entering and exiting the carriage smoothly and not jumping with the entry button;
4. Opening/closing of doors should be implemented, considering the operation of buttons;
5. The functionality of the door locking lock should be reproduced;
6. The elevator's location and immediate surroundings should correspond to the location of the actual vehicle;
7. The buttons for opening/closing the doors and the handle for opening the elevator underneath them should be placed a little higher;
8. The visible edge of the elevator cover should be added so that the edge hints at where to grab it;
9. On the front of the elevator platform, there is a safety device that automatically lowers upon contact with the platform; its functionality should be recreated;
10. The step under the door should extend and retract, adhering to reality;
11. The ergonomics of unscrewing the valves for manually lowering the elevator needs improvement.

Points (1), (2), (6), and (11) are seemingly minor recommendations. However, observations made during training and an exam on a real device indicated that their current representation in virtual reality could perpetuate behaviors that hinder work on a real object. An example is recommendations (1) and (2), which refer to the habit of turning the key. It is also worth noting that recommendation (11) was created in response to the surprise among trainees caused by the functionality of the real valves compared to the virtual model.

4.3. Fitting a Training Strategy

In addition to a virtual tool to support the didactic process, developing an appropriate training strategy is necessary to allow participants to acquire the required competencies actively. For this reason, four possible training strategies were analyzed:

1. Strategy 1 (ST1)—individual guidance of one trainee by one instructor—the instructor explains step by step what to do;
2. Strategy 2 (ST2)—the trainee uses a storyboard describing the steps after a prior observation of the instructor's demonstration—instructor support is limited;
3. Strategy 3 (ST3)—instructor adaptively supports on request—people observed once by the instructor and once by another participant;
4. Strategy 4 (ST4)—instructor adaptively supports on request—individuals observed the instructor once and another participant twice.

However, it should be noted that in each of the strategies highlighted above, the instructor discussed all the activities to be performed in the task realized in virtual reality (the group had a preview on a large screen) Once. For each strategy, an analysis was carried out on the execution times of all training activities in virtual reality. For this purpose, all volunteers were divided into four groups, and each group underwent training according to a different strategy.

It can be observed that there is a variation in the average times of completing activities using the virtual environment, as well as in the values of the standard deviation. Therefore, the following hypotheses can be put forward:

- The execution time of the first iteration of the tasks is longer if the trainee is to self-learn from the available materials and after only one observation of the activity;
- The time to complete the first iteration of the virtual exercise decreases with the number of observed iterations, and after two iterations of the observation of other trainees and one instructor, it reaches the level of individual guidance of one trainee by one instructor;
- The most time-disadvantageous case is to have trainees perform the exercise on their own after a single observation by the instructor (only with the support of training materials and limited assistance from the instructor);
- The variability in completion times as expressed by the standard deviation decreases most significantly after a single observation of other trainees; therefore, the prediction of exercise completion times becomes more precise at a more advanced stage of training.

The above hypotheses were confirmed statistically. Confirmation was obtained by constructing appropriate statistical parametric tests for the equality of the means of general populations and the equality of the standard deviations of general populations. Confirmatory results are shown in Table 4.

Table 4. Comparison of training strategies.

Statistical Parameters	ST1	ST2		ST3	ST4
Group	1	2		3	4
Sample size	12	13		12	11
Mean (mm: ss)	14:05	19:26		15:44	13:50
Std. deviation (mm: ss)	08:57	08:16		01:55	01:39
Hypothesis H0	$\mu_1 = \mu_2$	$\mu_2 = \mu_3$		$\mu_3 = \mu_4$	$\mu_1 = \mu_4$
t-statistics	1.49	1.45		2.42	0.09
Reject of H0?	yes	yes		yes	no
Critical t-quantile value			1.32		
Hypothesis H0	$\sigma_1 = \sigma_2$	$\sigma_2 = \sigma_3$		$\sigma_3 = \sigma_4$	$\sigma_1 = \sigma_4$
F-statistics	1.18	18.45		1.33	28.93
Critical F-quantile value			2.22		
Reject of H0?	no	yes		no	yes

Using observations made during the experiments, it was found that with successive iterations of the exercise, which other participants observe, a particular awareness of the activities to be performed develops. Therefore, increasing the group supervised by one instructor is possible with successive iterations. The maximal number of trainees supervised by one instructor is limited to three and optimistically to four. This is because it is necessary to keep in mind the required availability of an instructor for a given trainee in terms of content and the safety of operating the virtual environment. The balanced scenario indicated in Table 5 shows the feasibility of an elevator exercise for a person with disabilities by one instructor for 10 trainees within a (net training) time period of about 73 min (average times were increased by the value of the standard deviation).

Table 5. Proposal of a virtual reality training structure.

Stage	Instructor 1	Active Participants	Passive Participants	Estimated Net Time = Mean + Std. Deviation (min)
1	Demonstrates the operation of equipment and performs exercise activities in virtual reality	none	They observe, together on a large screen, the instructor's virtual activity and the instructor's physical activity	5
2	The instructor guides one trainee by explaining the tasks to him step by step	1—conducted individually by the instructor	They observe, together on a large screen, the virtual activity of the trainee, ask questions, and listen to concerns	23
3	Instructor adaptively supports participants in virtual reality, prompts upon request	2—conducted adaptively	They observe, divided into two subgroups, the virtual activity of one of the two trainees on dedicated screens, asking questions and listening to concerns	17
4	Instructor adaptively supports participants in virtual reality, prompts upon request	3–4—conducted adaptively	They observe, with divisions into three/four subgroups, each virtual activity of one of the trainees on dedicated screens, ask questions, and listen to concerns	14
5	Instructor adaptively supports participants in virtual reality, prompts upon request	4—conducted adaptively	They observe, with a division of four subgroups, each virtual activity of one of the trainees on dedicated screens, ask questions, and listen to concerns	14
			Total training time for ten participants (min)	73

5. Discussion

The results presented in the article were to verify the effectiveness of using a training program for conductors using VR technology and to answer four research questions formulated by project team members and presented in Section 3, "Methodology". Figure 6 shows the relationship between the results of the completed research tasks and the answers to the questions posed.

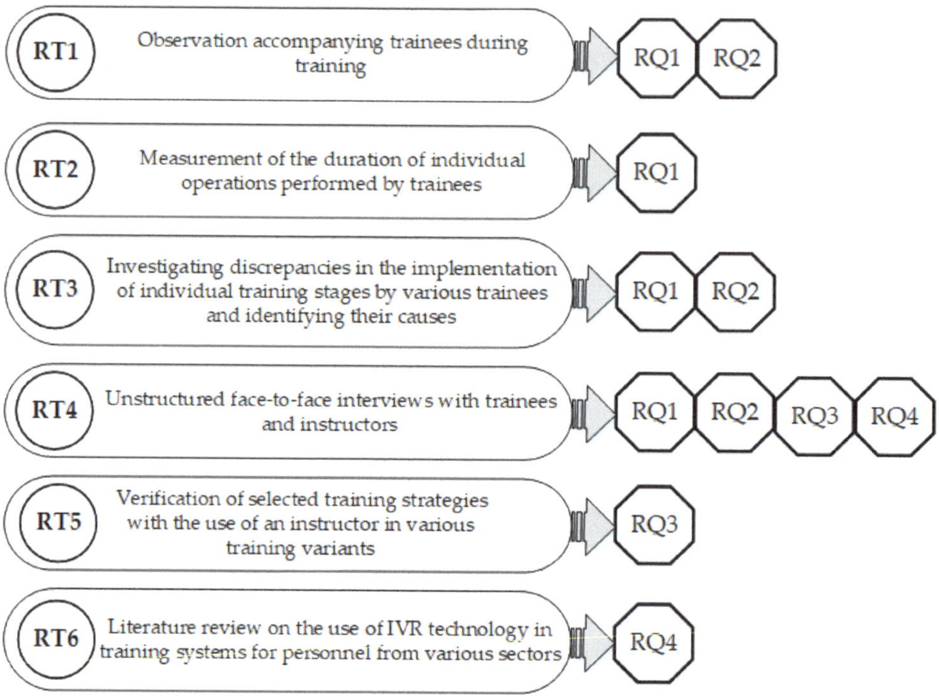

Figure 6. Assignment of the results obtained from various tasks, answering the formulated research questions.

The measurements of the execution times of individual training operations made it possible to identify the tasks in the training program that were the most time-consuming or characterized by highly diversified execution times. A significant time commitment could result from three premises: (1) a high level of precision or complexity in performing a given task, (2) an incorrect representation of reality in a virtual environment, and (3) training participants' predispositions. For this reason, it was necessary to conduct additional interviews with trainees and instructors. Thanks to these interviews, it was possible to identify the reasons for the recorded times of completing individual tasks. On the one hand, these interviews confirmed that the trainee's personal characteristics and skills may affect the trainee's level of immersion and the times of the operations performed. This result is consistent with the results reported by Janssen et al. [21]. On the other hand, it was possible to identify elements of the virtual world for which their mapping required improvement. The potential for further improvement of the developed training tool was also formulated.

An important element of the conducted research was the assessment of the effectiveness of the developed scenario and training tool. The experiment used two methods of evaluating training effectiveness, as in most studies described in the publications analyzed by Strojny and Dużmańska-Misiarczyk [36]. In our case, these were accompanying observations conducted by two experts, and the verification of skills was carried out based on a practical test conducted on a real object. In this way, it was possible to assess the level of participants' skills and identify the current limitations of the prepared VR training tool.

Thanks to this, it was possible to introduce improvements to increase trainees' immersion level and the effectiveness of training scenarios.

Based on the interviews and accompanying observations, the research team members determined the benefits of using the VR tool in the developed training scenarios. Most of them were consistent with the benefits described by Naranjo [35], namely the following:

- Increased activity of participants during the training and increased involvement in the implementation of the tasks assigned to them;
- Greater emotions accompanying the implementation of tasks, which had a positive impact on remembering the scope of performed operations;
- Ability to adapt scenarios to the needs of training participants.

At the same time, developing scenarios in a virtual environment allowed the development of many training scenarios and offered a comfortable form of training. As part of the experiment, the trainees performed tasks under the supervision of an instructor in a training room. However, the developed tool can also be used individually in a safe and comfortable space created by the user. This aligns with the positive aspects of VR highlighted by Xie et al. [4]. What primarily speaks in favor of the use of virtual reality in the case of the analyzed scenarios of elevator operations for people with disabilities is the lack of the need to shut down authentic wagons for training purposes or the implementation of training during breaks taken when disposing of these wagons for. Training participants also praised the possibility of making mistakes, based upon which they learned the correct behavior. In their statements, they emphasized the comfort of not having health risks and the pressure related to the costs of damage caused by wrong decisions. Researchers also identified boundary conditions for the training of train crew members by the Polish railway carriers. The average group of trainees is about ten to sixteen people. Thus, strategies for training were also analyzed. As a result, researchers found that one teacher can effectively use four VR goggles to teach fourteen students about lift operations. The time effort is half the time spent on a real object.

The main limitation of the conducted research is the age of the training participants. All volunteers were no more than 30 years old. The experiment confirms the effectiveness of VR technology training among young people with no or minimal professional experience. However, most have experience in games based on virtual reality. Nevertheless, the developed training tool should be used for people of different ages. Meanwhile, using VR tools can be a problem for senior staff. Therefore, the research should be supplemented with an experiment in which people aged 35–50 and 50–60 would also participate. Thanks to this, it will be possible to assess the effectiveness of the developed tool for various groups of railway personnel.

6. Conclusions

The research described in the article presents the results of verifying a training tool using IVR technology, which was developed as part of the project "Immersive training system for railway personnel using virtual reality technology". The research confirmed the higher efficiency of using VR tools in conductor training than traditional training forms. Training participants showed more significant involvement in the performance of training tasks and remembered the sequence of operations that had to be performed faster. Participants primarily emphasized the comfort of training in virtual reality, which was possible thanks to providing an educational function for mistakes made by participants. Critical results also concerned recommendations regarding changes to be made to both the scenario and the training tool. The introduced modifications and the developed training strategy will increase the effectiveness of the training conducted using the developed solution.

The results obtained in the research are of scientific importance and constitute the answer to the formulated research questions. The obtained results confirmed the benefits of using VR tools described by other researchers. It is worth noting that the innovation, in this case, was the development of IVR scenarios dedicated to the training of conductors. So far, such a solution has not been described in the literature. At the same time, various training strategies were verified, and on this basis, the most effective solution was selected. The practical contribution of the presented results is also important. Based on the conducted research, it was possible to develop recommendations regarding the changes that should be introduced in the developed training tool to better adapt to the real conditions of service operations. At the same time, the effectiveness of using this form of training concerning improving a selected group of employees was confirmed. Operators of passenger rail transport can, therefore, use the developed scenario and training tool.

As noted in Section 5, a limitation of the experiment conducted was the age of the volunteers who participated in the training. They were young people and were most familiar with the virtual reality environment. However, it is necessary to conduct a similar experiment on a group of people over 35. In this way, it will be possible to identify new potential for improving the developed tool and indicate recommended modifications. The results obtained in this age group will also allow for assessing the effectiveness of VR training in various groups of railway employees.

Author Contributions: Conceptualization, A.A.T., F.R. and A.J.-P.; methodology, A.A.T., F.R. and A.J.-P.; validation, A.A.T., F.R. and A.J.-P.; formal analysis, F.R.; investigation, A.A.T., F.R. and A.J.-P.; resources, A.A.T., F.R. and A.J.-P.; data curation, A.A.T., F.R. and A.J.-P.; writing—original draft preparation, A.A.T., F.R. and A.J.-P.; writing—review and editing, A.A.T.; visualization, A.A.T. and F.R. All authors have read and agreed to the published version of the manuscript.

Funding: This research was funded by the National Centre for Research and Development, grant number POIR.04.01.04-00-0141/19: "Immersive training system for railway personnel using virtual reality technology".

Institutional Review Board Statement: The study was conducted in accordance with the Declaration of Helsinki.

Informed Consent Statement: Informed consent was obtained from all subjects involved in the study. Volunteers were informed about the use of research results in the prepared publications.

Data Availability Statement: Not applicable.

Conflicts of Interest: The authors declare no conflict of interest.

References

1. Available online: https://Bulldogjob.Pl/for-Employers/Blog/Zmora-Rekrutacji-Naszych-Czasow-Rotacja-Pracownikow-Jak-Ja-Pokonac (accessed on 15 June 2023).
2. Szaciłło, L.; Jacyna, M.; Szczepański, E.; Izdebski, M. Risk Assessment for Rail Freight Transport Operations. *Eksploat. I Niezawodn.–Maint. Reliab.* **2021**, *23*, 476–488. [CrossRef]
3. Tubis, A.; Haładyn, S.; Kierzkowski, A.; Kisiel, T.; Restel, F.; Wolniewicz, Ł. Analysis of Human Errors in the Cargo Logistics Process in the Airport Zone. In Proceedings of the 16th International Conference on Probabilistic Safety Assessment and Management, PSAM 2022, Honolulu, HI, USA, 26 June–1 July 2022.
4. Xie, B.; Liu, H.; Alghofaili, R.; Zhang, Y.; Jiang, Y.; Lobo, F.D.; Li, C.; Li, W.; Huang, H.; Akdere, M.; et al. A Review on Virtual Reality Skill Training Applications. *Front. Virtual Real.* **2021**, *2*, 645153. [CrossRef]
5. Guo, K.; Zhang, L. Simulation-Based Passenger Evacuation Optimization in Metro Stations Considering Multi-Objectives. *Autom. Constr.* **2022**, *133*, 104010. [CrossRef]
6. Wang, Z.; Mao, Z.; Li, Y.; Yu, L.; Zou, L. VR-Based Fire Evacuation in Underground Rail Station Considering Staff's Behaviors: Model, System Development and Experiment. *Virtual Real.* **2023**, *27*, 1145–1155. [CrossRef]
7. Yu, S. Application of VR Intelligent Technology in Fire Evacuation Simulation of Urban Rail Transit Station. *Procedia Comput. Sci.* **2022**, *208*, 223–230. [CrossRef]
8. Vanichchanunt, P.; Tanmalaporn, T.; Suthamvijit, C.; Noisri, S.; Wuttisittikulkij, L.; Pongyart, W.; Paripurana, S. Virtual Reality for Railway Signaling System Training. In Proceedings of the 2023 20th International Conference on Electrical Engineering/Electronics, Computer, Telecommunications and Information Technology (ECTI-CON), Nakhon Phanom, Thailand, 9–12 May 2023; pp. 1–4.
9. Randeniya, N.; Ranjha, S.; Kulkarni, A.; Lu, G. Virtual Reality Based Maintenance Training Effectiveness Measures—A Novel Approach for Rail Industry. In Proceedings of the 2019 IEEE 28th International Symposium on Industrial Electronics (ISIE), Vancouver, BC, Canada, 12–14 June 2019; pp. 1605–1610.

10. Shen, Z.; Zou, J.; Chen, W.; An, X. VR Panoramic Video Technology and Its Application in Rail Transit Personnel Training in the 5G Era. In Proceedings of the 2020 5th International Conference on Mechanical, Control and Computer Engineering (ICMCCE), Harbin, China, 25–27 December 2020; pp. 2369–2373.
11. Sun, R.; Wu, Y.J.; Cai, Q. The Effect of a Virtual Reality Learning Environment on Learners' Spatial Ability. *Virtual Real.* **2019**, *23*, 385–398. [CrossRef]
12. Jensen, L.; Konradsen, F. A Review of the Use of Virtual Reality Head-Mounted Displays in Education and Training. *Educ. Inf. Technol.* **2018**, *23*, 1515–1529. [CrossRef]
13. Ryan, M.-L. *Narrative as Virtual Reality 2: Revisiting Immersion and Interactivity in Literature and Electronic Media*; JHU Press: Baltimore, MD, USA, 2015.
14. Walsh, K.R.; Pawlowski, S.D. Virtual Reality: A Technology in Need of IS Research. *Commun. Assoc. Inf. Syst.* **2002**, *8*, 20. [CrossRef]
15. Steuer, J. Defining Virtual Reality: Dimensions Determining Telepresence. *J. Commun.* **1992**, *42*, 73–93. [CrossRef]
16. Mandal, S. Brief Introduction of Virtual Reality & Its Challenges. *Int. J. Sci. Eng. Res.* **2013**, *4*, 304–309.
17. Lee, E.A.-L.; Wong, K.W. A Review of Using Virtual Reality for Learning. In *Transactions on Edutainment I. Lecture Notes in Computer Science*; Pan, Z., Cheok, A.D., Müller, W., El Rhalibi, A., Eds.; Springer: Berlin/Heidelberg, Germany, 2008; Volume 5080, pp. 231–241.
18. Mao, R.Q.; Lan, L.; Kay, J.; Lohre, R.; Ayeni, O.R.; Goel, D.P.; de Sa, D. Immersive Virtual Reality for Surgical Training: A Systematic Review. *J. Surg. Res.* **2021**, *268*, 40–58. [CrossRef] [PubMed]
19. Moorthy, K.; Munz, Y.; Adams, S.; Pandey, V.; Darzi, A. A Human Factors Analysis of Technical and Team Skills Among Surgical Trainees During Procedural Simulations in a Simulated Operating Theatre. *Ann. Surg.* **2005**, *242*, 631–639. [CrossRef] [PubMed]
20. Radianti, J.; Majchrzak, T.A.; Fromm, J.; Wohlgenannt, I. A Systematic Review of Immersive Virtual Reality Applications for Higher Education: Design Elements, Lessons Learned, and Research Agenda. *Comput. Educ.* **2020**, *147*, 103778. [CrossRef]
21. Janßen, D.; Tummel, C.; Richert, A.; Isenhardt, I. Towards Measuring User Experience, Activation and Task Performance in Immersive Virtual Learning Environments for Students. In *Immersive Learning Research Network*; Springer: Berlin/Heidelberg, Germany, 2016; pp. 45–58.
22. Adams, E. *The Designer's Notebook: Postmodernism and the 3 Types of Immersion*; Game Developer: London, UK, 2004.
23. Bjork, S.; Holopainen, J. *Patterns in Game Design*; Charles River Media: Needham, MA, USA, 2005.
24. Rogers, S. Virtual Reality: The Learning Aid of the 21st Century. *Forbes* **2019**. Available online: https://www.forbes.com/sites/solrogers/2019/03/15/virtual-reality-the-learning-aid-of-the-21st-century/?sh=16e98e8a139b (accessed on 1 June 2023).
25. Krokos, E.; Plaisant, C.; Varshney, A. Virtual Memory Palaces: Immersion Aids Recall. *Virtual Real.* **2019**, *23*, 1–15. [CrossRef]
26. Radhakrishnan, U.; Koumaditis, K.; Chinello, F. A Systematic Review of Immersive Virtual Reality for Industrial Skills Training. *Behav. Inf. Technol.* **2021**, *40*, 1310–1339. [CrossRef]
27. Tabbron, G.; Yang, J. The Interaction between Technical and Vocational Education and Training (TVET) and Economic Development in Advanced Countries. *Int. J. Educ. Dev.* **1997**, *17*, 323–334. [CrossRef]
28. Falck, A.-C.; Örtengren, R.; Högberg, D. The Impact of Poor Assembly Ergonomics on Product Quality: A Cost-Benefit Analysis in Car Manufacturing. *Hum. Factors Ergon. Manuf. Serv. Ind.* **2010**, *20*, 24–41. [CrossRef]
29. Bowyer, M.W.; Streete, K.A.; Muniz, G.M.; Liu, A.V. Immersive Virtual Environments for Medical Training. *Semin Colon Rectal Surg.* **2008**, *19*, 90–97. [CrossRef]
30. Backlund, P.; Engstrom, H.; Hammar, C.; Johannesson, M.; Lebram, M. Sidh—A Game Based Firefighter Training Simulation. In Proceedings of the 2007 11th International Conference Information Visualization (IV '07), Zurich, Switzerland, 4–6 July 2007; pp. 899–907.
31. Li, C.; Liang, W.; Quigley, C.; Zhao, Y.; Yu, L.-F. Earthquake Safety Training through Virtual Drills. *IEEE Trans. Vis. Comput. Graph.* **2017**, *23*, 1275–1284. [CrossRef]
32. Howard, M.C.; Gutworth, M.B.; Jacobs, R.R. A Meta-Analysis of Virtual Reality Training Programs. *Comput. Hum. Behav.* **2021**, *121*, 106808. [CrossRef]
33. Rupasinghe, T.D.; Kurz, M.E.; Washburn, C.; Gramopadhye, A.K. Virtual Reality Training Integrated Curriculum: An Aircraft Maintenance Technology (AMT) Education Perspective. *Int. J. Eng. Educ.* **2011**, *27*, 778–788.
34. Moreno, P.; Mayer, R.E. Virtual Reality and Learning: Cognitive and Motivational Effects of Students' Sense of Presence. *Proc. Hum.-Comput. Interact.* **2001**, 6567. Available online: https://citeseerx.ist.psu.edu/document?repid=rep1&type=pdf&doi=ac051f0f9452bc4730be682f4629d92f7f6b071d (accessed on 1 June 2023).
35. Naranjo, J.E.; Sanchez, D.G.; Robalino-Lopez, A.; Robalino-Lopez, P.; Alarcon-Ortiz, A.; Garcia, M.V. A Scoping Review on Virtual Reality-Based Industrial Training. *Appl. Sci.* **2020**, *10*, 8224. [CrossRef]
36. Strojny, P.; Dużmańska-Misiarczyk, N. Measuring the Effectiveness of Virtual Training: A Systematic Review. *Comput. Educ. X Real.* **2023**, *2*, 100006. [CrossRef]

applied sciences

Article

Receding Horizon Optimization for Cooperation of Connected Vehicles at Signal-Free Intersections under Mixed-Automated Traffic

Jian Gong [1], Weijie Chen [2,*] and Ziyi Zhou [3]

1 Intelligent Transportation System Research Center, Southeast University, Nanjing 210096, China; gongjian_reus@163.com
2 School of Civil and Transportation Engineering, Ningbo University of Technology, Ningbo 315211, China
3 School of Transportation, Southeast University, Nanjing 210096, China; zhouziyi1234@126.com
* Correspondence: wchechen@163.com

Abstract: This paper proposes a distributed coordination scheme for connected vehicles, including automated vehicles (AVs) and manual vehicles (MVs), at signal-free intersections. The cooperation issue of vehicles at an intersection is formulated into a multi-objective optimization problem that aims to eliminate conflicts and improve traffic mobility and fuel economy. For this purpose, the future trajectories of AVs and MVs are predicted by the respective car-following models, and are shared with neighboring vehicles in conflict relationships. The proposed scheme optimizes the sum of the performance of AVs within the cooperative zone in a prediction horizon. A distributed optimization algorithm in the receding horizon is presented to obtain the local optimal solutions, and is tested in simulations with different demand levels and penetration rates of AVs. The results show that the proposed scheme reduces travel time by 29.7–45.5% and 34.5–49.2%, and decreases fuel consumption by 27.6–35.3% and 21.6–29.9% under 70–100% penetration rates of AVs, compared to the no-control operation and fixed-time signal control strategy. In addition, a comparison simulation with the strategy of jointly optimizing the vehicle trajectory and signal timing is conducted to evaluate the relative merits of the proposed scheme.

Keywords: connected vehicles; automated vehicles; manual vehicles; signal-free intersections; distributed cooperation; receding horizon optimization

Citation: Gong, J.; Chen, W.; Zhou, Z. Receding Horizon Optimization for Cooperation of Connected Vehicles at Signal-Free Intersections under Mixed-Automated Traffic. *Appl. Sci.* **2023**, *13*, 11576. https://doi.org/10.3390/app132011576

Academic Editor: Juan-Carlos Cano

Received: 17 September 2023
Revised: 4 October 2023
Accepted: 20 October 2023
Published: 23 October 2023

1. Introduction

1.1. Background

Social progress and improvements in people's quality of life profit from the rapid development of urban transportation. Unfortunately, a sharp increase in vehicle ownership poses great challenges in traffic safety, mobility, and environmental sustainability [1]. Over the decades, the development of intelligent transportation systems (ITSs) has become a significant force in dealing with these challenges. In particular, emerging connected vehicles (CVs) technologies have aroused widespread attention due to their enormous potential in improving road capacity, driving safety, and fuel economy. Employing vehicle-to-everything (V2X) communication, CVs are able to cooperate in various traffic scenarios [2].

As bottlenecks of urban traffic, intersections have always been the focus of researchers and traffic engineers for improving traffic safety and smoothing traffic flow. Intersections can be categorized into two types, i.e., signalized and signal-free. Researchers have proposed various approaches to improve traffic efficiency at signalized intersections, e.g., fixed-time control, actuated control, adaptive control, and connected-vehicle-based methods [3,4]. These methods can improve traffic throughput at intersections to some extent, but unnecessary stopping and idling are still inevitable. For signal-free intersections, there are no signal lights or any other display devices to prompt drivers. Traditionally, drivers must observe and estimate the adequate vehicle clearance through visual judgment for safe crossing. Currently, a real-time and reliable CVs environment can be constructed using the V2X communication technology, which provides faster response and shorter vehicular gaps for AVs to pass through intersections. Significant efforts have been made to coordinate AVs at signal-free intersections under centralized or decentralized frameworks. The centralized frameworks globally decide the movements of all vehicles using a central controller, while under the decentralized frameworks, each vehicle is treated as a single agent and collaborates independently [5]. Two kinds of strategies have generally been presented for signal-free intersection coordination under the centralized or decentralized frameworks, i.e., heuristic strategies and trajectory planning strategies.

1.2. Literature Review

Heuristic strategies aim at adjusting the movements of vehicles to cross an intersection without collisions; they generally adopt the rule of first come, first served (FCFS). Typical heuristic strategies that have been proposed in the literature include resource reservation [6,7], fuzzy logic [8,9], and virtual platooning [10,11]. For instance, Dresner et al. [6] proposed a centralized resource reservation method to coordinate vehicles at a signal-free intersection with a single straight lane in each direction. This work was further extended in [7], which considered turning and improved the performance of the central controller. The central controller frequently interacted with vehicles and scheduled the time slots and space tiles, which suffered heavy communication burdens and the risk of deadlocks. Note that the existing heuristic strategies only regulate vehicles to pass through intersections without conflicts, but are incapable of further optimizing the trajectories of vehicles. In addition to collision avoidance, the trajectory planning strategies drive vehicles to a better traffic performance. For example, Lee et al. [12] presented a trajectory optimization algorithm to minimize the length of potential conflicting overlapping trajectories. Considering vehicle jitter, acceleration, and expected velocity, Dai et al. [13] formulated a multi-objective trajectory optimization problem. By using a Monte Carlo tree search, Xu et al. [14] planned the trajectories of vehicles through the global-optimal passing orders. Planning-based strategies have been shown to perform better than heuristic strategies under high traffic demand [15]. However, the trajectory planning strategies have a weakness in real-time practical implementation since the computation of the formulated problems suffers the curse of dimensionality.

Several recent studies on the cooperation of vehicles at signal-free intersections introduced the concept of multi-agent, i.e., consensus, control [16,17]. The coordinated vehicles are regulated to approach a desired state of motion to avoid collisions at intersections. For instance, Mirheli et al. [18] proposed a coordinated optimization trajectory algorithm by formulating a mixed-integer non-linear program (MILP), which aims at forcing vehicles to reach consensus. This approach pushed the distributed vehicle-level solutions toward global optimality, but the formulated MILP problems suffered a heavy computational burden. Xu et al. [11] presented a virtual consensus-based platoon control approach by transforming the two-dimensional vehicle cluster at the intersection into a one-dimensional virtual platoon in a virtual lane. However, this approach was not evaluated in the receding horizon framework. For those problems of vehicle collaboration at signal-free intersections, receding horizon control (RHC), also known as model predictive control (MPC) [19], has been utilized to dynamically capture predictions on vehicle states and future trajectories to optimize the coordination systems in real time. In the vehicle–intersection coordination framework proposed in [20], a conflict risk indicator was integrated into the MPC controller. Du et al. [21] formulated the corridor-level vehicle coordination problem in the RHC framework, where vehicle collision was avoided by using a consensus algorithm to force vehicles to track the reference speed.

Note that the existing research on the cooperation of vehicles at signal-free intersections is only confined to the fully connected AV environment. Since the mixed driving environment containing AVs and MVs will stay with us for an extended period, it is essential to establish control mechanisms for cooperating with both types of vehicles. In the current study, traffic signal lights still have to be used to regulate the mixed autonomy traffic flow at intersections [22,23]. To jointly optimize the signal timing and vehicle trajectories, a complex MILP is constructed [24]. However, MILP problems are NP-hard, in which the computational burden grows exponentially with the number of loading vehicles [25]. Moreover, introducing traffic lights still inevitably leads to stops and idling. Furthermore, some of the existing collision avoidance algorithms at signal-free intersections can only deal with the problems with just a few vehicles due to computation complexity [21,26]. The presented algorithms cannot be applied in a realistic traffic environment to coordinate vehicles globally under a receding horizon framework. In addition, the centralized coordination approaches suffer heavy computational burdens and struggle to handle high-density traffic. Whereas distributed coordination frameworks have the advantages of a lower communication burden, lower computational complexity, and better information protection. Nevertheless, the existing approaches on the distributed coordination of AVs at signal-free intersections are inadequate in some respects, e.g., a lack of restrictions on control inputs, vehicle states, and safety conditions [11], without fully considering the global conflict relationships [26,27], and losing sight of the mixed-automated vehicles environment [11,18,21,26,28]. To address the problem of the cooperation of AVs and MVs at signal-free intersections, we establish a distributed collaboration framework to address the above issues.

1.3. Contribution of This Work

In this paper, we develop a distributed cooperation scheme to coordinate AVs and MVs to pass through signal-free intersections. The core idea is to cooperatively regulate each AV to cross the intersection while keeping a safe distance with the vehicles in conflict relationships. Firstly, a conflict graph is constructed to describe the conflict relationships of the different approaching lanes. Based on the conflict graph, the communication topologies of vehicles are identified in real time. Then, the future trajectories of AVs and MVs are predicted by the respective car-following models and are shared with their neighboring vehicles in conflict relationships. By utilizing the predicted trajectory information of the conflicting MVs, AVs proactively optimize their trajectories to avoid conflicts while assisting the MVs to cross the intersection smoothly. In addition, the elimination of conflicts, traffic mobility, and fuel economy are synchronously considered in the proposed scheme. The intersection coordination issue is formulated into a multi-objective optimization problem aiming to minimize the weighted performance indexes of AVs in a prediction period. Finally, a receding horizon optimization algorithm is presented and tested with a typical four-phase intersection in simulations. The contributions of this study are listed below:

(1) Different from the existing literature only considering coordinating vehicles under a fully AV environment, this paper proposes a coordination scheme under which AVs and MVs can cooperatively pass through a signal-free intersection;
(2) The proposed scheme develops a practical application which concurrently considers the restrictions of control inputs, vehicle states, safety conditions, global conflict relationships, and the mixed-automated driving environment;
(3) A distributed multi-objective optimization algorithm is presented to synchronously eliminate the traffic conflicts and improve mobility and fuel economy in a receding horizon framework.

The remainder of this paper is structured as follows. Section 2 describes the problem statement. Section 3 details the methodology to formulate and solve the cooperative control problem and gives the receding horizon optimization algorithm. Simulations are conducted in Section 4; and Section 5 concludes this paper.

2. Problem Statement

We consider a typical symmetrical dual-lane intersection with four approaches and departures, as depicted in Figure 1. Each approach consists of a straight lane and a left-turn lane. Right-turn lanes are omitted here for simplicity since they have no impact on the collaboration results of the intersection. Here, we define a cooperative zone (CZ) with the scope of the four approaching lanes under the radius R. An intersection coordination unit (ICU) deployed in the center of the intersection assists the vehicles in sharing information in the CZ.

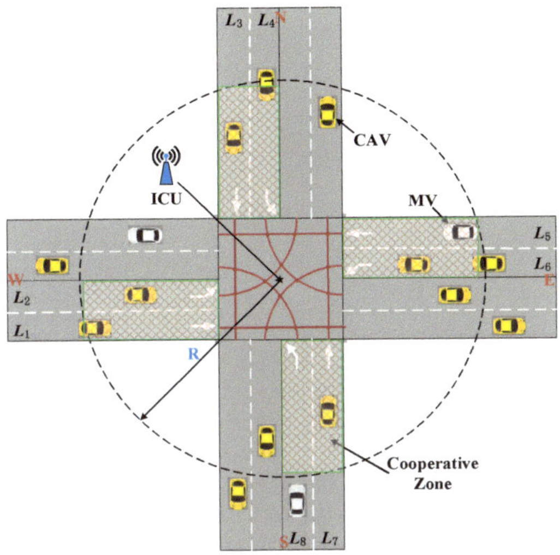

Figure 1. A scenario of connected vehicles passing through a signal-free intersection.

We assume that all vehicles are connected, and equipped with GPS or BDS, vehicle-to-infrastructure (V2I), and vehicle-to-vehicle (V2V) communication devices to exchange real-time information with the ICU and other vehicles. Each vehicle in the CZ transmits four necessary properties to the ICU, i.e., vehicle ID, vehicle type, origin-destination, and vehicle state (position and speed). (1) A vehicle ID will be automatically assigned to the vehicle when entering the CZ; (2) There are two vehicle types considered in this study: AVs and MVs. We consider high penetration rates of AVs (no less than 70%) in this paper to decrease traffic insecurity factors and alleviate drops in traffic efficiency caused by MVs; (3) The origin–destination of a vehicle means which edges the vehicle enters and departs from in the road network; (4) The vehicle state is the crucial information used to predict the future trajectory of the vehicle, which is transmitted to the correlative neighboring vehicles. At each time step, the ICU receives the information of the four properties of all vehicles in the CZ and produces a communication topology according to the conflict relationships of the vehicles. Through V2V communication, AVs exchange the information of the predicted trajectories with their neighboring vehicles based on the communication topology delivered from the ICU, and then optimize their movements. In addition, the behaviors of lane changing and overtaking are required to only operate before entering the CZ. To simplify the problem, we neglect communication delays, package losses, communication constraints, and computation delays, such that the states of all vehicles are updated simultaneously within a common global clock.

This paper aims to design a scheme for connected and mixed-automated vehicles to cooperatively pass through a signal-free intersection, as shown in Figure 1. The priority order for all vehicles passing through the intersection follows the "FCFS" rule. It is worth noting that only AVs in the CZ can be controlled. MVs and AVs beyond the CZ follow their own driving behaviors. Thus, we focus on regulating the longitudinal movements of AVs on the approaching lanes by utilizing the predicted trajectories of their neighboring vehicles. The coordination issue of connected and mixed-automated vehicles at a signal-free intersection will be addressed in the following section by optimizing the trajectories of AVs under a distributed framework.

3. Methodology

This section proposes a methodology for coordinating mixed-automated vehicles at a signal-free intersection and presents a detailed algorithm to address the cooperative driving problem.

3.1. Car-Following Models for AVs and MVs

This paper only considers longitudinal car-following models for two kinds of vehicles (AVs and MVs) on the approaching lanes. The movements of AVs follow the optimal control inputs designed by this study, and the Krauss model in [29] is adopted to describe the driving behavior of MVs.

The longitudinal movement dynamics of AVs can be modeled by the following discrete-time model in [30]:

$$\begin{cases} p_i(t+1) = p_i(t) + \delta v_i(t) + \frac{\delta^2}{2} u_i(t) \\ v_i(t+1) = v_i(t) + \delta u_i(t) \end{cases}, \quad i \in \Omega_A(t) \tag{1}$$

where δ is the discrete time interval, and $p_i(t)$ and $v_i(t)$ denote the position and speed of AV i at time step t, respectively. $u_i(t)$ is the control input to be designed at the control sample point t, and $\Omega_A(t)$ represents the time-varying set of AVs in the CZ at time step t.

The movements of MVs are not under control, and the Krauss model is introduced to estimate the longitudinal behavior of MVs. The Krauss model is a conflict-free model based on the safe distance with immediately preceding vehicles. Algorithm 1 is given to describe the updating strategy of the speed and position of MVs according to [29].

Algorithm 1 Trajectory-Updating Algorithm for MVs

Input: The speed of the immediately preceding vehicle v_p, the position and speed of ego vehicle p_m and v_m ($m \in \Omega_M(t)$), the vehicle gap with its immediately preceding vehicle g;
Output: Updated position and speed of the ego vehicle at the next step.
1: ***Repeat***
2: Acquire $v_p(t)$, $p_m(t)$, $v_m(t)$, and g.
3: $v_{safe} \leftarrow v_p(t) + \frac{g - v_m(t)\tau}{(v_p(t)+v_m(t))/2b_{max}+\tau}$
4: $\gamma_0 \leftarrow \min\left[v_m(t) + a_{max}, V_{max}, v_{safe}\right]$
5: $\gamma_1 \leftarrow \gamma_0 - \epsilon\{\gamma_0 - [v_m(t) - a_{max}]\}$
6: $v_m(t+1) \leftarrow v_{ran(\gamma_0,\gamma_1)}$
7: $p_m(t+1) \leftarrow p_m(t) + \delta v_m(t+1)$
8: ***Return*** $p_m(t+1)$ and $v_m(t+1)$;
9: $t \leftarrow t+1$
10: ***Until*** The ego vehicle departs from the CZ.

In Algorithm 1, the position and speed of the ego vehicle at the next time step are estimated according to the speed of its immediately preceding vehicle, the position and speed of the ego vehicle, and the vehicle gap between the two vehicles at the current time step. $\Omega_M(t)$ represents the dynamic set of MVs in the CZ at time step $t\delta$. v_{safe} denotes the maximum safe speed of the ego vehicle without collision with its immediately preceding vehicle at the next time step, which is computed by step 3 in Algorithm 1. τ is the reaction time of drivers, and ϵ is a constant. a_{max} and b_{max} are the maximum acceleration and braking deceleration, respectively. Steps 4–7 in Algorithm 1 show the updating rules of the Kruass model, where γ_0 denotes the desired speed, and γ_1 represents the speed with the maximum difference from the desired speed due to imperfect driving behaviors of drivers. $v_{ran(\gamma_0,\gamma_1)}$ denotes a random value for the speed picked from γ_0 and γ_1. Note that although this randomly selected updating strategy leads to different driving behaviors, the strong interactions between successive vehicles will decrease the freedom of individual drivers to choose their velocity.

3.2. Traffic Conflict Graph and Communication Topology

The complexity of the conflict types at intersections determines the difficulty level of traffic control. Typical intersections have 16 conflict points, as shown in Figure 1, which implies that vehicles from different approaching lanes may crash at these points. In addition, each vehicle is under a conflict relationship with its immediately preceding or following vehicle on the same lane, i.e., rear-end collision. We construct a conflict graph to describe the conflict relationships of the approaching lanes, as shown in Figure 2. Define the set of approaching lanes as $\mathscr{L}$, where $\mathscr{L} = \{L_1, L_2, L_3, L_4, L_5, L_6, L_7, L_8\}$, and the conflict set $\mathscr{C}_h, h \in \mathscr{L}$ denotes the set of lanes that has conflict relationships with lane h. Take the westbound straight lane L_1 as an example, $\mathscr{C}_{L_1} = \{L_1, L_3, L_6, L_7, L_8\}$, which means that the trajectories of vehicles on lane L_1 may conflict with those on lanes L_1, L_3, L_6, L_7, and L_8 at the intersection. Conversely, vehicles on lane L_1 have no risk of collision with other vehicles on the lanes outside the set $\mathscr{C}_{L_1}$.

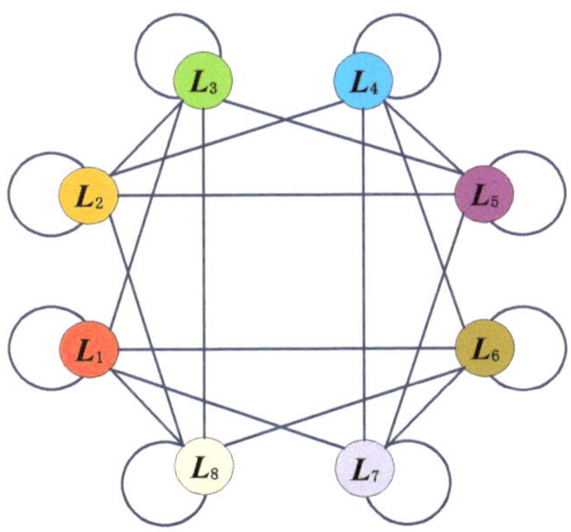

Figure 2. Traffic conflict graph for the intersection in Figure 1.

For eliminating collisions at intersections, a reasonable idea is to allow vehicles on the conflicting lanes to cross the intersection asynchronously while keeping a safe distance between each other. From the point of view of a single vehicle, no collision will occur if it retains a virtual safe car-following spacing with all potentially conflicting vehicles. We call these vehicles on the lanes in the conflict set the neighboring vehicles of the ego vehicle. The ego vehicle must exchange information with its neighbor vehicles in real time through V2V communication to maintain a virtual safe car-following spacing. Considering all vehicles in the CZ, the communication topology of vehicles can be modeled by a time-varying graph $\mathcal{G}_t = \{\mathcal{S}_t, \mathcal{E}_t\}$, where $\mathcal{S}_t = \{1, 2, \ldots, n_t\}$ is the set of nodes (vehicles), and $\mathcal{E}_t \subseteq \mathcal{S}_t \times \mathcal{S}_t$ is the set of edges (information flowing directions). The characteristic of graph $\mathcal{G}_t$ can be modeled by an adjacent matrix, which is defined as $\mathcal{A}_t = [a_{ij}(t)] \in \mathcal{R}^{n_t \times n_t}$ with

$$\begin{cases} a_{ij}(t) = 1, \text{if } \{j, i\} \in \mathcal{E}_t \\ a_{ij}(t) = 0, \text{if } \{j, i\} \notin \mathcal{E}_t \end{cases} \quad i, j \in \mathcal{S}_t \tag{2}$$

where $\mathcal{R}^{n_t \times n_t}$ represents the set of $n_t \times n_t$ real matrices, $\{j, i\} \in \mathcal{E}_t$ denotes that vehicle i can receive the information of vehicle j. Define the neighboring set of node i as $\mathcal{N}_i(t)$, where $\mathcal{N}_i(t) = \{j | a_{ij}(t) = 1, j \in \mathcal{S}_t\}$. In other words, node j is called a neighbor of node i if and only if $a_{ij}(t) = 1, j \in \mathcal{S}_t$. Correspondingly, we define a dual set $\mathcal{D}_i(t) = \{j | a_{ji}(t) = 1, j \in \mathcal{S}_t\}$, which represents that any node $j \in \mathcal{S}_t$ can receive the information of node i. Obviously, we have $\mathcal{N}_i(t) = \mathcal{D}_i$(t) under an undirected graph.

3.3. Distributed Cooperative Control Model for AVs

Considering the benefits of distributed frameworks, we will construct a distributed multiple-objective optimization model to address the cooperation issue of AVs and MVs at an intersection. In what follows, we will detail the objectives, necessary constraints, and problem formulation of this paper.

(1) ***Objectives***: We aim at regulating vehicles to cross the intersection without collisions by controlling all AVs in the CZ, while also improving traffic mobility and fuel economy. Thus, the objectives of this paper cover three parts, i.e., conflict elimination, traffic mobility, and fuel economy, which are detailed as follows.

First, the primary goal is to eliminate conflicts within the intersection area. For this purpose, we define the indexes of the virtual car-following tracking errors as

$$\mathscr{J}_{i1} = \sum_{k=0}^{N_p-1} \sum_{j\in\mathcal{N}_i(t), f\in\mathscr{C}_h} ||\bar{p}_{ih}(k+1|t) - \bar{p}_{jf}(k+1|t) + \mathscr{D}_{ij}||^2 \tag{3}$$

$$\mathscr{J}_{i2} = \sum_{k=0}^{N_p-1} \sum_{j\in\mathcal{N}_i(t), f\in\mathscr{C}_h} ||\bar{v}_{ih}(k+1|t) - \bar{v}_{jf}(k+1|t)||^2 \tag{4}$$

where N_p is the prediction horizon, $\bar{p}_{ih}$ and $\bar{p}_{jf}$ denote the distance to the stop line of vehicle i, $i \in \Omega_A(t)$ on lane h, $h \in \mathscr{L}$, and its neighboring vehicle j, $j \in \mathcal{N}_i(t)$ on lane f, $f \in \mathscr{C}_h$, respectively. $\bar{v}_{ih}$ and $\bar{v}_{jf}$ are the speeds of vehicles i and j, respectively. $\mathscr{D}_{ij}$ is the desired virtual position error between vehicles i and j, defined as

$$\mathscr{D}_{ij} = \begin{cases} l_i + d_{ij} + c_{ij}, & |\bar{p}_{ih}| \le |\bar{p}_{jf}| \\ -l_j - d_{ij} + c_{ij}, & |\bar{p}_{ih}| > |\bar{p}_{jf}| \end{cases} \tag{5}$$

where l_i and l_j denote the lengths of vehicles i and j, respectively, d_{ij} is the desired spacing between vehicles i and j, and c_{ij} is the compensation of distance within the intersection area, which is defined as the difference in the distance between vehicles i and j from the stop lines to the conflict point. $\mathscr{J}_{i1}$ indicates the performance of vehicle i in eliminating collisions with its neighboring vehicle j, $j \in \mathcal{N}_i(t)$ when crossing the intersection. Vehicle i and its neighboring vehicles are expected to arrive at the intersection with a desired virtual spacing. A small value of the virtual spacing may cause collisions of vehicles at the intersection, while a large value will reduce the traffic throughput. $\mathscr{J}_{i2}$ indicates that vehicle i is expected to keep a similar speed with its neighboring vehicle j to enhance safety when crossing the intersection.

Second, we expect smooth traffic flows that push each vehicle to run at a desired speed. It can also improve traffic mobility by regulating each vehicle on the same lane to approach an anticipated speed. Hence, we define the index of the variance of the desired speed as follows:

$$\mathscr{J}_{i3} = \sum_{k=0}^{N_p-1} ||\bar{v}_{ih}(k+1|t) - \mathscr{V}_h(k+1|t)||^2 \tag{6}$$

where $\mathscr{V}_h(k+1|t)$ is the desired speed defined as

$$\mathscr{V}_h(k+1|t) = \frac{\sum_{s\in\mathbb{S}_h(t)} \bar{v}_{sh}(k|t) + \mathscr{V}_{max}}{\mathbb{N}_h(t)+1} \tag{7}$$

where $\bar{v}_{sh}$ denotes the speed of vehicle s on lane h, $\mathbb{S}_h(t)$ is the set of vehicles on lane h, $\mathscr{V}_{max}$ is the maximum speed, and $\mathbb{N}_h(t)$ represents the number of vehicles on lane h. In Equation (7), the desired speed is defined as the average value of the speed of all the vehicles on the same lane and the maximum speed. This setting can propel vehicles on the same lane to run evenly and also increase the mean speed of vehicles in the network.

Accordingly, a smaller value of $\mathscr{J}_{i3}$ indicates that the speed of the vehicle is closer to the desired speed, which benefits smoother traffic flows and higher throughputs.

Third, we expect to reduce the fuel consumption of each AV by optimizing the local control cost. It is commonly believed that a lower cost of control inputs results in a higher fuel economy [30]. Hence, the index of the control cost is defined as follows

$$\mathscr{J}_{i4} = \sum_{k=0}^{N_p-1} ||u_i(k|t)||^2. \tag{8}$$

Combining Equations (3), (4), (6) and (8), we define an objective function with a weighted sum of the four indexes

$$\mathscr{J}_i = \omega_1 \mathscr{J}_{i1} + \omega_2 \mathscr{J}_{i2} + \omega_3 \mathscr{J}_{i3} + \omega_4 \mathscr{J}_{i4} \tag{9}$$

where ω_1, ω_2, ω_3, and ω_4 are weight factors to penalize the virtual position errors, relative speed errors, variances of the desired speed, and control cost, respectively. These weight coefficients can be adjusted according to actual traffic demand in practice.

(2) ***Constraints***: In what follows, we consider the constraints of the mobility characteristics of vehicles and the safety condition. The vehicles have to follow the car-following models according to their type, i.e., AVs or MVs. The trajectory of ego vehicle i, $i \in \Omega_A(t)$, is subject to satisfying the following constraints:

$$\begin{cases} \bar{p}_{ih}(k+1|t) = \bar{p}_{ih}(k|t) - \delta\bar{v}_{ih}(k|t) - \frac{\delta^2}{2}u_i(k|t) \\ \bar{v}_{ih}(k+1|t) = \bar{v}_{ih}(k|t) + \delta u_i(k|t), i \in \Omega_A(t), h \in \mathscr{L} \end{cases} \tag{10}$$

$$\begin{cases} \bar{p}_{jf}(k+1|t) = \bar{p}_{jf}(k|t) - \delta\bar{v}_{jf}(k|t) \\ \bar{v}_{jf}(k+1|t) = \mathscr{R}_a\{\min\{v_{jf}(k|t) + a_{max}, \mathscr{V}_{max}, v_{pf}(k|t) + \frac{g - v_{jf}(k|t)\tau}{(v_{pf}(k|t) + v_{jf}(k|t))/2b_{max} + \tau}\}, \\ \qquad (1-\epsilon)\min\{v_{jf}(k|t) + a_{max}, \mathscr{V}_{max}, v_{pf}(k|t) + \frac{g - v_{jf}(k|t)\tau}{(v_{pf}(k|t) + v_{jf}(k|t))/2b_{max} + \tau}\} \\ \qquad + \epsilon(v_{jf}(k|t) - a_{max})\},\ j \in \mathcal{N}_i(t), f \in \mathscr{C}_h \end{cases} \tag{11}$$

In addition, the trajectory of neighboring vehicle j is subject to either of the constraints (10) and (11) depending on its vehicle type, i.e., AV or MV, respectively.

In Equation (11), $\mathscr{R}_a$ is defined as a random selection function, v_{pf} denotes the speed of the immediately preceding vehicle, and the definitions of other parameters are given in Algorithm 1.

Considering the physical limitations of vehicles, road restrictions, and comfort of passengers, the speed and acceleration of AVs should be confined to a range. Denote $\mathscr{A}_{min}$, $\mathscr{A}_{max}$, $\mathscr{V}_{min}$, and $\mathscr{V}_{max}$ as the minimum acceleration, maximum acceleration, minimum speed, and maximum speed. The following constraints should be satisfied for each AV:

$$\mathscr{A}_{min} \leq u_i(k|t) \leq \mathscr{A}_{max}. \tag{12}$$

$$\mathscr{V}_{min} \leq v_{ih}(k+1|t) \leq \mathscr{V}_{max}. \tag{13}$$

To avoid rear-end collisions of vehicles on the same lane, we add a constraint to guarantee that vehicle i stays a safe distance from its immediately preceding vehicle q. Denote d_m as the minimum gap, and l_q is the length of vehicle q. The constraint of the safety condition is defined as follows:

$$\bar{p}_{ih}(k+1|t) - \bar{p}_{qh}(k+1|t) \geq d_m + l_q. \tag{14}$$

(3) ***Problem***: We aim to minimize the objective function (9) of all AVs in the CZ in a prediction horizon while guaranteeing that each AV satisfies the constraints (12)–(14). To this end, we construct a distributed multi-objective optimization model as follows:

$$\min_{u_i(0|t),\ldots,u_i(N_p-1|t)} \mathscr{J}_i(\bar{p}_{ih}, \bar{v}_{ih}, \bar{p}_{jf}, \bar{v}_{jf}, u_i) \tag{15}$$

subject to Equations (10)–(14), where the optimal control inputs of AVs are computed by minimizing the cost metric $\mathscr{J}_i$ at each time step.

Remark 1. *In the proposed distributed optimization control model, the performance index (9) and the inequality constraints in (12)–(14) are convex functions with respect to the decision variable $u_i(k|t)$. The equality constraint in (10) is an affine function of $u_i(k|t)$. Hence, the problem in (15) is a convex optimization problem. In addition, note that the feasible domain of the decision variable $u_i(k|t)$ is non-empty and the performance index function (9) is a coercive function during each time step. According to the Weierstrass lemma [31], feasible solutions to the optimization problem (15) can be always found in the feasible domain.*

Remark 2. *When analyzing the driving behavior of an MV approaching an intersection, it is important to consider two different situations. In the first situation, when the MV is not the leading vehicle in its lane, the Krauss model can be used to predict its trajectory. However, in the second situation, when the MV is the leading vehicle in its lane, its behavior is affected by various real-world factors such as visibility and the positions of conflicting vehicles in other lanes. To simplify this problem while characterizing the driving behavior of leading MVs, we propose using the Krauss model to simulate a virtual car-following behavior for the leading MV at the intersection. This involves assuming that a neighboring vehicle j of the leading MV m is closer to the intersection and can be treated as the virtual preceding vehicle of the leading MV. The virtual gap between vehicles m and j can be expressed as*

$$g_{mj} = \bar{p}_{mh} - \bar{p}_{jf} - l_j - d_{mj} + c_{mj}, \tag{16}$$

where the variables and parameters are defined in accordance with Equations (3) and (5). The gap g in the Krauss model for the leading MV m can be obtained as

$$g = \min_j g_{mj}. \tag{17}$$

In situations where there is no virtual preceding vehicle, the leading MV will accelerate to its maximum velocity, using maximum acceleration, and continue to cross the intersection.

Remark 3. *The idea of the proposed coordination scheme in this paper is inspired by the concept of consensus control of multi-agent systems [11,18,21,32]. Each AV is regulated to cross the intersection while staying a desired virtual safe distance from its neighboring vehicles in conflicting lanes while improving traffic mobility and fuel economy. In other words, each AV tends to reach a consensus with its neighboring vehicles by solving the local optimization problem (15). Therefore, through information sharing, the ego AV and its neighboring vehicles (if they are AVs) cooperatively optimize their trajectories, which pushes the distributed solutions towards the global optimality.*

Remark 4. *The current studies on vehicle consensus coordination at signal-free intersections, such as [11,18], are subject to a fully connected AV environment. All the vehicles must be controllable to achieve the objectives. The scheme presented in this paper can coordinate mixed-automated vehicles to cross the signal-free intersection. When potential conflicts occur between AVs and MVs, AVs use the predicted future trajectories of MVs to optimize their movements to avoid collisions. Note that if two or more MVs with conflict relationships approach the intersection simultaneously, they may slow down or even stop for safe crossing, resulting in a decrease in traffic efficiency. To alleviate this situation, we expect to apply our scheme under high penetration rates of AVs for potential future implementation.*

Remark 5. *The optimality of the problem solution in (15) cannot characterize the stability of the closed-loop system. The recently distributed MPC approaches focused on guaranteeing stability for the cooperation of vehicles on a one-dimension road, i.e., vehicular platoon [33,34]. In particular, the necessary assumptions were made for stability analysis that the vehicular platoon does not change, and all following vehicles in the platoon formed a spanning tree rooting at the leader. For the case of the cooperation of vehicles on a two-dimension road, i.e., intersection, most of the relevant distributed optimization algorithms (e.g., [18,21]), as well as this study, ignore the convergence analysis. The reasons for this are twofold: (1) the convergence plays a minor role in the performance of the cooperation of vehicles at intersections, while the optimally, e.g., safety, traffic mobility, and fuel economy, are the main concerns in practice; (2) the topology of vehicles on the road network is random, time-varying, and even disconnected, such that the stability of the system is difficult to analyze.*

3.4. Algorithm of Distributed Receding Horizon Optimization

In order to address the cooperative driving issue of mixed-automated vehicles at intersections, we present a distributed optimization algorithm under a receding horizon framework. The details are provided in Algorithm 2, which aims to find the optimal solutions for each AV in the CZ by solving (15). Figure 3 shows the basic procedure of the distributed receding horizon optimization scheme. The optimization problem is solved at each time step, and only the first element of the obtained control sequence is applied.

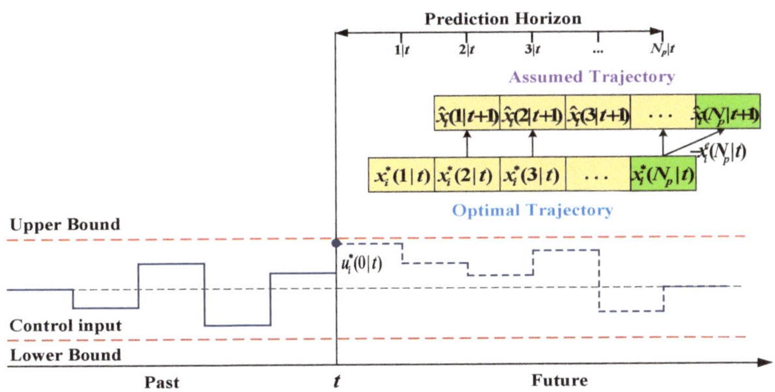

Figure 3. Illustration of the procedure of the distributed receding horizon optimization scheme.

In Algorithm 2, initialization is conducted in step 1 , which determines the start time of the algorithm, i.e., when ego AV i enters the CZ, and calculates initial N_p-step prediction states $\hat{x}_i(k+1|0) = [\hat{p}_{ih}(k+1|0), \hat{v}_{ih}(k+1|0)]^T, k = 0, \ldots, N_p - 1$, using (10). The communication topology of vehicles in the CZ at the current time step is identified in step 3. Based on the communication topology, ego AV i receives the information of N_p-step prediction states $\hat{x}_j(k+1|t), k = 0, \ldots, N_p - 1$, of neighboring vehicle $j, j \in \mathcal{N}_i(t)$ in step 4. Then, in step 5, ego AV i utilizes its current state and the predicted future N_p-step trajectories of neighboring vehicle $j, j \in \mathcal{N}_i(t)$ to solve the optimization problem (15). The first element of the obtained optimal control sequence is applied to AV i in step 6, i.e., $u_i(t) = u_i^*(0|t)$. The prediction of the trajectory of the neighboring vehicles is the key to solving the optimization problem (15) for the ego vehicle. Note that in the framework of distributed cooperation, the communication, computation, and control of all vehicles are assumed to be synchronous. Therefore, each AV i cannot receive the accurate N_p-step predicted trajectories of its neighboring vehicle j at the current step. To address this issue, we perform an operation to adopt the N_p-step assumed trajectories of neighboring vehicle j in the last shifted time window. Here, we take vehicle $i, i \in \Omega_A(t)$ as an example and illustrate the calculation process of the assumed trajectory, as depicted in Figure 3. Define $u_i^*(k|t), k = 0, \ldots, N_p - 1$ as the optimal control sequence, and $x_i^*(k+1|t) = [p_{ih}^*(k+1|t), v_{ih}^*(k+1|t)]^T, k = 0, \ldots, N_p - 1$

are the optimal states calculated by (10) in step 7. The assumed trajectory of vehicle i is computed in step 8 by removing the first state value of $x_i^*(k+1|t)$, $k=0,\ldots,N_p-1$, and adding the ending value $x_i^*(N_p|t) - x_i^e(N_p|t)$, where $x_i^e(N_p|t) = [v_{ih}^*(N_p|t)\delta, 0]^T$. The obtained assumed trajectory of vehicle i is transmitted to vehicle j, $j \in \mathcal{D}_i(t)$, and is regarded as the predicted trajectory of vehicle i in the next time window. In the same way, vehicle i receives the assumed trajectory of its neighboring vehicle j, $j \in \mathcal{N}_i(t)$. Note that if vehicle j is an MV, Equation (11) should be used to calculate the assumed trajectory of vehicle j. Similar approaches can be found in [33,35]. In the next shifted time window, the problem is reformulated and solved again until ego AV i departs from the CZ.

Algorithm 2 Algorithm of Distributed Receding Horizon Optimization

Input: The communication topology of vehicles in the CZ; the state of ego AV $i, i \in \Omega_A(t)$; N_p-step prediction states of its neighboring vehicle $j, j \in N_i(t)$

Output: Optimal trajectory of AV i.

1: *Initialization*: At time $t=0$ (the moment AV i enters the CZ), load the current state $\bar{x}_i(0|0)$; Assume $u_i(k|0)=0, k=0,\ldots,N_p-1$, and calculate the N_p-step prediction states $\hat{x}_i(k+1|0), k=0,\ldots,N_p-1$, using (10).

2: ***Repeat***

3: *Topology identification*: Acquire the communication topology of vehicles at the current step.

4: *Information reception*: Receive the information of N_p-step prediction states $\hat{x}_j(k+1|t)$, $k=0,\ldots,N_p-1$, of neighbor vehicle $j, j \in \mathcal{N}_i(t)$.

5: *Optimization solution*: At time $t>0$, according to the current state $\bar{x}_i(0|t)$ and received $\hat{x}_j(k+1|t), k=0,\ldots,N_p-1$, solve the optimal problem (15) and obtain the optimal control sequence $u_i^*(k|t), k=0,\ldots,N_p-1$.

6: *Implementation of optimal results*: Apply the first element of the optimal control sequence to AV i, i.e., $u_i(t)=u_i^*(0|t)$.

7: *Iteration of optimal states*: Calculate the optimal states $x_i^*(k+1|t), k=0,\ldots,N_p-1$, by using (10) and $u_i^*(k|t), k=0,\ldots,N_p-1$.

8: *Estimation of assumed states*: Calculate the N_p-step assumed states $\hat{x}_i(k+1|t+1)$ at time $t+1$ as

$$\hat{x}_i(k+1|t+1) = \begin{cases} x_i^*(k+2|t), & k=0,\ldots,N_p-2 \\ x_i^*(k+1|t) - x_i^e(k+1|t), & k=N_p-1 \end{cases},$$

then transmit $\hat{x}_i(k+1|t+1)$ to vehicle $j, j \in \mathcal{D}_i(t)$.

9: ***Return*** Optimal trajectory of AV i;

10: $t \leftarrow t+1$

11: ***Until*** Vehicle i departs from the CZ.

Remark 6. *Note that the N_p-step predicted trajectories of vehicles might not be entirely calculated since the vehicles may enter the intersection within the prediction horizon. The following operations are conducted to address this issue. (1) By detecting the predicted states of vehicle i, only the predicted information for the valid time steps before vehicle i passes through the stop line is retained, and the information of the remaining steps is set to null. (2) The formulated optimization problem for vehicle i is solved just to obtain the optimization control inputs in the valid time steps. (3) The assumed trajectory of vehicle i is calculated and then transmitted to its neighboring vehicle $j, j \in \mathcal{N}_i(t)$. (4) To maintain the integrity of the received assumed trajectory of vehicle i, the received null values are replaced by the desired assumed state values of the neighboring vehicles of vehicle j. (5) The optimization problem of vehicle j can be solved. The above operations guarantee that if a vehicle enters the intersection at a certain prediction step, the prediction information of the vehicle after this step will not affect the solution of optimization problem (15) for the ego vehicle and its neighboring vehicles.*

4. Simulation Results

In this section, simulations are conducted to evaluate the performance of the presented scheme for the cooperative driving of mixed-automated vehicles at an intersection. The proposed algorithm is coded in Python (Version 3.8.13) and run on a computer with Intel Core i5-9500 3 GHz CPU and 8 GB of DDR4-2666 RAM. The SUMO (Version 1.13.0) microscopic traffic simulator is used to construct the road network, generate traffic flows, and acquire performance measurements [36]. Vehicle information is transmitted between SUMO and our control model in Python through SUMO's TraCI interface.

4.1. Simulation Framework

We develop a simulation framework at a single intersection to test the effectiveness of the proposed algorithm in this paper. The test intersection has a typical symmetrical four-leg configuration. The length of each approach and departure section is approximately 193.6 m. Each approach includes two incoming lanes with the specific allowable movements, where the left lane only serves left-turn traffic, and the right lane serves through traffic. The north- and southbound approaches are main roads, and the east- and westbound approaches are minor roads. This distinction allows for the consideration of different traffic priorities and behaviors at the intersection. The radius of the cooperative zone is set as 100 m. The simulation duration is 1 h. Other simulation parameter settings can be seen in Table 1.

To investigate the sensitivity of the proposed algorithm, we consider three cases based on different traffic volumes. Table 2 summarizes the three cases tested in this study. Since our study focuses on environments with high penetration rates of AVs, for each case, four different penetration rates of AVs (i.e., 70%, 80%, 90%, 100%) are considered to evaluate the validity of the proposed algorithm. In addition, the simulation results in a 15 min period are compared to those under the no-control (NC) operation and fixed-time signal control (FSC) strategy.

Table 1. Parameter settings.

Parameter	Notation	Value
Common		
Radius of CZ (m)	R	100
Lane length (m)	L	193.6
Lane width (m)	W	3.2
Road speed limit (m/s)	V_{lim}	20
Vehicle length (m)	l	5
Minimum gap (m)	d_m	5
AVs		
Desired spacing (m)	d_{ij}	10
Maximum speed (m/s)	$\mathscr{V}_{max}$	18
Minimum speed (m/s)	$\mathscr{V}_{min}$	5
Maximum acceleration (m/s^2)	$\mathscr{A}_{max}$	3
Maximum deceleration (m/s^2)	$\mathscr{A}_{min}$	3
Prediction horizon (s)	N_p	5
Time interval (s)	δ	1
Weight	ω_1	2
Weight	ω_2	1
Weight	ω_3	1
Weight	ω_4	1
MVs		
Maximum speed (m/s)	V_{max}	18
Maximum acceleration (m/s^2)	a_{max}	3
Maximum deceleration (m/s^2)	b_{max}	4
Reaction time of drivers (s)	τ	0.5
Constant	ϵ	0.4

Table 2. Demand for different cases.

Case	North- and Southbound		East- and Westbound	
	Through Demand (veh/h/lane)	Left-Turn Demand (veh/h/lane)	Through Demand (veh/h/lane)	Left-Turn Demand (veh/h/lane)
1	250	125	150	75
2	300	150	200	100
3	350	175	250	125

4.2. Results

The proposed algorithm is tested under the different demand cases and penetration rates of AVs (PRAs) to investigate the performance of the scheme in this study. Table 3 shows the comparison of the mean travel time for the different demand and PRAs. The results show that the mean travel time increases slightly when the PRA decreases in cases 1 and 2. In contrast, the mean travel time for case 3 rises substantially with the reduction in PRA. On the other hand, by comparing the results vertically, it slightly differs in the mean travel time for the different cases under 80% to 100% PRA. Conversely, the mean travel time increases significantly for case 3 under 70% PRA. The results indicate that when the PRA is as low as 70% and the total traffic demand reaches 1800 vehicles per hour, there is a significant adverse effect on the traffic efficiency of the intersection. Compared to the results for under 100% PRA, the mean travel time increases by 28.9% under 70% PRA. Table 4 shows the results for fuel consumption under the different traffic demand and PRAs, respectively. Similarly, the differences in fuel consumption under different PRAs are less distinct for cases 1 and 2. There are even slight drops in fuel consumption when the PRA decreases, since the additional control cost for the desired performance indexes is reduced. The fuel consumption rises significantly with the decline in PRA in case 3. The results indicate that the proposed scheme has less apparent impact on fuel consumption with different PRAs under low demand, while the differences are distinct under high demand.

Table 3. Summary of mean travel time (s) for the three cases under different penetration rates of AVs.

Case	Penetration Rate of AVs			
	100%	90%	80%	70%
1	30.4	30.8	31.1	31.7
2	30.6	31.5	32.1	32.7
3	30.8	32.3	34.6	39.7

Table 4. Summary of mean fuel consumptions (mL) for the three cases under different penetration rates of AVs.

Case	Penetration Rate of AVs			
	100%	90%	80%	70%
1	31.8	31.6	31.4	31.5
2	33.4	33.3	33.5	33.7
3	34.7	35.3	36.4	38.8

Figure 4 shows the mean speed relative rates for case 3 in 15 min under different PRAs. The mean speed relative rate is defined as the ratio of the actual average speed to the desired speed. Figure 4 indicates that the mean speed under higher PRAs is generally closer to the desired speed than that under lower PRAs. In particular, during the 700 to 900 s period, the mean speed relative rate maintains a relatively high value under 90% and 100% PRA, while these drop dramatically under 70% and 80% PRA. This phenomenon occurs since more uncontrolled MVs are loaded onto the network during this period. Multiple

MVs slow down or even stop at the intersection for safe passing, resulting in reducing the mean speed of the road network. Figure 5 shows the number of stops for case 3 in 15 min under different PRAs. No stops and a few stops occur in the scenarios of 100% and 90% PRA, respectively. When the PRA drops to 70% and 80%, during the 700 to 900 s period, more vehicles stop at the intersection for safe passing, which leads to queuing. The above results indicate that the decrease in the PRA leads to a decline in the traffic efficiency of the intersection.

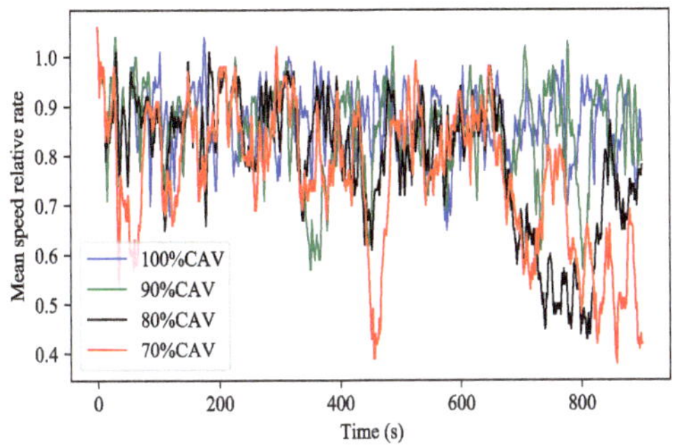

Figure 4. Comparison of the mean speed relative rates for case 3 in 15 min under different penetration rates of AVs.

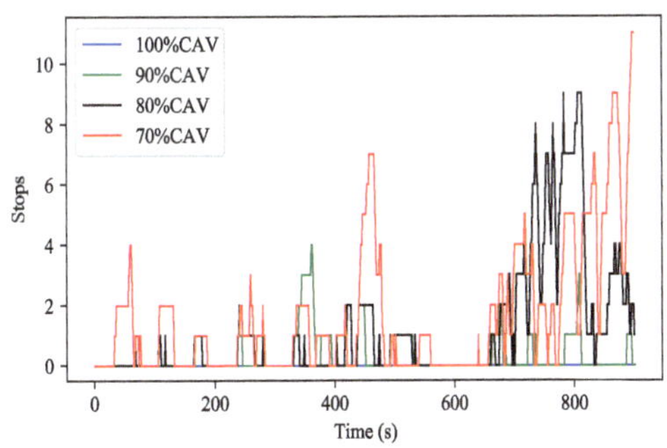

Figure 5. Comparison of the number of stops for case 3 in 15 min under different penetration rates of AVs.

To compare the performance of the proposed scheme with other methods used, we test the no-control (NC) operation and fixed-time signal control (FSC) strategy as benchmark baselines at the same intersection. The NC operation means allowing all vehicles to pass through the signal-free intersection by following the driving behaviors of MVs. The cycle time of the FSC system is set as 82 s. The green times of north- and southbound, and east- and westbound are chosen as 20 s and 15 s, respectively. The clearance time is 3 s. Figure 6 shows the trajectories of vehicles for case 3 in 5 min under the different strategies. It is observed that the vehicles with the proposed scheme in this paper adjust their trajectories after entering the CZ. Especially in the scenario of 100% PRA, the vehicles pass through the intersection smoothly. When the PRA is 70%, a few vehicles stop at the intersection. In contrast, the vehicles under NC or FSC frequently stop at the intersection, and lots of queues are formed. Figure 7 shows the speeds of vehicles for case 3 in 5 min under the different strategies. Corresponding to the results in Figure 6, the vehicles using the proposed scheme always pass the intersection without halting (speed greater than zero) under 100% PRA, and a small number of vehicles slow down to zero at the intersection under 70% PRA. When NC or FSC is adopted, more vehicles slow to a stop at the intersection. The above phenomena indicate that the proposed scheme in this paper notably outperforms NC and FSC in smoothing traffic flow and improving mobility. Figure 8 shows the acceleration curves of vehicles with respect to distance under the proposed scheme for case 3 in 5 min. We can easily observe that the vehicles adjust their acceleration significantly after entering the CZ. Notice that the maximum acceleration and deceleration of MVs are 3 m/s^2 and 4 m/s^2 in the scenario of 70% PRA, respectively, while the maximum acceleration and deceleration of AVs are both 3 m/s^2.

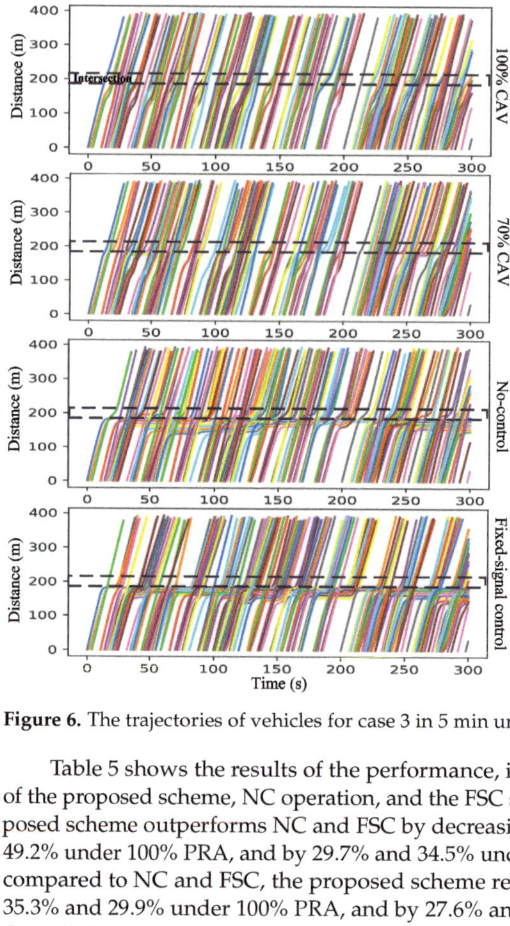

Figure 6. The trajectories of vehicles for case 3 in 5 min under the different strategies.

Table 5 shows the results of the performance, i.e., travel time and fuel consumption, of the proposed scheme, NC operation, and the FSC strategy. The results show that the proposed scheme outperforms NC and FSC by decreasing the mean travel time by 45.5% and 49.2% under 100% PRA, and by 29.7% and 34.5% under 70% PRA, respectively. In addition, compared to NC and FSC, the proposed scheme reduces the mean fuel consumption by 35.3% and 29.9% under 100% PRA, and by 27.6% and 21.6% under 70% PRA, respectively. Overall, the proposed scheme of this paper significantly improves traffic efficiency and fuel economy compared to NC and FSC.

Table 5. Summary of performance for case 3 under the proposed scheme, no-control operation, and the fixed-signal control strategy.

Performance	Proposed Scheme		NC	FSC
	100% AV	70% AV		
Travel time (s)	30.8	39.7	56.5	60.6
Fuel consumption (mL)	34.7	38.8	53.6	49.5

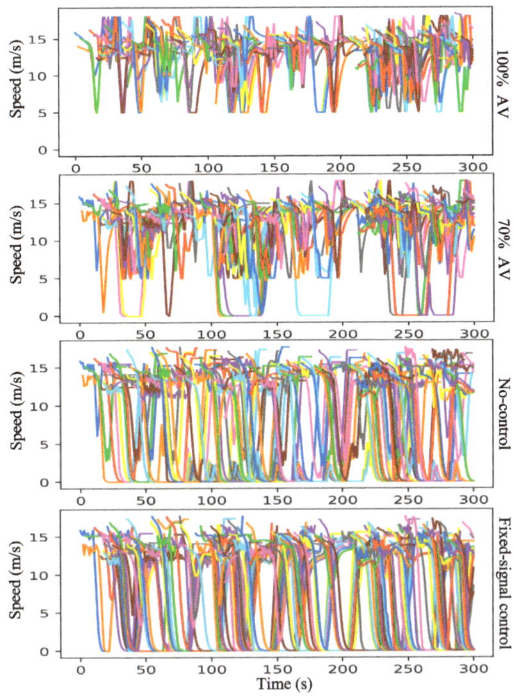

Figure 7. The speeds of vehicles for case 3 in 5 min under the different strategies.

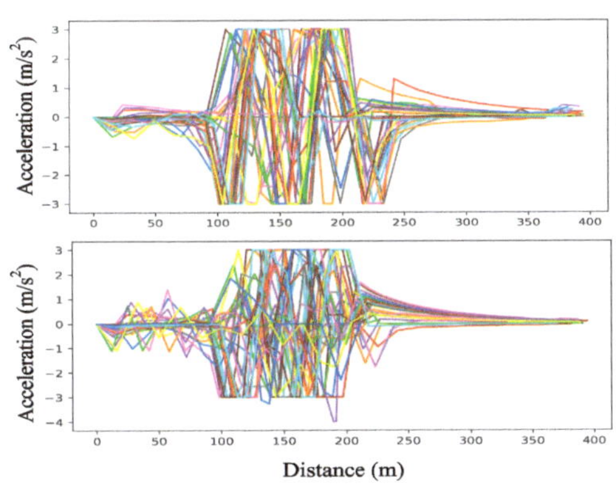

Figure 8. The acceleration of vehicles for case 3 in 5 min under the proposed strategy.

Figure 9 shows the mean speed of the road network with the different strategies for case 3 in 15 min, which reflects the traffic mobility to some extent. The vehicles with the proposed scheme in this study can always maintain a high mean speed in the scenario of 100% PRA. When the PRA is 70%, the mean speed declines, since more MVs are loaded onto the network and slow down at the intersection. In contrast, the mean speed of vehicles under NC or FSC is kept at a smaller value, reflecting the inefficiency of the traffic. Figure 10 shows the number of stops for case 3 in 15 min under the different strategies. It is easy to observe that almost no stops occur in the scenario of 100% PRA with the proposed scheme, and a small number of stops exist under 70% PRA. When NC or FSC is adopted, many vehicles stop near the intersection, leading to queuing. Figures 9 and 10 indicate that the proposed scheme in this paper can significantly improve mobility and smooth traffic flows compared to the NC operation and FSC strategy.

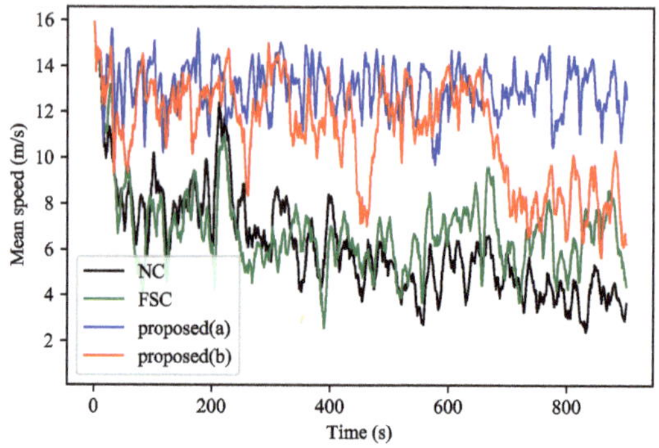

Figure 9. Comparison of the mean speed for case 3 in 15 min under the different strategies: (a) proposed scheme under 100% penetration rate of AVs; (b) proposed scheme under 70% penetration rate of AVs.

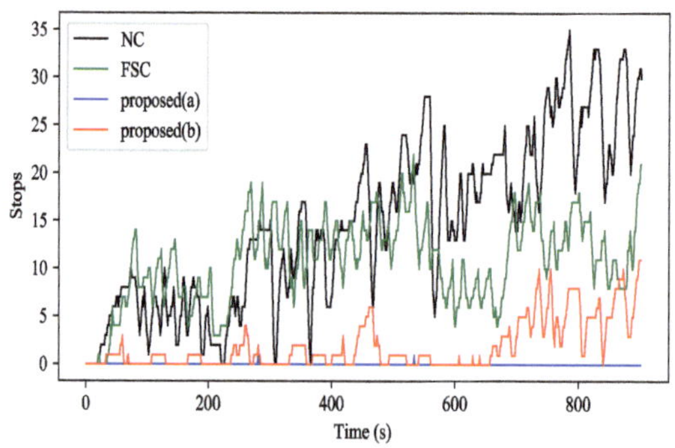

Figure 10. Comparison of the number of stops for Case 3 in 15 min under the different strategies: (a) proposed scheme under 100% penetration rate of AVs; (b) proposed scheme under 70% penetration rate of AVs.

Since few of the current state-of-practice strategies consider mixed-automated traffic at signal-free intersections, we have compared the results of the proposed scheme with a recently developed strategy that jointly optimized the signal timing plans with the trajectory of vehicles in Ramin et al. [24]. The paper developed a centralized mixed-integer non-linear program to coordinate the trajectories of AVs and signal timing plans under mixed-automated traffic. We adopt the same traffic parameters as [24] for comparative simulations under the two-levels of traffic demands, i.e., demand 1 and demand 2. Demand 1 denotes low traffic demand, with 300 veh/h/lane through demand and 24 veh/h/lane left-turn demand, and demand 2 represents high traffic demand, with 600 veh/h/lane through demand and 48 veh/h/lane left-turn demand. Table 6 shows the total delay between the scheme proposed in this paper and in [24] for the two demands under different PRAs in 5 min. The results indicate that the total delay increases with PRA reduction in both schemes. In addition, our proposed scheme performs better than [24] under high PRAs and low traffic demand. However, when the PRA drops below 70%, especially under high demand, the total delay in our scheme is higher than in [24]. This suggests that the role of signal timing optimization in [24] is more significant under high demand and a low PRA. Table 7 shows the computation performance between the proposed scheme in this paper and [24] for demand 2 under different PRAs. The parameters # Var. and # Con. denote the mean number of variables and constraints processed per second, and # Com. represents the mean number of communications required per second. The results show that the mean number of variables, constraints, and communications mostly decrease as PRA decreases in both schemes. In addition, the proposed scheme in this paper has fewer computation parameters to process per second than [24], which means that our scheme has an advantage in computational complexity.

Table 6. Comparison of total delays (s) between the scheme of this paper and Ramin et al. [24] for the two demands under different penetration rates of AVs.

PRA	Demand 1		Demand 2	
	This Study	**Ramin et al.** [24]	**This Study**	**Ramin et al.** [24]
100%	69.1	78.2	157.8	173.2
70%	164.2	191.6	1062.3	1057.0
50%	211.7	204.3	1552.1	1363.2
30%	265.6	228.3	2521.6	2000.1
0%	318.4	270.9	3256.2	2468.8

Table 7. Comparison of complexity parameters between the scheme of this paper and Ramin et al. [24] for demand 2 under different penetration rates of AVs.

PRA	# Var.		# Con.		# Com.	
	This Study	**Ramin et al.** [24]	**This Study**	**Ramin et al.** [24]	**This Study**	**Ramin et al.** [24]
100%	25.5	150.9	20.4	92.8	26.5	38.8
70%	20.3	130.3	16.2	88.5	21.6	33.0
50%	15.5	72.7	12.4	84.4	18.4	29.1
30%	10.4	51.3	5.2	82.0	15.1	25.3
0%	0	15.9	0	75.0	10.2	19.4

Var.: Mean number of variables per second. # Con.: Mean number of constraints per second. # Com.: Mean number of communications per second.

4.3. Discussion and Limitations

The existing literature on regulating mixed-automated traffic flow at intersections has focused on introducing traffic lights for right-of-way allocation, e.g., [23,24]. Our study attempts to optimize the behavior of AVs by predicting the trajectory of AVs and MVs, which leads all vehicles to pass through the signal-free intersection without collision and results in a satisfactory traffic performance. Compared to the strategies of jointly optimizing vehicle trajectories and signal timing, our scheme theoretically makes full use of right-of-way without idling caused by traffic lights. However, the increasing number of non-optimized MVs affects the performance of the proposed scheme. In other words, our scheme is sensitive to the PRA and traffic demand. Through the comparison simulation with [24], we find that our scheme performed better under a high PRA and low traffic demand. The results also imply that the application of signal lights and timing optimization for regulating mixed-automated traffic at intersections are more necessary and superior under a low PRA and high traffic demand. In addition, in terms of computational complexity, the proposed scheme in this paper has a certain computational demand since the optimization problem for each AV is solved at each step. Nevertheless, the proposed scheme still has significant advantages in computation compared MILPs jointly optimizing the vehicle trajectory and signal timing.

5. Conclusions

This paper developed a distributed cooperation scheme for coordinating vehicles at signal-free intersections in a mixed traffic environment of AVs and MVs. The traffic conflict graph was constructed to recognize the communication topologies of the vehicles in conflict relationships, assuming that all AVs and MVs were connected and shared information in the CZ. We formulated the coordination issue into a multi-objective optimization problem, which was solved by optimizing the sum of weighted indexes of the conflict elimination, traffic mobility, and fuel economy in a prediction horizon synchronously. The future trajectories of AVs and MVs, as the inputs of the optimization model, were predicted by using the vehicular longitudinal dynamic model and Krauss car-following model, respectively. To obtain the local optimal solutions, we presented a distributed receding horizon algorithm to calculate the optimal control inputs for each AV in the CZ. The simulations evaluated the effectiveness and superiority of the proposed scheme. The results showed that the proposed scheme outperformed the no-control operation by decreasing travel time by 29.7–45.5%, and reducing fuel consumption by 27.6–35.3% under 70–100% PRA. Compared to the fixed-time signal control strategy, the proposed scheme decreased travel time by 34.5–49.2% and reduced fuel consumption by 21.6–29.9% under 70–100% PRA. Additionally, according to the comparison results with the strategy of jointly optimizing the vehicle trajectory and signal timing, the proposed scheme performs better at high penetration rates of AVs and under not very dense traffic demand, and suffers less computational burden.

Though this study considers a typical dual-lane intersection with four approaches, it can be easily extended to on-ramps of expressways or other types of intersections according to the conflict relationships. In addition, advanced driver guidance systems for MVs are expected to be integrated into the intersection coordination framework in future studies. Based on the prompt information of the system, the drivers of MVs can actively manipulate the vehicles to cross the intersection safely and smoothly. In addition, the combination of artificial intelligence methods and other types of optimization methods with the proposed strategy of this paper will be further considered in the future. Furthermore, research on the distributed collaboration of multiple signal-free intersections will be studied under a mixed traffic environment.

Author Contributions: Methodology, J.G.; writing—original draft, J.G.; writing—review and editing, W.C. and Z.Z. All authors have read and agreed to the published version of the manuscript.

Funding: This work was supported by the National Natural Science Foundation of China, under grant 61833005.

Institutional Review Board Statement: Not applicable.

Informed Consent Statement: Not applicable.

Data Availability Statement: Measurement data is available from the authors upon request.

Conflicts of Interest: The authors declare no conflict of interest.

References

1. Alanazi, F. A systematic literature review of autonomous and connected vehicles in traffic management. *Appl. Sci.* **2023**, *13*, 1789. [CrossRef]
2. Sun, K.; Zhao, X.; Gong, S.; Wu, X. A cooperative lane change control strategy for connected and automated vehicles by considering preceding vehicle switching. *Appl. Sci.* **2023**, *13*, 2193. [CrossRef]
3. Antonio, G.P.; Maria-Dolores, C. Multi-agent deep reinforcement learning to manage connected autonomous vehicles at tomorrow's intersections. *IEEE Trans. Vehi. Technol.* **2022**, *71*, 7033–7043. [CrossRef]
4. Feng, Y.; Huang, S.E.; Wong, W.; Chen, Q.A.; Mao, Z.M.; Liu, H.X. On the cybersecurity of traffic signal control system with connected vehicles. *IEEE Trans. Intell. Transp. Syst.* **2022**, *23*, 16267–16279. [CrossRef]
5. Riostorres, J.; Malikopoulos, A.A. A Survey on the coordination of connected and automated vehicles at intersections and merging at highway on-Ramps. *IEEE Trans. Intell. Transp. Syst.* **2017**, *18*, 1066–1077. [CrossRef]
6. Dresner, K.; Stone, P. Multiagent traffic management: A reservation-based intersection control mechanism. In Proceedings of the 3rd International Joint Conference on Autonomous Agents and Multiagent Systems: (AAMAS), New York, NY, USA, 19–23 July 2004; pp. 530–537.
7. Dresner, K.M.; Stone, P. A Multiagent approach to autonomous intersection management. *J. Artif. Intell. Res.* **2008**, *31*, 591–656. [CrossRef]
8. Milanes, V.; Perez, J.; Onieva, E.; Gonzalez, C. Controller for urban intersections based on wireless communications and fuzzy logic. *IEEE Trans. Intell. Transp. Syst.* **2010**, *11*, 243–248. [CrossRef]
9. Onieva, E.; Milanes, V.; Villagra, J.; Perez, J.; Godoy, J. Genetic optimization of a vehicle fuzzy decision system for intersections. *Expert Syst. Appl.* **2012**, *39*, 13148–13157. [CrossRef]
10. Medina, A.I.M.; van de Wouw, N.; Nijmeijer, H. Cooperative intersection control based on virtual platooning. *IEEE Trans. Intell. Transp. Syst.* **2017**, *19*, 1727–1740. [CrossRef]

11. Xu, B.; Li, S.E.; Bian, Y.; Li, S.; Ban, X.J.; Wang, J.; Li, K. Distributed conflict-free cooperation for multiple connected vehicles at unsignalized intersections. *Transp. Res. C Emerg. Technol.* **2018**, *93*, 322–334. [CrossRef]
12. Lee, J.; Park, B. Development and evaluation of a cooperative vehicle intersection control algorithm under the connected vehicles environment. *IEEE Trans. Intell. Transp. Syst.* **2012**, *13*, 81–90. [CrossRef]
13. Dai, P.; Liu, K.; Zhuge, Q.; Sha, E.H.M.; Lee, V.C.S.; Son, S.H. Quality-of-experience-oriented autonomous intersection control in vehicular networks. *IEEE Trans. Intell. Transp. Syst.* **2016**, *17*, 1956–1967. [CrossRef]
14. Xu, H.; Zhang, Y.; Li, L.; Li, W. Cooperative driving at unsignalized intersections using tree search. *IEEE Trans. Intell. Transp. Syst.* **2019**, *21*, 4563–4571. [CrossRef]
15. Meng, Y.; Li, L.; Wang, F.Y.; Li, K.; Li, Z. Analysis of cooperative driving strategies for non-signalized intersections. *IEEE Trans. Veh. Technol.* **2018**, *67*, 2900–2911. [CrossRef]
16. Gong, J.; Cao, J.; Zhao, Y.; Wei, Y.; Guo, J.; Huang, W. Sampling-based cooperative adaptive cruise control subject to communication delays and actuator lags. *Math. Comput. Simul.* **2020**, *171*, 13–25. [CrossRef]
17. Lu, J.; Wang, Y.; Shi, X.; Cao, J. Finite-time bipartite consensus for multiagent systems under detail-balanced antagonistic interactions. *IEEE Trans. Syst., Man, Cyber., Syst.* **2021**, *51*, 3867–3875. [CrossRef]
18. Mirheli, A.; Tajalli, M.; Hajibabai, L.; Hajbabaie, A. A consensus-based distributed trajectory control in a signal-free intersection. *Transp. Res. C Emerg. Technol.* **2019**, *100*, 161–176. [CrossRef]
19. Zhang, X.; Fang, S.; Shen, Y.; Yuan, X.; Lu, Z. Hierarchical velocity optimization for connected automated vehicles with cellular vehicle-to-everything communication at continuous signalized intersections. *IEEE Trans. Intell. Transp. Syst.* 2023, *early access*.
20. Kamal, M.A.S.; Imura, J.I.; Hayakawa, T.; Ohata, A.; Aihara, K. A vehicle-intersection coordination scheme for smooth flows of traffic without using traffic lights. *IEEE Trans. Intell. Transp. Syst.* **2014**, *16*, 1136–1147. [CrossRef]
21. Du, Z.; HomChaudhuri, B.; Pisu, P. Hierarchical distributed coordination strategy of connected and automated vehicles at multiple intersections. *J. Intell. Transp. Syst.* **2018**, *22*, 144–158. [CrossRef]
22. Pourmehrab, M.; Elefteriadou, L.; Ranka, S.; Martin-Gasulla, M. Optimizing signalized intersections performance under conventional and automated vehicles traffic. *IEEE Trans. Intell. Transp. Syst.* **2019**, *21*, 2864–2873. [CrossRef]
23. Tajalli, M.; Hajbabaie, A. Traffic signal timing and trajectory optimization in a mixed autonomy traffic stream. *IEEE Trans. Intell. Transp. Syst.* **2021**, *23*, 6525–6538. [CrossRef]
24. Niroumand, R.; Tajalli, M.; Hajibabai, L.; Hajbabaie, A. Joint optimization of vehicle-group trajectory and signal timing: Introducing the white phase for mixed-autonomy traffic stream. *Transp. Res. C Emerg. Technol.* **2020**, *116*, 102659. [CrossRef]
25. Xu, H.; Cassandras, C.G.; Li, L.; Zhang, Y. Comparison of cooperative driving strategies for CAVs at signal-free intersections. *IEEE Trans. Intell. Transp. Syst.* **2021**, *23*, 7614–7627. [CrossRef]
26. Belkhouche, F. Collaboration and optimal conflict resolution at an unsignalized intersection. *IEEE Trans. Intell. Transp. Syst.* **2018**, *20*, 2301–2312. [CrossRef]
27. Bian, Y.; Li, S.E.; Ren, W.; Wang, J.; Li, K.; Liu, H.X. Cooperation of multiple connected vehicles at unsignalized intersections: Distributed observation, optimization, and control. *IEEE Trans. Ind. Electron.* **2019**, *67*, 10744–10754. [CrossRef]
28. Li, L.; Wang, F. Cooperative driving at blind crossings using intervehicle communication. *IEEE Trans. Veh. Technol.* **2006**, *55*, 1712–1724. [CrossRef]
29. Krauß, S. Microscopic Modeling of Traffic Flow: Investigation of Collision Free Vehicle Dynamics. Ph.D. Thesis, University of Cologne, Koln, Germany, 1998.
30. Gong, S.; Du, L. Cooperative platoon control for a mixed traffic flow including human drive vehicles and connected and autonomous vehicles. *Transp. Res. B Methodol.* **2018**, *116*, 25–61. [CrossRef]
31. Rudin, W. *Principles of Mathematical Analysis*; McGraw-Hill: New York, NY, USA, 1976; Volume 3.
32. Ishii, H.; Wang, Y.; Feng, S. An overview on multi-agent consensus under adversarial attacks. *Annu. Rev. Control* **2022**, *53*, 252–272. [CrossRef]
33. Zheng, Y.; Li, S.E.; Li, K.; Borrelli, F.; Hedrick, J.K. Distributed model predictive control for heterogeneous vehicle platoons under unidirectional topologies. *IEEE Trans. Control Syst. Technol.* **2017**, *25*, 899–910. [CrossRef]
34. Dunbar, W.B.; Caveney, D.S. Distributed receding horizon control of vehicle platoons: Stability and string stability. *IEEE Trans. Autom. Control* **2011**, *57*, 620–633. [CrossRef]
35. Dunbar, W.B.; Murray, R.M. Distributed receding horizon control for multi-vehicle formation stabilization. *Automatica* **2006**, *42*, 549–558. [CrossRef]
36. Lopez, P.A.; Behrisch, M.; Bieker-Walz, L.; Erdmann, J.; Flötteröd, Y.P.; Hilbrich, R.; Lücken, L.; Rummel, J.; Wagner, P.; Wießner, E. Microscopic traffic simulation using sumo. In Proceedings of the 21st International Conference on Intelligent Transportation Systems (ITSC), Maui, HI, USA, 4–7 November 2018; pp. 2575–2582.

modelling MDPI

Article

Supply-Driven Analysis for a Continuous Water Supply Network Based on a Pressure-Based Outflow at the House Outlets under Peak Withdrawal Scenarios

Conety Ravi Suribabu [1,*], Neelakantan Renganathan Thurvas [1] and Perumal Sivakumar [2]

1 Centre for Advanced Research in Environment, School of Civil Engineering, SASTRA Deemed University, Thanjavur 613401, Tamil Nadu, India; neelakantan@civil.sastra.edu
2 Department of Civil Engineering, North Eastern Regional Institute of Science and Technology (Deemed to be University), Nirjuli 791109, Arunachal Pradesh, India; psiva.nerist@gmail.com or siva_nerist@yahoo.co.in
* Correspondence: suribabu@civil.sastra.edu

Abstract: This research brings a new analysis method for a continuous water supply distribution network. The number of house service connections in different story buildings, rather than the nodal peak demand, shall be accounted for in the analysis. This work aims to consider the flow when pipes are opened in the house plumbing systems. The approach deviates from a traditional peak demand-based analysis of the water distribution network. The analysis gives the flow rate at each nodal point that could be observed in the different story buildings. The methodology is applied to a hypothetical network and shows how much flow and nodal pressure can occur when different percentages of consumers are in an active state. This study indicates that the peak demand-based sizing of the supply pipes could have a deficient capacity under real scenarios. The proposed analysis method will help to understand the actual behavior of the network.

Keywords: continuous water supply; house plumbing; peak demand; water distribution network

Citation: Suribabu, C.R.; Thurvas, N.R.; Sivakumar, P. Supply-Driven Analysis for a Continuous Water Supply Network Based on a Pressure-Based Outflow at the House Outlets under Peak Withdrawal Scenarios. *Modelling* **2023**, *4*, 515–528. https://doi.org/10.3390/modelling4040029

Academic Editors: Tomasz Nowakowski, Artur Kierzkowski, Agnieszka A. Tubis, Franciszek Restel, Tomasz Kisiel, Anna Jodejko-Pietruczuk and Mateusz Zając

Received: 23 September 2023
Revised: 26 October 2023
Accepted: 6 November 2023
Published: 8 November 2023

1. Introduction

The simulation of water distribution networks (WDNs) is performed in three ways for selected network geometry and hydraulic properties. They are demand-driven, pressure-driven, and volume-driven. The demand-driven analysis (DDA) results are valid if no pressure deficiency exists at the nodes. Pressure-driven analysis (PDA) can compute the pressure-dependent outflow for a designated minimum-maximum pressure range. Volume-driven analysis (VDA) computes the volume of water that could be supplied upon expected demand. Analysis based on these three approaches shall be required to better design WDNs and to understand the network behaviors when it is servicing under component failures and a shortfall in source water availability. Computational fluid dynamic (CFD) methods are commonly used in the assessment of hydrodynamic phenomena in pipes and connections [1].

The peak nodal demand is the main parameter for pipe sizing using DDA. The peak demand is calculated based on the forecasted population that needs to be served during peak water usage. The peak hourly demand is usually between 1.3 and 3.0 times the average hourly demand [2]. The entire design process is carried out using peak hourly demand. The volume of water required over a day is calculated based on per capita water usage. Its value varies from place to place, and different values are used for rural and urban communities. The analysis of domestic water consumption indicates 92 L/person/day for healthy urban lifestyles with restrictions on everyday activities and 175 L/person/day for unrestricted water usage. It is quantified according to estimated needs; the consumption rate is estimated as 227 L/person/day, and the ultimate level of consumption happens at the rate of 314 L/person/day [3]. The peak demand occurs between 6.00 a.m. and 7.00 a.m. and between 5.00 p.m. and 7.00 p.m. [4]. Residential water demand has shown high variability due to variations in water fixtures and appliances coupled with different routines and timing of withdrawal by the end users [5]. During peak time, water withdrawal may happen simultaneously from several fixtures. The flow from these fixtures occurs based on the pressure in the street main pipeline. The more pressure availability, the more volume will be delivered quickly, and vice versa. Hence, the pressure-dependent outflow from the fixtures at peak usage shall be accounted for in modelling. In the case of a flat terrain, the houses located near the source derive more benefits due to the availability of more pressure, whereas the houses located far away from the source are deprived of pressure, and less supply is possible. Due to these aspects, the simulation that considers the actual flow to the houses could provide a better analysis regarding meeting the demand, especially at peak times. An estimate of total outflow at peak usage of water shall be carried out, and the same shall be used in the analysis of WDN. This could provide valuable information on the pressure at several nodes and make it easier to manage pressure better. In the DDA and PDA, the demand for each node shall be set and analyzed. In VDA, the volume of water to be supplied at each node shall be set through outlet sizing storage at the outlet. The VDA is the most suitable method of analysis for intermittent water supply (IWS) systems, where water is collected in the underground storage system during supply hours [6]. In the continuous water supply (CWS) system, maintaining service pressure at peak hours is very important to meet the volume of water expected by consumers. In the present study, a new approach is formulated to simulate the network based on the number of fixtures kept open during peak hours in each house. The proposed approach analyzes the network based on the number of house service connections (HSCs) provided and the number of plumbing fixtures simultaneously active during peak hours. The analysis can be easily performed once the number of HSC incident to every node is known. The flow rate at each house at peak hours can be directly ascertained using this approach. The supply at each fixture happens only based on the available pressure, and the analysis procedure is named supply-driven analysis (SDA). The SDA can be implemented using the Emitter option available in the EPANET. This paper further presents a method for calculating the emitter coefficient under different scenarios.

2. Pressure Flow Relationship for House Service Lines

In a CWS system that directly connects with a house plumbing system (Figure 1), the flow rate for each HSC can be estimated using the arrangement shown in Figure 2. The outflow from the house plumbing system depends on the availability of pressure. Each house can have several outlets, like a water tap in the kitchen, a washing machine supply, a bathroom water shower, and a cistern flush in the lavatory. Simultaneous supply from all outlets provides the maximum possible withdrawal. It is to be noted that supply from these outlets is based on the available pressure. Hence, satisfying the required demand is related to time. There could be more than four fixtures in a house. Considering the worst scenario where five fixtures are opened simultaneously, this could be considered as peak demand. This peak demand can be estimated by considering the network between the connection at the main supply and any of the five plumbing fixtures of the house. The available static pressure in the water main in the street pushes the water to the service line based on the pressure available.

A house service pipeline connecting the house's internal pipe network can be treated as a separate system that can be simulated for various static pressures to obtain the relationship between pressure and supply. The prepared network layout consists of a reservoir that provides a static pressure head to the system, and a pipeline connecting the municipal main pipeline with the service line. The diameter of this pipeline can be taken as either 25 mm or 18 mm, and the diameter of the internal pipeline can be considered as either 18 mm or 12 mm. The diameter can be decided based on local practice. The minor loss coefficient for the service line can be taken as 1.8 [7]. The selection of house service pipe diameter and internal house pipe diameter can also be taken as actual values adopted in the respective region in the analysis. In the present work, a 18 mm diameter is considered for house service connections and a 12 mm diameter for raising pipes and internal pipes for single-story buildings (ground floor alone). In the case of two- and three-story buildings, 25 mm, 18 mm, and 12 mm diameters are considered for house service, raising, and internal pipes, respectively. The pipe network layout for a single-story building shown in Figure 2 is considered for the present study. The length of the house service pipeline and internal pipelines may vary from house to house. Hence, a reasonable length can be considered for the analysis. Usually, the plumbing fixture outlet level would have been positioned at different levels based on convenience and the ceiling height of the building. An average height of 1.5 m from the floor level is considered to obtain the flow vs. pressure head relationship. Table 1 gives the network details for a single-story building. The Hazen–Williams head loss equation is used, and the roughness coefficient for all pipes is taken as 130 for the network analysis. The static pressure will be provided by the source reservoir (R), which mimics the availability of static pressure in the pipeline. Water delivery at various outlets can be computed by connecting a reservoir at the respective outlets (R1 to R5). A minor loss coefficient of 1.8 is fixed to all pipes to account for losses due to bends, contractions, and fixtures. Two scenarios are considered to establish the relationship between supply and pressure head. In the first scenario, a maximum of five fixtures are considered to be active at peak time for all three types of buildings. In the second scenario, two fixtures are kept active for all three types of buildings. Through simulation with different static source pressure heads, the total supply for each house type is obtained. The EPANET 2.2 software [8] is used to analyze the network.

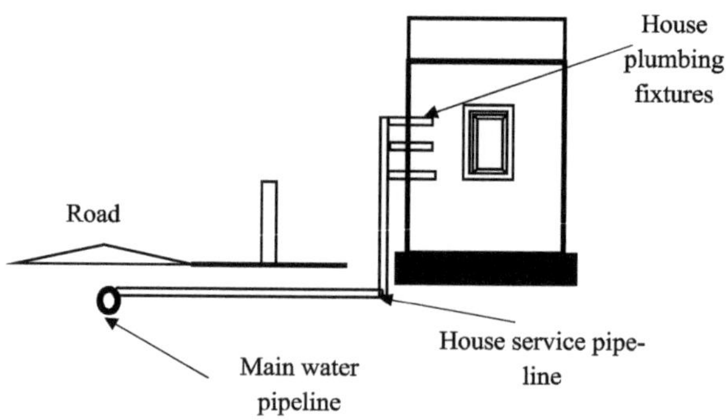

Figure 1. A typical house service connection and sample internal plumbing.

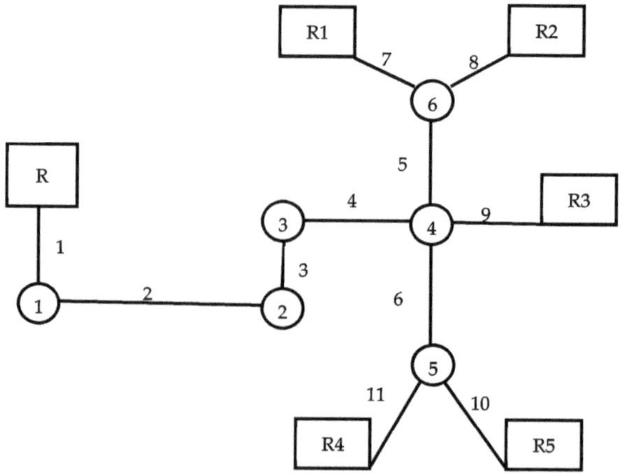

Figure 2. Prepared house service connection network for arriving pressure–flow relationships.

Table 1. Network details for the single-story house.

Reservoir/Node	Elevation (m)	Pipe Number	Length (m)	Diameter (mm)
R	Variable (5 m to 30 m)	1	1.00	18
R1	3.0	2	4.00	18
R2	3.0	3	2.00	18
R3	3.0	4	2.00	12
R4	3.0	5	3.00	12
R5	3.0	6	3.00	12
1	0.0	7	3.00	12
2	0.0	8	3.00	12
3	1.5	9	3.00	12
4	1.5	10	3.00	12
5	1.5	11	3.00	12
6	1.5	12	3.00	12

In the analysis, each pipe-connecting fixture (artificial reservoir) is provided with a check valve to prevent reverse flow in the pipeline. A plot is prepared between the flow and pressure head for all scenarios. The relationship between flow and pressure head is obtained as a power equation in MS Excel. Table 2 gives the equations for various scenarios obtained from the generated data. These equations are in the form of KP^n and are akin to the emitter equation used in EPANET (version 2.2) simulation software, in which K and n are the emitter coefficient and exponent, respectively. It is to be noted here that the obtained equations have different values of K and n. The K value for each node and each category of buildings can be set directly in the EPANET software, whatever the values. But no option is available to set a different value of n for each node. The software accepts one n for all nodes if the emitter option is activated for the simulation. To have a common exponent for double- and triple-story buildings, the emitter coefficient value for each case is corrected by multiplying with the correction coefficient as given in Table 2. The common n value corresponds to the single-story building when all five fixtures in the active state were used for all scenarios. Since there are three categories of buildings, it is not possible to set three K values for a node. Further, the emitter continues to be active when the node faces a negative pressure value during simulation. Under this circumstance, the emitter behaves as a source rather than a sink. To eliminate this situation and to handle three categories of buildings for supply, three small pipes of a 300 mm diameter, 0.1 m length, and roughness coefficient of 130 are connected with the demand node and with a dummy node at the other end, and a check valve is enabled for these artificial pipes (Figure 3). Now, in the fictitious nodes, emitter coefficients corresponding to single, double and triple stories are assigned. The elevation of all fictitious nodes shall be the actual demand node. No demand value shall be set to any nodes for this analysis. The network can be simulated after inputting the necessary data, as mentioned above. This basically requires the addition of a fictitious system to all nodes. Developments in hardware and software that are happening in the computing world have allowed to include all kinds of water supply information in modelling [9]. As the software is designed to handle more nodes and pipes for hydraulic analysis, the proposed approach can easily be adapted for any network size.

Table 2. Relationship between supply and pressure.

Type of Building	Number of Fixtures in Active State	Flow (Q) vs. Pressure Head (P) Relationship	R^2 Value	Correction Coefficient (Cc)	Adjusted Q vs. P Relationship
Single-story	5	$Q = 6.024P^{0.611}$	0.9987	0.971	$Q = 5.849P^{0.62}$
Double-story	5	$Q = 14.129P^{0.669}$	0.9955	1.176	$Q = 16.616P^{0.62}$
Triple-story	5	$Q = 15.287P^{0.670}$	0.9968	1.180	$Q = 18.039P^{0.62}$
Single-story	2	$Q = 5.206P^{0.612}$	0.9987	0.974	$Q = 5.071P^{0.62}$
Double-story	2	$Q = 10.775P^{0.671}$	0.9957	1.184	$Q = 12.758P^{0.62}$
Triple-story	2	$Q = 13.049P^{0.668}$	0.9975	1.172	$Q = 15.293P^{0.62}$

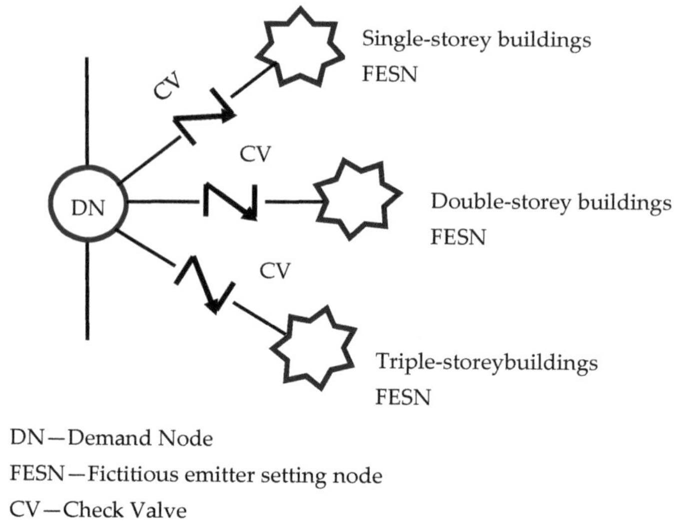

Figure 3. Fictitious component arrangement at each demand node for supply-driven analysis.

3. Hypothetical Network Preparation

A hypothetical network (Figure 4) is prepared to illustrate the proposed method of analysis of WDN. The network is supplied by one reservoir with a total head of 50 m. There are 15 demand nodes and 22 pipes. The elevation of all nodes is taken as 0 m. The geometrical properties of the network are presented in Table 3. The number of houses based on each type of story is assigned for each street, and then it is transferred to the end nodes of the pipe by taking half of the house service connections of each link. Table 4 provides node-wise house service connections in each type and gives each node's peak demand. The peak demand is taken as three times the average demand over a day. The Hazen–Williams roughness co-efficient is taken as 130 for all pipes. The purpose of peak demand is to show how the nodal pressure fares while the network is simulated using peak demand. A pressure of 30 m is required to supply water for the entire network. The per capita water requirement considered in this study is 200 liters per person. The population size per house is considered to be five for each floor.

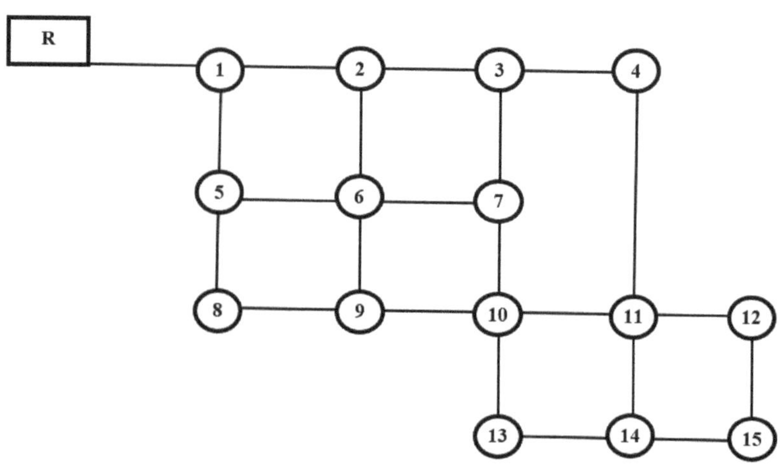

Figure 4. Layout of the hypothetical network.

Table 3. Hypothetical network details.

Pipe ID	Connecting Nodes	Length (m)	Pipe Diameter in mm	Number of Buildings in Each Type		
				Single Story	Double Story	Triple Story
1	R–1	50	250			
2	1–2	360	150	30	20	10
3	2–3	720	150	80	30	10
4	3–4	360	100	20	15	5
5	1–5	240	200	25	15	0
6	2–6	240	50	35	20	5
7	3–7	240	50	28	12	0
8	4–11	540	100	52	28	10
9	5–6	360	150	38	14	8
10	6–7	720	150	86	24	10
11	5–8	300	100	24	22	4
12	6–9	300	50	22	22	6
13	7–10	300	150	40	10	0
14	10–11	360	150	30	20	10
15	11–12	480	100	20	35	5
16	10–13	600	100	70	45	5
17	11–14	600	100	70	25	5
18	12–15	600	100	54	16	10
19	13–14	360	50	75	20	5
20	14–15	480	50	75	25	0
21	8–9	360	100	50	5	5
22	9–10	720	100	52	18	10

Table 4. Node-wise building type and nodal demand.

Node ID	Node-Wise Single-Story Building	Node-Wise Double-Story Building	Node-Wise Triple-Story Building	Calculated Demand for PF = 3 (LPM)
1	28	17	5	104
2	73	35	13	252
3	64	29	7	208
4	36	21	8	136
5	43	26	6	156
6	90	40	15	302
7	77	23	5	219
8	37	13	5	115
9	62	23	10	198
10	96	46	13	323
11	86	53	15	321
12	37	26	8	148
13	72	33	5	229
14	110	35	4	310
15	65	21	5	190

4. Results and Discussion

The network is simulated for peak demand, and it is found that the demand is met with the required pressure of 30 m at all nodes. According to this analysis, the network can supply all nodes when it is working under normal conditions, provided the withdrawal pattern is as per peak demand by all consumers. It can be seen from Figure 5 that when water is drawn as per peak demand/design demand, the pressure at all nodes is above 30 m, and demand will be supplied as per design value, provided that the normal condition exists. This shows that the network's pipe configuration is appropriate to supply the demand with the required service pressure of 30 m.

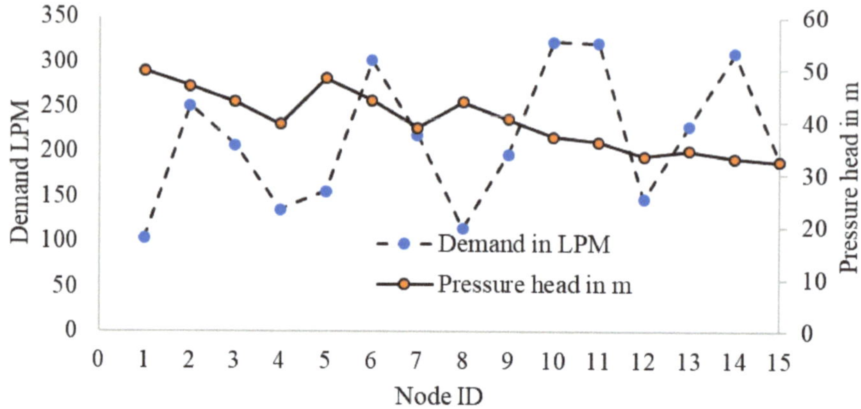

Figure 5. Plot showing the node-wise demand and corresponding pressure head for peak demand.

The CWS system, where the house connection service is linked directly to the internal plumbing system, will supply water whenever the internal plumbing outlets are opened. The flow through the system depends on the availability of pressure. The number of plumbing outlet fixtures could vary from house to house, and all its usage simultaneously occurs very rarely. If all outlets are opened simultaneously in every house, then the flow corresponding to that situation can be considered as the peak flow of the system. However, the pipe sizing in the WDN is not decided based on such flow rate. Hence, during its functioning, the supply variations among consumers are usually noticed, and consumers are happy in some regions, satisfied in a few regions, and unsatisfied in other regions, due to the large fluctuations in pressure and supply rate. To illustrate this scenario, the emitter option available is used as a unified plumbing arrangement that can show the supply based on the pressure available at that node.

The proposed SDA approach is applied considering certain critical scenarios where every user draws water simultaneously through their available plumbing outlets. Initially, the five and two plumbing fixtures are considered in active state for all houses. Rarely could simultaneous usage of plumbing fixtures, even at peak hours, by all consumers in a street occur. Hence, the simulation is also performed with 10%, 20%, 30%, 50%, and 75% of consumers in the active conditions for two fixtures alone for illustration. To simulate using the emitter expression developed for single-, double-, and triple-story buildings, each demand node is connected with the emitter node using a 0.1 m length and 300 mm diameter pipe having a Hazen–Williams roughness coefficient of 130, as shown in Figure 3. There will be three such fictitious systems to simulate these three types of dwellings. Each fictitious pipe is supported with a fictitious check valve to prevent reverse flow. Hence, each demand node will be connected with three fictitious pipe systems. In some cases, few nodes do not have three-story buildings; then, it is enough to connect two fictitious systems. Initially, the analysis is performed with 100% active condition for two and five fixtures in operational condition. The analysis clearly shows the deprived supply to several nodes when five and two fixtures are in 100% active condition. The pressure at node 1 alone is greater than 30 m, and the remaining nodes are experiencing pressure heads less than 30 m (Figure 6). The high supply rate of water in the advantaged nodes caused a deprived supply or extremely poor supply. Further, the supply variation between two fixtures and five fixtures in the active state is minimal. It is obvious that when more fixtures are open, the flow gets distributed due to a drop in pressure despite more paths being available at that instant, and a slightly higher flow may be observed compared to the flow when two fixtures are open. These two scenarios seldom occur in reality. As the simultaneous opening of fixtures in all houses is rare, this analysis provides a theoretical observation for municipal engineers. Estimating the probability of active consumers provides a better analysis. In the absence of such data, the simulation can be carried out for a selected percentage of the consumers who are in active condition. Figure 7 shows the variation of nodal pressure for a different proportion of consumers in an active state. The nodal pressure falls below 30 m for 12 nodes, and still, consumers receive water when 10% of house service connections (HSC) are in the active state. This scenario changes when the level of consumers' active state increases. The consumers living on the ground and first floors can cater to their needs. The consumers living on higher floors will be deprived of water supply due to poor pressure availability. The supply is possible only if the consumers at the advantaged nodes become inactive. The advantaged consumers always meet their requirements very quickly, and they close their fixtures as soon as the required volume is received. This could help to tide over to a certain extent for those consumers deprived at those moments.

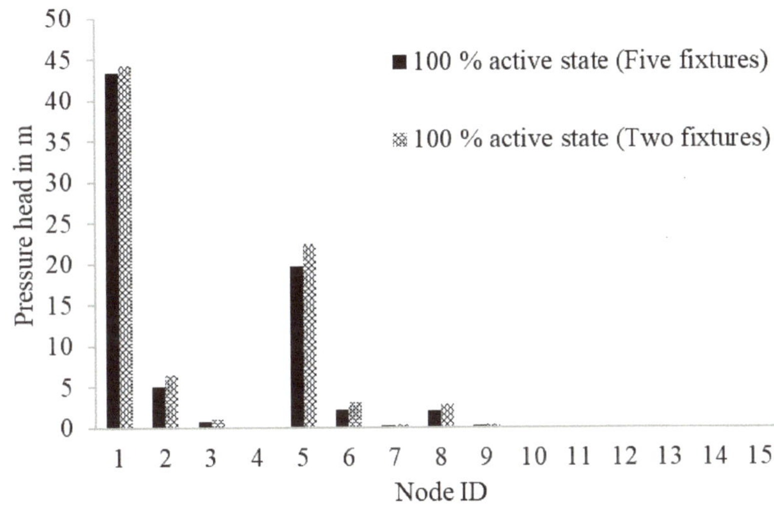

Figure 6. Nodal pressure head when two and five fixtures are in 100% active state.

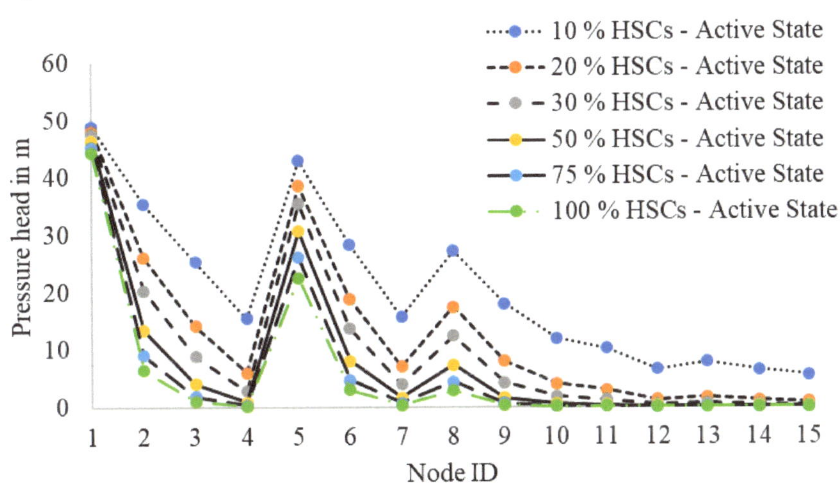

Figure 7. Observed nodal pressure head for various percentages of HSCs in active state.

It can be seen from Figure 8 that the consumers at nodes 1, 2, 5, and 6 receive the supply at a higher rate whenever they open their plumbing fixtures. The entire simulation is carried out with supply unit LPM. While comparing the corresponding design demand (three times the average demand) with 10% of house service connections under two fixture operating conditions, the supply-to-design demand (S/DD) ratio varies from 1.01 to 4.67 The variations observed for 20%, 30%, 50%, 5%, and 100% are (0.61 to 9.24), (0.37 to 13.75), (0.16 to 22.60), (0.07 to 33.39), and (0.04 to 43.90), respectively. The upper value of S/DD is increasing enormously while the active HSCs increase, and a reverse trend is observed for lower range values. This clearly shows, in the case of flat terrain, how consumers close to the source receive more water than those located remotely from the source. If this is visualized with respect to average demand, then the above-shown ranges shall be multiplied by three.

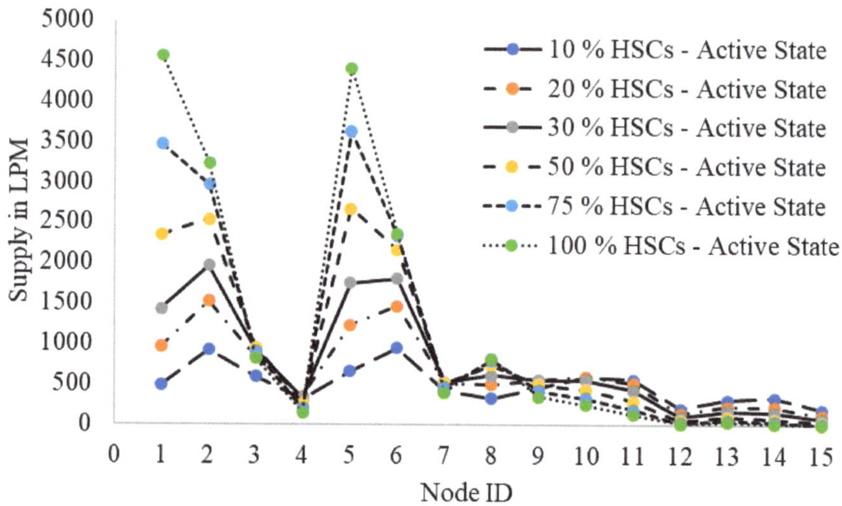

Figure 8. Observed supply at each node for various percentages of HSCs in active state.

More realistic results can be obtained if a more accurate relationship is developed based on field measurements. As the entire simulation analysis depends on the values of the emitter coefficient and exponent, the measurement of flow from the street main to the house service line under different fixtures in active state with varying static pressure in the main pipe could help to arrive at a more accurate relationship between pressure and flow. In developed countries, water billing based on the volume of water drawn can provide the usage of water for each service connection. But it will not reflect the peak usage of water. Hence, measuring the flow and pressure at selected intervals of time in the house service connection can help the modelers accurately simulate the networks. Further, laboratory-level investigation of typical house plumbing systems under different pressure heads and operation of fixtures could help better present the flow and pressure relationship. Using such relationships in the analysis of water distribution for CWS systems can replace the current practice of demand-based design. The simulated results for 10% active consumers denote the deficit in the pressure at several nodes like other cases considered in this analysis. The pipes 1 to 5, 7, 8, 11, and 15 to 22 are revised as 400, 350, 250, 200, 300, 200, 200, 200, 200, 200, 150, 150, 150, 150, 200, and 200 mm, respectively, in order to obtain a 30 m pressure head at all supply nodes. The diameter of the pipes carrying a large quantity of flow is increased and simulated by the proposed approach. It is noticed that the pressure at all demand nodes has reached above 30 m. Figure 9 shows the nodal pressure head for the initial and revised configurations of the network. Further, the S-DD ratio for the revised configuration has a higher value at all nodes than the initial configuration (Figure 10). The S-DD range for the revised pipe configuration (3.03–4.72) has become closer compared with the initial configuration (1.01–4.67). Further, it can be noticed from Figure 10 how the SDD ratio varies at each node for the initial and improved network configurations. The two curves become wider at the higher node numbers which are relatively far from the source. Since the pressure at all nodes is above 30 m for the revised configuration, supply to all floors is possible simultaneously when 10% HSCs are in the active condition. But, in the case of a network with the initial configuration, only partial supply at several nodes is possible. It can be understood that supply to the ground floor would be satisfied, but supply to higher floors may not be possible owing to inadequate pressure. It is evident from this study that an increase in the diameter of the pipes carrying more flow or posing more energy losses shall need to be revised suitably or optimally. In some cases, the head at the source shall be revised to maintain the service pressure. This study indicates that the peak factor (PF) varies from 9.09 to 14.15 with respect to average demand when 10% of HSCs are in the active condition for the revised configuration. The proposed analysis method can certainly be useful in the optimal design of WDNs where the sizing of pipes and fixing of the source head are the major objectives.

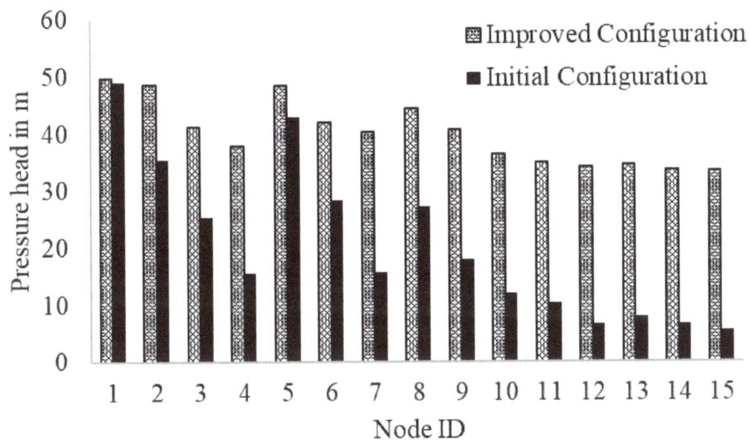

Figure 9. Nodal pressure head for initial and improved pipe configurations when 10% HSCs are in active state.

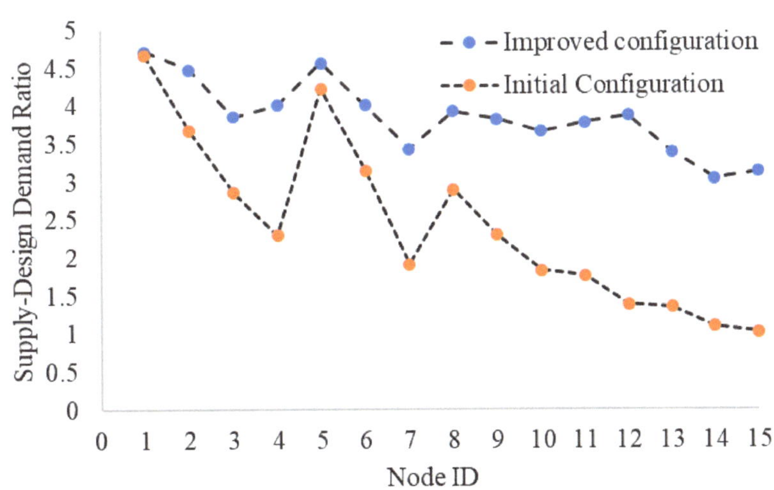

Figure 10. S/DD ratio for the initial and improved pipe configuration.

The approach described above provides the total supply for each type of building instead of the floor-wise supply rate. To know the floor-wise supply, the emitter nodes for each story shall be elevated to the level of the supply rather than following lumped supply quantification on the ground floor. To examine the effect of supply to different floors, the added fictitious emitter nodes are elevated to 1.5 m, 4.5 m, and 7.5 m, corresponding to the ground floor (GF), first floor (FF), and second floor (SF). The emitter equation corresponding to the GF is employed to simulate the supply. The simulation is performed for two fixtures in active condition for each house for different percentages of HSCs. Table 5 presents the floor-wise supply at each node. It can be seen from Table 5 that supply is possible to all floors when 10% of HSCs are in the active state. From 20% onwards, supply to certain floors does not occur due to inadequate pressure. In the case when 100% of HSCs are in the active state, the supply is only possible on the GF at all nodes, and no supply occurs on the first and second floors for most of the nodes except those nodes located near the source. Figure 11 shows the nodal supply for the initial and improved configurations. The nodal pressure at all the nodes is found to be greater than 30 m in the case of the improved network configuration, whereas in the initial configuration, the pressure varies from 15 m to 47 m. Though the initial configuration ensures supply to all floors when 10% of HSCs are in the active state, it is unable to supply water with the required pressure of 30 m. The total rate of supply for the initial and improved configurations is obtained as 5551.39 LPM and 7914.47 LPM, respectively, when 10% of HSCs are in the active state. The pipe size of the main pipe plays a vital role in maintaining pressure. Hence, the current practice of pipe selection, either by thumb rule or optimization, will not provide a practically feasible solution. Certainly, this study will draw a design practice followed for pipe sizing based on nodal demand.

Table 5. Floor-wise supply when two fixtures are in operation for various percentages of HSCs in the active state.

Node ID	Floor Wise Supply Status with Two Fixtures in the Active State																	
	10%			20%			30%			50%			75%			100%		
	GF	FF	SF	GF	FF	SF	GF	FF	SF	GF	FF	SF	GF	FF	SF	GF	FF	SF
1	Y	Y	Y	Y	Y	Y	Y	Y	Y	Y	Y	Y	Y	Y	Y	Y	Y	Y
2	Y	Y	Y	Y	Y	Y	Y	Y	Y	Y	Y	Y	Y	Y	Y	Y	Y	Y
3	Y	Y	Y	Y	Y	Y	Y	Y	Y	Y	Y	Y	Y	Y	**N**	Y	Y	**N**
4	Y	Y	Y	Y	Y	Y	Y	Y	Y	Y	Y	**N**	Y	**N**	**N**	Y	**N**	**N**
5	Y	Y	Y	Y	Y	Y	Y	Y	Y	Y	Y	Y	Y	Y	Y	Y	Y	Y
6	Y	Y	Y	Y	Y	Y	Y	Y	Y	Y	Y	Y	Y	Y	Y	Y	Y	Y
7	Y	Y	Y	Y	Y	Y	Y	Y	Y	Y	Y	**N**	Y	**N**	**N**	Y	**N**	**N**
8	Y	Y	Y	Y	Y	Y	Y	Y	Y	Y	Y	Y	Y	Y	Y	Y	Y	Y
9	Y	Y	Y	Y	Y	Y	Y	Y	Y	Y	Y	**N**	Y	**N**	**N**	Y	**N**	**N**
10	Y	Y	Y	Y	Y	Y	Y	Y	**N**	Y	**N**	**N**	Y	**N**	**N**	Y	**N**	**N**
11	Y	Y	Y	Y	Y	Y	Y	Y	**N**	Y	**N**	**N**	Y	**N**	**N**	Y	**N**	**N**
12	Y	Y	Y	Y	Y	**N**	Y	**N**	**N**	Y	**N**	**N**	Y	**N**	**N**	Y	**N**	**N**
13	Y	Y	Y	Y	Y	**N**	Y	Y	**N**	Y	**N**	**N**	Y	**N**	**N**	Y	**N**	**N**
14	Y	Y	Y	Y	Y	**N**	Y	**N**	**N**	Y	**N**	**N**	Y	**N**	**N**	Y	**N**	**N**
15	Y	Y	Y	Y	Y	**N**	Y	**N**	**N**	Y	**N**	**N**	Y	**N**	**N**	Y	**N**	**N**

Y—Supply reaching; N—Supply not reaching.

Table 6 gives the rate of supply per connection for the initial and upgraded network configurations. There is a large difference between the minimum and maximum supply to all floor levels in the case of the initial configuration. This difference appears to be smaller for the upgraded configuration. It is evident from this study that the rate of supply increases when the availability of pressure is higher. Further, it is to be noted that the flow rate per connection decreases when the floor levels increase. Figures 11 and 12 provide the rate of supply at each node for the initial and upgraded configurations when 10% of HSCs are in the active state. The supply of water to each floor varies for both configurations. The variation appears large in the initial configuration. It is clear from this study that the rate of supply depends on pressure availability, and proper water main and sub-main pipe sizing is important to provide a reliable supply of water to consumers. The proposed approach can be used to assess the performance of different design alternatives, such as the pipe diameter, number of connections, active number of fixtures, placement of valves, and source head. The use of the proposed approach for the design of continuous water supply system networks can help engineers and planners to optimize network performance, reduce costs, and ensure the reliable delivery of safe and clean water to users. The limitations of the proposed analysis are that (i) the analysis adopts a steady state condition, and (ii) the transient flow that occurs in the system due to the closure of various valves is not accounted. This method can be adopted in all topologies of dwellings.

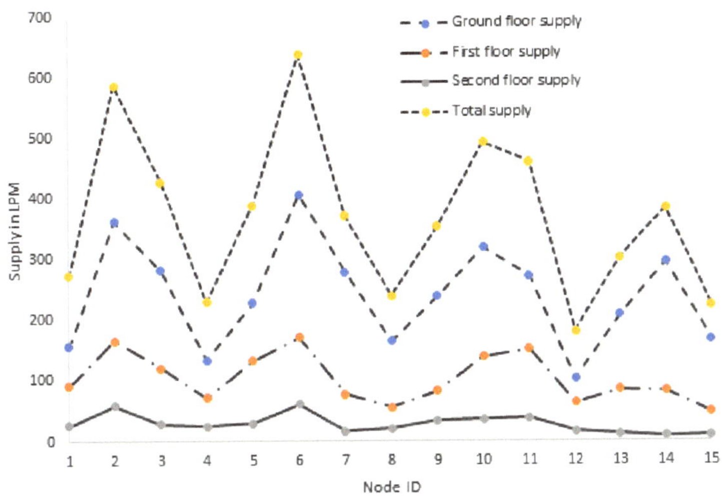

Figure 11. Nodal supply rate per connection for different level floors for the initial configuration.

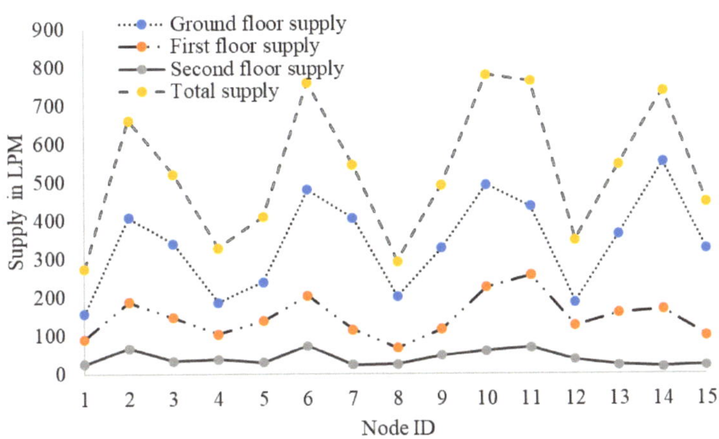

Figure 12. Nodal supply rate for different level floors for the improved configuration.

Table 6. Supply rate (LPM) per connection when 10% HSCs are in the active state.

Configuration	Ground Floor		First Floor		Second Floor	
	Maximum	Minimum	Maximum	Minimum	Maximum	Minimum
Initial configuration network	406.75	101.82	170.73	46.68	60.12	7.8
Upgraded configuration network	552.05	157.28	256.94	67.66	73.38	18.16

5. Conclusions

This paper presents the modeling method for the quantification of a pressure-based supply of water from the water main pipe to the house service pipe due to the opening of the fixtures at the house level. A new supply-driven concept is introduced for the analysis of WDN, where nodal demand is defined for the network in terms of the number of HSCs instead of traditional peak nodal demand. As the usual nodal peak demand estimated by multiplying the peak factor does not reflect the true peak demand, employing the proposed approach in the design of new WDNs can result in a better design and the identification of the nodes that could experience pressure deficit in the existing systems. The pressure and supply relationship obtained between the water main and HSCs can easily be adopted in the emitter tool available in the hydraulic simulation software (EPANET 2.2). The analysis using the proposed method indicates that the simultaneous opening of several plumbing fixtures could result in a major pressure deficit in the network, whereas the same network indicates that nodal demand is met with the required pressure when it is analyzed for peak demand. While analyzing with 10% active consumers in the system, close matching is found with peak demand-based analysis. This reinforces the need that certain pipes be increased its size, or the source head be increased owing to the peak supply that occurs when simultaneously opening the plumbing fixtures in the houses. The design of a water distribution network based on the projected population in the area helps to size the capacity of the storage system, whereas the projected number of house service connections and potential delivery of water when the fixtures are in the active state will be useful for the sizing of the pipes and pumps. The analysis depicts that the supply of large volumes of water at the high-pressure nodes in a relatively short time leads to a very high PF for which the main piping system is not designed. High friction loss in the main pipes that carry flow to nodes distant from the source creates pressure-deficient conditions. This study has also shown how an increase in the size of the pipe configuration of the network results in improved performance in terms of supply with required service pressure. The analysis additionally emphasizes the use of elevated fictitious nodes for each floor in evaluating the supply rate in the modeling. This paper further recommends additional theoretical and field research related to the pressure and supply of water at the various story buildings with respect to time that could help improve the pipe sizing design of water distribution networks.

Author Contributions: C.R.S. and P.S. participated in the conceptualization, methodology design, and result analysis for this research. N.R.T. was involved in the assessing application of the proposed method and the investigation of obtained results. All authors reviewed and edited the article. All authors have read and agreed to the published version of the manuscript.

Funding: This research received no external funding.

Data Availability Statement: All data, models, and code generated or used in this study are available in the manuscript.

Conflicts of Interest: The authors declare no conflict of interest.

References

1. Karpenko, M.; Bogdevičius, M. Investigation into the hydrodynamic processes of fitting connections for determining pressure losses of transport hydraulic drive. *Transport* **2020**, *35*, 108–120. [CrossRef]
2. Beal, C.D.; Stewart, R. A Identifying Residential Water End Uses Underpinning Peak Day and Peak Hour Demand. *J. Water Resour. Plan. Manag.* **2014**, *140*, 04014008. [CrossRef]
3. Crouch, M.L.; Jacobs, H.E.; Speight, V.L. Defining domestic water consumption based on personal water use activities. *AQUA—Water Infrastruct. Ecosyst. Soc.* **2021**, *70*, 1002–1011. [CrossRef]
4. Rizvi, S.; Rustum, R.; Deepak, M.; Wright, G.; Arthur, S. Identifying and analyzing residential water demand profile; including the impact of COVID-19 and month of Ramadan, for selected developments in Dubai, United Arab Emirates. *Water Supply* **2021**, *21*, 1144–1156. [CrossRef]
5. Bastidas Pacheco, C.J.; Horsburgh, J.S.; Attallah, N.A. Variability in Consumption and End Uses of Water for Residential Users in Logan and Providence, Utah, US. *J. Water Resour. Plan. Manag.* **2023**, *149*, 05022014. [CrossRef]

6. Suribabu, C.R.; Sivakumar, P.; Nivedita, S. Volume Driven Analysis for House Level Water Supply Assessment in an Intermittent Water Supply System. *ISH J. Hydraul. Eng.* **2022**, *29*, 459–467. [CrossRef]
7. Swamee, P.K.; Sharma, A.K. *Design of Water Supply Pipe Networks*; John Wiley & Sons, Inc.: Hoboken, NJ, USA, 2008.
8. Rossman, L.A.; Woo, H.; Tryby, M.; Shang, F.; Junke, R.; Haxton, T. *EPANET 2.2 User's Manual*; Water Infrastructure Division, Center for Environmental Solutions and Emergency Response, U.S. Environmental Protection Agency: Cincinnati, OH, USA, 2020.
9. Martínez-Solano, J.; Iglesias-Rey, P.L.; Pérez-García, R.; López-Jiménez, P.A. Hydraulic Analysis of Peak Demand in Looped Water Distribution Networks. *J. Water Resour. Plan. Manag.* **2008**, *134*, 504–510. [CrossRef]

applied sciences

MDPI

Article

A Novel Approach to Identify Industrial Logistics Decarbonization Opportunities: Method Development and Preliminary Validation

Philipp Miklautsch-Breznik *, Mario Hoffelner and Manuel Woschank

Chair of Industrial Logistics, Montanuniversitaet Leoben, 8700 Leoben, Austria
* Correspondence: philipp.milkautsch@unileoben.ac.at

Featured Application: This approach can assist industrial logistics professionals in obtaining a fresh viewpoint toward their decarbonization strategy.

Abstract: This article explores how different types of inventories affect the costs of decarbonizing transportation in manufacturing companies. For these companies, it is difficult to find affordable ways to reduce emissions from transportation given their resource scarcity. Additionally, they handle numerous inventory items that have varying transportation needs based on their order frequency and value, which necessitates the development of tailored inventory management strategies. One tool to do so efficiently is the ABC/XYZ analysis, which classifies items into nine different inventory categories. These groups have different economic importance and predictability, which impacts total logistics costs. Our literature analysis contends that lower-carbon transportation alternatives yield varying abatement costs contingent upon the specific inventory categories. Subsequently, we empirically validate this proposition through discrete-event simulations in two case studies involving Austrian manufacturing enterprises, employing combined road-rail transportation as an illustrative decarbonization measure. Statistical tests substantiate the significance of the XYZ dimension in influencing carbon emission abatement costs during the transition from road to rail transportation. In conclusion, our study offers a novel perspective on decarbonization efforts, underscoring the importance of leveraging established management tools to inform strategic decarbonization decisions. This research holds promise for catalyzing progress in overcoming entrenched challenges associated with decarbonization initiatives within industrial logistics.

Keywords: green logistics; decarbonization; climate change; industrial logistics; ABC analysis; XYZ analysis; simulation; transportation

Citation: Miklautsch-Breznik, P.; Hoffelner, M.; Woschank, M. A Novel Approach to Identify Industrial Logistics Decarbonization Opportunities: Method Development and Preliminary Validation. *Appl. Sci.* **2023**, *13*, 12277. https://doi.org/10.3390/app132212277

Academic Editor: Radu Godina

Received: 12 September 2023
Revised: 2 November 2023
Accepted: 6 November 2023
Published: 13 November 2023

1. Introduction

Decarbonizing logistics is, from a scientific perspective, crucial for mitigating climate change [1], and from an institutional perspective, crucial for achieving net-zero pledges [2]. Systematic approaches to developing strategies for logistics decarbonization from a freight owners' perspective already exist. For example, the 10C approach by McKinnon [3] consists of ten activities and includes possible options to consider. The field of developing carbon mitigation measures for transportation is well researched, from discussing different drivetrain technologies [4] and alternative fuels [5] in technical terms to developing green routing algorithms [6]. Regarding the application of greening measures in practice, generic total cost of ownership studies exist, but often only distinguish between factors that are not tangible from the perspective of the freight owners, e.g., the application scenarios in urban logistics [7], factors on the national level [8], or barriers for the implementation of specific measures [9,10]. Although all these contributions are important for the decarbonization progress, efficient and effective guidelines that freight owners can use to select decarbonization measures are missing. Most studies only rank the different measures by abatement costs. Although this is a common instrument to evaluate the efficiency of decarbonization measures and is highly important in selecting decarbonization measures [11–13], abatement costs only provide limited support for firms in identifying appropriate alternatives that can be ranked and prioritized based on their respective situation.

The challenge is to select a decarbonization measure that minimizes expenses and avoids disrupting operations [14,15]. This problem is particularly difficult to overcome when dealing with diverse goods that have varying value and volume compositions, which is the case in most industrial companies. To the best of our knowledge, there is currently no way for industrial companies to increase the effectiveness and efficiency of the process of identifying and selecting transport decarbonization measures for further evaluation.

Nevertheless, one positive result is that parallels can be drawn with another area of industrial engineering and management, that is, inventory management, which deals with many items and must prioritize them to manage inventory levels [16]. Therefore, instead of investigating each item on its own, a common approach is to classify goods with similar characteristics or importance to the firm's success. Inventory control tools for these purposes have been in use for economic reasons for many years. One of the best-known inventory control tools is the ABC/XYZ analysis, which classifies inventory items along the two dimensions, "value" and "order frequency," into nine types. For each of these types, different replenishment strategies are commonly applied to efficiently and effectively manage inventory levels [17].

With regards to these parallels, in this study, we introduce a novel approach to identify industrial logistics decarbonization measures that differentiate the transportation requirements of the transported goods in line with the ABC/XYZ analysis. Simply put, we propose that different transportation requirements come with different costs for lower-carbon transportation. From the inventory management perspective, higher transportation costs can be borne for some inventory items, as their storage is as expensive as their transportation [16]. If shipments in which such goods are transported are to be decarbonized, high abatement costs may have to be expected. For shipments of lower value or shipments with a stable demand, cost-effective decarbonization measures might be an option, leading to negative abatement costs. In the first part of this paper, we delineate this novel perspective from literature and define our proposition.

In the second part, we present the results of two simulation case studies, exemplarily testing the proposition using a promising decarbonization option, i.e., the shift from road to rail. We develop a discrete-event simulation using Python that mirrors the behavior of a logistics system using combined road–rail transportation. We decided to use this measure for several reasons, which are elaborated on in Section 4.1.

To summarize, the overarching goal of this paper is to introduce a novel perspective on carbon management for industrial logistics and provide the first evidence for its effectiveness. For researchers, this adds fertile ground for further research on the entrenched field of applying decarbonization measures in industrial logistics. As for practitioners, this paper offers an efficient way of investigating decarbonization measures to reduce their logistics' impact on climate change. Chances are high that ABC/XYZ analysis has already been implemented in a manufacturing company, which is why this approach proves useful for efficiently evaluating alternative and environmentally friendly transportation methods.

The paper is structured as follows: Section 2 presents the research methodology applied in this paper. Section 3 presents the results of a literature review on the selection of logistics decarbonization measures from an industrial company's perspective, the parallels to inventory management and the developed theory. Section 4 presents the argumentation for using combined road–rail transportation to evaluate the proposition and the method to do so. Section 5 outlines the case studies for the exemplary validation, the assumptions, and scenarios, as well as the simulation results. The paper concludes with a discussion of the limitations and implications in Section 6 and a summary of the results in Section 7.

2. Research Methodology

As the title of this paper suggests, this research is split into two consecutive parts. In the first part, the proposition is formulated. The initial motivation to do so emerged from the preceding research of the authors and discussion on decarbonization practices with industry experts. Throughout these talks, the authors recognized a repeating pattern where logistics managers stated that a specific decarbonization measure is only competitive for a certain type of good, i.e., a small portion of the inventory. Starting with this observation, we have gradually developed our theory with findings from the literature in a back-and-forth manner between literature and practice; we present it in the first part of this paper. This method of theory development is suggested to build theory from case studies [18] and applied by other researchers in supply chain management [19].

The second part of the paper deals with the validation of the developed proposition. However, since this proposition is very generic, it is not possible to test it in all its facets. For this reason, we have chosen a specific decarbonization measure to test the proposition. This measure is the shift from road transport to combined transport. We quantify the carbon abatement costs through a discrete-event simulation using primary data from two Austrian industrial companies.

The research process is visualized in Figure 1.

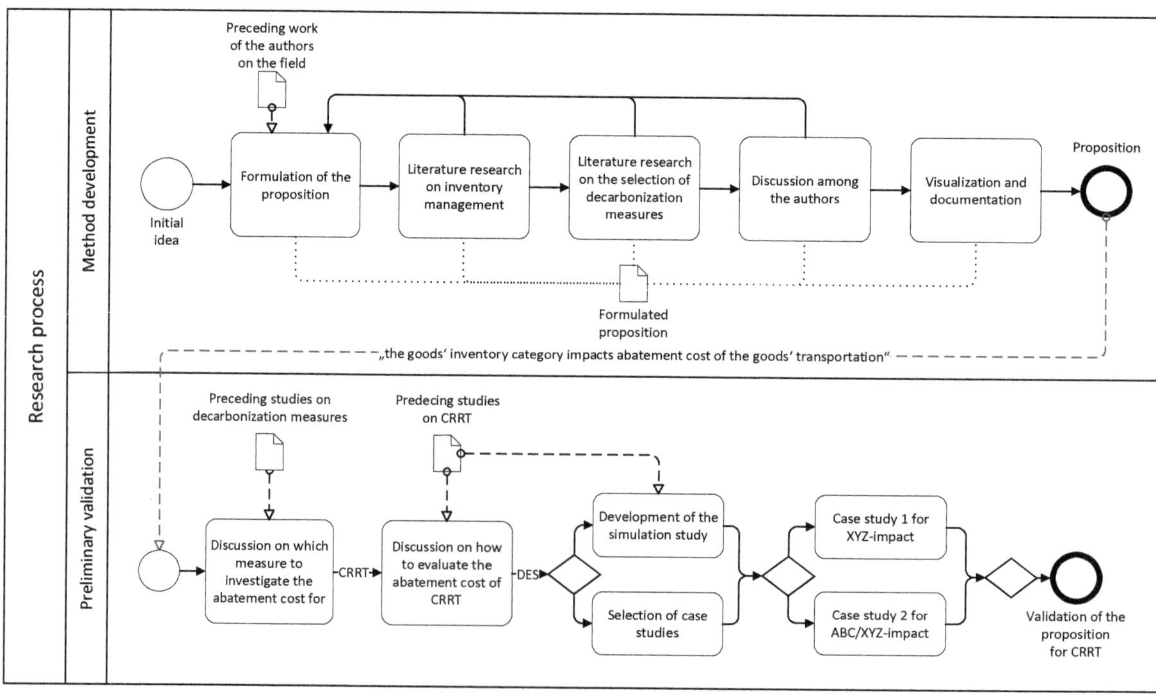

Figure 1. The research process of this study.

3. Theory Development

This section explains the challenges involved in selecting decarbonization measures from a manufacturing company's perspective and delineates our theory.

3.1. Selection of Decarbonization Measures

To decarbonize industrial logistics, researchers and practitioners discuss a vast number of different measures [20], which are commonly classified through the Avoid–Shift–Improve (ASI) approach [1,21,22]. As modern economic theories state [23], implementing all the measures is impossible in practice due to resource scarcity. As economic entities only have limited resources available, policymakers and managers who define decarbonization pathways need to prioritize decarbonization measures [24] for some criteria. On a macroeconomic level, policymakers may prioritize the highest-emitting economic sectors in their decision scope [25]. Delving deeper into the sectors, such as the building sector, scholars suggest prioritizing actions according to their decarbonization potential, respecting their detailed technical feasibility [26]. Other researchers promote prioritizing those technologies that are least invasive to the current environment, i.e., rooftop photovoltaics in combination with electric vehicles [27,28].

Although these prioritization efforts are all, from an environmentalist perspective, comprehensible, actual decarbonization decisions in the industry are rarely one-dimensional. For example, when investigating decisions on a microeconomic level, empirical research found that decarbonization alternatives in the building sector were evaluated and prioritized by 27 different criteria [29]. Similar research approaches are applied to other sectors, including energy generation. In one case, green hydrogen production alternatives were prioritized according to their efficiency and sustainability, considering the criteria capital cost, feedstock cost, operations and maintenance cost, hydrogen production, and CO_2 emission [30]. These criteria focus on cost, which already implies that prioritization based solely on the effectiveness of decarbonization is unlikely. Instead, decarbonization may only be possible economically, which is a theory supported by various authors (for instance, see [31,32]), not least for freight transport. Exemplarily, the factors hindering the adoption of electric trucks in the United States were researched, and the results indicated that the top causal factors are the business model and partnerships, product availability, and charging time [9]. A Delphi Study on factors affecting the adoption of alternative fuel-powered trucks in Germany found that cost and reliability factors are ranked highest among practitioners [10]. These criteria, in the end, all reflect the efficiency of electric trucks. Similarly, the chief impediments mentioned by industrial logistics experts when surveyed about green practices were mostly related to costs [14]. A study proposing a multi-criteria decision-making tool for industrial logistics practitioners defined eight criteria for evaluating decarbonization alternatives. The two criteria considered most relevant in the demonstration case study were "abatement cost" and "impact on logistics performance" [15].

In summary, the selection of decarbonization measures in industrial logistics practice prioritizes efficiency over effectiveness. As previously demonstrated in [15], the economic efficiency of decarbonization measures can thereby be well expressed in terms of the abatement costs AC. These are defined by:

$$AC_m = \frac{C_m - C_b}{G_m - G_b}, [C] = \text{EUR}, [G] = t\,CO_2\,e$$

describing the cost difference of a decarbonization measure m compared to a baseline situation b, concerning the greenhouse gas (GHG) emissions reduced by m. Simply put, AC indicates how much it costs to reduce one ton of GHG and, thus, to which extent that measure is competitive with other measures. Abatement costs are commonly reported in studies dealing with decarbonization pathways or options, for example, in the energy sector [11,12]. Nevertheless, AC starts to be frequently used in freight transport, as well. In a recent study by Chinese researchers, various decarbonization measures for sand and gravel transportation were analyzed using the ASI strategy. By calculating AC, it was found that switching to lower carbon modes of transport was the most competitive option [22]. The cross-sectoral study of Denmark's transition to fossil-free transportation revealed differences in abatement costs both within and between transport segments [13]. An analysis of American electric vehicle procurement incentives revealed that the utilization of electric vehicles influences the abatement costs related to them [33].

According to these studies, the efficiency of decarbonization measures, in terms of abatement costs, varies significantly between implementation scenarios characterized by, for example, weight, volume, frequency, origin, and destination. This emphasizes the challenge of identifying efficient decarbonization measures in the logistic network of a manufacturing company, which usually involves goods of varying priorities, suppliers, customers, and values. To the best of the authors' knowledge, it is still unknown how industrial firms, whose main competence lies outside the decarbonization of logistics, can identify efficient decarbonization measures effectively.

3.2. Inventory Management

Handling the above-mentioned challenge regarding the multiplicity of goods purchased, transported, and handled in a manufacturing company, is the responsibility of inventory management. Inventories frequently constitute a considerable portion of the total assets on the balance sheet of manufacturing companies; a 15–20% share is not uncommon [16]. Given that inventory holding costs can range up to 26%, inventories have a significant impact on a company's cost structure. The DuPont scheme incorporates inventories in the evaluation of the current asset in the return on investment (ROI) calculation and highlights the lever of inventory reductions: a 10% inventory reduction results in a 3.6% ROI increase [16].

Nevertheless, inventory management has been found to significantly affect both the economic success and sustainability of the company. Choices made on the inventory management level thereby influence a multitude of factors, e.g., the necessary distance, the frequency, or the mode of transportation. For instance, replenishment strategies establish the essential prerequisites for transportation processes in terms of frequency and volume, which, if optimized, can result in cost and environmental advantages [34]. Mode choice depends on lead time, volume, frequency, and costs [35]. Whereas fast modes enable priority shipments, they usually come with premiums. Slow modes enable efficient transportation of high volumes at one time and reduce the replenishment frequency but are less flexible [36]. Calibrating the inventory management strategy for the whole inventory is a complex task for a manufacturing company handling multiple products given the varying product requirements [37].

To effectively optimize inventory, organizations, therefore, commonly prioritize goods based on their relevance. One widely used tool for distinguishing highly relevant goods from less relevant ones is the ABC analysis. The analysis classifies articles, suppliers, or inventory movements according to their materiality for inventory management. The A category thereby comprises the highest-value elements that account for 70–80% of the value measured in monetary units. Usually, these elements only account for 5–10% of the number of elements, highlighting the most relevant ones for inventory management [38]. Regarding the inventory value, these elements are of utmost importance and provide the largest lever for improvements. Elements belonging to category B generally represent 25% in terms of quantity and 10–15% in terms of value. The rest of the elements, which are considered the least valuable, are categorized as C. These elements generally comprise 65% of the total quantity, but contribute only 10% to the overall value [38]. The ABC analysis is known for its ease of use and its clear graphical representation, making it a commonly used instrument among practitioners [16]. Nevertheless, it has been criticized frequently for its one-dimensionality, which is why researchers started to incorporate multi-criteria decision-making techniques in the field of inventory management [39]; they added, for example, non-financial criteria [40] and uncertainty [41] to the classical ABC analysis. One further dimension that has been used to detail the results of the ABC analysis is the demand fluctuation, which is investigated using the XYZ analysis [42]. Like ABC, XYZ classifies inventory elements into three categories but uses the coefficient of variation CV instead of the element's monetary value. X elements, thereby, have a stable demand and a $CV < 0.1$; Y elements show seasonal fluctuations having $0.1 \leq CV < 0.25$; and Z elements are identified by $CV \geq 0.25$, being vastly volatile and unpredictable [16].

3.3. The Developed Theory

As elaborated above, inventory management and its ABC/XYZ analysis is a crucial and extensively researched aspect of operations management. Manufacturing companies typically use distinct logistics strategies for each of the nine emerging cells [35], which differ by the transportation mode, the order frequency, and the lead time [43]. These factors thereby impact the transportation cost, inventory holding cost, and order cost, which, together, form the total logistics cost [35,44,45]. Thus, the ABC/XYZ category of an inventory item impacts the logistics strategy applied to it, which, in turn, impacts total logistics cost.

Thereby, a goods' ABC dimension primarily affects the inventory holding cost, as it represents the pecuniary value of the inventory item. The XYZ dimension, on the other side, impacts the required lead times, flexibility, and resulting vehicle utilization, as it represents the demand characteristics, thereby impacting transport and order costs.

Combining this with the findings of Section 3.1, that abatement costs for one decarbonization measure differ among transportation scenarios, we can, in turn, delineate that logistics strategies impact the abatement costs of specific measures.

Thereby, the abatement costs represent the total logistics cost difference between the conventional transportation technique and a lower-carbon alternative, divided by the mitigated carbon emissions. As the goods' category impacts the total logistics cost, we theorize that the goods' category also impacts the abatement costs of reducing carbon emissions from the goods' transport.

Figure 2 presents a graphical representation of the line of argumentation for this theory.

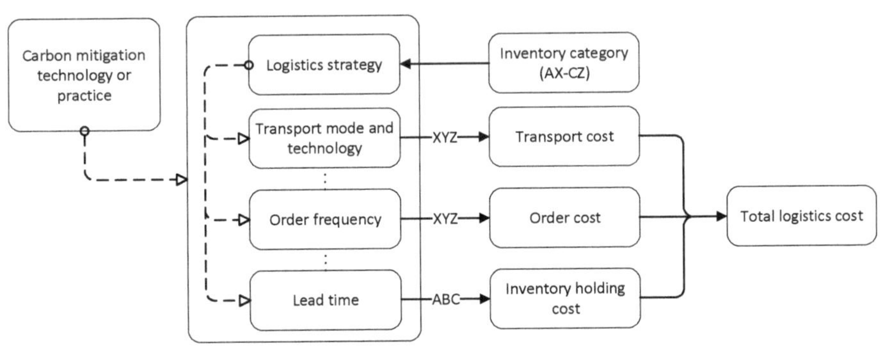

Figure 2. The impact of carbon mitigation practices on total logistics cost and the causal relationship with the inventory category.

This theory can be argued logically with different decarbonization measures, for example, vehicle selection. The truck class depends on several factors such as distance, speed, reliability, and flexibility. Items with a consistent demand (X items) may be transported in larger and slower trucks, while those with a volatile demand (Z items) require high-speed transportation in smaller trucks. A study by [13] found that abatement costs vary among various truck classes, which implies that the reduction of transportation emissions of X and Z items differ in this case. Additionally, there is evidence that the cost of reducing emissions for electric vehicles depends on their utilization [33], which suggests that the abatement costs for smaller, high-priority deliveries may be higher due to the need for low-utilization transport. Another example is the shift to lower-carbon modes of transportation, i.e., rail freight, which implies longer lead times. As elaborated above, these affect inventory holding costs. Thus, this measure is proposed to be more cost-effective for low-value goods (C) than for high-value goods (A), which implies different abatement costs for the transportation of A and Z items. The concept of differentiating abatement costs across inventory items is advocated by other researchers. For example, using different modes of transport for individual items across the entire range of products was found to identify products that are relatively cost-effective to abate, as well as more expensive-to-abate products [37].

To sum up, this theory leads us to our final proposition: The abatement costs of one carbon mitigation practice or technology differ significantly between the ABC/XYZ category of the good to which it is applied; simply put, "the goods' inventory category impacts abatement costs of the goods' transportation". Table 1 presents a conclusive summary of the literature on this topic, which is used to argue for this proposition.

Table 1. Literature summary.

Statement	References
Inventory management and its ABC/XYZ analysis are frequently applied in practice and well-researched	[16,34,35,37–41]
Inventory categories impact the selection of inventory strategies	[16,35]
Inventory strategies impact inventory cost	[16,35,43]
Inventory costs impact total logistics cost	[44,45]
Inventory strategies impact emissions	[46,47]
Implementing lower-carbon transportation technology, modes, or practices impacts total logistics cost	[13–15,20,33,48–50]
Selecting transport decarbonization measures on the product level can optimize total abatement cost	[37]
Transportation abatement costs differ across inventory categories	This study

Although our proposition has been carefully developed, it needs to be thoroughly evaluated. In the following section, we present an initial evaluation of the theory applied to one decarbonization measure, i.e., combined road–rail transportation.

4. The Methodology for the Proposition Validation

In-depth validation of the theory across all potential decarbonization measures exceeds the scope of a single article; therefore, in this study, we have opted to test the theory by way of exemplification using one specific decarbonization measure. In the upcoming sections, we present the rationale for opting for combined road–rail transportation (CRRT) as a model decarbonization measure, evaluating it through a discrete-event simulation approach. Additionally, we will highlight the critical facets of CRRT that are considered during the simulation, along with the description of the two case studies that were used for testing the proposition.

4.1. The Selection of Combined Road–Rail Transport as an Exemplary Decarbonization Measure

For the proposition validation, we selected CRRT as the exemplary measure, which is substantiated by several key factors. First, the emphasis on shifting towards lower carbon modes of transport as a central pillar of decarbonization literature [7] underscores the relevance and significance of rail transportation. As inland waterway transportation faces challenges regarding reliability [51,52], which is an important decisive factor for transportation users [53], rail transportation is the more common shifting alternative [21]. Since only a limited number of manufacturing firms have direct access to the rail network, intermodal transportation, particularly combined road–rail transport (CRRT), is promoted by various logistics service providers (see, e.g., [54–56]). The demonstrated cost-effectiveness of CRRT in mitigating emissions in China [22] and Europe [57,58] affirms its practicality and environmental viability.

Despite the potential advantages of CRRT, it presents significant challenges for shippers due to increased lead times and the involvement of multiple stakeholders in this type of transportation [59–62]. Therefore, this measure presents greater evaluation complexity compared to others, such as changing drivetrain technology. While the logistics system stays consistent with a change in drive technology, CRRT planning necessitates significantly extended replenishment times and consideration of rail network delays. For cost and emission assessments, it is necessary to consider two modes of transportation as well as load units and handling and storage procedures. This complexity creates challenges in conducting environmental and economic impact assessments. For these reasons, we opted to evaluate our proposition for CRRT, as it is a promising and hard-to-assess decarbonization measure. This selection aligns with the European Union's ambitious decarbonization goals, indicating that CRRT not only holds theoretical promise but also aligns with broader regional strategies for sustainable freight transport [63].

4.2. The Use of Discrete Event Simulation as the Evaluation Method

Because of the complexity of CRRT's impact assessment, simulation is a commonly applied methodology in researching intermodal transportation. Through simulation, researchers can gain a better understanding of mechanisms within the transportation system, reaching from the operations in transshipment hubs [64] to the utilization of transport corridors [65]. Simulation paradigms vary depending on the study's objective and encompass a multitude of possibilities, including agent-based [66], discrete-event [64], and system dynamics [67] approaches, as well as Monte Carlo simulation [49]. For the focal study, we developed a discrete-event simulation (DES) using Python and the Salabim library [68]. Within the simulation, the behavior of all elements and actors involved in CRRT, like consignees, shippers, hubs, trucks, trains, and load units, are modelled. We opted for DES because it is frequently applied to intermodal transportation [69–71] as well as the author's experience with DES.

4.3. The Case Studies Investigated for the Theory Validation

To validate our proposition, we use inbound shipment data from two Austrian industrial companies. The first case study is an Austrian electrical equipment manufacturer that provided us with nonfinancial data from its 2021 shipments over 11 months. Due to the attributes of the data, we cannot integrate the ABC part of the ABC/XYZ in this case study. For the validation of the ABC impact, we integrated a second case study, involving an Austrian cable manufacturer. The company provided us with shipment data from 2022, including the value of the goods.

For each case study, we combine all European inland road freight legs, including the final road leg for air or sea shipments, and road shipments as input for the DES. For each of the companies, we calculated costs and emissions from the status quo, i.e., road transportation, and compared it to two hypothetical shifting scenarios. Thereby, we were able to calculate hypothetical transportation abatement costs for each of the inventory categories.

Due to the numerous possibilities for planning, conducting, and controlling CRRT, we have made certain assumptions to ensure the simulation's effort is manageable. Therefore, the following constraints have been made:

- For competitive CRRT, the maximum number of transshipment operations during the main rail leg is set to one, i.e., no more than two rail services included in the main leg are allowed. The maximum allowed time for intermodal transportation is two days. To select the possible CRRT services, we search for the origin-destination pair on Routescanner.com (accessed on 11 September 2023) for each weekday and select the quickest connection that fulfils the requirements just mentioned. If there is no viable connection, we exclude the respective origin–destination pair from all scenarios. The rail services used are presented in the supplementary material.
- For pre- and post-haulage, we used the shortest possible routes from the suppliers to the origin terminal as well as from the destination terminal to the consignee. The distance and duration of these road haulages were acquired through the Openrouteservice.org (accessed on 11 September 2023) Distance API. We abstain from mentioning the precise addresses and distances due to confidentiality, but we can share that the average distances were 139 km and 169.5 km in the pre- and post-run for the first case, respectively, and 109.23 km and 193.82 km in the second case, respectively.

- The replenishment times differ between shipments from X, Y, and Z suppliers: X suppliers are given 7 calendar days replenishment time, Y shipments 5 days, and Z shipments 3 days. This reflects the predictability that is related to the different XYZ clusters. The replenishment time thereby determines how long before the planned arrival date the shipment is released by the supplier. For example, X suppliers are notified 7 calendar days in advance to send the shipment, either directly to the plant via road or to the first transshipment hub for intermediate storage.
- No further consolidation was considered. Shipment volumes and weights for intermodal shipments need to match the volumes and weights for the respective direct shipments that are given in the input data. It stands to reason that this is not the case with a real shift to rail as consolidation effects strengthen the business case for CRRT. However, we could not make any meaningful assumptions and would mix the evaluation with a second measure, which is why we decided against making any further assumptions on consolidation given the overall objective of the evaluation.
- To minimize cost, we aim to use 40-foot ISO containers. If the utilization of the 40-foot container is lower than 80% in terms of loading length, volume, or weight, we use 20-foot ISO containers instead. For competitive CRRT, each ILU utilization u_{ILU} needs to be larger than or equal to 80%.
- Combining the former two assumptions led us to discuss how to handle shipments that have $1 < u_{ILU} < 1.8$. Therefore, we introduced two dispatching modes for those shipments:
    a. In the first mode, we dispatch the whole shipment size to CRRT. This option constitutes the first shifting scenario, called "All ILUs on Rail" (ARA). In this scenario, the whole shipment, regardless of the shipment size, is dispatched to CRRT, meaning that one ILU is less utilized than 80%. This implies that some ILUs on rail are not well utilized, and the costs for renting, shipping, and handling the goods in this ILU are higher than for the better-utilized ones, but costs for direct road transport are obeyed. Before dispatching the ILUs via intermodal transportation, it is checked whether the ILU can reach the consignee on time with the available rail services. If this is not possible, the ILU is shipped directly by road transportation.
    b. This implies more cost for a low-utilized road transport. In this scenario, a minimum utilization of 80% is necessary for each load unit to be shipped via CRRT. If this utilization is not reached, the ILU is scheduled for direct road transportation. We call this scenario "Highly utilized ILUs on Rail" (URA).

Therefore, including the base case, three scenarios are presented for each case study, which are visualized in Figure 3.

Table 2 presents a summary of the input data that was used for the scenarios. The difference between the total and considered shipments is based on constraints 1 and 2 of the enumeration.

The emission data that is used is a combination of dynamic and static data. For the transportation legs, we use the EcoTransIT World (ETW) emission calculator as the shipments reflect real-world transports that have been conducted. ETW offers a comprehensive calculation methodology that accounts for real-world movements of ships and flights and integrates up-to-date emission factors, emission quantification standards, and traffic networks [72]. To address ILUs in the hubs, we use a static value of 30 kg CO_2e per transshipment activity, as suggested by the GLEC framework [73]. Emission intensities of transshipment activities differ between hubs and equipment used, but no detailed data was available to us regarding terminal-specific emission factors.

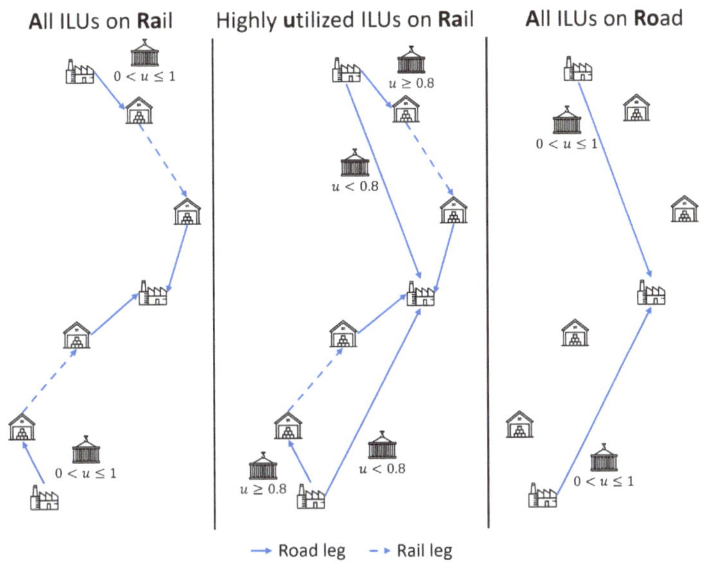

Figure 3. A schematic overview of the different dispatching mechanisms in the scenarios (from left to right the ARA, URA, and ARO scenarios).

Table 2. Descriptive statistics of the two cases under study.

	Case 1		Case 2	
	Total	*Considered*	*Total*	*Considered*
Number of suppliers	36	21	75	8
Number of products	n.a.	n.a.	359	10
Number of shipments	1215	275	1706	98
Shipment date range	February–December 2021	February–May 2021	January–December 2022	January–December 2022
Transport GHG emissions	548.51 t	247.25 t	305.60 t	153.01 t

4.4. Total Logistics Cost of Combined Road–Rail Transportation

Transportation costs generally comprise the cost components of each party involved in relocating goods [49]. Several studies have already defined the cost elements of intermodal transport and the comparison with unimodal transport. We use the cost functions defined in a recent case study on intermodal transportation [58] and detail them with other studies. The authors of [58] elaborate on the costs of intermodal transport services from a consignor's perspective and divide them into several parts. First, the pecuniary cost of transport PC includes the cost per kilometre and the cost of transshipment. Second, the monetary cost of transit time TT depends on the transport time, the value of the shipment, and the interest rate. Thirdly, the monetary cost of a delay D can be quantified, including fines and production downtime costs. Fourth, the cost of cargo loss C (i.e., damaged, expired, or stolen cargo) is a function of the value of the shipment, the fines, the production downtime costs, and the cost of reordering. Fifth, the study includes the cost of oversized cargo OC, which is highly dependent on the size of the cargo, and sixth, the social cost of transport SCT, the value of which depends on the method of quantification. Additional cost factors that have been considered in other studies are the expenses associated with storing a load unit at a terminal, the cost for the intermodal load units (ILU), incurred either as rent or depreciation, and the management and organization costs [74,75]. This study focuses on the perspective of the freight owner, which is why the total logistics costs are considered. Therefore, we have decomposed the aspects into six components that reflect the roles of all parties involved, including carriers, hub operators, railway operators, ILU owners, the organizing party, and the freight owner itself, including the transportation cost, order cost, and inventory holding cost. Respectively, these components are the cost for transport operations TC, the cost for hub operations HC, the cost for the ILU rent or depreciation $ILUC$, the cost for production downtime DC and the cost for capital CC, as well as the overhead cost for organizing the CRRT OC. TC includes cost for pre-, main, and post-haulage, and HC includes the cost of the origin and destination hub. Figure 4 illustrates the occurrence of these components in the transport chain. As the freight owner is the principal of the transport, the costs are borne directly or indirectly by the manufacturing company. As we are missing data on ILU thefts or damages, we neglect these cost components in the focal study.

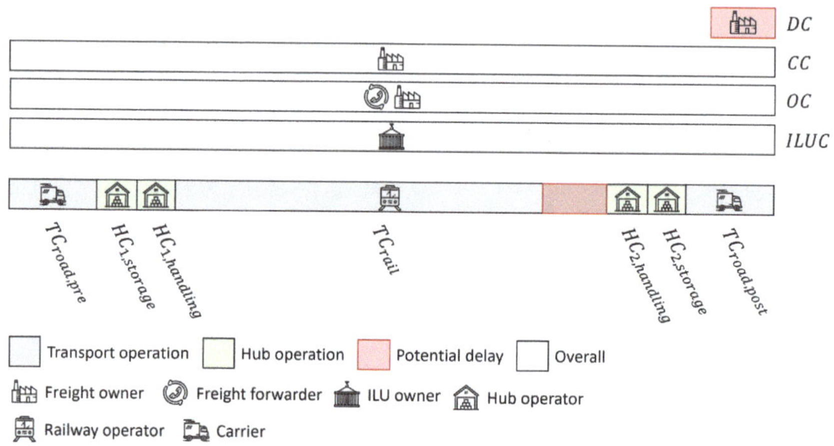

Figure 4. Cost structure of combined road–rail transportation from a freight owners' perspective (Icons from Freepik on Flaticon.com, accessed on 11 September 2023).

The cost data that was used in the simulation study is a combination of primary data and secondary data:

- TC_{road}: The input data from case study 1 were used to consider costs for direct truck transport. As a result, we developed a regressive function that depends on the truck utilization u to determine the freight charges per tonne-kilometre $TC_{road,direct,tkm}$. Specifically,

$$TC_{road,direct,tkm}(u) = 0.0582 + \frac{0.0244}{u}\frac{\text{EUR}}{\text{tkm}}$$

  Additional information on this can be found in Appendix A.
- TC_{rail}: To obtain pricing information for the main leg, we consulted the sales team at a railway operator for data related to a sample CRRT service. They provided us with a cost of 800 EUR for a single 60-foot wagon travelling an on-rail distance of 822 km and having a capacity of three 20-foot containers, which are commonly referred to as 20-foot equivalent units (TEU). Based on this information, we estimate the cost of transporting one 20-foot ILU per kilometre as:

$$TC_{rail,km} = \frac{800\text{EUR}}{3\text{TEU}*822\text{km}} = 0.32\frac{\text{EUR}}{\text{TEUkm}}$$

We are aware that rail transport pricing differs between operators, relations, and time, but regarding the efficiency of the research, we were not able to collect more detailed cost information empirically. Nevertheless, the 0.32 EUR/TEU used in this study is in the range of 0.46–1.35 EUR per forty-foot container provided by [75]. From this, we deduce that the exemplary price resulting from our research is representative of the European market.

- HC: For the hub cost, we use primary data from the involved terminals' homepages, if available. If not, the value of:

$$HC_{handling} = 48 \text{ EUR and } HC_{storage} = 0 \text{ EUR}$$

is used in line with [75]. For gateway movements, we use:

$$HC_{handling} = 27 \text{ EUR and } HC_{storage} = 0 \text{ EUR},$$

in line with [75] as no other data is available to us.
- $ILUC$: The ILU rental fee per day was retrieved from an Austrian container rental company, compared with data from [76,77], and set to

$$ILUC = 7.5\frac{\text{EUR}}{\text{day}}$$

- OC: Since actual road freight costs are included in the input data, the road OC is already included in the road freight charges.
- DC: As no data was available to us on the production delay, we initialized the simulation with:

$$DC = 500\frac{\text{EUR}}{\text{hour}}$$

- CC: In the first case, we cannot integrate CC in our evaluation due to the abundance of the shipment monetary values. For the second case, we compute the costs of capital by utilizing a literature-based interest of 12% from [62].

4.5. Distribution of Train Delay Times

Freight train delays negatively impact the reliability of intermodal freight networks and are an important, but under-researched element of combined transport [78]. Freight trains are typically less reliable regarding the planned arrival times than trucks, which is a planning challenge for freight owners. Reasons for freight train arrival delays can be grouped into two sets of factors [79]. The first set describes a deviation of the departure time from shunting yards or hubs. The distribution of these deviations in Swedish shunting yards has been investigated by [80]. In their paper, the authors find that the log-normal and the gamma distribution best approximate the departure delays and early departures, respectively. The second set describes delays during the journey of the train. Network capacity utilization, weather conditions, construction sites, and many more factors determine the actual arrival time of trains. To model the delays of freight trains in our simulation, we thus cannot rely on the departure time delays, as the influence of the journey would be neglected. As creating models that predict the arrival time by combining both sets of factors is its own body of research [79], we concentrate on approximating the actual arrival delay using a probability distribution. Some work in this field has already been done. For example, the delay time of arriving passenger trains was approximated for Chinese railway stations through exponential distribution [81]. Although passenger train punctuality is much higher than freight train punctuality in Europe (see, e.g., [82,83]), we argue that the nature of the factors influencing arrival delays are similar, and thus, the type of distribution approximating passenger train delays also approximates freight train delays. Therefore, we use the exponential distribution with the probability density function:

$$f(d) = \lambda e^{\lambda d}, d \in \mathbb{N}$$

to model the delay d of a freight train. For parameterizing, we use the punctuality information of the Austrian Railways, reporting that the share of combined wagonload traffic with less than 30-min delay in Austria in 2022 was 52% [82]. By calculating the cumulative distribution function:

$$F(d) = 1 - e^{-\lambda d}$$

we approximate $\lambda \approx 0.02531$ so that $F(29) \approx 0.52$. This results in a mean delay of $E = \frac{1}{\lambda} = 39.51$ min and a median delay of $m = \frac{\ln(2)}{\lambda} = 27.39$ min. In the simulation, we incorporate the delay by adding it to the scheduled lead time t_s, thereby defining the actual lead time t_a of a railway service as:

$$t_a = t_s + d$$

5. Validation Results

5.1. Results of the ABC/XYZ Analysis

In the first case study, the shipment data solely represents the supplier-level and not the product-level, which is why we conducted the XYZ analysis for the suppliers as suggested by [16]. The XYZ classification thresholds were set to $CV_{XY} = 0.3$ and to $CV_{YZ} = 0.6$. These values are in the range of the values presented by other literature, as elaborated in Table A1 in Appendix B. As values from the literature vary significantly, we established the thresholds to classify approximately 20% of suppliers as X, 50% as Y, and 30% as Z, as suggested by [16,38].

In the second case study, the shipment data contains weight and value data on the product-level. Thereby, each shipment represents the transport of a certain product from a supplier to the manufacturing plant. Due to the shipment data, we were able to conduct an ABC/XYZ analysis. Following the first case study, the XYZ thresholds were set to $CV_{XY} = 0.3$ and $CV_{YZ} = 0.6$. The ABC-thresholds were set to $s_{AB} = 0.8$ and $s_{BC} = 0.9$, whereby s is the share of the cumulative value, which aligns with other articles, as shown in Table A1.

Appendix B presents figures plotting the distribution curves of the XYZ and ABC/XYZ analysis of both case studies.

5.2. Results of the Simulation Study

The simulation of each scenario assessing the 275 shipments of the first case took about 5 min, and the scenarios of the second case about 1.8 min. The high-level results of the simulation study are comprehensively visualized in Table 3, outlining the total logistics cost of the scenarios for each case study along with the GHG emissions.

Table 3. Total logistics cost and GHG emissions resulting from the three scenarios for each case study.

	Case 1		Case 2	
	Total Logistics Cost EUR	*GHG Emissions t CO_2e*	*Total Logistics Cost EUR*	*GHG Emissions t CO_2e*
ARO	321,698.21	247.25	176,436.29	153.01
ARA	304,060.61	222.59	159,717.17	120.31
URA	308,854.52	224.32	166,241.26	131.04

In both cases, the ARA scenario is the best one from an economic and an environmental perspective. In the first case, costs are reduced by 5.8% and emissions by 9.97% when shifting all possible load units by rail. In the second case, this scenario mitigates cost by 9.48% and emissions by 21.37%. This is because the supplier structure in the second case is more consolidated. This can be seen from the statistics presented in Table 2, where indications for many small shipments are given, or in the ABC/XYZ visualization in Figure A3 in Appendix B, highlighting that, for example, a single product sourced by a single supplier accounts for more than 20% of the overall weight.

Interestingly, the URA scenarios result in slightly smaller cost and emission reductions, which was somehow surprising to us. For the first case, 3.99% cost and 9.27% emission reduction; and for the second case, 5.78% cost and 14.36% emission reduction are calculated.

Inventory holding costs were not included in the first case. Upon closer examination of the second case, these costs represent 1.23% of the total expenses in the ARO scenario, 8.25% in the ARA scenario, and 5.58% in the URA scenario. The differences between ARA and URA are due to a larger number of shipments routed via intermodal transport in the ARA scenario; this leads to longer overall lead times, which results in higher inventory holding costs.

According to a research project on the EU Combined Transport, as reported in 2015, the cost breakdown for a CRRT of a semitrailer from Germany to Italy with 435 km on the main rail leg is as follows: 3% from the cost of the load unit, 15% from road pre-haulage, 19% from road post-haulage, 3% from the exporting terminal, 6% from the importing terminal, and 55% from the rail leg [84]. The simulation in this study presents higher shares for pre- and post-haulages and a lower share for the rail leg due to the long pre- and post-carriage distances covered in our simulations. Specifically, in the URA scenario of Case 1, pre-haulage and post-haulage expenditures account for 33% and 32% of the total costs, respectively. As a result, rail transport expenses are comparatively low, at only 17%. Nevertheless, the expenses for terminals and ILU closely resemble those from the empirical report, implying the authenticity and comparability of our results.

To validate the proposition made in the first part of this article, we compare the costs and emissions of the ARA and URA scenarios to the ARO scenario by calculating abatement costs on the ABC/XYZ level available. Results are presented in Table 4.

Table 4. Transportation abatement cost (EUR per mitigated t CO_2e) on the ABC/XYZ levels for the evaluated scenarios, with cost savings in green and additional costs in red.

		Case 1			Case 2		
		X	*Y*	*Z*	*X*	*Y*	*Z*
ARA	*A*				−874.99	15.21	−891.09
	B				−989.48	−1169.40	−1251.47
	n.a.	−1510.92	−552.97	−1411.41			
URA	*A*				−849.64	−63.24	−892.56
	B						
	n.a.	−1514.00	−513.90	123.79			

On the left side of the table, the results of Case 1 are presented in the bottom lines of the ARA and URA scenarios as financial data was not available (*n.a.*). On the right side, the results of Case 2 are presented in rows separated according to the ABC category. The absence of C items in the study can be explained using the constraints mentioned in Section 4.3. Due to the small size of C-shipments, it is not possible to utilize a 20-foot container to a sufficient level. For the URA scenario, even the B shipments are too small, as only highly utilized load units are routed intermodally.

In Table 4, we have colored the cells according to their cost-effectiveness, with the green cells representing strongly negative abatement costs—cost savings accompanying emission savings—and the red cells representing positive abatement costs—cost increases associated with emission reductions.

In Case 1, abatement costs for X- and Z- shipments are similar, while AC for Y-shipments are still negative, but three times higher. The investigation of the URA scenario for the first case highlights the differences between the dispatching strategies. While the X- and Y-shipments evaluate to similar values, costs for Z-shipments significantly differ from the ARA scenario. If the utilization of load units on the train is focused, the Z shipments cause additional costs for badly utilized fallback road transportation, which impedes higher costs and emissions at a level that raises the abatement cost to a positive value. Comparing the results of Case 1 with Case 2, abatement costs at similar levels are observed. Interestingly, the abatement costs for the Y-shipments of Case 2 are higher in ARA than in URA. Despite the otherwise relatively similar values, the comparison of the cases shows that a generalization of the absolute abatement costs is hardly possible. However, this also shows that the abatement costs do differ between the ABC/XYZ classes, which supports our initial proposition and approach presented in Section 2.

To validate the proposition, we examined the results of the simulation runs of the second case study in more detail. Therefore, we conducted Kruskal–Wallis tests for three samples (shipments of the ARA scenario, the URA scenario, and both scenarios combined) and found that there is a significant difference in the abatement costs for at least two inventory groups in all samples. Delving deeper into the pairwise comparisons of the inventory categories' abatement costs, we found that abatement costs between the inventory groups BX-AY, AZ-AY, and AX-AY significantly differ for the ARA and Total samples, and between the AX-AY and AZ-AY groups for the URA sample. The methods used for the tests are described in Appendix C. Results show that, at least across the XYZ inventory groups, abatement costs differ significantly. Investigating the boxplots in Figure A4 shows that abatement costs are highest for the Y-shipments and similar for X and Z shipments. The influence of the ABC dimension was significant only in the ARA and total sample, indicating that the influence of inventory holding costs exists, but is not decisive in most cases.

To summarize, the high-level results of the simulated cases and scenarios indicate that abatement costs do pertain to the ABC/XYZ category. Further statistical examination of the second case study showed that there is a significant difference in the abatement costs of some inventory categories. These findings let us conclude that we can partially accept the developed proposition.

6. Discussion, Implications, and Limitations

This paper's approach is twofold. In the first part, the authors provide a novel perspective on industrial logistics' decarbonization measures by combining a well-known inventory management tool with transport decarbonization measures. From these considerations, it is proposed that transport abatement costs of a specific measure differ when it is applied across the nine different ABC/XYZ inventory categories. The second part of this research validates this proposition for an exemplary mitigation measure, combined road–rail transportation, while presenting some limitations. The following sections will discuss the implications and limitations of our findings for research as well as practice.

6.1. Implications for Practitioners from Simulating Combined Road–Rail Transportation

In the CRRT simulation, the rail scenarios denoted as ARA and URA, in both evaluated cases, yield an overall reduction in cost and emissions, resulting in negative carbon emission abatement costs. Consequently, despite being suboptimal due to extended pre- and post-carriage distances, the cost-effectiveness of CRRT was demonstrated. However, a more nuanced examination of the results by the ABC/XYZ classification unveils that certain inventory categories exhibit higher costs and/or emissions compared to the reference scenario.

Other research in this field finds that overall total logistics costs can be reduced when the transportation mode is not selected holistically, but product-wise [37]. This, in turn, supports our findings that total logistics costs for one carbon mitigation measure differ across inventory categories, which, in turn, leads to different abatement costs. Further, the costs associated with inventory holding significantly contribute to the total logistics expenses, comprising 1.23% in the reference scenario and escalating to a range of 5.58% to 8.25% in the rail scenarios. Nevertheless, this does not offset the cost savings realized from the rail segment. These two observations underscore the necessity of a comprehensive assessment encompassing total logistics costs, rather than focusing exclusively on transportation costs when evaluating mitigation strategies, which is a common practice. While transportation costs constitute the predominant component for freight owners, they do not comprise the entirety. Moreover, making overarching statements about the competitiveness of CRRT is unwarranted, underscoring the imperative to diligently scrutinize each CRRT application individually due to the intricate nature of the intermodal system. Differing shares of the cost elements in our study from preceding studies [84] highlight the presence of numerous parameters that influence the costs of the intermodal system. Our results also lead to the finding that the widely referenced break-even distance definition is not universally applicable to combined transport, which was also shown in other modelling studies [49].

In the ARA scenario, costs for direct road transportation are non-zero due to the incapacity of CRRT to deliver certain Z-shipments promptly to the consignee. This discrepancy arises from the stipulated replenishment time of three days for Z-shipments. In contrast, all Y-shipments, ordered five days in advance of their required delivery, can be feasibly transported via CRRT. This underscores the criticality of accurate production demand forecasting and the implementation of distinct dispatching strategies for varied inventory categories.

Despite the elevated abatement costs of AY inventory items in the ARA scenario of Case 2, the URA scenario registers higher total costs. This is attributed to the efficiencies achieved in B-shipments in the ARA scenario, which compensate for the inefficiencies in AY-shipments. This implies that, at least within Case 2, the 80% minimum utilization threshold may be set excessively high. If this level were lowered, the cost-effectiveness of CRRT could potentially be further enhanced.

Based on the insights derived from simulating CRRT, we posit that the marked disparities observed between cases, scenarios, and inventory categories underscore the imperative, first, to evaluate the total logistics costs holistically and, second, to optimize the dispatching strategy when transitioning to CRRT. Utilization levels and replenishment times emerge as pivotal determinants in total logistics costs when employing CRRT, potentially demarcating the boundary between cost savings and additional expenditures.

6.2. Implications for Practitioners from the Validation of the Approach

Regarding the proposition made, simulation results show that GHG abatement costs vary among different shipment categories (X, Y, and Z) in the URA scenario for Case 1, but less in the ARA scenario. For Case 2, a closer statistical investigation of the abatement costs on the shipment level showed that the abatement costs significantly differ between some inventory groups. Therefore, the more crucial ABC/XYZ dimension for the abatement costs was found to be the XYZ dimension.

The primary determinant in logistics is the logistics costs. When these costs vary among different alternatives, it implies that the likelihood of implementing certain alternatives is greater than others [14,15]. However, for decision-makers utilizing our developed method, the objective is not only to maximize economic savings but also to reduce emissions. To combine these two quantitative decision criteria, abatement costs can be calculated. We have demonstrated that the cost and emission impacts of shifting freight transport from road to combined road-rail transport differ depending on the transported goods. Briefly said, from a freight owners' perspective, GHG abatement costs for intermodal transportation vary, depending on the class of the inventory being transported. The most impactful variable, thereby, is the demand characteristics (order frequency and volume), described by the demand stability of the XYZ dimension. Therefore, we see that inventory holding costs, which are affected by the ABC dimension, are a less important criterion.

Nevertheless, our findings concurrently highlight the necessity for a nuanced assessment of the applicability of combined transport for different use cases to effectively exploit potential savings. The generalized assumption of break-even distances, dispatching strategies, or costs is not feasible, even within a single company. At the same time, the economic and environmental impact of a shift towards combined transport is difficult to evaluate in detail.

Since costs differ between some of the ABC/XYZ classes, and the ABC/XYZ analysis is a straightforward and well-known measure, this study provides practitioners with an efficient method for identifying potential promising use cases for combined transport. Even though the results differ between our simulation cases, a cost-effective reduction in emissions was demonstrated for the shift of X goods in every instance. For Y goods, the statistical tests showed a higher risk for positive abatement costs, and the Z goods varied across the case studies. This insight allows for minimal effort in identifying suitable goods for combined transport and provides a starting point for CRRT newcomers, decreasing the entry barrier to the intermodal system. Therefore, this approach aids industrial enterprises in efficiently decarbonizing their logistics.

6.3. Limitations and Further Research Directions

As with every research article, there are limitations to this study which create opportunities for further research. The main limitation of this study is the requirement of using only one decarbonization measure to validate the proposed method. The primary objective for future research is to evaluate the proposition for additional decarbonization strategies and suggest the most effective measures for each of the nine inventory categories. A large volume of literature is available that recommends and categorizes decarbonization techniques (see, e.g., [20] for a comprehensive list), which can be a useful starting point for future investigations. For example, a frequently discussed measure is the use of trucks with alternative drives. It would be interesting to investigate effective application scenarios and the differences in abatement costs among ABC/XYZ items. To test for these differences, we suggest collecting real-world data from case studies that have implemented GHG mitigation measures for a broad range of inventory items or conducting simulation-based studies. The first approach may be criticized for its comparability, while the second may be questioned for its validity. Nevertheless, data availability is a crucial issue for all approaches, which is why it is recommended that research institutions collaborate closely with industrial companies to obtain realistic data.

Due to missing data, we neglect costs for damaged or stolen cargo. These have been modelled in other studies [58] and might harm CRRT's competitiveness. Although prior research indicates that CRRT is cost-competitive only when the pre- and post-haul distances are less than 40 km and the primary rail haul is no less than 750 km [75], the scenarios evaluated in this simulation are beneficial in terms of cost. Further, we have not included external costs in our study, as these do not currently have to be paid. However, a recent European study shows how dramatically different the results could be if external costs had to be paid in transportation [75].

Regarding CRRT, many different parameters can be adjusted, and many different replenishment, order, dispatching, and routing strategies can be followed. As intermodal transportation is a highly complex system, we were not able to fully test all parameters but had to make some assumptions regarding how CRRT could be conducted. Although these assumptions were discussed with other researchers and practitioners, future studies might assess the impact of a change in load unit compositions, route selection, and many more operational aspects of CRRT.

We were not able to validate the calculated figures for the mean and median of the train delay for Austria due to an absence of data. An analysis of train delays for intermodal freight services departing in Luxembourg resulted in $E = 74.35$ min and $m = 12$ min [79]. Another study found the standard deviation of freight train delays on one Swedish railway line to be $\sigma = 64.19$ min [85]. Assuming the underlying data is subject to an exponential distribution means that $E = \sigma$. Comparing our computed results, $E = 39.51$ min and $m = 27.39$ min, and with these empirical data, we see that our approximation produces smaller delay times. On the other side, an older study utilizes an average anticipated delay time of $D = 30$ min [62]. To address the ambivalence in the results, the suitability of the exponential distribution should be further researched in future studies. During the literature review on train delay times, we found that the majority of studies dealt with real-time delay prediction for passenger trains, but almost no studies presented high-level metrics for freight trains [86]. Nevertheless, for simulations like ours and logistics network planning for shippers and logistics service providers, further research on this topic could be helpful.

7. Conclusions

Throughout this study, we delineated and preliminarily validated the proposition that the ABC/XYZ analysis can be used to cost-effectively diversify industrial logistics decarbonization measures. In detail, we propose that abatement costs of a specific low-carbon practice or technology differ when applied to transport goods of different ABC/XYZ categories. For the preliminary validation, we used two simulation case studies, assessing the impact of shifting goods from road to combined road-rail transportation. Results show that abatement costs differ between ABC/XYZ categories in some scenarios, but we could not determine a common pattern across the two studies, except that X-shipments were cost-efficient in all scenarios. For a more detailed examination, we conducted a Kruskal–Wallis test to compare the abatement costs from shipments of the nine inventory categories of the second case study. Results show that there is a significant difference in the resulting abatement costs across some inventory categories, which suggests that our proposition can be accepted partially for combined road–rail transportation. Thereby, the inventories' XYZ dimension impacts abatement costs more significantly than its ABC dimension. Nevertheless, the results of our tests refer to only two case studies, which is why more research is necessary to assess additional scenarios, supplier–consignee relationships, and industrial sectors. For researchers, this study provides a promising foundation for investigating abatement costs, which are a critical component of efforts to decarbonize transportation. For practitioners, the findings of this study indicate that the makeup of a company's inventory has a significant bearing on the costs of transporting goods in an ecologically responsible manner. This is vital for making knowledgeable choices in industrial logistics and harmonizing economic effectiveness with environmental responsibility.

Supplementary Materials: The following supporting information can be downloaded at: https://www.mdpi.com/article/10.3390/app132212277/s1.

Author Contributions: Conceptualization, P.M.-B.; methodology, P.M.-B.; software, P.M.-B. and M.H.; validation, P.M.-B. and M.H.; writing—original draft preparation, P.M.-B.; writing—review and editing, M.W. and P.M.-B.; visualization, P.M.-B.; supervision, M.W.; project administration, M.W. All authors have read and agreed to the published version of the manuscript.

Funding: This research belongs to the project "SME 5.0—A Strategic Roadmap Towards the Next Level of Intelligent, Sustainable and Human-Centred SMEs" (funded in the European Union's Horizon MSCA 2021 programme under the Marie Skłodowska-Curie grant agreement No. 10108648).

Institutional Review Board Statement: Not applicable.

Informed Consent Statement: Not applicable.

Data Availability Statement: Data is contained within the Supplementary Material.

Conflicts of Interest: The authors declare no conflict of interest.

Appendix A

To analyze the freight charges, we used visualizations to examine the input data. Our investigation revealed a clear non-linear correlation between truck utilization and freight charges per tonne-kilometre. Using scikit-learn, we created a Python script to fit our data to the regressive function $y(x) = a + \frac{b}{x}$. The code we used is accessible online at https://github.com/MUL-Chair-of-Industrial-Logistics/simple_regression (accessed on 11 September 2023) and outputs $a = 0.05816489075412676$ and $b = 0.02437902515302302$. The left subplot of visualizes the regression results.

Although we anticipated an association between transportation distance and freight charges per tonne-kilometre, we failed to identify a function that accurately models this relationship with acceptable residuals. The data point distribution is displayed in the right plot of Figure A1. This could be because the transportation data available to us primarily included long-distance shipments.

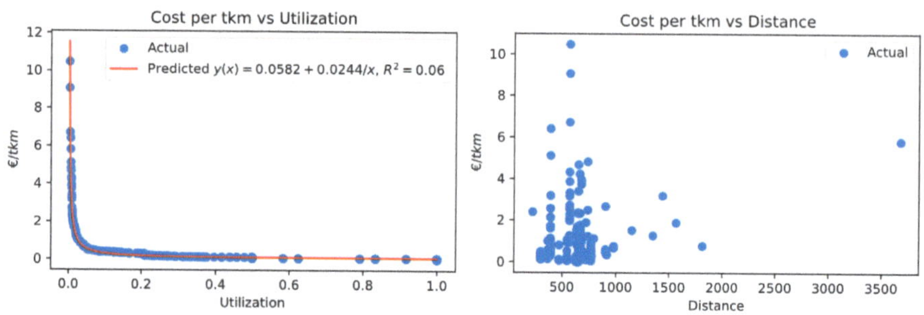

Figure A1. Graphical representation of the freight charges analysis.

Appendix B

The left plot of Figure A2 displays the characteristic curve of the XYZ-analysis plotted against the number of suppliers, while the right plot shows it against the cumulative weight.

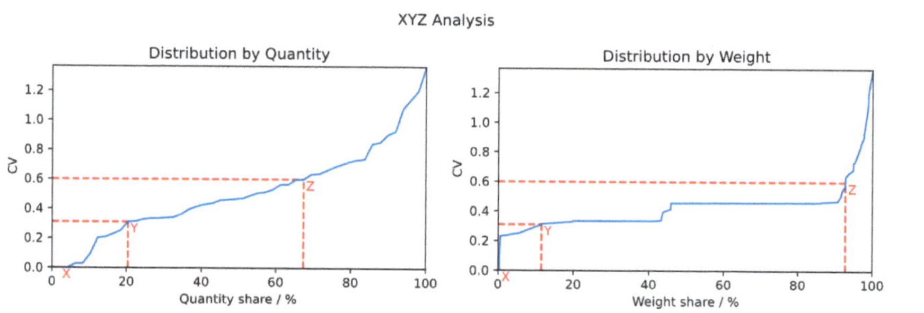

Figure A2. The characteristic curve of the XYZ analysis plotted against the number of suppliers (**left**) and the cumulative weight (**right**) for the first case study.

In Figure A3, the results of the ABC/XYZ analysis for the second case study are visualized.

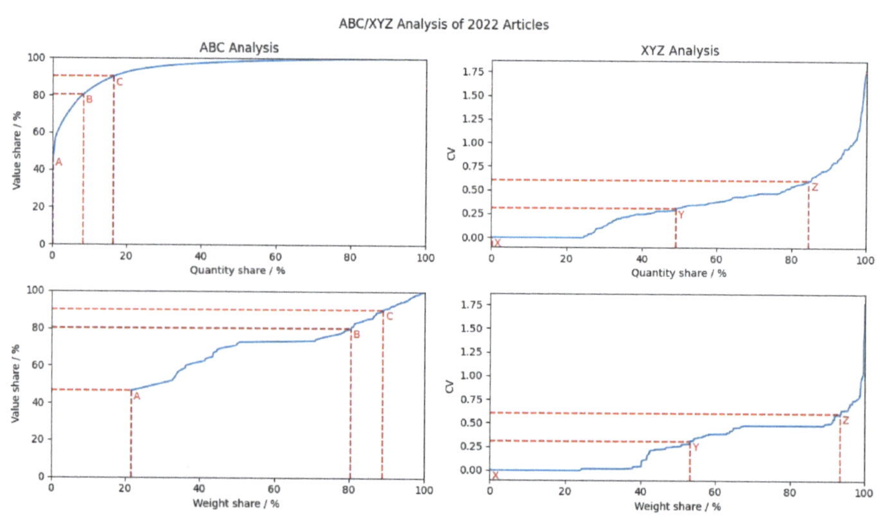

Figure A3. The characteristic curves of the ABC analysis (**left**) and the XYZ analysis (**right**) are plotted against the number of products (**top**) and the cumulative weight (**bottom**) for the second case study.

Table A1 presents the ABC/XYZ thresholds of this study in comparison to other references.

Table A1. Comparison of ABC/XYZ thresholds from the literature.

Reference	s_{AB}	s_{BC}	CV_{XY}	CV_{YZ}
This study	0.8	0.9	0.3	0.6
[16]	0.7–0.8	0.9–0.95	0.1	0.25
[38]	0.6–0.8	0.85–0.95		
[87]	0.8	0.95	0.5	1

Appendix C

To test the significance of the results, we calculated the abatement costs of each shipment shipped in the ARA and URA scenarios in the second case study. Then, we calculated the Kolmogorov–Smirnov and Shapiro–Wilk tests to test the abatement costs for normality using SPSS. We use an alpha value of $\alpha = 0.05$, which leads to a rejection of the hypothesis that the distribution of abatement costs is similar to a normal distribution. The results of the test for the shipments of the URA scenario, the ARA scenario, and both shipment data combined are presented in Table A2.

Table A2. Normality tests.

	Kolmogorov–Smirnov [a]			**Shapiro–Wilk**		
	Statistics	**df**	**Sig**	**Statistics**	**df**	**Sig**
abatement_cost_ara	0.209	55	<0.001	0.866	55	<0.001
abatement_cost_ura	0.267	32	<0.001	0.776	32	<0.001
abatement_cost_total	0.229	87	<0.001	0.852	87	<0.001

[a] Significance correction according to Lilliefors.

Due to the non-normality of the sample, we selected the Kruskal-Wallis – test for further investigation [88]. The test determines if all populations of one sample are identical [89]. In the focal case, the null hypothesis is that abatement costs are identical across the inventory categories, which implies that the alternative hypothesis is that abatement costs differ across inventory categories. Again, the alpha value was set to $\alpha = 0.05$. For all three samples (ARA shipments, URA shipments, and both shipments combined), the null hypothesis was rejected, and the alternative hypothesis was accepted, meaning that a statistically significant difference was found between at least two groups in all samples. Test results are displayed in Table A3.

Table A3. Kruskal–Wallis to test the null hypothesis.

	Kruskal–Wallis			
	Chi-Square	**df**	**Sig**	**H**
ARA	38.429	5	<0.001	reject
URA	19.176	2	<0.001	reject
Total	58.644	5	<0.001	reject

The pairwise group comparisons of the ARA and Total sample show significant differences ($sig < \alpha$) between the groups BX-AY, AZ-AY, and AX-AY. The group comparison of the URA scenario shows significant differences between the groups AX-AY and AZ-AY. These results are visualized in the boxplots shown in Figure A4, whereby SPSS visualizes outliers by circles (difference to 1st or 3rd quartile > 1.5 * interquartile range) and extreme outliers by asterisk (difference to 1st or 3rd quartile > 3 * interquartile range).

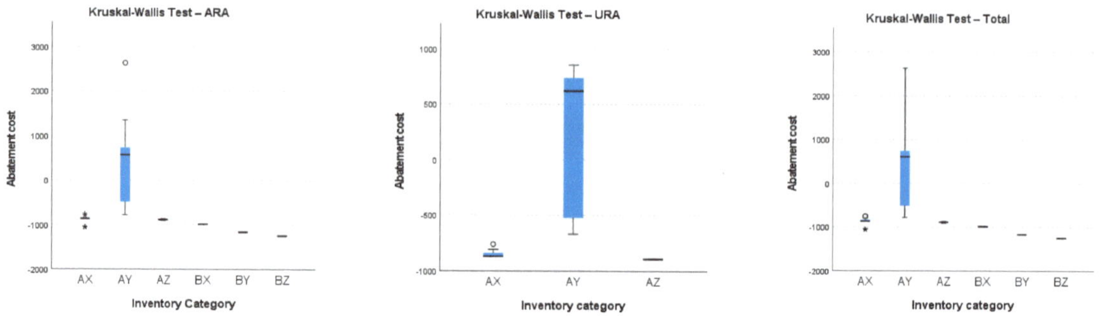

Figure A4. Boxplots of the abatement cost results for the ARA, URA and Total samples (left to right).

References

1. Jaramillo, P.; Kahn Ribeiro, S.; Newman, P.; Whitehead, J.; Hammer Strømman, A.; Ou, X.; Nugroho, S.B.; Lee, D.S.; Kajino, T.; Diemuodeke, O.E.; et al. Transport. In *Climate Change 2022: Mitigation of Climate Change: Contribution of Working Group III to the Sixth Assessment Report of the Intergovernmental Panel on Climate Change*; IPCC, Ed.; IPCC: Cambridge, UK, 2022.
2. Miklautsch, P.; König, A.; Woschank, M. Identifying a Country's Freight Transport-Intensive Economic Sectors and Their Logistics Emissions—Method Development and Exemplary Evaluation with Austria. *Sustainability* **2022**, *14*, 15050. [CrossRef]
3. McKinnon, A.C. *Decarbonizing Logistics: Distributing Goods in a Low Carbon World*; Kogan Page Limited: London, UK; New York, NY, USA, 2018; ISBN 9780749483807.
4. Cunanan, C.; Tran, M.-K.; Lee, Y.; Kwok, S.; Leung, V.; Fowler, M. A Review of Heavy-Duty Vehicle Powertrain Technologies: Diesel Engine Vehicles, Battery Electric Vehicles, and Hydrogen Fuel Cell Electric Vehicles. *Clean Technol.* **2021**, *3*, 474–489. [CrossRef]
5. Alonso-Villar, A.; Davíðsdóttir, B.; Stefánsson, H.; Ásgeirsson, E.I.; Kristjánsson, R. Technical, economic, and environmental feasibility of alternative fuel heavy-duty vehicles in Iceland. *J. Clean. Prod.* **2022**, *369*, 133249. [CrossRef]
6. Marrekchi, E.; Besbes, W.; Dhouib, D.; Demir, E. A review of recent advances in the operations research literature on the green routing problem and its variants. *Ann. Oper. Res.* **2021**, *304*, 529–574. [CrossRef]
7. Basma, H.; Rodríguez, F.; Hildermeier, J.; Jahn, A. *Electrifying Last-Mile Delivery: A Total Cost of Ownership Comparison of Battery-Electric and Diesel Trucks in Europe*; ICCT: Washington, DC, USA; Brussels, Belgium, 2022. Available online: https://theicct.org/wp-content/uploads/2022/06/tco-battery-diesel-delivery-trucks-jun2022.pdf (accessed on 8 August 2023).
8. Mojtaba Lajevardi, S.; Axsen, J.; Crawford, C. Comparing alternative heavy-duty drivetrains based on GHG emissions, ownership and abatement costs: Simulations of freight routes in British Columbia. *Transp. Res. Part D Transp. Environ.* **2019**, *76*, 19–55. [CrossRef]
9. Konstantinou, T.; Gkritza, K. Examining the barriers to electric truck adoption as a system: A Grey-DEMATEL approach. *Transp. Res. Interdiscip. Perspect.* **2023**, *17*, 100746. [CrossRef]
10. Anderhofstadt, B.; Spinler, S. Factors affecting the purchasing decision and operation of alternative fuel-powered heavy-duty trucks in Germany—A Delphi study. *Transp. Res. Part D Transp. Environ.* **2019**, *73*, 87–107. [CrossRef]
11. Bistline, J. Metrics for assessing the economic impacts of power sector climate and clean electricity policies. *Prog. Energy* **2021**, *3*, 43001. [CrossRef]
12. Cole, W.; Antonysamy, A.; Brown, P.; Sergi, B.; Mai, T.; Denholm, P. How much might it cost to decarbonize the power sector? It depends on the metric. *Energy* **2023**, *276*, 127608. [CrossRef]
13. Hagos, D.A.; Ahlgren, E.O. Exploring cost-effective transitions to fossil independent transportation in the future energy system of Denmark. *Appl. Energy* **2020**, *261*, 114389. [CrossRef]
14. Miklautsch, P.; Woschank, M. Decarbonizing Industrial Logistics. *IEEE Eng. Manag. Rev.* **2022**, *50*, 149–156. [CrossRef]
15. Miklautsch, P.; Woschank, M. The Missing Link: How to Bridge the Gap to Zero-Carbon Logistics: In Press. In *Erfolg Durch Nachhaltiges Energie- und Ressourcenmanagement*; Posch, W., Vorbach, S., Zsifkovits, H., Eds.; Nomos Verlagsgesellschaft: Baden-Baden, Germany, 2023.
16. Hoppe, M. *Bestandsoptimierung Mit SAP: Effektives Bestandsmanagement Mit SAP ERP und SAP SCM, 2.*; aktualisierte und erw. Aufl.; Galileo Press: Bonn, Germany, 2008; ISBN 9783836210829.
17. Ravinder, H.V.; Misra, R.B. ABC Analysis for Inventory Management: Bridging the Gap Between Research and Classroom. *AJBE* **2016**, *9*, 39–48. [CrossRef]
18. Eisenhardt, K.M. Building Theories from Case Study Research. *AMR* **1989**, *14*, 532. [CrossRef]
19. Timmer, S.; Kaufmann, L. Do Managers' Dark Personality Traits Help Firms in Coping with Adverse Supply Chain Events? *J. Supply Chain. Manag.* **2019**, *55*, 67–97. [CrossRef]
20. Miklautsch, P.; Woschank, M. A framework of measures to mitigate greenhouse gas emissions in freight transport: Systematic literature review from a Manufacturer's perspective. *J. Clean. Prod.* **2022**, *366*, 132883. [CrossRef]

21. Miklautsch, P.; Woschank, M. The adoption of industrial logistics decarbonization practices: Evidence from Austria. *Transp. Res. Interdiscip. Perspect.* **2023**, *21*, 100857. [CrossRef]
22. Zhu, Y.; Ma, H.; Sha, C.; Yang, Y.; Sun, H.; Ming, F. Which strategy among avoid, shift, or improve is the best to reduce CO_2 emissions from sand and gravel aggregate transportation? *J. Clean. Prod.* **2023**, *391*, 136089. [CrossRef]
23. Isler, O. Resource Scarcity. In *Encyclopedia of Evolutionary Psychological Science*; Shackelford, T.K., Weekes-Shackelford, V.A., Eds.; Springer International Publishing: Cham, Switzerland, 2021; pp. 6647–6649, ISBN 978-3-319-19649-7.
24. Brinken, J.; Trojahn, S.; Behrendt, F. Sufficiency, Consistency, and Efficiency as a Base for Systemizing Sustainability Measures in Food Supply Chains. *Sustainability* **2022**, *14*, 6742. [CrossRef]
25. Linton, S.; Clarke, A.; Tozer, L. Technical pathways to deep decarbonization in cities: Eight best practice case studies of transformational climate mitigation. *Energy Res. Soc. Sci.* **2022**, *86*, 102422. [CrossRef]
26. Yang, X.; Hu, M.; Zhang, C.; Steubing, B. Key strategies for decarbonizing the residential building stock: Results from a spatiotemporal model for Leiden, the Netherlands. *Resour. Conserv. Recycl.* **2022**, *184*, 106388. [CrossRef]
27. Liu, J.; Li, M.; Xue, L.; Kobashi, T. A framework to evaluate the energy-environment-economic impacts of developing rooftop photovoltaics integrated with electric vehicles at city level. *Renew. Energy* **2022**, *200*, 647–657. [CrossRef]
28. Chang, S.; Cho, J.; Heo, J.; Kang, J.; Kobashi, T. Energy infrastructure transitions with PV and EV combined systems using techno-economic analyses for decarbonization in cities. *Appl. Energy* **2022**, *319*, 119254. [CrossRef]
29. Too, J.; Ejohwomu, O.A.; Hui, F.K.P.; Herath, N.; Duffield, C. How Can Decarbonisation Alternatives Be Prioritised to Achieve Carbon Neutrality in Building Projects? In Proceedings of the Structures Congress 2023, New Orleans, LA, USA, 3–6 May 2023; Soules, J.G., Ed.; American Society of Civil Engineers: Reston, VA, USA, 2023; pp. 247–261, ISBN 9780784484777.
30. Xu, L.; Wang, Y.; Shah, S.A.A.; Zameer, H.; Solangi, Y.A.; Walasai, G.D.; Siyal, Z.A. Economic Viability and Environmental Efficiency Analysis of Hydrogen Production Processes for the Decarbonization of Energy Systems. *Processes* **2019**, *7*, 494. [CrossRef]
31. Cremer, A.; Berger, M.; Müller, K.; Finkbeiner, M. The First City Organizational LCA Case Study: Feasibility and Lessons Learned from Vienna. *Sustainability* **2021**, *13*, 5062. [CrossRef]
32. Ren, L.; Zhou, S.; Peng, T.; Ou, X. A review of CO_2 emissions reduction technologies and low-carbon development in the iron and steel industry focusing on China. *Renew. Sustain. Energy Rev.* **2021**, *143*, 110846. [CrossRef]
33. Woodley, L.; Rossetti, P.; Nunes, A. Targeted electric vehicle procurement incentives facilitate efficient abatement cost outcomes. *Sustain. Cities Soc.* **2023**, *96*, 104627. [CrossRef]
34. Žic, S.; Žic, J.; Đukić, G. Efficient planning and optimization of inventory replenishments for sustainable supply chains operating under (R, s, S) policy. *Sustain. Futures* **2023**, *5*, 100110. [CrossRef]
35. Zsifkovits, H.E. *Logistik*, 1st ed.; UTB GmbH; UVK Lucius: Stuttgart/München, Germany, 2012; ISBN 978-3825236731.
36. Rodrigue, J.-P. *The Geography of Transport Systems*, 5th ed.; Routledge: Abingdon, UK; New York, NY, USA, 2020; ISBN 9780429346323.
37. Drent, M.; Moradi, P.; Arts, J. Efficient emission reduction through dynamic supply mode selection. *Eur. J. Oper. Res.* **2023**, *311*, 925–941. [CrossRef]
38. Wannenwetsch, H. Analysen zur Kostenreduzierung in der Materialwirtschaft. In *Integrierte Materialwirtschaft, Logistik und Beschaffung*; Wannenwetsch, H., Ed.; Springer Berlin Heidelberg: Berlin/Heidelberg, Germany, 2014; pp. 31–63, ISBN 978-3-642-45022-8.
39. Flores, B.E.; Whybark, D. Multiple Criteria ABC Analysis. *Int. J. Oper. Prod. Manag.* **1986**, *6*, 38–46. [CrossRef]
40. Flores, B.E.; Whybark, D. Implementing multiple criteria ABC analysis. *J. Oper. Manag.* **1987**, *7*, 79–85. [CrossRef]
41. Chu, C.-W.; Liang, G.-S.; Liao, C.-T. Controlling inventory by combining ABC analysis and fuzzy classification. *Comput. Ind. Eng.* **2008**, *55*, 841–851. [CrossRef]
42. Clevert, D.-A.; Stickel, M.; Jung, E.M.; Reiser, M.; Rupp, N. Cost analysis in interventional radiology—A tool to optimize management costs. *Eur. J. Radiol.* **2007**, *61*, 144–149. [CrossRef]
43. Cardoso, J.M.; Leitão, P.; Geraldes, C.A.S. Enhancing Inventory Management Decisions in a Bakery: A Case Study. In *Flexible Automation and Intelligent Manufacturing: Establishing Bridges for More Sustainable Manufacturing Systems*; Silva, F.J.G., Ferreira, L.P., Sá, J.C., Pereira, M.T., Pinto, C.M.A., Eds.; Springer Nature: Cham, Switzerland, 2023; pp. 1049–1056, ISBN 978-3-031-38164-5.
44. Patil, R.A.; Patange, A.D.; Pardeshi, S.S. International Transportation Mode Selection through Total Logistics Cost-Based Intelligent Approach. *Logistics* **2023**, *7*, 60. [CrossRef]
45. Zeng, A.Z.; Rossetti, C. Developing a framework for evaluating the logistics costs in global sourcing processes. *Int. J. Phys. Distrib. Logist. Manag.* **2003**, *33*, 785–803. [CrossRef]
46. Fichtinger, J.; Ries, J.M.; Grosse, E.H.; Baker, P. Assessing the environmental impact of integrated inventory and warehouse management. *Int. J. Prod. Econ.* **2015**, *170*, 717–729. [CrossRef]
47. Becerra, P.; Mula, J.; Sanchis, R. Green supply chain quantitative models for sustainable inventory management: A review. *J. Clean. Prod.* **2021**, *328*, 129544. [CrossRef]
48. Darvish, M.; Archetti, C.; Coelho, L.C. Trade-offs between environmental and economic performance in production and inventory-routing problems. *Int. J. Prod. Econ.* **2019**, *217*, 269–280. [CrossRef]
49. Zgonc, B.; Tekavčič, M.; Jakšič, M. The impact of distance on mode choice in freight transport. *Eur. Transp. Res. Rev.* **2019**, *11*, 10. [CrossRef]

50. Miklautsch, P.; Woschank, M. Decarbonizing Industrial Logistics Through a GIS-Based Approach for Identifying Pareto-Optimal Combined Road-Rail Transport Routes. In *Towards a Smart, Resilient and Sustainable Industry*; Borgianni, Y., Matt, D.T., Molinaro, M., Orzes, G., Eds.; Springer Nature: Cham, Switzerland, 2023; pp. 372–383, ISBN 978-3-031-38273-4.
51. Hegyi, P.; Jaber, A.; Földes, D.; Csonka, B.; Silva, D.; Csiszár, C. Assessing the impact of climate change and weather conditions endangering inland waterway navigation. In Proceedings of the 2023 Smart City Symposium Prague (SCSP), Prague, Czech Republic, 25–26 May 2023; pp. 1–5, ISBN 979-8-3503-2162-3.
52. Seneviratne, S.I.; Zhang, X.; Adnan, M.; Badi, W.; Dereczynski, C.; Di Luca, A.; Ghosh, S.; Iskandar, I.; Kossin, J.; Lewis, S.; et al. Weather and Climate Extreme Events in a Changing Climate. In *Climate Change 2021—The Physical Science Basis*; IPCC, Ed.; Cambridge University Press: Cambridge, UK, 2021; pp. 1513–1766, ISBN 9781009157896.
53. Kotowska, I.; Mańkowska, M.; Pluciński, M. The Competitiveness of Inland Shipping in Serving the Hinterland of the Seaports: A Case Study of the Oder Waterway and the Szczecin-Świnoujście Port Complex. In *Integration as Solution for Advanced Smart Urban Transport Systems*; Sierpiński, G., Ed.; Springer International Publishing: Cham, Switzerland, 2019; pp. 252–263, ISBN 978-3-319-99476-5.
54. DHL. European Rail Transport Solutions. Available online: https://www.dhl.com/global-en/home/our-divisions/freight/rail-freight.html (accessed on 7 September 2023).
55. LKW WALTER. Combined Transport. Available online: https://www.lkw-walter.com/int/en/products-and-services/combined-transport (accessed on 7 September 2023).
56. DB Cargo. Combined Transport. Available online: https://www.dbcargo.com/resource/blob/5812214/8ecbe5b562224046295c483c1547c5f6/artikel-kombinierter-verkehr-pdf-data.pdf (accessed on 7 September 2023).
57. Miklautsch, P.; Woschank, M.; Heißenberger, J. Decarbonizing Construction Material Supply Chains: An Innovative Approach to Intermodal Transportation. In *Construction Logistics, Equipment, and Robotics*; Fottner, J., Nübel, K., Matt, D., Eds.; Springer Nature: Cham, Switzerland, 2023; pp. 123–136, ISBN 978-3-031-44020-5.
58. Rotaris, L.; Tonelli, S.; Capoani, L. Combined transport: Cheaper and greener. A successful Italian case study. *Res. Transp. Bus. Manag.* **2022**, *43*, 100792. [CrossRef]
59. Islam, D.M.Z.; Zunder, T.H. Experiences of rail intermodal freight transport for low-density high value (LDHV) goods in Europe. *Eur. Transp. Res. Rev.* **2018**, *10*, 24. [CrossRef]
60. Colicchia, C.; Creazza, A.; Dallari, F. Lean and green supply chain management through intermodal transport: Insights from the fast moving consumer goods industry. *Prod. Plan. Control* **2017**, *28*, 321–334. [CrossRef]
61. European Court of Auditors. *Rail Freight Transport in the EU: Still Not on The Right Track: Special Report No 08, 2016*; European Court of Auditors: Luxembourg, 2016.
62. Janic, M. Modelling the full costs of an intermodal and road freight transport network. *Transp. Res. Part D Transp. Environ.* **2007**, *12*, 33–44. [CrossRef]
63. European Commission. *Communication from the Commission to the European Parliament, the Council, the European Economic and Social Committee and the Committee of the Regions: Sustainable and Smart Mobility Strategy—Putting European Transport on Track for the Future*; Publications Office of the European Union: Brussels, Belgium, 2020.
64. Rosca, E.; Rusca, A.; Popa, M.; Rusca, F.; Olteanu, S. Modelling and Simulation for Intermodal Transportation in Logistics Chains. In *Modern Trends and Research in Intermodal Transportation*; Sładkowski, A., Ed.; Springer International Publishing: Cham, Switzerland, 2022; pp. 1–49, ISBN 978-3-030-87119-2.
65. Shibuya, K.; Shibasaki, R.; Kawasaki, T.; Tokuori, T. Stagnant logistics growth simulation on West African intermodal corridors. *Transp. Res. Interdiscip. Perspect.* **2023**, *21*, 100867. [CrossRef]
66. Kaddoura, I.; Masson, D.; Hettinger, T.; Unterfinger, M. An agent-based simulation approach to investigate the shift of Switzerland's inland freight transport from road to rail. *Transportation* **2023**, 1–22. [CrossRef]
67. Zhong, H.; Chen, W.; Gu, Y. A system dynamics model of port hinterland intermodal transport: A case study of Guangdong-Hong Kong-Macao Greater Bay Area under different carbon taxation policies. *Res. Transp. Bus. Manag.* **2023**, *49*, 100987. [CrossRef]
68. Van der Ham, R. Salabim. Available online: https://www.salabim.org/ (accessed on 7 September 2023).
69. Lopes, H.d.S.; Da Lima, R.S.; Leal, F.; Nelson, A.d.C. Scenario analysis of Brazilian soybean exports via discrete event simulation applied to soybean transportation: The case of Mato Grosso State. *Res. Transp. Bus. Manag.* **2017**, *25*, 66–75. [CrossRef]
70. Zum Felde, J.; Alias, C.; Goudz, A. Comparing generic and dedicated tools of discrete-event simulation for examining inland waterway transportation services. In Proceedings of the 2022 IEEE 6th International Conference on Logistics Operations Management (GOL), Strasbourg, France, 29 June–1 July 2022; pp. 1–10, ISBN 978-1-6654-6932-6.
71. Kogler, C.; Rauch, P. Discrete event simulation of multimodal and unimodal transportation in the wood supply chain: A literature review. *Silva Fenn.* **2018**, *52*, 9984. [CrossRef]
72. IVE mbH. EcoTransIT World—Methodology. Available online: https://www.ecotransit.org/en/methodology/ (accessed on 7 September 2023).
73. Smart Freight Centre. Global Logistics Emissions Council Framework for Logistics Emissions Accounting and Reporting: Version 2.0. Available online: https://www.smartfreightcentre.org/en/our-programs/global-logistics-emissions-council/calculate-report-glec-framework/ (accessed on 7 September 2023).
74. Crainic, T.G.; Kim, K.H. Chapter 8 Intermodal Transportation. In *Transportation*; Elsevier: Amsterdam, The Netherlands, 2007; pp. 467–537, ISBN 9780444513465.

75. Carboni, A.; Dalla Chiara, B. Range of technical-economic competitiveness of rail-road combined transport. *Eur. Transp. Res. Rev.* **2018**, *10*, 45. [CrossRef]
76. Ward, K. Renting a Container: A Complete Guide to Pricing, Options, and More. Available online: https://oatuu.org/renting-a-container-a-complete-guide-to-pricing-options-and-more/ (accessed on 29 August 2023).
77. Hellenic Shipping News Worldwide. 20ft Container Rental Price 2023: What's the Best Price to Pay [+One-Way Lease Rate]. Available online: https://www.hellenicshippingnews.com/20ft-container-rental-price-2023-whats-the-best-price-to-pay-one-way-lease-rate/ (accessed on 29 August 2023).
78. Kramarz, M.; Przybylska, E.; Wolny, M. Reliability of the intermodal transport network under disrupted conditions in the rail freight transport. *Res. Transp. Bus. Manag.* **2022**, *44*, 100686. [CrossRef]
79. Pineda-Jaramillo, J.; Bigi, F.; Bosi, T.; Viti, F.; D'ariano, A. Short-Term Arrival Delay Time Prediction in Freight Rail Operations Using Data-Driven Models. *IEEE Access* **2023**, *11*, 46966–46978. [CrossRef]
80. Minbashi, N.; Palmqvist, C.-W.; Bohlin, M.; Kordnejad, B. Statistical analysis of departure deviations from shunting yards: Case study from Swedish railways. *J. Rail Transp. Plan. Manag.* **2021**, *18*, 100248. [CrossRef]
81. Wang, P.; Zhang, Q. Train delay analysis and prediction based on big data fusion. *Transp. Saf. Environ.* **2019**, *1*, 79–88. [CrossRef]
82. Schienen-Control GmbH. Jahresbericht 2022: Ihr Recht am Zug. Schienen-Control. Available online: https://www.schienencontrol.gv.at/epaper/sc_jb_2022/html5.html (accessed on 27 July 2023).
83. Deutsche Bahn AG. Integrierter Zwischenbericht Januar-Juni 2022. Available online: https://zbir.deutschebahn.com/2022/fileadmin/pdf/DB_ZB22_d_web_06.pdf (accessed on 27 July 2023).
84. KombiConsult GmbH; Intermodality Ltd; PLANCO Consulting; Gruppo CLAS S.p.A. *Analysis of the EU Combined Transport: Contract No FV355/2012/MOVE/D1/ETU/ SI2.659386, Frankfurt/Main.* 2015. Available online: https://transport.ec.europa.eu/system/files/2016-09/2015-01-freight-logistics-lot2-combined-transport.pdf (accessed on 3 August 2023).
85. Corman, F.; Kecman, P. Stochastic prediction of train delays in real-time using Bayesian networks. *Transp. Res. Part C Emerg. Technol.* **2018**, *95*, 599–615. [CrossRef]
86. Tiong, K.Y.; Ma, Z.; Palmqvist, C.-W. A review of data-driven approaches to predict train delays. *Transp. Res. Part C Emerg. Technol.* **2023**, *148*, 104027. [CrossRef]
87. Scholz-Reiter, B.; Heger, J.; Meinecke, C.; Bergmann, J. Integration of demand forecasts in ABC-XYZ analysis: Practical investigation at an industrial company. *Int. J. Product. Perform. Manag.* **2012**, *61*, 445–451. [CrossRef]
88. Najmi, A.; Sadasivam, B.; Ray, A. How to choose and interpret a statistical test? An update for budding researchers. *J. Fam. Med. Prim. Care* **2021**, *10*, 2763–2767. [CrossRef]
89. Dodge, Y. Kruskal-Wallis Test. In *The Concise Encyclopedia of Statistics*; Dodge, Y., Ed.; Springer: New York, NY, USA, 2008; pp. 288–290, ISBN 978-0-387-31742-7.

drones

MDPI

Review

A Review of Icing Research and Development of Icing Mitigation Techniques for Fixed-Wing UAVs

Liang Zhou [1], Xian Yi [1,2,*] and Qinglin Liu [1]

1 Key Laboratory of Icing and Anti/De-Icing, China Aerodynamics Research and Development Center, Mianyang 621000, China; zlandtyt@163.com (L.Z.); aseity670@126.com (Q.L.)
2 National Key Laboratory of Aerodynamics Science and Technology for Aerospace Flight, China Aerodynamics Research and Development Center, Mianyang 621000, China
* Correspondence: yixian@cardc.cn; Tel.: +86-15892611314

Abstract: With the continuous expansion of Unmanned Aerial Vehicle (UAV) applications, the threat of icing on UAV flights has garnered increased attention. Understanding the icing principles and developing anti-icing technologies for unmanned aircraft is a crucial step in mitigating the icing threat. However, existing research indicates that changes in Reynolds numbers have a significant impact on the physics of ice accretion. Icing studies on aircraft operating at high Reynolds numbers cannot be directly applied to unmanned aircraft, and mature anti-icing/deicing techniques for manned aircraft cannot be directly utilized for UAVs. This paper firstly provides a comprehensive overview of research on icing for fixed-wing UAVs, including various methods to study unmanned aircraft icing and the identified characteristics of icing on unmanned aircraft. Secondly, this paper focuses on discussing UAV anti-icing/deicing techniques, including those currently applied and under development, and examines the advantages and disadvantages of these techniques. Finally, the paper presents some recommendations regarding UAV icing research and the development of anti-icing/deicing techniques.

Keywords: atmospheric icing; UAV; Reynolds number; aerodynamic penalties; ice mitigation techniques

Citation: Zhou, L.; Yi, X.; Liu, Q. A Review of Icing Research and Development of Icing Mitigation Techniques for Fixed-Wing UAVs. *Drones* **2023**, *7*, 709. https://doi.org/10.3390/drones7120709

Academic Editors: Tomasz Nowakowski, Artur Kierzkowski, Agnieszka A. Tubis, Franciszek Restel, Tomasz Kisiel, Anna Jodejko-Pietruczuk and Mateusz Zajac

Received: 15 November 2023
Revised: 11 December 2023
Accepted: 13 December 2023
Published: 18 December 2023

1. Introduction

Recently, unmanned aircraft systems have been among the most significant developments in aviation. The Unmanned Aircraft Systems Roadmap 2005–2030 [1] reports that over 250 UAV models are currently produced in 32 countries worldwide. Initially, drones saw military use, undertaking tasks such as aerial combat and military reconnaissance, because they were cost-effective, posed no threat to human life, and offered extended mission times and access to complex locations that manned aircraft could not reach. Predator, Phoenix, and Global Hawk drones have all found military applications. More recently, UAVs have demonstrated lengthy endurance capacity, increased safety, greater intelligence, and flexibility. Together, these qualities can effectively address the drawbacks of unsafe, less efficient, and costlier manual operations. Consequently, UAVs are increasingly being used across civil fields for a range of tasks—such as electricity inspections, disaster assessments, rescue operations, aerial photography and mapping, and logistics and delivery. Notably, during heavy rain floods in China's Henan Province in July 2021, the fixed-wing UAV "Pterodactyl" provided emergency communication assistance to the affected populace, facilitating emergency relief work in the disaster region [2]. Villeneuve [3] proposed the concept and design of using the GRIFF 135 heavy-lift rotary drone to spray anti-icing fluid on aircraft.

Currently, the drone industry is growing rapidly worldwide, with many countries introducing policies to meet national needs. However, drone icing is becoming a significant issue. Ice accretion on the leading edge of the wing alters the wing geometry and reduces the aerodynamic performance of the UAV, causing reduced lift, increased drag, and a higher risk of stalling [4]. The potential of flight icing (as seen in Figure 1 [5]) may render UAVs inoperable in icy weather and even result in crashes. During some military operations, 25% of UAVs experienced icing, negatively impacting mission success [6]. Three U.S. Army General Atomics MQ-1 Predator drones crashed due to icing in Afghanistan in 2001 and 2002, and the U.S. Army's Northrop Grumman Global Hawk drone also crashed due to icing during Operation Enduring Freedom in Afghanistan [7].

Figure 1. Ice accretion on fixed-wing UAV [5].

To mitigate the flight risk posed to UAVs by icing, current anti-icing strategies include grounding UAVs [8] or modifying their path planning [9]. Military missions, such as those in Kosovo, have suspended the use of Hunter drones from October to April [10], significantly inhibiting their operation in icing conditions. Existing anti-icing/deicing methods used for manned aircraft (e.g., anti-icing liquid, pneumatic type, electric-thermal type, etc.) are not entirely suitable for unmanned aircraft due to their complex structure, large mass, or high energy consumption. There is a need to develop innovative anti-icing/deicing strategies specifically for UAVs to mitigate the risk of icing. The United States, Canada, Russia, China, France, Turkey, and several other countries are developing anti-icing/deicing systems for UAVs, but further research is necessary, given that the application of such technology on UAVs is less developed and less widespread than on airplanes.

2. Methods of Studying Ice Accretion on a UAV

Research on icing has primarily focused on manned aircraft, wind turbines, and power lines. It was not until the early 21st century that UAVs garnered attention. Studies on icing in manned aircraft can be traced back to 1940, while research on icing in the domain of UAVs is relatively new and limited [11,12]. The early work on UAV icing, specifically discussed, was published by Siquig in 1990 [11], which is considered the first explicit discussion of UAV icing. With the increasing widespread use of UAVs, there is a growing demand for research in the field of UAV icing.

Compared to manned aircraft, UAVs are more susceptible to ice-related hazards due to three main reasons [13]. Firstly, the troposphere holds most of the water vapor in the atmosphere, experiencing complex airflow and weather phenomena like thunderstorms, fog, and low-temperature clouds. Most drones operate within the troposphere. Icing events for aircraft primarily occur during takeoff and landing phases, whereas many UAVs face the risk of icing throughout their entire flight process. Secondly, the smaller size of UAVs, including their wings and propellers, compared to larger aircraft means that ice of the same thickness has a greater impact on UAVs, increasing the probability of stalling in advance. Thirdly, limited energy resources and the absence of anti-icing/deicing capabilities in most UAVs restrict their ability to prevent icing. The Reynolds number, a crucial distinction between UAV and manned aircraft icing, is generally lower for UAVs, significantly influencing the icing process. Therefore, conducting research on UAV icing is of vital importance for safe UAV operations and designing effective anti-icing/deicing strategies. The rotation of the propeller of a rotary-wing drone creates additional flow complexity [14], resulting in different icing characteristics compared to fixed-wing drones. Due to limitations in length, this article only introduces research on icing of fixed-wing UAVs.

Experimental studies and numerical simulations are the primary methods employed in UAV icing research.

2.1. Experimental Studies

UAV icing tests are carried out mainly through flight and wind tunnel tests.

2.1.1. Flight Tests

Flight tests, the most effective method to obtain relevant data, are limited by test weather conditions, continuous monitoring and hence, are rare.

In 2013, Bottyán [15] performed UAV flight icing tests on two different UAVs (an ELBIT Skylark-I LE short-range UAV and a Northrop Grumman RQ-4 Global Hawk Block 30 high-altitude, long-endurance UAV) in the Carpathian Mountains in order to develop an ice accumulation model. In 2016, Williams et al. [16] used a prefabricated ice-like material glued to the wing of the Kahu-Hawk UAV for aerodynamics and stability tests. In 2017, Matiychyk et al. [17] conducted a natural icing flight test with an M-10-2 "Oko" UAV. The results showed that the leading edge of the wing, the leading edge of the tail, the video antenna, and the airspeed sensor's front surface were affected by icing. With a thickness of approximately 1.5 mm on the leading edge of the wing, the test showed that icing led to a 40% increase in drag and a 10% increase in energy consumption for drone operation, as per flight recorder analysis. Avery [18] performed three UAV flight tests in 2019 to confirm the icing vulnerability of small UAVs during low-altitude flights. One test revealed ice geometries formed on the UAV's surface due to the flight flow field, which confirmed the influence of pressure distribution on low-speed ice accretion physics. However, the lack of cloud signature sensors precluded the determination of atmospheric water content. In 2021, Siddique [19] conducted tests on the Sky Hunter fixed-wing UAV in three different weather conditions: windy, calm, and icy. Glaze ice was noticeable on the leading edge of the wing, while overflow water could be observed on the wing surface. The propeller was covered with a thick layer of ice. In comparison to calm wind conditions, the UAV experienced a significant increase in power consumption by approximately 240% during icy weather, resulting in a higher demand on battery power. In 2023, Han et al. [20] conducted a flight test to examine the characteristics of in-flight icing and its impact on fixed-wing UAV flight performance. During the flight, extensive ice structures were observed on almost all exposed surfaces of the UAV, such as the wings, body, stabilizers, pitot tubes, and propellers, with water backflow observed on the wing surface, as shown in Figure 2. When completing the same flight task, the power consumption during icing conditions was over 80% higher than when flying under normal non-icing weather conditions.

2.1.2. Lab Tests

A wind tunnel is an experimental device that artificially creates and controls airflow to simulate gas flow around an entity, observe physical phenomena, and measure the effect of airflow. It is a valuable tool for aerodynamic experiments and integral to vehicle research.

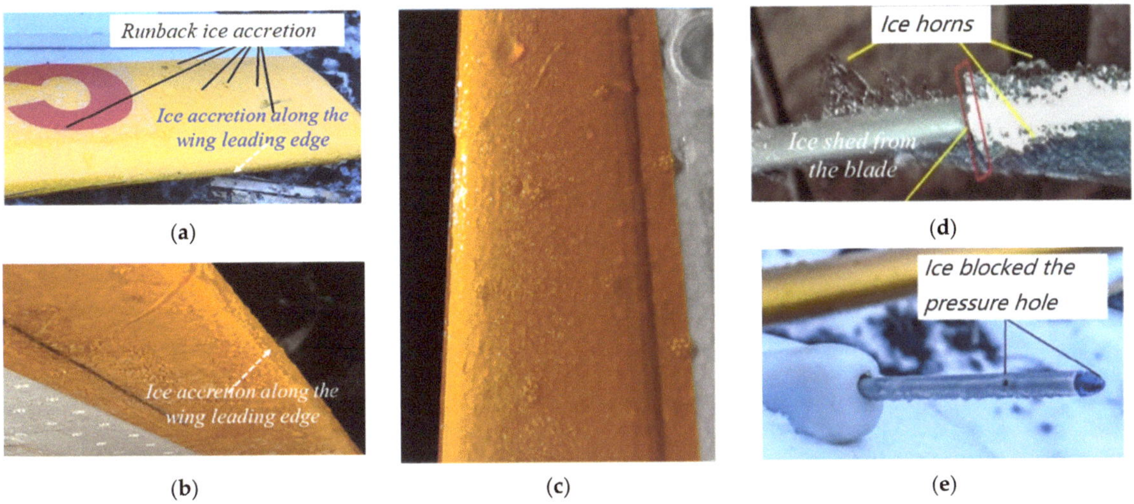

Figure 2. Snapshot images of the ice structures accreted on the UAV airframe surface ((**a**) wing upper surface; (**b**) wing lower surface; (**c**) vertical stabilizer; (**d**) propeller blade; (**e**) Pitot probe) acquired right after the UAV landed on the ground [20].

Avery and Jacob [21] conducted icing tests of cylinders at low Reynolds numbers in a cryogenic wind tunnel, comparing experimental and numerical results. Williams et al. [16] investigated the RG-15 airfoil in the NRC Altitude Icing Wind Tunnel in 2017. Oswald et al. [22] conducted an experimental study using 3D printed glaze shapes to analyze the icing performance of a medium-sized fixed-wing UAV with the RG-15 airfoil at the largest wind tunnel facility in the Karmen Institute, Belgium. The glaze shapes were derived from previous icing wind tunnel experiments, with their 2D profiles shown in Figure 3. Finally, the experimental results were compared with the FENSAP numerical results.

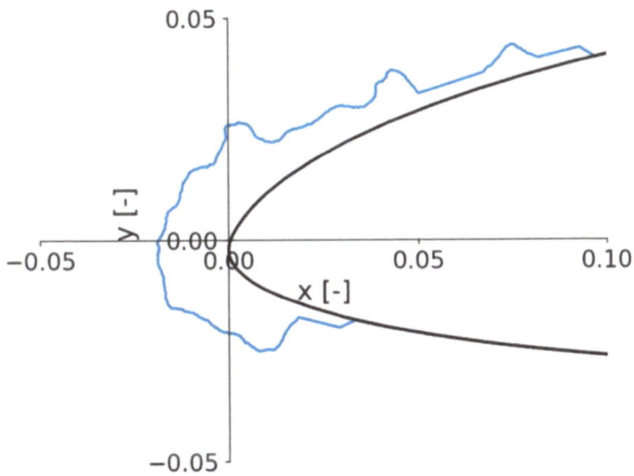

Figure 3. A 2D icing profile obtained from the experiment [22]. The blue line represents the ice shape, and the black line represents the clean airfoil.

Hann carried out three main experimental campaigns to gather ice shape data for UAV wings at low Reynolds numbers. In spring 2019, Hann [23] conducted experiments at the Cranfield icing wind tunnel on the RG-15 and NREL S826 airfoils. These tests yielded three different icing patterns: glaze, mixed, and rime ice, with Reynolds numbers around 10^5. In fall 2019, Hann et al. [24,25] performed additional icing studies on the RG15 airfoil at the VTT icing wind tunnel facility in Finland, focusing on glaze and mixed ice conditions. In 2020, Hann expanded the experimental study to the NREL S826 airfoil, conducted at the Norwegian University of Science and Technology wind tunnel, investigating three different icing conditions: glaze, mixed, and rime ice [26]. Li et al. [27] examined the dynamic ice accumulation and unsteady heat transfer processes of NACA0012 airfoils of thermoplastic materials and aluminum, respectively, under the same icing conditions in the icing research wind tunnel at Iowa State University.

2.2. Numerical Studies

Improved computer processing power and new computational fluid dynamics techniques have made numerical simulations a cost-effective and efficient way to study ice. Some countries have developed mature icing simulation software [28–30], including LEWICE in the United States, ONERA Icing Code in France, FENSAP-ICE in Canada, and so on. However, these existing icing simulation tools are validated for manned aviation under high Reynolds numbers [23], and further validation is needed to apply them for UAV icing simulation under low Reynolds numbers [31]. In order to better simulate the icing and anti-icing processes of UAVs, researchers have adopted two approaches; one is to verify the effectiveness of existing simulation tools and choose the more suitable one; the other is to develop new models.

2.2.1. Validation of Existing Simulation Tools

Hann [32] compared the panel-based LEWICE method to a modern CFD icing code, FENSAP-ICE. Both codes were used to generate three characteristic 2D ice shapes (rime, glaze, and mixed) at low Reynolds numbers on a NREL S826 wing. Figure 4 presents the simulation results for both codes. Rime results were the most similar, with both codes calculating a good match in the extent, direction, area, and overall shape of the ice, although three differences could be detected. Glaze and mixed ice results showed significant differences between the two icing codes. All three icing conditions in FENSAP-ICE exhibited leading-edge separation, with varying degrees of separation strength, weakest in rime and strongest in mixed ice. As a panel method, LEWICE is not designed to capture separation effects, which may fail to represent critical flow features adversely affecting the prediction of ice shapes. Therefore, LEWICE seems more appropriate for the study of UAV icing and aerodynamic losses under rime conditions.

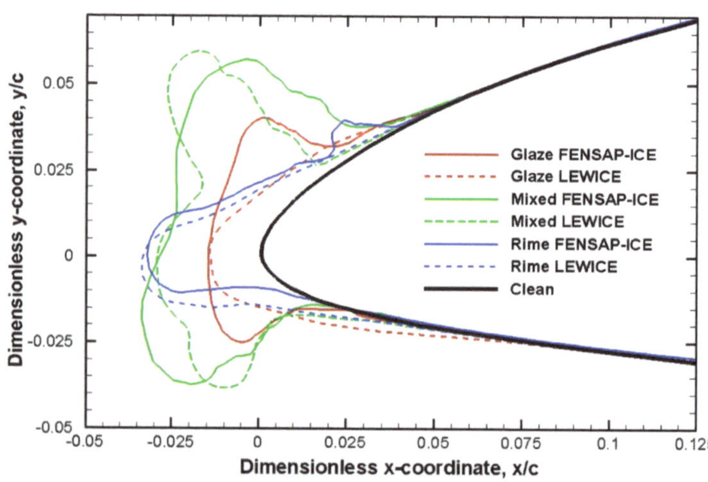

Figure 4. Ice shape results with LEWICE and FENSAP−ICE for the test cases [32].

Yirtici et al. [33] performed numerical icing simulations for three NREL S826 airfoils with different spreading ratios using the XFOIL [34] panel solver. XFOIL predicted larger ice shapes under frost conditions compared to LEWICE. In addition, the XFOIL solver predicted drag coefficients in good agreement with experimental data, but overpredicted lift coefficients in the low angle of attack to the stall region.

Oswald et al. [22] used the FENSAP-ICE flow solver module and the one-equation Spalart-Allmaras (SA) turbulence model for icing simulations of the RG-15 airfoil. The clean RG-15 airfoil was simulated in steady-state 2D RANS CFD with flow transition locations predicted by the software tool XFOIL. Flow transitions were assumed to be triggered by ice accumulation. Simulation parameters are derived from experimental conditions. Comparison of the numerical results with experimental data indicated that the SA turbulence model seemed to be of limited applicability for estimating ice-induced aerodynamic losses at low Reynolds numbers. Other higher-order turbulence models such as Reynolds stress and nonlinear eddy viscosity models should be investigated for numerical simulations.

Hann et al. [26] simulated the aerodynamic losses of the NREL S826 airfoil under icing using SA and Menter's k-ω SST turbulence models and compared them with experimental data. The simulations matched well with experimental data for clean and streamlined icing airfoils, but had limitations for complex ice shapes and stall prediction. Experimental stall behavior was not captured by either model. The SA model overpredicted lift values, while the k-ω SST model tended to underpredict the maximum lift angle and predict earlier stall.

The flow around UAVs operating at low Reynolds numbers ($O(10^5)$) is mainly laminar, leading to laminar separation bubbles (LSBs) that affect lift, drag, and pitch values. Muhammed and Virk [35] used k-ω SST and transitional k-ω SST turbulence models to numerically simulate and evaluate their ability to predict LSBs on the RG-15 airfoil at a Reynolds number of 1.07×10^5. The k-ω SST turbulence model predicted fully attached flow at lower angles of attack but failed to predict LSBs. On the other hand, the transitional k-ω SST turbulence model accurately predicted LSBs but with an earlier separation onset time.

2.2.2. Developing of New Models

Bottyán [15] developed an onboard icing estimation method with a simple ice accumulation model to examine the influence of airflow temperature, liquid water content (LWC), wing geometry, and airflow velocity on UAV wing icing.

Szilder and McIlwain [36] investigated the impact of Reynolds number on the icing process by using a morphogenetic icing method to derive an analytical model for UAV icing. The model was combined with a CFD solver (which could predict the Stanton number) and a droplet trajectory solver to simulate ice accretion on the NACA0012 2D airfoil in the Reynolds number range of 5×10^4 to 5×10^6. Some of the model's predictions were in close agreement with experimental results.

Avery and Jacob [21] developed a cylindrical volume icing model called ALRIA in Matlab to study UAV icing conditions. The ALRIA model was composed of droplet flow fields and thermal equilibrium sections, including low Reynolds number formulas for cylindrical components in the range of Re~3×10^5–8×10^5. During ALRIA's development, Avery enhanced the earlier work of Langmuir et al. [37] and derived a method to calculate droplet collection efficiency as a function of the stagnation point angle. Avery and Jacob subsequently conducted experiments, model simulations, and LEWICE3D simulations on chosen scenarios to compare their results. As shown in Figure 5, comparative analysis indicated that the ALRIA model was able to simulate ice accretion under dry and most wet icing conditions, but it performed poorly for large and thick ice accretions.

Flight experiments are costly and have limited validation. Numerical icing simulation software and wind tunnel tests have played an important role in solving aviation icing issues and should also be applied to UAV icing. However, there is still not enough work done in this new field. Additionally, existing experimental and numerical results often do not match, perhaps due to inaccurate simulations or errors in the tests. It is necessary to continue research on UAV icing.

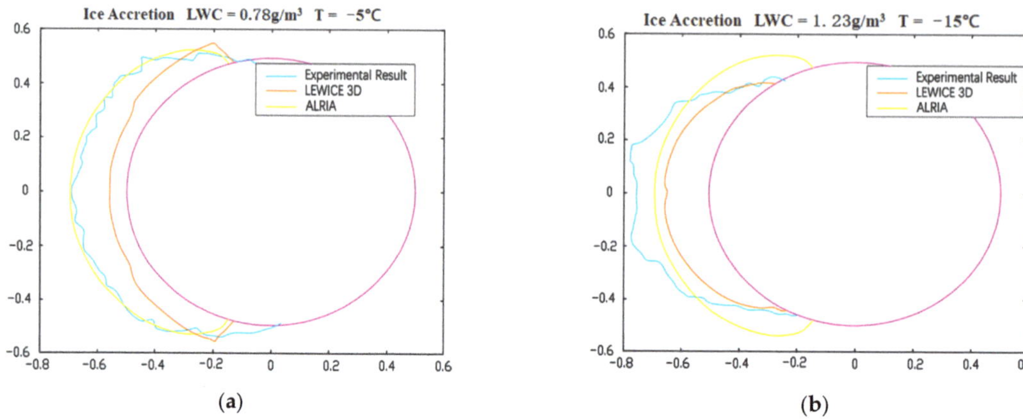

Figure 5. Comparison of experimental, model simulation, and LEWICE3D simulation results under given operating conditions: (**a**) −5 °C, 30.5 m/s, 0.78 g/m^3 for 5 min; (**b**) −15 °C, 30.5 m/s, 1.23 g/m^3 for 5 min [21]. The pink line represents the cross-section of the cylinder.

3. Advances in UAV Icing Research

Like manned aircraft, UAV icing can be affected by conditions such as free stream velocity, atmospheric temperature, LWC, median volume diameter of droplets (MVD), angle of attack, Reynolds number, and so on.

3.1. Typical Icing Characteristics of UAVs

Many of the characteristics of UAV icing are the same as for manned aircraft, as shown below. As the atmospheric temperature decreases, the ice changes from glaze to mixed ice and finally to rime [38]. The mass of ice increases with rising temperature [39]. Changes in the angle of attack cause changes in ice accretion position [16]. Increases in LWC and MVD lead to a greater amount of ice accretion, and the shape of ice shifts from streamlined to angular [38]. When investigating based on the FAR 25 Appendix C icing envelope, ice accretion is a function of temperature, LWC, and MVD. As the MVD value increases, the ice size decreases and eventually disappears due to the decrease in LWC [40]. When the LWC is large, a considerable amount of latent heat must be removed to freeze the impacting droplets completely, which is favorable for glaze. However, as the potential for latent heat removal increases at lower temperatures, the temperature also plays an important role here.

However, most unmanned drones have some typical characteristics in icing, given their differences from manned aircraft, such as lower Reynolds numbers, smaller flight speeds, lower flight altitudes, smaller sizes, and different body materials.

In low Reynolds number flows, a bubble can form at the leading edge of the UAV wing, which can cause laminar-to-turbulent flow transition. Low Reynolds numbers also impact surface heat transfer and supercooled droplet trajectory, which can, in turn, affect the airfoil icing process. Szilder and McIlwain [36] developed an icing analysis model for drones in 2011 to study the effect of Reynolds numbers on icing behavior. With increasing Reynolds numbers, while holding other parameters constant, ice transformed from feathery rime to a combination of glaze and rime and eventually to solely glaze. Flight conditions that resulted in glaze at high Reynolds numbers might result in rime at low Reynolds numbers. Additionally, as the Reynolds number increased, mixed ice conditions tended to occur at colder air temperatures and lower LWCs.

Szilder and McIlwain [36] also compared ice accretion at low and high Reynolds numbers for specific flight distances, which was crucial for understanding actual flight situations of UAVs. At a specific flight distance, as the Reynolds number increased, the relative ice accretion size significantly decreased while the actual ice mass increased. Low Reynolds numbers are advantageous for rime formation that lowers aerodynamic losses, but it also increases the relative thickness of ice, which raises aerodynamic losses. Therefore, the changes in aerodynamic characteristics during UAV flight are dependent on specific flight conditions.

The Reynolds number is a function of chord length and velocity. Bottyán et al. [15,41] observed in 2013 that when the ambient temperature was −4.2 °C, an increase in airspeed from 10 m/s to 130 m/s would cause icing behavior to change from less dangerous dry ice to more dangerous horn ice. In addition, it was observed that the ice accretion rate increases with speed, and the ice accretion thickness also increases with speed. However, the ice thickness at the stagnation point decreases and disappears completely at higher speeds. Hann and Johansen [24] studied the effects of chord length and velocity respectively on UAV icing based on the FENSAP-ICE software. The study revealed that airspeed had progressively increasing effect on rime, mixed, and glaze ice. Increasing airspeed slightly increased the relative ice thickness of rime, enlarged the droplet impact area but does not affect its streamlined ice shape. The impact of airspeed on mixed ice was different. As the airspeed increased, the morphology of mixed ice transitioned from rime-like to glaze-like, resulting in reduced ice thickness and significantly increased ice limit. The impact of airspeed on glaze was similar to that on mixed ice. At lower wind speeds, glaze on the airfoil had a streamlined shape, but it bore less resemblance to rime compared to low-speed mixed ice conditions. At higher speeds, the glaze formation became more complex, and the ice limit significantly increased.

The impact of chord length was consistent across the three types of ice formation, as shown in Figure 6. Firstly, the relative ice thickness significantly increased for smaller wing sections. Secondly, the relative ice limit was significantly increased. Thirdly, there was no significant impact of chord length variation on ice geometry. Greater relative ice thickness often indicates higher aerodynamic losses. This essentially means that unmanned aircraft are much more sensitive to icing than manned aircraft, with icing conditions rated as "trace" or "slight" for manned aircraft potentially being severe for small unmanned aircraft [42].

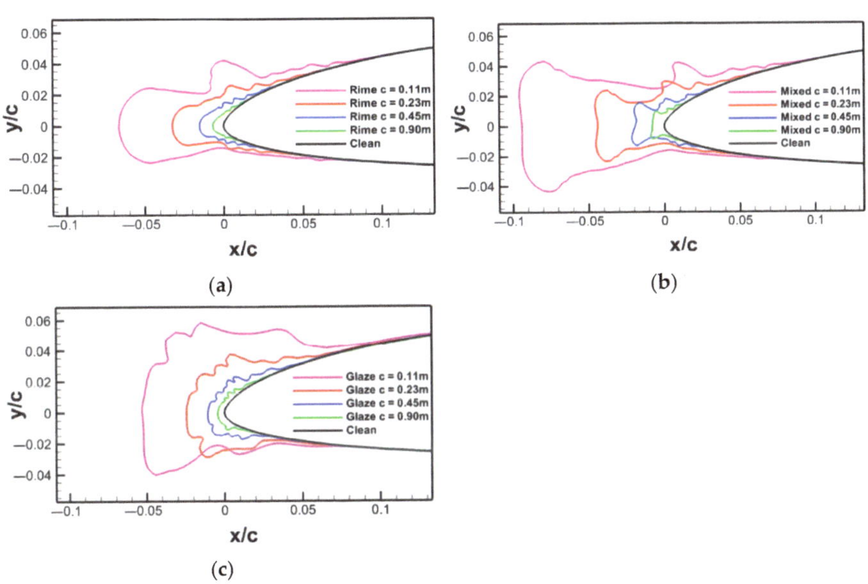

Figure 6. Ice shapes for the three meteorological icing conditions rime (**a**), mixed (**b**), and glaze (**c**) for the variation of chord length [24].

Flying altitude affects UAV icing through changes in natural environment, including lower temperature and atmospheric pressure at higher altitudes, varying liquid water content and droplet size. In addition, the body material of drones also plays a role in icing, with thermop materials releasing latent heat from melting much more slowly compared to aluminum. Li [27] found that the surface temperature near the leading edge of the wing of the UAV model using thermoplastic material is higher, the "heating" area is larger, the surface water flow on the wing surface is more obvious, and more complex stream ice structures are formed at downstream positions beyond the direct impact area of supercooled water droplets.

3.2. Changes in Aerodynamic Characteristics Caused by Icing

Reynolds number significantly affects the quantity, location, and geometry of ice accumulation, which are closely related to performance losses [4]. In 2007, Cistriani et al. [43] studied the aerodynamic performance of an iced UAV wing in a wind tunnel. Maximum lift was reduced by 30% compared to a clean wing. Hann et al. [44] demonstrated that icing increased drag, decreased lift, and lowered maximum angle of attack. Also, Hann et al. [32] analyzed aerodynamic performance penalties of the NREL S826 wing with rime, glaze, and mixed ice. Overall, mixed ice showed severe performance degradation followed by glaze and rime. Fajt [39], Hann [24], and Oswald [22,45] reported high aerodynamic losses of the RG-15 airfoil in glaze ice conditions, with drag coefficient increasing up to 170%. Szilder and Yuan reported severe aerodynamic losses of the SD7037 airfoil in glaze conditions [46]. Simulation results showed a maximum decrease in lift coefficient of 16%, drag coefficient increasing up to 300%, a 12% decrease in pitch moment, and an early stall at 9° (clean airfoil stalled at 11°). Hann et al. [26] conducted an analysis on the aerodynamic degradation of the NREL S826 airfoil at different Reynolds numbers during 2020. According to Figure 7, glaze and horn ice formations resulted in the most significant drag loss, with a corresponding 26% loss in lift and up to a 330% increase in drag coefficient at zero angle of attack. This substantial drag increase and changes in lift occurred due to the occurrence of large separation bubbles resulting from the horn geometry. The study found that the geometric shape of ice was the cause of flight performance degradation, with more complex ice formations causing more severe degradation in performance. The effect of Reynolds number on the results appeared to be relatively small, and higher Reynolds numbers led to greater lift and reduced drag levels.

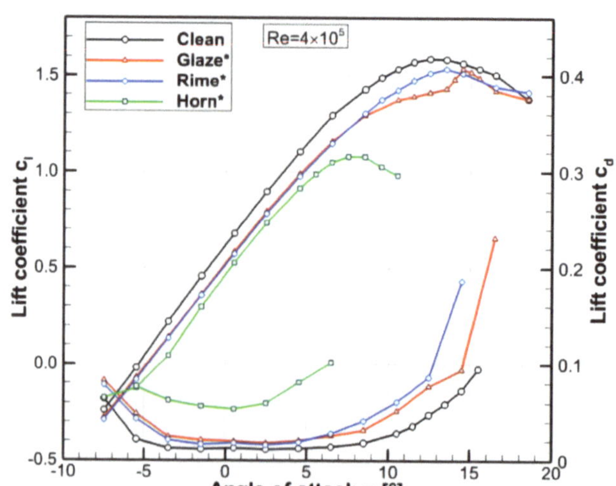

Figure 7. Experimental results for lift and drag of the LEWICE ice shapes for Re = 4×10^5 [26]. * indicates that the ice shape was calculated by LEWICE.

The laminar separation bubble (LSB) depends on the airfoil geometry and flow velocity, and can be observed within the Reynolds number range of 5×10^4 to 3×10^6 [47]. When formed, the LSB alters the behavior of the boundary layer, transitioning the flow from laminar pre-separation to turbulent post-reattachment. As the angle of attack increases and the separation bubble becomes smaller in length and thickness, the LSB moves upstream [35]. At low Reynolds numbers (defined as $Re < 10^6$), where the boundary layer is thicker and the LSB is larger in size [48], the significance of laminar flow characteristics and LSB becomes even more pronounced. It could potentially impact ice accretion and subsequently degrade aerodynamic performance [48]. Additionally, observations by Oo [49] revealed that flow at low Reynolds numbers experienced delayed reattachment, increased separation, and the possibility of secondary separation bubbles. Such intricate flow patterns at low Reynolds numbers may explain some of the "peculiar" phenomena encountered during UAV flights under icing conditions.

In numerical simulations, icing causes a decrease in lift and an increase in drag across the entire range of angle of attack values [26]. However, certain experiments reveal a sudden increase in lift when the angle of attack reaches a specific value after entering the stall zone. This lift increase could be due to the ice shape acting as a nose-down effect [50] or localized separation bubbles [51]. As Reynolds number decreases, the lower inertia of the boundary layer causes trailing edge separation to occur at a lower angle of attack, resulting in a more significant impact on lift. It explains why this anomalous behavior occurs at all Reynolds numbers, with the most prominent effect observed at low Reynolds numbers. Seifert and Richert [52] observed a lesser lift increase in their experiments. Similar effects at low Reynolds numbers were recorded by Jasinski et al. [53].

Ice-induced separation bubbles (ISBs) represent another type of airflow separation that affects the aerodynamic characteristics of an iced airfoil. ISBs form downstream of ice accretion at the leading edge, with their size increasing along with the angle of attack, while the starting position remains constant. High-frequency oscillations can be observed in the transitional region between ISBs and LSBs [54]. Both Oo [49] and Bragg [4] concluded that ISBs and LSBs exhibit similar characteristics. Oo observed an increase in lift and a decrease in drag for an iced airfoil compared to a clean airfoil (RG-15) under specific icing conditions. The increase in lift can be explained by the previous section. As for the decrease in drag, the authors attribute it to the relatively larger LSBs present on the clean airfoil compared to the ice-induced separation bubbles.

Appendix A summarizes the main findings of the UAV icing study cited in this article. Although research on UAV icing is currently insufficient, limited research progress indicates that icing significantly deteriorates the aerodynamic performance of UAVs, resulting in reduced lift, increased drag, and premature stall. The reduced maximum lift negatively impacts the stability of UAVs, whereas the heightened drag coefficient results in greater power consumption. These penalties in aerodynamic performance are likely to restrict the range and endurance of UAVs, thereby impacting the successful execution of their missions [55]. It further underscores the significance of developing anti-icing/deicing technologies for UAVs.

4. Existing Anti-Icing/Deicing Techniques for UAVs

The ice protection system (IPS) helps to minimize the negative impacts of icing on aircraft. Conventional aircraft deicing methods use coatings, chemicals, heat, and mechanics [56]. Chemical methods mainly utilize liquids, heat methods include gas heating and electric heating, and mechanical methods refer to using force to separate the ice from the substrate, including pneumatic and electric pulse methods. While these techniques are well established for aircraft, they are relatively new when it comes to UAVs [31]. The inherently limited payload and low available energy of UAVs necessitate lightweight and low-energy IPSs. Therefore, the mature anti-icing/deicing techniques used on aircraft cannot be directly applied to UAVs and require further assessment and improvement.

The pneumatic deicing technique [56] utilizes a high-pressure air source to rapidly expand expansion tubes on the wing surface to break up surface ice, which is then carried away by external airflow. This method is systematic and requires high-pressure gas sources, pressure valves, gas pipelines, and actuated expandable tubes, making it only suitable for large aircraft. However, the entire system is relatively heavy and burdensome for UAVs, and there is no available literature on its use for UAV ice protection.

The liquid anti-icing/deicing technique [57,58] utilizes liquid to reduce ice adhesion to aircraft surfaces or lower the freezing point of water on the surface. Although primarily used for ground anti-icing of aircraft, some aircraft carry liquid and have specific spraying devices to apply liquid for anti-icing/deicing during flight. The "weeping wing" system [59] on the RQ-1B drone [60] continuously pumps a deicing liquid film onto the wing, and liquid is also used for anti-icing/deicing on the MQ-9B UAV [61]. However, this method has a restricted effective time and requires significant liquid consumption, which is too heavy for drones. Therefore, the method is not ideal for anti-icing/deicing.

The hot air anti-icing/deicing technique [62] utilizes heated air from the engine to warm aircraft components and prevent icing. This method is well established and predominantly used for large passenger or military transport aircraft. However, UAVs typically rely on electricity as their power source and are incapable of providing a hot air supply. Consequently, this technique is not suitable for UAVs.

Coatings [63–65] primarily aim to reduce the adhesion strength of water or ice to surfaces, making it challenging for ice to form or adhere to the structure's surface, thus facilitating ice removal. Although coatings are energy-efficient, they cannot completely prevent icing. Their effectiveness is limited to delaying the freezing of water droplets on the coating surface [66,67]. Currently, researchers are striving to combine passive anti-icing techniques like coatings with active anti-icing/deicing methods such as heating to enhance overall anti-icing/deicing performance [68,69].

The electric anti-icing/deicing technique utilizes electric heating elements to convert electrical energy into heat energy. This process warms the aircraft's skin or components, thus achieving the desired anti-icing/deicing effect [70,71]. Anti-icing and deicing are two different modes of an electric heating system. The anti-icing mode applies continuous heat to the surface of a structure to prevent ice formation and is primarily used for components where ice formation is not allowed. The deicing mode allows ice to accumulate on the surface until a certain amount is reached, and then applies heat to melt it. The electric heating technique is lauded as an effective anti-icing/deicing method due to its high heating efficiency, precise control, and adaptable configuration. The Gulfstream Pilot Operating Handbook [72] suggests that the electric anti-icing system, operating continuously under icing conditions, should maintain a surface temperature of roughly 38 °C–54 °C. Consequently, a high power supply may be needed to satisfy enhanced anti-icing requirements [73,74]. For instance, a typical general aviation aircraft's anti-icing area, which utilizes a pad heater, is around 0.90 m^2, and necessitates around 23,250 W/m^2 of energy to attain the anti-icing temperature. This energy requirement is challenging for any aircraft to handle, let alone drones that have limited energy resources for anti-icing/deicing. The Chinese Wing Loong II drone uses an electric anti-icing system on its wings while simultaneously employing anti-icing coatings to decrease power usage [75]. However, with the emergence of fully electric aircraft, the electric anti-icing/deicing system will enter a new era of development.

Electric pulse deicing technique leverages the interaction between the magnetic field generated by the pulse coil circuit and the induced magnetic field of the metal skin surface. The interaction leads to high-frequency skin vibrations that remove ice from the surface [76,77]. Russia employed this technique successfully on IL-76 and IL-86 aircraft. In recent years, the Orion UAV was also equipped with an electric pulse system to manage icing hazards [78]. This system offers low energy consumption and high efficiency but has its limitations. Composite material skins pose a challenge for this technology because of the difficulty in generating induced magnetic fields.

Few anti-icing techniques are available for unmanned aerial vehicles, which face struggles in handling severe icing conditions. Table 1 lists the application and pros and cons of using traditional aircraft anti-icing technologies on UAVs.

Table 1. Application of conventional aircraft anti-icing/deicing techniques to UAVs.

Techniques	Pros	Cons	Applications
Pneumatic type	Simple operation and low cost	Large mass, easily damaged pneumatic sleeve	—
Liquid	No energy consumption	Large mass, short effective time	RQ-1B UAV MQ-9B UAV
Hot air	Air source available on aircraft	No air source on drones	—
Coating	No energy consumption	Short effective time	Wing Loong II UAV
Electric type	Good anti-icing/deicing effect and mature system	High energy consumption with overflow ice	Wing Loong II UAV
Electric pulse type	Low energy consumption and good deicing effect	Composite materials are not easy to generate magnetic fields	Orion UAV

5. Exploring Ice Mitigation Techniques for UAVs

Given that anti-icing/deicing techniques for drones are not yet sufficiently developed, scientists are actively working on developing ice mitigation techniques that are UAVs.

5.1. Airfoil Optimization

It seems that structural optimization offers a solution to minimize ice accretion on UAVs at the source.

Ghisu et al. [79] incorporated icing conditions into the optimization process for a general aviation airfoil. They considered the uncertainty of the ice accretion location and employed an adaptive polynomial chaos method for quantifying the uncertainty. The method computed how airfoil performance varied with the given changes in ice accretion location. The findings indicated that obtaining the optimal lift coefficient configuration came at the expense of a significant negative pitching moment and a 10% increase in drag coefficient in comparison to a clean wing. It is important to note that the ice shape in their optimization process remained fixed rather than occurring naturally from the icing process.

Li et al. [80] conducted a study in which they optimized a supercritical airfoil to consider the effects of icing. Their results demonstrated a 16.5% increase in the maximum lift coefficient for the icing airfoil at a drag coefficient of 3.7 during cruise conditions. However, the optimization process utilized a fixed ice shape and neglected variations in ice shape that corresponded to the airfoil and icing conditions. Dai et al. [81] proposed an objective optimization method to design ice-resistant airfoils, balancing performance between clean and icing conditions. The optimized configuration displayed considerable improvement in icing performance, indicating a 9.95% increase with a maximum lift coefficient of 4.133 during icing conditions.

Li et al. [40] utilized the Radial Basis Function—Enhanced Differential Evolutionary algorithm to perform two rounds of optimization on a specific UAV. In the initial bi-objective optimization, net lift-to-drag ratio and icing stall performance were considered. The results indicated that the optimal "cruise optimal" design improved the maximum lift coefficient for icing conditions by 18% compared to the traditional "icing optimal" design. During the second optimization, fluctuations in atmospheric icing conditions were taken into account since they significantly affect aerodynamic performance. Environmental parameters were regarded as uncertainties, and a Probability Density Function was assumed. To evaluate the maximum lift coefficient due to varying environmental parameters, the Nonintrusive Polynomial Chaos Expansion formula was utilized. Finally, the mean and variance of the resulting maximum lift coefficient were employed as objectives in the collection of data.

Figure 8 showed that the resulting "Polynomial Chaos Expansions optimal" airfoil had improved icing performance, with a 6.7% increase in maximum lift coefficient.

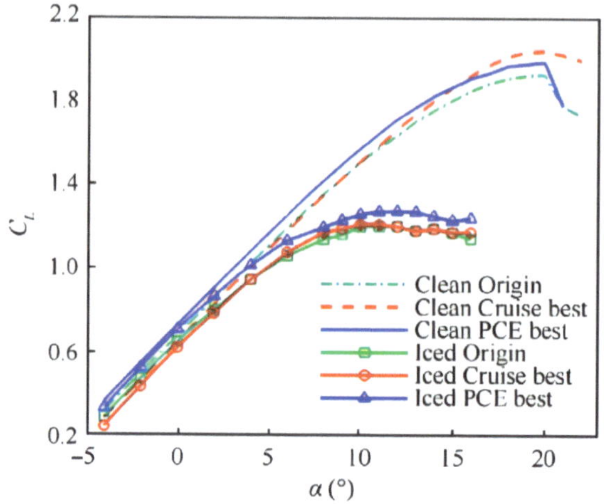

Figure 8. Airfoil optimization results [40].

5.2. Flight Optimization

Vukit [82] suggested that the most effective approach for avoiding the dangers of natural icing is through precise prediction of conditions that trigger icing and avoiding such conditions. According to several researchers, the weather research and forecasting (WRF) model is capable of accurately predicting meteorological conditions that lead to icing [83]. By combining the WRF model with proxy models [84], a safe envelope or graded safety zones icing in the target area can be generated. This offers crucial data support for optimizing UAV paths.

Narum et al. [85] developed an optimal path planning algorithm for a hybrid electric UAV with an electrical anti-icing/deicing system, using given meteorological forecasts as input data. Particle swarm optimization was used for energy consumption and overall time usage optimization. Multiple simulated datasets were used to test the algorithm. Path planning optimization considering meteorological forecasts for such platforms resulted in a 43% energy efficiency improvement or a 42% flight time reduction between two points, compared to standard straight-line cruising flights.

5.3. Coatings

Several methods have been suggested for minimizing ice adhesion, such as "air film isolation" represented by superhydrophobic (SHP) surfaces [86–88], "oil film isolation" represented by slippery liquid-infused porous surfaces (SLIPS) [89–91], and "water film isolation" represented by hydrophilic lubrication layers [92,93] and electrically heated coatings [94,95]. Although coatings have proved to be highly effective for anti-icing purposes, their insufficient durability remains a major drawback.

SHP surfaces have been able to passively reduce interfacial adhesion and promote droplet shedding, demonstrating excellent anti-icing performance. Ye et al. [96] created a hydrophobic coating system composed of fluorinated organic siloxane material for UAV icing protection that exhibited a contact angle $\theta \geq 115°$ according to experimental measurements. It is important to keep coating thickness within 20 nm to avoid compromising UAV endurance with excess weight. For possible application of SHP surfaces in anti-icing on metallic unmanned aircraft wings, Luo et al. [97] used electrostatic spraying and melt molding processes to make PEEK-PEEK/PTFE/k-SiO_2 composite coatings on the surface of 7075Al alloy, employing PTFE and PEEK as physically blended materials infused with modified hydrophobic silica (k-SiO_2) particles. The composite coatings exhibited SHP characteristics, retaining excellent mechanical strength even after 51 abrasion cycles and 50 peeling cycles with 3M tape. Moreover, in addition to remarkable self-cleaning properties, water droplets completely froze on the SHP coating at −15 °C within 940 s, which is 43.5 times slower than on a standard 7075Al alloy surface.

The concept of SLIPSs, initially introduced by Wong et al. [98], has gained popularity as a passive anti-icing solution. On SLIPSs, ice is isolated by a thin oil film retained within the porous structure, which substantially reduces ice adhesion through the "solid-oil-solid contact model". To demonstrate the potential for anti-icing, Kim et al. [89] developed aluminum-based SLIPSs that effectively prevent ice/frost accumulation by facilitating the removal of condensation water and achieving ultra-low ice adhesion. Li et al. [99] combined this surface with electrothermal technology to create a slippery liquid-infused porous electric heating coating (SEHC). Dynamic anti-icing tests revealed that SEHCs reduced energy consumption by 23% compared to dry porous electric heating coatings, and 11% compared to SHC porous heating coatings. Continuous high-speed jetting tests, static deicing/deicing cycles, and dynamic anti-icing tests confirmed that SEHCs exhibit superior durability compared to SHC porous electric heating coatings.

Similar to SHP surfaces and SLIPSs, ice on the water-based lubrication layer is also isolated by a water film within the structure. Inspired by ice skating, Chen et al. [92] used dopamine-modified hyaluronic acid to create anti-icing coatings with water-based lubrication layers under mild conditions. Moreover, thanks to the non-selective adhesion property of dopamine, this coating can be used on almost all types of solid surfaces. Dou et al. [93] grafted hydrophilic dihydroxymethyl propionic acid onto a polyurethane substrate to provide hydrophilicity to the coating, which absorbs moisture to isolate ice from the surface. Environmental tests conducted at −53 °C revealed that the water lubrication layer remained intact, and the ice adhesion strength persisted even after 30 deicing and icing cycles, confirming durability. Alternatively, electric heating coatings use Joule heating to melt ice, forming a water film that creates "water film isolation" conditions.

5.4. Electric Heating for Anti-Icing/Deicing

UAVs rely on IPSs that are lightweight and energy-efficient [31]. Electric systems are well suited for unmanned aircraft due to their maturity, nature, and ease of retrofitting onto existing airframes [100]. However, traditional electric heating systems consume a significant amount of energy, which necessitates measures to reduce energy consumption in order to make them suitable for unmanned aircraft. These measures primarily involve selecting materials with improved heating performance, optimizing processes, and implementing more reasonable designs. System optimization often involves a combination of multiple measures.

5.4.1. Selection of High-Quality Materials

Carbon-based materials, such as carbon nanotubes (CNTs) and graphene, which exhibit excellent conductivity and thermal conductivity, are ideal options for utilizing the conductor's Joule heating effect to de-ice unmanned aircraft [73,101].

In 2013, Buschhorn et al. [73] created a CNT-based IPS that only requires 1 kW/m^2 of power to prevent icing under mild icing conditions. Yao et al. found that a carbon nanotube network/carbon fiber composite structure could effectively de-ice within 15 s at a constant power density of 4.9 kW/m^2 [102]. Zhao developed an electric heating coating by incorporating multi-walled CNTs into polyurethane/paraffin composites, demonstrated a positive temperature coefficient effect, and could automatically control the maximum temperature during heating, thereby protecting the temperature-sensitive composite substrate and saving electricity [101].

In addition, the conductivity of graphene-modified carbon fiber/polyether ether ketone composites was around 0.1 S/m, resulting in a 35% increase of Joule heating-to-deicing energy conversion efficiency [103]. Glass fiber bundles coated with graphene displayed low electrical resistance (~1.7 Ω/cm) and rapid temperature rises (from −0.1 °C to 27.3 °C within 5 min) when exposed to a 10 V voltage [104].

Cortés et al. discovered that 3D printed circuits utilizing graphene nanosheets and carbon nanotube-reinforced resin are effective for anti-icing purposes [105]. Fluorinated decylated graphene nanoribbon films, a fresh design that merges the low polarizability fluorocarbons and the exceptional electrical conductivity of graphene nanoribbons, are anticipated to offer structural components effective anti-icing/deicing performance [106].

Hexagonal boron nitride is another exceptional heating material due to its high thermal conductivity, superior thermal and chemical stability, and mechanical strength, making it an excellent choice [107]. Since carbon-based materials are capable of shielding electromagnetic waves [108], anti-icing/deicing systems with high radio frequency (RF) transmission rates are necessary to ensure proper functioning of RF equipment, such as radar domes and antennas. In 2020, Hwang et al. proposed an RF transmission heating system for deicing, and its deicing effect was demonstrated by applying a voltage of 40 V to the system at a temperature of −20 °C [109].

5.4.2. Search for Cost-Effective Manufacturing Processes

The conductivity and thermal properties of carbon fibers allow them to serve as direct heating elements in electric heaters, eliminating the need for separate heating elements. This simplifies the integration of anti-icing/deicing systems into drones, reducing complexity and cost. However, numerous challenges arise from the techniques employed in manufacturing carbon-based heaters, including high costs, energy-intensive processes, limited lifespan, and additional weight on composite materials [110–113]. These factors consequently restrict the applications of carbon-based materials. Consequently, it is imperative to improve manufacturing technology.

Idris [114] employed extrusion printing technology to produce electrical contact points on carbon fibers during the manufacturing of heating devices. The results revealed that the heater was successfully produced, achieving high temperatures suitable for deicing. The proposed carbon fiber heater was estimated to increase costs by only USD 0.03/cm^2. Vertuccioa et al. [110] adopted expanded graphite to create flexible expanded graphite foils through a straightforward and eco-friendly solvent casting process with the addition of PVA. Karim et al. [104] first synthesized scalable amounts of graphene-based ink using the microfluidic exfoliation technique, then utilized the dip-drying solidification coating technique to manufacture highly conductive graphene-based glass fiber yarns. Both techniques are simple, scalable, and extremely suitable for industrial applications. Xu et al. [115] used a casting technique to produce laminated flexible pressure-sensitive adhesive graphene-based composite heaters with exceptional joule heating performance, thermal stability, electrical anisotropy, and mechanical properties. The heater was lightweight and environmentally friendly, enabling it to be used as a custom-shaped flexible deicing heater for various substrates. At a low power of roughly 2.1 W and a bias voltage of 6 V, the heating temperature peaked at 142.9 °C. In addition, the casting method employs simple equipment, has high production efficiency, yields uniform film performance, and is well established in industry.

5.4.3. Design of System Layout

UAVs possess restricted available energy, ensuring the careful design of the electric heating system and minimizing heat for the operating system.

Developed by the Norwegian University of Science and Technology's Autonomous Marine Operations and Systems Center and commercialized by UBIQ Aerospace [100], D•ICE is an electric IPS that implements a heating zone made of carbon fiber. Figure 9b [116] depicts the basic composition of D•ICE, exhibiting a continuously heated special heating zone located at the separation strip by the leading edge of the stagnation point, which is utilized to reduce the energy consumption required for deicing [117]. The strip boosts the aerodynamic force that acts on the ice, enhancing ice removal efficiency. The typical width of the separation strip is 2–3 mm. Since 2013, D•ICE has been integrated into various styles and sizes of fixed-wing UAVs, then tested in icing wind tunnels and during flight.

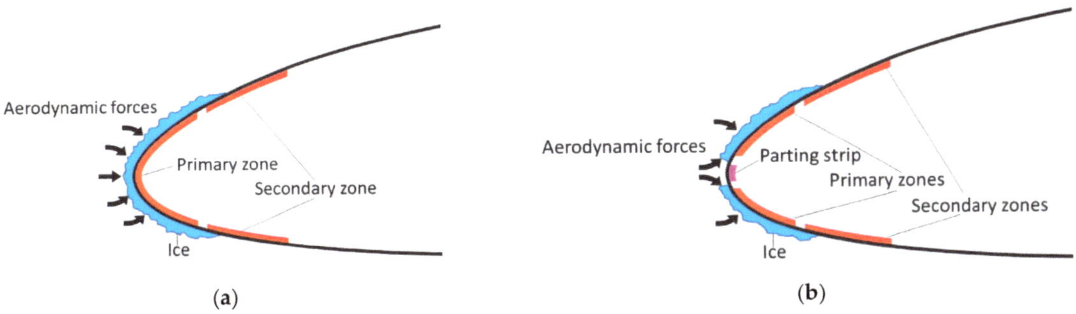

Figure 9. Schematic layout of the heating-zones for the (**a**) conventional deicing and (**b**) deicing with parting strip [116].

During autumn 2018, D•ICE was tested at the Cranfield icing wind tunnel to identify the minimum heat flux needed to maintain an ice-free surface under two different weather conditions by gradually decreasing the system's power supply [100]. In autumn 2019, the VTT icing wind tunnel facility tested two layouts of the D•ICE system (refer to Figure 9). These experiments were conducted for mid-size fixed-wing UAVs operating at low Reynolds numbers (Re = 8–9 $\times$ 10^5). By adopting separation strips, deicing energy requirements reduced by 50% when compared to traditional deicing systems. For separation strip systems, higher angles of attack resulted in shorter deicing times, reducing such duration by approximately 20–30%.

Roy et al. [118] designed and developed an integrated electrothermal anti-icing system using a thin etched foil heating film. The system was integrated into the leading edge of a CFRP laminate wing of a UAV. Simulated icing flight conditions included an airflow velocity of 102 m/s, an air temperature of −6.65 °C, an MVD of 20 µm for water droplets, an LWC of 0.78 g/m^3, and an exposure time of 600 s. The study investigated different IPS heat flux values ranging from 0 to 10 kW/m^2. Two IPS designs were considered, one with 5 heating film units and another with 15 units in the leading edge region. The results revealed that the 15 heating film IPS design with a film heat flux of 7.5 kW/m^2 achieved complete anti-icing functionality, maintaining a wing surface temperature distribution between 0–13 °C.

Sørensen et al. [119] developed a temperature-controlled anti-icing/deicing system for the airfoil of the X8 Skywalker UAV. The system structure is shown in Figure 10. The core of the system is a coating made of carbon nanomaterials graphene and carbon black, which provides anti-icing functionality. The thermocouple and humidity sensor are responsible for sensing the surface temperature of the airfoil leading edge and the environmental humidity, respectively. Power delivery to the coating is controlled by an Arduino microcontroller based on feedback from the thermocouple and humidity sensor. Different coating layouts were studied within a temperature range of −25 °C to +25 °C. Experiments showed that a layout in which the coating covers the entire length of the wing was preferred, and that the solution was capable of rapidly raising the surface temperature when needed (deicing) and maintaining a roughly constant temperature when needed, while keeping energy consumption within the weight and cost constraints imposed by the small size of the UAV. Sørensen and Johansen [120] investigated the heat flow produced by electric IPS when integrated into small UAVs. The temperature along the heat source was non-uniform under a fixed power supply of 10 kW/m^2. The results suggested that utilizing smart methods could be advantageous in reducing power consumption to a minimum.

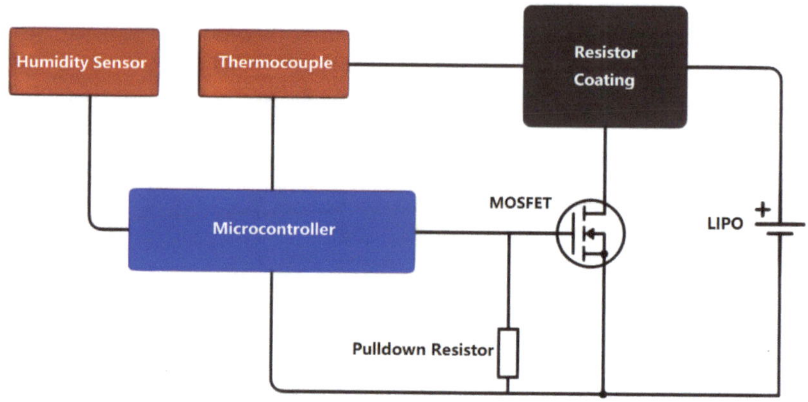

Figure 10. Schematic of ice protection system [119].

5.4.4. Combining with Coatings

When compared with traditional anti-icing/deicing systems, hybrid anti-icing/deicing systems that utilize hydrophobic coatings and electric heating systems offer higher energy efficiency, making them ideal for use in UAVs. There has been a significant rise in publications related to hybrid electric systems over the last few decades [121,122].

Zhao [95] employed multi-walled CNTs as heating materials, acrylic resin clear varnish as coating materials, and hydrophobic SiO_2 nanoparticles to create an SHP coating. Anti-icing experiments demonstrated that the multilayer coating not only accomplished low-temperature anti-icing (around 7 °C), but also had efficient electric heating and anti-icing performance. In comparison to conventional electric heating methods, the multilayer coating could reduce anti-icing energy consumption by up to 58%.

Bogoslov et al. [123] developed an integrated electrically heated anti-icing system (IE AIS) that relies on fluorescent film. The IE AIS features a 50 μm fluoroplastic film that functions as a passive anti-icing coating, and a 300 μm thick conductive fluoroplastic film that serves as the heating element. In to these components, the system also includes an electrical current and thermal insulation layer connected to a control system, alongside temperature-sensitive components specifically designed for aircraft parts. By making use of IE AIS, energy necessary for surface deicing could be decreased by 30% as compared to currently available electrically heated anti-icing systems.

Battelle [124] developed a technique that incorporates CNTs into both the barrier and paint layers. When an electric current runs through it, the system acts as a resistive heating coating and creates a highly conductive heating surface. This approach provides several benefits compared to traditional ice-removal systems. Firstly, it is lightweight and requires low power. Secondly, the method utilizes established coating application processes, enabling its direct application to the UAV fuselage.

Yan et al. [125] proposed an anti-icing solution that adds a surface insulation layer, metal heating layer, and bottom insulation layer under a biomimetic anti-icing structural layer to obtain a hybrid anti-icing skin. The biomimetic anti-icing layer, inspired by Qinling giant bamboo, is a multilayer non-uniform high and nanoscale structure. The metal heating layer is made of constantan, which has high resistivity and thermal conductivity and good ductility and bendability. Flight tests demonstrated the excellent performance of this hybrid anti-icing system.

Zhu [126] created a low-energy SHP electric heating anti-icing skin for a UAV by combining a polyimide SHP surface with an electric heating film. Figure 11a shows the production process. In order to optimize the energy utilization of the anti-icing skin, electric wire structures were designed for different regions of the airfoil model to generate differing power densities. The overall design, including the size and regions of the SHC electric heating skin, is depicted in Figure 11b. Region I, located at the front of the upper wing surface, is 90 mm long and responsible for anti-icing/deicing. Region II is 30 mm long and covers the leading edge of the wing, acting as a hot knife and requiring higher power density than the upper and lower wing surfaces. Region III, 40 mm long, is responsible for anti-icing/deicing the bottom of the wing. Anti-icing wind tunnel experiments demonstrated significant benefits for the SHC electric heating skin over ordinary electric heating skins, with reductions in energy consumption of 52%, 68.8%, and 89.9%, respectively, during three different icing conditions. Test conditions were very similar to those at the University of Quebec [70], with their anti-icing skin yielding reductions of just 13.3% and 34.2%, making the results from Zhu better than those from the University of Quebec.

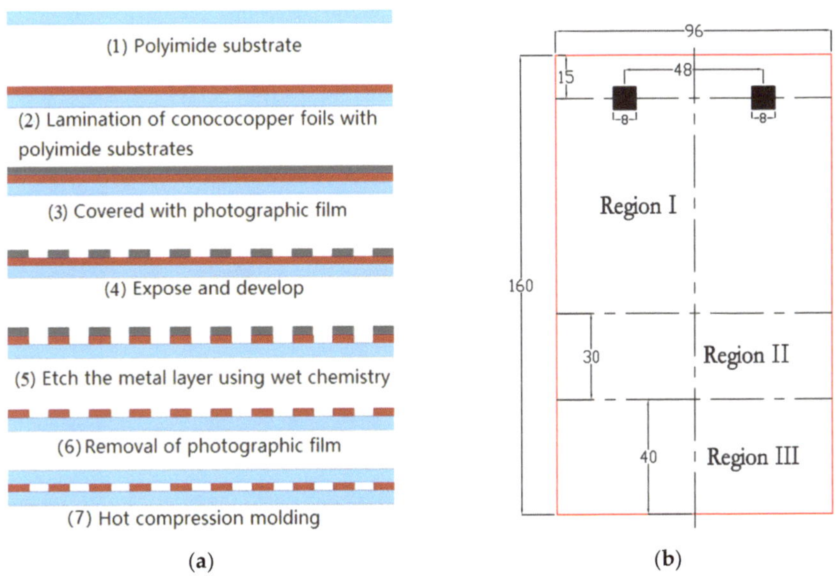

Figure 11. Electric heating skin's (**a**) production process and (**b**) overall size [126].

5.5. Piezoelectric Deicing

The piezoelectric deicing technique has also been developed for aviation [127]. When the current flows in one direction, the piezoelectric crystal bends. Thus, there are two ways for piezoelectric materials to achieve deicing: one is when high-frequency AC is applied to piezoelectric actuators [128], the substrate vibrates at a very high frequency, and the stress generated destroys the ice and substrate; the other is to let the piezoelectric substance work in the low-frequency band and excite the resonant frequency of the structure [129]. Piezoelectric systems have the advantages of small volume, light weight, arbitrary shape, and low energy consumption [130,131], but piezoelectric materials are prone to fatigue and have a relatively short service life. Overall, the piezoelectric technique is a promising deicing method.

Palacios [132,133] utilized piezoelectric patches to generate ultrasonic shear stress at high frequencies, studied the combination of hydrophobic coatings with ultrasonic deicing systems [134], and designed a piezoelectric deicing system [135] effective in promoting shedding ice layers ranging from 1.4 to 7.1 mm thick and under different icing conditions. Budinger [136] compared a ceramic deicing system with Langevin prestressed piezoelectric transducers to find an interesting compromise solution for an ultrasound-based piezoelectric deicing device.

Meanwhile, Venna et al. [129,137–139] used piezoelectric ceramics to excite low-frequency modes (below 1000 Hz) while Struggle et al. [140] performed similar experiments which proved successful deicing at 307 Hz. Villeneuve [128] designed a low-frequency piezoelectric deicing system for thinned Bell 206 main rotor blades and Bell 206 tail rotor blades, and demonstrated its deicing performance in an icing wind tunnel. For all structures, the power input of the actuator was less than 19 kW/m^2. Palanque [141] developed an electromechanical resonance deicing system based on low-frequency piezoelectric actuators, and proved that the prototype had an effective deicing performance with low peak power consumption (less than 10 kW/m^2) and fast deicing.

Surface acoustic wave (SAW) deicing employs piezoelectric material and shows promise as a deicing method due to its ability to provide localized heating, on-site control low power consumption, and system integration. It is achieved by applying high-frequency electrical signals to patterned interdigit electrodes on a piezoelectric substrate, which generates surface vibrations via the reverse piezoelectric effect [142,143]. Heat generation results from both the Joule heating of the interdigitated electrodes and the dielectric losses in the piezoelectric substrate [144,145]. The sketch of the SAW-induced melting process of glace ice aggregates is shown in Figure 12 [146].

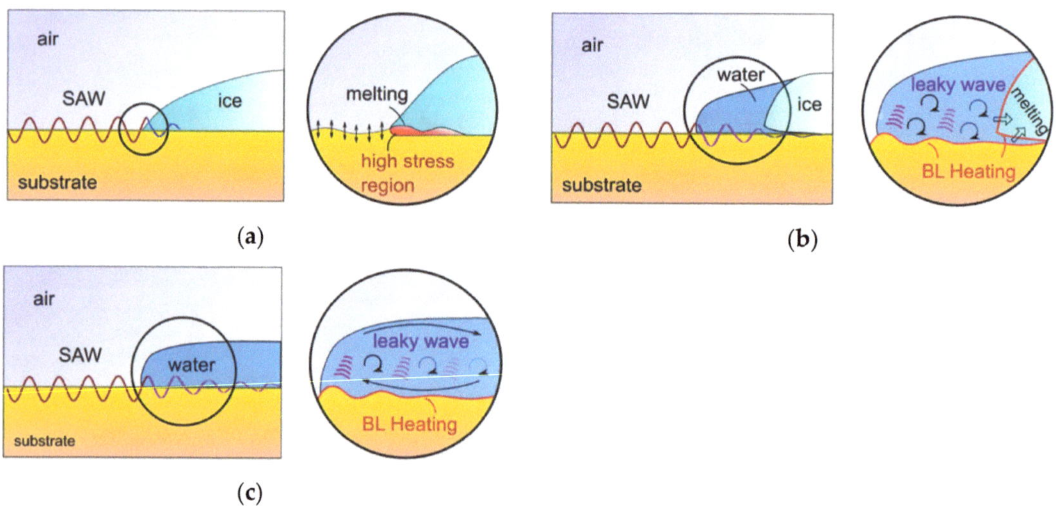

Figure 12. Sketch of the SAW-induced melting process of glace ice aggregates: (**a**) initial deicing process: the SAW interacts with the ice, creating regions of high stresses. A thin water film forms in the high-stress region; (**b**) melting process: the SAW efficiently leaks into the water-film. The ice melts at the well-defined interface with the water; (**c**) active anti-icing process: the ice is completely melted. (BL: Boundary layer) [146].

Zeng et al. [147] experimentally demonstrated that SAWs can reduce ice adhesion. Nampoothiri et al. [148] deiced microliter-sized droplets with low-power SAW actuation using interdigitated electrodes on a piezoelectric ($LiNbO_3$) substrate, suggesting the future development of efficient SAW-based deicing systems. Jacob et al. [146] demonstrated that energy-efficient deicing and prevention of ice formation could be achieved for larger volume piezoelectric substrates using high-frequency thickness-shear acoustic vibrations, with reduced ice adhesion. Jacob also attempted to transfer the deicing capability to low-thermal conductivity base materials.

5.6. Plasma Anti-Icing/Deicing

With the increasing research on active plasma flow control [149], attention has recently turned to the heating effect of the dielectric barrier discharge (DBD) actuator [150]. Especially, surface dielectric barrier discharge (SDBD) plasma has been found to be a promising approach for efficient anti-icing due to its high heating performance, low energy consumption, quick response, and high efficiency [151].

Some studies have shown that plasma actuators are effective anti-icing/deicing devices, and nanosecond dielectric barrier discharge (NSDBD) plasma actuators have been found to perform better than alternating current DBD plasma actuators [152–155]. Zhu et al. [156,157] investigated the heating performance of NSDBD and demonstrated that it can heat up quickly and has an ice-proof heating effect. Wei et al. [158,159] showed that NSDBD had excellent anti-icing performance, removing an average ice thickness of 3 mm within 4 s and preventing further ice formation in the plasma protection zone.

Furthermore, Niu et al. [160] simulated the anti-icing effect of the NSDBD plasma actuator located at the NACA0012 wing leading edge. The simulation results demonstrated that the plasma actuator produced high-temperature air that covered the anti-icing area of NACA0012, preventing ice formation. Niu observed that higher peak voltage and pulse frequency of the plasma actuator led to better anti-icing performance, and directional arrangement optimization of the plasma excitation device could also improve its anti-icing capability. Thus, Niu concluded that the NSDBD plasma anti-icing technique was effective and could be used to develop anti-icing systems for unmanned aircraft.

Gao [161] developed a set of nanosecond pulse (ns-) SDBD plasma anti-icing devices for a certain type of all-weather small UAV. The working principle of ns-SDBD plasma actuators is as shown in Figure 13 [162]. The device has 8 kV voltage output, 200 A maximum output current, and 20 kHz heavy-frequency working capability and was designed with aircraft load, energy consumption, and anti-icing effectiveness as the top priorities. The experiment demonstrated that the device had anti-icing performance under specific icing conditions. The team plans to continue developing more compact, lighter, and durable anti-icing devices in the future.

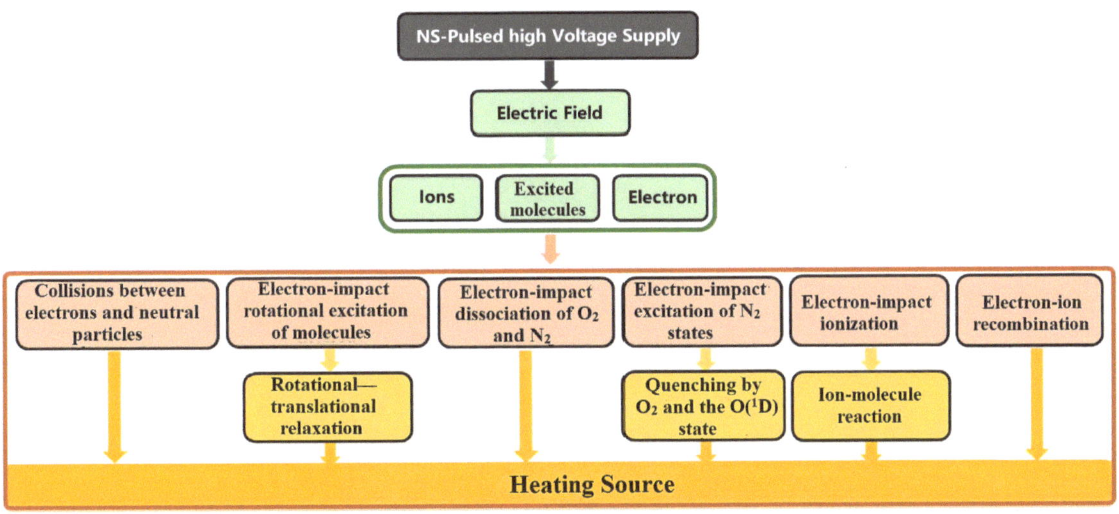

Figure 13. Ns-SDBD plasma heating mechanism [162].

5.7. Synthetic Thermal Jet for Anti-Icing/Deicing

Since the mid-1990s, synthetic jet technology, also known as zero-mass-flux jet technology, has become a popular topic in flow control research. The technology synthesizes jets by combining controlled vortex structures. Synthetic jets are considered one of the most promising active flow control methods because of their advantages such as zero mass flow rate, low energy consumption, quick response, and flexible control [163]. The piezoelectric synthetic jet actuator (PSJA) was first proposed by Glezer, Smith, and their research team at Georgia Institute of Technology in the 1990s, attracting researchers from all over the world due to its simple structure and high instantaneous pressure rise, making it potentially suitable as an ice breaker [164].

In 2013, Nikisha creatively proposed using a heated bottom synthetic jet actuator array to effectively control ice formation, studying the impact of parameters such as bottom heating, heating temperature, inflow temperature, droplet size, and distribution on ice formation [165–167]. Li designed a synthetic thermal jet actuator in 2015, based on a synthetic dual jet actuator [168,169]. The synthetic thermal actuator was to melt ice crystals on a cold surface in experiments on deicing principles, confirming the feasibility of deicing using synthetic thermal jets.

In 2017, Jiang [170] designed a novel electromagnetic synthetic dual jet actuator (as shown in Figure 14) for a specific UAV. By conducting fundamental experimental research on deicing an ice-covered wing using synthetic hot jets, it was found that the hot jet was more effective in reducing deicing time, by 25.0% up to 36.4%, when compared to pure heating deicing. Additionally, a 90° jet angle exhibited the most significant deicing effect, due to minimal momentum loss and enhanced heat transfer. Liu et al. added an impact rod structure to the traditional PSJA to improve its deicing performance in flight environments [171]. Furthermore, Gao et al. proposed a novel deicing strategy in which electric heating first eliminates the adhesion forces and then the PSJA discharge fractures the ice [172]. The PSJA operation only accounted for an insignificant 0.27% of the entire system's energy consumption and did not cause any surface bending.

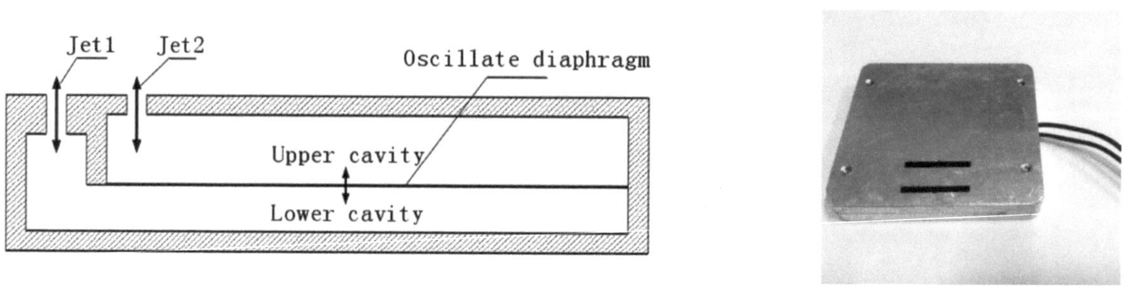

(a) (b)

Figure 14. The synthetic double jet actuator's (**a**) structural schematic diagram and (**b**) physical diagram [170].

In general, the above-mentioned methods each have advantages and disadvantages. Airfoil and flight optimization as well as coatings are energy-efficient but only offer reductions or delays in icing. Electric heating systems have great anti-icing/deicing capabilities but consume more energy. Piezoelectric deicing can result in alterations to the surface shape of the aircraft. Plasma and synthetic thermal jet methods are not yet mature and lack practical applications. However, these methods are not mutually exclusive and can be combined. Under a framework of airfoil optimization, multiple anti-icing/deicing measures can be integrated, such as coatings with electric heating, synthetic thermal jet with electric heating, and piezoelectric with coatings. In conclusion, additional research and optimization are necessary to develop an outstanding anti-icing/deicing system for unmanned aircraft.

6. Discussions

Scholars have made advancements in studying the icing phenomenon and anti-icing/deicing techniques for UAVs, but there is still much progress to be made in effectively addressing icing challenges.

Currently, the academic community lacks a comprehensive understanding of the UAV icing phenomenon, particularly in relation to the complex flow behavior at low Reynolds numbers, which remains unexplained accurately thus far. The most prominent methods for studying ice accretion are the icing wind tunnel experiment and numerical icing simulation. While existing icing numerical codes can reliably predict the accumulation of rime ice, they struggle to provide accurate forecasts for glaze and mixed ice due to the involvement of intricate physical factors. In addition, the icing wind tunnel tests on drones are not enough, and the mutual evidence between wind tunnel tests and numerical simulations is still insufficient. This lack of information not only makes it harder to understand icing on drones, but it also affects the assessment of anti-icing/deicing techniques. Considering the time-consuming and challenging nature of developing new icing numerical codes, as well as the existing codes' reliability, it is advisable to develop new models specifically for unmanned aircraft icing, apply them to the existing codes, and conduct extensive experiments to validate them.

On the contrary, the anti-icing/deicing system for UAVs must satisfy specific conditions such as low weight, minimal power consumption, and minimal aerodynamic design changes, due to constraints such as limited available energy, small size, light payload, and susceptibility to icing. However, some traditional anti-icing/deicing techniques cannot be used on drones, while others, although usable, cannot meet the above requirements. Therefore, developing an innovative approach becomes necessary when creating a novel anti-icing/deicing system for UAVs. Significantly improving the performance indicators of traditional techniques is not the only way. Developing new techniques is one approach, and considering the coupling of multiple ice mitigation techniques is also a good choice. In addition, the cost of anti-icing/deicing systems also needs to be considered.

7. Conclusions

The article provides a review of current knowledge regarding the icing phenomenon and anti-icing/deicing methods for fixed-wing UAVs. Previous studies show that UAV icing is similar to manned aircraft icing, but unique characteristics can arise due to the low Reynolds number, low flight speed, small size, low altitude, and different materials of most UAVs. In the limited research on UAV icing, it is still evident that icing poses severe hazards to UAVs. However, the existing anti-icing/deicing technology cannot meet the need of drones. The article discusses ice mitigation technology being explored and emphasizes the need for lightweight and energy-saving systems to reduce icing impact and ensure all-weather UAV operation.

Author Contributions: Conceptualization, L.Z. and X.Y.; methodology, L.Z. and X.Y.; investigation, L.Z. and Q.L.; data curation, L.Z.; writing—original draft preparation, L.Z.; writing—review and editing, L.Z. and X.Y.; supervision, X.Y. All authors have read and agreed to the published version of the manuscript.

Funding: The work reported in this paper is supported by National Natural Science Foundation of China (No. 12132019) & National Science and Technology Major Project of China (Project J2019-III-0010-0054).

Data Availability Statement: The data presented in this study are available on request from the corresponding author.

Conflicts of Interest: The authors declare no conflict of interest.

Appendix A

Appendix A provides a summary of the research section of the main text. It includes selected content and is not exhaustive. Table A1 presents research on the icing principles and characteristics of drones. Table A2 presents ice mitigation techniques directly relevant to drones in Section 5.

Table A1. A tabular summary for UAV icing research.

Researchers	Year	Research Method	Major Achievement or Key Findings
Bottyán et al. [15]	2013	Flight tests	A UAV ice accretion model was developed.
Williams et al. [16]	2017	Flight tests and lab tests	The ice shapes differed significantly from those on high Reynolds number airfoils.
Matiychyk et al. [17]	2017	Flight tests	The energy consumption of unmanned aerial vehicles increased after icing.
Siddique [19]	2021	Flight tests	
Han et al. [20]	2023	Flight tests	
Avery [18]	2019	Flight tests	The geometric shape of ice driven by the flight flow field appeared on the surface of the drone.
Avery and Jacob [21]	2022	Numerical simulation and lab tests	A cylindrical volume icing model called ALRIA was established and validated.
Oswald et al. [22]	2022	Numerical simulation	The Spalart Allmaras turbulence model was limited in its applicability to estimating aerodynamic losses caused by ice at low Reynolds numbers.
Hann et al. [24]	2019	Numerical simulation and lab tests	The influence of airspeed on frost, mixed ice, and glaze increased sequentially. The relative ice thickness and relative ice limit of smaller airfoils significantly increased.
Hann et al. [26]	2020	Numerical simulation and lab tests	Spalart Allmaras and Menter's k-ω SST model had limitations in simulating complex ice shapes and stall. More complex icing caused greater degradation of flight performance. As the Reynolds number increased, lift increased while drag decreased.
Li et al. [27]	2019	Lab tests	The surface water reflux on the wing surface using thermoplastic materials was more pronounced [25].
Hann [32]	2018	Numerical simulation	LEWICE and FENSAP-ICE showed good consistency in simulating frost ice, while glaze and mixed ice showed significant differences.
Yirtici et al. [33]	2020	Numerical simulation	The resistance coefficient predicted by XFOIL method was in good agreement with experimental data, while the lift coefficient was not.
Muhammed and Virk [35]	2023	Numerical simulation	Transition k-ω SST turbulence model could accurately predict LSB, but it predicted an earlier separation start time.
Szilder and McIlwain [36]	2011	Numerical simulation	A UAV ice accretion model was developed. As the Reynolds number increased, ice changed from frost ice to mixed ice, and finally became clear ice.
Fajt et al. [39]	2019	Numerical simulation	Rising temperature caused an increase in ice mass. Ice caused an increase in resistance.
Cistriani et al. [43]	2007	Lab tests	Ice caused a decrease in lift.
Hann et al. [44]	2017	Numerical simulation	Ice accumulation increased drag while reducing lift and maximum angle of attack.
Szilder et al. [46]	2015	Numerical simulation	
Oo et al. [49]	2020	Lab tests	At low Reynolds numbers, flow reattachment delayed and separation increased.

Table A2. A tabular summary for ice mitigation techniques of UAVs.

Subject	Researchers	Year	Research Method	Major Achievement or Key Findings
Airfoil optimization	Li et al. [40]	2022	RADE algorithm	"PCE" airfoil had improved icing performance, with a 6.7% increase in maximum lift coefficient.
Flight optimization	Narum et al. [86]	2020	Particle swarm optimization	A 43% energy efficiency improvement or a 42% flight time reduction between two points.
Coatings	Ye et al. [97]	2021	Lab tests	A hydrophobic coating system that exhibited a contact angle $\theta \geq 115°$ was created.
	Luo et al. [98]	2021	Lab tests	The PEEK-PEEK/PTFE/k-SiO_2 composite coatings for drone metal surfaces were created.
Electric heating combining with coatings	Hann et al. [117]	2021	Lab tests	The better design for layout of D•ICE was found.
	Roy et al. [119]	2021	Numerical simulation and lab tests	An integrated electrothermal anti-icing system using a thin etched foil heating film was designed and developed.
	Sørensen et al. [120]	2015	Numerical simulation and lab tests	A temperature-controlled anti-icing/deicing system for the airfoil of the X8 Skywalker UAV was designed.
	Yan et al. [127]	2023	Lab tests and flight tests	An anti-icing Skin with Micro-nano Structure was designed and tested.
	Zhu [128]	2018	Lab tests	A low-energy SHP electric heating anti-icing skin was created.
Plasma	Gao [164]	2021	Lab tests	A set of ns-SDBD plasma anti-icing device was created.
Synthetic thermal jet	Jiang [172]	2017	Lab tests	A novel electromagnetic synthetic dual jet actuator was designed.

References

1. Weatherington, D.; Deputy, U. *Unmanned Aircraft Systems Roadmap, 2005–2030*; U.S. Department of Defense: Washington, DC, USA, 2005.
2. Gao, H.; Yu, P. The Application Prospects of Drones in Emergency Communication. *Labor. Prot.* **2022**, 85–87. (In Chinese)
3. Villeneuve, E.; Karmouch, E.; Boulerice, X. Development of a small and transportable de-icing/anti-icing drone-mounted system. Part 1: System design. *Drone Syst. Appl.* **2022**, *10*, 155–177. [CrossRef]
4. Bragg, M.B.; Broeren, A.P.; Blumenthal, L.A. Iced-airfoil aerodynamics. *Prog. Aerosp. Sci.* **2005**, *41*, 323–362. [CrossRef]
5. How Is an UAV Affected by In-Flight Icing and Can We Simulate It Accurately? Available online: https://uavicinglab.com/2021/05/28/how-is-an-uav-affected-by-in-flight-icing-and-can-we-simulate-it-accurately/ (accessed on 15 January 2022).
6. Haulman, D.L. *US Unmanned Aerical Vehicles in Combat 1991–2003*; Air Force Historical Research Agency: Montgomery, AL, USA, 2003.
7. Botura, G.; Fahrner, A. *Icing Detection System–Conception, Development, Testing and Applicability to UAVs*; AIAA-2003-6637; Goodrich Corp: Akron, OH, USA, 2003.
8. Szilder, K.; McIlwain, S. In-flight icing of UAVs—The influence of flight speed coupled with chord size. *Can. Aeronaut. Sp. J.* **2012**, *58*, 83–94. [CrossRef]
9. Zhang, B.; Tang, L.; Roemer, M. Probabilistic weather forecasting analysis for unmanned aerial vehicle path planning. *J. Guid. Control. Dyn.* **2014**, *37*, 309–312. [CrossRef]
10. Koenig, G.; Ryerson, C.; Larsson, J.; Reehorst, A. Effect of Variable LWC on Ice Shape in the NASA-GRC IRT. In Proceedings of the 41st Aerospace Sciences Meeting and Exhibit, Reno, NV, USA, 6–9 January 2003.

11. Lynch, F.; Khodadoust, A. Erratum to "Effects of ice accretions on aircraft aerodynamics". *Prog. Aerosp. Sci.* **2002**, *38*, 669–767. [CrossRef]
12. Siquig, R.A. *Impact of Icing on Unmanned Aerial Vehicle (Uav) Operations*; Naval Environmental Protection Research Facility: Monterey, CA, USA, 1990.
13. Papadopoulos, C.; Ioannidou, S.; Panagiotou, P.; Yakinthos, K. Numerical investigation of the impact of tubercles and wing fences on the aerodynamic behaviour of a fixed-wing, tactical Blended-Wing-Body UAV platform. *IOP Conf. Ser. Mater. Sci. Eng.* **2022**, *1226*, 012015. [CrossRef]
14. Samad, A.; Villeneuve, E.; Blackburn, C.; Morency, F.; Volat, C. An Experimental Investigation of the Convective Heat Transfer on a Small Helicopter Rotor with Anti-Icing and De-Icing Test Setups. *Aerospace* **2021**, *8*, 96. [CrossRef]
15. Bottyán, Z. In-flight icing characteristics of unmanned aerial vehicles during special atmospheric condition over the carpathian-basin. *Landsc. Environ.* **2013**, *7*, 74–80.
16. Williams, N.B.A.; Brian, G.; Ol, M. The Effect of Icing on Small Unmanned Aircraft Low Reynolds Number Airfoils. In Proceedings of the 17th Australian International Aerospace Congress (AIAC), Melbourne, Australia, 26–28 February 2017.
17. Matiychyk, L.; Suvorova, N.; Tereshchenko, D.; Plakhotniuk, I.; Trachuk, K.; Komarova, K. Influence of Icing on Aircraft Performance of Unmanned Aerial Vehicle M-10-2 "Oko". *Proc. Natl. Aviat. Univ.* **2017**, *4*, 52–59. [CrossRef]
18. Avery, A.S. Ice Accretion on Small Unmanned Aircraft. Ph.D. Thesis, Oklahoma State University, Stillwater, OK, USA, 2019.
19. Siddique, M.A. An Experimental Study on the Effects of Adverse Weathers on the Flight Performance of an Unmanned-Aerial-System (UAS). Ph.D. Thesis, Iowa State University, Ames, IA, USA, 2021.
20. Han, N.H.; Siddique, M.A.; Zhang, Z.C.; Tian, L.C.; Hu, H.Y.; Hu, H. A flight-testing campaign to examine inflight icing characteristics and its effects on the flight performance of an Unmanned-Aerial-Vehicle. *Cold Reg. Sci. Technol.* **2023**, *207*, 103775. [CrossRef]
21. Avery, A.S.; Jacob, J.D. Ice accretion panel model for cylinders at low Reynolds numbers. *Bull. Atmos. Sci. Technol.* **2022**, *3*, 7. [CrossRef]
22. Oswald, J.W.; Enache, A.; Hann, R.; Glabeke, G.; Lutz, T. UAV Icing: Experimental and Numerical Study of Glaze Ice Performance Penalties on an RG-15 Airfoil. In Proceedings of the AIAA SCITECH 2022 Forum, San Diego, CA, USA & Virtual, 3–7 January 2022.
23. Hann, R. UAV Icing: Ice Accretion Experiments and Validation. In Proceedings of the International Conference on Icing of Aircraft, Engines, and Structures, Minneapolis, MN, USA, 17–21 June 2019.
24. Hann, R.; Johansen, T.A. UAV icing: The influence of airspeed and chord length on performance degradation. *Aircr. Eng. Aerosp. Technol.* **2021**, *93*, 832–841. [CrossRef]
25. Hann, R. Atmospheric Ice Accretions, Aerodynamic Icing Penalties, and Ice Protection Systems on Unmanned Aerial Vehicles. Ph.D. Thesis, Norwegian University of Science and Technology, Trondheim, Norway, 2020.
26. Hann, R.; Hearst, R.J.; Sætran, L.; Bracchi, T. Experimental and Numerical Icing Penalties of an S826 Airfoil at Low Reynolds Numbers. *Aerospace* **2020**, *7*, 46. [CrossRef]
27. Li, L.K.; Liu, Y.; Zhang, Z.C.; Hu, H. Effects of thermal conductivity of airframe substrate on the dynamic ice accretion process pertinent to UAS inflight icing phenomena. *Int. J. Heat Mass Transf.* **2019**, *131*, 1184–1195. [CrossRef]
28. Anderson, D.N. Rime-, Mixed-, and Glaze-Iced Evaluations of Three Scaling Laws. In Proceedings of the 36th Aerospace Sciences Meeting and Exhibit, Reno, NV, USA, 10–13 January 1994.
29. Wright, W.B.; Bidwell, C.S. Additional improvements to the NASA Lewis ice accretion code LEWICE. In Proceedings of the 33rd Aerospace Sciences Meeting and Exhibit, Reno, NV, USA, 9–12 January 1995.
30. Beaugendre, H.; Morency, F.; Habashi, W.G. FENSAPICE: Roughness effects on ice shape prediction. *J. Aircr.* **2003**, *40*, 239–247. [CrossRef]
31. Hann, R.; Johansen, T.A. *Unsettled Topics in Unmanned Aerial Vehicle Icing*; SAE EDGE Research Report EPR2020008; SAE International: Warrendale, PA, USA, 2020.
32. Hann, R. UAV Icing: Comparison of LEWICE and FENSAP-ICE for Ice Accretion and Performance Degradation. In Proceedings of the 2018 Atmospheric and Space Environments Conference, Atlanta, GA, USA, 25–29 June 2018.
33. Yirtici, O.; Körpe, D.S.; Ozgen, S. Aerodynamic Performance Losses due to Ice Formation on the UAV's Wing Profile. *J. Aeronaut. Space Technol.* **2020**, *13*, 207–215.
34. XFOIL Subsonic Airfoil Development System. Available online: http://web.mit.edu/drela/Public/web/xfoil/ (accessed on 11 November 2019).
35. Muhammed, M.; Virk, M. Steady and Time Dependent Study of Laminar Separation Bubble (LSB) behavior along UAV Airfoil RG-15. *Int. J. Multiphysics* **2023**, *17*, 55–76.
36. Szilder, K.; McIlwain, S. *In-Flight Icing of UAVs—The Influence of Reynolds Number on the Ice Accretion Process*; SAE Technical Paper 2011-01-2572; SAE International: Warrendale, PA, USA, 2011.
37. Langmuir, I. *A Mathematical Investigation of Water Droplet Trajectories*; Army Air Forces Headquarters, Air Technical Service Command: San Antonio, TX, USA, 1946.
38. Koenig, G.; Ryerson, C.; Kmiec, R. UAV Icing Flight Simulation. In Proceedings of the 40th AIAA Aerospace Sciences Meeting & Exhibit, Reno, NV, USA, 14–17 January 2002.
39. Fajt, N.; Hann, R.; Lutz, T. The Influence of Meteorological Conditions on the Icing Performance Penalties on a UAV Airfoil. In Proceedings of the 8th European Conference for Aeronautics and Space Sciences (EUCASS), Madrid, Spain, 1–4 July 2019.

40. Li, H.; Zhang, Y.; Chen, H. Optimization design of airfoils under atmospheric icing conditions for UAV. *Chin. J. Aeronaut.* **2022**, *35*, 118–133. [CrossRef]
41. Bottyán, Z. Estimation of In-Flight Icing Characteristics of UAVs during Different Meteorological Conditions. In Proceedings of the 8th International Conference on Intelligent Unmanned Systems (ICIUS 2012), Singapore, 22 October 2012.
42. Jeck, R.K. *A History and Interpretation of Aircraft Icing Intensity Definitions and FAA Rules for Operating in Icing Conditions*; Federal Aviation Administration Technical Report DOT/FAA/AR-01/91; SKYbrary: Anklam, Germany, 2001.
43. Cistriani, L. Falco UAV Low Reynolds Airfoil Design and Testing at Galileo Avionica. In *UAV Design Processes/Design Criteria for Structures*; Galileo Avionica Ronchi Dei Legionari (Italy) Simulators and Uav Business Unit; RTO: Neuilly-sur-Seine, France, 2007.
44. Hann, R.; Wenz, A.; Gryte, K.; Johansen, T.A. Impact of Atmospheric Icing on UAV Aerodynamic Performance. In Proceedings of the 2017 Workshop on Research, Education and Development of Unmanned Aerial Systems (RED-UAS), Linköping, Sweden, 3–5 October 2017.
45. Oswald, J.W. UAV Icing: Numerical and Experimental Study of Performance Penalties on an RG-15 Airfoil. Master's Thesis, University of Stuttgart, Stuttgart, Germany, 2021.
46. Szilder, K.; Yuan, W. The Influence of Ice Accretion on the Aerodynamic Performance of a UAS Airfoil. In Proceedings of the 53rd AIAA Aerospace Sciences Meeting, Kissimmee, FL, USA, 5–9 January 2015.
47. Hain, R.; Kähler, C.; Radespiel, R. Dynamics of laminar separation bubbles at low-Reynolds-number aerofoils. *J. Fluid Mech.* **2009**, *630*, 129–153. [CrossRef]
48. O'Meara, M.M.; Mueller, T.J. Laminar Separation Bubble Characteristics on an Airfoil at Low Reynolds Numbers. *AIAA J.* **1987**, *25*, 1033–1041. [CrossRef]
49. Oo, N.L.; Richards, P.J.; Sharma, R.N. Ice-Induced Separation Bubble on RG-15 Airfoil at Low Reynolds Number. *AIAA J.* **2020**, *58*, 5156–5167. [CrossRef]
50. Ericsson, L.E.; Reding, J.P. *Unsteady Airfoil Stall*; NASA Report CR-66787; NASA Center for Aerospace Information: Hanover, MD, USA, 1969.
51. Lissaman, P.B.S. Low-Reynolds-number airfoils. *Annu. Rev. Fluid Mech.* **1983**, *15*, 223–239. [CrossRef]
52. Seifert, H.; Richert, F. Aerodynamics of Iced Airfoils and Their Influence on Loads and Power Production. In Proceedings of the European Wind Energy Conference, Dublin, Ireland, 6–9 October 1997; pp. 459–460.
53. Jasinski, W.J.; Selig, M.S.; Bragg, M.B.; Shawn, C.N. Wind Turbine Performance Under Icing Conditions. *J. Sol. Energy Eng.* **1998**, *120*, 60–65. [CrossRef]
54. Oo, N.L.; Richards, P.; Sharma, R. Influence of an Ice-Induced Separation Bubble on the Laminar Separation Bubble on an RG-15 Airfoil at Low Reynolds Numbers. In Proceedings of the AIAA Aviation 2020 Forum, Virtual, 15–19 June 2020.
55. Muhammed, M.; Virk, M.S. Ice Accretion on Fixed-Wing Unmanned Aerial Vehicle—A Review Study. *Drones* **2022**, *6*, 86. [CrossRef]
56. Wang, J.; Ji, S.Y.; Yi, X.S.; Zhao, W.M. Research progress in aircraft anti-icing/deicing technology. *Aviat. Manuf. Technol.* **2015**, *495*, 30–32. (In Chinese)
57. Cornell, J.S.; Pillard, D.A.; Hernandez, M.T. Comparative measures of the toxicity of component chemicals in aircraft deicing fluid. *Environ. Toxicol. Chem.* **2000**, *19*, 1465–1472. [CrossRef]
58. Sapienza, R. Environmentally Benign Anti-Icing or Deicing Fluids. U.S. Patent 6,129,857, 2000.
59. *Unmanned Aerial Vehicle Reliability Study*; Technical Report; Office of the Secretary of Defense: Washington, DC, USA, 2003.
60. Peck, L.; Ryerson, C.; Martel, C. *Army Aircraft Icing*; Technical Report; Cold Regions Research and Engineering Laboratory: Hanover, HA, USA, 2002.
61. GA-ASI's MQ-9B RPAS Undergoes Cold Weather Validation Test. Available online: https://www.airforce-technology.com./news/gaasis-mq9b-cold-weather-validation (accessed on 27 February 2022).
62. Dong, W.; Zhu, J.; Zheng, M.; Chen, Y. Thermal analysis and testing of nonrotating cone with hot- air anti-icing system. *J. Propuls. Power* **2015**, *31*, 3. [CrossRef]
63. Boinovich, L.B.; Emelyanenko, A.M. Anti-icing Potential of Superhydrophobic Coatings. *Mendeleev Commun* **2013**, *23*, 3–10. [CrossRef]
64. Farzaneh, M.; Ryerson, C.C. Anti-icing and deicing techniques. *Cold Reg. Sci. Technol.* **2011**, *65*, 88–96. [CrossRef]
65. Shen, Y.Z.; Wu, Y.; Tao, J.; Zhu, C.L.; Chen, H.F.; Wu, Z.W.; Xie, Y.H. Spraying Fabrication of Durable and Transparent Coatings for Anti-Icing Application: Dynamic Water Repellency, Icing Delay, and Ice Adhesion. *ACS Appl. Mater. Interfaces* **2019**, *11*, 3590–3598. [CrossRef] [PubMed]
66. Lv, J.Y.; Song, Y.L.; Jiang, L.; Wang, J.J. Bio-Inspired Strategies for Anti-Icing. *ACS Nano* **2014**, *8*, 3152–3169. [CrossRef] [PubMed]
67. Varanasi, K.K.; Hsu, M.; Bhate, N.; Yang, W.S.; Deng, T. Spatial control in the heterogeneous nucleation of water. *Appl. Phys. Lett.* **2009**, *95*, 094101. [CrossRef]
68. Strobl, T.; Storm, S.; Thompson, D.; Hornung, M.; Thielecke, F. Feasibility study of a hybrid ice protection system. *J. Aircr.* **2015**, *52*, 2064–2076. [CrossRef]
69. Fortin, G.; Adomou, M.; Perron, J. *Experimental Study of Hybrid Anti-Icing Systems Combining Thermoelectric and Hydrophobic Coatings*; SAE Technical Paper 2011-38-0003; SAE International: Warrendale, PA, USA, 2011.
70. Ma, L.; Xiong, L.; Liu, L.; Yang, J. Experimental study on electro-thermal deicing technique for carbon fiber composite. *Acta Aeronaut. Astronaut. Sin.* **2012**, *33*, 54–61. (In Chinese)

71. Jiang, H.; Wang, H.T.; Liu, G.; Liu, J.P.; Zhang, X.N.; Chen, Y.Q.; Zhou, W.W. Light-weight, flexible, low-voltage electro-thermal film using graphite nanoplatelets for wearable/smart electronics and deicing devices. *J. Alloys Compd.* **2017**, *699*, 1049–1056. [CrossRef]
72. Rutherford, R.B. De-Ice and Anti-Ice System and Method for Aircraft Surfaces. U.S. Patent 6194685B1, 27 February 2001.
73. Buschhorn, S.T.; Kessler, S.S.; Lachmann, N.; Gavin, J.; Thomas, G.; Wardle, B.L. Electrothermal Icing Protection of Aerosurfaces Using Conductive Polymer Nanocomposites. In Proceedings of the 54th AIAA/ASME/ASCE/AHS/ASC Structures, Structural Dynamics, and Materials Conference, Boston, MA, USA, 8–11 April 2013; p. 1729.
74. Strehlow, R.H.; Moser, R. *Capitalizing on the Increased Flexibility that Comes from High Power Density Electrothermal Deicing*; SAE Technical Paper, 2009-01-3165; SAE International: Warrendale, PA, USA, 2009.
75. China's Domestically Produced Large-Scale Unmanned Aerial Vehicle Used for Artificial Weather Modification, the Wing Loong II, Successfully Completed Its Maiden Flight and Got First Place. Available online: http://www.iliema.cn/article-661981-1.html (accessed on 9 January 2022). (In Chinese).
76. Guang-Chao, L.I.; Jiang, H.E.; Lin, G.P. Electro-impulse de-icing (EIDI) technology study. *J. Aerosp. Power* **2011**, *26*, 1728–1735.
77. Zumwalt, G.; Friedberg, R. Designing an Electro-Impulse De-Icing System. In Proceedings of the 24th Aerospace Sciences Meeting, Reno, NV, USA, 6–9 January 1986.
78. Orion UAV Can Perform Ice Reconnaissance in the Arctic. Available online: https://www.airrecognition.com/index.php/news./defense-aviation-news/2021/october/7733-orion-uav-can-perform-ice-reconnaissance-in-the-arctic.html (accessed on 11 October 2021).
79. Ghisu, T.; Jarrett, J.P.; Parks, G.T. Robust design optimization of airfoils with respect to ice accretion. *J. Aircr.* **2011**, *48*, 287–304. [CrossRef]
80. Li, H.; Zhang, Y.; Chen, H. Optimization of supercritical airfoil considering the ice-accretion effects. *AIAA J.* **2019**, *57*, 4650–4669. [CrossRef]
81. Dai, J.Z.; Li, H.R.; Zhang, Y.F.; Chen, H.X. Optimization of multi-element airfoil settings considering ice accretion effect. *Chin. J. Aeronaut.* **2023**, *36*, 41–57. [CrossRef]
82. Vukits, T.J. Overview and Risk Assessment of Icing for Transport Category Aircraft And Components. In Proceedings of the 40th AIAA Aerospace Sciences Meeting and Exhibit, Reno, NV, USA, 14–17 January 2002.
83. Thompson, G. High Resolution Numerical Weather Model Forecasts of Icing at the Ground and in the Air. In Proceedings of the International Workshop on Atmospheric Icing of Structures, Reykjavík, Iceland, 23–28 June 2019.
84. Niu, J.; Sang, W.; Li, D.; Guo, Q.; Qiu, A.; Shi, M. Fast Prediction of Multiple Parameters Related to Iced Airfoil Based on POD and Kriging Methods. *J. Appl. Fluid Mech.* **2023**, *16*, 325–336. [CrossRef]
85. Hovenburg, A.; de Alcantara Andrade, F.A.; Hann, R.; Dahlin Rodin, C.; Johansen, T.; Storvold, R. Long range path planning using an aircraft performance model for battery powered sUAS equipped with icing protection system. *IEEE J. Miniaturization Air Space Syst.* **2020**, *1*, 76–89. [CrossRef]
86. Narum, E.F.L.; Hann, R.; Johansen, T.A. Optimal Mission Planning for Fixed-Wing UAVs with Electro-Thermal Icing Protection and Hybrid-Electric Power System. In Proceedings of the 2020 International Conference on Unmanned Aircraft Systems (ICUAS), Athens, Greece, 1–4 September 2020; pp. 651–660.
87. Subramanyam, B.S.; Kondrashov, V.; Ruhe, J.; Varanasi, K.K. Low ice adhesion on nano-textured superhydrophobic surfaces under supersaturated conditions. *ACS Appl. Mater. Interfaces* **2016**, *8*, 12583–12587. [CrossRef] [PubMed]
88. Guo, P.; Zheng, Y.; Wen, M.; Song, C.; Lin, Y.; Jiang, L. Icephobic/anti-icing properties of micro/nanostructured surfaces. *Adv. Mater.* **2012**, *24*, 2642–2648. [CrossRef] [PubMed]
89. Kim, P.; Wong, T.-S.; Alvarenga, J.; Kreder, M.J.; Adorno-Martinez, W.E.; Aizenberg, J. Liquid-infused nanostructured surfaces with extreme anti-ice and anti-frost performance. *ACS Nano* **2012**, *6*, 6569–6577. [CrossRef]
90. Liu, Q.; Yang, Y.; Huang, M.; Zhou, Y.; Liu, Y.; Liang, X. Durability of a lubricant-infused electrospray silicon rubber surface as an anti-icing coating. *Appl. Surf. Sci.* **2015**, *346*, 68–76. [CrossRef]
91. Ozbay, S.; Yuceel, C.; Erbil, H.Y. Improved icephobic properties on surfaces with a hydrophilic lubricating liquid. *ACS Appl. Mater. Interfaces* **2015**, *7*, 22067–22077. [CrossRef]
92. Chen, J.; Luo, Z.; Fan, Q.; Lv, J.; Wang, J. Anti-ice coating inspired by ice skating. *Small* **2014**, *10*, 4693–4699. [CrossRef]
93. Dou, R.; Chen, J.; Zhang, Y.; Wang, X.; Cui, D.; Song, Y.; Jiang, L.; Wang, J. Anti-icing coating with an aqueous lubricating layer. *ACS Appl. Mater. Interfaces* **2014**, *6*, 6998–7003. [CrossRef]
94. Zhao, Z.; Chen, H.; Liu, X.; Liu, H.; Zhang, D. Development of high-efficient synthetic electric heating coating for anti-icing/de-icing. *Surf. Coat. Technol.* **2018**, *349*, 340–346. [CrossRef]
95. Matsubayashi, T.; Tenjimbayashi, M.; Manabe, K.; Komine, M.; Navarrini, W.; Shiratori, S. Integrated anti-icing property of super-repellency and electrothermogenesis exhibited by PEDOT:PSS/cyanoacrylate composite nanoparticles. *ACS Appl. Mater. Interfaces* **2016**, *8*, 24212–24220. [CrossRef] [PubMed]
96. Ye, W.M.; Guo, K.; Qiu, W.P. Research and development of anti icing and hydrophobic coatings for unmanned aerial vehicles. *Mech. Electr. Inf.* **2021**, *672*, 18–20. (In Chinese)
97. Luo, H.Y.; Li, Y.; Huan, D.J.; Zhu, C.L.; Wang, J.X.; Zeng, D. Efficient Fabrication of Wear-Resistant PEEK Matrix Composite Coating with Superhydrophobicity for Self-Cleaning and Anti-Icing Applications. *Polym.-Plast. Technol. Mater.* **2021**, *60*, 1106–1121. [CrossRef]

98. Wong, T.S.; Kang, S.H.; Tang, S.K.Y.; Smythe, E.J.; Hatton, B.D.; Grinthal, A.; Aizenberg, J. Bioinspired self-repairing slippery surfaces with pressure-stable omniphobicity. *Nature* **2011**, *477*, 443–447. [CrossRef] [PubMed]
99. Liu, X.L.; Chen, H.W.; Zhao, Z.H.; Yan, Y.Y.; Zhang, D.Y. Slippery liquid-infused porous electric heating coating for anti-icing and deicing applications. *Surf. Coat. Technol.* **2019**, *374*, 889–896. [CrossRef]
100. Hann, R.; Borup, K.; Zolich, A.; Sorensen, K.; Vestad, H.; Steinert, M.; Johansen, T. *Experimental Investigations of an Icing Protection System for UAVs*; SAE Technical Paper 2019-01-2038; SAE International: Warrendale, PA, USA, 2019.
101. Zhao, Z.H.; Chen, H.W.; Liu, X.L.; Wang, Z.L.L.; Zhu, Y.T.; Zhou, Y.P. The development of electric heating coating with temperature controlling capability for anti-icing/de-icing. *Cold Reg. Sci. Technol.* **2021**, *184*, 103234. [CrossRef]
102. Yao, X.D.; Hawkins, S.C.; Falzon, B.G. An advanced anti-icing/de-icing system utilizing highly aligned carbon nanotube webs. *Carbon* **2018**, *136*, 130–138. [CrossRef]
103. Pan, L.; Liu, Z.; Kiziltaş, O.; Zhong, L.; Pang, X.; Wang, F.; Zhu, Y.; Ma, W.; Lv, Y. Carbon fiber/poly ether ether ketone composites modified with graphene for electro-thermal deicing applications. *Compos. Sci. Technol.* **2020**, *192*, 108117. [CrossRef]
104. Karim, N.; Zhang, M.; Afroj, S.; Koncherry, V.; Potluri, P.; Novoselov, K.S. Graphene-based surface heater for de-icing Applications. *RSC Adv.* **2018**, *8*, 16815–16823. [CrossRef]
105. Cortés, A.; Romate, X.F.S.; Jiménez-Suárez, A.; Campo, M.; Prolongo, M.G.; Ureña, A.; Prolongo, S.G. 3D printed anti-icing and de-icing system based on CNT/GNP doped epoxy composites with self-curing and structural health monitoring capabilities. *Smart Mater. Struct.* **2021**, *30*, 025016. [CrossRef]
106. Wang, T.; Zheng, Y.; Raji, A.-R.O.; Li, Y.; Sikkwma, W.K.A.; Tour, J.M. Passive anti-icing and active deicing films. *ACS Appl Mater Interfaces* **2016**, *8*, 14169–14173. [CrossRef] [PubMed]
107. Falin, A.; Cai, Q.; Santos, E.J.G.; Scullion, D.; Qian, D.; Zhang, R.; Yang, Z.; Huang, S.; Watanabe, K.; Taniguchi, T.; et al. Mechanical properties of atomically thin boron nitride and the role of interlayer interactions. *Nat. Commun.* **2017**, *8*, 15815. [CrossRef] [PubMed]
108. Thomassin, J.-M.; Jérôme, C.; Pardoen, T.; Bailly, C.; Huynen, I.; Detrembleur, C. Polymer/carbon based composites as electromagnetic interference (EMI) shielding materials. *Mater. Sci. Eng. R Rep.* **2013**, *74*, 211–232. [CrossRef]
109. Hwang, H.; Ma, K.Y.; Kim, J.W.; Yuk, D.; Hong, J.; Jung, J.H.; Yong, S.; Choi, J.; Kim, J.Y.; Shin, H.S. Radio-frequency-transmitting hexagonal boron nitride-based antiand de-icing heating system. *Nanoscale* **2020**, *12*, 21895–21900. [CrossRef] [PubMed]
110. Vertuccio, L.; De Santis, F.; Pantani, R.; Lafdi, K.; Guadagno, L. Effective de-icing skin using graphene-based flexible heater. *Compos. Part B Eng.* **2019**, *162*, 600–610. [CrossRef]
111. Ming, Y.K.; Duan, Y.G.; Zhang, S.Q.; Zhu, Y.; Wang, B.W. Self-heating 3D printed continuous carbon fiber/epoxy mesh and its application in wind turbine deicing. *Polym. Test.* **2020**, *82*, 106309. [CrossRef]
112. Galao, O.; Bañón, L.; Baeza, F.; Carmona, J.; Garcés, P. Highly conductive carbon fiber reinforced concrete for icing prevention and curing. *Materials* **2016**, *9*, 281. [CrossRef]
113. Zhao, H.; Wu, Z.; Wang, S.; Zheng, J.; Che, G. Concrete pavement deicing with carbon fiber heating wires Cold Reg. *Sci. Technol.* **2011**, *65*, 413–420.
114. Idris, M.K.; Qiu, J.F.; Melenka, G.W.; Grau, G. Printing electronics directly onto carbon fiber composites: Unmanned aerial vehicle (UAV) wings with integrated heater for de-icing. *Eng. Res. Express* **2020**, *2*, 025022. [CrossRef]
115. Xu, X.; Wang, F.; Mao, J. The flexible pressure-sensitive adhesive graphene-based composite heater based on the laminating structure for de-icing applications. *J. Mater. Sci. Mater. Electron.* **2021**, *32*, 13994–14005. [CrossRef]
116. Hann, R.; Enache, A.; Nielsen, M.C.; Stovner, B.N.; Beeck, J.V.; Johansen, T.A.; Borup, K.T. Experimental Heat Loads for Electrothermal Anti-Icing and De-Icing on UAVs. *Aerospace* **2021**, *8*, 83. [CrossRef]
117. SAE International. *Ice, Rain, Fog, and Frost Protection*; Aerospace Information Report AIR1168/4; SAE International: Warrendale, PA, USA, 2016.
118. Roy, R.; Raj, L.P.; Jo, J.-H.; Cho, M.-Y.; Kweon, J.-H.; Myong, R.S. Multiphysics anti-icing simulation of a CFRP composite wing structure embedded with thin etched-foil electrothermal heating films in glaze ice conditions. *Compos. Struct.* **2021**, *276*, 114441. [CrossRef]
119. Sørensen, K.L.; Helland, A.S.; Johansen, T.A. Carbon Nanomaterial-Based Wing Temperature Control System for In-Flight Anti-Icing and De-Icing of Unmanned Aerial Vehicles. In Proceedings of the IEEE Aerospace Conference Proceedings, Big Sky, MT, USA, 7–14 March 2015.
120. Sørensen, K.L.; Johansen, T.A. Thermodynamics of a Carbon Nano-Materials Based Icing Protection System for Unmanned Aerial Vehicle. In Proceedings of the IEEE Aerospace Conference, Big Sky, MT, USA, 5–12 March 2016.
121. Warwick, G. Nanotech promises energy-efficient anti-icing. *Aviat. Week Space Technol.* **2012**, *174*, 15.
122. Grinats, É.S.; Miller, A.B.; Potapov, Y.F.; Stasenko, A.L. Experimental and theoretical investigations of the processes of icing of nanomodified superhydrophobic and ordinary surfaces. *Vestn. Moskovsk. Gos. Obl. Univ. Ser. Fiz.-Mat.* **2013**, 3, 84–92.
123. Bogoslov, E.A.; Danilaev, M.P.; Mikhailov, S.A.; Pol'skii, Y.E. Energy efficiency of an integral anti-ice system based on fluoroplastic films. *J. Eng. Phys. Thermophys.* **2016**, *89*, 815–820. [CrossRef]
124. Battelle's Anti-Icing System for Remotely Piloted Aircraft (RPAs). Available online: http://www.battelle.org (accessed on 14 August 2013).
125. Yan, Z.X.; He, Y.; Yuan, W.Z.; Yuan, W.Z. Anti-icing Skin with Micro-nano Structure Inspired by Fargesia Qinlingensis. *Trans. Nanjing Univ. Aeronaut. Astronaut.* **2023**, 2, 115–123.

126. Zhu, B. Low Power Superhydrophobic Electrothermal Skin and Its Anti-icing Performance. Master's Thesis, Northwestern Polytechnical University, Xi'an, China, 2018. (In Chinese).
127. Deshpande, S.; Sæterdal, A.; Sundsbø, P.-A. Sea Spray Icing: The physical process and review of prediction models and winterization techniques. *J. Offshore Mech. Arct. Eng.* **2021**, *143*, 061601. [CrossRef]
128. Villeneuve, E.; Harvey, D.; Zimcik, D.; Aubert, R.; Perron, J. Piezoelectric deicing system for rotorcraft. *J. Am. Helicopter Soc.* **2015**, *60*, 1–12. [CrossRef]
129. Venna, S.V.; Lin, Y.J. In-Flight De-Icing Self-Actuating Wing Structures with Piezoelectric Actuators. In Proceedings of the American Society of Mechanical Engineers/International Mechanical Engineering Congress and Exposition, New Orleans, LA, USA, 17–22 November 2002; pp. 237–245.
130. Yao, S.J. Research on Wing Deicing Method Based on Piezoelectric Drivers. Master's Thesis, Nanjing University of Aeronautics and Astronautics, Nanjing, China, 2010.
131. Bai, T. Ice Detection and Deicing Method Based on Piezoelectric Materials. Ph.D. Thesis, Nanjing University of Aeronautics and Astronautics, Nanjing, China, 2015.
132. Palacios, J.; Smith, E.; Rose, J.; Royer, R. Ultrasonic de-icing of wind-tunnel impact icing. *J. Aircr.* **2011**, *48*, 1020–1027. [CrossRef]
133. Palacios, J.; Smith, E.; Rose, J. Investigation of an Ultrasonic Ice Protection System for Helicopter Rotor Blades. *Annu. Forum Proc.—AHS Int.* **2008**, *1*, 609–618.
134. Tarquini, S.; Antonini, C.; Amirfazli, A.; Marengo, M.; Palacios, J. Investigation of Ice Shedding Properties of Superhydrophobic Coatings on Helicopter Blades. *Cold Reg. Sci. Technol.* **2014**, *100*, 50–58. [CrossRef]
135. Overmeyer, A.; Palacios, J.; Smith, E. Ultrasonic de-icing bondline design and rotor ice testing. *AIAA J.* **2013**, *51*, 2965–2976. [CrossRef]
136. Budinger, M.; Pommier-Budinger, V.; Napias, G.; Silva, A.C. Ultrasonic ice protection systems: Analytical and numerical models for architecture tradeoff. *J. Aircr.* **2016**, *53*, 680–690. [CrossRef]
137. Venna, S.V.; Lin, Y.J. Development of Self-Actuating In-Flight De-Icing Structures with Power Consumption Considerations. In Proceedings of the American Society of Mechanical Engineers International Mechanical Engineering Congress and Exposition, Washington, DC, USA, 17–22 November 2003; pp. 45–53.
138. Venna, S.V.; Lin, Y.J. Mechatronic Development of Self-Actuating In-Flight De-Icing Structures. *IEEE/ASME Trans. Mechatron.* **2006**, *11*, 585–592. [CrossRef]
139. Venna, S.; Lin, Y.J.; Botura, G. Piezoelectric Transducer Actuated Leading Edge De-Icing with Simultaneous Shear and Impulse Forces. *J. Aircr.* **2007**, *44*, 509–515. [CrossRef]
140. Struggl, S.; Korak, J.; Feyrer, C. A Basic Approach for Wing Leading De-Icing by Smart Structures. In Proceedings of the SPIE Smart Structures and Materials + Nondestructive Evaluation and Health Monitoring, San Diego, CA, USA, 6–10 March 2011.
141. Palanque, V. Design of Low Consumption Electro-Mechanical De-Icing Systems. Ph.D. Thesis, Université de Toulouse ISAE-Supaero, Toulouse, France, 2022.
142. Tan, M.K.; Friend, J.R.; Yeo, L.Y. Interfacial Jetting Phenomena Induced by Focused Surface Vibrations. *Phys. Rev. Lett.* **2009**, *103*, 024501. [CrossRef]
143. Qi, A.; Yeo, L.Y.; Friend, J.R. Interfacial Destabilization and Atomization Driven by Surface Acoustic Waves. *Phys. Fluids* **2008**, *20*, 074103. [CrossRef]
144. Li, L.; Wu, E.; Jia, K.; Yang, K. Temperature Field Regulation of a Droplet Using an Acoustothermal Heater. *Lap Chip* **2021**, *21*, 3184–3194. [CrossRef]
145. Mehmood, M.; Nawaz Chaudhary, T.; Burnside, S.; Khan, U.F.; Yongqing Fu, R.; Chen, B. Coupling Mechanism of Kinetic and Thermal Impacts of Rayleigh Surface Acoustic Waves on the Microdroplet. *Exp. Therm. Fluid Sci.* **2022**, *133*, 110580. [CrossRef]
146. Jacob, S.; Pandey, S.; Moral, J.D.; Karimzadeh, A.; Gil-Rostra, J.; González-Elipe, A.R.; Borrás, A.; Winkler, A. Surface Acoustic Waves Equip Materials with Active Deicing Functionality: Unraveled Deicing Mechanisms and Application to Centimeter Scale Transparent Surfaces. *Adv. Mater. Technol.* **2023**, *8*, 2300263. [CrossRef]
147. Zeng, X.; Yan, Z.; Lu, Y.; Fu, Y.; Lv, X.; Yuan, W.; He, Y. Reduction of Ice Adhesion Using Surface Acoustic Waves: Nanoscale Vibration and Interface Heating Effects. *Langmuir* **2021**, *37*, 11851–11858. [CrossRef] [PubMed]
148. Nampoothiri, K.N.; Nath, A.; Satpathi, N.S.; Sen, A.K. Deicing of Sessile Droplets Using Surface Acoustic Waves. *Langmuir* **2023**, *39*, 3934–3941. [CrossRef] [PubMed]
149. Roth, J.R.; Sherman, D.M.; Wilkinson, S.P. Boundary Layer Flow Control with a One Atmosphere Uniform Glow Discharge Surface Plasma. In Proceedings of the 36th AIAA Aerospace Sciences Meeting and Exhibit, Reno, NV, USA, 12–15 January 1998. AIAA-98-0328.
150. Winkel, R.; Correale, G.; Kotsonis, M. Effect of Dielectric Material on Thermal Effect Produced by ns-DBD Plasma Actuator. In Proceedings of the 45th AIAA Plasma Dynamics and Lasers Conference, Atlanta, GA, USA, 16–20 June 2014. AIAA-2014-2119.
151. Li, Q.Q.; Hao, L.Y. Surface Dielectric Barrier Discharge Plasma and Its Applications. *High Volt. Technol.* **2016**, *42*, 1079–1090. (In Chinese)
152. Cai, J.; Tian, Y.; Meng, X.; Han, X.; Zhang, D.; Hu, D. An experimental study of icing control using DBD plasma actuator. *Exp. Fluids* **2017**, *58*, 102. [CrossRef]

153. Kolbakir, C.; Liu, Y.; Starikovskiy, A.; Miles, R.B.; Hu, H. An Experimental Investigation on The Thermal Effects of NS-DBD and AC-DBD Plasma Actuators for Aircraft Icing Mitigation. In Proceedings of the AIAA Science and Technology Forum and Exposition, Kissimmee, FL, USA, 8–12 January 2018. AIAA-2018-0164.
154. Liu, Y.; Hu, H. An experimental investigation on the unsteady heat transfer process over an ice accreting airfoil surface. *Int. J. Heat Mass Transf.* **2018**, *122*, 707–718. [CrossRef]
155. Liu, Y.; Kolbakir, C.; Hu, H. A Comparison Study on AC-DBD Plasma and Electrical Heating for Aircraft Icing Mitigation. In Proceedings of the AIAA Science and Technology Forum and Exposition, Kissimmee, FL, USA, 8–12 January 2018. AIAA-2018-0167.
156. Zhu, Y.; Starikovskaia, S. Fast gas heating of nanosecond pulsed surface dielectric barrier discharge: Spatial distribution and fractional contribution from kinetics. *Plasma Sources Sci. Technol.* **2018**, *27*, 124007. [CrossRef]
157. Zhu, Y.; Wu, Y.; Wei, B.; Xu, H.; Liang, H.; Jia, M.; Song, H.; Li, Y. Nanosecond-pulsed dielectric barrier discharge-based plasma-assisted anti-icing: Modeling and mechanism analysis. *J. Phys. D Appl. Phys.* **2020**, *53*, 145205. [CrossRef]
158. Wei, B.; Wu, Y.; Liang, H.; Zhu, Y.; Chen, J.; Zhao, G.; Xu, H. SDBD based plasma anti-icing: A streamwise plasma heat knife configuration and criteria energy analysis. *Int. J. Heat Mass Transf.* **2019**, *138*, 163–172. [CrossRef]
159. Wei, B.; Wu, Y.; Liang, H.; Chen, J.; Zhao, G.; Tian, M.; Xu, H. Performance and mechanism analysis of nanosecond pulsed surface dielectric barrier discharge based plasma deicer. *Phys. Fluids* **2019**, *31*, 091701. [CrossRef]
160. Niu, J.J.; Sang, W.M.; Zhou, F.; Li, D. Numerical investigation of an anti-icing method on airfoil based on the NSDBD plasma actuator. *Aircr. Eng. Aerosp. Technol.* **2021**, *93/4*, 592–606. [CrossRef]
161. Gao, X.L. Development of High-frequency ns-SDBD Plasma Generator and Preliminary Study on Anti-Icing Experiment. Master's Thesis, Chongqing University, Chongqing, China, 2021. (In Chinese).
162. Liu, Y.; Kolbakir, C.; Starikovskiy, A.Y.; Miles, R.; Hu, H. An experimental study on the thermal characteristics of NS-DBD plasma actuation and application for aircraft icing mitigation. *Plasma Sources Sci. Technol.* **2019**, *28*, 014001. [CrossRef]
163. Luo, Z.B.; Xia, Z.X. Advances in synthetic jet technology and applications in flow control. *Adv. Mech.* **2005**, *35*, 220–234. (In Chinese)
164. Smith, B.L.; Glezer, A. The formation and evolution of synthetic jets. *Phys. Fluids* **1998**, *10*, 2281–2297. [CrossRef]
165. Nagappan, N.; Golubev, V.V. On Icing Control Using Thermally Activated Synthetic Jets. In Proceedings of the 51st AIAA Aerospace Sciences Meeting including the New Horizons Forum and Aerospace Exposition, Grapevine, TX, USA, 7–10 January 2013.
166. Nagappan, N.; Golubev, V.V. Parametric Analysis of Icing Control Using Synthetic Jet Actuators. In Proceedings of the 21st AIAA Computational Fluid Dynamics Conference, San Diego, CA, USA, 24–27 June 2013.
167. Nagappan, N. Numerical Modeling of Anti-Icing Using an Array of Heated Synthetic Jets. Ph.D. Thesis, Embry-Riddle Aeronautical University, Daytona Beach, FL, USA, 2013.
168. Li, Y.J. Research on Airfoil Separate Flow Control and Airfoil Icing Control Using Dual Synthetic Jet Actuator. Master's Thesis, National University of Defense Technology, Changsha, China, 2015. (In Chinese).
169. Li, Y.J.; Luo, Z.B. An experimental investigation on the process of droplet icing/frosting and defrosting/deicing using dual synthetic jet. *J. Exp. Fluid Mech.* **2016**, *30*, 27–32. (In Chinese)
170. Jiang, H. Research on Anti-Icing/De-Icing Using the Heated Dual Synthetic Jet Technology. Master's Thesis, National University of Defense Technology, Changsha, China, 2017. (In Chinese).
171. Liu, X.C.; Liang, H.; Zong, H.H.; Xie, L.K.; Su, Z. Experimental investigation on de-icing by an array of impact rod-type plasma synthetic jets. *Plasma Sci. Technol.* **2023**, *25*, 035504. [CrossRef]
172. Gao, T.X.; Luo, Z.B.; Zhou, Y.; Yang, S.K. A novel de-icing strategy combining electric-heating with plasma synthetic jet actuator. *Proc. IMechE Part G J. Aerosp. Eng.* **2020**, *235*, 095441002094472. [CrossRef]

modelling

MDPI

Article

Controller Design for Air Conditioner of a Vehicle with Three Control Inputs Using Model Predictive Control

Trevor Parent, Jeffrey J. Defoe and Afshin Rahimi *

Department of Mechanical, Automotive & Materials Engineering, University of Windsor, 401 Sunset Ave., Windsor, ON N9B 3P4, Canada; parenttrev@gmail.com (T.P.); jdefoe@uwindsor.ca (J.J.D.)
* Correspondence: arahimi@uwindsor.ca

Abstract: Fuel consumption optimization is a critical field of research within the automotive industry to meet consumer expectations and regulatory requirements. A reduction in fuel consumption can be achieved by reducing the energy consumed by the vehicle. Several subsystems contribute to the overall energy consumption of the vehicle, including the air conditioning (A/C) system. The loads within the A/C system are mainly contributed by the compressor, condenser fan, and underhood aerodynamic drag, which are the components targeted for overall vehicle energy use reduction in this paper. This paper explores a new avenue for A/C system control by considering the power consumption due to vehicle drag (regulated by the condenser fan and active grille shutters (AGS)) to reduce the energy consumption of the A/C system and improve the overall vehicle fuel economy. The control approach used in this paper is model predictive control (MPC). The controller is designed in Simulink, where the compressor clutch signal, condenser fan speed, and AGS open-fraction are inputs. The controller is connected to a high-fidelity vehicle model in Gamma Technologies (GT)-Suite (which is treated as the real physical vehicle) to form a software-in-the-loop simulation environment, where the controller sends actuator inputs to GT-Suite and the vehicle response is sent back to the controller in Simulink. Quadratic programming is used to solve the MPC optimization problem and determine the optimal input trajectory at each time step. The results indicate that using MPC to control the compressor clutch, condenser fan, and AGS can provide a 37.6% reduction in the overall A/C system energy consumption and a 32.7% reduction in the error for the air temperature reference tracking compared to the conventional baseline proportional integral derivative control present in the GT-Suite model.

Keywords: model predictive control; vehicle air conditioning; fuel efficiency; fuel consumption optimization; vehicle energy efficiency; software-in-the-loop simulation

Citation: Parent, T.; Defoe, J.J.; Rahimi, A. Controller Design for Air Conditioner of a Vehicle with Three Control Inputs Using Model Predictive Control. *Modelling* **2024**, *5*, 117–152. https://doi.org/10.3390/modelling5010008

Academic Editors: Tomasz Nowakowski, Artur Kierzkowski, Agnieszka A. Tubis, Franciszek Restel, Tomasz Kisiel, Anna Jodejko-Pietruczuk and Mateusz Zając

Received: 21 November 2023
Revised: 20 December 2023
Accepted: 27 December 2023
Published: 3 January 2024

1. Introduction

The most effective method for A/C system control has been continuously evolving throughout the years as the state of the art for control design has also continued to evolve. Early implementations of controllers on A/C systems include on/off, proportional integral derivative (PID), and sliding-mode controllers. Høgh and Nielsen [1] presented a new and innovative control strategy that used a compressor and an electronic expansion valve for better control of the superheat level. The system efficiency was also analyzed to find the optimal superheat setpoint, which could reduce energy consumption. The refrigerant system was modeled to allow for model-based feedback and feedforward, which improved control of the superheat level. Koo et al. [2] discussed the use of a proportional–integral (PI)-controlled cascade structure to control the relative filling and evaporation pressure of a target system. They also introduced the super-twisting-algorithm-based 2-sliding mode controller (SMC) design for the target system. The 2-SMC design was found to be more effective than the classical first-order sliding mode controller (1-SMC) in handling plant parameter uncertainty. Liu et al. have worked on a survey [3] where they provided a comprehensive overview of different automatic control techniques that had been applied in A/C systems. The authors described the characteristics of A/C systems and discussed the variables that should be controlled. They used classification of the control approaches to explain the main features of these strategies used in A/C systems. Their paper clearly showed that, with the improvement in people's living standards and the emergence of global environmental and energy problems, there were increasing requirements for more advanced A/C control strategies. The authors summarized studies of control methods presented in the paper, and analyzed the development and new trends of A/C control approaches. Additionally, the paper discussed studies of PID control applied in A/C systems, including more advanced PID control methods to improve control system performance and to make full use of PID control advantages in the multi-input–multi-output (MIMO) A/C system control, and simplify the design and operation procedure of the control strategies. The above-mentioned controller designs are typically applied to the compressor to maintain the cabin at a specific temperature. These controllers are simple to design but most do not attempt to reduce the energy consumption of the system. Cvok et al. [4] proposed a method to optimize and control an electric vehicle heating, ventilation, and air conditioning (HVAC) system for better efficiency and thermal comfort. The authors used dynamic programming to find optimal control actions based on a simplified cabin model and static HVAC maps. They also designed a hierarchical control system that used an instantaneous optimization-based allocation algorithm and a superimposed cabin temperature controller. They verified their method by simulation for a cool-down scenario and compared it with the dynamic programming benchmark. They showed that their method could approach the benchmark by tuning the controller bandwidth and the allocation cost function. Kim et al. [5] presented the development of a mathematical dynamic model and observer-based set point control of a vapor compression cycle (VCC) system. The mean void fraction and moving boundary control volumes in a heat exchanger (HEX) were used for a lumped parameter model of an air conditioning system. A MIMO dynamic model of the VCC was presented in a nonlinear state space representation (SSR) and was reformulated into a linearized regular form for control design. A modified observer-based controller with SMC was introduced to attenuate the effects of observing approximation errors and external disturbances. The proposed control methodology ensured uniformly ultimately bounded tracking results of the desired low pressure and suction superheat. A simulation of controlling the VCC illustrated the effectiveness and performance in the presence of disturbances. Xie et al. [6] proposed an intelligent air conditioning system control strategy that can learn passengers' thermal comfort preferences. The proposed control algorithm directly and automatically decreased the cabin temperature to the passengers' preferred temperature without any manual adjustment.

More recent developments in sliding-mode controller applications show potential for energy savings related to the operation of the compressor and fans. However, these control methods still lack any consideration of vehicle drag and AGS control. For example, Huang et al. [7] presented an energy-saving controller for automotive air-conditioning/refrigeration (A/C-R) systems. The proposed controller consisted of two different time-scale layers. The outer or the slow time-scale layer, called a setpoint optimizer, is used to find the set points related to energy efficiency by using the steady-state model, whereas the inner or the fast time-scale layer is used to track the obtained set points. In the inner loop, thanks to its robustness, an SMC was utilized to track the set point of the cargo temperature.

In recent years, model predictive control applications for A/C systems have seen more research interest in academia. MPC lends itself well to A/C system control because it can directly consider constraints on inputs, states, and outputs in the control problem formulation. Liberati et al. [8] address the emerging challenge of managing smart microgrids for electric vehicle charging and various user loads while incorporating renewable energy sources and storage. Their work emphasizes the importance of advanced power systems and building automation control for real-world applications. The article introduces an economic model predictive control method that optimizes electric and heating resources in smart buildings, achieving nearly zero energy consumption and seamless participation in demand response programs. Simulation results, based on an Italian case study, showcase the approach's effectiveness in managing electric and thermal flexibility within a building. In a review by Borreggine et al. [9] on the state-of-the-art issues and solutions for MPC in power electronics, the authors reviewed the MPC techniques for power electronics and electrical drives, with a focus on finite control set MPC and explicit MPC. The paper discussed the advantages and drawbacks of both methods, such as flexibility, performance, computation effort, and steady-state error. The paper also presented several applications of MPC for different types of converters, loads, and powertrain configurations. The paper highlighted the potential of MPC for powertrain control, especially for hybrid and electric vehicles.

More recently, MPC applications have been applied to automotive A/C systems. Most of the current literature on automotive A/C system control focuses on minimizing the compressor power consumption and the tracking error of the discharge air (cooling requirement) in electric vehicles (EVs), as discussed in a review by Huang et al. [10]. Wang et al. [11] proposed a nonlinear MPC control for EVs to minimize evaporator blower power consumption but did not consider the AGS in the analysis. Pereira et al. [12] proposed an energy management system (EMS) for a fuel cell (FC) hybrid electric vehicle based on nonlinear model predictive control (NMPC) and a recurrent neural network (RNN). The NMPC allowed the formulation of control objectives not allowed by a linear MPC, such as maximum efficiency point tracking of the FC, while the RNN accurately predicted the FC nonlinear dynamics. The EMS was implemented on a low-cost development board and tested in real time on a hardware-in-the-loop test bench equipped with a real 3-kW FC stack. The experimental results demonstrated that the NMPC EMS was able to meet the vehicle's energy demand and operate the FC in its most efficient region. A comparative study was performed between the proposed NMPC, a linear MPC, and hysteresis band control. The results of this comparative study demonstrated that the NMPC provided better fuel economy and could reduce FC degradation. Xie et al. [13] presented an Intelligent Model Predictive Control (IMPC) strategy for the A/C–cabin coupled system that includes the influences of vehicle speed and external environment on the heat exchange with the cabin. The IMPC strategy integrates the vehicle speed previewer and the self-adaptor for passengers' thermal comfort. The IMPC has a more dynamic response of compressor speed to the car speed change and can automatically adjust cabin temperature, making it satisfy the thermal preference of the passenger with little control error of cabin temperature. The IMPC strategy saves more energy than other control strategies researched in their paper. Glos et al. [14] presented an application of NMPC algorithms for energy-efficient control of cabin temperature and air quality in fully electric vehicles (FEVs). The authors developed simplified dynamic models of the FEV cabin and HVAC system and formulated an optimal control problem that minimized power consumption while satisfying the user's comfort and safety requirements. The NMPC algorithms were implemented on a microcontroller and tested in a Processor in the Loop simulation. The paper showed that the NMPC approach could reduce the heat losses through ventilation and improve the range of the FEV. The paper also discussed the challenges and limitations of the NMPC implementation, such as computational complexity, state estimation, and actuator constraints. Kibalama et al. [15] presented an energy management strategy for the air conditioning system of an electrified vehicle, based on MPC. The authors used a nonlinear model of the vapor compression refrigeration system and formulated a constrained optimal control problem to minimize the energy consumption while tracking a desired evaporator pressure. They proposed a novel implementation method that reduced the computation time by approximating the value function from a global optimal solution obtained by dynamic programming (DP). They showed that their MPC approach achieved near-optimal performance on different drive cycles and thermal load requirements, and validated it on real-time hardware. The paper did not address the robustness of the controller to uncertainties or disturbances, or the extension to other thermal management systems.

A recent study by Göltz and Sawodny [16] investigated tracking performance and energy consumption of the A/C system using various control schemes, including MPC. However, this work mainly focuses on battery thermal management for battery electric vehicles (BEVs). Additionally, opening the AGS increases the cooling effort supported by the front-end airflow, allowing other components, such as the compressor, to reduce their working load and still satisfy the cooling requirements of the vehicle, as shown in a study by El-Sharkawy et al. [17]. This AGS airflow cooling comes at a cost. Shigarkanthi et al. [18] demonstrated that, as the AGS is opened, the drag force of the underhood vehicle airflow increases due to an increase in drag coefficient, concluding that the AGS has an impact on the fuel economy and power consumption of the vehicle. By controlling the condenser fan and AGS along with the compressor, it is expected that the energy used by the system can be further reduced but the cost of increased underhood drag must be considered.

Chen et al. [19] discuss the significance of real-time optimal control in improving the energy efficiency of cabin climate control systems in EVs. They introduced a novel linear-time-varying (LTV) MPC approach, which simplifies the optimization process and conducts a comprehensive parametric study to identify robust weighting factors. Comparative analysis reveals that the LTV-MPC outperforms rule-based and nonlinear economic MPC (NEMPC) controllers, achieving faster temperature regulation with significantly lower energy consumption while maintaining comparable computational efficiency.

Coppola et al. [20] focused on improving energy performance in platoons of autonomous connected vehicles. Their study presents an innovative Eco-Driving Control Architecture that utilizes NMPC for optimizing energy consumption while maintaining precise leader tracking. Analytical proofs of exponential stability and robust stability conditions are provided, demonstrating the effectiveness of the approach in various operational scenarios and parameter uncertainties within a Mixed Traffic Simulator (MiTraS) co-simulation platform. The results validate the architecture's ability to ensure eco-driving behavior for the entire vehicle platoon.

He et al. [21] introduced a stochastic model predictive controller (SMPC) for enhancing the energy efficiency of EV air conditioning systems. It employs a Markov-chain-based velocity predictor to anticipate future disturbances, and systematically quantifies the sensitivity of the electrified AC system to solar radiation, ambient temperature, and relative air flow speed. Comparative analysis with a rule-based controller and dynamic programming benchmark reveals that SMPC improves AC energy economy by 12%, reduces cabin temperature variation by over 50.4%, and enhances overall cabin comfort.

Huang et al. [22] presented a robust model predictive controller for enhancing temperature performance and reducing energy consumption in automotive air-conditioning/refrigeration (A/C-R) systems with three-speed and continuously-varying compressors. It includes a discrete MPC with a carefully chosen terminal weight for robustness under various conditions. The study also presents hybrid controllers that combine on/off control and discrete MPC, as well as a continuous MPC for systems with variable components, demonstrating energy savings of up to 23% compared to conventional on/off controllers in both experimental and simulation results.

Based on the reviewed literature, the existing work on the application of MPC in A/C for commercial vehicles does not consider the impact of the AGS actuator nor the power consumption due to underhood aerodynamic drag [16]. Additionally, some studies have not considered all possible control inputs in the overall design, such as missing AGS [11]. Furthermore, most studies, especially in recent years, focus on EVs as energy consumption and management has a much greater role in such vehicles [12–15]. This has left fossil-fueled energy management and A/C energy efficiency on the back burner in recent years. Additionally, the applications of nonlinear controllers, although more advanced and capable, require more computation and consideration compared to simpler alternatives.

Contribution: To the best of the authors' knowledge [23,24], the consideration of AGS control to reduce the overall vehicle power consumption and regulate the underhood airflow for cabin cooling has not been considered in the literature. Therefore, the work in this paper addresses a gap for non-electric vehicles, and additional control inputs and modeling enhancements not currently addressed in the literature. The starting point for this study is the modeling and control work by Zhang [25], where various control approaches were attempted for an internal-combustion-powered vehicle, with both the condenser fan and compressor clutch as actuators. Zhang [25] found that none of the approaches attempted was satisfactory and recommended MPC be attempted. In this paper, we apply MPC for the same vehicle, introduce the AGS as a new actuator into the control problem, and consider the additional underhood drag power consumption for the A/C system.

Several linear approximation MPCs have been formulated in the past, with the most popular being adaptive, as showcased by Dhar and Bhasin [26], and gain-scheduled model predictive controllers, as showcased by Hu et al. [27] for a commercial vehicle air brake system. These controller types will be discussed in more detail in Section 2.1. However, adaptive MPC is more promising for the application in this study due to its effectiveness for control designs where the structure of the optimization problem (number of states and constraints) remains the same across different operating points. Further discussions and rationales for the choice of adaptive MPC in this study are provided in Section 2.1. To address the identified gap in the literature and extend the work by Zhang [25] based on the system modeling improvements for the A/C system by the authors in [28], this paper has two main objectives:

Objective 1. In addition to controlling the compressor clutch only, determine the improvement in A/C system energy use and air-to-cabin reference temperature tracking that can be achieved by switching from PID (benchmark) to MPC.
Objective 2. Identify what further improvement, if any, can be achieved by including the condenser fan rotational speed and AGS position in the MPC formulation.

The remainder of this paper is organized as follows. Section 2 details background materials for the reader to understand the subsequent sections of the paper as well as the implemented methodology in this study. Section 3 includes a description of the problem formulation. Section 4 discusses the findings and implications of the results, as well as a discussion of the analysis. Finally, the conclusions of this study are offered in Section 5, followed by recommendations for future studies.

2. Materials and Methods

This section provides preliminary background on the theories used in this paper to help the reader follow the content more easily and to make the paper self-contained. Additionally, it provides the implemented methodology for the controller design and formulation based on the materials to help the reader follow the content with ease.

2.1. MPC Design

Model predictive control is an optimal control technique that leverages a precise mathematical model of a real system to proactively predict its future behavior. This predictive capability empowers MPC to anticipate potential system deviations and effectively optimize control inputs to achieve predefined performance objectives. By employing sophisticated optimization algorithms, MPC with the help of its more advanced derivatives can handle complex, nonlinear engineering systems with a high degree of accuracy and robustness, making it a valuable tool for a wide range of engineering applications. MPC's ability to incorporate constraints on both system states and control inputs ensures safe and reliable operation, even in the presence of uncertainties and disturbances. Its adaptability to various real-world scenarios, such as process control, robotics, and energy management systems, underscores its versatility and practical relevance. The combination of predictive modeling, optimization, and constraint handling renders MPC an essential asset in the pursuit of efficient and stable control solutions for complex engineering systems. A simple block diagram of an MPC is shown in Figure 1, where the dashed box includes the MPC components and their interactions, namely the Optimizer block and the Model block. The model block, also known as the prediction model, contains a simplified mathematical model of the real system (plant). The prediction model computes the predicted output, $\hat{y}$. The optimizer block contains an optimization algorithm used to determine the optimal control input u based on the error predicted by the prediction model. The details of the optimizer block in this work are outlined in Section 3.3. When the optimizer computes a control signal, it sends it to the prediction model and the behavior of the real system is predicted again. This process repeats several times over an optimization window, called the prediction horizon, to forecast the future behavior of the real system. When the entire control sequence is completed, only the immediate next control action, u^*, is sent to the real system with the states x and output y, designated as the plant block in Figure 1.

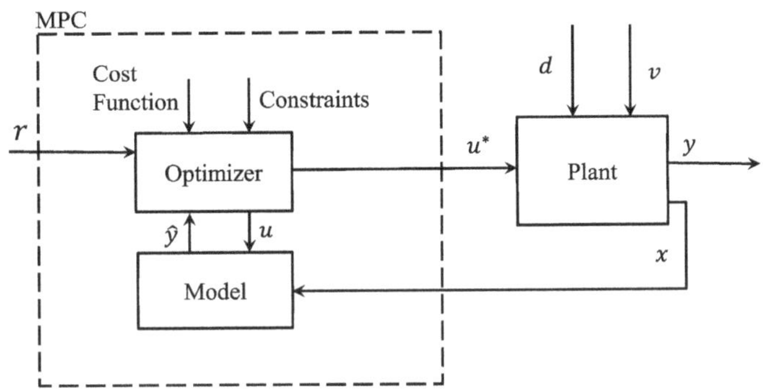

Figure 1. MPC block diagram.

The plant can be a real physical system, such as a physical vehicle, or a virtual representation of the system, such as a high-fidelity model of a physical vehicle. In this paper, the plant is a high-fidelity model, introduced in Section 2.3. The connection to the real system in this work is detailed in Section 3.1 based on the previous work by the authors [28].

2.1.1. Type Selection

Selecting the appropriate MPC technique based on the system's characteristics is an important task in the overall MPC design process. A linear time-invariant MPC method is suitable for a linear system with linear constraints and a quadratic cost function. This scenario results in a convex optimization problem with a unique global optimum. Numerous numerical methods and software can be employed to solve this well-studied problem. When moving to a nonlinear system, two methods are discussed: adaptive and gain-scheduled MPC controllers. These methods rely on linearization to approximate the nonlinear system around an operating point. While effective near this point, linearization's limitations outside this range are acknowledged.

Adaptive MPC involves computing a linear model on the fly as conditions change, updating the internal plant model for the MPC controller. This approach maintains the optimization problem structure across various operating points. Alternatively, gain-scheduled MPC involves offline linearization at multiple operating points, designing separate MPC controllers for each. While offering controller independence, this method requires managing controller switching and may demand more memory.

It is beneficial to use adaptive MPC when the optimization problem structure remains consistent across various conditions. When the structure changes, gain-scheduled MPC is preferable. However, a nonlinear system that resists accurate linearization requires nonlinear MPC. This advanced method employs a nonlinear plant model for precise predictions but solving the associated nonconvex optimization problem in real time can be challenging.

Therefore, adaptive and gain-scheduled MPC can be used for systems that can be approximated by linear models, and nonlinear MPC can be employed for highly nonlinear systems. The implementation efficiency of nonlinear MPC must be considered as it relies on specialized nonlinear solvers.

The A/C system used in this study was developed and published by the authors of this study in [28]. This model is an extension to a previously developed model by the industrial partner's collaborators where effects of underhood aerodynamics and modeling of system parameters using polynomial approximations are added for a more extensive and semi-analytical model of the system under study. The model in [28] requires an adaptive MPC for the control strategy development because it meets the following requirements for a choice of the adaptive MPC over a traditional MPC:

1. Time-varying dynamics: The system's dynamics in [28] change over time due to various factors such as varying operating conditions or system wear and tear.
2. Uncertain models: Some of the model parameters [28] are not precisely known as they are embedded in the model developed by the industrial partner where the exact values of the parameters are unknown and approximate values were used to build the semi-analytical model, and there is uncertainty about the model's accuracy.
3. Nonlinear systems: The system in [28] is highly nonlinear and for controlling systems with nonlinear behavior, where a linear model may not be sufficient, and an adaptive model helps to approximate the nonlinear dynamics better.
4. Disturbance rejection: As mentioned in item 2, some of the system parameters are not entirely known throughout the simulation as they are provided by the software-in-the-loop and depend on the driving cycle that the tests are being run over. When dealing with systems subject to disturbances that are not entirely predictable or change over time, adaptive MPC is the plausible choice.
5. Parameter variations: Some of the system parameters in [28] change over time, and their values are embedded in the software-in-the-loop provided by the industrial partner. In cases where there are abrupt changes or slow variations in system parameters, adaptive MPC can be useful.

Therefore, a variation of adaptive MPC, Successive Linearization Based Model Predictive Control (SLMPC) [29] (see Appendix A for detailed derivation), is chosen for the problem introduced in this paper due to its ability to reduce the linearization error caused by the system operating far from the trim state. In the adaptive MPC, a linear model of the system is computed on the fly as the operating conditions change and the internal prediction model is updated with each new linearized model. As the operating conditions change, only the model of the system is updated and the MPC formulation remains the same. As discussed in Section 2.3, the plant model for the A/C system used in this study has varying system parameters that are controlled by other subsystems in the vehicle, therefore making this system time-varying. The use of adaptive MPC ensures that, despite variations in system parameters, at every operating point, the linearization guarantees close agreement between the linearized model and the nonlinear plant model for the A/C system.

In addition to the type of MPC controller, there are parameters to be discussed in the design process for any MPC controller. These parameters are discussed next.

2.1.2. Design Parameters

For MPC design, there are essential parameters that require careful consideration to enhance both controller performance and computational efficiency. These parameters encompass the controller sample time, prediction horizon, control horizon, constraint handling, and weight assignment. Each parameter plays a pivotal role in optimizing the MPC algorithm's real-time performance while ensuring effective control of dynamic systems.

- **Sample time:** The choice of the sample time determines the controller's operational frequency, impacting its ability to respond to disturbances and setpoint changes. Balancing between rapid disturbance rejection and computational load, it is advisable to accommodate 10 to 20 samples within the open-loop system's rise time [30]. This ensures a responsive controller without overwhelming computational demands. Further details about this parameter in this study are provided in Section 3.4.2.

- **Prediction horizon:** The prediction horizon influences the extent to which the controller anticipates future plant behavior. A cautious prediction horizon encompasses a timeframe that captures significant system dynamics, preventing untimely responses to disturbances. To account for open-loop transient responses, selecting 10 to 20 samples is recommended [30]. Further details about this parameter in this study are provided in Section 3.4.2.
- **Control horizon:** Conversely, the control horizon determines the number of future control inputs that shape the predicted plant output. A smaller control horizon minimizes computational efforts but may compromise maneuverability. An optimal balance can be struck by setting the control horizon to 10–20% of the prediction horizon, ensuring a minimum of 2–3 steps to account for effective control [30]. Further details about this parameter in this study are provided in Section 3.4.2.
- **Constraints:** Constraints on inputs, input rate changes, and outputs can be imposed within the MPC framework. These constraints may be either hard or soft. The distinction lies in the feasibility of violation: hard constraints remain inviolable, while soft constraints can be breached within limits. Balancing the conflicting interests of inputs and outputs, it is advised to adopt soft constraints for outputs, eschewing simultaneous hard constraints on inputs and their rate of change. This minimizes the risk of infeasibility and ensures more adaptable control [30]. Further details about this parameter in this study are provided in Sections 3.3 and 3.4.3.
- **Weights:** Weight assignment is a critical aspect of optimizing control performance. Weights gauge the significance of objectives in MPC. These objectives involve tracking setpoints and achieving smooth control maneuvers. Achieving a harmonious performance balance necessitates cautiously weighing input rates and outputs relative to each other. Additionally, adjusting weights within these categories fine-tunes performance to the specific control objectives [30]. Further details about this parameter in this study are provided in Sections 3.3 and 3.4.4.

One can see the role of distinct MPC design parameters in shaping controller efficacy and computational efficiency. By strategically tuning these parameters, control engineers can harness MPC's potential to efficiently regulate dynamic systems while navigating the intricacies of real-time optimization.

2.2. Discrete State-Space General Form

MPC strategies commonly use discrete state-space formulations for the mathematical modeling of the system of interest. A continuous-time linear state-space model form must be discretized to form a discrete-time linear state-space model. Typically, MPC algorithms are designed in discrete time. A great amount of MPC designs have been developed using discrete-time approaches and a much smaller amount of continuous-time MPCs are present in the literature. In this work, a discrete-time MPC has been chosen due to the nature of the communication between the MPC and the real system, as discussed in Section 3.1.

The standard form of a continuous state-space system can be written as [31]:

$$\begin{aligned} \dot{x} &= Ax + Bu \\ y &= Cx + Du \end{aligned} \tag{1}$$

where A, B, C, and D are matrices describing the system dynamics. x is the state vector, u is the controlled input vector, and y is the output vector.

The continuous system in (1) can be written in a discrete state-space system as [32]:

$$\begin{aligned} x_{k+1} &= A_d x_k + B_d u_k \\ y_k &= C_d x_k + D_d u_k \end{aligned} \tag{2}$$

where A_d, B_d, C_d, and D_d are matrices describing the linearized system dynamics. k denotes the current timestep, x_k is the state vector, u_k is the controlled input vector, and y_k is the output vector. One can use forward Euler's method to convert (1) to (2).

Forward Euler's method is a simple discretization method for continuous-time systems. It is based on the approximation of the first derivative at $t = kT$ where T is the time step for the discretization. Having [31]

$$\dot{x}(t) = \frac{dx(t)}{dt} \approx \frac{1}{T}(x((k+1)T) - x(kT)) \tag{3}$$

Applying the approximate formula in (3) to the SSR system equation, we have [31]

$$\begin{aligned} &\tfrac{1}{T}(x((k+1)T) - x(kT)) \approx Ax(kT) + Bu(kT) \\ &x((k+1)T) \approx (I + TA)x(kT) + TBu(T) \end{aligned} \tag{4}$$

or [31]

$$\begin{aligned} x[k+1] &\approx (I + TA)x[k] + TBu[k] \\ x[k+1] &\approx A_d x[k] + B_d u[k] \end{aligned} \tag{5}$$

where $A_d = I + TA$ and $B_d = TB$ are the discrete system matrices for the discrete SSR of the system with sampling time T. The same approach can be followed to obtain C_d and D_d that similar to B_d would only have a multiplication by T. The sampling time for the discretization in this study was chosen as $T = 1$ s as detailed in Section 2.1.2. In this paper, the system equations of the dynamics of an automotive A/C system, as detailed in Section 2.3, are considered. These equations are nonlinear and, therefore, are linearized to conform to the standard continuous-time form in (1). These continuous-time equations are then discretized to conform to the discrete-time form in (2) using the forward Euler method. The details of the full process are described in Section 3.

2.3. *Introduction of A/C System*

A simple diagram of the automotive A/C system used in this study is depicted in Figure 2. The A/C system comprises four main components, the compressor, condenser, evaporator, and expansion valve. The controllable actuators of interest in this work are the compressor clutch, which switches the compressor (used to move the refrigerant throughout the cooling circuit) on and off, and the fan and AGS used to regulate the front-end airflow passing through the condenser. Compressor clutch π_c is not a continuous variable in practice and is a discrete binary variable as either engaged (with the value of 1) or disengaged (with the value of 0). In this study, π_c is initially treated as a continuous variable to be computed via optimization. When the MPC computes the continuous compressor clutch signal π_c, it rounds it to a value of either 0 or 1 prior to sending it to the GT-Suite model to ensure it is in the discrete form that GT-Suite expects (see Appendix B for a discussion on MPC relaxation and justification for applicability of this approach in this context).

The model used by the MPC is a semi-analytical nonlinear model of the A/C system by Parent et al. [28] (the authors of this study) as a prerequisite to run this study for controller design development. The dynamics of this model can be described by the heat exchanger pressures p_c for condenser and p_e for evaporator. The first system output of interest is the temperature of the evaporator air stream. The underlying continuous-time mathematical equations for the heat exchanger pressures and temperatures are [28]:

$$\begin{aligned} \dot{p}_c &= \frac{\left[Q_c \rho_a c_{p,air}(T_{air,in} - T_{air,out}) + \pi_c \dot{m}_{ref}(h_{in} - h_{out})\right]}{V\left[(1-\overline{\gamma})\frac{\partial(\rho_l h_l)}{\partial p} + \overline{\gamma}\frac{\partial(\rho_g h_g)}{\partial p} + (\rho_g h_g - \rho_l h_l)\frac{\partial \overline{\gamma}}{\partial p} - 1 + \frac{m_H c_H}{V}\left(\frac{\partial T_{wall}}{\partial p}\right)\right]} \\ \dot{p}_e &= \frac{\left[Q_e \rho_a c_{p,air}(T_{air,in} - T_{air,out}) + \pi_c \dot{m}_{ref}(h_{in} - h_{out})\right]}{V\left[(1-\overline{\gamma})\frac{\partial(\rho_l h_l)}{\partial p} + \overline{\gamma}\frac{\partial(\rho_g h_g)}{\partial p} + (\rho_g h_g - \rho_l h_l)\frac{\partial \overline{\gamma}}{\partial p} - 1 + \frac{m_H c_H}{V}\left(\frac{\partial T_{wall}}{\partial p}\right)\right]} \\ T_{air,out,c} &= T_{ref,c} + \left(T_{air,in,c} - T_{ref,c}\right)^{(-NTU_c)(K_{ca})} \\ T_{air,out,e} &= T_{ref,e} + \left(T_{air,in,e} - T_{ref,e}\right)^{(-NTU_e)(K_{ea})} \\ NTU &= \frac{\alpha A_s\left[1 - F_{fin}(1 - \eta_{FA})\right]}{\dot{m}_a c_{p,air}} \end{aligned} \tag{6}$$

where Q_c is the volume flow rate of the air through the condenser $[m^3/s]$, ρ_a is the air density $[kg/m^3]$, h_{in} and h_{out} are the inlet and outlet refrigerant enthalpies [joules], ρ_l, ρ_g are the densities $[kg/m^3]$ and h_l, h_g are the enthalpies [joules] for the refrigerant calculated at saturated liquid and vapor states, respectively. c_H is the specific heat of the HEX material [j/K·kg], $c_{p,air}$ is the specific heat of air at constant pressure [j/K·kg], $\overline{\gamma} \in [0,\ 1]$ is the mean void fraction of the refrigerant, V is the volume of the HEX [m^3], m_H is the material mass of the HEX [kg], T_{wall} is the wall temperature of the HEX [K], p_c and p_e are the refrigerant pressures in the condenser and evaporator [kPa], π_c is the compressor clutch engagement [on/off], $\dot{m}_{ref}$ is the refrigerant mass flow rate [kg/s], Q_e is the volume flow rate of the air from the blower fan [m^3/s], $T_{air,in}$ and $T_{air,out}$ are the inlet air temperatures to the condenser and evaporator [K], T_{ref} is the HEX refrigerant temperature [K], NTU is the number of transfer units of the air, and K_{ca} and K_{ea} act as multipliers on the Nusselt number. α is the convective heat transfer coefficient of the air [$W/m^2 \cdot K$], A_s is the external surface area of the HEX [m^2], $F_{fin} \in [0,\ 1]$ is the fraction of air-to-structure surface area on the fins, η_{FA} is the air-side fin efficiency, and $\dot{m}_a$ is the mass flow rate of the air [kg/s].

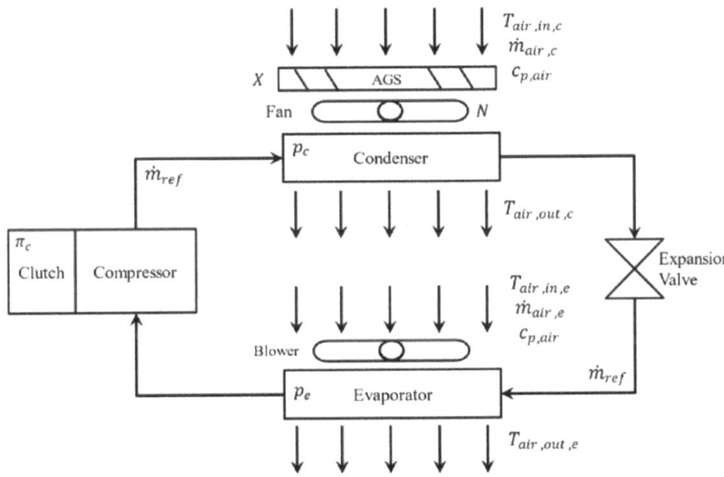

Figure 2. Simple A/C system diagram.

From the developments in [28], the definitions for Q_c and $T_{ref,c}$ in continuous time are as follows:

$$\begin{aligned} Q_c &= 0.07722\hat{N} + 0.01057v + 4.913 \times 10^{-5}v^2 + 0.2039X\hat{N} + 0.002671Xv \\ T_{ref,c} &= 8.4 \cdot 10^{-9}p_c^3 - 4.9 \cdot 10^{-5}p_c^2 + 0.12p_c - 34.9 \end{aligned} \tag{7}$$

where $\hat{N}$ is the normalized rotational speed of the condenser fan between zero and one, X is the open-fraction of the AGS between zero and one, and v is the vehicle speed [m/s]. The full list of fitted curves is available in Appendix A of [28] in Equations (17)–(20) and not further provided here for brevity.

Note that, in the system model in (6), there are terms that are updated from outside of the A/C system model and controller by other vehicle subsystems. These parameters are updated as the vehicle is operating and not controlled by the controller designed in this study for the A/C system. Hence, they are treated as system parameters in (6) that vary with time. The varying system parameters controlled by other subsystems in the vehicle dynamics are defined as follows:

$$\theta = \begin{bmatrix} \dot{m}_{ref} \\ Q_e \\ T_{air,in,c} \\ T_{air,in,e} \\ h_{in,c} \\ v \\ h_{in,comp} \end{bmatrix} = \begin{bmatrix} \theta_1 \\ \theta_2 \\ \theta_3 \\ \theta_4 \\ \theta_5 \\ \theta_6 \\ \theta_7 \end{bmatrix} \tag{8}$$

where $\dot{m}_{ref}$ is the mass flow rate of the refrigerant set by the compressor in GT-Suite [kg/s], Q_e is the volumetric flow rate of the air set by the evaporator blower model in GT-Suite [m^3/s], $T_{air,in,c}$ and $T_{air,in,e}$ are the air temperatures entering the condenser and evaporator [K], $h_{in,c}$ is the inlet enthalpy for the condenser [joules], v is the vehicle speed [m/s], and $h_{in,comp}$ is the inlet enthalpy for the compressor [joules]. All these parameters are measured and sent from the vehicle energy management (VEM) model to the MPC. It is important to note that this makes the system under study a nonlinear time-varying system, which would require an adaptive MPC type to handle the nonlinearities by linearizing the model at each operating point, as discussed in Section 2.1.1.

Since system energy consumption is of interest, the model includes the power consumption definitions in continuous time as [28]:

$$\begin{aligned} \dot{W}_{comp} &= \frac{\dot{m}_{ref}(h_{out}-h_{in})}{\eta_{comp}} \\ \dot{W}_{fan} &= \frac{Q_c \Delta p_t}{\eta_{fan}} \\ \Delta\dot{W}_{drag} &= \tfrac{1}{2}\rho_a A_{fr} v^3 \Delta C_d \\ g_2 = \dot{W}_{total} &= \dot{W}_{comp} + \dot{W}_{fan} + \Delta\dot{W}_{drag} \end{aligned} \tag{9}$$

where $\dot{W}_{comp}$ is the compressor's power consumption [j/s], $\dot{W}_{fan}$ is the fan's power consumption [j/s], and $\Delta\dot{W}_{drag}$ is power consumption associated with additional drag on the vehicle due to underhood airflow, which is set by the vehicle speed, fan rotational speed, and AGS open-fraction [j/s]. $\dot{W}_{total}$ is the total power consumption from A/C comprising of the previous three components [j/s]. $\dot{m}_{ref}$ is the mass flow rate of the refrigerant [m^3/s], h_{in} and h_{out} are the enthalpies of the refrigerant at the inlet and outlet of the compressor, respectively [j/kg], η_{comp} is the ratio of isentropic efficiency to volumetric efficiency of the compressor, Δp_t is the pressure rise required of the fan [Pa], η_{fan} is the fan's adiabatic efficiency, A_{fr} is the frontal area of the vehicle [m^2], ΔC_d is the increase in drag coefficient due to additional flow through the underhood at a given vehicle speed due to the AGS being open and/or the fan being on, and v is the vehicle speed [m/s].

With all the required background information discussed above, next, the problem definition is discussed in more detail.

3. Problem Definition

This section provides a step-by-step formulation of the problem studied in this paper.

3.1. Introduction

To conduct simulations to determine the desired MPC design parameters, the industrial partner supplied a high-fidelity model of the physical vehicle in GT-Suite v2020 (Build 3.0002), a multi-physics simulation software created by Gamma Technologies. This high-fidelity model is treated as a substitute for the real physical vehicle in this work (plant). GT-Suite v2020 (Build 3.0002) and MathWorks' Simulink 2021b have built-in cross-platform communication capabilities that allow signals to be sent to and from the two software packages. Therefore, MATLAB and Simulink are used for the MPC implementation. In the GT-Suite model, there are numerous independent cases set up for simulation, each representing a different standard drive cycle with time-series data loaded for each of the driving conditions, some of which are the system parameters in (8) required by the MPC from GT-Suite, such as the vehicle speed. The remaining elements of (8) are computed by the GT-Suite model and sent to the MPC, such as $\dot{m}_{ref}$ and Q_e. In this work, the MPC simulations are conducted for the SC03 drive cycle [33], which is a standard drive cycle used by the Environmental Protection Agency (EPA) to evaluate the vehicle fuel economy where the A/C system is active. The total duration of this drive cycle is 575 s.

3.2. System Equations

In this paper, an adaptive MPC, known as successive linearization-based MPC (SLMPC), is applied to the automotive A/C system illustrated in Figure 2, i.e., the A/C system is the plant. To define the elements of A, B, C, and D used in the prediction model described in (2), the system of nonlinear equations described in (6)–(9) must be linearized. These equations are linearized on the fly using the adaptive MPC approach described in Section 2.1, allowing the trim state to always be the current working state of the plant. The Jacobian matrices that map the system nonlinear equations in (6) to the linearized state-space representation A, B, C, D are calculated at every time step using the MATLAB *jacobian* function, where $jacobian(f, v)$ computes the Jacobian matrix of symbolic function f with respect to v. The (i, j) element of the result of $\partial f(i)/\partial v(j)$ is as follows:

$$\begin{cases} A = jacobian([f_1, f_2], [x_1, x_2]) \\ B = jacobian([f_1, f_2], [u_1, u_2, u_3]) \\ C = jacobian([g_1, g_2], [x_1, x_2]) \\ D = jacobian([g_1, g_2], [u_1, u_2, u_3]) \end{cases} \tag{10}$$

where f_1 is the definition of $\dot{p}_c$, f_2 is the definition of $\dot{p}_e$, g_1 is the definition of $T_{air,out,e}$ in (6), and g_2 is the definition of $\dot{W}_{total}$ in (9). x is defined in (27); u is defined in (28). Additionally, the collective terms g_1 and g_2 capture the system outputs y as defined in (29). Note that, due to the form of the definitions for f_1, f_2, g_1, and g_2, the full form of the Jacobians in (10) is not provided here as it would be infeasible to fit all the terms in a presentable format.

The numerical values of the state-space matrices are computed using a successive linearization approach [29], SLMPC, as a common approach adopted for MPC designs [34]. The concept of this approach is simple. At the beginning of each control iteration, the Jacobian matrices are evaluated at the current operating conditions of the complex system model, that being the current system states, controlled inputs, and system parameters. This will update the numerical values of the state-space matrix entries. By re-evaluating the state-space model at the current operating conditions, one ensures that the trim state is close to the current operating point. The intention of this is to reduce the linearization error caused by deviations from the trim state. Thus, the working condition of the complex system model will be treated as the trim state. Full theoretical development and proof of applicability for SLMPC can be found in [29]. This allows the controller to be initialized at the current system state at each time step, allowing accuracy to be maintained throughout the simulation since the controller will always work in the vicinity of the calculated trim state.

3.3. Quadratic Programming Formulation

Quadratic programming is an optimization process used to solve mathematical problems containing quadratic functions. The optimal solution to a quadratic programming problem is found by minimizing the following objective function, J [32]:

$$J = min\left[\frac{1}{2}X_{qp}^T H X_{qp} + f^T X_{qp}\right] \tag{11}$$

where X_{qp} is a vector containing the control objectives shown in (12), H is a symmetric matrix containing relative weightings for each control objective in the quadratic term of the cost function, and f is a vector containing relative weightings for each control objective in the linear term of the cost function. In (11), the elements of the vector X_{qp} are minimized to achieve the smallest possible value for the objective function.

For the MPC controllers, the controller is typically designed to track reference values for the states and outputs of the system while also considering the magnitude and rate of change of the controlled input values. In this case, the elements of X_{qp} are as follows [32]:

$$X_{qp} = \begin{bmatrix} \Delta u_0 \\ x_1' \\ \vdots \\ \Delta u_{h-1} \\ x_h' \\ y_1' \\ \vdots \\ y_h' \end{bmatrix}_{h(n_x+n_u+n_y)} \tag{12}$$

where the definitions of these terms are provided further below. This description of X_{qp} is for a discrete-time case since the approach used for the controller design in this study is also discrete-time due to the nature of the controller type. The dimension of the vector depends on the prediction horizon length h, the number of states n_x, number of inputs n_u, and the number of outputs n_y. The elements of the H matrix can be used to change the relative importance of each of the control objectives corresponding to the terms in the X_{qp} vector as outlined in (12).

The typical state-space equations can be modified to include the output and state reference tracking deviations. This will make forming the QP constraints much easier. The goal of the following modifications to the state-space model form is to explicitly show the output setpoint deviations, state setpoint deviations, and the rate of change of the controlled inputs since these are included in the X_{qp} vector of the objective function. In QP problem formulations, linear equality constraints containing the X_{qp} vector can be applied to the problem. These constraints can be used to mathematically define the elements of X_{qp}. This can be done by treating the discrete state-space model in (2) as an equality constraint and reorganizing the formulation to contain the elements of X_{qp} explicitly. The state-space formulation described in (2) is reorganized to explicitly include the elements of (12) using the approach outlined in [32]. As a special case of the general form of the objective function for QP in (11) when H is symmetric positive-definite, the objective function reduces to $J = min\left[\frac{1}{2}\left\|X_{qp} - X_{qp,s}\right\|_{H'}{}^2\right]$ where $H = H'TH'$ follows from the Cholesky decomposition of H and $f = -H'^T X_{qp,s}$. Hence, given the system in (2), the special case of the objective function in QP can be expanded as [32]:

$$\begin{cases} J = \sum\limits_{i=1}^{h} \left(\left\| y_{k+i} - y_{k+i,s} \right\|_{Q_y}^2 + \left\| x_{k+i} - x_{k+i,s} \right\|_{Q_x}^2 + \left\| u_{k+i-1} \right\|_{R_u}^2 + \left\| \Delta u_{k+i-1} \right\|_{R_{\Delta u}}^2 \right) \\ \qquad S.t. \ : Dynamic\ Model\ and\ Constraints \end{cases} \tag{13}$$

where the subscript s denotes a setpoint, i is the current step in the time horizon window, and I is the identity matrix. Note that subscript i is from 0 to the time horizon h as the dimensions of these vectors change with the size of the time horizon. To make the formulation easier, the system equations are written in

$$\begin{cases} x_{k+1} = A_d x_k + B_d \Delta u_k + B_d u_{k-1} \\ u_k = u_{k-1} + \Delta u_k \\ y_{k+1} = C_d x_{k+1} + D_d \Delta u_k + D_d u_{k-1} \end{cases} \tag{14}$$

where $\Delta u_k \triangleq u_k - u_{k-1}$. Rewriting the model in this form incorporates the rate of change of the controller inputs into the state-space model. The next step is to rewrite the state-space model in terms of setpoint deviations. So, the general form of the model equation becomes:

$$\begin{cases} x_{k+1} - x_{k+1,s} = A_d(x_k - x_{k,s}) + A_d x_{k,s} + B_d \Delta u_k + B_d u_{k-1} - x_{k+1,s} \\ u_k = u_{k-1} + \Delta u_k \\ y_{k+1} - y_{k+1,s} = C_d(x_{k+1} - x_{k+1,s}) + C_d x_{k+1,s} + D_d \Delta u_k + D_d u_{k-1} - y_{k+1,s} \end{cases} \tag{15}$$

Or in a compact form as

$$\begin{cases} \begin{bmatrix} x_{k+1} - x_{k+1,s} \\ u_k \end{bmatrix} = \begin{bmatrix} A_d & B_d \\ 0_{n_u \times n_x} & I_{n_u \times n_u} \end{bmatrix} \begin{bmatrix} x_k - x_{k,s} \\ u_{k-1} \end{bmatrix} \\ \quad + \begin{bmatrix} B_d \\ I_{n_u \times n_u} \end{bmatrix} \Delta u_k - \begin{bmatrix} x_{k,s} \\ 0_{n_u \times 1} \end{bmatrix} + \begin{bmatrix} A_d & B_d \\ 0_{n_u \times n_x} & I_{n_u \times n_u} \end{bmatrix} \begin{bmatrix} x_{k,s} \\ 0_{n_u \times 1} \end{bmatrix} \\ y_{k+1} - y_{k+1,s} = \begin{bmatrix} C_d & 0_{n_y \times n_u} \end{bmatrix} \begin{bmatrix} x_{k+1} - x_{k+1,s} \\ u_k \end{bmatrix} \\ \quad + \begin{bmatrix} 0_{n_u \times n_x} & D_d \end{bmatrix} \Delta u_k + \begin{bmatrix} 0_{n_u \times n_x} & D_d \end{bmatrix} \begin{bmatrix} x_k - x_{k,s} \\ u_{k-1} \end{bmatrix} \\ \quad - y_{k+1,s} + \begin{bmatrix} C_d & 0_{n_y \times n_u} \end{bmatrix} \begin{bmatrix} x_{k+1,s} \\ 0_{n_u \times 1} \end{bmatrix} \end{cases} \tag{16}$$

If we define

$$x'_k = \begin{bmatrix} x_k - x_{k,s} \\ u_{k-1} \end{bmatrix}, A' = \begin{bmatrix} A_d & B_d \\ 0_{n_u \times n_x} & I_{n_u \times n_u} \end{bmatrix}, B' = \begin{bmatrix} B_d \\ I_{n_u \times n_u} \end{bmatrix}, x'_{k,s} = \begin{bmatrix} x_{k,s} \\ 0_{n_u \times 1} \end{bmatrix}$$
$$y'_k = y_k - y_{k,s}, C' = \begin{bmatrix} C_d & 0_{n_y \times n_u} \end{bmatrix},$$
$$D\prime = \begin{bmatrix} 0_{n_u \times n_x} & D_d \end{bmatrix}, Q'_x = \begin{bmatrix} Q_x & 0_{n_x \times n_u} \\ 0_{n_u \times n_x} & R_u \end{bmatrix} \tag{17}$$

The form in (16) can be rearranged for time step 0 in the prediction horizon, where we initiate the dynamics with x_0, y_0, and u_{-1}, and given an equality constraint can be written in the form of $A_{eq} X_{qp} = B_{eq}$

$$\begin{cases} x'_1 = A' x'_0 + B' \Delta u_0 - x'_{1,s} + A' x'_{0,s} \\ y'_1 = C' x'_1 + D' \Delta u_0 + D' x'_0 - y_{1,s} + C' x'_{1,s} \end{cases}$$
$$\xrightarrow{A_{eq} X_{qp} = B_{eq}\ Form} \begin{cases} x'_1 - B' \Delta u_0 = A' \begin{bmatrix} x_0 \\ u_{-1} \end{bmatrix} - x'_{1,s} \\ y'_1 - C' x'_1 - D' \Delta u_0 = -y_{1,s} + C' x'_{1,s} + D' \begin{bmatrix} x_0 \\ u_{-1} \end{bmatrix} \end{cases} \tag{18}$$

And for any other time step between $i = 1$ up to $h - 1$ it can be written as follows:

$$\begin{cases} x'_{k+1} = A' x'_k + B' \Delta u_k - x'_{k+1,s} + A' x'_{k,s} \\ y'_{k+1} = C' x'_{k+1} + D' \Delta u_k + D' x'_k - y_{k+1,s} + C' x'_{k+1,s} \end{cases}$$
$$\xrightarrow{A_{eq} X_{qp} = B_{eq}\ Form} \begin{cases} x'_{k+1} - A' x'_k - B' \Delta u_k = A' x'_{k,s} - x'_{k+1,s} \\ y'_{k+1} - C' x'_{k+1} - D' \Delta u_k = -y_{k+1,s} + C' x'_{k+1,s} + D' x'_k \end{cases} \tag{19}$$

Hence, the objective function in (13) becomes:

$$\begin{cases} J = \sum_{i=1}^{h} \left(\|y'_{k+i}\|^2_{Q_y} + \|x'_{k+i}\|^2_{Q'_x} + \|\Delta u_{k+i-1}\|^2_{R_{\Delta u}} \right) \\ S.t.\ : Dynamic\ Model\ and\ Constraints \end{cases} \tag{20}$$

which can be written in compact form as:

$$\begin{cases} J = \sum_{i=1}^{h} X_{qp}^T H X_{qp} \\ S.t.\ : A_{eq} X_{qp} = B_{eq}\ and\ lb_X \leq X_{qp} \leq ub_X \end{cases} \tag{21}$$

where

$$A_{eq} = \begin{bmatrix} -B' & I_{n_{x^i}} & 0 & 0 & \cdots & \cdots & \cdots & \cdots & 0 \\ 0 & -A' & -B' & I_{n_{x^i}} & \ddots & \ddots & \ddots & \ddots & \vdots \\ \vdots & \ddots & \ddots & \ddots & \ddots & \ddots & \ddots & \ddots & \vdots \\ 0 & \cdots & 0 & -A' & -B' & I_{n_{x^i}} & 0 & \cdots & 0 \\ -D' & -C' & \cdots & \cdots & \cdots & 0 & I_{n_y} & \cdots & 0 \\ \vdots & \ddots & \ddots & \ddots & \ddots & \ddots & \ddots & \ddots & 0 \\ -D' & \cdots & \cdots & \cdots & \cdots & -C' & \cdots & 0 & I_{n_y} \end{bmatrix}_{h(n_x+n_u+n_y)\times h(n_x+2n_u+n_y)} \tag{22}$$

$$B_{eq} = \begin{bmatrix} A'\begin{bmatrix} x_0 \\ u_{-1} \end{bmatrix} - x'_{1,s} \\ A'x'_{1,s} - x'_{2,s} \\ \vdots \\ A'x'_{h-1,s} - x'_{h,s} \\ -y_{1,s} + C'x'_{1,s} + D'\begin{bmatrix} x_0 \\ u_{-1} \end{bmatrix} \\ \vdots \\ -y_{h,s} + C'x'_{h,s} + D'x'_h \end{bmatrix}_{h(n_x+n_u+n_y)\times 1} \tag{23}$$

with the lower and upper bounds for the entire length of the prediction horizon as

$$lb = \begin{bmatrix} lb_u - ub_u \\ \begin{bmatrix} lb_x - x_{1,s} \\ lb_u \end{bmatrix} \\ \vdots \\ lb_u - ub_u \\ \begin{bmatrix} lb_x - x_{h,s} \\ lb_u \end{bmatrix} \\ lb_y - y_{1,s} \\ \vdots \\ lb_y - y_{h,s} \end{bmatrix}_{h(n_x+n_u+n_y)\times 1} \tag{24}$$

$$ub = \begin{bmatrix} ub_u - lb_u \\ \begin{bmatrix} ub_x - x_{1,s} \\ ub_u \end{bmatrix} \\ \vdots \\ ub_u - lb_u \\ \begin{bmatrix} ub_x - x_{h,s} \\ ub_u \end{bmatrix} \\ ub_y - y_{1,s} \\ \vdots \\ ub_y - y_{h,s} \end{bmatrix}_{h(n_x+n_u+n_y)\times 1} \tag{25}$$

where lb_u and ub_u are the lower and upper bounds on the inputs, and lb_x, ub_x, lb_y, and ub_y are the lower and upper bounds for the states and outputs, respectively. Each element of the X_{qp} vector can be weighted individually, providing more significance to prioritizing certain control objectives over others. The weighting matrix H for the control objectives within the objective function (output reference tracking, state reference tracking, input value magnitude, and rate of change of the input commands) is [32]

$$H = \begin{bmatrix} \begin{bmatrix} R_{\Delta u} & 0 \\ 0 & Q'_x \end{bmatrix} & 0 & \cdots & \cdots & \cdots & 0 \\ 0 & \ddots & \ddots & \ddots & \ddots & \vdots \\ \vdots & \ddots & \begin{bmatrix} R_{\Delta u} & 0 \\ 0 & Q'_x \end{bmatrix} & 0 & \cdots & 0 \\ 0 & \cdots & 0 & Q_y & \cdots & 0 \\ \vdots & \ddots & \ddots & \ddots & \ddots & \vdots \\ 0 & \cdots & \cdots & \cdots & 0 & Q_y \end{bmatrix}_{h(n_x+2n_u+n_y)\times h(n_x+2n_u+n_y)} \tag{26}$$

$$Q'_x = \begin{bmatrix} Q_x & 0_{n_x \times n_u} \\ 0_{n_u \times n_x} & R_u \end{bmatrix}_{(n_x+n_u)\times(n_x+n_u)}$$

where $R_{\Delta u}$ is the weighting of the rate of change of the inputs, Q_x is the weighting of the state reference tracking deviations, R_u is the weighting of the input magnitudes, and Q_y is the weighting of the output reference tracking deviations. $R_{\Delta u}$, Q_x, R_u, and Q_y are diagonal matrices with the respective weights along the diagonal.

3.4. Implementation

The cabin cooling effort of the A/C system is controlled using three actuators: the compressor clutch, π_c, condenser fan, $\hat{N}$, and the AGS, X. The compressor clutch connects the compressor to the engine, enabling the compressor to operate and increase the pressure of the refrigerant leaving through the discharge port. As mentioned in Section 2.3, the compressor clutch π_c is not a continuous variable in practice and is a discrete binary variable. In this study, π_c is initially treated as a continuous variable to be computed via QP optimization. When the MPC computes the continuous compressor clutch signal π_c, it rounds it to a value of either 0 or 1 prior to sending it to the GT-Suite model to ensure it is in the discrete form that GT-Suite expects. Alternative methods exist that are better suited for dealing with mixed-variable systems like the one described. For instance, techniques such as hybrid predictive control models that incorporate both integer and continuous variables have been explored in the literature [35–38] and can be further investigated in future studies. These methods provide a more holistic approach to dealing with mixed-variable systems, allowing for more accurate and efficient control strategies in scenarios where variables exhibit both continuous and discrete behaviors.

The condenser fan and AGS regulate the flow of the ambient air stream entering the vehicle's front end, which passes through the condenser to exchange heat with the refrigerant. The fan's rotational speed is continuously controllable, and the AGS area open-fraction can continuously vary between fully closed and fully open. Note that some underhood airflow can still occur when fully closed due to the vehicle's motion and/or fan operation.

Based on the approach detailed in Section 2.2, using a time step of one second leaves only the A matrix changed due to the addition of the identity matrix, as shown in (3)–(5). By convention, the time step for MPC–plant interaction is the same as the one used within the MPC for future time predictions using the model up to the prediction horizon [32]. The time step of one second was selected after assessing the open-loop response of the air temperature output subjected to a unit step from each of the three controlled inputs for the continuous-time linear model. All other parameters, such as the external inputs, were held constant at the trim state selected based on an operating point observed for the GT-Suite (real system) model.

3.4.1. States, Inputs, and Outputs

For the simulations in this study, from the A/C system equations in (6)–(9) the state vector x is defined as:

$$x = \begin{bmatrix} p_c \\ p_e \end{bmatrix} = \begin{bmatrix} x_1 \\ x_2 \end{bmatrix} \quad (27)$$

where p_c and p_e are the condenser and evaporator refrigerant pressures. The input vector u is defined as:

$$u = \begin{bmatrix} \pi_c \\ \hat{N} \\ X \end{bmatrix} = \begin{bmatrix} u_1 \\ u_2 \\ u_3 \end{bmatrix} \quad (28)$$

where π_c is a continuous compressor clutch signal, $\hat{N}$ is the normalized rotational speed of the condenser fan between zero and one, and X is the open fraction of the AGS between zero and one.

The output vector, y, is defined as:

$$y = \begin{bmatrix} T_{air,out,e} \\ \dot{W}_{total} \end{bmatrix} = \begin{bmatrix} y_1 \\ y_2 \end{bmatrix} \quad (29)$$

where $T_{air,out,e}$ is the temperature of the air leaving the evaporator (entering the cabin) [K], and $\dot{W}_{total}$ is the total power consumption of the compressor [Watt] or [j/s] that is comprised of $\dot{W}_{comp}$, $\dot{W}_{fan}$, and $\Delta\dot{W}_{drag}$, as discussed in (9). The evaporator outlet air temperature is chosen as one of the outputs for the model because the controller's goal is to control the actuators to maintain this air temperature close to a reference or set value, thus regulating the cabin cooling performance. The other output, $\dot{W}_{total}$, is chosen because the second goal of this controller is to minimize the power consumption of the A/C system while satisfying the cooling objective.

It is important to note that, in the proposed formulation, feedback is taken into account by using the direct system outputs as the inputs for the controller. This means that no estimator or observer was used to obtain the state variables or the output variables of the system. In this study, it is assumed that the system outputs are measurable and noise-free, and that the system model is accurate. Therefore, a possible model mismatch is not considered.

However, in reality, these assumptions may not hold true. There may be measurement errors, nonlinearities, uncertainties, or disturbances that affect the system's performance and stability. In such cases, using an observer or an estimator could help to improve the feedback quality and robustness of the controller. This limitation can be further explored in future studies with the implementation of an observer or an estimator to handle potential system mismatches.

3.4.2. Sampling Time, Prediction Horizon, and Control Horizon

As per the recommendations and discussion provided in Section 2.1.2, the time unit of measurement for the governing equations used to derive the linearized model is in seconds. Thus, the units of the discretization time step are in seconds to maintain consistency. The rise time of the A/C system in this study is roughly 12.7 s, meaning a time step between 0.64 s to 1.3 s would fit 10 to 20 samples within the open-loop rise time, meaning a sampling or discretization time step of $T = 1$ s is satisfactory based on this best practice.

As per the recommendation for the choice of prediction horizon, in this study, a prediction horizon of 10 s or 10 time steps is chosen. This reduces the computational burden for the optimization process required during the prediction horizon by choosing the lower bound of the recommended range. However, in future studies, one can further optimize the prediction horizon. Additionally, a control horizon of two steps is chosen for the implementation of the adaptive MPC as it provides a sufficient balance between computation, efficiency, and performance for the MPC design.

It should be noted that the choices above were based on trial and error in the study to obtain satisfactory results for the industrial partner in this first phase of investigations for the implementation of an MPC controller for an A/C system in a vehicle. In future studies, additional investigations shall be pursued to further improve and optimize these choices for the controller design.

3.4.3. Constraints

As discussed in Section 3.3, constraints can be set on the optimization. The values of these three controlled inputs have physical limitations based on their range of operation. These are hard constraints that cannot be exceeded by any means. The mathematical number sets and limitations of these three controlled variables are provided by the industrial partner as:

$$\begin{aligned} \pi_c \in \mathbb{N}\, s.t. \quad & \pi_c = [0,1] := \left[l_{b_{u_1}},\, u_{b_{u_1}}\right] \\ \hat{N} \in \mathbb{Q}\, s.t. \quad & 0 \le \hat{N} \le 1 := \left[l_{b_{u_2}},\, u_{b_{u_2}}\right] \\ X \in \mathbb{Q}\, s.t. \quad & 0 \le X \le 1 := \left[l_{b_{u_3}},\, u_{b_{u_3}}\right] \end{aligned} \tag{30}$$

The states also have constraints but they are not due to physical limitations. The constraints on the states are imposed to prevent problems arising in the HEXs due to excessive or lack of pressure. The state constraints are as follows and are defined by the industrial partner:

$$\begin{aligned} 100 \text{ kPa} \le p_e \le 1300 \text{ kPa} := \left[l_{b_{x_1}},\, u_{b_{x_1}}\right] \\ 700 \text{ kPa} \le p_c \le 3000 \text{ kPa} := \left[l_{b_{x_2}},\, u_{b_{x_2}}\right] \end{aligned} \tag{31}$$

For the QP algorithm to run successfully, constraints must also be present for the outputs as well but, since the outputs do not have bounds for the application in this study, they are set to infinity as

$$\begin{aligned} -\infty\, K \le T_{air,out,e} \le +\infty\, K := \left[l_{b_{y_1}},\, u_{b_{y_1}}\right] \\ -\infty\, Watt \le \dot{W}_{total} \le +\infty\, Watt := \left[l_{b_{y_2}},\, u_{b_{y_2}}\right] \end{aligned} \tag{32}$$

Feasibility and infeasibility issues in MPC controllers with QP optimization can affect the stability and performance of the closed-loop system. If the QP problem has hard constraints, such as input or output limits, it may be impossible to satisfy them at some time steps, leading to infeasibility. If the QP solver detects infeasibility, it terminates immediately and returns an error in MATLAB. To avoid this situation, some methods have been proposed, such as using soft constraints, additional constraint checking horizon, terminal condition and cost, or disturbance preview [38].

In this study, the infeasibility problem shall not occur as the only hard constraints in the QP optimization in this study are on the inputs outlined in (30) and these constraints shall be met as, in any A/C system of a vehicle, there must be a combination of these inputs that provides some sort of long-term asymptotic stability and tracking for the cabin temperature even without the use of an MPC controller. This was primarily achieved by the baseline PID controlled in previous works and can be the bare minimum solution to this optimization problem that would exist at all times. Therefore, since there is no time limit on the settling time for the temperature tracking of the cabin, the control inputs and hard constraints in (30) and (31), as shown in simulation results for the ranges of p_c and p_e, shall not prevent the optimization problem from converging to a solution at any given time step and, hence, the infeasibility issue shall not arise.

3.4.4. Weights

Additionally, from Section 3.3, the elements of the X_{qp} vector defined in (12) can be weighted based on importance and contribution to the objective function through the weighting matrix H described in (26). As defined in (27)–(29), for this work, there are two states ($n_x = 2$), three controlled inputs ($n_i = 3$), and two outputs ($n_y = 2$). The specific weighting matrices inside (26) are defined as follows:

$$\begin{aligned} Q_x &= \begin{bmatrix} K_{p_c} & 0 \\ 0 & K_{p_e} \end{bmatrix} \\ Q_y &= \begin{bmatrix} K_{T_{air}} & 0 \\ 0 & K_{\dot{W}_T} \end{bmatrix} \\ R_u &= \begin{bmatrix} K_{\pi_c} & 0 & 0 \\ 0 & K_{\hat{N}} & 0 \\ 0 & 0 & K_X \end{bmatrix} \\ R_{\Delta u} &= \begin{bmatrix} K_{\Delta\pi_c} & 0 & 0 \\ 0 & K_{\Delta\hat{N}} & 0 \\ 0 & 0 & K_{\Delta X} \end{bmatrix} \end{aligned} \tag{33}$$

where K_{p_c} and K_{p_e} are the weightings for the state deviations, $K_{T_{air}}$ is the weighting for the air temperature tracking deviation, $K_{\dot{W}_T}$ is the weighting for the total power consumption, K_{π_c}, $K_{\hat{N}}$, and K_X are the weightings for the magnitude of the controlled inputs, and $K_{\Delta\pi_c}$, $K_{\Delta\hat{N}}$, and $K_{\Delta X}$ are the weightings for the rate of change of the controlled inputs. It is worth noting that the goal of the MPC is to reduce energy consumption, but the power consumption is modeled and weighted because energy is the time integral of power. Thus, reducing power consumption will also reduce energy consumption.

3.4.5. Scenarios

Two control input cases are considered: first, only controlling the compressor clutch and, second, adding the condenser fan speed and AGS position as additional controlled inputs. Thus, the u_k and Δu_k elements of the X_{qp} vector for the first case with clutch-only control are:

$$u_k = [\pi_c] \Delta u_k = [\Delta\pi_c] \tag{34}$$

and for the second case with full control of compressor clutch engagement, condenser fan speed, and open-fraction AGS position are:

$$u_k = \begin{bmatrix} \pi_c \\ \hat{N} \\ X \end{bmatrix}; \; \Delta u_k = \begin{bmatrix} \Delta\pi_c \\ \Delta\hat{N} \\ \Delta X \end{bmatrix} \tag{35}$$

It is important to note that, as the input vector changes in cases 1 and 2, the applied constraints in the optimization problem are also adjusted to only include the applied control inputs.

3.4.6. Summary

The overall system implementation is shown in Figure 3, where the structure of implemented SLMC is shown with notations that are the same as in Figure 1. Additionally, the environment under which each component is developed/operated is noted in red color.

To summarize this section and formalize the control problem investigated in this study, we have the continuous SSR system form in (1) that can be written in the discrete SSR system form in (2) with x defined in (27), u defined in (28), and y defined in (29). The matrices A, B, C, and D in (1) are the Jacobians calculated from (6)–(9) using (10). These matrices can be converted to their discrete form using Euler's method described in (3)–(5) to obtain A_d, B_d, C_d, and D_d in (2) with the sampling step of $T = 1$ s.

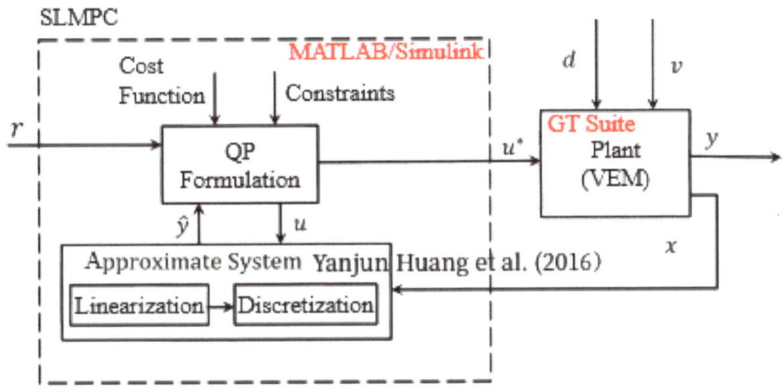

Figure 3. Successive Linearization Based MPC block diagram [22].

The goal of the optimization problem in SLMPC is to minimize the objective function in (11) in the general case and for this study in a simplified form in (13) and with constraints formed as (21), where the elements of X_{qp} and H are defined in general in (12)–(26) with specific definitions for this study in (33) subject to constraints in (24) and (25) in general form and for this study in (30)–(32).

Finally, the investigations are conducted for two scenarios, 1. clutch only as input for the MPC design as outlined in (34), and 2. full control input including compressor clutch engagement, condenser fan speed, and open-fraction AGS position as outlined in (35). The time or prediction horizon for the MPC design was set to 10 time steps, and since the time step for this study was chosen as $T = 1$ s it would equal a 10 s look ahead in time for the controller.

The results and analysis of the simulations with the VEM model, which was provided by the industrial partner in the GT-suite environment and is used as the plant for the simulations, are presented in Section 4, including some discussions on the observations made from the outcomes.

4. Results and Discussion

To evaluate the performance of the air temperature tracking, the root mean squared error (RMSE) between the actual air temperature from the GT-Suite model and the target reference temperature was calculated as:

$$RMSE = \sqrt{\frac{\sum_{i=1}^{n_{points}} \left(x_{actual,i} - x_{reference,i}\right)^2}{n_{points}}} \tag{36}$$

where $x_{actual,i}$ is the actual value from the GT-Suite model, $x_{reference,i}$ is the target reference value, and n_{points} is the total number of data points. The choice for the performance evaluation in (36) is to ensure that the tracking of the reference value is monitored and considered throughout the entire simulation, not just for a portion of it. That is why the summation in (36) is from 1 to the total number of data points or simulation outputs.

The performance of the GT-Suite model with the baseline PID control scheme is shown in Table 1, where $n_{\Delta\pi_c}$ is the clutch actuation frequency in the number of actuations per minute. These metrics will be referenced to the MPC results to compare the improvement of the new control scheme.

Table 1. Baseline VEM model performance metrics.

Model	$T_{eao}RMSE$ [C]	W_T [kJ]	$n_{\Delta\pi_c}$
Baseline GT-Suite	6.27	1.26×10^3	0

4.1. MPC Tuning Process

The weighting parameters of the MPC controller are set for the individual elements of the objective function as X_{qp}, defined in (12). The most crucial elements of (12) to be minimized are (i) the air temperature setpoint deviation corresponding to $K_{T_{air}}$ in (33), (ii) the instantaneous power consumption corresponding to $K_{\dot{W}_T}$ in (33), and (iii) the rate of change of the clutch command corresponding to $K_{\Delta\pi_c}$ in (33). The rate of change of the clutch command is critical to consider since the durability of the compressor can be a concern if a chattering phenomenon occurs, where the compressor clutch is constantly cycled between engaging and disengaging. Wear and tear on the compressor clutch can lead to reliability issues within vehicles and can increase expenses if the compressor requires maintenance more frequently. To quantify the actuation frequency of the clutch, the total number of clutch actuations throughout the cycle is recorded and divided by the 9.583 min (575 s) duration of the test cycle used in this study. With these three as the crucial weightings of interest for the MPC tuning, all other weightings in (33) are set to a value of one, while these three ($K_{T_{air}}$, $K_{\dot{W}_T}$, and $K_{\Delta\pi_c}$) are varied to ensure a reasonable contribution of various terms in the objective function.

There are optimal methods for tuning MPCs, as reported by Vazques et al. [39] using an artificial intelligence approach for real-time tuning of MPC for power converters and by Yamashita et al. [40] using a multi-objective tuning technique for MPC tuning in grinding circuits. However, in this study, a variation in the preliminary optimization method known as grid search with a systematic distribution of the grid points in the desired search space was considered as the focus of this study was more on the feasibility of implementing an MPC controller for the vehicle's A/C system rather than optimizing it. More advanced and optimal tuning techniques can be considered in future studies. A preliminary sensitivity assessment was conducted to determine the search space ranges for these three weighting values, $K_{\Delta\pi_c}$, $K_{T_{air}}$, and $K_{\dot{W}_T}$, to ensure the entire relevant search space is being explored. $K_{\Delta\pi_c}$, $K_{T_{air}}$, and $K_{\dot{W}_T}$ were systematically increased while all other weighting factors were held constant. For example, $K_{\Delta\pi_c}$ was increased until no further significant change was observed in the system's clutch actuation frequency.

To form the optimization search space, the valid weighting ranges for $K_{\Delta\pi_c}$ was identified as 1–1000 since values greater than 1000 presented no change to the clutch actuation frequency. The valid weighting range for $K_{T_{air}}$ was chosen between 1–50, as the values above this range provided no significant change to the air temperature's RMSE. Finally, the valid weighting range for $K_{\dot{W}_T}$ was chosen to be between 1 and 50 since integer values greater than 50 caused the compressor to shut off completely for the entire cycle, meaning the A/C system was inactive, and the air temperature's RMSE reached extreme and unrealistic values. Therefore, the optimization process for the MPC tuning could be described as

$$\begin{aligned} \text{maximize(or minimize)} \quad & f_{GS}\left(K_{\Delta\pi_c}, K_{T_{air}}, and\ K_{\dot{W}_T}\right) \\ \text{subject to} \quad & 1 \leq K_{\Delta\pi_c} \leq 1000 \\ & 1 \leq K_{T_{air}} \leq 50 \\ & 1 \leq K_{\dot{W}_T} \leq 50 \end{aligned} \tag{37}$$

where $f_{GS}(\cdot)$ is the function whose value needs to be maximized or minimized during the optimization process. A Latin Hypercube Sampling (LHS) approach outlined by Huntington and Lyrintzis [41] was used to determine the grid points (candidate solutions distribution) in the search space for the three MPC weightings within the defined search spaces in (37). A sample space of 300 unique combinations of weighting values was randomly generated in MATLAB using LHS to adequately fill the search space defined in (37). The MPC was simulated with the GT-Suite model in two different scenarios discussed in Section 3.4.5 for the SC03 cycle. The plant was operated with the MPC for each set of 300 weighting combinations to assess the system's behavior with the new control scheme.

As stated earlier, the three main metrics of interest are (i) the RMSE for the air temperature setpoint deviation, (ii) the total energy consumption throughout the cycle, and (iii) the clutch actuation frequency, each of which could be considered as the definition of $f_{GS}(\cdot)$ in (37). In a traditional approach to grid search optimization, one would minimize or maximize the value of $f_{GS}(\cdot)$ in (37) as one of the metrics outlined above. However, in this study, to provide a more insightful illustration of how the samples in the search space in (37) compare to one another against any of the above metrics, the results of the evaluation for all samples in the search space in (37) against all metrics above are shown in Figure 4. To further discuss the optimum points for this optimization approach, the points of interest (POIs) are introduced next to provide trade-offs and decision-making recommendations once the outcomes of all evaluations are obtained and visualized.

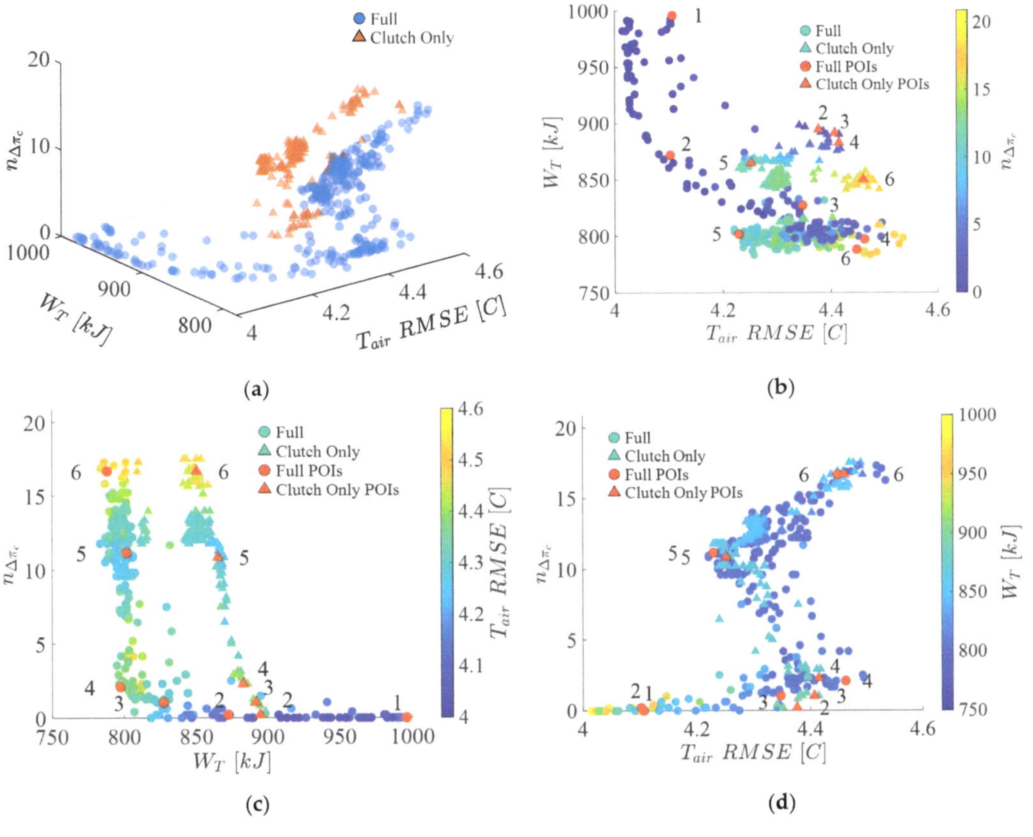

Figure 4. Evaluation of full control and clutch-only control MPCs for SC03 cycle: (**a**) 3D space with all three key metrics, (**b**) air temperature RMSE vs. total energy side profile, (**c**) total energy vs. clutch actuation frequency side profile, (**d**) air temperature RMSE vs. clutch actuation frequency side profile.

4.2. Points of Interest

First, we examine the three metrics of interest for both MPC configurations, as shown in Figure 4a. Four observations can be made here.

1. First, the upper bound of energy consumption is 1000 kJ, 21% lower than that for the baseline control scheme for the plant. Hence, all assessed MPC configurations significantly reduce A/C system energy consumption.

2. Second, in a similar way, the largest air temperature error is 4.6 °C, 27% less than that for the baseline controls. This suggests that the model being re-linearized at every operating point is key to good control performance, as the baseline controls do not include this.
3. Third, clutch actuation frequency ranges from 0 (same as baseline) to ~16 actuations per minute over the SC03 cycle, with higher clutch actuations, in general, being traded for lower energy consumption. This suggests there is no single optimal value of the QP weights but rather a trading relationship the control designer must choose within.
4. Finally, the difference between the total energy consumption using clutch-only MPC and full MPC is apparent, with clutch-only MPC yielding higher energy consumption. The impact of the inclusion of two extra actuators into the MPC framework is easier to establish by examining the same data in each of the three two-dimensional views in Figure 4b–d.

In Figure 4b, the data is projected onto the energy–temperature error plane. It is clear that including all actuators in the MPC framework lowers energy consumption for a given temperature error at a comparable or lower clutch actuation frequency. Giving the MPC the control ability over all actuators also enables higher-energy solutions with few to no clutch actuations, which is impossible with clutch-only control.

In Figure 4c, the data is projected onto the clutch actuation frequency–energy plane. The three-actuator MPC is shown to lower energy consumption by 50–100 kJ over the drive cycle at comparable numbers of clutch actuation frequency and again the higher-energy solutions with little to no clutch activation are apparent. In Figure 4d, the data is projected onto the clutch actuation frequency–temperature error plane. Here, it can be readily seen that the three-actuator MPC enables similar temperature errors and clutch actuation frequency but at reduced energy.

Some POIs are gathered in Table 2. When the clutch does not actuate in POI 1 for full control, the energy consumption is at its highest. However, the air temperature's RMSE is at its lowest. This scenario is most similar to the baseline control since the built-in compressor displacement control is the main driving force for the refrigerant control. As the compressor is active during the entire cycle, the energy consumption is high since the compressor is the primary energy consumption for the A/C system. Comparing this POI with the baseline performance, the energy consumption is reduced by 20.9% and the air temperature's RMSE is reduced by 34.6%. This demonstrates that, by just using MPC control of the compressor, fan, and AGS, there is a significant improvement to the reference temperature tracking and energy consumption.

Table 2. MPC performance comparison to baseline control's performance.

Model	Full Control			Clutch-Only Control		
	T_{eao}*RMSE* [C]	W_T [kJ]	$n_{\Delta\pi_c}$	T_{eao}*RMSE* [C]	W_T [kJ]	$n_{\Delta\pi_c}$
Baseline GT-Suite	6.27	1262.8	0	6.27	1262.8	0
POI 1	4.10	998.6	0	–	–	–
POI 2	4.10	872.1	0.20	4.38	894.7	0.20
POI 3	4.35	827.6	1.01	4.40	891.7	1.01
POI 4	4.46	797.3	2.03	4.41	882.9	2.03
POI 5	4.22	797.0	10.7	4.25	864.9	10.7
POI 6	4.44	788.2	16.2	4.46	850.4	16.2

For the clutch-only control results, the smallest clutch actuation frequency was found to be 0.2, meaning the clutch has at least one on/off actuation over the 575 s drive cycle. This can possibly be attributed to the fact that the baseline fan and AGS control inside the GT-Suite model are poor, typically running the fan when it is not needed for additional cooling and maintaining the AGS fully open regardless of the cooling requirement, causing the clutch to be more inclined to be the primary source of cooling regulation.

An interesting impact can be observed by increasing the compressor clutch actuation frequency by a small margin, from POI 1 to POI 2 and 3, which is performed by decreasing the weighting value of $K_{\Delta\pi_c}$ or increasing the weighting values of $K_{T_{air}}$ and $K_{\dot{W}_T}$. For full control, by increasing the actuation frequency from zero to 0.2, the energy consumption is decreased by 30.9% compared to the baseline. Similarly, from an actuation frequency of zero to 1.01, the energy consumption is reduced by 34.5% compared to the baseline. However, the RMSE for the air temperature is increased from 4.10 °C to 4.35 °C. POIs 5 and 6 demonstrate that, when the compressor clutch actuation frequency is above 2.03 for full control, no significant improvement to the energy consumption is achieved. However, in this region, the best air temperature's RMSE is found between a clutch actuation frequency of roughly 9.6–11.2. Comparing the POI results between the full control and clutch-only control clearly shows that, at the same clutch actuation frequency, the control of the fan and AGS provides an additional benefit to power consumption and air temperature tracking. The largest difference between the full and clutch-only control can be seen by comparing both cases at POI 3, where the full-control MPC uses 9.7% less energy compared to the clutch-only control for the same clutch actuation frequency. The weights associated with the POIs gathered in Table 2 are shown in Table 3.

Table 3. Weighting parameters for MPC simulations.

Model	Full Control			Clutch-Only Control		
	$K_{T_{air}}$	$K_{\dot{W}_T}$	$K_{\Delta\pi_c}$	$K_{T_{air}}$	$K_{\dot{W}_T}$	$K_{\Delta\pi_c}$
POI 1	1.0	35.3	956.4	–	–	–
POI 2	3.8	28.7	371.5	0.5	18.1	53.4
POI 3	7.2	35.6	837.1	0.8	20.0	223.5
POI 4	5.8	4.7	624.4	6.6	36.4	929.3
POI 5	22.4	3.1	551.4	16.4	35.4	563.3
POI 6	32.0	15.6	37.7	49.5	12.4	51.8

The system behavior for POIs 2 and 4 are shown in Figures 5 and 6, respectively, where they are both compared to the baseline system behavior. By investigating the behavior of the clutch-only MPC results shown in Figures 5 and 6, it can be immediately observed that the condenser fan rotational speed and AGS open fraction are controlled in the same manner as the baseline.

The impact of the full MPC on energy consumption can be recognized by comparing the condenser fan and AGS control to the baseline. For example, in POI 2, the full MPC is able to completely shut off the condenser fan for the majority of the cycle. In order to cause a sharp increase in air temperature, the clutch is disengaged, allowing the system to heat up quickly. However, when doing so, the full MPC controller drives the rotational speed of the condenser fan to its maximum value so as to not cause a significant overshoot in air temperature. This can be seen more often in POI 4, where the clutch actuation frequency is higher. In doing so, more energy is saved by disengaging the clutch and allowing the other two actuators to work toward satisfying the air temperature reference tracking objective. In POIs 5 and 6, the clutch actuation frequency is significantly higher than in POI 4, as seen in Table 2.

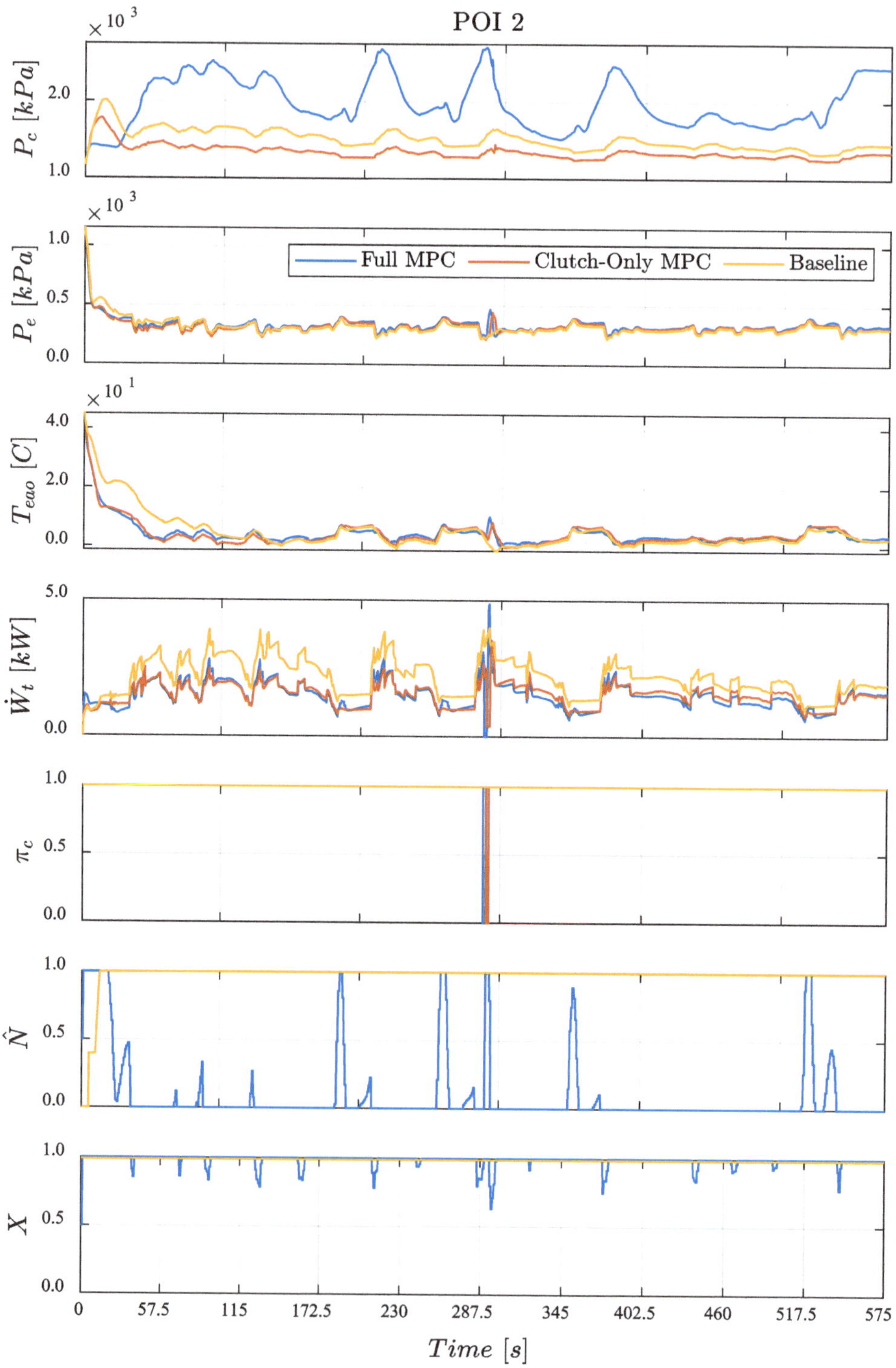

Figure 5. System behavior for baseline vs. clutch-only MPC vs. full MPC for POI 2.

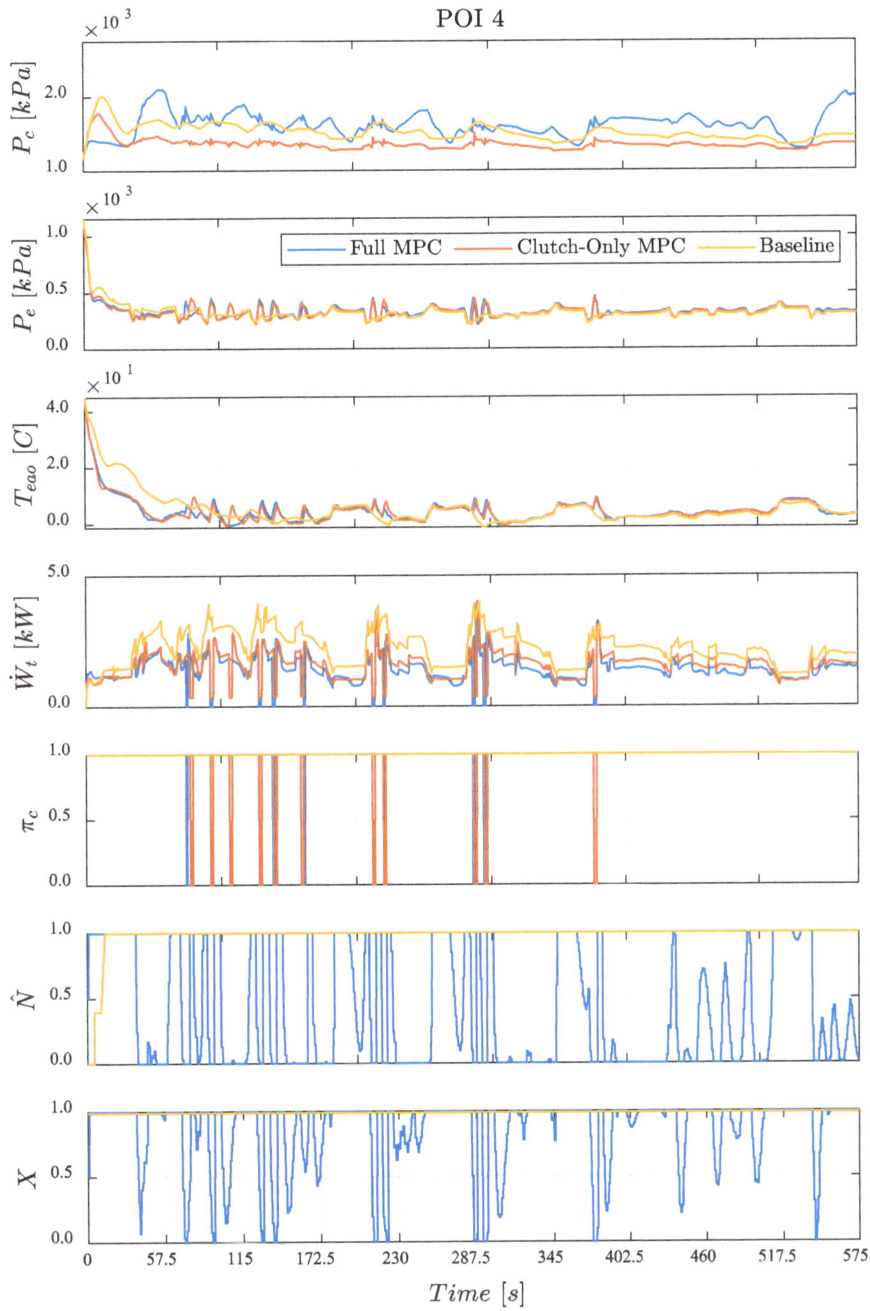

Figure 6. System behavior for baseline vs. clutch-only MPC vs. full MPC for POI 4.

The key takeaway is that there are multiple combinations of the weighting parameters which yield good performance, trading mainly between energy consumption and clutch actuation frequency. Choosing one is the designer's job, depending on the balance desired between energy use and clutch/compressor life longevity. For the vehicle of interest in this work, the fuel economy is 9.8 L/100 km when the A/C system is not active. In terms of fuel economy, the A/C system can typically use roughly 1.0 L/100 km, increasing the fuel consumption to 10.8 L/100 km [42]. In this work, the greatest observed energy savings by the full MPC is 37.6%, which would reduce the fuel economy number to 10.4 L/100 km, roughly yielding a 3.5% savings in fuel economy.

5. Conclusions and Future Work

In this paper, two MPC schemes were applied to the A/C system of a vehicle, that is, one controlling only the compressor clutch and one additionally controlling the condenser fan and AGS. The analysis showed that, by switching to MPC control, an improvement of 32.7% to the air temperature reference tracking and 37.6% to the energy consumption can be achieved. By disengaging the compressor when it is not needed, up to 37.6% in energy savings was recorded using the full-control MPC. However, achieving a steady-state solution at the air temperature setpoint of 4 °C was not achievable by controlling the fan and AGS alone, leading to a cycling behavior of the discrete compressor clutch. By increasing the actuation frequency of the compressor clutch from zero to 2.03, the energy benefit using the full-control MPC was improved from 20.9% to 36.9%. However, increasing the clutch actuation frequency further resulted in no further energy benefit. Nevertheless, the air temperature RMSE was improved by 5.4% by increasing the clutch actuation frequency from 2.03 to 10.7, at the expense of more wear and tear on the clutch itself.

From the analysis done in this work, it can be concluded that switching to an MPC scheme is worthwhile for A/C system control, even if there is no change in the components being actuated. This can be seen by the energy saving and air temperature tracking benefit of 32.6% and 32.2% provided by the clutch-only MPC. Further benefit can be obtained, especially in terms of clutch durability, by including all three actuators in the MPC scheme. The full-control MPC is able to provide a 34.5% energy benefit with a clutch actuation frequency of 1.01, whereas the clutch-only control provides a 32.7% energy benefit with a clutch actuation frequency of 16.2. The discrete number of points investigated in the Latin Hypercube Sampling did not permit formal identification of a Pareto front. However, the trades seen between the outputs strongly suggest such a front could be identified for any given vehicle; thus, optimal weightings could be chosen based on the designer's choice between clutch actuation frequency, energy consumption, and temperature error.

Author Contributions: Conceptualization, T.P., J.J.D. and A.R.; methodology, T.P., J.J.D. and A.R.; software, T.P. and A.R.; validation, T.P., J.J.D. and A.R.; formal analysis, T.P. and A.R.; investigation, T.P.; resources, J.J.D. and A.R.; data curation, T.P.; writing—original draft preparation, T.P.; writing—review and editing, J.J.D. and A.R.; visualization, T.P. and A.R.; supervision, J.J.D. and A.R.; project administration, J.J.D. and A.R.; funding acquisition, J.J.D. and A.R. All authors have read and agreed to the published version of the manuscript.

Funding: This research was funded by Mitacs Canada and Stellantis Canada, grant number IT25381.

Data Availability Statement: Data are contained within the article.

Acknowledgments: We thank Kevin Laboe, Nandan Sawkar, and Hamed Kharrati for supporting this work by providing insight, experimental data, and guidance.

Conflicts of Interest: The authors declare no conflicts of interest.

Appendix A. Successive Linearization Based MPC

Traditional MPC relies on discrete-time linear models, but most dynamic systems are nonlinear and continuous. To address this, one approach is to employ successive linearization. This involves linearizing the continuous nonlinear system at each sampling time around the current state, discretizing it, and using the resulting model to predict future values. A QP problem is then formulated to find a control sequence that minimizes a cost function. See Figure 3 for an illustration of this MPC approach based on successive linearization. The contents of this appendix are adapted from [29] for completeness of the presented work in this study.

Appendix A.1. Continuous Nonlinear System Linearization

Consider n_x, n_u, and n_o, as states, inputs, and outputs of a system, respectively resulting in state $\boldsymbol{x} \in R^{n_x}$, input $\boldsymbol{u} \in R^{n_u}$, and output $\boldsymbol{y} \in R^{n_o}$ vectors. The outputs y_i and states x_i would be expressed as $y_i = z_i(\boldsymbol{x}, \boldsymbol{u}), i = 1, .., n_o$ and $\dot{x}_i = f_i(\boldsymbol{x}, \boldsymbol{u}), i = 1, .., n_x$. The system dynamics in SSR form would be

$$\dot{\boldsymbol{x}} = F(\boldsymbol{x}, \boldsymbol{u}),\ \boldsymbol{y} = Z(\boldsymbol{x}, \boldsymbol{u}) \tag{A1}$$

Generally, the states and inputs in (A1) can be constrained as $\boldsymbol{x}_l \leq \boldsymbol{x} \leq \boldsymbol{x}_u$ with the lower and upper bounds as $\boldsymbol{x}_l, \boldsymbol{x}_u$, and $\boldsymbol{u}_l \leq \boldsymbol{u} \leq \boldsymbol{u}_u$ with lower and upper bounds as $\boldsymbol{u}_l, \boldsymbol{u}_u$, respectively. One can linearize (A1) by finding Jacobians at a set point as

$$A_c = \begin{bmatrix} \frac{\partial f_1}{\partial x_1} & \frac{\partial f_1}{\partial x_2} & \cdots & \frac{\partial f_1}{\partial x_n} \\ \frac{\partial f_2}{\partial x_1} & \frac{\partial f_2}{\partial x_2} & \cdots & \frac{\partial f_2}{\partial x_{n_x}} \\ \vdots & \vdots & \ddots & \vdots \\ \frac{\partial f_{n_x}}{\partial x_1} & \frac{\partial f_{n_x}}{\partial x_2} & \cdots & \frac{\partial f_{n_x}}{\partial x_{n_x}} \end{bmatrix} \tag{A2}$$

which can be written in compact form as $A_c(i,j) = \frac{\partial f_i}{\partial x_j}$. Following the same convention, the remaining SSR matrices can be formed as $B_c(i,j) = \frac{\partial f_i}{\partial u_j}, C_c(i,j) = \frac{\partial z_i}{\partial x_j}$, and $D_c(i,j) = \frac{\partial z_i}{\partial u_j}$. Let us denote the desired set point by $\boldsymbol{x}_{op}$, with time derivative $\dot{\boldsymbol{x}}_{op}$, and the control input $\boldsymbol{u}_{op}$. The partial derivatives in A_c, B_c, C_c, and D_c must be evaluated at $\boldsymbol{x} = \boldsymbol{x}_{\mathrm{op}}, \dot{\mathbf{x}} = \dot{\boldsymbol{x}}_{op}$, and $\boldsymbol{u} = \boldsymbol{u}_{\mathrm{op}}$. If we denote the linearized system state by $\boldsymbol{x}_L$, control by $\boldsymbol{u}_L$, and output by $\boldsymbol{y}_L$, the system evolution can be formulated as

$$\dot{\mathbf{x}}_{\mathbf{L}} = \dot{\mathbf{x}}_{\mathbf{op}} + A_c(\mathbf{x}_{\mathbf{L}} - \mathbf{x}_{\mathbf{op}}) + B_c(\mathbf{u}_{\mathbf{L}} - \mathbf{u}_{\mathbf{op}}). \tag{A3}$$

Gathering the constant terms of (A3) in a new term $\mathcal{K} = \dot{\boldsymbol{x}}_{op} - A_c \boldsymbol{x}_{op} - B_c \boldsymbol{u}_o \in R^{n_x}$, the SSR of the linearized system becomes

$$\mathbf{x}_{\mathbf{L}} = A_c \mathbf{x}_{\mathbf{L}} + B_c \mathbf{u}_{\mathbf{L}} + \mathcal{K}, \mathbf{y}_{\mathbf{L}} = C_c \mathbf{x}_{\mathbf{L}} + D_c \mathbf{u}_{\mathbf{L}}. \tag{A4}$$

Appendix A.2. Continuous Linearized System Discretization

The next step in this process is to discretize the continuous linearized system. To do so, let us multiply both sides of (A4) by $e^{-A_c t}$

$$e^{-A_c t}\mathbf{x}_{\mathbf{L}}(t) = e^{-A_c t} A_c \mathbf{x}_{\mathbf{L}}(t) + e^{-A_c t}(B_c \mathbf{u}_{\mathbf{L}}(t) + \mathcal{K}) \tag{A5}$$

and reorganize (A5) as

$$\frac{d}{dt}\left(e^{-A_c t}\mathbf{x}_{\mathbf{L}}(t)\right) = e^{-A_c t}(B_c \mathbf{u}_{\mathbf{L}}(t) + \mathcal{K}) \tag{A6}$$

The analytical solution for (A6) can be obtained as

$$\mathbf{x}_{\mathbf{L}}(t) = e^{A_c t}\mathbf{x}_{\mathbf{L}}(0) + \int_0^t e^{A_c(t-\tau)}(B_c \mathbf{u}_{\mathbf{L}}(\tau) + \mathcal{K})d\tau \tag{A7}$$

If we denote the discrete-time indices as $\boldsymbol{x}_L[k] = \boldsymbol{x}_L(kT_s)$, with the sampling time T_s, and considering $t = kT_s$ in (A7), we obtain

$$\begin{gathered} \mathbf{x}_{\mathbf{L}}[k+1] = e^{A_c T_s}\mathbf{x}_{\mathbf{L}}[k] + \\ \int_{kT_s}^{kT_s+T_s} e^{A_c(kT_s+T_s-\tau)}(B_c \mathbf{u}_{\mathbf{L}}(\tau) + \mathcal{K})d\tau \end{gathered} \tag{A8}$$

With the assumption that $u_L = u_L[k]$ does not change throughout a time step and A_c is invertible, (A8) becomes

$$\mathbf{x_L}[k+1] = e^{A_cT_s}\mathbf{x_L}[k] + A_c^{-1}\left(e^{A_cT_s} - I\right)(B_c\mathbf{u_L}[k] + \mathcal{K}) \tag{A9}$$

which gives (A9) as the exact solution of discretized (A4) with the system output as

$$\mathbf{y_L}[k] = C_c\mathbf{x_L}[k] + D_c\mathbf{u_L}[k]. \tag{A10}$$

The discrete-time SSR matrices can be defined as $A_d = e^{A_cT_s}$, $B_d = A_c^{-1}(e^{A_cT_s} - I)B_c$, $K = A_c^{-1}(e^{A_cT_s} - I)\mathcal{K}$, $C_d = C_c$, and $D_d = D_c$ to write (A9) and (A10) in a compact form. Since we are in the discrete-time domain from here on, we can denote $x_L[k]$ as x_D^k and $u_L[k]$ as u_D^k. Hence, the discrete-time linearized SSR yields

$$\mathbf{x_D^{k+1}} = A_d\mathbf{x_D^k} + B_d\mathbf{u_D^k} + \mathbb{K}, \mathbf{y_D^k} = C_d\mathbf{x_D^k} + D_d\mathbf{u_D^k}. \tag{A11}$$

Appendix A.3. Quadratic Programming Problem Derivation

In MPC formulation, we compute the system's future evolution within a finite horizon known as the prediction horizon. Predictions for a given set of control inputs are based on a discrete-time linear system model. The goal is to minimize a cost function to track a reference trajectory while adhering to constraints. At the current time step, k, the cost function typically involves the tracking error, represented by the difference between the output y_D^k and the desired reference trajectory r^k as $e^k = y_D^k - r^k$, as well as control inputs $\mathbf{u_D^k}$. The error's evolution over the prediction horizon commencing from time step k can be defined as

$$\begin{cases} \mathbf{e^k} = C_d\mathbf{x_D^k} + D_d\mathbf{u_D^k} - \mathbf{r^k} \\ \mathbf{e^{k+1}} = C_d\mathbf{x_D^{k+1}} + D_d\mathbf{u_D^{k+1}} - \mathbf{r^{k+1}} = C_dA_d\mathbf{x_D^k} + C_dB_d\mathbf{u_D^k} + C_d\mathbb{K} + D_d\mathbf{u_D^{k+1}} - \mathbf{r^{k+1}} \\ \mathbf{e^{k+2}} = C_d\mathbf{x_D^{k+2}} + D_d\mathbf{u_D^{k+2}} - \mathbf{r^{k+2}} = C_dA_d^2\mathbf{x_D^k} + C_dA_dB_d\mathbf{u_D^k} + C_dB_d\mathbf{u_D^{k+1}} + C_dA_d\mathbb{K} + C_d\mathbb{K} + D_d\mathbf{u_D^{k+2}} - \mathbf{r^{k+2}} \\ \vdots \end{cases} \tag{A12}$$

The error and control signals over the horizon, spanning n_p time steps, can be denoted as $\vec{e} \in \mathbb{R}^{n_p \cdot n_o}$ and $\vec{u} \in \mathbb{R}^{n_p \cdot n_u}$, respectively. Hence, once can write the system's behavior over the horizon as

$$\begin{aligned} \underbrace{\begin{bmatrix} \mathbf{e^k} \\ \mathbf{e^{k+1}} \\ \mathbf{e^{k+2}} \\ \vdots \\ \mathbf{e^{k+n_p-1}} \end{bmatrix}}_{\vec{\mathbf{e_k}}} &= \underbrace{\begin{bmatrix} C_d \\ C_dA_d \\ C_dA_d^2 \\ \vdots \\ C_dA_d^{n_p-1} \end{bmatrix}}_{P\in\mathbb{R}^{n_pn_o\times n_x}} \mathbf{x_D^k} + \\ &\underbrace{\begin{bmatrix} D_d & 0 & 0 & \cdots \\ C_dB_d & D_d & 0 & \cdots \\ C_dA_dB_d & C_dB_d & D_d & \cdots \\ \vdots & \vdots & \vdots & \ddots \\ C_dA_d^{n_p-2}B_d & C_dA_d^{n_p-3}B_d & C_dA_d^{n_p-4}B_d & \cdots \end{bmatrix}}_{H\in\mathbb{R}^{n_pn_o\times n_pn_u}} \cdot \\ &\underbrace{\begin{bmatrix} \mathbf{u_D^k} \\ \mathbf{u_D^{k+1}} \\ \mathbf{u_D^{k+2}} \\ \vdots \\ \mathbf{u_D^{k+n_p-1}} \end{bmatrix}}_{\vec{\mathbf{u_k}}} + \underbrace{\begin{bmatrix} 0 \\ C_d \\ C_d(I + A_d) \\ \vdots \\ C_d\left(I + \sum_{i=1}^{n_p-2} A_d^i\right) \end{bmatrix}}_{E\in\mathbb{R}^{n_pn_o\times n_x}} \mathbb{K} - \underbrace{\begin{bmatrix} \mathbf{r^k} \\ \mathbf{r^{k+1}} \\ \mathbf{r^{k+2}} \\ \vdots \\ \mathbf{r^{n_p-1}} \end{bmatrix}}_{\vec{\mathbf{r_k}}} \end{aligned} \tag{A13}$$

where P, H, and E denote the output prediction matrices, while $\mathbb{K}$ represents the constant term from (A11). The vectors $\vec{e_k}$, $\vec{u_k}$, and $\vec{r_k}$ encompass the combined error, control input, and reference vectors for the horizon commencing at time k. The concise expression for (A13) would then be as follows

$$\vec{\mathbf{e_k}} = P\mathbf{x_D^k} + H\vec{\mathbf{u_k}} + K \tag{A14}$$

where $K = E\mathbb{K} - \vec{r_k} \in R^{n_p \cdot n_o}$ collects all constant terms.

The cost function, which quantifies the controller's performance, is defined as

$$J = \frac{1}{2}\left(\vec{\mathbf{e}}_{\mathbf{k}}^T Q\vec{\mathbf{e_k}} + \vec{\mathbf{u_k}}^T R\vec{\mathbf{u_k}}\right) \tag{A15}$$

where $Q \in \mathbb{R}^{n_p n_o \times n_p n_o}$ and $R \in \mathbb{R}^{n_p n_u \times n_p n_u}$ are weight matrices for error and input, respectively. Substituting (A14) into (A15) yields

$$J = \frac{1}{2}\left(\left(Px_D^k + H\vec{u_k} + K\right)^T Q\left(Px_D^k + H\vec{u_k} + K\right) + \vec{u_k}T\vec{u_k}\right)$$

After eliminating terms unrelated to the control input $\vec{u_k}$, and combining the quadratic and linear components, we arrive at

$$J = \frac{1}{2}\vec{u}_k^T \underbrace{\left(H^T Q H + R\right)}_{G}\vec{u_k} + \underbrace{\left(x_D^{kT} P^T + K^T\right)}_{W^T} QH\vec{u_k} \tag{A16}$$

Utilizing matrices G and W, the objective function J becomes

$$J = \frac{1}{2}\vec{\mathbf{u}}_{\mathbf{k}}^T G\vec{\mathbf{u_k}} + W^T\vec{\mathbf{u_k}} \tag{A17}$$

Nevertheless, when control inputs carry a non-zero weight in the cost function, a steady-state output error may arise. In such cases, it might be more cost-effective to allow this error in J instead of raising the control inputs. A solution is to define the cost function in terms of control input increments rather than their absolute values. This approach minimizes the impact of control input values on J. This requires transforming the control vector elements as $u_D^k = u_D^{k-1} + \delta u_1, u_D^{k+1} = u_D^k + \delta u_2 = u_D^{k-1} + \delta u_1 + \delta u_2, \ldots, u_D^{k+n_p-1} = u_D^{k+n_p-2} + \delta u_{n_p} = u_D^{k-1} + \delta u_1 + \cdots + \delta u_{n_p}$ or, concisely, as

$$\underbrace{\begin{bmatrix} \mathbf{u_D^k} \\ \mathbf{u_D^{k+1}} \\ \vdots \\ \mathbf{u_D^{k+n_p-1}} \end{bmatrix}}_{\vec{\mathbf{u_k}}} = \underbrace{\begin{bmatrix} \mathbf{u_D^{k-1}} \\ \mathbf{u_D^{k-1}} \\ \vdots \\ \mathbf{u_D^{k-1}} \end{bmatrix}}_{\vec{\mathbf{u}}_{\mathbf{k-1}}} + \underbrace{\begin{bmatrix} I & 0 & \cdots \\ I & I & \cdots \\ \vdots & \vdots & \ddots \\ I & I & \cdots \end{bmatrix}}_{\Delta \in \mathbb{R}^{n_p n_u \times n_p n_u}} \underbrace{\begin{bmatrix} \delta \mathbf{u}_1 \\ \delta \mathbf{u}_2 \\ \vdots \\ \delta \mathbf{u_{n_p}} \end{bmatrix}}_{\vec{\delta \mathbf{u_k}}}$$

where Δ is a lower diagonal matrix with identity $I \in R^{n_u \times n_u}$ or a zero $0 \in R^{n_u \times n_u}$ matrix of the same size as its elements. Hence, (A14) can be rewritten as

$$\vec{\mathbf{e_k}} = P\mathbf{x_D^k} + H\left(\vec{\mathbf{u}}_{\mathbf{k}-1} + \Delta\vec{\delta\mathbf{u_k}}\right) + K. \tag{A18}$$

Since $\vec{u}_{k-1}$ is constant throughout the horizon, it can be included in K. The objective function J_Δ in δu is formulated as

$$J_\Delta = \frac{1}{2}\vec{\delta\mathbf{u_k}}_T^T G_\Delta \vec{\delta\mathbf{u_k}} + W_\Delta^T \vec{\delta\mathbf{u_k}} \tag{A19}$$

where $G_\Delta = \Delta^T G$ and $W_\Delta^T = \vec{u}_{k-1}^T G\Delta + W^T\Delta$.

A key aspect of MPC is its ability to handle prescribed constraints on both state variables and control inputs. To enforce constraints on the states, it is essential to establish the state vector's trajectory over the prediction horizon.

$$\underbrace{\begin{bmatrix} \mathbf{x_D^k} \\ \mathbf{x_D^{k+1}} \\ \mathbf{x_D^{k+2}} \\ \vdots \\ \mathbf{x_D^{k+n_p}} \end{bmatrix}}_{\vec{\mathbf{x}}_\mathbf{k}\in\mathbb{R}^{(n_p+1)n_x}} = \underbrace{\begin{bmatrix} I \\ A_d \\ A_d^2 \\ \vdots \\ A_d^{n_p} \end{bmatrix}}_{P_x\in\mathbb{R}^{(n_p+1)n_x\times n_x}} \mathbf{x_D^k} + \underbrace{\begin{bmatrix} 0 & 0 & 0 & \cdots \\ B_d & 0 & 0 & \cdots \\ A_dB_d & B_d & 0 & \cdots \\ \vdots & \vdots & \vdots & \ddots \\ A_d^{n_p-1}B_d & A_d^{n_p-2}B_d & A_d^{n_p-3}B_d & \cdots \end{bmatrix}}_{H_x\in\mathbb{R}^{(n_p+1)n_x\times n_p n_u}} \cdot \begin{bmatrix} \mathbf{u_D^k} \\ \mathbf{u_D^{k+1}} \\ \mathbf{u_D^{k+2}} \\ \vdots \\ \mathbf{u_D^{k+n_p-1}} \end{bmatrix} + \underbrace{\begin{bmatrix} 0 \\ I \\ I + A_d \\ \vdots \\ I + \sum_{i=1}^{n_p-1} A_d^i \end{bmatrix}}_{E_x\in\mathbb{R}^{(n_p+1)n_x\times n_x}} \mathbb{K} \quad \text{(A20)}$$

where $\vec{x_k}$ holds $n_p + 1$ state vectors from the time horizon, and P_x, H_x, and E_x are state prediction matrices. Hence, (A20) becomes

$$\vec{\mathbf{x_k}} = P_x\mathbf{x_D^k} + H_x\vec{\mathbf{u_k}} + E_x\mathbb{K} \quad \text{(A21)}$$

The state constraint $\mathbf{x}_1 \leq V\vec{\mathbf{x_k}} \leq \mathbf{x_u}$ can be written as $\boldsymbol{x}_1 - VP_x\boldsymbol{x}_D^k - VE_x\mathbb{K} \leq VH_x\vec{\boldsymbol{u}_k} \leq \boldsymbol{x}_u - VP_x\boldsymbol{x}_D^k - VE_x\mathbb{K} where\ V \in \mathbb{R}^{(n_p+1)n_c\times(n_p+1)n_x}$ picks n_c states with active constraints. If all states are constrained, then V is an identity matrix. With $\vec{\delta\boldsymbol{u}_k}$ and defining $U = VH_x$, the QP problem can be characterized as

$$\begin{aligned} &\underset{\vec{\delta u_k}}{\operatorname{argmin}} J = \tfrac{1}{2}\vec{\delta u_k}^T G_\Delta \vec{\delta u_k} + W_\Delta^T \vec{\delta u_k} \\ &\text{subject to } U_l \leq U\vec{\delta u_k} \leq U_u \\ &u_l \leq \vec{\delta u_k} \leq u_u \end{aligned} \quad \text{(A22)}$$

where $\boldsymbol{U}_l = \boldsymbol{x}_l - \boldsymbol{U}^k$ and $\boldsymbol{U}_u = \boldsymbol{x}_u - \boldsymbol{U}^k, \boldsymbol{U}^k = VP_x\boldsymbol{x}_D^k + VE_x\mathbb{K} + VH_x\vec{\boldsymbol{u}}_{k-1}$, while $u_l = \Delta^{-1}\left(u_l - \vec{\boldsymbol{u}}_{k-1}\right)$ and $u_u = \Delta^{-1}\left(\boldsymbol{u}_u - \vec{\boldsymbol{u}}_{k-1}\right)$. It is important to highlight that the constraints and bounds in (A22) can be either constant or time-varying.

In a typical QP solver, matrices G and W, in conjunction with constraint and bound information, are employed to determine an optimal solution. In closed-loop MPC, only the first control vector within the horizon is implemented on the system, while the subsequent control inputs are disregarded. Following the current time step, new system measurements are acquired. This iterative process, involving linearization, discretization, prediction matrix generation, and the application of the first control input, is repeated at each time step in a closed-loop manner.

Appendix B. Relaxation of MPC Formulation

The concept of using continuous variables to represent discrete ones is rooted in optimization and relaxation methods. In this approach, discrete variables are temporarily considered as continuous, simplifying the problem into a relaxation of the original. Solving this relaxed problem is more manageable. However, rounding the solution to the nearest integers does not always guarantee an optimal solution for the original problem because the relaxation can overestimate feasible solutions. Consequently, additional techniques like branch-and-bound algorithms are commonly employed to systematically explore and find optimal integer solutions.

In mathematical optimization and related fields, relaxation serves as a modeling strategy to approximate complex problems with more tractable ones. It allows for insights into the original problem. For example, in integer programming, linear programming relaxations enable rational solutions without integer constraints. In combinatorial optimization, Lagrangian relaxations address complex constraints, simplifying problem solutions. These relaxation techniques often complement branch-and-bound algorithms by helping to establish bounds.

Axehill et al.'s work [43] addresses the control of hybrid systems in Mixed Logical Dynamical (MLD) form, particularly those combining continuous and binary control signals. While binary states are not explicitly considered, this problem shares similarities with the finite alphabet control problem.

In standard linear MPC, the control problem is a QP task. In the hybrid context, binary variables are introduced, making it a Mixed Integer Quadratic Programming (MIQP) problem, often called Mixed Integer Predictive Control (MIPC). This shift from QP to MIQP transforms it into a computationally challenging non-convex problem. The contents of this appendix are adapted from [43] for completeness of the presented work in this study.

Appendix B.1. Notation

In this section, $\mathbb{S}^n$ represents symmetric matrices with n columns and $\mathbb{S}^n_{++}$ ($\mathbb{S}^n_{+}$) indicates symmetric positive (semi) definite matrices with n columns. The superscript $*$ denotes values of variables and functions at their optimum. The function $\text{diag}(\cdot)$ is defined such that, when applied to a vector, it forms a diagonal matrix with the vector's elements along the diagonal. When applied to a matrix, it creates a vector consisting of the matrix's diagonal elements. We frequently use the sets $\mathcal{T} = \{0, \ldots, N-1\}$ and $\mathcal{I} = \{1, \ldots, Nm\}$. Additional notation details can be found in [43].

Appendix B.2. Control Problem

For a system in the following form

$$\begin{cases} x(0) = x_0 \\ x(t+1) = Ax(t) + Bu(t), \forall t \in T \end{cases} \tag{A23}$$

where $A \in \mathbb{R}^{n\times n}$, $B \in \mathbb{R}^{n\times m}, x(t) \in \mathbb{R}^n$ is the state, and $x_0 \in \mathbb{R}^n$ is the initial state. We consider an MIPC problem over a prediction horizon of length N. We use an all-binary control input $u(t) \in \{0,1\}^m$ to keep derivations as simple as possible. It is worth noting that the results remain applicable even when combining real-valued and binary control signals. The objective function to minimize is given by:

$$\begin{aligned} &J_{\text{MPC}}(x(\cdot), u(\cdot)) \\ &= \tfrac{1}{2}\sum_{t=0}^{N-1} \left(\| x(t) \|^2_{Q_x} + \| u(t) \|^2_{Q_u}\right) + \tfrac{1}{2}\| x(N) \|^2_{Q_x} \end{aligned} \tag{A24}$$

where $\| v \|_Q^2 = v^T Qv, Q_x \in \mathbb{S}_+^n$ and where $Q_u \in \mathbb{S}_{++}^m$. At each time step, the system is additionally subjected to c linear inequality constraints represented as:

$$\begin{aligned} H_x x(t) + H_u u(t) + h \leq 0, \forall t \in \mathcal{T} \\ H_x x(N) + h \leq 0 \end{aligned} \tag{A25}$$

where $H_x \in \mathbb{R}^{c\times n}, H_u \in \mathbb{R}^{c\times m}$, and $h \in \mathbb{R}^c$. The MIPC problem can be expressed as an MIQP problem in two equivalent forms. In the first form, the equality constraints representing the system's dynamics are preserved, resulting in an MIQP problem given as:

$$\begin{aligned} \underset{\mathrm{x,u}}{\text{minimize}} \quad & J_{\mathrm{MIQP}_1}(\mathrm{x},\mathrm{u}) \\ \text{s.t.} \quad & [\mathrm{AB}]\left[\mathrm{x}^T\mathrm{u}^T\right]^T = b, \\ & [\mathrm{H_x H_u}]\left[\mathrm{x}^T\mathrm{u}^T\right]^T + \mathrm{h} \leq 0 \\ & \mathrm{u}_i \in \{0,1\}, \forall i \in \mathcal{I} \end{aligned} \tag{A26}$$

where $J_{\mathrm{MIQP}_1}(\mathrm{x},\mathrm{u}) = \frac{1}{2}\mathrm{x}^T\mathrm{Q_x x} + \frac{1}{2}\mathrm{u}^T\mathrm{Q_u u}$ for a suitable choice of A, B, H_x, H_u, h, b, and matrices Q_x and Q_u. The $x \in \mathbb{R}^{(N+1)n}$ and $u \in \mathbb{R}^{Nm}$ denote system states and control inputs, respectively. In the second form, the equality constraints in (A23) can be employed to remove the states as $\mathrm{x} = S_\mathrm{x} x_0 + S_\mathrm{u}\mathrm{u}$. This results, with a suitable S_x and S_u, in an MIQP problem similar to (A26) in a new form as

$$\begin{aligned} \underset{\mathrm{u}}{\text{minimize}} \quad & J_{\mathrm{MIQP}_2}(\mathrm{u}) \\ \text{s.t.} \quad & (\mathrm{H_x S_u} + \mathrm{H_u})\mathrm{u} + \mathrm{h} + \mathrm{H_x} S_\mathrm{x} x_0 \leq 0, \\ & \mathrm{u}_i \in \{0,1\}, \forall i \in \mathcal{I} \end{aligned} \tag{A27}$$

where $J_{\mathrm{MIQP}_2}(\mathrm{u}) = \frac{1}{2}\mathrm{u}^T(S_\mathrm{u}^T\mathrm{Q_x}S_\mathrm{u} + \mathrm{Q_u})\mathrm{u} + (S_\mathrm{x}x_0)^T\mathrm{Q_x}S_\mathrm{u}\mathrm{u} + \kappa$ and $\kappa = \frac{1}{2}(S_\times x_0)^T\mathrm{Q}_\times(S_\times x_0)$ is a constant. The optimal objective function values for the problems in (A26) and (A27) are equal, i.e., $J^*_{\mathrm{MPC}} = J_{\mathrm{MIQP}\ 1} = J^*_{\mathrm{MIQP}_2}$.

Appendix B.3. Relaxations

A simple method to relax the integer constraints substituting $u_i \in \{0,1\}$ with the relaxed constraint $u_i \in [0,1]$. This results in a QP problem, known as a QP relaxation. Following this approach, the QP relaxation for the problem in (A26) becomes:

$$\begin{aligned} \underset{\mathrm{x,u}}{\text{minimize}} \quad & J_{\mathrm{QP}_1}(\mathrm{x},\mathrm{u}) \\ \text{s.t.} \quad & [\mathrm{AB}]\left[\mathrm{x}^T\mathrm{u}^T\right]^T = b, \\ & [\mathrm{H_x H_u}]\left[\mathrm{x}^T\mathrm{u}^T\right]^T + \mathrm{h} \leq 0 \\ & 0 \leq \mathrm{u}_i \leq 1, \forall i \in \mathcal{I} \end{aligned} \tag{A28}$$

where $J_{\mathrm{QP}_1}(\mathrm{x},\mathrm{u}) = J_{\mathrm{MIQP}_1}(\mathrm{x},\mathrm{u})$. Hence, the QP relaxation of (A27) yields

$$\begin{aligned} \underset{\mathrm{u}}{\text{minimize}} \quad & J_{\mathrm{QP}_2}(\mathrm{u}) \\ \text{s.t.} \quad & (\mathrm{H_x S_u} + \mathrm{H_u})\mathrm{u} + \mathrm{h} + \mathrm{H_x} S_\mathrm{x} x_0 \leq 0 \\ & 0 \leq \mathrm{u}_i \leq 1, \forall i \in \mathcal{I} \end{aligned} \tag{A29}$$

where $J_{\mathrm{QP}_2}(\mathrm{u}) = J_{\mathrm{MIQP}_2}(\mathrm{u})$.

This relaxation is a semidefinite programming (SDP) problem and cannot generally form a QP. The SDP relaxation of (A26) yields

$$\begin{aligned} \underset{U,\mathrm{x},\mathrm{u}}{\text{minimize}} \quad & J_{\mathrm{SDP}_1}(U,\mathrm{x},\mathrm{u}) \\ \text{s.t.} \quad & [\mathrm{AB}]\left[\mathrm{x}^T\mathrm{u}^T\right]^T = b, \\ & [\mathrm{H_x H_u}]\left[\mathrm{x}^T\mathrm{u}^T\right]^T + \mathrm{h} \leq 0 \\ & U_{ii} = \mathrm{u}_i, \forall i \in \mathcal{I} \\ & U \succeq \mathrm{u}\mathrm{u}^T \end{aligned} \tag{A30}$$

where $J_{\mathrm{SDP}_1}(U,\mathrm{x},\mathrm{u}) = \frac{1}{2}\mathrm{x}^T\mathrm{Q_x}\mathrm{x} + \frac{1}{2}\mathrm{tr}(\mathrm{Q_u}U), U \in \mathbb{S}^{Nm}$ and where U_{ii} denotes matrix U's diagonal element i. The relaxation employed in (A30) is referred to as the equality-constrained SDP relaxation, which can equivalently be rewritten as

$$\begin{aligned} \underset{U(\cdot),x(\cdot),u(\cdot)}{\text{minimize}} \quad & J_{\mathrm{SDP}_{1'}}(U(\cdot),x(\cdot),u(\cdot)) \\ \text{s.t.} \quad & \text{(A23) and (A25)} \\ & U_{ii}(t) = u_i(t), \forall t \in \mathcal{T}, \forall i \in \{1,\ldots,m\} \\ & U(t) \succeq u(t)u(t)^T, \forall t \in \mathcal{T} \end{aligned} \tag{A31}$$

with

$$\begin{aligned} & J_{\mathrm{SDP}_{1'}}(U(\cdot),x(\cdot),u(\cdot)) \\ & = \frac{1}{2}\sum_{t=0}^{N-1}\left(x(t)^T Q_x x(t) + \mathrm{tr}(Q_u U(t))\right) \\ & + \frac{1}{2}x(N)^T Q_x x(N) \end{aligned} \tag{A32}$$

where $U \in \mathbb{S}^{Nm}$ in (A30) is substituted by $N\ U(t) \in \mathbb{S}^m$. Therefore, the number of variables (matrix elements) in (A31) increases linearly with the prediction horizon N, in contrast to the quadratic growth in (A30).

Likewise, the SDP relaxation of (A27) yields

$$\begin{aligned} \underset{U,\mathrm{u}}{\text{minimize}} \quad & J_{\mathrm{SDP}_2}(U,\mathrm{u}) \\ \text{s.t.} \quad & (\mathrm{H_x}S_\mathrm{u} + \mathrm{H_u})\mathrm{u} + \mathrm{h} + \mathrm{H_x}S_\mathrm{x}x_0 \leq 0 \\ & U_{ii} = \mathrm{u}_i, \forall i \in \mathcal{I} \\ & U \succeq \mathrm{u}\mathrm{u}^T \end{aligned} \tag{A33}$$

where $J_{\mathrm{SDP}_2}(U,\mathrm{u}) = \frac{1}{2}\mathrm{tr}((S_\mathrm{u}^T\mathrm{Q_x}S_\mathrm{u} + \mathrm{Q_u})U) + (S_\mathrm{x}x_0)^T\mathrm{Q_x}S_\mathrm{u}\mathrm{u} + \kappa$, $U \in \mathbb{S}^{Nm}$ and κ is a constant. It is important to note that the number of elements in U in (A32) increases quadratically with N. Further proof of the feasibility and quality of the relaxed solution can be found in [43].

Yang et al. [44] focused on developing real-time MPC for embedded systems, incorporating both continuous and discrete control inputs. They cast the online optimization in MPC as a MIQP and employed a branch and bound algorithm that handles multiple relaxed QPs. Due to embedded system constraints, they limited the number of relaxed problems and introduced custom heuristics to generate early feasible solutions.

For problems involving continuous variables in the relaxed states, an additional step is needed to revert the result to the feasible form of the original states, whether continuous or discrete. This process, referred to as "Rounding Heuristics", plays a crucial role. It involves determining how to round the solution, with an improperly rounded solution causing delays in the optimization process. The most naive rounding is to round obtained solution $x(t)$ to 1 $x(t) \geq 0.5$ in the relaxed solution where the range for $x(t)$ is $x \in [0,1]$ in continuous space or $x = 0\ or\ 1$ in discrete space. There is a need to reformulate the problem with appropriate weights if there are multiple discrete variables in the system's operation, as further discussed in [44].

References

1. Høgh, G.; Nielsen, R. Model Based Nonlinear Control of Refrigeration Systems. Master's Thesis, Aalborg University, Aalborg Municipality, Denmark, 2008.
2. Koo, B.; Yoo, Y.; Won, S. Super-Twisting Algorithm-Based Sliding Mode Controller for a Refrigeration System. In Proceedings of the 2012 12th International Conference on Control, Automation and Systems, Jeju, Republic of Korea, 17–21 October 2012; pp. 34–38.
3. Liu, J.; Zhou, H.; Zhou, X.; Cao, Y.; Zhao, H. Automative Air Conditioning System Control—A Survey. In Proceedings of the 2011 International Conference on Electronic & Mechanical Engineering and Information Technology, Harbin, China, 12–14 August 2011; pp. 3408–3412.
4. Cvok, I.; Škugor, B.; Deur, J. Control Trajectory Optimisation and Optimal Control of an Electric Vehicle HVAC System for Favourable Efficiency and Thermal Comfort. *Optim. Eng.* **2021**, *22*, 83–102. [CrossRef]
5. Kim, N.; Park, Y.; Son, J.E.; Shin, S.; Min, B.; Park, H.; Kang, S.; Hur, H.; Ha, M.Y.; Lee, M.C. Robust Sliding Mode Control of a Vapor Compression Cycle. *Int. J. Control Autom. Syst.* **2018**, *16*, 62–78. [CrossRef]
6. Xie, Y.; Liu, Z.; Liu, J.; Li, K.; Zhang, Y.; Wu, C.; Wang, P.; Wang, X. A Self-Learning Intelligent Passenger Vehicle Comfort Cooling System Control Strategy. *Appl. Therm. Eng.* **2020**, *166*, 114646. [CrossRef]
7. Huang, Y.; Khajepour, A.; Ding, H.; Bagheri, F.; Bahrami, M. An Energy-Saving Set-Point Optimizer with a Sliding Mode Controller for Automotive Air-Conditioning/Refrigeration Systems. *Appl. Energy* **2017**, *188*, 576–585. [CrossRef]
8. Liberati, F.; Giorgio, A.D.; Giuseppi, A.; Pietrabissa, A.; Habib, E.; Martirano, L. Joint Model Predictive Control of Electric and Heating Resources in a Smart Building. *IEEE Trans. Ind. Applicat.* **2019**, *55*, 7015–7027. [CrossRef]
9. Borreggine, S.; Monopoli, V.G.; Rizzello, G.; Naso, D.; Cupertino, F.; Consoletti, R. A Review on Model Predictive Control and Its Applications in Power Electronics. In Proceedings of the 2019 AEIT International Conference of Electrical and Electronic Technologies for Automotive (AEIT AUTOMOTIVE), Torino, Italy, 2–4 July 2019; pp. 1–6.
10. Huang, Y.; Wang, H.; Khajepour, A.; He, H.; Ji, J. Model Predictive Control Power Management Strategies for HEVs: A Review. *J. Power Sources* **2017**, *341*, 91–106. [CrossRef]
11. Wang, H.; Kolmanovsky, I.; Amini, M.R.; Sun, J. Model Predictive Climate Control of Connected and Automated Vehicles for Improved Energy Efficiency. In Proceedings of the 2018 Annual American Control Conference (ACC), Milwaukee, WI, USA, 27–29 June 2018; pp. 828–833.
12. Pereira, D.F.; Lopes, F.D.C.; Watanabe, E.H. Nonlinear Model Predictive Control for the Energy Management of Fuel Cell Hybrid Electric Vehicles in Real Time. *IEEE Trans. Ind. Electron.* **2021**, *68*, 3213–3223. [CrossRef]
13. Xie, Y.; Liu, Z.; Li, K.; Liu, J.; Zhang, Y.; Dan, D.; Wu, C.; Wang, P.; Wang, X. An Improved Intelligent Model Predictive Controller for Cooling System of Electric Vehicle. *Appl. Therm. Eng.* **2021**, *182*, 116084. [CrossRef]
14. Glos, J.; Otava, L.; Vaclavek, P. Non-Linear Model Predictive Control of Cabin Temperature and Air Quality in Fully Electric Vehicles. *IEEE Trans. Veh. Technol.* **2021**, *70*, 1216–1229. [CrossRef]
15. Kibalama, D.; Liu, Y.; Stockar, S.; Canova, M. Model Predictive Control for Automotive Climate Control Systems via Value Function Approximation. *IEEE Control Syst. Lett.* **2022**, *6*, 1820–1825. [CrossRef]
16. Göltz, S.; Sawodny, O. Design and Comparison of Model-Based Controllers for an Automotive Air Conditioning System in an Electric Vehicle. *Control Eng. Pract.* **2023**, *130*, 105376. [CrossRef]
17. El-Sharkawy, A.E.; Kamrad, J.C.; Lounsberry, T.H.; Baker, G.L.; Rahman, S.S. Evaluation of Impact of Active Grille Shutter on Vehicle Thermal Management. SAE Int. *J. Mater. Manuf.* **2011**, *4*, 1244–1254. [CrossRef]
18. Shigarkanthi, V.; Damodaran, V.; Soundararaju, D.; Kanniah, K. Application of Design of Experiments and Physics Based Approach in the Development of Aero Shutter Control Algorithm. *SAE Tech. Paper* **2011**. [CrossRef]
19. Chen, Y.; Kwak, K.H.; Kim, J.; Kim, Y.; Jung, D. Energy-efficient Cabin Climate Control of Electric Vehicles Using Linear Time-varying Model Predictive Control. *Optim. Control Appl. Methods* **2023**, *44*, 773–797. [CrossRef]
20. Coppola, A.; Lui, D.G.; Petrillo, A.; Santini, S. Eco-Driving Control Architecture for Platoons of Uncertain Heterogeneous Nonlinear Connected Autonomous Electric Vehicles. *IEEE Trans. Intell. Transport. Syst.* **2022**, *23*, 24220–24234. [CrossRef]
21. He, H.; Jia, H.; Sun, C.; Sun, F. Stochastic Model Predictive Control of Air Conditioning System for Electric Vehicles: Sensitivity Study, Comparison, and Improvement. *IEEE Trans. Ind. Inf.* **2018**, *14*, 4179–4189. [CrossRef]
22. Huang, Y.; Khajepour, A.; Bagheri, F.; Bahrami, M. Optimal Energy-Efficient Predictive Controllers in Automotive Air-Conditioning/Refrigeration Systems. *Appl. Energy* **2016**, *184*, 605–618. [CrossRef]
23. Parent, T. Developing a Compressor, Fan, and Active Grille Shutter Control Strategy for Air Conditioner Duty Cycles to Improve Overall Vehicle Power Consumption. Master's Thesis, University of Windsor, Windsor, ON, Canada, 2023.
24. Parent, T. Developing a Compressor, Fan, and Active Grille Shutter Control Strategy for Air Conditioner Duty Cycles to Improve the Overall Vehicle Power Consumption. Master's Thesis, Politecnico di Torino, Torino, Italy, 2023.
25. Zhang, Q.Z. Modeling, Energy Optimization and Control of Vapor Compression Refrigeration Systems For Automotive Applications. Ph.D. Thesis, Ohio State University, Columbus, OH, USA, 2014.
26. Dhar, A.; Bhasin, S. Indirect Adaptive MPC for Discrete-Time LTI Systems with Parametric Uncertainties. *IEEE Trans. Automat. Contr.* **2021**, *66*, 5498–5505. [CrossRef]
27. Hu, D.; Li, G.; Deng, F. Gain-Scheduled Model Predictive Control for a Commercial Vehicle Air Brake System. *Processes* **2021**, *9*, 899. [CrossRef]

28. Parent, T.; Defoe, J.J.; Rahimi, A. Nonlinear Modeling of an Automotive Air Conditioning System Considering Active Grille Shutters. *Modelling* **2023**, *4*, 70–86. [CrossRef]
29. Zhakatayev, A.; Rakhim, B.; Adiyatov, O.; Baimyshev, A.; Varol, H.A. Successive Linearization Based Model Predictive Control of Variable Stiffness Actuated Robots. In Proceedings of the 2017 IEEE International Conference on Advanced Intelligent Mechatronics (AIM), Munich, Germany, 3–7 July 2017; pp. 1774–1779.
30. Alhajeri, M.; Soroush, M. Tuning Guidelines for Model-Predictive Control. *Ind. Eng. Chem. Res.* **2020**, *59*, 4177–4191. [CrossRef]
31. Gajić, Z. *Linear Dynamic Systems and Signals*; Pearson Education: Upper Saddle River, NJ, USA, 2003; ISBN 978-0-201-61854-9.
32. Rawlings, J.B.; Mayne, D.Q. *Model Predictive Control: Theory and Design*; Nob Hill Pub: Madison, WI, USA, 2009; ISBN 978-0-9759377-0-9.
33. EPA SC03 Supplemental Federal Test Procedure (SFTP) with Air Conditioning. Available online: https://www.epa.gov/emission-standards-reference-guide/epa-sc03-supplemental-federal-test-procedure-sftp-air (accessed on 1 February 2021).
34. Dhar, A.; Bhasin, S. Adaptive MPC for Uncertain Discrete-Time LTI MIMO Systems with Incremental Input Constraints. *IFAC-PapersOnLine* **2018**, *51*, 329–334. [CrossRef]
35. De Prada, C.; Grossmann, I.; Sarabia, D.; Cristea, S. A Strategy for Predictive Control of a Mixed Continuous Batch Process. *J. Process Control* **2009**, *19*, 123–137. [CrossRef]
36. Karer, G.; Mušič, G.; Škrjanc, I.; Zupančič, B. Model Predictive Control of Nonlinear Hybrid Systems with Discrete Inputs Employing a Hybrid Fuzzy Model. *Nonlinear Anal. Hybrid Syst.* **2008**, *2*, 491–509. [CrossRef]
37. Rawlings, J.B.; Risbeck, M.J. Model Predictive Control with Discrete Actuators: Theory and Application. *Automatica* **2017**, *78*, 258–265. [CrossRef]
38. Takács, G.; Rohal'-Ilkiv, B. Stability and Feasibility of MPC. In *Model Predictive Vibration Control*; Springer: London, UK, 2012; pp. 253–285. ISBN 978-1-4471-2332-3.
39. Vazquez, S.; Marino, D.; Zafra, E.; Peña, M.D.V.; Rodríguez-Andina, J.J.; Franquelo, L.G.; Manic, M. An Artificial Intelligence Approach for Real-Time Tuning of Weighting Factors in FCS-MPC for Power Converters. *IEEE Trans. Ind. Electron.* **2022**, *69*, 11987–11998. [CrossRef]
40. Yamashita, A.S.; Martins, W.T.; Pinto, T.V.B.; Raffo, G.V.; Euzébio, T.A.M. Multiobjective Tuning Technique for MPC in Grinding Circuits. *IEEE Access* **2023**, *11*, 43041–43054. [CrossRef]
41. Huntington, D.E.; Lyrintzis, C.S. Improvements to and Limitations of Latin Hypercube Sampling. *Probabilistic Eng. Mech.* **1998**, *13*, 245–253. [CrossRef]
42. Mogro, A.E.; Huertas, J.I. Assessment of the Effect of Using Air Conditioning on the Vehicle's Real Fuel Consumption. *Int. J. Interact. Des. Manuf.* **2021**, *15*, 271–285. [CrossRef]
43. Axehill, D.; Vandenberghe, L.; Hansson, A. Convex Relaxations for Mixed Integer Predictive Control. *Automatica* **2010**, *46*, 1540–1545. [CrossRef]
44. Yang, L.; Katnik, A.; Ozay, N. Quickly Finding Recursively Feasible Solutions for MPC with Discrete Variables. In Proceedings of the 2019 IEEE Conference on Control Technology and Applications (CCTA), Hong Kong, China, 19–21 August 2019; pp. 374–381.

logistics

MDPI

Article

Examination of Logistics Simulation Demand Related to Enterprises: Focusing on a Hungarian County

Károly Szabó [1], László Szabó [1,*] and Richárd Kása [2]

[1] Department of Business Information Technology, Faculty of Finance and Accountancy, Budapest Business University, H 1149 Budapest, Hungary; szabo.karoly@uni-bge.hu
[2] Department of Management and Entrepreneurship, Faculty of Finance and Accountancy, Budapest Business University, H 1149 Budapest, Hungary; kasa.richard@uni-bge.hu
* Correspondence: szabo.laszlo4@uni-bge.hu; Tel.: +36-304449350

Abstract: *Background:* Despite the fact that some results can be found for the logistics simulation in the literature, there is a lack of an experiment based on inner company data. Therefore, the study aimed to determine the potential need for this kind of solution by Zala County enterprises. *Methods:* As a first step, the paper presents the existing literature with the help of a literature review. Afterwards, questionnaire sampling was conducted among local enterprises. The paper applies several statistical methods (e.g., descriptive statistics, SPSS, exploratory factor analysis, and structural equation modeling) to the sample for the validation of the reorganization demand. *Results:* The study obtained a total of 147 complete responses from the 1022 invitations. An overwhelming majority of the respondents indicated regular and significant delays in their logistics processes, as well as the need for a new simulation method. Based on the SEM model, it has been observed that IT solutions are being utilized in an inefficient manner, resulting in logistical system issues and operational damages. *Conclusion*: The paper successfully identified a research gap, the research of which can have not only theoretical but also many practical benefits. Hopefully, the results will generate other academic research in this field.

Keywords: logistics; Zala County; strategy; planning; simulation

Citation: Szabó, K.; Szabó, L.; Kása, R. Examination of Logistics Simulation Demand Related to Enterprises: Focusing on a Hungarian County. *Logistics* **2024**, *8*, 7. https://doi.org/10.3390/logistics8010007

Academic Editors: Tomasz Nowakowski, Artur Kierzkowski, Agnieszka A. Tubis, Franciszek Restel, Tomasz Kisiel, Anna Jodejko-Pietruczuk and Mateusz Zając

Received: 8 November 2023
Revised: 27 December 2023
Accepted: 29 December 2023
Published: 4 January 2024

1. Introduction

In contemporary times, supply chains have achieved a notable degree of efficiency. However, it is important to acknowledge that their operational effectiveness is not without limitations. In recent years, there has been an emergence of vulnerability, which has been brought to light due to a multitude of global events that have resulted in unparalleled disruptions within supply chains. The repercussions of these disruptions have been experienced in various sectors, ranging from the automotive industry to electronics. According to Pató and Herczeg [1], the source being referenced spans from pages 1 to 11.

Moreover, it is crucial to emphasize that prior studies have consistently identified a mounting difficulty in attaining punctual delivery and procurement as a result of escalating congestion in transportation networks [2]. The congestion observed in supply chains is frequently attributed to the growth of global trade and insufficient infrastructure, thereby intensifying the susceptibility of these supply chains. The issue not only results in the postponement of the transportation of unprocessed materials but also hampers the dissemination of finalized goods, thereby causing extended lead times and diminished levels of customer contentment.

Given the challenges, there is an increasing demand for supply chains that exhibit greater resilience. This may entail the integration of more comprehensive risk management strategies, the expansion of supplier diversification, or the implementation of digital technologies to enhance visibility and responsiveness. The objective is to establish supply chains that exhibit both efficiency and resilience, enabling them to endure disruptions without causing substantial disruptions to production or delivery processes. The preliminary examination of the literature clearly showed that there are certain results in the field of supply chain simulation, but so far no one has approached the simulation from an internal company point of view, which means a strong research gap this way. The main motivation of the thesis is essentially the primary analysis of this research gap with the local testing of a possible solution (a simulation based on inner company data), which possibly induces further research. The main engineering or scientific contribution of the results lies in the fact that the model created in this way can be tested in other geographical areas (either by other research teams in an open-source manner) or even further developed. Since the current paper aims to examine the necessity of this simulation (main question—is it even needed to develop a simulation based on inner company data?), the study determined the following research questions:

- How satisfied are the enterprises with their logistics processes, and how often can significant delays be experienced within the current (classical) intra-company logistics processes?
- What direct and indirect effects do the financial, market, and logistics system features of companies have on the way that company leaders view the external and internal logistics process difficulties that plague their operations?

These are the two most important questions or dimensions that have to be examined for a future reorganization implemented with a simulation tool based on inner company data. The system development and simulation results will be represented in the next research phase. In the current study, the definition of the concrete application demand and the examination of the possible reorganization possibilities were aimed at.

For the implementation of the research, Zala County was chosen (ZC in the following) in Hungary as the territory of the analysis. The choice of this area was based on its historical significance as well as its frequency in terms of logistics. However, recent investments have further enhanced its suitability for the study. These include the construction of the R76 and M70 roads, the establishment of the Zalaegerszeg test track, the planned construction of the M9 highway, and the ongoing construction of the logistics center or tank factory. These developments have significantly contributed to the county's logistics infrastructure, making it an ideal location for our research [3,4].

The structure of this paper is as follows: the next chapter conducts a literature review on basic supply chain operations and trends. Section 3 summarizes the process of hypothesis development and the research method used. The results obtained are presented and discussed in Section 4. Lastly, the concluding section highlights the theoretical and practical contributions of the study, along with its limitations and suggestions for future research.

2. Literature Review

2.1. An Overview of the Logistical Challenges Present in Contemporary Supply Chains

The efficient movement of goods and services from manufacturers to end consumers is crucially dependent on supply chains. Notwithstanding their complexity, these networks encounter various logistical challenges that may impede the smooth delivery of goods, escalate expenses, and affect customer contentment. The present study offers an all-encompassing analysis of the primary logistical obstacles encountered by supply chains in the contemporary, dynamic commercial landscape. The challenges that have been discussed pertain to diverse facets, including but not limited to fluctuations in demand, transportation and distribution, management of inventory, sharing of information, and sustainability. The objective of the analysis outlined in this paper is to augment our comprehension of these challenges and encourage additional research and the advancement of efficacious approaches to alleviate their consequences.

Demand variability poses a significant logistical challenge for supply chains, according to research. Variations in consumer demand may cause imprecise demand predictions, which can result in excessive inventory or insufficient stock. The "bullwhip effect" is a phenomenon that can disseminate throughout the supply chain, resulting in the amplification of demand fluctuations at every level [5]. It is a result of various factors, including but not limited to order batching, price fluctuations, promotional activities, and inaccuracies in forecasting. In order to address this challenge, it is imperative for supply chain managers to utilize sophisticated forecasting methodologies, establish close partnerships with customers, and execute agile replenishment tactics [6,7].

The optimization of transportation and distribution processes is of paramount importance in guaranteeing the punctual delivery of goods to consumers [8–11]. Supply chains frequently face logistical obstacles associated with transportation networks, such as congested roadways, restricted transportation capacity, and insufficient infrastructure. The challenges have the potential to result in setbacks, escalated expenditures, and subpar customer support. In addition, the emergence of electronic commerce has brought about novel intricacies, including obstacles pertaining to final delivery stages and the process of reverse logistics [12–15]. In order to tackle the aforementioned concerns, professionals in the field of supply chain management may utilize sophisticated analytical techniques and optimization models to streamline routing processes, enhance fleet management, and elevate delivery performance [16–18].

Effective inventory management is crucial for achieving a harmonious equilibrium between customer demand and supply chain expenses [19–22]. Supply chains often encounter difficulties in efficiently managing inventory levels. The maintenance of an excessive inventory can lead to the immobilization of working capital and a rise in holding costs. Conversely, inadequate inventory levels can result in stockouts and missed opportunities for sales. The challenges at hand are attributed to factors such as demand forecasting accuracy [23,24], lead time variability [25,26], supplier reliability [27–29], and seasonality [30,31]. The implementation of sophisticated inventory management strategies, such as just-in-time (JIT) and vendor-managed inventory (VMI), in conjunction with effective demand forecasting, can effectively mitigate these obstacles and optimize inventory levels [32–35].

The sharing of information plays a crucial role in the optimization of supply chain performance by facilitating coordination among its partners [36,37]. Insufficient information sharing and inadequate visibility across the supply chain frequently give rise to logistical challenges [38]. The absence of instantaneous data, isolated information systems, and apprehensions regarding cybersecurity can impede the dissemination of information, leading to procrastination, heightened response durations, and suboptimal decision-making. The utilization of integrated information systems, such as enterprise resource planning (ERP) and supply chain management (SCM) softwares, enables the smooth exchange of information, cooperative planning, and enhanced transparency [39,40]. The utilization of blockchain technology also holds the potential to significantly transform the field of supply chain management through its ability to offer enhanced transparency, heightened security measures, improved operational efficiency, and streamlined collaboration processes [41]. Nevertheless, it is crucial to acknowledge that the integration of blockchain technology within the framework of a supply chain presents certain obstacles. These include the requirement for standardization, the intricate nature of the technical aspects involved, and concerns pertaining to privacy and safeguarding of data [42].

The concept of sustainability has also gained significant importance in the domain of supply chain operations in recent times [43,44]. The logistical hurdles associated with sustainability encompass the reduction of carbon emissions, the minimization of waste, and the assurance of ethical sourcing practices [45–47].

The adherence to environmental regulations [48,49], the escalation of consumer expectations for sustainable goods [50], and the imperative to embrace circular economy principles present noteworthy obstacles for supply chains [51,52]. The challenges faced by companies can be tackled through the adoption of green logistics practices, alternative transportation modes, reverse logistics programs, and supplier collaboration for sustainable sourcing [53,54].

In summary, it can be said that an increasing pressure is present in the logistics area, but the presence of these global processes in Zala County has not been investigated so far. Accordingly, the paper seeks to find an answer to the research gap about how the global challenges are realized at the local level. How satisfied are local businesses with their logistics processes in light of the growing challenges, and what problems do they face on a daily basis?

2.2. *IT Support of Logistics Systems*

The utilization of IT is a crucial factor in augmenting the efficacy, perceptibility, and synchronization of logistics systems. Over the course of the last half-decade, noteworthy progress has been achieved in the realm of information technology solutions specifically designed for logistics, affording entities the opportunity to enhance the efficiency of their supply chain activities. This chapter offers a comprehensive survey of contemporary literature and investigations concerning IT assistance for logistics systems, with a particular emphasis on significant domains such as digital transformation, instantaneous tracking and supervision, data analysis, artificial intelligence, and blockchain innovation. The objective of this chapter is to underscore the potential advantages of IT implementation in logistics by scrutinizing the most recent advancements in this field. Additionally, this chapter endeavors to identify nascent challenges and delineate forthcoming research avenues.

Logistics information systems refer to the software and hardware components that are responsible for facilitating transportation and storage activities, as per their established definition. The objective of these systems is to provide customers with a service of superior quality. With the assistance of software, customers can obtain the appropriate goods in the correct quantity and at the optimal time while effectively managing costs and time [55]. According to the definition, the main objective of the information background supporting logistics processes is to ensure the provision of cost-effective services for production purposes. Hence, our attention will be directed towards the software component of the information technology systems. The primary objective of our research is to enhance the efficiency and effectiveness of processes by implementing software solutions, specifically through the use of decision support systems.

To illustrate the functioning of IT systems, we can provide an example wherein a closing set is provided. In this scenario, the system utilizes the given stock and future demand to estimate the quantity to be procured. Subsequently, the system generates a forecast based on this estimation. Subsequently, utilizing the estimation, it presents a decision-support proposal regarding the order, typically emphasizing the presence of a scarcity. Indeed, it is evident that the provided illustration is a reductionist portrayal that fails to consider numerous factors. The material management function in ERP systems is primarily carried out by the MRP modules. The operational structure of material requirements planning (MRP) systems is typically categorized into three primary priority domains:

- Production Plan: The central program that includes the periodical production plan(s) defined by the organization. The number of pieces to be produced is determined by the company's market forecasts and market needs.
- BOM—Bill of Materials (Installation tree)—the installation guide for the components that are required for the finished product. The BOM contains not only the order of installation of the components but also the required quantities.
- Inventory register: Shows the tracking of raw materials over time. It also includes the lead times required for material procurement as well as the production times.

In real terms, the aforementioned principles are further elaborated to the extent that material requirements planning (MRP) systems must possess the capability to accommodate orderable item sizes, commonly referred to as policy orders, as well as orders that have already been committed within the system [56,57].

Based on the existing literature, it can be argued that the fundamental operations of the systems have remained unchanged over time. Nevertheless, it is imperative to acknowledge that various modernization processes have emerged within the subject matter.

These advancements can be categorized into three distinct process lines. The three key areas of focus in this study are cloud-based solutions, customization, and artificial intelligence.

Regarding cloud-based solutions, it is noteworthy to mention that in the past, organizations predominantly relied on on-premises ERP applications and exhibited reluctance towards embracing cloud-based operations. As a result of the COVID-19 pandemic and other unforeseen circumstances, there has been a significant shift in opinion, leading to a growing preference among businesses for cloud-based enterprise resource planning (ERP) systems. This option presents a multitude of advantages, such as simplified execution, cost reduction, improved adaptability, the acquisition of novel functionalities, decreased dependence on internal IT assets, and seamless integration of supplementary capabilities [58]. The implementation of a cloud-based solution facilitated the opportunity for the subsequent development of a two-level ERP system. The utilization of cloud technology has facilitated the implementation of a Tier 1/Tier 2 ERP system, wherein the central company employs a Tier 1 full-application ERP, while subsidiaries utilize a Tier 2 ERP with additional modules integrated through cloud-based tools [59].

The second trend observed in 2023 pertains to the customization of enterprise resource planning (ERP) systems, a phenomenon closely linked to the growing adoption of artificial intelligence (AI) technologies. One notable trend that is gaining traction is the increasing prevalence of AI-driven assistive and conversational user interfaces, such as chatbots. These interfaces have the capability to comprehend user voice or text input and provide responses by leveraging customer and order data stored within the enterprise resource planning (ERP) system. Within the context of customization, it is crucial to emphasize that historically, enterprise resource planning (ERP) platforms presented challenges in terms of adapting to the unique requirements of individual businesses due to their intricate scripting languages. However, in recent times, organizations have been able to leverage cloud-based ERP platforms that are specifically designed for simplified configuration or "low-code" platforms. This further enhances the prospects of customized systems [60].

The third significant trend in the realm of enterprise resource planning (ERP) systems is the growing utilization of artificial intelligence (AI). In this current trajectory, it is evident that AI and machine learning (ML) capabilities have been incorporated into enterprise resource planning (ERP) systems primarily to function inconspicuously in order to meet the increasing need for customization and optimize various business operations. In contemporary times, there has been a notable shift in the approach of companies towards the integration of AI functionality within enterprise resource planning (ERP) systems. Previously, organizations would typically incorporate AI capabilities into specific ERP systems subsequent to their initial introduction. However, a growing trend has emerged wherein an expanding array of vendors now offer ERP softwares that inherently encompasses these integrated capabilities [61]. AI is predominantly employed in the field of data analysis. As organizations amass greater quantities of operational and customer data, they increasingly depend on AI to extract valuable business insights from this data. AI technologies have the capability to efficiently analyze extensive quantities of unstructured data, enabling the identification of patterns and the prediction of trends that would be challenging to discern through manual data analysis methods. While the analysis of ERP data has been feasible in the past, there is a growing emphasis on the ability to employ predictive analysis using these data by the year 2023, facilitated by artificial intelligence.

Software equipped with machine learning capabilities has the ability to analyze a maintenance company's machine repair data and make predictions regarding the occurrence of future failures. This can be illustrated with an example. The optimization of the maintenance schedule by the organization enables prompt repair or replacement of problematic parts. Building upon this concept, AI plays a pivotal role in the automation and enhancement of a diverse array of processes. One notable illustration can be found in the various opportunities presented by JIT systems. The utilization of artificial intelligence and machine learning techniques has the potential to enhance the efficiency of delivery and labor scheduling processes, leading to heightened productivity levels and decreased operational costs [62].

Based on the observed patterns, it can be posited that we are currently on the cusp of a novel paradigm in the realm of logistics software support or ERP systems. However, it is important to note that a significant advancement in this domain has not yet materialized. Upon reviewing the existing literature, it becomes evident that the present study aligns well with these advancements and can be effectively integrated with the ongoing AI application. In summary, it can be observed that the concrete modelling-simulation-beta testing has not yet been implemented. Consequently, the present research remains significant due to the existing gap that needs to be addressed in this particular domain [63].

The advent of digitalization has brought about a significant transformation in logistics systems, facilitating the automation and integration of diverse processes. Contemporary literature highlights the incorporation of digital technologies, including the IoT and cloud computing, within the realm of logistics operations. The utilization of IoT devices, sensors, and RFID technology enables the prompt acquisition of data, tracking of assets, and management of inventory. According to Kalkha et al. [64] and Trappey et al. [65], cloud-based platforms offer scalable infrastructure and facilitate smooth data sharing among supply chain partners, thereby promoting collaboration and improving the overall efficiency of the system.

The utilization of real-time tracking and monitoring systems has garnered considerable interest over the last half-decade. The progression in GPS technology, wireless communication, and data analytics have facilitated instantaneous monitoring of shipments, vehicles, and assets across the entire supply chain. Recent scholarly literature investigates the amalgamation of location-based services, mobile applications, and sensor networks for the purposes of inventory tracking and monitoring, transportation route optimization, and precise delivery estimations. The implementation of real-time monitoring has been found to have positive effects on supply chain responsiveness, customer service, and issue resolution [66–68].

In conclusion, the field of logistics systems has experienced noteworthy progress in IT support over the last half-decade. The potential for a revolutionary transformation of logistics operations exists through the integration of digitalization, real-time tracking and monitoring, data analytics, artificial intelligence, and blockchain technology. Successful implementation of a system is contingent upon addressing various challenges, including but not limited to data privacy and security, interoperability, and organizational change management. Subsequent investigations ought to concentrate on formulating comprehensive frameworks and assessing the effects of information technology implementation [69].

In connection with the outlined IT support, the main question is how much the IT infrastructure helps to alleviate the challenges described earlier in the paper at the local level.

Within this phenomenon, the long-term goal of the research team is to test new simulation tools as a possible solution. However, it is not possible to conduct a test without learning the current situation at the local level, the daily problems, the damages in logistics, and the development demands. As the examination of these dimensions gives the answer to that really important question, does a new simulation solution even have a right to exist? Is this long-term research goal even relevant? On the basis of the current literature, it is visible that the academic field has expanded a lot with several topics in the field, like current challenges in logistics and new trends in IT support. It is also visible that these global results have not been analyzed at the local level, which forms a research gap. Accordingly, the current study aims from one side to examine the local logistics problems, damages, and development demands. From another side, the paper tries to put the answers in a cause-and-effect framework, which helps in the more complex understanding of processes. The hypothesis development is presented in the next section on the basis of these literature experiences.

3. Hypothesis Development, Sampling and Methods

As mentioned, the paper seeks to investigate the effects caused in the supply chain. This inquiry pertains to the impact experienced by local companies as a consequence of these adverse effects, as well as the existence of any demand for reengineering. The current study was preceded by previous research [70] that aimed to define the variables to be measured and outline a research framework of the possible relations based on the responses of local companies. Through the answers and the literature review of this article, the following cause-and-effect framework was determined (See Figure 1).

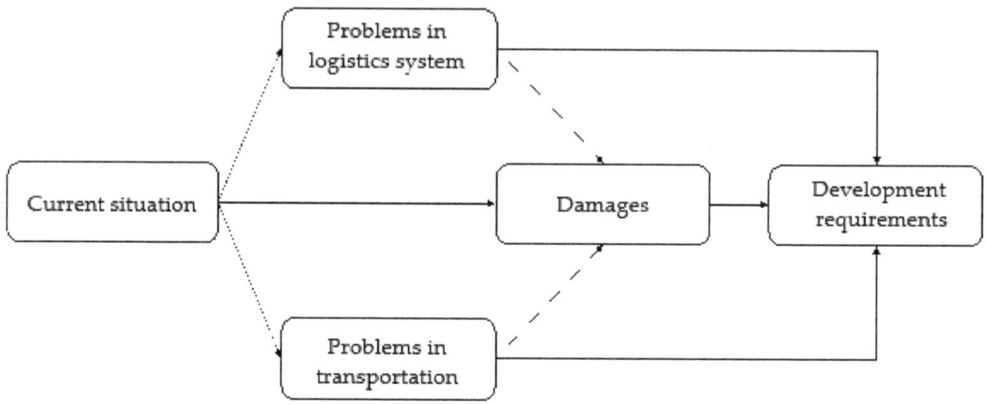

Figure 1. Research framework.

The paper aims to examine these aspects and the relationship between them. On the basis of the results of the previous study, the current understanding and adaptive review of the literature, and the implemented research framework, the following research questions and hypotheses were determined:

Q1: *How satisfied are the ZC enterprises with their logistics processes within the business operations?*

H1: *ZC enterprises are mostly dissatisfied with their logistics processes, which need reengineering.*

Q2: *How often are there significant delays in the intra-company logistics processes of ZC enterprises?*

H2: *There are regular and significant (perceptible for local companies) delays in the logistics processes of ZC enterprises.*

Q3: *What direct and indirect effects do the financial, market, and logistics system features of local companies have on the way that company leaders view the external and internal logistics process difficulties that plague their operations?*

H3: *Efficiency of local businesses' IT support and capacity directly influences their logistics processes at the system level, where low efficiency causes operative damages; to solve this problem, local companies mostly demand system reengineering solutions.*

Since the primary variables were already established through the use of qualitative methods in the prior work [70], the present study relied mostly on quantitative approaches. To back up our H1, H2, and H3 ideas, a questionnaire approach made perfect sense. For the examination of the hypotheses, an online, self-fulfilling enterprise questionnaire was developed, which will be shared through a chosen research frame. The questionnaire contains a mixture of closed (dichotomous and multiple-outcome) questions and a 5-point Likert scale. All of this was accomplished through a thorough and flexible analysis of qualitative and quantitative research techniques [70,71]. The results obtained will be analyzed with descriptive statistics in the case of H1 and H2, while for H3, a more complex method was determined. After the evaluation of several methods, the SEM model was chosen, as it gives the most broad picture of the cause and effect of a given territory. Within this, the study examines the results with the following values in the framework of an SPSS analysis: KMO (Kaiser-Meier-Olkin test) assesses the proportion of variance among variables that might be common variance; TVE (total variance explained) represents the cumulative proportion of variance explained (eigenvalues) by the retained factors; F-statistic is used in analysis of variance (ANOVA) to assess whether there are statistically significant differences among group means; the *p*-value is the probability of observing a test statistic as extreme as the one computed from the sample, assuming that the null hypothesis is true; the significance level, "Sig.," represents the chosen threshold for determining statistical significance. After establishing the methodology, our initial objective was to ascertain the particulars of the sampling, encompassing the research framework, query publication location, and methodology, among other factors. The questionnaire was organized in the following manner:

- data of the company;
- current situation of the company;
- problems of logistics processes;
- problems of logistics transportation;
- damages due to negative logistics effects;
- logistics developments.

In the vast majority of cases, the research topic also defines the target population to be selected. In this case, the "enterprise" in the title of the paper was the unit of analysis, which should be interpreted as the unit to be studied during the research. When selecting the target population, one of the most important attributes was territorial delimitation; namely, the research was conducted among ZC enterprises, the validation of which was partly explained earlier. (Number of enterprises are shown in Table 1).

Table 1. The number of registered enterprises in ZC.

Territorial Unit Name	Territorial Unit Level	Number of Registered Businesses (pcs)
Győr-Moson-Sopron	county	76,268
Vas	county	41,396
Zala	county	52,429
Western Transdanubia (Total)	region	170,093

Source: Own editing on the basis of the national statistics.

As a research frame, the Zala County Foundation for Enterprise Development database was targeted, which covers the entire entrepreneurial sphere of the county. As a database, at least 100 completion feedbacks were set as a goal to be achieved. The following filters were applied to the quantity:

- location in ZC;
- significant material flow (min. weekly order);
- having a logistics department at the operations level.

In order to evaluate the hypotheses, it is necessary to derive the factors of the model using the database variables, excluding the measured variables under consideration. In order to achieve this objective, exploratory factor analysis (EFA) was employed with Kaiser normalization and varimax rotation, as recommended in the existing literature [72–74]. Causal pathway models are constructed using structural equation modeling (SEM) based on the resulting factors from EFA [75]. In the context of structural equation modeling, researchers typically adhere to the overarching framework of the LISREL model. This entails constructing a causal model utilizing a causal multicausal regression approach and incorporating partial correlation techniques [76]. Pearson's zero-order linear correlation between the independent and dependent variables is decomposed into two additive components, with the exclusion of other factors. Consequently, the analysis involves examining the direct impact of the independent variable on the ultimate dependent variable, as well as the indirect impact of the independent variable on said variable through intermediary variables. Hence, the partial relationships observed in the model also indicate a causal relationship, which differs from the conventional regression approach. The construction of the model is as follows. In order to construct the models, an initial model was created for the entire sample through the utilization of exploratory regression analysis. Within this particular model, solely noteworthy causal relationships are identified and denoted by arrows to indicate their direction. The standardized β parameters are determined by their respective weights. Subsequently, all participants were eliminated from the dataset who had missing data for the variables of interest.

The models adhere to the structure outlined in the literature [77]. The calculations were conducted utilizing IBM AMOS software (version number–27.1.1.3). In the models, the input variables that serve as causes are connected to each other through covariances, indicating that they are not necessarily independent variables. These relationships also influence the development of the explained causal variables. The model's endogenous mediating variables are denoted by ei error terms. The standardized β parameters of the model represent the weights of the paths (links) connecting the variables. The determinant coefficients (R2) for each model represent the values assigned to the explained variables. The relationships among the models are subsequently elucidated through the utilization of the standard error (S.E.) while additionally providing the critical value (C.R.). The adequacy of the models is assessed using chi-square tests, which are deemed to be well-fitting if the *p*-value is less than 0.05. Furthermore, the average explanatory capability of the generated models was assessed by calculating the means of squared multiple correlations for the dependent variables. As an evaluation criteria, the study determined a minimum 0.200 correlation coefficient absolute value between the given variables on the basis of the literature on SEM modelling [77]. Above this absolute value, the given connection can be considered strong, and these connections formulate the path of the model to justify the H3 hypothesis.

The research was conducted in the ZC region, as mentioned. The selection of this area for the research was a rational decision based on the natural, geographical, and infrastructural resources available. Given the prevailing circumstances, ongoing development trends, and the abundance of available data, there is a significant opportunity to achieve precise and accurate outcomes in relation to the long-term objectives [78].

4. Results

The sample size for this research is denoted as n = 147, representing the number of fully completed questionnaires available for analysis. Given the nature of our research, which is focused on causal and explanatory relationships rather than providing a descriptive account of the situation in Hungary, representativeness was not prioritized in the sampling approach. Instead, a non-random sampling method was chosen with a smaller number of elements, as it aligns better with our research objectives [79,80].

4.1. Satisfaction of ZC Businesses with Their Logistics Processes

Based on the survey conducted among 147 companies, it was determined that a mere 20% of the respondents expressed complete satisfaction with their logistics processes. The majority of enterprises (67%) expressed partial satisfaction with the development of the daily material flow, whereas 13% of the enterprises reported complete dissatisfaction.

In addition to considering the satisfaction of enterprises, it is also important to investigate the perceived efficiency of logistics processes among the respondents. (For the present inquiry, we sought information regarding the primary determinants of the enterprise, such as market dynamics, financial condition, information technology and professional resources, competitive landscape, funding sources, availability of raw materials, and the efficacy of logistical operations).

As shown in Figure 2, 40% of respondents indicated that their logistics processes and operations are at an average level, 28% reported that these processes are below average, and 22% expressed that the logistics situation is exceptionally poor. In addition to assessing satisfaction levels pertaining to logistics processes and their operational efficacy, it is also pertinent to investigate the fundamental requirements for software support in addressing these challenges.

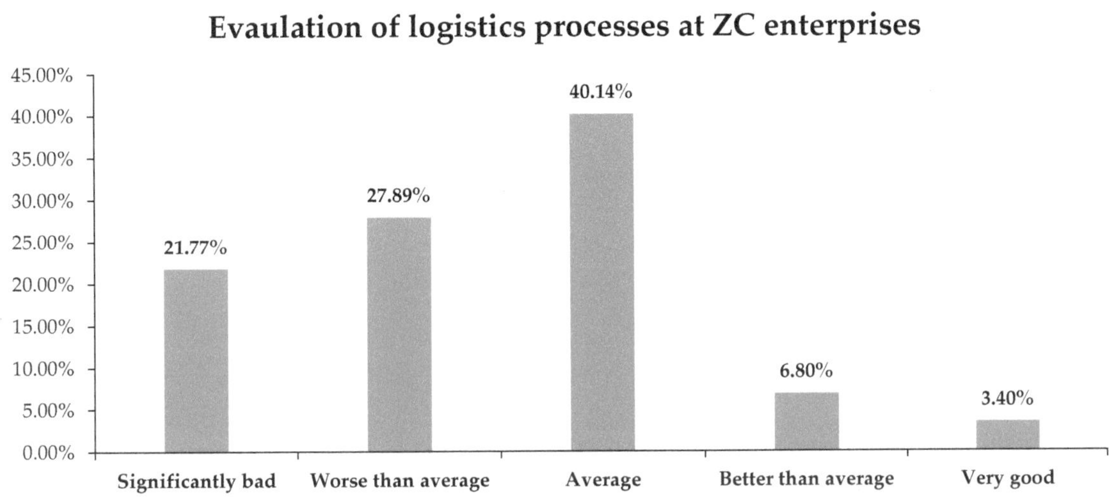

Figure 2. Evaluation of own logistics processes among local companies.

According to the survey data, a majority of 64.58% of the participants expressed their intention to utilize decision support software capable of facilitating logistics flows in order to restructure the logistics operations of the organization. This finding highlights a clear demand for a novel approach in the domain of logistics.

As a result, the results confirm the H1 hypothesis and establish it as a (T1) thesis. Specifically, 67% of the surveyed companies express partial satisfaction, while 13% report no satisfaction at all. Additionally, 49.66% perceive their logistics processes to be below average. Furthermore, 64.58% of respondents express a need for new decision support software to facilitate the reorganization of their logistics processes. Based on the findings of this study, it is appropriate to assert that a significant majority of the ZC enterprises express dissatisfaction with their current logistics processes, indicating a need for reengineering.

4.2. Problems of the Logistics Processes among ZC Enterprises

After the examination of basic demand and satisfaction, it was logical to explore the root causes of logistics problems. The question arose as to how significant these problems are at all. For this, the size and the risk of the delays perceived by businesses were measured.

As shown in Figure 3, the largest number of respondents (46.26%) indicated 1–5 times, but there were also such enterprises that indicated 75–100 cases. Overall, all of the companies indicated that there are at least 1–5 cases of delays. Based on this, it is visible that regular material flow level blockage is present in the case of local enterprises; only its significance is the question.

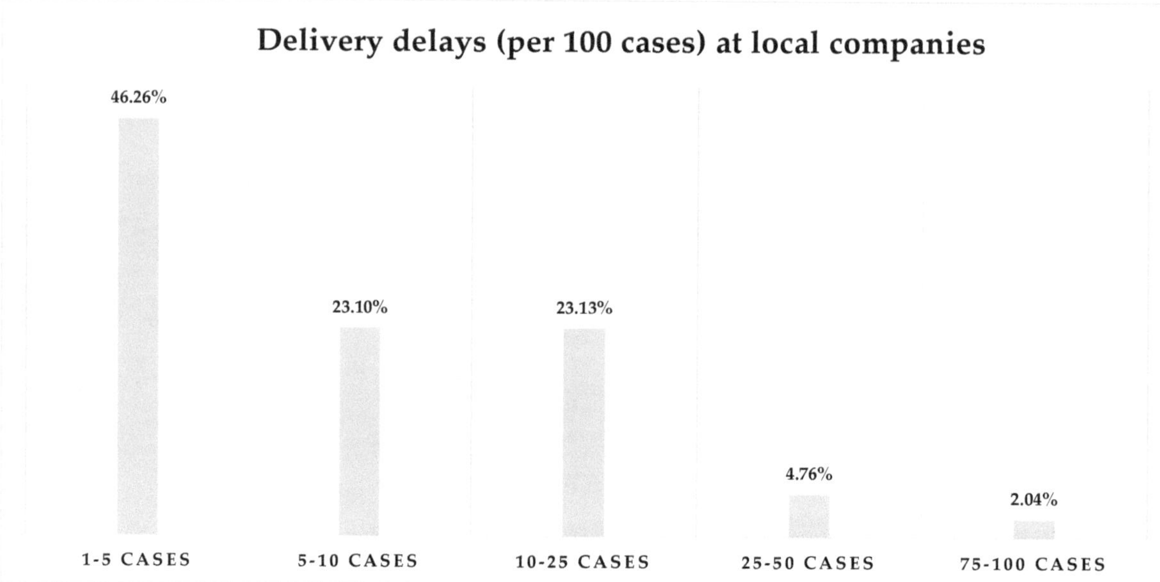

Figure 3. Frequency of delivery delays at ZC enterprises.

To assess the importance of the issue, the specific magnitude of the delays was surveyed. Within the participants, a notable proportion of 48.98% expressed that the delays encountered had a negligible impact. Conversely, the remaining 51% acknowledged that these delays exert a tangible influence on production processes and customer satisfaction. Out of the total number of enterprises, 14 of them, accounting for 9.5% of the sample, have been affected by significant delays. Based on the findings, it was observed that a significant majority of enterprises encounter frequent delays, averaging between 5 and 25 occurrences per 100 cases. Furthermore, it was determined that these delays have a notable impact in the majority of cases, accounting for 51% of instances. Consequently, the H2 hypothesis, which posits the existence of regular and significant delays in the logistics processes of businesses, was supported and subsequently transformed into a thesis (T2). It is important to highlight that these delays are caused by several problems, so the local companies could not find one definite cause of the result. Therefore, the main aim of the next section is partly to implement a cause-and-effect model to map the potential reasons.

4.3. Mapping the Logistics Processes of ZC Enterprises

In the second stage of the research, an examination to identify variables that exhibited a robust correlation was conducted, subsequently generating factors based on these variables. The database obtained through quantitative sampling was rendered more manageable through the implementation of factors. Additionally, given the successful validation of the H3 hypothesis, it became crucial to minimize the number of dimensions within the measured set of variables. Thirteen groups were created based on the correlation of observed variables and an a priori research framework. These groups are outlined in Table 2. The factors contain the answers from the questionnaire by grouping. The individual groups were combined during the SPSS analysis, according to which variables showed a strong correlation.

In the next part of the research, these groups of measurement variables were listed, then complex factors were created with the help of factor analysis, applying the principal component method, Kaiser normalization, and Varimax rotation. The obtained factors of the 5 EFA models are shown in Table 2.

Table 2. Factor analysis results of local companies' logistics processes.

Groups	Factors	No. of Measured Variables	KMO	Bartlett p	TVE	Min. Factor Load
Current situation	Economic situation	4	0.773	0.000	67.68	0.612
	IT support and capacity	2				0.817
Logistics problems	Logistics system problems	5	0.684	0.000	70.68	0.650
	Administrative problems	2				0.824
Transportation problems	External environmental problems	4	0.663	0.000	77.32	0.779
	Border crossing problems	2				0.917
	Organizational problems	2				0.705
	Problems with suppliers and labor shortages	2				0.688
Damages	Operative problems	4	0.840	0.000	74.45	0.699
	Strategic problems	4				0.633
Developments	Planning developments	4	0.703	0.000	75.14	0.767
	Market developments	3				0.641
	Search for new transportation routes	1				0.939

On the basis of the variance analysis, the most important lesson was that we must pay close attention to the following factors in order of importance for the operation of a completely optimal logistics system:

- building adequate production/service capacity—adequate infrastructure;
- supply of specialists—the existence of a sufficient number of qualified workers;
- IT supply—software support for company management is optimal.

The table presents the average disparities in the factors, categorized by the level of satisfaction that companies have with their logistical processes. The computation of significance values has been performed using ANOVA, and asterisks are used to indicate the locations where statistically significant differences exist.

Regarding the economic context, companies that expressed dissatisfaction with their logistics operations demonstrated a noteworthy negative mean (−0.665), whereas companies that reported partial satisfaction displayed a modest positive mean (0.169). This observation implies that there is potential for enhancement in this aspect. The observed difference between the groups was found to be statistically significant ($F = 6.339$, $p = 0.002$).

It is noteworthy that companies with higher levels of satisfaction with IT support and capacity exhibited a significantly positive mean value (0.564), indicating a favorable level of satisfaction. Conversely, companies with lower levels of satisfaction reported a negative mean value (−0.427), suggesting a less favorable level of satisfaction. The obtained results were found to be statistically significant ($F = 7.254$, $p = 0.001$), indicating a significant relationship between information technology (IT) and satisfaction with logistics.

The companies that expressed higher levels of satisfaction generally exhibited higher positive means, suggesting a lower occurrence of logistical problems. Conversely, the companies that expressed lower levels of satisfaction exhibited negative means, indicating a notable presence of significant problems. The observed disparities exhibited statistical significance in various areas, such as logistics system problems ($F = 29.171$, $p < 0.001$) and organizational problems ($F = 12.824$, $p < 0.001$).

Nevertheless, the levels of satisfaction did not exhibit a statistically significant disparity concerning matters pertaining to suppliers and labor shortages ($F = 0.764$, $p = 0.468$), indicating that this particular domain may pose challenges universally. The results indicate that there was a substantial and statistically significant disparity in operational issues ($F = 67.593$, $p < 0.001$). It was observed that companies with the highest level of satisfaction exhibited the most positive average (1.276), underscoring the significance of efficient operations in relation to overall satisfaction. In the 'Developments' category, companies that expressed lower levels of satisfaction with their logistics operations generally indicated positive averages, indicating a higher level of involvement in actively pursuing enhancements. The observed disparities in planning developments were found to be statistically significant (F = 15.178, $p < 0.001$).

The aforementioned findings emphasize the intricate relationship between different facets of logistics and satisfaction, serving as a strong foundation for enhancing logistics management. (Table 3 shows the concrete relationships values and results.)

Table 3. Anova results.

Groups	**Factors**	**Satisfaction with Logistics Processes**					
		Yes, Completely	**Partially, Some Processes Need Improvement**	**Not Satisfied at All**	**F**	**Sig.**	***p***
Current situation	Economic situation	−0.141	0.169	−0.665	6.339	0.002	**
	IT support and capacity	0.564	−0.083	−0.427	7.254	0.001	***
Logistics problems	Logistics system problems	0.983	−0.134	−0.804	29.171	0.000	***
	Administrative problems	0.447	−0.117	−0.070	3.759	0.026	*
	External environmental problems	0.503	−0.205	0.299	7.150	0.001	**
Transportation problems	Border crossing problems	0.239	−0.153	0.431	3.895	0.023	*
	Organizational problems	0.174	0.142	−1.005	12.824	0.000	***
	Problems with suppliers and labor shortages	0.139	0.003	−0.226	0.764	0.468	
Damages	Operative problems	1.276	−0.175	−1.037	67.593	0.000	***
	Strategic problems	0.556	−0.160	−0.016	6.156	0.003	**
Developments	Planning developments	−0.810	0.146	0.475	15.178	0.000	***
	Market developments	−0.342	0.021	0.414	3.458	0.034	*
	Search for new transportation routes	−0.442	0.008	0.633	7.209	0.001	**

***: $p < 0.001$; **: $0.001 \le p < 0.01$; *: $0.01 \le p < 0.05$.

As mentioned, the factors (e.g., economic situation) were implemented on the basis of the SPSS results. By this, the given factors contain everything valuable, connecting to the given topic, like the economic situation, the financial possibilities, the basic income, and the market situation, while the administrative problems the filing, track n' trace, and update problems. On the basis of these factors, the SEM model (with the method of structural equation modeling) of the logistics processes of ZC enterprises (see the Figure 4 below) was implemented for the examination of the H3 hypothesis.

If we study the model closer (in Figure 4), it is clearly visible that the model includes a wide range of logical cause-and-effect relationships, many of which were unknown before the research. The model has four vertical dimensions, which are the following:

- the current situation,
- the problem types,
- the caused damages
- and the needed developments for solving the problems.

The analysis reveals several discernible connections, such as the correlation between supplier and labor force issues and the prevailing economic conditions. In the event that we encounter notable challenges in the realm of suppliers and the labor force that surpass the average, it becomes apparent that our economic condition will concurrently undergo a decline. In the realm of border crossing, there exists a consistent negative correlation whereby companies experience heightened difficulties in crossing borders when faced with unfavorable economic conditions and inadequate IT support. (This outcome was also anticipated).

The model exhibited unexpected values and relationships. In this analysis, it is imperative to emphasize the correlation between IT support and external environmental issues, as well as the link between the economic situation and operational damages. Based on fundamental logic and empirical evidence, it was anticipated that a negative correlation would exist, implying that an improvement in the economic situation would result in reduced operational damages. Specifically, this expectation was particularly applicable to IT support, as it was hypothesized that higher levels of IT support would correspond to a decrease in external environmental issues. As observed in the present context, it is evident that a plausible rationale for the perceived economic advantage of larger companies lies in their greater financial resources. However, it is important to note that at this level, specifically beyond the small and medium-sized enterprise (SME) level, operational losses assume a considerably greater magnitude, if they occur. Despite the improved economic conditions, instances of operational destruction are perceived to be more significant in comparison to the majority of small and medium-sized enterprises (SMEs). This theory was partially validated through the reevaluation of the sample, however; further investigation is required to thoroughly understand the phenomenon over an extended period of time.

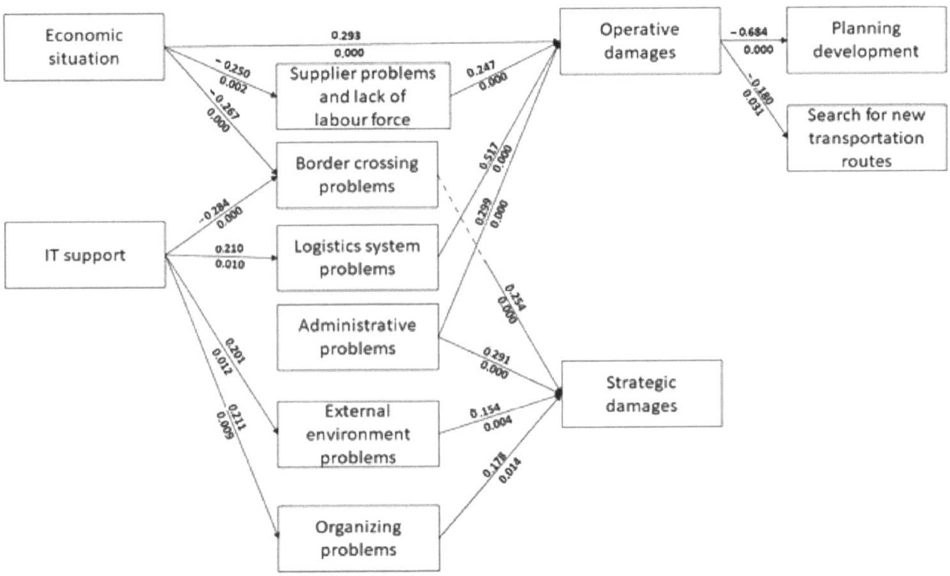

Figure 4. Structural equation model of the local companies' logistics processes.

Another aspect of the relationship pertained to the efficiency of IT support. Within the context of this section, the outcomes obtained unequivocally validated our hypotheses, as there was a robust positive correlation observed between the level of IT support and both logistics system issues and organizational challenges. Despite the availability of extensive IT support, a significant number of local companies fail to effectively utilize these resources. Consequently, despite the increased level of support, there is a corresponding increase in organizational or logistics system issues within the supply chain. This phenomenon is primarily manifested in companies characterized by intricate logistical operations yet lacking in sufficiently developed logistics software support. This practice was commonly observed in small and medium-sized enterprises (SMEs) and, to a lesser extent, in certain multinational corporations.

It is imperative to acknowledge that administrative problems constitute a distinct and autonomous factor within the model. The absence of a cause-and-effect relationship between these factors suggests that administrative problems can arise independently of the economic situation and IT support.

Additionally, it is crucial to emphasize the resultant consequences of these issues. Upon closer examination of the model, it becomes evident that this issue has both strategic and operational ramifications. The relationship between the second dimension, namely problems, and the third dimension, namely damages, exhibits a positive correlation. This implies that an increase in the magnitude of a problem is associated with a corresponding increase in the magnitude of the resulting damages. The relationship between the logistics system and operative damage is the most robust, thereby providing additional support for our hypothesis.

In terms of the dimension of damage, it is observed that four of the problems give rise to strategic damages. Anticipated challenges in the areas of organization, administration, and the external environment were expected due to the historical inclusion of these territories in strategic plans. In addition to the points, a noteworthy aspect of consideration was the issue surrounding border crossing difficulties. In terms of elucidation, it has been observed that this particular issue has exhibited a consistent recurrence within each designated timeframe. Owing to its regular manifestation, the resultant impact is perceived to be less immediate in nature, primarily lacking operational consequences but rather possessing a more profound strategic significance.

As evident from the preceding figure, it is observed that approximately fifty percent of the issues result in direct operational damages. The immediate occurrence of operational issues can be attributed to the logical consequences of both labor force shortages and supplier difficulties. Administrative issues can give rise to both operational and strategic challenges. The primary and most robust correlation identified earlier was the association between logistical system issues and operational impairments. At this stage of the study, a "relation route" was effectively identified within the model that provides support for hypothesis H3 (refer to the Figure 5 presented below).

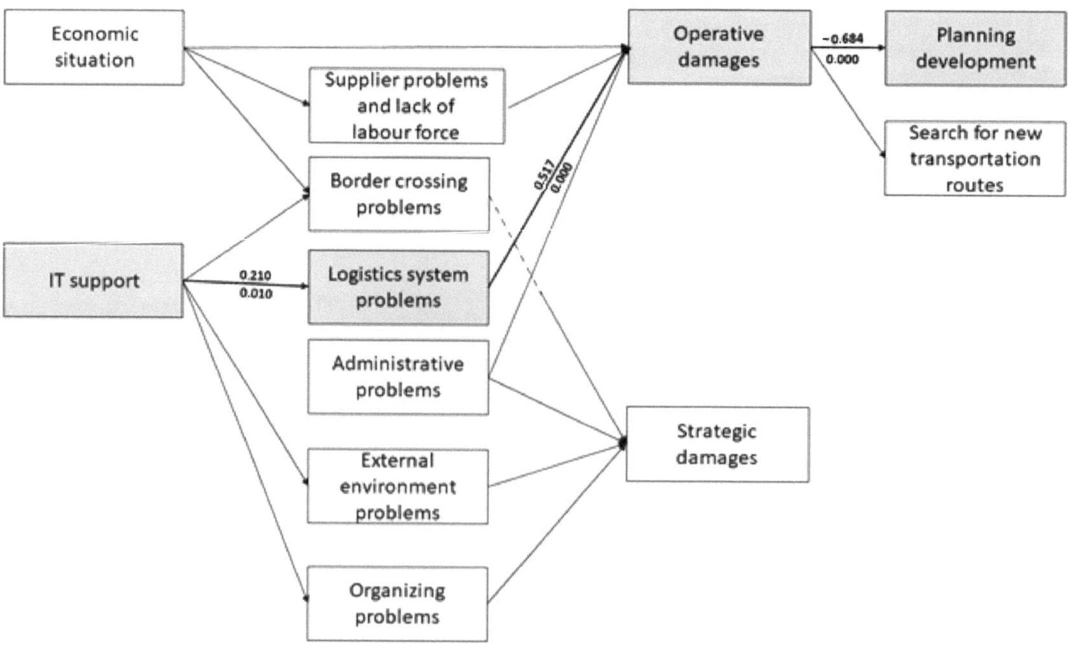

Figure 5. Relevant relations in connection with H3 hypothesis.

In order to analyze this route (shown in Figure 5 with grey colour), it can be posited that the effectiveness of IT support and capacity within local businesses has a direct impact on their logistical processes at a systemic level. It is observed that inadequate efficiency in this regard leads to operational disruptions in practice. Based on the analysis of the sample, a significant portion of the hypotheses were validated, with the exception of the inquiry regarding the demand for a solution.

In order to comprehensively analyze the subject matter, it is imperative to investigate the correlation between operational damages and the process of planning development.

Prior to conducting the research, it was hypothesized that a significant positive correlation would exist between the two variables. However, the obtained results revealed a contrary relationship. This implies that in cases where the damages incurred by the operative are substantial among the majority of respondents, there is a diminished requirement for planning development. Indeed, it is worth noting that the converse holds true as well, whereby a high demand for planning development corresponds to low damages.

Initially, we were taken aback by the direction of the relationship. However, upon conducting a thorough examination of the responses, we were able to formulate several plausible explanations for this phenomenon. One potential factor is that when a company experiences significant success, it is able to allocate greater resources towards strategic and operational planning. This leads to a paradox in which the company, despite minimizing operational costs, places considerable emphasis on planning. Simultaneously, the company actively seeks out novel approaches to enhance operational efficiency.

Indeed, the converse holds true in this scenario as well, indicating that in other types of companies, when the damages incurred are substantial, there is a lack of inclination towards seeking new avenues for development tools. Based on the responses, it is evident that these companies primarily experience operational losses due to their inability to allocate resources towards operational and strategic planning. The individuals' perception also exhibited a profound sense of despondency, which can be most accurately characterized as a morale marked by the belief that no solutions or interventions are effective. This implies that the efficiency of IT support is lacking, there are significant issues with the logistics system, and operational damages are prevalent. Consequently, under these circumstances, the need for planning becomes superfluous. Another subset of respondents in this study does not exhibit symptoms of depression. Instead, they lack knowledge about planning techniques or do not recognize or acknowledge their significance in crisis management. In both instances, the outcome is identical.

No causal relationship was observed between strategic damages and solution demand in the model. The primary explanation for this phenomenon lies in the fact that the responses provided by the companies predominantly reflect their perceptions of the presented issues, the extent of the damages incurred, and the appropriate course of action to be taken. The central concept under consideration in this instance is perception, which consequently leads to the logical inference that participants predominantly perceive practical (daily) consequences and prioritize them as their primary focus. Although the participants occupy decision-making positions, they tend to approach the questionnaire by primarily considering their daily business operations. However, this does not imply that they lack an understanding of strategic fit. Rather, they tend to have a clearer perception of these outcomes and processes when assessed through year-end or period results. A significant takeaway from this is that local companies should allocate greater resources towards the dissemination of their strategy during the implementation of routine processes.

To provide a concise overview of the model and its correlation, the local companies can be delineated based on the following characteristics:

- Companies have a given level of IT support and capacity in logistics, but they use the facilities in an inefficient way;
- This causes serious logistics problems at the system level;
- From these barriers to the everyday material flow, instant and direct operative damages are created;
- Well-prospered companies with proper planning (e.g., the PDCA approach) can handle these problems and keep the damages low.
- Despite this, these enterprises show a strong demand for planning and new approaches;
- The type of the company suffers from serious system problems and operative damages resulting from it but they do not pay attention to the planning and introduction of new tools.

Based on the findings acquired, it can be concluded that the utilization of the IT system is not optimized, resulting in logistical challenges and subsequent operational issues. The H3 hypothesis was partially supported, leading to its elevation to the status of a thesis (T3). The complete acceptance of the solution was contingent upon the confirmation of the demand, which was not definitively established. The investigation of this relationship will constitute a forthcoming undertaking within the context of our extensive research agenda.

5. Conclusions, Limitations and Future Research Direction

The primary objective of this study was to assess the fundamental demand for a new simulation tool based on internal company data. Within this, the paper examined the satisfaction levels of ZC companies with their logistics processes, determined the significance of any issues within these processes, and gained a general understanding of the utilization of IT tools, associated problems, damages, and potential solutions. The findings provide compelling evidence that there exists a notable level of dissatisfaction within the local logistics processes, while there is a strong need for new decision support software capable of facilitating the reorganization of logistics processes. As the SEM was implemented, it has been observed that the fundamental IT solutions are being utilized in an inefficient manner by the local companies, resulting in logistical system issues and operational damages. However, the solution sought for addressing this phenomenon lacks clarity, as it was anticipated that an alternative approach would be taken in this regard. The most important academic contribution of the results is that a new simulation research territory was validated on the local level. On the other hand, the current paper has some limitations. The most obvious one is the location. The research was conducted in Zala County, so a rightful question is: how do these processes develop in other areas? This can limit the generalizability of the results. Another limitation is the question of service logistics. In the study, we only dealt with hard-logistics processes, but an interesting question is: is it possible to implement the research in the field of information flow only? The third limitation is the resulting effects of logistics problems and risk management. These topics need a deeper analysis, as many of the respondents could not answer these aspects properly. The future research direction involves the expansion of the current survey on the national level as well as the acquisition of data and the implementation of sampling techniques for the simulation model (based on inner company data). After successful sampling, the next undertaking is to solve the model through simulation. It is anticipated that, over an extended period of time, the outcomes obtained will serve as the foundation for the development of a decision support system.

Author Contributions: Conceptualization, K.S. and R.K.; methodology, K.S. and R.K.; software, R.K.; validation, L.S and K.S.; formal analysis, K.S.; investigation, K.S., resources K.S. and L.S., data curation, K.S.; writing—original draft preparation, K.S.; writing—review and editing, K.S.; visualization, L.S.; supervision, L.S. All authors have read and agreed to the published version of the manuscript.

Funding: This research received no external funding.

Data Availability Statement: Data are contained within the article.

Conflicts of Interest: The authors declare no conflicts of interest.

References

1. Pató, B.S.G.; Herczeg, M. The Effect of the COVID-19 on the Automotive Supply Chains. *Stud. Univ. Babes-Bolyai Oeconomica* **2020**, *65*, 1–11. [CrossRef]
2. Gubán, M.; Kovács, G.; Kot, S. Simulation of Complex Logistical Service Processes. *Manag. Prod. Eng. Rev.* **2017**, *8*, 19–29. [CrossRef]
3. Szabó, L.; Szabó, K.; Gubán, M. Territorial examination of the logistics processes of enterprises. *Prosperitas* **2020**, *7*, 66–77. [CrossRef]
4. Szabó, K.; Szabó, L. Measurement of logistics processes for potential optimization purposes at enterprises based in a Hungarian county. *Prosperitas* **2023**, *10*, 4. [CrossRef]

5. Scukanec, A.; Rogic, K.; Babic, D. Bullwhip effect in supply chains. *Promet-Traffic Transp.* **2007**, *19*, 289–293.
6. Moon, K.L.K.; Lee, J.Y.; Lai, S.Y.C. Key drivers of an agile, collaborative fast fashion supply chain Dongdaemun fashion market. *J. Fash. Mark. Manag.* **2017**, *21*, 278–297. [CrossRef]
7. Shu, T.; Chen, S.; Lai, K.K.; Xie, C.; Wang, S.Y. A study of collaborative planning forecasting replenishment mechanism of agile virtual enterprises. In Proceedings of the 2006 IEEE International Conference on Management of Innovation and Technology, Singapore, 21–23 June 2006; Volume 1–2.
8. Arango, M.D.; Zapata, J.A. Multiobjective Model for The Simultaneous Optimization of Transportation Costs, Inventory Costs and Service Level in Goods Distribution. *IEEE Lat. Am. Trans.* **2017**, *15*, 129–136. [CrossRef]
9. Forgerini, F.L.; de Sousa, O.F. Flow optimization process in a transportation network. *Mod. Phys. Lett. B* **2021**, *35*, 2150113. [CrossRef]
10. Szentesi, S.; Illes, B.; Cservenak, A.; Skapinyecz, R.; Tamas, P. Multi-Level Optimization Process for Rationalizing the Distribution Logistics Process of Companies Selling Dietary Supplements. *Processes* **2021**, *9*, 1480. [CrossRef]
11. Tarapata, Z.; Nowicki, T.; Antkiewicz, R.; Dudzinski, J.; Janik, K. Data-Driven Machine Learning System for Optimization of Processes Supporting the Distribution of Goods Services—A case study. *Procedia Manuf.* **2020**, *44*, 60–67. [CrossRef]
12. Ekren, B.Y.; Stylos, N.; Zwiegelaar, J.; Kumar, V.; Turhanlar, E.E. Additive manufacturing integration in E-commerce supply chain network to improve resilience and competitiveness. *Simul. Model. Pract. Theory* **2023**, *122*, 102676. [CrossRef]
13. Jiang, Y.; Zhao, X.M.; Zhai, L.Y. Digital empowerment to improve the operational profitability in e-commerce supply chain. *Electron. Commer. Res. Appl.* **2023**, *58*, 101253. [CrossRef]
14. Sodhi, M.S.; Tang, C.S. Supply Chain Management for Extreme Conditions: Research Opportunities. *J. Supply Chain. Manag.* **2021**, *57*, 7–16. [CrossRef]
15. Wang, W.X.; Wang, S.Z.; Su, J.F. Integrated Production and Transportation Scheduling in E-Commerce Supply Chain with Carbon Emission Constraints. *J. Theor. Appl. Electron. Commer. Res.* **2021**, *16*, 2554–2570. [CrossRef]
16. Garcia, D.J.; You, F.Q. Supply chain design and optimization: Challenges and opportunities. *Comput. Chem. Eng.* **2015**, *81*, 153–170. [CrossRef]
17. Kassmann, D.; Allgor, R. Supply Chain Design, Management and Optimization. *Comput. Aided Chem. Eng.* **2006**, *21*, 101–106.
18. Sun, O.; Fan, N. A Review on Optimization Methods for Biomass Supply Chain: Models and Algorithms, Sustainable Issues, and Challenges and Opportunities. *Process Integr. Optim. Sustain.* **2020**, *4*, 203–226. [CrossRef]
19. Chen, Y.P.; Iyengar, G.; Wang, C. Robust Inventory Management: A Cycle-Based Approach. *M SOM-Manuf. Serv. Oper. Manag.* **2022**, *25*, 371–810. [CrossRef]
20. De Moor, B.J.; Gijsbrechts, J.; Boute, R.N. Reward shaping to improve the performance of deep reinforcement learning in perishable inventory management. *Eur. J. Oper. Res.* **2022**, *301*, 535–545. [CrossRef]
21. Khan, M.; Alshahrani, A.N.; Jacquemod, J. Digital Platforms and Supply Chain Traceability for Robust Information and Effective Inventory Management: The Mediating Role of Transparency. *Logistics* **2023**, *7*, 25. [CrossRef]
22. Panigrahi, R.R.; Mishra, P.C.; Samantaray, A.; Jena, D. Management of inventory for firms' efficiency—A study on steel manufacturing industry. *J. Adv. Manag. Res.* **2022**, *19*, 443–463. [CrossRef]
23. Nagashima, M.; Wehrle, F.T.; Kerbache, L.; Lassagne, M. Impacts of adaptive collaboration on demand forecasting accuracy of different product categories throughout the product life cycle. *Supply Chain. Manag.-Int. J.* **2015**, *20*, 415–433. [CrossRef]
24. Yani LP, E.; Aamer, A. Demand forecasting accuracy in the pharmaceutical supply chain: A machine learning approach. *Int. J. Pharm. Healthc. Mark.* **2023**, *17*, 1–23. [CrossRef]
25. Bandaly, D.; Satir, A.; Shanker, L. Impact of lead time variability in supply chain risk management. *Int. J. Prod. Econ.* **2016**, *180*, 88–100. [CrossRef]
26. Dominguez, R.; Cannella, S.; Ponte, B.; Framinan, J.M. On the dynamics of closed-loop supply chains under remanufacturing lead time variability. *Omega-Int. J. Manag. Sci.* **2020**, *97*, 102106. [CrossRef]
27. Islam, T.; Azeem, A.; Jabir, M.; Paul, A.; Paul, S.K. An inventory model for a three-stage supply chain with random capacities considering disruptions and supplier reliability. *Ann. Oper. Res.* **2022**, *315*, 1703–1728. [CrossRef] [PubMed]
28. Kamalahmadi, M.; Mellat-Parast, M. Developing a resilient supply chain through supplier flexibility and reliability assessment. *Int. J. Prod. Res.* **2016**, *54*, 302–321. [CrossRef]
29. Wang, C.; Li, M.C.; Cui, N.F. Improving reliability of assembly supply chains with disruption risk: Stakeholders' subsidy interplay. *Int. J. Prod. Res.* **2023**. [CrossRef]
30. Shukla, V.; Naim, M.M. Sensing endogenous seasonality in the case of a coffee supply chain. *Int. J. Logist.-Res. Appl.* **2018**, *21*, 279–299. [CrossRef]
31. Shukla, V.; Naim, M.M.; Thornhill, N.F. Rogue seasonality detection in supply chains. *Int. J. Prod. Econ.* **2012**, *138*, 254–272. [CrossRef]
32. Ganesan, S.; Wicaksono, H.; Valilai, O.F. Enhancing Vendor Managed Inventory with the Application of Blockchain Technology. In *Advances in System-Integrated Intelligence, SYSINT 2022*; Valle, M., Lehmhus, D., Gianoglio, C., Ragusa, E., Seminara, L., Bosse, S., Thoben, K.D., Eds.; Springer: Cham, Switzerland, 2023. [CrossRef]
33. Hong, X.P.; Wang, C.Y.; Xu, L.; Diabat, A. Multiple-vendor, multiple-retailer based vendor-managed inventory. *Ann. Oper. Res.* **2016**, *238*, 277–297. [CrossRef]

34. Lin, H.J. Investing in lead-time variability reduction in a collaborative vendor-buyer supply chain model with stochastic lead time. *Comput. Oper. Res.* **2016**, *72*, 43–49. [CrossRef]
35. Wen, C.L.; Wee, H.M.; Wu, S.M. Revisiting Lean Manufacturing Process with Vendor Managed Inventory System. In Proceedings of the 5th International Asia Conference on Industrial Engineering and Management Innovation (IEMI2014), Xi'an, China, 21–22 July 2014. [CrossRef]
36. Guan, Z.L.; Zhang, X.M.; Zhou, M.S.; Dan, Y.R. Demand information sharing in competing supply chains with manufacturer-provided service. *Int. J. Prod. Econ.* **2020**, *220*, 107450. [CrossRef]
37. Lee, T.S.; Zhou, Q. The Effect of Supply Information Sharing in a Supply Chain. In Proceedings of the Fourth International Conference on Operations and Supply Chain Management (ICOSCM 2010), Hong Kong and Guiangzhou, China, 25–31 July 2010.
38. Baah, C.; Agyeman, D.O.; Acquah, I.S.K.; Agyabeng-Mensah, Y.; Afum, E.; Issau, K.; Ofori, D.; Faibil, D. Effect of information sharing in supply chains: Understanding the roles of supply chain visibility, agility, collaboration on supply chain performance. *Benchmarking-Int. J.* **2022**, *29*, 434–455. [CrossRef]
39. Manavalan, E.; Jayakrishna, K. A review of Internet of Things (IoT) embedded sustainable supply chain for industry 4.0 requirements. *Comput. Ind. Eng.* **2019**, *127*, 925–953. [CrossRef]
40. Qureshi, M. Evaluating Enterprise Resource Planning (ERP) Implementation for Sustainable Supply Chain Management. *Sustainability* **2022**, *14*, 14779. [CrossRef]
41. Hsiao, S.J.; Sung, W.T. Blockchain-Based Supply Chain Information Sharing Mechanism. *IEEE Access* **2022**, *10*, 78875–78886. [CrossRef]
42. Xue, X.F.; Dou, J.P.; Shang, Y. Blockchain-driven supply chain decentralized operations—Information sharing perspective. *Bus. Process Manag. J.* **2021**, *27*, 184–203. [CrossRef]
43. Bratt, C.; Sroufe, R.; Broman, G. Implementing Strategic Sustainable Supply Chain Management. *Sustainability* **2021**, *13*, 8132. [CrossRef]
44. Nieuwenhuis, P.; Touboulic, A.; Matthews, L. Is Sustainable Supply Chain Management Sustainable? In *Sustainable Development Goals and Sustainable Supply Chains in the Post-Global Economy*; Yakovleva, N., Frei, R., Murthy, S.R., Eds.; Springer: Cham, Switzerland, 2019; Volume 7. [CrossRef]
45. Mridha, B.; Pareek, S.; Goswami, A.; Sarkar, B. Joint effects of production quality improvement of biofuel and carbon emissions towards a smart sustainable supply chain management. *J. Clean. Prod.* **2023**, *386*, 135629. [CrossRef]
46. Sarkar, B.; Sarkar, M.; Ganguly, B.; Cardenas-Barron, L.E. Combined effects of carbon emission and production quality improvement for fixed lifetime products in a sustainable supply chain management. *Int. J. Prod. Econ.* **2021**, *231*, 107867. [CrossRef]
47. Yadav, D.; Kumari, R.; Kumar, N.; Sarkar, B. Reduction of waste and carbon emission through the selection of items with cross-price elasticity of demand to form a sustainable supply chain with preservation technology. *J. Clean. Prod.* **2021**, *297*, 126298. [CrossRef]
48. Kalkha, H.; Khiat, A.; Bahnasse, A.; Ouajji, H. The Rising Trends of Smart E-Commerce Logistics. *IEEE Access* **2023**, *11*, 33839–33857. [CrossRef]
49. Wang, S.L.; Zhang, X. Influence of Environmental Regulation on Corporate Green Supply Chain Management: The Regulating Effect of Environmental Dynamism. *Front. Environ. Sci.* **2022**, *10*, 947022. [CrossRef]
50. Jaegler, A.; Goessling, T. Sustainability concerns in luxury supply chains: European brand strategies and French consumer expectations. *Bus. Strategy Environ.* **2020**, *29*, 2715–2733. [CrossRef]
51. Le, T.T. Linking big data, sustainable supply chain management and corporate performance: The moderating role of circular economy thinking. *Int. J. Logist. Manag.* **2023**, *34*, 744–771. [CrossRef]
52. Gkountani, V.A.; Tsoulfas, G.T.; Mouzakitis, Y. Mapping sustainability assessment methods in agri-food supply chains: A circular economy perspective. *Sci. Pap.-Ser. Manag. Econ. Eng. Agric. Rural. Dev.* **2022**, *22*, 361–368.
53. Shahparvari, S.; Soleimani, H.; Govindan, K.; Bodaghi, B.; Fard, M.T.; Jafari, H. Closing the loop: Redesigning sustainable reverse logistics network in uncertain supply chains. *Comput. Ind. Eng.* **2021**, *157*, 107093. [CrossRef]
54. Tavana, M.; Shaabani, A.; Santos-Arteaga, F.J.; Valaei, N. An integrated fuzzy sustainable supplier evaluation and selection framework for green supply chains in reverse logistics. *Environ. Sci. Pollut. Res.* **2021**, *28*, 53953–53982. [CrossRef]
55. Bikfalvi, P.; Dudás, L.; Hornyák, O.; Kulcsár, G.; Nehéz, K.; Tóth, T. *Logisztikai Informatika*; Nemzeti Tankönyvkiadó: Budapest, Hungary, 2009.
56. Goldston, J. The Evolution of ERP Systems: A Literature Review. *Int. J. Innov. Sci. Res. Technol.* **2020**, *5*, 9–17. [CrossRef]
57. Schubert, A. *Az Ellátási Lánc Információs Folyamatai/Information Processes of the Supply Chain (Műhelytanulmányok No. 85)*; Budapesti Corvinus Egyetem: Budapest, Hungary, 2007.
58. Elbahri, F.M.; Ismael Al-Sanjary, O.; Ali, M.A.M.; Ali Naif, Z.; Ibrahim, O.A.; Mohammed, M.N. Difference Comparison of SAP, Oracle, and Microsoft Solutions Based on Cloud ERP Systems: A Review. In Proceedings of the 2019 IEEE 15th International Colloquium on Signal Processing & Its Applications (CSPA), Penang, Malaysia, 8–9 March 2019; pp. 65–70. [CrossRef]
59. Liebetrau, F.; Mockel, A. Anomaly investigation of electric drive systems using big scattered test stand data on the example of a PMSM traction drive. In Proceedings of the 2021 IEEE 13th International Symposium on Diagnostics for Electrical Machines, Power Electronics and Drives (SDEMPED), Dallas, TX, USA, 22–25 August 2021; pp. 422–426. [CrossRef]
60. Monroy, C.; Domínguez-Martínez, E.; Taylor, B.; Marin, O.P.; Parise, E.; Reid, V.M. Understanding the causes and consequences of variability in infant ERP editing practices. *Dev. Psychobiol.* **2021**, *63*, e22217. [CrossRef] [PubMed]

61. Wilson, M.; Paschen, J.; Pitt, L. The circular economy meets artificial intelligence (AI): Understanding the opportunities of AI for reverse logistics. *Manag. Environ. Qual. Int. J.* **2022**, *33*, 9–25. [CrossRef]
62. Yathiraju, N. Investigating the use of an Artificial Intelligence Model in an ERP Cloud-Based System. *Int. J. Electr. Electron. Comput.* **2022**, *7*, 1–26. [CrossRef]
63. Bohács, G.; Kovács, G.; Rinkács, A. Production Logistics Simulation Supported by Process Description Languages. *Manag. Prod. Eng. Rev.* **2016**, *7*, 13–20. [CrossRef]
64. Fehér, P.; Kő, A.; Szabó, Z. Kapacitásmodellezés és az IT-architektúra-tervezés kihívásainak vizsgálata statisztikai és prediktív analitikai eszközökkel. *Statisztikai Szle.* **2016**, *94*, 1149–1164. [CrossRef]
65. Trappey, A.; Trappey, C.V.; Hsieh, A. An intelligent patent recommender adopting machine learning approach for natural language processing: A case study for smart machinery technology mining. *Technol. Forecast. Soc. Chang.* **2021**, *164*, 120511. [CrossRef]
66. Adelantado, F.; Ammouriova, M.; Herrera, E.; Juan, A.A.; Shinde, S.S.; Tarchi, D. Internet of Vehicles and Real-Time Optimization Algorithms: Concepts for Vehicle Networking in Smart Cities. *Vehicles* **2022**, *4*, 1223–1245. [CrossRef]
67. Chen, Y.T.; Sun, E.W.; Chang, M.F.; Lin, Y.B. Pragmatic real-time logistics management with traffic IoT infrastructure: Big data predictive analytics of freight travel time for Logistics 4.0. *Int. J. Prod. Econ.* **2021**, *238*, 108157. [CrossRef]
68. Fu, Y.; Ma, X.; Gao, K.; Li, Z.; Dong, H. Multi-Objective Home Health Care Routing and Scheduling with Sharing Service via a Problem-Specific Knowledge-Based Artificial Bee Colony Algorithm. *IEEE Trans. Intell. Transp. Syst.* **2023**. [CrossRef]
69. Wu, Z.; Zhou, R.; Goh, M.; Wang, Y.; Xu, Z.; Song, W. (DT4Smart) a digital twin-based modularized design approach for smart warehouses. *Int. J. Comput. Integr. Manuf.* **2023**, 1–22. [CrossRef]
70. Szabó, K.; Szabó, L.; Csanádi, Á. Strategic analysis of Zalaegerszeg Proving Ground. *J. Eng. Manag. Compet. (JEMC)* **2020**, *10*, 38–47. [CrossRef]
71. Lewin, K. Action Research and Minority Problems. In *Resolving Social Conflicts*; Harper and Row: New York, NY, USA, 1946.
72. Sajtos, L.; Mitev, A. *SPSS Kutatási és Adatelemzési Kézikönyv*; Alinea Kiadó: Budapest, Hungary, 2007.
73. Freedman, D.; Pisani, R.; Purves, R. *Statisztika*; Typotex: Budapest, Hungary, 2005.
74. Snedecor, G.W.; Cochran, W.G. *Statistical Methods*, 8th ed.; Iowa State University Press: Ames, IA, USA, 1989.
75. Füstös, L.; Kovács, E.; Meszéna, G.; Simonné Mosolygó, N. *Alakfelismerés. Sokváltozós Statisztikai Módszerek*; Új Mandátum Könyvkiadó: Budapest, Hungary, 2007.
76. In'nami, Y.; Koizumi, R. Structural Equation Modeling in Educational Research: A Primer. In *Application of Structural Equation Modeling in Educational Research and Practice*; Khine, M.S., Ed.; Sense Publishers: Rotterdam, The Netherlands, 2013; pp. 23–51.
77. Sebastian, N.; S. Nair, S.; P. Joseph, D. An overview of the pathway idea and its applications in statistical and physical sciences. *Axioms* **2015**, *4*, 530–553. [CrossRef]
78. Szabó, K.; Szabó, L. Possibilities of short supply chains in the sustainable logistics in Zala County. *J. Eng. Manag. Compet. (JEMC)* **2023**, *13*, 140–149. [CrossRef]
79. Hunyadi, L.; Mundruczó, G.; Vita, L. *Statisztika*; Aula Kiadó: Budapest, Hungary, 2000.
80. Mosteller, F.; Tukey, J.W. *Data Analisys and Regression*; Addison-Wesley: Reading, MA, USA, 1977.

applied sciences

MDPI

Article

Study on the Susceptibility of Drifting Snow in Ya'an–Qamdo Section of the Railway in Southwest China

Xue Zhou [1], Zhen Zhang [1,2,3], Weidong Yang [1,2,3] and Qingkuan Liu [1,2,3,*]

1 School of Civil Engineering, Shijiazhuang Tiedao University, Shijiazhuang 050043, China; stillwaterszx@163.com (X.Z.); zhangz@stdu.edu.cn (Z.Z.); yangwd@stdu.edu.cn (W.Y.)
2 State Key Laboratory of Mechanical Behavior and System Safety of Traffic Engineering Structures, Shijiazhuang Tiedao University, Shijiazhuang 050043, China
3 Innovation Center for Wind Engineering and Wind Energy Technology of Hebei Province, Shijiazhuang 050043, China
* Correspondence: qk@stdu.edu.cn

Abstract: To investigate the susceptibility of drifting snow along the Ya'an–Qamdo section of the railway, which is located in a high-altitude and cold plateau in Southwest China with scarce meteorological information, the Weather Research and Forecasting Model (WRF) is used in this paper to simulate the spatio-temporal distribution of meteorological data. According to the varying terrain, the railway section from Ya'an to Qamdo is divided into two regions along 100.8° E for double-layer nested simulation. The original land use data of the WRF model are used in region 1. Due to the increased number of mountains in region 2, the original data are replaced by the MCD12Q1v006 land use data, and the vertical direction layers are densified near the ground to increase simulation accuracy. The simulated results are compared with the observation data. It is found that after densification, the results have been significantly improved. The results obtained by the WRF model can accurately simulate the change trends of temperature, rainfall, and wind speed, and the correlation coefficients are relatively high, which verifies the accuracy of WRF for simulating complex terrain regions. The simulation results further indicate that approximately 300 km of the Ya'an–Qamdo railway may experience drifting snow. Among them, no drifting snow events occur in Ya'an County, and the areas with higher probability are located at the border between Luding County and Tianquan County, followed by Kangding area. The remaining areas have a probability of less than 10%. The WRF model demonstrates its capability in the drifting snow protection of railways with limited meteorological data.

Keywords: drifting snow; southwest railway; WRF; susceptibility; numerical simulation

Citation: Zhou, X.; Zhang, Z.; Yang, W.; Liu, Q. Study on the Susceptibility of Drifting Snow in Ya'an–Qamdo Section of the Railway in Southwest China. *Appl. Sci.* **2024**, *14*, 475. https://doi.org/10.3390/app14020475

Academic Editors: Artur Kierzkowski, Tomasz Kisiel, Agnieszka A. Tubis, Franciszek Restel, Mateusz Zając, Tomasz Nowakowski and Anna Jodejko-Pietruczu

Received: 18 November 2023
Revised: 28 December 2023
Accepted: 2 January 2024
Published: 5 January 2024

1. Introduction

Drifting snow is a theoretical study that integrates geography, meteorology, fluid mechanics, glaciology, and other disciplines. It is common in winter and spring seasons, and mostly occurs in Xinjiang, Inner Mongolia, Tibet, and Northeast China. It can cause snow accumulation on roads and block traffic. In order to reduce the harm of drifting snow, many scholars have studied its physical properties and prevention measures.

In terms of the fundamental physical properties of drifting snow, the two-phase flow theory proposed by Bagnold [1] provides a theoretical foundation for the study of wind and snow. Owen [2] obtained the calculation formula for the height of the saltation layer through a theoretical analysis of snow particle movement and field measurements of drifting snow. Kind [3] derived equations for calculating the concentration profile of snow particles by incorporating the theory of turbulent diffusion. Budd [4] analyzed the probability distribution of the equivalent particle size of snow particles in drifting snow and found that both gamma distribution and log-normal distribution could fit the effective particle size distribution of snow particles well. Schmidt [5] proposed a critical wind speed formula based on average snow particle size using field measurement data. Kind et al. [6] discovered that using high-density particles to simulate the salting process of snow particles can better fulfill the dynamic similarity requirements compared to using low-density particles. This is because high-density particles exhibit closer resemblance to actual snow particles in terms of snow pile shape and transport rate. These observations and empirical models provide valuable references for selecting key parameters in numerical simulations and wind tunnel experiments.

In terms of drifting snow control measures, Flaga et al. [7] conducted wind tunnel experiments to study the snow particle deposition of three different forms of large sports stadium roofs and found that the shape parameters of buildings have an important influence on the design of snow loads at different locations. Therefore, shape factors cannot be ignored in building load design. Qiang et al. [8] used artificial snow particles to simulate snowfall and snow drifting on flat roofs when they occur simultaneously in a low-temperature wind tunnel. They observed that the development of snow transport before reaching a saturated state followed a consistent pattern, regardless of whether there was concurrent snowfall or not. However, in cases where snowfall occurred concurrently with drifting, uneroded regions could be observed on the roof. Liu et al. [9] proposed equipment named "snow-wind combined experimental facility" to investigate snowdrifts around buildings and snow loads on building roofs; the experimental results prove the feasibility of this method, and the method can also be adopted on various shapes of building roofs. The wind tunnel tests using simulated material or moving snow into a low-temperature wind tunnel are not capable of accurately replicating the genuine properties and conditions of snow particles. Therefore, Li et al. [10] studied the snow accumulation in railway cuttings using a movable direct current low-temperature wind tunnel set up in the field. By adjusting the wind tunnel to simulate natural flow fields, they compared their experimental results with the detection results to verify the rationality of wind tunnel design and the accuracy of the experimental results. Due to the high cost of building low-temperature wind tunnels and the need for a large amount of complex preparation work to be carried out before experiments, some scholars have begun to use numerical simulations to study them as computing power improves. Previous researchers used computational fluid dynamics (CFDs) to study the effect of fence height, porosity, bottom clearance, and inclination angle on the snow accumulation around snow fences and their ability to prevent snow, which helps optimize their design [11–13]. Ma et al. [14] examined the effect of different forms of windbreaks on typical roadbed snow accumulation from aspects such as flow field, snow phase concentration, and sedimentation length. The results showed that the inclination angle of the windbreak was a key parameter affecting its effectiveness. Forward-leaning guide plates were more effective in preventing drifting snow during windy and snowy weather, and the angle between guide plates and the vertical axis should be controlled at around 45°.

Meteorological factors such as temperature, precipitation, wind speed, and wind direction have an important impact on the occurrence of drifting snow. The threshold wind-speed of dry snow is generally proportional to the temperature, with lower temperatures resulting in lower starting wind speeds [15]. However, when the temperature drops below −27.27 °C, the snow particles start to deteriorate, leading to an increase in the required starting wind speed [16]. Temperature is also a critical parameter affecting the shape of snow particles; −5 °C is the critical temperature at which snow crystals transition from plate-like to columnar [17]. Snow particles of different shapes undergo distinct development processes in a snowstorm flow. The number of suspended snow particles and the single-width snow transport rate of spherical, ellipsoidal, and cylindrical snow particles are two orders of magnitude higher than those of star-shaped and hexagonal prism-shaped snow particles [18]. At a height of 10 m, a wind speed of 3–8 m/s is sufficient to cause loose snow particles to jump and move [19]. When the angle between the line and the wind direction is greater than 55°, the most snow resistance is generated, and the closer the angle is to 90°, the greater the probability of snow resistance and the more severe the consequences [20]. These meteorological factors are interconnected and have various impacts on drifting snow.

The above research mainly focuses on the occurrence and aftermath of drifting snow. However, evaluating the susceptibility before its occurrence serves as the foundation for further research. Railway lines should avoid areas with high susceptibility or set different protection measures according to different occurrence probabilities along the way to save economic costs and improve protection efficiency. Most railways pass through high-altitude and cold regions with complex terrain and scarce meteorological information, making it difficult to judge the susceptibility of drifting snow along the railway line. The Weather Research and Forecasting Model (WRF) has been widely applied in regional simulations [21–23] and can partially address this issue. The model can decompose the simulated region into multiple grids with an accuracy of several kilometers. By providing boundary conditions, a series of atmospheric motion equations, including Newton's second law, the first law of thermodynamics, the continuity equation, and the state equation, can be solved to achieve weather forecasting. Therefore, the WRF model can provide meteorological data with a high resolution and continuous long time series at any location in the simulation area. Yang [24] analyzed the wind speed and wind direction characteristics during a storm and snow disaster in the Mayitas traffic corridor using the WRF dataset. In a separate study, Qi [25] employed WRF to investigate route selection schemes for the Keta Railway. Additionally, Luo et al. [26] utilized WRF to analyze snowfall, wind speed, and direction along the entire Keta Railway line and derived the probability of drifting snow occurrence for the entire route. Currently, the application of WRF is limited to the study of drifting snow on the Keta Railway, and there is a need to extend this research to include the study of drifting snow on the southwest railway, which holds significant importance.

Regarding the railway in Southwest China, Ma et al. [27] studied the distribution characteristics of extreme summer precipitation and the variation mechanism of total precipitation of the railway using daily rainfall observations and ERA5 reanalysis data in summer. The results showed that the extreme summer precipitation in these areas accounted for 30% of the total summer precipitation, and the extreme precipitation in the central and western regions was lower than that in the east, but with higher precipitation frequency. Due to the scarcity of observational data, some scholars use satellite data to study the temperature and precipitation changes in the region where the railway is located [28,29]. The WRF model can provide a more detailed spatiotemporal distribution of meteorological information. Gao et al. [30] evaluated the WRF model by examining the interannual variation, spatial structure, and 33-year time variation trend of surface temperature and precipitation. They concluded that the WRF model is capable of accurately analyzing significant weather phenomena on the Qinghai-Tibet Plateau. Chen et al. [31], Minder et al. [32], and Wrzesien et al. [33] have all verified the feasibility of using the WRF model to simulate precipitation in a complex topography area, and many scholars have used the WRF model to calculate southwest China [34–36]. Therefore, the above studies have demonstrated that applying the WRF model to the southwest region has credibility and can partially compensate for the lack of meteorological data.

Existing studies on Southwest China based on the WRF model mainly focus on climate, but there are few applications in terms of engineering in Southwest China. Due to the complex terrain and wide elevation difference in Southwest China, it is very difficult to conduct an engineering field investigation. Therefore, the application of the WRF model to combine meteorology and engineering can produce a large achievement with a low cost. In order to provide an example for the study of the susceptibility of drifting snow in areas lacking measured meteorological information, this paper takes the railway section from Ya'an to Qamdo in Southwest China as an object and studies its drifting snow susceptibility based on the WRF results. This research method can provide reference for places with the same terrain and lack of meteorological data and render guidance for railway drifting snow protection in southwest China. The structure of the remainder of this present study is as follows: Section 2 describes the geographical environment of a region which southwest railway passes through, the configuration of the WRF model, and the verification of the simulation results. Section 3 discusses the calculated results of temperature, precipitation, wind speed, wind direction, and the susceptibility of drifting snow on the railway. The conclusion of this study is finally summarized in Section 4.

2. Data and Methodology

2.1. Study Area

The southwest railway, as shown in Figure 1, is situated on the margin of Qinghai-Tibet Plateau and crosses various geomorphic units, such as a high mountain and gorge area, western Sichuan high mountain and plain area, and the core area of the Hengduan Mountains high mountain and gorge. The section starts from Ya'an and passes through Kangding, Litang, Dege, and other places before reaching Qamdo. The length of the railway is about 700 km, and the altitude varies greatly along the way, as shown in Figure 2 (the elevation data used come from https://www.earthdata.nasa.gov/ (accessed on 2 June 2023)). The complex terrain results in climatic features along the railway, and the spatial distribution of meteorological elements is significantly different.

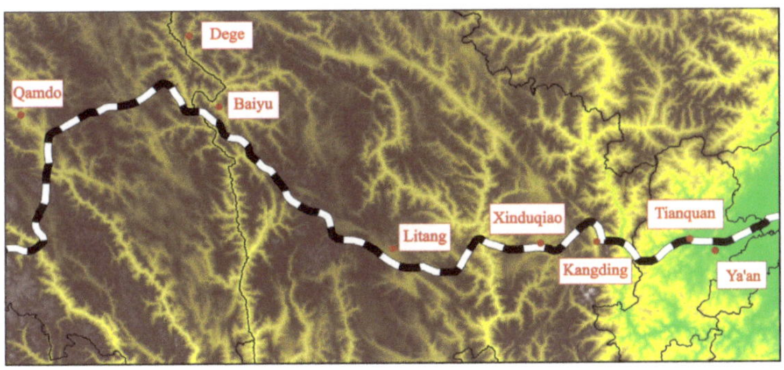

Figure 1. The railway of Ya'an–Qamdo section.

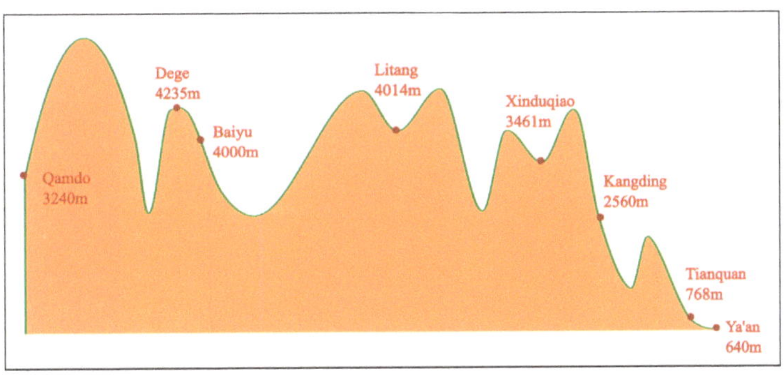

Figure 2. Elevation of Ya'an–Qamdo section.

Ya'an has a subtropical monsoon humid climate, with clear seasons, hot and rainy summers, and mild and humid winters. Tianquan area also has a subtropical monsoon climate, with warm winters and hot summers, and rain and heat occur simultaneously. Kangding to Xinduqiao area is divided into two parts by Zheduoshan Mountain. The eastern part has a warm climate with clear seasons, while the western part has cold and dry winters. Xinduqiao to Baiyu section is influenced by southwest monsoons and has cool and low-temperature summers and cold and dry winters. In Litang, Dege, and other places, the temperature is low and the annual precipitation is high and mainly in solid form.

2.2. WRF Setup and Experiment

2.2.1. Equations of WRF

In this paper, WRF (ARW) version 4.1.2 is employed, and a terrain following vertical coordinates is adopted, which is expressed as

$$\eta = \frac{p_{\mathrm{h}} - p_{\mathrm{ht}}}{\mu}, \tag{1}$$

where p_{h} is the static part of the air pressure, $\mu = p_{\mathrm{hs}} - p_{\mathrm{ht}}$, and p_{hs} and p_{ht} are the air pressure at the model surface and upper boundary, respectively.

Considering the horizontal pressure gradient error, the new variables are defined as the deviation from the reference state, and the atmospheric motion variables are expressed as follows

$$\left.\begin{array}{l} p = \overline{p}(\overline{z}) + p' \\ \phi = \overline{\phi}(\overline{z}) + \phi' \\ \alpha = \overline{\alpha}(\overline{z}) + \alpha' \\ \mu_{\mathrm{d}} = \overline{\mu}_{\mathrm{d}}(x,y) + \mu'_{\mathrm{d}} \end{array}\right\}, \tag{2}$$

where p is the corrected air pressure, ϕ is the geopotential height, $\phi = gz$, g is the gravitational acceleration, and z is the height of the position. α is the derivative of density, which is a non-conservative variable, and is a function of the vertical profile (x, y, η), $\mu_d = p_{\mathrm{dhs}} - p_{\mathrm{dht}}$, d represents dry air, and $p', \phi', \alpha', \mu'_{\mathrm{d}}$ are the original values of those variables without considering the horizontal pressure gradient error.

In the calculation process, the perturbation form of momentum equations of the WRF model can be written as:

$$\begin{array}{l} \frac{\partial U}{\partial t} + m_x\left(\frac{\partial Uu}{\partial x} + \frac{\partial Vu}{\partial y}\right) + \frac{\partial \Omega u}{\partial \eta} + \left(\mu_d \alpha \frac{\partial p'}{\partial x} + \mu_d \alpha' \frac{\partial \overline{p}}{\partial x}\right) + \\ (\alpha/\alpha_d)\left(\mu_d \frac{\partial \phi'}{\partial x} + \frac{\partial p'}{\partial \eta}\frac{\partial \phi}{\partial x} - \mu'_d \frac{\partial \phi}{\partial x}\right) = F_U, \end{array} \tag{3}$$

$$\begin{array}{l} \frac{\partial V}{\partial t} + m_y\left(\frac{\partial Uv}{\partial x} + \frac{\partial Vv}{\partial y}\right) + (m_y/m_x)\frac{\partial \Omega v}{\partial \eta} + \left(\mu_d \alpha \frac{\partial p'}{\partial y} + \mu_d \alpha' \frac{\partial \overline{p}}{\partial y}\right) + \\ (\alpha/\alpha_d)\left(\mu_d \frac{\partial \phi'}{\partial y} + \frac{\partial p'}{\partial \eta}\frac{\partial \phi}{\partial y} - \mu'_d \frac{\partial \phi}{\partial y}\right) = F_V, \end{array} \tag{4}$$

$$\begin{array}{l} \frac{\partial W}{\partial t} + (m_x m_y/m_y)\left(\frac{\partial Uw}{\partial x} + \frac{\partial Vw}{\partial y}\right) + \frac{\partial \Omega w}{\partial \eta} \\ -\frac{g}{m_y}(\alpha/\alpha_d)\left(\frac{\partial p'}{\partial \eta} - \overline{\mu}_d(q_v + q_c + q_r)\right) + \frac{g}{m_y}\mu'_d = F_w, \end{array} \tag{5}$$

And the potential temperature equation can be expressed as:

$$\frac{\partial \Theta}{\partial t} + m_x m_y\left(\frac{\partial U\theta}{\partial x} + \frac{\partial V\theta}{\partial y}\right) + m_y \frac{\partial \Omega \theta}{\partial \eta} = F_\Theta, \tag{6}$$

where U, V, and W are the momentum components controlling the three directions of the equation of atmospheric motion on the projection plane, respectively. Ω is the momentum of the vertical velocity field, Θ is the momentum of the potential temperature field, u, v, and w are velocity components in three directions, respectively, mx and my are magnification factors for map projections in x and y directions, respectively, $(m_x, m_y) = \frac{(\Delta x, \Delta y)}{(\Delta x_e, \Delta y_e)}$, and F_U, F_V, F_W, F_Θ are forced terms caused by model physical processes, turbulent mixing, spherical projection, and Earth rotation, respectively.

The mass conservation equation and the potential height equation are:

$$\frac{\partial \mu'_d}{\partial t} + m_x m_y\left(\frac{\partial U}{\partial x} + \frac{\partial V}{\partial y}\right) + m_y \frac{\partial \Omega}{\partial \eta} = 0, \tag{7}$$

$$\frac{\partial \phi'}{\partial t} + \frac{1}{\mu_d}\left[m_x m_y\left(U\frac{\partial \phi}{\partial x} + V\frac{\partial \phi}{\partial y}\right) + m_y \Omega \frac{\partial \phi}{\partial \eta} - m_y g W\right] = 0, \tag{8}$$

The static equation in the perturbation form is:

$$\frac{\partial \phi'}{\partial \eta} = -\alpha_d \mu_d' - \alpha_d' \overline{\mu}_d, \tag{9}$$

The diagnostic equation for complete air pressure (water vapor plus dry air) is:

$$p = p_0 (R_d \theta_m / p_0 \alpha_d)^{\gamma}, \tag{10}$$

where p_0 is the reference pressure (usually 105 Pa), R_d is the ratio constant of dry air to gas, $\theta_m \approx \theta(1 + 1.61 q_v)$, $\gamma = c_p / c_v$, and γ is the ratio of heat capacity c_p of dry air at constant pressure to heat capacity c_v of passenger.

2.2.2. Model Configuration

Ya'an, with a subtropical monsoon climate, is located in the western edge of Sichuan Basin. The area from Kangding to Litang belongs to the western Sichuan high mountain plain, with a wide and flat or hilly top surface, and mainly has a continental plateau-type and mountain-type climate. On the other hand, the area from Litang to Changdu belongs to the southeastern Tibetan Hengduan Mountains area, with high mountains, deep valleys, complex terrain, and mainly plateau climate. Since the terrain of Ya'an and Kangding is flatter and the climate is warmer than that of Litang Changdu, in order to adapt these differences, this paper divides the railway section from Kangding to Changdu into two regions with 100.8° E as the boundary and performs double-layer nested simulation. The outer grid is marked as d01, and the inner grid is marked as d02. Region 1 mainly includes Ya'an to Kangding and other places, and region 2 includes Litang to Qamdo and other places. The center points of the two regions are 30.03° N, 101.97° E and 30.6° N, 98.62° E, respectively. Figure 3 shows the specific simulation range.

The National Centers for Environmental Prediction (NCEP) Final Operational Global Analysis (FNL) (https://rda.ucar.edu/datasets/ds083.2/ (accessed on 13 February 2023)) with a horizontal resolution of 1° × 1° and a frequency of 6 h are selected as the initial conditions of the model. The simulation covers the period from 12 December 2016 to 28 February 2017, and the specific grid parameters of the two regions are shown in Table 1. In region 2, which has many mountains, the simulation accuracy is improved by using MCD12Q1v006 land use data with a horizontal resolution of 500 m instead of the original data. Moreover, to simulate the atmospheric processes within the boundary layer more precisely, the vertical direction is divided into 41 layers, with 15 layers densely distributed at a height of about 1.5 km near the ground. Table 2 shows the heights of each layer grid, and Figure 4 illustrates the elevated layers in both regions, the red line shows the position of the 15th layer.

Table 1. The grid parameters of two regions.

Parameters	Region 1		Region 2	
	d01	d02	d01	d02
Grid space	9 km	3 km	9 km	3 km
Grid points	100 × 100	88 × 88	100 × 100	148 × 130
ETA levels	32	32	41	41
Time step	27	27	27	27

Table 2. Vertical height of each mesh layer within 1.5 km.

Mesh layer	1	2	3	4	5
Height/m	4	30	64	93	131
Mesh layer	6	7	8	9	10

Table 2. *Cont.*

Height/m	183	234	283	554	946
Mesh layer	11	12	13	14	15
Height/m	1116	1181	1261	1367	1525

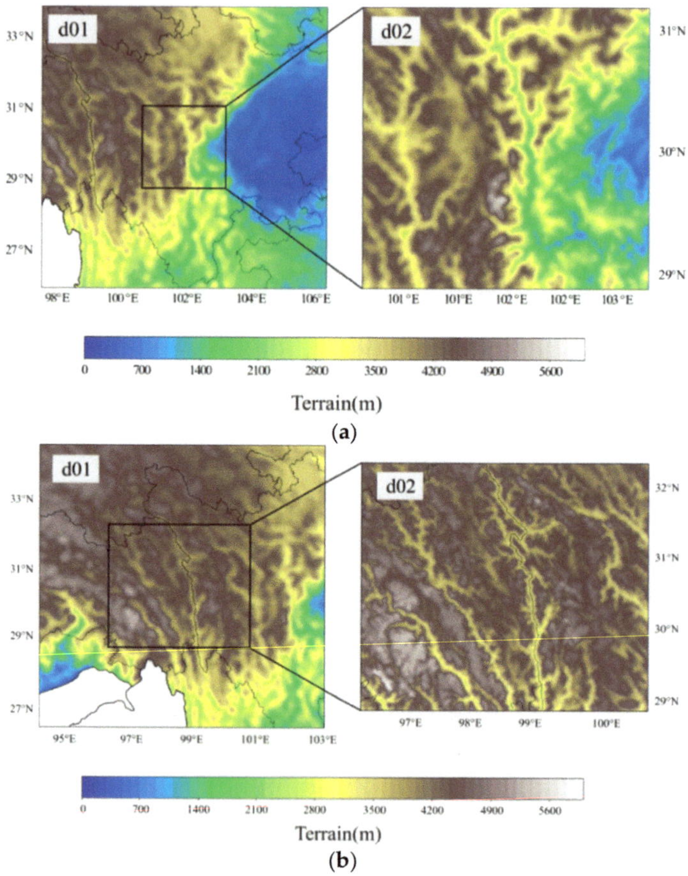

Figure 3. Simulation range for both regions. (**a**) The simulation range of region 1; (**b**) the simulation range of region 2.

Physical processes are crucial for determining the source and redistribution of momentum, heat, and water vapor in the atmospheric motion. These processes have a great impact on the overall dynamics of the atmosphere. Microphysics schemes mainly affect the precipitation simulation, among which the Purdue–Lin scheme and Thompson scheme are two relatively complex calculation schemes. These schemes involve variables such as rain, ice, snow, and graupel, making them suitable for high-resolution simulations. The planetary boundary layer scheme affects the wind simulation, among which the MRF scheme can provide ideal simulation results for the boundary layer conditions with good mixing, such as instability or convection, and the YSU scheme can simulate the turbulent motion and energy conversion in the atmosphere well and has wide applicability. A radiative parameterization scheme is used to calculate the surface heat budget. The land surface process is used to calculate the heat and water vapor fluxes at land–sea grids. The surface layer parameterization scheme is used to describe the surface characteristics and simulate the interaction between the surface and the atmosphere. Since the grid resolution in this paper is 9 km, the cumulus parameter scheme is not disabled. It is found that the simulation results of rainfall and wind speed using the Purdue–Lin scheme [37] and YSU scheme [38] are closer to the observed results, respectively. According to the characteristics of various schemes, this study combines the Purdue–Lin and Thompson microphysics schemes, as well as the YSU and MRF boundary layer schemes, which have significant impacts on rainfall and wind speed results, as shown in Table 3.

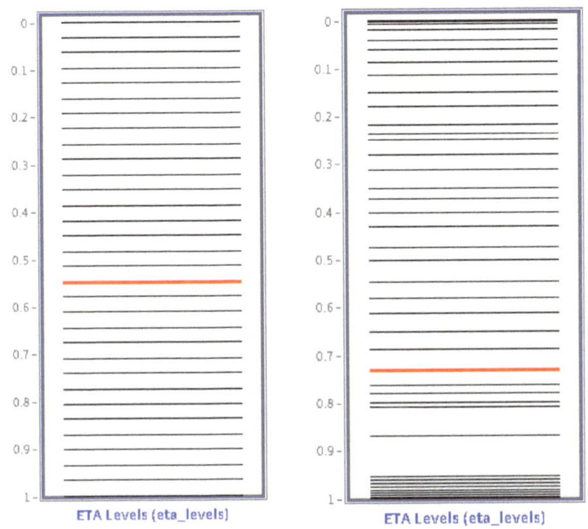

Figure 4. Vertical stratification of region 1 (**left**) and region 2 (**right**).

Table 3. Schemes of physical parameters for WRF simulation.

Scheme	Case1	Case2	Case3	Case4
mp_physics	Thompson	Purdue-Lin	Purdue-Lin	Thompson
bl_pbl_physics	YSU	MRF	YSU	MRF
ra_sw_physics	RRTM	RRTM	RRTM	RRTM
ra_la_physics	Dudhia	Dudhia	Dudhia	Dudhia
sf_surface_physics	Noah	Noah	Noah	Noah
sf_sfclay_physics	MM5	MM5	MM5	MM5
cu_physics	/	/	/	/

2.3. Verification of Simulation Results

2.3.1. Evaluation Statistics

To verify the accuracy of the simulation results, the Pearson correlation coefficient (R) is used to measure the linear correlation degree between the simulation sequence and the observation sequence. The closer R is to 1, the higher the correlation degree. The mean bias error (MBE) is used to verify the deviation between the simulation sequence and the observation sequence. The closer the result is to 0, the smaller the deviation. The root mean square error (RMSE) is used to verify the average deviation between the simulation sequence and the observation sequence. The closer the result is to 0, the better the simulation effect. The relevant calculation formulas are as follows:

$$R = \frac{\sum_{i=1}^{n} (X_i - \overline{X}_i)(X_0 - \overline{X}_0)}{\sqrt{\sum_{i=1}^{n} (X_i - \overline{X}_i)^2 \sum_{i=1}^{n} (X_0 - \overline{X}_0)^2}}, \quad (11)$$

$$\text{MBE} = \frac{1}{n}\sum_{i=1}^{n}(X_i - X_0), \tag{12}$$

$$\text{RMSE} = \sqrt{\frac{1}{n}\sum_{i=1}^{n}(X_i - X_0)^2}, \tag{13}$$

where X_i and X_0 are the simulated and observed value, $\overline{X}_i$ and $\overline{X}_0$ are the simulated and observed average values, and n is the number of records.

2.3.2. Verify Performance for WRF Model

The source of all the meteorological data used in this paper is from the China Meteorological Administration (https://data.cma.cn/ (accessed on 12 December 2022)). These kinds of data have advantages of high accuracy and stability, but the data of each station only represent the surrounding information. Taking the daily average wind speed from December 2016 to February 2017 as the verification data, the physical parameter scheme of case3 in Table 3, the original land use data of the WRF model, and the two vertical stratification methods mentioned above are used to calculate the meteorological information of Litang. Table 4 shows the evaluation index results of the two schemes. The scheme of vertical layers densified near the ground has increased the correlation coefficient by 77.72%, reduced the mean bias by 61.13%, and reduced the root mean square error by 59.04% compared with the non-density scheme. Figure 5 shows the comparison of the change trends of the simulation results and the measured results. It can be seen that the change trend of the calculation results with vertical layers densified near the ground is more consistent with the measured results. Therefore, near-surface encryption can significantly improve the simulation results of the mountain wind field.

Table 4. Evaluation results of different ETA levels.

ETA Levels	R	MBE	RMSE
32	0.10	7.23	7.69
41	0.44	2.81	3.15

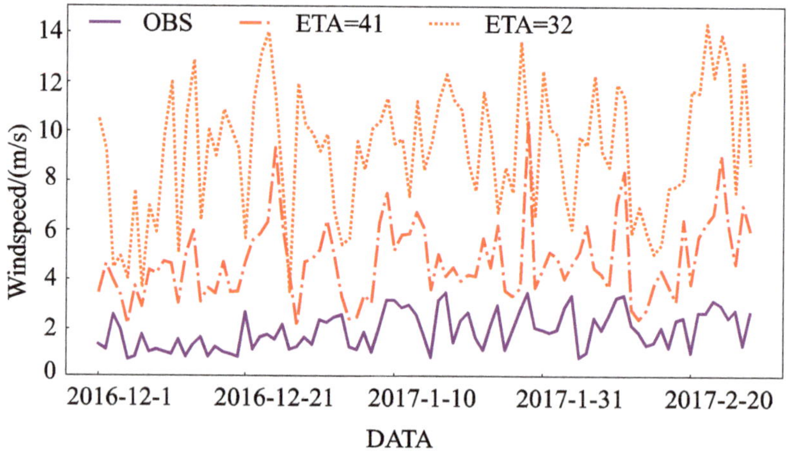

Figure 5. Wind speed variation trends at different ETA levels.

In order to validate the computed results of four cases presented in Table 3, the daily average wind speed results and precipitation results from December 2016 to February 2017 at Kangding were used for comparison. Table 5 shows the correlation coefficients of the two data. For rainfall, the Purdue–Lin scheme demonstrated a higher correlation coefficient compared to the Thompson scheme, while for wind speed, the YSU scheme outperformed the MRF scheme. Figure 6 displays the actual simulated and observed values. All four schemes accurately captured the changing trends, but the maximum and minimum wind speeds were higher than the actual results. Considering the comparison of correlation coefficients and the actual trend results, case3, shown in Table 3, appeared to be more reasonable. Therefore, this study adopted the third scheme for calculations in both region 1 and region 2.

Table 5. Comparison of results of different schemes.

Scheme	R		MBE		RMSE	
	Precipitation	Wind Speed	Precipitation	Wind Speed	Precipitation	Wind Speed
Case1	0.27	0.51	0.01	1.21	1.41	1.71
Case2	0.40	0.45	0.15	2.38	1.42	3.20
Case3	0.51	0.54	0.10	1.32	1.07	1.80
Case4	0.34	0.33	0.63	0.90	3.02	0.62

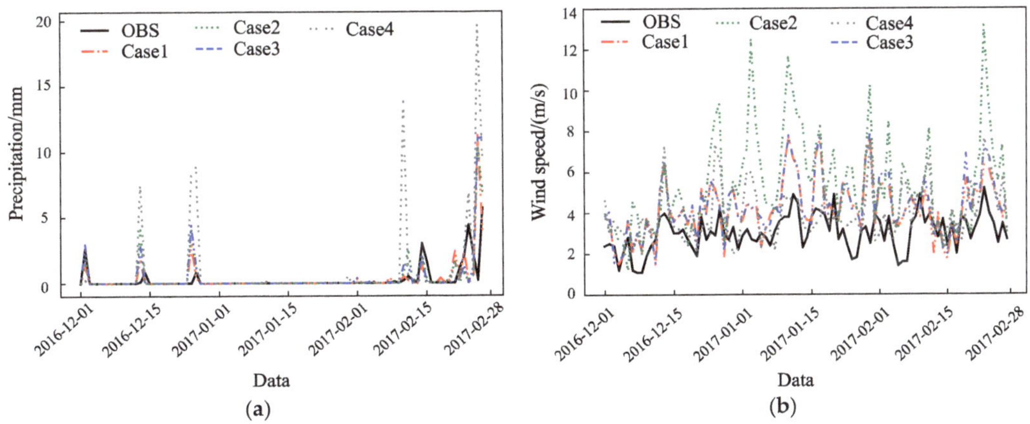

Figure 6. Comparison of simulation results in Kangding. (**a**) Comparison of precipitation result; (**b**) comparison of wind speed result.

To verify the accuracy of the calculation results of case3 comprehensively, this paper further validates the results of temperature, wind speed, and rainfall calculated by the WRF model. The simulated data are the WRF calculation results under the parameter settings in Tables 1 and 2 and case3 of Table 3. The validation data used to verify the calculation results of temperature and wind speed are the ground-based meteorological observation data of Kangding and Litang stations. Because there was no precipitation in the actual situation of the two meteorological stations in the observation data of 2022, the feasibility verification of the WRF model for rainfall used the daily average observation data of Litang from December 2016 to February 2017. The comparison results are shown in Table 6, and it can be found that the correlation coefficient of temperature is above 0.80, and the correlation coefficients of wind speed and rainfall are also above 0.50. In addition, the average deviation and root mean square error values are relatively small. By comparing the correlation coefficients presented in this paper with those published in the literature for temperature [39], wind speed [40], and rainfall [41] in similar terrain, it can be observed in Table 6 that the calculated correlation coefficients for temperature and wind speed in this study are slightly higher than those reported in previous studies. The correlation coefficients for rainfall closely align with the published values. Thus, the calculation results obtained in this study can be utilized to assess the susceptibility of drifting snow, providing valuable insights.

Table 6. Evaluation results of temperature, wind speed, and rainfall.

Meteorological Element	Station	R	MBE	RMSE	Comparison
Temperature	Kangding Litang	0.82 0.90	2.64 1.43	4.06 3.21	0.60–0.80
Wind speed	Kangding Litang	0.54 0.63	0.52 −1.69	2.26 2.37	0.51–0.56
Rainfall	Kangding Litang	0.54 0.58	1.32 1.21	1.80 1.52	0.50–0.60

We performed linear fitting on the calculation results. The closer the points in the figure are to the diagonal line (the black dash line in figure), the closer the simulation results are to the real results. The brighter the color of the points, the more points fall in that area. Figure 7 shows the fitting results of temperature and wind speed calculation data and their changes over time. It can be found that the model can capture the temperature changes well. The simulated temperature of Kangding station is concentrated within 0 to 5 °C, and the temperature of Litang station is concentrated within −5 °C to 5 °C. The WRF simulation results of wind speed for both places are concentrated in 1–2 m/s. The simulation results of Litang station are higher than the actual recorded values; however, the WRF model accurately captured the change trend and extreme values of wind speed at both stations. Figure 8 shows the comparison between the rainfall calculation results and the actual results. The WRF model successfully captured the rainfall events. In summary, although there is some error between the data simulated by the WRF model and the actual data, the correlation of the simulation results is good, and it can accurately capture the change trend of the data, which also verifies the effectiveness of the simulation method.

2.4. Occurrence Probability Determination Method

Vionnet et al. [42] analyzed the meteorological data detected when drifting snow occurred near the Alps and found that the main meteorological feature was that snow occurred when the wind speed reached a certain range, and this model was also applied by Guyomarc'h et al. [43]. Based on this discovery and previous studies, the drifting snow in this paper is defined as periods with precipitating snow with a wind speed in the range of 3–8 m/s [19]. In order to ensure that the precipitation is solid, the temperature is limited to less than 0 °C. Therefore, we calculated the probability by the percentage in the number of times that met the three conditions during the simulation period.

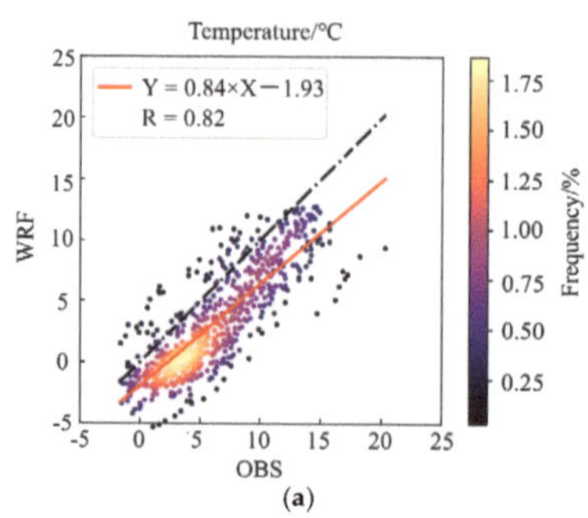

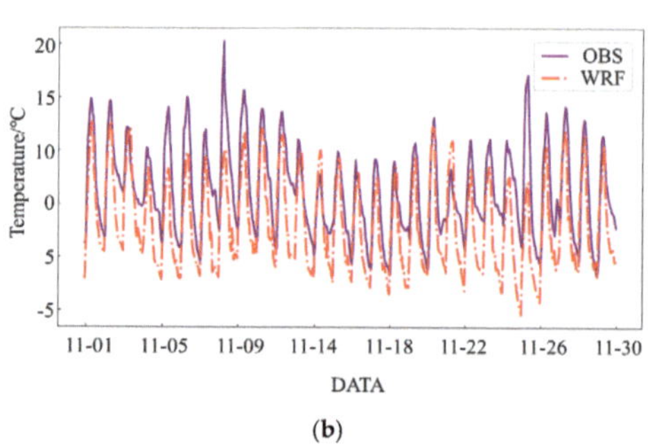

Figure 7. *Cont.*

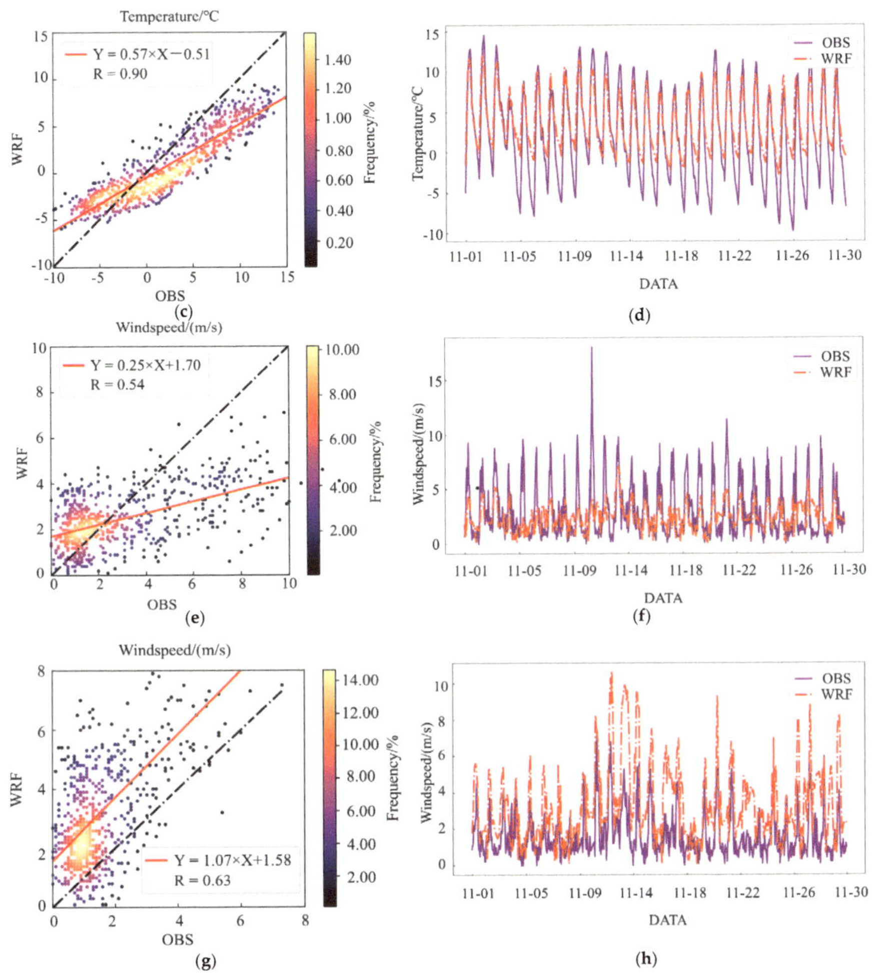

Figure 7. The results of the WRF fitting process and a comparison between the simulated and observed values. (**a**) The fitted line of temperature in Kangding; (**b**) the variation in temperature in Kangding; (**c**) the fitted line of temperature in Litang; (**d**) the variation in temperature in Litang; (**e**) the fitted line of wind speed in Kangding; (**f**) the variation in wind speed in Kangding; (**g**) the fitted line of wind speed in Litang; (**h**) the variation in wind speed in Litang.

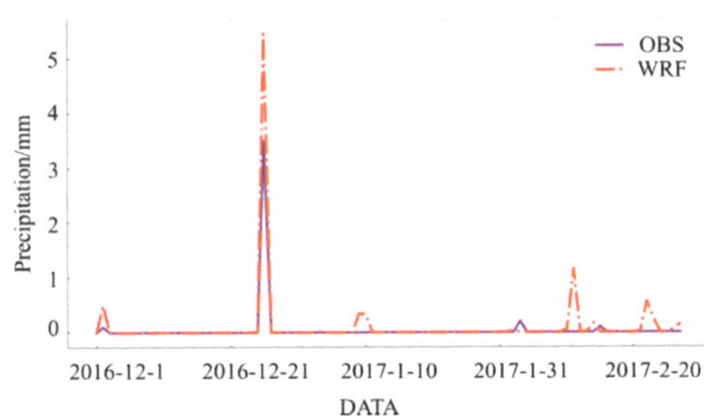

Figure 8. Comparison of daily rainfall in Litang.

3. Result and Discussion

3.1. Simulation Results of Temperature Distribution

The output results of the inner grid d02 are analyzed to obtain the monthly average temperature within the simulated area, as shown in Figure 9a–f. The results show that a minimum temperature of −14 °C in the study area is calculated by the WRF model, which ensures that the snow particles do not degrade due to excessively low temperatures. The temperature distribution around the railway is relatively regular. Region 1 bounded by Kangding is divided into two parts. The eastern side of Ya'an City has warmer temperatures than the western side, and the average monthly temperature stays above 0 °C, which matches the sub-tropical humid monsoon climate of Ya'an City that does not have harsh winters. On the other hand, most areas on the western side of Kangding are colder, with an average monthly temperature of around −2 °C. These temperatures change with time, and the average temperature in January is lower than that in December and February. In region 2, along the railway from Litang to Qamdo, the temperatures are frigid, always below 0 °C. Notably, the temperature in January is significantly lower than that in December and February.

3.2. Simulation Results of Precipitation Distribution

Snow is the material condition for the occurrence of drifting snow. Sufficient snow sources will increase the probability of drifting snow, and the amount of snowfall will also affect the intensity and duration of drifting snow. Figure 10 illustrates the monthly cumulative precipitation around the railway in winter. In region 1, precipitation is mainly concentrated in the Ya'an area east of Kangding and the area south of Kangding station. Precipitation is affected by temperature, and the eastern area has higher temperatures, so it has more precipitation than other areas. Notably, January exhibits the lowest average temperature and, consequently, the lowest level of precipitation. In February 2017, there was more precipitation in Kangding area, and the cumulative precipitation reached 500 mm in the southern part of Kangding. Conversely, the precipitation in region 2 is significantly lower than that in region 1. During the simulation period, only a few places have weak precipitation, and the areas with higher precipitation are far away from the railway. Moreover, the precipitation was highest in February 2017, which is similar to the result of region 1.

3.3. Simulation Results of Wind Field Distribution

Wind is the primary factor that affects the occurrence of drifting snow. The wind direction determines the movement direction of snow particles, and the wind speed determines the scale of drifting snow. Figure 11 illustrates the monthly average wind speed and direction distribution in the simulated area. The arrows in the figures represent the direction of wind speed. In region 1, the wind speed on the west side of 102° E is higher than that on the east side. In the three winter months, the wind speed in Kangding area is always higher than that in other surrounding areas. The average wind speed in January 2017 was higher than that in December 2016 and February 2017, with a maximum wind speed of 11.41 m/s. The wind direction around the railway is variable. In the section from Ya'an to Kangding, the wind direction is mainly eastward, parallel to the railway direction, with a small angle. On the west side of Kangding, the wind direction changes more frequently. Between 101.6° E and 102° E, it is mainly southwest wind. Between 101° E and 101.5° E, it is mainly south wind and southwest wind. In region 2, the whole area has high wind speed and higher than that in region 1, with a maximum wind speed of 12.33 m/s. The average wind speed in November is higher than that in the other two months. The wind direction is mainly southwest wind, forming a large angle with the railway.

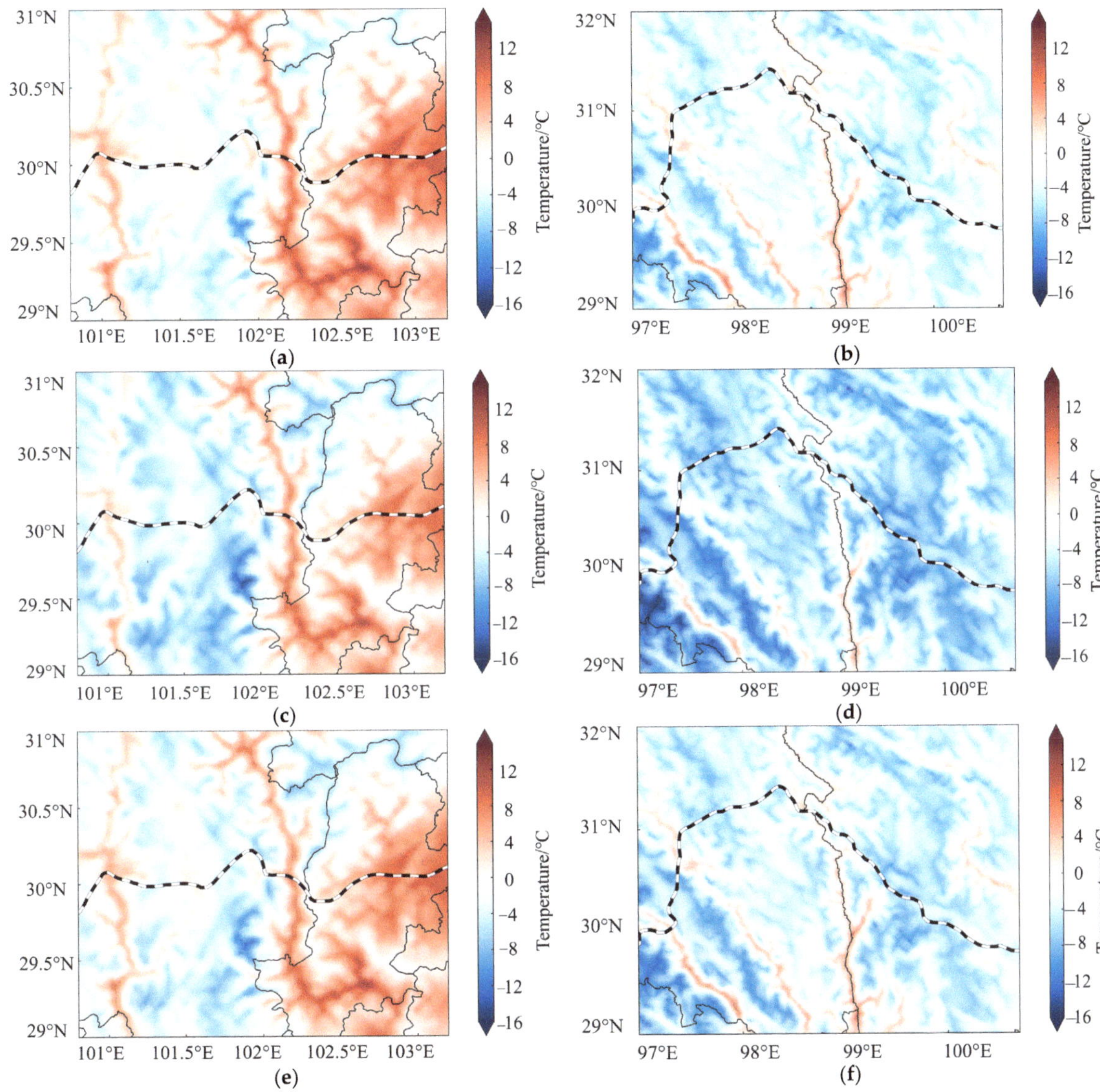

Figure 9. The distribution of average temperature in the simulation period. (**a**) Temperature of region 1 in December; (**b**) temperature of region 2 in December; (**c**) temperature of region 1 in January; (**d**) temperature of region 2 in January; (**e**) temperature of region 1 in February; (**f**) temperature of region 2 in February.

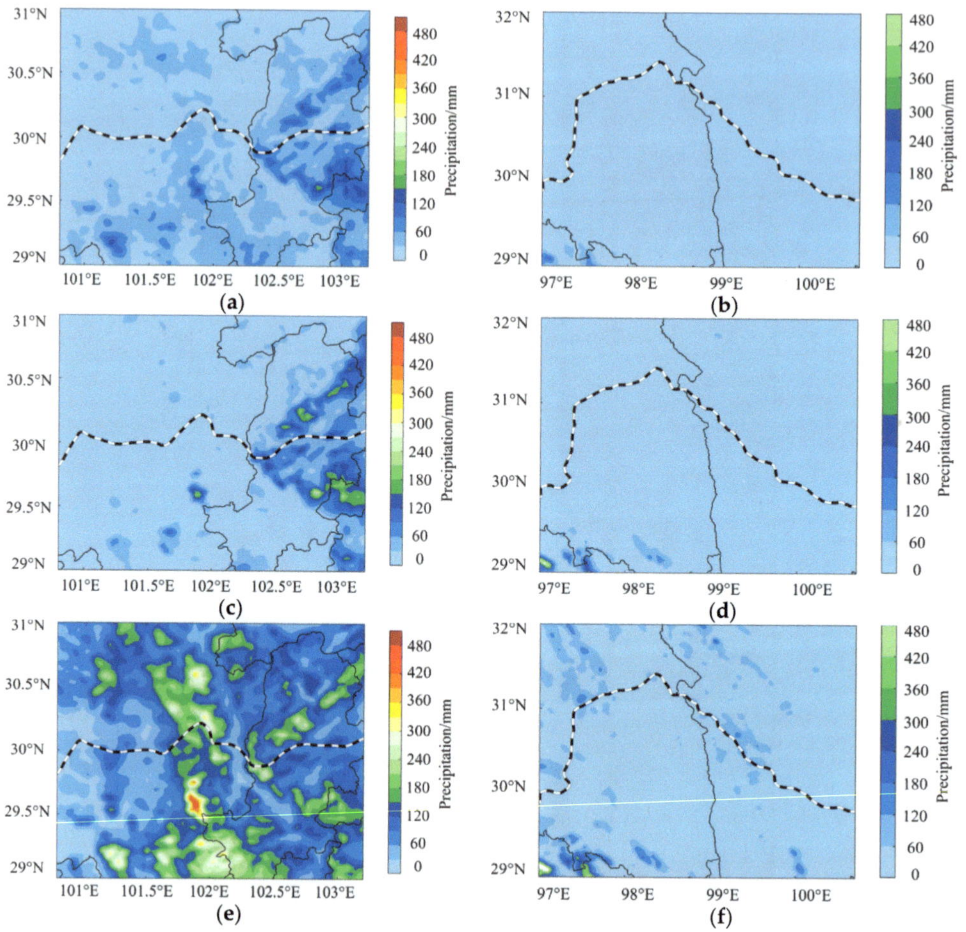

Figure 10. The distribution of cumulative precipitation in the simulation period. (**a**) Precipitation of region 1 in December; (**b**) precipitation of region 2 in December; (**c**) precipitation of region 1 in January; (**d**) precipitation of region 2 in January; (**e**) precipitation of region 1 in February; (**f**) precipitation of region 2 in February.

According to the different wind speeds and directions along the railway, five points, P1 (102° E, 30.13° N), P2 (101.6° E, 29.98° N), P3 (101.1° E, 30.03° N), P4 (99.50° E, 30.37° N), and P5 (97.7° E, 31.15° N), are selected as representatives to study the wind speed and direction distribution in their surrounding areas. The wind rose diagram and the wind speed probability density diagram are shown in Figure 12. In region 1, the maximum wind speed at P1 is 13.10 m/s, the primary dominant wind direction is south–southwest wind (SSW), accounting for 17.92% of the entire simulation period, the secondary dominant wind direction is southwest wind (SW), accounting for 16.82% of the entire simulation period, the wind speed is mainly concentrated around 2–5 m/s, and the wind speed in the probability interval of drifting snow occurrence (3–8 m/s) accounts for 61.82% of the total wind speed in the simulation period; at P2, the maximum wind speed is 17.91 m/s, the primary dominant wind direction is southwest wind (SW), accounting for 41.51% of the entire simulation period, the secondary dominant wind direction is west wind (E), accounting for 28.54% of the entire simulation period, the wind speed is mainly concentrated around 5–9 m/s, and the wind speed in the probability interval of drifting snow occurrence accounts for 56.39% of the total wind speed in the simulation period; at P3, the maximum wind speed is 16.7 m/s, the primary dominant wind direction is west–northwest wind (WNW), accounting for 22.51% of the entire simulation period, the secondary dominant wind directions are west wind and southwest wind, accounting for 18.81% and 18.39% of the entire simulation period, respectively, the wind speed is mainly concentrated around 6–10 m/s, and the wind speed in the probability interval of drifting snow occurrence accounts for 50.73% of the total wind speed in the simulation period.

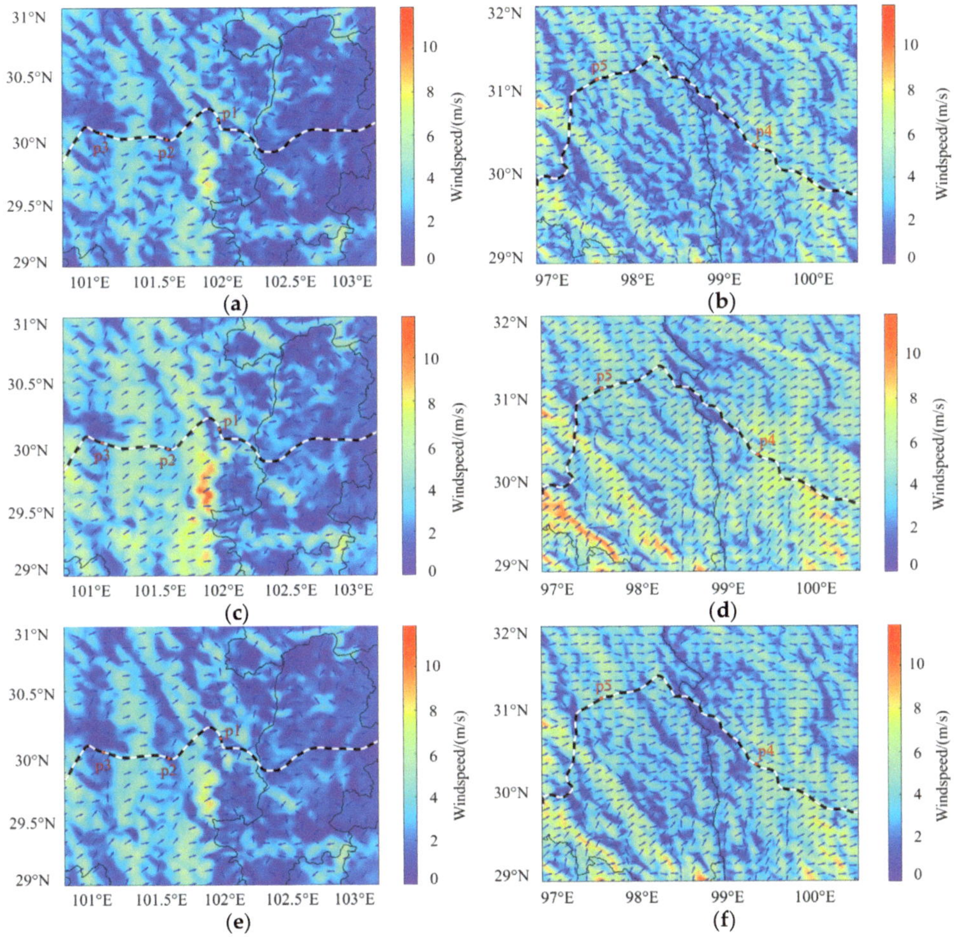

Figure 11. The distribution of average wind field in the simulation period. (**a**) Wind field of region 1 in December; (**b**) wind field of region 2 in December; (**c**) wind field of region 1 in January; (**d**) wind field of region 2 in January; (**e**) wind field of region 1 in February; (**f**) wind field of region 2 in February.

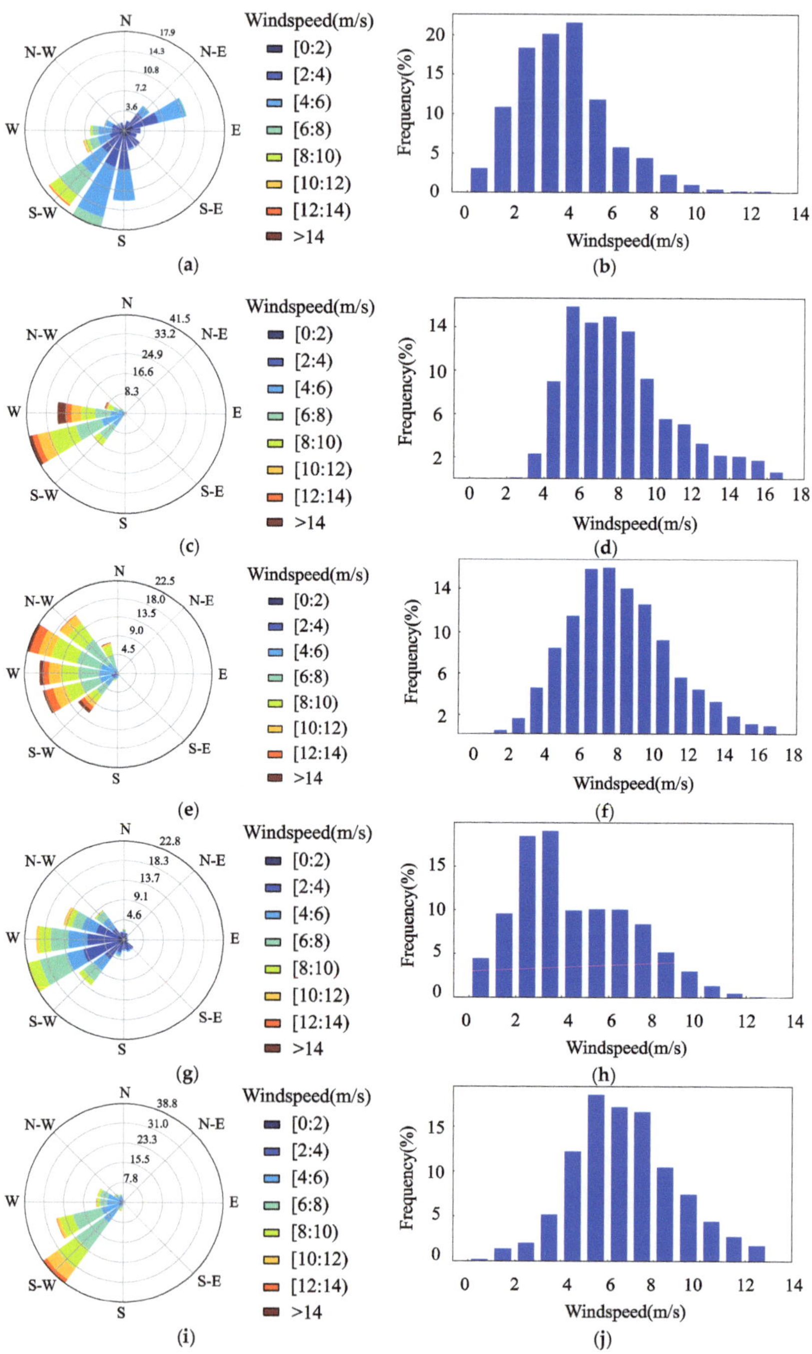

Figure 12. Wind rose diagram and the wind speed probability density graph. (**a**) the wind rose diagram of P1; (**b**) the wind speed probability density graph of P1; (**c**) the wind rose diagram of P2; (**d**) the wind speed probability density graph of P2; (**e**) the wind rose diagram of P3; (**f**) the wind speed probability density graph of P3; (**g**) the wind rose diagram of P4; (**h**) the wind speed probability density graph of P4; (**i**) the wind rose diagram of P5; (**j**) the wind speed probability density graph of P5.

In region 2, the maximum wind speed at P4 is 12 m/s, the primary dominant wind direction is south–southwest wind (SSW), accounting for 22.84% of the entire simulation period, the secondary dominant wind direction is west wind (W), accounting for 20.17% of the entire simulation period, the wind speed is mainly concentrated around 2–4 m/s, and the wind speed in the probability interval of drifting snow occurrence accounts for 55.59% of the total wind speed in the simulation period. At P5, the maximum wind speed is 14.90 m/s, the primary dominant wind direction is southwest wind (SW), accounting for 38.79% of the entire simulation period, the secondary dominant wind direction is southwest wind (SW), accounting for 26.81% of the entire simulation period, the wind speed is mainly concentrated around 4–8 m/s, and the wind speed in the probability interval of drifting snow occurrence accounts for 69.06% of the total wind speed in the simulation period. Since the climate of the study area is mainly influenced by the south branch jet stream of the westerly circulation, the wind direction is mostly west or southwest wind, which is consistent with the actual situation. Moreover, the proportion of wind speed in the interval of drifting snow occurrence is relatively large, which means a higher possibility of being affected by drifting snow.

3.4. Drifting Snow Susceptibility Judgment

As we can see in Figure 13, along the railway, the places with higher probability of drifting snow occurrence are located in Kangding County and Litang County, and there is a long railway line passing through the drifting snow susceptible areas in Litang, Batang, and Baiyu counties, with a total length of about 321 km that may experience drifting snow along the line. The entire area of Ya'an City has almost zero probability of drifting snow occurrence; about 12 km of railway in the adjacent area of Luding County and Tianquan County may experience drifting snow, with a probability of 15%; in Kangding County, there are two sections of railway that may experience drifting snow phenomenon, located at 101.9° E–102° E and 101.6° E–101.8° E, respectively, with lengths of 20 km and 23 km, respectively, and probabilities of 12% and 8%, respectively; about 22 km of railway in Yajiang County at 101.2° E–101.4° E may experience drifting snow, with a probability of about 4%; about 50 km of railway in Litang County at 99.8° E–100.2° E has a probability of 8%; about 87 km of railway passing through Batang County and Baiyu County at 99.15° E–99.7° E has a probability of 4% for drifting snow phenomenon; about 24 km of railway in Jomda County at 98.25° E–98.45° E has a probability of 6%; about 71 km of railway passing through Qamdo County and Chaya County at 97.50° E–97.90° E has a probability of 5%; about 12 km of railway in Chaya County and Basu County at 97.35° E–97.40° E has a probability of 6%. The railway located in Kangding County has a slightly higher probability of drifting snow phenomenon than other areas, and the length of the railway affected is larger, so the railway in this section needs to strengthen its protection against drifting snow.

3.5. Discussion of Occurrence Probability

As the WRF model relies on input boundary conditions, it can simulate diverse climatic scenarios by altering the input meteorological data to accommodate climate change. Climate variations, such as alterations in temperature and precipitation patterns, can exert influences on snow conditions. For instance, elevated temperatures can expedite snow melting rates, while shifts in precipitation can impact the formation and distribution of snow, ultimately affecting the occurrence of drifting snow. Given the WRF model's capacity to adjust to climate change, it can similarly accommodate shifts in drifting snow conditions due to climate variations. As a result, the derived probability of drifting snow occurrence based on the WRF model remains dependable.

This study utilizes temperature, snowfall, and wind speed information derived from WRF simulations to assess the probability of drifting snow around the railway. The calculated results have been validated against measured data and data from other published papers, which enhances the credibility of the findings in the paper. However, there are two errors in this paper: one is the calculation error of the WRF model itself, and the other is the error caused by the judgment model of drifting snow. If these two aspects are improved, the calculation results will be more accurate.

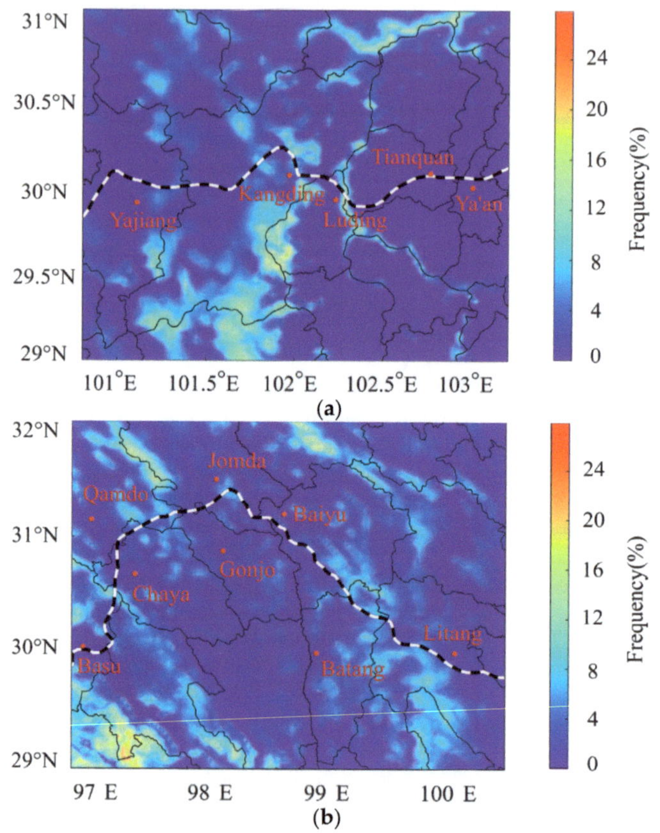

Figure 13. Probability of drifting snow along the railway. (**a**) The susceptibility of drifting snow in region 1; (**b**) the susceptibility of drifting snow in region 2.

The accuracy of the WRF calculation results is mainly affected by boundary conditions, parameter scheme combinations, and model parameters. The initial conditions are reanalysis data, that is, the historical weather information obtained by integrating radar satellite, aircraft, ship, and station observation data and numerical forecasting products using data assimilation technology. Therefore, the accuracy of the data used for assimilation will affect the accuracy of the reanalysis data and then affect the calculation results of the WRF model. The combination of different parameter schemes of the WRF model will produce different calculation results, and one model parameter of WRF may affect multiple meteorological variables, so the selection of model parameters needs to consider multiple meteorological variables. However, WRF has many parameter schemes and model parameters. With the current computing power, it is difficult to obtain the optimal parameter scheme combination and model parameters through thousands of long-time calculations, which will also affect the calculation results of WRF.

Meteorological information such as wind and snow are the main conditions that affect drifting snow occurrence, but the terrain around the railway, vegetation types, and the form of the roadbed of the railway also have a certain impact on it. Since the purpose of this paper is mainly to calculate the probability of drifting snow from the perspective of meteorological information, snow and wind data are mainly taken into account in the judgment model. However, terrain and other factors are not considered, so the occurrence probability calculated based on the drifting snow judgment model in this paper may be higher than the actual result, and the calculation result in this paper is more secure.

To reduce errors of possibility caused by various factors, improvements can be made in those aspects, and the data of temperature, precipitation, and wind speed in FNL can be replaced with observation data of weather stations to improve the accuracy of boundary conditions and further improve the accuracy of WRF simulation. Sensitivity analysis of the parameter scheme and model parameters can be carried out to find the optimal scheme and parameter under the condition of saving computing resources as much as possible. In addition to the main factors such as wind and snow, secondary factors such as terrain and vegetation are taken into account to establish a more comprehensive evaluation model of drifting snow susceptibility.

4. Conclusions

The railway from Kangding to Qamdo is divided into two regions for simulation in this paper. In region 2 with complex terrain, the vertical direction near the ground is densified to 15 layers at a height of about 1.5 km, and the MCD12Q1v006 land use data are used to replace the original data. The susceptibility of drifting snow along the railway is judged according to the simulation results. The conclusions are as follows:

(1) Among the simulations of temperature, rainfall, and wind speed, the simulation result of temperature is the best, followed by wind speed and rainfall. The WRF model can simulate the change trend of the three meteorological elements very accurately during the calculation period.

(2) For mountainous areas with complex terrain, using the method of increasing the number of layers in the vertical direction and densifying near the ground can significantly improve the simulation result of wind speed by the WRF model.

(3) The regional climate of the Ya'an to Qamdo railway is influenced by the south branch jet stream of the westerly circulation. The temperature is high in the east and low in the west, the rainfall is more in the east and less in the west, and the wind speed is high in the west and low in the east. The average wind speed is concentrated around 5–8 m/s, and the dominant wind direction in winter is mainly westerly and southwesterly.

(4) The WRF model can be applied to determine the susceptibility of drifting snow in areas lacking meteorological information: the total length of Ya'an–Qamdo section is about 700 km, of which about 321 km of railway may experience drifting snow. The railway is more likely to be affected by drifting snow disaster when passing through Kangding County, Luding County, and Tianquan County where they border each other; although there are places with higher probability of drifting snow occurrence in the simulation area, these places are far away from the railway and will not affect it.

In the term of railway protection with high susceptibility, the meteorological information calculated by WRF can be extracted and taken as the boundary condition of computational fluid dynamics (CFDs) so as to more accurately study the snow condition of various roadbeds and the protective effect of various forms of snow fence to achieve the best protection of the railway.

The WRF model is easy to access, has a fast calculation rate, and provides high reliability of results. It can be applied to any location lacking but requiring long time series and high-resolution meteorological information, creating significant economic value at a relatively low cost. Therefore, this computational method has enormous potential for applications in engineering, industry, and other fields.

Author Contributions: Conceptualization, Q.L.; Methodology, X.Z.; Data curation, W.Y.; Writing—original draft, X.Z.; Writing—review & editing, Z.Z. All authors have read and agreed to the published version of the manuscript.

Funding: This research was funded by Innovation research group project of Natural Science Foundation of Hebei Province of China (Grant No. E2022210078), Central Leading Local Science and Technology Development Fund Project (Grant No. 236Z5410G), High-end Talents Project of Hebei Province of China (Grant No. [2019] 63), and Funded by Science and Technology Project of Hebei Education Department (Grant No. QN2023093).

Institutional Review Board Statement: Not applicable.

Informed Consent Statement: Not applicable.

Data Availability Statement: The raw data supporting the conclusions of this article will be made available by the authors on request.

Conflicts of Interest: The authors declare no conflict of interest.

References

1. Bagnold, R.A. *The Physics of Blown Sand and Desert Dunes*; Methuen & Co., Ltd.: London, UK, 1941; Volume 265.
2. Owen, P.R. Saltation of uniform grains in air. *J. Fluid Mech.* **1964**, *20*, 225–242. [CrossRef]
3. Kind, R.J. Mechanics of aeolian transport of snow and sand. *J. Wind Eng. Ind. Aerodyn.* **1990**, *36*, 855–866. [CrossRef]
4. Budd, W.F. The drifting of nonuniform snow particles. *Stud. Antarct. Meteorol.* **1966**, *9*, 59–70.
5. Schmidt, R.A. Estimates of threshold windspeed from particle sizes in blowing snow. *Cold Reg. Sci. Technol.* **1981**, *4*, 187–193. [CrossRef]
6. Kind, R.J.; Murray, S.B. Saltation flow measurements relating to modeling of snowdrifting. *J. Wind Eng. Ind. Aerodyn.* **1982**, *10*, 89–102. [CrossRef]
7. Flaga, A.; Flaga, L. Wind tunnel tests and analysis of snow load distribution on three different large size stadium roofs. *Cold Reg. Sci. Technol.* **2019**, *160*, 163–175. [CrossRef]
8. Qiang, S.G.; Zhou, X.Y.; Kosugi, K.; Gu, M. A study of snow drifting on a flat roof during snowfall based on simulations in a cryogenic wind tunnel. *J. Wind Eng. Ind. Aerodyn.* **2019**, *188*, 269–279. [CrossRef]
9. Liu, M.M.; Zhang, Q.W.; Fan, F.; Shen, S.Z. Modeling of the Snowdrift in Cold Regions: Introduction and Evaluation of a New Approach. *Appl. Sci.* **2019**, *9*, 3393. [CrossRef]
10. Li, P.X.; Bai, M.Z.; Ding, L.S.; Cui, Z.Z.; Zhang, Z.C. Field measurement and outdoor wind tunnel test of snow-drifting: Snow distribution characteristics of railway subgrade and deposition mechanism. *J. Wind Eng. Ind. Aerodyn.* **2022**, *230*, 15. [CrossRef]
11. Basnet, K.; Constantinescu, G.; Muste, M.; Ho, H. Method to Assess Efficiency and Improve Design of Snow Fences. *J. Eng. Mech.* **2015**, *141*, 04014136. [CrossRef]
12. Li, P.X.; Bai, M.Z.; Li, J.X.; Qiu, S.M.; Wei, Z.J. Analysis of the influence of snow fences on snow redistribution under snow-drifting in railway cuttings. *Cold Reg. Sci. Technol.* **2022**, *196*, 103520. [CrossRef]
13. Nieto, P.J.G.; Diaz, J.J.D.; Castro-Fresno, D.; Munoz, F.B. Numerical simulation of the performance of a snow fence with airfoil snow plates by FVM. *J. Comput. Appl. Math.* **2010**, *234*, 1200–1210. [CrossRef]
14. Ma, W.; Luo, Y.; Li, F.; Sun, Y. The influence of wind deflectors on the snow distribution in road cuttings during snowdrift. *Cold Reg. Sci. Technol.* **2022**, *196*, 103505. [CrossRef]
15. Li, L.; Pomeroy, J.W. Estimates of threshold wind speeds for snow transport using meteorological data. *J. Appl. Meteorol.* **1997**, *36*, 205–213. [CrossRef]
16. Bychkova, V.I.; Kurbatova, M.M.; Zarochentsev, G.A.; Ignatov, R.Y. Blowing Snow Forecast Using Numerical Atmospheric Model Output Data. *Russ. Meteorol. Hydrol.* **2018**, *43*, 444–448. [CrossRef]
17. Libbrecht, K.G. Physical Dynamics of Ice Crystal Growth. *Annu. Rev. Mater. Res.* **2017**, *47*, 271–295. [CrossRef]
18. Huang, N.; Sang, J.B.; Han, K. A numerical simulation of the effects of snow particle shapes on blowing snow development. *J. Geophys. Res.-Atmos.* **2011**, *116*, D22206. [CrossRef]
19. Mellor, M. *Blowing Snow. Monogr. III-A3C*; US Army Cold Regions Research and Engineering Laboratory: Hanover, NH, USA, 1965.
20. He, B.H.; Bai, M.Z.; Liu, B.L.; Li, P.X.; Qiu, S.M.; Li, X.; Ding, L.S. Evaluation of Drifting Snow Susceptibility Based on GIS and GA-BP Algorithms. *Isprs Int. J. Geo-Inf.* **2022**, *11*, 142. [CrossRef]
21. Caldwell, P.; Chin, H.N.S.; Bader, D.C.; Bala, G. Evaluation of a WRF dynamical downscaling simulation over California. *Clim. Chang.* **2009**, *95*, 499–521. [CrossRef]
22. Chotamonsak, C.; Salathe, E.P.; Kreasuwan, J.; Chantara, S.; Siriwitayakorn, K. Projected climate change over Southeast Asia simulated using a WRF regional climate model. *Atmos. Sci. Lett.* **2011**, *12*, 213–219. [CrossRef]
23. Goncalves, M.; Barrera-Escoda, A.; Guerreiro, D.; Baldasano, J.M.; Cunillera, J. Seasonal to yearly assessment of temperature and precipitation trends in the North Western Mediterranean Basin by dynamical downscaling of climate scenarios at high resolution (1971–2050). *Clim. Chang.* **2014**, *122*, 243–256. [CrossRef]
24. Yang, X. Research on the Characteristics of Drifting Snow Disaster in Xinjiang Mayitas Traffic Corridor. *J. Railw. Eng. Soc.* **2018**, *35*, 1–6.

25. Qi, Y. Strategies of railway alignment selection for Karamay-Tacheng railway in Xinjiang considering snow drifting disasters. *J. Railw. Sci. Eng.* **2018**, *15*, 2813–2824.
26. Luo, Y.; Ma, W.-Y.; Sun, Y.-C. Prediction on wind speed and wind direction of traffic line in alpine region. *Gongcheng Lixue/Eng. Mech.* **2022**, *39*, 195–201. [CrossRef]
27. Ma, J.L.; Yao, X.P. Summer Extreme Precipitation in the Key Region of the Sichuan-Tibet Railway. *Adv. Atmos. Sci.* **2023**, *40*, 843–855. [CrossRef]
28. Lin, Z.Q.; Yao, X.P.; Du, J.; Zhou, Z.B. Refined Evaluation of Satellite Precipitation Products against Rain Gauge Observations along the Sichuan-Tibet Railway. *J. Meteorol. Res.* **2022**, *36*, 779–797. [CrossRef]
29. Zhang, J.P.; Zhao, T.B.; Zhou, L.B.; Ran, L.K. Historical Changes and Future Projections of Extreme Temperature and Precipitation along the Sichuan-Tibet Railway. *J. Meteorol. Res.* **2021**, *35*, 402–415. [CrossRef]
30. Gao, Y.; Xu, J.; Chen, D. Evaluation of WRF Mesoscale Climate Simulations over the Tibetan Plateau during 1979–2011. *J. Clim.* **2015**, *28*, 2823–2841. [CrossRef]
31. Chen, S.; Hamdi, R.; Ochege, F.U.; Du, H.; Chen, X.; Yang, W.; Zhang, C. Added Value of a Dynamical Downscaling Approach for Simulating Precipitation and Temperature Over Tianshan Mountains Area, Central Asia. *J. Geophys. Res.-Atmos.* **2019**, *124*, 11051–11069. [CrossRef]
32. Minder, J.R.; Letcher, T.W.; Skiles, S.M. An evaluation of high-resolution regional climate model simulations of snow cover and albedo over the Rocky Mountains, with implications for the simulated snow-albedo feedback. *J. Geophys. Res.-Atmos.* **2016**, *121*, 9069–9088. [CrossRef]
33. Wrzesien, M.L.; Durand, M.T.; Pavelsky, T.M.; Howat, I.M.; Margulis, S.A.; Huning, L.S. Comparison of Methods to Estimate Snow Water Equivalent at the Mountain Range Scale: A Case Study of the California Sierra Nevada. *J. Hydrometeorol.* **2017**, *18*, 1101–1119. [CrossRef]
34. Liu, L.; Zhang, X.Z.; Han, C.B.; Ma, Y.M. A performance evaluation of various physics schemes on the predictions of precipitation and temperature over the Tibet Autonomous Region of China. *Atmos. Res.* **2023**, *292*, 106878. [CrossRef]
35. Minola, L.; Zhang, G.F.; Ou, T.H.; Kukulies, J.; Curio, J.; Guijarro, J.A.; Deng, K.Q.; Azorin-Molina, C.; Shen, C.; Pezzoli, A.; et al. Climatology of near-surface wind speed from observational, reanalysis and high-resolution regional climate model data over the Tibetan Plateau. *Clim. Dyn.* **2023**. [CrossRef]
36. Zhou, X.; Ding, B.H.; Yang, K.; Pan, J.M.; Ma, X.G.; Zhao, L.; Li, X.; Shi, J.C. Reducing the Cold Bias of the WRF Model Over the Tibetan Plateau by Implementing a Snow Coverage-Topography Relationship and a Fresh Snow Albedo Scheme. *J. Adv. Model. Earth Syst.* **2023**, *15*, e2023MS003626. [CrossRef]
37. Zhi-Hui, Y.; Lian-Tang, D. Description of Microphysical Processes in WRF Model and Its Prediction Experiment. *Desert Oasis Meteorol.* **2007**, *1*, 1–6.
38. Xie, B.; Fung, J.C.H.; Chan, A.; Lau, A. Evaluation of nonlocal and local planetary boundary layer schemes in the WRF model. *J. Geophys. Res.-Atmos.* **2012**, *117*, D12103. [CrossRef]
39. Liu, L.; Ma, Y.; Menenti, M.; Zhang, X.; Ma, W. Evaluation of WRF Modeling in Relation to Different Land Surface Schemes and Initial and Boundary Conditions: A Snow Event Simulation Over the Tibetan Plateau. *J. Geophys. Res. Atmos.* **2019**, *124*, 209–226. [CrossRef]
40. Fan, Z.U.; Shuxian, F.; Yongwei, W.; Xi, W.U. Numerical simulation and error characteristics for wind speed on a wind farm over a complex underlying surface. *Trans. Atmos. Sci.* **2016**, *39*, 672–682.
41. Du, Y.; Xu, T.; Che, Y.; Yang, B.; Chen, S.; Su, Z.; Su, L.; Chen, Y.; Zheng, J. Uncertainty Quantification of WRF Model for Rainfall Prediction over the Sichuan Basin, China. *Atmosphere* **2022**, *13*, 838. [CrossRef]
42. Vionnet, V.; Guyomarc'h, G.; Bouvet, F.N.; Martin, E.; Durand, Y.; Bellot, H.; Bel, C.; Puglièse, P. Occurrence of blowing snow events at an alpine site over a 10-year period: Observations and modelling. *Adv. Water Resour.* **2013**, *55*, 53–63. [CrossRef]
43. Guyomarc'h, G.; Bellot, H.; Vionnet, V.; Naaim-Bouvet, F.; Déliot, Y.; Fontaine, F.; Puglièse, P.; Nishimura, K.; Durand, Y.; Naaim, M. A meteorological and blowing snow data set (2000–2016) from a high-elevation alpine site (Col du Lac Blanc, France, 2720 m a.s.l.). *Earth Syst. Sci. Data* **2019**, *11*, 57–69. [CrossRef]

applied sciences

MDPI

Article

Comparative Analysis of Deterministic Fundamental Diagrams Representative of Continuous and Interrupted Traffic Flow on Selected Regional Road in Croatia

Bojan Jovanović *, Marko Ševrović * and Grgo Luburić

Department of Transport Planning, Faculty of Transport and Traffic Sciences, University of Zagreb, 10000 Zagreb, Croatia; grgo.luburic@fpz.unizg.hr
* Correspondence: bjovanovic@fpz.unizg.hr (B.J.); msevrovic@fpz.unizg.hr (M.Š.)

Abstract: Since the inception of the traffic flow theory, numerous traffic flow models have been formulated by scholars in an effort to more accurately delineate the relationships between various traffic flow parameters. However, only a limited number of studies have explored the distinctions between fundamental traffic diagrams, which characterize continuous and interrupted traffic flow conditions. Addressing this research lacuna, we compared twelve "speed–density" and "flow–density" models fitted to empirical data collected under continuous and interrupted traffic flow conditions on a selected regional road in Croatia. The empirical data used to develop these models were extracted from video footage captured by an unmanned aerial vehicle on two representative road segments during characteristic peak and off-peak hours on workdays. Our analysis reveals that, depending on the selected traffic flow model and prevailing traffic flow conditions, the practical capacity of the observed regional road is estimated to be in the range from 799 to 2333 veh/h/lane. It was also discovered that the considered models reach practical capacity at a significantly different density under continuous and interrupted traffic stream conditions, i.e., between 37 and 129 veh/km/lane. The conducted *t*-tests underscore the need to employ distinct "speed–density" and "flow–density" regression functions for modeling continuous and interrupted traffic stream conditions.

Keywords: traffic flow theory; traffic flow modeling; fundamental traffic flow diagram; correlation analysis; regression analysis; unmanned aerial vehicles; speed–density relationship; flow–density relationship

Citation: Jovanović, B.; Ševrović, M.; Luburić, G. Comparative Analysis of Deterministic Fundamental Diagrams Representative of Continuous and Interrupted Traffic Flow on Selected Regional Road in Croatia. *Appl. Sci.* **2024**, *14*, 533. https://doi.org/10.3390/app14020533

Academic Editors: Tomasz Nowakowski, Artur Kierzkowski, Agnieszka A. Tubis, Franciszek Restel, Tomasz Kisiel, Anna Jodejko-Pietruczuk, Mateusz Zając and Arkadiusz Gola

Received: 24 July 2023
Revised: 30 November 2023
Accepted: 5 January 2024
Published: 8 January 2024

1. Introduction

Depending on the traffic stream conditions, traffic flows can be categorized as continuous, semi-continuous and periodically interrupted traffic flows. Continuous (uninterrupted) traffic stream conditions are present on open road sections without property access points and intersections. In a continuous traffic flow, vehicle speeds and traffic flow are influenced primarily by traffic density and secondarily by existing speed limits, roadway design elements and roadway environmental features. Under these conditions, vehicles do not stop until traffic density reaches its maximum value (congestion).

Semi-continuous traffic stream conditions are typically observed along freeway and motorway segments, particularly near various types of off-grade intersections. These conditions are marked by the additional interactions between vehicles in the main traffic flow and those involved in merging and diverging traffic flows, as well as between vehicles interacting along weaving road segments. In contrast to continuous traffic flow conditions, semi-continuous conditions are characterized by more frequent and substantial fluctuations in the speed, acceleration and deceleration of individual vehicles. Interrupted traffic flow conditions occur on road sections with unsignalized and signalized at-grade intersections, where two or more traffic flows cross-intersect each other. In these traffic stream conditions, the movements of individual vehicles in the conflicting traffic flows on signalized intersections are separated in time by traffic signals. Due to this fact, vehicles must stop their movement during each red phase to give way to vehicles in conflicting traffic streams. Vehicles also need to stop at minor approaches of unsignalized at-grade intersections in cases where they must yield to vehicles on the major intersection approaches. Due to the presence of periodic vehicle movements, interrupted traffic stream conditions differ significantly from continuous (uninterrupted) and semi-continuous traffic flows.

Over the past decades, numerous research works have attempted to describe and visualize the mathematical relationships between fundamental traffic flow parameters in different traffic flow regimes [1,2], utilizing diverse forms of traffic flow diagrams [3–6]. However, during this period, few researchers have specifically addressed the contrasts between the various functional forms of fundamental traffic flow diagrams suitable for representing the mathematical relationships between empirical traffic flow parameter values observed in both continuous and periodically interrupted traffic flow conditions. Therefore, in an attempt to reduce this research gap, this paper primarily aims to propose and compare optimal mathematical expressions for deterministic "speed–density" and "flow–density" fundamental diagrams suitable for describing the relationships between empirical values of relevant macroscopic traffic flow parameters in continuous and periodically interrupted traffic stream conditions on the ŽC5210 regional road within the Republic of Croatia. The purpose of this paper is to identify the differences in the expected ranges and in the relevant boundary and critical values of the empirical traffic flow parameter values representative of continuous and periodically interrupted traffic flow conditions on selected road sections of an undivided two-lane regional road and to define appropriate deterministic functions for describing the relationships between mean values of the observed traffic flow parameters. The research hypotheses are formulated as follows:

- H0 (Null Hypothesis): There is no statistically significant difference in the speed–density and flow–density functional relationships obtained for continuous and periodically interrupted traffic flow conditions on the representative road segments of the selected regional road, and therefore, the same mathematical formulation of "speed–density" and "flow–density" models can be used for modeling traffic flow characteristics on the observed regional road, both in continuous and interrupted traffic stream conditions.
- H1 (Alternative Hypothesis): There is a statistically significant difference in the speed–density and flow–density functional relationships obtained for continuous and periodically interrupted traffic flow conditions on the representative road segments of the selected regional road, and therefore, two alternative mathematical formulations of the "speed–density" and "flow–density" models need to be used to separately model the traffic flow characteristics under continuous and interrupted traffic stream conditions on the observed regional road.

The specific contributions of this research stem from (1) determining the calibrated "speed–density" equations and regression functions representative of modeling continuous and interrupted traffic stream conditions on regional (lower tier) roads in Croatia by fitting the selected macroscopic traffic flow models to empirical data collected from an open road segment and a road segment with signalized intersection from the regional road ŽC5210; (2) obtaining valuable insights into the distinctions between specific values of basic traffic flow parameters at boundary and critical points of "speed–density" and "flow–density" fundamental traffic flow diagrams of selected macroscopic traffic flow models, i.e., between the values of free flow speed, critical speed, critical density, jam density and practical road capacity representative of continuous and interrupted traffic stream conditions; (3) obtaining more realistic "speed–density" and "flow–density" relations due to the fact that empirical traffic flow density values are determined directly based on the actual number of vehicles present on the observed road segments in discrete time intervals, visible from the recorded aerial footage, i.e., they are not estimated based on occupancy or calculated indirectly by dividing empirical flow and speed values; (4) the possibility of applying the research findings to other geographic regions to identify the approximate functional forms of "speed–density" and "flow–density" fundamental diagrams, which could be used for modeling continuous and interrupted traffic stream conditions on other lower tier roads (regional and local roads) in the Republic of Croatia, as well as in comparable road categories in other countries. Regional and local road authorities could use these fundamental diagrams to make more informed decisions related to investment in road infrastructure, road traffic management, designing new and maintaining existing networks of regional and local roads. The aerial footage of observed road segments recorded in the scope of this research can be potentially used by researchers for the purpose of developing and integrating various algorithms and software based on artificial intelligence and machine learning, which would enable automatic video data processing, i.e., automatic detection and tracking of movements of individual vehicles in a traffic stream, automatic extraction of relevant traffic flow parameter values and automatic fitting of selected traffic flow models to collected empirical data.

The subsequent sections of this paper are organized in the following manner. The following chapter gives an overview of research related to this paper. Chapter 3 defines the macro and micro study areas within which drone-based aerial traffic flow video recording was conducted to collect the input data necessary for performing correlation and regression analysis between relevant traffic flow parameters. Chapter 4 provides detailed descriptions of the methodological steps employed for drone-based video recording of traffic flows on selected road sections, data processing, data filtering and analysis of the recorded aerial video files. Additionally, this chapter outlines the methodological procedures utilized for determining the empirical values of fundamental traffic flow parameters. It also highlights the approach taken to identify specific theoretical forms of fundamental traffic flow diagrams suitable for describing the real traffic flow characteristics on selected road sections. Chapter 5 delves into the results of the performed correlation and regression analysis carried out between the empirical values of observed traffic flow parameters and gives a comparative overview of suggested "speed–density" and "flow–density" models suitable for describing the traffic stream characteristics on observed road sections under continuous and interrupted traffic stream conditions. The conclusion gives a summary of the attained results and offers recommendations for conducting future research.

2. Literature Review

Since the inception of the traffic flow theory in the 1930s, a multitude of researchers have endeavored to develop mathematical models, which accurately represent the relationships between empirical traffic flow parameter values. During this period, many scientific papers, studies and projects were devoted to the development and validation of various deterministic [7–28] and stochastic representations [29–32] of basic traffic flow diagrams. The main goal of these efforts has been to represent the complex spatial and temporal characteristics of traffic flow more accurately.

Over the past 90 years, researchers have proposed various forms of fundamental traffic flow diagrams in their attempt to improve the initial Greenshields linear "speed–density" model developed in 1934 [7,8]. In 1959, Greenberg proposed a logarithmic form of the speed–density relationship [9]. A year later, Underwood suggested that the speed–density relationship can be described by the negative exponential function [10], while in 1961, Eddie presented the first two-regime traffic flow model based on two logarithmic functions [11]. In 1966, Dick developed the first traffic flow model specifically adapted for describing the relation between speed and density on urban roads [12]. In 1967, Drake suggested that the speed–density relationship can be represented by Gaussian distribution [13]. At the same time, May and Keller [33] and independently Pipes [34] developed generalized "speed–density" models by expanding the initial Greenshields linear model with the introduction of additional adjustment factors. Smulders, in 1990, proposed a combined linear–parabolic "speed–density" model [15]. In 1995, Del Castillo and Van Aerde independently presented new modified versions of the "speed–density" model [16,35]. In 2003, Chanut and Buisson developed a three-dimensional fundamental traffic flow diagram, based on which, it is possible to describe the differences present in a "flow–density" relationship at different percentages of heavy vehicles in a traffic stream [19]. Over the last two decades, an increasing number of researchers have suggested that the "speed–density" relationship can be more accurately described by different forms of logistic and sigmoid functions [21,23,36].

In their pursuit of finding the optimal deterministic and stochastic functional forms for the "speed–density", "flow–density" and "flow–speed" models, the majority of researchers have predominantly focused on studying vehicle interactions within uninterrupted traffic flow conditions, which has been regarded as the basic traffic flow condition representative of establishing the fundamental relations between the main traffic flow parameters. On the other hand, only a limited number of researchers have studied specific functional relationships between the macroscopic and microscopic traffic flow parameters under interrupted traffic flow conditions [37–44] and discussed the comparative differences between the values of traffic flow parameters measured in uninterrupted and interrupted traffic flow conditions [45–48]. In a 1996 study, Akcelik provided a comparative overview of the mathematical relations between speed, density and travel time, which can be used to describe the characteristics of traffic flow in uninterrupted and interrupted traffic stream conditions [45]. In 2005, Pueboobpaphan, Nakatsuji, Suzuki and Kawamura developed three distinct shock wave models, which can be used to convert vehicle speeds measured within uninterrupted traffic stream conditions on a road segment located prior to a signalized intersection into vehicle speeds present in the interrupted traffic stream conditions formed in the immediate vicinity of the signalized intersection [46]. In the same year, Pueboobpaphan et al. attempted to identify the relationship between vehicle speeds in uninterrupted and interrupted traffic stream conditions using the method of characteristics [47]. In a study conducted in 2021, Barka and Polis performed a calibration of the most commonly used volume-delay models, including BPR, Conical, Akcelik and modified Davison model, for continuous and interrupted traffic stream conditions in different road categories in the urban area of Thessaloniki in Greece [48].

In recent years, a growing number of researchers have used unmanned aerial vehicles (drones) to capture the aerial footage of traffic flows, based on which, it is possible to extract empirical values of relevant traffic flow parameters [49–55]. This innovative approach facilitates the analysis, comparison and assessment of prevailing traffic stream characteristics across various road network segments.

One of the main advantages of using unmanned aerial vehicles for traffic data collection stems from the fact that they allow data collection without significantly disrupting normal traffic flow. Therefore, by using drones, it is possible to ensure that the collected data remain representative of real-world conditions without introducing bias into samples, which are used to develop the "speed–density" and "flow–density" regression models. Furthermore, drones equipped with high-resolution cameras and advanced sensors can capture much finer details of complex traffic flow spatiotemporal characteristics, and aerial video footage recorded by drones can be used for numerous research purposes. With the further development and integration of various algorithms and software based on artificial intelligence and machine learning, aerial footage can be automatically processed and used to detect and track the movement of individual vehicles in a traffic stream, identify complex traffic flow patterns, extract the trajectory of vehicles, detect various traffic flow anomalies, predict potential traffic flow congestions and vehicle collisions. The ability to utilize these computational tools to process aerial video data far surpasses the limited possibilities of traditional traffic data collection methods. In addition to that, aerial video footage can be stored and used again at any time to conduct additional research. Regardless of the fact that the implementation of commercially available unmanned aerial vehicles is still limited by their relatively short battery life, there is no doubt that, in the future years, they will become fully accepted as one of the most advanced, accurate and reliable techniques for the collection of traffic data.

Quite a number of researchers have already implemented unmanned aerial vehicles for the purpose of capturing aerial video footage of traffic flows. For instance, Yahia and his team explored the potential of utilizing unmanned aerial vehicles for simultaneous traffic state estimation and incident detection on multiple road segments [56]. In a 2021 study, Ahmed, Ngoduy, Adnan and their colleagues proposed specific functional forms of fundamental traffic flow diagrams for describing the characteristics of heterogeneous and undisciplined traffic streams [57]. Specific functional forms of fundamental diagrams were developed based on empirical traffic data gathered from aerial footage captured by drones over representative road segments in the city of Karachi in Pakistan.

Recently, an increasing number of researchers suggested that the "speed–density" relationship can be more accurately described by compound forms of logarithmic, exponential, potential, logistic and sigmoid S-shaped multi-parameter functions [58–63]. Besides this, it should also be considered that various other types of mathematical models used in other fields, such as Morgan-Mercer-Flodin growth model, could be modified for the purpose of describing the relation between traffic flow speed and density [64].

3. Spatial and Temporal Scope of Research

The study area includes one macro- and two micro-observation zones (Figure 1). The macro-observation zone includes the urban road network of the city of Novalja in the Republic of Croatia, its bypass and the main radial roads in the vicinity of the city, including state road D106 and regional roads ŽC5210 and ŽC5151. The micro-observation zones include two road sections of regional road ŽC5210 with the following specific road design elements and traffic flow characteristics:

1. Road section A: The road section of regional road ŽC5210, spanning from a three-leg unsignalized intersection (connecting regional road ŽC5210 to Čiponjac South Street) to a four-leg roundabout (linking regional road ŽC5210 with state road D106), covering 310 m. The traffic flows were recorded on a straight section of this road, sufficiently distanced from adjacent intersections to ensure that all empirical traffic flow parameter values, representative of continuous traffic stream conditions, could be accurately extracted from the captured video files.
2. Road section B: The road section of regional road ŽC5210, situated near a signalized intersection (connecting regional road ŽC5210 to Caskin Way Street), 205 m in length. The traffic flows were recorded on a straight segment of the road, directly adjacent to the four-leg signalized intersection to ensure that the empirical traffic flow parameter values, indicative of interrupted traffic stream conditions, could be accurately extracted from the captured video files.

The reason behind selecting this particular geographic region and road category for this study stems from the following facts: (1) To date, there have been few studies in the world and in Croatia, which have developed the specific forms of "speed–density" and "flow–density" fundamental diagrams representative of modeling traffic flow characteristics on lower tier (regional and local) bi-directional undivided roads; (2) Due to high seasonal traffic flow fluctuations in the selected geographic region, i.e., higher traffic flow volumes on lower tier roads during the tourist summer season, it was possible to capture the different patterns of traffic flow dynamics present in various traffic flow regimes (free flow, normal and saturated traffic flow) and to obtain a sufficiently large empirical sample necessary for development and calibration of "speed–density" and "flow–density" fundamental diagrams, which would not be possible in other geographical regions of Croatia, where traffic flow on lower tier roads is typically very low during the whole year.

Figure 1. The research region with designated macro- and micro-observation zones. Source: Modified by authors based on maps and satellite imagery available in Google Maps, Geoportal DGU and Geoportal of Croatian public roads [65–67].

The empirical values of traffic flow parameters—essential for conducting statistical analysis to determine the optimal functional forms of "speed–density" and "flow–density" fundamental diagrams—were calculated on the basis of data obtained from the aerial photographs taken by unmanned aerial vehicles on two selected road sections of regional road ŽC5210 at representative 15 min intervals during peak hours (in the morning period from 7:00 to 9:00 a.m. and in the afternoon period from 15:00 to 17:00 p.m.) and off-peak hours (the period from 11:00 a.m. to 13:00 p.m.). Figure 1 shows the macro- and micro-observation zones situated within the study area, together with the indicated time periods of the aerial video recordings conducted and the size of the empirical samples collected on road sections A and B.

4. Methodology of Research

The basic steps of the methodology used for recording the aerial footage of selected road segments and for the processing, filtering and analysis of data extracted from captured video files are shown in Figure 2.

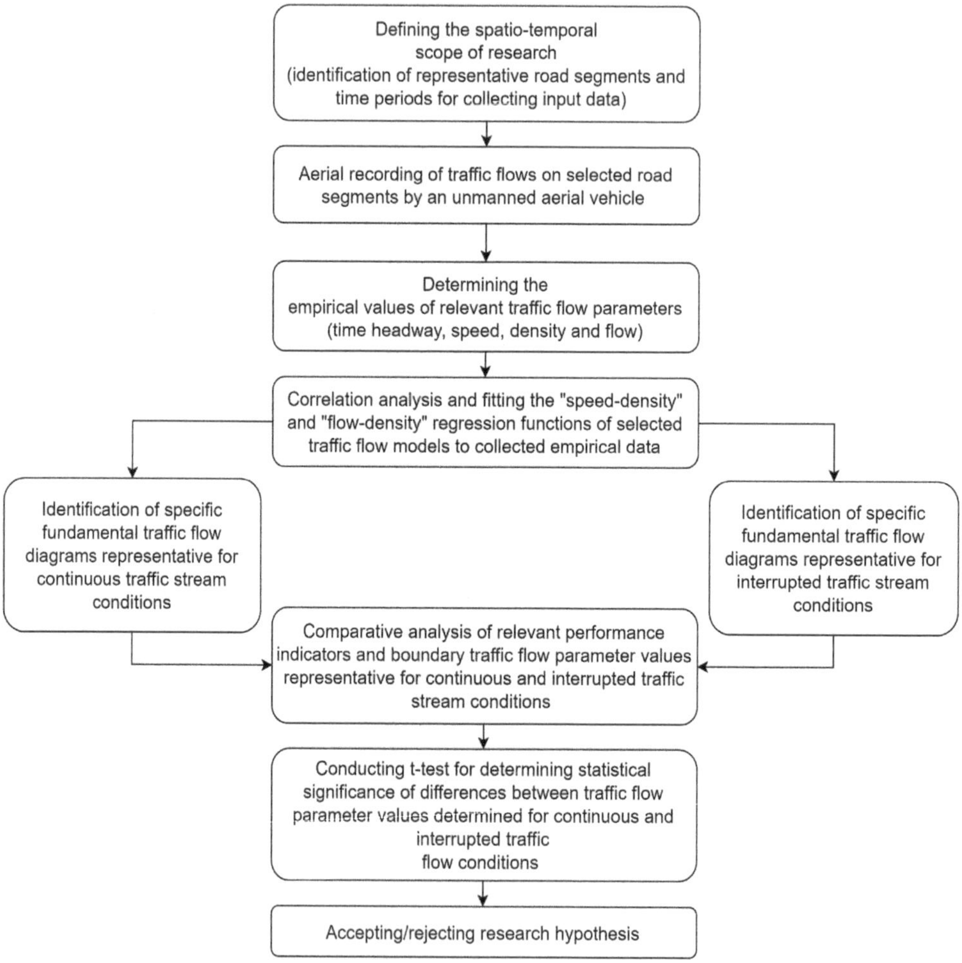

Figure 2. Flow chart of research methodology.

4.1. Defining the Spatio-Temporal Scope of Research

The empirical samples required to perform descriptive, correlation and regression analyses of the relevant macroscopic traffic flow parameters were formed based on data extracted from the aerial video files recorded on two selected road sections of regional road ŽC5210 in Croatia. To ensure the acquisition of empirical samples representative of continuous and interrupted traffic flows, aerial video recordings were conducted on one straight open road section without intersections and one road section near a four-leg signalized intersection. The recordings were conducted on workdays at representative five-, fifteen- and twenty-minute intervals during peak and off-peak hours to capture the different traffic flow regimes (free, normal, stable, unstable and saturated flow).

4.2. Capturing and Processing Aerial Video

The aerial recording of traffic flows on selected road segments was performed using an unmanned aerial vehicle MAVIC PRO equipped with advanced GPS/GLONASS satellite positioning systems, along with sensor systems enabling intelligent flight control, detection and avoidance of both stationary and moving obstacles during its operation. The drone was additionally equipped with a fully stabilized video camera, capable of capturing

12-megapixel photographs and recording aerial video footage in a very high 4K Quad HD resolution of 3840 × 2160 pixels. The aerial videos were captured at frame rates of 30 images per second. The recorded aerial video files were extracted from the drone's micro-SD card and transferred to a personal computer. They were then converted from their original MOV format into the MP4 format, ensuring compatibility for subsequent video data processing.

4.3. Determining the Empirical Values of Relevant Traffic Flow Parameters

Before reviewing the aerial video recordings, it was necessary to define the positions of entrance and exit reference lines separately for each observed road segment on ŽC5210 regional road. These lines were used for detection of vehicle entrance and exit times on each observed road segment, based on which, it was possible to calculate the empirical values of all relevant traffic flow parameters, including vehicle time headways, vehicle speeds, traffic density and traffic flow values. Vehicle time headways at defined entry/exit reference lines on each observed road section of ŽC5210 regional road were computed by subtracting the times at which two successive vehicles in a traffic stream have passed through each entry/exit reference line. Mathematically, this can be expressed using the following formulae:

$$th_{Ui} = t_{Ui+1} - t_{Ui}[s], \tag{1}$$

$$th_{Ii} = t_{Ii+1} - t_{Ii}[s], \tag{2}$$

where th_{Ui} and th_{Ii} represent vehicle time headways between two successive vehicles in a traffic stream determined on the entry and exit reference lines in seconds, respectively; t_{Ui} and t_{Ii} represent the arrival time of the i-th vehicle (leading vehicle) on the entry and exit reference line in seconds, respectively; t_{Ui+1} and t_{Ii+1} represent the arrival time of the first successive vehicle behind the i-th vehicle (following vehicle) on the entry and exit reference line in seconds, respectively. The velocity of each vehicle in the traffic stream was calculated by dividing the distance between the defined entry and exit reference lines, which represents the length of the observed road segment, by the difference between the exit times and the entry times recorded for each vehicle in the database, which represents the travel time of each vehicle through the observed road segment. This can be expressed using the following formula:

$$V_i = 3.6 \cdot v_i = 3.6 \cdot \frac{L_o}{\Delta t_i} = 3.6 \cdot \frac{L_o}{(t_{Ui} - t_{Ii})}[km/h], \tag{3}$$

where V_i represents the velocity of the i-th vehicle in the traffic stream, expressed in kilometers per hour; v_i represents the velocity of the i-th vehicle in the traffic stream, expressed in meters per second; L_o represents the length of the observed road segment in meters; and all other variables are as previously defined. In order to determine the empirical values of traffic density, the aerial videos were stopped at frames on which individual vehicles were present on the entry reference detection lines defined for each observed road segment. The current number of vehicles, which were present between the entry and exit reference lines at those times, was recorded in the database. The recorded number of vehicles between the entry and exit detection lines at discrete time intervals was then used to calculate traffic density using the following formula:

$$g_i = 1000 \cdot \frac{N_i}{L_0}[veh/km], \tag{4}$$

where g_i represents traffic density at discrete time interval i, expressed in the number of vehicles per kilometer, and N_i represents the number of vehicles on the observed road segment at discrete time interval i. All other variables are as previously defined. The traffic flow values were computed based on the previously determined empirical values of relevant macroscopic and microscopic traffic flow parameters, according to the following expressions:

$$q_i = V_i \cdot g_i = \frac{3600 \cdot Ni}{(t_{Ui} - t_{Ii})} = \frac{3600 \cdot Ni}{\Delta t} = \frac{3600}{t_h}[\text{veh/h}], \quad (5)$$

where q_i represents the traffic flow at discrete time interval i, expressed in vehicles per hour; Δt represents the travel time of the i-th vehicle in a traffic stream through the observed road segment, expressed in seconds; t_h represents the time headway in seconds; and all other parameters are as previously defined. All empirical traffic flow parameter values were stored in a database set up within the MS Excel platform. The traffic flow parameter values, computed based on data collected on two observed road sections of ŽC5210 regional road, were stored in two distinct data tables named "ZC5210_Road_Section_A" (Table S1) and "ŽC5210_Road_Section_B" (Table S2). Following a data filtering process, the values stored in relevant data columns were then extracted separately for each observed traffic stream and recorded in a comparative output table named "ZC5210_Comparative_Models" (Table S3). The comparative output table comprises the computed empirical values of relevant traffic flow parameters for a total of 1581 vehicles, i.e., a total of 4743 observations, including 1581 vehicle speed values, 1581 traffic density values and 1581 time headway values.

4.4. Comparative Analysis, Calibration and Fitting the Deterministic "Speed–Density" and "Flow–Density" Fundamental Traffic Flow Diagrams to Collected Empirical Data

The empirical values of traffic flow parameters were visualized on "speed–density" and "flow–density" scatterplots and then analyzed using the least-squares method to identify the best fitting regression functions, which could be effectively used for deterministic modeling of traffic flow characteristics under continuous and interrupted traffic stream conditions on the two observed road segments of ŽC5210 regional road.

In order to determine specific "speed–density" and "flow–density" fundamental traffic flow diagrams suitable for modeling traffic flow characteristics under continuous and interrupted traffic stream conditions, in the scope of this research, we comparatively analyzed, calibrated and fitted mathematical formulations of twelve selected traffic flow models. Based on the regression fitting procedure conducted using Python script created in the PyCharm integrated development environment [68], three distinct "speed–density" and three "flow–density" regression functions were established for each of the traffic flow models considered:

- "Speed–Density" and "Flow–Density" regression functions obtained using the overall statistical sample collected containing all empirical values of traffic flow parameters determined both under continuous and interrupted traffic stream conditions;
- "Speed–Density" and "Flow–Density" regression functions developed based on the partial empirical sample collected under continuous traffic stream conditions on the open road segment (road segment A); and
- "Speed–Density" and "Flow–Density" regression functions developed based on the partial empirical sample collected under interrupted traffic stream conditions on the road segment with signalized intersection (road segment B).

The obtained regression functions of proposed deterministic models were compared to highlight the differences among the deterministic fundamental traffic flow diagrams suitable for describing the continuous and interrupted traffic stream conditions on the selected undivided bi-directional two-lane road in the Republic of Croatia.

5. Results

5.1. Results of the Comparative and Regression Analysis of Selected "Speed–Density" Models Fitted to Empirical Data Contained in Overall Statistical Sample

The outcomes of the regression analysis conducted between the empirical values of speed and traffic density contained in the overall statistical sample collected are depicted in Figure 2 and Table 1. By comparing the values of Pearson coefficient (r) and coefficient of determination (r^2) obtained for the twelve "speed–density" models considered, it was discovered that Wang's five-parameter sigmoid model best fits the empirical data, with a Pearson correlation coefficient of 0.6191 and a coefficient of determination of 0.3833.

Table 1. Comparative overview of fundamental and calibrated "speed–density" equations of traffic flow models considered, obtained by fitting the "speed–density" regression functions to empirical data contained in the overall statistical sample collected from selected representative segments of regional road ŽC5210.

Model	Fundamental "Speed–Density" Equation	Calibrated "Speed–Density" Equation	R-Squared
Greenshields (1935) [7]	$v(g) = v_{max}\left(1 - \frac{g}{g_{max}}\right)$	$v(g) = 52.05\left(1 - \frac{g}{160.00}\right)$	0.2690
Greenberg (1959) [9]	$v(g) = v_c \ln\left(\frac{g_{max}}{g}\right)$	$v(g) = 18.02\ln\left(\frac{160}{g}\right)$	0.1563
Underwood (1961) [10]	$v(g) = v_{max}\exp\left(-\frac{g}{g_c}\right)$	$v(g) = 54.88\exp\left(-\frac{g}{69.05}\right)$	0.3760
Drake (1967) [13]	$v(g) = v_{max}\exp\left(-\frac{1}{2}\left(\frac{g}{g_c}\right)^2\right)$	$v(g) = 50\exp\left(-\frac{1}{2}\left(\frac{g}{40.75}\right)^2\right)$	0.3559
Modified Morgan (1975) [64]	$v(g) = \frac{v_{max} - v_{min}}{1+(m \cdot g)^n} + v_{min}$	$v(g) = \frac{49.84+4.32}{1+(0.018 \cdot g)^{1.72}} - 4.32$	0.3825
Wang (3P) (2011) [23]	$v(g) = \frac{v_{max}}{1+\exp(m(g-g_c))}$	$v(g) = \frac{74.14}{1+\exp(0.03(g-28.69))}$	0.3817
Wang (5P) (2011) [23]	$v(g) = v_c + \frac{v_{max} - v_c}{\left[1+\exp\left(\frac{g-g_c}{\theta_1}\right)\right]^{\theta_2}}$	$v(g) = \frac{74.69}{\left[1+\exp\left(\frac{g-35}{35.6}\right)\right]^{1.15}}$	0.3833
Kucharski and Drabicki (2017) [4]	$v(g) = \frac{v_{sw}}{\left[1+a\left(\frac{g}{g_o}\right)\right]^b}$	$v(g) = \frac{54.87}{\left[1+0.01\left(\frac{g}{60}\right)\right]^{86.45}}$	0.3611
Gaddam and Rao (5P) (2019) [58]	$v(g) = v_{max}\left\{\frac{\exp\left[-\left(\frac{g}{g_c}\right)^{1+\delta 1}\right] - \exp\left[-\left(\frac{g_{max}}{g_c}\right)^{1+\delta 1}\right]}{1-\exp\left[-\left(\frac{g_{max}}{g_c}\right)^{1+\delta 1}\right]}\right\}^{\delta 2}$	$v(g) = 44\left\{\frac{\exp\left[-\left(\frac{g}{57}\right)^3\right] - \exp\left[-\left(\frac{167}{57}\right)^3\right]}{1-\exp\left[-\left(\frac{167}{57}\right)^3\right]}\right\}^{0.8}$	0.3730
Gaddam and Rao (6P) (2019) [58]	$v(g) = v_{max}\left[1-\left(\frac{g}{g_{max}}\right)^{\gamma 1}\right]^{\gamma 2}\left[1+E\left(\frac{g}{g_{max}}\right)^{\gamma 3}\right]^{-1}$	$v(g) = 63.97\left[1-\left(\frac{g}{167}\right)^{0.14}\right]^{3.69}\left[1+492.5\left(\frac{g}{167}\right)^{0.40}\right]^{-1}$	0.3810
Fosu et al. (2020) [58]	$v(g) = v_c \ln\left(\frac{g_{max}+g_{min}}{g+g_{min}}\right)^{\phi}$	$v(g) = 25.88\ln\left(\frac{167}{g}\right)^{0.57}$	0.3123
Cheng et al. (2021) [63]	$v(g) = \frac{v_{max}}{\left[1+\left(\frac{g}{g_c}\right)^m\right]^{2/m}}$	$v(g) = \frac{56.13}{\left[1+\left(\frac{g}{83}\right)^1\right]^2}$	0.2681

The meaning of the parameters in the table is as follows: $v(g)$ represents speed as a function of traffic density, expressed in kilometers per hour; g—represents traffic density as an independent variable, expressed in the number of vehicles per kilometer; v_{max} denotes free-flow speed; g_{max} denotes jam density; v_c is the critical speed at road capacity; and g_c is the critical density at road capacity.

The second-best model is Morgan's modified logistic model, with correlation and determination coefficients of 0.6184 and 0.3825, respectively. Similar values of correlation and determination coefficients were also obtained for Wang's three-parameter sigmoid model ($r = 0.6178$ and $r^2 = 0.3817$), Gaddam and Rao's six-parameter model ($r = 0.6172$ and $r^2 = 0.3810$), Underwood's negative exponential model ($r = 0.6131$ and $r^2 = 0.3760$) and Gaddam and Rao's five-parameter model ($r = 0.6107$ and $r^2 = 0.3730$). Underwood's negative exponential model slightly outperformed Gaddam and Rao's six-parameter model, Kucharski and Drabicki's model, Drake's bell-curve model and Fosu's model. Greenshields' model achieved a determination coefficient of 0.2690 and therefore fitted empirical data better than Cheng's model and Greenberg's logarithmic model, which showed the weakest fit with the sample data, with the r-squared value being equal to 0.1563.

5.2. Results of the Comparative and Regression Analysis of Selected "Speed–Density" Models Fitted to Empirical Data Contained in Partial Statistical Samples Collected under Continuous and Interrupted Traffic Stream Conditions

In order to determine the specific functional forms of the "speed–density" models considered—representative of describing the relationship between the empirical values of speed and traffic density under continuous and interrupted traffic flow conditions—we conducted two separate regression analyses: the first one by fitting the traffic flow models considered to empirical data contained in the partial statistical sample collected on an open road segment (road segment A) and the second one by fitting the models considered to the observations included in the partial sample gathered on the road segment with signalized intersection (road segment B).

Table 2 gives a comparative overview of the specific mathematical formulations of the traffic flow models considered, which were identified as representative of modeling continuous and interrupted traffic stream conditions on the observed regional road, together with the correlation and determination coefficient values obtained. Based on the results obtained, it is evident that nine out of the twelve models proposed for modeling the continuous traffic flow condition achieved similar results in terms of correlation and determination coefficients. Most of the traffic flow models considered show an approximately equal determination coefficient value (r^2 between 0.28 and 0.27). The lowest determination coefficient value ($r^2 = 0.24$) was obtained for Greenberg's model. On the other hand, the results obtained for the models fitted to empirical data collected under interrupted traffic stream conditions show significantly lower correlation and determination coefficients compared to the models proposed for modeling continuous traffic stream conditions. The determination coefficient for regression models adjusted for interrupted traffic stream conditions ranges from 0.07 to 0.11, respectively.

Table 2. Comparative overview of specific mathematical formulations of the traffic flow models considered, obtained by fitting the "speed–density" regression functions to empirical data contained in partial statistical samples collected under continuous and interrupted traffic stream conditions.

Model	Regression Function for Continuous Traffic Stream Condition	R^2	Regression Function for Interrupted Traffic Stream Condition	R^2
Greenshields (1935) [7]	$v(g) = 53.53(1 - \frac{g}{160.00})$	0.29	$v(g) = 34.36(1 - \frac{g}{160.00})$	0.09
Greenberg (1959) [9]	$v(g) = 18.75\ln\left(\frac{160.00}{g}\right)$	0.24	$v(g) = 15.03\ln\left(\frac{160.00}{g}\right)$	0.06
Underwood (1961) [10]	$v(g) = 54.32\exp(-\frac{g}{100.54})$	0.28	$v(g) = 50.00\exp(-\frac{g}{58.59})$	0.07
Drake (1967) [13]	$v(g) = 50.00\exp\left(-\frac{1}{2}\left(\frac{g}{46.90}\right)^2\right)$	0.28	$v(g) = 50.00\exp\left(-\frac{1}{2}\left(\frac{g}{37.33}\right)^2\right)$	0.09
Modified Morgan (1975) [64]	$v(g) = \frac{51.57}{1+(0.02\cdot g)^{1.60}}$	0.28	$v(g) = \frac{30.47}{1+(0.02\cdot g)^{2.88}}$	0.09
Wang (3P) (2011) [23]	$v(g) = \frac{87.95}{1+\exp(0.02(g-21.77))}$	0.29	$v(g) = \frac{43.93}{1+\exp(0.02(g-58.21))}$	0.09
Wang (5P) (2011) [23]	$v(g) = \frac{66.93}{\left[1+\exp\left(\frac{g-41.94}{33.63}\right)\right]^{0.95}}$	0.28	$v(g) = \frac{31.32}{\left[1+\exp\left(\frac{g-45.98}{14.61}\right)\right]^{0.31}}$	0.15
Kucharski and Drabicki (2017) [4]	$v(g) = \frac{54.46}{1+\left[0.08\left(\frac{g}{60}\right)\right]^{7.42}}$	0.28	$v(g) = \frac{35.04}{1+\left[0.01\left(\frac{g}{60}\right)\right]^{45.35}}$	0.08
Gaddam and Rao (5P) (2019) [58]	$v(g) = 48.2\left\{\frac{\exp\left[-\left(\frac{g}{54.2}\right)^3\right]-\exp\left[-\left(\frac{167}{54.2}\right)^3\right]}{1-\exp\left[-\left(\frac{167}{54.2}\right)^3\right]}\right\}^{0.81}$	0.27	$v(g) = 30\left\{\frac{\exp\left[-\left(\frac{g}{74.8}\right)^3\right]-\exp\left[-\left(\frac{167}{74.8}\right)^3\right]}{1-\exp\left[-\left(\frac{167}{74.8}\right)^3\right]}\right\}^{0.66}$	0.10
Gaddam and Rao (6P) (2019) [58]	$v(g) = 53.75\left[\left(1-\left(\frac{g}{167}\right)\right)^{0.33}\right]^{4.4}\left[1+4.93\left(\frac{g}{167}\right)^0\right]^{-1}$	0.28	$v(g) = 33.97\left[1-\left(\frac{g}{167}\right)^{0.27}\right]^{3.65}\left[1+4.32\left(\frac{g}{167}\right)^0\right]^{-1}$	0.11
Fosu et al. (2020) [58]	$v(g) = 33.65\ln\left(\frac{167}{g}\right)^{0.55}$	0.25	$v(g) = 22.28\ln\left(\frac{167}{g}\right)^{0.36}$	0.07
Cheng et al. (2021) [63]	$v(g) = \frac{51.70}{\left[1+\left(\frac{g}{81.97}\right)^{1.54}\right]^{2/1.54}}$	0.28	$v(g) = \frac{30.36}{\left[1+\left(\frac{g}{64.67}\right)^{3.17}\right]^{2/3.17}}$	0.09

The obtained values of the determination coefficient (r^2) suggest that, depending on the traffic flow model considered, the amount of variability in the speed variable, which can be effectively explained by the density variable, varies from 15.63% to 38.33%. The R-squared values presented in Tables 1 and 2 indicate that the models considered can only explain the relatively small percentage of variability present in the measured vehicle speeds and flows. The low R-squared values obtained for "speed–density" and "flow–density" models fitted to the empirical data are a result of several factors.

The R-squared values obtained are especially low for models fitted to the empirical sample collected under interrupted traffic stream conditions. This can primarily be explained by the fact that interrupted traffic streams, especially near signalized intersections, have inherent variability due to stop-and-go movements. This introduces additional randomness in the data and therefore lowers the obtained values of the determination coefficient. Moreover, the scattering effect present in the "speed–density" and "flow–density" fundamental diagrams can be explained by numerous other contributing factors, including the impact of road geometry and the presence of specific road environment characteristics on the observed road segment, the impact of traffic flow structure, the presence of pedestrian flows near the signalized intersections and the variations in drivers' behavioral characteristics, just to name a few. On the other hand, it is also important to consider the fact that the models presented in this paper were developed based on the disaggregated empirical traffic flow parameter values, and therefore, it was expected that the R and R-squared values would be relatively low. By aggregating the measured speed and flow values into larger time intervals, it is possible to develop alternative mathematical formulations of "speed–density" and "flow–density" functions, which will have higher R and R-squared values; however, these models may be too simplistic for capturing the stochastic characteristics of real traffic flows, especially under interrupted traffic stream conditions.

Deterministic "speed–density" models fitted to empirical data contained in the statistical sample collected from the observed road segments of regional road ŽC5210 are presented on Figure 3, while Figure 4 presents comparative scatterplots with specific forms of "speed–density" regression functions suitable for describing the mathematical relation between mean traffic flow speed and mean traffic flow density under continuous and interrupted traffic stream conditions on the two observed segments of regional road ŽC5210 in the Republic of Croatia. The empirical observations collected on the open road segment and the road segment with signalized intersections—as well as the resulting regression functions suitable for modeling continuous and interrupted traffic stream conditions on the observed regional road—are represented in green and red colors on the scatterplots, respectively. The regression functions presented were obtained by fitting the traffic flow models considered to data contained in the statistical subsamples collected on road segments A and B. Based on the data shown in the comparative speed–density scatterplots, it is clear that all regression functions obtained by fitting the traffic flow models considered to the empirical sample collected on the open road segment are positioned above the regression functions determined based on the empirical sample collected on the road segment with signalized intersection for the entire range of possible density values. Based on the comparative analysis of these two groups of regression functions, it was possible to determine the absolute and relative differences among mean traffic flow speeds representative of continuous and interrupted traffic stream conditions at different traffic density levels, i.e., in different traffic flow regimes.

The fundamental traffic flow diagrams (Figures 5 and 6) representative of the observed road segments of regional road ŽC5210 were derived by transforming the proposed linear "speed–density" models into "flow–density" models. This transformation was conducted by applying the fundamental traffic flow equation, which establishes a direct relationship between the average vehicle flow and the product of average traffic flow speed and density.

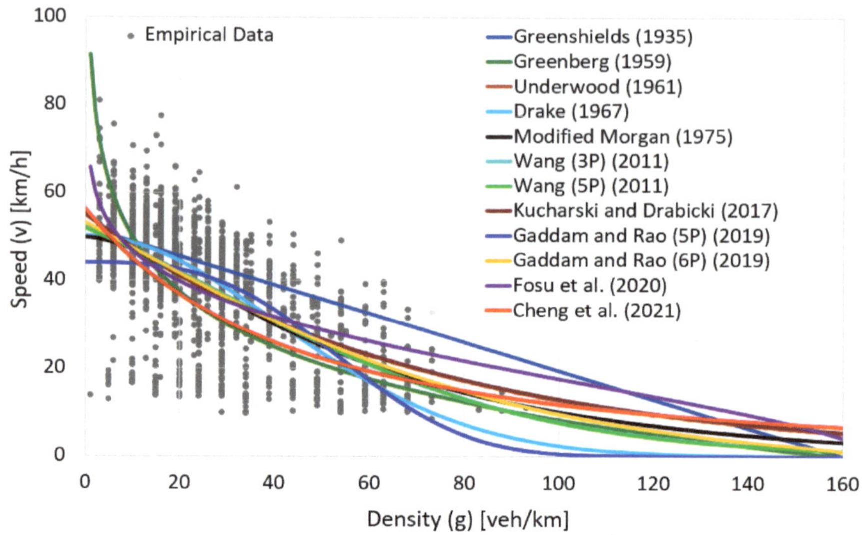

Figure 3. Deterministic "speed–density" models fitted to empirical data contained in the statistical sample collected from the observed road segments of regional road ŽC5210; (equations and references for considered traffic flow models are available in Tables 1 and 2). Source: Created by authors.

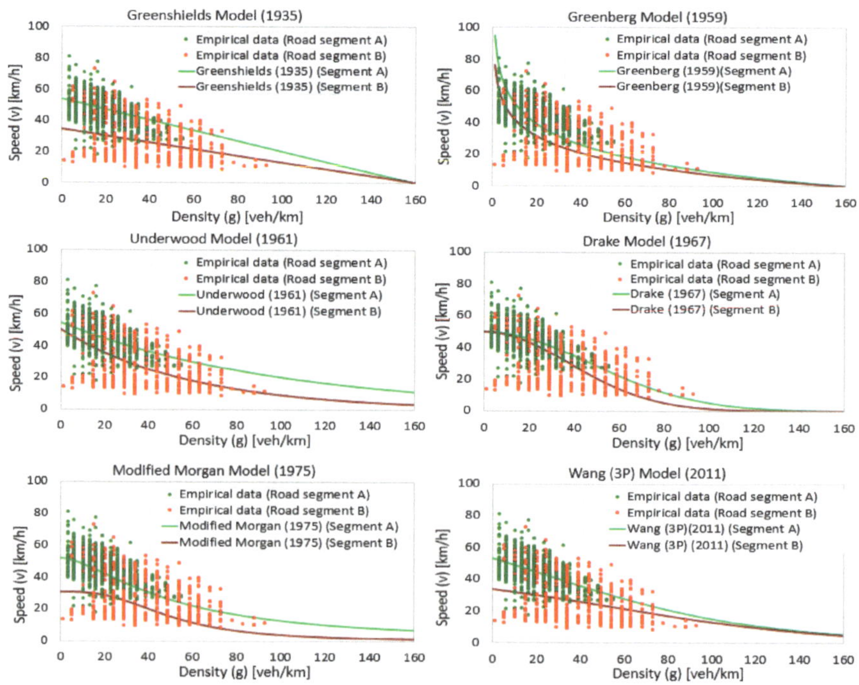

Figure 4. *Cont.*

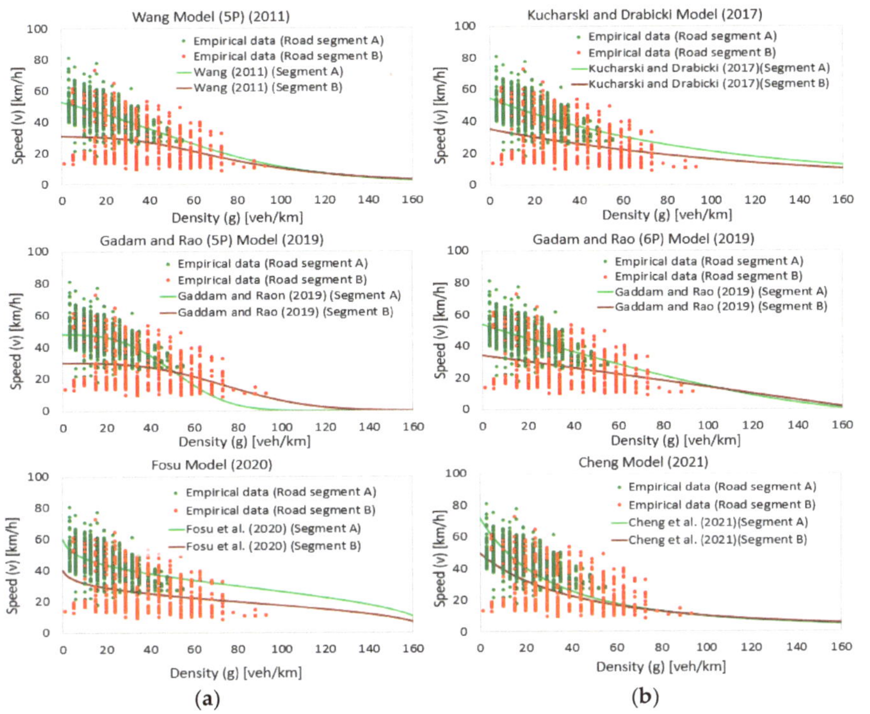

Figure 4. Comparative presentation of selected deterministic "speed–density" models fitted to the empirical data contained in the partial statistical sample collected on (**a**) Road segment A of ŽC5210 regional road (open road segment) and (**b**) Road segment B of ŽC5210 regional road (road segment with signalized intersection); (equations and references for considered traffic flow models are available in Tables 1 and 2). Source: Created by authors.

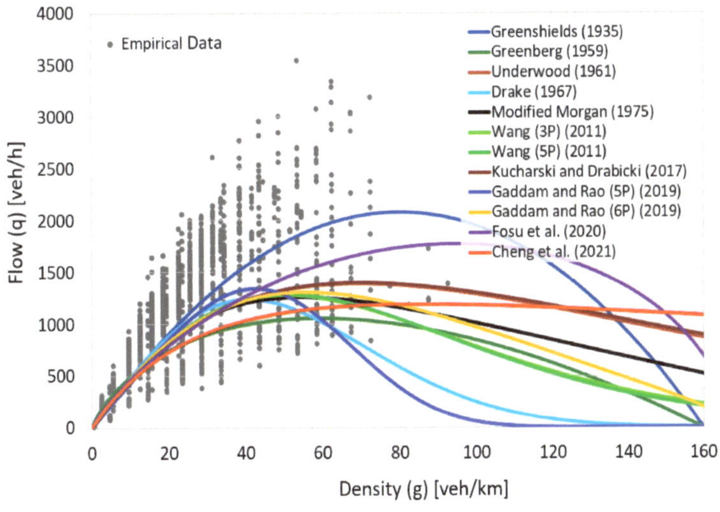

Figure 5. Selected deterministic "flow–density" models fitted to empirical data contained in the statistical sample collected on the observed road segments of regional road ŽC5210; (equations and references for considered traffic flow models are available in Tables 1 and 2). Source: Created by authors.

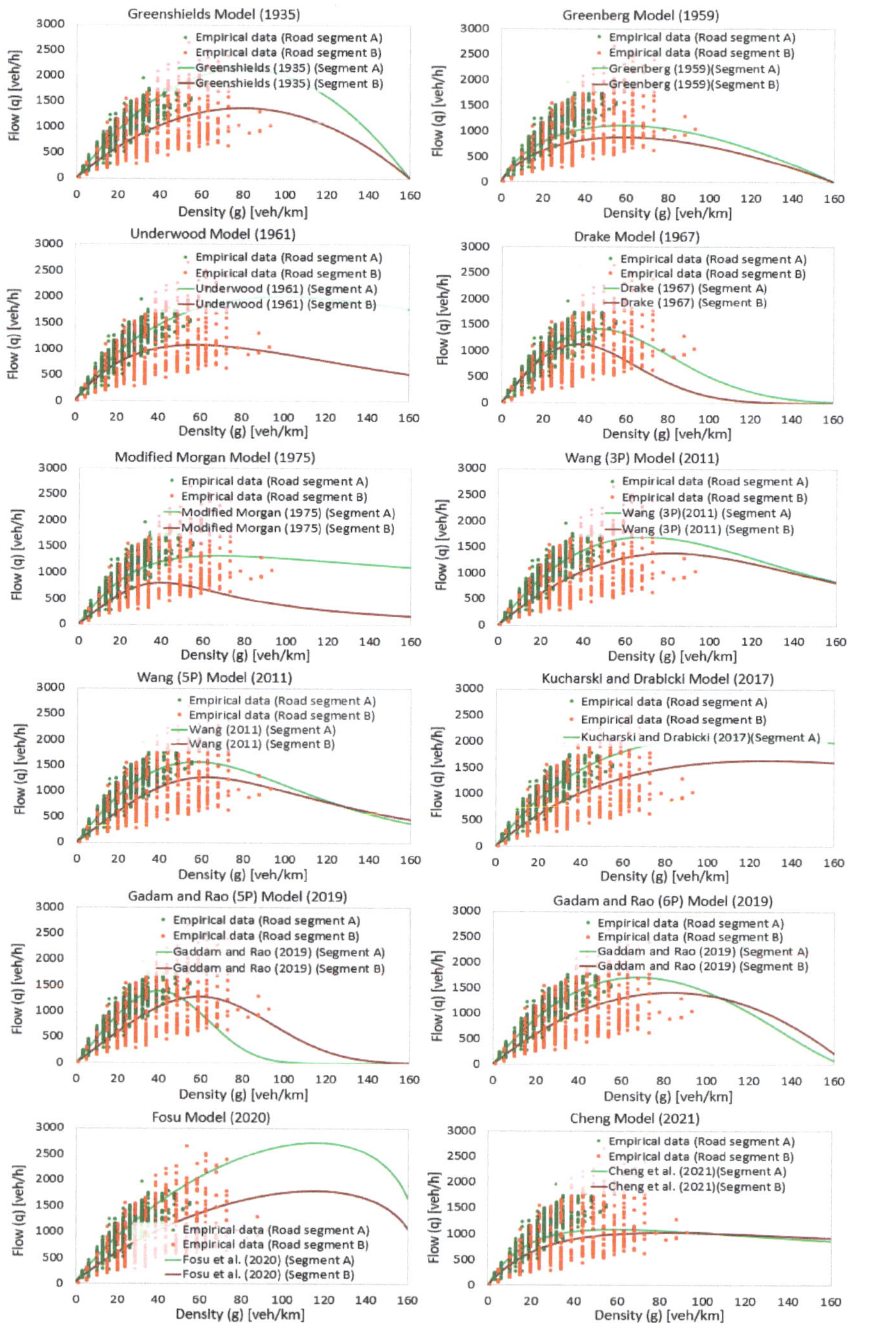

(**a**) (**b**)

Figure 6. Comparative presentation of selected deterministic "flow–density" models fitted to empirical data contained in the partial statistical sample collected on (**a**) Road segment A of ŽC5210 regional road (open road segment) and (**b**) Road segment B of ŽC5210 regional road (road segment with signalized intersection); (equations and references for considered traffic flow models are available in Tables 1 and 2). Source: Created by authors.

6. Discussion

6.1. Discussion on the Comparative Analysis of Selected "Speed–Density" Models Fitted to Empirical Data Collected under Continuous and Interrupted Traffic Stream Conditions

Based on the results obtained, it is evident that, for most of the models considered, the highest comparative difference among the mean traffic flow speeds representative of continuous and interrupted traffic stream conditions occurs in a free-flow regime. This difference is steadily reduced with the increase in traffic flow density, and it reaches its minimum in a congested traffic stream regime.

The difference among traffic flow speeds for continuous and interrupted traffic stream conditions computed by Greenshields' model ranges between 19.2 kph at free-flow speed (v_{max}) and 9.6 kph at critical (saturation) speed (v_c). In a saturated traffic stream regime, this difference is further reduced, and it becomes zero at jam density (g_{max}). The differences in predicted traffic flow speed obtained by Greenberg's model calibrated to empirical data collected in continuous and interrupted traffic stream conditions range from 18.9 kph at free flow to 3.7 kph at saturation density. In a congested regime, this difference continues to decrease, and it reaches zero at jam density.

The regression functions obtained by fitting the Underwood and Drake models suggest that the maximum comparative difference in the mean speed of continuous and interrupted traffic flow appears in the region between unsaturated and saturated traffic stream regime, i.e., when the traffic flow reaches practical road capacity. According to Underwood's model, the difference between continuous and interrupted traffic flow speed is equal to 4.3 kph in free-flow conditions, reaches its maximum of 12 kph at traffic density of 64 veh/km and then reduces back to 7.8 kph in a congested traffic flow regime. Even smaller differences are obtained using Drake's model, according to which the absolute speed difference predicted for continuous and interrupted traffic stream conditions is equal to 0 both at free-flow speed and at jam density, while the largest speed difference of 8.3 kph is reached at 57 veh/km. According to the modified Morgan logistic model, the differences between continuous and interrupted traffic flow speeds are reduced from 21.1 kph at free flow to 10.6 kph at critical density. In a congested traffic flow regime, the difference between these two speeds is further reduced to 5.9 kph.

Wang's three-parameter sigmoid model generates significantly different traffic flow speeds for continuous and interrupted traffic stream in free-flow conditions and approximately equal speed values in a saturated traffic flow regime, whereby the difference in the traffic flow speeds produced for continuous and interrupted traffic flow varies from 19.9 kph at free flow to 3.6 kph at saturation density. In a congested traffic flow regime, this difference is further reduced, and it reaches zero at jam density. Wang's five-parameter sigmoid model produces very similar results to Wang's three-parameter model over the entire range of possible density values, whereby the maximum difference between continuous and interrupted traffic stream speed does not exceed 1.5 kph. Kucharski and Drabicki's model shows similar differences between continuous and interrupted traffic flow speeds to Wang's three-parameter and five-parameter models in free-flow and normal traffic flow regimes and slightly larger differences in the saturated and congested traffic flow regime. The five-parameter model developed by Gadam and Rao produces somewhat unexpected results for the saturated and congested traffic stream regime. The difference in traffic flow speeds for continuous and interrupted traffic stream conditions computed by Gadam and Rao's five-parameter model ranges between 18.2 kph at free-flow speed and approximately 0 kph at critical (saturation) speed. Afterward, the traffic flow speed estimated for interrupted traffic stream conditions becomes higher than the speed estimated for continuous traffic stream conditions. On the other hand, an alternative version of Gadam and Rao's model with six parameters produces very similar results to Wang's three-parameter sigmoid model, whereby the difference in traffic flow speed estimated for continuous and interrupted traffic stream conditions decreases from 6.8 kph at free-flow speed to 1.76 kph at critical density, and it then reaches 0 at jam density. Fosu's logarithmic-potential form of the "speed–density" model generates very high differences, up to 42.5 kph, between continuous and interrupted traffic flow speeds in a free-flow regime. This difference is gradually reduced to 6.3 kph at saturation density, and it comes close to 0 at jam density. According to Cheng's model, the difference in the traffic flow speed produced for continuous and interrupted traffic stream conditions amounts to 21.34 kph in a free-flow regime and 6.7 kph at critical density. After reaching saturation density, the speeds estimated for continuous and interrupted traffic stream conditions become approximately equal.

If we consider the fact that, in a free-flow regime, the drivers are not restricted by the movements of other vehicles in the traffic flow, and therefore, they can drive at their desired speed, together with the fact that, in interrupted traffic stream conditions, vehicles are periodically slowed down and stopped due to the presence of signalized intersections along the road, it is logical to conclude that there must be a noticeable difference in the mean speed of continuous and interrupted traffic flow in a free-flow regime.

Following this reasoning, we argue that the "speed–density" regression functions obtained by fitting Greenberg's, Underwood's and Drake's models do not produce logical results. By taking into account the fact that Greenshields' model oversimplifies the speed–density relationship and therefore cannot be used for realistic representation of real traffic flow characteristics, which are non-linear in nature, we can come to the conclusion that only eight of the twelve traffic flow models considered are based on sound logical assumptions and are therefore practically applicable for describing the differences in mean traffic flow speed representative of continuous and interrupted traffic stream conditions. However, it is also important to emphasize here that the remaining models also have their limitations. For example, the mean traffic flow speed estimated by Wang's sigmoid models, Kucharski and Drabicki's, Fosu's and Cheng's models does not reach zero at jam density. If we also consider the fact that Gadam and Rao's five-parameter model produces somewhat unexpected results in a saturated and congested traffic flow regime, we can conclude that only modified Morgan's and Gadam and Rao's six-parameter models satisfy all logical assumptions based on the traffic flow theory.

6.2. Discussion on the Comparative Analysis of Selected "Flow–Density" Models Fitted to Empirical Data Collected under Continuous and Interrupted Traffic Stream Conditions

Based on the comparative analysis of fundamental traffic flow diagrams of the traffic flow models considered, we determined that the practical road capacity values representative of continuous and interrupted traffic stream conditions vary significantly depending on the selected model. The maximum practical capacity values, both for continuous and interrupted traffic stream conditions, are obtained using Fosu's model. According to Fosu's negative asymmetric "flow–density" curve, the practical road capacity of the observed regional road is reached at saturation densities of 96 and 116 veh/km/lane under continuous and interrupted traffic stream conditions, respectively. The practical road capacity (q_{max}) is equal to 2333 veh/h/lane and 1793 veh/h/lane in continuous and interrupted traffic flow, respectively. On the other hand, the lowest practical road capacity values for continuous and interrupted traffic stream conditions and, at the same time, the smallest mutual difference in practical road capacity representative of continuous and interrupted traffic stream conditions are predicted by Cheng's and modified Morgan's model, respectively. According to Cheng's model, the practical road capacity value is reached at a saturation density of 60 veh/km/lane, and it is equal to 1084 veh/h/lane in continuous traffic flow and 1030 veh/h/lane in interrupted traffic flow conditions. According to modified Morgan's model, the practical road capacity is reached at 69 veh/km/lane, and it is equal to 1331 veh/h/lane in continuous traffic conditions. In interrupted traffic stream conditions, the practical road capacity is reduced to 799 veh/h/lane, and it is achieved at a lower saturation density, i.e., at 40 veh/km/lane.

The practical road capacity value predicted by the remaining nine models ranges from 1104 to 2141 veh/h/lane for continuous traffic stream conditions and from 885 to 1640 veh/h/lane for interrupted traffic stream conditions. According to Greenshields' parabolic "flow–density" model, the practical road capacity is reached at a saturation density of 80 veh/km/lane, both in continuous and interrupted traffic stream conditions, and it is equal to 2141 veh/h/lane in uninterrupted and 1374 veh/h/lane in interrupted flows. Greenberg's model estimates much lower practical road capacity values, which are equal to 1104 veh/h/lane in continuous and 885 veh/h/lane in interrupted traffic stream conditions. Underwood's asymmetric "flow–density" parabolic curve reaches its practical road capacity at 2009 veh/h/lane in continuous traffic flow and at 1078 veh/h/lane in interrupted traffic flow conditions. However, these two practical capacity values are reached at significantly different density levels, i.e., at 101 veh/km/lane in continuous traffic stream conditions and at 59 veh/km/lane in interrupted traffic stream conditions. According to Drake's model, the practical roadway capacity in a continuous traffic flow is reached at 47 veh/km/lane, and it is equal to 1422 veh/h/lane, while the practical road capacity for an interrupted traffic flow amounts to 1132 veh/h/lane, and it is reached at a somewhat lower traffic density level (37 veh/km/lane). Wang's three-parameter and five-parameter "flow–density" sigmoid curves achieve maximum hourly traffic flows of 1699 and 1714 veh/h/lane at saturation densities of 68 and 69 veh/km/lane under continuous traffic stream conditions and practical capacities of 1381 and 1421 veh/h/lane at saturation densities of 81 and 83 veh/km/lane under interrupted traffic stream conditions. Kucharski and Drabicki's "flow–density" curve reaches a practical road capacity of 1388 veh/h/lane in continuous traffic stream conditions, at a density of 40 veh/km/lane. In interrupted traffic stream conditions, Kucharski and Drabicki's model predicts a relatively lower practical road capacity value of 1277 veh/h/lane, which is reached at a saturation density of 59 veh/km/lane. Gadam and Rao's five-parameter and six-parameter "flow–density" models reach the practical road capacity point at significantly different saturation densities, i.e., at 58 and 113 veh/km/lane under continuous and at 63 and 129 veh/km/lane under interrupted traffic stream conditions. Additionally, it is determined that Gadam and Rao's five-parameter model reaches practical road capacity at comparably higher traffic flow volumes, i.e., at 1562 veh/h/lane in continuous and at 1268 veh/h/lane in interrupted flows, while according to Gadam and Rao's six-parameter model, the practical road capacity is equal to 2099 veh/h/lane and 1640 veh/h/lane under continuous and interrupted traffic stream conditions, respectively.

Although Underwood's, modified Morgan's and Wang's three-parameter and five-parameter sigmoid models clearly produce much more realistic values of practical road capacity than Greenshields' and Greenberg's models, it is also evident that they are not able to reproduce the realistic hourly traffic flow volumes in a congested traffic flow regime. Drake's "flow–density" model produces realistic practical road capacity values, but it also incorrectly assumes that the vehicle flows in a free-flow regime are equal both under continuous and interrupted traffic stream conditions. In terms of the "flow–density" relationship, Gadam and Rao's six-parameter model produces the most realistic values for both continuous and interrupted traffic stream conditions.

The main limitation of modified Morgan's, Wang's, Fosu's and Cheng's models stems from the fact that they predict unrealistically high hourly traffic flow volumes at jam density, both in continuous and interrupted traffic stream conditions, which is not logically achievable in bumper-to-bumper traffic conditions. Underwood's model produces an even more unrealistic representation of the vehicle flows in a congested traffic stream regime, with hourly traffic flow volumes up to 1770 veh/h/lane for continuous traffic stream conditions and up to 521 veh/h/lane for interrupted traffic stream conditions.

Table 3 gives an overview of the relevant traffic flow parameter boundary values representative of continuous and interrupted traffic stream conditions, determined by fitting the "speed–density" and "flow–density" regression functions of the traffic flow models considered to empirical data collected on two selected segments of regional road ŽC5210, together with their comparative relative differences.

Table 3. Comparative overview of the relevant traffic flow parameter boundary values representative of continuous and interrupted traffic stream conditions, determined by fitting the traffic flow models considered to empirical data collected on two selected segments of regional road ŽC5210.

Continuous Traffic Stream Conditions				
Model	v_{max} [km/h]	v_c [km/h]	g_c [voz/km]	q_{max} [voz/h]
Greenshields (1935) [7]	53.5	26.8	80.0	2141
Greenberg (1959) [9]	∞	18.7	59.0	1104
Underwood (1961) [10]	54.3	19.9	101.0	2009
Drake (1967) [13]	50.0	30.3	47.0	1422
Modified Morgan (1975) [64]	51.6	19.3	69.0	1331
Wang (3P) (2011) [23]	53.4	24.6	69.0	1699
Wang (5P) (2011) [23]	53.7	25.2	68.0	1714
Kucharski and Drabicki (2017) [4]	48.2	34.7	40.0	1388
Gadam and Rao (5P) (2019) [58]	52.7	26.9	58.0	1562
Gadam and Rao (6P) (2019) [58]	54.5	18.6	113.0	2099
Fosu (2020) [58]	82.6	24.3	96.0	2333
Cheng (2021) [63]	72.3	18.1	60.0	1084
Interrupted Traffic Stream Conditions				
Model	v_{max} [km/h]	v_c [km/h]	g_c [voz/km]	q_{max} [voz/h]
Greenshields (1935) [7]	34.4	17.2	80.0	1374
Greenberg (1959) [9]	∞	15.0	59.0	885
Underwood (1961) [10]	50.0	18.3	59.0	1078
Drake (1967) [13]	50.0	30.6	37.0	1132
Modified Morgan (1975) [64]	30.5	20.0	40.0	799
Wang (3P) (2011) [23]	33.5	17.0	81.0	1381
Wang (5P) (2011) [23]	34.0	16.9	84.0	1421
Kucharski and Drabicki (2017) [4]	30.0	21.6	59.0	1277
Gadam and Rao (5P) (2019) [58]	30.9	20.1	63.0	1268
Gadam and Rao (6P) (2019) [58]	35.0	12.7	129.0	1640
Fosu (2020) [58]	42.5	15.5	116.0	1793
Cheng (2021) [63]	49.6	12.4	83.0	1030
Comparative Relative Differences [%]				
Model	v_{max} [km/h]	v_c [km/h]	g_c [voz/km]	q_{max} [voz/h]
Greenshields (1935) [7]	35.8%	35.8%	0.0%	35.8%
Greenberg (1959) [9]	-	19.8%	0.0%	19.8%
Underwood (1961) [10]	8.0%	8.2%	41.6%	46.4%
Drake (1967) [13]	0.0%	−1.1%	21.3%	20.4%
Modified Morgan (1975) [64]	40.9%	−3.5%	42.0%	40.0%
Wang (3P) (2011) [23]	37.3%	30.8%	−17.4%	18.7%
Wang (5P) (2011) [23]	36.8%	32.9%	−23.5%	17.1%
Kucharski and Drabicki (2017) [4]	37.7%	37.7%	−47.5%	8.0%
Gadam and Rao (5P) (2019) [58]	41.3%	25.2%	−8.6%	18.8%
Gadam and Rao (6P) (2019) [58]	35.7%	31.5%	−14.2%	21.9%
Fosu (2020) [58]	33.1%	33.7%	0.9%	34.2%
Cheng (2021) [63]	31.3%	31.3%	−38.3%	5.0%

The meaning of the parameters in the table is as follows: v_{max} denotes free-flow speed; v_c is the critical speed at road capacity; g_c is the critical density at road capacity and q_{max} represents road capacity (maximum flow).

Based on the data presented in Table 3, it is visible that the practical capacity of the observed regional road—determined by twelve calibrated traffic flow models fitted to empirical data collected under continuous traffic stream conditions—ranges from 1084 to 2333 veh/h/lane. On the other hand, according to the traffic flow models fitted to empirical data collected under interrupted traffic stream conditions, the practical road capacity of the observed road varies between 799 and 1793 veh/h/lane. Depending on the traffic flow model considered, the practical road capacity of the observed regional road in interrupted traffic stream conditions is 5%–46.4% lower than the practical road capacity determined for continuous traffic stream conditions.

The highest difference in practical road capacity values representative of continuous and interrupted traffic stream conditions (46.4%) is obtained based on Underwood's model. The flow–density model based on modified Morgan–Mercer–Flodin equation, as well as Fosu's and Greenshields' models, also produce significantly different practical road capacity values for continuous and interrupted traffic flow conditions. According to the modified Morgan–Mercer–Flodin equation, the practical road capacity of the observed regional road is reduced by 40% in interrupted traffic stream conditions, while Fosu's and Greenshields' models predict that the interrupted traffic flow has a 34.2% and 35.8% lower practical road capacity than the continuous traffic flow, respectively. The remaining traffic flow models suggest that the impact of interrupted traffic stream conditions on the reduction in roadway capacity is significantly lower, whereby the difference in the practical road capacity determined for continuous and interrupted traffic stream conditions ranges between 5% and 21.9%.

The regression functions of all the traffic flow models considered fitted to empirical data collected under continuous traffic stream conditions have free-flow speed values in the range of 48–82.6 kph. On the other hand, except for Greenberg's model, all the other models fitted to empirical data collected under interrupted traffic stream conditions show less pronounced variations in the free-flow speed parameter. According to Underwood's and Drake's models, the free-flow speed representative of modeling interrupted traffic stream conditions is equal to 50 kph; Cheng's model has a negligibly lower free-flow speed value (49.6 kph), while the remaining models show relatively lower values of free-flow speed, in the range of 30–40.3 kph. Depending on the traffic flow model selected, the relative difference in free-flow speed values representative of modeling traffic flow characteristics under continuous and interrupted traffic stream conditions can reach up to 41.3%.

The value of critical (saturation) speed, at which practical road capacity is reached, varies in the ranges of 18 kph–35 kph and 12 kph–31 kph for continuous and interrupted traffic stream conditions. The relative difference in the critical speed value determined under continuous and interrupted traffic flow ranges from −3.5% to 37.7%. The saturation traffic flow density value for continuous traffic flow conditions ranges between 40 and 117 veh/km/lane, while in interrupted traffic flow conditions, it ranges between 37 and 129 veh/km.

6.3. Discussion on the Results of the t-Tests Conducted

Lastly, *t*-tests were conducted to determine the statistical significance of the differences in empirical values of the main traffic flow parameters determined in continuous and interrupted traffic flow conditions. The results of these tests are presented in Table 4. Based on the results derived from the *t*-test for the speed variable, it is clear that the differences in vehicle speeds determined under continuous and interrupted traffic stream conditions are statistically significant. A derived *t*-statistic of 34.69 far exceeds the critical *t*-value of 1.96, leading to rejection of the null hypothesis. The positive *t*-statistic indicates that the mean speed of the first statistical sample, associated with continuous traffic stream conditions, is greater than the mean speed of the second statistical sample, related to interrupted traffic stream conditions. The resulting 95% confidence interval, spanning the range [17.97, 20.12], implies that there is a 95% probability that the actual difference in population mean speeds determined under continuous and interrupted traffic flow conditions on the observed regional road will fall within this interval.

Table 4. Comparative summary of t-test results for significance of the differences in empirical values of the main traffic flow parameters determined in continuous and interrupted traffic flow conditions on the observed road sections of regional road ŽC5210. Source: Created by authors.

Traffic Flow Parameter	t-Statistic	Critical t-Value	95% Confidence Interval	Hypothesis (H1)
Mean vehicle speed (V)	34.6934	1.96	[17.9685, 20.1221]	Accepted
Mean vehicle flow (q)	5.04610	1.96	[71.3028, 161.9837]	Accepted
Mean traffic density (g)	26.6395	1.96	[−20.6768, −17.8408]	Accepted

The outcomes of the t-test conducted for the vehicle flow variable also reveal a statistically significant distinction between hourly vehicle flow volumes determined in continuous and interrupted traffic stream conditions. The t-statistic in this case is 5.05, notably exceeding the critical t-value of 1.96 at the 5% significance level. The 95% confidence interval, ranging from 71.30 to 161.98 veh/h, implies that the actual difference in the means of traffic flow volumes determined under continuous and interrupted traffic stream conditions will lie between 71.3 and 161.98 veh/h in 95% of cases.

The results of the t-test conducted for the traffic density variable also confirmed the presence of statistically significant differences in the means of traffic density determined in continuous and interrupted traffic stream conditions. The negative t-statistic indicates that the density values determined on the open road segment are notably lower compared to density values determined on the road segment with signalized intersection. The results obtained also indicate that there is 95% certainty that the actual difference in the means of traffic density determined under continuous and interrupted traffic stream conditions will fall within the range from −20.67 veh/km to −17.84 veh/km.

7. Conclusions

In this paper, we investigated the comparative differences in fundamental traffic flow relations present under continuous and interrupted traffic stream characteristics on the two selected segments of regional road ŽC5210 in the Republic of Croatia to identify the specific functional forms of "speed–density" and "flow–density" fundamental diagrams suitable for modeling continuous and interrupted traffic stream conditions. The results and conclusions derived from this research can be summarized as follows:

- Based on the comparative analysis of R-squared values determined for the twelve traffic flow models considered fitted to empirical data collected under continuous and interrupted traffic stream conditions, it is revealed that Wang's five-parameter sigmoid model best fits the empirical data, with Pearson's correlation coefficient of 0.6191 and coefficient of determination of 0.3833.
- The results obtained indicate that the practical road capacity value representative of continuous and interrupted traffic stream conditions varies significantly depending on the selected model and prevailing traffic stream conditions. The practical capacity of the observed regional road, determined by twelve calibrated traffic flow models, ranges from 1084 to 2333 veh/h/lane under continuous and from 799 to 1793 veh/h/lane under interrupted traffic stream conditions. Depending on the traffic flow model considered, the practical road capacity under interrupted traffic stream conditions is 5–46.4% lower than under continuous traffic stream conditions, while the relative difference in free-flow speed values representative of modeling traffic flow characteristics under continuous and interrupted traffic stream conditions can reach up to 41.3%.
- The practical capacity of the observed regional road is reached at significantly different levels of traffic density when comparing continuous and interrupted traffic stream conditions. Depending on the selected model, the saturation traffic flow density value for continuous traffic flow conditions ranges between 40 and 117 veh/km/lane, while in interrupted traffic flow conditions, it ranges between 37 and 129 veh/km.

- The statistical significance of the differences in mean speed, mean hourly vehicle flow and mean traffic flow density values determined for continuous and interrupted traffic flow conditions by the proposed "speed–density" and "flow–density" models is confirmed by the *t*-tests conducted.
- We reject the Null Hypothesis (H0) and confirm the Alternative Hypothesis (H1) that there is a statistically significant difference in the speed–density and flow–density functional relationships obtained for continuous and periodically interrupted traffic flow conditions on the representative road segments of the selected regional road, and therefore, we conclude that two alternative mathematical formulations of the "speed–density" and "flow–density" models need to be used to separately model the traffic flow characteristics present under continuous and interrupted traffic stream conditions.

When interpreting the results and conclusions of the research conducted within this paper, it is necessary to consider the following methodological and spatio-temporal limitations and research constraints:

- The empirical samples used in this research to identify representative deterministic "speed–density" and "flow–density" models were gathered over relatively short periods and exclusively on the two representative road segments of ŽC5210 regional road. All of the models considered were fitted to an empirical sample, which did not include the empirical traffic flow parameter values measured in a congested traffic flow condition, i.e., at traffic flow densities higher than 93.00 veh/km.
- Twelve traffic flow models selected were fitted directly to disaggregated empirical data collected on the two observed road sections of regional road ŽC5210 in the Republic of Croatia. The impacts of higher levels of time aggregation of the observed traffic flow parameter empirical values on the shape of regression functions used to describe the relations between average traffic flow parameter values were not considered in this research.
- The impact of traffic flow structure, road environment characteristics, relevant road design and road infrastructure elements on the shapes of the obtained "speed–density" and "flow–density" regression functions was not considered in the scope of this research.

Based on the obtained results, derived conclusions and defined limitations of the research performed, the following suggestions and guidelines for future research are proposed:

- Future research should prioritize the collection of extensive statistical sample with a sufficient number of empirical values of traffic flow parameters measured in a congested traffic flow condition. Based on such sample, it would be possible to further examine the possibility of developing other specific non-linear regression models, which would enable more accurate descriptions of the mathematical relationship between average speed and density values under high-density (unstable) traffic flow conditions.
- In future research, it is also important to investigate the impact of higher levels of time aggregation of the observed traffic flow parameters, as well as the impact of traffic flow structure, road environment characteristics, relevant road design and infrastructure elements and other relevant factors, which possibly contribute to the scattering effect of empirical data visible in the "speed–density" and "flow–density" fundamental diagrams.

Supplementary Materials: The following supporting information can be downloaded at: https://www.mdpi.com/article/10.3390/app14020533/s1, Table S1: ZC5210_Road_Section_A; Table S2: ZC5210_
Road_Section_B; and Table S3: ZC5210_Comparative_Models. Tables S1 and S2 contain the input data used to perform the correlation and regression analysis and to develop the "speed–density"

and "flow–density" models representative of continuous and interrupted traffic stream conditions, which were obtained based on the processing and analysis of aerial video files recorded using an unmanned aerial vehicle on two selected road segments of ŽC5210 regional road in the Republic of Croatia. The tables include data on vehicle ID numbers, vehicle types, lengths of observed road segments and data on the empirical values of traffic density, time headways, vehicle speeds and vehicle flows, determined on the open road segment (Road Section A) and the road segment with signalized intersection (Road Section B) of the selected regional road. Table S3 contains the cumulative empirical sample of traffic flow parameter values collected on both observed road segments, together with the data on traffic flow parameter values, determined based on the developed "speed–density" and "flow–density" models.

Author Contributions: Conceptualization, B.J., M.Š. and G.L.; methodology, B.J.; validation, B.J., M.Š. and G.L.; formal analysis, B.J.; investigation, B.J. and M.Š.; resources, B.J., M.Š. and G.L.; data curation, M.Š.; writing—original draft preparation, B.J.; writing—review and editing, B.J., M.Š. and G.L.; visualization, B.J.; supervision, G.L.; project administration, M.Š. and G.L.; funding acquisition, M.Š. and G.L. All authors have read and agreed to the published version of the manuscript.

Funding: The article processing charges (APCs) for this paper were funded by the Research Fund of the Department of Transport Planning, Faculty of Transport and Traffic Sciences, University of Zagreb, Croatia.

Institutional Review Board Statement: Not applicable.

Informed Consent Statement: Not applicable.

Data Availability Statement: The input data used for conducting this research are available in Table S1: ZC5210_Road_Section_A; Table S2: ZC5210_Road_Section_B; and Table S3: ZC5210_Comparative_Models. Other data used to support the findings of this study are available from the corresponding author upon request.

Acknowledgments: This work is supported by the Department of Transport Planning, Faculty of Transport and Traffic Sciences, University of Zagreb, Croatia.

Conflicts of Interest: The authors declare no conflict of interest.

References

1. van Wageningen-Kessels, F.; Hoogendoorn, S.P.; Vuik, K.; van Lint, H. Traffic Flow Modeling: A Genealogy. *Transp. Res. E-Circ.* **2015**, *E-C195*, 1–16.
2. van Wageningen-Kessels, F.; van Lint, H.; Vuik, K.; Hoogendoorn, S. Genealogy of Traffic Flow Models. *EURO J. Transp. Logist.* **2015**, *4*, 445–473. [CrossRef]
3. Bramich, D.M.; Menéndez, M.; Ambühl, L. Fitting Empirical Fundamental Diagrams of Road Traffic: A Comprehensive Review and Comparison of Models Using an Extensive Data Set. *IEEE Trans. Intell. Transp. Syst.* **2022**, *23*, 14104–14127. [CrossRef]
4. Romanowska, A.; Jamroz, K. Comparison of Traffic Flow Models with Real Traffic Data Based on a Quantitative Assessment. *Appl. Sci.* **2021**, *11*, 9914. [CrossRef]
5. Dahiya, G.; Asakura, Y.; Nakanishi, W. Analysis of the Single-Regime Speed-Density Fundamental Relationships for Varying Spatiotemporal Resolution Using Zen Traffic Data. *Asian Transp. Stud.* **2022**, *8*, 100066. [CrossRef]
6. Zefreh, M.M.; Török, Á. Distribution of Traffic Speed in Different Traffic Conditions: An Empirical Study in Budapest. *Transport* **2020**, *35*, 68–86. [CrossRef]
7. Greenshields, B.D.; Thompson, J.T.; Dickinson, H.C.; Swinton, R.S. The Photographic Method of Studying Traffic Behavior. *Highw. Res. Board. Proc.* **1934**, *13*, 382–389.
8. Greenshields, B.D. Average Speed and Traffic Density as a Measure of Road Capacity. In Proceedings of the Highway Research Board, Washington, DC, USA, 30 November–3 December 1938.
9. Greenberg, H. An Analysis of Traffic Flow. *Oper. Res.* **1959**, *7*, 79–85. [CrossRef]
10. Underwood, R.T. Speed, Volume and Density Relationships. *Qual. Theory Traffic Flow.* **1960**, 141–188.
11. Edie, L.C. Car-Following and Steady-State Theory for Noncongested Traffic. *Oper. Res.* **1961**, *9*, 66–76. [CrossRef]
12. Dick, A.C. Speed/Flow Relationships within an Urban Area. *Traffic Eng. Control* **1966**, *8*, 393–396.
13. Drake, J.; Schofer, J.; May, A. A Statistical Analysis of Speed-Density Hypotheses. In Vehicular Traffic Science. *Highw. Res. Rec.* **1967**, *154*, 112–117.
14. Payne, H.J. Models of Freeway Traffic and Control. In *Mathematical Models of Public Systems*; Simulation Council: San Diego, CA, USA, 1971; Volume 1, pp. 51–61.
15. Smulders, S. Control of Freeway Traffic Flow by Variable Speed Signs. *Transp. Res. Part B Methodol.* **1990**, *24*, 111–132. [CrossRef]

16. Van Aerde, M.; Rakha, H. Multivariate Calibration of Single-Regime Speed-Flow-Density Relationships. In Proceedings of the Pacific Rim TransTech Conference, 1995 Vehicle Navigation and Information Systems Conference Proceedings, 6th International VNIS, A Ride into the Future, Seattle, WA, USA, 30 July–2 August 1995.
17. DelCastillo, J.M.; Benítez, F.G. On the Functional Form of the Speed-Density Relationship—I: General Theory. *Transp. Res. Part B Methodol.* **1995**, *29*, 373–389. [CrossRef]
18. Gartner, N.H.; Messer, C.J.L.; Rathi, A. *Traffic Flow Theory—A State-of-the-Art Report: Revised Monograph on Traffic Flow Theory*; Turner-Fairbank Highway Research Center: McLean, VA, USA, 2002; pp. 30–67.
19. Chanut, S.; Buisson, C. Macroscopic Model and Its Numerical Solution for Two-Flow Mixed Traffic with Different Speeds and Lengths. *Transp. Res. Rec. J. Transp. Res. Board.* **2003**, *1852*, 209–219. [CrossRef]
20. Helbing, D. Derivation of a Fundamental Diagram for Urban Traffic Flow. *Eur. Phys. J. B* **2008**, *70*, 229–241. [CrossRef]
21. MacNicholas, M.J. A Simple and Pragmatic Representation of Traffic Flow. In Proceedings of the Greenshields Symposium on the 75 Years of the Fundamental Diagram for Traffic Flow Theory, Woods Hole, MA, USA, 8–10 July 2008; pp. 161–177.
22. Transportation Research Board. *Highway Capacity Manual*, 5th ed.; Transportation Research Board: Washington, DC, USA, 2010.
23. Wang, H.; Li, J.; Chen, Q.-Y.; Ni, D. Logistic Modeling of the Equilibrium Speed–Density Relationship. *Transp. Res. Part A Policy Pract.* **2011**, *45*, 554–566. [CrossRef]
24. Horvat, R.; Kos, G.; Ševrović, M. Traffic Flow Modelling on the Road Network in the Cities. *Teh. Vjesn.-Tech. Gaz.* **2015**, *22*, 475–486. [CrossRef]
25. Elefteriadou, L.A. *Highway Capacity Manual*, 6th ed.; National Academies Press: Washington, DC, USA, 2016; ISBN 978-0-309-46019-4.
26. National Academies of Sciences; Engineering and Medicine. *Highway Capacity Manual*, 7th ed.; National Academies Press: Washington, DC, USA, 2022; ISBN 978-0-309-27566-8.
27. Gaus, A.; Muhammad, T.Y.S.; Ambo, U.S.H.; Liska, N. Mathematical Model of Traffic Speed and Capacity in the Archipelago Base. In Proceedings of the Workshop on Environmental Science, Society, and Technology (WESTECH 2020), Makassar, Indonesia, 16–17 October 2020; IOP Publishing: Bristol, UK, 2020.
28. Anupriya; Bansal, P.; Graham, D.J. Analytical Representations of the Fundamental Diagram of Traffic Flow for Highways: A Review of Theory and Empirics. *SSRN* **2022**, 1–38.
29. Qu, X.; Zhang, J.; Wang, S. On the Stochastic Fundamental Diagram for Freeway Traffic: Model Development, Analytical Properties, Validation, and Extensive Applications. *Transp. Res. Part B Methodol.* **2017**, *104*, 256–271. [CrossRef]
30. Ni, D.; Hsieh, H.K.; Jiang, T. Modeling Phase Diagrams as Stochastic Processes with Application in Vehicular Traffic Flow. *Appl. Math. Model.* **2018**, *53*, 106–117. [CrossRef]
31. Thonhofer, E.; Jakubek, S. Investigation of Stochastic Variation of Parameters for a Macroscopic Traffic Model. *J. Intell. Transp. Syst.* **2018**, *22*, 547–564. [CrossRef]
32. Bouadi, M.; Jia, B.; Jiang, R.; Li, X.; Gao, Z.-Y. Stability Analysis of Stochastic Second-Order Macroscopic Continuum Models and Numerical Simulations. *arXiv* **2022**, arXiv:2204.13937. [CrossRef]
33. May, A.D.; Keller, H.E. Non-Integer Car-Following Models. *Highw. Res. Rec.* **1967**, *199*, 19–32.
34. Pipes, L.A. Car Following Models and the Fundamental Diagram of Road Traffic. *Transp. Res.* **1967**, *1*, 21–29. [CrossRef]
35. Del Castillo, J.M.; Benitez, F.G. On the Functional Form of the Speed-Density Relationship—II: Empirical Investigation. *Transp. Res. Part B Methodol.* **1995**, *29*, 391–406. [CrossRef]
36. Liu, X.; Xu, J.; Li, M.; Wei, L.; Ru, H. General-Logistic-Based Speed-Density Relationship Model Incorporating the Effect of Heavy Vehicles. *Math. Probl. Eng.* **2019**, *2019*, 6039846. [CrossRef]
37. Amirgholy, M.; Nourinejad, M.; Gao, O.H. Optimal Traffic Control at Smart Intersections: Automated Network Fundamental Diagram. *Transp. Res. Part B Methodol.* **2020**, *137*, 2–18. [CrossRef]
38. Fourati, W.; Dabbas, H.; Friedrich, B. Estimation of Penetration Rates of Floating Car Data at Signalized Intersections. *Transp. Res. Procedia* **2020**, *52*, 228–235. [CrossRef]
39. Tilg, G.; Amini, S.; Busch, F. Evaluation of Analytical Approximation Methods for the Macroscopic Fundamental Diagram. *Transp. Res. Part C Emerg. Technol.* **2020**, *114*, 1–19. [CrossRef]
40. Satyajit, M.; Ankit, G. Speed Distribution for Interrupted Flow Facility under Mixed Traffic. *Phys. A Stat. Mech. Appl.* **2021**, *570*, 125798. [CrossRef]
41. Khan, M.A.; Ectors, W.; Bellemans, T.; Janssens, D.; Wets, G. Unmanned Aerial Vehicle-Based Traffic Analysis: A Case Study for Shockwave Identification and Flow Parameters Estimation at Signalized Intersections. *Remote Sens.* **2018**, *10*, 458. [CrossRef]
42. Kos, G.; Ševrović, M.; Jović, J. Harmonization and Synchronization Model of Interrupted Traffic Flows on Motorways. *Transport* **2018**, *33*, 364–369. [CrossRef]
43. Storani, F.; Di Pace, R.; Bruno, F.; Fiori, C. Analysis and Comparison of Traffic Flow Models: A New Hybrid Traffic Flow Model vs Benchmark Models. *Eur. Transp. Res. Rev.* **2021**, *13*, 58. [CrossRef]
44. Li, Y.; Arora, N.; Osorio, C. On the Fundamental Diagram of Signal Controlled Urban Roads. In Proceedings of the 11th Triennial Symposium on Transportation Analysis conference (TRISTAN XI), Mauritius Island, Mauritius, 19–25 June 2022.
45. Akcelik, R. Relating Flow, Density, Speed and Travel Time Models for Uninterrupted and Interrupted Traffic. *Traffic Eng. Control* **1996**, *37*, 511–516.

46. Pueboobpaphan, R.; Nakatsuji, T.; Suzuki, H.; Kawamura, A. Transformation between Uninterrupted and Interrupted Speeds for Urban Road Applications. *Transp. Res. Rec. J. Transp. Res. Board.* **2005**, *1934*, 72–82. [CrossRef]
47. Pueboobpaphan, R.; Yamamoto, F.; Nakatsuji, T.; Suzuki, H. Relationship between Uninterrupted and Interrupted Speed at Isolated Signalized Intersection. *J. East. Asia Soc. Transp. Stud.* **2005**, *6*, 2194–2209. [CrossRef]
48. Barka, R.E.; Politis, I. Fitting Volume Delay Functions under Interrupted and Uninterrupted Flow Conditions at Greek Urban Roads. *Eur. Transp. Trasp. Eur.* **2001**, *127*, 13–20. [CrossRef]
49. Meriem, B.; Stylianos, K.; Jeremy, T.; Zoi, C. Drones for Traffic Flow Analysis of Urban Roundabouts. *Int. J. Traffic Transp. Eng.* **2020**, *9*, 62–71.
50. Salvo, G.; Caruso, L.; Scordo, A. Urban Traffic Analysis through an UAV. *Procedia Soc. Behav. Sci.* **2014**, *111*, 1083–1091. [CrossRef]
51. Butilă, E.V.; Boboc, R.G. Urban Traffic Monitoring and Analysis Using Unmanned Aerial Vehicles (UAVs): A Systematic Literature Review. *Remote Sens.* **2022**, *14*, 620. [CrossRef]
52. Khan, M.A.; Ectors, W.; Bellemans, T.; Janssens, D.; Wets, G. Unmanned Aerial Vehicle–Based Traffic Analysis: Methodological Framework for Automated Multivehicle Trajectory Extraction. *Transp. Res. Rec. J. Transp. Res. Board.* **2017**, *2626*, 25–33. [CrossRef]
53. Salvo, G.; Caruso, L.; Scordo, A.; Guido, G.; Vitale, A. Traffic Data Acquirement by Unmanned Aerial Vehicle. *Eur. J. Remote Sens.* **2017**, *50*, 343–351. [CrossRef]
54. Khan, M.A.; Ectors, W.; Bellemans, T.; Janssens, D.; Wets, G. UAV-Based Traffic Analysis: A Universal Guiding Framework Based on Literature Survey. *Transp. Res. Procedia* **2017**, *22*, 541–550. [CrossRef]
55. Xue, C.; Jalayer, M.; Zhou, H. Validation of Traffic Simulation Model Output Using High-Resolution Video by Unmanned Aerial Vehicles. In Proceedings of the International Conference on Transportation and Development 2019, Alexandria, VA, USA, 9–12 June 2019; American Society of Civil Engineers: Reston, VA, USA, 2019; pp. 135–147.
56. Yahia, C.N.; Scott, S.E.; Boyles, S.D.; Claudel, C.G. Unmanned Aerial Vehicle Path Planning for Traffic Estimation and Detection of Non-Recurrent Congestion. *Transp. Lett.* **2022**, *14*, 849–862. [CrossRef]
57. Ahmed, A.; Ngoduy, D.; Adnan, M.; Baig, M.A.U. On the Fundamental Diagram and Driving Behavior Modeling of Heterogeneous Traffic Flow Using UAV-Based Data. *Transp. Res. Part A Policy* **2021**, *148*, 100–115. [CrossRef]
58. Fosu, O.G.; Akweittey, E.; Opong, J.M.; Otoo, M.E. Vehicular Traffic Models for Speed-Density-Flow Relationship. *J. Math. Model.* **2020**, *8*, 241–255.
59. Gaddam, H.K.; Rao, K.R. Speed–Density Functional Relationship for Heterogeneous Traffic Data: A Statistical and Theoretical Investigation. *J. Mod. Transp.* **2019**, *27*, 61–74. [CrossRef]
60. Péter, T.; Fazekas, S. Determination of Vehicle Density of Inputs and Outputs and Model Validation for the Analysis of Network Traffic Processes. *Period. Polytech. Transp. Eng.* **2014**, *42*, 53–61. [CrossRef]
61. del Castillo, J.M. Three New Models for the Flow–Density Relationship: Derivation and Testing for Freeway and Urban Data. *Transportmetrica* **2012**, *8*, 443–465. [CrossRef]
62. Ni, D.; Leonard, J.D.; Jia, C.; Wang, J. Vehicle Longitudinal Control and Traffic Stream Modeling. *Transp. Sci.* **2016**, *50*, 1016–1031. [CrossRef]
63. Cheng, Q.; Liu, Z.; Lin, Y.; Zhou, X.S. An S-Shaped Three-Parameter (S3) Traffic Stream Model with Consistent Car Following Relationship. *Transp. Res. Part B Methodol.* **2021**, *153*, 246–271. [CrossRef]
64. Yakasai, H.M.; Shukor MY, A. Predictive Mathematical Modelling of the Total Number of COVID-19 Cases for The United States. *Bioremediation Sci. Technol. Res.* **2020**, *8*, 11–16. [CrossRef]
65. Google. Google Maps. Available online: https://www.google.hr/maps (accessed on 20 June 2023).
66. State Geodetic Administration Geoportal of the State Geodetic Administration of the Republic of Croatia. Available online: https://geoportal.dgu.hr/ (accessed on 20 June 2023).
67. Croatian Roads Ltd. Geoportal of Croatian Public Roads. Available online: https://geoportal.hrvatske-ceste.hr/ (accessed on 20 June 2023).
68. JetBrains PyCharm. 2023. Available online: https://www.jetbrains.com/pycharm/ (accessed on 30 August 2023).

logistics

Article

A Bayesian Optimization Approach for Tuning a Grouping Genetic Algorithm for Solving Practically Oriented Pickup and Delivery Problems

Cornelius Rüther * and Julia Rieck

Operations Research Department, Institute for Business Administration and Information Systems, University of Hildesheim, Universitätsplatz 1, 31141 Hildesheim, Germany; rieck@bwl.uni-hildesheim.de
* Correspondence: ruether@bwl.uni-hildesheim.de

Abstract: *Background*: The Multi Depot Pickup and Delivery Problem with Time Windows and Heterogeneous Vehicle Fleets (MDPDPTWHV) is a strongly practically oriented routing problem with many real-world constraints. Due to its complexity, solution approaches with sufficiently good quality ideally contain several operators with certain probabilities.Thus, automatically selecting the best parameter configurations enhances the overall solution quality. *Methods*: To solve the MDPDPTWHV, we present a Grouping Genetic Algorithm (GGA) framework with several operators and population management variants. A Bayesian Optimization (BO) approach is introduced to optimize the GGA's parameter configuration. The parameter tuning is evaluated on five data sets which differ in several structural characteristics and contain 1200 problem instances. The outcomes of the parameter-tuned GGA are compared to both the initial GGA parameter configuration and a state-of-the-art Adaptive Large Neighborhood Search (ALNS). *Results*: The presented GGA framework achieves a better solution quality than the ALNS, even for the initial parameter configuration used. The mean value of the relative error is less than 0.9% and its standard deviation is less than 1.31% for every problem class. For the ALNS, these values are up to three times higher and the GGA is up to 38% faster than the ALNS. *Conclusions*: It is shown that the BO, as a parameter tuning approach, is a good choice in improving the performance of the considered meta-heuristic over all instances in each data set. In addition, the best parameter configuration per problem class with the same characteristics is able to improve both the frequency of finding the best solution, as well as the relative error to this solution, significantly.

Keywords: automatic algorithm configuration; black box optimization; Gaussian processes; machine learning model

Citation: Rüther, C.; Rieck, J. A Bayesian Optimization Approach for Tuning a Grouping Genetic Algorithm for Solving Practically Oriented Pickup and Delivery Problems. *Logistics* **2024**, *8*, 14. https://doi.org/10.3390/logistics8010014

Academic Editors: Tomasz Nowakowski, Artur Kierzkowski, Agnieszka A. Tubis, Franciszek Restel, Tomasz Kisiel, Anna Jodejko-Pietruczuk and Mateusz Zając

Received: 25 December 2023
Revised: 28 January 2024
Accepted: 29 January 2024
Published: 4 February 2024

1. Introduction

The less-than-truckload (LTL) transport sector typically involves large to medium-sized carriers. This sector is characterized by the fact that several transport requests are transported together in one vehicle. The freight of each request is transported from a specific starting point (pick-up customer) to a specific destination (delivery customer). As part of the vehicle routing, it must then be determined which customer is to be visited on which tour and in which order. The goods must be loaded and unloaded by employees at each customer. In order to ensure that employees and loading ramps are available and waiting times can be reduced, customer time windows are introduced as an addition. Since different types of goods are transported, e.g., palletized goods, but also pallet cages or cable drums, LTL carriers must have different types of vehicles in terms of capacity, speed or even pollutant emissions. Furthermore, carriers usually have several depot locations and vehicles are available at each depot from which they start and end their routes [1].

The described problem can be modeled as a Multi-Depot Pickup and Delivery Problem with Time Windows and Heterogeneous Vehicle Fleets (MDPDPTWHV). Due to the problem's complexity (it is $\mathcal{NP}$-hard), an extensive solution approach must be applied to achieve good solution quality (see Section 5.3). To do so, the Grouping Genetic Algorithm (GGA) presented in [1] has proven to be a good solution approach, which has shown promising results and solution quality.

Since the solution quality of a stochastic algorithm strongly depends on the *correct* combination of the executed operators, configuring the parameters that control the probabilities for the operator selection in the GGA is essential. The automatic optimization of these parameter configurations is defined as a Parameter Tuning Problem (PTP) (see Section 3.2). There are currently various solution approaches available for automated parameter configuration whose solution quality and suitability depends on the problem to be solved. In general, iterative parameter configurators, such as ParamILS or the Sequential Model-based Algorithm Configuration (SMAC), are state-of-the-art parameter tuning approaches [2], whereas model-based algorithms, such as SMAC, are very promising since they can be potentially executed more efficiently. Hence, an efficient configuration approach to tackle the PTP is sequential model-based Bayesian Optimization (BO), which was introduced in [3] and is a global optimization approach that is suitable for parameter tuning due to its convergence properties [4]. However, BO has not yet been applied as a parameter tuning approach in the context of meta-heuristics for solving vehicle routing problems (VRPs).

The main contributions of this paper are the introduction of a GGA for solving the MDPDPTWHV efficiently, which has been adopted from [1] and extended by new mutation and repair operators in order to enhance the solution quality for especially large and complex problem instances (see Sections 4.1 and 5.3). Moreover, an efficient configuration approach, based on Bayesian Optimization, to tackle the PTP is introduced and evaluated. The BO uses Gaussian Processes (GP) as probabilistic surrogate models for optimizing the function that predicts the utility of evaluating unknown parameter settings in the solution space (see Section 4.2.1). The results of the initial and tuned GGA are compared on 12 problem instance classes to a state-of-the-art Adaptive Large Neighborhood Search (ALNS) for the closely related Pickup and Delivery Problem with Time Windows (PDPTW). The classes differ regarding problem characteristics such as the distribution of customer locations (e.g., clustered or random) or size and position of time windows. Additionally, each class containing 1000 instances has been generated with 5 different numbers of depots (1, 4, 6, 8, and 9 depots) so that, in total, 6000 problem instances are used for the evaluation. It can be shown that the solution quality of the GGA can be improved by the implemented BO regarding both overall performance and for each problem instance class.

The paper is structured as follows. In Section 2, an extensive literature review of the related work regarding the Pickup and Delivery Problem contributions (cf. Section 2.1) and Parameter Tuning Problem methods (cf. Section 2.2) is given. The problems considered in this paper are defined in Section 3. In doing so, a new mathematical model formulation for the MDPDPTWHV is introduced in Section 3.1 and the Parameter Tuning Problem is specified in Section 3.2. The proposed methods for solving the studied problems are presented in detail in Section 4, whereas Section 4.1 details the concept of the implemented Grouping Genetic Algorithm; an explanation of the Bayesian Optimization approach with Gaussian Processes used as parameter configuration approach can be found in Section 4.2. The results, and a comprehensive discussion about them, are presented in Section 5. Finally, the paper closes with a short conclusion in Section 6, providing a summary, pointing out the limitations, and indicating avenues for future research.

2. Related Work

In recent decades, many approaches have been developed to efficiently solve Pickup and Delivery Problems (PDPs) with various (practically relevant) extensions or constraints. Due to the numerous contributions in this field over the years, Section 2.1 gives an overview of the solution methods that solve a PDP similar to MDPDPTWHV or that belong to the same class of meta-heuristics as the methods used in this paper. Extensive literature reviews are given in references [5,6].

All parameterizable approaches are to be optimized regarding the control parameters in order to enhance the solution quality. To do so, from the end of the last century, a number of approaches have been developed for automated parameter tuning. In Section 2.2, an overview of solution approaches in the field of automated parameter configuration is given. The focus is on parameter tuning approaches that were designed specifically for evolutionary algorithms (EAs). We aloso refer to important literature sources on parameter tuning in a more general context. An overview of parameter tuning methods for EAs is given in [7]. A more general overview of automatic parameter tuning for meta-heuristics can be found in [2]. It is worth mentioning that the contribution of this paper focuses on static parameter configuration. Other subjects, e.g., dynamic or adaptive parameter tuning, are not covered here (cf. [8]).

2.1. Pickup and Delivery Problems

In [9], a Pickup and Delivery Problem with Time Windows (PDPTW) is considered and several construction methods for an initial vehicle routing plan are developed. The presented methods differ, on the one hand, in whether the insertion is performed sequentially or in parallel, and, on the other hand, in which acceptance criterion (*First Insertion* or *Best Insertion*) is used. In another proposed approach, ref. [9] simply insert a request with a pickup and delivery node pair at the end of a route and transform this route into a feasible one via reasonable swap operators. A hill-climbing approach is applied as an acceptance criterion for each node swap. The presented methods are compared with the best known solutions of the PDPTW instances of [10]. In [11], the authors have developed an insertion heuristic for the PDPTW based on the time buffer before and after a service (attempted to be maximized for each insertion), in order to keep the most flexibility for the following insertions. As a second optimization criterion, ref. [11] uses the percentage of the length of intersecting paths, since a short vehicle routing plan intuitively contains few intersecting routes. The developed heuristic is evaluated using the PDPTW instances of [10]. A framework with operators for local neighborhood search is presented in [12] to solve several rich VRPs. Operators include swap, Or-Opt, 2-Opt, node relocation or string exchange. It should be noted that the approach presented can be easily embedded in a meta-heuristic, such as Simulated Annealing (SA) or Tabu Search (TS). The operators of the framework have been evaluated for the PDPTW on the [10] instances.

A reactive TS approach is presented by [13] in order to solve the PDPTW. Here, the tabu list stores which are positioned within a route are excluded for a customer node, while 'reactive' means that the parameters of the solution approach, such as the length of the tabu list, are adjusted depending on the current behavior of the procedure. The authors of ref. [13] apply three simple local neighborhood operators and an *Escape* operator, if none of the neighborhood operators generate a new solution. The approach is evaluated on self-generated problem instances. The authors of ref. [14] implement a SA approach for a PDPTW and one vehicle. To generate the start solution, several randomly generated solutions are considered and the best one is chosen. Furthermore, the generated solutions are used to choose the starting temperature. In order to improve the start solution, two randomly selected nodes are swapped at exactly the point when the time precedence specified for the upper, lower or middle time window value is not met. This approach is evaluated with self-generated problem instances and compared with two other implemented methods: a genetic algorithm and a hill-climbing approach.

The authors of [15] present a two-stage algorithm for solving PDPTW in which Simulated Annealing and Large Neighborhood Search (LNS) are implemented. In the first stage of the procedure, the number of vehicles required is first minimized. In the second stage, an LNS is presented to optimize the vehicle routing. The evaluation is carried out with the PDPTW instances of [10] and their best known solutions. In [16], the Pickup and Delivery Problem with Time Windows is investigated, whereby the objective function here is the weighted sum of the distance, the duration of the tour, and a penalty term for the non-served requests. The authors present an Adaptive Large Neighborhood Search (ALNS).

Several destroy (*Worst*, *Random*, and *Shaw Removal*) and repair operators (*Basic Greedy* and *Regret-k*) are used to determine the neighborhood under consideration. A simulated annealing approach is selected as the acceptance criterion for a new solution. The results of the ALNS are compared with the PDPTW problem instances from [10], among others. In addition, the authors present an extension of the ALNS in [17]. Here, they further develop destroy operators (e.g., *Cluster* and *Historical Node-Pair Removal*) and formulate these for Rich PDPTW. In [18], a slightly different PDP is presented: a PDP with transfer points to exchange cargo or people. The authors implement the ALNS of [16] and extend it with several destroy and repair operators to efficiently include transfer points. The approach is evaluated on 10 self-generated data sets, which are based on the practical application. The authors of ref. [19] present a PDP with heterogeneous vehicle types, whereby the vehicles have different configurations, e.g., to be able to carry wheelchair users. An ALNS is applied to several already calculated start solutions. The ALNS is adapted and extended on the basis of [16]. The approach is tested on Dial-a-Ride Problem (DARP) instances from the literature, among others.

In [20], an early Genetic Algorithm (GA) approach for solving the PDPTW is presented, whereby heterogeneous vehicles are also included in the considered formulation of the mathematical model. The authors choose a *Random Key* solution representation, in which each node is assigned a four-digit key. Based on the solution representation, a classic *Two-Point* crossover can be implemented. During the mutation, the vehicle index (which is the first digit of the four-digit key) is randomly changed for a pickup and delivery node pair. In order to minimize the number of vehicles, the vehicle with the smallest number of requests is deleted as in [1,21]. For each of the operators, the new individual is checked for feasibility and discarded if this is not the case. The GA is evaluated on self-generated problem instances. In [14], a GA for solving a PDPTW with one vehicle is developed. The authors use a path solution representation in which pickup and delivery nodes have the same expression, i.e., *duplicate gene representation*. The mutation operators implemented are a random exchange of two genes and a problem-specific operator that only permutes two randomly selected genes if the precedence with respect to the upper time window limits is not held. Since the implied order, given by the time windows, can become infeasible by conventional crossover operators, the authors of ref. [14] employ a special merge crossover. A steady-state GA variant is implemented, i.e., the generated children are inserted directly into the current population. The authors use self-generated test instances and compare two GA variants with different methods, such as hill-climbing, in their evaluation. A GA for Multi-Depot PDPTW with homogeneous vehicles is presented in [22]. The GA uses a path representation, i.e., the order of the customers is stored in the genotype. In order to increase diversification at the beginning, three variants are implemented to generate individuals for the initial population, whereby repair heuristics are used due to the solution representation. The genetic operators used include a one-point crossover and a mutation operator that randomly swaps two customers in the solution. The authors of [23] consider a slightly different route planning problem—a VRPTW with simultaneous delivery and pickup—and solve this problem with a GA. For this purpose, the authors use two parallel populations, for each of which different operators were developed. One population is responsible for diversification and the other for intensification. For the first population, there is only a recombination performed, which removes randomly selected nodes from the solution and re-inserts them again. To the second population, a crossover which copies entire routes from the parents and 11 mutation variants are applied. The selection of individuals is based on fitness and the best individuals are transferred to the new population using an elite mechanism. The evaluation is carried out with self-generated test instances and is additionally compared with CPLEX results. In reference [24], a Genetic Algorithm for PDPTW is presented, which also uses a path representation for the examined individuals. The selection of the parents is carried out using a tournament. Two crossover variants are implemented, each of which generates a child and exchanges either parts of the routes or the entire routes of the parents. The solutions can become infeasible, which means that a repair operator is required or the crossover is discarded in these cases. There are two additional mutation variants applied. The first variant dissolves a random vehicle and inserts the released requests into the remaining vehicles. The second variant selects a vehicle at random and tries to place each request on the associated route in a better position. The approach is evaluated on the PDPTW instances of [10] and compared with the best known solutions.

The authors of [21] also consider the PDP with Time Windows. As noted in the previous contributions, classical path representations can often lead to infeasible solutions as soon as genetic operators are executed. In order to eliminate this issue, and since the clustering of vehicle routing problems has a bigger impact on the solution quality, the authors of ref. [21] implement a Grouping Genetic Algorithm for the solution of the PDPTW. This means that the genotype uses a group-oriented representation. Thus, a gene is the set of all requests that are served by a vehicle. The presented group-oriented crossover is similar to a typical two-point crossover. During the mutation, a randomly selected gene is resolved and the requests from the associated vehicle are inserted into the remaining routes. For insertion, a classical double insertion heuristic is used, which operates on the phenotype or the route plan, respectively. The GGA is tested on the PDPTW data sets of [10], among others, using the best known solutions. In order to solve the Multi-Depot PDPTW with Heterogeneous Vehicle Fleets, the authors of ref. [1] adapt the GGA of [21] in order to consider the problem constraints, such as multiple depots and heterogeneous vehicles, properly. To do so, the group-oriented solution representation is adjusted and new mutation operators are introduced which take heterogeneous vehicles into account. The approach is evaluated on self-generated Multi-Depot PDPTWHV instances which are generated on the basis of the well-known [10] problem instances.

Pickup and Delivery Problems are widely investigated in several variants and solved with numerous approaches. However, contributions which tackle precisely this routing problem at hand cannot be found. Probably the most considered variant is the PDPTW which is closely related to the Multi-Depot Pickup and Delivery Problem with Time Windows and Heterogeneous Vehicles. For this problem, the GGA has proven its advantages compared to typical GA methods. Compared to existing contributions, the present paper introduces an adaption of the GGA presented in [1], which incorporates the problem constraints of the MDPDPTWHV through additional mutation operator types and a control parameter to manage when to use which type. In this way, the GGA becomes more efficient (cf. Section 4.1). Moreover, this contribution includes a new compact two-index model formulation of the MDPDPTWHV, with which the application of exact solvers (such as CPLEX) will be improved even for medium instance sizes due to the complexity reduction of the search space (cf. Section 3.1).

2.2. Parameter Tuning Problems

In [25], the Relevance Estimation and Value Calibration (REVAC), a population-based Estimation of Distribution Algorithm (EDA), is introduced for EAs. It is based on the principle of solving parameter tuning problems by estimating parameter relevance with normalized Shannon entropy. REVAC is an iterative algorithm starting with the assumption of a uniform distribution regarding the EDA and estimates the distributions of promising parameter values for each parameter within the configuration space. A different continuous black box optimization method is the Covariance Matrix Adaptation Evolution Strategy (CMA-ES) which is based on the concepts of self-adapting evolution strategies and is applied in [26]. In each of its iterations, a set of candidate configurations from a multivariate Gaussian distribution is sampled, whose co-variance matrix is adapted cumulatively. In [7], the authors compare REVAC and CMA-ES for tuning EA parameters. Here, they also combine the approaches with racing, e.g., Iterated F-Race from [27]. In Iterated F-Race, a small set of parameter configurations is sampled according to a probabilistic model and executed on several instances. Thereby, parameters are discarded as soon as enough statistical evidence is gathered against them. A Bayesian Optimization method to tune the parameter configuration of an EA solving a flow-shop scheduling problem is introduced in [28]. The authors show that the approach is able to improve the efficiency of the meta-heuristic by optimizing the six EA parameters considered.

The following literature focuses on parameter tuning for algorithms which do not necessarily belong to the EA class. In [29], an Iterative Local Search approach, called ParamILS, for the parameter configuration of algorithms is introduced, which basically chooses each parameter to be optimized step-by-step and individually. The ParamILS is evaluated on several configuration scenarios, e.g., parameter tuning on the CPLEX solver. In doing so, various benchmark instances are used, which are separated into training and test instances, in order to find sufficiently generalized algorithm parameters for a set of benchmark instances. While model-free parameter tuning approaches like ParamILS are relatively simple and can be applied out-of-the-box, model-based techniques (e.g., Bayesian Optimization) show very promising results for algorithm configuration. In [30], the Sequential Model-Based Optimization (SMBO) is extended for solving algorithm configuration problems with a large number of numerical as well as categorical parameters. The approach constructs a regression model in order to predict the performance of a certain parameter configuration on the algorithm to be optimized. After running the target algorithm with these parameters, the output is applied as additional data to enhance the model. The SMBO approaches developed are compared to ParamILS and a Genetic Algorithm, improving the performance of a local search and tree search algorithm and solving the Satisfiability Problem (SAT), as well as optimizing the performance of the Mixed-Integer Problem solver CPLEX. The irace package, a software package which provides several iterated racing approaches for improving parameter configuration, is introduced in [31]. In general, racing consists of three basic steps: (i) sampling new configurations regarding a certain distribution, (ii) performing racing and choosing the best configuration, and (iii) updating the sampling distribution in order to converge to an optimum. In [31], multiple variants of iterated racing methods are presented, e.g., an elitist racing procedure to ensure that the best configurations returned in racing are also the best ones over all racing rounds performed or a soft-restart approach in order to escape from local optima. They are compared regarding, e.g., an Ant Colony Optimization method solving the symmetric Travelling Salesman Problem. A heterogeneous vehicle routing problem with automatic guided ground and aerial vehicles is considered in [32]. The authors present a parameter tuning approach in order to improve the parameter configuration of the local search heuristics used to solve the routing problem. Thereby, a Bayesian Optimization approach and a Genetic Algorithm are compared for parameter tuning. It is shown that both approaches achieve solution quality enhancements; however, the BO is more efficient in doing so. In [33], a comprehensive study of the parameter configuration of the (meta-)heuristics for two Vehicle Routing Problems is considered, in which seven state-of-the-art parameter tuning approaches are compared. It can be shown that, overall, automated parameter configuration enhances the solution quality of the approaches considered. Moreover, there is no best known parameter configuration which provides the best solution quality over the entire benchmark (cf. [34]). For the Capacitated Arc Routing Problem, the authors of ref. [35] introduce a Bayesian Optimization approach in order to enhance the solution quality of a special meta-heuristic through automatic parameter configuration. The authors demonstrate that the tuned algorithm improves both the solution quality as well as the convergence speed.

Regarding the related work, Bayesian Optimzation has shown its strength in automatic parameter configuration in several fields of routing problems. Nevertheless, tuning a sophisticated GA for solving a high-complex routing problem is paramount, as the MDPDPTWHV has not yet been tackled using a BO approach in the literature. Hence, in this paper, we consider improving the solution quality of the GGA framework (cf. Section 4.2), i.e., for solving the Parameter Tuning Problem (cf. Section 3.2). However, in contrast to most of the former contributions, the focus in this paper is, additionally, to improve parameter configurations with the BO on different problem instance classes with regards to the data structure of each class (cf. Section 5.1).

3. Problem Definition

In this section, both problems considered in this paper are specified. In Section 3.1, a new compact two-index mathematical model formulation for the Multi-Depot Pickup and Delivery Problem with Time Windows and Heterogeneous Vehicle Fleets is introduced, whereas the Parameter Tuning Problem is described in detail in Section 3.2.

3.1. Mathematical Model for the Multi-Depot Pickup and Delivery Problem with Time Windows and Heterogeneous Vehicle Fleets

The pickup and delivery problem, which LTL carriers solve on a daily basis (cf. Section 1), is described on the basis of a smart two-index model formulation. The main advantage of this formulation compared to a typical three-index formulation is the significantly lower number of decision variables. Using this model, solutions with exact solvers can also be found for large, complex problem instances [1]. Moreover, the heterogeneity, in terms of capacity and speed, is modeled with a concept based on real-valued variables (cf. [36,37]). To do so, capacity and speed variables $w_i, u_i \in \mathbb{R}^{\geq 0}$ are introduced, the values of which are associated with the vehicle serving the customer i and forwarded to the following customer on the same route $v_i \in \mathbb{R}^{\geq 0}$. For solvers like CPLEX, this formulation provides support, since the relaxation of the problem is easier due to more real and less binary variables [38]. As a consequence, compared to [1], there is only the decision of whether an arc (i, j) is used or not with a binary variable x_{ij} model. The sets, parameters, and variables are listed in detail in Table 1.

Table 1. Parameters and variables for a compact two-index model formulation of the MDPDPTWHV.

indices and sets:	
$d \in \mathcal{D}$	set of depots, whereas each vehicle is modeled as an artificial depot with a start depot d_s and end depot $d_e = 2n + \delta + d_s$; δ is the number of depots (and vehicles) $\Rightarrow$ set of start depots $\mathcal{D}_s = \{0, \ldots, \delta - 1\}$, set of end depots $\mathcal{D}_e = \{2n + \delta, \ldots, 2n + 2\delta - 1\}$
$i, j \in \mathcal{N}$	set of customer nodes, $\mathcal{N} = \mathcal{N}_p \cup \mathcal{N}_d = \{\delta, \ldots, 2n + \delta - 1\}$
$i, j \in \mathcal{N}_p$	set of pickup locations, $\mathcal{N}_p = \{\delta, \ldots, n + \delta - 1\}$
$i, j \in \mathcal{N}_d$	set of delivery locations, $\mathcal{N}_d = \{n + \delta, \ldots, 2n + \delta - 1\}$
$i, j \in \mathcal{V}$	set of nodes, $\mathcal{V} = \mathcal{N} \cup \mathcal{D} = \{0, \ldots, 2n + 2\delta - 1\}$; in addition, $\bar{\mathcal{V}} = \mathcal{D}_s \cup \mathcal{N}$ and $\bar{\mathcal{V}} = \mathcal{N} \cup \mathcal{D}_e$ are declared
parameters:	
$[a_i, b_i]$	time window at node $i \in \mathcal{N}$ in which the service has to start
c_{ij} (d_{ij})	variable costs (distance) for travelling from i to $j \in \mathcal{V}$; $c_{ii} = \infty$
d_i	delivery demand at customer $i \in \mathcal{N}$, where $d_i > 0$, $\forall i \in \mathcal{N}_d$, and $d_i = 0$, $\forall i \in \mathcal{N}_p$; note that $d_{i+n} = p_i$, $\forall i \in \mathcal{N}_p$ holds
f_d	fixed costs for using a vehicle associated with depot $d \in \mathcal{D}$
κ_d	capacity of vehicle at depot $d \in \mathcal{D}$
$\bar{\kappa}$	maximum capacity of all vehicles, i.e., $\bar{\kappa} = \max_{d \in \mathcal{D}}\{\kappa_d\}$
l	load factor for vehicle loading measured in time per quantity unit
l^{const}	additional loading time to model preparations before service
M	big M for linearization of time window constraints
ν_d	reciprocal speed of the vehicle at depot $d \in \mathcal{D}$
$\bar{\nu}$	max. reciprocal speed of all vehicles, i.e., $\bar{\nu} = \max_{d \in \mathcal{D}}\{\nu_d\}$
p_i	pickup demand at customer $i \in \mathcal{N}$, where $p_i > 0$, $\forall i \in \mathcal{N}_p$, and $p_i = 0$, $\forall i \in \mathcal{N}_d$
s_i	service time at node $i \in \mathcal{V}$ holding $s_i = l \cdot (d_i + p_i) + l^{\text{const}}$, $\forall\, i \in \mathcal{N}$, and $s_d = 0$, $\forall\, d \in \mathcal{D}$
decision variables:	
L_i	current load of the visiting vehicle after serving node $i \in \mathcal{V}$
T_i	beginning of the service at node $i \in \mathcal{V}$
v_i	route number on which customer $i \in \mathcal{V}$ is assigned
w_i	capacity of the vehicle which serves customer $i \in \mathcal{V}$
u_i	reciprocal speed of the vehicle which serves customer $i \in \mathcal{V}$
x_{ij}	binary variable which indicates whether a vehicle uses the arc between nodes i and $j \in \mathcal{V}$ $(x_{ij} = 1)$ or not $(x_{ij} = 0)$

With the introduced notation, the model for the problem at hand has the following form:

$$\min \quad \sum_{i,j\in\mathcal{V}} c_{ij}x_{ij} + \sum_{d\in\mathcal{D}_s}\sum_{j\in\mathcal{N}_p} f_d x_{dj} \tag{1}$$

$$\text{s. t.} \quad \sum_{j\in\mathcal{N}\cup\mathcal{D}_e} x_{ij} = 1 \qquad \forall\, i \in \mathcal{N} \tag{2}$$

$$\sum_{i\in\mathcal{N}\cup\mathcal{D}_s} x_{ij} = 1 \qquad \forall\, j \in \mathcal{N} \tag{3}$$

$$\sum_{j\in\mathcal{N}_p} x_{dj} \le 1 \qquad \forall\, d \in \mathcal{D}_s \tag{4}$$

$$\sum_{j\in\mathcal{N}_p} x_{d_s j} - \sum_{i\in\mathcal{N}_d} x_{i d_e} = 0 \qquad \forall\, d_e, d_s \in \mathcal{D} \tag{5}$$

$$j x_{dj} - v_j \le 0 \qquad \forall\, d \in \mathcal{D}_s, j \in \mathcal{N}_p \tag{6}$$

$$v_j - j x_{dj} \le n(1 - x_{dj}) \qquad \forall\, d \in \mathcal{D}_s, j \in \mathcal{N}_p \tag{7}$$

$$v_i - v_j \le n(1 - x_{ij}) \qquad \forall\, i \in \mathcal{N}, j \in \bar{\mathcal{V}} \tag{8}$$

$$w_d = \kappa_d \qquad \forall\, d \in \mathcal{D}_s \tag{9}$$

$$w_j - w_i \le \bar{\kappa}(1 - x_{ij}) \qquad \forall\, i \in \bar{\mathcal{V}}, j \in \bar{\mathcal{V}} \tag{10}$$

$$u_d = \nu_d \qquad \forall\, d \in \mathcal{D}_s \tag{11}$$

$$u_j - u_i \le \bar{\nu}(1 - x_{ij}) \qquad \forall\, i \in \bar{\mathcal{V}}, j \in \bar{\mathcal{V}} \tag{12}$$

$$v_i - v_{i+n} = 0 \qquad \forall\, i \in \mathcal{N}_p \tag{13}$$

$$w_i - w_{i+n} = 0 \qquad \forall\, i \in \mathcal{N}_p \tag{14}$$

$$u_i - u_{i+n} = 0 \qquad \forall\, i \in \mathcal{N}_p \tag{15}$$

$$v_{d_s} - v_{d_e} = 0 \qquad \forall\, d_s, d_e \in \mathcal{D} \tag{16}$$

$$w_{d_s} - w_{d_e} = 0 \qquad \forall\, d_s, d_e \in \mathcal{D} \tag{17}$$

$$u_{d_s} - u_{d_e} = 0 \qquad \forall\, d_s, d_e \in \mathcal{D} \tag{18}$$

$$T_i + s_i + d_{ij}\, u_i - T_j \le M(1 - x_{ij}) \qquad \forall\, i \in \bar{\mathcal{V}}, j \in \bar{\mathcal{V}} \tag{19}$$

$$a_i \le T_i \le b_i \qquad \forall\, i \in \mathcal{V} \tag{20}$$

$$T_i + s_i + d_{i,i+n}\, u_i \le T_{i+n} \qquad \forall\, i \in \mathcal{N}_p \tag{21}$$

$$L_i - d_j + p_j - L_j \le \bar{\kappa}(1 - x_{ij}) \qquad \forall\, i \in \bar{\mathcal{V}}, j \in \bar{\mathcal{V}} \tag{22}$$

$$L_i \le w_i \qquad \forall\, i \in \mathcal{V} \tag{23}$$

$$x_{ij} \in \{0, 1\} \qquad \forall\, i, j \in \mathcal{V}, i \ne j \tag{24}$$

$$L_i, T_i, v_i, w_i, u_i \in \mathbb{R}^{\ge 0} \qquad \forall\, i \in \mathcal{V} \tag{25}$$

In the objective function (1), the variable costs and the fixed costs for the vehicles used are minimized. Constraints (2) and (3) ensure that each customer is served exactly once. Each vehicle may start, at most, once (4). Restrictions (5) model that each started vehicle must arrive at the corresponding end depot. The conditions (6) and (7) set the route index v_j to the index of the customer j served first. The unique route index v_j is passed by the inequalities (8) from customer i to j if j directly follows i. Capacity variables w_d are set by the conditions (9) of the vehicle's capacity associated with the depot $d \in \mathcal{D}$. This is passed from customer i to j by means of restrictions (10) if they are directly consecutive. The reciprocal speed u_i is set for the starting vehicles by restrictions (11). For a vehicle using arc (i, j), the reciprocal speed is passed from node i to j with constraints (12). Equations (13), (14), and (15) ensure the coupling of pickup and delivery on the same vehicle. Similarly, this is modeled for the corresponding start and end depots by (16), (17), and (18). If customer j is visited directly after i, the service start time at j must not be earlier than the service end time at i, plus the travel time between i and j due to (19). Further, the service start time has to start within the specified time window $[a_i, b_i]$ of customer i (20). The time precedence relationship between pickup and delivery nodes is modeled using inequalities (21) through the service start at node i and $i+n$. Restrictions (22) ensure that the load L_j is correctly updated after the customer i is visited. In addition, (23) guarantees that the capacity of the vehicle associated with the served customer i is not exceeded. Finally, the decision variables x_{ij} describing the transport over the arc (i,j) are defined as binary (24). All other variables (i.e., time variables S_i, T_i, loading variables L_i, variables for route indices v_i, capacity variables w_i, and (reciprocal) speed variables u_i) are restricted to non-negative real numbers (25).

3.2. Parameter Tuning Problem

The choice of parameters for algorithms is essential for their effectiveness and efficiency. The determination of suitable parameters for algorithms can be approached using the parameter tuning (or parameter configuration) problem [2]. The parameter tuning problem can be specified according to Eiben and Smit [7].

As Figure 1 demonstrates, different layers of the problem structure are considered and these are connected with a control and an information flow. The *application layer* describes the given optimization problem (here, the MDPDPTWHV, as in Section 3.1), which is solved by an algorithm (*algorithm layer*, here, e.g., the GGA from Section 4.1). How this particular algorithm is designed with respect to parameter configuration is described in the *design layer*. The quality of the parameter configuration chosen for an algorithm is evaluated in terms of solution quality.

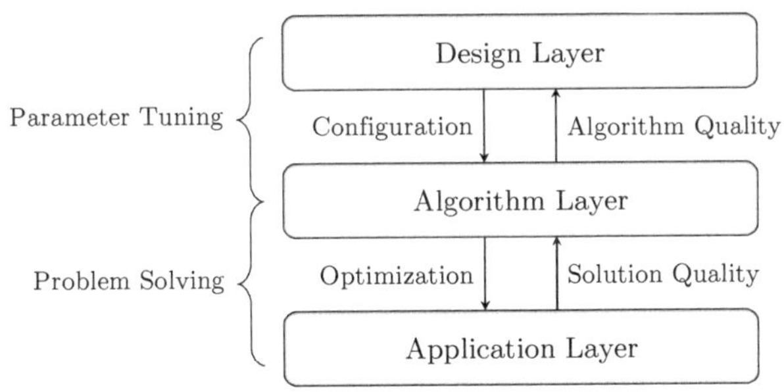

Figure 1. Control flow (**left**) and information flow (**right**) through the different layers in parameter tuning [7].

The PTP arises, among others, in the optimization of parameter settings for meta-heuristics [2]. Thus, many optimization methods such as a Genetic Algorithm depend on various known parameters that can be optimally chosen so that their performance can be improved in terms of finding good solutions [4]. Obviously, the performance of a meta-heuristic depends on the configuration of its parameters. Nevertheless, the configuration problem has not been formally considered in scientific publications for a long time [39]. The authors of ref. [40] describe how, until around 1999, parameter selection in the development of meta-heuristics was carried out manually. After evaluating various configurations, the one that delivered the best result was selected. The selection was thus often made on the basis of expert knowledge about the respective optimization problem. However, according to the *No Free Lunch Theorem*, there is no universal algorithm whose computations have the same solution quality for all optimization problems [34], so parameter tuning is by no means a one-time problem and it definitely makes sense to automate it [2].

It is worth pointing out that, besides the static parameter configuration, there are also dynamic, adaptive or self-adaptive approaches (cf. [8]). For these techniques, Machine Learning (ML) approaches are often reasonably applied due to their promising results. An overview of which meta-heuristic-related configuration problems can arise, and how ML techniques are applied to solve them for algorithms solving combinatorial optimization problems, can be found in [41]. A review of such learn-heuristics, especially for routing problems, is given by [42], while, in [43], a literature review for adaptive parameter configuration for evolutionary algorithms is given.

4. Proposed Methods

The methods to solve the considered problems, the MDPDPTWHV and the Parameter Tuning Problem, are presented in this section. In Section 4.1, a detailed overview of the Grouping Genetic Algorithm is given, followed by a comprehensive description of the Bayesian Optimization approach in Section 4.2.

4.1. Grouping Genetic Algorithm Framework

Genetic algorithms go back to [44]. They iteratively apply certain operators to a population of solutions so that, on average, each new generation tends to be better than the previous one, according to a predefined fitness criterion. The fitness criterion measures the ability of an individual to survive in a population and the algorithm strives to maximize fitness. The stop criterion is usually specified as a fixed number of generations or as a time limit for execution. In the end, the best solution found is returned. Genetic algorithms generally provide good results for complex routing problems (see, e.g., [14,21]), so it makes sense to focus on these.

In particular, the Grouping Genetic Algorithm presented in [1] produces adequate results for the MDPDPTWHV. Here, the authors show that the considered GA variants with population management regarding general replacement with elitism promise a good solution quality for the presented problem. However, the complexities of the problem instance's data structure can be various. Hence, in this paper, the GGA is extended by new mutation and repair operators in order to investigate which combination of operators is suitable for a given data structure. The GGA's general replacement version with binary tournament selection is used. In addition, the GGA of [1] has been enhanced through the permission of infeasible solutions by which the entire search space is investigated more effectively. An overview of the procedure of the GGA is given in Figure 2. For all terms and concepts of Genetic Algorithms, we kindly refer readers to [45].

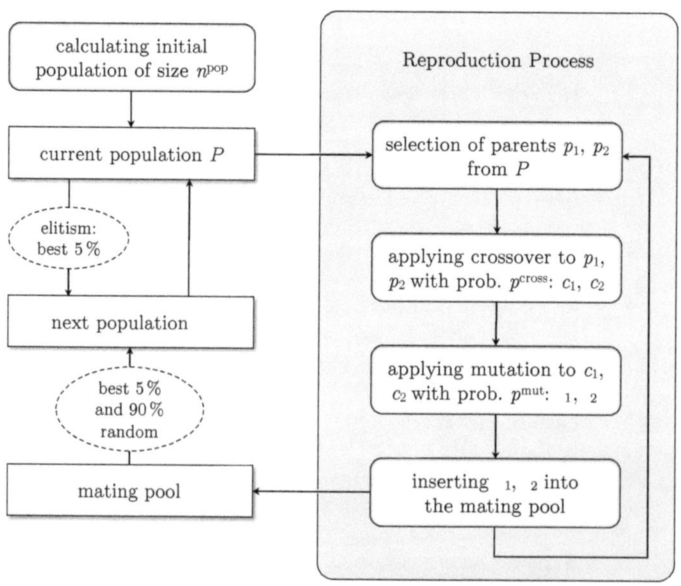

Figure 2. Grouping Genetic Algorithm framework with population management: *general replacement with elitism*.

The procedure of the GGA can be explained as follows: at the beginning, a population P of n^{pop} individuals is determined. It is assumed that each individual has a fitness value. This is equal to the sum of the variable and fixed costs of the vehicle routing solution (see objective function (1)) and also integrates a penalty term for non-served requests. The fitness value is to be minimized in the proposed GA variant and can be written as follows:

$$\sum_{i,j \in \mathcal{V}} c_{ij} x_{ij} + \sum_{d \in \mathcal{D}_s} \sum_{j \in \mathcal{N}_p} f_d x_{dj} + \gamma \sum_{i \in \mathcal{N}_p} \left(1 - \sum_{j \in \mathcal{N}} x_{ij}\right). \tag{26}$$

In general, for the penalty term in (26), the weight γ is to be chosen in such a way that the consideration of non-served requests results in the worst possiblesolution [46]. In order to achieve this constraint, γ is empirically chosen as follows:

$$\gamma := 2 \cdot \max_{i,j \in \mathcal{V}} \{c_{ij} + c_{ji}\} + \max_{d \in \mathcal{D}} \{f_d\}. \tag{27}$$

Please note that, if the GGA is implemented in a variant that only allows feasible solutions according serving all requests in an instance, γ in (27) has to be set to $\gamma := 0$. However, in preliminary studies, it has been found out that, the MDPDPTWHV allowing infeasible solutions in the manner of unserved requests, together with the fitness function (26) and penalty term (27), leads to an improved solution quality of the GGA.

Each individual is computed by using several sequential double-insertion heuristics that create routes one after another by inserting customers at their best possible position. To enhance the diversity of the initial population, 25% of the individuals are generated with *Best Insertion*, which starts the route with one request and inserts the best corresponding request in each iteration; 50% with *Random Insertion*, which chooses a random request for insertion in every iteration (cf. [1]); and 25% with a *Regret-k* heuristic (cf. [16]), which calculates a regret value to evaluate how much a greedy insertion will cost in the end and selects the request that has the highest costs (see Equation (30) in Section 4.1.5). A high level of diversity is achieved, as no duplicates are allowed within the population. The percentage values for applying the insertion heuristics considered have been determined in preliminary studies (cf. [1]).

As long as the maximum number of generations $\gamma^{\max}$ has not been met, two parents p_1, p_2 for generating offsprings are selected (cf. Section 4.1.2). With a probability of p^{cross}, one crossover operator variant is applied to p_1 and p_2, as in [1], in order to create two children c_1 and c_2 (cf. Section 4.1.3). Moreover, each child is modified by using one of the mutation operators with probability p^{mut} and so two mutated children $\tilde{c}_1$ and $\tilde{c}_2$ are generated (see Section 4.1.4). In case no operator is applied, the created offsprings are just clones of their ancestors. Generated individuals are stored in a mating pool from which the next population is selected. Finally, the best individual regarding its fitness is returned. Please note that $n^{\text{pop}}, \gamma^{\max}, p^{\text{cross}}$, and p^{mut} are parameters that are set manually. These parameters have have been determined in preliminary tests (cf. [1]).

The most difficult part of a GA is to find a good solution representation or genotype encoding, respectively, so that the crossover and mutation operators do not have to be too complex and decoding from genotype to phenotype is not time-consuming. For this reason, the approach of [1] is implemented as a Grouping GA in which each vehicle is represented by a gene and the pickup and delivery customers are grouped with their request index. A genotype encoding then stores the assignment of a request to the executing vehicle and can be represented by a vector of integers.

Figure 3 shows that the vehicles are represented by negative index values. All subsequent positive integers indicate the requests served by the corresponding vehicle. This encoding implies that a route for each vehicle (e.g., route 2 in Figure 3) has to be calculated, whenever the genotype decoding is necessary, e.g., for the determination of the fitness value. Since this is very time-consuming and not necessarily deterministic with regard to the applied heuristics, the corresponding phenotype of an individual is also be stored (cf. [1]), i.e., the vehicle routing solution. Thus, it is less likely that good features of an individual are removed from the phenotype when applying crossover or mutation operators. Please note that the phenotype must always be modified together with the genotype during the solution process.

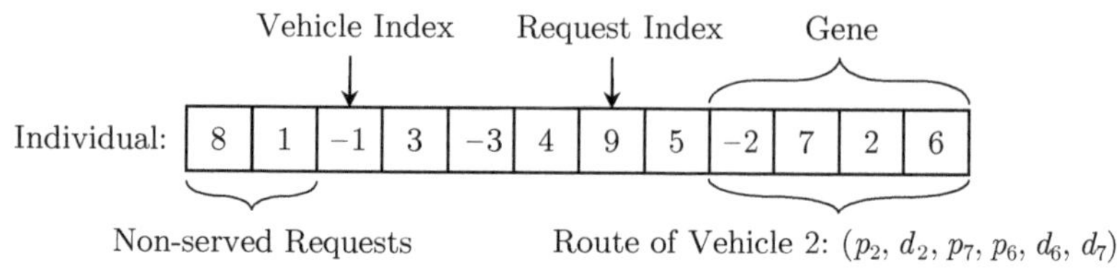

Figure 3. Genotype encoding for the Grouping Genetic Algorithm.

The following Sections 4.1.1–4.1.5 describe in detail the operators extended compared to [1] and those whose control parameters are to be optimized, i.e., mutation and repair operators. The selection and crossover operators, as well as population management, are adopted from the GGA of [1].

4.1.1. Population Management

Due to the promising results of the GGA with general replacement as population management [1], this variant was implemented in the present framework. Hence, a mating pool of 1.5 times the generation size is filled with children, whereas duplicates are prevented. In order to keep the best solutions from the last generation, elitism is applied by copying the best 5% of the generation into the next. Subsequently, the best 5% of individuals from the mating pool are selected and the rest of the next generation is randomly filled with individuals from the mating pool (cf. Figure 2).

4.1.2. Selection

In order to select parents without duplicates for offspring generation, binary tournament is used, since this selection approach has provided the most promising results. Here, two individuals, called a *tournament*, are chosen randomly from the population and the best individual is used as the first parent. The second parent is selected analogously.

4.1.3. Crossover Operator

The implemented group-oriented crossover variants are adopted from the GGA of [1]. A simplified illustration of how the crossover works for a representation with nine requests distributed over three possible vehicles is given in Figure 4. Two crossover points (both of them between two genes) are randomly chosen for parent 1. Then, all genes between the two crossover points (inner genes) are transferred with a probability of 50% and, otherwise, the genes outside of them (outer genes) are chosen at a randomly chosen insertion point into parent 2 to generate a child. This requires double-served requests and double-used vehicles coming from parent 2 to be deleted from the child. In doing so, non-served requests have to be re-inserted into the child by repair operators (see Section 4.1.5). In Figure 4, the inner genes are not transferred into the child. The outer genes mean that vehicles 1 and 2 have to be eliminated from the child. In addition, requests 2 and 7 must be removed. Finally, requests 5 and 9 remain and have to be re-inserted by a repair operator again. For a detailed explanation, we kindly refer readers to [1].

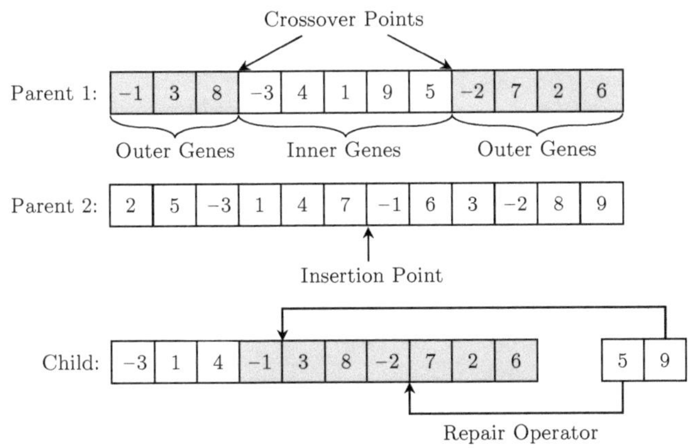

Figure 4. Crossover operator when transferring genes from parent 1 to parent 2.

4.1.4. Mutation Operators

Contrary to [1], the mutation operators are basically classified in *vehicle-based* and *request-based* mutations. The selection of the operator class is carried out by a specific probability value $p^{\text{choice}}(n)$ that models the likelihood for choosing request-based mutation in generation n:

$$p^{\text{choice}}(n) = 0.1 \cdot \exp\left(\ln(8)\,\frac{n}{\gamma^{\max}}\right). \tag{28}$$

In this way, the value p^{choice} of Equation (28) prefers vehicle-based operators in the beginning of the Genetic Algorithm and request-based operators at the end (the value p^{choice} starts at 0.1 in the first generation, $n = 0$, and ends at 0.8 in the last generation, $n = \gamma^{\max}$). In this manner, the trade-off is controlled between solution quality and computational time, since the vehicle-based mutation has a faster and broader investigation in the search space, while request-based mutation has a slower and more accurate search behaviour, which, in general, describes *exploration vs. exploitation*. Please note that, due to the choice of probability p^{choice}, vehicle-based and request-based mutation are not equally weighted.

The vehicle-based mutation is executed through selecting a vehicle index. The corresponding vehicle is deleted from the genotype and phenotype. Afterwards, all requests served by the erased vehicle must be reinserted, which is carried out by a repair operator (see Section 4.1.5), such that the solution holds all constraints. Here, a new vehicle may also have to be introduced. A simplified scheme of the mutation operator is given at the top of Figure 5. As in [1], the following four vehicle selection mechanisms are applied in order to tackle the problem of heterogeneity:

Costs/Number-of-Request Ratio. The ratio of route costs and the number of served requests is calculated for each vehicle and used to generate a roulette wheel. By spinning the roulette wheel, a vehicle index is selected. In particular, expensive routes with a small number of requests are preferred for removal.

Number of Requests. The vehicle with the minimum number of requests is used for deletion.

Random Vehicle Index. Here, a random vehicle index is selected, thus all vehicles are equally likely to be chosen.

Random Genotype Position. The variant chooses a random position within the genotype and selects the respective vehicle index. Hence, vehicles with many requests are preferred.

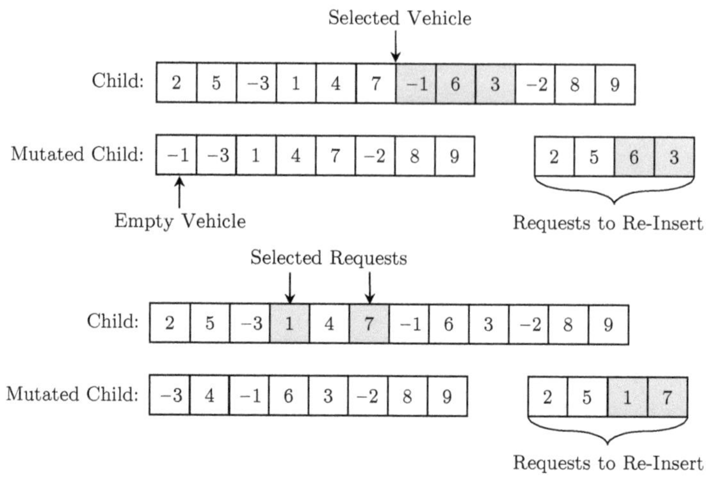

Figure 5. Mutation operators and mutated child to be repaired (using an additional empty vehicle if needed) with vehicle-based mutation (**top**) and request-based mutation (**bottom**).

Request-based mutation is a newly added concept. Particularly if instances have many requests per vehicle or if in the course of the search the solution space must be examined more precisely for intensification, it is useful to remove individual requests from the genotype or phenotype, respectively. Basically, a number of requests to be released is randomly determined and then eliminated from the individual with one of the mechanisms described below. A simplified procedure is shown at the bottom of Figure 5.

Historical Request Pair. This concept is adopted from the approach of [17] and uses a matrix $(h_{ij})_{i,j\in\mathcal{R}}$ of weights for all requests $\mathcal{R}$. The matrix stores how promising a combination of two requests on the same vehicle is. To do so, the elite is evaluated after each generation and a matrix element h_{ij} is increased by one if the requests i and j are served together on one vehicle. The matrix $(h_{ij})_{i,j\in\mathcal{R}}$ can be interpreted as long-term memory. Since combinations with high quality in a certain generation do not necessarily have to be globally optimal, the matrix is adjusted during each generation through reducing h_{ij} by a factor $\tau \in (0,1)$, i.e., $h_{ij} := \tau h_{ij}$. To determine the requests to be eliminated, a score σ_i is specified in which h_{ij} values are summed up if request j is served with i on the same vehicle in the considered individual. Requests with low σ_i are non-promising combinations on the same vehicle and, consequently, are removed from the individual.

Similarity Measure. The variant is derived from the similarity-measure-based removal operator of [16]. Here, the first request is chosen randomly and, additionally, those requests are removed which are *similar* with respect to the measure $\tilde{\rho}_{ij}$ from Equation (29), i.e., those that have the smallest values. Please note that $i, j \in \mathcal{R}$ and request i have pickup customer i and delivery customer $i+n$ (see Section 3.1).

$$\begin{aligned}\tilde{\rho}_{ij} =&\, \alpha_1(c_{ij} + c_{i+n,j+n}) + \alpha_2(|a_i - a_j| + |a_{i+n} - a_{j+n}|) \\ &+ \alpha_3(|b_i - b_j| + |b_{i+n} - b_{j+n}|) + \alpha_4(|p_i - p_j|) \end{aligned} \tag{29}$$

Thus, those requests are defined as similar whose pickup and delivery nodes, start and end time windows, and demands to be transported are close to each other, since, for these, the exchange to another vehicle offers potential for direct insertion and, therefore, an improvement in the fitness value. For comparability, all values are scaled to the interval $[0,1]$. The weights $\alpha_k, k = 1,\ldots,4$ are elements of $\mathbb{R}^{\geq 0}$.

Additionally, a so-called *swap operator* with a probability of 20% is applied, after one of the six mutation variants is taken into account. This operator tries to swap used and free vehicles to minimize the sum of fixed costs. In order to keep the search in the solution space broad, the choice of vehicles to be swapped is guided by a roulette wheel.

4.1.5. Repair Operator

When applying crossover or mutation operators, the individual has often to be repaired due to removed requests. During repair, requests need to be re-inserted. Here, several insertion methods are used that determine the best insertion positions for each request with respect to a cost value. Contrary to [1], a *Regret-k* insertion method beside the greedy insertion is implemented. The approaches are quite similar to the double insertion heuristics used for generating the initial population. However, the proposed repair operators are parallel insertion heuristics, i.e., the possible insertion positions are evaluated for all available vehicles. If there are requests that cannot be served by available vehicles, a new (empty) vehicle has to be introduced whenever feasible.

Greedy Insertion. For each request, the insertion costs for all possible positions in each vehicle are calculated and the position with the lowest costs is chosen (cf. [1]).
Regret-k Insertion. For each request r, the difference in how much worse it is to insert a request in the k-th best vehicle v_k instead of the best one v_1 is calculated, while the differences from the best to the k-th best vehicle are summed up. The request $\tilde{r}$ with the maximum value is supposed to be inserted into its best position in the current iteration, since, later on, the insertion costs increase (cf. [16]). Formally, this can be written as in (30):

$$\tilde{r} = \arg\max_{r} \left\{ \sum_{l=1}^{k} (\Delta_r^{v_l} - \Delta_r^{v_1}) \right\}. \tag{30}$$

The Regret-k insertion is applied for $k = 2, 3, 4, \delta$. Additionally, in order to avoid repeated solutions from one individual to another, a tabu list can be applied which includes the vehicle in which a request should preferably not be inserted again.

Both the two new request-based mutation variants, as well as the Regret-k repair operator (with possible tabu list), effect the GGA in the manner that the exploration of a known area in the solution space will be improved. This is very important in the context of the MDPDPTWHV in comparison to related, less complex vehicle routing problems, since, with higher problem complexity, the challenge of finding a sufficiently good solution with only simple operators intensifies. Hence, the rough searching intended by deleting an entire vehicle leads, rather, to new areas than does investigating a complex smaller part of the solution space.

4.2. *Bayesian Optimization for Tuning the Parameters of the Grouping Genetic Algorithm*

As depicted in Figure 1, the Parameter Tuning Problem is the optimization problem which is embedded in the top layer above the solution approach for solving the origin problem. Due to this fact, the considered origin problem's solution algorithm has to be executed reasonably often, according to the PTP, in order to find a sufficient algorithm parameter configuration (see Section 3.2). Hence, a common drawback of many parameter tuning approaches for meta-heuristics (see, e.g., the methods in Section 2) is the considerable number of evaluations of the objective function required to determine the best parameter configuration [28].

As the performance of a Genetic Algorithm is sensitive to the parameter setting and the structure of a problem instance, the evaluation of specific parameter configurations can be very time-consuming. Moreover, optimizing the parameters of meta-heuristics is a *black box* problem, since the relationship between the parameter configurations of the algorithm and its performance cannot be measured metrically [2].

Altogether, *Bayesian Optimization* is a reasonable choice as a parameter tuning algorithm, since this method manages to cope with few evaluations of the objective function to be optimized, which also need not be known [28,47]. Furthermore, BO is well applicable to functions defined in a solution space with dimension $D \leq 20$ (number of parameters) and tolerates noise with respect to the function evaluation [48], which is typical behaviour for stochastic methods like a GA. Last but not least, the results of [28] show that BO is, in general, a good approach in solving the Parameter Tuning Problem for EAs. For a detailed overview of Bayesian Optimization, we kindly refer readers to [47–49].

Let $\varphi : \mathbb{X} \to \mathbb{R}$ be an unknown *black box function* that cannot be modeled in closed form, i.e., for an argument $\boldsymbol{x} \in \mathbb{X} \subseteq \mathbb{R}^D$, the function value $\varphi(\boldsymbol{x})$ can only be determined by evaluating φ. In the Parameter Configuration Problem, φ describes the *utility* of a parameter vector $\boldsymbol{x} \in \mathbb{X}$, which can be, e.g., the mean of the objective values over all the considered instances. The goal of Bayesian Optimization is to find a configuration $\boldsymbol{x}^* = \arg\min_{\boldsymbol{x}\in\mathbb{X}} \varphi(\boldsymbol{x})$ describing the global minimum of φ. The method iteratively learns a probabilistic model that estimates the utility function φ by known function values $\varphi(\boldsymbol{x})$ of different points $\boldsymbol{x} \in \mathbb{X}$ in the parameter space. For this purpose, an *a priori probability distribution* $P(\varphi)$ over the utility function φ and an *acquisition function* $a_{P(\varphi)} : \mathbb{X} \to \mathbb{R}$, that quantifies the utility of evaluating the function φ at each parameter configuration $\boldsymbol{x}$, are required [50]. Let n data points in the set $\mathcal{D}_n = (\boldsymbol{x}_i, y_i)_{i=1,\dots,n}$ with $y_i = \varphi(\boldsymbol{x}_i) + \varepsilon_i$ be known for function φ. Then, Bayesian Optimization iteratively repeats the following three steps [47]:

1. Find the next configuration $\boldsymbol{x}_{n+1}$ whose evaluation is most promising, i.e., which maximizes the acquisition function under the condition of the known data points $\mathcal{D}_n$:

$$\boldsymbol{x}_{n+1} = \arg\max_{\boldsymbol{x}\in\mathbb{X}} a_{P(\varphi|\mathcal{D}_n)}(\boldsymbol{x}) \,.$$

2. Determine $y_{n+1} = \varphi(\boldsymbol{x}_{n+1}) + \varepsilon_{n+1}$ with possible noise ε_{n+1} and add $(\boldsymbol{x}_{n+1}, y_{n+1})$ to the previous data points: $\mathcal{D}_{n+1} = \mathcal{D}_n \cup \{(\boldsymbol{x}_{n+1}, y_{n+1})\}$.
3. Update the probability model $P(\varphi \mid \mathcal{D}_{n+1})$ and, consequently, the acquisition function $a_{P(\varphi|\mathcal{D}_{n+1})}$.

Here, the update of the probabilistic model is carried out in step (3) via the *Bayes' Theorem* (31), where

$$P(\varphi \mid \mathcal{D}_{n+1}) \propto P(\mathcal{D}_{n+1} \mid \varphi) P(\varphi) \,. \tag{31}$$

The theorem states that the posteriori probability distribution $P(\varphi \mid \mathcal{D}_{n+1})$ for the new data points $\mathcal{D}_{n+1}$ is proportional to the a priori model $P(\varphi)$ multiplied by the similarity of the data points $\mathcal{D}_{n+1}$ to the assumed model for φ.

In condition (31), $P(\mathcal{D}_{n+1} \mid \varphi)$ consequently expresses how likely the data are under the model assumptions that are known a priori. If the assumption is that the utility function φ is very smooth and noise-free, data with high variance should be recognized as less likely than data that hardly deviate from the mean [47].

Figure 6 shows an iteration of Bayesian Optimization. The black box function $\varphi(x)$ to be optimized is drawn here in blue and the mean estimated function $m(x)$ is dashed in black. At each point $x \in \mathbb{X}$ the estimated utility function value $m(x)$ with a variance of $k(x, x)$, which is described by the confidence interval in y-direction (gray), can be determined depending on the already evaluated data points $\mathcal{D}_n$ (red) and the used a priori model $P(\varphi)$. These data are incorporated into the acquisition function (shown below), which determines a score for evaluating the possible x vectors. Here, the selected value x_{n+1} receives a high score because the knowledge about the range containing x_{n+1} is not yet enough for a sufficiently good estimation, as the variance is high.

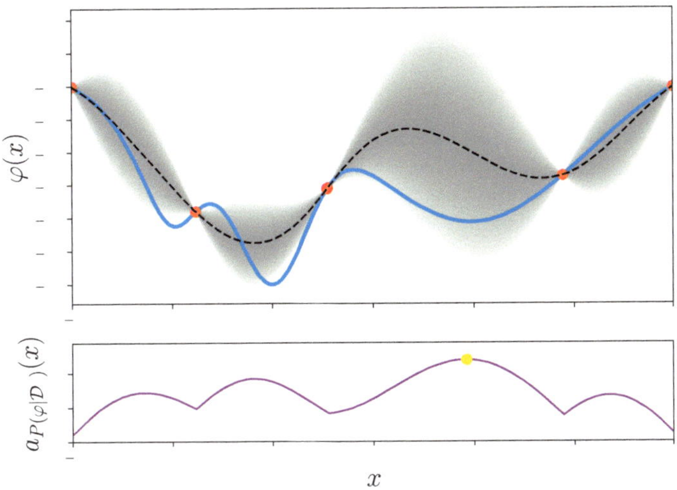

Figure 6. Bayesian Optimization with Gaussian Processes as a priori model (**top**) and acquisition function $a_{P(\varphi|\mathcal{D}_n)}$ (**bottom**).

The solution quality of Bayesian Optimization significantly depends on the choice of the a priori probability distribution (see Section 4.2.1) and on the evaluation of the utility by a surrogate model, i.e., the acquisition function (see Section 4.2.2). Both components are described in the following subsections.

4.2.1. Gaussian Processes

For a priori distributions, possible models are *Gaussian Processes*, *Random Forests* or *Bayesian Neural Networks* [50]. Gaussian Processes are most commonly chosen in the literature [48] and also provide promising results in [28,35,50]. Therefore, a Gaussian Process (GP) is used in the presented Bayesian Optimization approach. In the following, some insights into the GP used here are given. For a detailed description of Gaussian Processes, we kindly refer readers to [51,52].

For a Gaussian Process, a *covariance function* (also called a *kernel*) has to be defined. It can be assumed that the utility function φ for the problem at hand is non-smooth (cf. Section 4.1). This is because different function values $\varphi(x)$ can be obtained due to the uniform distributed randomly chosen operators, even if φ is evaluated several times with the same parameters $\boldsymbol{x}$. Thus, kernel functions coming from the class of *Automatic Relevance Determination* (ARD) *Matérn* kernels are appropriate for the Parameter Tuning Problem presented, since they can model non-smooth functions [53]. In the BO approach used in this paper, the ARD Matérn $\frac{5}{2}$ kernel is applied, which is defined as follows:

$$k_{\mathrm{M52}}(\boldsymbol{x},\boldsymbol{x}') = \theta_0\left(1+\sqrt{z}+\frac{z}{3}\right)\exp\left(-\sqrt{z}\right) \tag{32}$$

with $z = 5r^2(\boldsymbol{x},\boldsymbol{x}')$, while the function r^2 is defined as $r^2(\boldsymbol{x},\boldsymbol{x}') = \sum_{d=1}^{D}\frac{(x_d-x'_d)^2}{\theta_d^2}$. Since the influence of the individual parameters on the solution quality is not known a priori, the parameter $\theta_d = 1$ is set for all $d = 0,\ldots,D$ within $r^2(\boldsymbol{x},\boldsymbol{x}')$ in the presented approach.

Due to the stochastic nature of the GGA to be optimized, the solution for a given parameter configuration is expected to be subject to noise [2]. To model this noise, the kernel function (33) is additionally used [51]:

$$k_{\mathrm{Noise}}(\boldsymbol{x},\boldsymbol{x}') = \begin{cases}\sigma_n^2 & ,\boldsymbol{x}=\boldsymbol{x}' \\ 0 & ,\text{sonst}\end{cases}. \tag{33}$$

The parameter σ_n^2 describes the variance of the modeled noise. Altogether, the covariance function $k(x, x')$ (34), considered for the Gaussian Process in the BO approach at hand, is composed of (32) and (33) (cf. defining GP kernels in [51]):

$$k(x, x') = k_{\text{M52}}(x, x') + k_{\text{Noise}}(x, x'). \tag{34}$$

In order to purposefully design the choice of the next parameter configuration to be evaluated in step (1), a reasonable acquisition function must be defined. This function makes use of the model assumed by the Gaussian Process to predict the utility for evaluating certain parameter settings $x \in \mathbb{X}$.

4.2.2. Acquisition Function

The acquisition function $a_{P(\varphi)}$ is designed to estimate, based on the current data $\mathcal{D}_n$, which areas of the parameter space $\mathbb{X}$ to investigate in order to improve the model's prediction of the Gaussian Process. Therefore, the acquisition function should be constructed to achieve a high value at parameters $x \in \mathbb{X}$, where the predicted utility function value $\varphi(x)$ is low or the prediction's quality is poor, i.e., there is too little information available on that part of the considered parameter space for a sufficiently accurate estimation [47]. These two cases describe the classical problem of *exploitation vs. exploration*. This is tackled in several acquisition functions that can be found in the literature, such as *Probability of Improvement*, *GP Upper Confidence Bound*, and *Expected Improvement* (EI) [53]. Since EI automatically controls the balance between exploitation and exploration, and achieves sufficiently good results in BO approaches [28], it is also used in the approach at hand. With respect to the currently observed data $\mathcal{D}_n$, the acquisition function with EI is defined as follows:

$$a^{\text{EI}}_{P(\varphi|\mathcal{D}_n)}(x) = (\varphi(x^*) - m(x))\Phi(z) + \sigma(x)\phi(z) \tag{35}$$

with $z = \frac{\varphi(x^*) - m(x) - \xi}{\sigma(x)}$, where x^* is the best parameter configuration so far, i.e., $x^* = \arg\min_{x \in \mathcal{D}_n} \varphi(x)$, and $\sigma^2(x)$ describes the predicted variance $k(x, x)$. The function ϕ represents the density function and Φ the cumulative distribution function of the normal distribution $\mathcal{N}(m(x), k(x, x))$. By the acquisition function $a^{\text{EI}}_{P(\varphi|\mathcal{D}_n)}$ in (35), $x \in \mathbb{X}$ are preferred if $m(x) < \varphi(x^*)$, i.e., x could reduce the utility function value, or if there is a high variance $\sigma^2(x)$, i.e., the utility function value for x cannot be estimated with sufficient quality based on the current data $\mathcal{D}_n$. To influence exploitation and exploration in addition to the intrinsic control by the two summands of (35), a parameter $\xi \geq 0$ can be selected. The larger ξ is, the more exploration is preferred. Therefore, it is recommended to start with a sufficiently large ξ and to let it approach zero over the Bayesian Optimization process [47].

To approximate the best parameter configuration determined by the acquisition function $x_{n+1} = \arg\max_{x \in \mathbb{X}} a^{\text{EI}}_{P(\varphi|\mathcal{D}_n)}(x)$, which has to be evaluated next for φ, $a^{\text{EI}}_{P(\varphi|\mathcal{D}_n)}$, is evaluated on a sufficiently large number of random parameter configurations $x \in \mathbb{X}$. The best configuration is chosen as x_{n+1}.

5. Results and Discussion

In this section, the results of the Bayesian Optimization approach regarding the parameter configuration for the Grouping Genetic Algorithm are presented. In doing so, the most promising GGA variant of [1] (described in Section 4.1) is compared to a state-of-the-art algorithm developed to solve a related Pickup and Delivery Problem, the Adaptive Large Neighbourhood Search (ALNS) from [16]. In addition, an approach which randomly selects several parameter configurations and chooses the parameters with best fitness values of the GGA is applied in order to show that the BO performs the optimization of the parameter configuration in a goal-oriented manner.

Please note that the parameter configuration of the ALNS has not been optimized by the Bayesian Optimization approach, since the GGA, with its initial and optimized parameter configuration, is supposed to be compared to a state-of-the-art algorithm regarding the considered Pickup and Delivery Problem. Moreover, the ALNS does not offer the feature to tune the parameter configuration used, since it adjusts the parameters adaptively during the optimization process (cf. [16]).

At first, we describe how the benchmark data sets are created (see Section 5.1). Then, the results for the initial and parameter optimized GGA compared to the ALNS are presented (see Section 5.2). Finally, the different parameter configurations, as well as improvements in each problem class, are highlighted (see Section 5.3).

5.1. Data Generation

The evaluation of the meta-heuristics is performed using self-generated problem instances. These were created with a single-depot data set generator and on the basis of practical circumstances and experience (cf. [1]), so that several practically oriented single-depot problem instances with different depot positions are built and layered on top of one another for the generation of multi-depot instances. In this way, data sets with 1200 instances each, and 1, 4, 6, 8, and 9 depots, are constructed. According to our cooperation partner, these data sets represent the structures in reality. Each of the data sets contains 12 problem classes including 100 instances created with the same parameters in the data set generator. Thus, 60 classes with different characteristics regarding, e.g., number of depots, time window size, vehicle capacity or the constellation of the customer positions, are provided [1]. All data sets are available in [54].

5.2. Parameter Configuration

The initial parameters used for the GGA have been determined in preliminary studies during the development of the solution approach. They are displayed in Table 2.

For optimizing the parameters (see Table 2), the Bayesian Optimization is applied on the probabilities for mutation and repair operators, since tuning parameters such as n^{pop}, $\gamma^{\max}$, ω, and λ particularly affect the number of individuals to be generated. Therefore, previous investigations have shown that optimizing these parameters results in the intuitive solution of setting these values to their maximum. Furthermore, the choice for crossover and mutation rates is reasonable with respect to the problem at hand. Finally, due to the similarity of the crossover operators, optimizing the selection probabilities of these variants will not have a huge impact on the solution quality.

Table 2. Initial parameters for evaluating the GGA variants.

Param.	Value	Description
n^{pop}	50	population size
p^{cross}	1.0	crossover rate
p^{mut}	0.3	mutation rate
$\gamma^{\max}$	250	maximum number of generations
ω	1.5	relative mating pool size with respect to population size (i.e., $n^{\text{pop}} = 50$ results in 75 individuals)
λ	0.05	proportion of the elite within the population
p_i^{cross}	0.5	probability of applying one of the crossover variants, $i = 1, 2$
$p_1^{\text{mut}_v}$	0.4	probability of applying vehicle-based mutation *Costs/Number-of-Request Ratio*
$p_2^{\text{mut}_v}$	0.4	probability of applying vehicle-based mutation *Number of Requests*
$p_3^{\text{mut}_v}$	0.1	probability of applying vehicle-based mutation *Random Vehicle Index*
$p_4^{\text{mut}_v}$	0.1	probability of applying vehicle-based mutation *Random Genotype Position*
$p_1^{\text{mut}_r}$	0.6	probability of applying request-based mutation *Historical Request Pair*
$p_2^{\text{mut}_r}$	0.4	probability of applying request-based mutation *Similarity Measure Selection*
p_1^{repair}	0.55	probability for using the *Greedy Insertion* Repair Operator
p_2^{repair}	0.25	probability for using the *Regret*-2 Repair Operator
p_3^{repair}	0.10	probability for using the *Regret*-3 Repair Operator
p_4^{repair}	0.05	probability for using the *Regret*-4 Repair Operator
p_5^{repair}	0.05	probability for using the *Regret*-δ Repair Operator

In order to determine the best configuration of parameters to be optimized, the BO is performed with 20 iterations. Beforehand, for modeling a sufficiently good a priori probability distribution, the utility function values are determined for 10 randomly drawn parameter combinations. In this way, the Bayesian Optimization is applied to 10 randomly chosen problem instances (i.e., the training data) from each class to identify parameters with which the GGA has the best solution quality regarding the mean of the objective value over these instances. Since the GGA is a stochastic meta-heuristic, each instance is additionally evaluated five times to obtain a stable parameter configuration. In preliminary studies, it was established that a larger set of training data yields worse GGA results, as well as an increased computational time for the BO approach. Hence, the number of instances used for training data is chosen with respect to efficiency. Since the selection of training instance data sets is also a combinatorial optimization problem, it is evident that the manual tests for finding the best training sets represent a limitation in this study. Altogether, the described procedure means, according to Figure 1, that the BO, as the parameter tuning approach optimizing the design layer, executes the GGA (associated with the algorithm layer) 150 times in total for each instance.

Finally, the quality of the optimized parameters is tested on the multi-depot data sets by evaluating the GGA with the best found parameters class-wise over all problem instances (which can be seen as the test data).

To find parameter configurations with the random optimization, 30 iterations are performed in which a parameter configuration is chosen randomly and applied to the problem instances known as training data. Here, the GGA is also evaluated five times on each instance. The parameter configurations that obtained the best fitness are used to evaluate the GGA class-wise for all problem instances of a multi-depot data set as in the BO approach.

5.3. Evaluation and Discussion

For evaluating the considered algorithms, they were compared with respect to their computational time and relative error. The latter was calculated for each problem instance $\mathcal{I}$ and each approach $A \in \mathcal{A} = \{\text{GGA}_{\text{BO}}, \text{GGA}_{\text{Rand}}, \text{ALNS}\}$ using

$$\epsilon_{\mathcal{I}}(A) := \frac{f_{\mathcal{I}}(A) - f_{\mathcal{I}}^*}{f_{\mathcal{I}}^*}, \tag{36}$$

where $f_{\mathcal{I}}^*$ describes the objective function value of the best known solution for the problem instance $\mathcal{I} \in \mathcal{I}_{\mathcal{D}}$ (cf. [1]), i.e., $f_{\mathcal{I}}^* := \min_{A \in \mathcal{A}} f_{\mathcal{I}}(A)$. Furthermore, $\mu_\epsilon(A)$ specifies the mean value and $\sigma_\epsilon(A)$ the standard deviation of the relative error (36) for an approach A over all instances $\mathcal{I}$ of a data set $\mathcal{I}_{\mathcal{D}} = \{1\text{D}, 4\text{D}, 6\text{D}, 8\text{D}, 9\text{D}\}$.

It is worth mentioning that the best known solution $f_{\mathcal{I}}^*$ is not necessarily the optimal solution of the instance $\mathcal{I}$. Due to the complexity of the optimization problem of Section 3.1 (the MDPDPTWHV is $\mathcal{NP}$-hard), it is natural in the field of combinatorial optimization that exact solvers can only guanrantee the exact solution for small instances. Therefore, this is an evident limitation of the results: only heuristic solutions can be compared to each other due to the large amount and sizes of the problem instances.

In Table 3, the mean values and standard deviations of the relative error with respect to (36) are displayed for all five data sets. In doing so, the initial and optimized parameter configurations can be compared. First of all, it is worth mentioning that the solution quality of the GGA, even with the initial parameter setting, is better than that of the ALNS (with initial parameters), since $\mu_\epsilon^{\text{init}}$ is smaller for all data sets. Respecting the parameter configuration, it can be seen that both the mean value μ_ϵ as well as standard deviation σ_ϵ decrease for the GGA from initial to optimized parameters for all data sets. This shows that optimizing the parameter configuration for the GGA further improves the solution quality in comparison to the ALNS. Moreover, the GGA becomes more stable in finding sufficiently good solutions, which is shown by the smaller standard deviation σ_ϵ of the relative error. Since the corresponding mean and standard deviation values for the ALNS also increase, the GGA finds more best known solutions than it did initially. In addition, the mean value of the optimized GGA's computational time $t_{\text{cpu}}^{\text{opt}}$ is up to 38% faster than that of the ALNS. Finally, the parameter configurations tuned by the BO result in better and more stable solutions for the GGA than the ones of the randomized optimization. Especially when the complexity of the problem instances increases, i.e., with larger numbers of depots, the gap between the mean value $\mu_\epsilon^{\text{opt}}$ and the standard deviation $\sigma_\epsilon^{\text{opt}}$ of the relative error becomes larger. Hence, parameter configuration with the proposed Bayesian Optimization approach is purposeful and reasonable.

Table 3. Mean value and standard deviation of the relative error for the initial and optimized GGA parameter configuration in comparison to the ALNS, as well as mean computational time.

	Approach		$\mu_\epsilon^{\text{init}}$	$\sigma_\epsilon^{\text{init}}$	$\mu_\epsilon^{\text{opt}}$	$\sigma_\epsilon^{\text{opt}}$	$t_{\text{cpu}}^{\text{opt}}$
1D	GGA	BO	0.79%	1.48%	0.89%	1.31%	15.49 s
		Random			0.99%	1.39%	14.67 s
	ALNS		1.08%	1.48%	1.99%	1.95%	17.49 s
4D	GGA	BO	0.48%	1.06%	0.61%	1.08%	17.48 s
		Random			0.66%	1.12%	17.45 s
	ALNS		0.86%	1.42%	1.50%	1.91%	19.10 s
6D	GGA	BO	0.63%	1.34%	0.66%	1.11%	34.82 s
		Random			0.80%	1.24%	33.15 s
	ALNS		1.01%	1.51%	2.00%	2.15%	48.07 s
8D	GGA	BO	0.79%	1.49%	0.72%	1.19%	66.77 s
		Random			0.88%	1.23%	55.56 s
	ALNS		0.97%	1.40%	2.18%	2.24%	100.05 s
9D	GGA	BO	0.84%	1.55%	0.78%	1.25%	83.64 s
		Random			0.91%	1.30%	73.14 s
	ALNS		1.00%	1.48%	2.26%	2.23%	135.94 s

Nevetheless, the results in Table 3 also show that there is a slight trade-off in computational time when improving solution quality. The slightly worse random parameter configuration is a bit faster with regards to the mean of the computatonal time $t_{\text{cpu}}^{\text{opt}}$. However, comparing this value with the corresponding result of the PDPTW state-of-the-art solution approach, ALNS, it can be seen that that this effect is less severe. Finally, it is worth pointing out that the various mutation and repair operators are reasonable for solving the MDPDPTWHV, as they tackle the complexity of the considered Pickup and Delivery Problem in the right manner, since the initial solutions $\mu_\epsilon^{\text{init}}$ in Table 3 also show better results for the GGA than for the ALNS.

The fact that the solution quality of the GGA is improved by the parameter configuration carried out by the BO approach can also be seen in each problem class. In Figure 7, for the four-depot data set, the frequency of how often an approach found the best solution within each of the 12 problem classes is depicted. It can be observed that parameter tuning improves the solution quality of the GGA per class, since the frequency of the best solution found increases from the initial (top) to the optimized (bottom) parameter configuration. Similar effects can be shown for the other four data sets.

In Figure 8, the probability distribution for selecting one of the (vehicle-/request-based) mutation or repair operators is displayed for each optimized problem instance class for the four-depot data set. Additionally, the initial parameter configuration is depicted. Here, it becomes clear, at least for the four-depot data set configurations, that the selection probabilities of the operators have to be varied over the problem classes in order to enhance the solution quality within each class. This is due to the different data structures across the different classes. Similar effects regarding the selection probabilities can be seen in the results for the other multi-depot data sets.

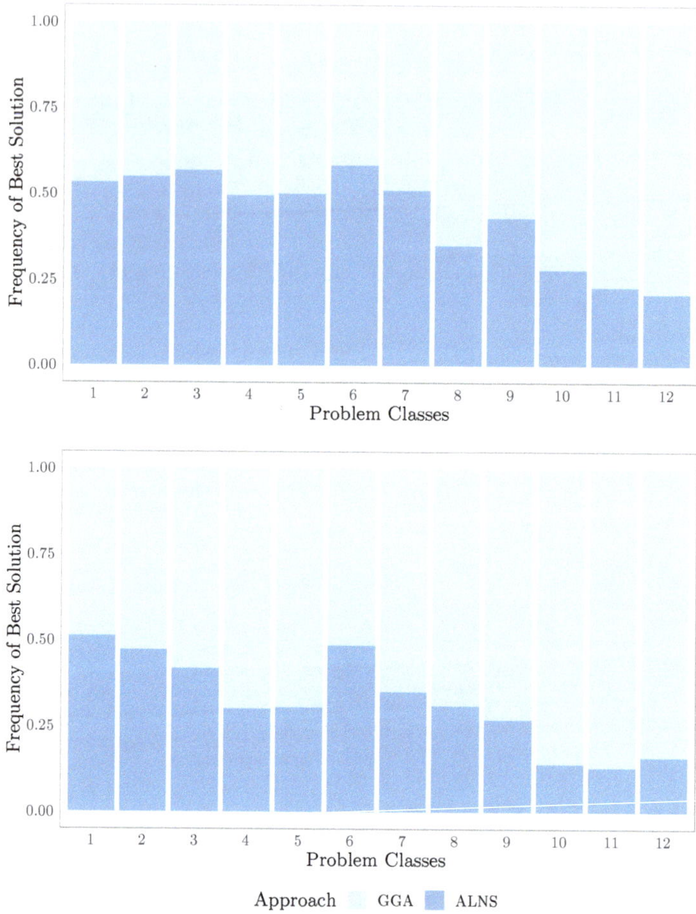

Figure 7. Frequencies of best solution found by the approaches within the 12 problem classes with 4 depots: initial parameter configuration (**top**) and optimized parameters (**bottom**).

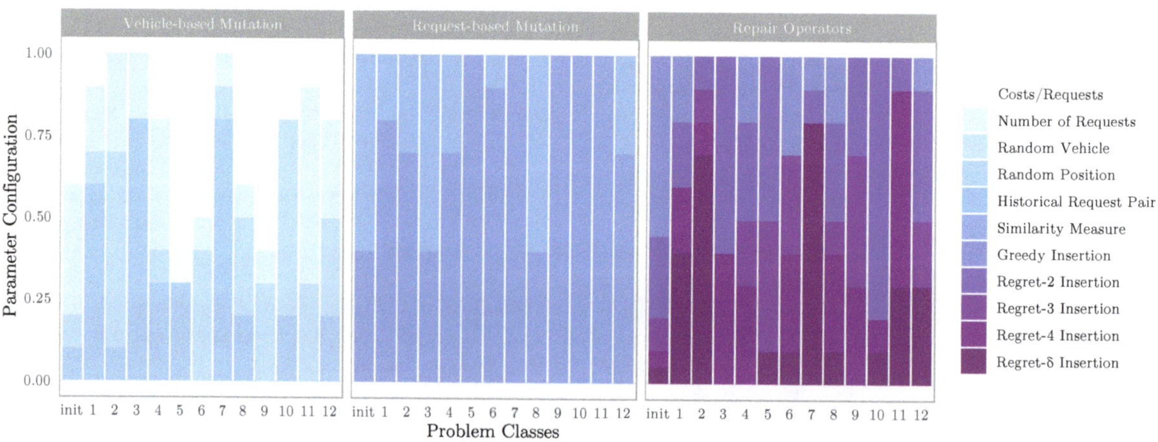

Figure 8. Different parameter configurations for each problem class compared to the initial parameter setting (*class* init) for the GGA with binary tournament evaluated on the 4-depot data set.

6. Conclusions

In this paper, a Bayesian Optimization approach with Gaussian Processes was considered for the parameter tuning of a Grouping Genetic Algorithm solving practically oriented Pickup and Delivery Problems. In order to evaluate improvements in solution quality for the optimized parameter configurations, the BO found the best parameters for each of the five data sets (with 1, 4, 6, 8, and 9 depots) and considered problem classes (with 100 instances each). Thus, 60 parameter configurations were determined, which were then adopted to solve all instances of the same corresponding class.

It could be shown that the BO is able to improve the solution quality of the considered GGA significantly, while the computational time of the GGA is kept low in comparison to a state-of-the-art solution approach (see Table 3). Additionally, the BO results in better parameter configurations for the GGA than randomly choosing parameters, which can be seen as brute-force parameter configuration in comparison. Thus, this is a purposeful approach for improving the solution quality of the GGA. Moreover, the existence of various mutation and repair operators could be proven to be appropriate, since different structures within the instances have to be tackled in different ways (see Figure 8). This resulted in a solution method for carriers in the LTL market that reliably generates a good solution for the MDPDPTWHV in a short time. The result is a considerable competitive advantage over the competition, who only solve the routing problems at hand with the help of less effective methods.

A limitation of this paper is that only one parameter tuning approach was used to improve a GGA solving the MDPDPTWHV. The purpose of our work was to show that optimizing parameters of different problem instance classes is useful due to the complexity of the pickup and delivery problem to be solved. Nevertheless, it is worth comparing Bayesian Optimization with other methods for automatic parameter configuration, such as SMAC or ParamILS (see Section 2.2).

The Genetic Algorithm itself also has limitations. The search process of the GGA strongly depends on the quality of the initial solutions or the request sequence within the associated representation. The generation of the initial population is therefore an important process. In addition, good (feasible) solutions are particularly characterized by the fact that all customer locations are served and the capacity of the vehicles is highly filled. Starting from such a good solution, the application of an operator can quickly lead to an infeasible solution, because meeting the customer time windows then requires the use of another vehicle that is not available. Moreover, even if the probabilities for a certain problem class are determined as best as possible on average, there can always be outlier instances that do not react well to the corresponding choice.

In future research, the proposed BO approach should be compared to other metaheuristic parameter tuning methods. In doing so, characteristics such as solution quality, convergence properties, and computational efficiency, as well as scalability, should be analyzed. Applying dynamic or adaptive parameter tuning techniques in comparison to static methods should also be considered (cf. [42,43]). Furthermore, the knowledge that similar instances (within a class) can be solved efficiently in the same manner is supposed to be included in an a priori parameter selection approach. To do so, key performance indicators which identify similar instances have to be developed in order to define problem classes in a general way. Then, the best parameter configuration can be determined for each class and be applied on new classified instances. The classification task can be carried out by, e.g., neural networks. For each new problem instance that occurs due to daily changing customers and changing demands, LTL carriers can then immediately determine the corresponding class and apply the appropriate algorithm for the solution. A basic idea for such an approach using a BO for the parameter tuning can be found in [55]. Finally, how the BO approach behaves for the automated parameter configuration of solution approaches for other routing problems should be checked in practice. For example, a VRPTWHV could be solved or even an MDVRPTWHV (even with the GGA introduced in this paper).

Author Contributions: Conceptualization, C.R.; methodology, C.R. and J.R.; software, C.R.; validation, C.R.; formal analysis, C.R.; investigation, J.R.; data curation, C.R.; writing—original draft preparation, C.R.; writing—review and editing, C.R. and J.R.; visualization, C.R.; supervision, J.R.; project administration, J.R.; funding acquisition, J.R. All authors have read and agreed to the published version of the manuscript.

Funding: This research was funded by the BMBF project "L2O: Learning to Optimize" grant number 01IS20013A.

Data Availability Statement: The benchmark instances used for the results of this paper are available online at https://www.uni-hildesheim.de/fb4/institute/bwl/betriebswirtschaft-und-operations-research/, accessed on 23 December 2023.

Conflicts of Interest: The authors declare no conflicts of interest. The funders had no role in the design of the study; in the collection, analyses, or interpretation of data; in the writing of the manuscript; or in the decision to publish the results.

References

1. Rüther, C.; Rieck, J.A. Grouping Genetic Algorithm for Multi Depot Pickup and Delivery Problems with Time Windows and Heterogeneous Vehicle Fleets. In *Evolutionary Computation in Combinatorial Optimization (EvoCOP)*; Paquete, L., Zarges, C., Eds.; Springer: Cham, Switzerland, 2020; pp. 148–163.
2. Huang, C.; Li, Y.; Yao, N.X. A Survey of Automatic Parameter Tuning Methods for Metaheuristics. *IEEE Trans. Evol. Comput.* **2020**, *24*, 201–216. [CrossRef]
3. Jones, D.R.; Schonlau, M.; Welch, W.J. Efficient Global Optimization of Expensive Black-Box Functions. *J. Glob. Optim.* **1998**, *13*, 455–492. [CrossRef]
4. Mockus, J. Bayesian Global Optimization. In *Encyclopedia of Optimization*; Floudas, C.A., Pardalos, P.M., Eds.; Springer: Berlin/Heidelberg, Germany, 2001; pp. 123–127.
5. Berbeglia, G.; Cordeau, J.-F.; Gribkovskaia, I.; Laporte, G. Static Pickup and Delivery Problems: A Classification Scheme and Survey. *Top* **2007**, *15*, 1–31. [CrossRef]
6. Parragh, S.N.; Doerner, K.F.; Hartl, R.F. A Survey on Pickup and Delivery Problems—Part II: Transportation Between Pickup and Delivery Locations. *J. Betriebswirtschaft* **2008**, *58*, 81–117. [CrossRef]
7. Smit, S.K.; Eiben, A.E. Comparing Parameter Tuning Methods for Evolutionary Algorithms. In Proceedings of the IEEE Congress on Evolutionary Computation, Trondheim, Norway, 18–21 May 2009; pp. 399–406.
8. Drake, J.H.; Kheiri, A.; Özcan, E.; Burke, E.K. Recent advances in selection hyper-heuristics. *Eur. J. Oper. Res.* **2019**, *285*, 405–428. [CrossRef]
9. Hosny, M.I.; Mumford, C.L. Constructing Initial Solutions for the Multiple Vehicle Pickup and Delivery Problem with Time Windows. *J. King Saud Univ. Comput. Inf. Sci.* **2012**, *24*, 59–69. [CrossRef]
10. Li, H.; Lim, A. A Metaheuristic for the Pickup and Delivery Problem with Time Windows. *Int. J. Art. Intell. Tools* **2001**, *12*, 160–167.
11. Lu, Q.; Dessouky, M.M. A New Insertion-Based Construction Heuristic for Solving the Pickup and Delivery Problem with Time Windows. *Eur. J. Oper. Res.* **2006**, *175*, 672–687. [CrossRef]
12. Irnich, S. A Unified Modeling and Solution Framework for Vehicle Routing and Local Search-Based Metaheuristics. *INFORMS J. Comp.* **2008**, *20*, 270–287. [CrossRef]
13. Nanry, W.P.; Barnes, J.W. Solving the Pickup and Delivery Problem with Time Windows using Reactive Tabu Search. *Transport. Res B-Meth.* **2000**, *34*, 107–121. [CrossRef]
14. Hosny, M.I.; Mumford, C.L. The Single Vehicle Pickup and Delivery Problem with Time Windows: Intelligent Operators for Heuristic and Metaheuristic Algorithms. *J. Heuristics* **2010**, *16*, 417–439. [CrossRef]
15. Bent, R.; Hentenryck, P.V. A Two-Stage Hybrid Algorithm for Pickup and Delivery Vehicle Routing Problems with Time Windows. *Comp. Oper. Res.* **2006**, *33*, 875–893. [CrossRef]
16. Ropke, S.; Pisinger, D. An Adaptive Large Neighborhood Search Heuristic for the Pickup and Delivery Problem with Time Windows. *Trans. Sci.* **2006**, *40*, 455–472. [CrossRef]
17. Pisinger, D.; Ropke, S. A General Heuristic for Vehicle Routing Problems. *Comput. Oper. Res.* **2007**, *34*, 2403–2435. [CrossRef]
18. Masson, R.; Lehuédé, F.; Péton, O. An Adaptive Large Neighborhood Search for the Pickup and Delivery Problem with Transfers. *Trans. Sci.* **2013**, *47*, 344–355. [CrossRef]
19. Qu, Y.; Bard, J.F. The Heterogeneous Pickup and Delivery Problem with Configurable Vehicle Capacity. *Trans. Res. Part C Emerg. Tech.* **2013**, *32*, 1–20. [CrossRef]
20. Jung, S.; Haghani, A. Genetic Algorithm for a Pickup and Delivery Problem with Time Windows. *Transp. Res. Record* **2000**, *1733*, 1–7. [CrossRef]
21. Pankratz, G. A Grouping Genetic Algorithm for the Pickup and Delivery Problem with Time Windows. *OR Spectr.* **2005**, *27*, 21–41. [CrossRef]

22. Alaia, E.B.; Dridi, I.H.; Borne, P.; Bouchriha, H. A Comparative Study of the PSO and GA for the m-MDPDPTW. *Int. J. Comput. Commun. Control* **2018**, *13*, 8–23. [CrossRef]
23. Wang, H.F.; Chen; Y.Y. A Genetic Algorithm for the Simultaneous Delivery and Pickup Problems with Time Window. *Comp. Ind. Eng.* **2012**, *62*, 84–95. [CrossRef]
24. Crèput, J.-C.; Koukam, A.; Kozlak, J.; Lukasik, J. An evolutionary approach to pickup and delivery problem with time windows. In Proceedings of the Computational Science (ICCS 2004), Kraków, Poland, 6–9 June 2004; Bubak, M., van Albada, G.D., Sloot, P. M. A., Dongarra, J., Eds.; Springer: Berlin/Heidelberg, Germany, 2004; pp. 1102–1108.
25. Nannen, V.; Eiben, A.E. Relevance Estimation and Value Calibration of Evolutionary Algorithm Parameters. In Proceedings of the IJCAI International Joint Conference on Artificial Intelligence, Hyderabad, India, 6–12 January 2007; pp. 975–980.
26. Hansen, N. The CMA evolution strategy: A comparing review. In *Towards a New Evolutionary Computation: Advances in the Estimation of Distribution Algorithms*; Lozano, J.A., Larranaga, P., Inza, I., Bengoetxea, E., Eds.; Springer: Berlin/Heidelberg, Germany, 2006; pp. 75–102
27. Balaprakash, P.; Birattari, M.; Stützle, T. Improvement strategies for the F-race algorithm: Sampling design and iterative refinement. In *Hybrid Metaheuristics*; Bartz-Beielstein, T., Blesa Aguilera, M. J., Blum, C., Naujoks, B., Roli, A., Rudolph, G., Sampels, M., Eds.; Springer: Berlin/Heidelberg, Germany, 2007; pp. 108–122.
28. Roman, I.; Ceberio, J.; Mendiburu, A.; Lozano, J. A. Bayesian Optimization for Parameter Tuning in Evolutionary Algorithms. In Proceedings of the IEEE Congress on Evolutionary Computation (CEC), Vancouver, BC, Canada, 24–29 July 2016; pp. 4839–4845.
29. Hutter, F.; Hoos, H.H.; Leyton-Brown, K.; Stützle, T. ParamILS: An Automatic Algorithm Configuration Framework. *J. Artif. Int. Res.* **2009**, *36*, 267–306. [CrossRef]
30. Hutter, F.; Hoos, H.H.; Leyton-Brown, K. Sequential model-based optimization for general algorithm configuration. In *Learning and Intelligent Optimization*; Coello, C.A., Eds.; Springer: Berlin/Heidelberg, Germany, 2011; pp. 507–523.
31. López-Ibáñez, M.; Dubois-Lacoste, J.; Pérez Cáceres, L.; Birattari, M.; Stützle, T. The Irace Package: Iterated Racing for Automatic Algorithm Configuration. *Oper. Res. Perspect.* **2016**, *3*, 43–58. [CrossRef]
32. Ramasamy, S.; Mondal, M.S.; Reddinger, J.-P.F.; Dotterweich, J.M.; Humann, J.D.; Childers, M.A.; Bhounsule, P.A. Heterogenous Vehicle Routing: Comparing Parameter Tuning Using Genetic Algorithm and Bayesian Optimization. In Proceedings of the International Conference on Unmanned Aircraft Systems (ICUAS), Dubrovnik, Croatia, 21–24 June 2022; pp. 104–113.
33. Rasku, J.; Musliu, N; Kärkkäinen, T. On Automatic Algorithm Configuration of Vehicle Routing Problem Solvers. *J. Veh. Rout. Alg.* **2019**, *2*, 1–22. [CrossRef]
34. Wolpert, D.H.; Macready, W.G. No Free Lunch Theorems for Optimization. *IEEE Trans. Evol. Comput.* **1997**, *1*, 67–82. [CrossRef]
35. Huang, C.; Yuan, B.; Li, Y.; Yao, X. Automatic Parameter Tuning using Bayesian Optimization Method. In Proceedings of the IEEE Congress on Evolutionary Computation (CEC), Wellington, New Zealand, 10–13 June 2019; pp. 2090–2097.
36. Rieck, J.; Zimmermann, J. A New Mixed Integer Linear Model for a Rich Vehicle Routing Problem with Docking Constraints. *Ann. Oper. Res.* **2010**, *181*, 337–358. [CrossRef]
37. Furtado, M.G.S.; Munari, P.; Morabito, R. Pickup and Delivery Problem with Time Windows: A New Compact Two-index Formulation. *Oper. Res. Lett.* **2017**, *45*, 334–341. [CrossRef]
38. Feillet, D. A Tutorial on Column Generation and Branch-and-Price for Vehicle Routing Problems. *4OR* **2010**, *8*, 407–424. [CrossRef]
39. Calvet, L.; Juan, A.A.; Serrat, C.; Ries, J. A Statistical Learning Based Approach for Parameter Fine-tuning of Metaheuristics. *Stat. Oper. Res. Trans.* **2016**, *40*, 201–224.
40. Eiben, A.E.; Hinterding, R.; Michalewicz, Z. Parameter Control in Evolutionary Algorithms. *IEEE Trans. Evol. Comput.* **1999**, *3*, 124–141. [CrossRef]
41. Karimi-Mamaghan, M.; Mohammadi, M.; Meyer, P.; Karimi-Mamaghanb, A.M.; Talbi, E.-G. Machine learning at the service of meta-heuristics for solving combinatorial optimization problems: A state-of-the-art. *Eur. J. Oper. Res.* **2022**, *296*, 393–422. [CrossRef]
42. Hussein, A.A.; Yassen, E.T.; Rashid, A. Learnheuristics in routing and scheduling problems: A review. *Samarra J. Pure Appl. Sci.* **2023**, *5*, 60–90. [CrossRef]
43. Aleti, A.; Moser, I. A Systematic Literature Review of Adaptive Parameter Control Methods for Evolutionary Algorithms. *ACM Comput. Surv.* **2016**, *49*, 1–35. [CrossRef]
44. Holland, J.H. *Adaptation in Natural and Artificial Systems*; University of Michigan Press: Ann Arbor, MI, USA, 1975.
45. Eiben, A.E.; Smith, J.E. *Introduction to Evolutionary Computing*, 2nd ed.; Natural Computing Series; Springer: Berlin/Heidelberg, Germany, 2015.
46. Michalewicz, Z.; Fogel, D.B. *How to Solve It: Modern Heuristics*, 2nd ed.; Springer: Berlin/Heidelberg, Germany, 2004.
47. Brochu, E.; Cora, V. M.; de Freitas, N. A Tutorial on Bayesian Optimization of Expensive Cost Functions, with Application to Active User Modeling and Hierarchical Reinforcement Learning. *arXiv* **2010**, arXiv:1012.2599.
48. Frazier, P.I. A Tutorial on Bayesian Optimization. *arXiv* **2018**, arXiv:1807.02811.
49. Shahriari, B.; Swersky, K.; Wang, Z.; Adams, R.P.; de Freitas, N. Taking the Human Out of the Loop: A Review of Bayesian Optimization. *Proc. IEEE* **2016**, *104*, 148–175. [CrossRef]
50. Klein, A.; Falkner, S.; Bartels, S.; Hennig, P.; Hutter, F. Fast Bayesian Hyperparameter Optimization on Large Datasets. *Electron. J. Stat.* **2017**, *11*, 4945–4968. [CrossRef]

51. Rasmussen, C.E.; Williams, C.K.I. *Gaussian Processes for Machine Learning*; Series Adaptive Computation and Machine Learning; MIT Press: Cambridge, MA, USA, 2006.
52. Duvenaud, D.K. Automatic Model Construction with Gaussian Processes. Ph.D. Thesis, University of Cambridge, Cambridge, UK, 2014.
53. Snoek, J.; Larochelle, H.; Adams, R.P. Practical Bayesian Optimization of Machine Learning Algorithms. *arXiv* **2012**, arXiv:1206.2944.
54. Betriebswirtschaft und Operations Research Homepage. Available online: https://www.uni-hildesheim.de/fb4/institute/bwl/betriebswirtschaft-und-operations-research/ (accessed on 23 December 2023).
55. Rüther, C.; Chamurally, S.; Rieck, J. An a-priori parameter selection approach to enhance the performance of genetic algorithms solving pickup and delivery problems. In *International Conference on Operations Research*; Trautmann, N., Gnägi, M., Eds.; Lecture Notes in Operations Research; Springer: Cham, Switzerland, 2022; pp. 66–72.

energies

Article

Cooperative Application of Onboard Energy Storage and Stationary Energy Storage in Rail Transit Based on Genetic Algorithm

Deshi Kong [1,*] and Masafumi Miyatake [2]

1 Green Science and Engineering Division, Sophia University, Tokyo 102-8554, Japan
2 Department of Engineering and Applied Sciences, Sophia University, Tokyo 102-8554, Japan; miyatake@sophia.ac.jp
* Correspondence: d-kong-7m1@eagle.sophia.ac.jp

Abstract: The transition towards environmentally friendly transportation solutions has prompted a focused exploration of energy-saving technologies within railway transit systems. Energy Storage Systems (ESS) in railway transit for Regenerative Braking Energy (RBE) recovery has gained prominence in pursuing sustainable transportation solutions. To achieve the dual-objective optimization of energy saving and investment, this paper proposes the collaborative operation of Onboard Energy-Storage Systems (OESS) and Stationary Energy-Storage Systems (SESS). In the meantime, Non-dominated Sorting Genetic Algorithm-II (NSGA-II) is applied to optimize the ESS capacity and reduce its redundancy. The simulation is programmed in MATLAB. The results show that the corporation of OESS and SESS offers superior benefits (70 kWh energy saving within 30 min operation) compared to using SESS alone. Moreover, the OESS plays a significant role, emphasizing its significance in saving energy and investment, therefore presenting a win–win scenario. It is recommended that the capacity of OESS be designed to be two to three times that of SESS. The findings contribute to the ongoing efforts in developing more sustainable and energy-efficient transportation solutions, with implications for the railway industry's investment and broader initiatives in energy saving for sustainable urban mobility.

Keywords: rail transit; ESS; SESS; OESS; SMES; Lithium-ion battery; regenerative braking; NSGA-II; energy recovery; energy management

1. Introduction

This research aims to achieve multi-objective optimization of energy consumption and cost in rail transit by coordinating OESS's and SESS's capacity. Therefore, NSGA-II is applied and works well, as expected.

In general, the pantograph-catenary is the primary energy supply for a train's operation in rail transit [? ?]. To improve the diversity and stability of energy supply in emergencies, renewable energy sources like photovoltaic power have also been introduced in rail transit [?]. On the other hand, as a supplement to the primary energy supply system, one key area of focus is the recovery and efficient utilization of RBE in railway transit by ESS. Regarding the construction of ESS, there are three types: power-density ESS, energy-density ESS, and hybrid energy-storage systems (HESS). The rated output power divided by the storage device's volume yields the power density (W/kg or W/liter). Power-density ESS like SMES and supercapacitors appropriate for high discharge current have quick response power quality applications. Similarly, stored energy divided by the volume refers to the energy density (Wh/kg or Wh/liter). Energy-density ESS like Li-ion holds higher energy density than conventional batteries, excelling in space reduction, mobility increase, and operating time extension [?]. Ratniyomchai introduced the application of ESS in electrified railways, especially batteries, flywheels, electric double-layer capacitors, and

Citation: Kong, D.; Miyatake, M. Cooperative Application of Onboard Energy Storage and Stationary Energy Storage in Rail Transit Based on Genetic Algorithm. *Energies* **2024**, *17*, 1426. https://doi.org/10.3390/en17061426

Academic Editors: Artur Kierzkowski, Tomasz Kisiel, Franciszek Restel, Tomasz Nowakowski, Agnieszka A. Tubis, Anna Jodejko-Pietruczuk and Mateusz Zając

Received: 17 February 2024
Revised: 4 March 2024
Accepted: 11 March 2024
Published: 15 March 2024

HESS. The storage and reuse of RBE is managed by energy-storage devices depending on the purpose of each system [? ?]. By lowering the frequency of battery charge and discharge and controlling battery peak current, Li introduced HESS with Superconducting Magnetic Energy Storage (SMES) and batteries in electric buses, extending battery life [?].

The efficient management and storage of energy have become integral to sustainability goals. In general, according to the allocation of ESS, there are also three types: SESS, OESS, and both. Unlike OESS, SESS is not confined to the train itself but is distributed strategically along the railway network. SESS plays a crucial role in capturing and storing the excess energy generated during regenerative braking events to improve efficiency and reliability in urban rail systems [?]. Lamedica demonstrates the utilization of Particle Swarm Optimization (PSO) in developing an optimal siting and sizing methodology for designing a SESS tailored for railway lines and maximizing the investment's financial return [?]. On the other hand, Andres Ovalle proposed the optimal energy storage sizing formulation, taking the characteristics of different modern battery and supercapacitor technologies into account, and the objective function to minimize the trade-off between energy-storage capacity and charging rates based on a real-time simulation [?]. At the same time, Dupré developed a methodology that divides installation into stages with different budgets and periods to obtain optimum installation of ESSs in a railway line, balancing energy saving and profit [?]. Also, simulations in different conditions managed by different algorithms have been conducted by many authors, all of which prove that installing SESS will lead to energy saving for the railway system [? ? ? ? ?]. Also, real examples have been commercially applied around the world [? ? ? ?]. This paper proposes the utilization of a hybrid energy-storage system (HESS) combining SMES and conventional batteries in rail transit.

SMES is a high-power-density energy-storage technology that relies on the principle of superconducting magnets. SMES takes advantage of the unique properties of superconducting materials, which exhibit zero resistance at low temperatures. A strong magnetic field is generated by passing an electric current through a superconducting coil, allowing for the storage of electrical energy. In addition, it exhibits a fast response in milliseconds. It is mainly applied for network stability applications [? ?]. While in rail transit, thanks to its unique characteristics, SMES is well-suited for recovering the intermittent and random nature of RBE. However, the negative impact of strong magnetic fields on the environment and their high cost are the main obstacles to deploying SMES devices.

Lithium-ion battery (LIB) presents a rechargeable characteristic. Electrical energy is stored and released through the embedding and de-embedding of lithium ions in its chemical reactions. This innovative technology has gained widespread adoption, positioning itself as one of the most prevalent energy-storage solutions available in commercial markets today. On the other hand, the high-energy-density characteristic distinguishes Lithium-ion batteries (LIBs) from SMES, which allows them to store more electrical energy in a light and small form. Consequently, LIB has become the preferred energy-storage technology for electric vehicles and mobile devices. Furthermore, successful commercialization and significant global investments have led to a considerable reduction in its cost. Although LIBs offer high energy density, they can still be relatively heavy compared to alternative ESS technologies like ultracapacitors [?]. In railway transit applications, weight directly impacts energy consumption and vehicle performance.

By utilizing a combination of several energy-storage technologies, the HESS system effectively delivers and controls energy. The two complementary technologies that constitute HESS's core are LIB and SMES. While LIB is distinguished by its high energy density and wide applicability, SMES is recognized for its quick reaction in milliseconds, exceptional power density of up to 2000 W/kg, and extended life expectancy. The HESS system may efficiently utilize the advantages of both LIB and SMES to accomplish efficient energy delivery and storage. Meanwhile, SMES can be regarded as a buffer for temporary energy storage to reduce the load of the battery, thus reducing the energy exchange frequency of the battery and extending its lifespan. One challenge is the specification of power-density

ESS and energy-density ESS. Integrating both effectively requires careful design. In the meantime, higher capacity does not always lead to more significant energy savings [?]. Optimal sizing is crucial.

A consistent power-supply infrastructure is frequently absent in remote or isolated regions. Under such circumstances, trains may face challenges relying directly on an external power source. OESS serves as a solution, enabling trains to provide energy independently in areas lacking power supply. OESS enables trains to capture RBE immediately and store it in real time. It also allows the train to utilize this energy as soon as needed. Batteries and supercapacitors are commonly applied in OESS [?]. However, it is regarded that high input and output power models are only sometimes feasible for battery energy systems to operate at [?]. In the meantime, OESS introduces an extra burden on the train along with more energy consumption [? ?]. In addition, the converter is a massive burden on driving range and design as well [?]. Ahmad proposed a chopper topology that reduces mass and volume with high chopper efficiency [?]. Miyatake investigated electric double-layer capacitors as an OESS due to their high energy density. A mathematical model is formulated using a widely applicable optimization technique known as sequential quadratic programming, which can determine optimal acceleration/deceleration and current commands at each sampling point, maintaining fixed transfer time and distance [?]. Wu introduced OESS with on-route constraints to model the real-world scenario. Based on the proposed MILP model, degradation of the OESS influences discharge/charge strategy, and the energy consumption is reduced by 11.6% with the introduction of OESS recovering RBE [?]. González-Gil considered Lithium-ion battery (LIB) and nickel-metal hydride (NiMH) batteries as viable options for OESS [?]. In addition, Pulazza proposed that the energy transmission congestion resulting from renewable energy can be managed by installing OESS, which proves the advantage of the installation of OESS [?]. Similarly, different types of OESS are also applied in commercial operation [? ? ? ? ? ? ?].

On the other hand, the application of OESS improves the efficiency of train power delivery because the energy does not need to be delivered through the catenary to a SESS but is embedded directly in the train itself. It is important to note that this does not mean that SESS is useless. Combination applications of SESS and OESS are usually installed in smart grids, microgrids, wind farms, etc. [? ? ?]. The cooperation between SESS and OESS is just part of this paper's proposal. They can work together to optimize the recovery and utilization of RBE throughout the rail transit system. OESS can capture energy quickly on the train, while SESS can distribute stored energy more evenly throughout the rail network to be shared and utilized when needed. Considering that the weight of OESS influences the energy consumption of the train, LIB is adopted as OESS. From a cost–benefit perspective, due to the introduction of OESS, the quantity and capacity of all the SESSs will be decreased, compared with the case only equipped with SESS [?], leading to substantial gains in energy savings and electricity cost reduction [?].

Besides the concern of SESS and OESS, capacity optimization of ESS is of great significance. Pang applied NSGA-II to address the capacity configuration of Energy-Storage Systems (ESS) in rail transit, considering two objectives: minimizing braking resistor energy consumption and configuration cost [?]. Similarly, Mundra also takes advantage of NSGA-II, achieving dual-objective optimization for the peak-to-average ratio of the total energy demand and electricity usage charge in smart grid [?]. On the other hand, Qayyum adopted PSO to minimize the nano grid energy trading cost while meeting energy demand [?]. Meanwhile, Li utilized Improved Particle Swarm Optimization (IPSO) to balance system economy and stability in the distribution grid [?]. On the other hand, to enhance the coordination between Transit-Oriented Development (TOD) and station-area land use in developing a potential city transit, Pishro employed Multiple Linear Regression (MLR) to establish Node–Place–Ridership–Time (NPRT) equations. This approach surpasses the accuracy of the earlier Node–Place (NP) and Node–Place–Ridership (NPR) models, delivering more precise outcomes [?]. Similarly, Pishro develops eight Multiple Linear Regression (MLR) equations for each hub by combining mathematical techniques and machine learning. It yields valuable insights that guide decision-making and facilitate the development of transportation systems [?]. Boukerche proposed machine-learning (ML)-based methods for building traffic-prediction models that are less restrictive to the prediction task as they require less prior knowledge of the relationships between different traffic modes and can better fit the nonlinear features in the traffic data [?]. Hitachi created and implemented an AI-driven hybrid railway traffic-management system to aid in the automation of the intricate process of planning train schedules [?]. Essien suggested a novel urban traffic-prediction model using deep learning. The model integrates insights from tweets along with traffic and weather data to enhance accuracy and reliability in predicting urban traffic patterns [?].

All in all, this paper adopts HESS configuration as SESS, combining high-power-density ESS, SMES, and high-energy-density ESS, LIB. To explore the combination of SESS and OESS in rail transit energy management, OESS utilizes LIB. To optimize the ESS capacity, minimize redundancy, and balance trade-offs between multi-objectives, cost, and energy consumption, NSGA-II is applied [?]. This is because of its ability to achieve a notably enhanced distribution of solutions and improved convergence closer to the Pareto-optimal front across various problem scenarios. The Parallel Computing Toolbox has been introduced to save computation time.

2. Methodology

2.1. Topology

A practical simulation of the SESS in rail transit involves designing a specific network configuration, as depicted in Figure **??**. In this setup, three trains with OESS travel from Station A to Station E, while another three trains with OESS travel in the opposite direction, from Station E to Station A. The entire operation adheres to a predefined timetable, outlined in Figure **??**. There are two substations strategically placed along the railway line. Each substation is equipped with a SESS comprising a SMES and a set of LIBs in parallel.

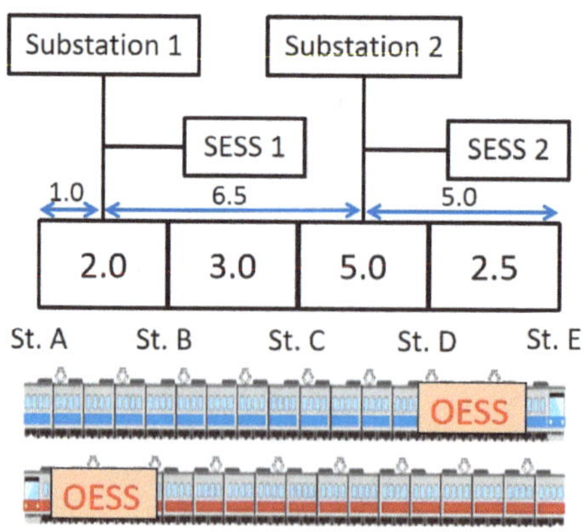

Figure 1. Topology of the Railway.

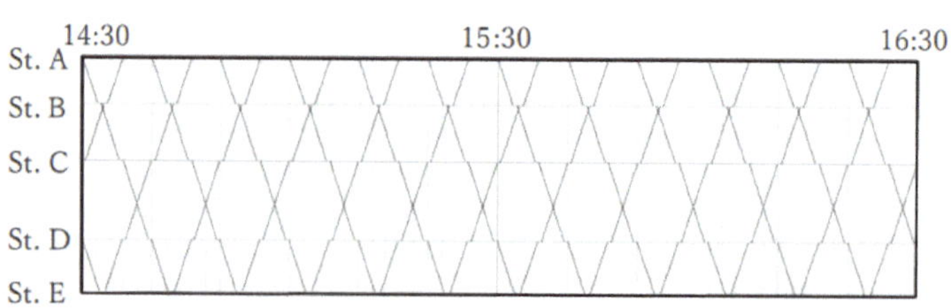

Figure 2. Timetable.

The simulation encompasses modeling the feeding system, trains, and substations. It is worth mentioning that the feeding system presents the connected nodes adopting the term "tie nodes". This approach offers a significant benefit as it eliminates the necessity for a comprehensive modification of the circuit topology whenever there is a change in the feeding system, as depicted in Figure **??**. The tie nodes can make the different node voltages identical to connect the feeders for different directions under the substations. This technique enables simple modeling of feeding circuits and avoids using the conductance matrix. On the other hand, the train model is constructed by voltage, current, and notch. The notch simulates a driver's or Automatic Train Operation's (ATO) command to accelerate or decelerate. The whole operation is based on a timetable. In the meantime, the substation model is represented by current and voltage. It is detailed in Reference [**?**].

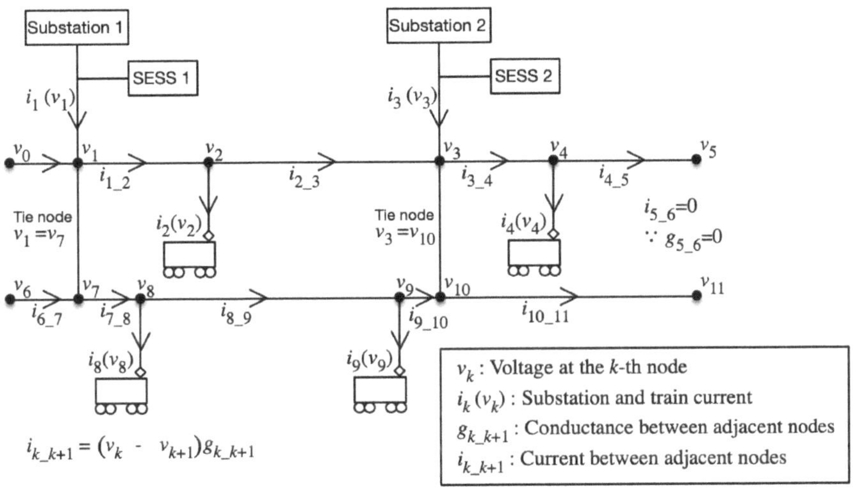

Figure 3. Circuit Topology.

2.2. NSGA-II

Non-dominated Sorting Genetic Algorithm-II (NSGA-II) is a multi-objective optimization algorithm for solving problems with multiple objective functions. It is an extension of the genetic algorithm that aims to find a non-dominated (non-inferior) solution in the objective space and form a Pareto front, i.e., no other solution can outperform this set of solutions on all objectives. Meanwhile, the fast non-dominated sorting approach is introduced with significant improvement over the complexity of other algorithms in the same category. On the other hand, by incorporating elitism, NSGA-II ensures that the best solutions are carried over to the next generation, improving the convergence towards the true Pareto-optimal front [**?**].

The operation of NSGA-II follows the flowchart Figure **??**. First, a set of potential solutions to the optimization problem is generated. These solutions are usually represented as chromosomes in a genetic algorithm context. Then, for each solution in the population, the objective functions are calculated, which evaluate how 'fit' or 'good' the solution is. Next, the best solutions from the current population to parents are selected to create a new generation of solutions. Crossover combines two parents to produce children for the next generation. Mutation introduces random changes to some of the children to maintain genetic diversity. Finally, the algorithm selects the best solutions based on their fitness, and these become the new population for the next iteration until there is no significant improvement between generations or when a satisfactory solution is found. However, NSGA-II also meets some challenges. In this paper, SESS and OESS capacity allocation are the solutions. To further optimize the energy management of the transportation system and discover the balance between investment and energy consumption, NSGA-II is applied.

NSGA-II involves sorting individuals based on non-domination, with a time complexity of O(MN^2), where N is the population size, and M is the number of objectives. This can become computationally expensive for large populations or high-dimensional problems. In the meantime, storing and maintaining the non-dominated fronts over generations can increase memory usage. Therefore, a server equipped with high computation ability and large RAM is of great necessity. On the other hand, the algorithm aims to discover and maintain a diverse set of solutions that approximate the entire Pareto front. However, due to the complexity of the search space and the inherent characteristics of the optimization problem, the algorithm might converge to or focus on specific regions of the Pareto front, forming what is known as local Pareto fronts [**?**]. One solution combines multiple algorithm runs with different random seeds or initial conditions. The ensemble of runs can provide more comprehensive coverage of the Pareto front, reducing the risk of converging to local Pareto fronts.

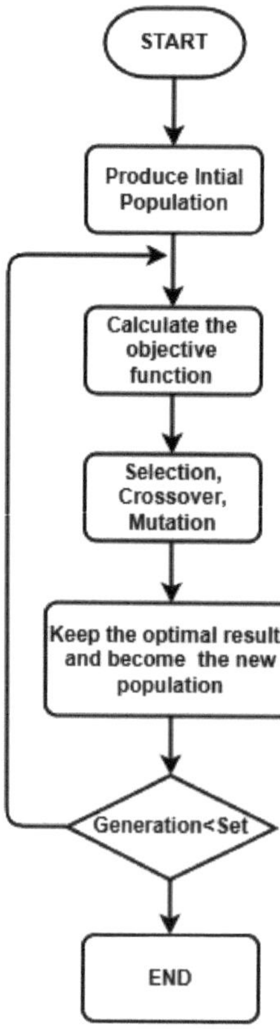

Figure 4. Flowchart of NSGA-II.

The objective function F consists of two sub-objectives: cost minimum (represented by $f1$) and energy consumption minimum (represented by $f2$), shown in Equation (1). In Equation (2), the composition of cost function $f1$ is associated with the number of battery modules in each SESS (K_i) and each OESS (K_p), as well as the fixed costs for different components like SMES ($cost_{sc}$), DCDC converters ($cost_{dcdc}$), battery ($cost_{bt}$), and the cost of electricity ($cost_e$). On the other hand, the sub-objective ($f2$) is a measure of the total energy consumption aggregating the energy supplied across substation (E_{sub}), SESS (E_{sess}), and OESS (E_{oess}) associated with the train system. The energy supplied by the substation is mainly for the train's operation. The energy supplied from SESS and OESS is calculated by the difference in energy capacity between the beginning and end of the operation, shown in Equations (3) and (4). In addition, the cost of electricity is based on the energy consumption in all substations, as shown in Equation (5).

$$Objective \quad : \quad F(K_i, K_p, E_{sub}, E_{sess}, E_{oess}) \quad = \quad [f_1(K_i, K_p), f_2(E_{sub}, E_{sess}, E_{oess})] \tag{1}$$

$$Sub - Objective : \begin{cases} f_1(K_i, K_p) = \sum_{i=1}^{N}(cost_{sc} + 2 \times cost_{dcdc} + cost_{bt} \times K_i + cost_e) \\ \qquad + \sum_{p=1}^{M}(cost_{bt} \times K_p) \\ \\ f_2(E_{sub}, E_{sess}, E_{oess}) = E_{sub} + E_{sess} + E_{oess} \end{cases} \tag{2}$$

$$E_{sess} = \sum_{i=1}^{N}(Esinitial_i - Esend_i) \tag{3}$$

$$E_{oess} = \sum_{p=1}^{M}(Eoinitial_p - Eoend_p) \tag{4}$$

$$cost_{alle} = E_{sub} \times cost_e \tag{5}$$

K_i: Amount of battery modules in SESS i
K_p: Amount of battery modules in OESS p
i: Order of substations
p: Order of trains
E_{sub}: Energy consumption in all substations
E_{sess}: Energy consumption in all SESS
$Esinitial_i$: Energy stored in SESS i in the beginning
$Esend_i$: Energy stored in SESS i in the end
E_{oess}: Energy consumption in all OESS
$Eoinitial_p$: Energy stored in OESS p in the beginning
$Eoend_p$: Energy stored in OESS p in the end
N: Amount of substations
$cost_{sc}$: Cost of SMES
$cost_{dcdc}$: Cost of DCDC converter
$cost_{bt}$: Cost of a battery module
$cost_{alle}$: Total electricity fee
$cost_e$: Cost of electricity fee per kWh
M: Quantity of trains

Table **??** outlines the predetermined parameters on which the algorithm runs.

Table 1. NSGA-II Parameter Settings.

Setting	Value
Amount of Population	30
Amount of Variable	5
Variable Range	[1100]
Iteration Count	40
Selection Strategy	Binary Tournament
Crossover Strategy	Simulated Binary
Crossover Factor	20
Crossover Probability	100%
Mutation Strategy	Polynomial
Mutation Factor	20
Mutation Probability	20%

Table **??** [**?**], Tables **??** and **??** provide specifics of the predetermined parameters for train, costs, SESS, and OESS. Specifically, the operating period is set at 30 min, from 2 p.m. to 2:30 p.m., under the timetable outlined in Figure **??**. This timetable dictates when trains travel between stations, ensuring that the simulation accurately reflects the timing and scheduling constraints of an actual railway system. Within 30 min of simulation, three trains with OESS travel from Station A to Station E, while another three trains with OESS operate in the reverse way. The schedule influences the operation of these substations, as they must manage the energy flow to and from the trains according to their arrival and departure times. These parameters play a crucial role in the simulation, influencing the capacity allocation, cost assessment, and energy consumption within the specified time frame.

Table 2. Specifications for Train.

Setting	Value
Mass	310.4 (tons)
Occupant Capacity	60%
Braking Mechanism	Regenerative braking
Terminal velocity in Constant Torque	50 (km/h)
Terminal velocity in Constant Output	80 (km/h)
Max Acceleration Rate	2.5 (km/h/s)
Max Deceleration Rate	3 (km/h/s)
Nominal Voltage	1500 (V)
Max Regenerative Current	3000 (A)
Starting/Cutoff Voltages for Regenerative Current Restriction	1750 (V), 1800 (V)

Table 3. Initial Investment.

Expenditure	Value
SMES Module	50,000,000 (JPY) [**?**]
Battery Module	645,337 (JPY)
DCDC Converter Module	4,302,250 (JPY)
Electricity Cost	15.65 (JPY/kWh) [**?**]

Table 4. Specifications for SESS and OESS.

Setting	Value
Inductance of SMES (L)	0.1 (H)
Maximum Superconductor Current Limit (Isc_max)	3000 (A)
Initial Superconductor Current (Isc)	2625 (A)
Minimum Superconductor Current Limit (Isc_min)	2250 (A)
DCDC Converter Efficiency	95%
Battery Output Power (P)	500 (W)
Battery Capacity	10 (Ah)
Initial Battery State of Charge (SOC)	50%
SOC Limit	30–80%

It is worth mentioning that the deceleration corresponds to a stop in the power supply to the motors. However, as the train continues to move with inertia, the axle reacts on the motor rotor to produce an induced current due to electromagnetic induction, i.e., the motor becomes a generator. At the same time, the torque generated by the rotation of the drive motor acts on the wheels to slow down the vehicle, which is called regenerative braking [?]. Utilizing the electrical energy generated by electromagnetic induction is called regenerative braking energy recovery. In the meantime, due to the regenerative energy output, the pantograph voltage increases. To avoid unpredictable damage to the catenary-connected appliances, which cannot withstand voltages exceeding the rated value, the regenerative current restriction is conducted in accordance with the catenary voltage, shown in Figure **??** and Table **??**. In addition, the RBE is first consumed by auxiliary applications. If the auxiliary application cannot consume all RBE, then the OESS stores most of the RBE to power the train in the next acceleration. The excess RBE that OESS cannot store transmits and stores in nearby SESS. Calculation of the distance between the train and SESS induces the nearest SESS. By sizing capacity and allocating priority of OESS and SESS based on the train's energy demand and operational state, utilization of RBE is fully optimized, energy consumption decreases, and the system's economic benefit is enhanced.

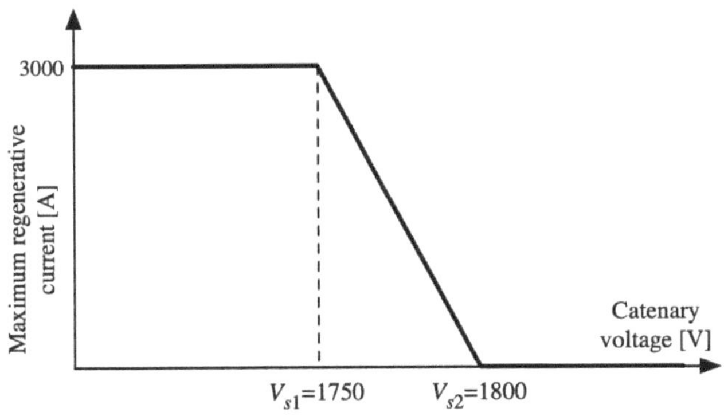

Figure 5. Regenerative Current Output Control.

3. Results and Discussion

In this study, MATLAB is employed as the simulation tool for modeling rail transit scenarios. Meanwhile, the Parallel Computing Toolbox and a server equipped with an AMD Ryzen Threadripper 3970X 32-Core CPU and 64 GB of RAM are applied to save simulation time. The cases under examination are categorized into five parts.

3.1. Case A

Neither SESS nor OESS is introduced into the rail transit system in Case A. The topology is depicted in Figure **??**, and the predefined parameters for substations, trains, and capital costs are detailed in Tables ??–??. The operation period is 30 min from 14:00 to 14:30, according to the timetable shown in Figure ??. The simulation results are shown in Figure ?? and Table ??. The cost only consists of the electricity fee.

Table 5. Case A: Simulation Result of Railway System without ESS.

Energy Consumption	Cost	Capacity of All SESSs	Capacity of All OESSs
1196.28 (kWh)	18721.84 (JPY)	0 (kWh)	0 (kWh)

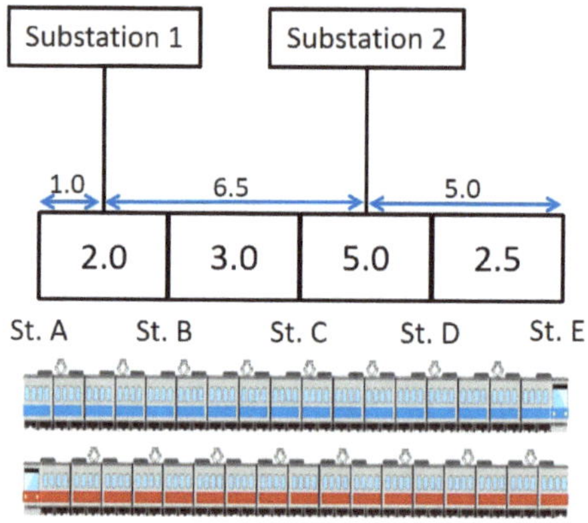

Figure 6. Topology of the Railway without ESS.

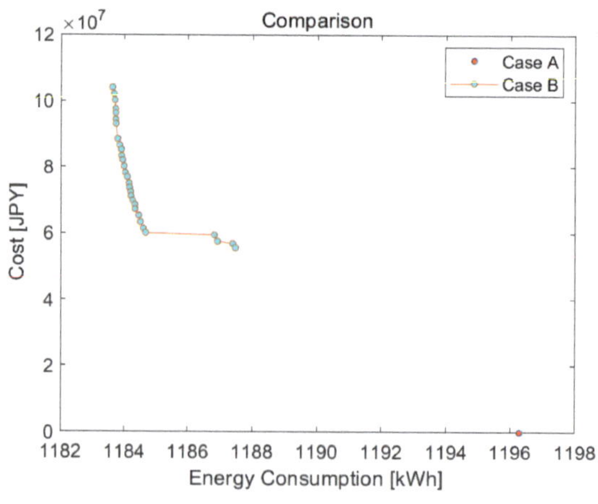

Figure 7. Cases Comparison: A and B.

3.2. Case B

In Case B, only SESS is integrated, and the capacity of all LIBs is tailored through customization using NSGA-II. The energy consumption varies from 1183.63 to 1187.47 kWh. The cost varies from 1.04×10^8 to 5.56×10^7 JPY. They are shown in Figure ??. On the other hand, the capacity distribution is depicted in Figure ??. Clearly, with the integration of SESS, the more the investment increases, the more energy is saved. However, energy savings are not apparent with a double investment. It was analyzed that there was a transmission loss from the regenerative braking device to SESS. At the same time, the SMES's capacity is fixed and limited due to its complicated construction to optimize its capacity.

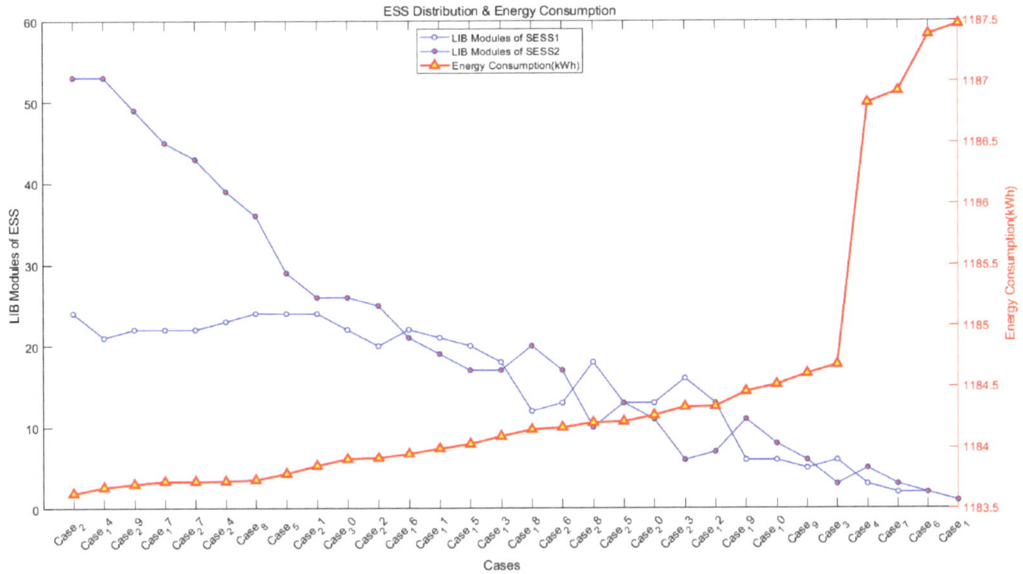

Figure 8. Case B: Capacity Distribution and Energy Consumption.

3.3. Case C

In Case C, both SESS and OESS are implemented in rail transit. Simultaneously, the capacities of LIBs for both SESS and OESS are optimized for uniformity through NSGA-II. Energy consumption varies from 1054.72 to 1111.30 kWh. And the cost varies from 1.84×10^9 to 5.94×10^7 JPY. They are depicted in Figure ??. Meanwhile, the relationship between capacity distribution and energy consumption is presented in Figure ??. Compared with case B, with the introduction of OESS, there is approximately 76.2 kWh of energy saving when the investment is around 5.94×10^7 JPY. And with more investment increases, there are further energy savings.

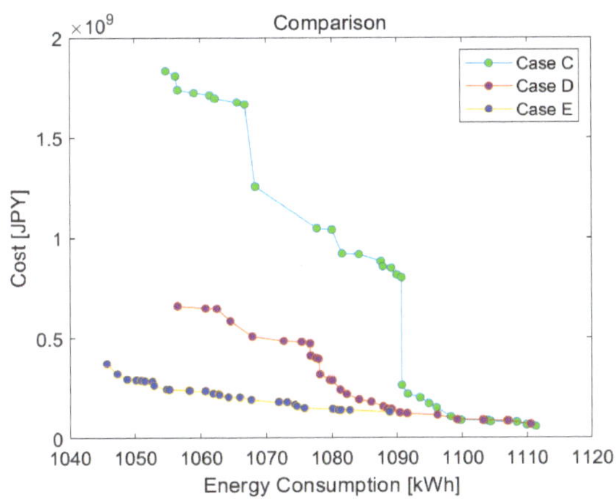

Figure 9. Pareto front cases comparison: C,D,E.

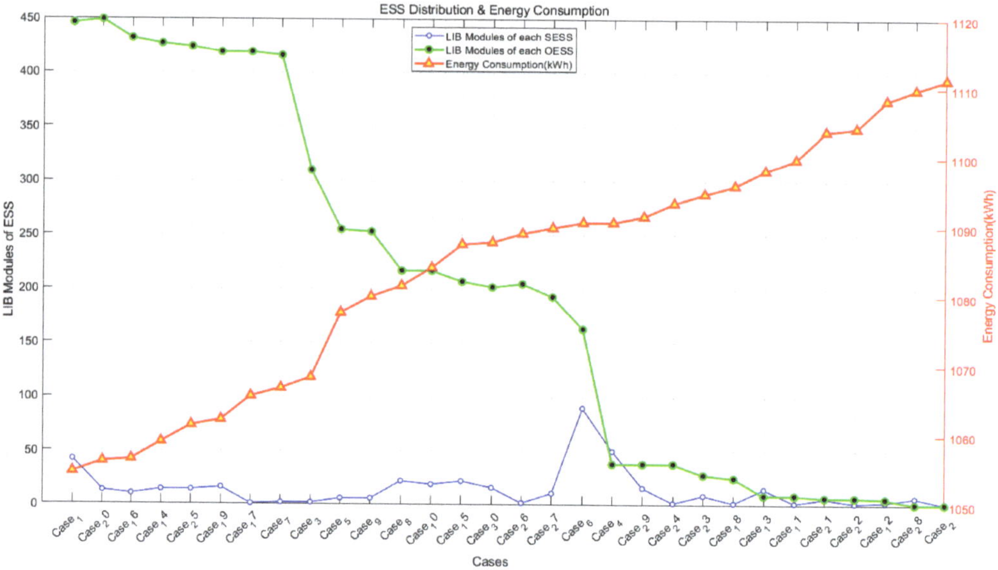

Figure 10. Case C: Capacity Distribution and Energy Consumption.

3.4. Case D

In Case D, both SESS and OESS are installed. However, due to the difficulty of equipment manufacturing, the SESS capacities may vary between stations, while OESS capacities remain uniform across all trains. Consequently, only SESS LIB capacities undergo customization, while OESS LIB capacities are optimized for uniformity. Energy consumption varies from 1056.48 to 1110.50 kWh. And the cost varies from 6.60×10^8 to 7.05×10^7 JPY. They are depicted inFigure **??**. Simultaneously, the relationship between capacity distribution and energy consumption is presented in Figure **??**.

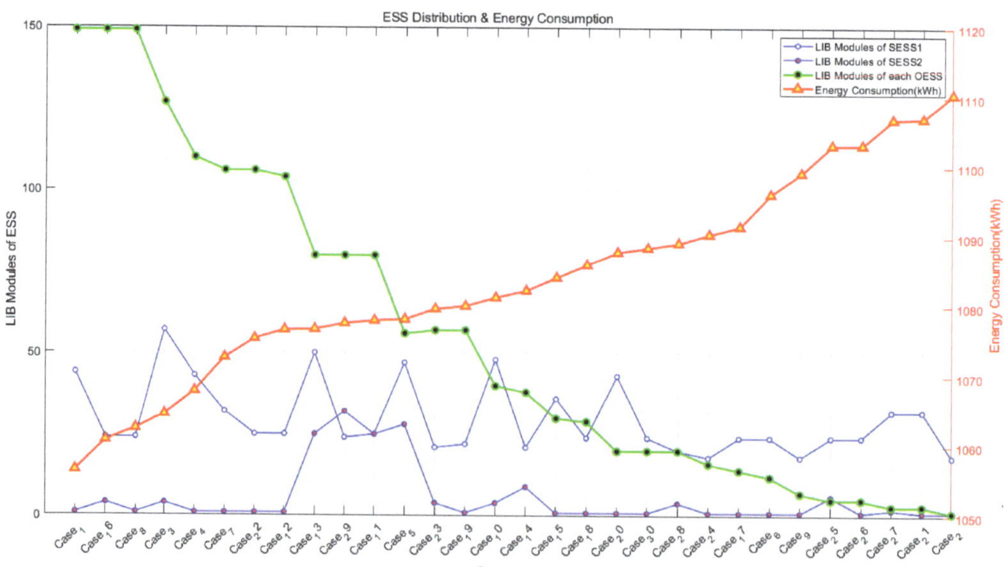

Figure 11. Case D: Capacity Distribution and Energy Consumption.

3.5. Case E

In Case E, similarly, both SESS and OESS are installed, but as a comparison, the capacities of LIBs for both SESS and OESS undergo customization using NSGA-II. Energy consumption varies from 1045.70 to 1088.83 kWh. And the cost varies from 3.74×10^8 to 1.33×10^8 JPY shown in Figure **??**. On the other hand, the relationship between capacity distribution and energy consumption is presented in Figure **??**. With the application of SESS and OESS, energy consumption decreased. It is noticeable that compared with OESS, the capacities of SESSs vary relatively flat in different cases. It further proves that the capacities of OESSs play a significant role in saving energy and investment.

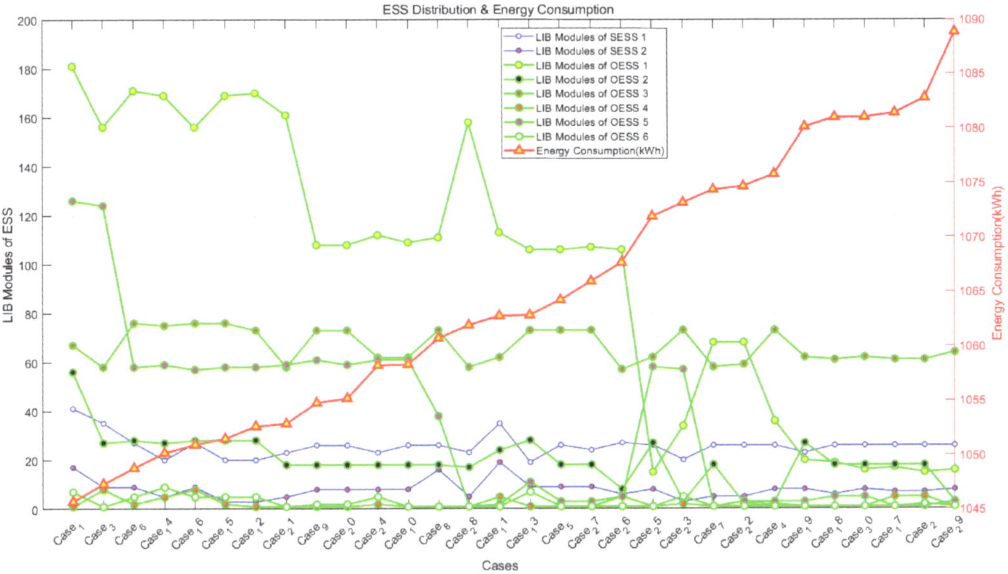

Figure 12. Case E: Capacity Distribution and Energy Consumption.

3.6. Cases Comparison

Table **??** summarizes the case comparison in configuration, impact on energy saving, and cost. As illustrated in 3.2 Case B, introducing SESS (Capacity Customization) has a limited impact on energy saving. On the contrary, the cost increased significantly, which is not a beneficial solution for rail transit construction.

Table 6. Cases Configuration and Impact Comparison.

Case	Configuration	Impact on Energy Saving	Impact on Cost
A	No SESS or OESS	Baseline for comparison	Baseline for comparison
B	Only SESS integrated	Limited energy savings	Increased due to SESS investment
C	SESS and OESS with uniform capacities	Notable energy savings	Highest cost suggesting redundancy
D	SESS with customized capacities and uniform OESS	Notable energy savings	A balance in cost
E	SESS and OESS with customized capacities	Notable energy savings	Lowest cost. Not practical in manufacturing train

Comparing Cases C, D, and E reveals that including OESS alleviates the rail transit's reliance on SESS capacity for energy management. Moreover, despite the increased investment, more energy savings are achieved compared to Case B. It is noteworthy that when energy consumption is similar in Cases C, D, and E, Case E incurs the highest cost among the three. This suggests potential redundancy in the Energy-Storage Systems (ESS) capacity. Conversely, Case D, which pragmatically optimized SESS and OESS capacity, presents a more reasonable outcome.

Moreover, the depicted Figure **??** demonstrates that with the introduction of the utilization rate of RBE into energy-cost Pareto front of Case D, as cost rises, the utilization rate of RBE also increases in a trend, therefore promoting efficient energy management. Additionally, Figure **??** represents the SESS1, SESS2, and OESS's capacity proportion in each optimal non-dominated solution of Case D, named Case 1 to Case 30. It is clear that the capacity of SESS1 and SESS2 have a minimal impact on the RBE utilization rate. Instead, the OESS predominantly plays a crucial role in recovering RBE. Therefore, by optimizing the capacities and configurations of SESS and OESS, rail transit systems can significantly reduce energy consumption and operational costs. In particular, the uniform capacity of each OESS should be two to three times the average capacity of each SESS.

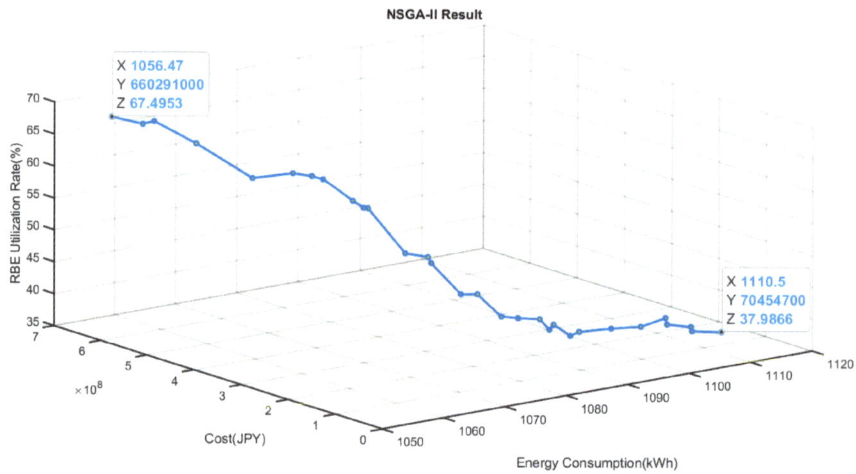

Figure 13. Case D: Cost Energy Regenerative Braking Energy Utilization rate.

4. Conclusions

Through the capacity optimization of LIB using NSGA-II, the integration of SESS proves effective in recovering RBE. Nearly 30 kWh is saved within 30 min of operation. In the meantime, the HESS configuration of SESS, combining SMES and LIB, ensures rapid response and high power exchange while providing a relatively large energy storage capacity to store regenerative braking energy. Furthermore, the collaborative implementation of both SESS and OESS emerges as a superior strategy for recycling. This approach not only achieves significant energy savings (approximately 100 kWh) but also demonstrates a mutually beneficial outcome for both energy consumption and investment efficiency within 30 min of operation.

Particularly noteworthy is the vital role played by OESS in this integrated system. The capacity of OESS is recommended to be two to three times that of SESS. The OESS emphasizes its significance in realizing a win–win scenario for optimizing energy consumption and investments in the rail transit system.

5. Possible Directions for Future Studies

The findings of our study on the coordination of SESS and OESS in rail transit for achieving multi-objective optimization in energy and cost suggest several promising avenues for future research. In this section, we outline challenges and potential directions for further investigation based on the insights gained from our study.

5.1. Optimization of SMES Capacity

Unlike conventional battery-based ESS, SMES presents unique challenges due to its complexity. A possible exploration is optimizing SMES's capacity within the rail transit operations. By delving deeper into determining optimal SMES capacity parameters, researchers can refine energy management strategies and maximize cost savings. Future studies could employ advanced optimization techniques to systematically evaluate the impact of varying SMES capacities on system performance metrics, such as energy efficiency and operational costs.

5.2. Application and Comparison of Different Energy-Storage Systems (ESSs)

Comparative studies involving diverse Energy-Storage Systems (ESSs) offer another avenue for exploration. By evaluating the performance of different ESS technologies, such as lithium-ion batteries, flywheels, and compressed air energy storage, researchers can identify the most suitable solutions for specific rail transit applications. Future research could involve comprehensive assessments of each ESS type under varying operating conditions, considering factors such as energy capacity, response time, and lifecycle costs. Furthermore, conducting sensitivity analyses and scenario-based simulations can provide valuable insights into the robustness and resilience of different ESS technologies in real-world deployment scenarios.

5.3. Validation Experiment

Validation experiments represent a critical step toward confirming the efficacy and feasibility of proposed coordination strategies in practical settings. Future research efforts could focus on conducting field trials and pilot implementations to empirically validate the performance of the developed models and methodologies. Researchers can gain access to real-world data and infrastructure by collaborating with industry partners and transit agencies, facilitating the validation process.

Author Contributions: Conceptualization, D.K. and M.M.; methodology, D.K.; software, D.K.; validation, D.K. and M.M.; formal analysis, D.K.; investigation, D.K.; resources, D.K. and M.M.; data curation, D.K.; writing—original draft preparation, D.K.; writing—review and editing, M.M.; visualization, D.K.; supervision, M.M.; project administration, D.K.; funding acquisition, M.M. All authors have read and agreed to the published version of the manuscript.

Funding: This research received no external funding.

Data Availability Statement: Data is contained within the article.

Acknowledgments: The authors would like to thank all the colleagues in Miyatake's laboratory for their kind help.

Conflicts of Interest: The authors declare no conflicts of interest.

Abbreviations

The following abbreviations are used in this manuscript:

RBE	Regenerative Braking Energy
ESS	Onboard Energy-Storage Systems
ESS	Stationary Energy-Storage Systems
NSGA-II	Non-dominated Sorting Genetic Algorithm-II
HESS	Hybrid Energy-Storage Systems
SMES	Superconducting Magnetic Energy Storage
PSO	Particle Swarm Optimization
LIB	Lithium-ion Battery
NiMH	Nickel-Metal Hydride
IPSO	Improved Particle Swarm Optimization
ATO	Automatic Train Operation
L	Inductance
Isc	Initial Current in Superconductor
Isc_max	Upper Limit of Current
Isc_min	Lower Limit of Current
SOC	Initial State of Charge for Battery
K_i	Quantity of battery modules in each SESS
K_p	Quantity of battery modules in each OESS
i	Order of substations
p	Order of trains
E_{sub}	Energy consumption in all substations
E_{sess}	Energy consumption in all SESS
$Esinitial_i$	Energy stored in SESS i in the beginning
$Esend_i$	Energy stored in SESS i in the end
E_{oess}	Energy consumption in all OESS
$Eoinitial_p$	Energy stored in OESS p in the beginning
$Eoend_p$	Energy stored in OESS p in the end
N	Quantity of substations
$cost_{sc}$	Cost of SMES
$cost_{dcdc}$	Cost of DCDC converter
$cost_{bt}$	Cost of battery per module
$cost_{alle}$	All cost of electricity fee
$cost_e$	Cost of electricity fee per kWh
M	Quantity of trains

References

. Song, Y.; Lu, X.; Yin, Y.; Liu, Y.; Liu, Z. Optimization of Railway Pantograph-Catenary Systems for Over 350 km/h Based on an Experimentally Validated Model. *IEEE Trans. Ind. Inform.* **2024**. [CrossRef]
. Liu, Z.; Song, Y.; Gao, S.; Wang, H. Review of Perspectives on Pantograph-Catenary Interaction Research forHigh-Speed Railways Operating at 400 km/h and Above. *IEEE Trans. Transp. Electrif.* **2024**. [CrossRef]
. Shen, X.; Wei, H.; Wei, L. Study of trackside photovoltaic power integration into the traction power system of suburban elevated urban rail transit line. *Appl. Energy* **2020**, *260*, 114177. [CrossRef]
. Chen, H.; Cong, T.N.; Yang, W.; Tan, C.; Li, Y.; Ding, Y. Progress in electrical energy storage system: A critical review. *Prog. Nat. Sci.* **2009**, *19*, 291–312. [CrossRef]
. Ratniyomchai, T.; Hillmansen, S.; Tricoli, P. Recent developments and applications of energy storage devices in electrified railways. *IET Elect. Syst. Transp.* **2014**, *4*, 9–20. [CrossRef]
. Ghaviha, N.; Campillo, J.; Bohlin, M.; Dahlquist, E. Review of application of energy storage devices in railway transportation. *Energy Proc.* **2017**, *105*, 4561–4568. [CrossRef]
. Li, J.; Zhang, M.; Yang, Q.; Zhang, Z.; Yuan, W. SMES/Battery Hybrid Energy Storage System for Electric Buses. *IEEE Trans. Appl. Supercond.* **2016**, *26*, 1–5. [CrossRef]
. González-Gil, A.; Palacin, R.; Batty, P. Sustainable urban rail systems: Strategies and technologies for optimal management of regenerative braking energy. *Energy Convers. Manag.* **2013**, *75*, 374–388. [CrossRef]
. Lamedica, R.; Ruvio, A.; Palagi, L.; Mortelliti, N. Optimal siting and sizing of wayside energy storage systems in a DC railway line. *Energies* **2020**, *13*, 6271. [CrossRef]
. Ovalle, A.; Pouget, J.; Bacha, S.; Gerbaud, L.; Vinot, E.; Sonier, B. Energy storage sizing methodology for mass-transit direct-current wayside support: Application to French railway company case study. *Appl. Energy* **2018**, *230*, 1673–1684. [CrossRef]

. Roch-Dupre, D.; Gonsalves, T.; Cucala, A.P.; Pecharroman, R.R.; López-López, Á.J.; Fernandez-Cardador, A. Multi-stage optimization of the installation of Energy Storage Systems in railway electrical infrastructures with nature-inspired optimization algorithms. *Eng. Appl. Artif. Intell.* **2021**, *104*, 104370. [CrossRef]
. Teymourfar, R.; Asaei, B.; Iman-Eini, H. Stationary super-capacitor energy storage system to save regenerative braking energy in a metro line. *Energy Convers. Manag.* **2012**, *56*, 206–214. [CrossRef]
. Kobayashi, H.; Asano, J.; Saito, T.; Kondo, K. A Power Control Method to Save Energy for Wayside Energy Storage Systems in dc-Electrified Railway System. *Electr. Eng. Jpn.* **2016**, *196*, 56–66. [CrossRef]
. Ciccarelli, F.; Iannuzzi, D.; Kondo, K.; Fratelli, L. Line-voltage control based on wayside energy storage systems for tramway networks. *IEEE Trans. Power Electron.* **2015**, *31*, 884–899. [CrossRef]
. Kobayashi, H.; Kondo, K.; Iannuzzi, D. A theoretical analysis on dynamic and static characteristics of control strategies for wayside energy storage system in DC-electrified railway. In Proceedings of the IEEE International Conference on Electrical Systems for Aircraft, Railway, Ship Propulsion and Road Vehicles International Transportation Electrification Conference (ESARS-ITEC), Nottingham, UK, 7–9 November 2018.
. Ceraolo, M.; Giglioli, R.; Lutzemberger, G.; Conte, M.; Pasquali, M. Use of electrochemical storage in substations to enhance energy and cost efficiency of tramways. In Proceedings of the AEIT Annual Conference, Palermo, Italy, 3–5 October 2013.
. Konishi, T.; Morimoto, H.; Aihara, T.; Tsutakawa, M. Fixed energy storage technology applied for DC electrified railway. *IEEJ Trans. Elect. Electron. Eng.* **2010**, *5*, 270–277. [CrossRef]
. Gillespie, A.J.; Johanson, E.S.; Montvydas, D.T. Energy storage in Pennsylvania: SEPTA's novel and innovative integration of emerging smart grid technologies. *IEEE Veh. Technol. Mag.* **2014**, *9*, 76–86. [CrossRef]
19. Increasing Energy Efficiency Optimized Traction Power Supply in Mass Transit Systems. Available online: https://docplayer.net/89948259-Siemens-com-mobility-increasing-energy-efficiency-optimized-traction-power-supply-in-mass-transit-systems.html (accessed on 26 December 2023).
. Radcliffe, P.; Wallace, J.S.; Shu, L.H. Stationary applications of energy storage technologies for transit systems. In Proceedings of the IEEE Electrical Power & Energy Conference, Halifax, NS, Canada, 25–27 August 2010.
. Hsu, C.S.; Lee, W.J. Superconducting magnetic energy storage for power systems application. In Proceedings of the 1992 IEEE Conference Record of the Industrial and Commercial Power Systems Technical Conference, Pittsburgh, PA, USA, 4–7 May 1992.
. Sutanto, D.; Cheng, K.W.E. Superconducting magnetic energy storage systems for power system applications. In Proceedings of the 2009 International Conference on Applied Superconductivity and Electromagnetic Devices, Chengdu, China, 25–27 September 2009.
. Zhang, B.; Lu, S.; Peng, Y.; Wu, C.; Meng, G.; Feng, M.; Liu, B. Impact of On-Board Hybrid Energy Storage Devices on Energy-Saving Operation for Electric Trains in DC Railway Systems. *Batteries* **2022**, *8*, 167. [CrossRef]
. Arboleya, P.; Bidaguren, P.; Armendariz, U. Energy Is On Board: Energy Storage and Other Alternatives in Modern Light Railways. *IEEE Electrif. Mag.* **2016**, *4*, 30–41. [CrossRef]
. Wang, Y.; Wang, B.; Zhang, T.; Nazaripouya, H.; Chu, C.C.; Gadh, R. Optimal energy management for Microgrid with stationary and mobile storages. In Proceedings of the IEEE/PES Transmission and Distribution Conference and Exposition (T&D), Dallas, TX, USA, 3–5 May 2016.
. Ceraolo, M.; Lutzemberger, G. Stationary and on-board storage systems to enhance energy and cost efficiency of tramways. *J. Power Sources* **2014**, *264*, 128–139. [CrossRef]
. Stana, G.; Brazis, V.; Prieditis, M. Simulation-based comparisons of mobile and stationary energy storage systems applied for electric transport. In Proceedings of the 20th European Conference on Power Electronics and Applications (EPE'18 ECCE Europe), Riga, Latvia, 17–21 September 2018.
. Pavlovský, M.; Guidi, G.; Kawamura, A. Assessment of coupled and independent phase designs of interleaved multiphase buck/boost DC-DC converter for EV power train. *IEEE Trans. Power Electron.* **2014**, *29*, 2693–2704. [CrossRef]
. Ahmad, H.J.; Hagiwara, M. A Compact High-Power Noninverting Bidirectional Buck-Boost Chopper for Onboard Battery Energy Storage Systems. *IEEE Trans. Power Electron.* **2022**, *37*, 1722–1735.
. Miyatake, M.; Matsuda, K. Energy saving speed and charge/discharge control of a railway vehicle with on-board energy storage by means of an optimization model. *IEEJ Trans. Elect. Electron. Eng.* **2009**, *4*, 771–778. [CrossRef]
. Wu, C.; Zhang, W.; Lu, S.; Tan, Z.; Xue, F.; Yang, J. Train speed trajectory optimization with on-board energy storage device. *IEEE Trans. Intell. Transp. Syst.* **2019**, *20*, 4092–4102. [CrossRef]
. Pulazza, G.; Zhang, N.; Kang, C.; Nucci, C.A. Expansion planning model coordinated with both stationary and transportable storage systems for transmission networks with High RES penetration. In Proceedings of the 2020 IEEE International Conference on Environment and Electrical Engineering and 2020 IEEE Industrial and Commercial Power Systems Europe (EEEIC/I&CPS Europe), Madrid, Spain, 9–12 June 2020.
. Daoud, M.I.; Abdel-Khalik, A.S.; Elserougi, A.; Ahmed, S.; Massoud, A.M. DC bus control of an advanced flywheel energy storage kinetic traction system for electrified railway industry. In Proceedings of the 39th Annual Conference of the IEEE Industrial Electronics Society (IECON), Vienna, Austria, 10–13 November 2013.
. Sitras HES Hybrid Energy Storage System for Rail Vehicles. 2014. Available online: https://assets.new.siemens.com/siemens/assets/api/uuid:aa06262d-3ebf-4bcb-9566-bd6b0db960a5/productinformation-sitras-hes-e.pdf (accessed on 1 January 2024).

. Alstom Wins Largest Light Rail Contract in Australia to Deliver Next Generation Trams for Melbourne, Victoria. 2022. Available online: https://www.alstom.com/sites/alstom.com/files/2022/04/21/20222104_PR_NGT_Melbourne__EN.pdf (accessed on 1 January 2024).
. Steiner, M.; Scholten, J. Energy storage on board of DC fed railway vehicles. In Proceedings of the 2004 IEEE 35th Annual Power Electronics Specialists Conference, Aachen, Germany, 20–25 June 2004.
. Steiner, M.; Scholten, J. *Improving Overall Energy Efficiency of Traction Vehicles*; Bombardier Transportation: Mannheim, Germany, 2006.
. Devaux, F.O.; Tackoen, X. Overview of Braking Energy Recovery Technologies in the Public Transport Field. Energy Recovery Working Paper. Available online: https://www.stib-mivb.be/irj/go/km/docs/resource/tickettokyoto/sites/default/files/downloads/T2K_ER_overview_technologies_web.pdfview_technologies_web.pdf(stib-mivb.be) (accessed on 1 January 2024).
. Steiner, M.; Klohr, M.; Pagiela, S. Energy storage system with ultracaps on board of railway vehicles. In Proceedings of the 2007 European Conference on Power Electronics and Applications, Aalborg, Denmark, 2–5 September 2007.
. Winter, M.; Passerini, S. Lithium ion batteries as key component for energy storage in automotive and stationary applications. In Proceedings of the IEEE 33rd International Telecommunications Energy Conference (INTELEC), Amsterdam, The Netherlands, 9–13 October 2011.
. Hayajneh, H.S.; Bashetty, S.; Salim, M.N.B.; Zhang, X. Techno-economic analysis of a battery energy storage system with combined stationary and mobile applications. In Proceedings of the 2018 IEEE Conference on Technologies for Sustainability (SusTech), Long Beach, CA, USA, 11–13 November 2018.
. Jiang, X.; Chen, J.; Zhang, W.; Wu, Q.; Zhang, Y.; Liu, J. Two-step optimal allocation of stationary and mobile energy storage systems in resilient distribution networks. *J. Mod. Power Syst. Clean Energy* **2021**, *9*, 788–799. [CrossRef]
. Moura, P.; Sriram, U.; Mohammadi, J. Sharing mobile and stationary energy storage resources in transactive energy communities. In Proceedings of the 2021 IEEE Madrid PowerTech, Madrid, Spain, 28 June–2 July 2021.
. Tao, P.; Ronghua, Z.; Luqing, J.; Zhongping, Y.; Fei, L.; Kaiqi, S. Capacity Configuration Method of Urban Rail Energy Storage System Based on NSGA-II and Simplified Energy Storage Model. In Proceedings of the 2023 IEEE 3rd International Conference on Industrial Electronics for Sustainable Energy Systems (IESES), Shanghai, China, 26–28 July 2023.
. Mundra, P.; Arya, A.; Gawre, S.; Mehroliya, S. Independent Demand Side Management System Based on Energy Consumption Scheduling by NSGA-II for Futuristic Smart Grid. In Proceedings of the IEEE Hybrid and Distributed Smart Grid Conference (IEEE-HYDCON), Hyderabad, India, 11–12 September 2020.
. Qayyum, F.; Jamil, H.; Jamil, F.; Kim, D. Predictive Optimization Based Energy Cost Minimization and Energy Sharing Mechanism for Peer-to-Peer Nanogrid Network. *IEEE Access* **2022**, *10*, 23593–23604. [CrossRef]
. Li, Y.; Lv, Q.; Zhang, Z.; Dong, H. Optimal Allocation of Energy Storage Capacity of Distribution Grid Based on Improved Particle Swarm Optimization Algorithm. In Proceedings of the 2022 3rd International Conference on Advanced Electrical and Energy Systems (AEES), Lanzhou, China, 23–25 September 2022.
. Amini Pishro, A.; Yang, Q.; Zhang, S.; Amini Pishro, M.; Zhang, Z.; Zhao, Y.; Postel, V.; Huang, D.; Li, W. Node, place, ridership, and time model for rail-transit stations: A case study. *Sci. Rep.* **2022**, *12*, 16120. [CrossRef]
. Amini Pishro, A.; L'hostis, A.; Chen, D.; Amini Pishro, M.; Zhang, Z.; Li, J.; Zhao, Y.; Zhang, L. The Integrated ANN-NPRT-HUB Algorithm for Rail-Transit Networks of Smart Cities: A TOD Case Study in Chengdu. *Buildings* **2023**, *13*, 1944. [CrossRef]
. Azzedine, B.; Wang, J. Machine learning-based traffic prediction models for intelligent transportation systems. *Comput. Netw.* **2020**, *181*, 107530.
. Hitachi Website. Available online: https://www.hitachi.com/rev/archive/2021/r2021_05/pdf/05a10.pdf (accessed on 28 February 2024).
. Essien, A.; Petrounias, I.; Sampaio, P.; Sampaio, S. A deep-learning model for urban traffic flow prediction with traffic events mined from twitter. *World Wide Web* **2021**, *24*, 1345–1368. [CrossRef]
. Deb, K.; Pratap, A.; Agarwal, S.; Meyarivan, T.A.M.T. A fast and elitist multiobjective genetic algorithm: NSGA-II. *IEEE Trans. Evol. Comput.* **2002**, *6*, 182–197. [CrossRef]
. Murata, R.; Miyatake, M.; Akiba, T.; Tajima, M. Study on a feeding circuit model formulated to use multipurpose solvers for multi-train simulators. In Proceedings of the 2016 International Conference on Electrical Systems for Aircraft, Railway, Ship Propulsion and Road Vehicles International Transportation Electrification Conference (ESARS-ITEC), Toulouse, France, 2–4 November 2016.
. Sato, Y.; Sato, M. Using Dominated Solutions at Edges to the Diversity and the Uniformity of Non-dominated Solution Distributions in NSGA-II. *SN Comput. Sci.* **2022**, *3*, 432. [CrossRef]
. J-TREC Website. Available online: https://www.j-trec.co.jp/company/070/02/jtr02_98-103.pdf (accessed on 1 March 2024).
. Zhu, J.; Zhang, H.; Yuan, W.; Zhang, M.; Lai, X. Design and cost estimation of superconducting magnetic energy storage (SMES) systems for power grids. In Proceedings of the 2013 IEEE Power & Energy Society General Meeting, Vancouver, BC, Canada, 21 July 2013.
. TEPCO Website. Available online: https://www.tepco.co.jp/ep/corporate (accessed on 1 June 2021).
. Khodaparastan, M.; Mohamed, A.A.; Brandauer, W. Recuperation of regenerative braking energy in electric rail transit systems. *IEEE Trans. Intell. Transp. Syst.* **2019**, *20*, 2831–2847. [CrossRef]

logistics

MDPI

Article

Investigating the Crash Protection Performance of a Medical Carrier Bag for Drone Transport

Fraser McLeod [1,*], Tom Cherrett [1], Andy Oakey [1], Katherine Theobald [1], Tim Waters [1], Matt Grote [1], John Armstrong [1], Jack Denny [1] and Alex Murray [2]

[1] Faculty of Engineering and Physical Sciences, University of Southampton, Southampton SO17 1BJ, UK; t.j.cherrett@soton.ac.uk (T.C.); a.oakey@soton.ac.uk (A.O.); k.theobald@soton.ac.uk (K.T.); t.p.waters@soton.ac.uk (T.W.); m.j.grote@soton.ac.uk (M.G.); j.armstrong@soton.ac.uk (J.A.); jack.denny@soton.ac.uk (J.D.)

[2] Motion Robotics Limited, Southampton SO30 3DS, UK; alex.murray@motion-robotics.co.uk

* Correspondence: f.n.mcleod@soton.ac.uk

Abstract: *Background:* Drone transport regulations in Europe require a crash-protected container (CPC) to be used for the carriage of dangerous goods. With increasing interest in the use of drones for medical logistics, the motivation behind this research was to investigate whether the existing approved medical carriers could also pass as CPCs. To date, there has been little practical experimentation on or theoretical research into the crash protection performance of medical containers. *Methods:* Addressing this gap, this paper reports findings from a series of drop test experiments to investigate the crashworthiness of a standard medical carrier bag used by the National Health Service (NHS) in the UK. Th drop tests were performed from heights of up to 122 m using standard medical carriers containing bags of dyed saline to examine the robustness of the carrier and whether it could contain any leakages, a key requirement for transporting dangerous goods. *Results:* The tests found that the medical carrier failed on some drops, with the zipped lid being identified as the main weakness. *Conclusions:* A new understanding of the carrier's terminal velocity, impact acceleration, and failure mechanisms were gained and subsequent strengthening and waterproofing remedial measures recommended. New insights and practical recommendations are provided relating to performing formal drop tests and how to conduct these using a drone.

Keywords: drone logistics; crash protection; drop testing; dangerous goods

Citation: McLeod, F.; Cherrett, T.; Oakey, A.; Theobald, K.; Waters, T.; Grote, M.; Armstrong, J.; Denny, J.; Murray, A. Investigating the Crash Protection Performance of a Medical Carrier Bag for Drone Transport. *Logistics* **2024**, *8*, 31. https://doi.org/10.3390/logistics8010031

Academic Editor: Mladen Krstić

Received: 7 December 2023
Revised: 14 February 2024
Accepted: 4 March 2024
Published: 15 March 2024

1. Introduction

The transport of infectious substances and other dangerous goods (DG) has specific regulations depending on the mode of transport [1]. For transport using remotely piloted aircraft (drones), one of the requirements in European Union law is the use of an approved crash-protected container (CPC) [2]. This has also been adopted in the United Kingdom (UK), where the national aviation authority (the Civil Aviation Authority (CAA)) is responsible for all guidance and approvals related to the carriage of DG by drones [3]. In association with the CAA, the UK's Vehicle Certification Agency (VCA) specified a drop testing procedure for a CPC, originally published in April 2022 [4] and amended in January 2024 [5]. Containers meeting the test requirements may, at the discretion of the CAA, be approved as suitable for the carriage of DG using drones. As the CPC regulations and drop testing procedure are recent, there has been no prior reported practical experimentation or theoretical research into determining the crash protection performance (and failure mechanisms) of a standard medical carrier bag or on conducting drop tests from a drone. Without such knowledge, it remains challenging to propose modifications to improve the crashworthiness of standard medical carriers necessary for regulatory approval. This paper directly addresses this research gap by describing drop tests of an insulated carrier bag that is commonly used by the UK National Health Service (NHS) for various medical logistics uses, including the movement of patient diagnostic samples, blood stock for transfusion, and aseptic medicines carried in intravenous (IV) bags.

Most patient diagnostic samples are transported to testing laboratories via road, but there is growing interest in using drones in specific cases, particularly where the terrain is challenging, resulting in long and unreliable journey times, or where the transported materials must be received urgently [6]. Where the items to be transported are classed as dangerous goods, it is important that the container used does not allow any contents to be released in the worst-case scenario of the drone crashing or the container becoming detached and falling to the ground during flight.

Over the last 10 years, many medical drone logistics applications have been reported worldwide [7–16]. Often, these have been bespoke use cases or trials in rural areas where the logistics of transportation by road is difficult and where air transport can significantly reduce the journey times by crossing over geographical features such as water, mountains, or forests. Drones have also been used for humanitarian and emergency relief logistics, delivering urgently needed medical supplies. The related literature has focused on identifying opportunities for using drones for medical logistics, or for other types of cargo transportation, and on identifying the various barriers to their wider implementation, including legal, financial, medical, technical, and logistical issues, but particularly relating to safety and the sharing of airspace with crewed aircraft. Another strand of research covers public attitudes towards the use of drones for cargo deliveries and how those may shape future trends in use, where noise, privacy, safety, and security tend to be the main concerns [17–19]. The attitudes of healthcare workers have also been surveyed, with most expressing positive attitudes towards drone deliveries [20,21].

Drone crashes can occur for several reasons, including mechanical faults, GPS errors, battery failure, communications failure, poor weather, magnetic interference, errors made by an autopilot system or by a human controller, or malicious interference [22–24]. As safety is a primary concern, drone flight paths are usually designed to be as far away from buildings and people as possible. Tall buildings in urban environments are not only hazards to be avoided but can induce wind turbulence, making drone navigation more difficult [25]. Nevertheless, with most hospitals situated in urban areas, there is demand for medical drone use in cities. A forerunner in the field of urban medical logistics by drone is Matternet, who have been flying drones (quadcopters with a 20 km range, 36 kph average speed, and 2 kg maximum payload) between hospitals and testing laboratories in Switzerland since 2017, with thousands of flights having been made [8].

The consequences of any drone crash will depend on the drone's construction, size, weight, and speed at impact. This has been studied for crashes into people ('ground risk') [26,27] and into other flying aircraft ('air risk') [28]. The analysis and quantification of the potential risks and consequences of a drone crash are complex. In [29], a bespoke software model was developed to predict the risk of a fatality of an individual on the ground due to a drone crash based on dynamic population densities, which allows drone flight paths to be planned to avoid areas of higher risk. However, this model does not include any additional risks that may be involved due to the specific case of a drone carrying DG cargo. Studies of drone travel impacts on medical cargoes have focused on the effects of drone vibrations on cargo quality assurance [8,30,31], with no studies found on the impacts of the potential release of DG into the surroundings as a result of a drone crash. The proposed methods for avoiding a drone crash or mitigating against it include the use of a parachute or airbag on detecting a fall [19,32] and the use of on-board technology to diagnose imminent failure and to redirect drones to a safe spot to crash-land [33].

The ramifications on the drone platform design in terms of scale and payload capacity could be significant if bespoke CPCs are mandated, in addition to the potential economic considerations for consignors and consignees of having to procure such carriage systems. Of interest here, in the context of medical logistics, is whether the existing approved packaging solutions could withstand such catastrophic failures. This study aimed to determine whether a standard carrier bag, approved for general transport use by the UK

Medicines and Healthcare Products Regulatory Agency (MHRA), could be approved as a crash-protected container for transport using drones. If such approval could be obtained, it would enable the use of drones for the transport of infectious substances, and it would avoid the need for the procurement of bespoke containers and associated packaging systems.

1.1. Relevant Regulations and Background to the Test Procedure

The overarching principles governing the safe transport of dangerous goods (DG) by air are laid out in Annex 18 to the Convention on International Civil Aviation (known as the Chicago Convention) [34]. These principles have been expanded into the 'Technical Instructions for the Safe Transport of Dangerous Goods by Air' (known as the Technical Instructions; TI), published biennially by the International Civil Aviation Organisation (ICAO) [35]. These regulations provide detailed requirements for the transport of each of the 3000+ substances classified as DG, covering aspects such as packing instructions, quantity limits, loading, documentation, training of personnel, leak/spill procedures, and emergency response. International civil aviation operations must comply with the regulations, and domestic operations are also encouraged to comply as a best practice [36]. The regulations apply to DG carried by all aircraft types, including drones (although national aviation authorities may grant alleviations for drone operations to deviate from the TI, provided an equivalent level of safety is achieved) but have developed over the years exclusively from the perspective of crewed aviation, in particular the operation of large, fixed-wing, airliner-type aircraft that are typically used to transport the vast majority of air freight. The interpretation and application of these regulations to drone operations are an emerging legislative arena, still under development in different regions around the world [37].

In general, the purpose of the TI is to protect packages containing DG from damage, leaks, or spills during routine aircraft operations [37]. However, in 2021, the European Union Aviation Safety Agency (EASA) published legislation detailing that a crash-protected container (CPC) would be required for drones operating in the Specific Category when carrying DG [2], with this legislation also adopted in the UK to support inter-operability. In the absence of a CPC, such that there may be a high risk to third parties in the event of an accident, the carriage of DG is only permissible by drones in the Certified Category [3]. The inference of this CPC requirement is that DG should be prevented from escaping into the surroundings not only during routine drone operations but also in situations up to and including the worst-case scenario of a drone crash (i.e., above and beyond the expectations for crewed aircraft operations). This is because, with the exception of drones in the Certified Category, drones are not certified to the same rigorous airworthiness standards as crewed aircraft [36].

Following this legislation, a working group, including some of the authors of this paper, was created to establish a test regime for CPCs, to enable drones to carry DG [38]. A formal test procedure was specified that required containers carrying a payload to be dropped from height without resulting in leakage of the contents [4], where the payload may be one of (i) packaging conforming to the TI; (ii) a fine, inert powder or granular material placed directly into the container without the use of lining or inner packaging; (iii) a liquid poured directly into the container without the use of lining or inner packaging.

The key requirements that must be met in the CPC test procedure are as follows:

- A set of three identical containers must be provided for testing.
- The gross mass of the test container will dictate the maximum gross mass of the approval.
- Tests must be conducted in ambient conditions (ideally, air temperature 5–25 °C), with windspeed < 10 knots (5.14 m/s) and no significant precipitation.
- The container can be elevated to drop height by any suitable means, with the drop height being the maximum height above ground level authorised for a particular drone operation, but where it can be demonstrated that terminal velocity is reached from a lower height, this height (plus 10%) may be used instead.

- If an aircraft is used as the drop platform, it should ideally be stationary (i.e., hovering), but where forward flight is unavoidable, the airspeed should be minimised as far as possible.
- The container must be released in its normal transport orientation.
- The container must strike the surface of the designated impact zone, where the surface must be reasonably level, smooth and flat, and constructed of concrete, asphalt, or another material of similar hardness that would conform to public highway standards.
- A container shall be deemed to have passed if:
  - There is no major disruption to its structure and no visible hole/gap through which contents might escape (even if they have not escaped), and;
  - Its contents are solids and there is no leakage from the container, or;
  - Its contents are liquids and there is no leakage from the container, either immediately or (at least) eight hours after the test.
- All three containers must pass. If one container fails, the test is failed overall. A pass for any remaining container cannot be carried forward to a new test, i.e., retesting must be performed using three untested containers.
- For the CPC approval application, a test report shall be produced detailing the conduct of the test and including photographs of the containers at all stages of testing, and wherever possible, a video of the whole test sequence.

A consequence of testing with payload type (i) is that the use of TI-compliant packaging introduces a requirement for a rigid outer packaging, as stipulated in the TI in Packing Instruction (PI) 650 for DG classified as UN 3373 (Biological substance, Category B), as is the case for patient diagnostic samples. The TI-compliant package can either be placed inside another container designed to provide crash protection (i.e., a separate CPC), or the outer packaging of the package itself can be relied on to satisfy the drop test requirements (i.e., the outer packaging effectively forms the CPC). In either case, any approval is limited to the specific packaging configuration tested, so any modifications to the packaging would require separate testing.

The tests for proving compliance with the requirements for rigid outer packaging do not appear to be clearly defined. The TI do specify a 'stack test' (no leakage or distortion liable to reduce strength or cause instability in stacks of packages when a force is applied to the top surface of a package equivalent to a 3 m high stack of identical packages for 24 h), but this seems more relevant to multiple packages stacked in an aircraft cargo hold rather than single packages carried by drones (e.g., as an underslung load). This highlights the need for drone-specific provisions to be explicitly included in future versions of the TI.

Testing with payload types (ii) or (iii) above has the advantage that approval then allows carriage within the approved CPC of any TI-compliant package configuration for solids or liquids, respectively, up to the tested payload mass.

The test procedure indicates that the drop height should typically be the operational height specified by the competent authority (121.9 m (400 feet), in the UK), with provision to drop from a lower height where it can be demonstrated that the lower drop height, plus a 10% safety margin, is sufficient for the loaded container to attain terminal velocity.

1.2. Containers for Crash Protection

In this section, we highlight the desired attributes needed for this application and the types of containers and packaging materials that might be most suitable. Here, the main requirement is that, after a fall from height, the container should not be damaged in such a way that its contents can escape from it. Although not a requirement of the CPC test procedure, it would be desirable if the contents remained undamaged and salvageable. It is also desirable that the container be as light as possible, to reduce the overall weight and so it is able to carry a wide range of packaged items, as opposed to a bespoke container and packaging arrangements designed for a specific application.

In general, the key to a good crash (shock) protection system is its ability to absorb the sudden impact energy and release it relatively slowly. Viscoelastic materials such as rubber and neoprene are well suited to shock absorption as they behave both like a viscous (liquid) material, deforming under load and transmitting forces in all directions, and like an elastic material, returning to its original shape when the load is removed; however, viscoelastic materials are notoriously temperature- and frequency-dependent [39]. The distribution of force across the impact zone is also important and, in some applications, a thin rigid layer of a metal or composite material can be used to distribute the impact across a large area of soft, energy-absorbing material.

In general transportation, a wide variety of materials are used to protect against shock and vibrations in normal transit conditions. Bubble wrap is commonly used to provide cushioning and is very light, with density as low as 11 kg/m^3 [40]. Similarly, 'air bags' or 'air pillows' use air inside a soft plastic cover for cushioning, and some can be deflated and reinflated for reuse. Protective foam materials include expanded polystyrene, expanded polypropylene, expanded polyethylene, polyurethane, and low-density polyethylene films. Most of these materials are versatile, as they can be moulded, cut, or glued together to form complex shapes, tailored to the items being transported. The typical forms seen in practice include loose-fill beads, endcaps, egg-box shapes, or 'foam-in-place', where polyurethane foam is injected to fill gaps within the container [41]. Most packaging foams are so-called 'closed cell' foams, which provide a good barrier against liquids and gases ('open cell' foams are more sponge-like) [40]. Styrene–acrylonitrile is a semi-rigid foam which had better shock absorbance than polystyrene foam in drop tests [40], with a density of about 16 kg/m^3. Polyolefin foams combine several materials to provide good flexibility and protection from multiple impacts, with typical densities from 16 to 32 kg/m^3, and are available in expanded (mouldable) and extruded forms [40]. Sorbothane, a proprietary polyurethane, may be a suitable material to use due to its ability to absorb up to 80% of impact force across a wide range of operating temperatures (−29 °C to 72 °C) [42].

There are a wide range of commercially available 'flight cases' that are designed to prevent damage to sensitive equipment in transit. A common type of these is a 'rack case', with standard 19″ rails, designed for electronic equipment such as computers or sound systems, often constructed using acrylonitrile butadiene styrene (ABS), a hard thermoplastic, for the outer case and aluminium supports internally, some of which are designed to act as shock absorbers [43]. Some other flight cases use foam linings for cushioning, which have the advantage of being relatively light [44]. Most of these cases are designed for the normal shocks and vibrations expected in transit and are not necessarily designed to be crash-resistant. The first CAA approval of a CPC was announced in September 2023 by Viking Drone Packaging [45]. A patent application from the same company, published in February 2023, suggests that the approved container and its related packaging system were designed to carry up to 150 medical sample tubes [46]. Although the patent application mentions the possibility of easily adapting the container to carry different products, the CPC test procedure states that any modifications would be subject to new drop tests and separate CAA approval.

2. Materials and Methods

The type of container (bag) used in all the drop tests undertaken in this research goes by the manufacturer name of 'Versapak' (Figure 1) [47]. This type of bag is widely used in medical logistics as an insulated pathology sample carrier. A medium-sized Versapak was used (external dimensions: 460 mm × 305 mm × 255 mm; internal dimensions 400 mm × 225 mm × 195 mm (=18 L); empty weight 2.15 kg). The bag has a zipped lid; four rubber pads on the bottom corners, each 80 mm × 52 mm × 7 mm; and covered foam padding inside on all six faces, with each pad having a thickness of 45 mm. The main materials used in the construction of the bag are PVC, nylon thread, foam padding, insulating foil, and stiffening materials.

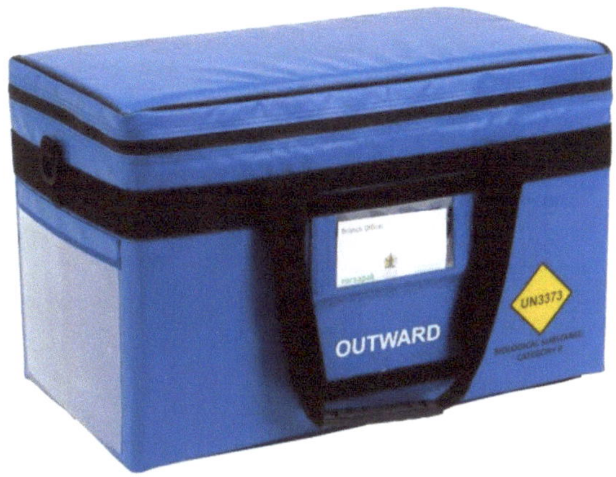

Figure 1. A medium-sized Versapak carrier bag.

The sample ('specimen') carried within each Versapak was a 500 mL IV bag [48], with dimensions of 300 mm × 120 mm × 30 mm and weighing 0.5 kg, containing saline injected with red dye to allow for the identification of any leakage. This type of specimen was chosen as the aim was to test whether the Versapak and its accompanying packaging would be able to contain a significant spillage if the IV bag burst on impact. Each IV bag was wrapped in two layers of cling film. Each specimen was carried alongside a frozen cold pack (168 mm × 87 mm × 35 mm, 0.45 kg) [49] made with high-density polyethylene, which is known for its strength and high impact resistance [50]. The cold pack was used in the tests as it would often be required in practice for temperature control [51]. Each cold pack was wrapped in one layer of bubble wrap, which is common practice to prevent the medical cargo from coming into direct contact with the cold surface and potentially freezing. Only one IV bag was used in these tests to simulate the delivery of a single treatment of aseptic medicine, though it should be noted that, in practice, multiple treatments can be carried in a single consignment [51]. Each of the drop tests used a new Versapak and IV bag, except for one practice drop test, which used a repaired Versapak (Section 2.1).

2.1. Practice Drop Testing

Before dropping the bags from a drone, three practice drop tests were conducted in June 2023 using a disused lift shaft inside a closed University of Southampton building. The purpose of these tests was to learn how the bags and their contents would perform when dropped from a height of approximately 49 m (160 feet), where that height represents half of that used in the drone drop tests. Each bag was dropped by hanging it via its handles from a pole, extended into the open lift shaft and pushing it off, slowly, using another pole. The bottom of the lift shaft was made of concrete, with a raised section (ledge) going from side to side, of a 0.59 m height, a 0.69 m depth, 0.4 m from the lift entrance (front), and 1.04 m from the back wall (Figure 2). The aim was to land the bags onto a flat surface, but this proved to be difficult to achieve due to the raised section (see Results (Section 3)). Video recordings were taken of all the practice drop tests.

The packaging configuration used for each practice drop test was:

- Practice drop test 1—The specimen was placed inside the Versapak on top of, but not attached to, the cold pack (Figure 3).
- Practice drop test 2—A made-to-measure 'soft' foam block was used to fill the interior of the Versapak, with a cut-out made in the foam to house the specimen and cold pack (Figure 4). According to the foam supplier [52], the 'soft' foam had a nominal hardness in the range of 100–140 newtons and a density in the range of 31.4–34.7 kg/m^3.
- Practice drop test 3—This used the same packaging configuration as in the first practice drop test using the same Versapak from the first test but using a new specimen.

The Versapak, damaged from the first test, was repaired using gaffer tape to provide added strength, particularly around the zip (Figure 5). This tape seal would also have affected the air pressure release on impact.

Figure 2. Bottom of lift shaft with second dropped bag near back wall.

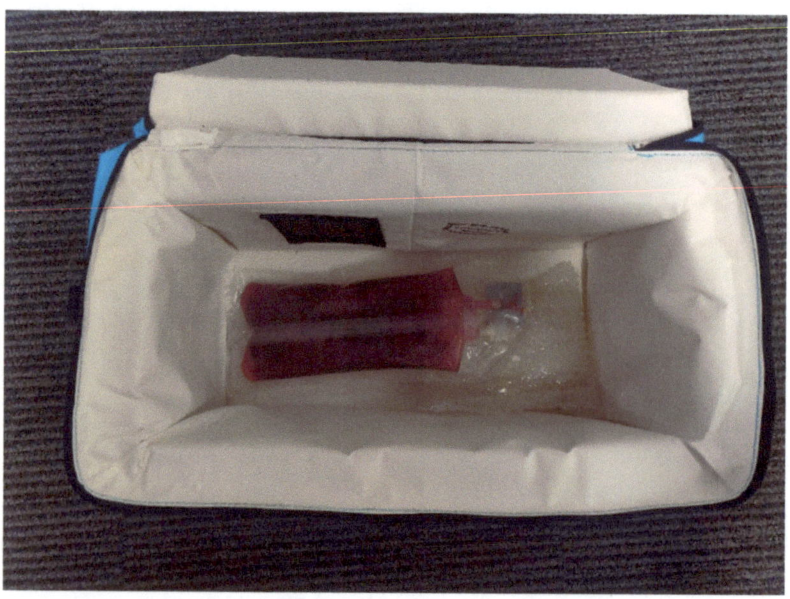

Figure 3. IV bag on cold pack inside Versapak as used in practice drops 1 and 3.

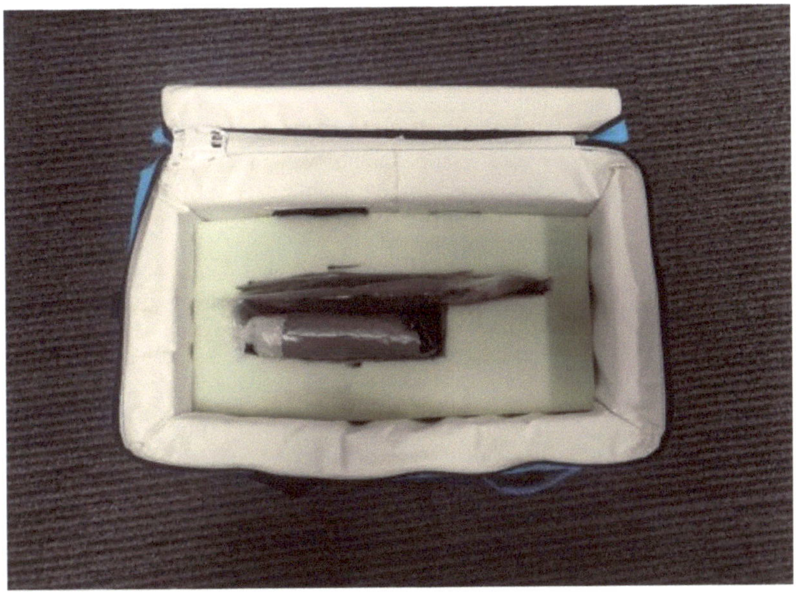

Figure 4. Foam packaging used in practice drop 2.

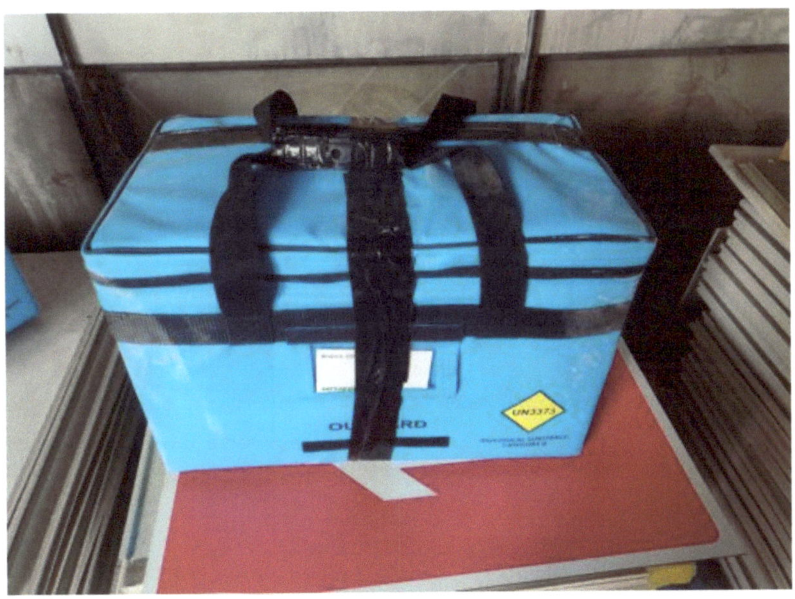

Figure 5. Taped container used in practice drop 3.

2.2. Main Drop Testing by Drone

The main drop tests, by drone, took place at the Snowdonia Aerospace Centre, Llanbedr Airfield in North Wales, in July 2023. Eight drop tests were made over three separate days. Drop tests 2 and 5 are not described here as one test was invalidated when the bag landed on grass to the side of the hardstanding, and the other was for a separate test configuration. Video recordings were taken of all the drop tests, both from the ground and air, the latter using a GoPro 360 camera mounted onto the drone. Three packaging configurations were tested to assess the level of crash protection needed:

(1) **Drop tests 1 and 4**—Foam protection was used as for practice drop 2 (Figure 4), with the specimen placed inside a polythene 'freezer bag' (26 × 35 cm) with tie handles (not tied), inside a polythene bag with a zip seal (30 × 40 cm), alongside a frozen cold pack wrapped in bubble wrap. Drop test 1 used the same 'soft' foam as described for practice drop test 2, while drop test 4 used a slightly firmer 'medium' foam, described by the supplier [52] as having a nominal hardness in the range of 115–155 newtons and a density in the range of 31.4–34.7 kg/m^3;

(2) **Drop tests 3, 6, and 7**—The IV bag was wrapped in cling film and then placed inside a polythene 'freezer bag' with tie handles (not tied), inside a polythene bag with a zip seal, inside a 'leak-proof' polythene liner (610 × 915 mm, 125 micron (500 gauge) thickness) [53], inside a second identical leak-proof liner (Figure 6), alongside a frozen cold pack, wrapped in bubble wrap, and using duct tape to seal the bags.

(3) **Drop test 8**—As for drop tests 3, 6, and 7 but not using the freezer bag or zip bag (the simplest configuration).

Figure 6. Specimen and packaging used in drop tests 3, 6, 7 (IV bag, freezer bag, zip-seal bag, two leak-proof liners) (cold pack not shown).

The wrapped specimen was placed inside the Versapak bag. It was not fixed within the bag so could move during the drop tests, especially vertically, in the bag's normal orientation, as, while the package nearly filled the Versapak in terms of length and width, it only filled about one quarter of its height (Figure 7). In practice, this would mean that the container could be filled with four packages of this size.

Each Versapak containing a packaged specimen was held underneath a 'Titan' hexacopter drone, designed, manufactured, and operated by Motion Robotics Ltd. (Southampton, UK) [54], using a specially designed remote-controlled harness grab/release mechanism on the underside of the drone using two clamps that close/open to grab/release a harness attached to the bag (Figure 8). The drone specification was a height of 0.682 m, a frame size of 0.49 m × 0.7 m, a rotor diameter of 0.762 m, a payload capacity of 15 kg, and a maximum take-off weight of 38 kg. The total weight of the container, specimen, packaging, and harness was approximately 4.57 kg, with some negligible variations for the different packaging configurations.

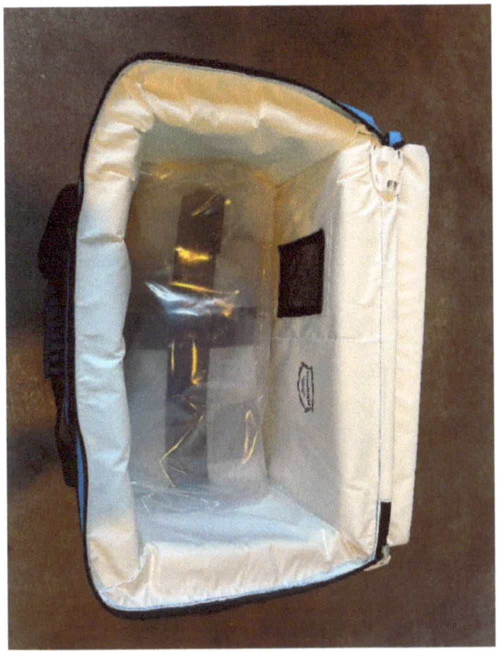

Figure 7. Wrapped specimen inside container.

Figure 8. Hexacopter drone with bag to be mounted via attached harness.

One of the VCA test requirements was that the drops should be made from at least a minimum height that ensured that terminal velocity would be achieved, plus a 10% safety margin. All drops were made from a height of 98 m, where this height was determined as explained in Appendix A. The drone was positioned above or to the side of the runway, making allowances for any lateral drift expected due to prevailing wind conditions, aiming to drop the bag onto the runway and not onto the softer grassy areas on either side. A stationary hover position was achieved for all the drops.

Following the test procedure requirements, each dropped container was visually inspected for damage on the runway and then placed on a plastic tray and removed to a nearby hangar for a more detailed inspection after at least eight hours, the purpose of this delay being to ensure that any leaks could be observed. The test procedure requires that the container be stored in an orientation that would allow liquid to escape (e.g., if there was a hole at top, then it should be placed top-down). In these tests, containers were stored upside down as the zipped lid was deemed to be the most likely source of any leakage, in the absence of any obvious holes appearing. The other test requirements include the recording of relative humidity and wind speed at the time of each drop, as well as providing photographic or video evidence of the various stages of the drop testing, that is, before, during, and immediately after each drop and at the bag inspection eight or more hours afterwards.

2.3. Measuring Devices

The shock levels experienced during impact were measured for drop tests 1 and 4, using a MEMS data logging triaxial shock accelerometer (the MSR 175) [55]. This sensor was configured to its maximum 6.4 kHz sampling frequency and measured acceleration in the range of ± 200 g, with a trigger level of 10 g. The sensor was secured within the internal pocket of the Versapak.

The acceleration levels experienced during the drop were measured for drop tests 3 and 7. This used a different triaxial MEMS data logging accelerometer (the Axivity AX6) [56], configured to its maximum sampling frequency of 1.6 kHz and measured acceleration in the range of ± 16 g. The sensor was secured within the internal pocket of the Versapak.

The drone's altitude was measured using readings from the flight controller's three barometers, fed through an extended Kalman filter. This information was fed back to the ground control station and relayed to the pilot during flight. During the drop tests, flight involved manually flying to an altitude of 98 m, followed by a short, automated flight, to a set waypoint above the runway at 98 m. The flight controller then notified the ground control station that the desired position and altitude had been achieved. The drone was also allowed to stabilise at this new waypoint for 10–15 s before initiating the release of the payload.

The wind speeds were measured using a digital anemometer at three locations around the site. The average and maximum gusting wind speeds were measured at all three points immediately prior to the commencement of each drone operation, and an average value between these locations was taken.

3. Results

3.1. Practice Drops Down a Lift Shaft

For all three practice bag drops, there was lateral movement of the bag during the drop, either towards the front or back wall, which caused the bags to the strike walls before landing. This was likely caused by circulating air effects. The drop time was estimated, in each case, from the video recordings, to be about 3.0 s, giving an estimated average drop speed of 16.3 m/s.

3.1.1. Practice Drop Test 1

During the drop, the bag moved laterally towards the front wall of the lift shaft and struck a lift shaft entrance ledge (Figure 9) about two-thirds of the way down, with the bag then bouncing to and glancing off the back wall before hitting the rear top edge of the ledge at the bottom of the lift shaft and ending near the back wall. The impacts with the lift shaft entrance ledge and the ledge at the bottom of the lift shaft were both observed to be significant. The bag lid was partially opened following the drop test but the IV bag was undamaged, despite the two impacts.

Figure 9. Bag about to hit lift shaft entrance ledge (left side of figure).

3.1.2. Practice Drop Test 2

During the drop, the bag moved laterally towards the back wall of the lift shaft, glancing off it near the bottom of the drop, with the bag landing behind the ledge, near the back wall. The bag lid partially opened on impact, but the IV bag was undamaged.

3.1.3. Practice Drop Test 3

This bag did not touch the side walls on the way down. The bag struck the front edge of the ledge at the bottom of the lift shaft before falling to the floor below and bouncing back onto the ledge. The taped bag lid did not open but the IV bag was ruptured inside and had leaked its contents, although no liquid escaped outside.

3.1.4. Lessons from Practice Drop Tests

The practice drop tests indicated that the bag's zip is its weakest point, with its teeth being forced open on impact. Zip-strengthening measures could be considered, such as a Velcro zip cover to keep the lid closed if the zip fails. If the zip is liable to fail, it would be desirable to ensure that the specimen packaging is leak-proof.

3.2. Main Drop Tests with the Drone

The main drop test results are described here, relating to shock and acceleration measurements, estimation of the time to reach terminal velocity, and crash performance. The drop times, from bag release at the drop height of 98 m to hitting the runway below, ranged between 5.69 s and 5.94 s, giving mean drop speeds ranging between 16.5 m/s and 17.2 m/s. In all cases, the IV bag ruptured on impact, splitting along its long edge and leaking into the packaging as described below. The frozen cold pack was undamaged in all cases.

3.2.1. Shock and Acceleration Measurements

The acceleration measurements during impact (shock) are illustrated for drop test 4 (Figure 10). This shows the measured acceleration values for the three orthogonal directions (x, y, z) and for the 'resultant acceleration', given by $\sqrt{(x^2 + y^2 + z^2)}$. The $\pm$200 g maximum limit, configured for the sensor, was reached in all three orthogonal directions from which it is concluded that the resultant acceleration during impact exceeded 346 g. Similar results and conclusions were obtained for drop test 1 and would be expected for the other drop tests as well (not measured).

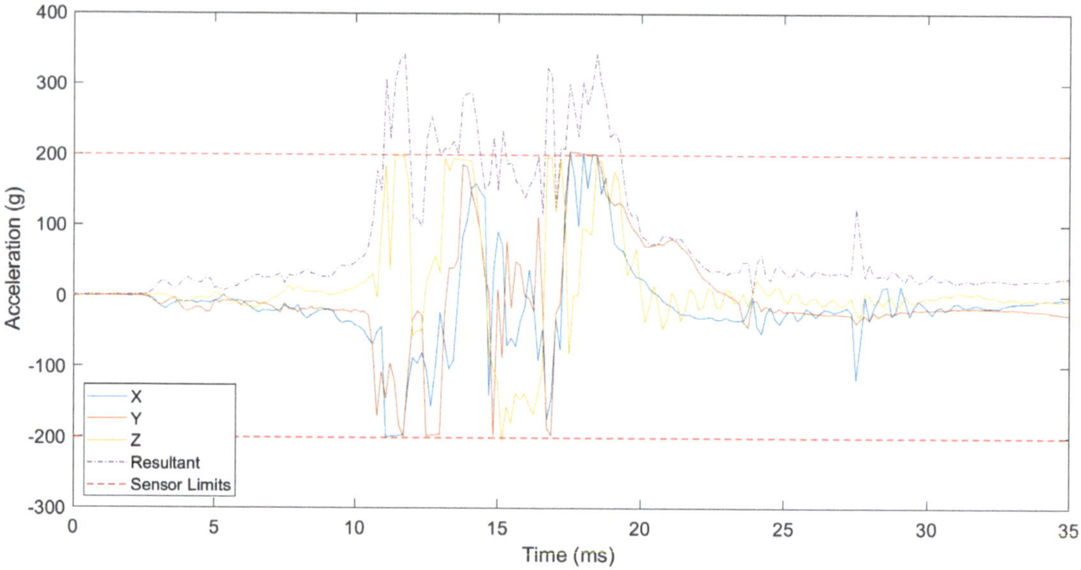

Figure 10. Acceleration levels during impact for drop test 4.

The resultant acceleration measurements recorded for drop tests 3 and 7 (Figure 11) were used to check whether the containers reached terminal velocity (zero acceleration) before hitting the ground, as required by the VCA's drop test procedure. These measurements were low-pass-filtered (smoothed) using a zero-phase 8th-order Butterworth filter [57] with a cut-off frequency of 5 Hz. The results indicated that zero acceleration was reached at 4.26 s for bag 3 and at 4.21 s for bag 7. As the drop times were 5.69 s and 5.68 s, respectively, the results confirmed that terminal velocity was attained for these drop tests and that the chosen drop height of 98 m, which had included a 10% safety margin (Appendix A), had been appropriate. For comparison, the measured acceleration curves are shown alongside the theoretical acceleration curve for the descent with the longest theoretical time to reach terminal velocity (5.71 s), where the container falls end face (smallest surface area, 0.077775 m^2) downwards, as explained in Appendix A. In practice, from the video footage of the drops, it was observed that the containers initially fell bottom side down (surface area 0.1173 m^2), this being the orientation at release, and then 'tumbled' (rotated slowly in varying directions) during the drops. The observed variations in the measured accelerations are assumed to be due to this tumbling motion during descent, resulting in angular accelerations and changes in the frontal area and hence drag forces.

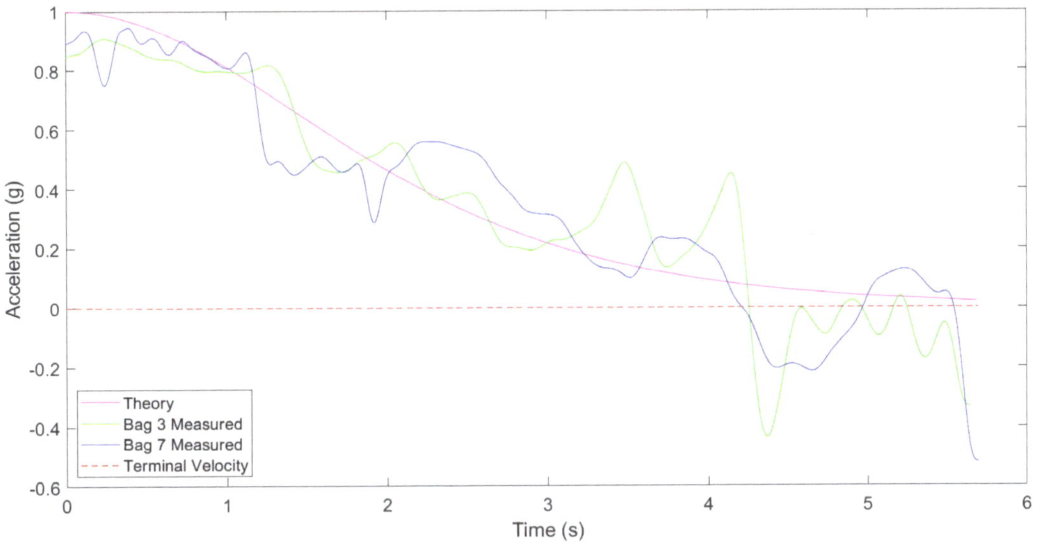

Figure 11. Measured and theoretical accelerations during drops.

3.2.2. Crash Performance

For drop tests 1 and 4, the added foam protections ('soft' and 'medium') did not prevent the IV bags from rupturing (Figure 12). The polythene freezer bag and polythene zip bag were both ineffective in preventing leakage into the container. Although both containers remained intact, without any tears or holes observed, they both leaked contents onto the runway and thus failed the test procedure. Examination of the containers revealed that liquid had escaped through the zip, suggesting that this may be an area for consideration in the design of a CPC.

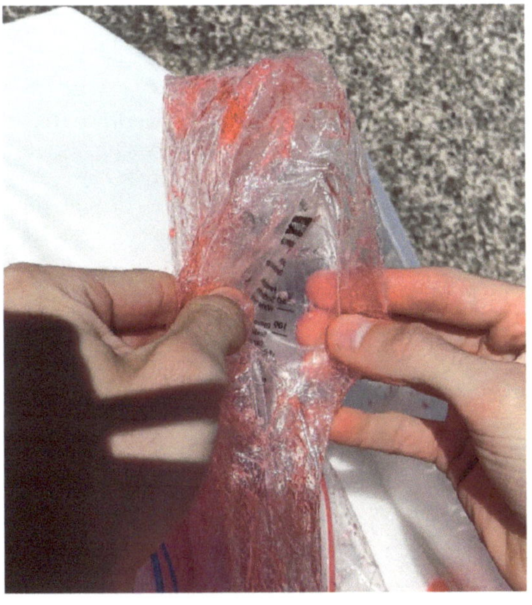

Figure 12. IV bag split down long edge (drop test 1).

In drop tests 3, 6, 7, and 8, the use of two leak-proof polythene liners was largely effective in containing the liquid that escaped from the ruptured IV bag in each case, aside from a few spots of liquid that escaped into one of the Versapaks (on drop test 7). In three of the four tests, liquid escaped from the inner liner into the outer liner due to the inner liner being punctured. In the other test, all the liquid remained inside the inner liner.

Some relatively minor deformations of each container were observed due to the landing impact. On one of the drop impacts, the zip was forced open, but the lid remained closed; the zip remained intact for the others. No liquid had escaped from any of the containers after the eight-hour observation period.

3.2.3. Lessons from Main Drops

Despite being dropped from twice the height of the practice drops, the Versapak containers were damaged less in the main drops (from a drone) than in the practice drops (down a lift shaft). This was likely due to landing on the flat surface of the airfield runway as compared to colliding with walls and landing on the edge of a ledge at the bottom of the lift shaft. Nevertheless, the bag zip was forced open in one of the main drop tests, which is a test failure despite no liquid contents escaping the container. As also observed in the practice drops, some strengthening and waterproofing of the zip would be needed for this type of container to pass the test requirements. General strengthening of the container may also be considered if the container is to be defined as 'rigid outer packaging', although it may be preferable to add a layer of rigid outer packaging to the contents instead (e.g., a cardboard box). The use of at least two 'leak-proof' liners seems necessary for this application as the inner liner was punctured on three of the drops. Further discussion of potential design refinements can be found in Section 4.5.

Gusts of wind of up to 7.2 m/s were experienced during the drop tests, which led to lateral bag movements (drifts) of up to 26 m. This made positioning the drone to land bags on the 50 m wide runway quite difficult to achieve, and, in one case, the test failed when the expected drift did not occur as the wind subsided when the bag was dropped, resulting in the bag landing on the grass to the side of the runway. Ideally, the tests would be undertaken in less windy conditions, but this can be difficult to achieve given the substantial operational logistics (e.g., booking of facilities, gaining permissions, number of people involved).

4. Discussion

In addition to the primary objective of testing the potential suitability of the standard medical carrier bag for carrying DG using a drone, the drop tests also provided useful information for further research and future testing, including how to estimate the terminal velocity and determine a suitable drop height. Here, we suggest improvements to the methods used and discuss how the current regulations (drop testing procedure) may affect the future testing needed to support the use of drones for carrying DG.

4.1. Sensors and Monitoring

Based on the measurements from the accelerometers during the drop, presented in Figure 11 alongside the theoretical estimates from Appendix A, there was good agreement between theory and practice in terms of the height needed and the time taken to reach terminal velocity for the given container dimensions and weight. This suggests that the prediction model (Appendix A) may be adequate for determining the drop height for other container sizes and weights. This prediction model could be improved through the measurement of the drag forces (e.g., in a wind tunnel) to determine the range of drag coefficients of the container at multiple orientations, as opposed to the assumed fixed value of 2.1, based on the approximation of the container as a cuboid. Validation of the model and additional improvements could be achieved through improvements in the measurements taken during the drop, such as recording the acceleration in the rotational degrees of freedom, in addition to the orthogonal directions, to provide greater insight into the tumbling movement of the container during drops. Better control and recording of the sensor orientation could also provide additional understanding. During this trial, acceleration sensors were secured within the internal pocket of the Versapak, which gave readings for the container acceleration. It may be of interest, in future research, to affix sensors to the specimen as well, especially where the specimen can move within the container.

To gain a better understanding of the severity of the impact, a sensor with a greater range could be used, as accelerations in this trial exceeded the 200 g range of the sensor used.

4.2. Regulatory Aspects

When testing a CPC using TI-compliant packaging (i.e., payload type (i) in Section 1.1), the test procedure states that any approval issued to a successfully tested CPC shall be limited to carriage of the specific packaging configuration tested. At present, with no CAA approval having been received yet for the container used in these drop tests, it is unclear how such an approval will be worded to specify what minor variations in packaging will be permitted, if any. Any approval will only cover payloads up to, but not exceeding, the tested payload mass. In this case, it may be advisable to undertake tests using the maximum expected payload mass. If feasible, it would be useful to have a categorisation of different packaging systems and rank them in terms of their crashworthiness so that it would not be necessary to test them all. If, for example, a lower-ranked packaging system successfully passed the test procedure, then approval could cover all higher-ranked packaging systems as well.

As medical logistics are often not routine, there may be many variations in the DG payloads to be transported, and it would not be feasible to test them all individually. If the test procedure approval is overly restrictive, it may severely limit the types of goods and packaging that can be used, whilst adding costs to the testing requirements. The test procedures offering the greatest freedom are those where the payloads are unpackaged solids or liquids poured directly into the container, as approval then covers any TI-compliant packaging configuration. The Versapak bags used in these tests are used in medical practice to transport a wide variety of types of samples, medicines, or equipment, most not DG. Non-DG items may have to be carried separately from DG, and the test procedure does not apply to them.

The drop test procedure for payload type (i) requires the outer packaging to be rigid, where this refers either to the package inside the container or the container itself. This requirement, stipulated in PI 650 (relevant to UN 3373 substances), appears to be appropriate for the operation of large, crewed airliners, where containers may be stacked on top of each other in the cargo hold and thus must be rigid. For the type of application here, where a single container is held underneath a drone, no stacking is involved, and rigidity of the container seems less appropriate, especially as it may not be a desirable property for a CPC if it does not provide good impact absorption. More drone-specific regulation concerning DG packaging would be desirable.

The latest drop test procedure (version 2) was published by the VCA in January 2024. This added some further information and clarifications, which were partly a result of the drop tests reported here and our discussions of these with the VCA. The added parts included advice to consult with the VCA, before testing, if unsure whether the proposed CPC design, contents, or drop height were acceptable; the statement that a set of three successful tests are required; and the inclusion of a full description of (a) the container design and construction, (b) all packaging used, (c) the impact area, (d) any terminal velocity calculations used.

4.3. Operational Aspects

Undertaking these drop tests was operationally challenging due to the potential risks involved with dropping an object from height, especially in windy conditions. With the container drifting laterally up to 26 m in these tests, accurately hitting a hardstanding area of limited width (50 m here) can be difficult and potentially dangerous. Ideally, drop tests should be made in less windy conditions, as stated in the test procedure, to ensure a safe landing in the desired area. The test requires landing on a hard flat surface, so landing elsewhere, such as on a grassy verge, would constitute a test failure. In practice, many different landing conditions could be envisaged, with differing impacts on a dropped CPC. Lower levels of damage would likely occur when landing on water or soft ground or where cushioned by vegetation, while greater levels of damage would likely occur if landing on an uneven or sharp surface such a building roof, wall, or fence, as was observed here when dropping bags down a lift shaft. The impact angle would also be a factor. The effect of the landing conditions on crash impacts can be considered as further research.

The drop tests were made from lower than the maximum permitted drone operating height (121.9 m) to minimise these risks; however, informal feedback has indicated that dropping from a lower height (98 m, here), based on terminal velocity calculations, should only be undertaken when dropping from the maximum operating height is impractical. A perceived concern here is the potential reaction from the public if tests are not made from the maximum operating height. Nevertheless, we would argue that dropping from the minimum height guaranteed to achieve terminal velocity is a valid and safer test method, especially if higher drone operating heights are permitted in future. In this case, guidance on suitable drop heights for different payload weights and container shapes and sizes would be useful, and the formulae presented in Appendix A may be useful as a basis for providing an online calculating tool.

Future designs of drone cargo systems may benefit from international standardisation to help reduce the needs for such tests, using technologies similar to the large commercial aircraft standardised air freight containers and pallets, known as 'unit load devices' [58].

If using a cold pack, its position relative to the IV bag may affect the damage to the IV bag, depending on how the bag lands. This has not been investigated here but could be of interest in further research.

An arguable weakness of the CPC drop test procedure is that it assumes the container falls to the ground in isolation, whereas a drone crash could involve the impact of the drone and container in combination. This situation is likely to alter the dynamics and magnitude of the forces on the container at impact (compared to container impact in isolation), due to factors such as different terminal velocities for drone/container combinations; powered flight into terrain at various velocities; or the size/mass/construction of the drone impacting the container as it hits the ground. That said, it would be impractical to insist on testing drone/container combinations in the CPC drop test procedure due to the damage to drones that would be incurred and the many different possible combinations that would need to be tested and approved. Instead, a pragmatic approach has been adopted in designing the CPC drop test procedure based on dropping the container in isolation, which is viewed by the experts at the VCA as adequate to ensure the integrity of containers in combination or isolation, although the veracity of this view has not been established exhaustively.

4.4. Subsequent Testing

Following the main drop tests reported in this paper, an opportunity arose to drop three Versapak containers, containing the same specimens, from the maximum permitted operating height (121.9 m) onto a concrete surface using a fixed-wing drone travelling at a speed of 33 m/s. These additional drops did not follow the test requirement of waiting eight hours to inspect the containers, but inspections immediately after crash landing indicated only minor surface damage (scratches) to the containers, with no liquid escaping from the inner liners into the containers.

4.5. Further Development of Medical Crash Protected Containers

The reported drop tests have indicated elements of the standard bag design that would require modifications to allow for their use as a CPC. In particular, the zip needs to be considered to ensure that it remains closed during and after impact. Consideration could also be given to using a waterproof zip to potentially limit or negate the need for the use of leak-proof liners in the packaging design. Consideration could also be given to increasing the rigidity of the container to reduce the amount of deformation it undergoes during impact which may reduce the likelihood of the zip failing. An additional consideration for the development of a medical crash-protected container could be to reduce the impact force by reducing the terminal velocity. This could be achieved through the incorporation of wings or fins to increase the drag coefficient of the bag. Structural dynamic models of the container could be developed to examine the impact behaviour, to further assist with strengthening the carrier design. CPC design should bear in mind that an increase in the mass of the carrier may result in a reduction in the payload capacity or in the range of the drone.

The implementation of any of these design considerations may increase the cost of the carrier in comparison with the current standard version. If so, this would likely result in the development of a specific (non-standard) medical CPC, as these changes would have limited or no benefit in business-as-usual ground transport, and users would not pay extra for these where not needed. In this case, the non-standard carrier (CPC) would only be purchased for specific use cases involving drone logistics.

5. Conclusions

Prior to this study, the crashworthiness and potential failure mechanisms of a standard medical carrier were not known. New experimental findings from a programme of drop tests provide useful empirical evidence that can guide the container manufacturer for the design of a modified version of the container for future testing, approval, and use as a CPC. These modifications relate to the container's strength and its water tightness, where its zipped lid was identified as its main weak point. Strengthening of the container may also allow it to qualify as the rigid outer packaging layer in a TI-compliant packaging configuration, although this may not be necessary where the transported goods already have a rigid outer packaging layer. Although liners can be used as part of the packaging to prevent leakage during testing, the test approval then only applies to that specific packaging configuration, which may be overly restrictive in practice. The research and development of a carrier bag that could be used for both general transport and for transporting dangerous goods using drones are a recommendation of this study, as this would be a highly desirable asset for healthcare services worldwide. The research has also provided useful insights into the practical test requirements and how to conduct the tests, as well as informing the VCA in amending the drop test procedure.

Author Contributions: Conceptualisation, T.C.; methodology, T.C., K.T., A.O., F.M., J.A. and A.M.; software, T.W.; validation, F.M., K.T., A.O. and J.D.; formal analysis, K.T., T.W. and F.M.; investigation, T.C., K.T., A.O., F.M., J.A. and A.M.; resources, T.C., K.T., A.O., F.M., J.A. and A.M.; data curation, A.O. and K.T.; writing—original draft preparation, F.M., M.G., K.T. and A.O.; writing—review and editing, T.C., T.W., J.A. and J.D.; visualisation, K.T.; supervision, T.C. and T.W.; project administration, T.C.; funding acquisition, T.C. All authors have read and agreed to the published version of the manuscript.

Funding: This research was funded by the UK Department for Transport, as part of the Solent Future Transport Zones project.

Data Availability Statement: The data and video footage created during this study were not publicly available at the time of writing for privacy reasons. They may be available on request later, subject to approval having been obtained.

Conflicts of Interest: Author Alex Murray was employed by the company Motion Robotics Limited. The remaining authors declare that the research was conducted in the absence of any commercial or financial relationships that could be construed as a potential conflict of interest.

Appendix A Determination of Drop Height

The terminal velocity (V_t) of a falling object is given using Formula (A1):

$$V_t = \sqrt{\frac{2mg}{\rho A C_d}} \tag{A1}$$

where m = mass (4.57 kg), g = acceleration due to gravity (9.81 m/s^2), ρ = density of air (1.225 kg/m^3), A = projected area of the object, C_d = drag coefficient [59]. A drag coefficient of 2.1 was assumed, based on the values given for different types of common shapes, where the container (Versapak) shape approximates that of a cuboid (rectangular box) [60]. The projected (cross-sectional) area of the container is dependent on how it falls and will vary as it tumbles. The minimum projected area is the area of the end face of the container, that is, 0.077775 m^2 (=0.305 m × 0.255 m), while the maximum is 0.1987 m^2 calculated using Formula (A2):

$$\sqrt{d^2\,h^2 + d^2 w^2 + h^2 w^2} \tag{A2}$$

where w, d, and h are the three dimensions of the container [61]. This gave a terminal velocity ranging between 13.24 m/s for the slowest possible descent (A = 0.1987 m^2) and 21.17 m/s for the fastest possible descent (A = 0.077775 m^2).

The relationship between velocity (v) and time (t) for a falling object is given using Formula (A3) [62]:

$$v = V_t \tanh\left(\frac{gt}{V_t}\right) \tag{A3}$$

Plotting this relationship for the fastest and slowest descents (Figure A1) indicated that the time to reach 99% of terminal velocity ranged between 3.57 s for the slowest possible descent and 5.71 s for the fastest possible descent. [Note: 'terminal velocity' is a theoretical upper limit that is never reached, hence the use of 99% here].

The height needed to achieve terminal velocity was calculated, via numerical integration, to range between 35 m, for the slowest possible descent, and 89 m, for the fastest possible descent (Figure A2). This led to the decision to undertake drop tests from a height of 98 m (=89 m + 10%), where a 10% added safety margin was used, as required by the drop test procedure [4].

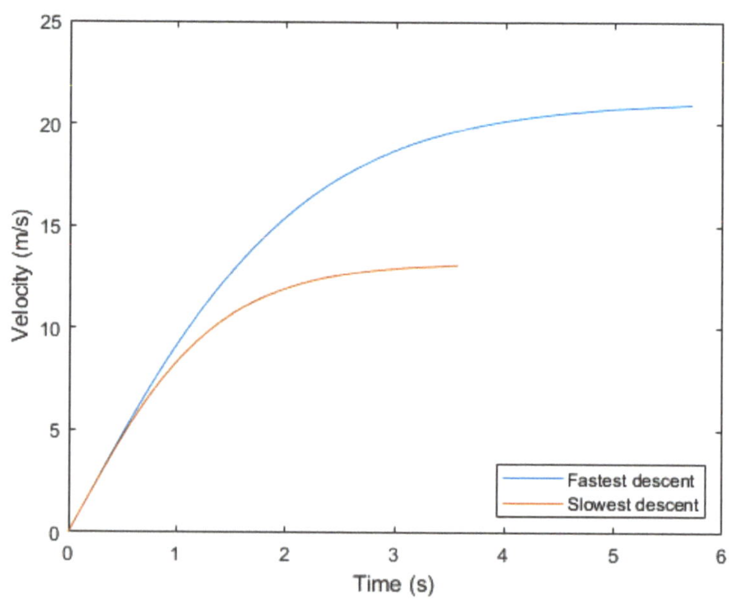

Figure A1. Velocity/time curves, with end point representing 99% of terminal velocity.

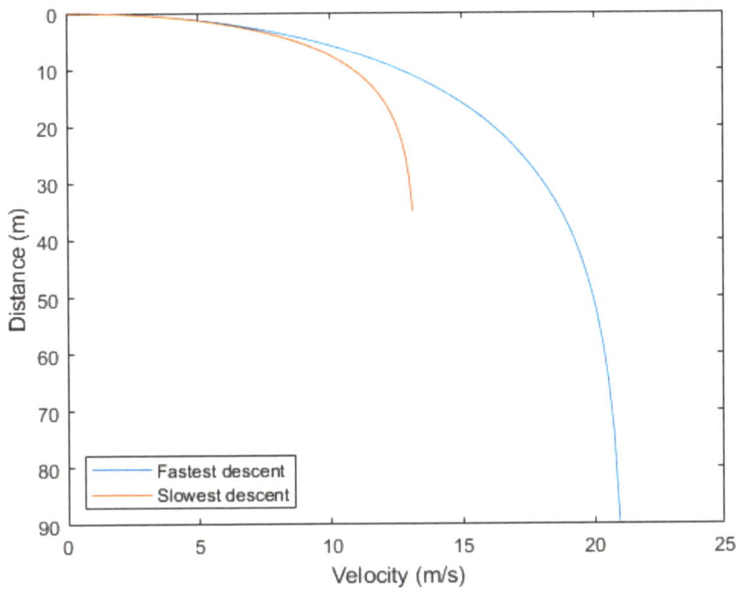

Figure A2. Distance (height)/velocity curves, with end point representing 99% of terminal velocity.

References

1. Sánchez, J.H.; Gouveia, S.; Cameselle, C. Transport of High-Risk Infectious Substances: Packaging for the Transport of Category A Infectious Specimens in Spain. *Int. J. Environ. Res. Public Health* **2022**, *19*, 12989. [CrossRef]
2. European Union Aviation Safety Agency. *Easy Access Rules for Unmanned Aircraft Systems (Regulations (EU) 2019/947 and (EU) 2019/945)*; European Union Aviation Safety Agency: Cologne, Germany, 2022.
3. Civil Aviation Authority. CAP2555: Guidance on the Carriage of Dangerous Goods as Cargo for UAS/RPAS Operators in the Specific Category. 2023. Available online: https://publicapps.caa.co.uk/modalapplication.aspx?appid=11&mode=detail&id=12257 (accessed on 9 February 2024).
4. Test Procedure (Crash Protected Containers for Dangerous Goods Carried by Remotely Piloted Aircraft Systems), Version 01. Available online: https://dc346c11-9b06-4f11-8326-4d9532cfa24c.usrfiles.com/ugd/dc346c_55a117d638ab4062a5531cef5fbb872b.pdf (accessed on 2 November 2023).
5. Vehicle Certification Agency. *Test Procedure (Crash Protected Containers for Dangerous Goods Carried by Remotely Piloted Aircraft Systems)*; Version 02; Vehicle Certification Agency: Leatherhead, UK, 2024.
6. Flemons, K.; Baylis, B.; Khan, A.Z.; Kirkpatrick, A.W.; Whitehead, K.; Moeini, S.; Schreiber, A.; Lapointe, S.; Ashoori, S.; Arif, M.; et al. The use of drones for the delivery of diagnostic test kits and medical supplies to remote First Nations communities during COVID-19. *Am. J. Infect. Control* **2022**, *50*, 849–856. [CrossRef] [PubMed]
7. Rana, K.; Praharaj, S.C.; Nanda, T. Unmanned Aerial Vehicles (UAVs): An Emerging Technology for Logistics. *Int. J. Bus. Manag. Invent.* **2016**, *5*, 86–92.
8. Poljak, M.; Šterbenc, A. Use of drones in clinical microbiology and infectious diseases: Current status, challenges and barriers. *Clin. Microbiol. Infect.* **2020**, *26*, 425–430. [CrossRef]
9. De Silvestri, S.; Capasso, P.J.; Gargiulo, A.; Molinari, S.; Sanna, A. Challenges for the Routine Application of Drones in Healthcare: A Scoping Review. *Drones* **2023**, *7*, 685. [CrossRef]
10. Aurambout, J.-P.; Gkoumas, K.; Ciuffo, B. Last mile delivery by drones: An estimation of viable market potential and access to citizens across European cities. *Eur. Transp. Res. Rev.* **2019**, *11*, 30. [CrossRef]
11. Kamat, A.; Shanker, S.; Barve, A.; Muduli, K.; Mangla, S.K.; Luthra, S. Uncovering interrelationships between barriers to unmanned aerial vehicles in humanitarian logistics. *Oper. Manag. Res.* **2022**, *15*, 1134–1160. [CrossRef]
12. Lin, C.A.; Shah, K.; Mauntel, L.C.C.; Shah, S.A. Drone delivery of medications: Review of the landscape and legal considerations. *Am. J. Health Syst. Pharm.* **2018**, *75*, 153–158. [CrossRef]
13. Rejeb, A.; Rejeb, K.; Simske, S.; Treiblmaier, H. Humanitarian Drones: A Review and Research Agenda. *Internet Things* **2021**, *16*, 100434. [CrossRef]
14. Sah, B.; Gupta, R.; Bani-Hani, D. Analysis of barriers to implement drone logistics. *Int. J. Logist. Res. Appl.* **2021**, *24*, 531–550. [CrossRef]
15. NHS Drone Delivery Trial. Available online: https://skyports.net/nhs-trials/ (accessed on 19 October 2023).

16. Wankmüller, C.; Kunovjanek, M.; Mayrgündter, S. Drones in emergency response–evidence from cross-border, multi-disciplinary usability tests. *Int. J. Disaster Risk Reduct.* **2021**, *65*, 102567. [CrossRef]
17. Leong, M.K.; Koay, K.Y. Towards a unified model of consumers' intentions to use drone food delivery services. *Int. J. Hosp. Manag.* **2023**, *113*, 103539. [CrossRef]
18. Kunze, O. Drones in Freight Transport. In *International Encyclopedia of Transportation*; Vickerman, R., Ed.; Elsevier: Oxford, UK, 2021; pp. 374–381.
19. Aydin, B. Public acceptance of drones: Knowledge, attitudes, and practice. *Technol. Soc.* **2019**, *59*, 101180. [CrossRef]
20. Comtet, H.E.; Johannessen, K.A. The Moderating Role of Pro-Innovative Leadership and Gender as an Enabler for Future Drone Transports in Healthcare Systems. *Int. J. Environ. Res. Public Health* **2021**, *18*, 2637. [CrossRef]
21. Sham, R.; Siau, C.S.; Tan, S.; Kiu, D.C.; Sabhi, H.; Thew, H.Z.; Selvachandran, G.; Quek, S.G.; Ahmad, N.; Ramli, M.H.M. Drone Usage for Medicine and Vaccine Delivery during the COVID-19 Pandemic: Attitude of Health Care Workers in Rural Medical Centres. *Drones* **2022**, *6*, 109. [CrossRef]
22. Kumar, R.; Singh, J.; Yadav, P.; Semwal, N.; Yadav, S.; Bhorey, A.; Bhaumik; Dhawan, H. Parachute deployment system for safe recovery of a drone. *Mater. Today Proc.* **2023**. [CrossRef]
23. Baldini, A.; Felicetti, R.; Ferracuti, F.; Freddi, A.; Iarlori, S.; Monteriù, A. Real-time propeller fault detection for multirotor drones based on vibration data analysis. *Eng. Appl. Artif. Intell.* **2023**, *123*, 106343. [CrossRef]
24. Krichen, M.; Adoni, W.Y.H.; Mihoub, A.; Alzahrani, M.Y.; Nahhal, T. Security Challenges for Drone Communications: Possible Threats, Attacks and Countermeasures. In Proceedings of the 2022 2nd International Conference of Smart Systems and Emerging Technologies (SMARTTECH), Riyadh, Saudi Arabia, 9–11 May 2022; pp. 184–189. [CrossRef]
25. Gianfelice, M.; Aboshosha, H.; Ghazal, T. Real-time Wind Predictions for Safe Drone Flights in Toronto. *Results Eng.* **2022**, *15*, 100534. [CrossRef]
26. Koh, C.H.; Low, K.H.; Li, L.; Zhao, Y.; Deng, C.; Tan, S.K.; Chen, Y.; Yeap, B.C.; Li, X. Weight threshold estimation of falling UAVs (Unmanned Aerial Vehicles) based on impact energy. *Transp. Res. Part C Emerg. Technol.* **2018**, *93*, 228–255. [CrossRef]
27. Svatý, Z.; Nouzovský, L.; Mičunek, T.; Frydrýn, M. Evaluation of the drone-human collision consequences. *Heliyon* **2022**, *8*, e11677. [CrossRef]
28. Lu, X.; Liu, X.; Li, Y.; Zhang, Y.; Zuo, H. Simulations of airborne collisions between drones and an aircraft windshield. *Aerosp. Sci. Technol.* **2020**, *98*, 105713. [CrossRef]
29. Pilko, A.; Sóbester, A.; Scanlan, J.P.; Ferraro, M. Spatiotemporal Ground Risk Mapping for Uncrewed Aircraft Systems Operations. *J. Aerosp. Inf. Syst.* **2023**, *20*, 126–139. [CrossRef]
30. Johannessen, K.A.; Wear, N.K.S.; Toska, K.; Hansbø, M.; Berg, J.P.; Fosse, E. Pathologic Blood Samples Tolerate Exposure to Vibration and High Turbulence in Simulated Drone Flights, but Plasma Samples Should be Centrifuged after Flight. *IEEE J. Transl. Eng. Health Med.* **2021**, *9*, 1–10. [CrossRef] [PubMed]
31. Oakey, A.; Waters, T.; Zhu, W.; Royall, P.G.; Cherrett, T.; Courtney, P.; Majoe, D.; Jelev, N. Quantifying the Effects of Vibration on Medicines in Transit Caused by Fixed-Wing and Multi-Copter Drones. *Drones* **2021**, *5*, 22. [CrossRef]
32. Ansari, K.I.B.H.; Shendge, A.; Pakhrani, K.; Singh, R. Design and development of an auto-inflatable airbag as the failsafe system of unmanned aerial vehicle. *Mater. Today Proc.* **2023**, *77*, 983–990. [CrossRef]
33. NASA Develops Technology to Help Drones Crash-Land Safely during Emergencies. Available online: https://geoawesomeness.com/nasa-develops-technology-help-drones-crash-land-safely-emergencies/ (accessed on 19 October 2023).
34. International Civil Aviation Organization. *Annex 18 to the Convention on International Civil Aviation-The Safe Transport of Dangerous Goods by Air*, 4th ed.; International Civil Aviation Organization: Montreal, QC, Canada, 2011.
35. International Civil Aviation Organization. *Technical Instructions for the Safe Transport of Dangerous Goods by Air (Doc 9284)*; International Civil Aviation Organization: Montreal, QC, Canada, 2023.
36. Grote, M.; Cherrett, T.; Oakey, A.; Royall, P.G.; Whalley, S.; Dickinson, J. How Do Dangerous Goods Regulations Apply to Uncrewed Aerial Vehicles Transporting Medical Cargos? *Drones* **2021**, *5*, 38. [CrossRef]
37. Grote, M.; Oakey, A.; Pilko, A.; Smith, A.; Cherrett, T. Drones-the scope for integration into multi-modal urban logistics services. In *The Routledge Handbook of Urban Logistics*; Monios, J., Budd, L., Ison, S., Eds.; Routledge: Abingdon, UK, 2023; pp. 72–90.
38. Civil Aviation Authority. CAP2383: Using Drones to Carry Dangerous Goods Using Crash Protected Containers, BEIS RPF2 Project Outcomes. 2022. Available online: https://publicapps.caa.co.uk/modalapplication.aspx?appid=11&mode=detail&id=11607 (accessed on 9 February 2024).
39. Shu, Z.; You, R.; Zhou, Y. Viscoelastic Materials for Structural Dampers: A Review. *Constr. Build. Mater.* **2022**, *342*, 127955. [CrossRef]
40. Selke, S.E.M.; Culter, J.D. Foams, Cushioning, and Distribution Packaging. In *Plastics Packaging*; Carl Hanser Verlag: Munich, Germany, 2016; pp. 339–352.
41. Mills, N.J. Chapter 12-Product packaging case study. In *Polymer Foams Handbook*; Mills, N.J., Ed.; Butterworth-Heinemann: Oxford, UK, 2007; pp. 281–306.
42. Data Sheets (Performance Curves). Available online: https://www.sorbothane.com/technical-data/technical-data-sheet/ (accessed on 19 October 2023).
43. Pulse (ABS-4U 19 inch Rack ABS Flight Case–4U). Available online: https://www.pulse-audio.co.uk/product/19-inch-rack-abs-flight-case-4u/ (accessed on 20 October 2023).

44. 19 Inch Rack Cases. Available online: https://flightcaseuk.com/ (accessed on 20 October 2023).
45. Viking Drone Packaging World's First to Be Certified as a Crash Protected Container. Available online: https://vikingpack.co.uk/viking-drone-packaging-worlds-first-to-be-certified-as-a-crash-protected-container/ (accessed on 20 October 2023).
46. Crash Protected Package for Transport by Remotely Piloted Aircraft Systems. Available online: https://patentscope.wipo.int/search/en/detail.jsf?docId=WO2023014721&_cid=P11-LNENHB-26244-1 (accessed on 20 October 2023).
47. Versapak (Insulated Pathology Sample Carrier (Medium)). Available online: https://www.versapak.co.uk/new-medium-insulated-medical-carrier-pathology (accessed on 20 October 2023).
48. Sodium Chloride (Saline) Bag 500 mL. Available online: https://www.henryschein.co.uk/gb-en/medical-gb/p/pharmaceuticals/solvents-and-solutions/sodium-chloride-saline-bag-500ml/pha258 (accessed on 23 August 2023).
49. Briquettes (Cold Packs). Available online: https://www.emballiso.com/en/products-packaging-isothermal/accumulators-cold-industry-transport (accessed on 20 October 2023).
50. What is HDPE? Available online: https://www.acmeplastics.com/what-is-hdpe (accessed on 20 October 2023).
51. Oakey, A. Investigating the Scope for Integrating Uncrewed Aerial Vehicles into Mixed-Mode Fleets to Support National Health Service Logistics Operations. Ph.D. Thesis, University of Southampton, Southampton, UK, 28 September 2023.
52. GB Foam Direct. Available online: https://www.gbfoamdirect.co.uk/ (accessed on 20 October 2023).
53. Anti Spillage Liner for Medical Carriers. Available online: https://www.versapak.co.uk/anti-spillage-liner-for-medical-carriers (accessed on 20 October 2023).
54. Motion Robotics. Available online: https://www.motion-robotics.co.uk/ (accessed on 20 October 2023).
55. Transport Shock Data Logger MSR175. Available online: https://www.msr.ch/en/product/transportation-shock-data-logger-msr175/ (accessed on 20 October 2023).
56. Axivity AX6. Available online: https://axivity.com/product/ax6 (accessed on 20 October 2023).
57. Butterworth Filter Design. Available online: https://www.electronics-tutorials.ws/filter/filter_8.html (accessed on 20 October 2023).
58. Döppner, D.A.; Schoder, D.; Siejka, H. Big Data and the Data Value Chain: Translating Insights from Business Analytics into Actionable Results-The Case of Unit Load Device (ULD) Management in the Air Cargo Industry. In Proceedings of the European Conference on Information Systems, Research-in-Progress Papers, Paper 7, Münster, Germany, 26–29 May 2015. Available online: https://aisel.aisnet.org/ecis2015_rip/7 (accessed on 7 December 2023).
59. Drag Force and Terminal Speed. Available online: https://openstax.org/books/university-physics-volume-1/pages/6-4-drag-force-and-terminal-speed (accessed on 20 October 2023).
60. Drag Coefficient. Available online: https://www.engineeringtoolbox.com/drag-coefficient-d_627.html (accessed on 20 October 2023).
61. Projected Area of a Cuboid. Available online: https://demonstrations.wolfram.com/ProjectedAreaOfACuboid/ (accessed on 20 October 2023).
62. Newton's 2nd Law, Drag. Available online: http://electron6.phys.utk.edu/PhysicsProblems/Mechanics/1-Simple%20point%20particles/2nd%20law-drag.html (accessed on 20 October 2023).

energies

Article

Operational Analysis of an Axial and Solid Double-Pole Configuration in a Permanent Magnet Flux-Switching Generator

Manuel Garcia Neto [1], Francisco Ferreira da Silva [2] and Paulo José da Costa Branco [2,*]

[1] Department of Electrical and Computer Engineering (DEEC), Instituto Superior Técnico, University of Lisbon, 1049 Lisbon, Portugal; manuel.neto@tecnico.ulisboa.pt

[2] Mechanical Engineering Institute (IDMEC), Instituto Superior Técnico, University of Lisbon, 1049 Lisbon, Portugal; francisco.ferreira.silva@tecnico.ulisboa.pt

* Correspondence: pbranco@tecnico.ulisboa.pt

Abstract: There are two main beneficial characteristics that doubly salient permanent magnet (PM) electrical machines present for aircraft applications: armature windings and PMs excitation sources placed on the stator side (maintenance and thermal management), and having a clear-cut rotor without PMs or excitation windings (vulnerable at high speeds due to associated centripetal mechanical stresses). Within this framework, a doubly salient permanent magnet (DSPM) generator was conceived by optimizing the stator size and rotor structure to minimize the torque ripple and maximize the root-mean-square (RMS) voltage value per turn of each generator phase. Firstly, a comparison between the 2D and 3D finite element method (FEM) models is made considering the results of 3D finite element analysis (FEA) as our benchmark in order to understand the accuracy of the 2D results against our benchmark model, the 3D one. A multi-objective design strategy based on a 2D FEA is made, it is set to have characteristics closest to optimal for a Boeing 767 turbine, that is, the necessary electromotive force for a required power of 90 kW at 3000 rpm, feeding a simplified Boeing 767 electrical power distribution system. The results show that the machine could not deliver the required power at 3000 rpm since the 2D FEA demonstrates that the 2D model gives optimistic results when compared with the 3D FEM model. However, with a 3D FEA of the machine feeding the aircraft load, it was seen that the machine's efficiency is 92%, suggesting that this machine can be a plausible solution.

Keywords: switched flux generator; oscillatory torque; linked flux; torque density

Citation: Neto, M.; da Silva, F.F.; Branco, P.J.d.C. Operational Analysis of an Axial and Solid Double-Pole Configuration in a Permanent Magnet Flux-Switching Generator. *Energies* **2024**, *17*, 1698. https://doi.org/10.3390/en17071698

Academic Editors: Artur Kierzkowski, Tomasz Kisiel, Agnieszka A. Tubis, Franciszek Restel, Mateusz Zając, Tomasz Nowakowski and Anna Jodejko-Pietruczuk

Received: 28 February 2024
Revised: 28 March 2024
Accepted: 29 March 2024
Published: 2 April 2024

1. Introduction

Some generators of aircraft power generation systems are synchronous machines with permanent magnets (PMs) [1,2]. A higher power density and efficiency and being more compact and lighter due to the lack of any slip-ring-based external excitation are some advantages [3]. However, one emerging machine topology is the flux-switching machine, or, more specifically, doubly salient machine, due to its higher reliability at higher rotor speeds due to its simple rotor design [4]. This mechanical characteristic is the main reason for studying a DSPM topology generator suitable for integration into aircraft fuselage since it must endure all atmospheric conditions with the utmost reliability. The electromechanical torque will be our study benchmark for the optimization process to obtain the highest possible output power of the generator, optimizing only the 2D geometry of the generator. Nevertheless, other points of the study are the reduction in the vibrations and electromagnetic forces transmitted to PMs due to being placed in a stator iron core [5] and the reduction in the temperature decaying flux effect in PMs due to the better temperature dissipation [6].

Still, the current aircraft electrical generation systems can be separated into two kinds of technology: variable-speed constant-frequency systems and constant-speed constant-frequency systems [1,7]. The integrated drive generator (IDG) is commonly denominated as a constant-speed constant-frequency system. Most of the civil aircraft use this technology [7], which comprises a continuously set variable-speed (CVT) gearbox and generator. The integration of a CVT becomes a useful resource in this area since it gives the possibility to offer solutions with constant-frequency AC generation systems [8]. A lower reliability and efficiency and higher weight are the main drawbacks of [1,8]. The growing demand for the electrification of aircraft motivates the development of new solutions. A new trend in research involves exploring the potential of integrating generators within the structure of aircraft turbines, particularly for high-range aircraft, known as aircraft embedded generation systems. In the work of [2], it was verified that, inside the low-pressure turbine casing, there exists an enclosed space capable of accommodating a generator without compromising the aerodynamics of the turbine. Some advantages over the IDG can be depicted: being lighter, more efficient and reliable, less costly, and having easier integration on aircraft fuselage systems. However, the increase in the complexity of designing the turbines can be one drawback.

Consequently, the research aim is to integrate the DSPM generator into the aircraft fuselage, assuming an IDG solution since it is a mature technology. Nevertheless, the possibility of integrating the generator inside the aircraft turbine as an embedded generation solution is investigated.

2. Doubly Salient Permanent Magnet (DSPM) Machine and Methodology of Study Overview

The starting point of this project is the DSPM obtained in [9], which is presented in Figure 1. This machine is now going to be called the DSPM S4R5. Also, the methodology used to study the different DSPM geometries will be reviewed. The finite element analysis (FEA) will be our benchmark study of this machine using a distributed-parameter simulation environment finite element method (FEM) program. More specifically, for all multiphysics 2D and 3D FEM simulations, the software used was Comsol Multiphysics version 6.1. So, it will be depicted to understand how the FEM models are built. Finally, an overview of the optimization process is presented.

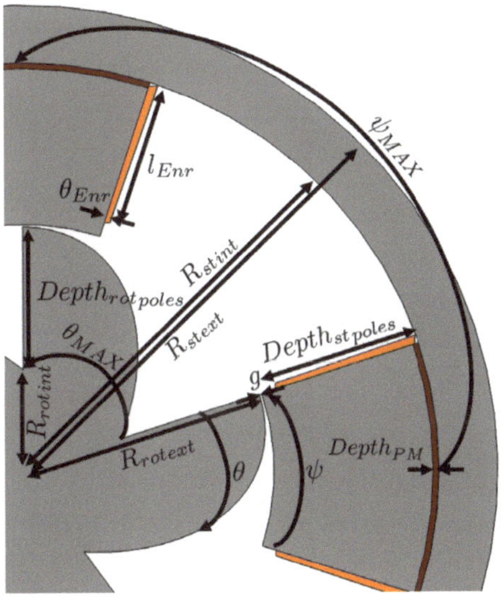

Figure 1. Presentation of the measures relating to the 3D geometry of the initial concept using a 2D cut plane.

2.1. Geometry

Figure 1 shows the parameters relative to the 2D plot of the DSPM S4R5 geometry. Table 1 presents the machine dimensions of Figure 1. $Depth_{DSPM}$ corresponds to the active length of the machine or stack iron length. The 3D plot is shown in the next section.

Table 1. Geometric parameter values.

Parameter	Description	Dimension
$R_{st_{ext}}$	External radius of ring stator	324 [mm]
$R_{st_{int}}$	Internal radius of ring stator	284 [mm]
$R_{rot_{ext}}$	External radius of rotor	170 [mm]
$R_{rot_{int}}$	Internal radius of rotor	70 [mm]
$Depth_{st_{poles}}$	Depth of stator poles	112.4 [mm]
$Depth_{rot_{poles}}$	Depth of rotor poles	100 [mm]
$Depth_{PM}$	Depth of PMs	4 [mm]
g	Airgap distance	1.6 [mm]
θ	Rotor pole angle	36°
ψ	Stator pole angle	36°
θ_{MAX}	Rotor pole pitch	$\frac{360°}{Rot_{np}} = 72°$
ψ_{MAX}	Stator pole pitch	$\frac{360°}{Est_{np}} = 90°$
l_{Enr}	Winding length	100 [mm]
θ_{Enr}	Winding aperture angle	0.04 [rad]
$Depth_{DSPM}$	DSPM total length	80 [mm]

The main geometric feature of the machine is the shape of the rotor teeth, which is given by a quadratic Bézier curve. In a Bézier curve, the initial and final points belong to it, while the intermediate control points (for a quadratic curve, there is only one intermediate control point) determine the curvature of the curve [10]. The minimum airgap value considered is 1.6 mm as a minimum mechanical safety margin. The coils have 20 turns each for the initial geometry's number of turns (N_t). Nevertheless, during the geometry optimization process, one turn is assumed for the sake of simplicity.

2.2. Materials

The materials corresponding to the different components of the machine are presented in Figure 2. The PMs are highlighted in Figure 2 by zooming in on one of them.

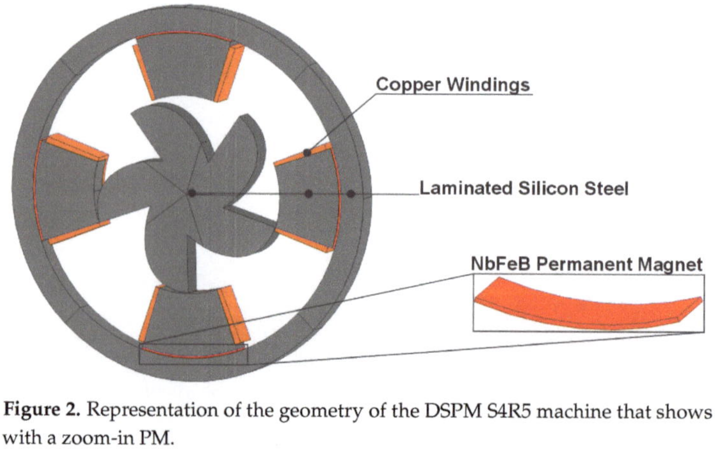

Figure 2. Representation of the geometry of the DSPM S4R5 machine that shows the materials used with a zoom-in PM.

For the iron core, Silicon Steel NGO 35JN200, which has a 0.35 mm lamination thickness, is considered. Its density (ρ_{Fe}) is 7600 kg/m^3, with a B-H characteristic curve presented in Figure 3 as assumed in [9] and its electrical conductivity (σ_{Fe}) of 1 S/m being considered as isotropic conductivity to reduce computational time.

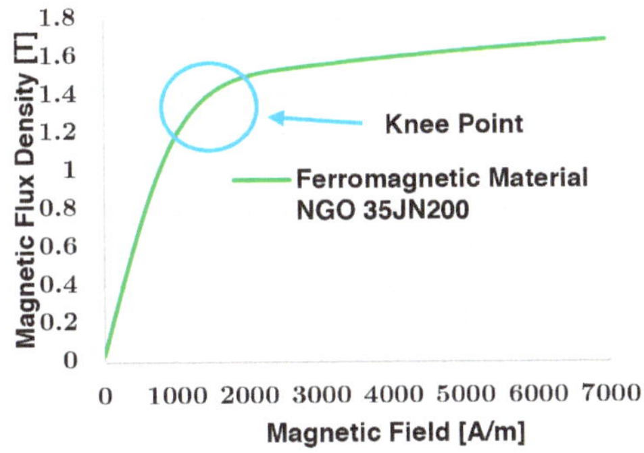

Figure 3. B-H curve of Silicon Steel NGO 35JN200.

The NdFeB magnets present a B-H curve given by a slope with a remanent flux density ($B_{r_{PM}}$) of 1.4 T with a coercive magnetic field of 1.08 MA/m. Their relative magnetic permeability (μ_{PM}) is around 1.03. The electrical conductivity of the PM is assumed to have a value of 1 S/m to mitigate simulation errors of induced currents. Also, the density of the PMs (ρ_{PM}) is 7600 kg/m^3.

The copper windings are considered to have a relative magnetic permeability and relative electrical permittivity similar to the air. However, their electrical conductivity (σ_{copper}) has a value of 60 MS/m. Figure 4 shows the coil numbering and the permanent magnet orientation.

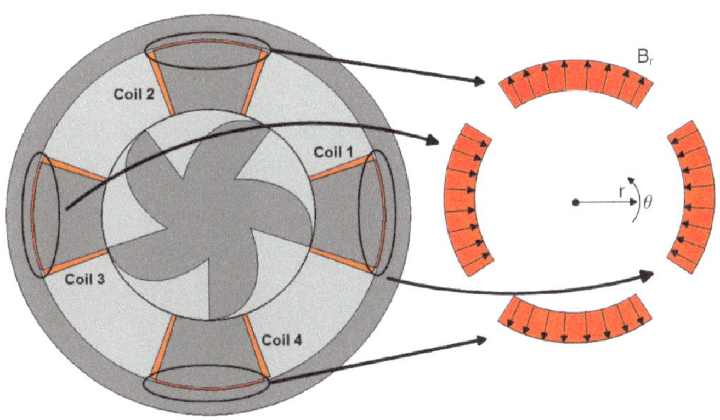

Figure 4. Two-dimensional plane of DSPM S4R5 for illustration of PM flux direction and coil numbering.

2.3. Generator Specifications

The pretended specifications for this aircraft are based on [9,11]. So, the needed electrical apparent power output is at least 90 kVA at a speed of 3000 rpm. As detailed later, values such as phase voltage and current will depend on the final design achieved and the AC/DC/AC power converters feeding the electric load typical of a 767 aircraft.

The aircraft load is simulated as a load fed by an AC/DC/AC converter. Therefore, for simplification, the rectifier (AC/DC converter) is considered to have properly dimensioned LC filters to have a two-level square current wave as phase current during the optimization process.

The assumed value of nominal current density for the coils is 5 [A/mm^2], with a filling factor of 0.5 to ensure the thermal limits and a normal slot factor [12].

2.4. FEM Study of DSPM S4R5 Machine

The equations that govern the FEM models are presented in Equations (1) and (2), where $\vec{H}$ is the magnetic field vector, $\vec{B}$ is the magnetic flux density vector, and $\vec{J}$ is the current density vector.

$$\begin{aligned} \text{Formulation } V_m : \nabla \times \vec{H} = 0 &\implies \vec{H} = -\nabla V_m \\ \vec{B} = \mu\vec{H} &\implies \vec{B} = -\mu\nabla V_m \\ \nabla \cdot \vec{B} = 0 &\implies -\nabla \cdot (\mu\nabla V_m) = 0 \end{aligned} \tag{1}$$

$$\begin{aligned} \text{Formulation } \vec{A} : \nabla \cdot \vec{B} = 0 &\implies \nabla \times \vec{A} = \vec{B} \\ \vec{H} = \frac{1}{\mu}\vec{B} &\implies \vec{H} = -\frac{1}{\mu}\nabla \times \vec{A} \\ \nabla \times \vec{H} = \vec{J} &\implies \nabla \times (-\frac{1}{\mu}\nabla \times \vec{A}) = \vec{J} \end{aligned} \tag{2}$$

The magnetic vector potential equations are considered for the stator side since currents are applied to the coils, and the magnetic scalar potential equations are neglected on the rotor side since the eddy currents in the rotor are neglected.

2.4.1. Magnetic Flux and Torque Calculation

The study of magnetic flux (Ψ) and torque (T) is relevant since these components relate to the output power, input power, and mechanical vibrations, respectively. Therefore, the study process of these quantities considers an angular variation in the rotor position of θ_{MAX}, which corresponds to an electrical cycle.

Magnetic Flux Calculation

For the 3D model, an FEA is performed assuming a stationary condition, while, for 2D geometries, the FEA considers a time-dependent study since our work verified that the time-dependent study is faster than a stationary one in this condition.

The induced voltage calculation in a stationary situation for 3D geometries is obtained by running the gradient of flux to the position of the rotor since the rotor speed is assumed to be constant. This avoids a time-consuming, time-dependent study.

Electromagnetic Torque Calculation: Arkkio's Method

The software Comsol 6.1 can perform torque computation through Maxwell's stress tensor or by Arkkio's method [13], characterized by a better precision than normal Maxwell's stress tensor. Like magnetic flux computation, the study of torque (T) will be time-dependent for 2D geometry and stationary for 3D geometry, remembering that the rotor rotates an electrical cycle.

2.4.2. Iron Losses Determination

Modeling of Iron Losses

The iron losses model assumed builds on Bertotti's model, where Equation (3) presents the equation that governs the model. The variables P_{Fe}, f, B_p, K_h, K_f, and K_{exc} are the iron losses [W], the frequency [Hz], the peak value of the signal of the magnetic flux density [T], and the dimensionless coefficients of hysteresis, eddy current, and excess losses, respectively.

$$P_{Fe} = K_h f B_p^2 + K_f f^2 B_p^2 + K_{exc} f^{1.5} B_p^{1.5} \tag{3}$$

Iron Losses' Data Map

The majority of power loss data of the Silicon Steel 35JN200 is extracted from the power loss manufacturer's table. However, the remaining missing data are obtained by extrapolating the manufacturer's data by fitting those data with the iron losses model Equation (3). The Berttoti coefficients are obtained by the data fitting for different frequencies. Figure 5 shows 3D graphs to better visualize the values of losses in iron with the level of induction, as well as the frequency relative to the extrapolation of the range of the most significant harmonic components of the magnetic flux density evolution.

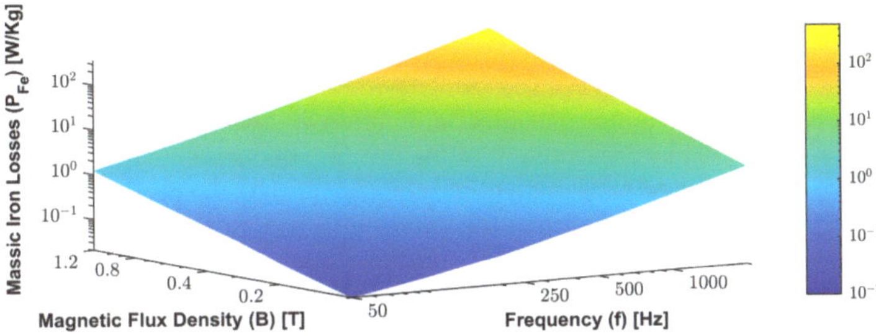

Figure 5. Iron losses 3D graphic representation for the frequency and magnetic flux density values range where the most significant harmonics will be present.

Determination of Iron Losses

As a first step, acquiring $\vec{B}$ in the entire mesh and assuming two electrical cycles is necessary for accurate fast Fourier transform (FFT) computation. With those harmonic values of magnetic flux density, it is possible to obtain the massic losses in the machine's core by interpolating those values with the power loss data shown in Figure 5. Moreover, the data obtained from FEM software also provide two other matrices: the identification of the element nodes and their geometric coordinates. This allows us to compute the volume of each element for either 2D geometries or 3D geometries. After obtaining the total volume of each element, knowing the value of massic losses in each element and its density, the total value of iron losses is given by Equation (4), where k is relative to the mesh elements. P_{Fe_t}, P_{Fe_k}, Vol_{t_k}, ρ_{Fe} are the iron losses in the core [W], the massic iron losses in each mesh element [W/kg], the volume of each element [m^3], and the density of the material that corresponds the element [kg/m^3], respectively.

$$P_{Fe_t} = \sum_{k=1}^{n} P_{Fe_k} \times \text{Vol}_{t_k} \times \rho_{Fe} \quad (4)$$

2.4.3. Electrical Equivalent Circuit Study

The equivalent electric circuit is based on a first-order state-space model as shown in Equation (5), where the state variable is the phase current. The matrices $[R]$, $[L]$, $[N]$, $[v]$, and $[i]$ are the resistance matrix, inductance matrix, $[N]$ matrix, voltage vector, and current vector, respectively. Those first three matrices are represented in Equation (6). For the sake of simplicity, the phase current is considered the output variable, and the rotor speed (ω) and voltage are considered the input variables. This analysis considers one phase per pole, so a four-phase system.

$$[v] = -[R][i] - [L][\dot{i}] + [N]\omega \quad (5)$$

$$[R] = \begin{bmatrix} R_s & 0 & \dots & \dots \\ 0 & Rs & 0 & \dots \\ \dots & 0 & \dots & \dots \\ \dots & \dots & \dots & Rs \end{bmatrix}$$
$$[L] = \begin{bmatrix} L_{11} & M_{12} & \dots & M_{1n} \\ M_{21} & L_{22} & M_{23} & \dots \\ \dots & M_{32} & \dots & \dots \\ M_{n1} & \dots & \dots & L_{nn} \end{bmatrix} [N] = \begin{bmatrix} N_1 \\ N_2 \\ \vdots \\ N_n \end{bmatrix} \tag{6}$$

Firstly, we start to determine the parameters of the equivalent circuit model using the FEM model. The FEM simulation software determines the resistive value of the windings (R_s) by the quotient of the product between the number of turns in the winding (N_t) and the average length of the coil ($l_{winding}$) by the product between σ_{copper} and the cross-section area of the wire of the winding (S_{turn}). Recalling the axial symmetry in 3D geometry, the resistive value in 3D is multiplied by two.

Using FEA results, the inductance and N maps are obtained by the gradient of each phase's linked flux to the current and rotor position, respectively. However, taking into account that the study is based on only the phase current of coil 1, the inductance maps obtained will reflect one phase, phase 1, which corresponds to the first column of $[L]$. The column vector $[N]$ is given by the derivative of the linked flux of the coils to the rotor position. Equation (7) shows $[L]$ and $[N]$ matrices.

$$\begin{cases} L_{11} = \frac{\partial \Psi_1}{\partial i_1} \\ L_{21} = \frac{\partial \Psi_2}{\partial i_1} \\ L_{31} = \frac{\partial \Psi_3}{\partial i_1} \\ L_{41} = \frac{\partial \Psi_4}{\partial i_1} \end{cases} \quad \begin{cases} N_1 = \frac{\partial \Psi_1}{\partial \theta_r} \\ N_2 = \frac{\partial \Psi_2}{\partial \theta_r} \\ N_3 = \frac{\partial \Psi_3}{\partial \theta_r} \\ N_4 = \frac{\partial \Psi_4}{\partial \theta_r} \end{cases} \tag{7}$$

Afterward, since there is symmetry between electrical cycles, only one electric cycle was simulated using a stationary study. Visual observation of the linked flux evolution of phase 1 indicates that the maximum and minimum values of linked flux correspond to the location of the d and q axes, respectively. To corroborate the location of the *dq* axis, one sets out the value of B_r to 1.4 T and makes the stationary study once again to obtain torque evolution since the electromagnetic torque is higher when the rotor is aligned with the d-axis than when it is aligned with the q-axis.

2.5. Geometry Optimization

2.5.1. Optimization Objectives

The objective functions considered are the minimization of useless torque ($f_1(x) = Min(T_{useless})$) and the maximization of torque density ($f_2(x) = Max(\frac{\overline{T}}{m})$). Since our study focuses on maximizing the output electric power and is made at stationary conditions, the average value of input electromechanical torque is a good gauge for maximization. However, without belittling, minimizing the vibrations transmitted to the PMs is also considered to diminish the oscillatory component of the electromechanical torque. These objectives are shown in Equation (8).

$$f(x) = \begin{cases} Min(T_{useless\,RMS}), T_{useless} = (T - \overline{T}) \\ Max(\frac{\overline{T}}{m}), m \rightarrow \text{DSPM S4R5 machine mass} \end{cases} \tag{8}$$

2.5.2. Design Variables

This thesis will study the effect of geometric dimensions in electromechanical torque. So, to maximize the freedom degree of the optimization, it will consider almost all geometric independent variables as decision variables as exposed in Figure 1, which are R_{stext}, R_{rotext}, $Depth_{st_{poles}}$, $Depth_{rot_{poles}}$, $Depth_{PM}$, g, θ, ψ, l_{Enr}, θ_{Enr}, $Weight_{PC_A}$, $Weight_{PC_B}$, $Weight_{PC_C}$, r_{PC_B}, and θ_{PC_B}.

2.5.3. Constraints

According to Figure 1, it is possible to observe that there are geometrical constraints that cannot be violated so that there are no intersections between the different DSPM machine elements. Therefore, the geometrical constraints are enumerated below.

1. $R_{st_{ext}} - (Depth_{st_{poles}} + g + R_{rot_{ext}}) - 8\ [\text{mm}] > 0$
2. $Est_{ext} - (Depth_{st_{poles}} + g + R_{rot_{ext}}) - 40\ [\text{mm}] < 0$
3. $l_{Enr} - Depth_{st_{poles}} < 0$
4. $\psi_{Max} - \psi - 2\theta_{Enr} \geq 0$
5. $r_{PC_B} - R_{rot_{ext}} > 0$
6. $r_{PC_B} - (R_{rot_{ext}} + g + Depth_{st_{poles}} - Depth_{PM}) < 0$

The limits of the design variables are presented in Table 2.

Table 2. Lower and upper limits of design variables used to optimize DSPM S4R5 machine.

Design Variable	Lower and Upper Limits
R_{stext}	240–290 [mm]
R_{rotext}	135–155 [mm]
$Depth_{st_{poles}}$	70–90 [mm]
$Depth_{rot_{poles}}$	50–110 [mm]
$Depth_{PM}$	4–50 [mm]
g	1.95–2.2 [mm]
θ_{Enr}	0.05–0.45 [rad]
l_{Enr}	55–70 [mm]
ψ	0.25–0.65 [rad]
θ	0.5–0.8 [rad]
r_{PC_B}	170–220 [mm]
θ_{PC_B}	−0.1–0.1 [rad]
$Weight_{PC_A}$	0.2–0.6
$Weight_{PC_B}$	0.2–0.7
$Weight_{PC_C}$	0.4–1

2.5.4. Optimization Program: Genetic Algorithm

To perform the optimization, the NSGA-II (Non-Sorting Dominant Genetic Algorithm) algorithm is used, which is based on the basic idea of the evolution of populations throughout generations, considering a sufficient number of generations to acquire the Pareto front or curve [14], where, next, the optimization algorithm is briefly described.

Considering the operation principle of the NSGA-II algorithm, the optimization process starts evaluating the constraints violated to study only the valid geometries. The admissible geometries are studied in a stationary study of an electrical cycle, and its results are sorted as described previously to make the next generation. This process is repeated to the next generations until the Pareto curve converges. The number of generations and the number of individuals considered are at least 150 [15] and 75, respectively. Figure 6 illustrates a flowchart of the whole process considering the NSGA-II algorithm and electromagnetic study algorithm to visualize the parallelism of the processes in the optimization.

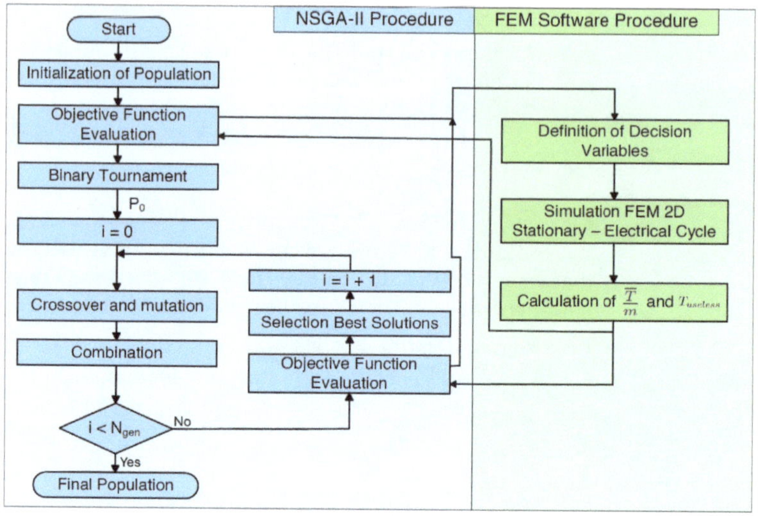

Figure 6. NSGA-II and FEM procedure for DSPM S4R5 machine optimization. Adapted from [16].

3. Results

3.1. Comparison of Flux and Torque Calculations Based on 2D and 3D Geometries at No Load and Lumped Parameters Study

Figure 7 illustrates, on the left side, the DSPM S4R5 and, on the right side, the axial symmetry of the DSPM S4R5 model used to reduce computational time. The PMs are represented in blue to highlight their visualization. Also, Figure 8 shows the corresponding 2D geometry.

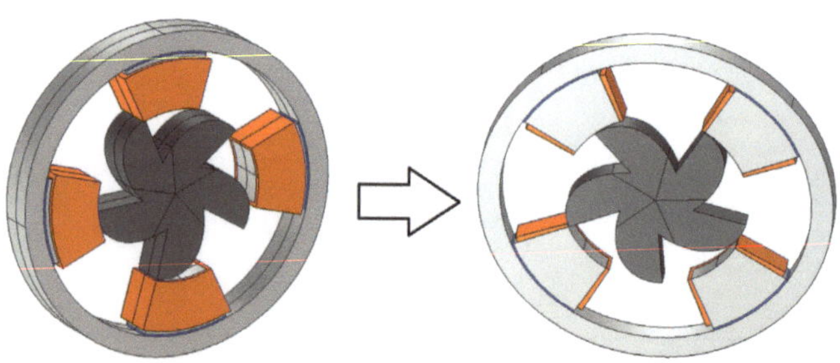

Figure 7. Illustration of 3D complete geometry and a half-cross-cut plane of 3D entire geometry.

Figure 8. Two-dimensional geometric representation of DSPM S4R5.

Figure 9 shows that the 3D geometry has a higher average and lower peak-to-peak amplitude values of linked flux than its 2D counterpart. As 3D geometry presents lower magnetic reluctance from the standpoints of the PMs, this means a rise in their magnetic load and, in turn, in their total magnetic flux. The constraints imposed on leakage paths by the 2D geometry compel the magnetic flux to cross from the stator into the rotor. In short, as 2D geometry gives more optimistic results in output voltage than 3D geometry, 2D geometry can only be used in the preliminary phase of designing machines with large diameter–length ratios.

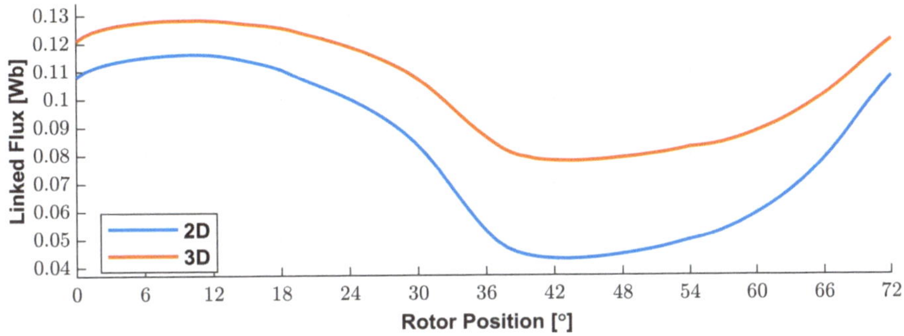

Figure 9. Linked flux results for 2D geometry and 3D geometry.

The linked flux evolution, as an image of the induced voltage, is studied to analyze the voltage quality. So, the computation of the weighted total harmonic distortion (WTHD) of linked flux is performed as shown in Figure 9. The WTHD is given by Equation (9) [17].

$$\mathrm{WTHD} = \frac{\sqrt{\sum(\frac{V_n}{n})^2}}{V_1}, n \rightarrow \text{harmonic order} \tag{9}$$

The WTHD for 2D geometry is approximately 4.17%, and the WTHD for the 3D one is approximately 4.25%, thus demonstrating that the 2D geometry has a good approximation to the 3D geometry in no-load conditions. This means that the voltage quality can be analyzed using 2D geometry to reduce computational time.

On the other hand, as described in Section 2.4.1, Figure 10 is obtained regarding the electromechanical torque evolution.

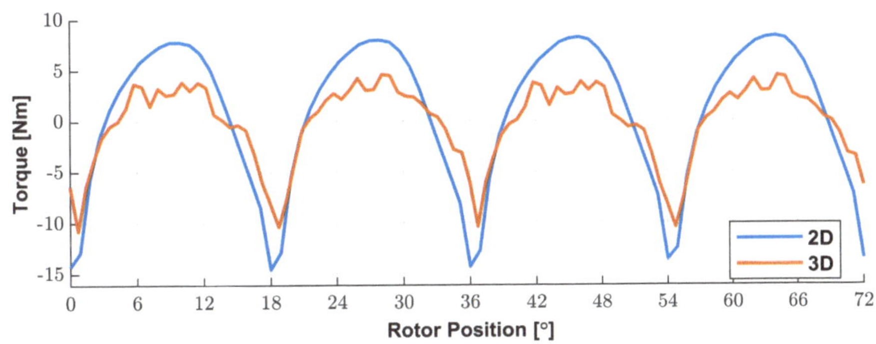

Figure 10. Torque results for 2D geometry and 3D geometry.

The 3D results present a certain unexpected ripple in torque. The torque data were not processed to maintain the peaks since our study verified that smoothing the data imposes a deviation in peaks. The ripple effect is caused by the inaccuracy of the mesh and stationary study. As in 2D geometry, the torque amplitude is higher, which means that 2D geometry gives more conservative results on oscillatory torque than its 3D counterpart. So, once more, 2D must be used only in a preliminary design phase. To obtain more accurate results, 3D geometry must be used.

As the study of magnetic flux was already made, a study of the lumped-element parameters of the equivalent electric model of the machine is performed, as described in Section 2.4.1. The stator coil resistance is 7.2 mΩ. We recall that the windings have only 20 turns to a total active cross-section of 543 mm^2, meaning a value of 27.15 mm^2 of the cross-section of each turn.

Figures 11 and 12 show the preponderance of the effects of the rotor position in inductance fluctuation compared to the effects of phase current. The peak-to-peak value of 300 μH due to the rotor position effect and a neglectable peak-to-peak value due to the phase current effect justify the previous sentence. Therefore, the core saturation effect is not visible. The self-inductance is about five times greater than the adjacent mutual inductances and around seven times higher than the opposite mutual inductance. Their geometrical arrangement gives the differences between mutual inductance values, either mean or oscillatory values. Relative to the parameter N, these maps relate the EMF produced by the DSPM S4R5 machine with the rotor position and the current apart from rotor speed as displayed in Equation (5). Moreover, it is highlighted that the decaying effect due to the current is similar for the N maps to the linked flux maps.

Figures 11 and 12 show each coil's inductance and N maps. The yellow and blue colors represent each map's higher and lower values.

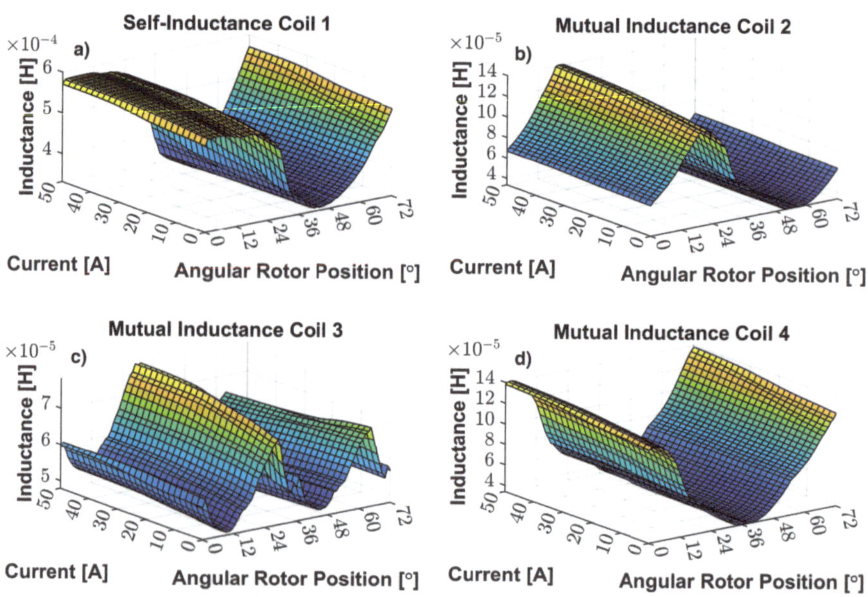

Figure 11. Inductance maps: (**a**) self-inductance of coil 1; (**b**) mutual inductance of coil 2 with regard to phase current of coil 1; (**c**) inductance of coil 3 with regard to phase current of coil 1; (**d**) inductance of coil 4 with regard to phase current of coil 1.

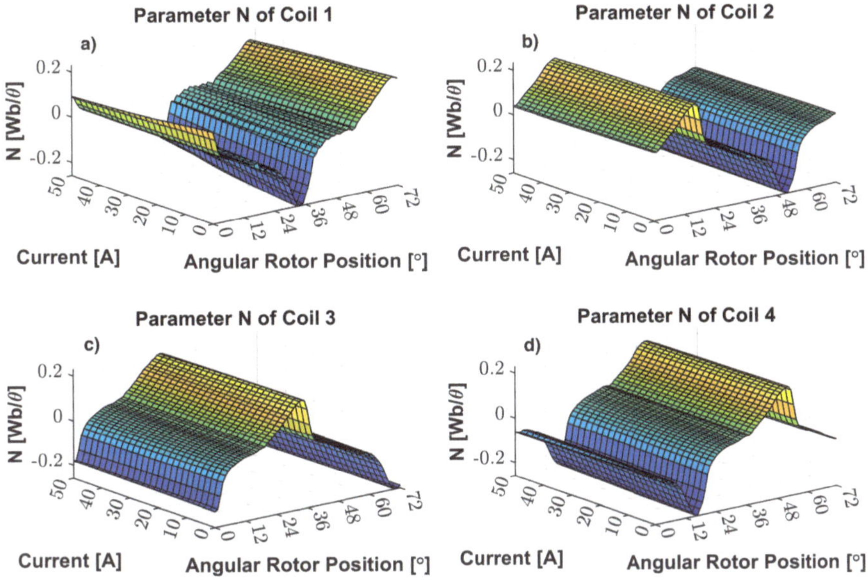

Figure 12. *N* Parameter map: (**a**) *N* of coil 1; (**b**) *N* of coil 2; (**c**) *N* of coil 3; (**d**) *N* of coil 4.

3.2. Iron Loss Calculations at No Load

The method discussed in Section 2.4.2 applied to 3D geometries will be our benchmark since the 3D geometry is the most accurate model. Therefore, the loss values obtained for 2D and 3D geometry are 362W and 284W, respectively.

Thus, the 2D geometry results present an oversize of about 45% concerning the most accurate value, 3D geometry. Once more, this means that the 2D model enables a rough estimation of iron losses in the preliminary phase of the design of machines with a large diameter–length ratio. Due to the high discrepancy in results, the 3D geometry must be used for improved accuracy.

3.3. Phase Current Shape Study

This study is based on the results of [18–20], where four current waveforms are seen as a sinusoidal waveform, a square waveform, and two rectified sinusoidal waveforms.

The different kinds of waveforms of current assumed to be studied are shown in Figure 13 for only one phase current with an amplitude of 110 A. Also, it is highlighted that the number of points considered to acquire the results is enough to represent the fast transitions of the waveforms with the required accuracy. For the sinusoidal waveform, the power angle δ that provides the biggest torque mean value is $0°$.

Most current waveforms in Figure 13 derive from a cosine wave, where the blue one is the proper one. To replicate a current waveform given by a power electronic converter, the waves in the orange and violet lines are the cosine waves considering firing angles. However, as the theory of conventional switched reluctance machines mentions, the higher output power is given when the range given by firing angles is in the interval where the inductance of the machine is decreasing. So, a square wave is obtained (green color).

Table 3 shows the average reaction torque value results, the standard deviation of the torque to represent the oscillatory torque, and the ratio between them.

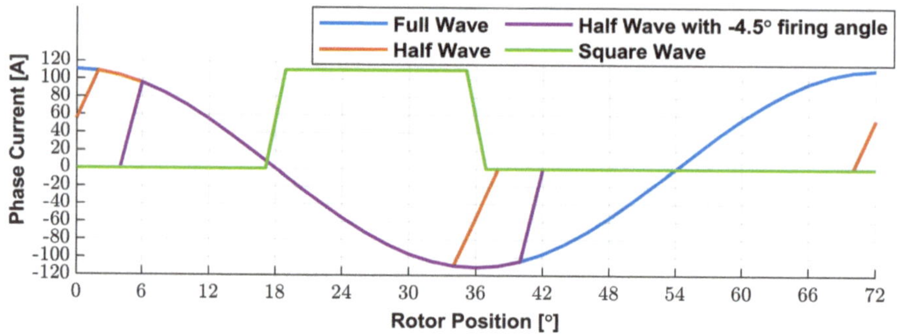

Figure 13. Different current waveforms used to obtain the best current waveform regarding input torque mean value and oscillatory torque.

Table 3. Current waveform results in terms of electromagnetic torque mean value and oscillatory torque.

Current Waveform	$\overline{T}$ [Nm]	σ_T [Nm]	$\frac{\sigma_T}{\overline{T}}$
Cosine wave	−0.42	40.7	97.1
Half-cosine	−0.57	20.4	35.8
Half-cosine with −4.5° firing angle	−0.86	21.1	24.5
Square wave	−20.1	11.1	0.553

The best waveform is the square waveform, which occurs when imposed at the moment, and a misalignment of the rotor and stator poles occurs since it has the lowest ratio. So, the square current waveform will be used in the optimization process. Also, the phase current waveform can significantly improve the electromechanical torque mean value and, in turn, the delivery of active power.

3.4. Optimization of the DSPM S4R5 Machine Results

The best front obtained by the optimization process is illustrated in the left part of Figure 14. The selected point is in the middle part of the Pareto curve. The region above the red line represents an output power equal to the minimal power requirements of 90 kW, considering 3000 rpm. The right part of Figure 14 illustrates the geometry representative of the selected point.

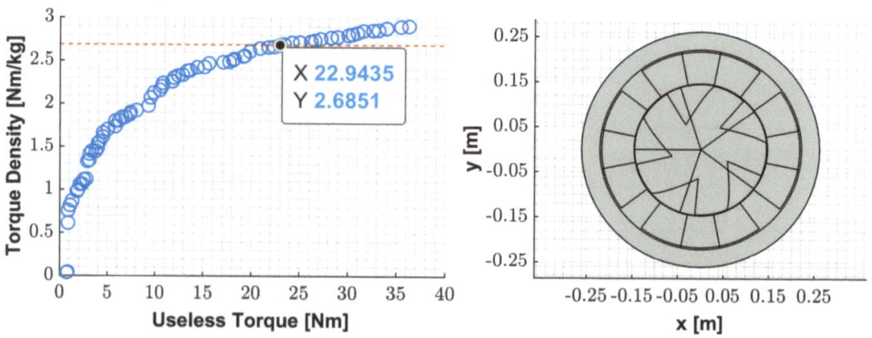

Figure 14. Pareto curve obtained for the optimization process (in **left** part) and the optimal geometry selected from the point shown in the Pareto curve (in **right** part).

The geometry obtained has a decrease in its total volume, concerning the initial geometry, of around 35%, by a reduction in its radius to about 19.4% relative to the initial geometry. The most relevant difference is the rotor teeth design, approximating a rectangular shape like the classical switched reluctance machines' rotors. The main geometrical parameters of the machine are exposed in Table 4.

Table 4. Geometric parameter values of the optimized DSPM S4R5 machine.

Parameter	Dimension
$R_{st_{ext}}$	261.3 [mm]
R_{rotext}	144.5 [mm]
$Depth_{st_{poles}}$	74.8 [mm]
$Depth_{rot_{poles}}$	80.1 [mm]
$Depth_{PM}$	4 [mm]
g	2.04 [mm]
θ_{Enr}	0.445 [rad]
l_{Enr}	61.9 [mm]
ψ	0.395 [rad]
θ	0.8 [rad]
r_{PC_B}	208.8 [mm]
θ_{PC_B}	0.065 [rad]
$Weight_{PC_A}$	0.292
$Weight_{PC_B}$	0.598
$Weight_{PC_C}$	0.717

Moreover, to visualize the behavior of this new optimal geometry with nominal load, Figures 15 and 16 illustrate the magnetic flux and torque evolution along an electrical cycle for the 3D geometry.

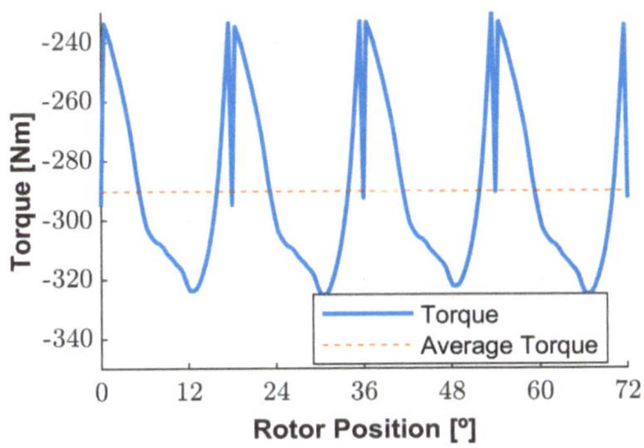

Figure 15. Torque evolution of the results of optimized DSPM S4R5.

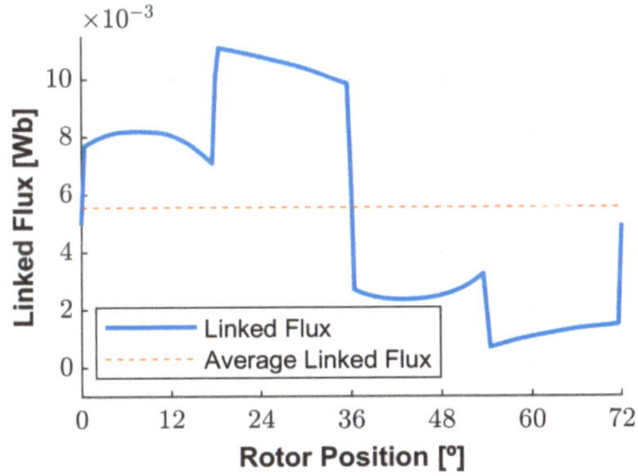

Figure 16. Magnetic flux evolution of the results of optimized DSPM S4R5.

The torque presents a sharp evolution when the stator poles are almost at the right and when right after the alignment, which can impose mechanical stress on the material. However, as the peaks are at around 10% of the nominal mean value of the electromechanical torque, the mechanical stress is not harsh. The linked flux evolution presents discontinuity zones due to the imposed switching load currents, translating into voltage peaks.

Figures 17 and 18 show the results for the 3D geometry of torque and linked flux in the same load condition as the 2D geometry. Once more, these results are our benchmark results.

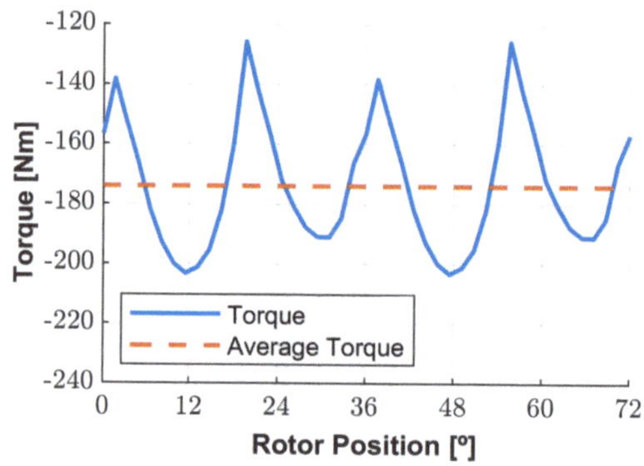

Figure 17. Electromagnetic torque evolution of the 3D geometry of the optimized DSPM S4R5 in same load condition as optimization process.

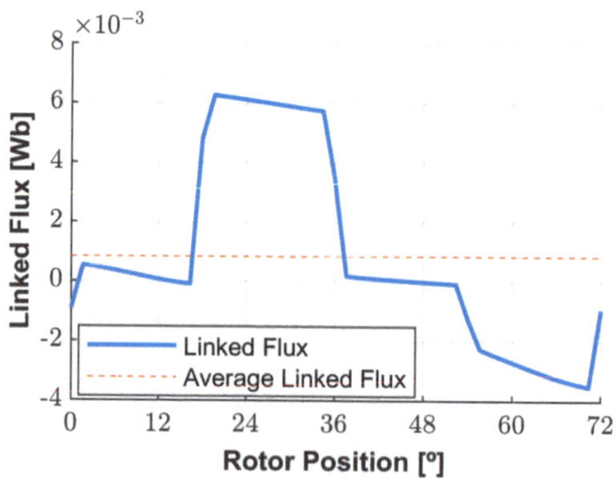

Figure 18. Magnetic flux evolution of the 3D geometry of the optimized DSPM S4R5 in same load condition as optimization process.

The mean value of linked flux and electromechanical torque decreased. The average value of torque has a decrease of around 46.5%, meaning that the selection of the geometries of the optimization process must take that into account. The average of the electromechanical power is about 47.75 kW. Also, the average value of linked flux decreased its value by 82.8%, and the maximum value decreased by about 45%, indicating that the 2D geometry gives errors above 10% compared to 3D geometry for the linked flux values of large-diameter–length ratio machines in load conditions. In short, this confirms that the 2D model must be oversized for high-diameter–length ratio machines to attain the requirements in the most accurate 3D model.

The voltages, currents, and total electric instantaneous power are presented in Figures 19–21. The indexes 1, 2, 3, and 4 shown in the legends are relative to coils 1, 2, 3, and 4. It is recalled that the voltages were obtained without considering the resistive voltage drops of coils since, for one turn, these components were around three orders of magnitude below the output voltage value. Nevertheless, we recall that the generator voltage distortion does not affect the load since it is considered an electronic interface with respective AC filters between the generator and load that must guarantee that the load voltage has low WTHD.

As visualized in Figure 18, the voltages are highly distorted as linked flux due to the imposition of pulsed currents. The transient zones of the commutation of the current cause voltage peaks. So, the peaks seen in the power are derived from that fact.

The difference between the average output power and average input power could occur due to the lack of accuracy of the results. During the work of this dissertation, the torque results were more accurate than the voltage results. So, our benchmark is the electromechanical torque. This indicates that, for studying electrical power, it is necessary to use a simpler model: the electric equivalent model of the DSPM S4R5.

Although knowing that the Joule losses and iron losses can be studied, the efficiency of the machine is given by the quotient of the difference between the total mechanical power and the studied losses by the proper total mechanical power. As visualized in Figures 20 and 21, the four phases are switched on by a quarter of the electrical period. So, relative to the Joule losses, this means that it can be computed for a single coil, presenting a value of 722 W. The iron losses computed as mentioned in Section 2.4.2 have a value of 3.05 kW. So, the efficiency of the optimized machine is around 92%, neglecting mechanical losses. This efficiency value is acceptable since the optimization process did not consider this aspect.

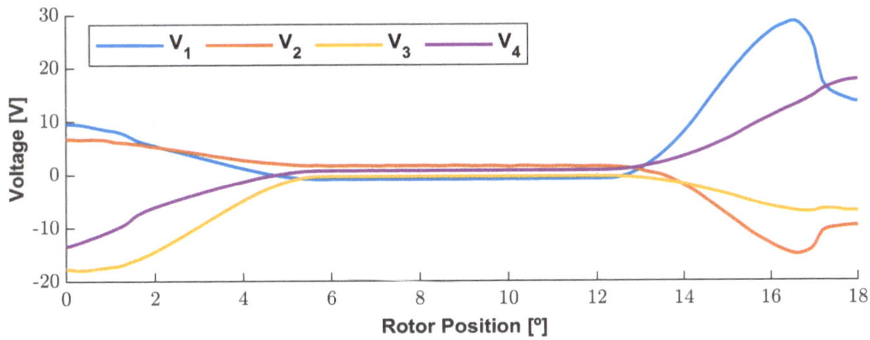

Figure 19. Evolution of the generator voltages imposed by the ideal phase current waveforms.

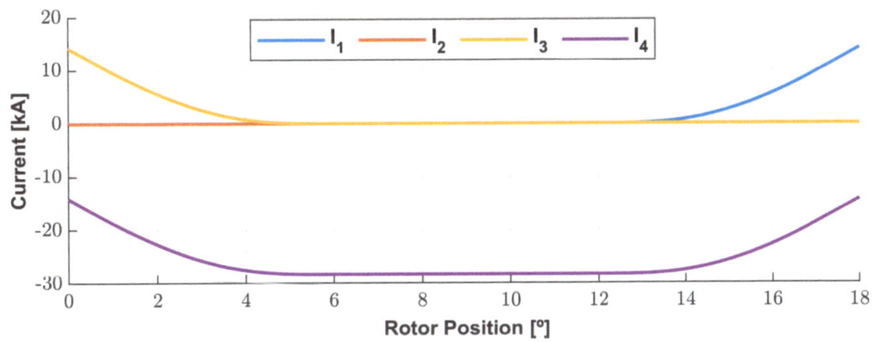

Figure 20. Evolution of the generator ideal phase current waveforms.

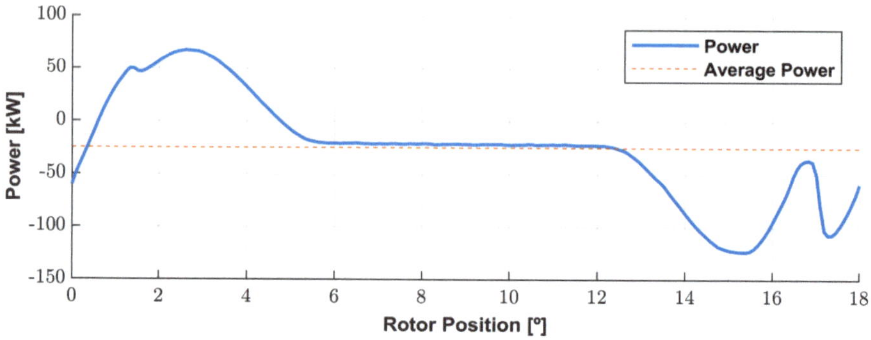

Figure 21. Evolution of the total generator instantaneous power feeding the ideal phase current waveforms.

3.5. Comparison of Initial and Optimized Geometries at No Load and Lumped Parameters Study

The initial geometry is recalled to visualize its differences to the optimized geometry, being the first one illustrated in Figure 22 and the last in Figure 23.

Figure 22. Illustration of the initial 3D geometry.

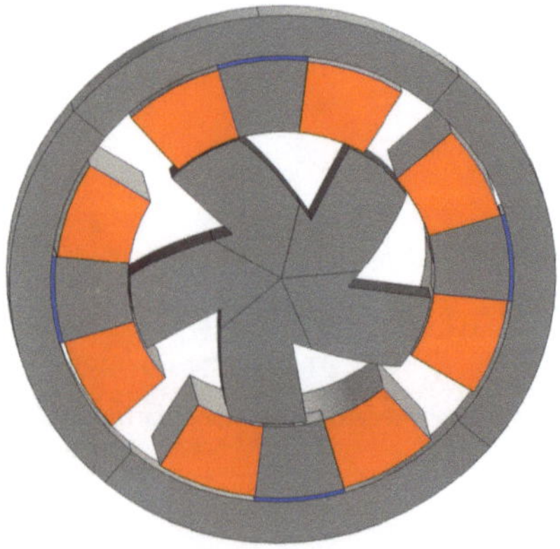

Figure 23. Illustration of the optimized 3D geometry.

To compare with [9], the number of turns per coil is now assumed to be 20. Figure 24 shows the evolution of the linked flux for the initial and optimized machines in order to compare both. The major difference remains in the mean value since the optimized geometry presents a reduction of 42.9% relative to the initial one. Besides that, it is observable that the peak-to-peak value of the optimized geometry is about 60 mWb compared to 50 mWb. This translates into a higher output voltage considering the same rotor speed for both geometries.

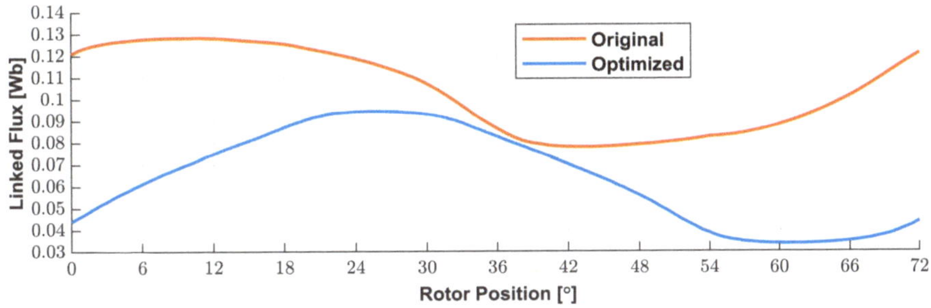

Figure 24. Linked flux results for initial geometry and optimized geometry for no-load condition.

Once more, using the WTHD in linked flux waveform to compare the voltage waveform quality, the WTHD value is 1.1% compared to 4.2%. This means that the voltage quality of the optimized DSPM S4R5 machine is greater than the original one.

Figure 25 illustrates the electromagnetic torque for the initial and optimized 3D machines for comparison purposes. The evolution in the lower peaks is less rough in the new geometry than in the original one since there is just one abrupt variation in those peaks instead of two, as in the initial geometry. This helps to minimize the mechanical stress imposed on the shaft. Once more, the results present a significant ripple because they are not filtered since the smoothing would distort the peak values.

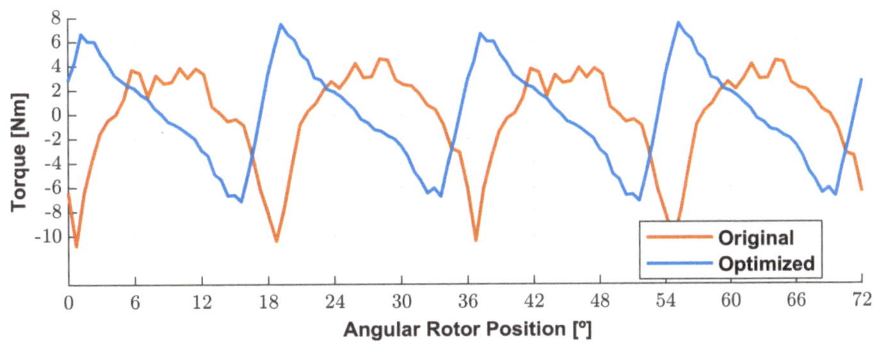

Figure 25. Torque results for initial geometry and optimized geometry for no-load condition.

For the comparison of iron losses of the original and optimized DSPM S4R5 machines, the iron losses calculation was made by an element method as described in Section 2.4.2, obtaining 286 W for the original geometry and 204 W for the optimized geometry. Thus, it can be verified that there is a reduction of 28.67% in core losses compared with the original geometry. As the rotor speed of both machines is 3000 rpm, the difference between them, according to Equation (3), is the magnetic flux density in the core of both machines. Since the harmonic spectrum of linked flux reflects the behavior of each harmonic of magnetic flux density, the WTHD is a measure of the higher harmonics amplitude relative to the fundamental one. So, as the WTHD of the linked flux of the optimized geometry is lower than the original geometry, the decaying of the harmonic components for the optimized geometry is higher than the initial geometry. This means that the main difference between the machines is the lower harmonic component of the optimized one.

The lumped parameters representation of the optimized machine is studied, as referred to in Sections 2.4.1 and 3.1. The stator coil resistance has an ohmic value of 0.62 mΩ.

Figures 26 and 27 show each coil's inductance and *N* maps. Once more, the inductance of the machine is higher in the aligned position than in the misaligned position. Figure 26 shows, for coil 1, its self-inductance and the mutual inductances with coils 2, 3, and 4. By making a comparison between Figures 11 and 26, it can be observed that the self- and mutual inductances of the optimized machine have lower values than the original one. The superiority of self-inductance over mutual inductance is similar to what was seen in Section 3 since there is a proportionality that is around four times greater for the mutual inductance of adjacent phases and around five times greater for the inductance of the opposite phase. The *N* parameter tends to be a more sinusoidal wave than the initial geometry, as verified by the reduction in the value of WTHD, which implies a non-oversizing of the filters of the converter. The peak values are similar between the geometries. So, considering constant rotor speed, the induced voltage generated is expected to be the same for both geometries in no-load conditions. Once more, the yellow and blue colors represent each map's higher and lower inductance values. In essence, the optimization process studied allows for improving the overall electric performance of the DSPM S4R5.

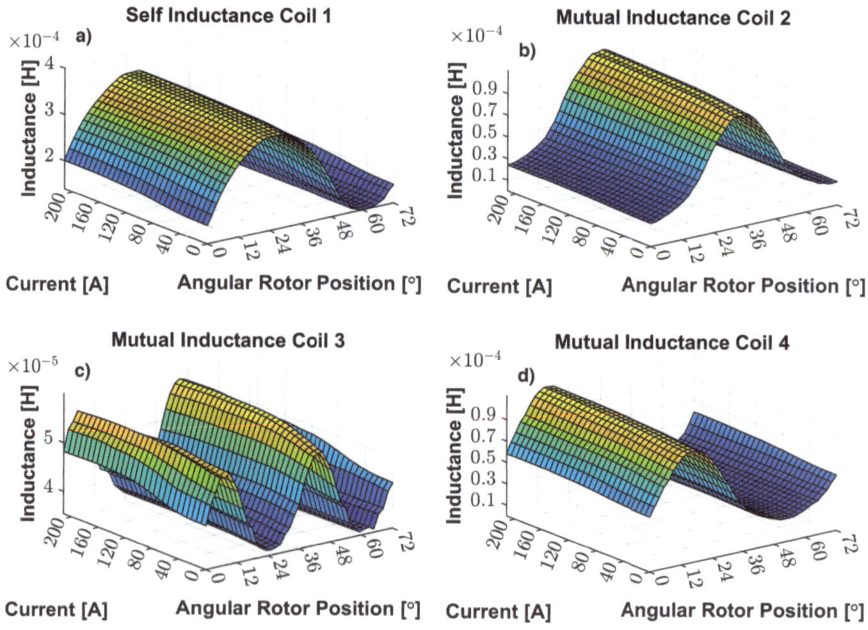

Figure 26. Inductance maps: (**a**) self-inductance of coil 1; (**b**) mutual inductance of coil 2 with regard to phase current of coil 1; (**c**) inductance of coil 3 with regard to phase current of coil 1; (**d**) inductance of coil 4 with regard to phase current of coil 1.

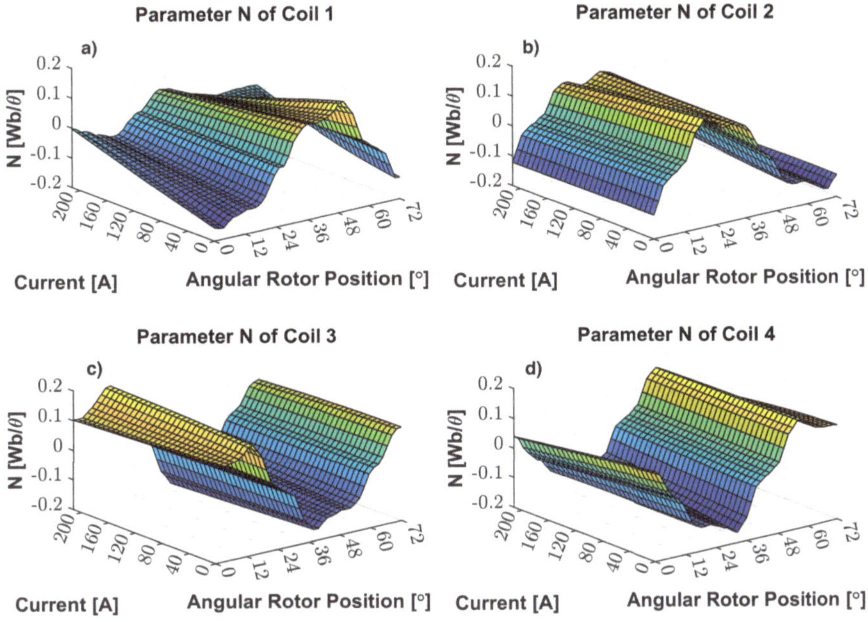

Figure 27. *N* Parameter map: (**a**) *N* of coil 1; (**b**) *n* of coil 2; (**c**) *n* of coil 3; (**d**) *n* of coil 4.

3.6. Loading Analysis of the Optimized Geometry

3.6.1. Aircraft Equivalent Electrical System

To study the optimized generator in an electric circuit, the equivalent circuit of the aircraft is considered, as depicted in [11]. The generator is modeled by a controlled current source, which is the control signal given by the equivalent electric system. The voltage measurement of each controlled current source gives the system feedback.

3.6.2. Electrical Response of the Generator with Non-Commanded AC/DC Converter

The simulation of the equivalent electric system of the aircraft studied in [11] and the equivalent model proceeded in a lumped-parameter circuit software using the methodology described in Section 2. Figures 28 and 29 illustrate the evolution of the voltages and the currents of each coil. The voltage waveform of each coil is highly distorted due to the no-load situation since the AC and DC filters were not properly adjusted for this generator. Also, the voltage waveforms between coils are not symmetrical. Figure 30 represents the generator output's instantaneous power evolution. The average output power is around 4 kW, well below the output power obtained in the ideal case. The voltage decaying is abrupt just after the peak, implying an abrupt output power decrease. So, the lack of control implies a degradation in the electric performance of the generator, which results in a significant reduction in its output power. This shows the need to design a control system to improve the DSPM performance.

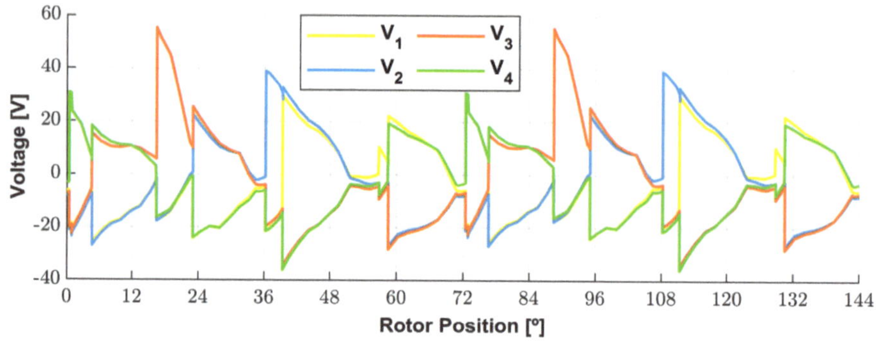

Figure 28. Evolution of the generator voltages imposed on the AC/DC converter of the equivalent electric aircraft system.

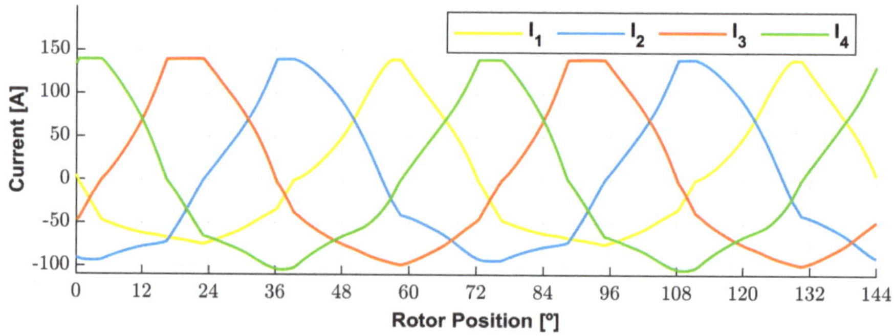

Figure 29. Evolution of the generator currents feeding the AC/DC converter of the equivalent electric aircraft system.

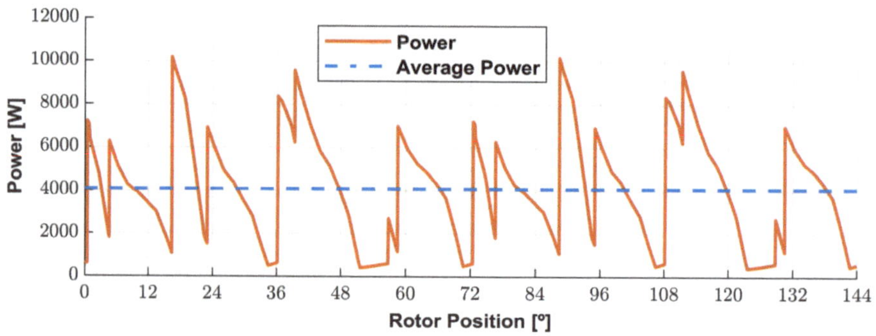

Figure 30. Evolution of the generator output instantaneous power feeding the AC/DC converter of the equivalent electric aircraft system.

4. Discussion

This study has two main sections. The first one studied the initial geometry of the DSPM S4R5 in no-load and load conditions, and the second one studied the optimization process and the optimized geometry in load conditions and also viewed the no-load response.

First, the results of the 2D geometry were compared with the 3D geometry to understand what quantities can be studied in 2D. The 2D geometry did not provide enough accurate results for linked flux, torque, and iron losses, allowing us to neglect the 3D model study. However, the 2D WTHD had a neglectable deviation (−0.08%) concerning the 3D model, so the voltage quality can be studied in 2D. Also, the shape of the electromechanical torque evolution is accurate enough in the 2D geometry study compared to the 3D one. The overestimation in iron losses of the 2D model compared to the 3D one (+45%) can also provide robust results in the design of the machine. In short, the 2D model can be used in the preliminary phase of the design and in the optimization of the process of large-diameter–length ratio machines since it gives more robust results in terms of oscillatory torque. So, an important contribution of this work is that it shows a methodology of study by 2D FEA to minimize the computation time of large-diameter–length ratio machines where the end-winding effects are significant.

Finally, regarding the optimized geometry, an optimization process, a study of the 3D geometry in load conditions, a study in no-load conditions, and a study of the electrical response to the aircraft simplified electrical system of [11] were made. The main conclusions of this work are as follows: for large-diameter–length ratio machines, the preliminary 2D study must be oversized to the total electric power output to comply with the requirements of the application being the oversize factor, dependent on the proper diameter–length ratio; the optimized generator obtained presented a plausible performance for ideal phase currents in terms of oscillatory torque (peaks around 10%), induced voltage, WTHD of linked flux, iron losses (decreased around 29%), and efficiency (92%), which allows us to confirm that the DSPM topology can be a possible solution to integrate into an aircraft power generation system.

Author Contributions: Conceptualization, M.G.N., F.F.d.S., and P.J.d.C.B.; methodology, M.G.N., F.F.d.S., and P.J.d.C.B.; investigation, M.G.N.; writing—original draft preparation, M.G.N.; writing—review and editing, F.F.d.S. and P.J.d.C.B.; supervision, F.F.d.S. and P.J.d.C.B.; project administration, P.J.d.C.B. All authors have read and agreed to the published version of the manuscript.

Funding: This work was supported by FCT, through IDMEC, under LAETA, project UID/EMS/50022/2020.

Data Availability Statement: The original contributions presented in the study are included in the article, further inquiries can be directed to the corresponding author.

Conflicts of Interest: The authors declare no conflicts of interest.

References

1. Madonna, V.; Giangrande, P.; Galea, M. Electrical power generation in aircraft: Review, challenges, and opportunities. *IEEE Trans. Transp. Electrif.* **2018**, *4*, 646–659. [CrossRef]
2. Cavagnino, A.; Li, Z.; Tenconi, A.; Vaschetto, S. Integrated generator for more electric engine: Design and testing of a scaled-size prototype. *IEEE Trans. Ind. Appl.* **2013**, *49*, 2034–2043. [CrossRef]
3. Rucker, J.E.; Kirtley, J.L.; McCoy, T.J. Design and analysis of a permanent magnet generator for naval applications. In Proceedings of the IEEE Electric Ship Technologies Symposium, Philadelphia, PA, USA, 27 July 2005; pp. 451–458.
4. Farahani, E.F.; Baker, N.J.; Mahmouditabar, F. An Innovative H-Type Flux Switching Permanent Magnet Linear Generator for Thrust Force Enhancement. *Energies* **2023**, *16*, 5976. [CrossRef]
5. Fernández, D.; Martinez, M.; Reigosa, D.; Guerrero, J.M.; Alvarez, C.M.S.; Briz, F. Permanent magnets aging in variable flux permanent magnet synchronous machines. *IEEE Trans. Ind. Appl.* **2020**, *56*, 2462–2471. [CrossRef]
6. Haavisto, M.; Tuominen, S.; Kankaanpää, H.; Paju, M. Time dependence of demagnetization and flux losses occurring in sintered Nd-Fe-B permanent magnets. *IEEE Trans. Magn.* **2010**, *46*, 3582–3584. [CrossRef]
7. Wheeler, P.; Bozhko, S. The more electric aircraft: Technology and challenges. *IEEE Electrif. Mag.* **2014**, *2*, 6–12. [CrossRef]
8. Lando, J.L. Fixed Frequency Electrical Generation System and Corresponding Control Procedure. U.S. Patent 7,064,455, 20 June 2006.
9. Esteves, A.N.F. Design and Optimization of an Axial and Solid Double-Inner-Rotor Configuration in PM Flux Switching Generators. Master's Thesis, Instituto Superior Técnico, Lisboa, Portugal, 2021.
10. Reimers, D. Bezier Curves. 2011. Available online: https://ctan.math.utah.edu/ctan/tex-archive/macros/latex/contrib/lapdf/bezinfo.pdf (accessed on 27 February 2024).
11. Silva, J.L.D. Analyses of Solutions for the Attenuation of Cogging Torque in Flux Switching Electrical Generators with Permanent Magnets in the Stator. Master's Thesis, Instituto Superior Técnico, Lisboa, Portugal, 2016.
12. Di Tommaso, A.O.; Genduso, F.; Miceli, R.; Nevoloso, C. Fast procedure for the calculation of maximum slot filling factors in electrical machines. In Proceedings of the 2017 Twelfth International Conference on Ecological Vehicles and Renewable Energies (EVER), Monte Carlo, Monaco, 11–13 April 2017; pp. 1–8. [CrossRef]

13. Arkkio, A. Analysis of Induction Motors Based on the Numerical Solution of the Magnetic Field and Circuit Equations. Ph.D. Thesis, Helsinki University of Technology, Espoo, Finland, 1987.
14. Deb, K.; Pratap, A.; Agarwal, S.; Meyarivan, T. A fast and elitist multiobjective genetic algorithm: NSGA-II. *IEEE Trans. Evol. Comput.* **2002**, *6*, 182–197. [CrossRef]
15. Gibbs, M.S.; Maier, H.R.; Dandy, G.C.; Nixon, J.B. Minimum number of generations required for convergence of genetic algorithms. In Proceedings of the 2006 IEEE International Conference on Evolutionary Computation, Vancouver, BC, Canada, 16–21 July 2006; pp. 565–572.
16. Peixoto, I.S.; da Silva, F.F.; Fernandes, J.F.; Vaschetto, S.; Branco, P.C. A Distributed Equivalent-Permeability Model for the 3D Design Optimization of Bulk Superconducting Electromechanical Systems. *IEEE Trans. Appl. Supercond.* **2023**, 33 .[CrossRef]
17. Barbie, E.; Rabinovici, R.; Kuperman, A. Analytic formulation and optimization of weighted total harmonic distortion in single-phase staircase modulated multilevel inverters. *Energy* **2020**, *199*, 117470. [CrossRef]
18. Chen, H.; Tang, T.; Han, J.; Ait-Ahmed, N.; Machmoum, M.; Zaïm, M.E.H. Current waveforms analysis of toothed pole Doubly Salient Permanent Magnet (DSPM) machine for marine tidal current applications. *Int. J. Electr. Power Energy Syst.* **2019**, *106*, 242–253. [CrossRef]
19. Hua, W.; Cheng, M. A new model of vector-controlled doubly-salient permanent magnet motor with skewed rotor. In Proceedings of the 2008 International Conference on Electrical Machines and Systems, Wuhan, China, 17–20 October 2008; pp. 3026–3031.
20. Bi, Y.; Fu, W.; Niu, S.; Huang, J. Comparative Analysis of Consequent-Pole Flux-Switching Machines with Different Permanent Magnet Arrangements for Outer-Rotor In-Wheel Direct-Drive Applications. *Energies* **2023**, *16*, 6650. [CrossRef]

infrastructures

MDPI

Article

Multi-Context Point Cloud Dataset and Machine Learning for Railway Semantic Segmentation

Abderrazzaq Kharroubi [1,*], Zouhair Ballouch [1,2], Rafika Hajji [2], Anass Yarroudh [1] and Roland Billen [1]

1 Unité de Recherche SPHERES, Geomatics Unit, University of Liège, 4000 Liège, Belgium; zouhair.ballouch@student.uliege.be (Z.B.); ayarroudh@uliege.be (A.Y.); rbillen@uliege.be (R.B.)
2 College of Geomatic Sciences and Surveying Engineering, Hassan II Institute of Agronomy and Veterinary Medicine, Rabat 10101, Morocco; r.hajji@iav.ac.ma
* Correspondence: akharroubi@uliege.be

Abstract: Railway scene understanding is crucial for various applications, including autonomous trains, digital twining, and infrastructure change monitoring. However, the development of the latter is constrained by the lack of annotated datasets and limitations of existing algorithms. To address this challenge, we present Rail3D, the first comprehensive dataset for semantic segmentation in railway environments with a comparative analysis. Rail3D encompasses three distinct railway contexts from Hungary, France, and Belgium, capturing a wide range of railway assets and conditions. With over 288 million annotated points, Rail3D surpasses existing datasets in size and diversity, enabling the training of generalizable machine learning models. We conducted a generic classification with nine universal classes (Ground, Vegetation, Rail, Poles, Wires, Signals, Fence, Installation, and Building) and evaluated the performance of three state-of-the-art models: KPConv (Kernel Point Convolution), LightGBM, and Random Forest. The best performing model, a fine-tuned KPConv, achieved a mean Intersection over Union (mIoU) of 86%. While the LightGBM-based method achieved a mIoU of 71%, outperforming Random Forest. This study will benefit infrastructure experts and railway researchers by providing a comprehensive dataset and benchmarks for 3D semantic segmentation. The data and code are publicly available for France and Hungary, with continuous updates based on user feedback.

Keywords: railways; LiDAR; dataset; semantic segmentation; machine learning

Citation: Kharroubi, A.; Ballouch, Z.; Hajji, R.; Yarroudh, A.; Billen, R. Multi-Context Point Cloud Dataset and Machine Learning for Railway Semantic Segmentation. *Infrastructures* **2024**, *9*, 71. https://doi.org/10.3390/infrastructures9040071

Academic Editor: Lucía Díaz-Vilariño

Received: 22 February 2024
Revised: 21 March 2024
Accepted: 28 March 2024
Published: 9 April 2024

1. Introduction

Railway infrastructure plays a pivotal role in modern transportation systems, necessitating continuous innovation and precise management and maintenance [1]. The integration of advanced technologies, such as mobile LiDAR (Light Detection and Ranging), into railway applications has introduced transformative capabilities to the industry. To fully leverage these technologies, data processing methods, like semantic segmentation, are essential for understanding the complex and dynamic railway environment. Semantic segmentation, a fundamental component of computer vision, enables the categorization of each pixel in an image or point in a point cloud into predefined classes, facilitating scene understanding. In this context, machine learning has emerged as a promising approach. Deep neural networks, known for their ability to extract intricate patterns and representations from extensive datasets, offer the potential to automate critical tasks within the railway domain [2,3].

However, the effective application of machine learning in the railway environment depends on a crucial factor: the availability of annotated data. Unlike urban contexts, where several annotated datasets and benchmark are available [4–8], the railway environment lacks comprehensive benchmarks and annotated datasets tailored specifically to railway applications. The lack of annotated data presents a significant challenge for machine learning models development and evaluation. To train and validate semantic segmentation models effectively, labelled data, representing the diverse range of objects and structures found along railway tracks, including tracks, signals, poles, overhead wires, vegetation, and more, are required [9,10].

Moreover, as railways turn to machine learning (ML), there is a rising need for models that are accurate and easy to interpret. Using complex models, especially in safety-critical areas like railways, raises concerns about interpretation. Traditional ML models are vital for building trust, especially in tasks like semantic segmentation, in the railway environment. Additionally, the challenge extends beyond data labelling; it also involves the establishment of benchmarks that enable researchers and practitioners to objectively assess model performance. Benchmarks serve as standardized metrics for the quantitative evaluation and comparison of novel algorithms. In the absence of a railway-specific benchmark, the assessment of model performance becomes subjective and ad hoc, hindering progress and the adoption of machine learning solutions in railway applications.

In response to this challenge, we introduce the Rail3D dataset, which aims to address the shortage of annotated data specifically tailored to railway applications. Rail3D comprises extensive point cloud data collected across diverse railway contexts in Hungary, France, and Belgium, covering approximately 5.8 km. This dataset not only mitigates the lack of annotated data but also serves as a foundational benchmark for semantic segmentation. The public availability of Rail3D represents a significant step toward fostering collaboration and innovation in the railway industry. We invite researchers, practitioners, and enthusiasts to explore, enhance, and contribute to this dataset, advancing the development of deep learning solutions that enhance the safety, efficiency, and sustainability of railway infrastructure on a global scale.

In the railway sector, where decision-makers might not be machine learning experts, the clarity and reliability of models are paramount [10]. Benchmarking LightGBM, Random Forest, and the deep-learning-based KPConv allows us to assess the effectiveness of both traditional machine learning and advanced deep learning techniques. This comparative analysis is vital for selecting a model that not only delivers high performance but is also interpretable and aligns with safety-critical standards in railway applications. Through this benchmarking process, we aim to identify a model that combines accuracy and efficiency, ensuring it can be effectively used and trusted in the context of railway infrastructure management.

The main objective of this paper is to develop a large scale and multi-context 3D point cloud dataset for railway semantic segmentation tasks and assess models' performance on it. The contributions of our work are as follows:

- Present a multi-context railway point cloud dataset for semantic segmentation: We introduce Rail3D, a unique dataset for railway environments. Its special because it covers a wide range of railway scenes from three countries, helping to fill the gap in available data for training machine learning models in this field.
- Evaluate three state-of-the-art semantic segmentation models: We compare three state-of-the-art models for understanding railway scenes: KPConv, LightGBM, and Random Forest. This helps find the best tool for the task, considering what is most important in railways: accuracy and safety.
- Assess the generalization capabilities of the best performing model: We check how well the best model works in different places. This is important to make sure the model is reliable and can be used in other real-world railway settings.

The rest of the paper is organized as follows: Section 2 presents current semantic segmentation techniques, with a special focus on railway applications and available datasets. Section 3 outlines the specifics of the subset we curated to develop Rail3D, explains how we annotated the data, and introduces the three baseline methods selected for our analysis. Section 4 details the experiments we conducted and their outcomes. Section 5 discusses what we found and ideas for future research.

2. Related Works

The availability of datasets for training has so far played a key role in the progress and comparison of ML models. These allow scene understanding across several tasks, including classification, semantic/instance/panoptic segmentation, and object detection, etc. In the following subsections, we summarize the state of the art of semantic segmentation, we investigate the existing application in railway contexts, and present the available datasets.

2.1. Semantic Segmentation

In the dynamic field of 3D point cloud semantic segmentation, significant advancements have been made through various approaches [11]. Model-driven methods, including RANSAC [12], Hough Transform, and region growing, continue to provide a solid foundation for segmenting point clouds, offering robustness in fitting geometric shapes and handling locally homogeneous properties. Meanwhile, knowledge-driven methods [13,14], incorporating explicit rules and matching strategies, are often used to enhance segmentation results in tandem with other approaches. Rule-based algorithms follow a set of predefined rules to categorize points, offering simplicity and computational efficiency [15]. Matching-based methods, on the other hand, excel in accuracy and robustness by matching the point cloud to a database of 3D models. These methods are particularly useful in scenarios where precise object recognition is essential, but they can be computationally intensive and require extensive model databases. Knowledge-driven methods often complement data-driven and model-driven approaches by refining segmentation results, ensuring that the identified objects align with domain-specific knowledge or predefined rules.

Finally, data-driven methods [11] have gained popularity, using machine and deep learning techniques to extract features and classify points. These data-driven methods span projection-based strategies, including multi-view [16–21], and spherical image approaches [22–25], as well as volumetric [26–29] and lattice-based representations [30–32], with hybrid representations combining their strengths. Point-based methods have made significant strides, introducing neural network architectures like PointNet and PointNet++ [33,34], point convolution networks [35,36], and Graph Neural Networks [37–40] to capture complex relationships and dependencies within point clouds. Recently, transformer-based methods [41–44] have emerged as state-of-the-art, using attention mechanisms to model global dependencies in point clouds. Figure 1 summarize these semantic segmentation methods.

The most recent advances in 3D semantic segmentation include self-supervised learning, multi-task learning, transformer networks, foundation models, and 3D-LLMs. Self-supervised learning methods can train models without the need for labelled data [45]. This can be helpful for reducing the cost and effort of training 3D semantic segmentation models, especially for large-scale applications. Multi-task learning methods train models to perform multiple tasks simultaneously, such as object detection and scene understanding [46]. This can help 3D semantic segmentation models to learn better representations of the data, which can lead to improved performance on all the tasks. Transformer networks are a type of neural network architecture that has been shown to be effective for 3D semantic segmentation [41]. They can learn long-range dependencies in point clouds, which can be helpful for better results. Foundation models, such as SAM [47], are Large Visual Models (LVMs) that have been trained on a massive dataset of images and text. Foundation models can be used to train 3D semantic segmentation models in a variety of ways, such as by training a downstream model on top of the foundation model. Finally, the 3D-LLMs [48] is a new type of large language model that is specifically designed to understand and reason 3D data. 3D-LLMs can be used to perform a variety of 3D-related tasks, such as 3D captioning, 3D questions answering, and 3D-assisted dialog.

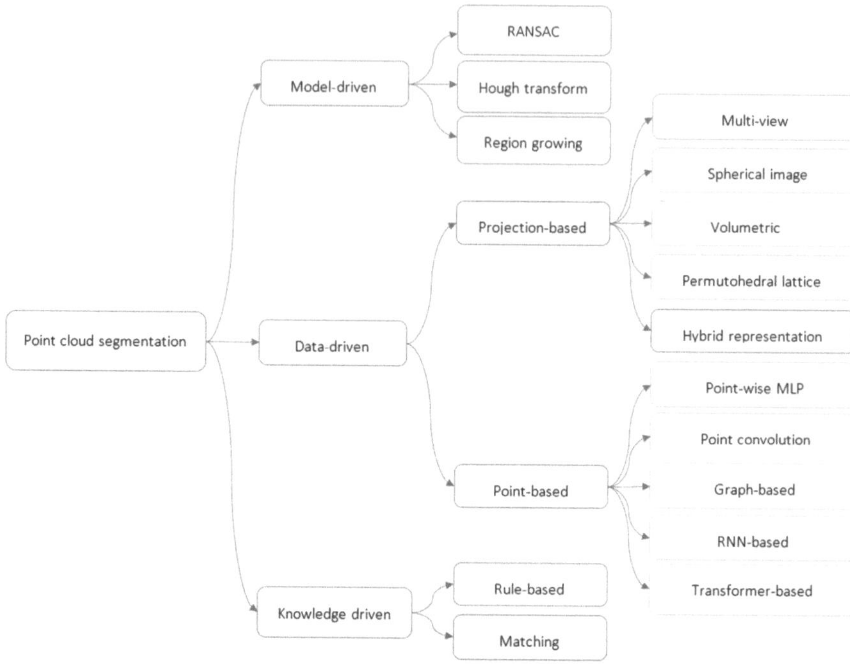

Figure 1. Summary of 3D semantic segmentation methods.

2.2. Semantic Segmentation for Railways

Even if semantic segmentation has gained progress, particularly in urban environment, applications in the railway context remain limited. Arastounia [49] developed an automated method to recognize key components of railroad infrastructure from 3D LiDAR data. The used dataset covers a 550 m section of Austrian rural railroad and contains thirty-one keys element with only XYZ coordinate information. The method segmented rail tracks, contact cables, catenary cables, return current cables, masts, and cantilevers based on their physical characteristics and spatial relationships. Recognition involved analyzing local neighborhoods, objects shape, topological relationships, and employing algorithms like KD tree's structure [50] and FLANN's nearest neighbor search [51] for computational efficiency. The approach achieved 96.4% accuracy and an average of 97.1% precision at point level. Despite the high metrics, this methodology cannot be generalized, the parameters need to be adjusted, and it remains particularly sensitive to occlusion. Additionally, the used data is not publicly available.

Heuristic methods have a long history, but learning-based approaches are becoming more prevalent. The latter aim to lessen reliance on parameters and enhance generalization capabilities. Chen et al. [52] proposed a method for automatically classifying railway electrification assets using mobile laser scanning data. They used a multi-scale Hierarchical Conditional Random Field (HiCRF) model to capture spatial relationships, improving classification accuracy compared to local methods. The model achieved an overall accuracy of 99.67% for ten objects classes. The authors in [53] proposed a deep learning-based methodology for the semantic segmentation of railway infrastructure from 3D point clouds. PointNet++ and KPConv were evaluated in four diverse scenarios, exhibiting a remarkable ability to adapt to varying data quality and density conditions. Notably, it achieved a mean accuracy of 90% when assessed on a 90 km long railway dataset used for both training and testing. They also evaluated the pros and cons of their approach, identifying the impact of intensity values as input features on segmentation results. Their best-performing tested architecture achieves a mean Intersection over Union (mIoU) of 74.89%. Furthermore, the methodology demonstrated its versatility by maintaining robust performance across different assets, such as rails, cables, and traffic lights, with an F1-score above 90% in all cases. Bram et al. [54] used PointNet++, SuperPoint Graph, and Point Transformer as three main deep learning models to perform semantic segmentation of railways catenary arches. The models were trained and assessed on the Railways Catenary Arches dataset with 14 labelled classes. PointNet++ achieved the best performance with over 71% in Intersection over Union (IoU).

The common disadvantage of these studies is that they do not evaluate their methods on the same dataset, in the same context, or with the same number of classes. This makes it almost impossible to make a fair comparison between all these studies, which do not, in general, make their data publicly available.

2.3. *Railways Dataset*

In the context of railway scene understanding, several image datasets have been developed. RailSem19 [55] offers 8500 annotated sequences from a train's perspective, including crossings and tram scenes. FRSign (French Railway Signaling) [56] contains over one hundred thousand images annotated over six French railway traffic lights with acquisition details (time, date, sensors information, and bounding boxes). RAWPED [57,58] is dedicated to pedestrian detection in railway scenes. GERALD [59] features five thousand images and annotations for German railway signals. OSDaR23 [60], a multi-sensor dataset, captures various railway scenes with an array of sensors. These datasets are vital for developing and testing algorithms for railway scene analysis, promoting safety and automation in the industry. Compared to image datasets, there is a shortage of open 3D point cloud datasets. To date, we found the following:

- WHU-Railway3D: Introduced by Wuhan University of Science and Technology in 2023, this is a new railway point cloud dataset. It covers 30 km, contains four billion points, and is annotated using 11 classes. Its great advantage is that it covers three different environments, including urban, rural, and plateau. Each environment covers approximately 10 km: the urban railway dataset spans 10.7 km, captured using Optech's Lynx Mobile Mapper System in central China; the rural railway dataset, covering approximately 10.6 km, was collected through an MLS system with dual HiScan-Z LiDAR sensors; and, lastly, the plateau railway dataset, spanning around 10.4 km, was obtained using a Rail Mobile Measurement System equipped with a 32-line LiDAR sensor [61].
- Catenary Arch Dataset: Covering an 800 m stretch of railway track near Delft, Netherlands, this dataset captures fifteen catenary arches in high detail [62]. The data collection process involved the use of a Trimble TX8 Terrestrial Laser Scanner (TLS). This dataset provides the XYZ coordinates of points but does not include color information, intensity, or normal for the points. It is referenced within the Rijksdriehoeksstelsel coordinate system using meters as units. The number of points per arch varies from 1.6 million to 11 million, and each point is manually labelled into 14 distinct classes [62].
- OSDaR23: The Open Sensor Data for Rail 2023 dataset offers a comprehensive multi-sensor perspective of the railway environment, recorded in Hamburg, Germany during September 2021 [60,63]. It uses a railway vehicle equipped with an array of sensors, including lidar, high and low resolution RGB cameras, an InfraRed camera, and a radar. The data were acquired through a Velodyne HDL-64E lidar sensor operating at a 10 Hz frequency. These lidar data are further enriched with annotations of various object classes (20), such as trains, rail tracks, catenary poles, signs, vegetation, buildings, persons, vehicles, etc.
- Others dataset: Several other datasets have been used for railway scene understanding tasks, but they are not publicly available or irrelevant. They include the Austrian rural railroad [49], UA_L-DoTT [64], Vigo dataset [2,53], TrainSim [65], INFRABEL-5 [66], Railway SLAM Dataset [67], MUIF [68] by the Italian Railways Network Enterprise (RFI), and MOMIT [69].

Existing 3D point cloud datasets for railway scene understanding are often restricted in size, variety, and semantic class, limiting their practical applicability. Our Rail3D dataset addresses these limitations by providing a comprehensive collection of railway environments from Hungary, France, and Belgium. This multi-context approach includes diverse railway landscapes, with various railway types, track conditions, and densities. Table 1 summarizes the datasets and their key specifications.

Table 1. Statistics overview of the existing railways point cloud datasets.

Dataset	Year	Additional Fields	Points/Sequences	Classes	Country
Catenary Arch	2021	Intensity	55 million	14	Netherlands
WHU-Railway3D	2023	Intensity, Number Of Returns	4 billion	11	China
OSDaR23	2023	Intensity	45 sequences	20	Germany
Rail3D	2024	RGB, Intensity	288 million	09	France, Hungary, Belgium

3. Materials and Methods

This section details the process of labelling the dataset and the methods of benchmarks. We begin by describing the data specifications and then the used classes and how the annotation process was performed. Then, we describe the three methods for semantic segmentation in railway environments.

3.1. Dataset Specification

The Rail3D dataset is the first multi-context point cloud dataset for railways semantic understanding. It is composed of three distinct datasets from Hungary, France, and Belgium, providing a diverse representation of railways. For this, we annotated three datasets named HMLS, SNCF, and INFRABEL. The firsts two datasets are publicly available, while the third one was provided by Infrabel, the Belgian railway infrastructure manager, under a confidentiality agreement. The datasets were acquired using different LiDAR sensors, ensuring a wide range of point densities and acquisition conditions. This diversity is crucial for developing robust and generalizable models for railway scene understanding. The specifications of each dataset are detailed in the subsections below.

3.1.1. Hungarian MLS Point Clouds of Railroad Environment

This LiDAR dataset represents a collection acquired by the Hungarian State Railways using the mobile mapping system (MMS)—Riegl VMX-450 high-density. These LiDAR scans were conducted from a railroad vehicle recording data at a rate of 1.1 million points per second. It exhibited high precision, with an average 3D range accuracy of 3 mm and a maximum threshold of 7 mm, ensuring high-quality data capture. The positional accuracy of the acquired point clouds averaged 3 cm, with a maximum threshold of 5 cm. This dataset not only contains georeferenced spatial information in the form of 3D coordinates but also incorporates intensity and RGB data, enhancing its utility for diverse applications. The dataset adheres to the Hungarian national spatial reference system, designated as EPSG:23700. Originally, three distinct datasets were thoughtfully selected, each representing different topographical regions within Hungary. The original data can be found in [70].

3.1.2. French MLS Point Clouds of Railroad Environment

The SNCF LiDAR dataset is a valuable resource provided by SNCF Reseau, the French state-owned railway company. This dataset comprises approximately 2 km of rail-borne LiDAR data, representing a non-annotated subset of a more extensive collection. The geospatial reference system employed for this dataset is RGF93-CC44/NGF-IGN69, ensuring its alignment with regional mapping standards and geographic referencing. The dataset is primarily presented in compressed LAZ format with a total of 16 tiles publicly available under the Open Database License (ODbL).

3.1.3. Belgian MLS Point Clouds of Railroad Environment

For several years, Infrabel, the Belgian railway infrastructure manager, was using LiDAR technology to acquire data over the country's railways network. The lidar acquisition process involves mounting a Z + F 9012 lidar sensor on the front part of a train (EM202 vehicle) and recording the point cloud data as the train travels along the tracks. The train collects data for every railway line in Belgium at least twice a year. This is highly relevant for 3D change detection, which we will be looking at in our next studies. In addition to LiDAR, four cameras are used to capture the colors, two at the front and two at the back. However, for this application, we only used the point cloud without intensity and without colors. The point clouds were encoded in LAS format and the coordinates were in Lambert Belge 72. We chose three different areas in Belgium: Brussels, mid-way between Brussels and Ghent, and the south of Ghent. Each of these contexts presented a different challenge, either in terms of occlusion, complexity, or unbalanced classes.

As a summary, we provide, in Table 2, an overview of the Rail3D datasets, detailing key metrics such as total points, total length, attributes, and the number of tiles per set (for train, validation, and test). Overall, for each railway scene, the training data accounted 60%, the validation data accounted for about 20%, and the test data accounted for about 20%.

Table 2. Description of datasets in numbers of points, length (in meter), attributes, and number of tiles per set (train, validation, and test).

Dataset	Total Points	Total Length	Attributes	Train	Validation	Test
HMLS	104,958,796	2140 m	X, Y, Z, R, G, B, Intensity	17	6	6
SNCF	143,313,588	1600 m	X, Y, Z, R, G, B, Intensity	10	3	3
INFRABEL	39,648,006	2087 m	X, Y, Z, Intensity	12	4	5

3.2. *Classes Typology*

To choose class labels, we combined literature reviews [2,53] with insights from railway industry experts from Infrabel. This approach helped us to identify classes critical for railway infrastructure applications, ensuring our labels reflect both academic and practical considerations. This resulted in the following list of classes and their corresponding objects as illustrated in Figure 2:

- Ground (label 1): represents various ground surfaces, including rough ground, cement, and asphalt.
- Vegetation (label 2): covers all types of vegetation present in the railway environment, such as trees and median vegetation.
- Rail (label 3): refers to the physical track structure.
- Poles (label 4): includes all types of poles found along the railway, such as catenary poles and utility poles.
- Wires (label 5): covers various overhead wires, including catenary wires, power lines, and communication cables.
- Signaling (label 6): represents all types of railways signaling equipment, such as traffic lights and traffic signs.
- Fence (label 7): includes several types of fences and barriers used along the railways, such as security fences, noise insulation panels, and guardrails.
- Installation (label 8): refers to larger structures that are part of the railway infrastructure, such as boxes and passenger cabins.
- Building (label 9): covers all types of buildings located within the railway corridor, such as houses, warehouses, and stations.

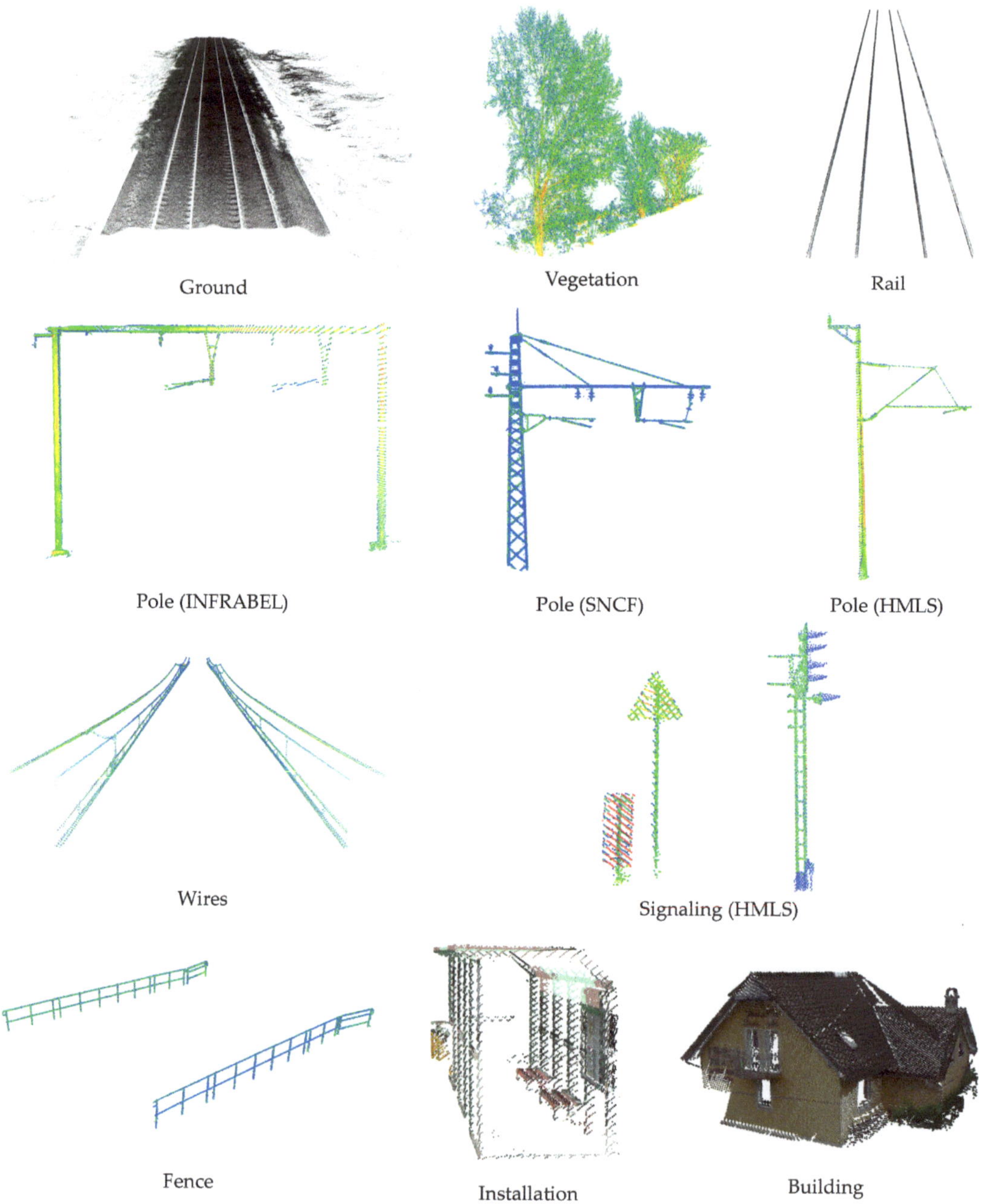

Figure 2. Illustration of each object class from the three datasets.

3.3. Data Annotation

The quality of the annotation is a determining factor for a good dataset, as well as for all the successive tasks. As shown by Silvia et al. (2023)'s results in [71], the annotation differs from one person to another, even for the same scene and with the same class definition. This has a significant impact on the results of deep learning semantic segmentation models, with a relationship of $R^2 = 0.765$ between the points labelled in concordance and the

F1-score. Even if the discordances do not exceed 1% in general, we tried to perform an entirely manual annotation, without using any automatic algorithms. The annotation was performed on CloudCompare [72] (version 2.13.0) using the QCloudLayers tool and was controlled by a second and a third annotator. They were given guidelines to help them with the annotation process, but it was up to them to make the final decision on the class label for each point. Table 3 details points per class, revealing dominant vegetation and ground classes. The minority classes, like signals and buildings, may pose challenges for the semantic segmentation models.

Table 3. The classes repartition of each subset.

Dataset	Ground	Vegetation	Rail	Poles	Wires	Signaling	Fences	Installation	Building
HMLS	69,712,961	32,208,451	1,157,524	667,108	430,949	74,687	359,907	74,272	272,937
SNCF	51,214,064	85,506,927	2,592,131	1,390,768	2,187,299	110,467	224,576	87,356	0
INFRABEL	33,698,796	1,381,599	2,041,258	654,757	1,735,242	24,926	51,100	60,328	0

In Figures 3–5, we show samples of our annotation. Each color code represents a class, as indicated in the legend.

Figure 3. Examples of our annotation of the Hungarian MLS dataset, displayed in colors (**left**) and corresponding classes (**right**). The legend is valid for the rest of the Figures 4 and 5.

Figure 4. Examples of our annotation of the SNCF dataset, displayed in colors (**left**) and corresponding classes (**right**).

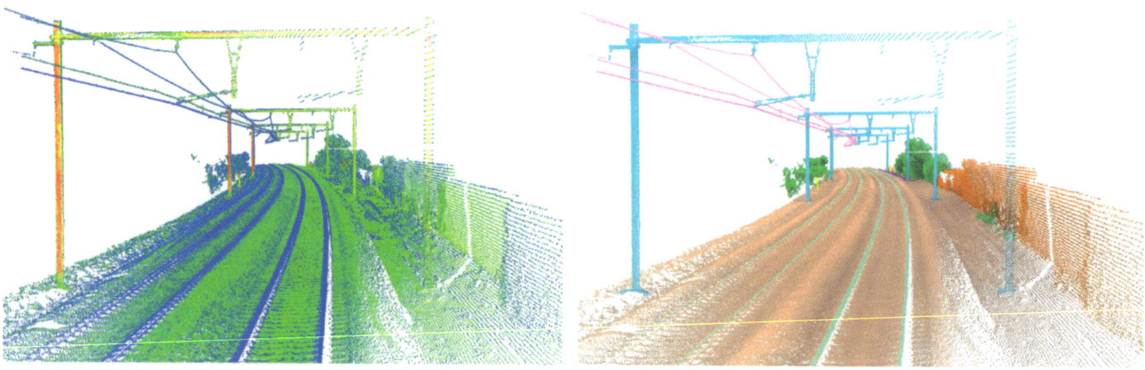

Figure 5. Point cloud from the INFRABEL dataset, displayed in intensity (**left**) and corresponding labels (**right**).

3.4. Baseline Methods

In this subsection, we explore three baseline models for semantic segmentation of our Rail3D dataset: KPConv, LightGBM, and 3DMASC. KPConv, known for its novel convolution operation tailored for point clouds, stands out for its efficiency in semantic segmentation tasks and it shows high performance over several datasets [6,8,73]. LightGBM, a gradient-boosting framework, is chosen for its accuracy and high efficiency, particularly when dealing with handcrafted features from point clouds [74]. 3DMASC, the most recent one, offers an accessible and explainable approach for point cloud classification, managing diverse datasets effectively [75]. While KPConv has shown exceptional performance in various benchmarks, LightGBM and 3DMASC are well-established in the literature, offering robust solutions for point cloud classification. Our choice of these models is based on their proven capabilities and the recent advancements they represent in the field.

3.4.1. Deep Learning Baseline

To evaluate semantic segmentation models using our datasets, we investigated the Kernel Point Convolution (KPConv) [36]. KPConv offers a novel convolution operation specifically designed for point clouds. It departs from traditional grid-based convolutions by utilizing kernel points to define local support regions. These kernel points possess learnable weights and capture the spatial relationships between points within their area of influence. Deformable KPConv further enhances this capability by adapting the location of kernel points according to the point cloud's geometry (Figure 6). We opted for KPConv due to its ability to efficiently handle point cloud data and its demonstrated success in semantic segmentation tasks [7].

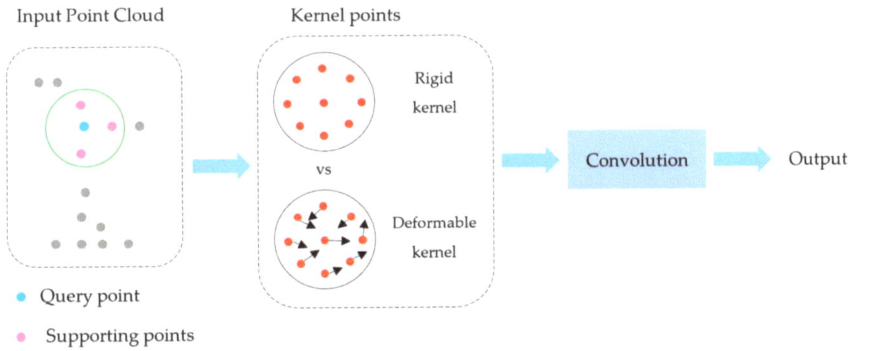

Figure 6. KPConv defines a set of kernel points to aggregate local features and performs convolution on supporting points, which are selected from a range search around the query point. (Illustration adapted from [76]).

The KP-FCNN model is built as a fully convolutional network designed for point cloud semantic segmentation. Its encoder has five layers; each layer has two convolutional blocks. The first block in each layer reduces the size of the input, except for the very first layer. These blocks are like ResNet blocks but use KPConv for the convolution process, along with batch normalization and leaky ReLU activation. After the encoding process, global average pooling collects the features, which are then passed through fully connected and SoftMax layers, like how traditional image CNNs work. The decoder part of the model uses nearest neighbor upsampling to improve the detail of point-wise features, using skip connections to combine features from both the encoder and decoder at different points. The KP-FCNN architecture is detailed in the bellow illustration in Figure 7.

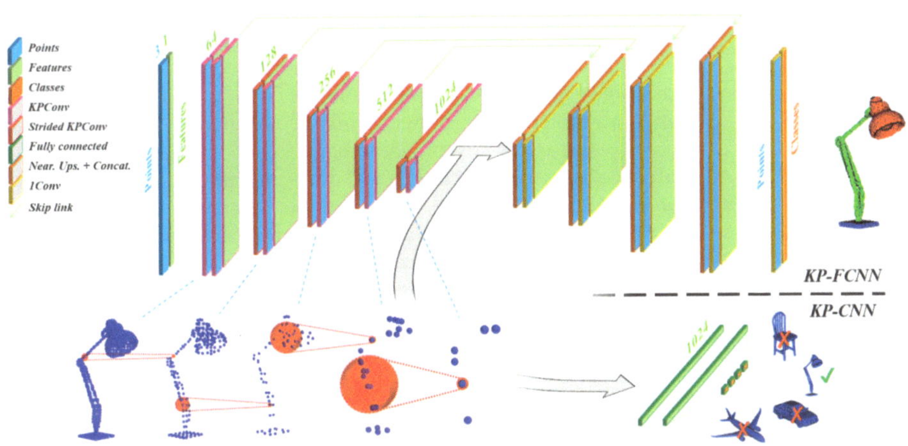

Figure 7. Illustration of the Kernel Point Convolution network architecture [77] used as a baseline method in our study. The top illustration is for semantic segmentation; the bottom one is for the classification task.

To handle the large size of point clouds, the model works with smaller portions of the point cloud, shaped into spheres with a set radius. This method helps manage the large amount of data effectively. Our use of the model follows the original training guidelines, including how big the input spheres are and how the kernel points are arranged. However, we simplified the input features to just a constant value (equal to 1) that represents the shape of the data points. This simplification, along with using a sphere-based method for testing and a voting system to decide the final segmentation, helps keep the segmentation accurate even with the large size and complexity of railway point clouds.

3.4.2. Machine Learning Baseline

The extraction of meaningful information from raw 3D point cloud data is crucial for the effective application of machine learning algorithms in classification tasks. We opted into the use of handcrafted features, which are carefully selected descriptors that encapsulate the geometric, spectral, and local dimensional characteristics of the point cloud. The process of feature selection and extraction is important for enhancing the capability of machine learning models to conduct accurate classifications. By selecting the right handcrafted features, the classification algorithms (Random Forest and LightGBM) can achieve a more robust understanding and interpretation of the complex railway assets. We followed the general steps:

- Feature identification: this selection process aims to identify a set of informative attributes that effectively capture the characteristics of the points and their surrounding context. These attributes can be derived from various sources including the point cloud's geometry, spectral information, and local dimensionality. Common geometrical features include eigenvalues of the neighborhood covariance matrix which help identify local shapes like lines, planes, contours, edge, and volume. Additionally, point density and verticality estimations can be informative for classification. In the case of lidar data, features based on multiple returns, such as the number of returns or return ratio, are valuable for distinguishing ground, buildings, and vegetation. Height variations (e.g., Z range) and radiometric information like backscattered intensity further enhance classification, especially when dealing with objects that share similar geometries.
- Feature extraction: it involves analyzing a point's local neighborhood to capture its geometric properties. The choice of neighborhood type, whether spherical, cylindrical, or cubic, along with the decision to compute descriptive features at single or multiple scales significantly influences the classification outcome. Multi-scale analysis has shown greater descriptive power, capturing scene elements of varying sizes and the geometric variations of objects more effectively. While optimal parameterization, including neighborhood type, number of scales, and scale radius, is crucial, automatic scale selection based on minimizing redundancy and maximizing classification accuracy is an active area of research.
- Feature selection: high-dimensional data can harbor irrelevant or redundant features that inflate model complexity and hinder learning. Selection methods aim for an optimal feature subset, maximizing informativeness while minimizing redundancy. Filter-based methods rank features using relevance scores (e.g., Fisher's index or information gain index) but may miss feature synergies. Wrapper and embedded methods integrate the classifier, using accuracy (wrapper) or feature importance (embedded) for selection [75].
- Classification: this involves assigning labels to individual points within a 3D point cloud. Common approaches include instance-based methods like k-nearest neighbors (kNN) and Support Vector Machines (SVMs), as well as ensemble methods like Random Forests (RFs) and gradient-boosting algorithms like LightGBM. RFs remain popular due to their ease of use, robustness to overfitting, and ability to provide feature importance metrics. However, these individual point classifiers, including RFs, struggle to incorporate spatial context beyond what is captured in the feature vector. Contextual classification methods address this by explicitly modelling relationships between neighboring points during training. While techniques like Conditional Random Fields (CRFs) can achieve higher accuracy by leveraging spatial context, they are

computationally expensive and may not perfectly capture complex relationships due to training data limitations. Our tests leverage the strengths of LightGBM, a powerful gradient-boosting algorithm, and Random Forest, while indirectly incorporating contextual knowledge through carefully chosen feature descriptors.

Our first chosen 3D point cloud classification approach relies on handcrafted features and the LightGBM Classifier, a highly efficient and accurate gradient-boosting framework. We have carefully outlined a set of steps to improve classification performance (Figure 8). Given an annotated point clouds set, we extracted features, focusing on attributes like spatial distribution and local context. We used established features based on eigenvalues, as summarized in Table 4. To manage the unordered and varying density of 3D point clouds, we utilize a multi-scale and efficient spherical neighborhood for feature computation. Then, a concatenation step of features was performed before training the classifier. Based on the features' importance results and the performance metrics, we can even use the model directly or once it has been retrained again with only the most important features.

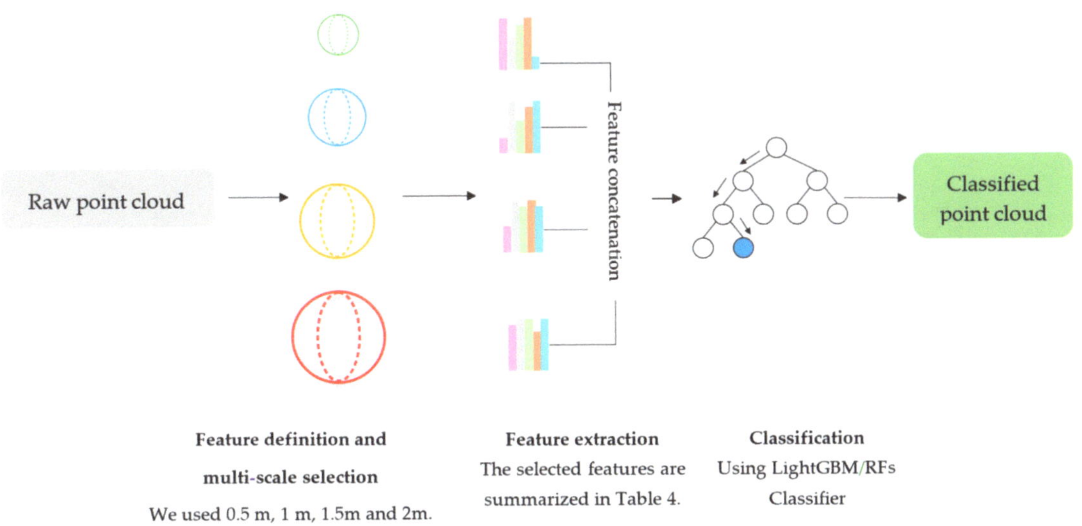

Figure 8. Multi-scale neighborhood selection, feature extraction, feature concatenation, and classification using LightGBM and Random Forest.

Table 4. Used handcrafted features [75], each one calculated at a multi-scale sphere diameter: 0.5 m, 1 m, 1.5 m, and 2 m.

Features Category	Acronym	Description
Neighborhood features	PCA1	PCA: λ 1/(λ 1 + λ 2 + λ 3)
	PCA2	PCA: λ 2/(λ 1 + λ 2 + λ 3)
	PCA3	PCA: λ 3/(λ 1 + λ 2 + λ 3)
	SPHERICITY	(λ 3/λ 1)
	LINEARITY	(λ 1–λ 2)/λ 1
	PLANARITY	(λ 2–λ 3)/λ 1
	VERTICALITY	$\left|\frac{\pi}{2} - angle(e_i, e_z)\right|_{i\epsilon(0,2)}$
	ZRANGE	Zmax—Zmin
	ZMAX	Zmax—z
	ZMIN	Z—Zmin
	ANISOTROPY	Anisotropy measure
	DIP	Computed from PCA
	NORMDIP	Dip Angle (point cloud must have Normal)

The eigenvalues $\lambda_1 \geq \lambda_2 \geq \lambda_3 \in \mathbb{R}$ and the corresponding eigenvectors $e_1, e_2, e_3 \in \mathbb{R}^3$ of the neighborhood covariance matrix, defined by equation bellow. Where p is a given point and $\overline{p}$ is the centroid of the neighborhood N (the sphere points) [78]:

$$Cov\,(N) = \frac{1}{|N|}\sum\nolimits_{p \in N} (p - \overline{p})(p - \overline{p})^T.$$

The DIP feature extracted from principal components analysis (PCA) refers to the orientation or inclination of the surface at a point in the point cloud, relative to a vertical reference. Specifically, the DIP measures how much a surface "dips" away from the horizontal plane. NORMDIP (Normal Dip Angle) refines the concept of DIP by considering the angle of the surface's inclination (or dip) relative to the normal (perpendicular) vector of the point cloud's surface at that point. This feature requires that the point cloud has Normal calculated for each point, which represent the perpendicular direction to the surface at each point.

The use of multi-scales is essential for capturing the variability within railway scene point clouds. This method, in contrast with single-scale approaches, enhances classification performance by adapting to the diverse sizes and shapes present in the data. The selection of scales was systematically parameterized to account for object diversity and neighborhood configurations, optimizing feature extraction across different scales. For feature extraction, we utilized eigenvalues from the neighborhood covariance matrix of each point to evaluate the point cloud geometry, focusing on attributes such as linearity, planarity, sphericity, etc. (see Table 4). Feature selection was conducted using embedded strategies, notably leveraging Random Forest-based metrics to identify the most informative features (feature importance). The classification phase employed LightGBM, chosen for its gradient-boosting capabilities and efficiency in processing the selected features. The parameters for LightGBM were determined through a grid search.

The second model is 3DMASC [75], which represents an accessible and explainable method for point cloud classification. It manages several attributes, scales, and point clouds, enabling the processing of diverse point cloud datasets, including those with spectral and multiple return attributes. The feature extraction process incorporates classical 3D data multi-scale descriptors, novel spatial variation features, and dual-cloud features, ensuring a comprehensive capture of both geometrical and spectral information. Subsequent steps in scale selection and feature fusion contribute to enhanced classification performance. The method is based on Random Forest as a classification model and CloudCompare for feature extraction, selection, importance, and classification. These are calculated either based on a spherical neighborhood or the k-nearest neighbors. Table 4 showed a summary of the features we used for this test. In Random Forest classification, feature importance is determined theoretically by evaluating how each feature contributes to reducing uncertainty or impurity in the model's predictions across all decision trees within the forest. This process involves calculating the decrease in impurity, often measured using the Gini impurity or entropy, attributed to each feature at every split point in each tree. The model aggregates these decreases across all trees to assign an importance score to each feature. The more a feature decreases impurity on average across all trees, the more important it is. This method allows us to identify which features play a crucial role in the model's decision-making process, highlighting their relative importance in predicting the target label.

4. Experiments and Results

This section evaluates the three chosen state-of-the-art chosen methods (KPConv, LightGBM, and 3DMASC). The first subsection describes the implementation of each method to ensure replicability. The second subsection discusses the metrics used for evaluation as well as the results of the different tests performed in our study.

4.1. Implementation

These experiments required the creation of python code and the implementation of the original experiments. For 3DMASC, from feature extraction to inference, it was performed using the 3DMASC's CloudCompare Plugin and following the training procedure in [75]. Whereas, for KPConv, we utilize PyTorch original implementation. For the training experiment we used the same procedure as in [77]. For LightGBM, we used CloudCompare Plugin for feature extraction and feature importance. Then, we created Python scripts using the LightGBM framework (version 4.3.0) for classification. The code is available at https://github.com/akharroubi/Rail3D, accessed on 14 December 2023. We adopted the training procedure in [79]. All these codes were accessed on 14 January 2024 and experiments were conducted on a workstation with a NVIDIA GeForce RTX 3090 graphics card and an i9-10980XE CPU @ 3.00 GHz with 256 GB RAM.

For the first experiment, we used KPConv without any changes to the network architecture [77]; only the hyperparameters were adapted. We used 15 kernel points with an input sphere radius of 2.5 m and a first subsampling grid size of 0.06 m to capture fine details. The convolution radius was set to 2.0 m, with a deformable convolution radius of 4.0 m for flexible feature learning. The kernel point influence was 'linear', and the aggregation mode was 'sum'. We initialized the feature's dimension at 128 and employed batch normalization with a momentum of 0.02. The model was trained for a maximum of 400 epochs, with a learning rate of 1×10^{-2}, a batch size of 5, epoch steps of 500, and a validation size of 50, optimizing for both accuracy and efficiency. In contrast to the original article, however, we did not add any input features (only X, Y, and Z coordinates), except for a constant feature equal to one, which encoded the geometry of the input points.

For the LightGBM experiment, we used the original training strategy as in [79]. When selecting a method for neighborhood extraction from point clouds, we considered cylindrical, nearest neighbor (k-NN), and spherical approaches. We chose the spherical neighborhood method. This decision was informed by research indicating that cylindrical neighborhoods can be more advantageous in extracting useful features from point clouds [78]. The spherical method was selected for its ability to uniformly capture the local structure around a point in all directions, providing a more balanced and comprehensive understanding of the point's immediate environment. This approach ensures that the features extracted are representative of the point's spatial context. For the parameters, we set the boosting_type to 'gbdt' for gradient-boosting decision tree, with num_leaves at 100 and max_depth at 30 to allow for complex model learning. The learning_rate was set to 0.05, with n_estimators at 100 for the number of boosting rounds. We also configured subsample_for_bin at 200, subsample at 0.8 for the subsample ratio of the training instance, and min_child_samples at 20 to ensure sufficient data for each tree. The class_weight was set to "balanced" to address class imbalance. We used the dataset split detailed in Table 2 for training, validation, and testing across all models, ensuring consistency in model evaluation and comparison.

For the 3DMASC experiment, we used the Random Forest method with a spherical neighborhood for feature extraction and four scales (0.5 m, 1 m, 1.5 m, and 2 m). We set the max_depth of the trees to 30 and limited the max tree count to 150, allowing the model to learn detailed features without becoming overly complex. The min sample count was set to 10, ensuring that each leaf had enough samples to make reliable decisions. The 3DMASC plugin utilizes Random Forests to inherently assign importance scores to each of the 60 features (10 features at 4 scales) employed in the analysis. To enhance the interpretability of these scores, the plugin addresses redundancy through a two-step selection process. First, a bivariate pre-selection based on information gain (IG) and Pearson's correlation removes highly correlated features, resulting in a set with clearer importance scores. Second, the plugin implements an embedded backward selection using the RF feature importance. This process iteratively removes features with the lowest importance while monitoring the impact on model performance via Out-of-Bag (OOB) scores. The selection stops when removing a feature significantly decreases the OOB score, identified through a sliding window approach. This approach within 3DMASC combines filter-based pre-selection for interpretability with embedded selection to capture feature interactions within the RF model [75].

4.2. Evaluation

We performed both qualitative and quantitative evaluations of the semantic segmentation results. For the class-level evaluation, we used the Intersection over Union (IoU) as a metric. As a global indicator, we employed the overall accuracy (OA) and mean of the Intersection over Union (mIoU) [80]. The following equations were used to compute each of these metrics:

$$\text{IoU} = \frac{TP}{TP + \text{FN} + FP} \tag{1}$$

$$\text{OA} = \frac{TP + \text{TN}}{TP + \text{TN} + FP + \text{FN}} \tag{2}$$

$$\text{meanIoU} = \frac{\sum IoU_n}{N} \tag{3}$$

where *TP*, *TN*, *FP*, and *FN* refer to true positives and negatives, false positives, and negatives, respectively. *N* represents the total number of classes.

According to Table 5, which shows the results on HMLS subset, LightGBM demonstrates an overall accuracy (OA) of 0.94 positioned between the 3DMASC (RF) and KPConv approaches, which achieved OAs of 0.93 and 0.97, respectively. LightGBM also achieves a mean Intersection over Union (mIoU) of 0.60, demonstrating noteworthy proficiency in precisely delineating diverse semantic classes; it displays competitive efficacy across various categories, particularly in classes like Vegetation and Wires. However, KPConv outperforms both RF and LightGBM, particularly in classes like Ground, Rail, and the minority classes.

Table 5. Quantitative experimental results on the HMLS dataset.

Approach	OA	mIoU	Ground	Vegetation	Rail	Poles	Wires	Signaling	Fences	Installation	Building
KPConv	0.97	0.86	0.95	0.91	0.94	0.93	0.99	0.96	0.90	0.13	0.99
3DMASC	0.93	0.57	0.82	0.89	0.46	0.72	0.96	0.09	0.33	0.08	0.80
LightGBM	0.94	0.60	0.83	0.91	0.48	0.75	0.97	0.20	0.30	0.15	0.85

The results on SNCF subset show that the LightGBM method demonstrates an overall accuracy (OA) of 0.93 in the semantic segmentation task, positioned between the Random Forest (RF) and KPConv approaches, which achieved OAs of 0.91 and 0.97, respectively. LightGBM also achieves a mean Intersection over Union (mIoU) of 0.67, highlighting notable performance in accurately segmenting various semantic classes. It exhibits competitive performance across various semantic classes, with notable strengths in classes such as Rail. However, KPConv outperforms both RF and LightGBM method, particularly in classes like Ground and Signaling in Table 6.

Table 6. Quantitative experimental results on the SNCF dataset.

Approach	OA	mIoU	Ground	Vegetation	Rail	Poles	Wires	Signaling	Fences	Installation
KPConv	0.97	0.81	0.92	0.96	0.79	0.96	0.99	0.67	0.97	0.23
3DMASC	0.91	0.64	0.62	0.90	0.82	0.77	0.89	0.08	0.94	0.08
LightGBM	0.93	0.67	0.68	0.92	0.86	0.78	0.88	0.16	0.85	0.20

On the Infrabel subset, the LightGBM method demonstrates an overall accuracy (OA) of 0.97, confirming its position between the Random Forest (RF) and KPConv approaches, which achieved OAs of 0.96 and 0.99, respectively. LightGBM also achieves a mean Intersection over Union (mIoU) of 0.71, indicating significant performance in accurately segmenting various semantic classes. It exhibits competitive performance across a variety of semantic classes, with notable strengths in classes such as Vegetation and Wires. However, KPConv outperforms both the RF and LightGBM methods, particularly in classes like Poles and Installation in Table 7.

Table 7. Quantitative experimental results on the INFRABEL dataset.

Approach	OA	mIoU	Ground	Vegetation	Rail	Poles	Wires	Signaling	Fences	Installation
KPConv	0.99	0.84	0.99	0.84	0.95	0.97	0.99	0.40	0.69	0.89
3DMASC	0.96	0.70	0.96	0.84	0.85	0.88	0.99	0.22	0.56	0.28
LightGBM	0.97	0.71	0.97	0.86	0.87	0.89	0.98	0.25	0.63	0.26

Feature importance analysis revealed that the Normal dip (Dip Angle) at scales of 2 m emerged as the one of the most important features, indicating their significant role in the classification accuracy. Following these, verticality at 2 m scale was also identified as key contributor to the model's performance. These findings, visualized in our feature importance Figure 9, underscore the value of these specific features in enhancing the semantic segmentation of railway scene point clouds.

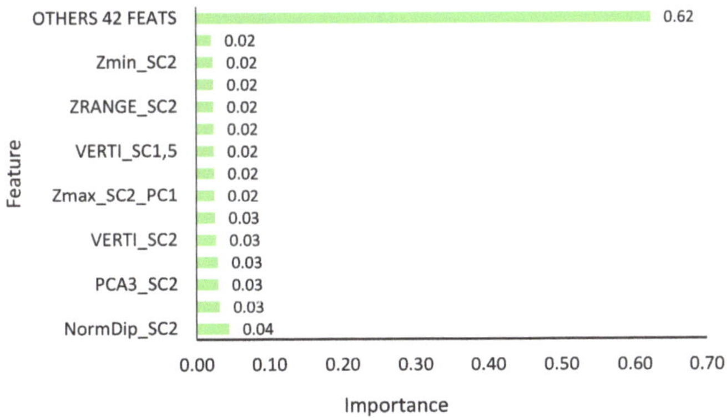

Figure 9. Feature importance selected using the LightGBM results on the Infrabel subset.

Since KPConv achieved the best performance results, we evaluated its generalization over other datasets. We conducted a series of additional experiments; we trained the model using only HMLS and assessed its performance on SNCF and INFRABEL datasets. Subsequently, we trained a larger model (KPConv**) utilizing all three datasets and evaluated its performance on the test subset. The results demonstrated a significant improvement in model generalizability when trained on a multi-context dataset. This highlights the importance of incorporating diverse data from various railway environments to develop models that can effectively handle real-world scenarios and adapt to different contexts. The results of these tests are summarized in the table below. The KPConv model trained only on HMLS gave results of 0.61 in mIoU on SNCF and 0.55 in mIoU on INFRABEL. This is mainly because the railway contexts, in general, are quite similar. But, the results were much improved when we used the whole dataset. We reached 0.76 mIoU for SNCF and 0.64 mIoU for INFRABEL. The bad result given by KPConv on the Fences class is due to its misclassification as a building even though it does not exist in INFRABEL. Also, KPConv** gave a result of almost 0 IoU in the Signaling class, as this feature was classified into the Pole class. Table 8 summarize these findings.

KPConv training was monitored by tracking both training loss and accuracy (Figures 10 and 11). The training loss indicates the model's ability to learn from the provided data, decreasing as the model fits the training examples. Conversely, accuracy reflects the model's performance in correctly classifying points during training. A continuously decreasing loss accompanied by a rising accuracy towards a plateau suggests the model is effectively learning the semantic relationships within the point cloud data. Meanwhile, the loss ideally exhibits a continuous decrease as the model learns from the data; occasional spikes or fluctuations may occur.

Table 8. Quantitative results of KPConv trained on HMLS and tested on SNCF and INFRABEL. KPConv** was trained on all the dataset and tested on the test subset.

Dataset	Model	OA	mIoU	Ground	Vegetation	Rail	Poles	Wires	Signaling	Fences	Installation
SNCF	KPConv	0.95	0.61	0.88	0.94	0.61	0.84	0.89	0.26	0.81	0.24
	KPConv**	**0.97**	**0.76**	**0.91**	**0.96**	**0.78**	**0.97**	**0.99**	**0.83**	**0.98**	**0.47**
INFRABEL	KPConv	0.98	0.55	0.98	0.76	0.86	0.75	0.97	**0.27**	0	0.34
	KPConv**	**0.99**	**0.64**	**0.99**	**0.82**	**0.94**	**0.88**	**0.99**	0	**0.68**	**0.45**

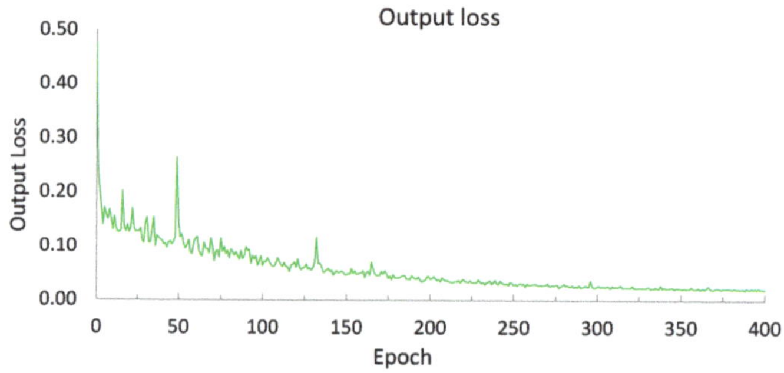

Figure 10. KPConv training loss on Rail3D Dataset (model named KPConv**).

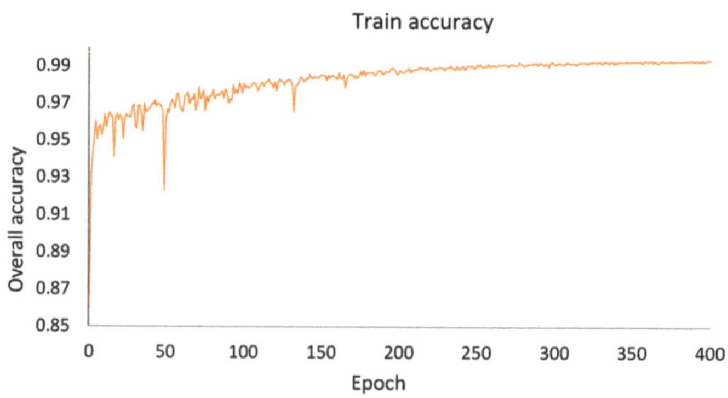

Figure 11. KPConv training accuracy on Rail3D dataset (model named KPConv**).

In the following figure we visually show samples of the point cloud, ground truth inputs, and the results of KPConv, LightGBM, and 3DMASC, successively.

More qualitative results are presented in Appendix A for the HMLS subset, in Appendix B for the SNCF, and in Appendix C for Infrabel. Each include scenes, with the ground truth, KPConv predictions, KPConv errors, 3DMASC predictions, and 3DMASC errors. We have not visually identified any significant differences between 3DMASC and LightGBM, even though the latter gives better quantitative results. Figure 12 highlights the problem of the misclassification of a fence next to railway tracks, which was classified as a building. This problem is due to the imbalance between these two classes, given that buildings are a minority but also that this type of fence can only be found in the INFRABEL dataset. All three approaches faced the same difficulty, with KPConv being the most accurate; visually, we cannot see difference between the 3DMASC and LightGBM results.

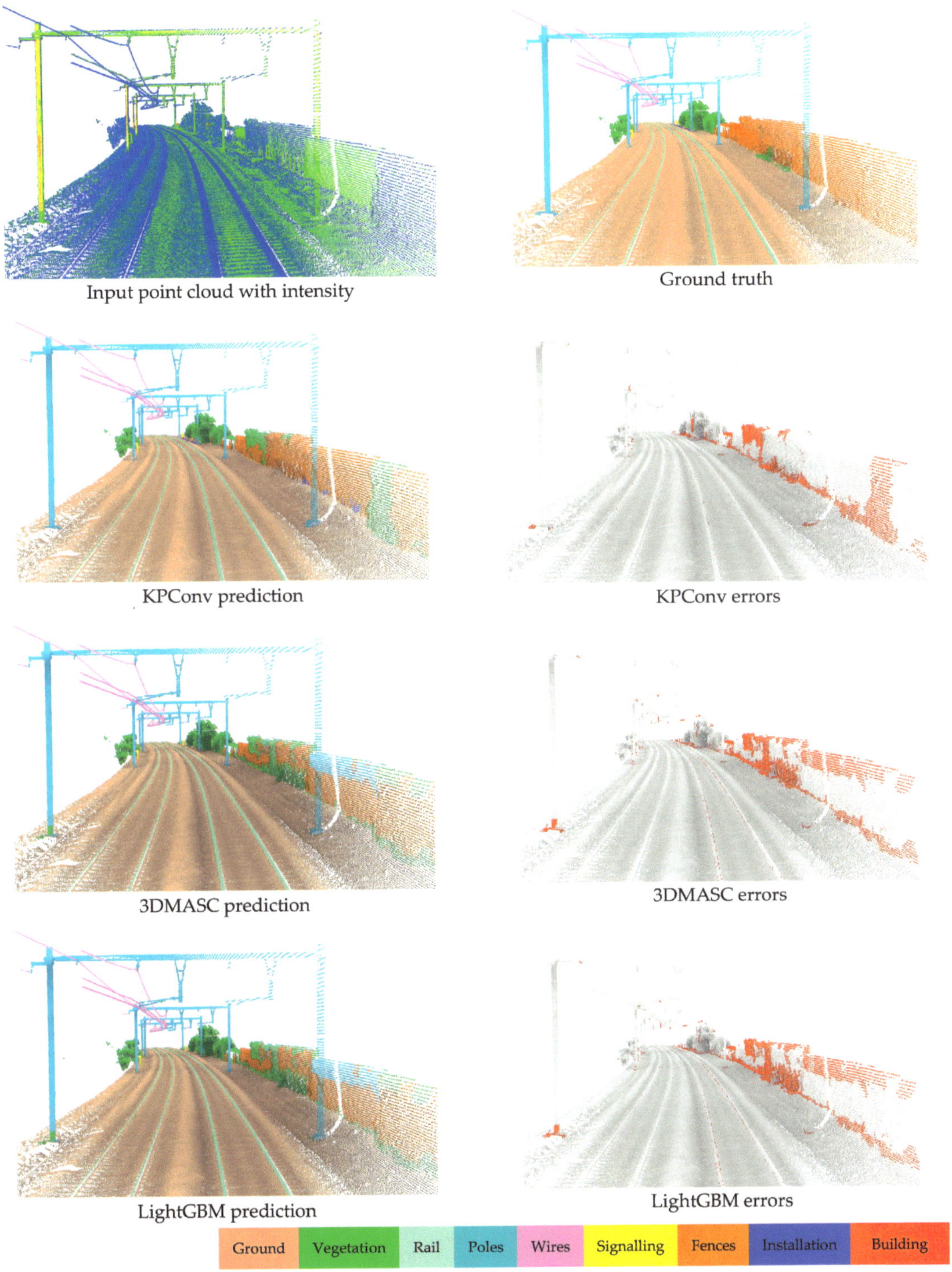

Figure 12. Input point cloud, with KPConv, 3DMASC, and LightGBM predictions. Each is associated with an error relative to the ground truth. The same class colors are used in the previous figures.

5. Discussion

This work addresses two complementary aspects: firstly, the proposal of three subset representing different contexts of railway environments and, secondly, the introduction of three baselines to addressing semantic segmentation challenges. Regarding the proposed dataset, it presents the first comprehensive dataset for semantic segmentation in railway environments, covering three distinct railway contexts from Hungary, France, and Belgium. These datasets provide rich resources for applications in railway environments. With a significant annotation of over 288 million points, the created datasets stand out in both size and diversity compared to existing datasets. This extensive volume of annotated points enhances the dataset's ability to cover a wide range of situations and variations within railway environments. Furthermore, the datasets developed across these railway environments ensure effective learning for machine learning models, confirming their generalization to different contexts. This was confirmed by the results obtained in Table 8.

The comparison between KPConv-, 3DMASC-, and LightGBM-based results provides valuable insights about the performance of different classification methods for large-scale rail borne point cloud data. The results of the comparison indicate that, while KPConv may excel in capturing intricate local geometric features, LightGBM highlights promising performance in handling large-scale datasets with its computational efficiency. 3DMASC, although a reliable method, may exhibit limitations in dealing with the complexity and unbalanced nature of railways point cloud data. Furthermore, LightGBM offers superior computational efficiency and faster training times compared to traditional Random Forests. The performance of different methods was assessed through a combination of qualitative and quantitative evaluations. Across Tables 5–7, the LightGBM method consistently gave good performance in semantic segmentation, with overall accuracies (OAs) ranging from 0.93 to 0.97 and mean Intersection over Union (mIoU) values from 0.60 to 0.71. Notably, the LightGBM approach highlights efficient results within a brief timeframe, 100 time faster than 3DMASC, emphasizing its practical applicability. This is because LightGBM is designed to be highly parallelizable while 3DMASC is not. Parallelization allows LightGBM to divide the training task into multiple smaller tasks, which can be run on multiple cores or CPUs simultaneously. While demonstrating competitive accuracy, particularly excelling in classes like Rail, Vegetation, and Wires, the machine learning method consistently falls slightly behind the performance of KPConv. The latter's superior accuracy, especially in specific classes, suggests areas for potential refinement in the handcrafted feature method, highlighting the ongoing challenge of balancing speed and precision in semantic segmentation tasks. The qualitative results obtained using various methods have confirmed the quantitative findings. The traditional ML methods yielded qualitative results closely aligned with the ground truth across all three subsets. This underscores the good selection of geometric features, indicating a positive impact on differentiating railway objects. While color and intensity are often valuable features that can enhance classification performance, we opted to exclude them from our study for two primary reasons. Firstly, the color data acquired exhibited inferior quality due to variations in acquisition times and seasons, leading to heterogeneity between contexts. Secondly, the intensity values lacked calibration across the three contexts, as different lidar sensors were employed. To validate our choice of relying exclusively on spatial coordinates (x, y, and z) as input, we conducted a test using KPConv that verified the adverse effect of color and intensity on classification accuracy within our specific case study (refer to Figure 13). One solution to correct this sky-colored points effect would be to perform a pre-processing step, as suggested in [81].

In addition, misclassification occurred where signaling poles were mistakenly categorized as catenary poles. This is due to the strong resemblance between the two and the lack of use of intensity or a specific color or geometric shape for these signs. In other cases, as shown in Figure 12, LightGBM and 3DMASC were unable to extract the Fence class correctly, given the strong similarity with the Building class. However, KPConv was able to classify it correctly. This led to a further experiment on KPConv to assess its generalizability. The use of a multi-context dataset, encompassing railways from three different countries, was shown to be crucial for boosting the generalizability of semantic segmentation models. By training on data from diverse railway environments, the model develops a more robust representation of railway structures and can better adapt to unseen contexts. This was demonstrated in our experiments, where the KPConv model trained on our multi-context dataset achieved significantly higher mIoU scores (0.76 and 0.64) on the SNCF and INFRABEL datasets compared to the model trained on the HMLS dataset alone (0.61 and 0.55). This suggests that our dataset can generalize to unseen railway environments, making it a valuable resource for developing robust semantic segmentation models for railway applications.

Figure 13. The poor quality of the colors led to poor classification (e.g., in vegetation). (**Top**): input. (**Middle**): predictions with colors. (**Bottom**): predictions without colors.

6. Conclusions

In this paper, we proposed the first multi-context point cloud dataset called Rail3D. This dataset covers three countries—Hungary, France, and Belgium—with a total length of almost 5.8 km and approximately 288 million points. It covers the nine most relevant classes for railway applications: Ground, Vegetation, Rail, Poles, Wires, Signaling, Fence, Installation, and Building. The LightGBM machine learning model for semantic segmentation demonstrated promising performance on three distinct railway datasets. It achieves significantly higher mean Intersection over Union (IoU) scores compared to 3DMASC but still falls behind the KPConv's performance. The best performing model, a fine-tuned KPConv, achieved a mean Intersection over Union (mIoU) of 86%. Meanwhile, LightGBM achieved a mIoU of 71%, outperforming Random Forest with 70%. The Rail3D dataset, publicly available for France and Hungary at https://github.com/akharroubi/Rail3D, accessed on 14 December 2023, will overcomes limitations in existing datasets, fostering the training of more generalizable learning models. We recommend evaluating our developed dataset using new machine learning methods to assess their performance. Future work will focus on the correction of sky-colored points and experimentation on the contribution of this pre-processing step in the railway environment. It also involves the exploration of the use of semantic information within Rail3D for change detection task.

Author Contributions: Conceptualization, A.K. and R.B.; Data curation, A.K.; Funding acquisition, A.K. and R.B.; Investigation, A.K. and Z.B.; Methodology, A.K., Z.B. and R.B.; Supervision, R.B. and R.H.; Validation, R.B. and R.H.; Visualization, A.K.; Writing—original draft, A.K.; Writing—review and editing, A.K., Z.B. and A.Y. All authors have read and agreed to the published version of the manuscript.

Funding: Abderrazzaq Kharroubi is an Aspirant of the Fonds de la Recherche Scientifique FNRS.

Data Availability Statement: The data used in this study are publicly available at https://github.com/akharroubi/Rail3D (accessed on 28 November 2023).

Acknowledgments: This study is part of the first author's Ph.D. thesis. The original data used in this study are publicly available at https://ressources.data.sncf.com/explore/dataset/nuage-points-3d for the French data and at https://data.mendeley.com/datasets/ccxpzhx9dj for the Hungarian data (accessed on 28 November 2023).

Conflicts of Interest: The authors declare no conflicts of interest.

Appendix A

Input

Figure A1. *Cont.*

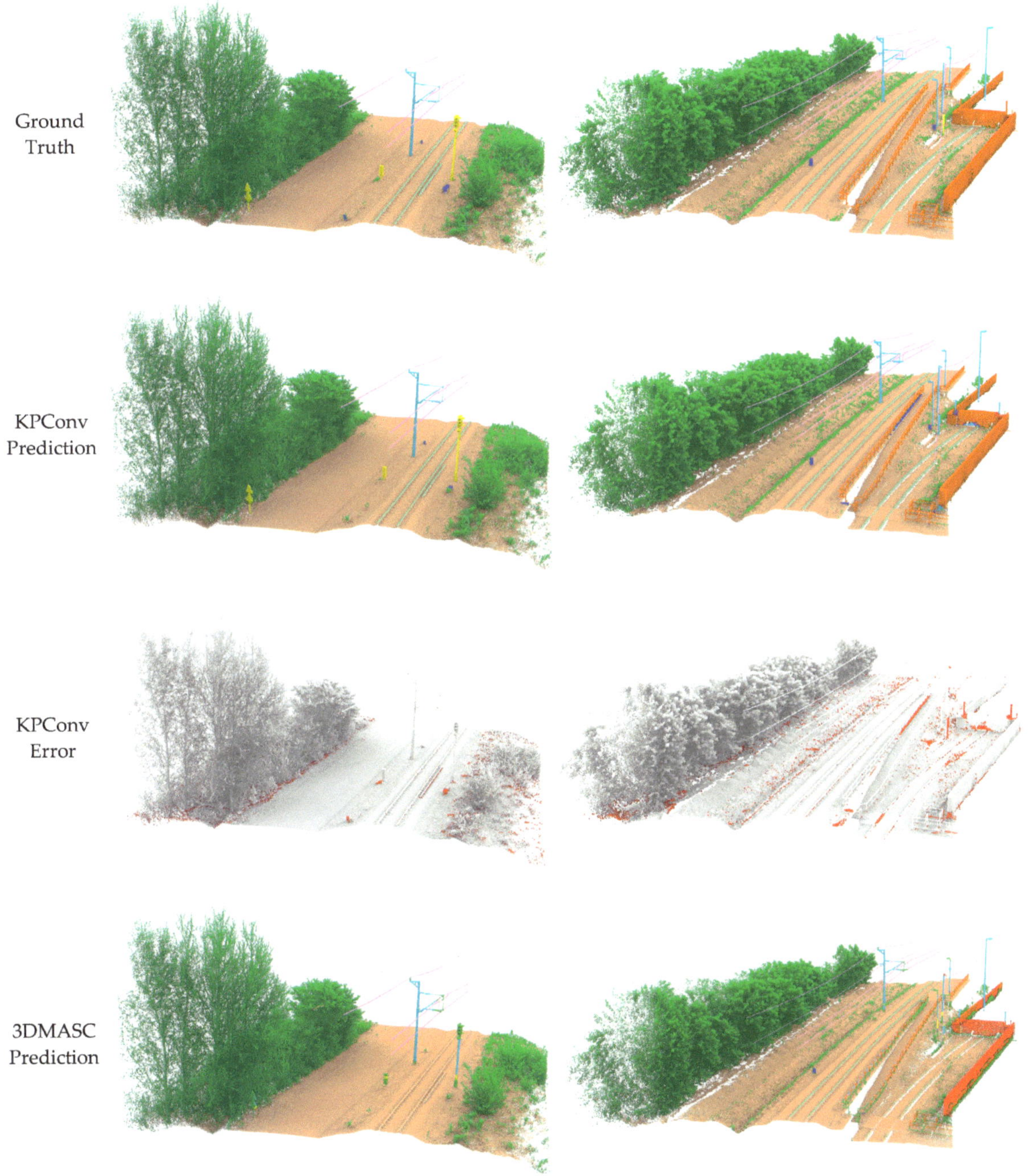

Figure A1. *Cont.*

3DMASC
Error

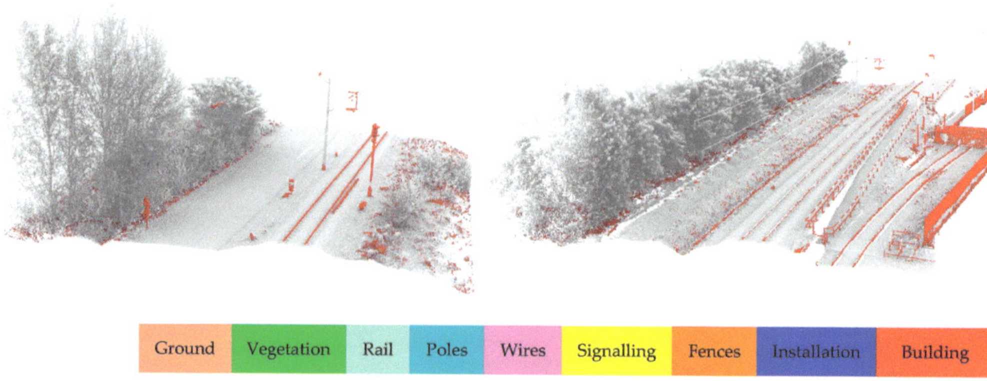

Figure A1. Qualitative results on HMLS subset.

Appendix B

Input

Ground
Truth

Figure A2. *Cont.*

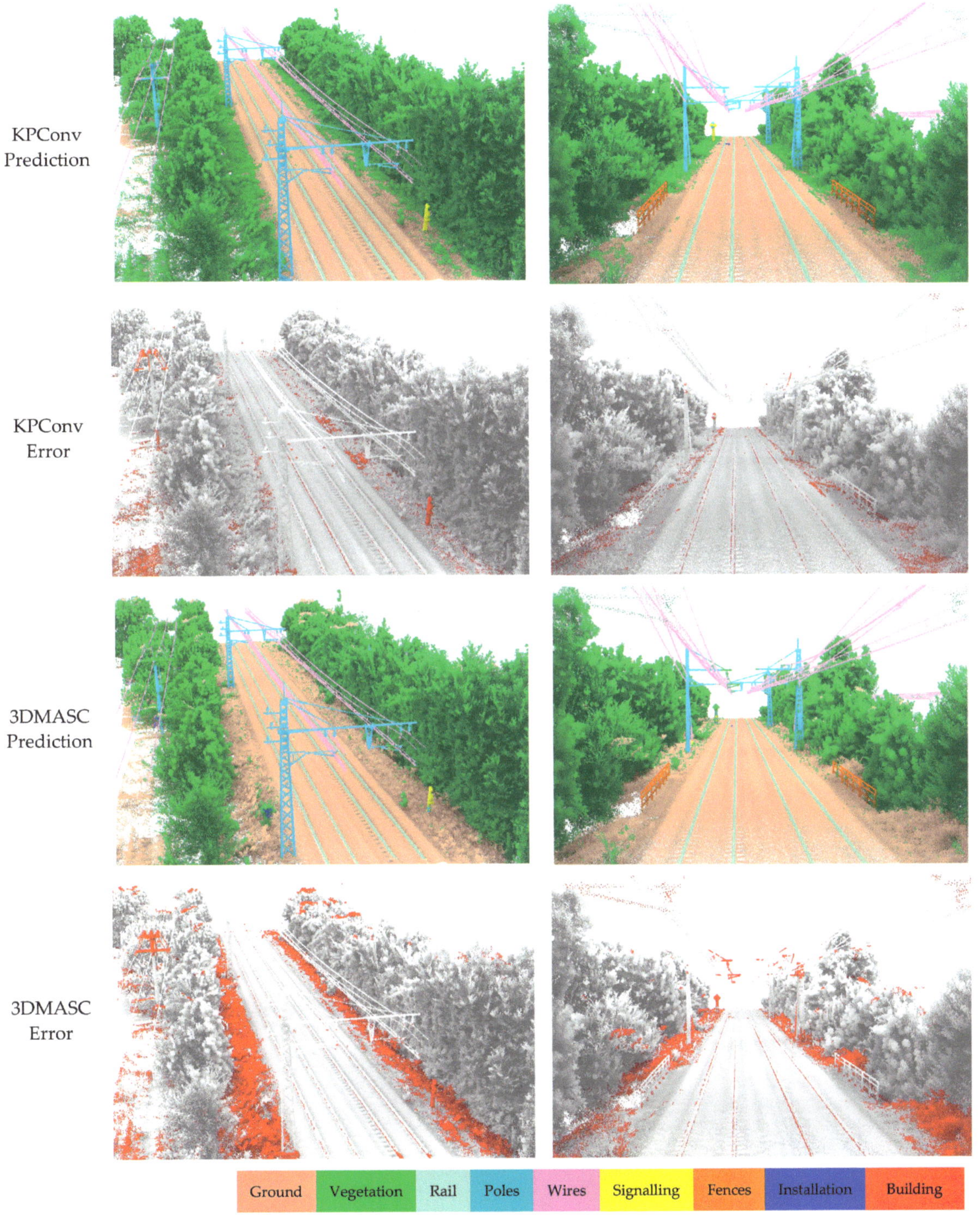

Figure A2. Qualitative results on SNCF subset.

Appendix C

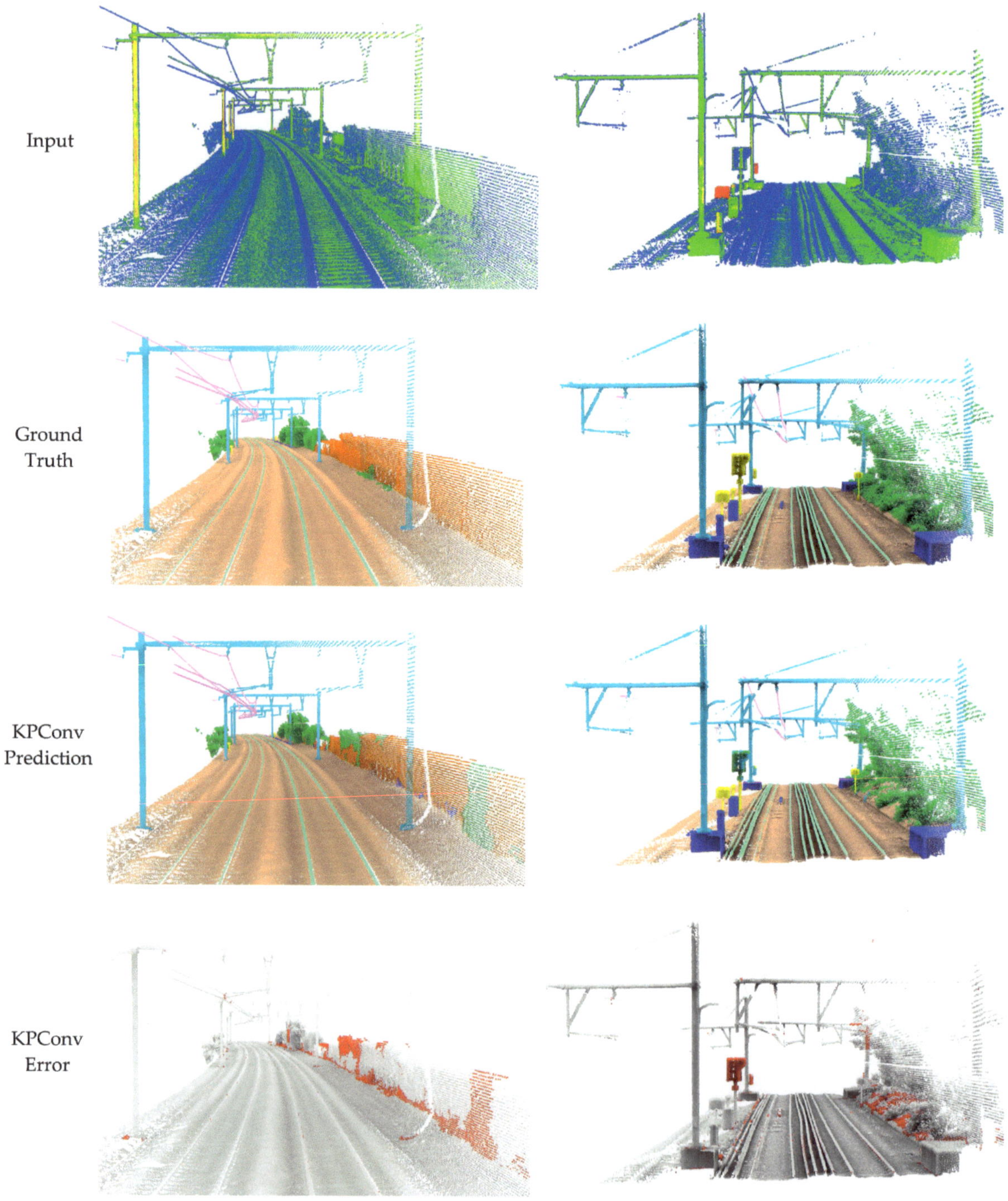

Figure A3. *Cont.*

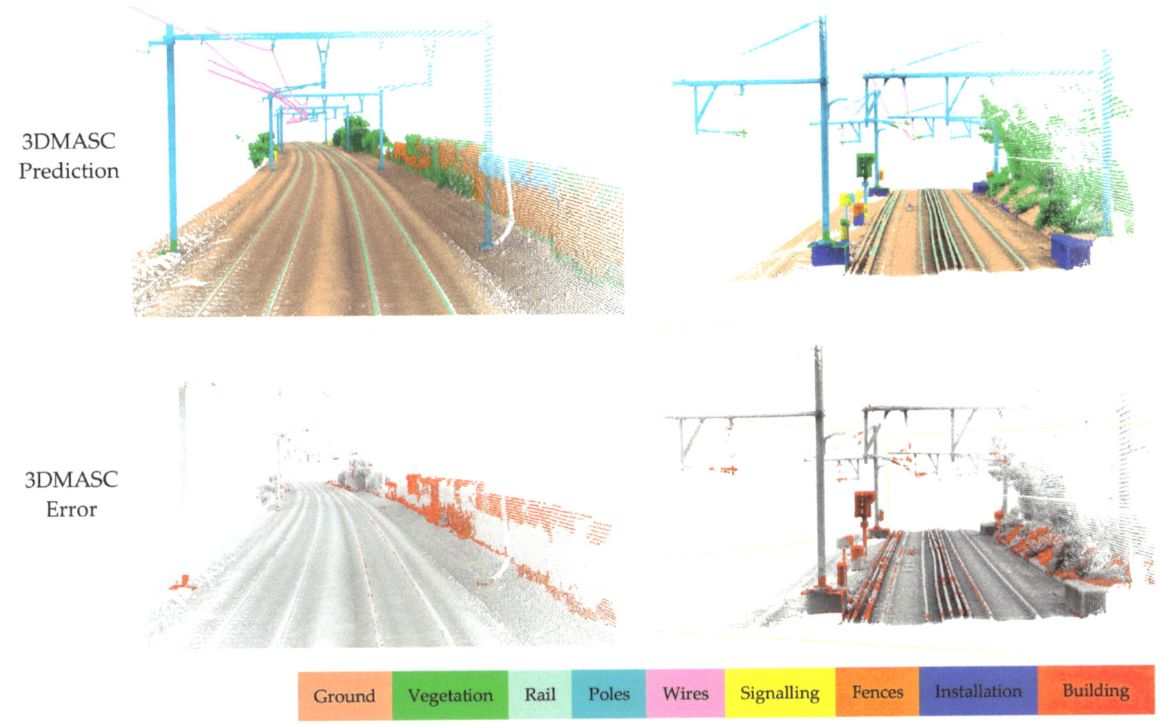

Figure A3. Qualitative results on INFRABEL subset.

References

1. Soilán, M.; Sánchez-Rodríguez, A.; Del Río-Barral, P.; Perez-Collazo, C.; Arias, P.; Riveiro, B. Review of laser scanning technologies and their applications for road and railway infrastructure monitoring. *Infrastructures* **2019**, *4*, 58. [CrossRef]
2. Lamas, D.; Soilán, M.; Grandío, J.; Riveiro, B. Automatic point cloud semantic segmentation of complex railway environments. *Remote Sens.* **2021**, *13*, 2332. [CrossRef]
3. Chen, X.; Chen, Z.; Liu, G.; Chen, K.; Wang, L.; Xiang, W.; Zhang, R. Railway overhead contact system point cloud classification. *Sensors* **2021**, *21*, 4961. [CrossRef]
4. Roynard, X.; Deschaud, J.-E.; Goulette, F. Paris-Lille-3D: A Point Cloud Dataset for Urban Scene Segmentation and Classification. Available online: http://caor-mines-paristech.fr/fr/ (accessed on 14 December 2023).
5. Tan, W.; Qin, N.; Ma, L.; Li, Y.; Du, J.; Cai, G.; Yang, K.; Li, J. Toronto-3D: A Large-scale Mobile LiDAR Dataset for Semantic Segmentation of Urban Roadways. Available online: https://www.cloudcompare.org (accessed on 14 December 2023).
6. Behley, J.; Garbade, M.; Milioto, A.; Quenzel, J.; Behnke, S.; Stachniss, C.; Gall, J. SemanticKITTI: A Dataset for Semantic Scene Understanding of LiDAR Sequences. April 2019. Available online: http://arxiv.org/abs/1904.01416 (accessed on 14 December 2023).
7. Lytkin, S.; Badenko, V.; Fedotov, A.; Vinogradov, K.; Chervak, A.; Milanov, Y.; Zotov, D. Saint Petersburg 3D: Creating a Large-Scale Hybrid Mobile LiDAR Point Cloud Dataset for Geospatial Applications. *Remote Sens.* **2023**, *15*, 2735. [CrossRef]
8. Roynard, X.; Deschaud, J.-E.; Goulette, F. Paris-Lille-3D: A Large and high-quality ground truth urban point cloud dataset for automatic segmentation and classification. *arXiv* **2017**, arXiv:1712.00032. Available online: http://arxiv.org/abs/1712.00032 (accessed on 14 December 2023). [CrossRef]
9. Zhang, R.; Wu, Y.; Jin, W.; Meng, X. Deep-Learning-Based Point Cloud Semantic Segmentation: A Survey. *Electronics* **2023**, *12*, 3642. [CrossRef]
10. Oh, K.; Yoo, M.; Jin, N.; Ko, J.; Seo, J.; Joo, H.; Ko, M. A Review of Deep Learning Applications for Railway Safety. *Appl. Sci.* **2022**, *12*, 572. [CrossRef]
11. He, Y.; Yu, H.; Liu, X.; Yang, Z.; Sun, W.; Mian, A. Deep Learning Based 3D Segmentation: A Survey. *arXiv* **2021**, arXiv:2103.05423. Available online: http://arxiv.org/abs/2103.05423 (accessed on 14 December 2023).
12. Schnabel, R.; Wahl, R.; Klein, R. Efficient RANSAC for Point-Cloud Shape Detection. *Comput. Graph. Forum* **2007**, *26*, 214–226. [CrossRef]

13. Truong, Q.H. Knowledge-Based 3D Point Clouds Processing. 2014. Available online: https://theses.hal.science/tel-00977434 (accessed on 14 December 2023).
14. Ponciano, J.-J.; Roetner, M.; Reiterer, A.; Boochs, F. Object Semantic Segmentation in Point Clouds—Comparison of a Deep Learning and a Knowledge-Based Method. *ISPRS Int. J. Geo-Inf.* **2021**, *10*, 256. [CrossRef]
15. Alkadri, M.F.; Alam, S.; Santosa, H.; Yudono, A.; Beselly, S.M. Investigating Surface Fractures and Materials Behavior of Cultural Heritage Buildings Based on the Attribute Information of Point Clouds Stored in the TLS Dataset. *Remote Sens.* **2022**, *14*, 410. [CrossRef]
16. Su, H.; Maji, S.; Kalogerakis, E.; Learned-Miller, E. Multi-View Convolutional Neural Networks for 3D Shape Recognition. Available online: http://vis-www.cs.umass.edu/mvcnn (accessed on 14 December 2023).
17. Hamdi, A.; Giancola, S.; Ghanem, B. MVTN: Multi-View Transformation Network for 3D Shape Recognition. Available online: https://github.com/ajhamdi/MVTN (accessed on 14 December 2023).
18. Dai, A.; Nießner, M. 3DMV: Joint 3D-Multi-View Prediction for 3D Semantic Scene Segmentation. *arXiv* **2018**, arXiv:1803.10409. Available online: http://arxiv.org/abs/1803.10409 (accessed on 14 December 2023).
19. Kundu, A.; Yin, X.; Fathi, A.; Ross, D.; Brewington, B.; Funkhouser, T.; Pantofaru, C. Virtual Multi-view Fusion for 3D Semantic Segmentation. *arXiv* **2020**, arXiv:2007.13138. Available online: http://arxiv.org/abs/2007.13138 (accessed on 14 December 2023).
20. Boulch, A.; Guerry, J.; Le Saux, B.; Audebert, N. SnapNet: 3D point cloud semantic labeling with 2D deep segmentation networks. *Comput. Graph.* **2018**, *71*, 189–198. [CrossRef]
21. Yang, Y.; Wu, X.; He, T.; Zhao, H.; Liu, X. SAM3D: Segment Anything in 3D Scenes. *arXiv* **2023**, arXiv:2306.03908. Available online: http://arxiv.org/abs/2306.03908 (accessed on 14 December 2023).
22. Wu, B.; Wan, A.; Yue, X.; Keutzer, K. SqueezeSeg: Convolutional Neural Nets with Recurrent CRF for Real-Time Road-Object Segmentation from 3D LiDAR Point Cloud. *arXiv* **2017**, arXiv:1710.07368. Available online: http://arxiv.org/abs/1710.07368 (accessed on 14 December 2023).
23. Wang, Y.; Shi, T.; Yun, P.; Tai, L.; Liu, M. PointSeg: Real-Time Semantic Segmentation Based on 3D LiDAR Point Cloud. *arXiv* **2018**, arXiv:1807.06288. Available online: http://arxiv.org/abs/1807.06288 (accessed on 14 December 2023).
24. Karara, G.; Hajji, R.; Poux, F. 3D point cloud semantic augmentation: Instance segmentation of 360° panoramas by deep learning techniques. *Remote Sens.* **2021**, *13*, 3647. [CrossRef]
25. Ando, A.; Gidaris, S.; Bursuc, A.; Puy, G.; Boulch, A.; Marlet, R. Marlet. RangeViT: Towards Vision Transformers for 3D Semantic Segmentation in Autonomous Driving. *arXiv* **2023**, arXiv:2301.10222. Available online: http://arxiv.org/abs/2301.10222 (accessed on 14 December 2023).
26. Xu, Y.; Tong, X.; Stilla, U. Voxel-based representation of 3D point clouds: Methods, applications, and its potential use in the construction industry. *Autom. Constr.* **2021**, *126*, 103675. [CrossRef]
27. Fang, Z.; Xiong, B.; Liu, F. Sparse point-voxel aggregation network for efficient point cloud semantic segmentation. *IET Comput. Vis.* **2022**, *16*, 644–654. [CrossRef]
28. Ye, M.; Wan, R.; Xu, S.; Cao, T.; Chen, Q. DRINet++: Efficient Voxel-as-point Point Cloud Segmentation. *arXiv* **2021**, arXiv:2111.08318.
29. Li, H.; Guan, H.; Ma, L.; Lei, X.; Yu, Y.; Wang, H.; Delavar, M.R.; Li, J. MVPNet: A multi-scale voxel-point adaptive fusion network for point cloud semantic segmentation in urban scenes. *Int. J. Appl. Earth Obs. Geoinf.* **2023**, *122*, 103391. [CrossRef]
30. Hang, S.; Jampani, V.; Sun, D.; Maji, S.; Kalogerakis, E.; Yang, M.-H.; Kau, J. SPLATNet: Sparse Lattice Networks for Point Cloud Processing. *arXiv* **2018**, arXiv:1802.08275. Available online: http://arxiv.org/abs/1802.08275 (accessed on 14 December 2023).
31. Rosu, R.A.; Schütt, P.; Quenzel, J.; Behnke, S. LatticeNet: Fast Point Cloud Segmentation Using Permutohedral Lattices. *arXiv* **2019**, arXiv:1912.05905. Available online: http://arxiv.org/abs/1912.05905 (accessed on 14 December 2023).
32. Rosu, R.A.; Schütt, P.; Quenzel, J.; Behnke, S. LatticeNet: Fast Spatio-Temporal Point Cloud Segmentation Using Permutohedral Lattices. *arXiv* **2021**, arXiv:2108.03917. Available online: http://arxiv.org/abs/2108.03917 (accessed on 14 December 2023). [CrossRef]
33. Qi, C.R.; Su, H.; Mo, K.; Guibas, L.J. PointNet: Deep Learning on Point Sets for 3D Classification and Segmentation. *arXiv* **2016**, arXiv:1612.00593. Available online: http://arxiv.org/abs/1612.00593 (accessed on 14 December 2023).
34. Qi, C.R.; Yi, L.; Su, H.; Guibas, L.J. PointNet++: Deep Hierarchical Feature Learning on Point Sets in a Metric Space. *arXiv* **2017**, arXiv:1706.02413. Available online: http://arxiv.org/abs/1706.02413 (accessed on 14 December 2023).
35. Wu, W.; Qi, Z.; Fuxin, L. PointConv: Deep Convolutional Networks on 3D Point Clouds. *arXiv* **2018**, arXiv:1811.07246. Available online: http://arxiv.org/abs/1811.07246 (accessed on 14 December 2023).
36. Thomas, H.; Qi, C.R.; Deschaud, J.-E.; Marcotegui, B.; Goulette, F.; Guibas, L.J. KPConv: Flexible and Deformable Convolution for Point Clouds. *arXiv* **2019**, arXiv:1904.08889.
37. Zeng, Z.; Xu, Y.; Xie, Z.; Wan, J.; Wu, W.; Dai, W. RG-GCN: A Random Graph Based on Graph Convolution Network for Point Cloud Semantic Segmentation. *Remote Sens.* **2022**, *14*, 4055. [CrossRef]
38. Jiang, T.; Sun, J.; Liu, S.; Zhang, X.; Wu, Q.; Wang, Y. Hierarchical semantic segmentation of urban scene point clouds via group proposal and graph attention network. *Int. J. Appl. Earth Obs. Geoinf.* **2021**, *105*, 102626. [CrossRef]
39. Landrieu, L.; Simonovsky, M. Large-scale Point Cloud Semantic Segmentation with Superpoint Graphs. *arXiv* **2017**, arXiv:1711.09869. Available online: http://arxiv.org/abs/1711.09869 (accessed on 14 December 2023).

40. Li, G.; Müller, M.; Thabet, A.; Ghanem, B. DeepGCNs: Can GCNs Go as Deep as CNNs? *arXiv* **2019**, arXiv:1904.03751. Available online: http://arxiv.org/abs/1904.03751 (accessed on 14 December 2023).
41. Lu, D.; Xie, Q.; Wei, M.; Gao, K.; Xu, L.; Li, J. Transformers in 3D Point Clouds: A Survey. *arXiv* **2022**, arXiv:2205.07417. Available online: http://arxiv.org/abs/2205.07417 (accessed on 14 December 2023).
42. Zhao, H.; Jiang, L.; Jia, J.; Torr, P.; Koltun, V. Point Transformer. *arXiv* **2020**, arXiv:2012.09164. Available online: http://arxiv.org/abs/2012.09164 (accessed on 14 December 2023).
43. Lai, X.; Liu, J.; Jiang, L.; Wang, L.; Zhao, H.; Liu, S.; Qi, X.; Jia, J. Stratified Transformer for 3D Point Cloud Segmentation. *arXiv* **2022**, arXiv:2203.14508. Available online: http://arxiv.org/abs/2203.14508 (accessed on 14 December 2023).
44. Zhou, J.; Xiong, Y.; Chiu, C.; Liu, F.; Gong, X. SAT: Size-Aware Transformer for 3D Point Cloud Semantic Segmentation. *arXiv* **2023**, arXiv:2301.06869. Available online: http://arxiv.org/abs/2301.06869 (accessed on 14 December 2023).
45. Fei, B.; Yang, W.; Liu, L.; Luo, T.; Zhang, R.; Li, Y.; He, Y. Self-supervised Learning for Pre-Training 3D Point Clouds: A Survey. *arXiv* **2023**, arXiv:2305.04691. Available online: http://arxiv.org/abs/2305.04691 (accessed on 14 December 2023).
46. Lin, X.; Luo, H.; Guo, W.; Wang, C.; Li, J. A Multi-task Learning Framework for Semantic Segmentation in MLS Point Clouds. In *Artificial Intelligence and Security*; Lecture Notes in Computer Science (Including subseries Lecture Notes in Artificial Intelligence and Lecture Notes in Bioinformatics); Springer Science and Business Media Deutschland GmbH: Berlin/Heidelberg, Germany, 2022; pp. 382–392. [CrossRef]
47. Kirillov, A.; Mintun, E.; Ravi, N.; Mao, H.; Rolland, C.; Gustafson, L.; Xiao, T.; Whitehead, S.; Berg, A.C.; Lo, W.-Y.; et al. Segment Anything. *arXiv* **2023**, arXiv:2304.02643. Available online: http://arxiv.org/abs/2304.02643 (accessed on 14 December 2023).
48. Hong, Y.; Zhen, H.; Chen, P.; Zheng, S.; Du, Y.; Chen, Z.; Gan, C. 3D-LLM: Injecting the 3D World into Large Language Models. *arXiv* **2023**, arXiv:2307.12981. Available online: http://arxiv.org/abs/2307.12981 (accessed on 14 December 2023).
49. Arastounia, M. Automated recognition of railroad infrastructure in rural areas from LIDAR data. *Remote Sens.* **2015**, *7*, 14916–14938. [CrossRef]
50. Vosselman, G.; Klein, R. Visualisation and structuring of point clouds. In *Airborne and Terrestrial Laser Scanning*; Vosselman, M.G., Maas, H.G., Eds.; CRC Press (Taylor & Francis): Boca Raton, FL, USA, 2010; pp. 45–81.
51. Muja, M.; Lowe, D.G. Fast Approximate Nearest Neighbors with Automatic Algorithm Configuration. In *VISAPP 2009, Proceedings of the Fourth International Conference on Computer Vision Theory and Applications, Lisboa, Portugal, 5–8 February 2009*; Ranchordas, A., Araújo, H., Eds.; INSTICC Press: Lisbon, Portugal, 2009; Volume 1, pp. 331–340.
52. Chen, L.; Jung, J.; Sohn, G. Multi-Scale Hierarchical CRF for Railway Electrification Asset Classification from Mobile Laser Scanning Data. *IEEE J. Sel. Top. Appl. Earth Obs. Remote Sens.* **2019**, *12*, 3131–3148. [CrossRef]
53. Grandio, J.; Riveiro, B.; Soilán, M.; Arias, P. Point cloud semantic segmentation of complex railway environments using deep learning. *Autom. Constr.* **2022**, *141*, 104425. [CrossRef]
54. Ton, B.; Ahmed, F.; Linssen, J. Semantic Segmentation of Terrestrial Laser Scans of Railway Catenary Arches: A Use Case Perspective. *Sensors* **2023**, *23*, 222. [CrossRef]
55. Zendel, O.; Murschitz, M.; Zeilinger, M.; Steininger, D.; Abbasi, S.; Beleznai, C. RailSem19: A Dataset for Semantic Rail Scene Understanding. Available online: www.wilddash.cc (accessed on 14 December 2023).
56. Harb, J.; Rébéna, N.; Chosidow, R.; Roblin, G.; Potarusov, R.; Hajri, H. FRSign: A Large-Scale Traffic Light Dataset for Autonomous Trains. *arXiv* **2020**, arXiv:2002.05665. Available online: http://arxiv.org/abs/2002.05665 (accessed on 14 December 2023).
57. Toprak, T.; Aydın, B.; Belenlioğlu, B.; Güzeliş, C.; Selver, M.A. Railway Pedestrian Dataset (RAWPED). *Zenodo* **2020**. Available online: https://zenodo.org/records/3741742 (accessed on 14 December 2023).
58. Toprak, T.; Belenlioglu, B.; Aydin, B.; Guzelis, C.; Selver, M.A. Conditional Weighted Ensemble of Transferred Models for Camera Based Onboard Pedestrian Detection in Railway Driver Support Systems. *IEEE Trans. Veh. Technol.* **2020**, *69*, 5041–5054. [CrossRef]
59. Leibner, P.; Hampel, F.; Schindler, C. GERALD: A novel dataset for the detection of German mainline railway signals. *Proc. Inst. Mech. Eng. Part F J. Rail Rapid Transit.* **2023**, *237*, 1332–1342. [CrossRef]
60. Tagiew, R.; Klasek, P.; Tilly, R.; Köppel, M.; Denzler, P.; Neumaier, P.; Klockau, T.; Boekhoff, M.; Schwalbe, K. OSDaR23: Open Sensor Data for Rail 2023. *arXiv* **2023**, arXiv:2305.03001. Available online: http://arxiv.org/abs/2305.03001 (accessed on 14 December 2023).
61. WHU-Railway3D: A Diverse Dataset and Benchmark for Railway Point Cloud Semantic Segmentation. Available online: https://github.com/WHU-USI3DV/WHU-Railway3D (accessed on 13 March 2024).
62. Ton, B. *Labelled High Resolution Point Cloud Dataset of 15 Catenary Arches in The Netherlands*; 4TU.ResearchData: Delft, The Netherlands, 2022. [CrossRef]
63. Tagiew, R.; Köppel, M.; Schwalbe, K.; Denzler, P.; Neumaier, P.; Klockau, T.; Boekhoff, M.; Klasek, P.; Tilly, R. Open Sensor Data for Rail 2023. *arXiv* **2023**, arXiv:2305.03001. [CrossRef]
64. Eastepp, M.; Faris, L.; Ricks, K. UA_L-DoTT: University of Alabama's large dataset of trains and trucks. *Data Brief* **2022**, *42*, 108073. [CrossRef]
65. D'Amico, G.; Marinoni, M.; Nesti, F.; Rossolini, G.; Buttazzo, G.; Sabina, S.; Lauro, G. TrainSim: A Railway Simulation Framework for LiDAR and Camera Dataset Generation. *arXiv* **2023**, arXiv:2302.14486. Available online: http://arxiv.org/abs/2302.14486 (accessed on 14 December 2023). [CrossRef]
66. Fayjie, R.; Vandewalle, P. Few-shot learning on point clouds for railroad segmentation. *Electron. Imaging* **2023**, *35*, 100-1–100-5. [CrossRef]

67. Wang, Y. *Railway SLAM Dataset*; IEEE Dataport: Piscataway, NJ, USA, 2022. [CrossRef]
68. Corongiu, M.; Masiero, A.; Tucci, G. Classification of railway assets in mobile mapping point clouds. In *International Archives of the Photogrammetry, Remote Sensing and Spatial Information Sciences—ISPRS Archives*; International Society for Photogrammetry and Remote Sensing: Hanover, Germany, 2020; pp. 219–225. [CrossRef]
69. Riquelme, J.L.A.; Ruffo, M.; Tomás, R.; Riquelme, A.; Pagán, J.I.; Cano, M.; Pastor, J.L. 3D Point Cloud of a Railway Slope—MOMIT (Multi-Scale Observation and Monitoring of Railway Infrastructure Threats) EU Project—H2020-EU.3.4.8.3.—Grant Agreement ID: 777630. *Zenodo* **2020**. Available online: https://zenodo.org/records/3777996 (accessed on 14 December 2023). [CrossRef]
70. Cserep, M. Hungarian MLS Point Clouds of Railroad Environment and Annotated Ground Truth Data. *Mendeley Data* **2022**. Available online: https://data.mendeley.com/datasets/ccxpzhx9dj/1 (accessed on 28 November 2023). [CrossRef]
71. González-Collazo, S.M.; Balado, J.; González, E.; Nurunnabi, A. A discordance analysis in manual labelling of urban mobile laser scanning data used for deep learning based semantic segmentation. *Expert Syst. Appl.* **2023**, *230*, 120672. [CrossRef]
72. Girardeau-Montaut, G. CloudCompare. 12 July 2023. Available online: https://www.cloudcompare.org/ (accessed on 14 December 2023).
73. De Gélis, I.; Lefèvre, S.; Corpetti, T. Siamese KPConv: 3D multiple change detection from raw point clouds using deep learning. *ISPRS J. Photogramm. Remote Sens.* **2023**, *197*, 274–291. [CrossRef]
74. Sevgen, E.; Abdikan, S. Classification of Large-Scale Mobile Laser Scanning Data in Urban Area with LightGBM. *Remote Sens.* **2023**, *15*, 3787. [CrossRef]
75. Letard, M.; Lague, D.; Le Guennec, A.; Lefèvre, S.; Feldmann, B.; Leroy, P.; Girardeau-Montaut, D.; Corpetti, T. 3DMASC: Accessible, explainable 3D point clouds classification. Application to Bi-Spectral Topo-Bathymetric lidar data. *ISPRS J. Photogramm. Remote Sens.* **2024**, *207*, 175–197. Available online: https://hal.science/hal-04072068 (accessed on 14 December 2023). [CrossRef]
76. Li, Y.; Fan, C.; Wang, X.; Duan, Y. SPNet: Multi-Shell Kernel Convolution for Point Cloud Semantic Segmentation. *arXiv* **2021**, arXiv:2109.11610. Available online: http://arxiv.org/abs/2109.11610 (accessed on 14 December 2023).
77. Thomas, H.; Qi, C.R.; Deschaud, J.-E.; Marcotegui, B.; Goulette, F.; Guibas, L.J. KPConv: Flexible and Deformable Convolution for Point Clouds. In Proceedings of the IEEE/CVF International Conference on Computer Vision (ICCV), Seoul, Republic of Korea, 24 October–2 November 2019.
78. Thomas, H.; Deschaud, J.-E.; Marcotegui, B.; Goulette, F.; Le Gall, Y. Semantic Classification of 3D Point Clouds with Multiscale Spherical Neighborhoods. *arXiv* **2018**, arXiv:1808.00495. Available online: http://arxiv.org/abs/1808.00495 (accessed on 14 December 2023).
79. Ke, G.; Meng, Q.; Finley, T.; Wang, T.; Chen, W.; Ma, W.; Ye, Q.; Liu, T.-Y. LightGBM: A Highly Efficient Gradient Boosting Decision Tree. 2017. Available online: https://github.com/Microsoft/LightGBM (accessed on 14 December 2023).
80. Grandini, M.; Bagli, E.; Visani, G. Metrics for Multi-Class Classification: An Overview. *arXiv* **2020**, arXiv:2008.05756. Available online: http://arxiv.org/abs/2008.05756 (accessed on 14 December 2023).
81. González, E.; Balado, J.; Arias, P.; Lorenzo, H. Realistic correction of sky-coloured points in Mobile Laser Scanning point clouds. *Opt. Laser Technol.* **2022**, *149*, 107807. [CrossRef]

drones

MDPI

Article

Digital Battle: A Three-Layer Distributed Simulation Architecture for Heterogeneous Robot System Collaboration

Jialong Gao, Quan Liu, Hao Chen, Hanqiang Deng, Lun Zhang, Lei Sun and Jian Huang *

College of Intelligence Science and Technology, National University of Defense Technology, Changsha 410073, China; gaojialong@nudt.edu.cn (J.G.); liuquan@nudt.edu.cn (Q.L.); chenhao@nudt.edu.cn (H.C.); denghanqiang16@nudt.edu.cn (H.D.); zhanglun@nudt.edu.cn (L.Z.); sunlei17a@nudt.edu.cn (L.S.)
* Correspondence: huang_jian@nudt.edu.cn

Abstract: In this paper, we propose a three-layer distributed simulation network architecture, which consists of a distributed virtual simulation network, a perception and control subnetwork, and a cooperative communication service network. The simulation architecture runs on a distributed platform, which can provide unique virtual scenarios and multiple simulation services for the verification of basic perception, control, and planning algorithms of a single-robot system and can verify the distributed collaboration algorithms of heterogeneous multirobot systems. Further, we design simulation experimental scenarios for classic heterogeneous robotic systems such as unmanned aerial vehicles (UAVs) and unmanned ground vehicles (UGVs). Through the analysis of experimental measurement data, we draw several important conclusions: firstly, the replication time characteristics and update frequency characteristics of entity synchronization in our system indicate that the replication time of entity synchronization in our system is relatively short, and the update frequency can meet the needs of multirobot collaboration and ensure the real-time use and accuracy of the system; secondly, we analyze the bandwidth usage of data frames in the whole session and observe that the server side occupies almost half of the data throughput during the whole session, which indicates that the allocation and utilization of data transmission in our system is reasonable; and finally, we construct a bandwidth estimation surface model to estimate the bandwidth requirements of the current model when scaling the server-side scale and synchronization-state scale, which provides an important reference for better planning and optimizing of the resource allocation and performance of the system. Based on this distributed simulation framework, future research will improve the key technical details, including further refining the coupling object dynamic model update method to support the simulation theory of the coupling relationship between system objects, studying the impact of spatiotemporal consistency of distributed systems on multirobot control and decision making, and in-depth research on the impact of collaborative frameworks combined with multirobot systems for specific tasks.

Keywords: heterogeneous robot systems; distributed simulation; communications middleware; collaborative mechanism

Citation: Gao, J.; Liu, Q.; Chen, H.; Deng, H.; Zhang, L.; Sun, L.; Huang, J. Digital Battle: A Three-Layer Distributed Simulation Architecture for Heterogeneous Robot System Collaboration. *Drones* **2024**, *8*, 156. https://doi.org/10.3390/drones8040156

Academic Editors: Tomasz Nowakowski, Artur Kierzkowski, Agnieszka A. Tubis, Franciszek Restel, Tomasz Kisiel, Anna Jodejko-Pietruczuk and Mateusz Zając

Received: 19 February 2024
Revised: 11 April 2024
Accepted: 16 April 2024
Published: 18 April 2024

1. Introduction

Unmanned and autonomous systems play an essential role in various fields—whether industrial, civil, or military—and experts of unmanned and autonomous systems have become integral members across diverse sectors, encompassing industrial, civil, and military applications, prompting experts from these domains to explore the multifaceted potentialities of task collaboration within heterogeneous robot systems. Heterogeneous robot systems refer to an assembly of robots that are diverse in terms of their functionalities, structures, and capabilities and are designed to work collaboratively towards achieving complex objectives [1]. This consortium of assorted robotic platforms often includes a mix of unmanned aerial vehicles (UAVs) and unmanned ground vehicles (UGVs), among others, enabling a richer interaction and broader operational scope compared to homogeneous systems. The synergy within these heterogeneous systems strives to harness the distinct strengths and capabilities of each robot type, enhancing the overall efficiency, adaptability, and resilience of the collective in executing coordinated tasks or missions across varying environments. These fields continue to consider how to research the various possibilities of task collaboration within heterogeneous robot systems.

However, before the effective deployment of multirobot systems to perform complex tasks can be achieved, the research community still faces numerous fundamental technical challenges. These challenges, as listed in [2], encompass areas such as big data, the Internet of Things, task complexity, autonomous machine learning, scalability and heterogeneity trade-offs, coalition formation and task allocation, human in the loop, transfer learning, unified frameworks, communication constraints, and connectivity uncertainty. In addition, there are complex application challenges outlined in [3], including adaptive heterogeneous architecture and modeling methods for robot swarm systems; distributed perception and cognition of high-dimensional situations [4]; intelligent decision making and planning of robot swarm systems that can be guided, trusted, and evolved; and autonomous collaborative control of robot swarm systems.

In recent years, significant advancements have been made in the fields of robot operating systems, unmanned system task planning, formation control, ad hoc network communication, and intelligent algorithms. The collaborative sensing of robot clusters primarily encompasses collaborative positioning technology [5,6], collaborative target recognition, multisource sensor collaborative optimization, and collaborative target tracking [7]. Intelligent decision making and planning in cluster systems involve traditional methods such as the Hungarian algorithm, auction algorithm, particle swarm algorithm, and corridor-based methods [8], as well as collaborative task networks constructed using machine learning models [9]. The cooperative control of robot clusters has also seen extensive theoretical research on fundamental modes like consistency control, formation control [10], and formation tracking control [11].

When robots perform complex tasks, whether operating independently or collaborating in clusters, they require a simulation architecture for integrated testing and validation. Distributed simulation technology, owing to its characteristics, holds the potential to serve as the simulation verification scheme that most closely resembles the actual working environment of heterogeneous multirobot systems engaged in collaborative task algorithms. Past developments in distributed system technology have encompassed research on synchronization techniques based on message dialogue [12], the implementation of flexible access technologies founded on active services [13], and the provision of reliable operational support guarantees [14]. These advancements have positioned distributed simulation as a viable approach to accurately simulate and verify collaborative task algorithms in robot clusters [15–17], facilitating effective evaluation and testing in environments closer to real-world conditions. Through distributed simulation, it becomes possible to validate the collaborative behavior and performance of robots in complex environments while also assessing the robustness and reliability of robot systems. This contributes significant support and assurance for the rapid advancement and application of robot technology. In addition, there are many simulation systems that can also provide UAV operation simulation training when considering the way people control the loop [18–20].

Based on the theoretical work of predecessors, we propose an open system simulation architecture. In this paper, we first list some relevant technical achievements that inspired our work and describe the progress they have made (see Section 2). Next, we focus on introducing the architecture of the three-layer network simulation verification platform we propose for implementing collaborative algorithms of heterogeneous robot systems (see Section 3). In the experimental section (refer to Section 4), we delineate the experimental procedures and scenario designs. In the discussion Section 5, we analyze network characteristics and provide reference suggestions for building similar architectures. Finally, in Section 6, we offer some shallow insights into the study of collaborative networks.

2. Related Work

Over the past 25 years, robot simulation platforms have evolved almost simultaneously with robotics (see Table 1). Initially, research and development enterprises of robot hardware platforms provided simulation modules that could be integrated into Simulink, enabling users to facilitate secondary development. Meanwhile, the complexity of robot structures increased, leading to a wider range of tasks. As robot diversity expanded, the modeling complexity also grew higher. Consequently, a single module could no longer meet user demands, and independent simulation software began to emerge in collaborations between enterprises and universities [21–24].

Subsequently, as feature engineering and other multisensor fusion technologies advanced, simulation platforms not only had to provide basic dynamic simulation but also had to simulate digital signals from multiple sensors, such as visual cameras (even infrared), LiDAR (light detection and ranging) in [23,25,26], and SONAR (sound navigation and ranging) in [27], allowing for a comprehensive and realistic representation of the robot's perception capabilities in the virtual environment. Fast-forward to the past five years, and with the progress of artificial intelligence technology, simulation platforms have found greater utility in providing training samples. This utilization leverages efficient training mechanisms offered by emerging learning algorithms, thus enabling robots to perform better on many nonlinear problems [28,29].

Table 1. A representative series of simulation platforms for robot systems.

First Published Year	Institution	Software *	Tool	Render	Dynamics	Sensors	Vehicles
1998	Swiss Federal Institute of Technology in Lausanne [21]	webots	OpenGL		ODE	IMU, RGBD	Robots
2001	Simon Fraser University [30]	Stage	FLTK	OpenGL	–	IMU, RGBD	Robots
2007	University of California and University of Pittsburgh [31]	USARSim	Unreal Engine 2.0		ODE	IMU, RGB	Robots
2008	Coppelia Robotics	V-REP	OpenGL		ODE, Bullet Physics, Newton Dynamics, Vortex Physics Engine	IMU, RGB, LiDAR, sonar, GPS	Wheeled robots, tracked robots, four-wheel drive vehicles, aircraft
2010	Carnegie Mellon University [32]	OpenRave	OSG		IKFast	IMU, RGB	Robotic arm
2011	Free University of Brussels [33]	ARGoS	Qt-OpenGL		ODE Chipmunk	IMU, RGB	Robots
2012	Universitat Jaume I [27,34]	UWSim	OSG	osgOcean	Bullet	IMU, RGB, sonar	Underwater robots
2013	Technical University of Darmstadt	Hector	Gazebo(ROS)	OpenGL	ODE	IMU, RGB	Drones
2016	University of Zurich (ETH Zurich) [35]	RotorS	Gazebo(ROS)	OpenGL	ODE	IMU, RGBD	Drones
2016	OpenAI [22]	OpenAI-Gym	Gym		Mujoco [36]	IMU	Multijoint robot
2017	Inter, Toyota, etc. [37]	CARLA	Unreal Engine		PhysX	IMU, RGBD	Ground vehicles
2017	MicroSoft [23]	Airsim	Unreal Engine		PhysX fastsim	IMU, RGBD, Segment, LiDAR	Drones, ground vehicles
2019	MIT [38,39]	FlightGoggle	Unity		-	IMU, RGBD, Segment	Drones, ground vehicles
2020	University of Texas and Stanford University [40]	robosuite	mujoco-py	OpenGL	Mujoco	IMU, RGBD	Robots
2020	National University of Defense Science and Technology, Beihang University, etc. [41,42]	XTDrone	Gazebo(ROS)	OpenGL	ODE	IMU, RGBD	Drones, robots
2021	University of Zurich and ETH Zurich [43]	Flightmare	Unity		-	IMU, RGBD, Segment	Drones
2021	Stanford University [44]	IGibson	Unity		-	IMU, RGBD, Segment	Drones
2021	Beihang University and Central South University [45]	Rflysim	Unreal Engine		CopterSim	IMU, RGBD	UAV
2021	Nvidia [24]	Issac Gym	Issac Gym		PhysX	IMU	Bipedal robot, robotic arm, etc.
2022	University of Hong Kong [25]	MARSIM	ROS	OpenGL	-	IMU, LiDAR (HD)	Quadrotor
2023	Institute of Automation, Chinese Academy of Sciences [26]	NeuronsGym	Unity3D		-	IMU, LiDAR	Mecanum wheeled robot
2023	Cosys-Lab (FTI) of University of Antwerp, Chinese Academy [46]	COSYS-Airsim	Unreal Engine		-	IMU, LiDAR, RGB	Drones, ground vehicles

Software *: All software versions are publicly available as of December 2023.

These systems provide platform support for research, learning, testing, and validation of control, perception, navigation, and adversarial tasks. Driven by application requirements, the development of robot simulation technology has been continuously progressing. It has evolved from simulating single objects to multiple heterogeneous objects, from simple static environments to complex environments, and from cooperative collaboration to adversarial games. In addition, the scale of software has been growing increasingly large.

The previous approach of simulating and accessing multiple controllers on a single computing terminal is no longer sufficient to meet the demands. A distributed simulation architecture is needed to address the challenges of scalability. Therefore, our task is to propose a scalable distributed multirobot collaborative simulation architecture to meet the growing demands for functionality and complexity.

Furthermore, the distributed nature of the simulation architecture possesses an inherent advantage: it more accurately reflects the communication characteristics present in real-world working conditions, regardless of whether the multirobot system employs centralized or distributed collaborative algorithms. This advantage underscores the significance of utilizing a distributed network structure for deploying multirobot systems.

3. Architecture for Digital Battle

Drawing on the design experience of control and perception modules from related works, and taking into account the coordination of multirobot systems at the information level as well as the coupling relationships in kinematics, dynamics, and other aspects, we formally propose a three-tiered distributed simulation architecture tailored for heterogeneous multirobot collaboration.

Taking into consideration the potential complexity associated with the collaborative operation of heterogeneous robot systems, we adopted three fundamental principles in our design approach:

i The simulation process and outcomes must adhere to the laws of physics, ensuring their consistency across distributed simulation nodes;

ii The capabilities of each robot's motion platform are determined by the structural design, accessory selection, and control implementation;

iii The collaborative algorithms employed in heterogeneous robot systems should be validated through authentic distributed verification methods.

According to these three designing principles, we propose an architecture to organize models and algorithms by three key layers (Figure 1):

(1). Distributed virtual simulation network layer.

It is mainly aimed at maintaining a unified virtual world for every distributed participant. As most massively online role-playing games have employed, a popular technical solution is a client–server(C/S) network, where the clients keep a local copy of the virtual world, and the server keeps the authority of the simulation with the responsibility for four services, viz. physical interaction, scenario management, synchronization service, and session management.

(2). Perception and control subnetwork layer.

We retained many of the advantages of Airsim's solution [23]. This subnetwork connects the virtual avatar with its bonded "brain" (i.e., the one that possesses the control). The virtual avatar could be a motion platform with actuator and sensor units, while the "brain" could be a mature hardware with an RTOS (real-time operating system), a set of algorithms, or even a person, i.e., the subnetwork could work either in the artificial intelligence mode or the human-in-the loop mode.

(3). Collaborative communication service network layer.

As has been proven to be a success in real-time distributed embedded systems, the idea of a data distributed service (DDS) [47] is also adopted here to provide a basic information transmission mechanism. It is believed to support most of the possible collaborations between the upper compute node of heterogeneous robot systems.

As a distributed system, the startup process of simulation is a bit cumbersome. There are two primary constraints on the startup order: (1) The distributed virtual simulation network server should launch before other clients participate. (2) The perception and control subnetwork can only establish the connection between the avatar and the "brain" after the virtual character (i.e., the avatar) has been created in the virtual world. In addition, the collaborative communication service network can work whenever a "brain" is online. Next, more details about these three key layers' services are described.

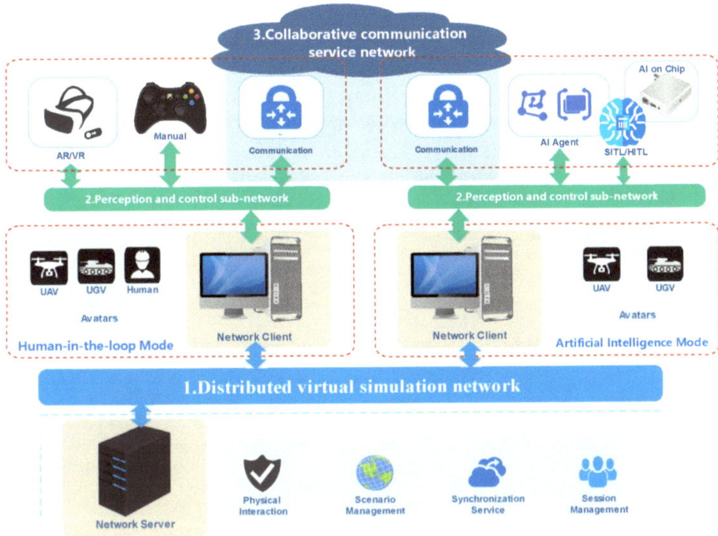

Figure 1. The architecture of the distributed simulation platform with three key layers of the network.

3.1. Distributed Virtual Simulation Network

To assure all the participants feel like staying in the same virtual world, the distributed virtual simulation network provides a series of easy-to-use C/S services. Among these services, session management only deals with the long-term connection mechanism for virtual clients and servers (Figure 2), while scenario management, physical interaction, and the synchronization service work together to coordinate processing related to simulation content on distributed platforms.

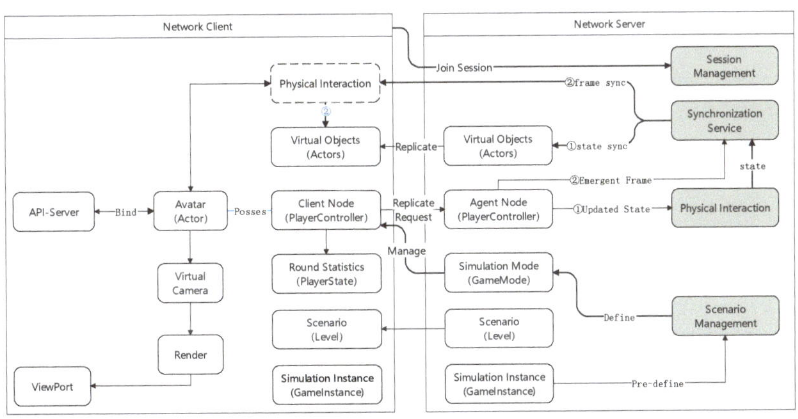

Figure 2. Architecture of the distributed virtual simulation network.

3.1.1. Session Management

From the perspective of distributed simulation, session management is the functional basis for the server–client architecture connection [48]. The essential functions provided mainly include providing an available port when creating simulation services on the server side. The client discovers and joins the server side in the network, providing role selection and publicity of virtual robot system objects for specific simulation tasks.

The entire distributed simulation session is divided into three stages (Figure 3): (1) At the beginning After creating and joining a simulation session, the server controls all participating clients' entry into the their local simulation map synchronously, generates corresponding virtual objects at the starting position set in the scenario management, and obtains control permissions over them. (2) During the process. During the simulation, session management provides continuous connection support for data synchronization between the server and the client. This synchronized data contain individual dynamic objects in the scene, as well as robot system entities controlled by the client. (3) At the end. After the distributed simulation task ends, the clients exit the server-managed session and end all data services. All these features are customizations of and upgrades on the mechanics of the game engine's multiplayer sessions.

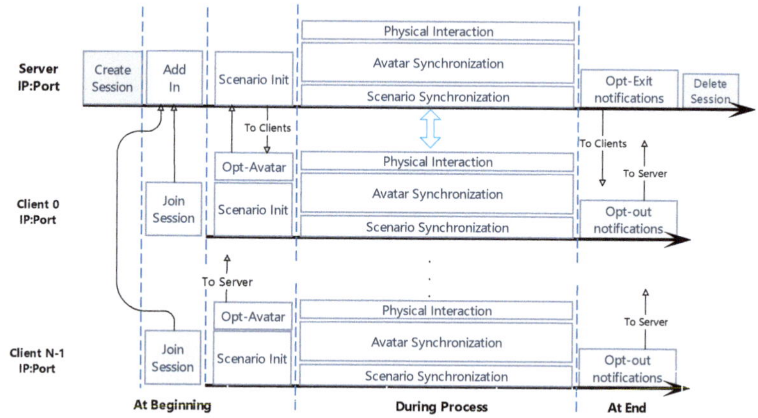

Figure 3. The entire distributed simulation session. The session mainly organizes the server and clients in long-term connections and automatically delivers data for scenario synchronization, avatar synchronization, and physical interaction.

3.1.2. Scenario Management

Like most simulation systems and virtual games, the scenario usually defines a global coordinate system in virtual space (either in a local three-dimensional Cartesian coordinate system or using spherical coordinates of the Earth), initializes various entity objects in that coordinate system, and manages every simulation-involved object, e.g., landscape, buildings, roads, nonplayer characters (NPCs), robotics under control, and even the wind field.

As shown in Figure 4, scenario management is mainly a matter of storage and calculation.

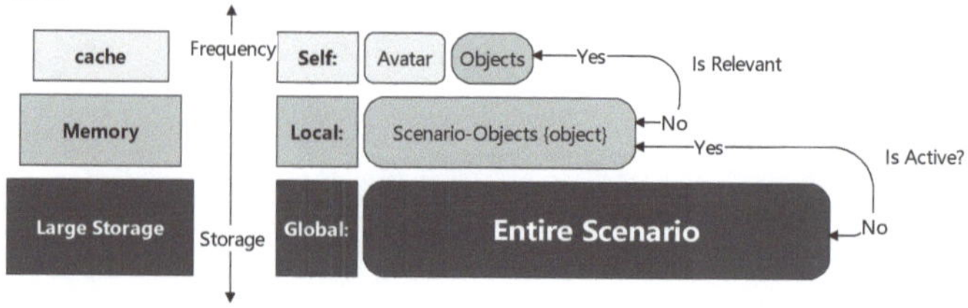

Figure 4. Scenario management plans storage based on local dependencies and activity.

The difference is that the server updates a global scenario, while clients only act on local space.

(1). On the server.

Based on the fundamental understanding that the server possesses superior parallel computing and memory read/write capabilities, it is responsible for maintaining global scene information. When receiving a request from a client to update object properties in the scene, the server first checks for potential conflicts, then determines the multicast node list based on the activity level of each client involved in the update, and finally distributes the update results to the respective clients.

(2). On the client.

The client-side scene management loads local object data according to the specific requirements of each client. There are three main categories for handling object data: (1) objects closely related to the controlled objects are stored in a cache for efficient access and support dynamic calculations at higher frequencies; (2) objects that are not directly affected by the client, but require awareness, are kept in memory and synchronized with the server's replication commands; (3) unrelated objects are set to an inactive state and remain stored on the hard disk as a resource.

3.1.3. Physical Interaction

Inspired by Mujoco's state-of-the-art performance in contact dynamics [36], we believed it is worthwhile to provide an interaction mechanism to support similar calculation on our distributed platform. Commonly, interaction calculation has three key steps:

Step 1.

Since a certain interaction must be triggered by some specific condition, we might as well define the conditional function $\mathbf{C}$ as in Equation (1). For arbitrarily selected objects obj_i and obj_j with generalized state parameters X_i and X_j, the positiveness of function $\mathbf{C}$ can indicate whether the interaction conditions are met. Mathematically, function $\mathbf{C}$ is a binary relationship.

$$c_{i,j} = \mathbf{C}(X_i, X_j) \tag{1}$$

A graph $\mathbf{G} = (V, E)$ can describe the interaction relations within objects [49]. The vertex set $V \triangleq \{obj_i\}$ and the edge set $E = \{e_{i,j}\}$ construct an adjacency matrix, with each element $e_{i,j}$,

$$e_{i,j} = \begin{cases} 1, & c_{i,j} >= 0 \\ 0, & else \end{cases} \tag{2}$$

(For those discrete event-driven interactions that can be linearly separated into pairs, skip directly to step 3.)

Step 2.

For those relatively complicated cases, instead of listing various kinds of interaction models in detail, we use standard Hamiltonian systems as a theoretic representative of multi-rigid-body dynamics [50]. Suppose there exists a connected subgraph $G_l = (V_l, E_l)$ (i.e., $V_l \subseteq V$, $E_l \subseteq E$, and the derived distance matrix D_l of G_l, s.t., $\forall d_{i,j} \in D_l$, there is $0 < d_{i,j} \leq l$). This subgraph G_l illustrates the object set V_l as isolated from other objects by distance l. The edges in E_l and the corresponding constraints determine a tangent bundle (i.e., configuration space) spanned by the objects in V_l, e.g., along with the constraint, [51]

$$\boldsymbol{\psi}(\mathbf{p}, \mathbf{q}, t) = \mathbf{0} \tag{3}$$

The constrained forces, subject to the D'Alembert–Lagrange principle, define the virtual displacements ω and constrained forces $\mathbf{Q}^C$ where there would be $\omega^T \mathbf{Q}^C = 0$, and also with the assumption that multi-rigid-body problem is subject to the Hamilton's principle of least action.

$$\delta \mathbf{S}(\mathbf{p}, \mathbf{q}, t) \triangleq \delta \int_t^{t+T} (\mathbf{p} \cdot \mathbf{q} - H(\mathbf{p}, \mathbf{q}, t))dt = 0 \tag{4}$$

The motion equation is the cotangent bundle in Equation (5), where $\mathbf{F}^u$ refers to the combined force of the input driving force and the possible dissipative force [52–54].

$$\dot{\mathbf{q}} = \frac{\partial H}{\partial \mathbf{p}}, \dot{\mathbf{p}} = -\frac{\partial H}{\partial \mathbf{q}} + \mathbf{Q}^C(\mathbf{p}, \mathbf{q}, \mathbf{t}) + \mathbf{F}^u \tag{5}$$

If the expression in Equation (5) is separable, the explicit differential expression could be constructed easily [55], but for those implicit issues, the construction method depends on the area of the problem and is beyond the scope of this article.

Step 3.

Suppose we already have an explicit expression pattern Equation (6) for every object involved in the interaction subgraph $G_I = (V_I, E_I)$. The evolution of the isolated interacted system can be figured out and updated for every distributed client.

$$X_i(t+T) = \mathrm{F}(X_i(t), \mathbf{Q}^C, \mathbf{F}^u, T) \tag{6}$$

3.1.4. Synchronization Service

A synchronization service is the key to maintaining limited consistency for each node in a distributed virtual network. Figure 2 shows two paths for synchronization, namely, emergent frame and state update.

(1). Emergent frame.

In objects with coupling relationships (e.g., constraints on the state), if there are more than two related clients, emergent frame is the event-driven way to reconstruct the motion model on the related clients. This specific method is based on the remote procedure call (RPC) mechanism, and the server triggers the response of the relevant client to reconstuct the motion model (i.e., step 2 in the physical interaction).

(2). State update.

It is a cycle-adaptive multicast method. The client will detect whether the properties of the local object change for every time interval and send a synchronization request to the server if they change. The server accepts the request and confirms that it is valid and then packages the object of the application attribute change in the next state synchronization frame and synchronizes it to the relevant client according to the multicast list. The selection of time intervals is positively correlated with the number of occurrences in other client local zones and does not exceed the upper limit allowed by the network. Assuming that all properties with amount D are recognized as needing to be synchronized to other $n-1$ clients within the synchronization allowable time T_s, the supremum of the state update network traffic as stream S_v can be inferred as

$$\sup S_v = \frac{(n-1)D(n)}{T_s} \tag{7}$$

3.2. Perception and Control Subnetwork

The perception and control network is a one-to-one network abstracted from Airsim [23]. Sensor data, actuator signals, perception data, and the desired state can either use the Mavlink protocol to connect the algorithm hardware and the virtual client (Figure 5) through the serial port or use remote procedure call (RPC) locally to realize the interoperability between the algorithm software and the virtual client or access the user through the peripheral of human–computer interaction.

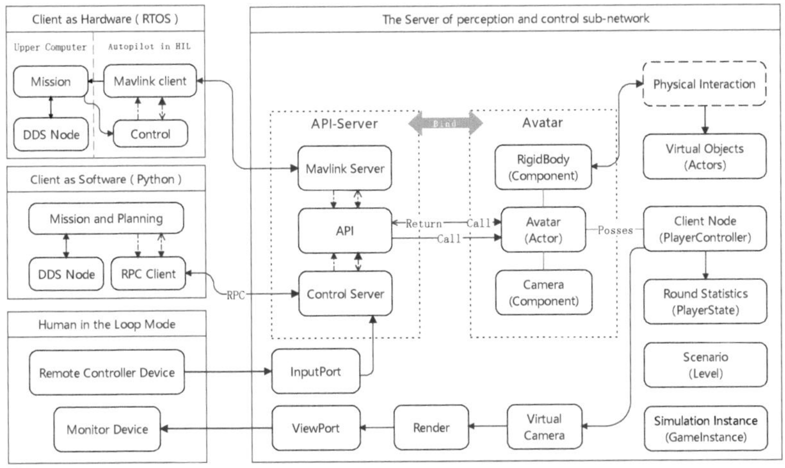

Figure 5. Architecture of the perception and control subnetwork.

Unlike the other two-layer networks, the perception and control subnetwork is an independent connection mechanism, which only requires the baud rate of the local hardware interface or the read/write rate of interprocess communication to meet the bandwidth requirements of state data and control instructions.

3.2.1. Artificial Intelligence Mode

The simulation verification subject for this platform's design is the robot system's planning and control algorithm. For clients in artificial intelligence mode, algorithm access methods are mainly divided into hardware in the loop and software in the loop.

Since the update frequency of sensor data and control data is fixed, it may as well be set up where m_i is the size of the ith datum and f_i is the corresponding update frequency. Thus, the average traffic on the perception and control subnetwork is

$$S_c = \sum_{i=1}^{n} f_i \cdot m_i \tag{8}$$

(1). Hardware in the loop.

Thanks to the contribution of the open-source community, we can directly integrate Mavlink support into the hardware in the loop and enable autonomous driving hardware to drive the corresponding robot system in the virtual environment. The hardware in the loop is usually transmitted through the serial communication interface at a negotiated baud rate and the actual effective bandwidth appropriate to Equation (8).

(2). Software in the loop.

The simulation of software in the loop, based on the interprocess communication mode, connects the input and output of the artificial intelligence algorithm with the request and service form of remote process call, and the artificial intelligence algorithm is responsible for environmental perception, task decision making, action planning, and specific control node function implementation and the organizational framework and is not limited by the three-layer distributed simulation system architecture of this paper, as long as it meets the process communication rate constraints of the local perception-control loop.

3.2.2. Human-in-the-Loop Mode

The human-loop mode provides two possibilities for use. One is to directly send the input of the remote control handle of the human as a command signal to the API end of the robot system, which is a form of human–computer interaction with direct control; second, the client operator exists in the digital space as an avatar, which is aimed at verifying the collaborative mode of human–machine symbiosis.

The direct control mode is mainly connected to the operation control in the serial port mode of the controller simulator, and the human–computer symbiosis mode is realized through OpenXR access to VR or AR human–computer interaction devices.

3.3. *Collaborative Communication Service Network*

Different from the distributed virtual simulation network's primary–secondary structure and perception and the control subnetwork's one-to-one structure, the collaborative communication service network is a decentralized and completely flat distributed structure. For this design purpose, we use FastDDS (version >= 2.14) as the communication middleware for this layer of the network.

However, from the perspective of simulation construction, it is worth noting that the collaborative communication service network and the distributed virtual simulation network share the physical bandwidth of the network, but generally speaking, the application's use of bandwidth is closely integrated with the collaborative problem itself, and we can only give an estimate of the bandwidth usage margin.

Assuming that the actual maximum bandwidth available to the network is W and the distributed virtual simulation network utilizes S_v bandwidth in Equation (7), thus, we have,

$$Margin = W - S_v \tag{9}$$

3.3.1. Basic Communication Mechanism

As in Figure 6, DDS enables all intelligent algorithms of robot systems in a virtual scene to connect to the same publish/subscribe network. In the collaborative data space of the network, each topic can be regarded as a communication channel, and the nodes related to the topic in the network will subscribe to the topic corresponding to the channel and publish collaborative information to it as needed. The specific publishing process is as follows: first, encode the collaborative information according to the standardized JSON format, then serialize it into a string, and finally, publish it to the topic. It is possible to transform a data-centric DDS network into a task-centered collaborative network through the above mechanism.

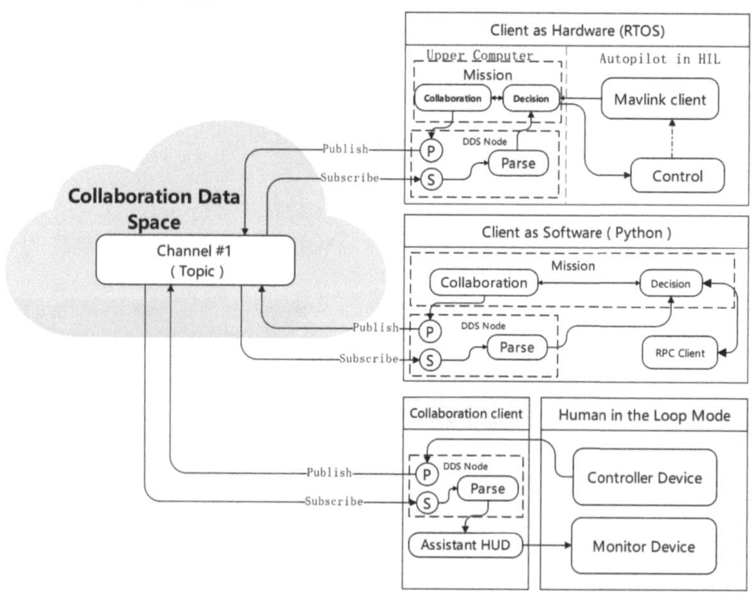

Figure 6. Architecture of the collaborative communication service network.

3.3.2. Distributed Collaboration Process

This article provides an interop protocol process based on the above communication mechanism. Think of a collaborative task simply as a diagram of activity (Figure 7) between two entities that request and respond together (see another work in [56]). The public topic refers to the channel that all intelligent systems will subscribe to, mainly for requests and responses for collaborative tasks. When the requester and responder of a task match, create a dedicated task channel for continuous collaborative communication requirements.

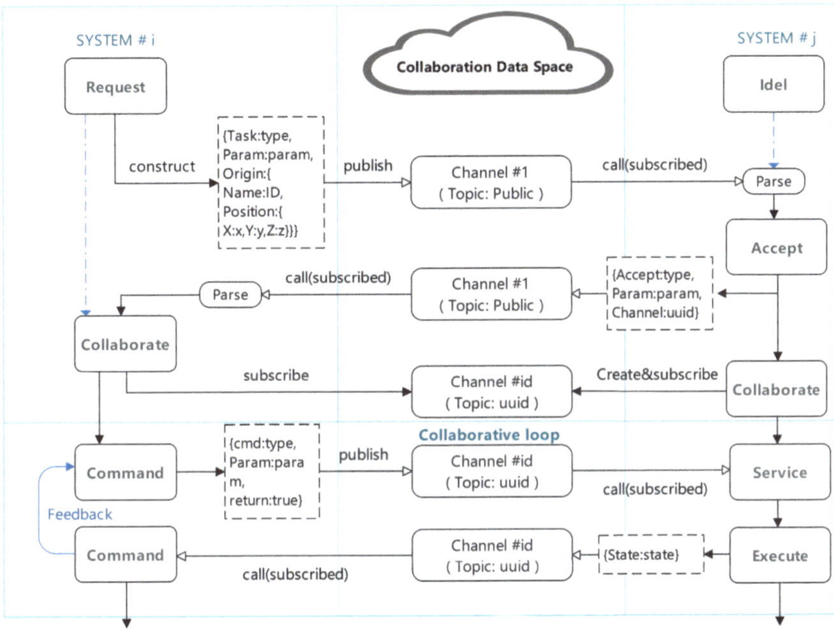

Figure 7. A diagram of activity for the distributed collaboration process.

In some special cases, it may not be necessary to form a dedicated topic channel, and when the coordination requirements are matched and confirmed, the task parameters will be sufficient to guide the jth cooperative system to complete the corresponding operation.

4. Example and Experiments

Based on the three-layer distributed simulation architecture, the performance test process is roughly divided into four steps:

1. Design a virtual scene, define the coordinate system in the scene, place buildings and roads, and set the initial generation position of the robot system;
2. Connect the computing platforms to the network, and specify a server and multiple clients among them, as well as setting the corresponding avatars (UGV or UAV) with their collaborative tasks.
3. Start the simulation service and record the network stream frames at the same time with an improved network profiler;
4. Analyze the distributed data service and compare the theoretical estimation.

Note: Any entity or logical object will be instantiated as a subclass of the actor when the experiment is implemented. Therefore, actors will be used as a general term when analyzing and describing results.

4.1. Scenario Description

The scene map covers an area of about 1 square kilometer, and the mission area is a village located in the desert hills, surrounded by mountains on three sides, with residential buildings scattered and arranged, with an average floor height of 20 m, and a two-lane road running through it. According to preliminary reconnaissance, the command post of a group of terrorist organizations is roughly located in the area enclosed by the black-and-blue wire frame in Figure 8.

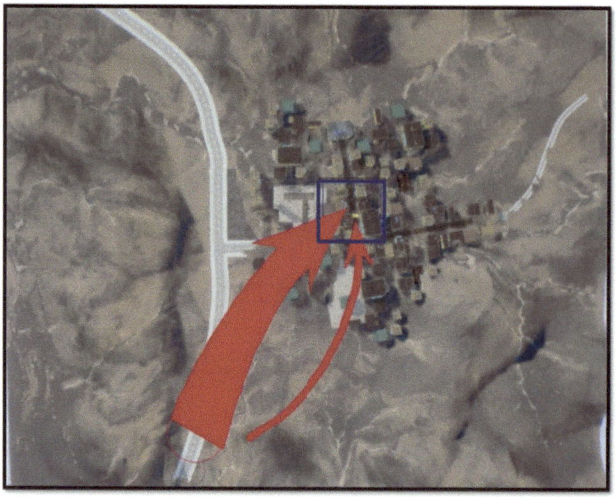

Figure 8. Scenario: The mission is to eliminate the enemy forces in the blue square area.

The enemy is in a rough position surrounded by a quadrilateral, and troops are deployed in the area as follows: a small building (commander on the second floor), 2 guard towers (snipers), and small groups of enemy troops (machine gunners and riflemen) patrolling the ground. The objective of the mission is to eliminate the enemy forces in the area as quickly as possible.

4.2. Collaborative Tasks

The cyclic execution process of two typical types of collaborative tasks (Figure 9) is as follows:

(1). Target-Detection–Bomb-Strike Loop:

Within this loop, the reconnaissance quadcopter is responsible for detecting potential targets, and once a target is confirmed, the attack quadcopter executes the bomb-strike action. The reconnaissance quadcopter employs its sensor system to identify and locate the target and subsequently communicates the relevant information to the attack quadcopter. Upon receiving the information, the attack quadcopter promptly responds, launching a strike on the target and carrying out the bomb attack.

(2). Bomb-Attack–Ammunition-Supply Loop:

In this loop, after executing a bomb strike, the attack quadcopter requires replenishment of ammunition in order to continue its mission. The armored unmanned vehicle serves as the provider of ammunition, transporting the required ammunition to a designated location based on the needs of the attack quadcopter. Once the ammunition resupply is completed, the attack quadcopter can proceed with the next round of bomb strikes.

Through the implementation of such cyclic execution processes, effective coordination between the reconnaissance quadcopter and the attack quadcopter, as well as between the attack quadcopter and the armored unmanned vehicle, is achieved, facilitating efficient and synchronized combat operations.

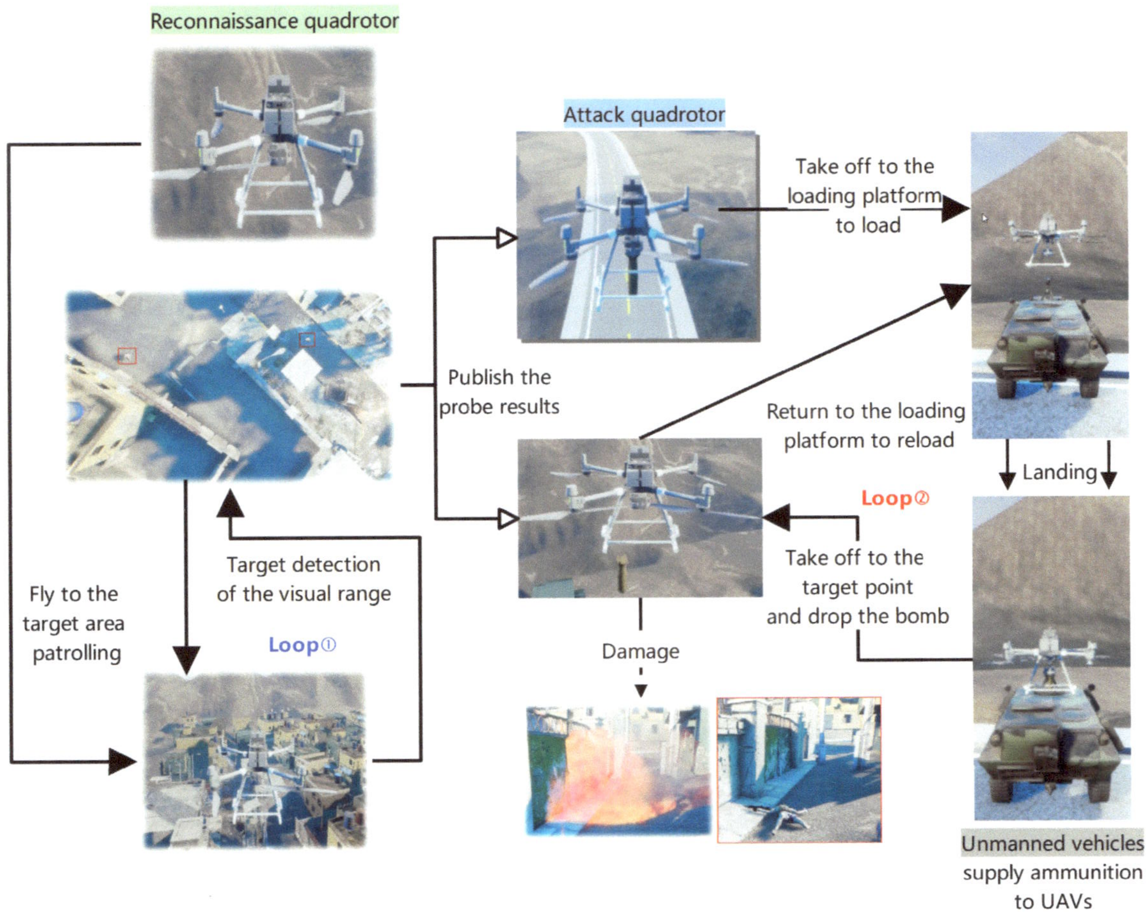

Figure 9. The cyclic execution process of two typical types of collaborative tasks. Loop ①: The cycle of "target detection–bomb strike" between the reconnaissance quadcopter and the attack quadcopter. Loop ②: The cycle of "bomb attack–ammunition supply" between attack quadcopter and armored unmanned vehicles.

4.3. Sampling Software

In this subsection, we explore a valuable addition to the software (Figure 10): a new feature module that enables users to output network traffic and performance information in batches to JSON files. Although this enhancement does not change the usage method, it greatly simplifies subsequent data processing, making analysis and optimization more efficient.

Figure 10. The Network Profiler interface for data analysis and transforming the file format.

(1). Introduction to relevant functions:

Network Profiler is an independent tool designed to display network traffic and performance information captured by the engine during game runtime. This tool is highly effective in identifying areas of high bandwidth consumption in multiplayer games, allowing users to view the percentage of bandwidth occupied by various actors, PRCs, and attributes, thereby assisting in optimizing network performance.

(2). Recording and analyzing sessions:

Before using Network Profiler, users need to record the relevant data for analysis. Recording can be achieved by enabling the engine's process-tracking feature, typically by compiling the engine into a debug build or using an editor build for nondebug configurations. Additionally, users can control the recording of network data by adding the command line parameter "networkprofiler=true" at engine startup or using other command line instructions.

The recorded data will be saved to the specified path, allowing users to open the file in the standalone tool for further analysis. It is important to note that temporary files will be renamed according to a specific scheme when the analysis session ends, facilitating accurate tracking and data processing.

5. Discussion

From the previous Introduction, our framework provides virtual simulation session management in C/S architecture, a basic perception and control logic framework, and a flexible configuration collaboration mechanism. We compare these three features with existing software in Table 2. We compared simulation software for multirobot collaboration and found that the highlight of our work was the ability to manage state synchronization and coupling calculations for multiple clients using a session mechanism. Therefore, we discuss the characteristics of entity synchronization, bandwidth characteristics in the time domain, and the relationship between bandwidth and the number of clients in combination with experimental measurement data.

Table 2. Comparative analysis of the key characteristics of popular multiheterogeneous robot simulation software programs.

Software	Last Update	Feature			
		C/S Architecture	Session Management	Perception and Control Logic	Coordination Mechanism
ARGoS	2022	✓		✓	✓
Stage	2020	✓		✓	✓
Webots	2021	✓		✓	✓
USARSim	2013			✓	
CoppeliaSim (V-REP)	2022	✓		✓	✓
Digital Battle	2024	✓	✓	✓	✓

5.1. Actor–Synchronous Feature

This part of the experiment mainly examines different actors' actual replication frequencies and replication times during synchronization. One UAV and two UGVs were set up for each of the three clients, and nonplayer characters that could track moving actors were also set up in the scene. The drone just received the hovering instruction while the UGVs are driven along the road by joystick.

The results of the data visualization are shown in Figure 11. The plot records the distribution box plot of the replication frequency of each actor that triggered the overall data change of the simulation at the beginning, during the process, and at the end of the simulation. Since the drone was only controlled to hover, the state change was moderated, and the update frequency was slow (merely 10 Hz). The UGVs were constantly moving, so the update frequency was high (always more than 75 Hz). Nonplayer characters were stationary or patrolling for a period because they could not see moving vehicles, so the median was very close to the upper quartile (about 50 Hz). In addition, object name sorting was performed from the smallest to the largest amount of data synchronized throughout the simulation.

For the resulting data of the same experiment, we also analyzed the replication time distribution when different actor states were synchronously managed to replicate to remote clients, and the visualization results are shown in the ridge-line plot (Figure 12). Actors representing motion entities had a shorter replication time overhead because their state descriptions were relatively single and were usually state vectors of rigid bodies. They could usually complete the replication of states within 0.03 milliseconds, which could meet the time interval allowed for frame synchronization. The peaks in distribution roughly indicated several independently updated states or events for that actor.

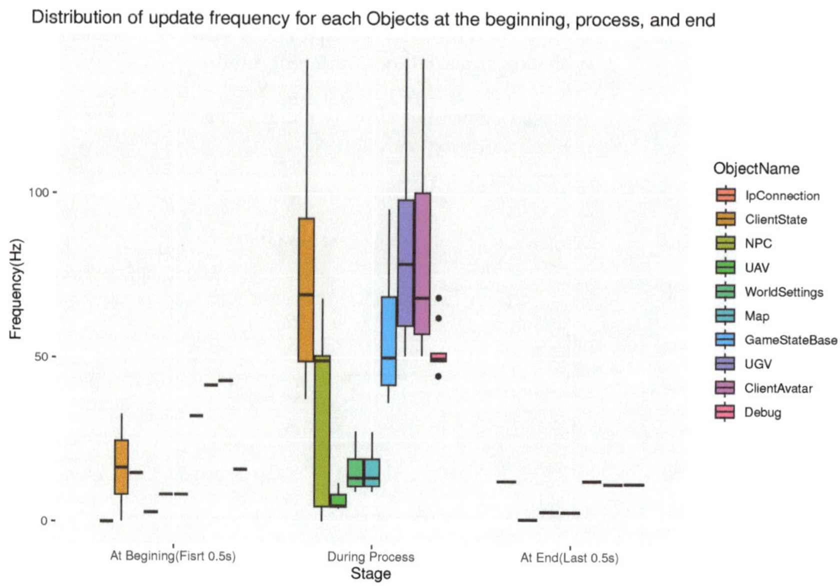

Figure 11. The replication (update) frequency for each actor varies at the beginning, during the process, and at the end. Among all synchronized actors, their update frequencies are within the range of 0 to 150 Hz.

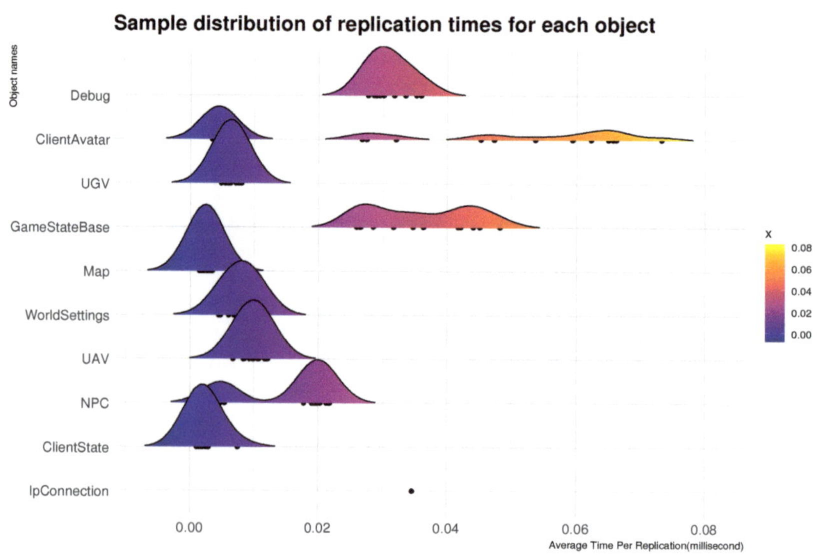

Figure 12. Replication time ridge plot of different actor states synchronized. The replication time varies from 0 to 0.08 milliseconds.

5.2. Time Domain Analysis

In this part of the experiment, the goal was to test the data bandwidth and total throughput of as many clients as possible based on existing lab resources and to examine the correlation between data frames and state data. We built a server (Intel i9 with 32 GB memory) and invited 12 students to participate. They provided their laptops as clients (computer models and configurations were diverse) and their wisdom to design collaborative tasks, and we finally formed 15 distributed client nodes (some students were single-machine multiopen clients) to complete the distributed simulation experimental process. The duration of the simulation was less than 240 s.

Figure 13A provides an intuitive picture of the bandwidth occupied by the data frame, where the packet was initially used to complete the configuration work required for initialization on the distributed network. The data frame monolith was significant, but the synchronization frequency was relatively low. In the intermediate process, the data frame was mainly used to synchronize the dynamic properties of the actor between the various clients, so its monolith was small, but it was updated relatively frequently. At the end of the simulation, the number of entities was reduced, and the bandwidth consumption of synchronized data frame rate was also reduced. The above situation corresponds to the stage characteristics shown in Figure 3. In addition, the median bandwidth occupied by data frames was 8 KB/s, and the maximum bandwidth usage was about 25 KB/s. The results of Figure 13B reveal a positive correlation between the size of the data frame and the amount of the attribute data in the frame. Therefore, we can use a rough estimate of the amount of data that need to be synchronized. Furthermore, the results of Figure 13B can be verified bidirectionally by Figure 13C, and the bandwidth occupied by attribute synchronization fluctuated below 50% but occasionally exceeded 87.5% throughout the simulation time, which confirms the coefficient of 2.13 obtained by linear fitting $y = 0.0096 + 2.13x$ in Figure 13B.

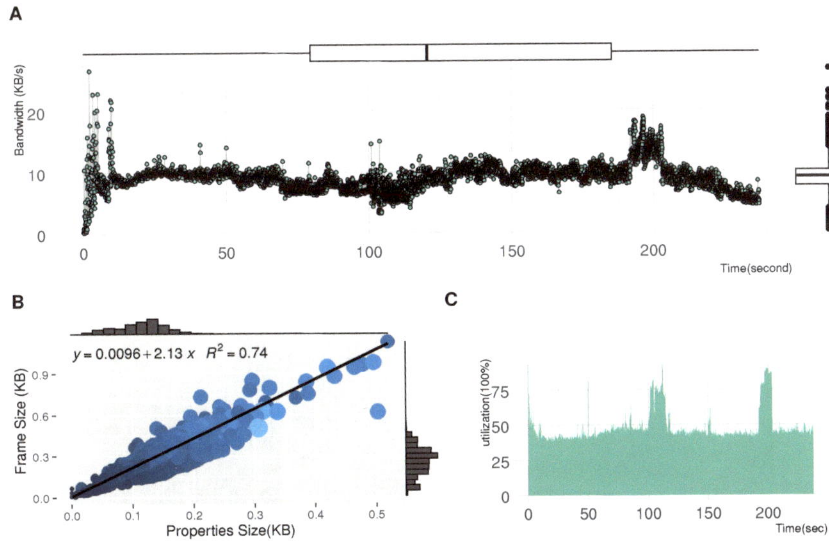

Figure 13. (**A**) Bandwidth distribution over the time domain. (**B**) The size correlation between the frame and the properties data. (**C**) Time distribution of attribute data on bandwidth usage.

At the same time, we also made a chord graph of the total data throughput of the server and 15 clients in Figure 14. Through comparison, it can be found that the data throughput on the server side reaches almost half of the total throughput, and the throughput of each client is different. By observing the total amount of data input and output of each end, it can also be found that the input amount of the client is slightly greater than the output, while the output of the server side is slightly larger than the input.

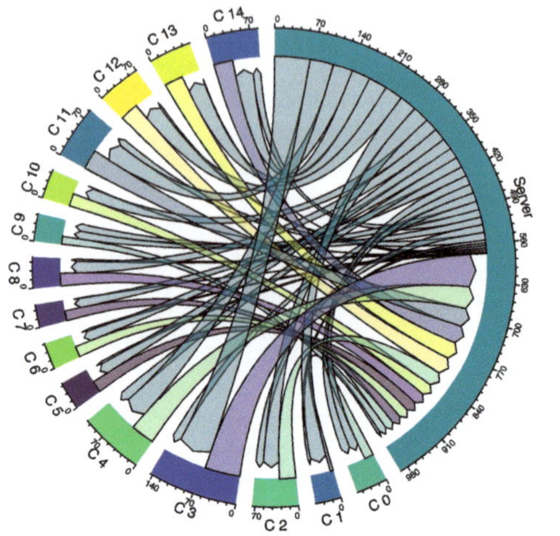

Figure 14. A chord diagram for throughput within the server and 15 clients.

5.3. Scale and Bandwidth

Based on several experiments, we analyze the relationship between the physical object scale of the robot system (equivalent to the client node scale of the distributed virtual simulation network within this framework) and the communication bandwidth in the network.

In the process of data analysis, because the theoretical value of maximum bandwidth is almost impossible due to the existence of synchronization strategy, but assuming that the distribution of bandwidth has specific mean point characteristics, we select the mean bandwidth as the fitted sample point to obtain the distribution relationship of bandwidth with the scale of clients and the total number of attribute values, as shown in Figure 15.

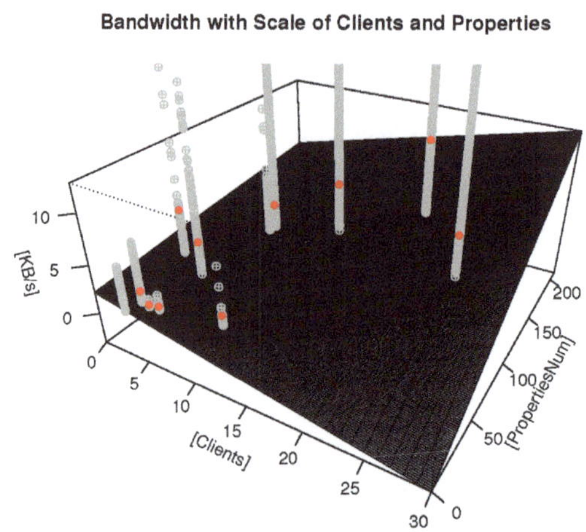

Figure 15. The bandwidth fits with scale of clients and properties. The gray points refer to the bandwidth from experiment data grouped by the scale of clients and properties, while the red points are the mean value in the summaries of each group. Furthermore, the light-blue surface fit the red points as in Equation (10).

$$S_{mean} = a \cdot n \cdot D_n - b \cdot n + c \cdot D_n + d \tag{10}$$

Among the parameters, a stands for average size of each property, b represents the average output of each client, c is a relative ratio of the volume of properties maintained by the server, and d could be some necessary command and debug data. As a numeric result, we obtained $a = 0.002083$ KB/s, $b = 0.1818$ KB/s, $c = 0.013199$ KB/s, $d = 2.257655$ KB/s. Compared with the form of Equation (7), we can derive a similar form based on Equation (10).

$$S = 0.002083 \cdot n \cdot (D_n - 87) + 0.013199 D_n + 2.257655 \tag{11}$$

Thus, its parameters can be symbolized as

$$S = p_u \cdot n(D_n - \bar{D}) + cD_n + d \tag{12}$$

where the p_u represents property replication traffic related to client scale, and $\bar{D}$ is the part of the property that is independent of the number of clients.

6. Conclusions and Future Work

In this study, we compared simulation software for multirobot collaboration and identified unique highlights of our work. Our work is to successfully leverage the session mechanism to manage state synchronization and coupling computation across multiple clients. Through the analysis of experimental measurement data, we drew the following important conclusions:

1. We discussed the replication time characteristics and update frequency characteristics when the entities are synchronized. We found that in our system, the replication time of entity synchronization is relatively low, and the update frequency can meet the needs of multirobot collaboration, ensuring the real-time use and accuracy of the system;
2. We analyzed the bandwidth usage of data frames throughout the session. We observed that the server side took up almost half of the data throughput during the entire session, which indicates the rational allocation and utilization of data transfer in our system;
3. We built a bandwidth estimation surface model to estimate the bandwidth requirements of the current model when scaling the server-side scale and synchronization-state scale. This model provides us with an important reference to better plan and optimize the resource allocation and performance of the system.

In future research, we will improve the following key technical details based on this distributed simulation framework:

1. Although the dynamic model update method of coupled objects was given in this architecture, efforts are still needed to further refine the method to support the simulation theory of the system coupling relationship between objects;
2. The synchronous management adopted in this architecture has already partially improved the communication delay problem, but it is still necessary to study the influence of the spatiotemporal consistency problem of the distributed system on multirobot control and decision making;
3. Because the default collaborative network in this study is a simple event-based communication mechanism, it only integrates data services and designs and implements the basic collaboration framework, and it does not consider its impact on the overall distributed network communication bandwidth; therefore, the in-depth study of the collaborative framework will be further promoted in the future in combination with the specific tasks of multirobot system collaborative work.

7. Patents

This section outlines a novel distributed simulation method and system devised specifically for multi-UAV (unmanned aerial vehicle) systems. The methodology bifurcates the distributed simulation tasks inherent in multi-UAV systems into three fundamental concepts: virtual simulation, control perception, and task coordination. These concepts are underpinned by the establishment of three key networks: the distributed virtual simulation network, the perception control subnetwork, and the DDS-based collaborative communication service network.

The distributed virtual simulation network serves as the backbone for deploying virtual simulation tasks across distributed nodes. Its comprehensive server plays a pivotal role by providing a spectrum of essential services, including rule arbitration, map scene management, real-time state synchronization, multipoint session management, intelligent simulation, and real-time situational awareness. Conversely, the client segment within the distributed virtual simulation network furnishes local computing capabilities encompassing dynamics, collision interaction, damage simulation, and graphic rendering.

The system architecture is meticulously crafted to operationalize the aforementioned method, leveraging its inherent advantages such as simplicity, wide applicability, and high integration.

Furthermore, it is pertinent to note the patent details:

Classification: H04L41/042; H04L47/783; H04L9/40; H04L67/08; H04L67/59.

Application Number: 202310416953.5.

Author Contributions: Conceptualization, J.G. and J.H.; methodology and writing—original draft, J.G.; writing—review and editing, Q.L., H.C., H.D., L.Z., L.S. and J.H. All authors have read and agreed to the published version of the manuscript.

Funding: This research received no external funding.

Data Availability Statement: Not applicable.

Acknowledgments: While developing this architecture and the corresponding platform, the open-source community provided the model objects, and all enthusiasts contributed to robot systems that lay a good foundation for the design and implementation of our architecture. We also thank Huang for allowing us to conduct system performance testing in the multirobot system simulation course, as well as the 12 students who provided the hardware equipment and took the time to complete the test.

Conflicts of Interest: The authors declare no conflicts of interest.

References

1. Tselegkaridis, S.; Sapounidis, T. Simulators in Educational Robotics: A Review. *Educ. Sci.* **2021**, *11*, 11. [CrossRef]
2. Rizk, Y.; Awad, M.; Tunstel, E.W. Cooperative Heterogeneous Multi-Robot Systems: A Survey. *ACM Comput. Surv.* **2019**, *52*, 29. [CrossRef]
3. Xiang, J.; Dong, X.; Ding, W.; Suo, J.; Shen, L.; Xia, H. Key technologies for autonomous cooperation of unmanned swarm systems in complex environments. *Acta Aeronaut. Astronaut. Sin.* **2022**, *43*, 527570.
4. Tubis, A.A.; Poturaj, H.; Dereń, K.; Żurek, A. Risks of Drone Use in Light of Literature Studies. *Sensors* **2024**, *24*, 1205. [CrossRef] [PubMed]
5. Ding, Y.; Xiong, Z.; Xiong, J.; Cui, Y.; Cao, Z. OGI-SLAM2: A hybrid map SLAM framework grounded in inertial-based SLAM. *IEEE Trans. Instrum. Meas.* **2022**, *71*, 2519014. [CrossRef]
6. Xun, Z.; Huang, J.; Li, Z.; Xu, C.; Gao, F.; Cao, Y. CREPES: Cooperative RElative Pose EStimation towards Real-World Multi-Robot Systems. *arXiv* **2023**, arXiv:2302.01036.
7. Wang, J.; Wu, Y.; Chen, Y.; Ju, S. Multi-UAVs collaborative tracking of moving target with maximized visibility in Urban Environment. *J. Frankl. Inst.* **2022**, *359*, 5512–5532. [CrossRef]
8. Wang, Z.; Xu, C.; Gao, F. Robust Trajectory Planning for Spatial-Temporal Multi-Drone Coordination in Large Scenes. In Proceedings of the 2022 IEEE/RSJ International Conference on Intelligent Robots and Systems (IROS), Kyoto, Japan, 23–27 October 2021; pp. 12182–12188.
9. Raziei, Z.; Moghaddam, M. Adaptable automation with modular deep reinforcement learning and policy transfer. *Eng. Appl. Artif. Intell.* **2021**, *103*, 104296. [CrossRef]
10. Badrno, H.; Baradarannia, M.; Bagheri, P.; Badamchizadeh, M.A. Distributed Predictive Consensus Control of Uncertain Linear Multi-agent Systems with Heterogeneous Dynamics. *Iran. J. Sci. Technol. Trans. Electrical Eng.* **2022**, *47*, 255–267. [CrossRef]
11. Li, W.; Zhang, H.; Gao, Z.; Wang, Y.; Sun, J. Fully Distributed Event/Self-Triggered Bipartite Output Formation-Containment Tracking Control for Heterogeneous Multiagent Systems. *IEEE Trans. Neural Netw. Learn. Syst.* **2022**, *34*, 7851–7860. [CrossRef]
12. Yang, S.M.; Kim, K. Implementation of the conversation scheme in message-based distributed computer systems. *IEEE Trans. Parallel Distrib. Syst.* **1992**, *3*, 555–572. [CrossRef]

13. Xu, J.; Romanovsky, A.; Randell, B. Concurrent exception handling and resolution in distributed object systems. *IEEE Trans. Parallel Distrib. Syst.* **2000**, *11*, 1019–1032. [CrossRef]
14. Chen, J.; Huang, L. Supporting Dynamic Service Updates in Pervasive Applications. In Proceedings of the 2011 IEEE 35th Annual Computer Software and Applications Conference, Munich, Germany, 18–22 July 2011; pp. 273–278. [CrossRef]
15. Calderón-Arce, C.; Brenes-Torres, J.C.; Solis-Ortega, R. Swarm Robotics: Simulators, Platforms and Applications Review. *Computation* **2022**, *10*, 80. [CrossRef]
16. Cho, W.J.; Kim, S.; Kim, Y.; Moon, Y.H. Advanced Co-Simulation Platform for UAV Simulations under Virtual Wireless Network Environments. *IEEE Access* **2022**, *10*, 95498–95508. [CrossRef]
17. Phadke, A.; Medrano, F.A.; Sekharan, C.N.; Chu, T. Designing UAV Swarm Experiments: A Simulator Selection and Experiment Design Process. *Sensors* **2023**, *23*, 7359. [CrossRef] [PubMed]
18. Koç, D.; Seçkin, A.Ç.; Satı, Z.E. Evaluation of Participant Success in Gamified Drone Training Simulator Using Brain Signals and Key Logs. *Brain Sci.* **2021**, *11*, 1024. [CrossRef]
19. Covaciu, F.; Iordan, A.-E. Control of a Drone in Virtual Reality Using MEMS Sensor Technology and Machine Learning. *Micromachines* **2022**, *13*, 521. [CrossRef]
20. Lee, D.; Kim, H.; Yoon, H.; Lee, W. Usability Comparison between 2D and 3D Control Methods for the Operation of Hovering Objects. *Drones* **2023**, *7*, 520. [CrossRef]
21. Michel, O. Cyberbotics Ltd. Webots™: Professional Mobile Robot Simulation. *Int. J. Adv. Robot. Syst.* **2004**, *1*, 5. [CrossRef]
22. Brockman, G.; Cheung, V.; Pettersson, L.; Schneider, J.; Schulman, J.; Tang, J.; Zaremba, W. OpenAI Gym. *arXiv* **2016**, arXiv:1606.01540.
23. Shah, S.; Dey, D.; Lovett, C.; Kapoor, A. Airsim: High fidelity visual and physical simulation for autonomous vehicles. In Proceedings of the Field and Service Robot, Results of the 11th International Conference, Zurich, Switzerland, 12–15 September 2017; pp. 621–635.
24. Makoviychuk, V.; Wawrzyniak, L.; Guo, Y.; Lu, M.; Storey, K.; Macklin, M.; Hoeller, D.; Rudin, N.; Allshire, A.; Handa, A.; et al. Isaac Gym: High Performance GPU-Based Physics Simulation for Robot Learning. *arXiv* **2021**, arXiv:2108.10470.
25. Kong, F.; Liu, X.; Tang, B.; Lin, J.; Ren, Y.; Cai, Y.; Zhu, F.; Chen, N.; Zhang, F. MARSIM: A Light-Weight Point-Realistic Simulator for LiDAR-Based UAVs. *IEEE Robot. Autom. Lett.* **2022**, *8*, 2954–2961. [CrossRef]
26. Li, H.; Liu, S.; Ma, M.; Hu, G.; Chen, Y.; Zhao, D. NeuronsGym: A Hybrid Framework and Benchmark for Robot Tasks with Sim2Real Policy Learning. *arXiv* **2023**, arXiv:2302.03385.
27. Gwon, D.H.; Kim, J.; Kim, M.H.; Park, H.G.; Kim, T.Y.; Kim, A. Development of a side scan sonar module for the underwater simulator. In Proceedings of the 2017 14th International Conference on Ubiquitous Robots and Ambient Intelligence (URAI), Jeju, Republic of Korea, 28 June–1 July 2017; pp. 662–665. [CrossRef]
28. Kumar, A.; Li, Z.; Zeng, J.; Pathak, D.; Sreenath, K.; Malik, J. Adapting Rapid Motor Adaptation for Bipedal Robots. In Proceedings of the 2022 IEEE/RSJ International Conference on Intelligent Robots and Systems (IROS), Kyoto, Japan, 23–27 October 2022; pp. 1161–1168. [CrossRef]
29. Jiang, Y.; Zhang, T.; Ho, D.; Bai, Y.; Liu, C.K.; Levine, S.; Tan, J. SimGAN: Hybrid Simulator Identification for Domain Adaptation via Adversarial Reinforcement Learning. In Proceedings of the 2021 IEEE International Conference on Robotics and Automation (ICRA), Xi'an, China, 30 May–5 June 2021; pp. 2884–2890. [CrossRef]
30. Vaughan, R.T. Massively multi-robot simulation in stage. *Swarm Intell.* **2008**, *2*, 189–208. [CrossRef]
31. Carpin, S.; Lewis, M.; Wang, J.; Balakirsky, S.; Scrapper, C. USARSim: A robot simulator for research and education. In Proceedings of the 2007 IEEE International Conference on Robotics and Automation, Rome, Italy, 10–14 April 2007; pp. 1400–1405. [CrossRef]
32. Diankov, R. Automated Construction of Robotic Manipulation Programs. Ph.D. Thesis, Carnegie Mellon University, Robotics Institute, Pittsburgh, PA, USA, 2010.
33. Pinciroli, C.; Trianni, V.; O'Grady, R.; Pini, G.; Brutschy, A.; Brambilla, M.; Mathews, N.; Ferrante, E.; Di Caro, G.; Ducatelle, F.; et al. ARGoS: A modular, multi-engine simulator for heterogeneous swarm robotics. In Proceedings of the 2011 IEEE/RSJ International Conference on Intelligent Robots and Systems, San Francisco, CA, USA, 25–30 September 2011; pp. 5027–5034. [CrossRef]
34. Prats, M.; Pérez Soler, J.; Fernandez, J.; Sanz, P. An open source tool for simulation and supervision of underwater intervention missions. In Proceedings of the 2012 IEEE/RSJ International Conference on Intelligent Robots and Systems, Vilamoura-Algarve, Portugal, 7–12 October 2012; pp. 2577–2582. [CrossRef]
35. Furrer, F.; Burri, M.; Achtelik, M.; Siegwart, R. *Rotors—A Modular Gazebo Mav Simulator Framework*; Springer: Berlin/Heidelberg, Germany, 2016; pp. 595–625.
36. Todorov, E.; Erez, T.; Tassa, Y. MuJoCo: A physics engine for model-based control. In Proceedings of the 2012 IEEE/RSJ International Conference on Intelligent Robots and Systems, Vilamoura-Algarve, Portugal, 7–12 October 2012; pp. 5026–5033. [CrossRef]
37. Dosovitskiy, A. CARLA: An Open Urban Driving Simulator. In Proceedings of the Conference on Robot Learning, Mountain View, CA, USA, 13–15 November 2017.

38. Guerra, W.; Tal, E.; Murali, V.; Ryou, G.; Karaman, S. FlightGoggles: Photorealistic Sensor Simulation for Perception-driven Robotics using Photogrammetry and Virtual Reality. In Proceedings of the 2019 IEEE/RSJ International Conference on Intelligent Robots and Systems (IROS), Macau, China, 3–8 November 2019; pp. 6941–6948. [CrossRef]
39. Guerra, W.; Tal, E.; Murali, V.; Ryou, G.; Karaman, S. FlightGoggles: A Modular Framework for Photorealistic Camera, Exteroceptive Sensor, and Dynamics Simulation. *arXiv* **2021**, arXiv:1905.11377v2.
40. Zhu, Y.; Wong, J.; Mandlekar, A.; Mart'in-Mart'in, R. robosuite: A Modular Simulation Framework and Benchmark for Robot Learning. *arXiv* **2020**, arXiv:2009.12293.
41. Xiao, K.; Tan, S.; Wang, G.; An, X.; Wang, X.; Wang, X. XTDrone: A Customizable Multi-rotor UAVs Simulation Platform. *arXiv* **2020**, arXiv:2003.09700.
42. Xiao, K.; Ma, L.; Tan, S.; Cong, Y.; Wang, X. Implementation of UAV Coordination Based on a Hierarchical Multi-UAV Simulation Platform. *arXiv* **2020**, arXiv:2005.01125.
43. Song, Y.; Naji, S.; Kaufmann, E.; Loquercio, A.; Scaramuzza, D. Flightmare: A Flexible Quadrotor Simulator. In Proceedings of the Conference on Robot Learning, Virtual, 16–18 November 2020.
44. Li, C.; Xia, F.; Martin-Martin, R.; Lingelbach, M.; Srivastava, S.; Shen, B.; Vainio, K.; Gokmen, C.; Dharan, G.; Jain, T.; et al. IGibson 2.0: Object-Centric Simulation for Robot Learning of Everyday Household Tasks. *arXiv* **2021**, arXiv:2108.03272.
45. Wang, S.; Dai, X.; Ke, C.; Quan, Q. RflySim: A Rapid Multicopter Development Platform for Education and Research Based on Pixhawk and MATLAB. In Proceedings of the 2021 International Conference on Unmanned Aircraft Systems (ICUAS), Athens, Greece, 15–18 June 2021; pp. 1587–1594. [CrossRef]
46. Jansen, W.; Verreycken, E.; Schenck, A.; Blanquart, J.E.; Verhulst, C.; Huebel, N.; Steckel, J. COSYS-AIRSIM: A Real-Time Simulation Framework Expanded for Complex Industrial Applications. In Proceedings of the 2023 Annual Modeling and Simulation Conference (ANNSIM), Hamilton, ON, Canada, 23–26 May 2023; pp. 37–48.
47. Schlesselman, J.; Pardo-Castellote, G.; Farabaugh, B. OMG data-distribution service (DDS): Architectural update. In Proceedings of the IEEE MILCOM 2004. Military Communications Conference, Monterey, CA, USA, 31 October–3 November 2004; Volume 2, pp. 961–967. [CrossRef]
48. Statzer, J.M. AdvancedSessionsPlugin. 2017. Available online: https://github.com/mordentral/AdvancedSessionsPlugin (accessed on 11 February 2021).
49. Wang, K.; Ju, H. An explicit modelling method of joint-space inertia matrix for tree-chain dynamic system. *Int. J. Non-Linear Mech.* **2022**, *144*, 104033. [CrossRef]
50. Müller, M.; Macklin, M.; Chentanez, N.; Jeschke, S.; Kim, T.Y. Detailed Rigid Body Simulation with Extended Position Based Dynamics. *Comput. Graph. Forum* **2020**, *39*, 101–112. [CrossRef]
51. Udwadia, F.E. Constrained Motion of Hamiltonian Systems. *Nonlinear Dyn.* **2016**, *84*, 1135–1145. [CrossRef]
52. Celledoni, E.; Leone, A.; Murari, D.; Owren, B. Learning Hamiltonians of constrained mechanical systems. *J. Comput. Appl. Math.* **2022**, *417*, 114608. [CrossRef]
53. Huang, Z.; Chen, J.; Zhang, Z.; Tian, Q. Lie group variational integral algorithm for multi-rigid body dynamics simulation. *J. Dyn. Control* **2022**, *20*, 8. (In Chinese) [CrossRef]
54. Borisov, A.V.; Mamaev, I.S. 1 Rigid Body Equations of Motion and Their Integration. In *Rigid Body Dynamics Hamiltonian Methods, Integrability, Chaos*; Higher Education Press: Beijing, China, 2018; pp. 19–78.
55. Kang Feng, Q.M. *Symplectic Geometric Algorithms for Hamiltonian Systems*; Zhejiang Science and Technology Press: Hangzhou, China, 2003.
56. Deng, H.; Huang, J.; Liu, Q.; Zhou, C.; Gao, J. BGSD: A SBERT and GAT-based Service Discovery Framework for heterogeneous distributed IoT. *Comput. Netw.* **2023**, *220*, 109488. [CrossRef]

applied sciences

MDPI

Article

How Does the Built Environment Affect Mechanical Parking Space Planning: A Case Study in Xi'an City

Yuejiao Wang [1] and Weijia Li [2,3,*]

[1] Key Laboratory of Road and Traffic Engineering, Ministry of Education, Tongji University, Shanghai 200092, China; 2111233@tongji.edu.cn
[2] School of Architecture, Chang'an University, Xi'an 710064, China
[3] Engineering Research Center of Collaborative Planning of Low-Carbon Urban Space and Transportation, Universities of Shaanxi Province, Xi'an 710061, China
* Correspondence: weijiali@chd.edu.cn

Abstract: Mechanical parking lots and spaces are known as the "energy saver" of urban space because of their small footprint, high efficiency, and environmental protection. However, the location and number of mechanical parking lots and space planning have become an important part of effectively exerting the function of mechanical parking lots. In order to explore the planning problem of mechanical parking lots, this study used the gradient boosting decision tree–Shapley additive explanations (GBDT-SHAPs) to measure the non-linear impact of the urban built environment on the mechanical parking spaces ratio and extract the optimal threshold of key variables. The results show that land use mix and distance to Bell Tower (CBD) are two key variables affecting mechanical parking space planning, and both have a non-linear relationship with the built environment. The threshold values are 0.83 and 7 km. The results will provide urban and transport planners with strategies for planning mechanical parking lots and spaces.

Keywords: mechanical parking spaces ratio; GBDT; built environment; SHAP value; non-linear relationship

Citation: Wang, Y.; Li, W. How Does the Built Environment Affect Mechanical Parking Space Planning: A Case Study in Xi'an City. *Appl. Sci.* **2024**, *14*, 4666. https://doi.org/10.3390/app14114666

Academic Editors: Tomasz Nowakowski, Artur Kierzkowski, Agnieszka A. Tubis, Franciszek Restel, Tomasz Kisiel, Anna Jodejko-Pietruczuk and Mateusz Zając

Received: 6 April 2024
Revised: 24 May 2024
Accepted: 25 May 2024
Published: 29 May 2024

1. Introduction

The problem of parking and traffic operation efficiency caused by a rapid increase in mobility is very serious. Limited land resources and an increase in car ownership have exacerbated the conflict between parking supply and demand [1]. According to the parking survey in Xi'an in 2022, there are about 3.347 million car parking spaces in the city, and the average parking space is about 0.85, which is lower than the 1.1–1.3 recommended by the national standard (http://zygh.xa.gov.cn/web_files/zygh/file/2023/10/19/202310191543486971221.pdf, accessed on 2 April 2024). Moreover, improper parking supply and layout planning can result in a 30–50% inefficient operation [2], wasting road resources, reducing traffic efficiency, and increasing environmental pollution.

Compared with the traditional flat parking lots, the mechanical parking lots are smarter. Modern mechanical parking systems use the central controller, data collection, analysis, and processing, and, ultimately, generate intelligent monitoring instructions and send them to the drive system, so as to achieve intelligent vehicle flow planning, parking access, and monitoring. The Smart Parking System (SPS), which uses IoT, AI, and multi-agent systems, can better organize parking lots and manage traffic [3]. The core concept of the system is to automatically allocate parking spaces without human intervention. SPS uses IoT technologies and devices (e.g., sensors) to assist drivers. The mobile application allows users to receive real-time parking information about the selected parking area. They can find the nearest available parking space, which can be booked online in advance, thereby reducing traffic congestion, air pollution, and related health risks [4,5]. Moreover, the mechanical parking space has the advantages of saving space, improving parking efficiency, environmental protection, and energy saving. It can reduce the occupation of land, effectively solve the problem of parking space shortage and low parking efficiency, and thus reduce urban traffic congestion and environmental pollution. Therefore, mechanical parking spaces have important advantages and development potential in urban parking management. The proper planning and design of mechanical parking lots and spaces is the key to effectively relieving traffic congestion and reducing environmental impact. Therefore, it is necessary to carry out a reasonable planning layout of mechanical parking spaces and to determine the location and number of mechanical parking spaces by discussing the influence of factors on the construction of mechanical parking spaces, so as to provide theory and application for parking planning and urban planning.

In order to reveal which factors have a greater impact on the planning and construction of mechanical parking lots and mechanical parking spaces, and to explore what built environments are needed to be equipped with mechanical parking lots and how many mechanical parking spaces are needed, so as to provide a theoretical and practical basis for the planning and construction of mechanical parking lots and parking spaces. Traditional linear regression and non-linear regression have constraints on the trends of the model, which may bias the predictive accuracy of the model and the degree of influence of the independent variable on the dependent variable, and thus bias the planning guidance. Therefore, the gradient boosting decision tree (GBDT) method was used to study the non-linear relationship between the mechanical parking ratio and the built environment. It has been shown that there is a non-linear relationship between the built environment and parking demand [1], and that there is a relationship between parking demand and parking supply [6]; therefore, there should be some relationship between the built environment and parking supply. The effect of built environment factors on the mechanical parking space ratio is quantitatively analyzed from global and local perspectives. The main contributions of this study are as follows: on the one hand, there are few studies on mechanical parking planning. Some studies have analyzed the internal organization, control strategy, site selection and cost, mechanical structure [7–12], and the design of the device and intelligent system optimization of the single facilities, while there is a lack of regional planning studies that provide theoretical and practical support for the spatial layout planning of mechanical parking. On the other hand, very few studies have addressed the relationship between car parks and the built environment, especially mechanical parking lots, and there is a lack of analysis of the impact of factors.

There are six sections as follows. Section 2 discusses relevant research in the field of parking lots and built environments. Section 3 introduces the models used in this research, including the GBDT and SHAP models. Section 4 presents the study area and data. Section 5 presents the results. Section 6 provides the discussion and conclusion.

2. Related Works

In the field of transport, parking research is becoming increasingly important. The reasonable setting of parking space is an important means to effectively alleviate the problem of parking difficulties and reduce the impact on dynamic traffic flow and environmental pollution. The built environment can influence parking planning by influencing parking demand, which can provide decision guidance for parking planning. Therefore, this paper summarizes two aspects. The first aspect is the analysis of the basic theory and research methods of the parking and built environment; the second aspect is the research method used for an analysis of the relationship between parking and built environment.

2.1. Parking and Built Environment

The 5D built environment indicator system mainly includes density, diversity, design, destination accessibility, and distance to transit [13]. Population density and parking are the sixth and seventh "D" in "7D" [14], which is developed from the "5D" theory. Several research studies have analyzed the relationship between the built environment and travel behavior [15], such as mode choice [16] and parking behavior [17–19].

Parking demand is influenced by built environment factors, such as land use [20], location conditions [21], and so on. For parking demand prediction, macro prediction methods mainly include the parking generation rate model, travel attraction model, traffic volume parking demand model, and multiple regression analysis prediction model [22,23]. With the deepening of parking surveys and demand analysis, parking demand prediction models are gradually enriched with survey methods [24] and a consideration of influencing factors such as traffic demand allocation [25] and parking behavior [26,27]. With the development of data collection and storage technology, the observation scale of parking surveys is becoming smaller and smaller, the collection of continuous parking data is becoming easier and easier, and some scholars have proposed the use of time series and machine learning methods for the short-term prediction of parking demand [28]. Parking cost is the most important variable influencing parking demand, followed by location and road density [1]. There is a non-linear influence of the built environment on the demand and shows an obvious threshold effect [12]. Chen has proposed that the construction index (18.92%) and location (15.23%) are the most important built environment factors influencing parking demand [29]. The parking supply is often predicted by parking demand by considering the supply and demand ratio or the supply and demand uncertainty. It is difficult to choose an appropriate supply and demand ratio. Exploring the relationship between the built environment and mechanical parking space might be a new thought for parking supply analysis and prediction.

2.2. The Research Method of the Relationship between Parking and Built Environment

Traditional methods such as Ordinary Least Squares (OLSs) [30], the logit method [1] h, and Geographical Weighted Regression (GWR) [31] have been used to dig out the relationship between parking-related variables and built environment indicator systems. Machine learning, which has higher predictive performance, has been widely used in the transportation field [21,32–34]. Due to higher predictive accuracy and operational efficiency, lots of researchers have used the Gradient Boosting Decision Tree (GBDT) model to analyze the influence of the built environment on travel-related variables [33,35,36], or to predict parking-related variables [21,37].

Machine learning algorithms are black box models, i.e., their inner workings are difficult to understand and explain [38]. The interpretability of machine learning has attracted much attention. Partial dependence plots (PDPs) and accumulated local effects (ALE) plots could describe how features affect the prediction of a machine learning model on average. The ALE method has three advantages: Firstly, the graph created by the ALE method is unbiased. As ALE can deal with the correlation between features, the graph produced is not affected by the joint effect. Secondly, ALE is faster than PDP and requires fewer computations than PDP, and thirdly, the interpretation of ALE graphs is very clear and unambiguous. By removing the effects of correlated variables, one can easily interpret the feature variables and model results [39].

One of the difficulties in using machine learning methods is the interpretation of their results [40]. To overcome this problem, Shapley additive explanations (SHAPs), which are based on the classical Shapley values, are used to provide local explanations and give detailed feature importance and attribution at the level of individual observations. This method has brought great potential for understanding machine learning models of spatial data, as each observed value is geographically referenced and allows spatial visualization of any estimated feature attribution. The SHAP value of y_i for sample i is calculated by the sum of the predicted mean value of all samples and the SHAP value of sample x_{ij} (($f(x_{ij})$, which is the individual SHAP value. j is the feature j). When $f(x_{ij}) > 0$, it means that the feature enhances the prediction value and plays a positive role, and vice versa [41]). The implementation of local SHAP as SHAP is provided in software such as Python 3.12.3 and R 4.3.3. Machine learning methods such as GBDT, xgboost, and LightGBM can call the SHAP command in the scikit-learn package.

Although the built environment and parking research methods use the advanced GBDT algorithm [29], there is almost no research on mechanical parking and the interpretability of variables and almost no research on the interpretation of local features. To address the unresolved issue in parking planning, the mechanical parking rate was selected as the dependent variable, and the variables based on the "7D" built environment indicator system were used as the independent variable. The GBDT model was used to establish the non-linear relationship to predict the mechanical parking spaces ratio, and then to interpret the predictions from both global and local perspectives and to reveal the contribution of each variable with the SHAP value. This research will provide a theoretical basis for the design and planning of mechanical parking spaces and lots.

3. Materials and Methods

3.1. GBDT Method (Gradient Boosting Decision Tree)

Among machine learning models, the GBDT model is good at representing non-linear relationships and does not require prior assumptions about the relationships between variables. At the same time, it can provide the relative importance of the variables in the results, which is conducive to revealing the results. Moreover, it can deal with the problem of multi-collinearity within variables. GBDT method uses bagging and adding regularization to avoid noise and outlier data, which makes GBDT more robust [36]. Thus, GBDT is now widely used for regression and classification prediction [42].

GBDT is an advanced ensemble learning model based on the CART algorithm. It uses weak classifiers (CART trees) to construct an ensemble model and trains multiple decision tree models through iterations and overlays to achieve the goal of improving prediction accuracy. In each iteration, based on the previous iteration, the loss function (Equation (1)) and the pseudo-residuals (Equation (2)) are calculated. A new decision tree is constructed, and then all the generated decision trees are weighted and fused (Equation (4)) according to the weights of the decision trees (Equation (3)) by gradient descent. The GBDT model combines the decision tree with the idea of the ensemble in an effective way and improves the prediction accuracy of the model. The following steps are usually included.

$$L_i^{(t)} = \left(y_i - f^{(t-1)}\right)(x_i))^2 \quad (1)$$

$$r_i^{(t)} = y_i - f^{(t-1)}(x_i) \quad (2)$$

$$y^{(t)} = learning\_rate \quad (3)$$

$$F_i(x) = \sum_{t=1}^{m} y^{(t)} f^{(t)}(x_i) \quad (4)$$

where $L_i^{(t)}$ denotes the loss function value of the ith sample of the tth decision tree; $r_i^{(t)}$ denotes the pseudo residual of the ith sample; y_i is the actual value of the ith sample, which is the prediction of the $t-1^{\text{th}}$ decision tree for the ith sample; $y^{(t)}$ denotes the weight of the tth tree, and learning_rate is the learning rate. $F_i(x)$ denotes the final dependent prediction for the ith sample, and $f^{(t)}(x_i)$ denotes the dependent prediction result for the ith sample of the tth tree on the observation sample x. The prediction results of all trees are weighted and summed according to the weights to obtain the final prediction result.

For more information, see Li et al. [38].

3.2. Shapley Additive Explanation (SHAP)

The SHAP is an ex-post model interpretation approach, the core idea of which is to compute the marginal contribution of features to the model output, and then to interpret the "black box model" both globally and locally. The SHAP constructs an additive interpretation model, where all features are considered "contributors". For each prediction sample, the model generates a prediction value, and the SHAP value is the value assigned to each feature in the sample.

In the SHAP interpretation model, the contribution of the feature element i to the predicted value $f(x)$ is attributed to the Shapley value of the feature element, which is denoted as $\varnothing_i(f,x)$. The formula is [33]:

$$\varnothing_i(f,x) = \sum_{S\subseteq N} \frac{|S|!(M-|S|-1)!}{M!} \times [f_x(S\cup\{i\}) - f_x(S)] \tag{5}$$

where N is the set of all features; S is an order-dependent permutation subset of N, and f is the marginal contribution of feature i to the subset S; M is the total number of input features.

4. Study Area and Data

The study area is the center city of Xi'an, Shannxi province. Xi'an land space overall planning (2021–2035) (http://zygh.xa.gov.cn/web_files/zygh/file/2022/11/08/202211081852155156437.pdf, accessed on 4 April 2024) defined the central urban area of Xi'an is about 766 square kilometers, including seven districts. The total number of parking lots is 2506, of which 216 have mechanical parking spaces. In order to explore the influence of built environment factors on mechanical parking spaces and provide a theoretical basis for the construction of mechanical parking lots, this study selected the proportion of the mechanical parking spaces ratio as the dependent variable, and the range of independent variables is [0, 1]. A value of 0 means that all parking spaces are ordinary parking spaces, 0.5 indicates that mechanical parking spaces account for 50% of the total parking spaces, and 1 indicates that all parking spaces are mechanical parking spaces. The distribution of parking lots (yellow dots) and parking lots including mechanical parking spaces (red dots) in the center city of Xi'an are shown in Figure 1.

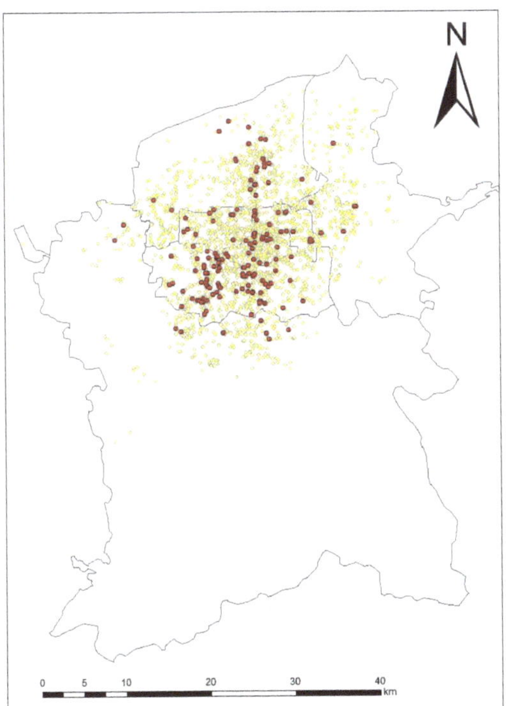

Figure 1. The distribution of parking lots.

The survey data of parking lots in Xi'an obtained in this study include the basic characteristics of parking lots, such as latitude and longitude coordinates, number of parking spaces and mechanical parking spaces, cost, and architecture type. The analysis of spatial data must take into account the scale effect and the zoning effect. In order to analyze the built environment around the parking lots, it is necessary to choose the appropriate size of the analysis unit. The whole modeling process requires that the spatial analysis unit should not be too large in order to ensure the validity of the measurement of the built environment on the one hand and to highlight the intrinsic variability of the factors of the built environment on the other hand. In reference to the selection of buffer zones in the study on the built environment and motor vehicle travel behavior [34], this paper finally selects the 500 m buffer zone of the parking lots as the research unit. Ten indicators were selected from the "7D" system to construct the independent variable system. Three parking lot characteristic indicators are cost, area, and architecture type.

The open source population density raster data are obtained from the World Population Data website (https://hub.worldpop.org/ (accessed on 20 May 2024)); Point of Interest (POI) data are obtained using the Python program, which is used to access the Gaud API to crawl, and the coordinate transformation plug-in of ArcGIS 10.2 is used to transform the GCJ-02 coordinate system of Gaud into the WGS-84 coordinate system; the road network data intercepts the road network within the scope of the study through the open source map website (OpenStreetMap, OSM) and converts the dual-line road network into a single-line network. The dual-line roads specify the direction of vehicle travel. For an intersection, there is one intersection regardless of the number of lanes. Without this treatment, multiple intersections are counted based on lane crossings.

The Kolmogorov–Smirnov test (K-S test) is a non-parametric hypothesis test in statistics that is used to test whether a single sample follows a certain distribution or whether two samples follow the same distribution. The K-S test can not only assess the degree of fit of the sample data to the theoretical distribution, which can help us to better understand the characteristics and regularity of the data distribution, but also determine whether the distribution of features is the same in the training set and the test set. If the K-S test value is greater than 0.05, the data follow a normal distribution. Through the statistical analysis of the indicators, the K-S test shows that the significance of each variable is greater than 0.05. The sample follows a normal distribution and has good quality. The multi-collinearity test of the explanatory variables is carried out using SPSS 26.0. The variance inflation factor (VIF) of the variables is less than 10. Therefore, there are no strong collinearity variables, and all independent variables can be included in the model. Table 1 gives the summary of the variables.

Table 1. Summary of variables.

Variables	Description	Min	Max	Mean	Std.
Mechanical parking space ratio	The ratio of mechanical parking spaces to total parking spaces	0	1	0.05	0.18
Land use mix	Entropy index of POI land use type in 500 m circular buffer zone of research object	0	0.92	0.76	0.1
Bus stop density	The number of bus stops in the 500 m circular buffer zone of the research object	0	16.56	3.74	2.77
Transit stop density	The number of transit stops in the 500 m circular buffer zone of the research object	0	7.65	1.05	1.74
Network density	The road length in the 500 m circular buffer zone of the research object	0	21.77	7.95	3.40
Intersection density	The number of interactions in the 500 m circular buffer zone of the research object	0	61.11	11.64	9.27
Population density	The population density in the 500 m circular buffer zone of the research object	0	17.16	1.34	2.56

Table 1. *Cont.*

Variables	Description	Min	Max	Mean	Std.
Distance to CBD	The distance from the center point of the buffer zone to CBD (Bell Tower)	0.21	29.51	7.92	4.29
Fee	Parking fee	0	10	2.11	1.48
Area	The parking lot is located in Xi'an (Inside the Ming City Wall, Ming City Wall-Second Ring, Second Ring-Third Ring, and Third Ring Outside).	1	4	2.94	0.91
Architecture type	1. Hotel; 2. Public; 3. Temporary; 4. Business office; 5. Relics scenic spot; 6. Housing; 7. Comprehensive business; 8. Hospital; 9. Other	1	9	5.76	2.71

5. Results

5.1. The Result of the Prediction Model

A total of 2506 buffer areas ($\pi \times 500 \times 500$) are used to construct models for predicting the mechanical parking spaces ratio. Python is used to conduct the prediction using the GBDT model. First, all samples are divided into 80% training and 20% test groups. Then, the GBDT model is fitted and the parameters are selected using the grid search method. The model parameters were determined using the five-fold cross-validation method. The data set was randomly divided into five subsets of equal sample size. Four of them were selected as the training set and the remaining one as the test set. The depth of the tree is 3 and the number of trees is 150. The prediction R^2 of the GBDT and OLS models are 37% and 11%. The GBDT model achieves a higher prediction accuracy.

5.2. Global Interpretability

The relative importance of the variables obtained in Python is shown in Figure 2. The relative importance of a variable is the predicted contribution of an independent variable to the dependent variable relative to other independent variables. The higher the relative importance, the greater its effect on the dependent variable. Land use mix (46.2%) and distance to CBD (38.2%) (Bell Tower in Xi'an) are two of the most important variables affecting the mechanical parking spaces ratio. Followed by the variables architecture type (5.4%), network density (4.2%), and fee (3.2%). Chen et al. [29] proposed that proximity to the city center is the second most important factor and that intersection density has the lowest impact on parking demand, which is consistent with our study.

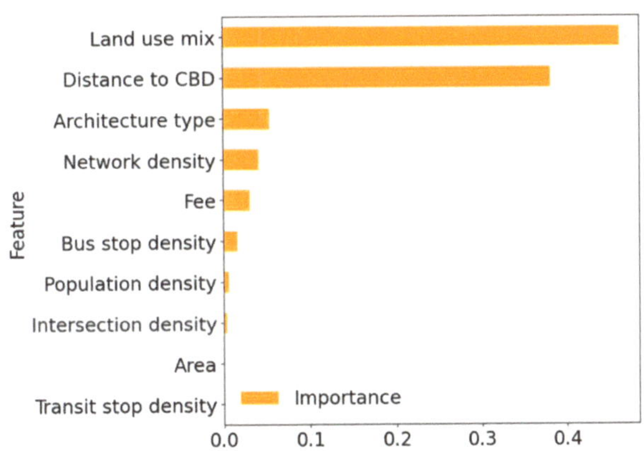

Figure 2. The relative importance of each variable.

The ALE plots show a non-linear and threshold effect of factors on the MPS ratio. The blue line indicates the value of the independent variable. The denser the blue line, the higher the frequency of that value. Figure 3a shows that the land use mix is positively associated with the MPS ratio. When the land use mix is less than 0.83, there is no obvious variation in the MPS ratio. While it exceeds 0.83, its effects are faced with a sharp increase. The distance to CBD is negatively associated with the MPS ratio, as shown in Figure 3b. It shows a threshold which is about 7 km. When the MPS ratio is less than 7 km, it faces a large slope. When it exceeds 7 km, the influence is minimal. The reason is that the land resources in the city center are scarce, and mechanical parking spaces are needed to improve the level of land use, meet parking demand, and improve turnover efficiency. Different architecture types have different effects on the MPS ratio (Figure 3c). The hotel, public, and temporary areas need more mechanical parking spaces. The network density has an extreme effect (Figure 3d). The higher or lower the density, the greater the impact on the parking rate. The road network in the city center is dense and the shortage of land requires mechanical parking spaces. The road network outside the suburbs is sparse, and some large attractions, such as parks or commercial centers, have a high demand for parking. See from Figure 3e, the influence of bus stops on MPS ratios is positive and the threshold is 5 units/km^2. The higher the density, the higher the ratio, mainly because the higher the density, the more likely it is to be in the city center, and, therefore, the higher the requirement for the ratio. See from Figure 3f, the influence of fees on MPS ratios is positive and the threshold is 2.5 yuan/h. The higher the cost, the higher the rate. The main reason is that, on the one hand, the parking fee in the city center is high, and it is more likely to build mechanical parking spaces. On the other hand, the cost of a mechanical parking lot is higher than that of an ordinary parking lot. The effects of population density, intersection density, area and transit stop density are close to zero.

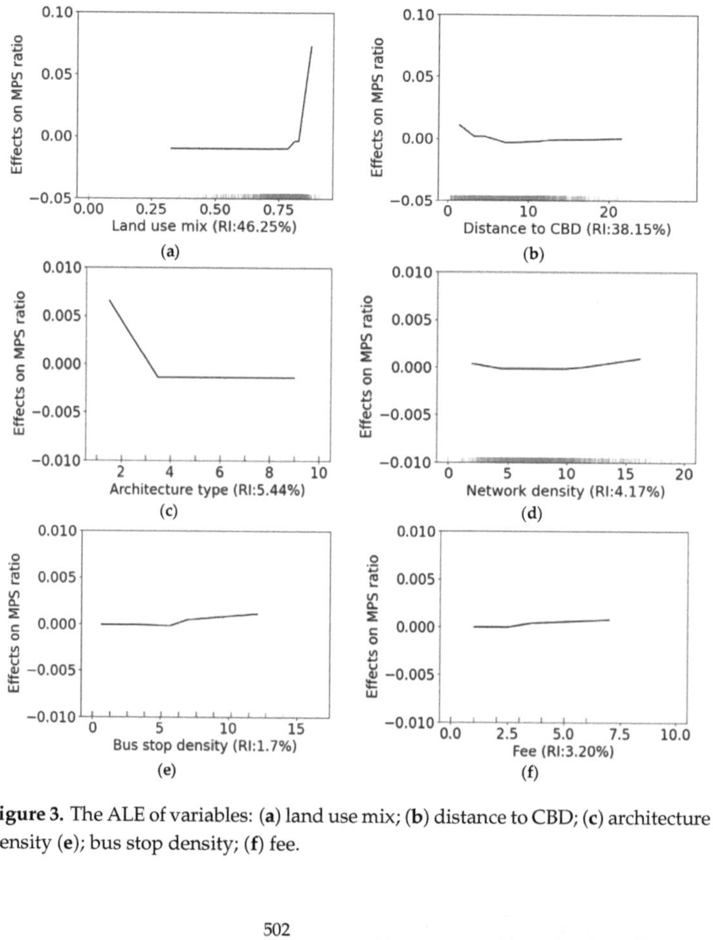

Figure 3. The ALE of variables: (**a**) land use mix; (**b**) distance to CBD; (**c**) architecture type; (**d**) network density (**e**); bus stop density; (**f**) fee.

The SHAP tree explainer is used to explain the model globally and locally. The order of the influence of the SHAP value on variables is given from large to small in Figure 4. It is almost the same as the importance of the variables obtained from the GBDT simulation. Land use mix, distance to CBD, and architecture type are three variables with large influence and have the best global interpretability to MPS ratio.

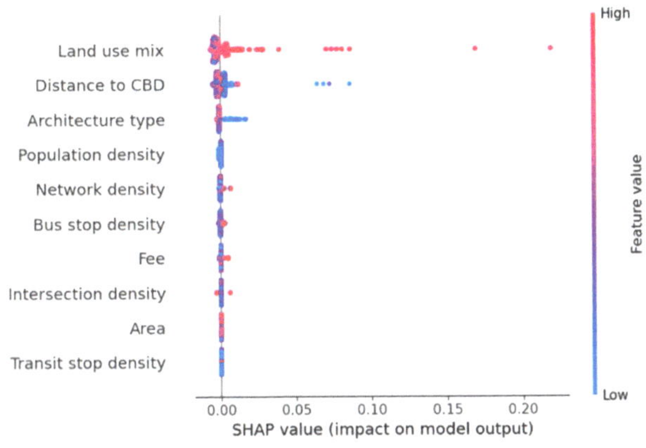

Figure 4. The SHAP values of variables.

5.3. Local Interpretability

The local interpretability of the model shows the prediction contribution of each parking lot buffer with a different spatial coordinate. Two buffers are randomly selected and show the SHAP value of each variable, which is shown in Figure 5. The base value is the same (0.04787), which is the average of all samples. The red bar indicates the positive influence and the blue color means the negative influence. As shown in Figure 5a, the variables with positive SHAP values are population density and distance to CBD. The remaining variables have a negative influence. The total SHAP value is 0.05, which is higher than the mean value. Figure 5b shows another sample of which variables have different influences. These two samples show that the local interpretability of variables is different from the global interpretability, and the local interpretability varies across buffers.

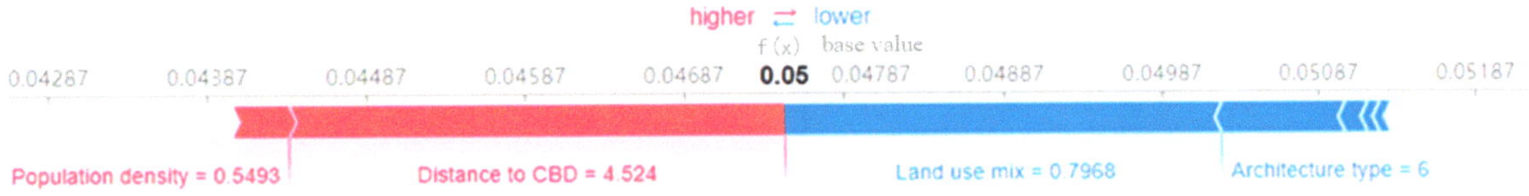

(**a**)

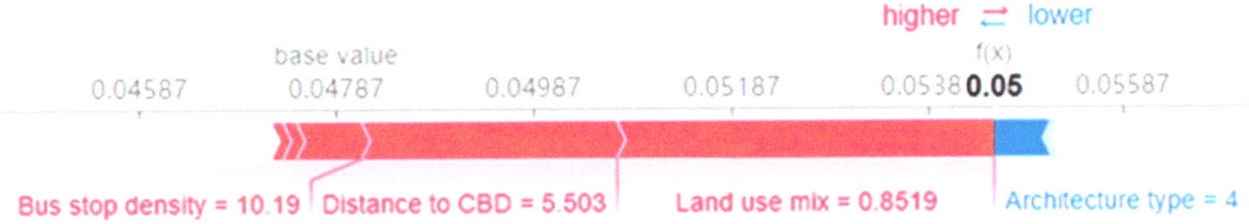

(**b**)

Figure 5. Local SHAP values are randomly selected from two buffers: (**a**) Buffer 1; (**b**) Buffer 2.

Figure 6 shows the local SHAP values of six different variables. Figure 6a,b,e,f shows variables of the distance to CBD, land use mix, bus stop density, and fee. These four figures show the influence of variables on the MPS ratio has strong spatial aggregation. Figure 6c,d show architecture type and population density. The influence is more dispersed in space. These graduated symbol maps can suggest the MPS ratio that is dominated by that particular variable. The local SHAP values of different variable distributions are also different. This is because different variables have different influences on the model at different spatial positions. These figures show the different contributions of each variable, which can help urban and traffic planners generate tailored strategies for parking lots and mechanical parking spaces planning.

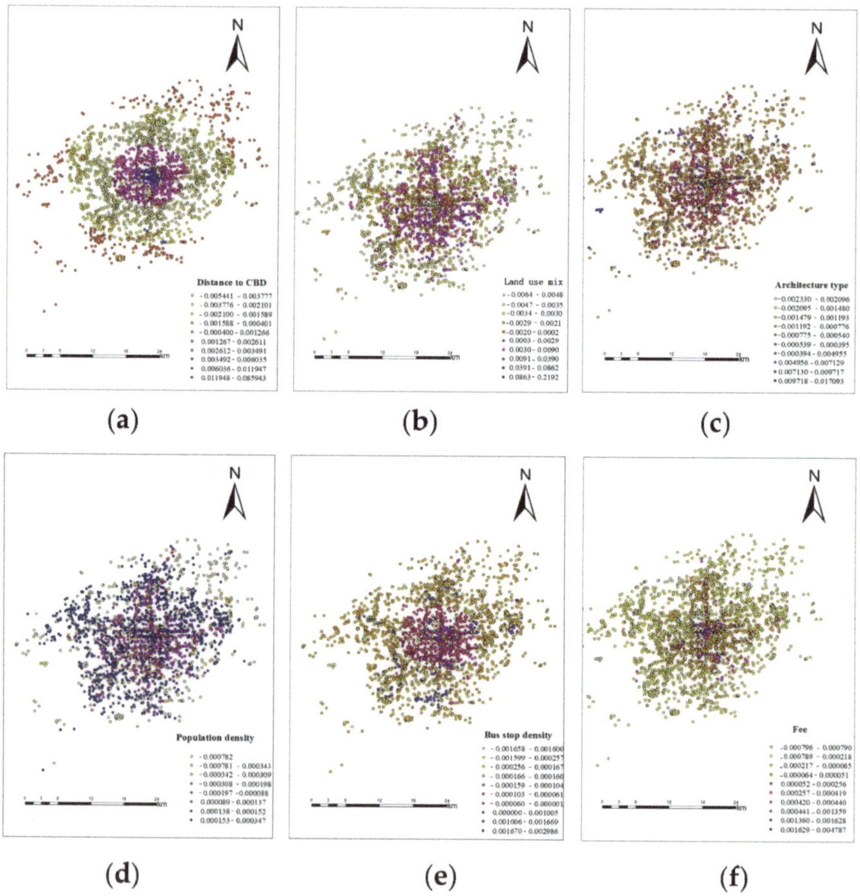

Figure 6. Contribution of different variables obtained from SHAP values over space: (**a**) distance to CBD; (**b**) land use mix; (**c**) architecture type; (**d**) population density; (**e**) bus stop density; (**f**) fee.

6. Discussion and Conclusions

The non-linear relationship between the mechanical parking spaces ratio (MPS ratio) and the built environment is constructed by using the GBDT model. Moreover, the ALE plot is used to simulate the influence of variables on the mechanical parking spaces ratio and to determine the influence threshold. In addition, the SHAP value is used to reveal the local feature of built environment variables on the MPS ratio. The local feature means the influence of variables on a sample which is a buffer for a parking lot. Due to the different spatial locations of the different buffers, the built environment around them is different, which makes the effect of the factors on the MPS ratio show different effects. Based on these studies, some valuable insights are provided for the planning of mechanical parking lots and spaces. The results make three contributions to the parking analysis. First, the results reveal the non-linear influence of built environment variables on the mechanical parking spaces ratio. Second, the relative importance of variables on the mechanical parking spaces ratio helps planners make the mechanical parking lots and spaces planning. Third, the use of SHAP values to describe the spatially local impacts of the built environment can reveal the spatial variability of variable impacts.

From Figure 2, the land use mix and the distance to CBD play an important role in the planning of mechanical parking spaces and lots. The construction and maintenance costs of the mechanical parking spaces are higher than those of conventional parking spaces. Therefore, the mechanical parking spaces or parking lots in the necessary locations can not only meet the parking demand but also save costs. In this study, it is more necessary to build mechanical parking spaces or parking lots in an area where the land use mix is higher than 0.83 and the distance from the city center is less than 7 km. However, according to the local effect diagram in Figure 6, the construction of mechanical parking spaces or parking lots can also be considered for areas where the variables have a positive effect. For building types, the impact of the construction of mechanical parking spaces or parking lots can be considered when planning parking lots for building types with a higher impact, such as hotels with higher impact.

According to the local effect diagram in Figure 6, the variables in some areas have a positive impact on the mechanical parking spaces ratio; therefore, it is necessary to build a mechanic parking space for the area with a large value of the variable. The variables in some areas have a negative effect on the mechanical parking spaces ratio, which requires the construction of mechanical parking spaces for areas where the built environment variables have lower values. According to the results of the study, shown in Figure 6a, it can be seen that the positive effect of the variable distance to CBD on the construction of mechanical parking is mainly evident in the city center as well as in the periphery. Thus, the closer to the city center, the more mechanical parking spaces are needed to meet the parking demand. However, the ALE graph shows that, although the distance to the CBD has the second highest impact, the impact trend is not very large. This conclusion also reflects that the CBD does not advocate a lot of parking lots, as this will increase parking demand and thus traffic congestion in the city center area. At the same time, the periphery can be a park-and-ride site and provide a reference for the installation of park-and-ride sites. In addition, as shown in Figure 6f, the higher the parking fees, the more mechanical parking spaces there will be; moreover, mainly because of the high construction costs of mechanical parking garages, only high charges can cover normal operating and maintenance costs. This finding can also provide a basis for charging for mechanical parking lots. Above all, planners should make mechanical parking planning schemes more responsive to the built environment of the area and the travel characteristics of residents (the built environment influences the travel characteristics of residents). This allows the mechanical parking system to play its proper role. Thus, the results of this study provide theoretical and practical references for car park planning and parking space allocation.

This study still has some limitations and issues that need to be discussed and solved in the future. As the empirical analyses are based on a single city, whether the impact of built environment factors on parking demand is equally significant in different cities has not been verified. In the future, we can pay attention to the corresponding problems of different cities and different subregions and explore the changes of the influence mechanism in different regions; moreover, we can also pay attention to whether different cities and regions have the same significant conclusions, in order to better serve the urban construction and to improve the travel environment and improve travel satisfaction.

Author Contributions: Conceptualization, W.L. and Y.W.; methodology, W.L.; software, W.L. and Y.W.; validation, W.L.; formal analysis, W.L.; investigation, W.L.; resources, W.L.; data curation, W.L. and Y.W.; writing—original draft preparation, W.L. and Y.W.; writing—review and editing, W.L.; visualization, W.L.; supervision, W.L.; project administration, W.L.; funding acquisition, W.L. All authors have read and agreed to the published version of the manuscript.

Funding: The research was funded by the Education Department of Shaanxi Provincial Government [grant number 22JE004].

Institutional Review Board Statement: Not applicable.

Informed Consent Statement: Not applicable.

Data Availability Statement: The raw data supporting the conclusions of this article will be made available by the authors upon request.

Conflicts of Interest: The authors declare no conflicts of interest.

References

1. Peng, T.; Chen, J.; Huang, Y. Examining the relationship between built environment and urban parking demand from the perspective of travelers. *J. Clean. Prod.* **2023**, *385*, 255175014. [CrossRef]
2. Arnott, R.; Williams, P. Cruising for parking around a circle. *Transp. Res. Part B-Methodol.* **2017**, *104*, 357–375. [CrossRef]
3. Zulfiqar, H.; Rivano, H.; Le Mouël, F. A survey on smart parking systems in urban cities. *Concurr. Comput.-Pract. Exp.* **2023**, *35*, 3229–3253. [CrossRef]
4. Belkhala, S.; Benhadou, S.; Boukhdir, K.; Medromi, H. Smart Parking Architecture based on Multi Agent System. *Int. J. Adv. Comput. Sci. Appl.* **2019**, *10*, 378–382. [CrossRef]
5. Rocco, G.; Pipino, C.; Pagano, C. An Overview of Urban Mobility: Revolutionizing with Innovative Smart Parking Systems. *Sustainability* **2023**, *15*, 13174. [CrossRef]
6. Zong, F.; Zeng, M.; Yu, P. A parking pricing scheme considering parking dynamics. In *Transportation*; Springer: Berlin/Heidelberg, Germany, 2023.
7. Zhang, Y.N.; Huang, Q.B.; Cui, Y.Z. Modeling and Simulation of Virtual Prototype of the Lift-Sliding Mechanical Parking Equipment Based on Solid Works. In Proceedings of the International Conference on Modeling, Simulation and Optimization, Beijing, China, 26–27 December 2009.
8. Zhang, J.S.; Yang, H.; Chen, Z.; Yu, T.; Liu, H. Probability distribution and recommended values of passenger car external dimensions in mechanical parking garage design. *J. Asian Archit. Build. Eng.* **2022**, *21*, 1942–1958. [CrossRef]
9. Qin, H.M.; Yang, X.; Wu, Y.; Gaun, H.; Wang, P.; Shahinpoor, N. Analysis of parking cruising behaviour and parking location choice. *Transp. Plan. Technol.* **2020**, *43*, 717–734. [CrossRef]
10. Xu, B.; Yu, B.; Tao, Z. Strength design of a mechanical parking system. In Proceedings of the International Conference on Mechanical Structures and Smart Materials (ICMSSM 2013), Xiamen, China, 16–17 November 2013.
11. Tazaki, Y.; Okuda, H.; Suzuki, T. Parking Trajectory Planning Using Multiresolution State Roadmaps. *IEEE Trans. Intell. Veh.* **2017**, *2*, 298–307. [CrossRef]
12. Amari, A.; Moussaid, L.; Tallal, S. New Parking Lot Selection Approach Based on the Multi-Criteria Decision Making (MCDM) Methods: Health Criteria. *Sustainability* **2023**, *15*, 938. [CrossRef]
13. Ewing, R.; Cervero, R. Travel and the Built Environment. *J. Am. Plan. Assoc.* **2010**, *76*, 265–294. [CrossRef]
14. Wang, Z.; Song, J.; Zhang, Y.; Li, S.; Jia, J.; Song, C. Spatial Heterogeneity Analysis for Influencing Factors of Outbound Ridership of Subway Stations Considering the Optimal Scale Range of "7D" Built Environments. *Sustainability* **2022**, *14*, 16314. [CrossRef]
15. Gan, Z.; Yang, M.; Feng, T.; Timmermans, H.J. Examining the relationship between built environment and metro ridership at station-to-station level. *Transp. Res. Part D-Transp. Environ.* **2020**, *82*, 102332. [CrossRef]
16. Yin, C.Y.; Shao, C.F.; Wang, X.Q. Built Environment and Parking Availability: Impacts on Car Ownership and Use. *Sustainability* **2018**, *10*, 2285. [CrossRef]
17. Meng, F.Y.; Du, Y.; Chong, Y.; Wong, S.C. Modeling heterogeneous parking choice behavior on university campuses. *Transp. Plan. Technol.* **2018**, *41*, 154–169. [CrossRef]
18. Christiansen, P.; Engebretsen, Ø.; Fearnley, N.; Hanssen, J.U. Parking facilities and the built environment: Impacts on travel behaviour. *Transp. Res. Part A-Policy Pract.* **2017**, *95*, 198–206. [CrossRef]
19. Jiang, Z.; Mondschein, A. Analyzing parking sentiment and its relationship to parking supply and the built environment using online reviews. *J. Big Data Anal. Transp.* **2021**, *3*, 61–79. [CrossRef]
20. Aydinoglu, A.C.; Iqbal, A.S. Determining Parking Demand and Locating Parking Areas Using Geographic Analytics Methods. *J. Urban Plan. Dev.* **2021**, *147*, 05020035. [CrossRef]
21. Liu, Q.; Chen, J.W.P.; Xiao, Z. How does parking interplay with the built environment and affect automobile commuting in high-density cities? A case study in China. *Urban Stud.* **2016**, *54*, 3299–3317. [CrossRef]
22. Zeng, C.; Zhou, X.; Yu, L. Parking Generating Rate Prediction Method Based on Grey Correlation Analysis and SSA-GRNN. *Sustainability* **2023**, *15*, 13016. [CrossRef]
23. Tang, X.; Ma, J.; Zhou, S.; Shan, T. Parking Demand Prediction Method of Urban Commercial-Office Complex Buildings Based on the MRA-BAS-BP Algorithm. *J. Adv. Transp.* **2022**, *2022*, 2529912. [CrossRef]
24. Lee, Y.W. Establishment of an on-street parking demand estimation model through a survey on parking during day and night. *Asia Life Sci.* **2015**, 471–484.

25. Han, Y.; Shan, J.; Wang, M. Optimization design and evaluation of parking route based on automatic assignment mechanism of parking lot. *Adv. Mech. Eng.* **2017**, *9*, 168781401771241. [CrossRef]
26. Kadkhodaei, M.; Shad, R.; Ali Ziaee, S.; Kadkhodaei, M. Prediction model for drivers' tendency to perpetrate a double parking violation on urban trips. *Transp. Policy* **2023**, *141*, 331–339. [CrossRef]
27. Leurent, F.; Boujnah, H. A user equilibrium, traffic assignment model of network route and parking lot choice, with search circuits and cruising flows. *Transp. Res. Part C-Emerg. Technol.* **2014**, *47*, 28–46. [CrossRef]
28. Wang, T.; Li, S.; Li, W. A Short-Term Parking Demand Prediction Framework Integrating Overall and Internal Information. *Sustainability* **2023**, *15*, 7096. [CrossRef]
29. Chen, J.; Liu, K.; Di, J.; Peng, T. Nonlinear Model of Impact of Built Environment on Urban Parking Demand. *J. Transp. Syst. Eng. Inf. Technol.* **2021**, *21*, 197–203.
30. Gabbe, C.J.; Pierce, G.; Glowers, G. Parking policy: The effects of residential minimum parking requirements in Seattle. *Land Use Policy* **2020**, *91*, 104053. [CrossRef]
31. Liu, Q.; Zhu, M.J.; Xiao, Z.P. Workplace Parking Provision and Built Environments: Improving Context-Specific Parking Standards Towards Sustainable Transport. *Sustainability* **2019**, *11*, 1142. [CrossRef]
32. Ma, X.W.; Yin, Y. Short-Term Prediction of Bike-Sharing Demand Using Multi-Source Data: A Spatial-Temporal Graph Attentional LSTM Approach. *Appl. Sci.* **2022**, *12*, 1161. [CrossRef]
33. Liu, K.L.; Chen, J.; Li, R. Nonlinear Effects of Community Built Environment on Car Usage Behavior: A Machine Learning Approach. *Sustainability* **2022**, *14*, 6722. [CrossRef]
34. Liu, C.M.; Wang, Y.; Li, W. An Urban Built Environment Analysis Approach for Street View Images Based on Graph Convolutional Neural Networks. *Appl. Sci.* **2024**, *14*, 2108. [CrossRef]
35. Zhang, W.J.; Zhao, Y.; Cao, X.; Lu, D.; Chai, Y. Nonlinear effect of accessibility on car ownership in Beijing: Pedestrian-scale neighborhood planning. *Transp. Res. Part D-Transp. Environ.* **2020**, *86*, 102445. [CrossRef]
36. Yang, W.Y. The nonlinear effects of multi-scale built environments on CO_2 emissions from commuting. *Transp. Res. Part D-Transp. Environ.* **2023**, *118*, 103736. [CrossRef]
37. Ye, X.; Wang, J.; Wang, T.; Yan, X.; Ye, Q.; Chen, J. Short-Term Prediction of Available Parking Space Based on Machine Learning Approaches. *IEEE Access* **2020**, *8*, 174530–174541. [CrossRef]
38. Li, W.J.; Kockelman, K.M. How does machine learning compare to conventional econometrics for transport data sets? A test of ML versus MLE. *Growth Chang.* **2022**, *53*, 342–376. [CrossRef]
39. Apley, D.W.; Zhu, J.Y. Visualizing the effects of predictor variables in black box supervised learning models. *J. R. Stat. Soc. Ser. B-Stat. Methodol.* **2020**, *82*, 1059–1086. [CrossRef]
40. Lee, C.; Lee, S. Exploring the Contributions by Transportation Features to Urban Economy: An Experiment of a Scalable Tree-Boosting Algorithm with Big Data. *Land* **2022**, *11*, 577. [CrossRef]
41. Yu, C.C.; Deng, Y.; Qin, Z. Traffic volume and road network structure: Revealing transportation-related factors on PM2.5 concentrations. *Transp. Res. Part D-Transp. Environ.* **2023**, *124*, 103935. [CrossRef]
42. Ding, C.; Wang, D.; Ma, X.; Li, H. Predicting Short-Term Subway Ridership and Prioritizing Its Influential Factors Using Gradient Boosting Decision Trees. *Sustainability* **2016**, *8*, 1100. [CrossRef]

energies

MDPI

Article

Innovative, Three-Dimensional Model for Time-Dependent, Mechanical Battery Module Behaviour Due to Cell Volume Change

Tolga Bozalp [1,2,*], Shraddha Suhas Kulkarni [2], Holger Opfer [1] and Thomas Vietor [2]

1 Volkswagen AG, Industriestraße Nord, 38239 Salzgitter, Germany
2 Institute for Engineering Design, Technische Universität Braunschweig, Hermann-Blenk-Straße 42, 38108 Braunschweig, Germany
* Correspondence: tolga.bozalp@volkswagen.de

Abstract: Battery cells experience volume changes due to intercalation and ageing processes, which may pose a challenge when integrating cells into a battery module. This study presents an innovative, numerical model, which spatially resolved predicts the time-dependent, overall mechanical behaviour of battery modules caused by volume changes in built-in cells. An already self-developed battery module model, which statically describes the three-dimensional (3D), mechanical behaviour in a 0D simulation environment, is extended by the time dimension for dynamic modelling. The existing model abilities and features are maintained, such as the inclusion of multiple size scales from the cell to module level as well as the automatized model building process for the investigation of different module designs in regard to the number and arrangement of foam pads and multiple other design parameters. The validation of the predication abilities against those of complex, commercial software solutions, which use Finite Element Analysis (FEA) in a 3D model environment, have shown good agreement regarding sensitivity, robustness and numerical stability, revealing the impact and interdependencies of model parameters as well as the numerical limits of the model. In this study, the potential of the novel model regarding computational time and resources is underlined, making it a useful and effective tool for fast optimization studies.

Keywords: battery module; pouch cell; lithium ion; simulation; dynamic; time-dependent; swelling; breathing; multi-scale; ageing; mechanical

Citation: Bozalp, T.; Kulkarni, S.S.; Opfer, H.; Vietor, T. Innovative, Three-Dimensional Model for Time-Dependent, Mechanical Battery Module Behaviour Due to Cell Volume Change. *Energies* **2024**, *17*, 2941. https://doi.org/10.3390/en17122941

Academic Editors: Artur Kierzkowski, Tomasz Kisiel, Agnieszka A. Tubis, Franciszek Restel, Mateusz Zając, Tomasz Nowakowski and Anna Jodejko-Pietruczuk

Received: 21 May 2024
Revised: 3 June 2024
Accepted: 9 June 2024
Published: 14 June 2024

1. Introduction

It has become generally known that lithium-ion battery cells show electrothermal in addition to mechanical behaviour. Here, reversible and irreversible volume change in cells can be distinguished between each other. On the one hand, intercalation and extraction processes while charging and discharging are expressed through a reversible increasing and decreasing of the cell thickness [1–5]. Even small lithium-ion cells show that the volume expansion of their electrode sandwich can be as much as 1.5%, when charging with low current rates from 0% to 100% SOC (state of charge) [4]. On the other hand, lithium-ion cells experience an irreversible volume expansion over their cycle and calendar life due to internal degradation processes [5–9]. Post-mortem analyses show that they include processes such as lithium plating or the growth of the solid electrolyte interphase (SEI) [6,7]. For example, in automotive applications like battery-driven electric vehicles, an increased number of large-size, high-energy lithium-ion cells are electrically connected and integrated into a mechanical housing structure [10]. This time-dependent, geometrical behaviour of the cells must be considered in the early development phase of the constructive design of a battery module. The most common measurement to consider this challenge is the application of foam pads between the cells of the module. The assessment of the mechanical design of a battery module by experimental testing is a long-lasting process, which consumes a lot of financial resources. A design development without any analytical simulation tool, which maps the spatially resolved, time-dependent force displacement of a battery module according to physical laws, is inefficient and demands multiple optimization loops, which multiplies the time and cost of the product development phase.

There have been publications in recent years which have investigated the topic of swelling modelling with different approaches on the cell as well as module level. Generally, the previously proposed models can be categorized on the basis of the included number of dimensions in 1D, 2D and 3D simulation models. On the cell level, the mechanical swelling model is either single-domain and standalone or part of a coupled, multiphysical model. As a standalone model, it has been shown that a simple phenomenological model can effectively predict global swelling forces over cycles for prismatic cells, using an experimentally calibrated, stress–strain-based constitutive model [11]. Modelling a cell two-dimensionally also enables the prediction of cell swelling and stress, spatially solved as a position within electrode layers. Therefore, a 2D mechanical model as part of a multiphysical cell model to predict the electrochemical–mechanical coupled behaviour for prismatic cells has also been developed [12]. The calculation of swelling is conducted as a function of position within the electrode layer and time [12]. For swelling prediction on the cell level, a homogenization scheme is applied and solved by finite element software and then validated against a detailed cell model also using the same software [12]. A similar modelling approach was used in a publication which investigated the ageing and swelling behaviour of cylindrical cells [13]. Using simplified 2D geometries in finite element software, a swelling simulation study was conducted to support the experimental ageing results [13]. In a different publication, it was also shown by a similar modelling technique, which couples the electrochemical and mechanical behaviour, that it is possible to model the influence of the cell casing on the swelling behaviour [14]. When modelling the cell swelling three-dimensionally, there are also two different approaches, standalone and coupled. Similar to the previously described 1D approach [11], there is also a publication where the global swelling force of a prismatic cell due to internal degradation processes is measured in cycle life tests, and a model is derived by fitting a 3d model in finite element software to the measurement data, which describe the desired free swelling of the cell [15]. On the other side, there have been numerous publications which have focused on the 3D simulation of battery cell swelling as part of a multiphysical cell model, where different physical processes are coupled with their mechanical behaviour. A 3D mechanical model as part of a multiphysical cell model to predict the mechanical pouch cell behaviour of two interacting cells under overcharge conditions was developed by fluid–structure coupling simulation using finite element software [16]. Another model which couples the electrothermal and mechanical cell behaviour during cycling was also presented previously. The model fully couples the three physical domains in finite element software and three-dimensionally spatially resolves it in dependence of time [17]. In a different previously published paper, the methodology and results of a battery cell model were presented, which is not only multiphysical, coupling electrothermal–chemical–mechanical behaviour, but also multi-scale [18]. In this time-dependent model, even the porous microstructures based on the images of the surface morphologies of anode and cathode surfaces are included in finite element software [18]. Therefore, mechanical stress in particles is also simulated during cycling [18].

On the battery module level, there have also been some publications examining the effects of cell breathing or swelling. These simulation models can also be divided according to the number of spatial dimensions included in 1D as well as 3D. Numerous studies have mainly focused on the 3D simulation of the swelling of battery modules. In the course of the experimental characterization of possible mechanical inhomogeneities within prismatic modules due to breathing and swelling, appropriate 3D simulations, considering cells and other built-in components, were conducted, using finite element software in order to study the effects on stress distribution and end plate deformation on the module level [19]. Similar approaches were also used in further simulation studies of battery module swelling [15,20], while in all 3D studies, the degree of details differs. On the one hand, simplified 3D geometries for module components are examined for rough estimations of module elongations [15]. On the other hand, a different model includes geometries closer to reality, in order to calculate exact mechanical stresses and their distributions on module housing components [20]. Studies presenting 1D model approaches for the swelling behaviour of battery modules show that their results can also agree well with more complex and time-consuming 3D simulations. These models resemble each other in the regard that they simplify the geometry of module components drastically to an equivalent spring stiffness one at a time [11,21,22]. In the next step, the stiffnesses of all cells in the module are summed up to one cumulative, equivalent stiffness [11,21,22]. If other components are considered, such as foam pads, the same procedure is conducted for them as well [11,21]. All equivalent component springs are connected in series afterwards. Applying a force equilibrium equation then allows for the calculation of the global resulting displacements and forces at the border region of the considered components, when a load is applied. In a previous publication, the global swelling force on the module end plate was calculated by applying a free swelling force on the equivalent spring for the cells at first, which acted on the equivalent spring for the pads, behind which an ideal stiff end plate was located [11]. In the next step, the force equilibrium was applied. A different publication introduced a model in dependence of state of charge and state of health (SOH) by applying the force equilibrium between the module frame and cell stack stiffness, in order to predict module deformation and average pressure for the cells inside [21]. Even though no publication gives specific details about calculation power and time, it can be assumed that 1D simulation models offer a potential advantage in regard to calculation time and resources, while on the other hand, they lack spatial and temporal resolution.

A substantial development in that regard was made in a previously published paper which introduced a novel method for the three-dimensional simulation of battery module swelling [23]. Within the framework of the study, a newly developed simplified 0D mathematical model was presented, which is able to predict the three-dimensional, mechanical battery module behaviour. In comparison to previous publications in the field of the simplified mathematical modelling of battery modules based on mass–spring systems [11,21,22], the new model is able to predict global module deformations not one- but three-dimensionally. The biggest achievement of the model is the tremendous calculation time advantage in comparison to complex 3D simulations, which use commercial finite element software, without sacrificing much degree of detail by still modelling the module three-dimensionally. The main input for the model is a cell swelling model, which is implemented in the module model [23]. The module model then calculates the module force–displacement behaviour over its lifetime statically. A successful validation against experimental test data was also conducted.

In this study, the previously published static, mathematical model for the three-dimensional simulation of the force–displacement behaviour of battery modules is extended by the time dimension for application in dynamic modelling. Consequently, the newly developed model distinguishes itself from previous publications in the field of the simplified mathematical modelling of battery modules based on mass–spring systems [11,21,22] not only by modelling the force–displacement behaviour three-dimensionally but also now is capable of dynamic modelling for time-dependent behaviour, which is a clear distinguishing factor in comparison to the previous publication, which only applied static modelling [23]. The key model features as well as the automatic model building process, which comfortably and quickly allows us to investigate different module designs, are kept for the novel model as well. The mathematical model as well as the fundamental technical mode of operation are described, at first; in the next step, its prediction abilities are validated against those of complex software solutions using Finite Element Analysis (FEA) by means of a systematic sensitivity and robustness study. In addition, the ability to predict the effects of design variations as well as the robustness and numerical stability of the model are investigated. After a presentation and discussion of the simulation and validation results, at the end, a conclusion is drawn, and an outlook is given.

2. Simulation Model

The newly developed dynamic model for the three-dimensional simulation of the mechanical battery module behaviour due to cell volume change is described in the first part of this chapter. Consequently, an overview of the complex battery module model is given, which is used for the validation of the simplified model. Since in this study the focus is placed on dynamic modelling, in addition to the applied geometrical and material parameters, the temporal and numerical parameters of the simulation need to be investigated and are therefore described as well.

2.1. *Simplified Dynamic Modelling by Novel Mathematical Approach*

As an alternative to an FEA model, another 3D mathematical model for the mechanical behaviour of battery modules is developed in the commercial software MATLAB/Simulink R2022b. This model is a further development of an existing static, mathematical model [23]. It is intended to map the overall mechanical force–displacement behaviour of battery modules due to cell volume change in a spatially and also time-resolved manner over their cycle and service life, considering their individual components, such as cells, foam pads and housing parts. By the inclusion of time, dynamic loads can be applied. The density and mass of components are also considered, so that dynamic effects, such as impact loads, can be mapped. On the other hand, damping effects are neglected in the novel model.

The automatized model building process by intelligent algorithms is maintained by adopting it into the Simulink simulation environment. By simply providing the algorithm with the data it needs via predefined constant blocks, such as geometrical and design information about the module as well as its components and their mechanical material behaviour, the algorithm processes the information to build up a mathematical equations system, which represents the time-dependent force–displacements of the defined module. In the same step, a geometrical discretization of the module and its components is conducted, similar to complex finite element software. In comparison, though, the number of elements, in which the module is geometrically discretized, is higher by several orders of magnitude in commercial software. The model and consequently the fundamental mathematical equations are based on physical laws and realized as a mass–spring system with time-dependent behaviour, which can be described in matrix notations as follows in Equation (1):

$$\{F\} = [M]\cdot\left\{\ddot{U}\right\} + [K]\cdot\{U\} \tag{1}$$

The global mass and stiffness matrices are represented by [M] and [K], while {U} and $\left\{\ddot{U}\right\}$ stand for the global displacement and acceleration vectors. The global force vector {F} is equated to the sum of the products, as described in Equation (2). In explicit notation, considering each vector and matrix entry, the equation can be described as follows:

$$\begin{Bmatrix} F_1 \\ \vdots \\ F_n \end{Bmatrix} = \begin{bmatrix} M_{11} & \cdots & M_{1n} \\ \vdots & \ddots & \vdots \\ M_{n1} & \cdots & M_{nn} \end{bmatrix} \cdot \begin{Bmatrix} \ddot{u}_1 \\ \vdots \\ \ddot{u}_n \end{Bmatrix} + \begin{bmatrix} K_{11} & \cdots & K_{1n} \\ \vdots & \ddots & \vdots \\ K_{n1} & \cdots & K_{nn} \end{bmatrix} \cdot \begin{Bmatrix} u_1 \\ \vdots \\ u_n \end{Bmatrix} \tag{2}$$

The battery module is geometrically discretized by mass nodes and interconnecting springs with stiffnesses according to this modelling approach. All masses and spring stiffnesses can be written into a mass and stiffness matrix. The displacement and acceleration of each mass node as well the forces which act on them are accordingly represented by vectors. By setting up Equation (2), a mathematical equation system is formed for describing the mechanical behaviour of each node as a result of the acting forces. By determining the movement of each mass node, it is possible to describe the mechanical behaviour of the entire module. The equations generated by the algorithm are subsequently solved with the Newmark-beta method [24]. The numerical integration method is paramount for the model and follows a general procedure, consisting of a series of iteration steps to solve the following by Newman-defined relations [24]:

$$\dot{u}_{n+1} = \dot{u}_n + (1-\gamma)\cdot\ddot{u}_n\cdot h + \gamma\cdot\ddot{u}_{n+1}\cdot h \quad (3)$$

$$u_{n+1} = u_n + \dot{u}_n\cdot h + \left(\frac{1}{2} - \beta\right)\cdot\ddot{u}_n\cdot h^2 + \beta\cdot\ddot{u}_{n+1}\cdot h^2 \quad (4)$$

While h stands for the time interval and u for the displacements, the two parameters γ and β were introduced as part of the method. Basically, they allow the user to have influence on the numerical integration process by adjusting them. While the parameter γ affects the accuracy of the results, the parameter β has influence on the stability of the numerical integration procedure [24]. The parameter γ is fixed, since an artificial damping is introduced into the investigated system, if it differs from the value of 0.5 [24]. In contrast, the parameter β may vary and should lie between 0 and 0.25 [24]. The selection of the parameter β not only influences the acceleration during a time interval but also shows a general correspondence for some values [24]. Specifically, consistent variation in acceleration during a time interval can be determined for at least four β-values [24], which are shown in Table 1.

Table 1. Influence of numerical stability parameter β on acceleration during time interval [24].

Numerical Stability Factor β	Value	Acceleration During Time Interval
β_1	1/4 = 0.250	Uniform
β_2	1/6 = 0.167	Linear variation
β_3	1/8 = 0.125	Step function
β_4	0	Double pulses

Analogous to complex FEA software Abaqus 2023 Hotfix 1, the nodes of the elements are shared by adjusting elements. To simulate multiple components of the battery module, such as cells, compression pads and housing and their interaction, additional air layers with very low stiffness and negligible thickness are modelled around each component. This helps to capture the individual deformations of the components similar to multi-body simulations.

The jelly stack inside the cell is modelled as a homogeneous body, on which the swelling forces are applied, leading to the deformation of the cell casing and subsequently other components, causing neighbouring elements to collide with each other. This is mainly due to the very low stiffness and thickness of air layers. For the calculation of the force, when two components come in contact, the following Equation (5) is used [25]:

$$F_c = -\omega_n\cdot g_n\cdot e_n \quad (5)$$

Here, F_c stands for the contact force, ω_n for the penalty factor, g_n for the gap function and e_n for normal direction vectors.

Furthermore, the model offers the possibility to restrict the displacement and rotation of user-defined nodes. Therefore, the fixation of the module at its corners by screws according to a specific mounting position inside a battery system or test jig is applicable. Also, the fixation of components by adhesive joints, such as between cell cans and the adjacent module frame, can be implemented and investigated.

Besides the previously mentioned input data, the algorithm needs information about the type of investigated battery module, since the simulation model is suitable for modules, which consists of pouch or prismatic cells, also considering a parameterizable metal case of the cells. When it comes to parametrizing the mechanical material behaviour of the other module components, there are different options. Simplified material behaviour can be entered, such as the Young's Modulus (YM) for cell and its jelly stack and module housing components, as well as more complex force–deflection curves (FDCs) for foam pads. Due to the wide range of selection options in pre-processing, the developed model offers maximum flexibility for modelling battery modules of different geometries, arrangements, materials, cells and other properties. A geometric visualization of the configured battery module is also available for checking the input data. Due to the model's dynamic nature and in order to calculate the time-dependent displacements and velocities, it is required to specify a time step size and a total simulation time. Figure 1 gives an overview of the relations of the processes of the newly developed dynamic simulation model. As already mentioned, a necessary data and information set about the module and its components is needed as an input for the algorithm to build up the model. This information is only needed once at the beginning of the simulation. On the other hand, since the simulation is time-dependent, for the mechanical state of each cell, information about its respective jelly stack needs to be provided for each time step, because after model build-up, the simulation starts running and processing the information. The momentary, spatially resolved Young's Modulus and dimensions of each cell jelly stack can be provided by submodels, which are connected to the module simulation model.

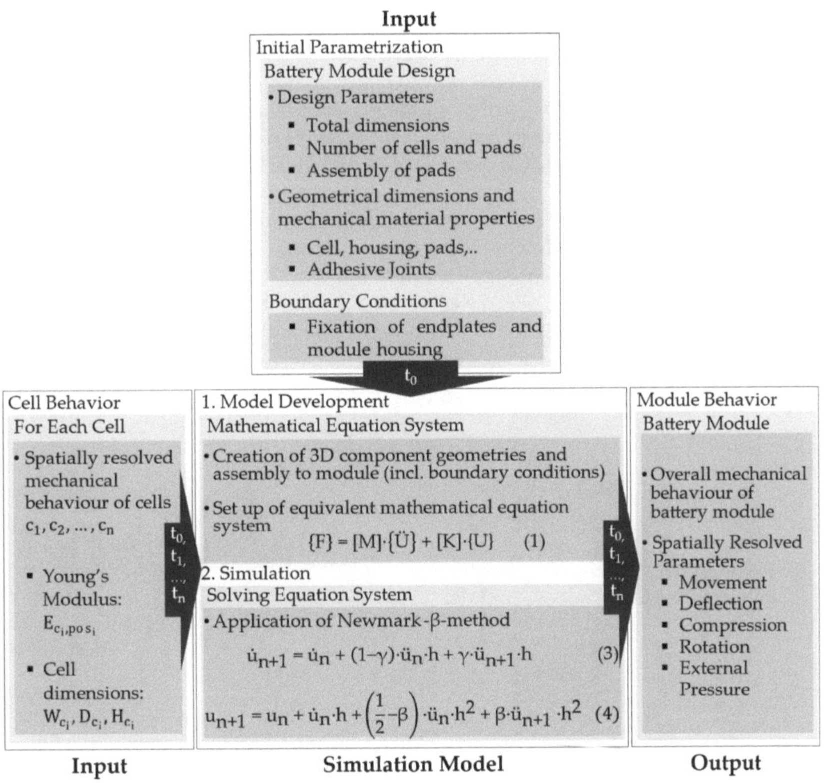

Figure 1. An overview of the processes for the novel dynamic simulation model (updated).

After defining the model parameters and input data, all components of the battery module are created as 3D volume elements with linear interpolation so that the complex model can be simplified, using a mathematical system of equations. The simulation model then solves the equation system applying the Newmark-beta method, in order to map the swelling behaviour of an entire battery module. Once the equation system is solved, the results of each time step are outputted, which includes spatially resolved displacements, movements, compressions, deflections and much more information about the current state of the module at each time step. Finally, the visualization of the deformed module geometry with the display of displacements is implemented for a meaningful interpretation of the simulation results.

2.2. Complex Dynamic Modelling by Commercial FEA Software

In order to validate the previously described simplified model, a complex model of a battery module is created by the use of the commercial software Abaqus FEA. A direct comparison of the results allows for not only a qualitative but also quantitative assessment of the newly developed simplified model. The steps for the model build-up follow the common state of the art, which were already used in a previous publication regarding the static model for mechanical battery module behaviour [23]. Mainly, the steps consist of modelling the geometries of each component at first, before their material behaviour is parametrized. All components, such as cells, jelly stacks, pads, module frame and end plates are assembled together, afterwards. The meshing as part of the geometrical discretization is conducted in the next step. Here, it must be noted that the number of elements used is significantly higher in contrast to the previously described mathematical model. After contact definitions, boundary conditions and loads are set, the simulation time and time step size have to be defined. In order to have comparability between the models, it is important to set up the same total simulation time in order to see if dynamic effects are mapped accordingly. A dynamic, explicit solver is configured in the complex model, in order to simulate with high accuracy by the application of small time step sizes. Finally, after all the parametrization is finished, the simulation is started. The investigated battery module is also identical to the one from the previous paper [23]. The investigated module consists of 24 pouch cells with multiple foam pads, which are all integrated into a single-part mechanical housing. End plates are welded onto the open ends of the housing. By setting limiting boundary conditions, the module is mounted at its corners. In the following Figure 2, a schematic representation of the 3D model in the Abaqus FEA software environment is shown. Additionally, a section cut is placed, and a cross-sectional view of the module is pictured. Meanwhile, Figure 3 shows the model after the meshing process is finished, and the high number of elements can be reviewed.

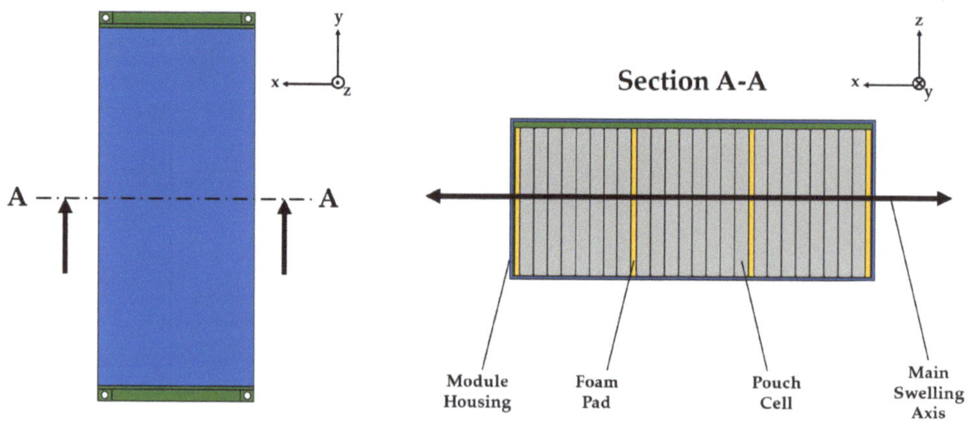

Figure 2. The view of the investigated battery module in the complex 3D simulation environment as the top (**left**) and sectional (**right**) view [23,26]. Redrawn with permission from Bozalp et al., Proceedings of the 11th International Conference on Simulation and Modeling Methodologies, Technologies and Applications SIMULTECH; published by SIMULTECH, 2021. Redrawn with permission from Bozalp et al., Proceedings of the International Conference and Exhibition SIA Paris 2021 Powertrain & Power Electronics; published by SIA, 2021.

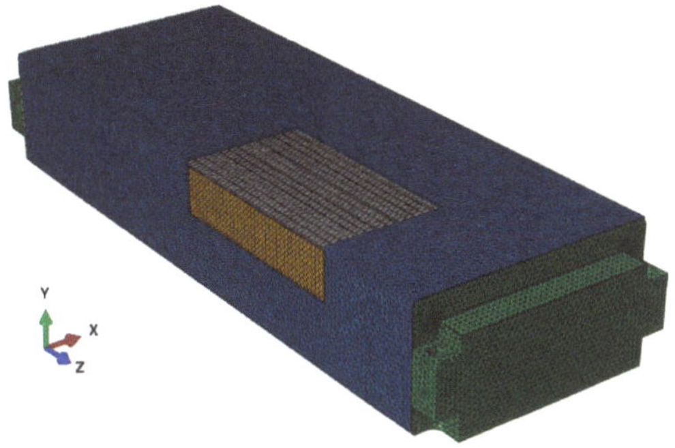

Figure 3. The view of the investigated battery module in the complex 3D simulation environment after the meshing process with the cut-off module frame [23,26]. Redrawn with permission from Bozalp et al., Proceedings of the 11th International Conference on Simulation and Modeling Methodologies, Technologies and Applications SIMULTECH; published by SIMULTECH, 2021. Redrawn with permission from Bozalp et al., Proceedings of the International Conference and Exhibition SIA Paris 2021 Powertrain & Power Electronics; published by SIA, 2021.

3. Model Validation by Study of Sensitivity, Robustness and Numerical Stability

A modified local sensitivity study is conducted not only to validate the simplified model regarding its prediction ability for the dynamic swelling behaviour of the investigated battery module but also to prove its robustness. For this reason, a systematic simulation study will be conducted in this section. The effect of the stiffness of different battery module components on the overall module force–displacement behaviour is studied. Therefore, the focus is placed on the module components along the main axis of cell swelling, which have the highest impact on module deformation (Figure 2). These include the stiffnesses of the jelly stack of the cells, the module housing, which consists of multiple components, such as the top, bottom and end plate, and foam pads. In Figure 4, a schematic overview of the local sensitivity study is pictured.

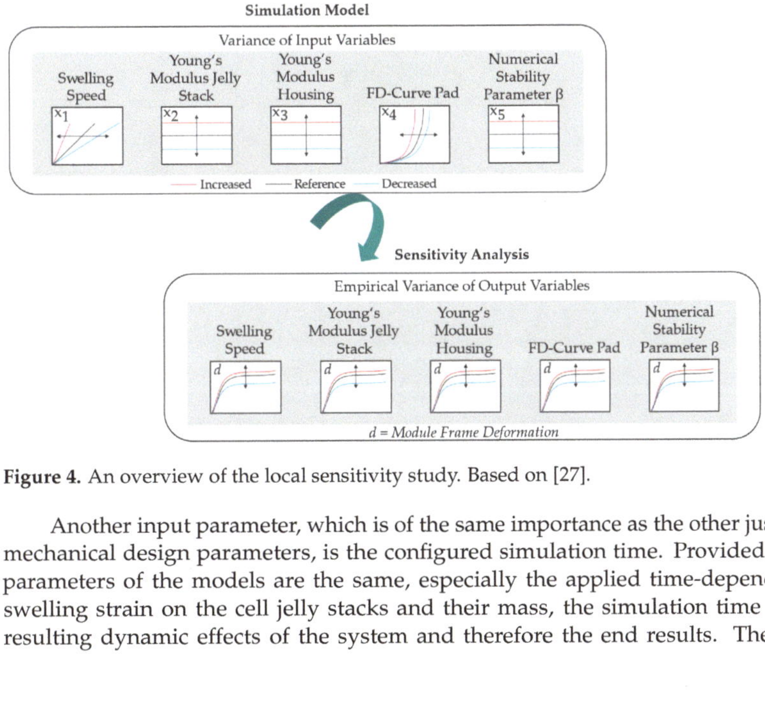

Figure 4. An overview of the local sensitivity study. Based on [27].

Another input parameter, which is of the same importance as the other just described mechanical design parameters, is the configured simulation time. Provided that all the parameters of the models are the same, especially the applied time-dependent, linear swelling strain on the cell jelly stacks and their mass, the simulation time defines the resulting dynamic effects of the system and therefore the end results. The numerical stability of the model is also examined in this chapter. For this reason, the impact of the previously described numerical stability parameter β on the overall module results is investigated by taking on the values of Table 1 one at a time while investigating the other input variables. Since preliminary investigations have revealed that the simplified model does not converge for the border limit operation value of β_4, the focus of this investigation is placed on β_1 to β_3.

In view of the above, the model of the investigated pouch battery module is set up with the simplified mathematical model in Simulink as well as with the commercial software Abaqus FEA. In the first step, this is conducted by using the reference geometrical and material properties for each component and the overall module. It is also ensured that both the simplified mathematical and complex model have the same input parameters regarding the total simulation time as well as the temporal evolvement of the cell jelly stack swelling ratio. After model development is completed, one investigated input parameter is systematically changed by a specifically defined value around the reference centre value for each simulation case. An overview of the applied Design of Experiments (DoE) is shown in Table 2. While 0 stands for the reference value of an input parameter, −1 and +1 represent the expression of a characteristic. The investigated input parameter is changed by a fixed value. While +1 means that a fixed value is added to the reference value, −1 stands for the opposite. During the analysis of the mechanical material input parameters, additionally for each test also, the numerical stability parameter β is changed, in order to investigate its effect on the results.

Table 2. Design of Experiments (DoE) for model validation study. Increase of reference value by fixed value (+1) represented in green and decrease (−1) in orange. Numeric stability parameter is investigated in each test case by use of β_1 (light blue), β_2 (light green) or β_3 (light yellow).

Test	Swelling Speed	YM Jelly Stack	YM Housing	FDC Pad	Parameter β
1	0	0	0	0	β_1
2	−1	0	0	0	β_1
3	+1	0	0	0	β_1
4	0	0	0	0	β_2
5	0	0	0	0	β_3
6	−1	0	0	0	β_2
7	−1	0	0	0	β_3
8	+1	0	0	0	β_2
9	+1	0	0	0	β_3
10	0	−1	0	0	β_1
11	0	−1	0	0	β_2
12	0	−1	0	0	β_3
13	0	+1	0	0	β_1
14	0	+1	0	0	β_2
15	0	+1	0	0	β_3
16	0	0	−1	0	β_1
17	0	0	−1	0	β_2
18	0	0	−1	0	β_3
19	0	0	+1	0	β_1
20	0	0	+1	0	β_2
21	0	0	+1	0	β_3
22	0	0	0	−1	β_1
23	0	0	0	−1	β_2
24	0	0	0	−1	β_3
25	0	0	0	+1	β_1
26	0	0	0	+1	β_2
27	0	0	0	+1	β_3

The aim of the first three simulations is to map the ability of the simplified model for dynamic behaviour over time. Secondly and equally important, the first simulations are needed to determine the optimal total simulation time and time step size for the simplified model. In general, the trade-off between the time step size and the accuracy as well as stability of a numerical simulation is well known. While a small time step size provides high accuracy and stability, it extends the computing time. Therefore, a small time step size and total simulation need to be chosen for the simulations in this study. But due to varying model properties and settings, such as the number of elements or simplified physical equations, as well as a different solver type and method, it is necessary for the numerical stability of the simplified model to apply a time step size different to the complex FEA, while the total simulation is set as identical. For the following Tests 4–27, the same determined total simulation time and time step size are applied for each model itself.

The results of the simulations will be evaluated at defined points on the battery module housing. Specifically, the deformations of the defined characteristics points are examined to assess the quality of the spatially resolved prediction of the mechanical battery module behaviour. In Figure 5, it is schematically shown which points on the battery module housing are investigated.

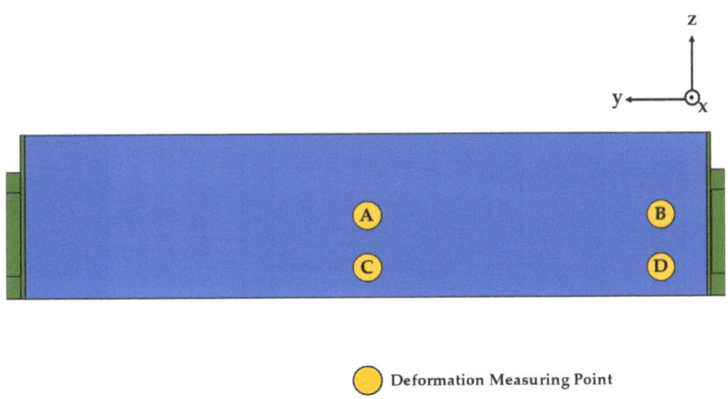

Figure 5. Illustration of deformation measuring points on battery module housing [23,26]. Redrawn with permission from Bozalp et al., Proceedings of the 11th International Conference on Simulation and Modeling Methodologies, Technologies and Applications SIMULTECH; published by SIMULTECH, 2021. Redrawn with permission from Bozalp et al., Proceedings of the International Conference and Exhibition SIA Paris 2021 Powertrain & Power Electronics; published by SIA, 2021.

4. Results of Model Validation

In this chapter, the results of the previously described systematic sensitivity, robustness and numerical stability analysis of the simplified model are presented. The impact of each investigated parameter is presented in a separate subchapter one at a time.

4.1. Variation in Swelling Speed

In this subchapter, the simplified model is examined regarding its ability to map the dynamic evolvement of the mechanical force–displacement behaviour of the investigated battery module spatially resolved over time during cell jelly stack swelling. In the first step, the examined components of the module, represented as the input parameters x_2 to x_4, are simulated with a fixed set of mechanical properties and stiffnesses in Test 1. This set up also represents the reference module design and test. The results of the simplified model in Test 1 are shown in Figure 6.

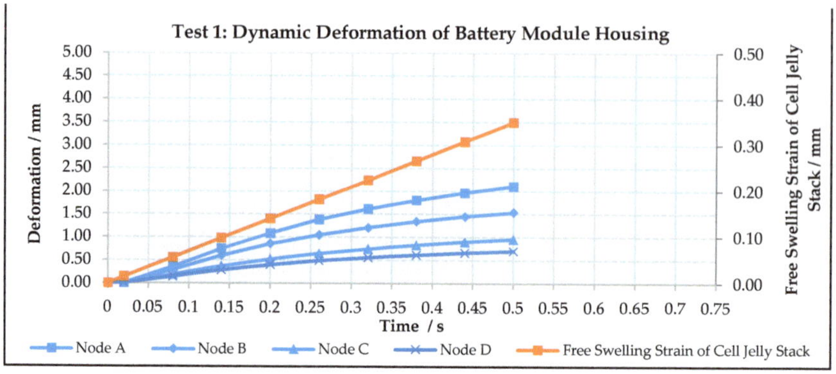

Figure 6. Temporal and spatial evolvement of battery module housing deformation in response to cell jelly stack swelling (Test 1).

The swelling speed as input parameter x_1 is varied in Tests 2 and 3 as the only model parameter. The results of these simulations are shown in Figure 7. For each assessed node, there is a diagram, which shows a comparison of the results of the simplified and the complex model with changing simulation time.

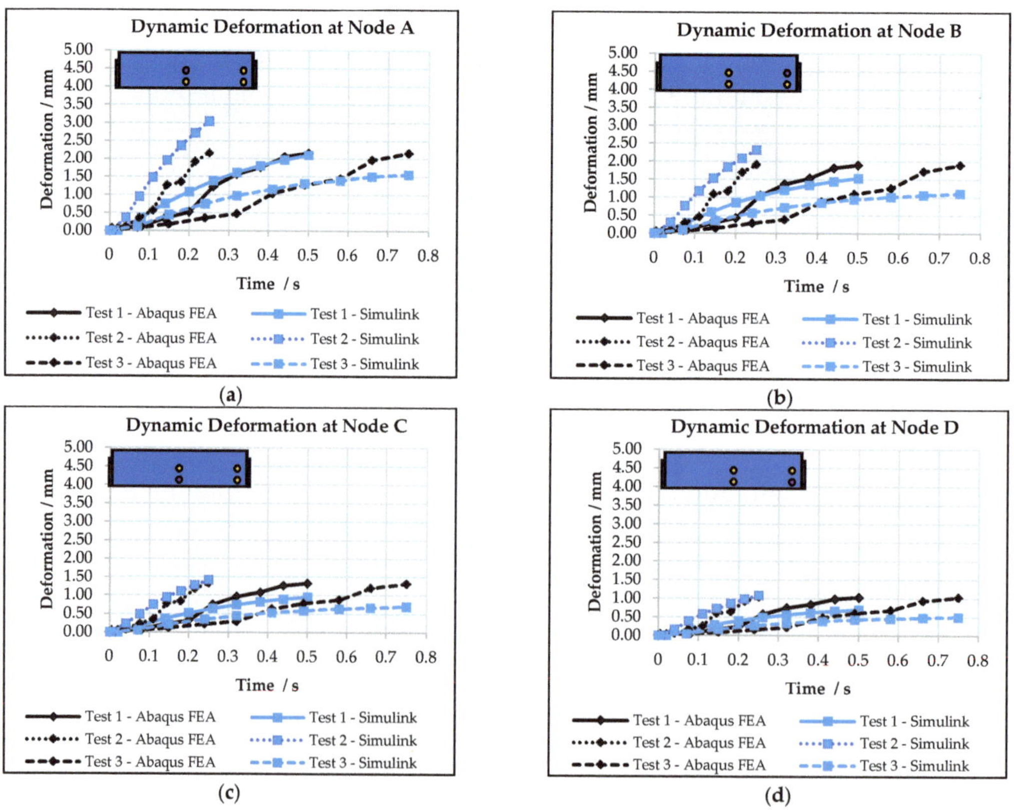

Figure 7. Comparison of time-dependent deformations of battery module housing at different nodes in dependence of swelling speed of cell jelly stacks (Test 1–3). Considered deformation point on the module housing is highlighted in purple: (**a**) node A (**b**) node B (**c**) node C (**d**) node D.

The results show that the simulation time of 0.5 s shows the best agreement between Simulink and Abaqus FEA regarding the total predicted displacements, when the cells reach their total swelling strain. Therefore, for the following sensitive analyses, this simulation time and time step size are used.

4.2. Variation in Young's Modulus of Cell Jelly Stack

In Tests 10 to 15, the impact of the cell jelly stack stiffness on the overall mechanical behaviour of the module during cell jelly stack swelling is presented. The Young's Modulus of the jelly stack of the cells is reduced and then increased by a fixed value around the reference centre value (13 MPa). Figure 8 shows a comparison of the deformation results at the end of the simulation time of 0.5 s for the different nodes on the battery module housing. In each diagram, it is also demonstrated how the numerical stability parameter β effects the prediction of the simplified model.

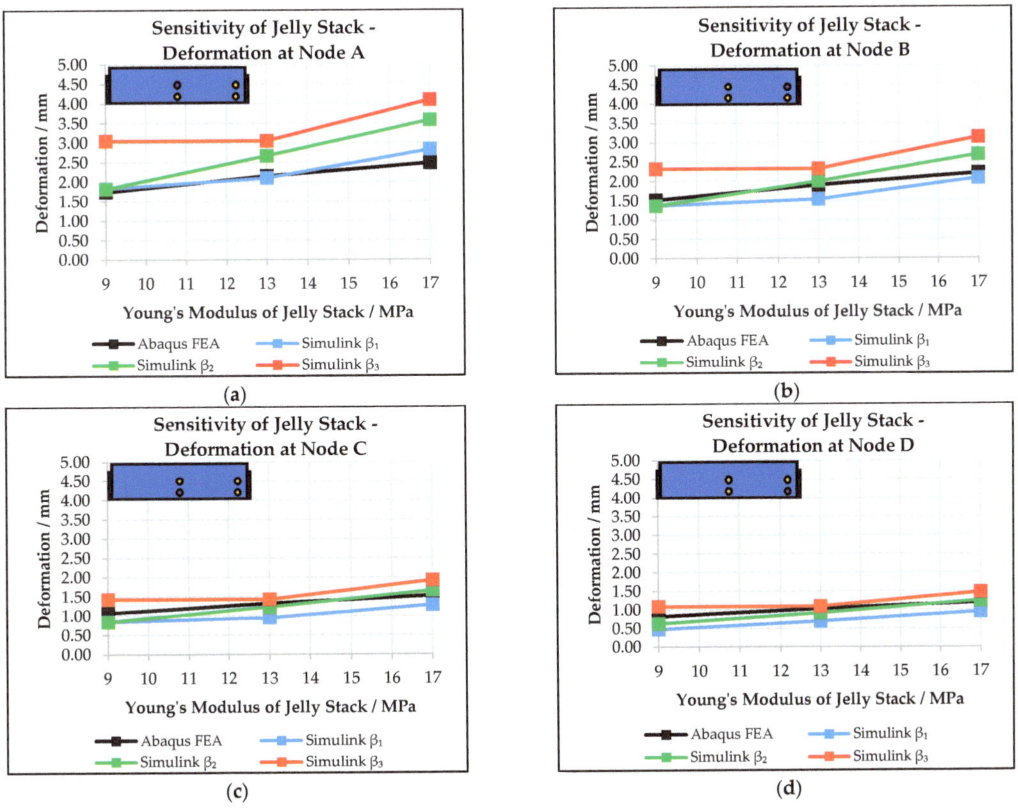

Figure 8. Comparison of time-dependent deformations of battery module housing at different nodes in dependence of Young's Modulus of cell jelly stacks and β-parameter. Considered deformation point on the module housing is highlighted in purple: (**a**) node A (**b**) node B (**c**) node C (**d**) node D.

4.3. Variation in Young's Modulus of Battery Module Housing Components

In the next step, it is shown how the stiffness of the battery module housing, expressed by its Young's Modulus, influences its deformation due to the built-in, swollen cells (Tests 16 to 21). The Young's Modulus of the module housing is reduced and then increased by a fixed value around the reference centre value (69 MPa). Also, in this case, the simulations are conducted with the simplified model first and then validated against the complex model, using finite element software Abaqus FEA. The impact of the numerical stability parameter β in the simplified model is again included. These results are shown in the following Figure 9.

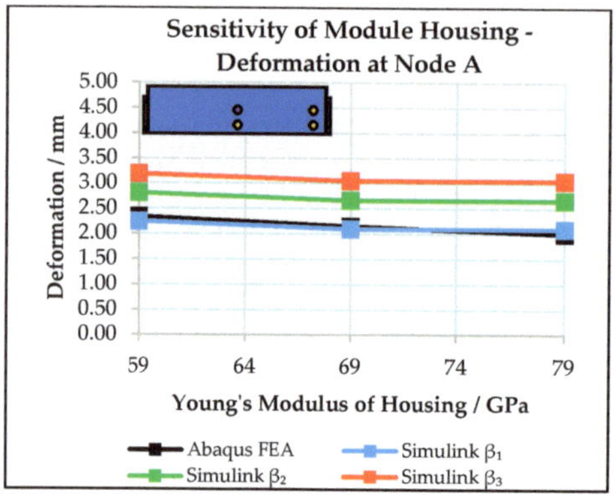

(**a**)

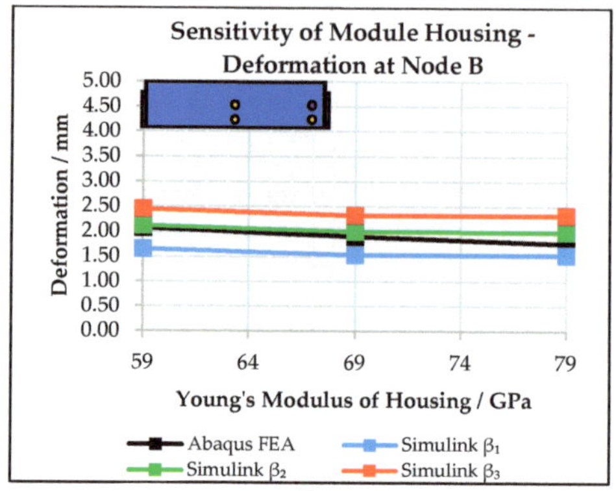

(**b**)

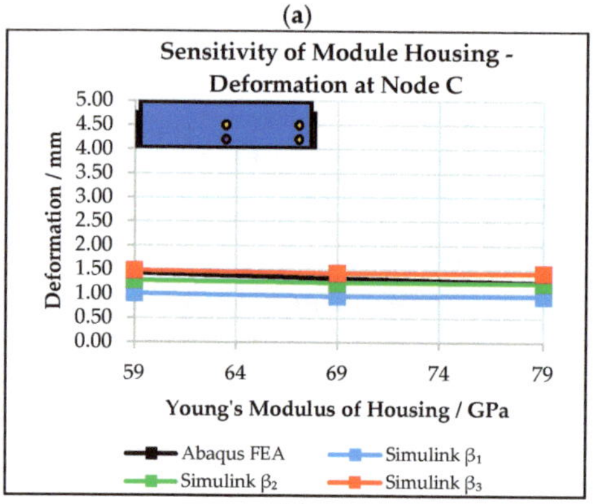

(**c**)

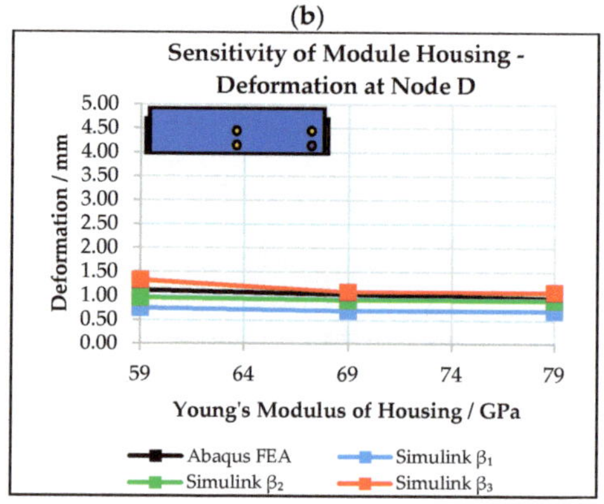

(**d**)

Figure 9. Comparison of time-dependent deformation of battery module housing at different nodes in dependence of Young's Modulus of battery module housing and β-parameter. Considered deformation point on the module housing is highlighted in purple: (**a**) node A (**b**) node B (**c**) node C (**d**) node D.

4.4. Variation in Force–Deflection Curve of Foam Pad

At last, the sensitivity study focuses on the impact of the foam pad, which has the main function of offering space for cell swelling, while its increase in reaction forces is moderate. Therefore, the force–displacement curve of the applied pad is influential for the mechanical behaviour of the module. For the last Tests 22 to 27, the force–displacement curve of the pad of the reference module from Test 1 is adopted. While in Tests 22–24 a curve for a softer pad is used (Pad Stiffness 1), the pad of Tests 25–27 has a steeper increase in reaction force (Pad Stiffness 3). These results are graphically summarized in Figure 10. When changing the β-parameter for the soft pad from its reference value, the simulation becomes instable, and the calculation is aborted in the simplified model.

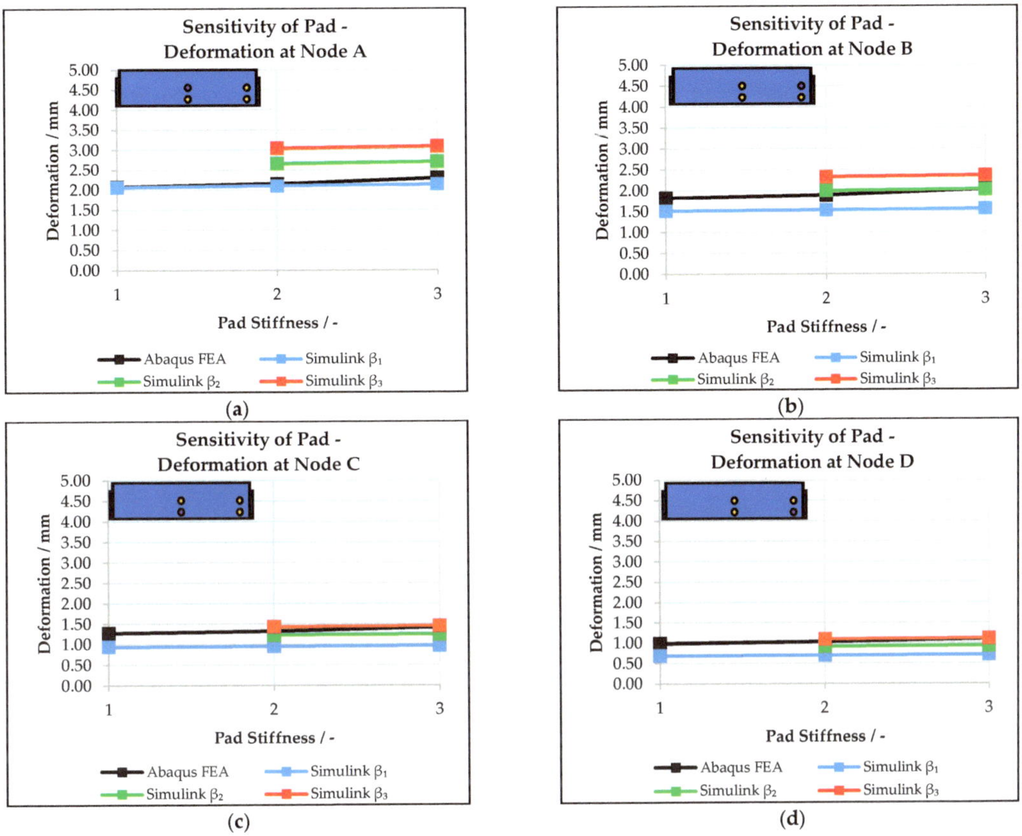

Figure 10. Comparison of time-dependent deformation of battery module housing at different nodes in dependence of pad force–deflection curve and β-parameter. Considered deformation point on the module housing is highlighted in purple: (**a**) node A (**b**) node B (**c**) node C (**d**) node D.

5. Discussion

In this chapter, the previously shown results of the sensitive analyses are interpreted and discussed. For this reason, the outcomes of each investigated input parameter are analyzed one at a time and in the same order as previously presented, starting from Test 1 as the reference module. Subsequently, putting the partial analyses of all tests together, an overall assessment of the novel model is conducted.

In the first step of Test 1, the reference setting of the module design was configured, and its swelling behaviour was simulated. The results, presented in Figure 6, clearly show that the novel model not only has the ability of resolving the overall module spatially but also temporally. While in the first time steps until 0.02 s each cell jelly stack swells by 0.014 mm, the module housing does not show any deformation, which is mainly due to the foam pads, which absorb the forces during that time and doing so, largely restrain them from being passed on to the module frame. Afterwards, the forces act on the inside of the module housing, and a deformation is effective on the module outside. The deformations are distributed inhomogeneously over the housing area in the main swelling direction, while the biggest deformation can be observed in the centre region (Node A). This proportionally large area has a lower structural stiffness compared to the edge and corner regions (Node B and D) and therefore gives in more easily, even though the material stiffness is parametrized the same for the whole housing components. When comparing the results of the simplified model from Simulink with different simulation times and therefore different swelling speeds of the cell jelly stacks, the slope of the module housing deformations also develops accordingly to the speed (Figure 7). A comparison of the results of Test 1 with those of the complex model reveals that they match well after firstly overpredicting them. On the other hand, when comparing the results of Tests 2 and 3, where the simulation time is reduced and elongated, the simplified model firstly also overpredicts the deformations (Figure 7). While in Test 2, the overpredicted deformations keep increasing, in Test 3, at first, they level but end up underestimating the deformations towards the end of the simulation time. Consequently, the complex model predicts, independent of the swelling speed, the same total module housing deformations, while in the simplified model, the deformations increase with higher swelling speeds. One possible explanation for this behaviour in the simplified model is the non-consideration of damping effects in its main underlying Equation (2). For example, vibrations or structural responses in the model are not damped and therefore can build up over the simulation time, leading to higher total deformations. On the other hand, accuracy differences between the model results in general are due to the different time step sizes used in each model. Specifically, the time step size in the simplified model is greater than the one used in Abaqus FEA by a factor of nearly 800, allowing for much shorter total computation times (Figure 11). Since there is a general correspondence between time step size as well as number and model prediction accuracy, this in turn implies that deviations in the results are to be expected. Furthermore, the novel model has more simplifications to it, such as a significantly lower number of volume elements, which would also explain deviations and lower accuracy. In total, the number of elements used in the complex model is higher by a factor of nearly 2300 (Figure 11). Not only does the number of elements differ but also their shape. While the complex model uses different kind of element shapes, depending on the component and its geometric complexity, in order to simulate with best possible accuracy, the simplified model only uses one particular shape for the reasons of simplification and speed in the model build-up phase. Due to the listed model parameters, which are known for having a strong influence on the results, further analyses are recommended to fully clarify their effects on the results of the simplified model.

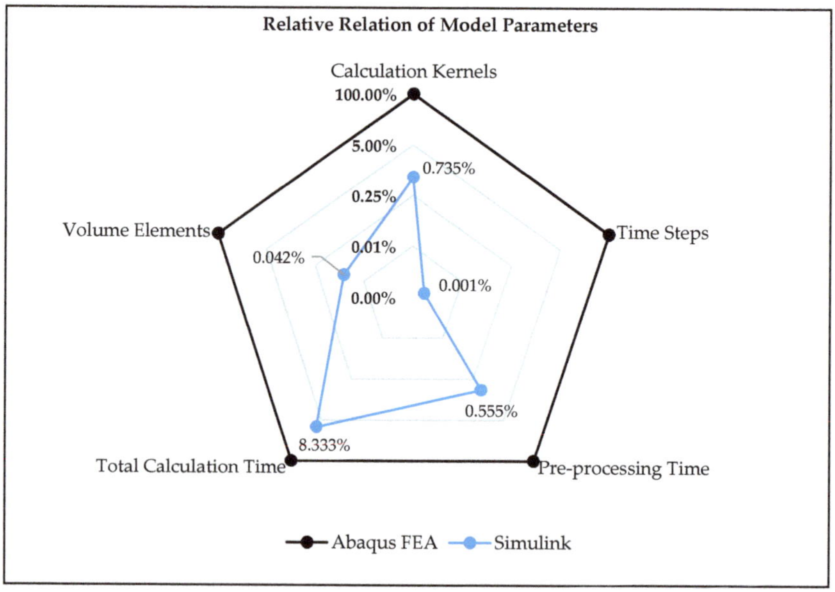

Figure 11. Relative relations between simplified and complex model parameters on average.

After the simulation time of 0.5 s was identified as the one with the best matching of the results, the following sensitive analyses were conducted, applying the determined simulation time. In general, the sensitive analyses show that the novel model is robust and able to predict the effect of design changes on the mechanical stiffnesses of the investigated components, when the cells reach their maximum swelling state at the end of the simulation time. As seen in Figures 8–10, the model maps the slope and direction of the deformation gradient correctly, when changing the stiffnessor force–displacement curve of an investigated design input parameter. Considering the reference β-parameter, it is noticeable that the highest match of the model results is provided for Node A, while the other nodes follow the general trend well, when changing the stiffness of a parameter but keeping an off-set. The previously described model parameter differences, such as element number, size and shape, have a major influence on the displacement of nodes, possibly explaining this behaviour. Another factor which needs to be considered when it comes to the accuracy of the novel model is the numerical stability factor β. The acceleration during the time intervals, expressed by the parameter, impacts the calculated displacement of the model nodes and consequently the accuracy of the predictions, which is most likely also due to the non-consideration of damping effects. In Figures 8–10, it is shown that a change in the parameter influences the simplified model results in comparison to the ones of the complex model. Especially, nodes B, C and D can reach a much better accuracy by adopting the parameter β. In the considered range for the input parameters, all the simulated test cases finished without any instability in the numeric procedure except for the investigated ones of the soft pad. Here, a variation in the β-parameter from its reference value leads to instabilities in the numerical solving procedure, ending with a numerical abortion. This is another good example for the influence and effectiveness of the parameter to actively control the numerical stability of a simulation.

6. Conclusions

Within the framework of this study, an existing simplified mathematical model, which statically describes the overall three-dimensional, mechanical behaviour of battery modules under cell swelling conditions, was successfully further developed by adding the time dimension to it. By the expansion of the fundamental physical equations that are underlying the model, the module behaviour is not only spatially but also temporally resolved, enabling it to map time-dependent, dynamic effects. In the field of the simplified mathematical modelling of battery modules based on mass–spring systems, the novel model shows significant progress that has been made by creating a model which distinguishes itself from previous publications by mapping the three-dimensional force–displacement behaviour and at the same time enables dynamic modelling for time-dependent behaviour, which is also a clear step forward in comparison to the previous publication applying static modelling [23]. Furthermore, the included algorithm, which automatically builds up a mathematical equation system representing the investigated module by simply provided module parameters, such as the number of cells and pads and their arrangement, was also further developed. In doing so, the maximum flexibility for modelling battery modules of different designs is offered. New features added to it include the spatial fixation of individual or a group of nodes, in order to consider design properties, such as adhesive joints between components or the mounting conditions of the module in a battery system as well. Besides the geometrical configuration of all the module components, such as cells, jelly stacks, pads and module housing parts, their mechanical material behaviour is also parametrized by adding information, such as Young's Modulus or even force–deflection curves.

The ability of the model to predict the overall time-dependent mechanical behaviour of battery modules was investigated for an exemplary battery module design. For validation purposes, a module, consisting of 24 pouch cells and multiple foam pads, was modelled with the novel mathematical model as well as in complex finite element software Abaqus FEA. In the process of conducting the first small study regarding the optimal simulation time and time step size, the simplified model's ability to spatially and temporally resolve the force–displacement behaviour with the biggest deformations to be observed in the centre region (Node A) was revealed. In doing so, the novel proved its ability to map structural inhomogeneous stiffness distributions, e.g., of the module frame. A subsequent local sensitivity study demonstrated that the novel model is also able to predict the global force–displacement behaviour well, mapping the slope and direction of the deformation gradient accurately, when changing the stiffness or force–displacement curve of the investigated design input parameter. In this process, this study affirms that numerical stability parameter β needs to be chosen carefully in order to predict the spatially resolved deformations of the module. Due to the fact that the novel model simplifies and reduces not only the geometrical but also temporal discretization, it reveals its potential regarding the total computation time and capacity needed. Even though in comparison to the complex model, the novel model does not even need 1% of the number of calculation kernels and 10% of the total calculation time on average, the examined deformation results show a good agreement between the two models, which can make it a useful and effective tool in the early development phase for fast optimization studies.

To the best knowledge of the authors, it is the first time that a simulation model allows for the mapping of time-dependent mechanical battery module behaviour three-dimensionally under cell swelling conditions, which is based on a simplified mathematical approach in a 0D simulation environment. Even though the novel model in its current state shows good results and has brought significant improvements to the mathematical simulation of mechanical battery module behaviour, there is still optimization potential left to increase the accuracy and structural model behaviour. Especially the impact and interdependencies of model parameters as well as the demonstrated numerical limits of the model need further investigations.

7. Outlook

In view of the above, multiple model parameters need further examination regarding their impact and interdependencies. For example, the effect of the non-consideration of damping processes in the main underlying equation needs to be investigated in following studies, possibly explaining if undamped vibrations and structural responses in the model build up over simulation time lead to higher total deformations. Another aspect which needs further examination is the potential increase in accuracy by raising the number of time steps and volume elements used in the simplified model.

Moreover, the numerical boundaries and instabilities of the model need to be explored more extensively. In the current state, there are already exemplary model limits demonstrated, such as during the sensitivity analysis of the soft pad. In this context, the influence of the numerical stability parameter β on the numerical procedure as well as on the overall results is another research aspect for further studies of the model. The effect of the time step size on the prediction accuracy of the novel model as well as on its numerical stability also needs further examination and is highly recommended for future work.

The results, opinions and conclusions expressed in this publication are not necessarily those of Volkswagen Aktiengesellschaft.

Author Contributions: Conceptualization, T.B. and H.O.; methodology, T.B. and S.S.K.; validation, T.B.; formal analysis, T.B. and S.S.K.; investigation, T.B.; resources, T.B. and H.O.; writing—original draft preparation, T.B. and S.S.K.; writing—review and editing, T.B. and S.S.K.; visualization, T.B.; supervision, T.V. and H.O. All authors have read and agreed to the published version of the manuscript.

Funding: The support of the German Research Foundation (DFG) and the Open Access Publication Fund of Technische Universität Braunschweig is acknowledged by the authors.

Data Availability Statement: For confidentiality reasons, there will be no further data or information provided. This also applies to the developed models and its codes.

Conflicts of Interest: The authors declare no conflicts of interest.

References

1. Jiang, W.; Li, H.; Wang, S.; Wang, S.; Wang, W. Dynamic Volumography of Cylindrical Li-Ion Battery Cells by Watching Its Breath During Cycling. *CCS Chem.* **2023**, *5*, 1308–1317. [CrossRef]
2. Lee, J.H.; Lee, H.M.; Ahn, S. Battery dimensional changes occurring during charge/discharge cycles—Thin rectangular lithium ion and polymer cells. *J. Power Sources* **2003**, *119–121*, 833–837. [CrossRef]
3. Yuan, C.; Hahn, Y.; Lu, W.; Oancea, V.; Xu, J. Quantification of electrochemical-mechanical coupling in lithium-ion batteries. *Cell Rep. Phys. Sci.* **2022**, *3*, 101158. [CrossRef]
4. Oh, K.; Siegel, J.B.; Secondo, L.; Kim, S.U.; Samad, N.A.; Qin, J.; Anderson, D.; Garikipati, K.; Knobloch, A.; Epureanu, B.I.; et al. Rate dependence of swelling in lithium-ion cells. *J. Power Sources* **2014**, *267*, 197–202. [CrossRef]
5. Mohtat, P.; Lee, S.; Siegel, J.B.; Stefanopoulou, A.G. Reversible and Irreversible Expansion of Lithium-Ion Batteries Under a Wide Range of Stress Factors. *J. Electrochem. Soc.* **2021**, *168*, 100520. [CrossRef]
6. Willenberg, L.K.; Dechent, P.; Fuchs, G.; Sauer, D.U.; Figgemeier, E. High-Precision Monitoring of Volume Change of Commercial Lithium-Ion Batteries by Using Strain Gauges. *Sustainability* **2020**, *12*, 557. [CrossRef]
7. Louli, A.; Ellis, L.; Dahn, J. Operando Pressure Measurements Reveal Solid Electrolyte Interphase Growth to Rank Li-Ion Cell Performance. *Joule* **2019**, *3*, 745–761. [CrossRef]
8. Cannarella, J.; Arnold, C.B. Stress evolution and capacity fade in constrained lithium-ion pouch cells. *J. Power Sources* **2014**, *245*, 745–751. [CrossRef]
9. Simunovic, Z.; Dressler, R.; Alter, E.D.; Trussler, S.; Harlow, J.; Johnson, M.; McMonigle, C.; Fisher, M.; Metzger, M. Laser Scanning Method for High-Resolution Thickness Mapping of Lithium-Ion Pouch Cells. *J. Electrochem. Soc.* **2023**, *170*, 120532. [CrossRef]
10. Dörnhöfer, A. Konzept zur Betriebsfestigkeitsanalyse von Hochvoltspeichern. In *Betriebsfestigkeitsanalyse Elektrifizierter Fahrzeuge*, 1st ed.; Springer: Berlin/Heidelberg, Germany, 2019; pp. 39–68.
11. Govindaraj, A.L. Phenomenological Swelling Model of Battery Module. Master's Thesis, KTH Royal Institute of Technology, Stockholm, Sweden, 2022.
12. Gupta, P.; Gudmundson, P. A multi-scale model for simulation of electrochemically induced stresses on scales of active particles, electrode layers, and battery level in lithium-ion batteries. *J. Power Sources* **2021**, *511*, 230465. [CrossRef]
13. Blazek, P.; Westenberger, P.; Erker, S.; Brinek, A.; Zikmund, T.; Rettenwander, D.; Wagner, N.P.; Keckes, J.; Kaiser, J.; Kazda, T.; et al. Axially and radially inhomogeneous swelling in commercial 18650 Li-ion battery cells. *J. Energy Storage* **2022**, *52*, 104563. [CrossRef]
14. Chen, Y.; Yang, L.; Guo, F.; Liu, D.; Wang, H.; Lu, J.; Zheng, J.; Yu, X.; Li, H. Mechanical-electrochemical modeling of silicon-graphite composite anode for lithium-ion batteries. *J. Power Sources* **2022**, *527*, 231178. [CrossRef]
15. Knutsson, A.; Trombati, J. Finite Element Modelling of Cell Swelling for Traction Batteries. Master's Thesis, Chalmers University of Technology, Gothenburg, Sweden, 2021.
16. Sun, Y.; Lu, H.; Jin, Y. Experimental and Numerical Study on Mechanical Deformation Characteristics of Lithium Iron Phosphate Pouch Battery Modules under Overcharge Conditions. *Energy Fuels* **2021**, *35*, 15172–15184. [CrossRef]
17. Mei, W.; Duan, Q.; Lu, W.; Sun, J.; Wang, Q. An investigation on expansion behavior of lithium ion battery based on the thermal-mechanical coupling model. *J. Clean. Prod.* **2020**, *274*, 122643. [CrossRef]
18. Hong, Y.; Zheng, J.; Deng, D.; Wu, Q.; Shen, W. Heterogeneous aging of large-scale flexible lithium-ion batteries based on micro-heterostructures and simulation. *Int. J. Energy Res.* **2020**, *44*, 12112–12125. [CrossRef]
19. Zhong, X.; Yang, L.; Li, N.; Chu, Z.; Chen, J.; Zhu, S.; Song, W.; Ai, S.; Chen, H. In-situ characterizations and mechanism analysis of mechanical inhomogeneity in a prismatic battery module. *J. Power Sources* **2022**, *548*, 232053. [CrossRef]
20. Li, Y.; Wei, C.; Sheng, Y.; Jiao, F.; Wu, K. Swelling force in Lithium-Ion power batteries. *Ind. Eng. Chem. Res.* **2020**, *59*, 12313–12318. [CrossRef]
21. Hahn, S.; Theil, S.; Kroggel, J.; Birke, K.P. Pressure Prediction Modeling and Validation for Lithium-Ion Pouch Cells in Buffered Module Assemblies. *J. Energy Storage* **2021**, *40*, 102517. [CrossRef]
22. Cai, T.; Mohtat, P.; Stefanopoulou, A.G.; Siegel, J.B. Li-ion Battery Fault Detection in Large Packs Using Force and Gas Sensors. *IFAC-Pap.* **2020**, *53*, 12491–12496. [CrossRef]
23. Bozalp, T.; Siddiqui, M.A.R.; Opfer, H.; Vietor, T. Novel Method for the Three-Dimensional Simulation of Mechanical Ageing of Battery Modules. In Proceedings of the 11th International Conference on Simulation and Modeling Methodologies, Technologies and Applications SIMULTECH, Paris, France, 7–9 July 2021.
24. Newmark, N.M. A Method of Computation for Structural Dynamics. *ASCE J. Eng. Mech. Div.* **1959**, *85*, 67–94. [CrossRef]
25. Kim, N. Finite Element Analysis for Contact Problems. In *Introduction to Nonlinear Finite Element Analysis*, 1st ed.; Springer: New York, NY, USA, 2014; pp. 367–426.

26. Bozalp, T.; Opfer, H.; Wurzenberger, J.C. Innovative Method for Coupled, Multi-Physical Simulation of Battery Modules. In Proceedings of the International Conference and Exhibition SIA Paris 2021 Powertrain & Power Electronics, Paris, France, 9–10 June 2021.
27. Siebertz, K.; Van Bebber, D.; Hochkirchen, T. Sensitivitätsanalyse. In *Statistische Versuchsplanung*, 2nd ed.; Springer: Berlin/Heidelberg, Germany, 2017; pp. 415–430.

energies

MDPI

Article

Analysis of Stray Current Leakage in Subway Traction Power Supply System Based on Field-Circuit Coupling

Shan Lin [1], Zhixi Tang [2,*], Xia Chen [1], Xuehua Liu [2] and Yunsheng Liu [2]

[1] Guangzhou Metro Design & Research Institute Co., Ltd., Guangzhou 510010, China; linshan@gmdi.cn (S.L.); chenxia@gmdi.cn (X.C.)
[2] School of Electrical and Automation Engineering, East China Jiaotong University, Nanchang 330013, China; 9920230054@jgsu.edu.cn (X.L.); liuyunsheng@ecjtu.edu.cn (Y.L.)
* Correspondence: tangzhixi@ecjtu.edu.cn

Abstract: In a rail transit system, there is a constant leakage of current from the subway rails to the earth, and these stray currents have complex propagation paths and a wide range of influence. Since no stray current collection devices are installed at subway depots, some of the stray current leaking from the mainline will converge at the depot, seriously corroding the structural reinforcement and buried metal of the station, thereby jeopardizing the normal operation of subway trains and passenger safety. In this paper, a field-circuit coupling method is proposed to analyze the current leakage and distribution law of the subway mainline and depot. It is found that the failure of the gauge block at the mainline will trigger the maximum leakage of rail current. Additionally, it is observed that the stray current distribution at the depot is mainly influenced by the operating status of the one-way conduction device (OWCD) and the change of rail potential. These results validate the applicability and effectiveness of the field-circuit coupling method proposed in this paper and provide new technical support for the study of stray current leakage distribution in subways.

Keywords: subway; stray current; leakage; distribution; field-circuit

Citation: Lin, S.; Tang, Z.; Chen, X.; Liu, X.; Liu, Y. Analysis of Stray Current Leakage in Subway Traction Power Supply System Based on Field-Circuit Coupling. *Energies* **2024**, *17*, 3121. https://doi.org/10.3390/en17133121

Academic Editors: Artur Kierzkowski, Tomasz Kisiel, Agnieszka A. Tubis, Franciszek Restel, Mateusz Zając, Tomasz Nowakowski and Anna Jodejko-Pietruczuk

Received: 20 May 2024
Revised: 11 June 2024
Accepted: 18 June 2024
Published: 25 June 2024

1. Introduction

In rail transit systems, rails are usually used as the return path for subway traction current [1]. However, since the rails cannot be completely insulated from the ground, there exists a partial current leakage into the earth, which is called stray current. Subway stray current has the characteristics of a complex propagation path and a wide influence range [2]. If no stray current collection device is installed at a subway depot, some of the stray current leaking from the mainline will converge at the depot, causing serious corrosion to the structural reinforcement of the station and the buried metal [3]. This situation jeopardizes the normal operation of the subway train and the safety of the passengers. In-depth exploration and analysis of the subway stray current leakage situation can help implement corresponding measures to ensure the safe operation of the subway system, prolong the service life of the facilities, and reduce maintenance costs, thus making a contribution to the safety and sustainable development of urban rail transit. Therefore, it is of great significance to accurately analyze the current leakage situation and distribution law in the mainline and depot.

Ref. [4] evaluates the stray current leakage on a long subway line by considering the variability of the subway structure, splitting the return current system into several differential units, modeling the resistive network, and discussing the uniform boundary conditions for solving the equations. Ref. [5] proposes a coupling simulation model during the dynamic operation of multiple trains, analyzes the coupling effect of the mainline and the depot, and quantitatively calculates and analyzes the stray current and rail potentials by using the constructed resistive network model, and evaluates the coupling effect. Ref. [6] uses finite element software to build a subway tunnel model based on the open excavation method, explores the mechanism of track transition resistance on stray current, and accurately analyzes the distribution law of subway stray current under many different track transition resistances. Ref. [7] couples a single subway line with the power system structure, builds a finite element model, and studies the stray current distribution characteristics under different soil structure conditions and traction conditions. Ref. [8] proposes a simulation model of field-circuit coupling, which accurately analyzes the diffusion distribution of the stray current by regionally equating the spatial and conductor structures in the DC subway return system and modeling the stray current diffusion field using the direct boundary element method.

The resistive network method used by the above-mentioned scholars is based on Kirchhoff's circuit laws and Ohm's law, and differential equations are established to solve for the stray current distribution. The equivalent circuit model is often built according to the structure of the ground network; there are the "rail–earth" two-layer equivalent model [4], the "rail–SCCM–earth" three-layer equivalent model [9], and the "rail–SCCM–buried metal-earth" four-layer equivalent model [10], as well as the equivalent model of the subway depot [11], and so on. In related studies, loop resistance (transition resistance, rail longitudinal resistance, buried metal longitudinal resistance, etc.) [12], traction load [13], train position [14], and substation distance [8] are analyzed as variable covariates for correlation analysis, and the changes of the actual system covariates are simulated by raising the transition resistance and shortening the substation distance.

However, the resistive network model can only be calculated based on the equivalent of the set resistance value, which makes it difficult to accurately reflect the diffusion distribution of stray current around the line. The advantage of the finite element method, on the other hand, lies in the modeling of complex three-dimensional diffusion scenarios [7,15–17] of stray currents or the calculation of dynamic processes in complex structures to accurately simulate real physical systems. When using finite element for the dynamic simulation of stray current, each simulation moment needs to manually modify the size and location of the equivalent current source of the train, and then carry out the repetitive dissections of the whole model, which leads to a greatly extended time for the dynamic simulation of the stray current distribution.

In summary, to accurately and efficiently analyze the stray current distribution, a field-path coupling method is proposed in this paper. The field model of stray current leakage distribution is established by the finite element method to accurately obtain the stray current circulation path and distribution, and then the convenience of the equivalent calculation of the resistance network is utilized, and the equivalent resistance network of the traction power supply system is coupled to obtain the dynamic simulation of the stray current and simulate the leakage distribution of the stray current in the surrounding soil.

The main contributions of this paper can be summarized as follows:

(1) In this paper, a field-circuit coupling method is proposed.
(2) In this paper, two scenarios are modeled, analyzed, and discussed for the mainline and depot of the subway line, respectively.
(3) This paper analyzes the current leakage and distribution law at the mainline and depot of the subway.
(4) The methodology presented in this paper will facilitate the design of stray current mitigation measures for high-risk areas in both the mainline and the subway depot.

This paper describes the principle of finite element field and resistor network in Section 2, the coupling modeling of two scenarios of mainline and depot in Section 3, and the current leakage distributions and laws of the subway mainline and depot in Section 4, which validates the model proposed in this paper. The results are compared and discussed in Section 5, and conclusions are drawn in Section 6.

2. Modeling Fundamental

2.1. The Finite Element Field

Considering that stray current distribution is a constant electric field problem, the soil is simplified and can be regarded as uniformly distributing the stray current in the nearby soil. Therefore, the physical field current module is used in the finite element modeling; i.e., Maxwell's equations are solved using the "AC/DC module" [18]. The governing equations of the model are

$$\nabla \cdot \boldsymbol{J} = 0 \tag{1}$$

$$\nabla \times \boldsymbol{E} = 0 \tag{2}$$

The constitutive relationship is

$$\boldsymbol{J} = \sigma \boldsymbol{E} \tag{3}$$

The boundary condition is

$$\boldsymbol{E}_{1t} = \boldsymbol{E}_{2t} \tag{4}$$

$$\boldsymbol{J}_{1n} = \boldsymbol{J}_{2n} \tag{5}$$

where $\boldsymbol{E}$ is the electric field strength, $\boldsymbol{J}$ is the current density, γ is the soil conductivity, $\boldsymbol{E}_t$ is the tangential component of the electric field strength, and $\boldsymbol{J}_n$ is the normal component of the current density.

2.2. The Resistor Network

An equivalent resistive network is used to represent the subway traction return system. The scope of equivalence mainly includes the overhead contact line, rails, stray current collection mat (SCCM), earth, track-bed, rail-to-SCCM transition resistance, and SCCM-to-earth transition resistance [19]. All the symbols are used in Figures 3 and 6.

(1) Overhead contact line equivalent

The overhead contact line of length L is equivalent to n resistors in series, so for each resistor, the equivalent length is also $l = L/n$. The resistance value of each resistor is $R_C \cdot l$, where R_C is the unit resistance of the overhead contact line. The current source is used as an equivalent of the train, extracting current from the overhead contact line nodes and injecting it into the rail nodes.

(2) Rail equivalent

The rails are the main trajectory of the DC current returning to the substation. A rail of length L is equivalent to n resistors in series, so for each resistor, the equivalent length is $l = L/n$. The resistance of each resistor is therefore $R_G \cdot l$, where R_G is the unit resistance of the rail.

(3) SCCM equivalent

The SCCM is used as an absorber of leakage current from the rail. An SCCM of length L is equivalent to n resistors in series, so for each resistor, the equivalent length is $l = L/n$. The resistance value of each resistor is therefore $R_P \cdot l$, where R_P is the unit resistance of the grid.

(4) Earth equivalent

The earth can provide many paths for stray currents in DC subway systems, especially in subway depots. Horizontal resistive networking is used as an equivalent of the earth. A rectangular X-Y coordinate system is established in this paper by choosing an area of length X and width Y covering the inner zone of the depot, the outer zone of the depot, and part of the mainline [20]. The earth in this area is divided into *q-row* and *p-column* units along the x and y directions, respectively. Therefore, the equivalent length of a single unit is $a = X/q$ and the equivalent width is $b = Y/p$, as shown in Figure 1, where h is the depth of the earth unit. According to the conductor resistance equation, the equivalent resistance of the earth unit is R_x in the *x-direction* and R_y in the *y-direction*.

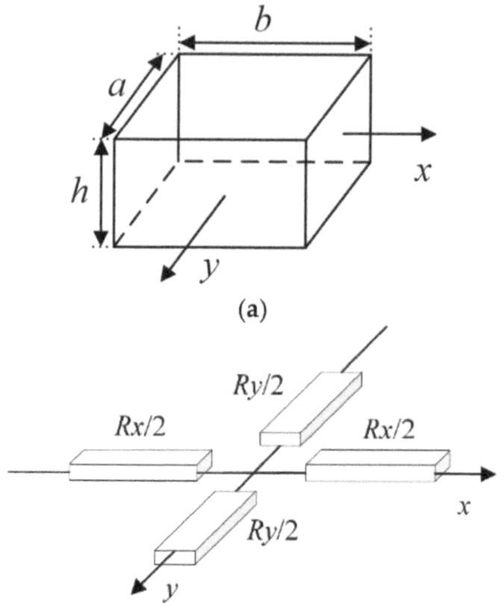

Figure 1. Equivalent of the earth unit. (**a**) The earth unit; (**b**) equivalent in x and y directions.

(5) Track-bed equivalent

To calculate the transition resistance, the equivalent resistance of the track-bed needs to be calculated first, as shown in Equation (6):

$$R_d = \frac{\rho_1 L_y}{S} \quad (6)$$

R_d is the resistance of the track-bed; ρ_1 is the resistivity of the track-bed concrete; L_y is the distance from the surface of the track-bed to the SCCM; and S is the area of the track-bed in the transverse direction.

(6) Rail-to-SCCM transition resistance equivalent

The rail-to-SCCM transition unit resistance R_g is calculated as shown in Equation (7):

$$R_g = (\frac{d_1 \cdot R_f}{L + d_1} + R_d) \cdot L \quad (7)$$

R_g is the transition resistance of the interval; R_f is the equivalent resistance of a single set of fasteners; and R_d is the resistance of the track-bed in that interval.

(7) SCCM-to-earth transition resistance equivalent

The SCCM-to-earth transition unit resistance $R_{g'}$ is calculated as shown in Equation (8):

$$R_{g'} = \frac{\rho_2 l}{s} = \frac{\rho_2 l}{L d_2} \quad (8)$$

where ρ_2 is the transition resistivity of the SCCM-to-earth; l is the depth of the SCCM from the earth; and d_2 is the width of the SCCM.

(8) OWCD equivalent

The OWCD is equivalent to a diode that cuts off when a reverse voltage is applied.

2.3. Field-Circuit Coupling

Dynamic processes in complex structures can be calculated using finite elements [21], which are capable of accurately modeling real physical systems. The resistor network is suitable for reducing spatial problems to planar problems [22], and it can transform the 3D current electric field problem into a 2D distributed parameter problem of a circuit. However, the resistance network does not consider the effect of grounding resistance and it cannot obtain the distribution of current and voltage on any wire.

Therefore, this paper combines these two methods and proposes a field-circuit coupling method, which can utilize the characteristics of the resistive network model to calculate the stray current distribution simply and conveniently, and it also can take advantage of the software simulation model to accurately simulate the distribution of electric fields and currents within the earth environment.

3. Modeling of Field-Circuit Coupling

According to the actual operation of a Guangzhou subway line from station A to station B, the traction current and distance data are collected when the train is traveling, as shown in Figure 2. The total length of the section from station A to station B is 1050 m, and the time of train travel is 82 s. When the train has been traveling for 20 s, it is 84 m away from station A, it is in the acceleration stage, and the positive traction current is 3390 A. When the train has been traveling for 43 s, it is 538 m away from station A, in a state of uniform speed, and the traction current is 67 A; after 43 s then the train enters the deceleration stage, and the traction current becomes negative. When the train has been traveling for 55 s, it is 836 m away from station A, it is in the deceleration stage, and the traction current is reversed to a maximum of −3177 A.

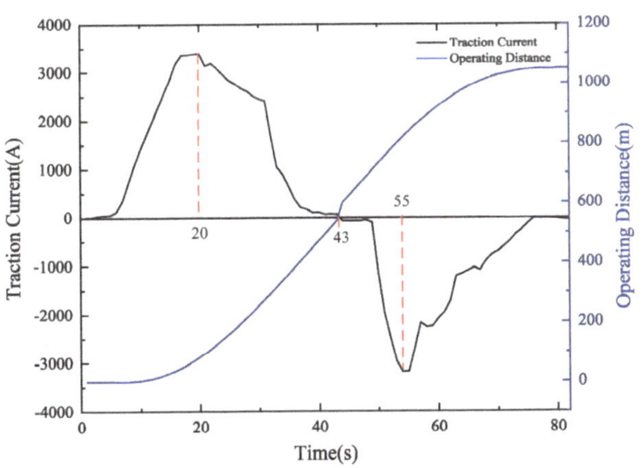

Figure 2. Train operating data.

To verify the correctness and efficiency of the field-circuit coupling method proposed in this paper, this paper analyzes and discusses a total of 2 scenarios for the mainline and the depot of the subway, respectively.

3.1. Modeling of Current Leakage at the Mainline

3.1.1. Modeling of Field-Circuit Coupling

The field-circuit coupling method is used to analyze the mainline, and the overall modeling is shown in Figure 3. In the model, finite element modeling is used for the middle section of the mainline within a range of 10 m, and the rest of the traction power system section is modeled by the resistive network, to analyze the leakage of current and the distribution law at the mainline.

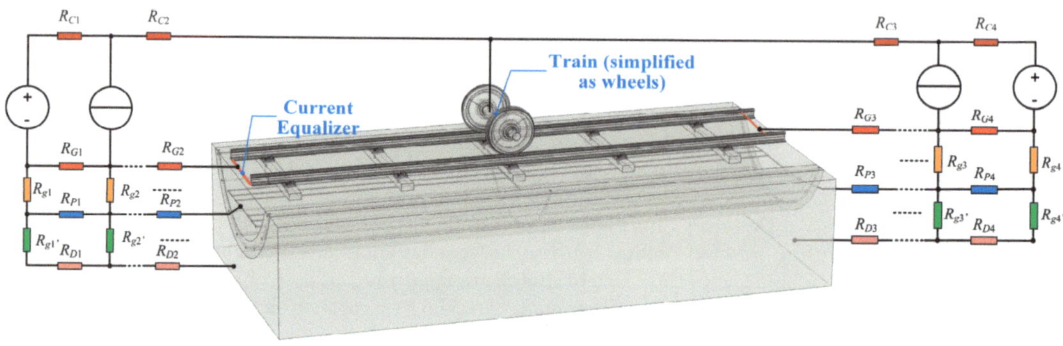

Figure 3. Modeling of field-circuit coupling at the mainline.

3.1.2. Modeling of Finite Element Field

A finite element solution model containing wheels, rails, fasteners, sleepers, track-bed, SCCM, and shield tunnel structure is established, as shown in Figure 4. Among these, the longitudinal structural reinforcement is uniformly distributed along the shield tunnel, the transverse reinforcements are welded to all the longitudinal structural reinforcements every 5 m, and three longitudinal reinforcements of the SCCM are arranged under each rail. The line was constructed using 60 kg/m rails with 1435 mm standard gauges. The wheelset is located directly above the rail and serves as the port of the model to which the DC current excitation is applied. The material properties in the model other than fasteners are shown in Table 1.

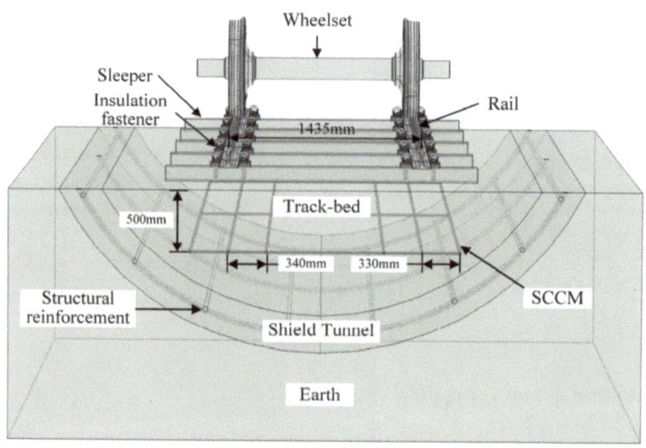

Figure 4. Finite element model.

Table 1. Simulation model material parameters.

Project	Relative Permittivity	Conductivity (S/m)
Rail	1	3×10^6
SCCM	1	2.8×10^6
Reinforcing bar	1	2.9×10^6
Tunnel concrete	1	2×10^{-3}
Track-bed concrete	1	2×10^{-3}
Earth	1	2×10^{-2}

The fastener, as the only rail-to-ground insulation device, is the main source of current leakage at the mainline [23], so the fastener is modeled separately in this paper. At present, elastic bar type III fasteners are widely used in the subway, and the fastener structure in the model is shown in Figure 5. The materials and properties of the type III fastener are shown in Table 2. The fastener is made up of gauge blocks, type III elastic strips, iron tie-plates, under-rail pads, elastic pads, bolts, gaskets, and insulated bushing. Of these, gauge blocks, under-rail pads, and elastic pads are the main insulating devices. Gauge blocks adjust the rail gauge and also isolate the rails from the elastic strips. The under-rail pad reduces the vibration generated when the train passes by, and at the same time avoids the contact edge between the rail and the iron tie-plates. Elastic pads are mainly used to dampen vibration and maintain electrical insulation between the iron tie-plates and concrete sleepers.

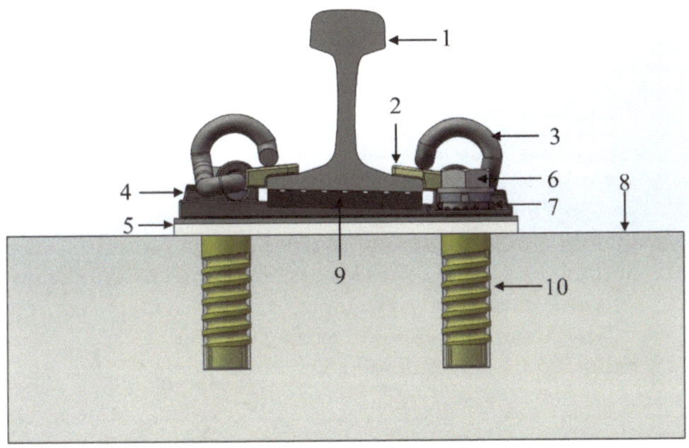

Figure 5. The structure of an elastic bar type III fastener. 1—rail, 2—gauge block, 3—III-type elastic strip, 4—iron tie—plate, 5—elastic pad, 6—bolt, 7—gasket, 8—sleeper, 9—under-rail pad, 10—insulated bushing.

Table 2. Material and parameters of type III fastener.

Structure	Material	Relative Permittivity	Conductivity (S/m)
Elastic strip	60Si2Mn	1	2.38×10^{6}
Gauge block	Glass fiber-reinforced polyamide	1.63	1×10^{-10}
Iron tie-plate	QT450-10	1	2×10^{6}
Under-rail pad	Thermoplastic polyester elastomer	5	1.587×10^{-12}
Elastic pad	Thermoplastic polyester elastomer	5	1.587×10^{-12}
Gasket	Q235-A steel	1	3×10^{6}
Bolt	S45C steel	1	4.59×10^{7}
Insulated bushing	Glass fiber-reinforced polyamide	1.63	1×10^{-10}

3.1.3. Modeling of Resistive Networks

The establishment of a bilateral power supply "rail–SCCM–earth" three-layer resistance network model, based on the circuit Kirchhoff's law and Ohm's law calculations to analyze the distribution of stray currents, sets the direction of the current in the reflux system to the right, as shown in Figure 6.

I is the traction current; I_1 and I_2 are the traction currents supplied to the train by the traction substation. R_C is the overhead contact line unit resistance; R_G is the rail unit resistance; R_P is the SCCM unit resistance; R_D is the earth unit resistance; R_g is the rail-SCCM transition unit resistance; $R_{g'}$ is the SCCM–earth transition unit resistance. I_{G1} and I_{G2} are the train currents flowing in the left and right rails, respectively.

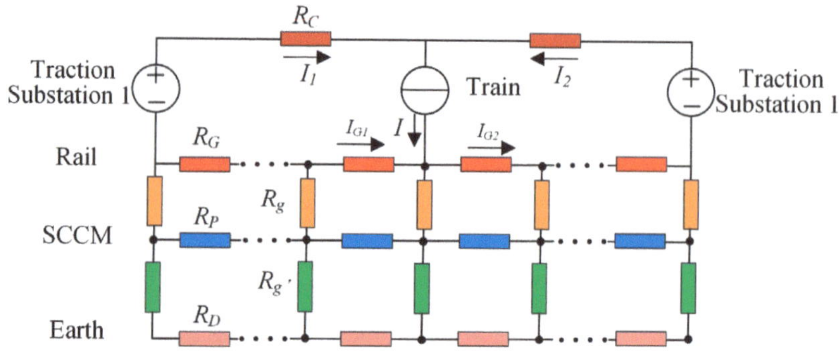

Figure 6. "Rail–SCCM–earth" resistance network model.

3.2. *Modeling of Current Leakage at the Depot*

3.2.1. Modeling of Field-Circuit Coupling

Using the field-coupled method to analyze the subway depot, finite element simulation software is used to establish the finite element solution model including the train, the parking inspection depot, and the line rails outside the depot, a small portion of the mainline rails, a small portion of the mainline SCCM, the one-way conduction device, and the earth. The direct grounding point current obtained from the Simulink resistance network model is used as input to couple with the earth domain finite element simulation model to obtain the field-circuit coupling model of the stray current distribution in the depot, and the overall modeling at the depot is shown in Figure 7.

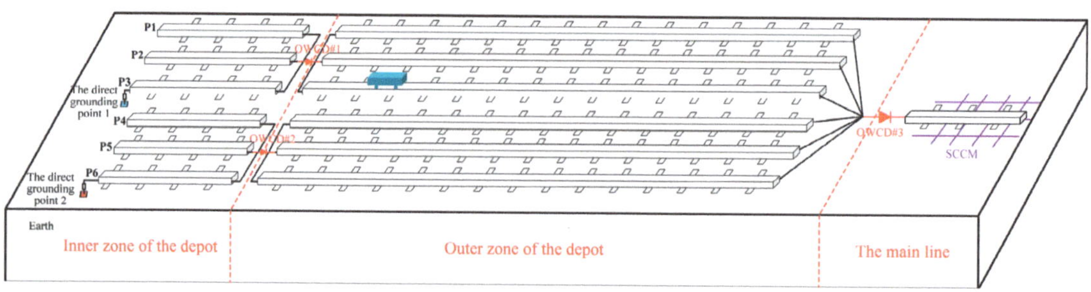

Figure 7. Field-circuit coupling model at depot.

3.2.2. Modeling of Finite Element Field

The volume of earth is 1400 m × 500 m × 100 m, of which the size of depot is 1200 m × 500 m × 100 m; the size of part of the mainline is 200 m × 500 m × 100 m. The SCCM of the mainline is located at 2 m below the rails of the mainline, and the coordinates of the direct grounding point #1 and #2 are $x = 100$, $y = 180$, $z = 0$; $x = 100$, $y = 450$, and $z = 0$, respectively. Where the earth is set to a constant conductivity factor of 2×10^{-2} S/m, the conductivity of metals (e.g., rail, SCCM, train, etc.) is set to 3×10^{6} S/m, and the conductivity of the sleeper is 1×10^{-5} S/m. The model is shown in Figure 8.

The direct grounding points #1 and #2 are used as the input boundaries for modeling stray currents in the depot. The lower surface of the earth model is set to 0 potential to simulate the potential at earth infinity, to study the distribution of the corresponding direct grounding point current in the earth domain when the train is running at different positions.

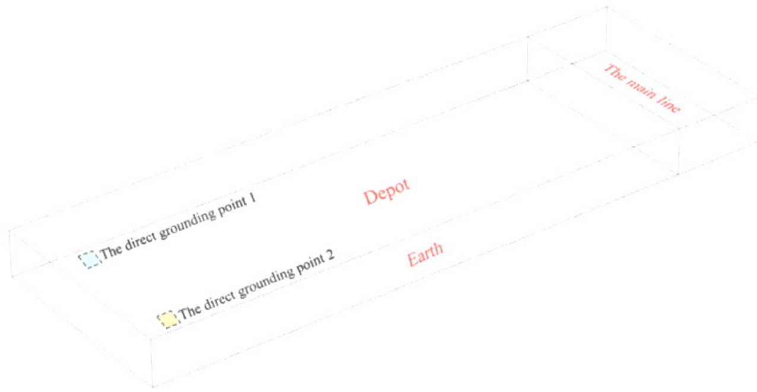

Figure 8. Regional division.

3.2.3. Modeling of Resistive Networks

When dealing with the depot where the ground rail line is closely arranged, it is crucial to consider the boundary conditions when utilizing the resistance network equivalent model. When the boundary conditions are complex, the results of the analytical expression of the stray current distribution are obtained with a huge error. Therefore, in this paper, the analysis has been carried out using resistive equivalence in Section 2, and the calculation of the stray current distribution in the depot has been implemented using Simulink 2019b. Build the equivalent circuit model of traction power supply to the depot, transfer the finite element calculation data to Simulink, and re-feed the waveform and amplitude information to the GUI interface after simulation. The design structure is shown in Figure 9.

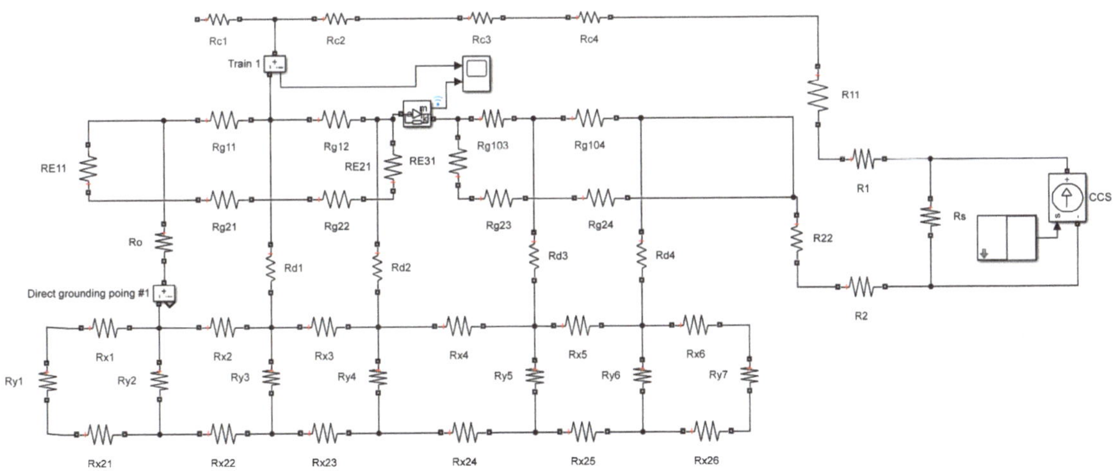

Figure 9. Simplified simulation model of the depot.

The simulation software platform is used to set the parameters of each component and run the data results and current waveforms. After repeated parameter calculations and adjustments to correct the test, accurate data on depot stray currents are obtained. These data serve as the basis for constructing the field-circuit coupling model.

4. Analysis of Results

4.1. Analysis of Current Leakage under Different Insulation Properties of Fasteners at the Mainline

The fastener, as the only rail-to-ground insulation device, is the main source of current leakage on the mainline. This section analyzes the impact of the fastener at the mainline on the current leakage of the rail under 3 insulation conditions.

4.1.1. Fastener Insulation Intact

The distribution of current when the fastener insulation is intact is shown in Figure 10a,b. Figure 10a shows the current density of the fastener; the maximum value is 8.1×10^{-4} A/cm^2, located in the direct connection between the gauge block and rail; the minimum value is 2.2×10^{-11} A/cm^2, mainly concentrated in the elastic pad and rail between the sleeper. As the fastener insulation is intact, the conductivity of each insulating part meets the design requirements, and the overall current density remains low. The current leakage mainly has two paths, as shown in Figure 10b. One is through the gauge block, and the second is through the under-rail pad, circulating to the iron tie-plate, elastic pad, sleeper, and track-bed. Since the conductivity of the gauge block at room temperature is higher than that of the under-rail pad, the current flow arrows of the gauge block in Figure 10b are more intensive.

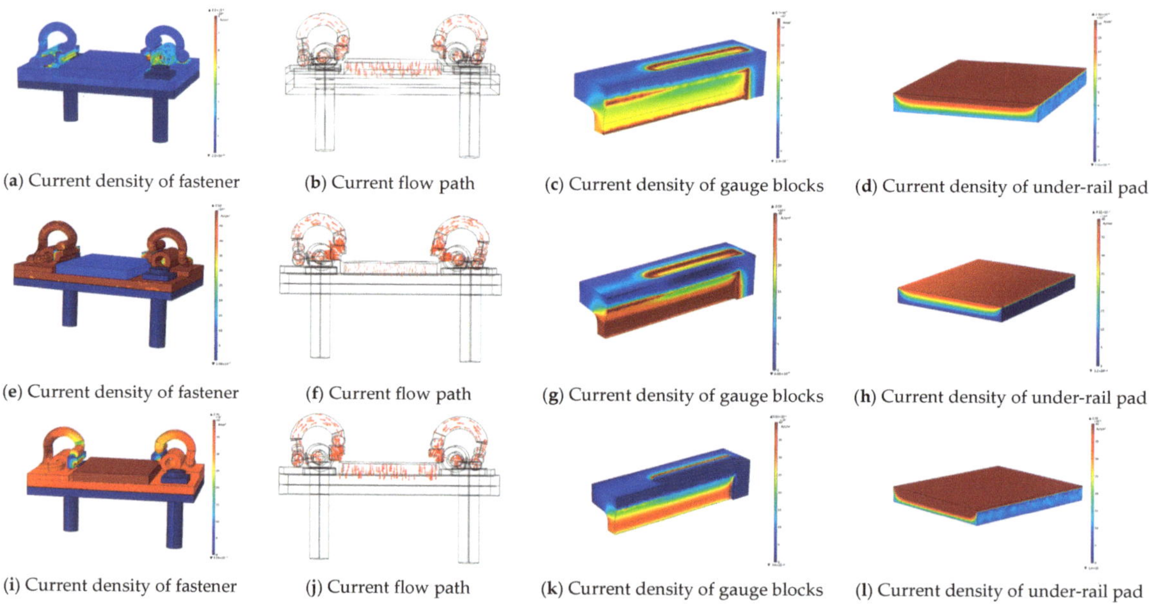

(**a**) Current density of fastener (**b**) Current flow path (**c**) Current density of gauge blocks (**d**) Current density of under-rail pad
(**e**) Current density of fastener (**f**) Current flow path (**g**) Current density of gauge blocks (**h**) Current density of under-rail pad
(**i**) Current density of fastener (**j**) Current flow path (**k**) Current density of gauge blocks (**l**) Current density of under-rail pad

Figure 10. Finite element simulation of fastener under 3 insulation conditions.

Since the insulating materials are in direct contact with the rail, analyzing the current density of the gauge block and the under-rail pad helps in understanding the distribution of rail leakage. The current density distributions of the gauge block and the under-rail pad are shown in Figure 10c,d when the train has been running for 20 s. The current density modes of the gauge block and the under-rail pad are shown in Figure 10c,d. Figure 10c shows the current density mode distribution of the gauge block, with a maximum value of 8.1×10^{-4} A/cm^2. Figure 10d shows the distribution of the current density mode of the under-rail pad, and the maximum value is 2.58×10^{-8} A/cm^2. It can be seen that the current density mode of the gauge block is much larger than that of the under-rail pad, which indicates that when the insulating property of the rail fasteners is intact, the current leaking from the rails is mainly conducted from the gauge block to the iron tie-plate. Therefore, the stray current density of the fastener leakage is roughly equivalent to the current density mode of the gauge block.

4.1.2. Gauge Block Insulation Failure

The surface resistance of the gauge block decreases due to humidity, water leakage, pollution, and other factors. At the same time, the train running back and forth will bring long-term stress and vibration to the gauge block, which will lead to ruptures and internal insulating property damage and accelerate aging. Considering these factors together, the insulating performance of the gauge block will be seriously degraded, and it is difficult to ensure rail-to-ground insulation. According to stray current protection regulations [24], the DC conductivity threshold of the gauge block is 1×10^{-8} S/m, which is higher than this threshold, indicating that it cannot meet the normal rail-to-ground insulation. When the conductivity of the gauge block decreases to 1×10^{-5} S/m, it indicates a serious failure of its insulating effect. Therefore, in this paper, the conductivity of the fastener gauge block at 535 m from station A in the mainline is changed from 1×10^{-10} S/m to 1×10^{-5} S/m, to simulate the serious failure of the insulating effect of the gauge block at this location and analyze its effect on the current leakage of the rail.

Figure 10e,f demonstrate the fastener current density distribution during the 20 s of train operation. As shown in the fastener current density graph in Figure 10e, the current densities at the gauge block, iron tie-plate, and elastic strip are larger, and the maximum value is located on the inner side of the gauge block, which is 0.59 A/cm^2. When the gauge block insulation is severely compromised, it exacerbates the current leakage into the gauge block, with a corresponding increase in current density at the iron tie-plate and elastic strip. The flow path is depicted based on the ratio of current density values in each part of the fastener, as illustrated in Figure 10f. It shows that the gauge block becomes the main path for current leakage, and the current density arrows become dense, while the current density arrows at the under-rail pad become sparse.

When the gauge block insulation fails, the current density mode of the gauge block and the under-rail pad is shown in Figure 10g,h. As can be seen from the figure, the maximum value of the gauge block current density mode is 0.59 A/cm^2, and the maximum value of the under-rail pad current density mode is 4.11×10^{-8} A/cm^2. The significant difference of five orders of magnitude between the two values indicates that when the gauge block fails, its current density mode is primarily responsible for the stray current density of the rail leakage.

4.1.3. Under-Rail Pad Insulation Failure

The under-rail pad is exposed to the air, contaminated and affected by underground water seepage, or subjected to long-term overload stress leading to aging and cracking, resulting in a reduction of surface resistance and a decline in the internal insulating properties, affecting the level of insulation between the rail and the earth. According to stray current protection regulations, the DC conductivity threshold is 1×10^{-8} S/m, and when the conductivity drops to 2×10^{-6} S/m, the insulation of the under-rail matting fails seriously. Therefore, in this paper, the conductivity of the fastener under-rail pad at 535 m from station A in the model line is changed from 1.587×10^{-12} S/m to 2×10^{-6} S/m to simulate the effect of the serious failure of the under-rail pad's insulating performance on the leakage current of the rails at this location.

The distribution of fastener current density when the train has been running for 20 s is shown in Figure 10i,j. Figure 10i shows that the current density at the under-rail pad, iron tie-plate, and elastic strip is larger, of which the current density at the upper surface of the under-rail pad is the largest, reaching a maximum value of 0.39 A/cm^2. The current density at the gauge block is small. The current path is plotted according to the ratio of current density values in each part of the fastener, as shown in Figure 10j. After the failure of the under-rail pads, the under-rail pad becomes the main path of current leakage, and their current density arrows are denser. The conductivity of the gauge block is smaller compared to the elastic pad, so some of the current will flow through the iron tie-plate-elastic strip and then converge at the gauge block.

When the insulation of the under-rail pad fails, the current density modes of the gauge block and under-rail pad are depicted in Figure 10k,l. It can be seen that the maximum value of the gauge block current density mode is 5.63×10^{-6} A/cm^2, and the maximum value of the under-rail pad current density mode is 0.39 A/cm^2. The difference between the two values is 5 orders of magnitude, so when the under-rail pad fails, it is mainly believed that the under-rail pad current density mode is used as a leakage current density of the rail.

4.2. Analysis of Current Leakage under Different Positions of Trains at the Depot

This section analyzes the stray current distribution law in the depot of the train in 4 different positions, as shown in Table 3, modeled using the field-circuit coupling method proposed in this paper.

Table 3. Parameters of different operating states and simulation results.

Mode	Train Position	Traction Current	State	Direct Grounding Points #1 and #2 Current
1	inner zone of the depot (x = 200, y = 180)	280 A	acceleration	65 A; −64 A
2	outer zone of the depot (x = 600, y = 180)	420 A	acceleration	0 A; −0.02 A
3	Mainline (x = 1300, y = 250)	1500 A	acceleration	0.012 A; −0.012 A
4	inner zone of the depot (x = 600, y = 180) outer zone of the depot (x = 1200, y = 250)	420 A −420 A	acceleration deceleration	Current at Train 1: 0.54 A Current at Train 2: −0.54 A

4.2.1. The Train in the Inner Zone of the Depot

For safety, trains departing from inside the garage have lower speeds and lower traction currents. In this paper, for example, a train on the P3 parking line is assumed to run from the end of P3 (x = 100 m, y = 180 m) to the depot entrance (x = 380 m, y = 180 m) assuming that the traction current is kept at 300 A.

To make the calculation clearer and easier to analyze, the data are logarithmically processed, as shown in Figure 11a. When the train is located at x = 200 m, y = 180 m at the inner zone of the depot area, the maximum value of stray current is 1.39 A/m^2, located in the direct grounding point #1; the minimum value of stray current is −1.58 A/m^2, located in the direct grounding point #2. At the end of the depot, the stray currents are relatively large and are concentrated at the direct grounding points #1 and #2. Outside the depot, the stray currents in the earth are small. From the figure, it can be seen that the stray currents flow in the earth diverge at the end of P3 but collect at the end of P6. It can be inferred that this is because the running rails of P3 and P6 are directly grounded at the end, and OWCD#2 is in conduction at the same time, which provides a return path with better conductivity, allowing the current to mainly leak from the ground at P3 and be absorbed by the ground at P6.

4.2.2. The Train in the Outer Zone of the Depot

As the train runs in the outer area of the depot, it begins to speed up and the traction current increases. As an example, a train on P3 in front of the depot is assumed to keep the traction current at 450 A. The train runs from the entrance of the depot (x = 380 m, y = 180 m) to the exit of the depot (x = 1200 m, y = 250 m).

When the train is located at x = 600 m and y = 180 m, the same logarithmic treatment is applied to the data, and the distribution of stray current in the depot is shown in Figure 11b. When the train is located in the outer zone of the depot, the maximum value is 7.38×10^{-3} A/m^2, which is located at the direct grounding point #2, and the minimum value is -1.29×10^{-3} A/m^2. The value of the stray current on the earth's surface is generally small, with most of the stray currents being concentrated at the direct grounding point #2. It can be seen that when the train operates in the outer zone of the depot, the stray currents in the earth are pooled into the direct grounding point #2 inside the depot.

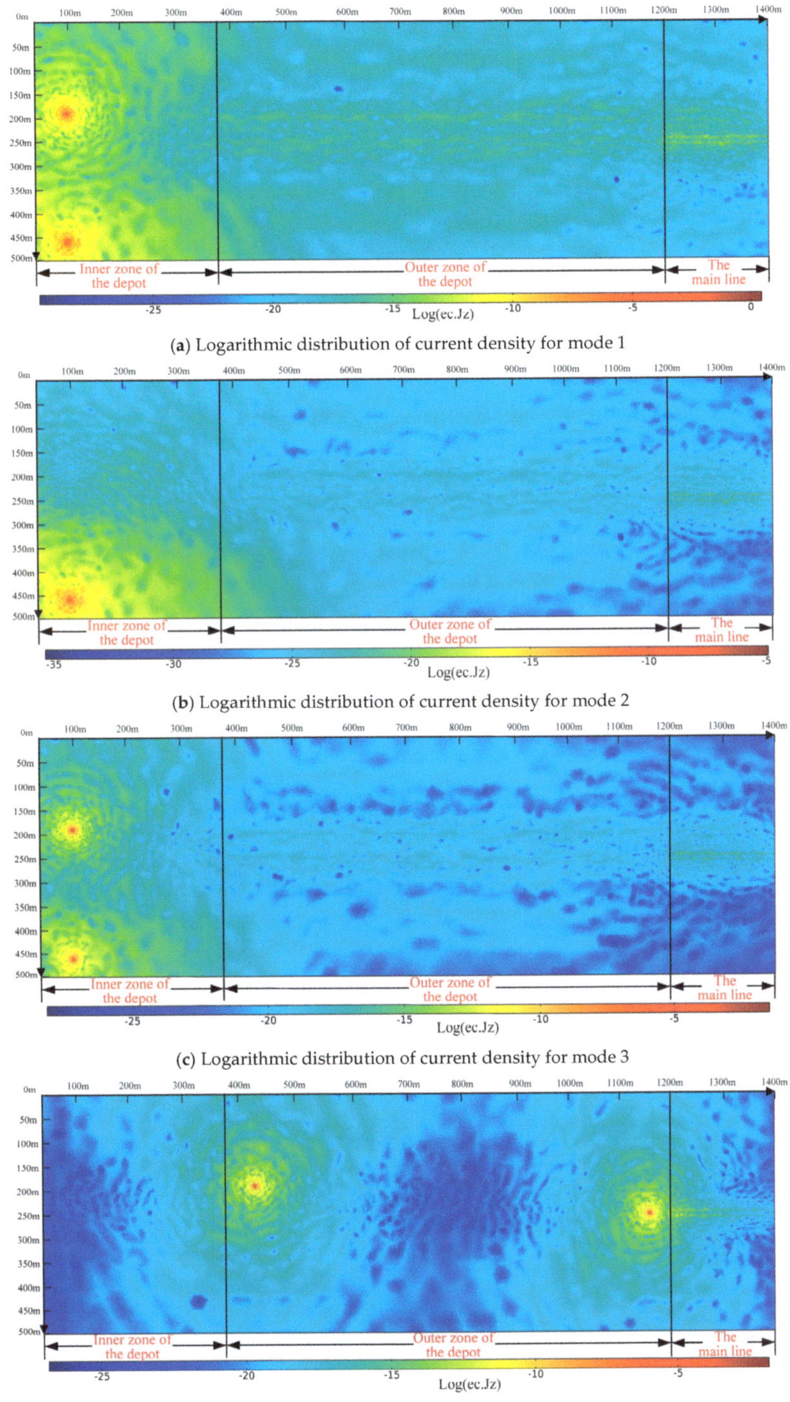

(a) Logarithmic distribution of current density for mode 1

(b) Logarithmic distribution of current density for mode 2

(c) Logarithmic distribution of current density for mode 3

(d) Logarithmic distribution of current density for mode 4

Figure 11. Distribution of stray currents on the earth surface of the depot under different positions of the train.

4.2.3. The Train on the Mainline

The speed of the train is further increased when it runs on the mainline, and the traction current gets even larger. Taking a train at P3 in front of the depot as an example, assuming that the traction current is kept at 1500 A, the train runs from the exit of the depot (x = 1200 m, y = 250 m) to the first section of the partial mainline (x = 1400 m, y = 250 m).

When the train is located at x = 1300 m and y = 250 m, the data are also logarithmically processed, and the stray current distribution in the depot is shown in Figure 11c. The maximum value is 0.03 A/m^2 and the minimum value is -5.37×10^{-3} A/m^2 when the train is located at the mainline. The cathode of OWCD#3 is connected to the rails of the mainline and is subjected to a reverse voltage up to the point. The current on the mainline cannot enter the depot through OWCD#3, but stray currents leak into the earth pass through the structural reinforcement of the depot, converge at the direct grounding points #1 and #2, and flow into the depot rails. It can be seen that the currents on the earth's surface mainly converge at the direct grounding points #1 and #2 in the inner zone of the depot. Since the OWCD#3 cutoff, the stray current in the mainline propagates mainly through the earth to the direct grounding point, so the currents on the earth's surface in the outer zone of the depot are low.

4.2.4. Two Trains in the Depot

To simulate a realistic stray current distribution in the depot, two trains are considered as running in different states. Train 1 is located at x = 400 m, y = 180 m, with a traction current of 420 A, in the acceleration state; Train 2 is located at x = 1200 m, y = 250 m, with a traction current of -420 A, in the deceleration state.

The same logarithmic process is applied to the data, and the stray current distribution in the depot is shown in Figure 11d. When there are two trains in the depot, the maximum value of stray current is 0.17 A/m^2, which is located directly under train 1; the minimum value is -0.15 A/m^2, which is located directly under train 2. Since the leakage current at traction train 1 is positive and the leakage current at braking train 2 is negative, the energy of the braking train is fed back to the traction train. In this case, all OWCDs are cut off, the magnitude of stray currents is greatly reduced, and the stray currents are mainly distributed on the ground from the vicinity of the traction train to the braking train. It can be seen that the leakage current dissipates at traction train 1 and converges at braking train 2. Since all the unidirectional conductors are reverse cutoff, there is no current dissipation or convergence at the direct grounding points #1 and #2.

5. Discussion

5.1. Discussion of the Situation at the Mainline

The stray current leakage of the fastener under three insulation conditions is shown in Figure 12. The current density mode of the fastener gauge block at 535 m from station A is chosen as the stray current leaking from the rail, and the current leaking from the rail when the train runs from station A to station B is obtained. When the fastener insulation is intact and the train has been running for 20 s, the leakage current tends to be close to 0. However, when the gauge block insulation fails due to aging or other factors, causing serious insulation failure, the leakage current value increases dramatically. During a 20-s run of the train, the positive traction current is the largest, with the current density reaching 0.59 A/cm^2. Conversely, during a 55-s run, the traction current is negatively the largest, with the current density reaching 0.56 A/cm^2 at its maximum value. When the under-rail pad board surface is covered in dirt or experiences aging cracking, its internal insulation performance declines. During a 20-s train run, the positive traction current is large, with the current density reaching up to 0.39 A/cm^2. In a 55-s run of the train, the negative traction current is large, with the current density reaching up to 0.36 A/cm^2.

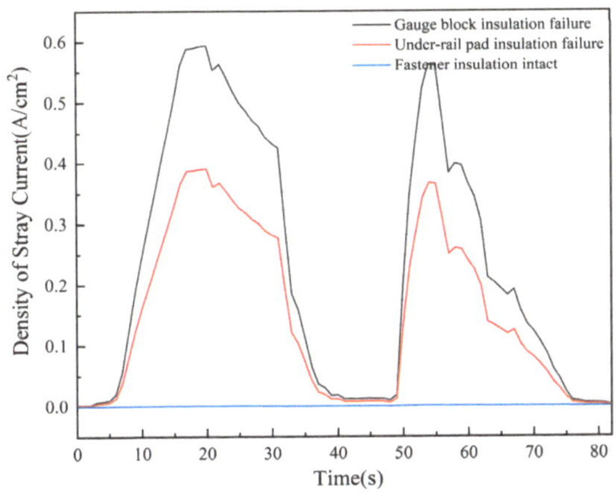

Figure 12. Comparison of current density under 3 insulation conditions for fastener.

5.2. Discussion of the Situation at the Depot

The stray currents are mainly concentrated at direct grounding point #2 when the train is located in the inner zone of the depot, as compared to when the train is located in the outer zone of the depot. Since the OWCD of the parking line P1-P3 is subjected to a counter voltage and cutoff, the current flow path is cut off by the OWCD, resulting in small stray currents at direct grounding point #1.

While the train is located in the mainline, the current primarily leaks through grounding points #1 and #2. Compared with the train located in the depot, the leakage current from the rail is reduced. This reduction is due to the OWCD withstanding the reverse voltage, preventing the return current on the mainline from flowing through the OWCD into the depot. However, some stray currents leak to the earth through the structural reinforcement of the depot, converging at direct grounding points 1 and 2 before flowing into the depot rail.

6. Results

The work fulfills the stated objective of providing a field-circuit coupling method to analyze stray current distribution on earth surfaces. The main conclusions of this paper are as follows:

(1) This paper proposes a field-circuit coupling method and analyzes the leakage current and its distribution law in two scenarios: subway mainline and the depot. The applicability and effectiveness of the proposed field-circuit coupling method are verified through these analyses.

(2) This paper builds a model of current leakage from fasteners on the subway mainline. By modeling the structure of type III fasteners, it studies and analyzes the influence of fasteners on rail current leakage under three different insulating conditions. It is found that the failure of the gauge block results in the greatest degree of current leakage from the rails.

(3) This paper builds a current leakage model of the grounding point at the subway depot and investigates four different actual subway depot line layouts and grounding situations. Simulation results indicate that the characteristics of stray current distribution in the depot are primarily influenced by the operational status of the OWCD, which is in turn affected by changes in rail potential. Additionally, when multiple trains are running in the depot, the complex operational status of the OWCD leads to a more varied and intricate distribution of stray current.

Author Contributions: Software, X.L. and Y.L.; Resources, S.L. and X.C.; Writing—original draft, Z.T.; Writing—review & editing, Z.T. All authors have read and agreed to the published version of the manuscript.

Funding: This research was funded by Natural Science Foundation of Jiangxi grant number 20224BAB204053.

Data Availability Statement: The original contributions presented in the study are included in the article, further inquiries can be directed to the corresponding authors.

Conflicts of Interest: Authors Shan Lin, Xia Chen were employed by the company Guangzhou Metro Design & Research Institute Co., Ltd. The remaining authors declare that the research was conducted in the absence of any commercial or financial relationships that could be construed as a potential conflict of interest.

References

1. Yang, S.; Liu, C. Discrete modeling and calculation of traction return-current network for 400 km/h high-speed railway. *Proc. Inst. Mech. Eng. Part F-J. Rail Rapid Transit* **2023**, *237*, 445–457. [CrossRef]
2. Chen, L.; Huang, K.; Wei, K.; Gao, D.; Peng, Q.; Lin, S. Study on the Potential Distribution Characteristics of Substation Grounding Grid Considering Stray Current. In Proceedings of the 2022 IEEE 6th Conference on Energy Internet and Energy System Integration (EI2), Chengdu, China, 11–13 November 2022; pp. 1284–1289.
3. Li, C.; Du, Q.; Guo, Y.; Liu, Y.; Yang, F.; Chen, L.; Zhang, X.; Huang, G.; Gao, G.; Wu, G. Modeling of Stray Currents from Metro Intruding into Power System Considering the Complex Geological Conditions in Modern Megacities. *IEEE Trans. Transp. Electrif.* **2023**, *9*, 1653–1663. [CrossRef]
4. Zhou, Q.; Lin, S.; Lin, X.; Wang, A. A Uniform Model for Stray Current of Long-Line DC Metro Systems. *IEEE Trans. Transp. Electrif.* **2022**, *8*, 2915–2927. [CrossRef]
5. Zheng, Z.; Du, G.; Huang, W.; Wang, J.; Fan, M.; Zhu, Z. Coupling Effect of Reflux Safety Parameters in Main line and Depot of Multitrain DC Traction Power Systems. *IEEE Trans. Transp. Electrif.* **2022**, *8*, 527–541. [CrossRef]
6. Song, C.; Li, Z.; Wang, H.; Liu, X. Analysis of influence law of rail transition resistance on subway stray current. In Proceedings of the 2022 7th Asia Conference on Power and Electrical Engineering (ACPEE), Hangzhou, China, 15–17 April 2022; pp. 1631–1635.
7. Gao, H.; Wei, R.; Liu, J.; Wang, Z.; Zhong, H.; Cao, Y.; Tang, M.; Zhang, J. Analysis of Stray Current Intrusion into Grid Substation under Multiple Factors. In Proceedings of the 2022 4th International Conference on Control and Robotics (ICCR), Guangzhou, China, 2–4 December 2022; pp. 188–192.
8. Liu, W.; Pan, Z.; Tang, Y.; Zhou, L.; Bhatti, A.A.; Niu, J. A Field-Circuit Coupling Diffusion Model for Stray Current of DC Metro System. *IEEE Trans. Transp. Electrif.* 2023, *early access*.
9. Mariscotti, A.; Fichera, F.; Ogunsola, A. Evaluating stray current from DC electrified transit systems with lumped parameter and multi-layer soil models. In Proceedings of the IEEE Eurocon 2013, Zagreb, Croatia, 1–4 July 2013; pp. 1187–1192.
10. Wang, C.; Li, W.; Wang, Y.; Xu, S.; Fan, M. Stray Current Distributing Model in the Subway System: A review and outlook. *Int. J. Electrochem. Sci.* **2018**, *13*, 1700–1727. [CrossRef]
11. Liu, W.; Zhou, L.; Pan, Z.; Bhatti, A.A.; Huang, X.; Zhang, J. Dynamic diffusion model of stray current in DC traction power supply system. *IEEE Trans. Power Deliv.* **2023**, *38*, 2170–2182. [CrossRef]
12. Vranesic, K.; Serdar, M.; Lakusic, S.; Kolar, V.; Mariscotti, A. Dynamic Stray Current Measuring Methods in Urban Areas. *Balt. J. Road Bridge Eng.* **2022**, *17*, 146–170. [CrossRef]
13. Zhu, J.; Mao, C.; Wang, Z.; Chen, J.; Wang, D.; Luan, L.; Qiao, Y.; Wang, H. A novel suppression method for grounding transformer against earth current from urban rail transit. *IEEE Trans. Ind. Electron.* **2020**, *68*, 11651–11662. [CrossRef]
14. Du, G.; Wang, J.; Jiang, X.; Zhang, D.; Yang, L.; Hu, Y. Evaluation of Rail Potential and Stray Current with Dynamic Traction Networks in Multitrain Subway Systems. *IEEE Trans. Transp. Electrif.* **2020**, *6*, 784–796. [CrossRef]
15. Vranesic, K.; Lakusic, S.; Serdar, M.; Alar, V. Detrimental effect of stray current on rails and fastening systems in urban railway tracks. *Constr. Build. Mater.* **2023**, *400*, 132645. [CrossRef]
16. Chang, W.C.; Kuo, C.C.; Lin, W.C.; Hsieh, M.N. Simulation of Stray and Core Shielding Loss for Power Transformer Based on 2D/3D FEM. *IEEE Access* **2023**, *11*, 16943–16950. [CrossRef]
17. Jabbehdari, S.; Mariscotti, A. Distribution of stray current based on 3-Dimensional earth model. In Proceedings of the 2015 International Conference on Electrical Systems for Aircraft, Railway, Ship Propulsion and Road Vehicles (ESARS), Aachen, Germany, 3–5 March 2015; pp. 1–6.
18. Lin, S.; Liu, X.; Zhang, X.; Cai, Z. Research on Stray Current Distribution in a Multitrain Subway System using Finite Element Modeling. *Int. J. Electrochem. Sci.* **2021**, *16*, 211250. [CrossRef]
19. Kunpeng, L.; Xiaobin, C.; Hao, S.; Chao, X.; Yunyun, Z.; Wei, L.; Haiou, S.; Zhixin, H. The Equivalent Circuit and Influencing Factors of Subway Rail Fastener Track-to-earth Resistance. In Proceedings of the 2018 IEEE Transportation Electrification Conference and Expo, Asia-Pacific (ITEC Asia-Pacific), Bangkok, Thailand, 6–9 June 2018; pp. 1–5.

20. Lin, S.; Huangfu, Y.; Zhou, Q.; Wang, A. Evaluation and Analysis Model of Stray Current in the Metro Depot. *IEEE Trans. Transp. Electrif.* **2021**, *7*, 1780–1794. [CrossRef]
21. Cai, Z.; Zhang, X.; Cheng, H. Evaluation of DC-Subway Stray Current Corrosion with Integrated Multi-Physical Modeling and Electrochemical Analysis. *IEEE Access* **2019**, *7*, 168404–168411. [CrossRef]
22. Ni, Y.; Yu, K.; Zhuo, C.; Zeng, X.; He, F.; Leng, Y.; Han, W. Study on distribution characteristics of metro stray current. In Proceedings of the 2021 IEEE International Conference on Electrical Engineering and Mechatronics Technology (ICEEMT), Qingdao, China, 2–4 July 2021; pp. 19–25.
23. Vranešić, K.; Lakušić, S.; Serdar, M. Influence of Stray Current on Fastening System Components in Urban Railway Tracks. *Appl. Sci.* **2023**, *13*, 5757. [CrossRef]
24. Mariscotti, A. Stray Current Protection and Monitoring Systems: Characteristic Quantities, Assessment of Performance and Verification. *Sensors* **2020**, *20*, 6610. [CrossRef] [PubMed]

infrastructures

MDPI

Article

Optimization Models for the Maintenance Management of Tropical Paved and Unpaved Roads

Taciano Oliveira da Silva [1], Heraldo Nunes Pitanga [2], Emerson Cordeiro Lopes [1], Laura Carine Pereira Ribeiro [1], Gustavo Henrique Nalon [1,*], Klaus Henrique de Paula Rodrigues [1], José Carlos Lopes Ribeiro [1] and Khaled Ksaibati [3]

1 Department of Civil Engineering, Federal University of Viçosa, Viçosa 36570-900, Brazil; taciano.silva@ufv.br (T.O.d.S.); emerson.lopes@ufv.br (E.C.L.); laura.ribeiro@ufv.br (L.C.P.R.); klaus@ufv.br (K.H.d.P.R.); jcarlos.ribeiro@ufv.br (J.C.L.R.)

2 Department of Transportation and Geotechnics, Federal University of Juiz de Fora, Juiz de Fora 36036-330, Brazil; heraldo.pitanga@ufv.br

3 Department of Civil and Architectural Engineering, University of Wyoming, Laramie, WY 82071, USA; khaled@uwyo.edu

* Correspondence: gustavo.nalon@ufv.br

Abstract: The degradation of paved and unpaved roads stands as a critical concern in contemporary infrastructure management. When faced with limited budgets, it is important to identify the optimal combination of road preservation strategies to minimize the lifecycle cost of the road network. Specific studies are necessary to improve the maintenance management systems and analyze the behavior of road surface deformation. To narrow these knowledge gaps, this study investigates a management system that focuses on the application of optimization techniques for managing both paved and unpaved tropical roads. Probabilistic deterioration models were constructed using the Markovian process, resulting in precise degradation curves in the context of 18 unpaved road segments in the Zona da Mata County of Minas Gerais (Brazil), along with 88 paved roads located in Minas Gerais. An optimization algorithm was proposed for the prediction of maintenance resources for unpaved and paved roads, emphasizing the cost-effectiveness of preventive and minor rehabilitation treatments over reconstruction. Comparisons between the maintenance costs of unpaved and paved roads indicated that the full rehabilitation costs of paved roads were approximately 10 times higher per kilometer compared to those of unpaved roads. The models effectively captured the trend wherein a major treatment leads to minor additional treatments being necessary for the subsequent several years in both scenarios. The findings of this study provide future directions for the optimized allocation of resources in the management of transportation infrastructures.

Keywords: transportation infrastructures; pavement management system; tropical roads; probabilistic deterioration models

Citation: Silva, T.O.d.; Pitanga, H.N.; Lopes, E.C.; Ribeiro, L.C.P.; Nalon, G.H.; Rodrigues, K.H.d.P.; Ribeiro, J.C.L.; Ksaibati, K. Optimization Models for the Maintenance Management of Tropical Paved and Unpaved Roads. *Infrastructures* **2024**, 9, 100. https://doi.org/10.3390/infrastructures9070100

Academic Editors: Tomasz Nowakowski, Artur Kierzkowski, Agnieszka A. Tubis, Franciszek Restel, Tomasz Kisiel, Anna Jodejko- Pietruczuk and Mateusz Zając

Received: 28 May 2024
Revised: 20 June 2024
Accepted: 24 June 2024
Published: 27 June 2024

1. Introduction

1.1. Research Context

The degradation of the existing road network is a pressing concern in contemporary infrastructure management. The existing road network has aged, leading to the appearance of different types of distresses such as cracking, raveling and potholing on the road surface. In the Brazilian state of Minas Gerais (MG), which is the focus of the present study, a total of 271,160 km of roadway must be maintained by federal, state, or local entities [1].

The deterioration level of roads is directly related to different factors, such as the volume of traffic and the maintenance activities undertaken over their lifespan. Nowadays, the costs of maintenance and rehabilitation activities are rising dramatically [2]. The lack of maintenance procedures has resulted in irreparable harm to economic growth rates, as poorly maintained roads contribute to delays, road accidents and increased vehicle operating costs. Without proper and timely maintenance interventions, the condition of the pavement rapidly deteriorates from "good" to "poor" over a relatively short period, necessitating higher funding for rehabilitation [3].

Over the past decade, maintenance management systems have significantly improved due to technological advances. Numerous researchers have dedicated efforts to elaborating on maintenance management systems and investigating the road surface deformation behavior. They include interesting strategic and systematic processes aimed at maintaining and enhancing the road network [4–6].

1.2. Literature Review

1.2.1. Maintenance Management of Paved and Unpaved Roads

Pavements are designed to last for a certain period. Throughout their life cycle, they begin in optimal conditions, declining until they reach failure [7]. The evaluation of the pavement deterioration level is important in assisting those responsible for selecting maintenance activities in their decision-making process. Deterioration should be estimated based on functional and structural factors [8]. Maintaining paved roads in appropriate conditions requires special attention. Depending on the structural health of the road, varying degrees of maintenance activities are necessary. Different types of paved road maintenance can be used [9,10]:

- Asphalt overlays: roads undergo restoration through resurfacing procedures. Rather than removing the deteriorating pavement, it serves as a base onto which a new layer is applied.
- Patching: this involves filling potholes or excavated areas in asphalt pavement. The swift repair of such issues helps to prevent further deterioration and the need for costly repairs. Without prompt patching, water infiltration can lead to more extensive and serious infrastructure issues.
- Reshaping: when a road surface is too damaged to be smoothed, the unpaved sub-base can be reworked to eliminate large potholes and restore a flattened crown.
- Slab replacement: pavement sections that are chipped, cracked, or uneven are replaced, avoiding the need to replace the entire stretch.
- Smoothing: reworking the surface of a road without delving too deeply into the sub-base. Smoothing procedures address minor damage resulting from regular use.

In addition, it is important to highlight that reconstruction is carried out when the paved roads have deteriorated beyond cost-effective maintenance. Reconstruction is usually carried out in phases to minimize traffic disruptions.

On the other hand, unpaved roads are constructed when the volume of commercial traffic (gross weight exceeding 3 tons) is low [11]. In this case, it is necessary to identify and characterize aspects related to the ride comfort, users' safety, vehicle performance and transportation costs, based on subjective and objective evaluations [12–14]. Several agencies and local researchers have suggested different types of subjective rating methods for representing unpaved road conditions, such as PASER, Unpaved-PASER, Ride Quality Rating Guide (RQRG) and Gravel Roads Rating System (GRRS) [15]. The GRRS offers standardized evaluations for seven different types of distresses: potholes, rutting, washboards, loose aggregate, dust, crown and roadside drainage [16].

Five primary factors affecting the quality of an unpaved road can be listed: traffic load, subgrade quality, construction practices, water and maintenance programs. According to Kabindra, Pramen and Thomas [17], the following types of maintenance programs are generally recommended for unpaved roads:

- Routine maintenance: this involves mitigating blockages and damage on unpaved roads through regularizing surfaces or patching methods. It also involves maintaining bridges and drainage structures, in addition to implementing erosion control procedures.
- Periodic maintenance: employing maintenance practices to address various issues that arise, where the extent of distress does not warrant full-scale rehabilitation or reconstruction services.

- Rehabilitation: applying a thin unpaved overlay to improve the durability and functionality of the road.

1.2.2. Optimization Methodologies for Road Management Systems

Life cycle costs analysis (LCCA) systems have been widely used to investigate the overall economic effectiveness of various alternatives for the design of pavements and their maintenance and rehabilitation (M&R). The LCCA enables the appropriate assessment of the lifetime impacts of their practices [18–20]. In addition, various statistical models and machine learning algorithms have been elaborated on to improve the performance of transportation systems, based on innovative strategies for traffic modeling, estimation and monitoring [21]. Accurate systems for traffic evaluation are vital tools that support decision-making by commuters and transportation system officials, particularly in managing road conditions effectively [22,23].

According to previous literature [16,24], the management systems can be considered a multi-objective optimization approach due to different factors: (i) engineers or decision-makers want to maximize the network road conditions within a limited budget; (ii) preventive and minor rehabilitation treatments are more cost-effective than reconstruction (iii) the budget must exceed a certain threshold to maximize societal benefits; and (iv) the best combination of road segments for rehabilitation should include those with more traffic volume. In some works, this system can be conceptualized as a two-level optimization process, as it connects decision-makers at two levels [25]. This scenario can be exemplified by two sets of objective functions: one aligned with governmental objectives and the other driven by the priorities of city governors or the company's stakeholders.

The Management System (MS) can be associated with the integrated coordination of various activities involved in the design, construction and maintenance processes related to road infrastructures, allowing them to provide acceptable conditions at minimal costs. The MS encompasses various methods aiding decision-makers in identifying optimized strategies for constructing, evaluating and maintaining roads in an acceptable condition over a specified period within a designated budget [26]. The optimization approaches encompass mathematical programming techniques aimed at minimizing or maximizing various factors to enhance the road management systems [27–29].

1.2.3. Pavement Management System (PMS)

The implementation of a Pavement Management System (PMS) is particularly crucial in situations where road conditions are poor and resources are limited. A robust PMS aims to delineate maintenance strategies that minimize costs while maximizing the return on invested resources, thereby extending the service life of roads and properly allocating limited budgets [30]. Furthermore, an effective PMS must continually update itself by exploring alternative strategies, identifying the optimal workflow, and leveraging feedback mechanisms to enhance overall decision-making processes [31].

The following procedures typically guide the customization of a pavement management process to meet the unique needs of each local agency: defining the roadway network and gathering inventory data, collecting condition data, predicting road conditions, selecting treatments, reporting results, choosing pavement management tools and ensuring ongoing updates [32].

Predicting road conditions requires a comprehensive understanding of pavement performance and the distinctions between analyzing it at the network and project levels. At the network level, the entire network is evaluated to identify the appropriate types of maintenance and the road sections to be repaired. At the project level, only the selected sections are analyzed to determine the most appropriate treatment procedures [33].

The road performance models are derived from various pavement characteristics (e.g., functional conditions, width, length, rut) and enable the prediction of pavement parameters arising from traffic loads and climatic conditions (e.g., deflections, stresses, strains, moisture content, temperature, etc.) [34–36].

1.2.4. Unpaved Roads Management System (URMS)

According to Saha, Ksaibati and Atadero [8], an Unpaved Roads Management System (URMS) is a strategic and systematic process for maintaining and upgrading the road network. When funding is limited, the URMS identifies the best combination of projects for road preservation, providing the most benefit to society in terms of the overall lifecycle cost of the road network. A URMS consists of various stages, with the "prediction of conditions step" being the most intricate. To predict the paved road conditions, it is necessary to develop deterioration models based on the road's conditions over the years. The established models facilitate the prediction of key responses of the paved road, based on empirical, mechanistic or mechanistic–empirical information calibrated with field-observed data.

URMSs lack standardized procedures for determining surface condition ratings or deterministic approaches to modeling surface deterioration. Evaluating roads and establishing the pavement condition index (PCI) are essential for assessing the infrastructure conditions. However, the typical PCI methodology is not applicable to unpaved roads, as it primarily focuses on detailed examinations of distresses of paved roads, such as cracking and potholes [37].

To determine future needs based on road conditions, a URMS applies either probabilistic or deterministic models. In Brazil, deterministic models are commonly utilized, based on expert opinion curves derived from field staff experience. However, these deterministic techniques do not properly account for uncertainties associated with pavement deterioration due to traffic and weather. In contrast, probabilistic models take these uncertainties into consideration, resulting in more precise curves [8].

1.3. Research Gaps and Original Contributions of the Present Study

In situations where funding is limited, the MS provides the optimal combination of projects for road preservation, reaching the greatest benefits in terms of the overall lifecycle cost of road infrastructures [38]. Since the growth of the road network has not kept pace with the considerable increase in transport demand in Brazil, efficient pavement management systems (PMSs) for tropical roads are necessary.

When confronting budget constraints, determining the optimal preservation procedures becomes pivotal for minimizing the overall costs throughout the lifecycle of paved and unpaved roads. The previous studies of Rejani et al. [39] only focused on the pavement maintenance and rehabilitation of urban roads, using the Highway Development and Management System (HDM-4) to minimize the total costs for a target road roughness. Therefore, specific studies are still needed to formulate robust management systems for both paved and unpaved roads. According to previous works [40,41], the inherent limitations of existing approaches highlight the need to develop computational methods that can effectively analyze the residual conditions of different infrastructure elements. Moreover, a research gap exists in standardized procedures for the surface condition exploration of unpaved roads. The conventional PCI methodology is inapplicable to unpaved roads, given its predominant focus on detailed examinations of distresses specific to paved roads.

To narrow these knowledge gaps, the present paper proposes a methodology for identifying the best combination of initiatives for road preservation within a specified budget. In MG, approximately 90% of roads are unpaved, and most roads require an efficient management system. Therefore, the present work presents innovative optimization techniques for managing unpaved and paved roads using similar structures and simple datasets. Maintenance resources were predicted for 18 unpaved road segments in the Zona da Mata County (ZMC) of MG, along with 88 paved roads located in MG.

2. Materials and Methods

This section presents the methods used for data collection, the roadway segmentation process, the procedures for the characterization of road conditions, the deterioration models applied in this study and the algorithm proposed to identify the best combination of projects for road preservation. It is important to mention that the present model does not consider political factors but rather purely the life cycle costs related to unpaved and paved roads.

2.1. Road Segmentation and Classification

Four different types of data were collected to develop this research: roadway segmentation data, traffic counts, road width values and road conditions information. Roadway segmentation was conducted by defining segments that start and end at points of new construction, intersections with other roads or other alterations along the road.

Data from unpaved roads were directly collected in the field, which imposes some challenges regarding accessibility and terrain conditions. For example, the average width of the sections was determined using tape measurements. Rutting was obtained using a standardized ruler and straightedge. The average daily traffic volume was determined using the methodology proposed in the DNIT manual for traffic studies. Data regarding paved roads were obtained from the "Rodovias" website managed by the government of the state of MG.

Based on the segmentation process and using the ArcGIS software, a total of 18 unpaved road segments were delineated in the ZMC of MG, alongside 88 paved roads distributed throughout MG. Figure 1 shows a paved road in MG awaiting maintenance to prolong its service life, while Figure 2 illustrates an unpaved road in poor condition in MG.

Figure 1. Paved road located in MG. Reproduced from Salomão et al. [42], as permitted under the Creative Commons Attribution License Agreement.

The unpaved road conditions were characterized by seven distinct parameters: cross-section/crown, roadside drainage, rutting, potholes, loose aggregate, dust and corrugation. The Pavement Surface Evaluation and Rating (PASER) manual describes that the surface of unpaved roads can be rated on a five-point scale, where five represents an excellent condition and one indicates a failed condition for unpaved roads. For this paper focusing on MG roads, we proposed transforming this five-point scale into a nine-point scale, following the Gravel Roads Rating System (GRRS) guidelines [15]. This approach aimed to reduce errors when human raters need to assess the infrastructure conditions. Then, a Pavement Serviceability Index (PSI) was calculated, following the recommendations of the RQRG created by the Wyoming Technology Transfer Center (WYT2/LTAP) [43]. This nine-points scale was applied for unpaved roads, based on the Overall Road Condition Index (ORCI). This methodology resulted in the ride quality, and the extreme values nine and one represent the best and worst conditions, respectively.

Figure 2. Unpaved road located in the ZMC of MG.

On the other hand, paved roads were assessed using the AASHTO methodology, which ranks conditions from 0 to 100, with 100 representing a newly constructed road and 0 indicating a failed road. According to this rating system, the primary variables are the distresses rutting, fatigue cracking, longitudinal cracking, transverse cracking, patches, potholes, raveling, bleeding and roughness. The evaluation results were reported using the AASHTO guide to road classification, categorizing roads as "Excellent", "Good", "Fair", "Poor", or "Failed" [44]. Therefore, a five-point scale was used for the paved roads investigated in this research.

2.2. Deterioration Models

Huntington and Ksaibati [15] developed a deterioration model for roads in Wyoming (USA), which served as a foundation in the current study for elaborating the deterioration model for the Brazilian roads.

An average deterioration was estimated in the present research, based on overall road conditions. At first, the obtained pavement condition database was appropriately organized according to the defined condition states. Then, the first aspect of the data quantification process involved categorizing the pavements. The subsequent step encompassed generating data matrices for each pavement classification.

The deterioration evolution of unpaved and paved roads was predicted through the Markovian chains. After defining the number of conditions (k) to be considered in the statistical evaluation, the transition matrix with dimensions $k \times k$ was used to evaluate the transition probability of states [45], as indicated in Equation (1).

$$P\Delta t = \begin{bmatrix} p_{11} & p_{12} & p_{13} & \cdots & p_{1k} \\ 0 & p_{22} & p_{23} & \cdots & p_{2k} \\ 0 & 0 & p_{33} & \cdots & p_{3k} \\ \vdots & \vdots & \vdots & & \vdots \\ 0 & 0 & 0 & \cdots & 1 \end{bmatrix} \tag{1}$$

The unit value presented in this matrix ($P\Delta t$) corresponds to the maximum input value for each event. The probability value of p_{kk} was defined considering that there was no further condition state after the k state, so it is a limited value [46]. The transition probability from a state i to a state j, for a generic time interval (Δt), is represented by p_{ij}, with $i = j = 1, \ldots, k$, as indicated in Equation (2) [47].

$$p_{(\Delta t)ij} = \text{Prob}\ (X_{t+\Delta t} = j | X_t = i) \quad (2)$$

Then, Markovian chain techniques and regression analyses were conducted using the R statistical software (version 3.5). Using the probabilistic Markovian process and beginning with the data matrix, the transition matrix was calculated to derive the probability distribution matrix. Subsequently, this matrix was used to determine the expected age of the roads based on their conditions, resulting in the development of a statistical model for road deterioration.

2.3. Algorithm Proposed to Identify the Best Combination of Projects for Road Preservation

Five treatment options were proposed for the unpaved roads: general maintenance, chemical and mechanical stabilization, preventive rehabilitation, medium rehabilitation and paving procedures (GM, 1-R, 2-R, 3-R and 4-R, respectively). In addition, six treatment options were proposed for paved roads: general maintenance, preventive rehabilitation, minor rehabilitation, medium rehabilitation, major rehabilitation and reconstruction (GM, 1-R, 2-R, 3-R, 4-R and 5-R, respectively). The characteristics of these treatment options were discussed in Section 3. The decision trees for the unpaved and paved roads are illustrated in Figures 3 and 4, respectively.

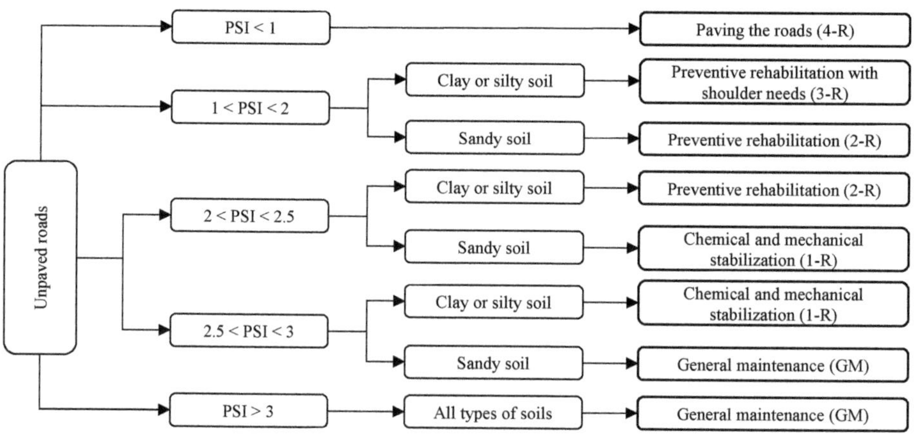

Figure 3. Decision tree for the unpaved roads.

Hence, it is evident that the proposed URMS and PMS take into account not only the cost factor but also local conditions, including the ORCI and the average daily traffic (ADT) data. The objective of the developed model was to maximize a function that considers three independent parameters: the road condition, ADT, and cost-factor, as indicated in Equations (3) and (4).

$$\text{Maximize} \sum_{i=1}^{n} \frac{ORCI \times ADT_i}{Cost - factor_{\ i}\ (treatment\ type)} \times (x_i) \quad (3)$$

$$\text{Subjected to} \sum_{i=1}^{n} TreatmentCost_{\ i} \times (x_i) \leq Budget\ , x_i \in \{0,1\} \quad (4)$$

where ADT_i expresses the average daily traffic for road i, Cost-Factor$_i$ is the function of the treatment type, x_i is an integer equal to one if the project is selected and zero if it is not selected and ORCI represents the minimum index among the distress indices. This is a combinatorial optimization problem where one selects a collection of projects of maximum value while satisfying some weight constraints.

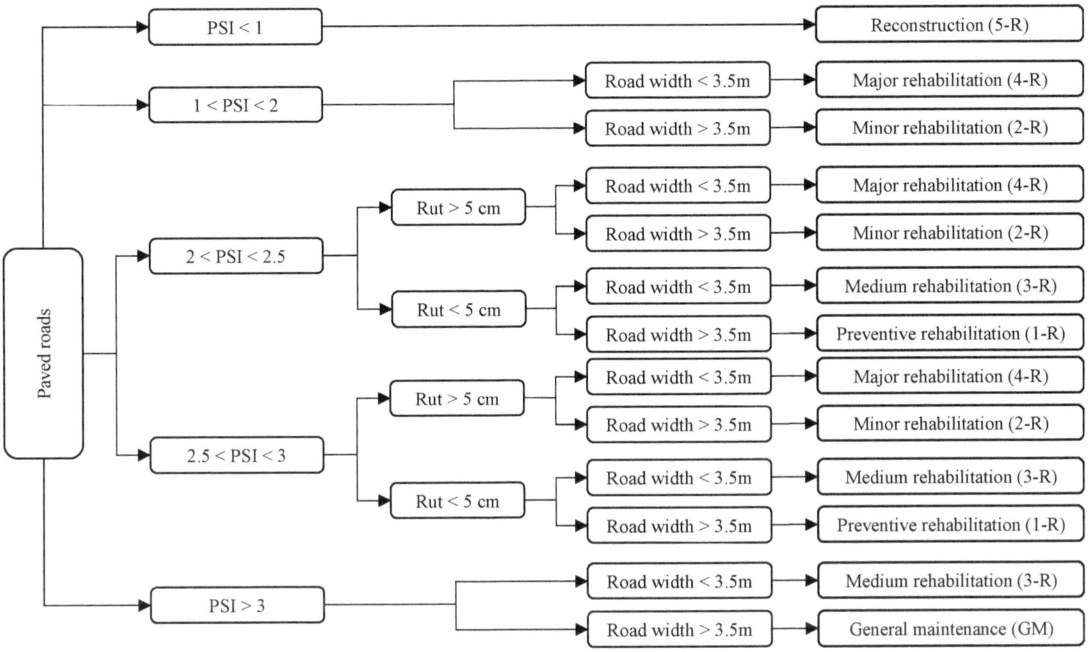

Figure 4. Decision tree for the paved roads.

3. Results and Discussion

The collected information for unpaved and paved roads is presented in the comprehensive database summarized in Tables 1 and 2, respectively. The information provided in these tables is also graphically represented in Figures 5 and 6, respectively. A detailed analysis of the traffic counts reveals significant patterns in vehicle usage. The unpaved road segments experienced traffic volumes of less than 400 vehicles per day, indicating lighter usage. In contrast, a notable portion of the paved road segments, approximately 20%, handle a heavier traffic load, with daily vehicle counts exceeding 400. The road widths of the segments varied significantly, ranging from 3 to 7 m. The collective length of the road segments was substantial, with the 18 unpaved segments and 88 paved segments together spanning a total distance of 15,236 km. This extensive network underscores the importance of understanding traffic patterns and road characteristics for the effective management of Brazilian roads.

Table 1. Dataset obtained for unpaved roads of MG.

Segment	1	2	3	4	5	6	7	8	9	10	11	12	13	14	15	16	17	18
PSI	1	1.8	2.81	2.51	1.5	0.88	1.63	0.89	1.8	1.03	1.19	3.06	1.22	1.63	0.9	0.93	1.22	3.24
Length (km)	1.94	0.91	0.62	0.65	1.02	0.13	0.27	1.52	0.57	1.18	1.00	2.65	0.26	0.56	0.13	0.90	0.60	1.30
Soil	Sand	Silt	Clay	Sand	Silt	Clay	Silt	Silt	Sand	Silt	Clay	Silt	Silt	Clay	Clay	Sand	Sand	Silt

The Markovian process enabled the determination of the following matrices: the transition probability data, transition probability matrix and probability distribution matrix. The results presented in Tables 3 and 4 represent the transition probability data obtained from the unpaved and paved roads datasets, respectively.

Table 2. Dataset obtained for paved roads of MG.

Segment	**1**	**2**	**3**	**4**	**5**	**6**	**7**	**8**	**9**	**10**	**11**	**12**	**13**	**14**	**15**	**16**	**17**	**18**
PSI	1	2	3	3	2	3	2	1	2	3	2	2	2	2	2	2	1	2
Length (km)	35	9	6.5	45	10	32	17	23	7	103	291	3	60	26	75	33	11	97
Width (m)	7.0	7.0	7.0	7.0	7.0	3.8	6.5	3.0	7.2	7.2	7.2	7.2	7.2	7.2	7.2	3.7	6.6	4.0
Rut (cm)	0.4	0.4	0.4	0.4	0.5	0.51	0.53	0.45	0.77	0.43	0.51	0.51	0.51	0.51	0.5	0.48	0.5	0.54
Segment	**19**	**20**	**21**	**22**	**23**	**24**	**25**	**26**	**27**	**28**	**29**	**30**	**31**	**32**	**33**	**34**	**35**	**36**
PSI	3	3	2	2	3	1	2	3	2	2	2	2	1	2	2	2	1	2
Length (km)	50	6.5	91	6.5	5	43	105	19	21	28	15	14	11	29	10	177	247	268
Width (m)	6.5	8.0	3.7	3.7	4.2	4.2	3.6	3.6	3.6	8.0	4.5	4.1	6.6	6.6	6.6	4.2	4.1	4.1
Rut (cm)	0.51	0.48	0.77	0.77	0.77	0.48	0.5	0.45	0.51	0.51	0.43	0.5	0.54	0.54	0.54	0.45	0.45	0.77
Segment	**37**	**38**	**39**	**40**	**41**	**42**	**43**	**44**	**45**	**46**	**47**	**48**	**49**	**50**	**51**	**52**	**53**	**54**
PSI	2	3	2	1	2	2	1	3	3	2	2	3	4	4	4	3	2	4
Length (km)	91	109	17	49	160	49	6	124	59	69	17	239	831	209	823	341	41	670
Width (m)	3.8	3.8	80	8.0.	8.0	8.0	8.0	3.8	4.4	3.4	3.3	3.8	3.8	5.5	5.5	5.0	5.0	3.7
Rut (cm)	0.45	0.5	0.76	0.77	0.25	0.43	0.4	0.48	0.25	0.76	0.77	0.56	0.57	0.53	0.54	0.53	0.5	0.4
Segment	**55**	**56**	**57**	**58**	**59**	**60**	**61**	**62**	**63**	**64**	**65**	**66**	**67**	**68**	**69**	**70**	**71**	**72**
PSI	4	4	4	3	3	3	2	3	2	3	3	2	4	4	3	2	4	3
Length (km)	338	246	11	547	311	890	417	457	77	268	73	202	154	875	324	137	946	140
Width (m)	5.2	5.2	5.2	5.2	5.0	5.0	5.0	5.0	5.0	5.0	5.0	5.0	5.0	5.0	5.0	5.0	5.0	5.0
Rut (cm)	0.55	0.51	0.4	0.4	0.57	0.58	0.57	0.81	0.54	0.58	0.86	0.86	0.79	0.58	0.84	0.99	0.76	0.84
Segment	**73**	**74**	**75**	**76**	**77**	**78**	**79**	**80**	**81**	**82**	**83**	**84**	**85**	**86**	**87**	**88**		
PSI	4	2	3	4	3	3	3	2	4	2	1	3	4	1	3	3		
Length (km)	48	179	30	27	78	212	79	4	25	151	142	256	198	135	86	15		
Width (m)	3.7	3.4	4.7	4.0	4.0	4.0	4.0	4.0	4.0	4.0	4.0	4.0	4.0	4.0	4.0	4.0		
Rut (cm)	0.55	0.49	0.45	0.25	0.77	0.77	0.25	0.25	0.25	0.5	0.76	0.48	0.48	0.58	0.55	0.58		

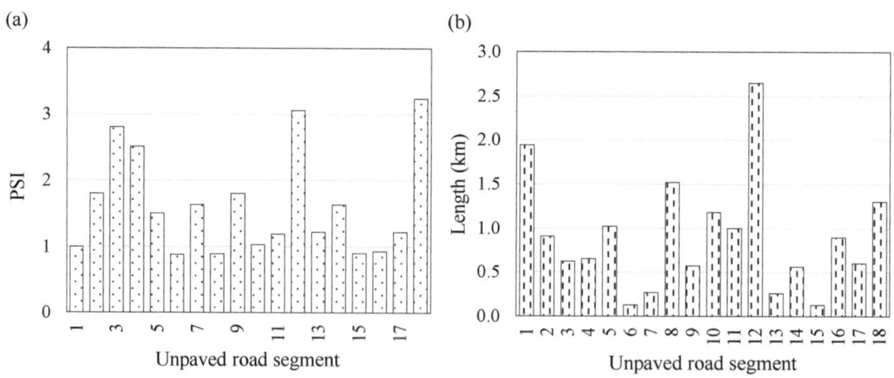

Figure 5. PSI (**a**) and length (**b**) values obtained for unpaved roads of MG.

Table 5 shows the transition probability matrix for the unpaved roads. A transition matrix of a Markovian process contains information on how the deterioration process moves from one state to another in terms of probabilities. According to the obtained data presented in Table 5, it was possible to observe that after 1 year, the percentage of roads that stay at level 9 was represented by the element 0.333. Similarly, the 0.481 element represents the percentage of roads that deteriorate from level 9 to level 8, and 0.185 represents the percentage of roads that deteriorate from level 9 to level 7.

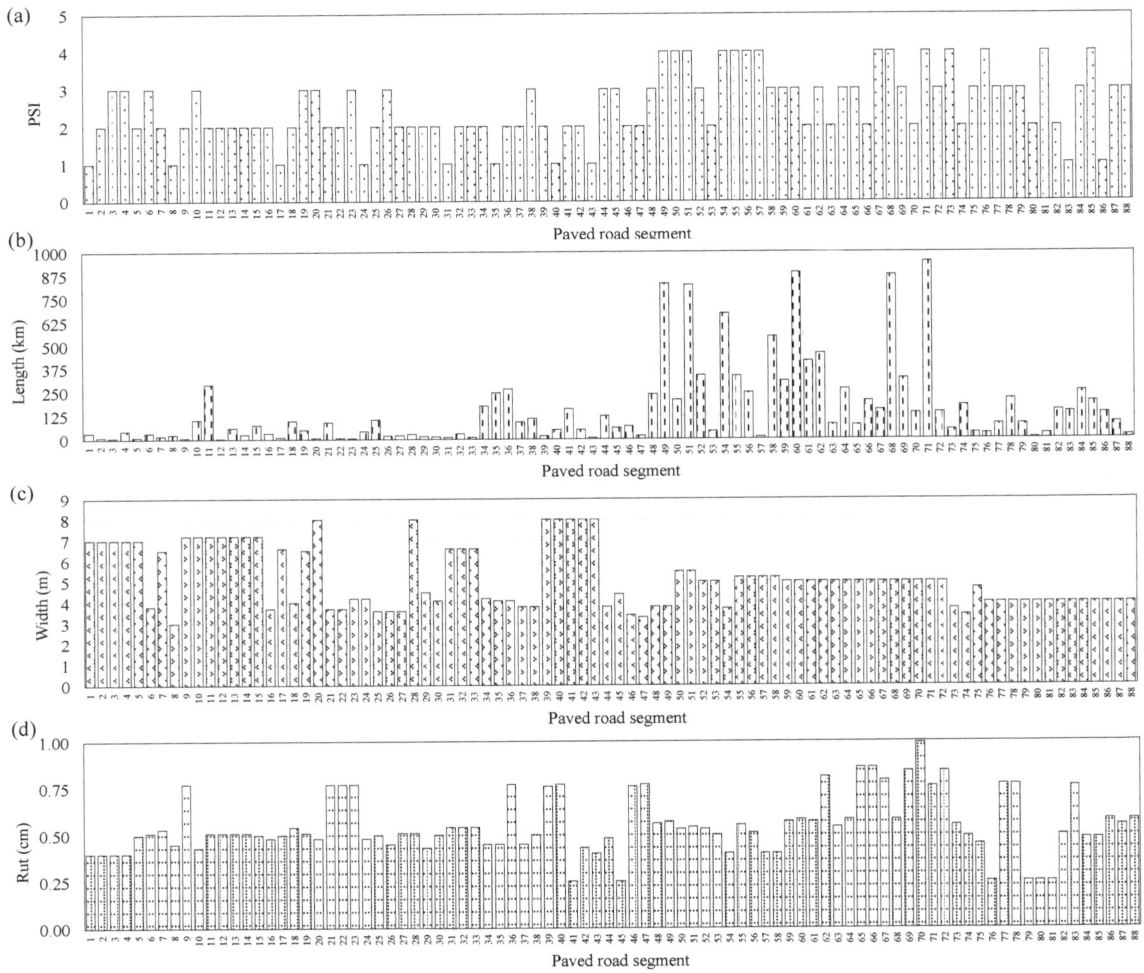

Figure 6. PSI (**a**), length (**b**), width (**c**) and rut (**d**) values obtained for paved roads of MG.

Table 3. Transition probability data obtained from the unpaved roads dataset.

RQRG	9	8	7	6	5	4	3	2	1
9	9	13	5	---	---	---	---	---	---
8	---	1	10	3	---	---	---	---	---
7	---	---	---	11	1	---	---	---	---
6	---	---	---	4	10	1	---	---	---
5	---	---	---	---	7	12	3	---	---
4	---	---	---	---	---	2	6	---	---
3	---	---	---	---	---	---	1	9	---
2	---	---	---	---	---	---	---	3	3
1	---	---	---	---	---	---	---	---	2

The deterioration models were derived from regression analyses applied to the mean data obtained from the Markovian distribution. Using these models, it was possible to determine the time taken to transition from an excellent condition to a failed condition. Equations (5) and (6) show the deterioration models obtained for the unpaved and paved roads, respectively. In these equations, URCI indicates the unpaved roads condition index, PCI indicates the pavement condition index and AGE is the age (in years).

Table 4. Transition probability data obtained from the paved roads dataset.

PCI	5	4	3	2	1
5	25	17	---	---	---
4	---	57	31	---	---
3	---	---	147	51	---
2	---	---	---	52	11
1	---	---	---	---	2

Table 5. Transition Probability Matrix from the unpaved roads dataset after 1 year.

RQRG	9	8	7	6	5	4	3	2	1
9	0.333	0.481	0.185	---	---	---	---	---	---
8	---	0.071	0.714	0.214	---	---	---	---	---
7	---	---	---	0.917	0.083	---	---	---	---
6	---	---	---	0.267	0.667	0.067	---	---	---
5	---	---	---	---	0.318	0.545	0.136	---	---
4	---	---	---	---	---	0.250	0.750	---	---
3	---	---	---	---	---	---	0.100	0.900	---
2	---	---	---	---	---	---	---	0.500	0.500
1	---	---	---	---	---	---	---	---	1.000

Table 6 represents the transition probability matrix for paved roads of MG. These probabilities are assumed to be constant, i.e., the transition matrix is stationary. According to the data reported in Table 6, it can be observed that the number 0.595 represents the percentage of roads that stay at level five after 1 year. Similarly, the element 0.405 represents the percentage of roads that deteriorate from level five to level four.

Table 6. Transition Probability Matrix from the paved roads dataset after 1 year.

PCI	5	4	3	2	1
5	0.595	0.405	---	---	---
4	---	0.648	0.352	---	---
3	---	---	0.742	0.258	---
2	---	---	---	0.825	0.175
1	---	---	---	---	1.000

The probability distribution derived from the application of the Markovian probability process for unpaved roads is presented in Table 7. In this table, states of maximum probability were highlighted in red. For example, comparing the obtained data from year 1 to year 5, there was a probability of 0.004 for the unpaved road segments to stay at level nine and a probability of 0.037 for these segments to be at level two after 5 years. In year 5, the highest probability value (0.327) indicated the unpaved road segments of MG at level five.

Table 8 represents the probability distribution derived from the application of the Markovian probability process to the paved road dataset. Again, states of maximum probability are shown in red. Comparing the distribution from year 1 to year 5, there was a probability of 0.075 for the paved road segments to stay at level five and a probability of 0.024 for the segments to be at level one. However, the highest probability value in year 5 was 0.415, which indicated paved roads at level three.

Table 7. Probability distribution obtained from the unpaved roads dataset using the Markov process.

RQRG		9	8	7	6	5	4	3	2	1	Total
Year	1	0.333	0.481	0.185	0	0	0	0	0	0	1
	2	0.111	0.194	0.451	0.230	0.011	0	0	0	0	1
	3	0.037	0.067	0.272	0.413	0.185	0.021	0.001	0	0	1
	4	0.012	0.023	0.123	0.312	0.351	0.134	0.041	0.001	0	1
	5	0.004	0.007	0.049	0.173	0.327	0.246	0.152	0.037	0	1

Table 8. Probability distribution obtained from the paved roads dataset using the Markov process.

PCI		5	4	3	2	1	Total
Year	1	0.595	0.405	0	0	0	1
	2	0.354	0.503	0.143	0	0	1
	3	0.211	0.469	0.283	0.037	0	1
	4	0.126	0.389	0.375	0.104	0.006	1
	5	0.075	0.303	0.415	0.183	0.024	1

$$URCI = 0.035 \times (AGE)^2 - 1.082 \times (AGE) + 9.107 \quad (5)$$

$$PCI = 0.005 \times (AGE)^2 - 0.268 \times (AGE) + 4.463 \quad (6)$$

These results suggest that unpaved roads tended to deteriorate more rapidly over time. The good alignment between the experimental findings derived from the methodology outlined in this paper and the results of previous studies underscores the robustness of the proposed methodology. According to previous literature, the faster deterioration of unpaved roads is affected by permeability, materials, construction, traffic, environment and drainage factors. Among them, the most important factor is the permeability and weather [33]. As a result of these factors, tropical roads experience faster deterioration, and unpaved road surfaces are more permeable compared to paved surfaces.

Details related to the five treatment options proposed in this methodology for the unpaved roads (GM, 1-R, 2-R, 3-R and 4-R) are summarized in Table 9, while Table 10 shows the six possible treatment options (GM, 1-R, 2-R, 3-R, 4-R and 5-R) proposed for the paved roads. The estimated costs were provided in BRL because it is the currency used in MG, thus reflecting the local economic context.

Table 9. Treatment types and costs for unpaved roads.

Treatment Type	Details and Applications	Estimated Cost per km	
		BRL	USD
GM (General maintenance)	Biannual maintenance procedures	600.00	110.62
1-R (Chemical and mechanical stabilization)	Conventional stabilizing agents	20,000.00	3687.32
2-R (Preventive rehabilitation)	DNIT recommendations	3180.00	586.28
3-R (Preventive rehabilitation with shoulder needs)	DNIT recommendations	71,387.00	13,161.32
4-R (paving the roads)	Addition of pavement	400,000.00	73,746.31

Note: All amounts in BRL have been converted to USD based on the exchange rate from 19 June 2024, where USD 1 = BRL 5.4240.

In Table 11 and Figure 7, the results provided by the optimization algorithm for unpaved roads are organized into four distinct scenarios, as follows: (i) conditions under current circumstances (referred to as the "current condition"), (ii) conditions following four cycles of 3 months without any maintenance (termed "do nothing"), (iii) conditions after applying treatments without budgetary constraints (designated as "no budget constraints") and (iv) conditions after applying treatments necessary for maintaining the road's condition within a limited budget (referred to as "limited budget"). The rationale behind the selection of a 3-month period stems from the observed tendency for unpaved roads to require the next level of maintenance every 3 months [24].

Table 10. Details of treatment types proposed in this methodology and costs for paved roads.

Treatment Types	Details and Application	Estimated Cost per km	
		BRL	USD
GM (general maintenance)	Asphalt patching Pothole repair Crack sealing Road striping	1875.00	345.69
1-R (preventive rehabilitation)	Chip seal Micro-surface Thin overlay (lower than 5 cm)	51,800.00	9550.15
2-R (minor rehabilitation)	Preparation of the surface Seal coat and thick overlay (higher than 5 cm) 1-R plus shoulder or widening requirements	100,000.00	18,436.58
3-R (medium rehabilitation)	Applied to good-condition roads with shoulder needs 2-R plus shoulder or widening requirements	150,000.00	27,654.87
4-R (major rehabilitation)	Applied to narrow roads with shoulder or widening needs	200,000.00	36,873.16
5- R (reconstruction)	Reconstruction of the road	300,000.00	55,309.73

Note: All amounts in BRL have been converted to USD based on the exchange rate from 19 June 2024, where USD 1 = BRL 5.421299.

Table 11. Unpaved roads treatment decisions obtained with the optimization algorithm.

Segment	Current Condition				Do Nothing		No Budget Constraints		Limited Budget	
	PSI	Length L (km)	Soil Type	PSI × L (km)	PSI	PSI × L (km)	PSI	PSI × L (km)	PSI	PSI × L (km)
1	1.00	1.94	Sand	1.940	0.92	1.783	4.10	7.954	0.92	1.783
2	1.80	0.91	Silt	1.638	1.72	1.564	4.00	3.640	1.72	1.564
3	2.81	0.62	Clay	1.742	2.73	1.692	3.90	2.418	3.90	2.418
4	2.51	0.65	Sand	1.632	2.43	1.578	2.51	1.632	2.51	1.632
5	1.50	1.02	Silt	1.530	1.42	1.446	4.00	4.080	1.42	1.446
6	0.88	0.13	Clay	0.114	0.80	0.104	4.10	0.533	0.80	0.104
7	1.63	0.27	Silt	0.440	1.55	0.418	4.00	1.080	1.55	0.418
8	0.89	1.52	Silt	1.353	0.81	1.228	4.10	6.232	0.81	1.228
9	1.80	0.57	Sand	1.026	1.72	0.979	4.00	2.280	1.72	0.979
10	1.03	1.18	Silt	1.215	0.95	1.119	4.00	4.720	0.95	1.119
11	1.19	1.00	Clay	1.190	1.11	1.108	4.00	4.000	1.11	1.108
12	3.06	2.65	Silt	8.109	2.98	7.892	3.06	8.109	3.06	8.109
13	1.22	0.26	Silt	0.317	1.14	0.296	4.00	1.040	4.00	1.040
14	1.63	0.56	Clay	0.913	1.55	0.867	4.00	2.240	1.55	0.867
15	0.90	0.13	Clay	0.117	0.82	0.106	4.10	0.533	0.82	0.106
16	0.93	0.90	Sand	0.837	0.85	0.763	4.10	3.690	0.85	0.763
17	1.22	0.60	Sand	0.732	1.14	0.683	4.00	2.400	1.14	0.683
18	2.34	1.30	Silt	3.042	0.92	1.190	4.00	5.200	4.00	5.200

Table 12 and Figure 8 present similar results for four different scenarios considered in the optimization algorithm for the paved roads. In this case, the "do nothing" scenario reflects conditions after 12 months without any maintenance. The reason for this 12-month period is that the typical lifespan of paved roads without maintenance is longer than that of unpaved roads [24].

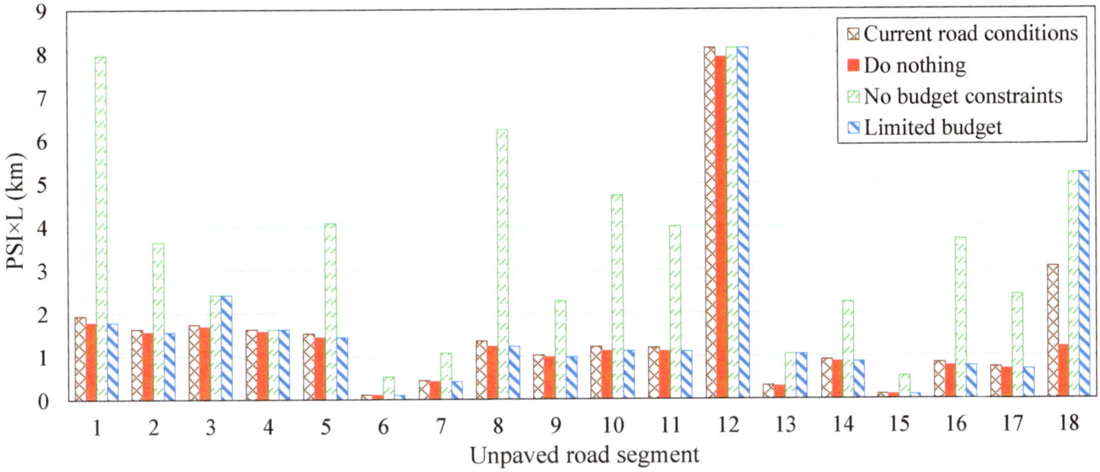

Figure 7. Unpaved roads treatment decisions obtained with the optimization algorithm.

Table 12. Paved roads treatment decisions obtained with the optimization algorithm.

Paved Road	Current Condition					Do Nothing		No Budget Constraints		Limited Budget	
	PSI	Length L (km)	Width (m)	Rut (cm)	PSI × L (km)	PSI	PSI × L (km)	PSI	PSI × L (km)	PSI	PSI × L (km)
LMG-633	1	35	7.0	0.40	35	0.96	33.4	4.30	150.5	0.96	33.4
LMG-820	2	9	7.0	0.40	18	1.86	16.7	4.00	36.0	4.00	36.0
MG-050	3	6.5	7.0	0.40	19.5	2.80	18.2	3.90	25.4	2.80	18.2
MG-123	3	45	7.0	0.40	135	2.80	126.0	3.90	175.5	2.80	126.0
MG-123/BR-262	2	10	7.0	0.50	20	1.86	18.6	4.00	40.0	4.00	40.0
MG-129	3	32	3.8	0.51	96	2.80	89.6	4.00	128.0	2.80	89.6
MG-129/BR-120	2	17	6.5	0.53	34	1.86	31.6	4.00	68.0	1.86	31.6
MG-164	1	23	3.0	0.45	23	0.95	21.9	4.30	98.9	0.95	21.9
BR-460	3	79	4.0	0.25	237	2.80	221.2	3.90	308.1	2.80	221.2
BR-462	2	4	4.0	0.25	8	1.86	7.4	4.00	16.0	1.86	7.4
BR-464	4	25	4.0	0.25	100	3.75	93.8	4.00	100.0	4.00	100.0
BR-474	2	151	4.0	0.50	302	1.86	280.9	4.00	604.0	1.86	280.2
BR-482	1	142	4.0	0.76	142	0.95	134.9	4.30	610.6	0.96	135.6
BR-491	3	256	4.0	0.48	768	2.80	716.8	3.90	998.4	2.80	716.8
BR-494	4	198	4.0	0.48	792	3.75	742.5	4.00	792.0	4.00	792.0
BR-496	1	135	4.0	0.58	135	0.95	128.3	4.30	580.5	0.95	128.3
BR-497	3	86	4.0	0.55	258	2.80	240.8	4.00	344.0	2.80	240.8
BR-499	3	15	4.0	0.58	45	2.80	42.0	4.00	60.0	2.80	42.0

Comparisons between the maintenance costs of unpaved and paved roads indicated that the full rehabilitation costs of county paved roads were approximately 10 times higher per kilometer compared to those for unpaved roads, considering the difference in maintenance treatment frequency, which is approximately four times higher for unpaved roads (every 3 months) compared to that for paved roads (every 12 months). The models accurately depicted the pattern where major treatments result in a minimal need for further treatments over the following several years in both situations.

It is important to highlight that other countries may have different optimization priorities. For instance, engineers might aim to maximize the overall network road conditions given an unrestricted budget. In the present study, the options for maximizing network road conditions and minimizing costs were thoroughly examined to showcase the potential of the proposed methodology.

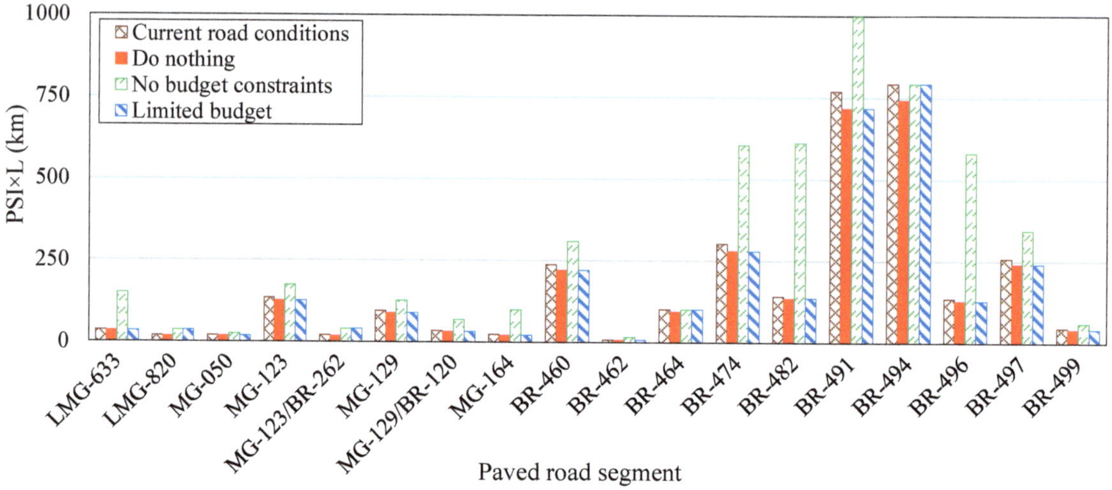

Figure 8. Paved roads treatment decisions obtained with the optimization algorithm.

4. Conclusions

In the present research, optimization techniques for the management of unpaved and paved roads of MG were applied, enabling the evaluation of the best combination of projects for road preservation, considering different constraints. The following conclusions were derived from the present paper:

- Based on the transition probability data, transition probability matrices and probability distribution matrices, it was possible to quantify the probability of unpaved and paved roads being in different condition levels after various periods.
- The Markovian probability distribution matrix revealed that an MG unpaved road that is initially at the highest condition level (nine) would have a higher probability of being at level eight after 1 year (48.1%) and a higher probability of being at level five after 5 years (32.7%). Similarly, a paved road that is initially at the highest condition level (5) would be more likely to remain at level 5 (59.5%) after 1 year and more likely to remain at level 3 (41.5%) after 5 years.
- The regression analysis models developed in this research enabled the determination of the time required for unpaved and paved roads of MG to go from being in excellent condition to being in a failed condition, based on the age of the roads.
- The results of the optimization algorithm for unpaved and paved roads were organized into four distinct scenarios: (i) conditions under the current circumstances (referred to as "current condition"), (ii) conditions without any maintenance ("do nothing"), (iii) conditions after applying treatments without budget constraints ("no budget constraints") and (iv) conditions after applying treatments required to maintain the road condition within a limited budget ("limited budget"). These results showed that the best results were achieved without budget constraints.
- Different preservation projects were evaluated in order to reach the maximization of the overall condition of paved and unpaved roads at minimal costs. Comparisons between the maintenance costs of unpaved and paved roads suggested that the full rehabilitation costs of paved roads were around 10 times higher per kilometer compared

to those of unpaved roads, considering the difference in the maintenance treatment frequency, which is approximately four times higher for unpaved roads (every 3 months) compared to that for paved roads (every 12 months).

- The models effectively captured the trend wherein a major treatment leads to minor additional treatments being needed for the subsequent several years in both scenarios.

The findings of the present study provide future directions for the optimized allocation of resources for the management of roads in MG. Further research is recommended to investigate the implementation of the proposed methodology in other regions and identify minor changes in some procedures that may be needed to reflect distinct local conditions.

Author Contributions: Conceptualization, T.O.d.S., H.N.P. and L.C.P.R.; methodology, T.O.d.S., H.N.P., L.C.P.R., J.C.L.R. and K.K.; software, J.C.L.R. and K.K.; validation, E.C.L., G.H.N. and K.H.d.P.R.; investigation, L.C.P.R., J.C.L.R. and K.K.; resources, T.O.d.S. and H.N.P.; data curation, L.C.P.R.; writing—original draft preparation, L.C.P.R.; writing—review and editing, T.O.d.S., H.N.P., E.C.L., G.H.N. and K.H.d.P.R.; visualization, E.C.L., G.H.N. and K.H.d.P.R.; supervision, T.O.d.S., H.N.P., J.C.L.R. and K.K.; project administration, T.O.d.S.; funding acquisition, T.O.d.S. and H.N.P. All authors have read and agreed to the published version of the manuscript.

Funding: This research received no external funding.

Data Availability Statement: The data are contained within the article.

Acknowledgments: The authors wish to express their gratitude for the support from the Civil Engineering Department at Universidade Federal de Viçosa, as well as for the scholarship provided by the National Council for Scientific and Technological Development—CNPq.

Conflicts of Interest: The authors declare no conflicts of interest.

References

1. CNT. *Transport Yearbook 2022*; National Transport Confederation: Brasília, Brazil, 2022.
2. Babashamsi, P.; Md Yusoff, N.I.; Ceylan, H.; Md Nor, N.G.; Salarzadeh Jenatabadi, H. Evaluation of Pavement Life Cycle Cost Analysis: Review and Analysis. *Int. J. Pavement Res. Technol.* **2016**, *9*, 241–254. [CrossRef]
3. Bonin, G.; Polizzotti, S.; Folino, N.; Rossi, C.; Teltayev, B. Development of a Road Asset Management System in Kazakhstan. In *Transport Infrastructure and Systems*; Taylor & Francis Group: London, UK, 2017; Volume 1, pp. 537–547.
4. Chen, W.; Zheng, M. Multi-Objective Optimization for Pavement Maintenance and Rehabilitation Decision-Making: A Critical Review and Future Directions. *Autom. Constr.* **2021**, *130*, 103840. [CrossRef]
5. Khahro, S. Defects in Flexible Pavements: A Relationship Assessment of the Defects of a Low-Cost Pavement Management System. *Sustainability* **2022**, *14*, 16475. [CrossRef]
6. Ngene, B.U.; Bassey, D.E.; Busari, A.A.; Bamigboye, G.O.; Nworgu, A.T. Influence of GIS on Sustainable Pavement Maintenance: A Comparative Review. *IOP Conf. Ser. Mater. Sci. Eng.* **2021**, *1036*, 012039. [CrossRef]
7. Elhadidy, A.A.; Elbeltagi, E.E.; El-Badawy, S.M. Network-Based Optimization System for Pavement Maintenance Using a Probabilistic Simulation-Based Genetic Algorithm Approach. *J. Transp. Eng. Part. B Pavements* **2020**, *146*, 04020069. [CrossRef]
8. Saha, P.; Ksaibati, K.; Atadero, R. Developing Pavement Distress Deterioration Models for Pavement Management System Using Markovian Probabilistic Process. *Adv. Civ. Eng.* **2017**, *2017*, 8292056. [CrossRef]
9. Brower, A. Six Types of Road Maintenance. *Milling Tools*, 31 May 2018.
10. Al Shehhi, A.H.O.; Jokhio, G.A. The Environmental and Economic Impacts of the Use of Recycled Asphalt During the Preventive Maintenance of Roadways in the UAE. In *BUiD Doctoral Research Conference 2023*; Springer: Berlin/Heidelberg, Germany, 2024; pp. 162–171.
11. Rohith, D.; Srikanth, K.; Chowdary, V.; Prasad, C. Performance Evaluation of Gravel Road Sections Sealed Using Open Graded Premix Surfacing with Bitumen and Bitumen Emulsion as the Binders. *Transp. Dev. Econ.* **2016**, *2*, 13. [CrossRef]
12. Mbiyana, K.; Kans, M.; Campos, J.; Håkansson, L. Literature Review on Gravel Road Maintenance: Current State and Directions for Future Research. *Transp. Res. Rec. J. Transp. Res. Board* **2023**, *2677*, 506–522. [CrossRef]
13. Saeed, N.; Dougherty, M.; Nyberg, R.G.; Rebreyend, P.; Jomaa, D. A Review of Intelligent Methods for Unpaved Roads Condition Assessment. In Proceedings of the 2020 15th IEEE Conference on Industrial Electronics and Applications (ICIEA), Kristiansand, Norway, 9 November 2020; pp. 79–84.
14. Žuraulis, V.; Sivilevičius, H.; Šabanovič, E.; Ivanov, V.; Skrickij, V. Variability of Gravel Pavement Roughness: An Analysis of the Impact on Vehicle Dynamic Response and Driving Comfort. *Appl. Sci.* **2021**, *11*, 7582. [CrossRef]
15. Huntington, G.; Ksaibati, K. Visual Assessment System for Rating Unsealed Roads. *Transp. Res. Rec. J. Transp. Res. Board* **2015**, *2474*, 116–122. [CrossRef]
16. WYDOT. *Transportation Asset Management Plan*; WYDOT: Cheyenne, WY, USA, 2014.

17. Shrestha, K.K.; Shrestha, P.P.; Kandie, T.K. A Road Maintenance Management Tool for Rural Roads in Kenya. In Proceedings of the Construction Research Congress 2014; American Society of Civil Engineers: Reston, VA, USA, 13 May, 2014; pp. 289–298.
18. Santos, J.; Ferreira, A.; Flintsch, G. A Multi-Objective Optimization-Based Pavement Management Decision-Support System for Enhancing Pavement Sustainability. *J. Clean. Prod.* **2017**, *164*, 1380–1393. [CrossRef]
19. Asghari, V.; Hsu, S.-C.; Wei, H.-H. Expediting Life Cycle Cost Analysis of Infrastructure Assets under Multiple Uncertainties by Deep Neural Networks. *J. Manag. Eng.* **2021**, *37*, 04021059. [CrossRef]
20. Omidian, P.; Khaji, N. A Total Life-Cycle Cost–Resilience Optimization Framework for Infrastructures Management Using Different Retrofit Strategies. *Sustain. Resilient Infrastruct.* **2023**, *8*, 675–698. [CrossRef]
21. Harrou, F.; Zeroual, A.; Hittawe, M.; Sun, Y. *Road Traffic Modeling and Management*; Elsevier: Amsterdam, The Netherlands, 2022; ISBN 9780128234327.
22. Harrou, F.; Zeroual, A.; Hittawe, M.M.; Sun, Y. Traffic Congestion Detection: Data-Based Techniques. In *Road Traffic Modeling and Management*; Elsevier: Amsterdam, The Netherlands, 2022; pp. 141–195.
23. Harrou, F.; Zeroual, A.; Hittawe, M.M.; Sun, Y. Recurrent and Convolutional Neural Networks for Traffic Management. In *Road Traffic Modeling and Management*; Elsevier: Amsterdam, The Netherlands, 2022; pp. 197–246.
24. Saha, P.; Ksaibati, K. Developing an Optimization Model to Manage Unpaved Roads. *J. Adv. Transp.* **2017**, *2017*, 9474838. [CrossRef]
25. Ruuska, S.; Miettinen, K.; Wiecek, M.M. Connections Between Single-Level and Bilevel Multiobjective Optimization. *J. Optim. Theory Appl.* **2012**, *153*, 60–74. [CrossRef]
26. Wolters, A.; Zimmerman, K.; Schattler, K. *Implementing Pavement Management Systems for Local Agencies: Implementation Guide*; University of Illinois Urbana-Champaign: Champaign, IL, USA, 2011.
27. Sánchez Soliño, A. Optimizing Performance-Based Mechanisms in Road Management: An Agency Theory Approach. *Eur. J. Transp. Infrastruct. Res.* **2015**, *15*, 465–481. [CrossRef]
28. Hamdi; Hadiwardoyo, S.P.; Correia, A.G.; Pereira, P. New Optimization Strategies of Pavement Maintenance: A Case Study for National Road Network in Indonesia Using Integrated Road Management System. *AIP Conf. Proc.* **2017**, *1855*, 040015.
29. Kazama, V.S.; Corte, A.P.D.; Robert, R.C.G.; Sanquetta, C.R.; Arce, J.E.; Oliveira-Nascimento, K.A.; DeArmond, D. Global Review on Forest Road Optimization Planning: Support for Sustainable Forest Management in Amazonia. *Ecol. Manag.* **2021**, *492*, 119159. [CrossRef]
30. Shon, H.; Cho, C.-S.; Byon, Y.-J.; Lee, J. Autonomous Condition Monitoring-Based Pavement Management System. *Autom. Constr.* **2022**, *138*, 104222. [CrossRef]
31. Dong, J.; Meng, W.; Liu, Y.; Ti, J. A Framework of Pavement Management System Based on IoT and Big Data. *Adv. Eng. Inform.* **2021**, *47*, 101226. [CrossRef]
32. Pulugurta, H.; Shao, Q.; Chou, Y.J. Pavement Condition Prediction Using Markov Process. *J. Stat. Manag. Syst.* **2009**, *12*, 853–871. [CrossRef]
33. El-Maaty, A. Temperature Change Implications for Flexible Pavement Performance and Life. *Int. J. Transp. Eng. Technol.* **2017**, *3*, 1. [CrossRef]
34. Wu, S.; Muhunthan, B. A Mechanistic-Empirical Model for Predicting Top-down Fatigue Cracking in an Asphalt Pavement Overlay. *Road Mater. Pavement Des.* **2019**, *20*, 1322–1353. [CrossRef]
35. da Silva, T.O.; Pitanga, H.N.; Silva, N.A.B.; Cardoso, H.S.; Lopes, E.C.; de Paula Rodrigues, K.H.; Nalon, G.H.; Borges, P.R. Effects of Short-Term Aging on the Mechanical and Structural Performance of Hot Mix Asphalt Mixtures: A Case Study Complemented by Statistical Analysis. *Road Mater. Pavement Des.* **2024**, *25*, 1–14. [CrossRef]
36. Gkyrtis, K.; Plati, C.; Loizos, A. Mechanistic Analysis of Asphalt Pavements in Support of Pavement Preservation Decision-Making. *Infrastructures* **2022**, *7*, 61. [CrossRef]
37. Amin, S.R. The Pavement Performance Modeling: Deterministic vs. Stochastic Approaches. In *Numerical Methods for Reliability and Safety Assessment*; Springer International Publishing: Cham, Switzerland, 2015; pp. 179–196.
38. Leandro, R.; Beja, I.; Bessa, I.; Andrade, L.; Vasconcelos, K.; Bernucci, L.; Bittencourt, T.; Bortes, A.; Neto, A.; Florêncio, H.; et al. Development of the Asphalt Pavement Deterioration Model Using Instrumentation and Weigh in Motion System. *Rev. ANTT* **2015**, *2*, 1–12. (In Portuguese)
39. Rejani, V.U.; Janani, L.; Venkateswaralu, K.; Sunitha, V.; Mathew, S. Strategic Pavement Maintenance and Rehabilitation Analysis of Urban Road Network Using HDM-4. *Int. J. Pavement Res. Technol.* **2023**, *16*, 927–942. [CrossRef]
40. Gong, C.; Cheng, M.; Peng, Y.; Ding, W. Seepage Propagation Simulation of a Tunnel Gasketed Joint Using the Cohesive Zone Model. *Tunn. Undergr. Space Technol.* **2024**, *147*, 105726. [CrossRef]
41. Gong, C.; Xie, C.; Zhu, H.; Ding, W.; Song, J.; Ge, Y. Time-Varying Compressive Properties and Constitutive Model of EPDM Rubber Materials for Tunnel Gasketed Joint. *Constr. Build. Mater.* **2024**, *433*, 136734. [CrossRef]
42. Salomão, P.E.A.; Santos, J.A.G.; Ferreira, R.d.S.; Gonçalves, B.B.; de Carvalho, P.H.V.; Starich, R. Impactos Ambientais Gerados Pela Construção e Operação de Rodovias. *Res. Soc. Dev.* **2019**, *8*, e278101368. [CrossRef]
43. Ksaibati, K.; Huntington, G.; Evans, B.; Jones, J.; Harman, M.; Hellrung, D.; Meena, E. Mitigating the Impact of Oil and Gas Traffic on Southeastern Wyoming County Roads. In *T2 Roads on the Range*; Department of Civil and Architectural Engineering, University of Wyoming: Laramie, WY, USA, 2012.
44. AASHTO. *Pavement Management Guide*; AASHTO: Washington, DC, USA, 2001.

45. Ortiz-García, J.J.; Costello, S.B.; Snaith, M.S. Derivation of Transition Probability Matrices for Pavement Deterioration Modeling. *J. Transp. Eng.* **2006**, *132*, 141–161. [CrossRef]
46. Stewart, W. *Probability, Markovian Chains, Queues, and Simulation: The Mathematical Basis of Performance Modeling*; Princeton University Press: Princeton, NJ, USA, 2009.
47. Yang, J. Road Crack Condition Performance Modeling Using Recurrent Markovian Chains and Artificial Neural Networks. Master's Thesis, University of South Florida: Tampa, FL, USA, 2004.

logistics

MDPI

Review

Comprehensive Review of Robotized Freight Packing

German Pantoja-Benavides [1], Daniel Giraldo [1], Ana Montes [1], Andrea García [2], Carlos Rodríguez [1,*], César Marín [2] and David Álvarez-Martínez [1]

[1] School of Engineering, University of los Andes, Bogota 111711, Colombia; gf.pantoja10@uniandes.edu.co (G.P.-B.); ds.giraldoh@uniandes.edu.co (D.G.); a.montesf@uniandes.edu.co (A.M.); d.alvarezm@uniandes.edu.co (D.Á.-M.)

[2] Integra S.A., Pereira 660003, Colombia; vgarcia@integra.com.co (A.G.); cmarin@integra.com.co (C.M.)

* Correspondence: crodrigu@uniandes.edu.co

Abstract: *Background*: This review addresses the emerging field of automated packing cells, which lies at the intersection of robotics and packing problems. Integrating these two fields is critical for optimizing logistics and e-commerce operations. The current literature focuses on packing problems or specific robotic applications without addressing their integration. *Methods*: To bridge this gap, we conducted a comprehensive review of 46 relevant studies, analyzing various dimensions, including the components of robotic packing cells, the types of packing problems, the solution approaches, and performance comparisons. *Results*: Our review reveals a significant trend towards addressing online packing problems, which reflects the dynamic nature of logistics operations where item information is often incomplete. We also identify several research gaps, such as the need for standardized terminologies, comprehensive methodologies, and the consideration of real-world constraints in robotic algorithms. *Conclusions*: This review uniquely integrates insights from robotics and packing problems, providing a structured framework for future research. It highlights the importance of considering practical robotic constraints. It proposes a research structure that enhances the reproducibility and comparability of results in real-world scenarios. By doing so, we aim to guide future research efforts and facilitate the development of more robust and practical automated packing systems.

Keywords: automated systems; load management; logistics; materials handling; palletizing

Citation: Pantoja-Benavides, G.; Giraldo, D.; Montes, A.; García, A.; Rodríguez, C.; Marín, C.; Álvarez-Martínez, D. Comprehensive Review of Robotized Freight Packing. *Logistics* **2024**, *8*, 69. https://doi.org/10.3390/logistics8030069

Academic Editors: Tomasz Nowakowski, Artur Kierzkowski, Agnieszka A. Tubis, Franciszek Restel, Tomasz Kisiel, Anna Jodejko-Pietruczuk and Mateusz Zając

Received: 27 May 2024
Revised: 11 June 2024
Accepted: 18 June 2024
Published: 8 July 2024

1. Introduction

The fourth industrial revolution, characterized by the convergence of cutting-edge technologies, has ushered in a new era of logistics and packing processes. Logistics 4.0 has introduced innovative technologies such as automation, big data analytics, twin simulation, human-robot interaction, and autonomous vehicles. Within this context, 3D robotic packing has emerged as a cutting-edge solution, revolutionizing traditional packing methodologies and presenting novel opportunities for the industry [1].

In the dynamic world of e-commerce and logistics, efficient packing processes are crucial to optimizing shipping and storage operations. As the global e-commerce market continues to soar, with an increasing number of consumers opting for online shopping, the demand for efficient packing operations has never been more critical. E-commerce companies face immense pressure to deliver products to customers quickly and accurately, making 3D robotic packing an invaluable asset in this competitive landscape.

While human operators have traditionally handled packing processes, the rise of robotic technology has opened new horizons for automation in the packing domain. Robotic and human packing have several advantages and disadvantages that are mentioned below:

Working time: Robots can work uninterrupted for longer than humans, increasing productivity. While operators have fixed schedules, need resting time, and may get sick, robots can work 24 h a day, seven days a week, and have scheduled maintenance.

Costs: Operators have costs associated with their salaries and health insurance [2]; meanwhile, robots have costs related to their acquisition and maintenance. The cheaper option is case-dependent, and a financial analysis may be needed.

Space: Robots generally require more space to operate than human operators. Large space warehouses imply more (variable and fixed) costs.

Flexibility: Although robots' flexibility has remarkably improved in the last years, human operators are still more flexible considering gripping and possible positions of the items within the container, mainly when irregular items are considered.

Integrality of goods: Robots are less likely to make mistakes than human operators, which can reduce the risk of damaged goods. Additionally, the programmability of robots enables the incorporation of regulatory constraints to ensure compliance with safety guidelines and standards. For example, robots can be programmed to avoid stacking poisons on top of food items [3].

The motivation for conducting this study stems from two key elements. First, automated packing cells emerge from the intersection of two fields: robotics and packing problems. No review specifically addresses the integration of packing and robotics in the literature. Existing reviews are focused either on packing problems [4,5] or on specific robotic applications, such as in the food industry [6–8] and leather handling [9]. Second, the dynamic nature of logistics and e-commerce necessitates exploring how robotic packing systems can adapt to the challenges within e-commerce operations. Therefore, this study aims to examine and organize the developments in robotic packing systems.

This study offers the following contributions. First, it is the first review to integrate robotics and packing problems by exploring robotic packing cells. Second, this study identifies research opportunities within this integration, providing guidance for future research in the field. Third, it proposes a research structure that will facilitate the comparison of proposed methodologies, enhancing reproducibility and the practical application of appealing methods in real-world scenarios.

Therefore, this manuscript aims to fill this research gap to indicate the advances, challenges, and prospects related to 3D packing with robots. Moreover, since this integration has boomed in recent years, publications related to this topic need a standard structure that allows researchers to compare their methodologies properly. For this reason, this manuscript proposes a structure for researching robotic packing.

This paper is organized as follows. Section 2 states the scope of this review and describes the methodological procedure used to compile the studies considered in this review. Section 3 provides a description of the components of robotic packing cells. Section 4 presents the classification and constraints of the packing problems. Section 5 describes the solution methods for solving the packing problems. Section 6 indicates how studies tend to compare the performance of the proposed approaches. Section 7 shows research gaps and possible directions for closing them. Finally, Section 8 presents the conclusions.

2. Scope and Methodology

Several research lines in automated activities related to handling 3D items with robots exist. The scope of this review exclusively focuses on the research line concerning the packing of 3D items within containers using robots. This scope includes different grades of human/robot interaction, i.e., collaborative robots (cobots) are considered. The research lines not covered in this review are picking (or order picking), rearrangement, and assembly.

The review methodology to establish the state of the art of 3D robotic packing was based on the ones proposed by [4,10]. First, the Scopus and Web of Science (WoS) databases were searched for the most relevant articles published in leading journals and refereed conference proceedings. Additionally, we expanded our search to include IEEE Xplore for engineering and technological advancements, Google Scholar for grey literature and articles not indexed in primary databases, and ProQuest for dissertations and theses. We combined specific and broad search terms, incorporating Boolean operators and truncation to refine our search results. The keywords covered various aspects of 3D robotic packing and related technologies. The primary keywords included: "3D robotic packing", "robotic packaging", "automated packing systems", "industrial robotics", and "Logistics 4.0". These words were relevant to finding studies specifically focused on the packing process involving robotic systems in three dimensions, mentioned within the industrial sector. One of the example search queries in databases was ("3D robotic packing" OR "robotic packaging" OR "automated packing systems") AND ("industrial robotics" OR "Logistics 4.0"). Relevant publications were those that did meet the scope and contributed significant information. This first inquiry identified the initial studies that approached 3D robotic packing. Then, a branching of manuscripts was performed using a breadth pruning approach with the cited references within the publications, i.e., the references of all considered publications were reviewed; if deemed relevant, they were included in the review, and their references are subsequently analyzed. This approach helps uncover additional relevant publications that might not have been found in the initial database search. This iterative procedure continued until no more suitable studies were found. Finally, a forward-looking search strategy was used to identify new research advances and incorporate these newly identified studies into the review, ensuring the literature review is up-to-date and comprehensive. This procedure was used to identify which publications cited the set of suitable publications. Figure 1 shows the stepwise elimination procedure, particularly those involving robotics and automation; several vital aspects are typically examined to ensure efficient and effective operation, such as Packing Algorithms, Robotic Systems and Components, Sensor Systems, Software Architecture, Packing Strategies, Physical Constraints, Environmental and Safety Constraints, Performance Metrics, Cost Considerations, and Application-Specific Requirements. These factors help ensure the 3D packing literature review is efficient, reliable, and suitable.

This methodology ensures the literature review is up-to-date and allows obtaining the main body of robotic 3D packing. After applying this methodology, 46 studies were left. Table 1 shows that most publications were research articles and that conference papers also play a significant role in knowledge dissemination, fostering discussion and collaboration among researchers. Figure 2 shows that robotic-packing-related publications have been booming in recent years, evidencing the influence of Logistic 4.0 on packing problems.

Table 1. Composition of main body of 3D robotic packing according to the publication format.

Publication Format	Number of Articles	Percentage	References
Research paper	29	63.04%	[2,11–38]
Conference paper	15	32.61%	[39–53]
Doctoral thesis	1	2.17%	[54]
Master's thesis	1	2.17%	[55]
Total	46	100%	

It is worth mentioning that the methodology used to collect relevant information only considered research publications. Therefore, the findings and insights of this review are biased toward these publications. This disclaimer is mentioned since most studies in 3D robotic packing are found in patents rather than research publications.

After collecting the main body of 3D robotic packing, the publications were analyzed regarding the physical layout, the packing problem, the solution methodologies for the packing problem, and the performance assessment shown in Figure 3. Therefore, the first step consisted of extracting the physical features of the layout of the packing cells. This overview involved elements such as the robot, conveyor, vision system, and end-effector gripper. Second, the studies were considered in the packing problem dimension to gain information about the problem types (according to state-of-the-art typologies) and constraints. Third, the studies were classified according to the solution methodology (mathematical models, heuristic algorithms, metaheuristics, machine learning models, hybrid algorithms, and preset patterns). Finally, the performance assessment was performed by classifying the manuscripts according to the instance types used (real test instances, classic data sets, and other test instances) and the comparison used (previous methodologies, diverse algorithm versions, and manual procedures).

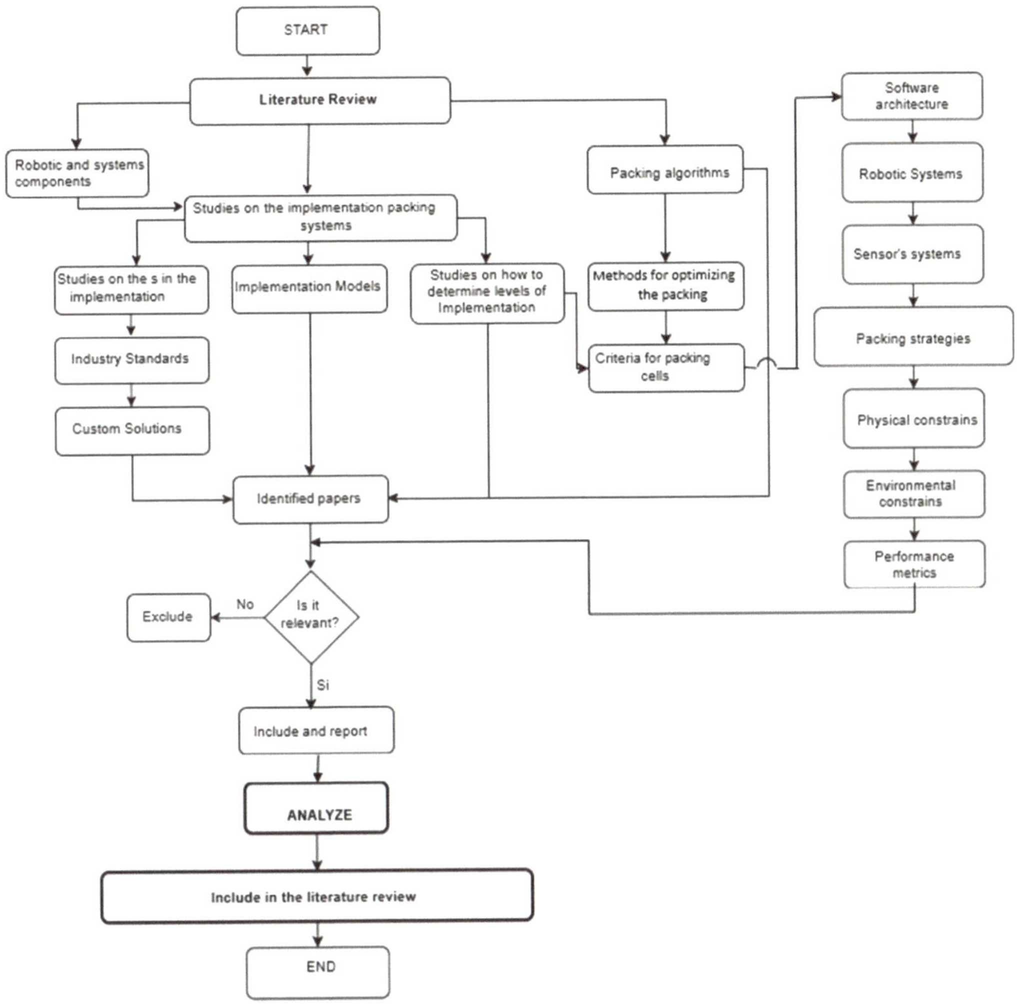

Figure 1. Flow chart for stepwise elimination procedure.

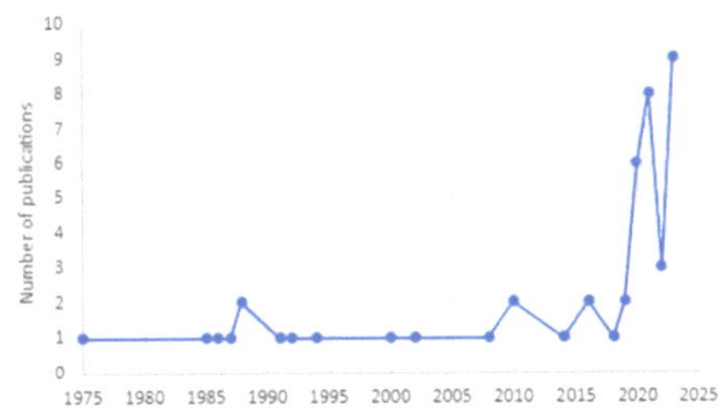

Figure 2. Distribution of the reviewed papers per year of publication from 1975 to 2023.

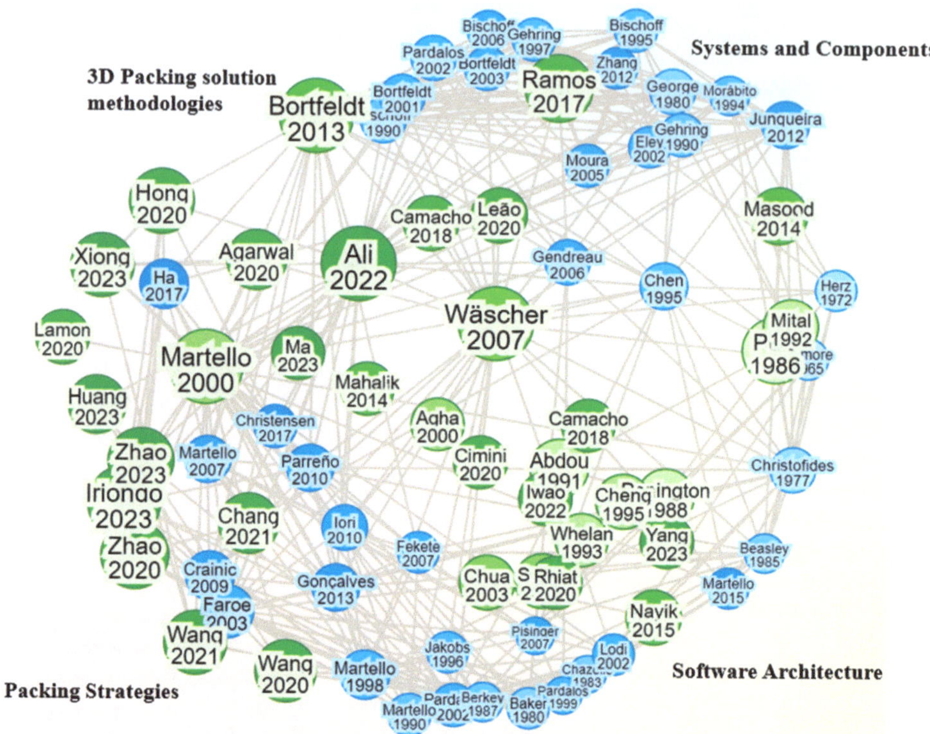

Figure 3. Literature review map based on physical features, packing problem dimension, solution methodology, and architecture [1–9,11–62].

3. Robotic Packing Cell

This section describes the general layout and the components identified in the reviewed studies. For this purpose, it is pertinent to define a robotic packing cell in general terms. In this manuscript, an automated packing cell consists of a configuration in which several items, equipment, areas, and at least one robot are grouped to perform a packing activity. The general structure of a cell consists of a picking area, objects to be packed, a robot, and a packing area. In the picking area, the objects to be packed lie (or arrive) before the packing procedure. This area may include a structure such as a table or a conveyor belt. The packing robot takes the objects from the picking area and positions them somewhere in the packing area. The packing area is usually a container, such as a box or a pallet.

Regarding the items to be packed, this study considers three-dimensional objects of any shape (regular and irregular). The most popular shape is the cuboid because it is the standard shape (even for containers) in logistics operations. Additionally, it is worth mentioning that homogeneous and heterogeneous scenarios are considered, i.e., cases in which there is only one type of item and cases in which there are several types of items.

Packing patterns can be built based on the available information about the to-be-packed items. The information is related to, among other features, their dimensions, shape, and order in the sequence of appearance (for example, when using a conveyor belt). It is considered an offline framework if all the information is known a priori to the packing activity. Otherwise, the framework is considered to be an online one. Packing methodologies use the items' information to find packing patterns; therefore, the packing pattern's utilization is expected to increase as more items' information is available.

Notably, some robotic packing cells incorporate perception systems located at one or more positions within the cell. These systems are generally used to obtain information on the current state of the items (e.g., dimensions, position, and orientation) and the packing pattern (changes in the packing pattern throughout its construction).

The packing cell components identified in the reviewed studies are detailed below. Section 3.1 indicates the main features of the packing robots. Section 3.2 describes the identified process-specialized areas within the robotic packing cell. Section 3.3 presents the sensor or perception systems. Section 3.4 mentions the software architecture that integrates algorithms and components of the packing cells. Finally, in Section 3.5, some special situations are discussed.

3.1. Robots

Robots used in the packing process are diverse; however, they can be grouped according to several dimensions. This section aims to group the packing robots considering the available features from the reviewed studies.

The most common robot manufacturers found in the studies were KUKA [11,12,39–42], followed by Universal Robot [13–16,43] and IBM [17,18,55]. Most studies used mass-manufactured robots except for the one developed by [44]; they presented a robot prototype called UPSarm to pack boxes within truck containers.

Generally, packing robots have a fixed position within the working area; however, there were two studies in which the robot was on a mobile platform. Integrating a mobile platform allows the robot to increase its reach area, enabling it to perform more complex tasks. However, this integration could also mean increased task completion time due to the navigation (transportation) time. The studies that included a mobile platform were the ones developed by [40,45]. They used the mobile platform that allowed the robot to move from the picking to the packing area. Moreover, the robot could position itself differently within the picking/packing areas to facilitate the grasping/placement processes.

Most of the reviewed studies consider a single-arm robot; however, the studies developed by [19,20] have a different setup. The former contemplates a dual-arm industrial robot (Nextage) for a grocery packing environment. The latter uses two robots (UR3) to pack a bending item.

Packing robots are equipped with a device for grasping and placing cargo. All studies consider an end-effector device, which in this study is generically called a gripper. Robots have a maximum load capacity that limits the cargo's weight the robots can handle. The reviewed studies generally used robots with a maximum load of less than 10 kg. For example, KUKA robots have a maximum load capacity ranging from 7 kg to 14 kg. Robots from Universal Robots have a maximum load capacity ranging from 3 kg (for the UR3) to 10 kg (for the UR10). The robots with the highest maximum load capacity in the reviewed studies are the STEP SR20E [30] and the ER20-robot [37], with a maximum load capacity of 20 kg. It is worth mentioning that within the robot's maximum load capacity, the weight of the gripper must be considered. Four broad types of grippers were found in the reviewed studies (Table 2).

Table 2. Number of studies according to the type of gripper used.

Gripper Type	Number of Studies	References
Suction	25	[2,12,13,15,16,19,23–25,30–32,34,37,39,41–43,46,47,49–51,54,55]
Lateral paddles	11	[11,14,17–20,27–29,38,40]
Upper and lower paddles	2	[21,44]
Electromagnets	1	[45]
Not specified	8	[22,26,33,35,36,48,52,53]

First, the suction or vacuum type grasps the items through vacuum generation (generally by an air compressor) and releases them by removing the vacuum. These grippers take the items from the top (although it is possible to grasp them from a side surface). The maximum weight the gripper can hold is limited by the vacuum generation and the strength of the suction cups (besides the cargo weight limit of the robots). The suction grippers could be preferred because they reduce collision chances with the packed items as long as the gripper's size is smaller than the top surface of the item.

The second most used gripper type uses lateral paddles (or fingers). These grippers work as clamps to grasp and release the items from the sides. Therefore, the weight of the items is limited by the closing force of the paddles. It is worth mentioning that this force may damage items in situations where the items are fragile. A disadvantage of these grippers is that their use is more collision-prone or tends to leave unwanted gaps in the packing pattern.

The third gripper type found in the reviewed studies consists of upper and lower paddles. This gripper type is unusual since only two studies considered them. The first one was developed by [21], who used it to avoid the disadvantage of lateral paddle grippers. The lower finger could slide back to place boxes in the desired locations. The other study was developed by [44], who used a gripper with one upper palm and two lower paddles. The lower paddles could extend (to grasp the boxes) and retract (to release the boxes).

Finally, the fourth gripper type uses an electromagnet at the wrist of the robot. It was found in the study developed by [45]. This gripper requires the items to be magnetic or attached to a magnetically attracted material. The authors used the latter strategy by placing a ferromagnetic sheet on the upper surface of the items (boxes). It is worth mentioning that the authors used this strategy to simulate a suction gripper.

3.2. Process-Specialized Areas

A general layout consists of mainly two process-specialized areas related to picking and packing. These areas must be within the robot's working space to manipulate the items. The picking area is where the small items lie before they are packed into the container. The placement area is where the container is located and where the packing pattern is built.

Picking areas of the reviewed studies can be classified according to two criteria: the item's availability for picking and obstacles within the picking area. The former refers to whether all items are available for picking or not. The latter is associated with the presence or absence of obstacles, such as the container walls in which the items lie.

Table 3 indicates that most studies have focused on cases in which there are no obstacles and not all items can be reached by the robot; i.e., items arrive at the picking area one at a time or in groups. This choice likely stems from the prevalence of this particular real-world scenario. This scheme for the picking area includes staging areas [17,18], preparation tables [22,44], conveyors, and mockup conveyors (items are placed into a picking position) [23,51]. On the other hand, the least used format is the one with obstacles, and the robot can only reach some items [24].

Table 3. Number of studies according to the picking area classification.

	With Obstacles (Walled Container)	Without Obstacles (Unwalled Container)
All items are available	4 [12,15,40,42]	7 [13,14,19,20,31,36,41]
Some items are available	1 [24]	27 [2,11,16–18,21–23,25,26,28–30,32,34,37,38,43–48,51,52,54,55]
Not specified	6 [27,33,35,39,49,50]	

Some studies have focused on the scenario in which all items are available for picking with and without obstacles. In the former case, researchers have used box-type containers [12,15,40,42]. Meanwhile, in the latter case, researchers have used tables, other pallets, and the floor in which all items lay close to the robot.

Packing areas are generally associated with containers where the packing pattern is built. Table 4 shows that the most common placement containers are pallets. These containers are flat, portable platforms without physical walls or roofs. Therefore, the stability of the packing pattern only depends on the stability provided by the stacked items. Moreover, the absence of walls makes them less collision-prone for the robot than all containers. The second most common placement containers are boxes, distinguished for having walls. The walls can provide stability (or contact points) to the packing pattern; however, they can cause some collisions with the robot in the packing process.

Table 4. Number of studies according to the placement containers.

Placement Containers	Number of Studies	References
Pallet	19	[2,16–18,21–23,28,32,34,36,39,43,45,47,48,51,54,55]
Box	17	[11–15,19,20,24,29–31,38,40–42,49,50]
Crucible	2	[25,46]
Cart	2	[26,37]
Truck container	1	[44]
Not specified	5	[27,33,35,52,53]

The less common placement containers are crucibles [25,46], wheeled carts [26,37], and truck containers [44]. These containers have walls like boxes but differ in shape (crucibles) and access for the robot. Truck containers are packed through one side while carts are packed through one side and from the top.

Besides picking and packing areas, some studies consider other process-specialized areas. Some reviewed studies contemplated the Buffer area associated with the buffer constraint (see Section 4.2.3). In this area, the items can be temporarily stored during the packing process [16,27,41,44].

Other studies contemplated areas used to measure the items [17,18] or prepare them [23,44]. In the former, the items' features, such as dimensions and weight, are determined manually or with other devices. In the latter, the items are adjusted to a suitable picking orientation for the packing robot.

Finally, some studies included a temporal storage area where items could be placed if they would not fit in the packing area [2,28]. The placement of an item in this area generally marks the end of the packing process, and a new container has to be used.

3.3. Sensor Systems

Several reviewed studies incorporated sensor systems to identify the items' position, size, and orientation. RGB-D (Red, Green, Blue + Depth) sensors were the most common systems. The two most popular RGB-D sensor manufacturers in the reviewed studies were Intel Real Sense and Asus. The following models of Intel Real Sense were found: SR300 [13,41,42], D435 [14,15,24], and D435i [16,43]. The only model of Asus found in the reviewed studies is the Xtion Pro Live [29,45].

Researchers have also used light beams (including lasers) to determine more precisely the items' dimensions [17,18] and surface heights [21,25,46].

The sensor systems are located in several sections of the physical layout of the packing cell. Five sections were found in the reviewed studies: on the packing robot, in the general work environment, and in the picking, packing, and detailing areas. It is worth clarifying that sensors in the work environment can simultaneously obtain information about picking and packing areas and even the robot [14,29].

The number of sections in which sensor systems were located varied among studies; however, the maximum number of areas with sensors was three. Table 5 shows in which and how many layout sections the reviewed studies used sensor systems. This table shows that sensor systems are usually located in two sections and have been mainly located to monitor the items and the picking and packing areas.

Table 5. Number of studies according to the layout's section in which sensor systems were used.

Layout Section	Number of Studies			
	1 Section	2 Sections	3 Sections	Total
Packing	2 [20,21]	10 [13–16,24,25,41–43,46]	3 [12,30,37]	15
Picking	4 [26,33,39,47]	7 [11,13,15,16,41–43]	3 [12,30,37]	14
Item detailing	3 [17,18,28]	3 [24,25,46]	2 [12,30]	8
Robot	2 [45,50]	1 [11]	1 [37]	4
Work environment	1 [29]	1 [14]	0	2
Not specified	19 [2,19,22,23,27,31,34–36,39,40,44,48,49,51–55]			

Ref. [47] proposed a particular sensor system in the picking area to monitor the incoming items on a conveyor. The system had two sensors in the picking area, one for the last item on the conveyor and the other for some items behind it. This scheme can be used to obtain information on future items that will be packed. This information can improve the packing pattern's utilization.

Some studies used sensor systems that focused on the item. These sensors were used to gain more accurate information on the items, like their dimensions [17,18] or their surface. In the last case, the sensor was generally located between the picking and packing areas. This information was used along with the information obtained with other sensors that focused on the packing [24,25,46] and picking [12,30] areas to improve the efficiency of the packing process or to guarantee vertical stability. For example, the items' bottom and the packing pattern's top surfaces can be scanned to identify the best positioning options. This strategy is advantageous when irregular items are considered.

Table 5 shows that several studies did not use or specify the focus of the perception systems. The main reason for not using a sensor system was that several studies did not need them because they focused on simulations. Computational simulations allow working with perfect information; therefore, all information related to the items' position, dimension, and orientation (at least in the picking area) is known. This situation makes sensors unnecessary since sensors are essentially used to gather information.

3.4. *Software Architecture*

A software architecture is a high-level design and organization of a software system. The architecture is essential to the software development process because it shows all the components, their tasks, and how they are integrated. The software architecture improves the understanding of a system with several components, as happens in applications of 3D robotic packing. Unfortunately, several reviewed studies did not clearly explain all the architecture used in the study. However, this situation has been improving with the inclusion of diagrams [12,14,16] and more detailed descriptions [15,29].

3.5. *Special Cases*

Two studies contemplated the human-robot interaction (collaborative robot or cobot) in the packing process. Ref. [29] used a cobot for packing irregular items within a box, and ref. [45] used a cobot for box palletizing. The concept of collaborative robots started to gain traction in recent years, considerably later than noncollaborative robots. This situation explains the low number of studies that consider cobots.

Robots and cobots have several advantages and disadvantages in packing applications. Robots generally have a larger payload capacity than cobots due to the safety constraints integrated into cobots. These constraints may also limit, in some cases, the working speeds. Additionally, the packing process is generally repetitive and predictable (a robot-favored scenario), at least in the homogeneous case, i.e., when the items are identical. However, some packing tasks can have higher complexity, making cobots a better option due to their flexibility. For example, when irregular items are handled, the packing pattern's stability is highly complex, and human input can facilitate this process.

Another special case that is worth mentioning is the use of simulations. Some studies used simulations, others used physical experiments, and others used both. Physical and simulated environments have features that can make them a better option for an application depending on the study's objective. Physical environments are more accurate since they are subjected to natural forces that may impact the result of a process. On the other hand, simulations depend on models representing those forces, and some use simplifications that may distort the results. These models can take high computation times if not simplified enough, making the simulation slower than reality. However, simulations can use elementary models that allow many iterations to be carried out in a fraction of a physical experiment's time.

It is worth mentioning that simulations and physical environments are not mutually exclusive; a packing application may benefit from both schemes. Both schemes can be used to validate and optimize the robotic packing system. For example, an application may use a simulation environment to plan and design the automated packing system, validate, optimize algorithms (including the robot kinematics), and analyze the impact of changes within the packing cell. On the other hand, the physical environment can be used to verify the actual system's functionality and identify elements to improve.

Simulators found in the reviewed studies are URSARSim [39], MATLAB-Robotics Toolbox [48], Klamp't [49], V-Rep [40], Blender [42], pyBullet [14,31,53], Unity [36], and Bullet [32]. Besides these simulators, some studies have developed simulation setups [22,50].

4. Packing

This section explains the most relevant aspects of the packing considered in the reviewed studies. The reviewed studies are organized according to cutting and packing classifications (Section 4.1) and their constraints (Section 4.2).

4.1. Typologies

Packing problems form part of the broader area called Cutting and Packing (C&P). These problems are considered to have fundamental similarities in their structure: there is a set of small items that are assigned to a set of large objects (containers) and satisfy the containment and non-overlapping constraints [4,56,57]. The former constraint states that the small items must entirely lie within the large objects. The latter constraint indicates that the small items must not overlap.

Ref. [56] proposed a typology whose purpose was to unify several versions of C&P problems that appeared in the literature with different names but with the same structure. This typology has four criteria (dimensionality, kind of assignment, assortment of large objects, and assortment of small items). It classifies C&P problems with four entrances, one for each criterion. For example, 1/V/I/R is a one-dimensional problem in which the whole set of small items must be assigned to a subset of identical large objects. The small items consist of a few item types, each with many items.

Ref. [57] proposed an improved typology based on the one presented by [56]. The newer typology has five criteria (dimensionality, kind of assignment, assortment of small items, assortment of large objects, and shape of small items). It has been broadly accepted by the scientific community, possibly due to the use of names commonly used in the jargon of the C&P community. Considering only the kind of assignment and the assortment of small items criteria, the authors classified the pure C&P problems into broad groups called Basic Problem Types. Moreover, if, besides the previous criteria, the assortment of large objects criterion is considered, then the C&P problems are classified into more specific groups called Intermediate Problem Types. Finally, if all criteria are considered, the C&P problems are classified into Refined Problem Types.

This typology contemplates that some problems do not fit the two, four, or five criteria groups and constitute exceptional cases or variations. The studies considered in this review are classified according to the typology proposed by [57] to determine the trends and gaps related to the addressed packing problems.

4.1.1. Basic Problem Types

The Basic Problem Types are obtained by identifying the type of assignment and the assortment of small items criteria. The first criterion distinguishes between output maximization and input minimization. The former implies that the quantity of large objects (containers) is insufficient to accommodate all small items; therefore, the objective is to maximize the utilization of the containers. On the other hand, the latter means there are enough large objects to accommodate all small items; the objective is to minimize the number of containers used. The second criterion indicates whether the small items are identical, can be grouped into a few types with each type having a significant demand, or are grouped into several types with each type having a low demand.

The classification into the Basic Problem Types is shown in Figure 4. It is worth noting that several reviewed studies did not clearly state the packing problem they were tackling. Therefore, the classification was performed according to the information in the methodology description, the application context, and the performed tests.

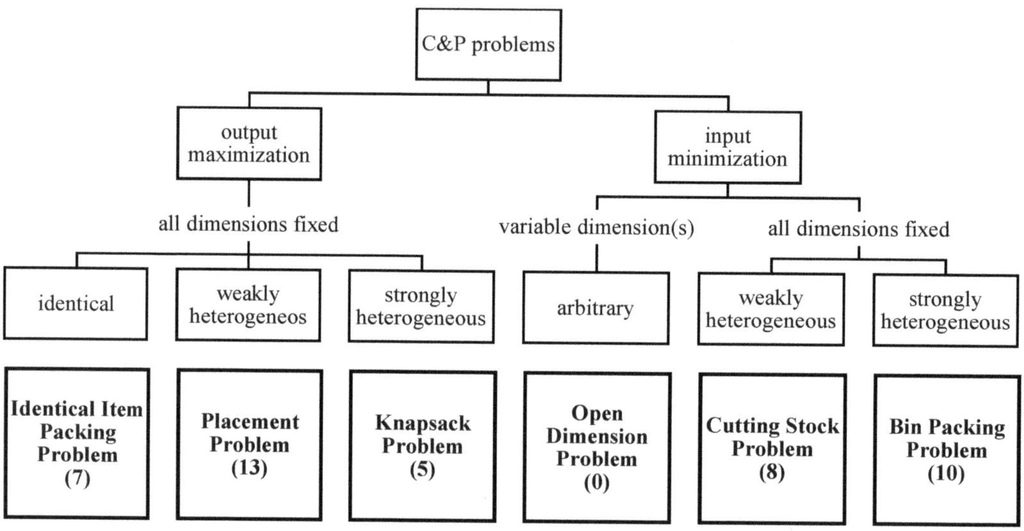

Figure 4. Classification of the reviewed studies into the Basic Problem Types proposed by [57]. This top-down hierarchy diagram has three levels. First, C&P problems are divided into two branches according to the type of assignment criterion. Then, each branch is divided into three branches according to the assortment of small items criterion. Below the six boxes, the names of the basic problem types and the number of classified studies are shown: Identical Item Packing Problem [22,29,34,41,42,48,55], Placement Problem [12,15,17,18,26,28,35–37,43,51–53], Knapsack Problem [14,30,32,39,49], Open Dimension Problem, Cutting Stock Problem [11,16,19,21,27,40,45,47], and Bin Packing Problem [2,13,23–25,31,33,44,46,54].

Two studies were classified as Bin Packing Problems and shared a particular feature regarding the container. In the studies developed by [25,46], the container was a crucible allowed to be filled more than its top as long as the small items (polycrystalline silicon nuggets) were stable. This situation could be interpreted as an Open Dimension Problem; however, the authors of this review consider it insufficient to catalog them as such. Therefore, these studies are deemed to have containers of fixed dimensions. Two studies were not classified within the Basic Problem Types shown in Figure 4. These studies were developed by [20,50], who focused on packing a single item. These studies are of input minimization since all items are packed. However, the assortment of small items is unclear to this manuscript's authors. Therefore, these studies are considered special cases or variations of packing problems.

In real applications in logistic centers and warehouses, the robotic packing cells tend to focus on packing all items, i.e., input-minimization. However, Figure 4 shows that several studies are framed under the output-maximization scheme. This situation occurs because some studies limit their experiments to the packing of a single container, and the packing process ends when no more items can be placed within the container. Therefore, these problems are classified as output-maximization problems.

4.1.2. Intermediate Problem Types and the Other Criteria

Following the typology proposed by [57], the reviewed studies are further cataloged into the Intermediate Problem Types by considering the assortment of large objects criterion. Table 6 shows the Intermediate Problem Types found in the reviewed studies according to the type of assignment criterion. This table indicates that the 3D packing problems associated with robotic packing tend to address single (IIPP, SLOPP, and SKP) or identical large objects (SSSCSP and SBSBPP). The study developed by [23] was the only study that addressed a weakly heterogeneous assortment of large objects. They solved MBSBPP instances for a logistics company implementing a GRASP algorithm.

Table 6. Classification of the reviewed studies into the Intermediate Problem Types proposed by [57].

Type of Assignment	Intermediate Problem Type	Number of Studies	References
Output maximization	Identical Item Packing Problem (IIPP)	7	[22,29,34,41,42,48,55]
	Single Large Object Placement Problem (SLOPP)	13	[12,15,17,18,26,28,35–37,43,51–53]
	Single Knapsack Problem (SKP)	5	[14,30,32,39,49]
Input minimization	Single Stock Size Cutting Stock Problem (SSSCSP)	8	[11,16,19,21,27,40,45,47]
	Single Bin Size Bin Packing Problem (SBSBPP)	9	[2,13,24,25,31,33,44,46,54]
	Multiple Bin Size Bin Packing Problem (MBSBPP)	1	[23]

The scope of this review established the dimensionality (first criterion of the improved typology) to 3D. Therefore, the remaining criterion (shape of small items) is addressed next. This criterion distinguishes between regular and irregular items (see Table 7). The former includes, among others, cylinders, balls, and boxes; meanwhile, the latter indicates arbitrary shapes. Within the studies that addressed regular-shaped items, boxes are by far the most common, possibly due to their vast applications in the logistics industry. In fact, only the study developed by [50] considered cylinders besides boxes. They proposed a placement-position search technique to pack the objects near the corner of the container.

Table 7. Number of studies according to the shape of small items.

Shape of Small Items	Number of Studies	References
Regular	32	[2,11,12,16–18,21–23,26,28,30,32–37,39–45,47,48,50–52,54,55]
Irregular	14	[13–15,19,20,24,25,27,29,31,38,46,49,53]

4.1.3. Online and Offline Packing Problems

Finally, packing problems can be classified based on the availability of item information, distinguishing between offline and online problems [4]. In offline problems, all item information is available beforehand, while in online problems, the information is incomplete. Table 8 shows that more studies have considered the online version of packing problems than the offline version. This trend indicates that research on automated packing cells focuses more on scenarios where there may be uncertainty in the dimensions of the items to be packed. Such scenarios are frequently encountered in logistics and transportation centers driven by e-commerce, where different items must be packed, grouped, and shipped to various destinations.

Table 8. Number of studies according to the type of problem regarding information availability.

Type of Problem	Number of Studies	References
Offline	18	[11,13,14,17–19,22,23,34,36,39–41,48–50,53,55]
Online	28	[2,12,15,16,20,21,24–33,35,37,38,42–47,51,52,54]

It is worth mentioning that the study developed by [27] introduced two problem variations called Nondeterministically Ordered Packing (NDOP) and Quasi-Online Packing (QOP). The former assures that a packing plan exists for every possible order of arrival of a given set of items. The latter consists of packing the items one at a time as they arrive, i.e., it is embedded in an online packing framework.

4.2. Constraints

Ref. [58] comprehensively reviewed constraints applied in the container loading process and classified them into five categories (container-related constraints, item-related constraints, cargo-related constraints, positioning constraints, and load-related constraints). These constraints have been accepted and generalized to other packing applications. However, ref. [59] proposed a more practical classification with two groups. The first one is the safety constraints related to the integrity of the cargo and the safety of the operators. The second one is the logistic constraints associated with operational decisions that do not depend on the physical features of the items and containers. The constraints found in the reviewed studies are detailed according to this last classification.

4.2.1. Safety Constraints

Regarding the safety constraints, none of the studies considered the weight limit constraint. The study developed by [17] was the only study that assessed the weight distribution constraint by conditioning that heavy items must be placed at the bottom of the container and light items on top of the heavy ones. This constraint is suitable to the study's context: grocery packing.

The orientation constraint is more relevant for boxes than for other shapes because arbitrary shapes can be symmetric [29,50] or can have multiple orientations [14,15]. For this reason, this constraint is discussed considering the studies that only handled boxes. The most common approach was only to allow two orientations obtained by a 90° rotation around the vertical axis, consistent with how the robot picks the boxes up. These two orientations are limited because grasping or placing boxes from a different surface than the top can cause difficulties associated with the grippers and the robot kinematic. For example, in some cases, the robot may push the items instead of grasping them from one of the item's sides, or the robot could collide with the structure in which the items lie.

Other approaches for the orientation constraint are discussed below. The study developed by [48] was the only one that used a single orientation for the pieces. They palletized boxes grabbed from a conveyor belt with a simple packing approach that exploits the circumstance of identical boxes and a single orientation of the boxes. An interesting strategy regarding orientation management consisted of allowing the packing algorithm to consider several (or all six) possible orientations of the boxes; however, the boxes were manually pre-oriented in a previous area before the robot took them [23,26]. An unusual strategy was presented by [33], who limited the number of orientations by the degrees of freedom of the robotic arm and imposed that the box's largest dimension cannot be aligned with the arm's vertical axis. Thus, the two box's orientations are not considered.

The stacking constraint refers to avoiding damaging the packed boxes and protecting the goods within the boxes. The studies developed by [2,54] were the only studies that considered toxicity and fragility constraints. The former is associated with how suitable a box is near others, considering its contents' toxicity. The latter refers to the load-bearing strength (pressure or weight that a box supports on top) and whether the contents will be damaged in the event of a fall from the top.

The physical positioning constraint is an unusual constraint found in three studies. Ref. [50] proposed a placement-position search technique to place an item in the corner of the container. Refs. [25,46] tackled the robotic packing of polycrystalline silicon nuggets into crucibles, which adopted a packing construction based on the dimensions of the nuggets. First, the bed layer consisted of small gravel-sized nuggets. Then, an alternate filling process was followed until the crucible was full. This process involved placing large nuggets near the crucible's wall and smaller nuggets in the inner part of the crucible.

The last safety constraint is associated with the stability of the small items in the packing pattern. This constraint distinguishes vertical (or static) and horizontal (or dynamic) stability [59]. The former addresses the stability of the items when the container is not moving, i.e., the stability of the packing pattern (when in construction and finished) subdued to the gravity force. The latter addresses the stability of the items when the container is moving, i.e., the stability in the transportation in which the items are subdued to accelerations in different directions.

None of the reviewed studies addressed the horizontal stability constraint; however, there are different approaches concerning the vertical stability constraint. Some studies did not consider the vertical stability constraint because it was not needed in the study's application. For example, ref. [54] pointed out that the pallet (container) can be secured with strapping and wrapping machines in industrial applications. Moreover, refs. [20,50] packed a single item; therefore, the stability constraint was unnecessary. However, it is worth mentioning that in the study developed by [20], the object did not easily fit within the container, but it was flexible. Therefore, they used another robot to guarantee the position and stability of the item in the packing process. Another situation where the stability constraint is not required consists of packing only one layer of items, i.e., without stacking items above others [29,40].

The vertical stability can indirectly be considered by using layer construction. This strategy can be used when the boxes (small items) are identical [22] or when they share the same height [28]. Furthermore, besides the stability of the boxes (individually), some studies use layer construction to improve the stability of the whole packing by alternating packing layers. The alternating layers are obtained by different packing layers [52] or by packing a rotated packing layer [21,34].

Considering the studies that handled boxes, unfortunately, some of them did not clearly state the strategy used to incorporate the vertical stability constraint; however, the following strategies were spotted: first, the full-support constraint, i.e., all the area of the base of a box is supported by the other boxes or the container's floor [23]. Another strategy to incorporate vertical stability when the items are boxes is a more technical approach (since it is based on physics). It seeks to guarantee that the center of mass of each box lies within the convex hull of support points [30,47]. Finally, the last found strategy was using heightmaps, a two-dimensional representation of the height of a three-dimensional surface [16,43].

Vertical stability is challenging to incorporate into the solution methodologies when handling irregular items due to the estimation of stable poses of these items in the landscape that forms other items. The complexity of these conditions has made researchers resort to various approaches to consider the vertical stability constraint with irregular items. Ref. [49] used force and torque balances that required a convex programming solver to validate whether an arrangement is stable. Ref. [31] developed an algorithm that creates a belief space representation of the world and updates it according to the observations taken. This procedure allows the algorithm to estimate the changes in the packing pattern.

The most usual strategy to consider vertical stability when handling irregular items is constantly monitoring the packing pattern. This monitoring can be performed by raster methods [19], line representation of heights [25,46], or heightmaps [14,15]. It is worth noting that the monitoring requires sensors or cameras that perceive depth.

4.2.2. Logistic Constraints

Regarding the logistic constraints, the reviewed studies only considered complexity constraints, i.e., constraints that reflect technological and human limitations. These constraints in the scope of this review are associated with robot-packable patterns that include guillotine packing patterns and collision avoidance. Several studies did not consider the collision avoidance constraint or did not clearly state the strategy to achieve it [24,26,32]. Other studies indirectly considered it by limiting the rotation of the boxes to the vertical axis [12], grasping several items with the gripper at once (instead of one) [34], performing guillotine packing patterns [18], using different types of grippers [19], or using a large robot and trajectory path with top-down loading motions that was hardly going to cause a collision [23,25,46].

Researchers who have tackled robotic 3D packing have used several strategies to handle the collision avoidance constraint. When the items are boxes, the most common strategy is oversizing them to overcome the uncertainty associated with the exact size and position of the item, the vision system, and the picking point [16–18,35]. This strategy tends to leave gaps between the packed items. Another strategy consists of packing in a far-to-near fashion from the robot. Zhao et al. in [30,47] achieved this packing method by implementing a reward function for collision-free trajectories. Another strategy is to modify the trajectory path by creating artificial (dummy) points the robot has to reach [22,28].

When irregular items are handled, the most common strategy for avoiding collisions is using heightmaps in which possible placement areas are validated for collision considering the nearby landscape [14,15,49].

A strategy for collision avoidance suitable for regular and irregular items consists of sampling the trajectory to obtain several steps. Then, each step was subdued to a collision check in which possible collisions between the robot (including the carrying item) and environment obstacles (packed items and container) were detected. This approach was used by [11,27,44].

The scope of this review also addresses approaches that contemplate human-robot interaction. In these cases, collision detection is crucial in protecting the integrity of the people near the working robot. Therefore, this constraint is managed with robust perception systems that constantly monitor the robot's movements and the environment [29,45].

It is worth mentioning that some studies allowed (or even wanted) collisions instead of avoiding them to successfully achieve a desired packing pattern. Ref. [20] used the collisions between the flexible item to be packed and the container to perform the bending of the item. Shome et al. in [41,42] allowed collisions when the items were pushed to the desired position.

4.2.3. Automatic Environment Constraints

Besides the safety and logistic constraints, some additional constraints appear in robotic packing cells. The most resounding constraints are buffering, deletion, and repacking. The first one allows items to be temporarily stored in a holding area before being packed into the container. This technique enables the packing decision to be taken later. Therefore, the packing utilization and flexibility of the packing process may improve at the expense of increasing the task's time and the problem's complexity. The utilization may increase because there are more positioning options, and a better option may be chosen. On the other hand, the time may increase because more robot movements are required to move items from the picking area to the buffer area and from the buffer area to the packing area.

The second constraint consists of not packing some incoming items or removing some packed items from the packing pattern. Generally, the removed items are packed in a subsequent container. Finally, the last constraint allows the packed items (typically the last ones) to be rearranged.

None of the studies considered the deletion constraint. The buffer constraint was considered only by four studies. Refs. [16,27,43] adopted a *k*-buffering, i.e., a variable size buffer. Although these methodologies allow a considerable value for *k*, it is worth mentioning that it is limited by the robot's reach and the space and location of the buffer in real applications. Ref. [43] realized a sensibility analysis of the buffer size (maximum number of items in the holding area). They showed how the container's utilization increases as the buffer's size increases. The other study that contemplated this constraint was the one developed by [44], who used a buffer of 12 boxes in a container loading framework.

The repacking constraint was only considered by [16]. They used a staging area as the buffer and a temporary storage area for repacking procedures. When their methodology detected an arrangement that improved the packing density, the items that needed repositioning were unpacked from the pallet to the staging area and repacked into a better position.

5. Solution Methodologies

Finding optimal solutions to the 3D packing problem with robots is challenging due to the complexity and combinatorial nature of the task. This complexity is increased in an online framework due to incomplete information about the items. Furthermore, since 3D robotic packing is related to real-time application, all systems must provide fast results. Therefore, researchers have focused more on efficient solutions, i.e., high-quality solutions obtained in a reasonable time. This situation is evidenced in that the preferred methodologies are heuristics and algorithms based on machine learning (see Table 9).

Table 9. Solution methodologies for the packing problem.

Methodologies	Number of Studies	References
Heuristic	28	[2,11,13,17–19,21,25,27–29,33,34,37–39,42,44–53,55]
Machine learning	11	[12,14–16,30–32,35,36,43,47]
Mathematical model	2	[26,54]
Predetermined pattern	5	[22,41,52,54,55]
Metaheuristic	3	[23,24,40]
Matheuristic	1	[33]

Heuristics are the most preferred methodology to solve the packing problem within robotic packing cells. These methodologies are characterized by being fast (low execution time) and their simple structure. In the case of robotic packing cells, heuristics are closely related to how the packing task develops. The packing task consists of iteratively adding items into a container while ensuring some practical constraints until the container is complete or until there are no more items to pack. This structure also describes the procedure that constructive heuristics follow.

The most relevant aspects in which heuristics differ are the selection of the item to pack (including their orientation) and the position in which the items will be packed. These decisions can vary in the order they are taken. In the offline framework, some researchers prefer first to choose a position and then choose the item to pack [39], or vice versa. However, in an online framework, item selection usually is not necessary, and the heuristic focuses on choosing the best position for the standby item (into the cargo space or buffer).

The item and position selections are based on different prioritization criteria and representations. The most known prioritization criterion is associated with the Deepest Bottom Left (DBL) strategy, in which items are placed starting at the deepest bottom left corner, and the packing pattern is built towards the opposite corner [19]. However, there are more complex prioritization criteria, such as choosing the positions closer to the container's corners [23].

Regarding space representation, researchers have used techniques such as maximal spaces [23], residual spaces [39], and height maps [16]. The first two relate to cuboid sub-containers, which are created or modified as more items are placed within the container. The last one uses a 2D grid, where each cell represents the height of the packed items.

Heuristics are scenario dependent; i.e., the satisfactory performance on a scenario does not necessarily transfer to other scenarios. For this reason, researchers have been focusing on developing more flexible (intelligent) algorithms that can adapt to a much more comprehensive range of scenarios. Therefore, machine-learning approaches have gained popularity among researchers. These approaches return fast solutions; however, they take considerable time in their training [30]. Moreover, these methods require a deep understanding of the problem to extract useful information that the algorithms can use.

Machine Learning approaches can be used for boxes and irregular-shaped items. In the former case, for example, ref. [12] formulated the packing problem as a Markov Decision Process (MDP) and used the actor-critic architecture. The goal was to maximize the space utilization ratio of the container, achieving a trade-off between exploration and exploitation. This actor-critic architecture contains a neuronal network that predicts the probability of each action and a value function that evaluates the state. In the latter case, ref. [31] developed the Sample Algorithm (SA) to solve problems with partially observed states and deterministic action effects. The objective is to minimize the entropy of the packed grocery items. Moreover, ref. [30] proposed a machine-learning algorithm for packing arbitrarily shaped items with partial observations.

Other less-used methods are discussed next. Mathematical programming techniques aim to find optimal solutions (minimizing the waste or maximizing the utilization) by deciding the items' positions within the container considering practical constraints. Generally, solving a problem through mathematical programming is computationally demanding, and, depending on the problem size, it can take considerable time.

Ref. [54] proposed a mixed integer programming model for solving the 3D palletizing problem with diverse cardboard box sizes, considering the pallet's capacity and box dimension, orientation, and sequence. The objective function is related to minimizing the operation time. Ref. [26] formulated a 3D stacking problem including practical-robotic constraint and vertical stability of the packages in an online framework. This formulation was solved using dynamic programming in an MPC (Model Predictive Control) framework with adaptative windows.

Predetermined patterns are found when the packed items are boxes and computed before the packing process begins. These patterns consist of several layers that usually repeat until the container is full. These patterns appear when there is a single box type in the whole container [41] or at least in each layer [52].

Metaheuristics leverage their ability to explore complex search spaces and find high-quality solutions. Several metaheuristic approaches solve offline packing problems. However, these algorithms are often not adaptable to online packing problems due to incomplete item information. Only two studies consider a metaheuristic approach. Ref. [23] used a Greedy Randomized Adaptative Search Procedure (GRASP) algorithm to solve the MBS-BPP in an offline framework. This algorithm did not consider additional constraints related to the robot; however, in [60] they analyzed the feasibility of the packing patterns due to the inverse kinematic of the manipulator. Ref. [24] employed a Differential Evolution (DE) algorithm known for its effectiveness in other optimization problems. The algorithm is based on iteratively moving objects around in the container until a local optimum is reached. The algorithm uses a differential update rule to move objects; i.e., each object's movement is based on the others in the container.

Finally, an interesting approach is using mathematical models and metaheuristics, i.e., matheuristic. Ref. [33] proposed a matheuristic consisting of two rolling-horizon mixed integer linear programming (MILP) and heuristics (First Fit and Best Fit) to obtain an optimal or near-optimal solution.

6. Instances and Performance Comparisons

The methodologies proposed in the reviewed studies to solve robotic 3D packing need to be evaluated to identify in which scenarios they perform better. Moreover, methodologies should be compared to determine their strengths and limitations in a given scenario. The evaluations and comparisons enable the validity and credibility of research findings. Furthermore, these procedures enhance research transparency and encourage knowledge sharing and peer validation, improving the state-of-the-art solutions for 3D packing problems.

Methodologies are tested on a set of instances whose origin may vary. Table 10 shows that in 3D robotic packing, the most common instance type, comes from real-life applications, followed by randomly generated instances and instances obtained from other studies (data sets). This situation indicates that several studies are application-dependent. Furthermore, some real-life applications are not detailed enough to be used by other studies (probably due to confidentiality agreements), which limits the comparisons between methodologies.

Table 10. Number of studies according to the type of instances.

Type of Instances	Number of Studies	References
Real-life instances	19	[2,12,14–16,18,20,21,23–25,28,30,38,41,46,47,51,55]
Randomly generated	12	[13,19,26,31,33–36,42,45,49,54]
Data set	8	[11,27,29,32,37,39,43,53]
Not specified	7	[17,22,40,44,48,50,52]

The most appropriate way to compare the performance of methodologies is by comparing their results over the same instances. Some instances of the online framework of the packing problem that have been proposed are CUT-1, CUT-2, and RS [32]. In CUT-1 and CUT-2, items are generated by sequentially "cutting" the bin into pre-defined item types; therefore, the items can be arranged into a perfect packing pattern. Meanwhile, in the RS, the sequence of items is randomly generated. These instances are relatively new but widely used [16,43,47].

Ref. [61] proposed some that have been used by other researchers in the robotic 3D packing community [11,33]. These instances consist of various-size cuboid items divided into nine classes. The first five classes consist of a single bin size and five item types. The items' dimensions are randomly generated according to uniform distributions. The sixth, seventh, and eighth classes have different bin sizes, and the item's dimensions are obtained with uniform distributions. The last class is generated by cutting the bin into smaller parts.

Irregular-shape instances can be found in some challenges, such as the one organized by Amazon (Amazon Picking Challenge). These challenges aim to foster innovation and advancements in robotic systems capable of picking and packing items from shelves or bins. Participants are asked to develop autonomous robots that can navigate warehouse environments, identify and grasp specific items, and place them accurately into containers or onto conveyors [27,62].

Since the most common instance type was not classic data sets (widely used instances by other researchers), it is worth analyzing how the reviewed studies compared their methodologies. Table 11 shows that most studies do not compare their methodologies or compare them with other versions of the methodologies by changing some procedures or parameters. This situation is concerning since, in this way, it is hard to identify the strengths or limitations of methodologies. This concern is deepened because few studies compare their methodologies with previous works.

Table 11. Number of studies according to the type of comparison.

Benchmark	Number of Studies	References
Algorithm	17	[11–14,16,23,32,33,35–37,41,43,49,54]
Previous work	5	[30,31,39,46,53]
Human	4	[2,25,29,45]
Not specified	18	[15,17–22,24,26–28,34,38,40,42,44,48,50–52,55]

The robotic 3D packing pretends to replace or complement the packing operation performed by humans. Therefore, some studies compared their methodologies to manual operation in terms of efficiency [32,45].

7. Final Remarks

The number of publications on robotic 3D packing is limited but has increased in recent years (Figure 2). This situation indicates that this field is booming, and more publications are expected in the coming years. The following notes may guide upcoming studies focusing on unexplored areas.

7.1. Limitations

As stated in Section 2, this study considers publications but does not include patents. Given these developments' novelty and real-world applications, the number of patents related to robotic packing cell systems is substantial. Additionally, this review is biased toward the problems addressed in the publications, which may only partially reflect the impact across some industries where robotic packing cells can be used. Many studies focus on issues encountered in e-commerce, where the heterogeneity of objects to be packed is high. However, some manufacturing industries have low heterogeneity; homogeneous cases (where objects are identical) can occur. While homogeneous packing problems are less challenging from a packing perspective than heterogeneous ones, they still represent significant practical applications in specific industries. This review may only partially capture these homogeneous scenarios, emphasizing the need for further research to encompass a broader range of industrial applications.

7.2. Managerial Implications

Integrating robotic systems can significantly enhance efficiency and accuracy in logistics operations, which is crucial for meeting e-commerce demands. Managers should invest in flexible robots that adapt to dynamic environments with incomplete item information, improving responsiveness. While robotic systems have higher initial costs, they can reduce long-term operational expenses by minimizing labor costs and increasing productivity. Careful cost-benefit analyses are necessary to justify these investments. These analyses should contemplate scaling operations to meet varying demand levels. Additionally, robots require more space than human workers, necessitating optimized warehouse layouts.

Strategic planning is critical when considering the implementation of robotic systems. The type of packing problem to be solved, the nature of the merchandise and containers, and the system layout must be carefully considered. Identifying the most suitable robots for the task and designing a layout that includes a perception system, picking zone, packing zone, merchandise, and robot are crucial steps. These elements must be well-organized to optimize the layout distribution. By addressing these factors, organizations can significantly enhance their operational performance and gain a competitive edge in the logistics and e-commerce sectors.

7.3. Research Opportunities

More research is needed on the layout of packing cells and the development of equipment selection guides. Current studies often do not specify the reasons for choosing specific robots, particularly in simulations where various robots could be selected. Additionally, the structures that allow robots greater reachability, such as mobile platforms or reel systems, need further exploration. This research would help set up packing cells properly, optimizing robot performance and efficiency.

Further investigation is required into the maximum velocities and accelerations that suction-type grippers can handle while ensuring object retention. While suction-type grippers are preferred for their simplicity and suitability for common warehouse materials, more information on their performance limits is needed [52]. This research would help design more reliable and efficient grippers for various packing applications.

Research is necessary to develop a clear strategy for determining the number and positions of cameras and sensors in packing cells. Some studies use multiple sensors for the same area. However, the impact of varying the number of cameras still needs to be discovered [13,15]. Accurate perception systems are crucial for assessing objects' shape, position, and orientation. This research would improve the reliability and replicability of robotic packing systems.

More attention should be paid to the software architecture of robotic packing cells, integrating all modules, such as packing optimization, perception systems, and robot kinematics. A well-designed software architecture, including a Fault Recovery Module, is essential for detecting, responding to, and overcoming operational failures. This research would facilitate the development and operational efficiency of robotic packing cells.

Further research is needed to find the best human-robot configurations for packing activities. Despite increasing implementation in various industries, only some studies have considered cobots in robotic packing. Exploring schemes where human operators assist in picking or packing tasks would enhance the effectiveness of cobots in avoiding collisions and ensuring proper handling. This research would optimize human-robot collaboration in packing processes.

Research should address the bias toward handling regular shapes in packing problems, as fewer publications focus on irregular shapes [56,57]. While recent studies have started to close this gap, more attention is needed on multiple container systems, particularly heterogeneous ones. Sensor systems must recognize container dimensions, positions, and orientations accurately. This research would enhance the versatility and application of robotic packing cells in various industries.

More studies should address automatic environment constraints such as buffering, repacking, and deletion in robotic packing. Future research should focus on k-bounded packing spaces, where algorithms decide which container to use based on physical limitations. This research would improve the efficiency and adaptability of robotic packing cells in dynamic environments.

Comparative research on algorithm performance is necessary, using classic database instances found in ESICUP and OR-Library for offline frameworks. There are no classic databases for online frameworks, likely due to recent interest in this area. Establishing a repository for these instances would facilitate performance comparison and improve the robustness of packing algorithms. This research would standardize algorithm evaluation and enhance the development of effective packing solutions.

7.4. Structure for Future Research

For effective communication in future studies on robotic packing, the packing problem must be clearly stated, including the objective function and constraints, and classified according to the most recent typologies. The equipment and layout used in the study must be clearly described. The architecture of the robotic packing software should be included and explained, detailing all algorithms used in the system (packing, perception systems, robot operation). Videographic evidence is suggested to facilitate understanding, showing one or several operation cycles of the system [16,23,31,47].

8. Conclusions

The growing research interest in robotic 3D packing is evidenced by the increasing number of publications, indicating that this area is booming and likely to see continued expansion. The reviewed studies highlight a significant focus on online packing problems, emphasizing the need to address scenarios where item information is incomplete, a common challenge in dynamic logistics and e-commerce environments.

This review uniquely integrates insights from robotics and packing problems, providing a structured framework bridging the gap between these traditionally separate fields. Integrating robotic systems into logistics operations can significantly enhance efficiency and accuracy, which is crucial for meeting the demands of e-commerce. Managers should consider investing in flexible robots capable of adapting to dynamic environments and incomplete item information, supported by careful cost-benefit analyses and strategic planning.

While this review has identified several research gaps, it is important to note their practical implications. The need for standardized terminologies, comprehensive methodologies, and the consideration of real-world constraints in robotic algorithms are not just academic concerns but are crucial for the effective implementation of robotic 3D packing systems. Future research should focus on developing equipment selection guides, exploring the maximum capabilities of suction-type grippers, and establishing clear strategies for sensor placement and software architecture. The review also highlights the potential for improved human-robot collaboration through more extensive research and implementation of cobots. There is a need to explore the best human-robot configurations for packing activities to enhance efficiency and safety.

Furthermore, the reviewed studies reveal a bias toward handling regular shapes, such as boxes and single-container systems. Future research should address the challenges of packing irregular shapes and multi-container systems, particularly in heterogeneous environments. The practical applications and industrial relevance of the findings and methodologies discussed in this review underscore the importance of continued research and development in robotic 3D packing. By providing a structured approach to integrating robotics with packing problems, this review aims to facilitate the development of more robust and practical automated packing systems for various industrial applications.

Author Contributions: Conceptualization, D.Á.-M.; methodology, G.P.-B., D.G., A.M., A.G. and D.Á.-M.; validation, C.M., C.R. and D.Á.-M.; formal analysis, D.Á.-M.; investigation, D.Á.-M.; resources, D.Á.-M.; data curation, D.Á.-M.; writing—original draft preparation, G.P.-B. and D.G.; writing—review and editing, D.Á.-M.; visualization, D.Á.-M.; supervision, C.R. and D.Á.-M.; project administration, A.M.; funding acquisition, C.M., C.R. and D.Á.-M. All authors have read and agreed to the published version of the manuscript.

Funding: This research was made possible thanks to the funding from the Patrimonio Autónomo Fondo Nacional de Financiamiento para la Ciencia, la Tecnología y la Innovación Francisco José de Caldas. The APC was funded by Universidad de los Andes.

Conflicts of Interest: Authors Andrea García and César Marín were employed by the company Integra S.A. The remaining authors declare that the research was conducted in the absence of any commercial or financial relationships that could be construed as a potential conflict of interest.

References

1. Cimini, C.; Lagorio, A.; Romero, D.; Cavalieri, S.; Stahre, J. Smart logistics and the logistics operator 4.0. *IFAC-PapersOnLine* **2020**, *53*, 10615–10620. [CrossRef]
2. Mital, A. Comparison of manual and automated palletizing of mixed size and weight box containers. *Int. J. Ind. Ergon.* **1992**, *9*, 65–74. [CrossRef]
3. Agha, H.N.; Decamp, W.H.; Shell, R.L.; Hall, E.L. Robotic Palletizing of Fixed-and Variable-Size/Content Parcels. In *Handbook of Industrial Automation*; CRC Press: Boca Raton, FL, USA, 2000.
4. Ali, S.; Ramos, A.G.; Carravilla, M.A.; Oliveira, J.F. On-line three-dimensional packing problems: A review of off-line and on-line solution approaches. *Comput. Ind. Eng.* **2022**, *168*, 108122. [CrossRef]

5. Leao, A.A.S.; Toledo, F.M.B.; Oliveira, J.F.; Carravilla, M.A.; Alvarez-Valdés, R. Irregular packing problems: A review of mathematical models. *Eur. J. Oper. Res.* **2020**, *282*, 803–822. [CrossRef]
6. Chua, P.Y.; Ilschner, T.; Caldwell, D.G. Robotic manipulation of food products—A review. *Ind. Robot. Int. J.* **2003**, *30*, 345–354. [CrossRef]
7. Mahalik, N. Advances in Packaging Methods, Processes and Systems. *Challenges* **2014**, *5*, 374–389. [CrossRef]
8. Nayik, G.A.; Muzaffar, K.; Gull, A. Robotics and Food Technology: A Mini Review. *J. Nutr. Food Sci.* **2015**, *5*, 4.
9. Whelan, P.F.; Batchelor, B.G. Flexible packing of arbitrary two-dimensional shapes. *Opt. Eng.* **1993**, *32*, 3278–3287. [CrossRef]
10. Webster, J.; Watson, R.T. Analyzing the Past to Prepare for the Future: Writing a Literature Review. *MIS Q.* **2002**, *26*, xiii–xxiii.
11. Agarwal, M.; Biswas, S.; Sarkar, C.; Paul, S.; Paul, H.S. Jampacker: An Efficient and Reliable Robotic Bin Packing System for Cuboid Objects. *IEEE Robot. Autom. Lett.* **2021**, *6*, 319–326. [CrossRef]
12. Xiong, H.; Ding, K.; Ding, W.; Peng, J.; Xu, J. Towards reliable robot packing system based on deep reinforcement learning. *Adv. Eng. Inform.* **2023**, *57*, 102028. [CrossRef]
13. Wang, F.; Hauser, K. Dense Robotic Packing of Irregular and Novel 3D Objects. *IEEE Trans. Robot.* **2022**, *38*, 1160–1173. [CrossRef]
14. Huang, S.; Wang, Z.; Zhou, J.; Lu, J. Planning Irregular Object Packing via Hierarchical Reinforcement Learning. *IEEE Robot. Autom. Lett.* **2023**, *8*, 81–88. [CrossRef]
15. Iriondo, A.; Lazkano, E.; Ansuategi, A.; Fernandez, A.; Maurtua, I. Dynamic mosaic planning for a robotic bin-packing system based on picked part and target box monitoring. *Int. J. Adv. Manuf. Technol.* **2023**, *125*, 1965–1985. [CrossRef]
16. Yang, S.; Song, S.; Chu, S.; Song, R.; Cheng, J.; Li, Y.; Zhang, W. Heuristics Integrated Deep Reinforcement Learning for Online 3D Bin Packing. *IEEE Trans. Autom. Sci. Eng.* **2023**, *21*, 939–950. [CrossRef]
17. Puls, F.M.; Tanchoco, J.M. Robotic implementation of pallet loading patterns. *Int. J. Prod. Res.* **1986**, *24*, 635–645. [CrossRef]
18. Penington, R.A.; Tanchoco, J.M.A. Robotic palletization of multiple box sizes. *Int. J. Prod. Res.* **1988**, *26*, 95–105. [CrossRef]
19. Iwao, K.; Koyama, K.; Wan, W.; Nishi, T.; Harada, K. Robotic Grasp/Motion Planning for Efficient Packing Assuming Multiple Grippers. *Res. Sq.* **2022**. [CrossRef]
20. Ma, W.; Zhang, B.; Han, L.; Huo, S.; Wang, H.; Navarro-Alarcon, D. Action Planning for Packing Long Linear Elastic Objects into Compact Boxes with Bimanual Robotic Manipulation. *IEEE/ASME Trans. Mechatron.* **2023**, *28*, 1718–1729. [CrossRef]
21. Seger, B. Application of an Industrial Robot for Palletising of Boxes. *Ind. Robot. Int. J.* **1975**, *2*, 10–13. [CrossRef]
22. Lim, S.; Yu, S.; Han, C.; Kang, M. Palletizing simulator using optimized pattern and trajectory generation algorithm. *Mech. Syst. Signal Process.* **2010**, *24*, 1487–1502.
23. Camacho, G.A.; Alvarez, D.; Cuellar, D. Heuristic Approach for The Multiple Bin-Size Bin Packing Problem. *IEEE Lat. Am. Trans.* **2018**, *16*, 620–626. [CrossRef]
24. Hong, Y.D.; Kim, Y.J.; Lee, K.B. Smart pack: Online autonomous object-packing system using rgb-d sensor data. *Sensors* **2020**, *20*, 4448. [CrossRef] [PubMed]
25. Sujan, V.A.; Dubowsky, S.; Ohkami, Y. Robotic Manipulation of Highly Irregular Shaped Objects: Application to a Robot Crucible Packing System for Semiconductor Manufacture. *J. Manuf. Process.* **2002**, *4*, 1–15. [CrossRef]
26. Chang, Q.; Sebghati, A. Efficient Algorithm Based on Adaptive Window–Model Predictive Control for Automatic Stacking in Warehouse Center. *IEEE Access* **2021**, *9*, 94813–94825. [CrossRef]
27. Wang, F.; Hauser, K. Robot Packing with Known Items and Nondeterministic Arrival Order. *IEEE Trans. Autom. Sci. Eng.* **2021**, *18*, 1901–1915. [CrossRef]
28. Abdou, G.; Lee, E. Physical model for robotics palletization. *Comput. Ind.* **1991**, *16*, 255–266. [CrossRef]
29. Balatti, P.; Leonori, M.; Ajoudani, A. A flexible and collaborative approach to robotic box-filling and item sorting. *Robot. Auton. Syst.* **2021**, *146*, 103888. [CrossRef]
30. Zhao, H.; Pan, Z.; Yu, Y.; Xu, K. Learning Physically Realizable Skills for Online Packing of General 3D Shapes. *ACM Trans. Graph.* **2023**, *42*, 165. [CrossRef]
31. Adu-Bredu, A.; Zeng, Z.; Pusalkar, N.; Jenkins, O.C. Elephants Don't Pack Groceries: Robot Task Planning for Low Entropy Belief States. *IEEE Robot. Autom. Lett.* **2022**, *7*, 25–32. [CrossRef]
32. Zhao, H.; Zhu, C.; Xu, X.; Huang, H.; Xu, K. Learning practically feasible policies for online 3D bin packing. *Sci. China Inf. Sci.* **2021**, *65*, 112105. [CrossRef]
33. Ojha, A.; Agarwal, M.; Singhal, A.; Sarkar, C.; Ghosh, S.; Sinha, R. A generalized algorithm and framework for online 3-dimensional bin packing in an automated sorting center. In Proceedings of the 2021 Seventh Indian Control Conference (ICC), Virtually, 20–22 December 2021; pp. 135–140.
34. Masood, S.H.; Khan, H.A. Development of pallet pattern placement strategies in robotic palletisation. *Assem. Autom.* **2014**, *34*, 151–159. [CrossRef]
35. Verma, R.; Singhal, A.; Khadilkar, H.; Basumatary, A.; Nayak, S.; Singh, H.V.; Kumar, S.; Sinha, R. A Generalized Reinforcement Learning Algorithm for Online 3D Bin-Packing. *arXiv* **2020**, arXiv:2007.00463.
36. Murdivien, S.A.; Um, J. BoxStacker: Deep Reinforcement Learning for 3D Bin Packing Problem in Virtual Environment of Logistics Systems. *Sensors* **2023**, *23*, 6928. [CrossRef] [PubMed]
37. Shuai, W.; Gao, Y.; Wu, P.; Cui, G.; Zhuang, Q.; Chen, R.; Chen, X. Compliant-based robotic 3D bin packing with unavoidable uncertainties. *IET Control Theory Appl.* **2023**, *17*, 2241–2258. [CrossRef]

38. Zhang, R.; Chen, X.; Wan, Z.; Wang, M.; Xiao, X. Deep Learning-Based Oyster Packaging System. *Appl. Sci.* **2023**, *13*, 13105. [CrossRef]
39. Schuster, M.; Bormann, R.; Steidl, D.; Reynolds-Haertle, S.; Stilman, M. Stable stacking for the distributor's pallet packing problem. In Proceedings of the IEEE/RSJ 2010 International Conference on Intelligent Robots and Systems, IROS 2010—Conference Proceedings, Taipei, Taiwan, 18–22 October 2010; pp. 3646–3651.
40. Rhiat, A.; Aggoun, A.; Lachere, R. Combining Mobile Robotics and Packing for Optimal deliveries. *Procedia Manuf.* **2020**, *44*, 536–542. [CrossRef]
41. Shome, R.; Tang, W.N.; Song, C.; Mitash, C.; Kourtev, H.; Yu, J.; Boularias, A.; Bekris, K.E. Towards Robust Product Packing with a Minimalistic End-Effector. In Proceedings of the 2019 International Conference on Robotics and Automation (ICRA), Montreal, QC, Canada, 20–24 May 2019; pp. 9007–9013.
42. Shome, R.; Tang, W.N.; Song, C.; Mitash, C.; Kourtev, H.; Yu, J.; Boularias, A.; Bekris, K.E. Tight Robot Packing in the Real World: A Complete Manipulation Pipeline with Robust Primitives. *arXiv* **2021**, arXiv:1903.00984.
43. Song, S.; Yang, S.; Song, R.; Chu, S.; Li, y.; Zhang, W. Towards Online 3D Bin Packing: Learning Synergies between Packing and Unpacking via DRL. In Proceedings of the 6th Conference on Robot Learning, Atlanta, GA, USA, 6–9 November 2023; Liu, K., Kulic, D., Ichnowski, J., Eds.; PMLR: New York, NY, USA, 2023; Volume 205, pp. 1136–1145.
44. Cheng, H.H.; Penkar, R. Stacking irregular-sized packages by a robot manipulator. *IEEE Robot. Autom. Mag.* **1995**, *2*, 12–20. [CrossRef]
45. Lamon, E.; Leonori, M.; Kim, W.; Ajoudani, A. Towards an Intelligent Collaborative Robotic System for Mixed Case Palletizing. In Proceedings of the 2020 IEEE International Conference on Robotics and Automation (ICRA), Paris, France, 31 May–31 August 2020; pp. 9128–9134.
46. Sujan, V.A.; Dubowsky, S. Application of a model-free algorithm for the packing of irregular shaped objects in semiconductor manufacture. In Proceedings of the 2000 ICRA. Millennium Conference. IEEE International Conference on Robotics and Automation, San Francisco, CA, USA, 24–28 April 2000; Symposia Proceedings (Cat. No.00CH37065). Volume 2, pp. 1545–1550.
47. Zhao, H.; She, Q.; Zhu, C.; Yang, Y.; Xu, K. Online 3D Bin Packing with Constrained Deep Reinforcement Learning. In Proceedings of the AAAI Conference on Artificial Intelligence, Vancouver, BC, Canada, 2–9 February 2021; Volume 35, pp. 741–749.
48. Xu, Y.; Liu, Y.; Hao, L.; Cheng, H. Design of palletizing algorithm based on palletizing robot workstation. In Proceedings of the 2016 IEEE International Conference on Real-Time Computing and Robotics (RCAR), Angkor Wat, Cambodia, 6–9 June 2016; pp. 609–613.
49. Wang, F.; Hauser, K. Stable Bin Packing of Non-convex 3D Objects with a Robot Manipulator. In Proceedings of the 2019 International Conference on Robotics and Automation (ICRA), Montreal, QC, Canada, 20–24 May 2019; pp. 8698–8704.
50. Esaki, K.; Kimura, N.; Ito, K. Placement-position search technique for packing various objects with robot manipulator. In Proceedings of the 2016 IEEE International Conference on Systems, Man, and Cybernetics (SMC), Budapest, Hungary, 9–12 October 2016; pp. 246–251. [CrossRef]
51. Mazouz, A.K.; Shell, R.L.; Hall, E.L. Expert System For Flexible Palletizing Of Mixed Size And Weight Parcels. In *Intelligent Robots and Computer Vision VI*; Casasent, D.P., Hall, E.L., Eds.; SPIE: Bellingham, WA, USA, 1988; Volume 0848, pp. 556–565.
52. Dzitac, P.; Mazid, A.M. An Efficient Control Configuration Development for a High-speed Robotic Palletizing System. In Proceedings of the 2008 IEEE Conference on Robotics, Automation and Mechatronics, Chengdu, China, 21–24 September 2008; pp. 140–145.
53. Pan, J.H.; Hui, K.H.; Gao, X.; Zhu, S.; Liu, Y.H.; Heng, P.A.; Fu, C.W. SDF-Pack: Towards Compact Bin Packing with Signed-Distance-Field Minimization. In Proceedings of the 2023 IEEE/RSJ International Conference on Intelligent Robots and Systems (IROS), Detroit, MI, USA, 1–5 October 2023; IEEE: New York, NY, USA, 2023; pp. 10612–10619.
54. Tsai, D.-M. Modeling and Analysis of Three-Dimensional Robotic Palletizing Systems for Mixed Carton Sizes. Ph.D. Thesis, Iowa State University, Ames, IA, USA, 1987.
55. Mertens, P. An Industrial Palletizing System for Industrial Robots. Ph.D. Thesis, Virginia Polytechnic Institute and State University, Blacksburg, VA, USA, 1985.
56. Dyckhoff, H. A typology of cutting and packing problems. *Eur. J. Oper. Res.* **1990**, *44*, 145–159. [CrossRef]
57. Wäscher, G.; Haußner, H.; Schumann, H. An improved typology of cutting and packing problems. *Eur. J. Oper. Res.* **2007**, *183*, 1109–1130. [CrossRef]
58. Bortfeldt, A.; Wäscher, G. Constraints in container loading—A state-of-the-art review. *Eur. J. Oper. Res.* **2013**, *229*, 1–20. [CrossRef]
59. Ramos, A.G.; Oliveira, J.F. Cargo stability in the container loading problem-state-of-the-art and future research directions. In Proceedings of the Operational Research: IO2017, Valença, Portugal, 28–30 June 2018; Springer International Publishing: Cham, Switzerland, 2018; pp. 339–350.
60. Álvarez-Martínez, D.; Camacho, G. Modelling the kinematic properties of an industrial manipulator in packing applications. In Proceedings of the 2018 IEEE 14th International Conference on Control and Automation (ICCA), Anchorage, AK, USA, 12–15 June 2018; pp. 1028–1033.

61. Martello, S.; Pisinger, D.; Vigo, D. The Three-Dimensional Bin Packing Problem. *Oper. Res.* **2000**, *48*, 256–267. [CrossRef]
62. Rennie, C.; Shome, R.; Bekris, K.E.; De Souza, A.F. A dataset for improved rgbd-based object detection and pose estimation for warehouse pick-and-place. *IEEE Robot. Autom. Lett.* **2016**, *1*, 1179–1185. [CrossRef]

applied sciences

MDPI

Article

Commuter Experience: An Assessment of Metro-Train Comfort Amidst Operational Vibroacoustic Conditions

Krzysztof Kowalczyk and Michał Opala *

Faculty of Transport, Warsaw University of Technology, Koszykowa 75, 00-662 Warsaw, Poland
* Correspondence: michal.opala@pw.edu.pl

Abstract: This study investigates the impact of different passenger rolling stock structures and train configurations on vibrations and noise generated during operation. Vibroacoustic measurements were performed during acceleration, constant speed, and braking phases to analyze the effects using statistical analysis according to the relevant standards, revealing a statistically significant relationship between the equivalent noise level and vibration dose. In the context of the expanding Warsaw metro network, which is adding new lines and modern rolling stock, trends in the development of metro rolling stock were analyzed using five different types of metro vehicles, from the oldest to the newest designs. Vehicle performance was ranked in the context of ride comfort using standards and combined vibration and noise measures. The research results allowed for a collective comparative assessment of the construction of individual types of passenger rolling stock in terms of vibroacoustic phenomena, thanks to which it is possible to assess the impact of modern solutions and the justification for investing in modern rolling stock. Newer trains generally record significantly lower vibration and noise levels. The difference between the oldest and the most modern vehicle types is 57% for the vibration acceleration level (0.08 ms^{-2} RMS) and 66% for the noise level (2.2 dB L_{Aeq}).

Keywords: metro; vibration; noise; measurements; statistical analysis; comfort

Citation: Kowalczyk, K.; Opala, M. Commuter Experience: An Assessment of Metro-Train Comfort Amidst Operational Vibroacoustic Conditions. *Appl. Sci.* **2024**, *14*, 6137. https://doi.org/10.3390/app14146137

Academic Editor: Gino Iannace

Received: 8 June 2024
Revised: 28 June 2024
Accepted: 10 July 2024
Published: 15 July 2024

1. Introduction

Rolling stock, a crucial part of railway transportation systems with a limited operational lifespan, requires replacement to keep services efficient. Within today's budget constraints, railroad and transit operators inevitably face decisions on the most suitable equipment for specific services. This raises questions about what equipment is best suited to replace existing ones, and various agencies and operators employ diverse metrics to determine the optimal rolling stock configurations [1]. In the realm of rolling stock performance, it is widely acknowledged that passengers prioritize clean, comfortable vehicles that facilitate easy boarding and alighting. The challenge lies in determining the relative importance of these factors, including comfort level. Estimating the effects of improving rolling stock on ride quality is possible by relating these improvements to fares and In-Vehicle Time (IVT) metrics. Improved ride quality can increase passenger satisfaction, which may result in higher ridership levels or longer journey times [2,3]. For example, if a 10% improvement in passenger satisfaction leads to a 2% increase in ridership or a 5% decrease in the perceived In-Vehicle Time (IVT), this could be used to estimate the impact on fare revenue or operational costs. Transportation planning also faces environmental challenges, particularly in terms of noise and vibration associated with railways. Vibrations originating at the vehicle–track interface can affect passengers and infrastructure, contributing to structural noise. Vibrations can cause fatigue, muscle strain, headaches, dizziness, nausea, and other health issues when experienced for extended periods or at high intensities [4,5]. In railway environments, vibrations are generated by train movement on tracks, track irregularities, wheel–rail interaction forces, and pantograph–catenary interaction [6,7]. The discomfort caused by vibration is influenced by frequency content (low frequencies generally cause more discomfort), amplitude, duration, and pattern. Additionally, vibrations within the track can accelerate wear on components such as sleepers, pads, and fastenings.

Noise exposure can lead to temporary or permanent hearing loss, sleep disturbances, stress, anxiety, and other adverse health effects [4,5]. In railway environments, noise sources include rolling stock engines, wheelsets, brakes, air conditioning systems, passenger interactions, track-borne noise (e.g., rail joint clatter), pantograph–catenary interaction, and structure-borne vibration transmission from the train to its surroundings. The discomfort caused by noise is influenced by factors such as sound pressure level, frequency content, duration, and pattern.

The combined effect of vibration and noise on comfort can be assessed using subjective measures like annoyance ratings or objective physiological responses (e.g., heart rate variability). Several studies have investigated the relationship between these two factors in various transportation contexts, including railways. These investigations often involve controlled laboratory experiments where participants are exposed to different vibration and noise levels while completing tasks related to their comfort perception or performance under such conditions [8].

Examining the specific case of the Warsaw metro, we analyzed vibrations and noise across various metro train types, ranging from older models to the most modern ones (Figure 1). The evaluation analyses vibration and noise levels at different positions within each train's interior. The goal is to identify trends in vibroacoustics and assess how each type performs in terms of generating, transmitting, and mitigating vibrations. While there have been extensive studies on vibroacoustic phenomena in railways, there is still a gap in understanding how these factors collectively influence overall passenger comfort. Additionally, there is a need for convenient rating scales to compare different train types. The implications of these findings are substantial for transportation planning and operation, emphasizing the importance of considering multiple factors to improve the overall passenger experience.

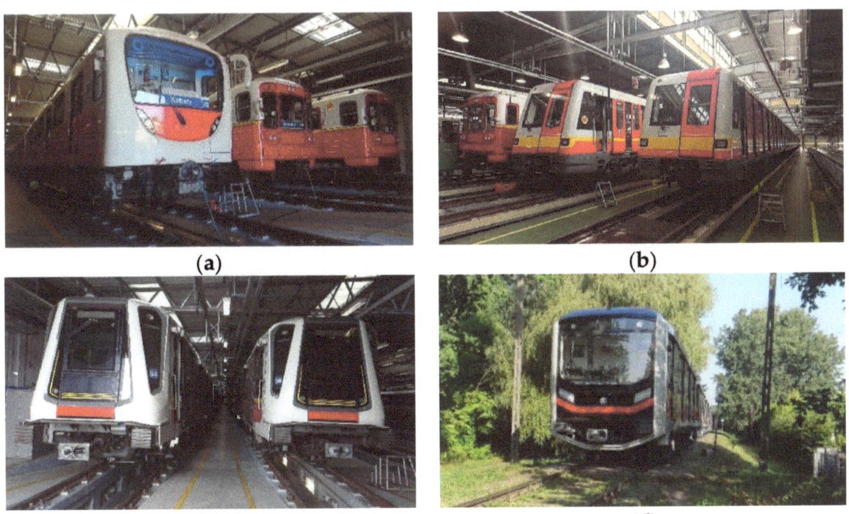

(a) (b) (c) (d)

Figure 1. (**a**) Set of the newer type 81 series compared to the version with an old front wall, (**b**) Alstom Metropolis 98b, (**c**) Siemens Inspiro, and (**d**) Škoda Varsovia.

Vibrations and noise in metro trains originate from two sources: external factors, such as poor infrastructure maintenance causing wear and noise during train passage, and internal components of the metro vehicle itself. The first group includes issues like railway turnouts or curved lines in tunnels, contributing to increased noise both inside and outside the train. The second group involves vibrations and noise generated directly by individual components within the metro vehicle. Key contributors to internal noise are the converter and inverter. The converter, audible even when the train is stationary, changes voltage from the third rail for electrical equipment, often operating at 110 V. The inverter controls voltage for acceleration and braking, noticeable in specific train models like Metropolis 98b. The pneumatic system, controlling brakes and doors, produces increased noise and vibrations, particularly evident in Russian-made trains. Wheelset axle issues, including failures, defects, and damage to bearings, also impact vibrations and noise in metro vehicles. Rail vehicle body vibrations are categorized into stiff and flexible modes, affecting ride comfort. Stiff modes, like bouncing and rolling, occur at around 1 Hz, while flexible modes, impacting human balance, occur in the 4 to 12 Hz range. Minimizing stiff modes' impact with springs and shock absorbers does not directly address flexible modes, which require distinct designs [9–11]. Modern lightweight bodies, while reducing weight, may compromise structural stiffness. Relying solely on rigid modes for comfort analysis might overlook critical phenomena, especially considering structural flexibility's impact on acceleration levels in the vehicle floor [12,13]. Proper wagon design and second-stage suspension assembly selection are crucial to prevent excessive vibrations [14–18]. Noise and vibration levels are further influenced by the drive system and bogie running gear characteristics, contributing to the overall vibroacoustic climate. This paper considers rail vehicles used in the Warsaw metro. In Table 1, basic parameters regarding the rolling stock construction are presented.

Table 1. Selected design parameters of metro vehicles.

Parameter	**81**	**Metropolis**	**81 New**	**Inspiro**	**Varsovia**
Drive engines	Serial	Asynchronous	Serial	Asynchronous	Asynchronous
Start	Resistor-based	Inverter-based	Resistor-based	Inverter-based	Inverter-based
I-st degree suspension	Coil springs	Metal–rubber springs, conical	Coil springs	Metal–rubber springs conical	Metal–rubber springs
II-nd degree suspension	Coil springs and hydraulic dampers	Pneumatic (air cushion) and hydraulic dampers (vertical and horizontal)	Coil springs and hydraulic dampers	Pneumatic (air cushion) and hydraulic dampers	Pneumatic (air cushion) and hydraulic dampers (vertical and horizontal)
Structure of the coach body	Steel, welded	Aluminum alloy and steel, riveted	Steel, welded	Aluminum alloy, welded	Aluminum alloy
Compressor	Piston-based	Screw compressor	Piston-based	Screw compressor	Piston-based
Ventilation of the passenger compartment	Free flow	Forced	Forced	Forced	Forced
Brake	Electro-dynamic + pneumatic	Electro-dynamic with recuperation and pneumatic	Electro-dynamic + pneumatic	Electro-dynamic with recuperation	Electro-dynamic with recuperation and electro-pneumatic

In series 81 vehicles, each train wagon is driven. In Metropolis trains, wagons 1 and 6 have non-powered bogies; in Inspiro trains, bogies in wagons 2 and 5 are not powered; and in the latest Škoda trains, non-powered bogies are used in wagons 3 and 4. Additionally, only Inspiro trains use non-rotational bogies.

The Warsaw metro uses ballast-less (concrete) and sleeper-less tracks. The concrete foundation (acts as a ballast) on the M1 line uses B25-class concrete, appropriately formed with channels. For part of the length, the rails are placed in the lagging. S60E1 and S49E1 rails are used along with a rail attachment node. Both parts are connected with anchor bolts, fixed (glued) in the concrete base using epoxy concrete. On the Kabaty–Wilanowska section, the track surface is fastened with a K-type fastening with a triple ring and a Łpa2 clamp.

To the north of the Wilanowska station, Skl-3 spring clips, FC104-type vibration-isolating pads (made of cork elastomer), and FC585-type under-rail spacers are used. Thanks to these measures, the track is more flexible. The switches used are a modification of standard railway turnouts S60–1:9–190 and crossings S60–1:4444, fastened with 24/30 mm anchor bolts. Welding of contactless tracks with a track welding machine was replaced with thermite welding, and the glued-compressed joint welded into the track was replaced with a joint made on-site in the track. Preventive rail grinding was also introduced, which involved removing the decarburized layer and improving the profile of the rail head on the wheel running surface. These treatments not only reduce noise and improve travel comfort, but also extend the service life and save energy consumption [19].

Apart from the elements shaping the passenger's surroundings, it is essential to highlight those exclusively tied to individual attributes. The level of comfort experienced during the journey, whether positive or negative, can be impacted by factors such as weight, height, body structure, overall physical and mental health, and the passenger's gender and age. The interpretation of comfort can vary among passengers due to factors like body posture, the unique dynamic response of the individual's body, and even personal expectations [20–23]. The current literature predominantly focuses on the separate evaluation of different factors, often neglecting the combined impact of these factors.

The conclusions from the literature seem consistent, in that subjective assessment of the vibration level is mainly independent of the simultaneous noise occurring within the range of typical values for means of transport. However, subjective noise assessment simultaneously depends on the vibration level [24–27]. If an overall assessment of stimuli is required, noise dominates the assessment with less influence from vibration levels. The overall annoyance resulting from the combined effect of vibration and noise is greater than that of just one type of stimulus.

Several models of the total nuisance caused by noise and vibration can be found in the literature. Psychophysical models focus on the direct relationship between physical stimuli, such as the level of noise and vibration, and people's psychological reactions, such as the level of discomfort or irritation. These models rely on experiments and measurements to understand how different levels of noise and vibration are perceived by humans. Perceptual models go a step further and consider not only sensory aspects but also psychological and cognitive factors. They include psychophysics elements and additionally consider factors such as attention, expectations, and individual differences in perception. These models focus on a more comprehensive understanding of how people evaluate noise and vibration in the context of their experience and psychology.

The combined effect of sound and vibration on human perception can be described using various mathematical models, depending on the specific application or research problem. One standard method for analyzing this interaction is to combine the individual effects of noise and vibration using a weighted sum or product.

The presented assessment of the metro vibroacoustic environment employed three distinct approaches. The initial approach is grounded in international standards addressing vibration comfort and noise in rail vehicles individually. The second approach utilizes psychophysical models [16,24], and the third one uses a relative measure proposed by the authors. This measure considers the combined effects of vibrations and noise, relying on objective measurements.

2. Research Methods

The noise and vibration acceleration measurements took place on the same track, specifically a section of 1108 m from A3 Imielin station to A4 Stokłosy station. However, due to different standards and requirements for measuring noise and vibration comfort, an additional 7592 m (from A3 Imielin station to A11 Politechnika station) was measured for vibrations. The tests were conducted every day from 8:00 p.m. to 11:00 p.m. to increase the results' repeatability and ensure similar measurement conditions. The measurement sites for vibration and noise were two selected wagons of each type of metro train. The measurements took place in the first (control) wagon and the third (middle) wagon. The detailed arrangement of measurement points and their numbering are shown in the drawings below (Figures 2 and 3).

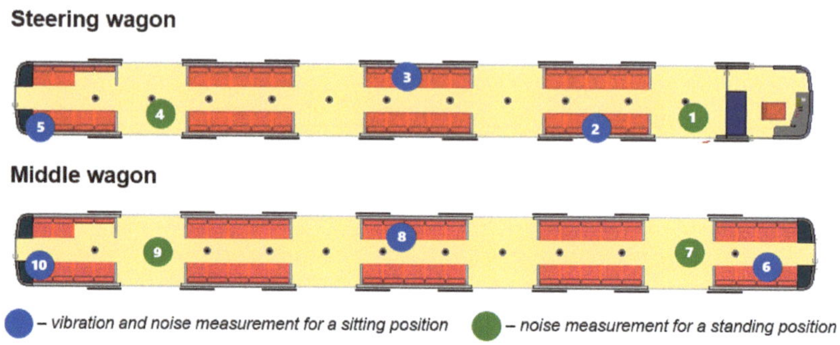

Figure 2. Places and numbering of measurement points in metro cars. Own study based on [28].

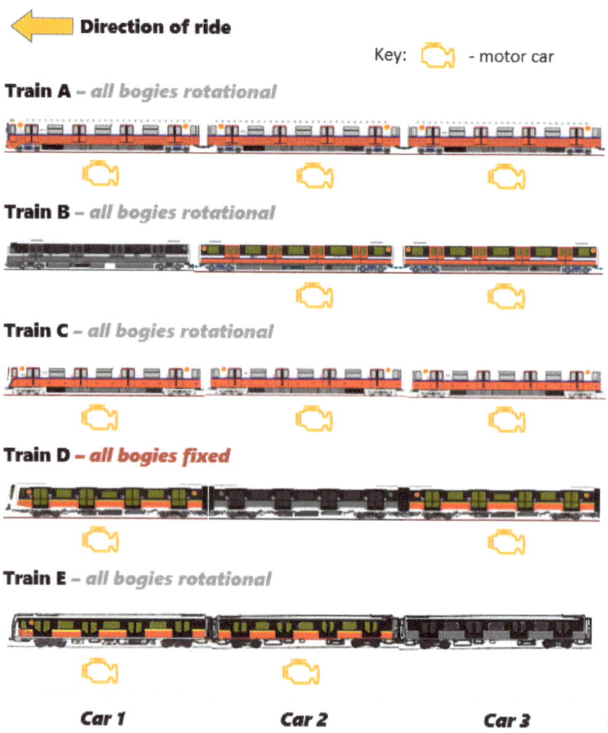

Figure 3. List of tested metro trains, taking into account the location of motor cars [28,29].

Vibration measurements were carried out using a triaxial seat acceleration sensor (10 mV/ms^{-2}), which was placed non-invasively on a passenger seat during the tests. The sensor was connected to a laptop computer using a USB cable and a NI-9233 measurement card with USB housing. Noise was measured using a portable recorder TASCAM DR-44WL at a height of 1.2 m for the sitting position and at a height of 1.6 m for the standing position [30].

Measurements were performed in five train sets of different types, which were further marked with letters as trains A, B, C, D, and E. Two wagons were selected for vibration and noise measurement for each type of train; a total of ten different wagons in regular operation were taken into account.

Assessment Basics

ISO 2631-1 and EN 12299 [31,32] standards are available for evaluating ride comfort for rail vehicle passengers. However, they have some differences in their approach. ISO standards are globally recognized and widely used across industries. EN 12299, despite being younger than ISO 2631-1, has gained significant popularity due to its focus on ride comfort for rail passengers and its statistical method that provides a more precise evaluation of vibration levels. ISO 2631-1's sensitive vertical and lateral vibration ranges are 4–12 Hz and 0.6–2 Hz, respectively. On the EN 12299, the sensitive ranges are 4–16.5 Hz for vertical vibration and 0.6–2 Hz for lateral vibration. Applicable frequency ranges are 0.5–80 Hz for ISO 2631-1 and 0.4–100 Hz for EN 12299. One significant difference between the two standards is that EN 12299 does not take into account vibration dose value (VDV) metrics. VDV is a measure of the cumulative exposure to vibration over a certain period, which can be important for assessing long-term effects on passengers' comfort and health. While there is no universal consensus on the impact of exposure time on riding comfort for shorter periods, everyday experience shows that more prolonged exposure to a harmful factor is more significant than shorter ones. VDV is also said to be a preferred measurement for exposure to jolts, shocks, and intermittent vibration because it is sensitive to peaks in acceleration levels. In the presented paper, the aVDV metric is the same as the metric used in Stevens' psychophysical model, used for comparison purposes; therefore, it seemed convenient to consider this measure.

According to the ISO 2631-1 document, the RMS acceleration value was calculated from the following formula:

$$a_w = \left[\frac{1}{T}\int_0^T a_w^2(t)\, dt \right]^{0.5} \left(\mathrm{ms}^{-2}\right), \tag{1}$$

where $a_w(t)$ is the acceleration weighted as a function of time and T is the measurement time(s). If the crest factor given by Formula (2) exceeded the value of 9, then the vibration dose was analyzed.

$$Crest = \frac{ZP}{RMS} \tag{2}$$

where ZP is the peak acceleration value of $a_w(t)$. The amount of vibration dose VDV is assessed according to the following relationship (3):

$$VDV = \left[\int_0^T a_w^4(t)dt \right]^{0.25} \left(\mathrm{ms}^{-1.75}\right) \tag{3}$$

where T is the measurement duration. VDV is a coefficient for which the maximum values affecting the riding comfort of passengers have not yet been unanimously determined. The directive of the European Parliament specifying vibration exposure limits for employees in an eight-hour working period [33] is calculated according to Formula (4):

$$VDV(8) = VDV \cdot \left[\frac{T_e}{T_m}\right]^{0.25} \left(\mathrm{ms}^{-1.75}\right) \tag{4}$$

where *VDV(8)* is the daily exposure (8 h), *VDV* is the frequency-weighted vibration dose, T_e is the exposure period (8 h), and T_m is the *VDV* measurement period (15 min).

VDV(8) evaluates the effects of vibration from an occupational health perspective and is not informative for comfort assessment (compulsory for employers/workers according to EU Directive 89/391/EEC, Table 2). It is given for comparison purposes.

Table 2. Worker vibration exposure limits for an eight-hour reference period [33].

	Arm–Hand, RMS	Whole Body, RMS	Whole Body, VDV
Vibration threshold value	$2.5\ m/s^2$	$0.5\ m/s^2$	$9.1\ m/s^{1.75}$
Maximum dose value	$5\ m/s^2$	$1.15\ m/s^2$	$21\ m/s^{1.75}$

In the case of noise, coefficients were used, such as the equivalent sound pressure level corrected by the A filter [30,34], used for tests carried out at a standstill or while driving at a constant speed (or during acceleration/braking), described by Equation (5):

$$L_{pAeq,T} = 10\log_{10}\left(\frac{1}{T}\int_0^T \frac{[p_a(t)]^2}{{p_0}^2}dt\right) = 20\log_{10}\left(\frac{1}{T}\int_0^T \frac{p_a(t)}{p_0}dt\right)(\mathrm{dB}) \quad (5)$$

where T is the measurement time (s), $p_a(t)$ is the RMS sound pressure value with A filter (Pa), and p_0 is the reference sound pressure level (p_0 = 20 μPa). The maximum *A*-weighted sound pressure level in a given time interval was determined using fast time weighting F. It is used when examining noise during vehicle acceleration and braking.

$$L_{pAFmax} = \max(p_a(t))\ (\mathrm{dB}) \quad (6)$$

3. Results and Discussion

The first part of the analysis is devoted to vibrations, the second to noise, and the last to the overall assessment of the vibroacoustic climate. Measurements at individual points of a given train were performed at intervals while riding in the same section. According to the standards, the noise should be measured separately in the three different phases (acceleration, constant speed, and braking), while the vibration measurement results should refer to all phases combined.

3.1. Vibration

The highest average RMS acceleration of 0.17 ms^{-2} was achieved by train C. In five of the six measurement points (2, 3, 5, 8, and 10), this train achieved the highest values of this measure (Figure A1). In most of these points, train A, with congruent technical specifications to train C, recorded similar parameters, thus taking second place among the highest average RMS acceleration values obtained by the train. The next pair was formed by the latest metro cars, trains D and E, which achieved comparable values at all measurement points, which were the lowest in each case. Train E achieved the lowest average RMS acceleration value of 0.06 ms^{-2}, and train D had the second smallest—0.07 ms^{-2}.

The weighted RMS values of accelerations in the one-third-octave bands (Figure 4) did not exceed the permissible values in any of the cases. The highest values were achieved for the vertical direction in the frequency range of 10–60 Hz for the constructions with coil springs in second-stage suspensions. In contrast, the designs with pneumatic springs have lower acceleration levels. The traction rod between train D's bogie and car body had no adverse effect on the acceleration level relative to other vehicles. In contrast, the other vehicles without traction rods had higher acceleration levels in the horizontal plane direction. In the vertical direction, the train equipped with a traction rod had the lowest acceleration levels measured at the car's body ends, compared to other vehicles. The levels at the car body center were higher in the 8–25 Hz frequency range but still comparable to B and E trains and less than A and C trains. The absence of a traction rod requires using a traction–torsion device at the bogie pivot, which can generate additional stiffness parallel to the second-stage springing at a level higher than the basic stiffnesses (pneumatic springs only) for the transverse and vertical guides. This may impact passenger comfort, especially with poorer track quality.

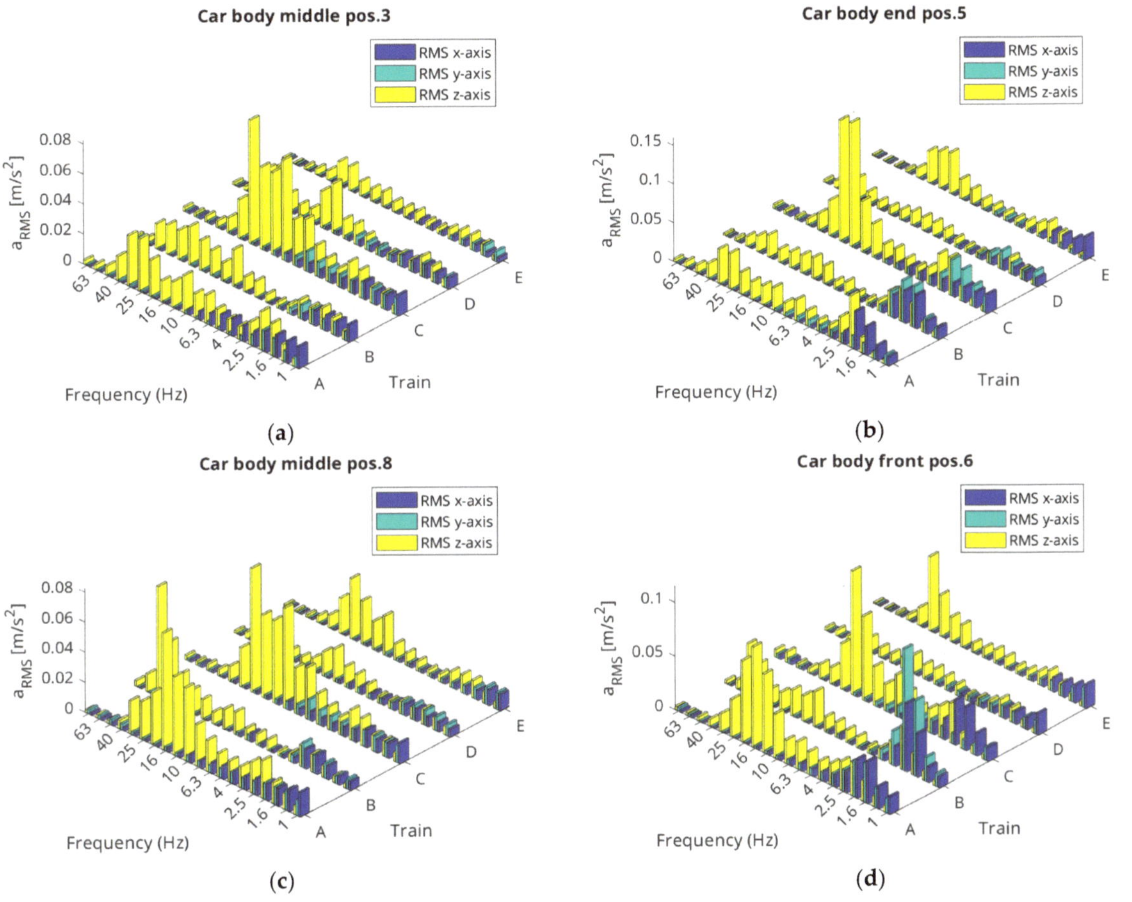

Figure 4. Frequency analysis in the one-third-octave bands for individual measurement positions (**a**) 3, (**b**) 5, (**c**) 8, and (**d**) 6.

The vibration dose results for the eight-hour work period aligned with VDV vibration doses (Figure A2) did not exceed the threshold values in any tested cases.

Crest factor values for the "x" direction were consistent across train types, with train D exhibiting the highest value of 13.1 and train B the lowest at 8.9 (Figure A3a). Train B's lower crest factor in the first three measurement points was attributed to the absence of a power bogie in its first wagon, reducing vibrations during acceleration and braking. Train D, equipped with fixed bogies, demonstrated higher crest factor values, particularly in curves due to increased resistance. The "z" direction crest factor differentiation was more pronounced (Figure A3b), with train E recording the highest average value of 40.1. In contrast, train C had the lowest at 10.9, correlating with its lowest values at specific measurement points.

Train D achieved the highest crest factor values in the middle and end of both tested wagons, i.e., at points 3 (16.1), 5 (10.2), 8 (23), and 10 (16.1). Trains A and E recorded similar values at points 3, 5, and 6. The C train has the most similar values at each point. Therefore, the difference between the extreme measured coefficient values was the smallest and amounts to 4, while the largest was 17.2 for train D. The lowest average value of the crest factor in z-direction was achieved by train C (10.9). The average crest factor of train A may be puzzling because, despite the similarities in design to train C, it is over 78% higher and amounts to 19.5. In contrast to the "x" direction, train D had the second-lowest average crest factor score of 13.7, and train B had the highest crest factor value of 21.6.

3.2. Acceleration Phase—Noise

The equivalent sound pressure level (L_{Aeq}) during acceleration in the tested trains ranged from 53.6 dB (point 2) to 70.1 dB (point 7), as illustrated in Figure A4. The highest average value was achieved by train A (66.3 dB) and the lowest by train D (58 dB) (Table 3).

Table 3. Measured noise values (dB) during acceleration (a), braking (b), and riding at a constant speed (c).

	Measuring Point	Riding Phase	 Train A		 Train B		 Train C		 Train D		 Train E	
			L_{Aeq}	L_{pAFmax}	L_{Aeq}	L_{pAFmax}	L_{Aeq}	L_{pAFmax}	L_{Aeq}	L_{pAFmax}	L_{Aeq}	L_{pAFmax}
First wagon	1	a	66.3	70.9	59.3	63.6	66.1	72.5	55.3	58.9	62	66.1
		c	75.1	78.4	67.8	73.1	75.0	79.6	67.7	72.4	70.1	72.7
		b	69.6	77.7	58.0	68.1	68.5	79.5	63.4	71.7	55.5	63.4
	2	a	63.8	67.9	62.3	66.7	59.1	64.0	53.6	56.7	61.5	66.2
		c	73.1	77.8	69.6	73.4	72.9	66.7	67.7	71.3	69.9	73.4
		b	64.0	74.1	58.8	68.2	60.6	67.0	60.5	63.8	62.9	63.0
	3	a	63.1	68.6	61.9	66.0	58.3	65.9	55.9	59.8	61.5	66.5
		c	65.0	68.4	60.8	64.2	64.7	67.5	58.8	63.5	68.8	72.7
		b	60.8	69.8	53.1	60.8	53.6	63.1	58.6	63.9	54.5	63.3
	4	a	66.2	69.9	60.0	64.4	63.1	67.9	56.8	62.5	61.1	65.9
		c	75.8	80.1	71.7	77.6	73.4	78.1	69.9	74.2	68.9	71.8
		b	69.0	78.1	57.9	66.5	69.2	77.1	64.3	71.0	62.8	68.0
	5	a	64.4	69.8	60.3	62.4	62.3	67.1	57.7	62.8	62.4	65.4
		c	67.9	71.0	61.3	64.4	65.0	71.1	63.3	67.3	71.3	74.0
		b	59.0	69.0	51.3	58.6	61.2	71.4	61.4	63.1	62.9	67.8
Third wagon	6	a	65.9	71.1	63.0	67.4	64.7	70.0	57.7	61.1	64.4	69.5
		c	71.9	74.2	65.0	67.8	65.7	68.3	63.3	67.4	71.6	74.4
		b	64.9	71.7	57.1	64.5	61.0	70.4	62.4	65.5	62.1	70.2
	7	a	70.1	77.2	63.5	68.4	68.0	73.6	61.0	67.7	66.6	73.2
		c	75.8	79.7	75.0	80.6	71.8	75.9	71.0	75.2	74.9	77.8
		b	71.4	81.8	64.7	72.9	65.7	74.8	64.4	67.5	67.9	71.4
	8	a	67.2	72.0	63.1	67.4	63.7	71.2	58.8	66.0	66.2	73.1
		c	71.8	73.6	63.9	65.7	67.7	70.9	61.9	66.2	74.0	77.3
		b	66.5	73.4	58.9	65.5	61.8	70.2	57.0	61.5	66.9	68.0
	9	a	68.2	73.2	65.4	71.0	64.9	70.9	61.2	66.8	66.2	71.8
		c	79.6	83.1	73.5	78.1	71.5	76.2	71.1	75.2	72.2	75.3
		b	71.0	78.4	68.2	75.7	67.3	75.5	64.0	67.0	66.2	70.3
	10	a	67.9	72.8	64.5	69.1	61.1	66.7	61.7	66.6	65.8	72.2
		c	67.5	69.5	68.7	71.2	62.2	64.7	64.0	66.4	72.5	75.4
		b	63.5	71.3	59.3	65.4	61.5	67.2	61.2	64.6	66.2	69.8

These trains recorded extreme values in all measurement places (train A, the highest; train D, the lowest), creating a range for each point. The exception to this rule is the last measurement point because the lowest value of 61.1 dB was recorded for train C. Train A's results are attributed to resistor-based starting. Interestingly, in each case, the average noise level measured in the first wagon was lower than in the third.

Slightly higher values were recorded for the maximum sound level (L_{pAFmax}). On average, it ranged from 62.9 dB for train D (Table 3), through 66.6 dB (train B) and 69 dB (trains C and E), to 71.3 dB for train A. The distribution was similar to L_{Aeq}-measured L_{pAFmax} values at individual measurement points. Thus, to a predominant extent, trains A and D, again, shaped the range of this coefficient—train A recorded the maximum values and vehicle D the minimum. Again, each vehicle had, on average, higher values measured at points in the third wagon. Trains A, B, C, and E achieved similar values at points 2 and 3. They ranged from 64 dB to 68.6 dB. This is interesting because train D achieved values not exceeding 59.8 dB at both points.

3.3. Constant Speed Phase—Noise

The average values of the L_{Aeq} parameter oscillated around 70 dB (Figure A5, Table 3). Three trains achieved values lower than this figure—train D 65.9 dB, train B 67.7 dB, and train C 69 dB. The last two exceeded the 70 dB limit, train E by 1.7 dB and train A by 2.7 dB. As expected, the lowest L_{Aeq} values were obtained in each of the trains at points 3, 5, 6, 8, and 10, i.e., places away from the bogies. Their values range from 58.8 dB to 74 dB and are, on average, higher in the third wagon. In 70% of the measurement locations, train D achieved the lowest values. In the case of train A, the free flow of ventilation in the passenger compartment contributes to the increase in noise level. In 60% of the measurement sites, this train achieved the highest values. Interesting data comes from train E. The differences between neighboring measurement places are the smallest for this train and amount to a maximum of 3.3 dB (points 6 and 7), which indicates the steady level of soundproofing for this type of vehicle.

3.4. Braking Phase—Noise

The average values of the equivalent sound pressure level (L_{Aeq}) during braking ranged from 58.7 dB for train B to 66 dB for train A (Figure A6). Train B recorded the lowest values at six points, while train A achieved the highest values by five times. Vehicle E was the only one to record extreme values at three different points, i.e., at point 1, it achieved the lowest value of all trains, amounting to 55.5 dB, and at points 5 and 10, it achieved the highest values of 62.9 dB and 66.2 dB, respectively. Train D, this time, achieved the lowest values at three points located in the third car. Train A obtained the highest maximum sound level (L_{pAFmax}) for the braking period, reaching 81.8 dB at point 7, while train D achieved the lowest average value of 66 dB.

3.5. Comparative Assessment of the Metro Vehicles Structure from the Point of View of Recorded Vibrations and Noise

In order to compare the vibroacoustic climate, two different methods were used; the first is Stevens' psychophysical model [16] with root-sum-square (RSS) representation [24], and the second is the authors' combined relative measure. Stevens' psychophysical law states that the subjective response increases exponentially with the objective physical magnitude of the stimulus. This can be expressed mathematically as follows:

$$\psi = k\varphi^n \tag{7}$$

where ψ is the subjective quantity, k is a constant, φ is the objective quantity of the stimulus, and n is the exponent dependent on the stimulus. The stimulus can be any continuum that can affect the human senses, such as the sound intensity, magnitude of vibration, temperature, or luminescence of light. When it comes to noise and vibration, the subjective quantity can be defined as follows:

$$\psi_s = k_s \varphi_s^{n_s}\ , \tag{8}$$

$$\psi_v = k_v \varphi_v^{n_v} \tag{9}$$

where ψ_s and ψ_v represent the subjective magnitude of noise and vibration, k_s and k_v are constants for noise and vibration, and $\varphi_s^{n_s}$ and $\varphi_v^{n_v}$ are the physical magnitude of noise and vibration with their exponents, respectively. Using this relationship, the subjective magnitude of noise and vibration can be written in logarithmic form as an equation.

$$\log_{10}\psi_s = \log_{10}k_s + n_s \log_{10}\varphi_s \tag{10}$$

$$\log_{10}\psi_v = \log_{10}k_v + n_v \log_{10}\varphi_v \tag{11}$$

Assuming $L_{Aeq} \propto 20\log_{10}\varphi_s$, $a_{VDV} \propto \varphi_v$ (or $a_{RMS} \propto \varphi_v$), and $L_{Aeq} = L_{AE} - 10\log_{10}\left(\frac{T}{T_0}\right)$, the follow is true:

$$\psi_s = k_s 10^{0.05 n_s\ L_{Aeq}}, \tag{12}$$

$$\psi_v = k_v (a_{VDV})^{n_v}. \tag{13}$$

The root-sum-square model (RSS) assumes [24] that the total discomfort caused by combined noise and vibration can be predicted by taking the square root of the discomfort caused by noise alone and vibration alone, $\psi_c = \sqrt{\psi_s^2 + \psi_v^2}$, where ψ_c means total discomfort, ψ_s is the discomfort caused by the noise itself, and ψ_v is discomfort caused by the vibrations themselves. The lower the value of this indicator, the lower the overall nuisance and the better the comfort. A formula was used to evaluate the obtained measurement range:

$$\psi_c = \left[(0.217 \cdot 10^{0.039\ L_{Aeq}})^2 + (245 \cdot (a_{VDV})^{1.04})^2 \right]^{0.5}. \tag{14}$$

The second type of comparison was made using the relative measure $R_{ij}(x,\ y)$ [–], where x is the reference value and y is the average value of the j-th indicator for the i-th train, determined for a set of n measurement points in a given train. The reference value $x = A_{ij}^{max}$ is the maximum value from the set of values of a given indicator $\overline{S}_{ij}$ for all trains. It follows that when $x = y$, then $R_{ij}(x,\ y) = 0$; in any other case, $R_{ij}(x,\ y) > 0$ when the indicator's value is lower than the reference level x. The value $R_{ij}(x,\ y) = 0$ is the most unfavorable; it means that a given indicator $\overline{S}_{ij}$ has taken the maximum value (in the case of the vibration and noise level measures used, the higher the value of a given indicator, the greater the vibration and noise level). The final assessment compares the F_i values, which are the average value of all relative measures $R_{ij}(x,\ y)$. The higher the value of this indicator, the lower the overall nuisance and the better the comfort (a higher value means that the values of individual indicators are further from the unfavorable values).

$$\overline{S}_{ij} = \frac{1}{n}\sum_{k=1}^{n} s_{ijk}\ ;\quad S_{ijk} = \{s_{ij1} \ldots s_{ijn}\} \tag{15}$$

$$A_{ij}^{max} = max\{\overline{S}_{ij}\ :\ j = 1 \ldots m\} \tag{16}$$

$$R_{ij} = 1 - \frac{\overline{S}_{ij}}{A_{ij}^{max}} \tag{17}$$

$$F_{ci} = \frac{1}{m}\sum_{j=1}^{m} R_{ij} \tag{18}$$

It is interesting to note the consistency of the results between the individual vibration and noise measurements (international standards) and the combined vibroacoustic measurement (F_c, Steven's metrics), as presented in the paper. There appears to be a general agreement between the two sets of results, with train D being the most comfortable according to individual and combined metrics while train A was the least comfortable for both metrics. However, it should be noted that the production dates of each train type vary significantly: A-train-type production started in 1976, and D-train-type production started in 2012; hence, there is a vast technological gap between their constructions, and they can be considered extreme cases. On the other hand, according to individual metrics, train B, produced in 1997, subsequently reached the second lowest noise level (L_{Aeq}) and third lowest vibration level (VDV and RMS). Train E reached the fourth lowest noise level (L_{Aeq}), first lowest RMS level, and second lowest VDV level. It would not be obvious to rank these trains based solely on their performance in the international standards until we use combined metrics. However, the combined metric might not always lead to significant variations from what one would expect based on the individual measurements alone due to the strong physical correlation between vibration and noise. They are closely related phenomena in rail transport, as they often have common sources and can interact. In the presented case of noise and vibration measurements, a statistically significant relationship (p-value = 0.018, alpha = 0.05) can be observed between the equivalent noise level and vibration dose (Figure 5a) with a moderate positive linear relationship (Pearson correlation coefficient PCC = 0.43). Compared to VDV, L_{Aeq} is less sensitive to peaks and more focused on overall energy, whereas VDV emphasizes high-intensity peaks due to its fourth power integration. RMS provides an average power measure and is less sensitive to peaks than both VDV and L_{Aeq}, focusing on the effective value of the signal. Each metric is designed to capture cumulative effects in their respective domains, with varying sensitivities to fluctuations and peaks in the data.

Table 4. Final evaluation of the results obtained.

Tested Indicator	 **Train A**	 **Train B**	 **Train C**	 **Train D**	 **Train E**
Average crest factor CF, longitudinal direction, variable speed	10.2	9.0	9.5	13.1	9.7
Average crest factor CF, vertical direction, constant speed	19.5	21.6	10.9	13.6	40.1
RMS	0.138	0.109	0.169	0.066	0.059
VDV	1.456	0.937	1.376	0.663	0.853
L_{Aeq}, acceleration	66.3	62.3	63.1	58.0	63.8
L_{pAFmax}. acceleration	71.3	66.6	69.0	62.9	69.0
L_{Aeq}. constant speed	72.4	67.7	69.0	65.9	71.4
L_{Aeq}. braking	66.0	58.7	63.0	61.7	62.8
L_{pAFmax}. braking	74.5	66.6	71.6	66.0	67.5
L_{Aeq}, on average	68.2	62.9	65.0	61.8	66.0
L_{pAFmax}. on average	73.8	68.3	70.8	66.3	70.3
F_c1 (excluding CF)	0.03	0.16	0.04	0.24	0.18
F_c2 (including CF)	0.10	0.21	0.14	0.26	0.17
ψ_c	375.5	237.1	349.5	169.4	223

The cross-correlation between the signals is very low in the range of measured frequencies (sampling rate 2000 Hz for vibration and 48 kHz for noise). At the same time, a high correlation was observed between the vibration acceleration signals themselves, regardless of the measurement location and wagon type, and a low correlation between the acoustic signals themselves (Table 5).

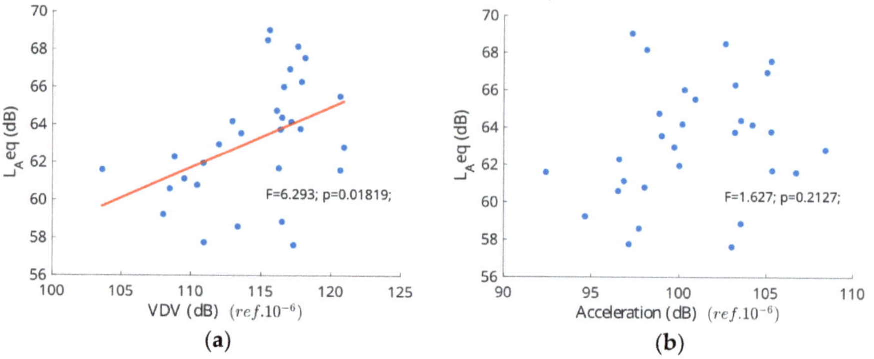

Figure 5. Relationship between the magnitude of vibration and the mean noise level using the observations from all trains: (**a**) statistically significant relationship for the vibration dose value (p-value = 0.018, alpha = 0.05, PCC= 0.43) and (**b**) statistically insignificant relationship for weighted RMS acceleration.

Table 5. Normalized cross-correlation for sound and vibration signals at three locations across the wagon (3—wagon center; 2, 5—wagon ends; V—vibration, S—sound measurement).

	Vibration–Sound		
location	S-2	S-3	S-5
V-2	0.02	0.06	0.07
V-3	0.02	0.06	0.07
V-5	0.01	0.05	0.05
	Vibration–Vibration		
location	V-2	V-3	V-5
V-2	1.00	0.91	0.87
V-3	0.91	1.00	0.87
V-5	0.87	0.87	1.00
	Sound–Sound		
location	S-2	S-3	S-5
S-2	1.00	0.11	0.11
S-3	0.11	1.00	0.10
S-5	0.11	0.10	1.00

Only two wagons in each train were tested, with train B being the only one with a steering wagon. The newest train, train E, achieved values comparable to train C in every noise category. Despite considerable design similarities between the A and C series of trains, trains C performed better than trains A. Trains A recorded the highest average values for both vibrations and noise. Noteworthy is the significantly increased CF value for the vertical vibration signal of train E. Considering that this is a new vehicle and the other indicators are at acceptable levels, this may mean that the dynamic systems in the new vehicle, such as suspension systems or braking systems, adapt and undergo specific changes during the initial period of operation. Initial factory settings and calibrations can be refined during subsequent operations.

From the acoustics point of view, the condition for admitting any rail vehicle into operation is that the vehicle only meets standards related to external noise emission. Issues related to assessing noise inside rail vehicles, except noise in the workplace, are often ignored [35,36]. The technical specifications for interoperability relating to the "Rolling Stock—Noise" subsystem were adopted as a reference point for permissible noise levels in rail vehicles [37]. Some studies suggest even higher values for an acoustically comfortable metro interior space [38].

4. Summary

While individual factors and their impact on passenger comfort have been extensively studied, their combined effect needs to be more pronounced. Each approach offers unique advantages and limitations. For instance, international standards provide a solid foundation for evaluating vibration comfort but may only capture part of the full scope of passenger perceptions and preferences. The advantage of psychophysical models is that they provide insights into how humans perceive and respond to vibrations and noise in transportation environments. This information can be valuable for designing vehicles that minimize discomfort and improve passenger experience. However, they rely on subjective data from human participants, which can vary between individuals and may not represent the entire population. Additionally, psychophysical tests can be time-consuming and expensive to conduct. The combined relative measures also offer advantages and limitations. The advantage of the combined metric is that it provides an objective measure of the overall nuisance resulting from the combined effect of vibration and noise, making it helpful in comparing trains systematically and quantitatively. However, using combined measures requires careful consideration when selecting appropriate comparison criteria, as different factors may not be directly comparable due to varying units or scales used in measuring them.

The highest average values of the equivalent and maximum sound pressure levels, corrected by filter A, occurred during the constant speed phase. The observation is consistent for all the measured vehicles (five train types, two wagons in each train; Table 3). The signals were recorded while driving at the normal service speed at the same track section between two stations. A possible explanation is that trains traveling through the tunnel in the opposite direction influenced the noise level measured. This section of the tunnel is constructed in such a way that the distance between the track axes is approximately 4.7 m. Comfort limits for vibrations (ISO 2631-1) and noise (TSI standard) were consistently within acceptable ranges. Passenger-generated noise and driver techniques significantly impact recorded differences in results. During measurements on trains E, where drivers were still improving their skills, the operation was not consistently smooth. Vibration and noise values during constant speed and braking depend on the measurement location, with higher values above the bogie and lower values between them. During acceleration, noise levels in a given wagon remain similar in different measurement locations. However, the first carriage is usually quieter than the third. The manufacturer's use of rolling bogies positively influences vibration reduction during speed changes, though it has negligible influence when traveling at a constant speed.

Despite receiving positive evaluations in either vibration or noise categories, a specific vehicle might have a less favorable overall rating for vibroacoustic nuisance when considering the combined effects of vibration and noise. Noise levels in metro vehicles exhibit variation based on the train type. Vehicle D emerges as the quietest during constant speed and acceleration, while train B achieves the best average result during braking. Using non-rotational bogies in vehicle D did not impede its excellent vibration results, recording the lowest average vibration dose.

Based on data from all the trains, a statistically significant relationship can be observed between the equivalent noise level and vibration dose with a moderate positive linear relationship (PCC = 0.43).

Newer trains generally record significantly lower vibration and sound levels (Table 4). The difference between the oldest (A) and the most modern (E) vehicle type is 57% for the vibration acceleration level (0.08 ms^{-2} RMS) and 66% for the noise level (2.2 dB L_{Aeq}). The difference is due to several factors. Newer models feature more advanced suspension systems, superior acoustic insulation, improved aerodynamics, and more efficient and quieter propulsion and braking systems, all contributing to reduced noise and vibration levels.

Author Contributions: Conceptualization, M.O. and K.K.; methodology, M.O.; software, M.O.; validation, K.K.; formal analysis, K.K. and M.O.; investigation, K.K. and M.O.; resources, K.K.; data curation, K.K.; writing—original draft preparation, K.K.; writing—review and editing, M.O.; visualization, K.K.; supervision, M.O.; project administration, M.O.; funding acquisition, M.O. All authors have read and agreed to the published version of the manuscript.

Funding: This paper was co-financed under the research grant of the Warsaw University of Technology, supporting the scientific activity in the disciplines of Civil Engineering, Geodesy, and Transport.

Institutional Review Board Statement: Not applicable.

Informed Consent Statement: Not applicable.

Data Availability Statement: The original contributions presented in the study are included in the article, further inquiries can be directed to the corresponding author.

Conflicts of Interest: The authors declare no conflicts of interest. The funders had no role in the design of the study; in the collection, analyses, or interpretation of data; in the writing of the manuscript; or in the decision to publish the results.

Appendix A

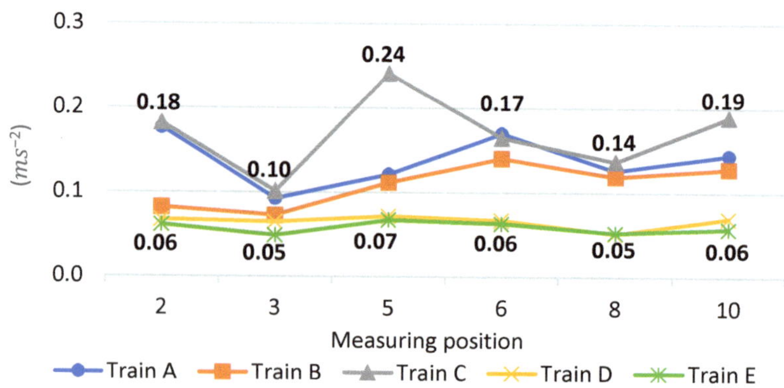

Figure A1. Frequency−weighted acceleration RMS.

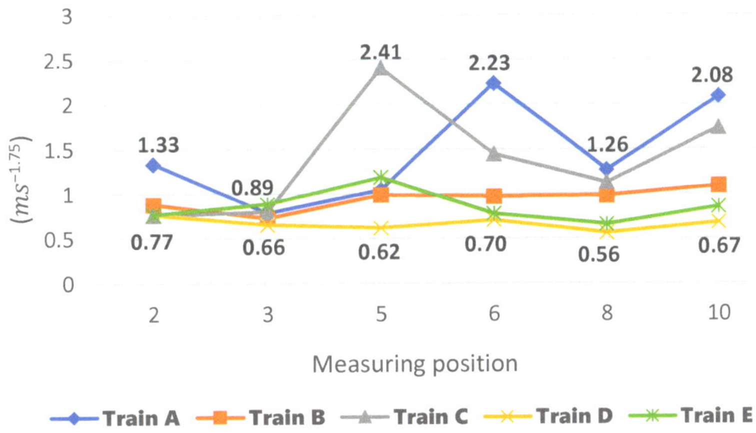

Figure A2. Vibration dose value VDV.

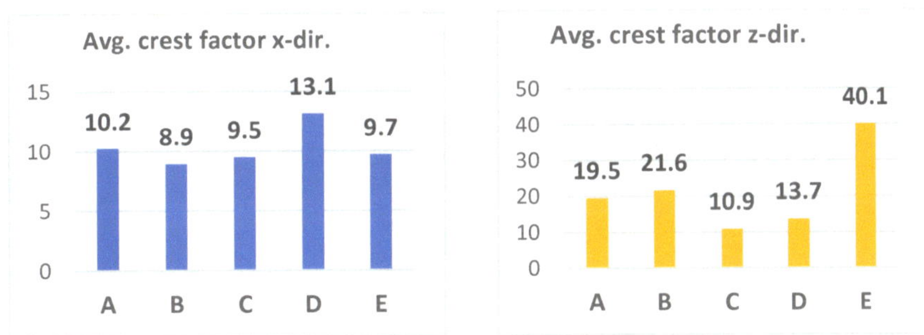

(**a**) (**b**)

Figure A3. Average crest factors for (**a**) x-direction and (**b**) z-direction.

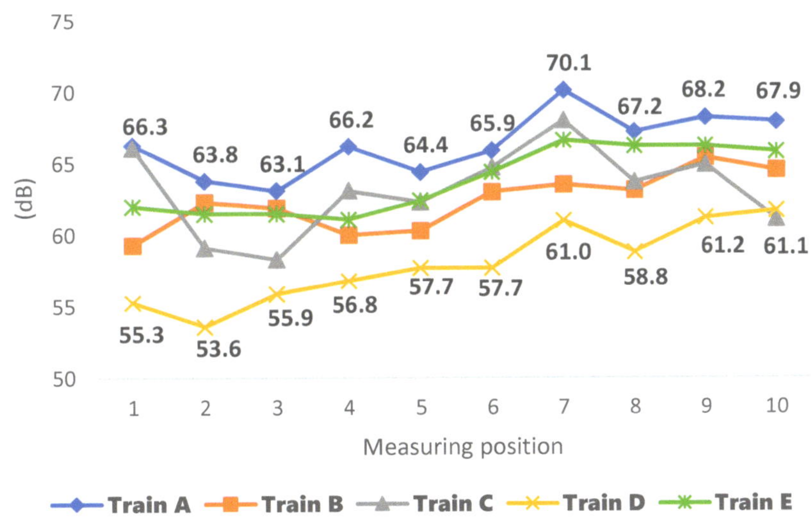

Figure A4. L_{Aeq} during acceleration.

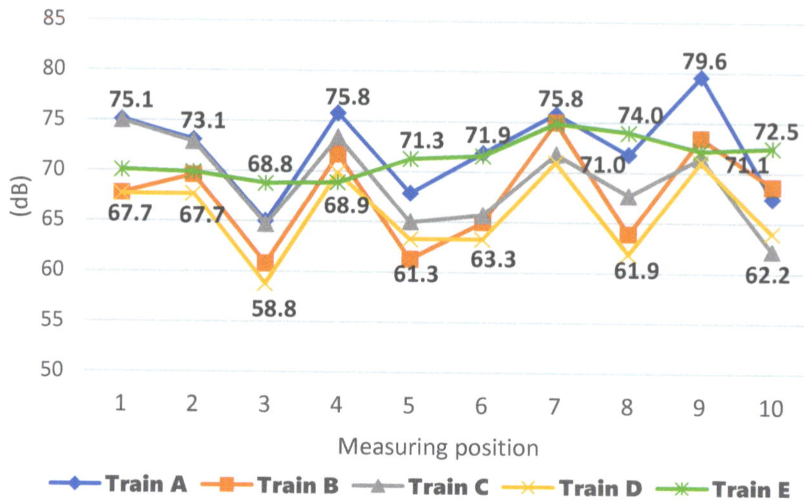

Figure A5. L_{Aeq} during ride with a constant speed.

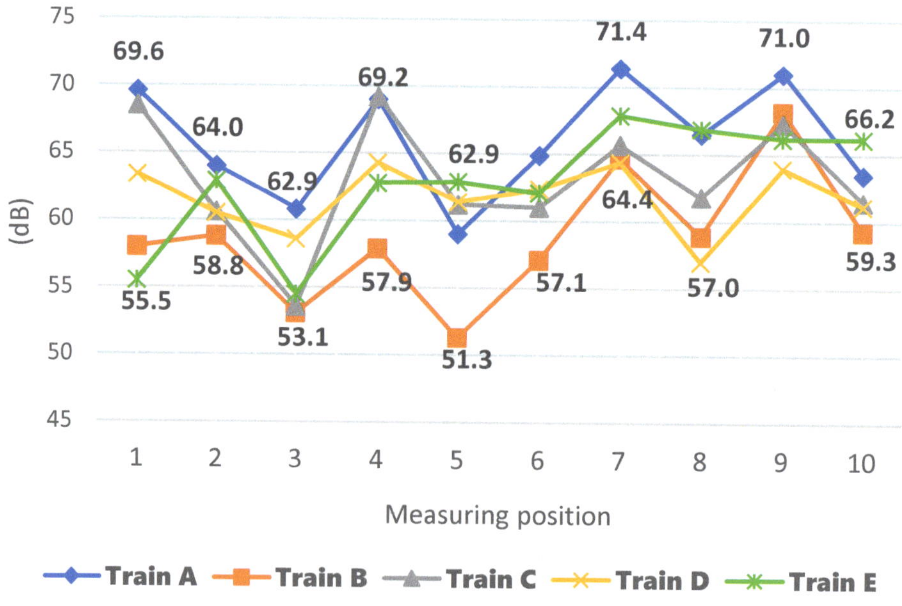

Figure A6. L_{Aeq} during braking.

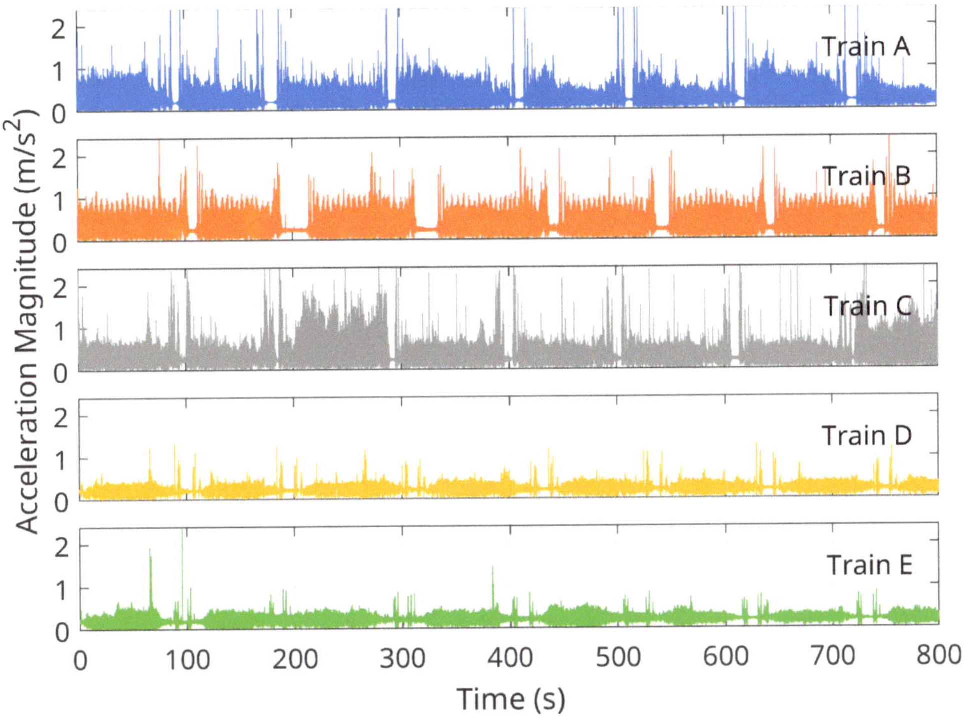

Figure A7. Acceleration magnitude measured at position 8 (middle wagon center) for 8 subsequent stations travel cycle (acceleration-constant speed-braking).

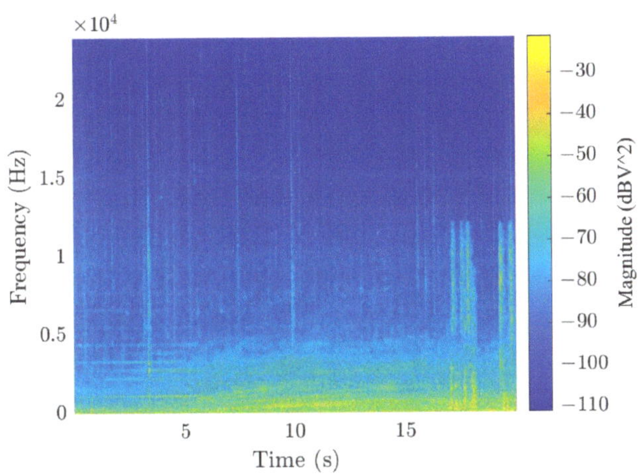

Figure A8. Noise spectrogram for acceleration phase, Train D, location 2.

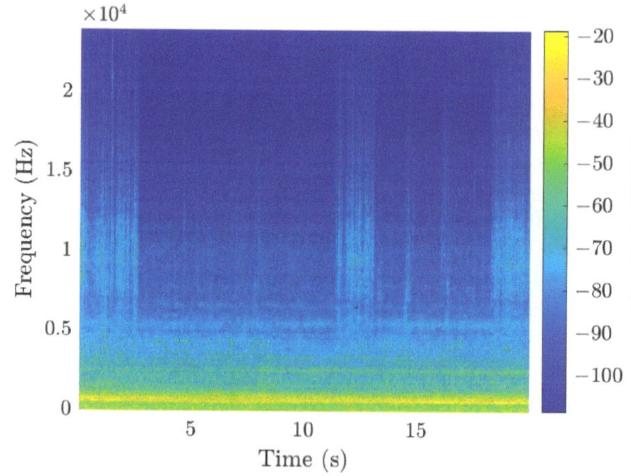

Figure A9. Noise spectrogram for constant speed phase, Train D, location 2.

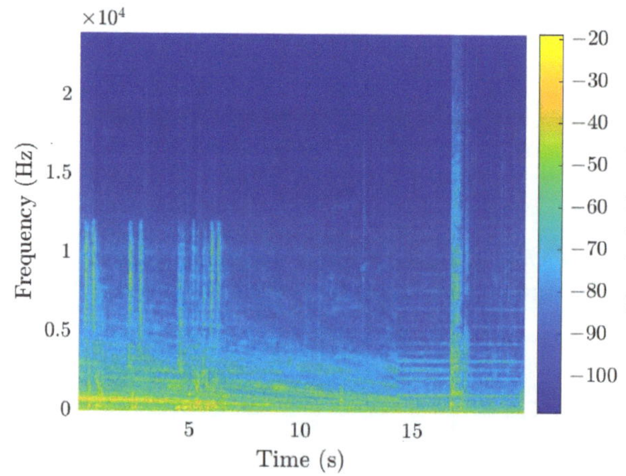

Figure A10. Noise spectrogram for braking phase, Train D, location 2.

References

1. Liu, R.; Song, A.C.; Nelson, D.O.; Liu, R.; Song, A.C.; Nelson, D.O. A Comprehensive Approach for Rolling Stock Planning: Combining Train Performance Simulation and Life-Cycle Cost Analysis. *J. Transp. Res. Forum* **2004**, *43*, 105–120. [CrossRef]
2. Paulley, N.; Balcombe, R.; Mackett, R.; Titheridge, H.; Preston, J.; Wardman, M.; Shires, J.; White, P. The Demand for Public Transport: The Effects of Fares, Quality of Service, Income and Car Ownership. *Transp. Policy* **2006**, *13*, 295–306. [CrossRef]
3. Wardman, M. Public Transport Values of Time. *Transp. Policy* **2004**, *11*, 363–377. [CrossRef]
4. Crocker, M.J.; Arenas, J.P. *Engineering Acoustics: Noise and Vibration Control*, 1st ed.; Wiley: Hoboken, NJ, USA, 2020; ISBN 978-1-118-49642-8.
5. Griffin, M.J. *Handbook of Human Vibration*; Elsevier: Amsterdam, The Netherlands, 1990; ISBN 978-0-12-303040-5.
6. Bruni, S.; Bucca, G.; Carnevale, M.; Collina, A.; Facchinetti, A. Pantograph–Catenary Interaction: Recent Achievements and Future Research Challenges. *Int. J. Rail Transp.* **2018**, *6*, 57–82. [CrossRef]
7. Liu, Z.; Song, Y.; Gao, S.; Wang, H. Review of Perspectives on Pantograph-Catenary Interaction Research for High-Speed Railways Operating at 400 Km/h and Above. *IEEE Trans. Transp. Electrific.* **2023**. [CrossRef]
8. Nering, K.; Kowalska-Koczwara, A.; Stypuła, K. Annoyance Based Vibro-Acoustic Comfort Evaluation of as Summation of Stimuli Annoyance in the Context of Human Exposure to Noise and Vibration in Buildings. *Sustainability* **2020**, *12*, 9876. [CrossRef]
9. Dumitriu, M.; Mazilu, T.; Apostol, I.I. Scale Models to Verify the Effectiveness of the Methods to Reducing the Vertical Bending Vibration of the Railway Vehicles Carbody: Applications and Design Elements. *Appl. Sci.* **2023**, *13*, 2368. [CrossRef]
10. Morales, A.L.; Chicharro, J.M.; Palomares, E.; Ramiro, C.; Nieto, A.J.; Pintado, P. Experimental Analysis of the Influence of the Passengers on Flexural Vibrations of Railway Vehicle Carbodies. *Veh. Syst. Dyn.* **2022**, *60*, 2825–2844. [CrossRef]

11. Popprath, S.; Benatzky, C.; Bilik, C.; Kozek, M.; Stribersky, A.; Wassermann, J. Experimental Modal Analysis of a Scaled Car Body for Metro Vehicles. In Proceedings of the Thirtheenth International Congress on Sound and Vibration (ICSV13), Vienna, Austria, 2–6 July 2006; 8p.
12. Diana, G.; Cheli, F.; Collina, A.; Corradi, R.; Melzi, S. The Development of a Numerical Model for Railway Vehicles Comfort Assessment through Comparison with Experimental Measurements. *Veh. Syst. Dyn.* **2002**, *38*, 165–183. [CrossRef]
13. Yan, L.; Chen, Z.; Zou, Y.; He, X.; Cai, C.; Yu, K.; Zhu, X. Field Study of the Interior Noise and Vibration of a Metro Vehicle Running on a Viaduct: A Case Study in Guangzhou. *Int. J. Environ. Res. Public Health* **2020**, *17*, 2807. [CrossRef] [PubMed]
14. Dižo, J.; Steišunas, S.; Blatnický, M. Vibration Analysis of a Coach with the Wheel-Flat Due to Suspension Parameters Changes. *Procedia Eng.* **2017**, *192*, 107–112. [CrossRef]
15. Sharma, S.K.; Sharma, R.C.; Lee, J.; Jang, H.-L. Numerical and Experimental Analysis of DVA on the Flexible-Rigid Rail Vehicle Carbody Resonant Vibration. *Sensors* **2022**, *22*, 1922. [CrossRef] [PubMed]
16. Sun, Y.; Gong, D.; Zhou, J. Study on Vibration Reduction Design of Suspended Equipment of High Speed Railway Vehicles. *J. Phys. Conf. Ser.* **2016**, *744*, 012212. [CrossRef]
17. Weidenfeld, S.; Schmitz, M.-T.; Sanok, S.; Henning, A.; Aeschbach, D.; Elmenhorst, E.-M. Effects of Railway Rolling Noise on Perceived Pleasantness. *Transp. Res. Part D Transp. Environ.* **2024**, *126*, 103995. [CrossRef]
18. Dižo, J.; Blatnický, M.; Harušinec, J.; Suchánek, A. Assessment of Dynamics of a Rail Vehicle in Terms of Running Properties While Moving on a Real Track Model. *Symmetry* **2022**, *14*, 536. [CrossRef]
19. Graff, M. Metro w Warszawie. *TTS Tech. Transp. Szyn.* **2008**, *14*, 25–35.
20. Wawryszczuk, R.; Kardas-Cinal, E.; Lejk, J.; Sokołowski, M. Methods of Passenger Ride Comfort Evaluation—Tests for Metro Cars. *Sensors* **2023**, *23*, 5741. [CrossRef] [PubMed]
21. Wawryszczuk, R.; Kardas-Cinal, E. Analysis of Ride Comfort in Selected Types of Rail Vehicles. *J. KONBiN* **2021**, *51*, 157–183. [CrossRef]
22. Palomares, E.; Morales, A.L.; Nieto, A.J.; Félix, M.; Chicharro, J.M.; Pintado, P. Is the Standard Ride Comfort Index an Actual Estimation of Railway Passenger Comfort? *Veh. Syst. Dyn.* **2022**, *61*, 2811–2824. [CrossRef]
23. Peng, Y.; Zhou, J.; Fan, C.; Wu, Z.; Zhou, W.; Sun, D.; Lin, Y.; Xu, D.; Xu, Q. A Review of Passenger Ride Comfort in Railway: Assessment and Improvement Method. *Transp. Saf. Environ.* **2022**, *4*, tdac016. [CrossRef]
24. Huang, Y.; Griffin, M.J. The Discomfort Produced by Noise and Whole-Body Vertical Vibration Presented Separately and in Combination. *Ergonomics* **2014**, *57*, 1724–1738. [CrossRef] [PubMed]
25. Huang, Y.; Griffin, M.J. The Effects of Sound Level and Vibration Magnitude on the Relative Discomfort of Noise and Vibration. *J. Acoust. Soc. Am.* **2012**, *131*, 4558–4569. [CrossRef] [PubMed]
26. Maigrot, P.; Parizet, É.; Marquis-Favre, C. Annoyance Due to Combined Railway Noise and Vibration: Comparison and Testing of Results from the Literature. *Appl. Acoust.* **2020**, *165*, 107324. [CrossRef]
27. Aladdin, M.F.; Jalil, N.A.A.; Guan, N.Y.; Rezali, K.A.M. Perturbation Effect of Noise on Overall Feeling of Discomfort from Vertical Whole-Body Vibration in Vibro-Acoustic Environment. *Int. J. Ind. Ergon.* **2021**, *83*, 103136. [CrossRef]
28. Transport 131, Artwork. Available online: http://Transport131.Com.Pl/Projj_491.Html (accessed on 1 July 2024).
29. United Tire, Artwork. Available online: https://Unitedtire.Com/Blog/View/Top-5-Reasons-a-Check-Engine-Light-Comes-On (accessed on 1 July 2024).
30. *PN-EN ISO 3381:2005*; Railway Applications. Acoustics. Noise Measurement inside Railbound Vehicles. ISO: Geneva, Switzerland, 2005.
31. *ISO 2631-1*; Mechanical Vibration and Shock. Evaluation of Human Exposure Whole-Body Vibration. Part 1: General Requirements. International Organization for Standardization: Geneva, Switzerland, 1997.
32. *EN 12299:2009*; Railway Applications—Ride Comfort for Passengers—Measurement and Evaluation. The Stationery Office Limited: Norwich, UK, 2009.
33. *EU Directive 2002/44/EC of the European Parliment and of the Council of 25 June 2002*; The Stationery Office Limited: Norwich, UK, 2002.
34. *ISO 1996-1*; Acoustics—Description, Measurement and Assessment of Environmental Noise—Part 1: Basic Quantities and Assessment Procedures. ISO: Geneva, Switzerland, 2016.
35. Tomaszewski, F.; Orczyk, M. Comparison of noise inside selected types of rail vehicles during rides. *Pojazdy Szyn.* **2014**, *3*, 1–7. (In Polish) [CrossRef]
36. Orczyk, M. Ocena Hałasu Na Stanowisku Motorniczego Wybranych Typów Tramwajów (Assessment of Noise at the Working Site of a Tram Driver of Selected Types of Trams). *Pr. Nauk. Politech. Warsz. Transp.* **2017**, *119*, 293–300.
37. *EU Commission Decision of 4 April 2011 Concerning the Technical Specifications of Interoperability Relating to the Subsystem Rolling Stock; Noise of the Trans-European Conventional Rail System (Notified under Document C(2011) 658) (Text with EEA Relevance) (2011/229/EU), CELEX1*; The Stationery Office Limited: Norwich, UK, 2015.
38. Ma, M.; Ran, W.; Wu, J.; Li, M.; Qu, X. Evaluating the Impact of Metro Interior Noise on Passenger Annoyance: An Experimental Study. *Int. J. Environ. Res. Public Health* **2022**, *19*, 5041. [CrossRef] [PubMed]

logistics

Article

Examining the Intention to Adopt an Online Platform for Freight Forwarding Services in Thailand: A Modified Unified Theory for Acceptance and Use of Technology (UTAUT) Model Approach

Nattakorn Pinyanitikorn [1], Walailak Atthirawong [2] and Wirachchaya Chanpuypetch [1,3,*]

1 Multidisciplinary and Interdisciplinary School, Chiang Mai University, Chiang Mai 50200, Thailand; nattakorn_pi@cmu.ac.th
2 School of Science, King Mongkut's Institute of Technology Ladkrabang, Bangkok 10520, Thailand; walailaknoi@gmail.com
3 Faculty of Agro-Industry, Chiang Mai University, Samut Sakhon 74000, Thailand
* Correspondence: wirachchaya.c@cmu.ac.th

Abstract: *Background*: The freight forwarding industry is undergoing digital transformation through the implementation of online platforms designed to enhance operational efficiency and transparency. Despite these benefits, the adoption of these platforms has been slower than anticipated due to customer concerns and industry-specific challenges. *Methods*: This study investigates the factors influencing the intention to adopt and the actual use of online platforms for freight forwarding services among business customers in Thailand. A modified Unified Theory for Acceptance and Use of Technology (UTAUT) model, incorporating perceived risk, serves as the theoretical framework. Survey data were collected from 400 respondents in managerial or higher-level positions involved in freight shipping within Thai firms and analyzed using a structural equation model (SEM). *Results*: The analysis reveals that performance expectancy, effort expectancy, social influence, and facilitating conditions positively influence adoption intention, while perceived risk negatively impacts it. Firm size moderates the effect of social influence, with a stronger impact observed in larger enterprises. *Conclusions*: The findings offer practical insights for Thai freight forwarders, suggesting strategies to improve customer acceptance and encourage the adoption of online platforms. Addressing the identified factors could lead to improved efficiency and greater integration of digital technologies in the logistics industry.

Keywords: freight forwarding services; online platforms; UTAUT; logistics; Thailand

Citation: Pinyanitikorn, N.; Atthirawong, W.; Chanpuypetch, W. Examining the Intention to Adopt an Online Platform for Freight Forwarding Services in Thailand: A Modified Unified Theory for Acceptance and Use of Technology (UTAUT) Model Approach. *Logistics* **2024**, *8*, 76. https://doi.org/10.3390/logistics8030076

Academic Editors: Tomasz Nowakowski, Artur Kierzkowski, Agnieszka A. Tubis, Franciszek Restel, Tomasz Kisiel, Anna Jodejko- Pietruczuk and Mateusz Zając

Received: 17 May 2024
Revised: 12 July 2024
Accepted: 29 July 2024
Published: 31 July 2024

1. Introduction

The role of freight forwarding services is crucial in enabling international trade by overseeing the intricate process of moving goods across borders for shippers. Acting as intermediaries between exporters/importers and carriers, freight forwarders offer various services, including transportation planning, customs clearance, cargo consolidation, and documentation handling. This industry is characterized by its high fragmentation, with a diverse range of players such as global logistics integrators, regional firms, and local small and medium enterprises (SMEs) competing based on factors like network coverage, specialized knowledge, and customer relations [1–3].

In the current business landscape, the rapid growth of e-commerce and digital technologies has posed challenges for numerous industries, including logistics and shipping [2,4]. To address the changing demands of customers, businesses are innovating their models and improving their value propositions. In the freight forwarding sector, exemplified by companies like DHL and UPS, online platforms have emerged to streamline services for their partners and customers [5]. These platforms are disruptive, leveraging digital capabilities to provide instant quotations, aggregate capacity, enable real-time tracking, and offer self-service shipment management solutions [2,6].

The digitalization of the freight forwarding industry has given rise to online platforms that provide shippers with real-time quotes, booking, tracking, and document management capabilities [6,7]. These platforms leverage cloud computing, API integration, and data analytics to streamline manual processes, enhance transparency and shipment visibility, and improve operational efficiency [8]. However, despite the potential benefits of these technologies, the adoption of online platforms remains limited in some contexts [9–11]. While global forwarders have introduced online booking and tracking tools in the market, most local firms have been slow to adopt them due to factors such as lack of customer awareness, legacy processes, and resource constraints [12,13]. This variation in adoption rates highlights the differing levels of digital readiness and technology acceptance.

In Thailand, freight forwarders are planning to transition their customers from traditional booking methods through salespersons to an online platform. However, there is no government support for developing a central platform for freight forwarding or other logistics services. As a result, private sectors must individually develop and heavily invest in their platform applications to support their businesses. This initiative aims to improve the service experience and enhance their competitiveness in a highly competitive marketplace.

However, the adoption of online-based freight forwarding service platforms in Thailand remains an innovative and relatively new concept. Despite this, Thailand does not face significant barriers regarding the quality of telecommunication and internet networking, which are sufficient to provide wide-area coverage. Within the Thai industry environment, the primary customers of freight forwarding services are business companies, including both large multinational corporations (MNCs) and, predominantly, small and medium enterprises (SMEs). Therefore, understanding the expectations and concerns of these customer groups regarding the adoption of online-based freight forwarding services within the context of Thai culture is essential. This knowledge is crucial for developing niche platforms that meet customer needs and facilitate the transition to online services.

Several theoretical models have been proposed to explain user acceptance of new technologies; they include the Technology Acceptance Model (TAM), the Theory of Planned Behavior (TPB), and Innovation Diffusion Theory (IDT), and the Unified Theory of Acceptance and Use of Technology (UTAUT) [14]. Among these, the UTAUT, proposed by Venkatesh et al. [15], integrates elements from various prominent models in the context of information systems [14,15]. Compared to other models, the UTAUT offers a more holistic approach by incorporating a wider range of factors and their interactions, making it highly effective in predicting technology acceptance. Additionally, the UTAUT accounts for the moderating effects of gender, age, experience, and voluntariness of use, providing a nuanced understanding of different user groups [15]. While models like TAM focus primarily on perceived usefulness and ease of use [16], the UTAUT's broader scope and consideration of social and contextual factors make it a robust and versatile tool for analyzing technology adoption across various settings. Consequently, the UTAUT has been widely applied in recent research to test factors influencing technology acceptance and adoption across various contexts.

Therefore, the primary aim of this study is to examine the factors that influence firms' intentions to use an online platform for freight forwarding services in Thailand. A significant contribution of this study is the analysis of variables within a modified UTAUT model to determine the effects of various factors related to the adoption of online platforms in the Thai context. This study introduces perceived risk as an additional factor to examine its relationship within the proposed model. Furthermore, interesting moderating variables are suggested to investigate different perspectives among user groups that may affect their adoption intentions for using information technology in the freight forwarding industry.

The paper is organized as follows: Section 2 reviews the literature and develops the hypotheses of this study. Section 3 describes the methodology. Section 4 presents the data analysis and results. Section 5 provides research discussion and implications. And Section 6 concludes this study.

2. Background Literature and Hypotheses

2.1. The Unified Theory of Acceptance and Use of Technology (UTAUT)

The Unified Theory of Acceptance and Use of Technology (UTAUT) was developed by Venkatesh et al. in 2003 [15]. The UTAUT posits that that four key factors influence the behavioral intention to adopt a technology: performance expectancy (perceived usefulness), effort expectancy (ease of use), social influence (subjective norms), and facilitating conditions (organizational and technical support). The model also includes four moderating variables—gender, age, experience, and voluntariness of use—which are hypothesized to affect the relationships between the predictors and behavioral intention [15].

The UTAUT has been widely applied to study technology adoption across various contexts, including the acceptance of E-commerce platforms [17], work from home technologies during the COVID-19 pandemic [18], implementation of omnichannel marketing in purchasing [19], mobile banking in China [20], E-payment in Yemen [21], student ICT adoption in Ghana [22], customer usage of logistics technologies such as buy-online-and-pickup-in-store (BOPS), smart lockers and drone delivery [23], sustainable usage intention of blockchain in logistics and supply chain management companies [24–26], logistics platform resource integration [27], material procurement in construction companies in India [28], and cloud computing services [29]. These studies have generally confirmed the predictive power of the model while emphasizing the need to adapt it based on the specific technology, regional, and industrial context.

Various studies have proposed modified UTAUT models to examine specific contexts, often by adding or removing independent and moderating variables to the model. In this review, we focus on the adoption of technology within the logistics industry. Additionally, we review the factors used in related contexts, considering the countries involved in the technology adoption. This review aims to identify the most suitable factors for analysis. The related literature on modified UTAUT models for testing the intention to adopt technology in various contexts is summarized in Table 1.

Table 1. The related literature on modified UTAUT models for testing the intention to adopt technology.

Source	Year	Technology	Context	PE	EE	SI	FC	PP	HM	PR	TR	PC	CO	PV	AT	HB	IN	AG	GE	EX	VU	ED	FS	IS
[30]	2013	Security as a service	Germany, Austria, and Switzerland	✓	✓					✓	✓				✓								✓	✓
[29]	2016	Cloud computing services	Germany	✓	✓	✓	✓			✓	✓							✓	✓	✓				
[31]	2019	Big data in supply chain management	Brazil	✓	✓	✓	✓				✓													
[32]	2020	Enterprise resource planning (ERP)	Developing and Asian countries	✓	✓	✓	✓															✓	✓	
[33]	2020	Online technology during COVID-19 quarantine time	Thailand	✓	✓	✓	✓			✓	✓													
[34]	2022	M-payment	Retailers in Egypt	✓	✓	✓	✓			✓	✓							✓	✓			✓		
[35]	2022	Cashless payment system	Thailand	✓	✓	✓				✓	✓													
[36]	2022	Digital health information system	China	✓	✓	✓	✓	✓		✓		✓	✓											
[37]	2022	E-marketplace	Small businesses in India	✓	✓	✓	✓							✓				✓		✓				
[38]	2022	E-retail platforms	Sub-Saharan Africa	✓	✓	✓			✓	✓	✓							✓	✓	✓				
[39]	2022	Fintech	Port's users in Ghana and sub-Saharan Africa	✓	✓	✓	✓		✓					✓		✓		✓						
[26]	2023	Blockchain adoption	Logistics companies in Ho Chi Minh City, Vietnam	✓	✓	✓	✓													✓				
[40]	2023	Mobile pocket software	A logistics company in Indonesia	✓	✓	✓	✓		✓	✓	✓		✓											
[41]	2023	Fintech service	Jeddah, Saudi Arabia	✓	✓	✓	✓			✓	✓	✓					✓							
[42]	2023	Imported spirit flash delivery applications	China			✓				✓						✓	✓					✓		

Note: UTAUT independent variables: PE = performance expectancy; EE = effort expectancy; SI = social influence; FC = facilitating condition. UTAUT moderating variables: AG = age; GE = gender; EX = experience; VU = voluntariness of use. Modified UTAUT independent variables: PP = perceived playfulness; HM = hedonic motivation; PC = privacy concern; PR = perceived risk; TR = trust; CO = perceived cost; PV = perceived vulnerability; AT = attitude; IN = innovativeness; HB = habit. Modified UTAUT moderating variables: ED = educational qualification; FS = firm size; IS = industrial sector.

This study aims to explore perceived risk as an additional influencing factor in the intention to adopt an online platform for freight forwarding services, integrating it into the UTAUT model in alignment with similar recent studies [29,30,33–35,38,40–42]. Additionally, we investigate the potential moderating effects of firm size and user generation on the relationship between the direct factors and the intention to adopt an online platform [29,30,32,34,37,38]. Consequently, the factors influencing the intention to adopt online freight forwarding services are comprehensively reviewed in the following section to develop research hypotheses.

2.2. Development of Influencing Factors toward the Intention to Use an Online Platform for Freight Forwarding Services

The traditional UTAUT model posits that performance expectancy, effort expectancy, and social influence positively influence behavioral intention to use, while facilitating conditions positively influence actual use behavior. We have modified the research model by incorporating an additional factor, perceived risk. This factor is expected to negatively impact the intention to use an online platform for freight forwarding services. This section details the hypotheses developed based on these relationships and how they influence the intention to use the platform.

2.2.1. Performance Expectancy (PE)

According to [15], performance expectancy (PE) refers to the belief that a specific technology will improve efficiency and effectiveness of tasks [21]. In this context, PE focuses on users' perceptions of convenience, service-tracking enhancements, and quality of service offered by an online platform for freight forwarding services [20,21,43]. While PE might not capture all aspects of perceived value (e.g., cost saving), it remains a critical factor influencing user adoption [17]. Previous research supports a positive relationship between perceived usefulness (similar to PE) and intention to use technological systems [2,32,44]. Therefore, we hypothesized that a higher perceived PE (convenience, service tracking, quality of service) will lead to a stronger intention to use an online platform for freight forwarding services. This hypothesis will be further tested empirically in this study.

H1: *Performance expectancy has a positive influence on the intention of Thai firms to use an online platform for freight forwarding services.*

2.2.2. Effort Expectancy (EE)

Effort expectancy (EE) reflects the perceived difficulty in using a technology. Users who find a technology difficult to learn or use may become discouraged and discontinue use [15]. In the context of online freight forwarding platforms, EE refers to the perceived ease of using the platform's interface, navigating functionalities, and completing tasks like booking shipments or tracking goods. Previous research has shown that perceived ease of use (similar to EE) positively impacts the intention to use technological systems [31,45]. Studies by Alwahaishi and Snášel (2013) [46,47] and Sarfaraz (2017) [48] also confirm the significant influence of EE on users' intention to adopt ICT technologies. Therefore, the following hypothesis explores the relationship between EE and intention to use an online platform for freight forwarding services.

H2: *Effort expectancy has a significant positive influence on the intention of Thai firms to use an online platform for freight forwarding services.*

2.2.3. Social Influence (SI)

Social influence (SI) reflects the effect of referent opinions from important others on individual user behavior [49,50]. In the context of Thai firms, SI can be shaped by recommendations from industry peers, logistics associations, or trusted business partners [2,3]. According to SI theory, users are more likely to adopt a technology if those they respect recommend it [15,51]. Therefore, we proposed the following hypothesis:

H3: *Social influence has a positive influence on the intention of Thai firms to use an online platform for freight forwarding services.*

2.2.4. Perceived Risk (PR)

One critical factor not explicitly addressed in the original UTAUT model is perceived risk (PR). In freight forwarding, the stakes are high, with substantial financial and operational implications tied to logistics efficiency and reliability. PR directly influences trust, which is crucial for users when adopting new technologies in this sector [52]. Additionally, the logistics industry frequently handles sensitive data, such as shipment details, customer information, and financial transactions [53]. The PR associated with data security and privacy can significantly affect companies' willingness to adopt online freight forwarding platforms. Therefore, PR plays a vital role in adoption decisions within this domain. PR refers to the uncertainties and potential negative outcomes a firm perceives from adopting a new system or service [34,54–56].

In the context of online platforms, common risk concerns include data security and privacy breaches, system reliability and uptime, integration and interoperability issues, and loss of control over critical business processes [34,35,54,56]. Several empirical studies have demonstrated that PR can negatively impact the adoption intentions of various technologies, such as e-procurement [28], enterprise resource planning (ERP) [32], and blockchain in supply chains [24,25]. These findings suggest that including PR as an additional factor in the UTAUT model could provide a more comprehensive understanding of the factors influencing the adoption of online freight forwarding platforms. Therefore, the fourth hypothesis was proposed as follows:

H4: *Perceived risk has a negative influence on the intention of Thai firms to use an online platform for freight forwarding services.*

2.2.5. Facilitating Conditions (FCs)

Facilitating conditions (FCs) refer to the resources and knowledge that users need to effectively use a technology [15]. They refer to the belief that there is an organizational and technical infrastructure in place to support the system [57]. If users lack these resources and knowledge, they may not continue using the technology [49]. Alwahaishi and Snášel (2013) [47] studied the factors affecting the acceptance and use of mobile internet as an ICT application in a consumer context and found strong support for FCs based on empirical data. Similarly, FCs have been found to influence consumer adoption of cloud computing services in Germany [29], blockchain adoption in logistics companies in Vietnam [26], online shopping in Bangladesh [58], and E-payment in Yemen [21]. Additionally, studies have shown a significant relationship between FCs and the actual use of internet banking [44]. Most research confirms that intention to use positively impact the actual use of technologies, such as adopting big data in supply chain management [31], and material procurement in construction companies in India [28]. Based on these findings, two hypotheses regarding the FC factor were formulated as follows:

H5: *Facilitating conditions have a positive influence on the intention of Thai firms to use an online platform for freight forwarding services.*

H6: *Facilitating conditions have a positive influence on the actual use of an online platform for freight forwarding services by Thai firms.*

2.2.6. Intention to Use (IU) and Actual Use (AU)

In the traditional UTAUT model, intention to use (IU) is widely recognized as a strong predictor of actual use (AU). Given its close alignment with behavior, many studies incorporating intentions into their frameworks focus on measuring behavioral adoption intentions. Individuals tend to be more engaged when their intention towards using a technology is positive, significantly impacting usage behavior with the intention to continue use, indicating a proactive effort and planning to use the technology. Previous studies have shown a positive influence of IU on AU, such as in the use of ICT in tourism [43] and E-government services [59]. Based on these observations, the seventh hypothesis was formulated as follows:

H7: *Intention to use has a positive influence on the actual use of an online platform for freight forwarding services by Thai firms.*

2.2.7. Moderators

Additionally, this study aims to explore how firm size and user generation moderate the relationship between the main factors and the intention to use an online platform for freight forwarding services. Existing studies indicate that firm size may have specific characteristics that affect the adoption of information technology, particularly in small and medium-sized enterprises [9,11,37,60–63]. Furthermore, we refine the traditional UTAUT model by adjusting the respondents' age to consider generation or age group, as this factor may influence the intention to use information systems. Therefore, the following hypotheses were developed to be tested with empirical data in the context of Thailand. Eight sub-hypotheses were proposed to test the moderating effect of firm size (H8a–d) and user generation (H9a–d) on the relationships between intention to use (IU) and the proposed UTAUT predictors: performance expectancy (PE), effort expectancy (EE), social influence (SI), and perceived risk (PR).

H8a: *The impact of performance expectancy on intention to use is moderated by firm size.*

H8b: *The impact of effort expectancy on intention to use is moderated by firm size.*

H8c: *The impact of social influence on intention to use is moderated by firm size.*

H8d: *The impact of perceived risk on intention to use is moderated by firm size.*

H9a: *The impact of performance expectancy on intention to use is moderated by user generation.*

H9b: *The impact of effort expectancy on intention to use is moderated by user generation.*

H9c: *The impact of social influence on intention to use is moderated by user generation.*

H9d: *The impact of perceived risk on intention to use is moderated by user generation.*

2.3. Proposed Structural Model

Based on the literature review, this study proposes a modified UTAUT-based model to analyze the factors influencing the intention to adopt an online platform for freight forwarding services by Thai firms. This model expands the UTAUT framework by integrating perceived risk as an additional significant inhibitor of adoption intention, alongside the four primary predictors: performance expectancy, effort expectancy, social influence, and facilitating conditions.

Furthermore, the model includes two moderating variables, firm size and user generation, to explore their potential impacts on the relationships between the predictors and the intention to adopt online platforms for freight forwarding services. Firm size is suggested as a moderator based on previous studies [30,32], which indicates that larger firms may have distinct considerations and capabilities regarding technology adoption compared to small and medium-sized enterprises (SMEs) [9,60]. User generation is also incorporated as a moderator based on UTAUT's proposition that younger individuals may exhibit greater openness to new technologies than older generations. The proposed modified UTAUT-based model for this study is illustrated in Figure 1.

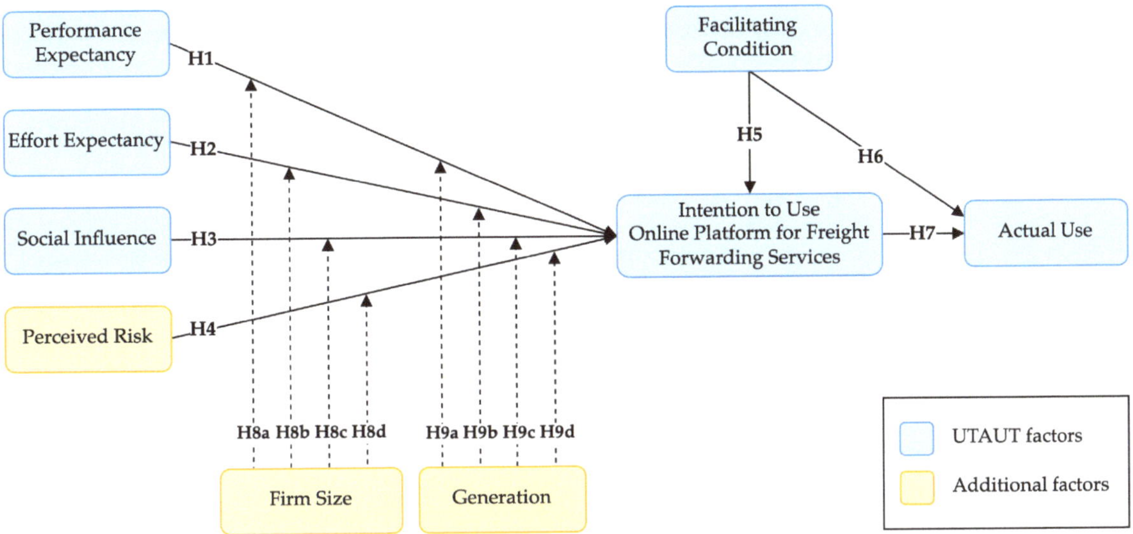

Figure 1. The proposed modified UTAUT-based model.

3. Methodology

3.1. Participants

This study utilized a quantitative, cross-sectional survey design to collect data for testing the hypothesized relationships in the research model. The target population comprised employees in Thai firms responsible for making decisions or managing processes related to the use of freight forwarding services for their organizations' cross-border shipping needs. This included roles such as supply chain managers, logistics managers, import/export managers, and shipping executives.

To recruit participants, a purposive sampling approach was employed through two main channels. Firstly, an industry database maintained by the Thai International Freight Forwarders Association (TIFFA) was used to identify potential respondents based on their job titles and company profiles. To mitigate bias, the population within this group was proportionally divided into strata based on the key demographic variable of firm size. Participants from each stratum were then selected using sampling techniques to reduce selection bias. Secondly, the customer databases of five leading freight forwarders operating in Thailand were utilized to reach out to key decision-makers in their client organizations. The participating forwarders were selected based on their market share, service offerings, and willingness to support the research.

The sample size was determined to be 400, which is widely recognized and applied across various research fields for its statistical accuracy of $\pm 5\%$ and is often considered the most cost-effective sample size [64,65]. According to the literature [66], a minimum sample size of 200 is typically required for robust SEM analysis.

3.2. Measures and Questionnaire

This study employed a survey questionnaire, which was presented in the Thai language for accessibility and structured into two main sections. The first section gathered demographic information and usage experience of participants, including their position, age group, average use of freight forwarding shipment services, and the size of their company. The second part consisted of Likert-scale questions concerning factors influencing the intention to adopt an online platform for freight forwarding services. Respondents rated their opinions using a five-point Likert scale, ranging from "strongly agree" (5) to "strongly disagree" (1).

Before distribution, the questionnaire underwent content validity testing using the Index Objective Congruence (IOC) to ensure the accuracy of its content [67]. Three experts from selected freight forwarding companies reviewed the questionnaire. Based on their feedback, minor modifications were made to improve the wording and flow of the questions.

To assess the questionnaire's reliability, it was pretested with a sample of 30 logistics professionals to ensure clarity, comprehensiveness, and reliability. Cronbach's alpha coefficient (α) was used to measure the questionnaire's reliability, with a value greater than 0.7 indicating good reliability [68].

The final survey instrument was implemented on an online survey platform. Data collection took place from February to April 2024, during which rigorous quality checks were implemented to ensure the validity of responses. Survey forms with suspicious response styles or incomplete answers were excluded from the analysis. The measurement is the questionnaire used in this research is presented in Table 2.

Table 2. Measurement items.

Constructs	Items	Source
Performance Expectancy (PE)	PE1: The use of an online platform for freight forwarding services will enhance our logistics performance compared to paper-based processes. PE2: Utilizing an online platform of freight forwarding services enables us to compete our logistics activities more rapidly. PE3: The use of an online platform for freight forwarding services improves the quality of our logistics activities. PE4: Using an online platform for freight forwarding services simplifies our logistics activities. PE5: Utilizing an online platform of freight forwarding services enhances our efficiency in logistics activities.	[3,15,69,70]
Effort Expectancy (EE)	EE1: Interacting with an online platform for freight forwarding services would be clear and understandable. EE2: It would be easy for us to become proficient at using an online platform for freight forwarding services in our business. EE3: Learning to operate an online platform for freight forwarding services is straightforward for our staffs. EE4: An online platform for freight forwarding services is compatible with our company.	[15,69]
Social Influence (SI)	SI1: The top management of our company believes that using an online platform for freight forwarding services is essential for digital transformation. SI2: Staff involved in freight booking believe that using an online platform for freight forwarding services should rely more on digitized documents. SI3: Using an online platform for freight forwarding services enhances commercial satisfaction for supply chain partners/customers.	[3,15,69]
Perceived Risk (PR)	PR1: Utilizing digital or paperless documents through an online platform for freight forwarding services may result in the potential loss of important information of our company. PR2: Using an online platform for freight forwarding services may lead to a possible loss of time for our staffs compared to traditional processes. PR3: Using an online platform for freight forwarding services may result in a potential loss of performance accuracy. PR4: Implementing an online platform for freight forwarding services may incur the possible cost of training staff. PR5: Using an online platform for freight forwarding services may lead to a potential loss of cooperation between our company and the freight forwarder.	[34,35,55]

Table 2. *Cont.*

Constructs	Items	Source
Facilitating Conditions (FCs)	FC1: Our company possesses the necessary information technology facilities to effectively utilize an online platform for freight forwarding services. FC2: Our company has the requisite knowledge to effectively utilize an online platform for freight forwarding services. FC3: There is technical support available online to assist with any difficulties encountered in using an online platform for freight forwarding services.	[15,69]
Intention to Use (IU)	IU1: I believe our company would adopt an online platform for freight forwarding services if it were available in the market. IU2: Our company is seriously considering the adoption of an online platform for freight forwarding services in the near future. IU3: Our company has a plan to utilize an online platform for freight forwarding services as a part of our digital transformation strategy.	[15,71]
Actual Use (AU)	AU1: Our company is prepared to utilize an online platform for freight forwarding services as a significant step towards digital transformation. AU2: Our company uses an online platform for freight forwarding services to partially or fully replace traditional freight forwarding services. AU3: Our company use an online platform for freight forwarding services to manage logistics.	[15,32]

3.3. Data Analysis: Structural Equation Modeling (SEM)

In this study, we followed a two-step approach to structural equation modeling (SEM) analysis, as recommended by [72]. Firstly, we conducted confirmatory factor analysis (CFA) to assess the reliability and validity of the measurement scales used in our proposed model. This step allowed us to evaluate the adequacy of the measurement model by examining goodness-of-fit criteria, including the minimum discrepancy divided by degrees of freedom (CMIN/df), the Comparative Fit Index (CFI), Tucker–Lewis Index (TLI), Root Mean Square Error of Approximation (RMSEA), and Standardized Root Mean Square Residual (SRMR) [73].

After validating the measurement model, we estimated the structural model using maximum likelihood estimation. This step involved examining path coefficients to determine the strength and significance of relationships between latent variables in our model. We also evaluated the overall fit of the structural model using fit indices.

Data analysis was performed using the IBM SPSS and AMOS 26.0 statistical packages. SPSS was utilized for data preparation and descriptive statistics, whereas AMOS 26.0 was employed for CFA and SEM analysis. This study included a sample of 400 respondents who are employees in Thai firms responsible for decisions or processes related to the use of freight forwarding services. The two-step SEM analysis provided valuable insights into the concurrent influences among the factors in our conceptual model, helping us to understand the complex relationships between variables and their implications for theory and practice.

4. Results

4.1. Demographic Structure of Respondents

The main part of this research study involved distributing an online questionnaire. A total of 400 complete responses were collected, with all respondents having experience using freight forwarding services in the past three months. The majority of the respondents held positions as import or export managers (43.3%) and logistics managers (35.5%). In terms of firm size, 45.3% were from small enterprises, while 35.5% were from medium-sized enterprises. Regarding generational categories (age group), the majority were Gen Y or millennials (30–43 years old), accounting for 65.5% of all respondents. Table 3 presents key aspects of the demographic structure of the respondents.

Table 3. Analysis of the demographic characteristics.

Categories	Dimensions	Frequency (*n*)	Percentage (%)
Position	Chief/managing director	29	7.3%
	Import/export manager	173	43.3%
	Logistics manager	142	35.5%
	Other management levels or equivalent	56	14.0%
Generation	Gen X (over 43 years old)	61	15.3%
	Gen Y or Millennials (30–43 years old)	262	65.5%
	Gen Z (29 years old or less)	77	19.3%
Average usage shipment	Less than 5 TEUs a month	100	25.0%
	5–20 TEUs a month	140	35.0%
	21–50 TEUs a month	84	21.0%
	More than 50 TEUs a month	76	19.0%
Firm size	Small (less than THB 50 million/year)	181	45.3%
	Medium (THB 51–300 million/year)	142	35.5%
	Large (more than THB 300 million/year)	77	19.3%
Use of freight forwarding service (past 3 months)	Yes	400	100.0%
	No	0	0%
	Total	400	100%

4.2. Assessment of the Measurement Model

Before analyzing the measurement model, descriptive statistics were computed for the 28 observed variables related to the acceptance of adopting an online platform for freight forwarding services in Thai firms. Each statement in the questionnaire, detailed in Table 2, was rated on a 5-point Likert scale ranging from "strongly agree" (5) to "strongly disagree" (1). The survey, conducted with 400 respondents, indicated that the degree of agreement for each observation item was higher than the moderate value or scale "3", except for the item measuring "Perceived Risk (PR)". This particular variable highlighted barriers or negative outcomes in the adoption of an online platform for freight forwarding services, where a lower score indicated a more positive outlook. The scale was recoded such that "strongly disagree" corresponded to a score of 5 and "strongly agree" to 1.

In preparation for structural equation model (SEM) or confirmatory factor analysis (CFA), assessing the normality of the data is crucial [74]. Typically, multivariate normality is tested, and for sample sizes larger than 300, normality can be evaluated using histograms and by examining the absolute values of skewness and kurtosis. In SPSS, these values should ideally approximate zero in a normal distribution [75,76]. For this study, skewness and kurtosis values within the range of ± 1.96 indicate normal distribution [77]. Detailed descriptive statistics of the variables and results of normality testing are presented in Table 4.

To confirm the reliability and validity of the structural model, individual questionnaire responses were analyzed. Factor loading (FL) values were calculated for each question to determine the Average Variance Extracted (AVE) and Construct Reliability (CR) values. The results are presented in Table 5, where it can be observed that all FL values exceed 0.5, all AVE values exceed 0.5, and all CR values exceed 0.6 (with CR values for all variables exceed 0.80) [78]. Cronbach's alpha values presented in Table 5 meet the minimum acceptable value of 0.70 for all the variables [68]. Therefore, the internal consistency and reliability of the scales used in this study are considered satisfactory.

Subsequently, each item within every factor underwent rigorous evaluation using reduction techniques to confirm its alignment with the intended constructs. Confirmatory factor analysis (CFA) is a statistical method essential for examining relationships among latent variables which is integral to SEM methodology and crucial for validating measurement models in path or structural analyses. CFA estimates latent variables based on the correlated variation within the dataset. Before assessing the structural model in SEM, researchers ensure that the measured variables accurately reflect the intended constructs or factors.

Table 4. Descriptive statistics of variables (n = 400).

Constructs	Items	Mean	Standard Deviation	Skewness	Kurtosis
Performance Expectancy (PE)	PE1	3.96	0.678	−0.291	0.115
	PE2	3.97	0.705	−0.299	−0.061
	PE3	3.96	0.778	−0.477	−0.037
	PE4	3.95	0.810	−0.452	−0.256
	PE5	3.99	0.771	−0.317	−0.461
Effort Expectancy (EE)	EE1	4.06	0.786	−0.697	0.328
	EE2	4.09	0.701	−0.298	−0.362
	EE3	4.08	0.762	−0.742	0.607
	EE4	4.03	0.665	−0.088	−0.565
Social Influence (SI)	SI1	4.04	0.650	−0.477	0.852
	SI2	3.99	0.761	−0.357	−0.109
	SI3	4.03	0.791	−0.746	0.482
Perceived Risk (PR)	PR1	3.61	0.851	−0.727	1.004
	PR2	3.74	0.819	−0.080	−0.616
	PR3	3.99	0.883	−0.415	−0.739
	PR4	3.70	0.716	−0.379	0.076
	PR5	3.64	0.733	−0.273	−0.124
Facilitating Condition (FC)	FC1	3.99	0.629	−0.296	0.496
	FC2	4.06	0.669	−0.366	0.251
	FC3	3.96	0.747	−0.629	0.526
Intention to Use (IU)	IU1	4.04	0.582	−0.156	0.529
	IU2	4.05	0.694	−0.297	−0.184
	IU3	4.10	0.623	−0.198	.037
Actual Use (AU)	AU1	3.75	0.589	−0.691	0.927
	AU2	3.75	0.588	−0.914	1.257
	AU3	3.73	0.633	−1.013	1.217

The structural equation model (SEM) in this study demonstrates a strong fit across several key goodness-of-fit criteria. The CMIN/df ratio, which measures the discrepancy between the model and observed data adjusted for degrees of freedom, is reported at 3.306, indicating a good fit, as values below 5 are generally considered acceptable. The Root Mean Square Error of Approximation (RMSEA) stands impressively low, at 0.044, well below the threshold of 0.08, suggesting a very close fit of the model to the data. The Standardized Root Mean Square Residual (SRMR) is also low, at 0.032, indicating excellent fit where values closer to zero denote better model fit. The Non-normed Fit Index (NNFI) or Tucker–Lewis Index (TLI) is reported at 0.779, slightly below the preferred threshold of 0.90 but still acceptable, suggesting the model provides a reasonable fit relative to a baseline model. Similarly, the Comparative Fit Index (CFI), at 0.833, though also slightly below 0.90, indicates a good overall fit of the model to the data. These findings collectively indicate that the SEM in this study exhibits a robust fit according to most criteria, suggesting it accurately represents the relationships among the variables under investigation. However, the slightly lower values of NNFI and CFI suggest that there may still be room for improvement, such as refining the model specification or addressing potential measurement issues. The results, alongside the cut-off values denoting a good fit, are presented in Table 6.

Table 5. Determining the reliability and validity of the model.

Constructs	Items	FL	AVE	CR	Cronbach's α
Performance Expectancy (PE)	PE1	0.789	0.651	0.903	0.863
	PE2	0.836			
	PE3	0.850			
	PE4	0.785			
	PE5	0.771			
Effort Expectancy (EE)	EE1	0.867	0.608	0.859	0.782
	EE2	0.790			
	EE3	0.844			
	EE4	0.585			
Social Influence (SI)	SI1	0.792	0.647	0.845	0.723
	SI2	0.749			
	SI3	0.867			
Perceived Risk (PR)	PR1	0.851	0.679	0.914	0.877
	PR2	0.770			
	PR3	0.779			
	PR4	0.875			
	PR5	0.841			
Facilitating Condition (FC)	FC1	0.785	0.639	0.841	0.711
	FC2	0.855			
	FC3	0.754			
Intention to Use (IU)	IU1	0.810	0.721	0.886	0.801
	IU2	0.819			
	IU3	0.915			
Actual Use (AU)	AU1	0.789	0.714	0.882	0.799
	AU2	0.866			
	AU3	0.877			

Table 6. CFA statistics of model fit.

Goodness-of-Fit Indexes	Result Model	Cut-Off for Good Fit	Source
CMIN/df	3.306	Good ≤ 3 and acceptable < 5	[79,80]
RMSEA	0.044	Excellence < 0.01, good > 0.01–0.05, medium > 0.05 to 0.08, and poor > 0.1	[81–84]
SRMR	0.032	<0.08	[83]
NNFI or TLI	0.779	0 = poor fit Close to 1 = very good fit	[73,85]
CFI	0.833	0 = poor fit Close to 1 = very good fit	[83,86]

4.3. Structural Model and Hypotheses Testing

This step involved testing the structural model to elucidate the relationships between constructs in the research model. Path analysis, a common form of SEM analysis, was employed to investigate the direct impact and causal relationships among variables. The significance of each hypothesized structural path was assessed using standardized path coefficients (β) and p-values.

Table 7 displays the path values, critical ratio (C.R.), standard error (S.E.), and significance level (p-value). In the analysis, the absolute value of the standardized path coefficient (β) indicates the magnitude of influence, with positive values indicating a positive influence and negative values indicating a negative influence. The critical ratio (C.R.) for the independent variable is a z statistic, equal to the ration of the parameter estimate to its standard deviation. The p-value represents the probability of the statistical test of C.R.

Table 7. Results of coefficient path analysis.

Hypothesis	Path		β	S.E.	C.R.	Result
H1	Performance Expectancy (PE)	→ Intention to Use (IU)	0.478 ***	0.078	3.925	Supported
H2	Effort Expectancy (EE)	→ Intention to Use (IU)	0.168 *	0.034	2.442	Supported
H3	Social Influence (SI)	→ Intention to Use (IU)	0.262 ***	0.065	3.321	Supported
H4	Perceived Risk (PR)	→ Intention to Use (IU)	−0.211 ***	0.022	−4.451	Supported
H5	Facilitating Conditions (FCs)	→ Intention to Use (IU)	0.458 ***	0.073	3.611	Supported
H6	Facilitating Conditions (FCs)	→ Actual Use (AU)	0.442 ***	0.059	4.579	Supported
H7	Intention to Use (IU)	→ Actual Use (AU)	0.663 ***	0.104	6.753	Supported

Note: *** $p < 0.001$; * $p < 0.05$.

Six paths were found to be significant at $p < 0.001$, while one path was significant at $p < 0.05$. All paths were considered significant, as the C.R. exceeded 1.96 and the p-value was less than 0.05. The notation "***" and "*" denotes p-values less than 0.001 and 0.05, respectively.

The analysis of influencing factors revealed that hypotheses H1–H7 are supported. Perceived risk (PR) was found to have a significant negative impact on the intention to use (IU) ($\beta = -0.211$, $p < 0.001$), supporting the fourth hypothesis (H4). This finding contradicts various existing literature on the acceptance of online platforms [55,56,87], which suggests that PR reduces the willingness to use an online platform for freight forwarding services. Organizations are aware of potential loss of time and important information in services, as well as of the capabilities of their staff.

Additionally, the effect of performance expectancy (PE) on the intention to use (IU) was found to be positive and significant ($\beta = 0.478$, $p < 0.001$), supporting the first hypothesis (H1). Firms intending to use an online platform for freight forwarding services believe that the platform will simplify and enhance efficiency in their activities compared to paper-based processes. Similarly, the result indicates that effort expectancy (EE) was positively and significantly related to the IU ($\beta = 0.168$, $p < 0.05$), confirming the second hypothesis (H2). This result suggests that firms are making efforts to understand and try new information systems for their business.

Furthermore, the third hypothesis (H3) revealed that social influence (SI) has a positive and significant impact on the intention to use (IU) ($\beta = 0.262$, $p < 0.001$). Digital transformation can enhance satisfaction among supply chain partners, thereby supporting H3.

Moreover, the fifth hypothesis (H5) suggested that facilitating conditions (FCs) have a positively significant effect on the intention to use (IU) ($\beta = 0.458$, $p < 0.001$), thus H5 was supported. Similarly, FCs also positively influenced actual use (AU), confirming the sixth hypothesis (H6) ($\beta = 0.442$, $p < 0.001$). This indicates that support from the company, including managerial support to enhance knowledge for digital transformation and information technology facilities, promotes the use of an online platform in their activities.

Finally, the intention to use (IU) was found to have a positively significant effect on actual use (AU) ($\beta = 0.663$, $p < 0.001$), supporting H7. The staff of the firm's intention to use an online platform for freight forwarding services leads to actual use of the platform and prompts digital transformation in their logistics processes. The structural model and path coefficients is illustrated in Figure 2.

4.4. Analysis of Moderating Effects

The moderating effects of firm size and user generation on the structural model were examined using multigroup analysis. Hypotheses five and six predicted that organization size and generation would moderate the effect of variables (performance expectancy (PE), effort expectancy (EE), social influence (SI), and perceived risk (PR)) on the intention to use of an online platform for freight forwarding services, respectively. Each moderating variable was split into two groups and analyzed using the critical ratios approach [73].

For the first moderating variable, "firm size," respondents were divided into small and medium-sized enterprises (SMEs) ($n = 323$) and large-sized enterprises ($n = 77$) for examination. For the second moderating variable, "generation," respondents were split into Generation X (age over 43 years old, $n = 61$) and Generation Y or Millennials and Generation Z (age less than 43, $n = 339$) for comparison based on generation context.

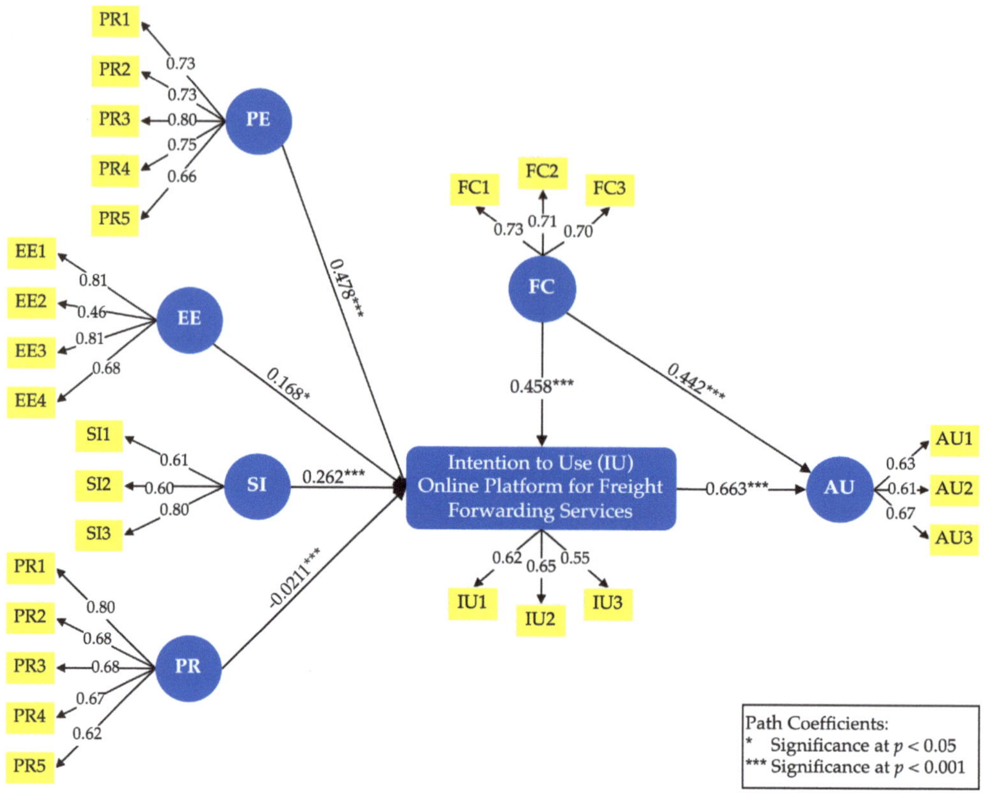

Figure 2. Structural model and path coefficients.

The effects of the two moderating variables are presented in Table 8. Estimates are unstandardized coefficients. z-scores indicate the significance of difference between path coefficients. The symbols ***, **, and * indicate significance at $p < 0.001$, 0.01, and 0.05, respectively.

Table 8 presents the results, with z-scores indicating the significance of the difference between path coefficients. The analysis reveals that social influence (SI) has a significant impact on the intention to use (IU) in both firm sizes, with the effect being significantly stronger among large organizations. Therefore, hypothesis H5c was confirmed with a z-score greater than 1.96 (z-score = 2.187, $p < 0.01$). However, the other factors investigated, including performance expectancy (PE), effort expectancy (EE), and perceived risk (PR), showed no significant differences between SMEs and large enterprises, as indicated by z-scores less than 1.96. This means hypotheses H8a, H8b, and H8d were rejected.

Regarding user generation or age group as a moderating variable, Generation Y and Generation Z were regrouped into the same category due to their similar information technology skills and knowledge compared to previous generations. The results indicate that there were no significant differences in any of the constructs between Generation X and Generation Y or Z respondents. Consequently, all six hypotheses, H9a–H9d, were rejected, as the z-scores were less than 1.96. In other words, the effect of PE, EE, SI, and PR on adoption intention is not moderated by user generation.

Table 8. Effects of moderating variables.

Firm Size			SME Estimate (n = 323)	Large Enterprise Estimate (n = 77)	z-Score	Moderation
H8a	Performance Expectancy (PE)	→ Intention to Use (IU)	0.572 ***	0.179	−1.412	No
H8b	Effort Expectancy (EE)	→ Intention to Use (IU)	0.154 *	0.242	0.314	No
H8c	Social Influence (SI)	→ Intention to Use (IU)	0.201 *	0.785 ***	2.187 **	Yes
H8d	Perceived Risk (PR)	→ Intention to Use (IU)	−0.234 ***	−0.053	1.623	No
Generation			**Generation X Estimate (n = 61)**	**Generation Y and Z Estimate (n = 339)**	**z-Score**	**Moderation**
H9a	Performance Expectancy (PE)	→ Intention to Use (IU)	0.810	0.364 **	−0.617	No
H9b	Effort Expectancy (EE)	→ Intention to Use (IU)	−0.777	0.198 **	0.522	No
H9c	Social Influence (SI)	→ Intention to Use (IU)	1.701	0.177 *	−0.732	No
H9d	Perceived Risk (PR)	→ Intention to Use (IU)	0.518	−0.252 ***	−0.805	No

Note: *** $p < 0.001$; ** $p < 0.01$; * $p < 0.05$.

5. Discussion and Implications

5.1. Discussion of Results

This study aimed to analyze the factors influencing the adoption an online platform for freight forwarding services, leading to actual use behavior in Thailand. The main variables observed were performance expectancy (PE), effort expectancy (EE), social influence (SI), facilitating condition (FC), and perceived risk (PR), based on the modified UTAUT model. A survey was conducted with 400 participants from Thai firms that have experience using freight forwarding services.

Out of the fifteen hypotheses tested, eight were accepted (H1–H7), while the rest were rejected (H8a, H8b, H8d, H9a–H9d). H8c was accepted based on the moderating effect tests for firm size.

The findings of this study confirm the significant positive impact of performance expectancy (PE) on the intention to use an online platform for freight forwarding services in Thailand. This result aligns with [15], suggesting that PE influences continuance intention. Additionally, this study is consistent with previous research on the use of new technologies in logistics activities [41,42,47,88], indicating that performance expectancy, effort expectancy, and risk perception significantly influence users' intention to use new technologies. This implies that a higher perceived degree of PE by users leads to a greater intention to use an online platform.

Effort expectancy (EE) was also found to significantly affect the intention to use an online platform for freight forwarding service positively. This indicates that a high perceived degree of EE by user leads to a greater intention to use the platform. As explained by [15], EE reflects the perceived difficulty in using a new technology. The results of this study are consistent with previous research showing that perceived ease of use (similar to EE) affects user satisfaction [42,55] and continued use [42], as well as intention to use [47].

Social influence (SI) was found to have a significant positive effect on the intention to use an online platform for freight forwarding services. This suggests that a higher perceived degree of SI among users leads to a greater intention to use the technology. SI reflects the effect of referent opinion on individual user behavior [89]. These results are consistent with previous studies [31,90], which also found that SI significantly influences the intention to continue using online platforms or information technologies for related logistics and supply chain management activities. Additionally, Lee and Song [56] found that performance expectancy and social influence positively influence behavioral intention, a finding that aligns with the results of this study.

The results of this study confirmed that perceived risk (PR) significantly influenced the intention to use an online platform for freight forwarding services, with a negative impact. This implies that a higher perception of associated risks among users leads to a lower level of intention to use the technology. Digital risk represents unexpected breakdowns, contagions, and interruptions between the company and business partners in the supply chain [91]. Similarly, Lee and Song [56] studied the effects of trust and PR on user adoption of a new technology service and found that trust and PR are direct antecedents of intention to use, which is consistent with the findings of this study. Abrahão et al. [92] investigated the intention to use ICT services based on the UTAUT and found that 76% of behavioral intentions were explained by performance expectancy, effort expectancy, social influence, and perceived risk, which is also supported by the findings of this study.

The results confirmed that facilitating conditions (FCs) significantly affected the intention to use and actual use of an online platform for freight forwarding services in Thai firms, with a positive impact. This indicates that a higher favorability of FCs as perceived by users leads to a higher level of adoption intention toward the technology. FCs refer to users having the necessary resources and knowledge to use the technology, as explained in [15]. In an organizational context, FCs are defined as one's belief that an organizational and technical infrastructure exists to support the system [57,88]. Similarly, Alwahaishi and Snášel [46] empirically investigated how FCs influence the acceptance and use of ICT applications. In logistics companies, FCs positively influence the behavioral intention to use blockchain [26]. Alalwan et al. [44] found similar results supporting that there is a statistically significant relationship between facilitating conditions and the actual use of online financial services, which is also similar to some of the facilities needed to use e-services. For SMEs in European nations, digitalization adoption is also influenced by FCs [93], as supported by this study regarding our participants' demographic (Table 3), where 80.8% of respondents are SMEs.

In the proposed modified UTAUT model, the impact of the main factors was not moderated by firm size and generation, except for social influence (SI). Firm size was found to moderate SI, suggesting that the relationship between SI and intention to use was significant for both SMEs and particularly for large-sized enterprises. Large enterprises often rely on high-potential emerging technologies or technological innovations to enhance their competitiveness, outperform competitors [94,95], and ensure business sustainability [96,97].

The results confirmed that the intention to use significantly influenced the actual use in the case of applying an online platform for freight forwarding services in Thailand. This finding is consistent with several other studies [59,98,99].

5.2. Practical Implications

In response to the findings suggesting practical implications for freight forwarding companies, the action plan focuses on several key strategies aimed at improving the adoption of online platforms. The first strategy emphasizes highlighting the performance benefits of these platforms, such as enhanced logistics efficiency and reduced reliance on paper-based processes. To achieve this, the marketing team will develop comprehensive marketing materials, including videos and case studies. Concurrently, the customer success team will organize workshops and webinars to demonstrate firsthand how these platforms can optimize logistics operations.

The second strategy centers on ensuring robust IT support and infrastructure. This involves establishing a dedicated IT support team to provide immediate assistance to customers encountering platform-related issues. Additionally, the IT department will focus on developing and reinforcing the necessary IT infrastructure to ensure seamless platform functionality, particularly catering to the needs of SMEs lacking extensive IT resources.

A crucial aspect of the plan involves leveraging social influence to foster platform adoption. This strategy begins with identifying and partnering with prominent freight forwarding firms willing to serve as early adopters and demonstrate successful platform implementations. This initiative aims to create positive role models within the industry. Moreover, the marketing team will collaborate to document and promote these success stories through various channels, thereby building confidence and interest among potential users.

Addressing perceived risks associated with online platforms constitutes the fourth strategy. Immediate measures include implementing stringent data security protocols to safeguard customer information. Concurrently, the customer success team will actively collect and disseminate positive testimonials from satisfied customers, fostering trust and mitigating apprehensions surrounding platform usage. Furthermore, enhancing customer support services will ensure prompt resolution of any concerns, further bolstering customer confidence in the platform's reliability.

Lastly, streamlining the user experience stands as the fifth strategy to facilitate platform adoption. This involves refining the platform's user interface through rigorous user testing and iterative design improvements, overseen by the UX/UI design team. Integrating artificial intelligence (AI) tools will further enhance user interaction by providing instant and continuous support. Simultaneously, the training department will develop comprehensive user tutorials and guides within two months to streamline onboarding processes and minimize the learning curve for new users.

In conclusion, by adhering to this structured action plan, freight forwarding companies can effectively implement the identified practical implications. This holistic approach not only addresses performance benefits and IT infrastructure requirements but also harnesses social influence, mitigates perceived risks, and enhances user experience. Through these concerted efforts, companies can achieve heightened efficiency, customer satisfaction, and competitive advantage in the marketplace.

5.3. Theoretical Implications

Based on the comprehensive discussion and practical implications derived from this study on factors influencing the adoption of online platforms for freight forwarding services in Thailand, the following theoretical implications can be formulated:

(1) Extended application of UTAUT model: This study reinforces the robustness of the modified UTAUT model in predicting technology adoption behaviors within the logistics and supply chain management context. It underscores the predictive power of variables such as performance expectancy, effort expectancy, social influence, perceived risk, and facilitating conditions across diverse organizational settings.

(2) Insights into technology adoption drivers: By identifying and validating factors influencing adoption intentions, such as perceived performance benefits and ease of use, this research provides deeper insights into the cognitive and motivational aspects that shape technology adoption decisions. This contributes to refining theoretical frameworks on technology acceptance by highlighting the nuanced interplay of these factors.

(3) Contextual understanding in logistics: This study enriches the understanding of technology adoption in the logistics industry, emphasizing industry-specific challenges and opportunities. It underscores the importance of organizational readiness, social influences, and perceived risks as critical factors requiring tailored strategies for successful implementation of new technology platforms.

(4) Implication for risk management strategies: This study's findings on the negative impact of perceived risk on adoption intentions emphasize the critical role of effective risk management strategies in technology adoption initiatives. This prompts further exploration into strategies for mitigating risks and integrating them into adoption frameworks to alleviate user concerns and enhance adoption rates.

(5) Generalizability and transferability of findings: The research findings contribute to the generalizability of technology adoption theories across various contexts, providing insights applicable to similar industries and technological innovations. This supports broader applications of theoretical frameworks in understanding and predicting technology adoption behaviors across different organizational environments.

These theoretical implications not only advance academic understanding of technology adoption processes but also offer practical guidance for stakeholders in the logistics sector.

They can help in devising effective adoption strategies, improving operational efficiencies, and fostering growth driven by innovation.

6. Conclusions

6.1. Research Conclusions

This study's findings significantly contribute to understanding the factors influencing the adoption an online platform for freight forwarding services among business customers in the Thai market. The research concludes that performance expectancy (PE), effort expectancy (EE), social influence (SI), perceived risk (PR), and facilitating conditions (FCs) are critical determinants of adoption intention. Additionally, this study reveals a moderating effect of firm size on the relationship between social influence (SI) and intention to use (IU), indicating that larger firms may be more influenced by social factors in their adoption decisions.

These findings have several implications for both theory and practice. Theoretically, this study extends the UTAUT model by incorporating perceived risk (PR) as an additional main factor and firm size as a moderating variable, providing a more comprehensive understanding of technology adoption in the logistics industry in the Thai context. Practically, the results can guide freight forwarding service providers in developing tailored strategies to promote the adoption of online platforms, such as emphasizing the performance benefits, simplifying the user experience, addressing social influences, mitigating perceived risks, and enhancing facilitation conditions.

Overall, this research contributes valuable insights that can inform the development and implementation of an online platform for freight forwarding services, ultimately enhancing the efficiency and competitiveness of the logistics industry in Thailand.

6.2. Limitations and Future Research Direction

This study has several limitations that present opportunities for future research. Firstly, the cross-sectional design, while useful for capturing a snapshot of adoption perceptions, restricts understanding of how these decisions evolve over time. Longitudinal studies could provide valuable insights into the dynamic nature of adoption decisions. Secondly, the focus on the Thai market may limit the generalizability of the findings. Replicating this study in other countries and cultures would enhance the external validity of the results.

Another limitation is the sample composition, which included more respondents from SMEs than large enterprises. This imbalance could have influenced the results, reflecting the predominance of small and medium-sized companies in the Thai market. It suggests the need for a more balanced sample in future studies to enhance the robustness of findings regarding moderator effects. While full measurement invariance testing was not conducted due to limitations in the Thai industry environment and time constraints, future research should consider employing larger sample sizes with balanced group distributions and conducting thorough measurement invariance testing to strengthen the validity of findings in multigroup analysis (MGA).

Additionally, due to resource constraints including a limited time and budget, a comprehensive statistical power analysis was not conducted. Addressing this limitation in future research would ensure that sample sizes are adequately determined to minimize the risk of Type II errors [100]. While the UTAUT model identifies several key factors influencing technology acceptance, it may overlook context-specific variables that are crucial in certain settings. Future research could explore additional moderators, such as industry sector and technology readiness, to further refine the understanding of adoption drivers and barriers within specific context. To investigate indirect effects and more intricate interactions between components, future research should include a mediating variable of intention to use between exogenous variables. This approach will allow for a more thorough comprehension of the phenomenon being studied.

Author Contributions: Conceptualization, N.P. and W.C.; methodology, N.P., W.A. and W.C.; validation, W.A. and W.C.; formal analysis, N.P.; investigation, N.P.; writing—original draft preparation, N.P. and W.C.; writing—review and editing, N.P., W.A. and W.C.; visualization, W.A. and W.C.; supervision, W.A. and W.C. All authors have read and agreed to the published version of the manuscript.

Funding: This research received no external funding.

Data Availability Statement: The data presented in this study are available on request from the first author. The data are not publicly available due to ethical restrictions.

Acknowledgments: This research was partially supported by Chiang Mai University.

Conflicts of Interest: The authors declare no conflicts of interest.

References

1. Song, D. A Literature Review, Container Shipping Supply Chain: Planning Problems and Research Opportunities. *Logistics* **2021**, *5*, 41. [CrossRef]
2. Raza, Z.; Woxenius, J.; Vural, C.A.; Lind, M. Digital Transformation of Maritime Logistics: Exploring Trends in the Liner Shipping Segment. *Comput. Ind.* **2023**, *145*, 103811. [CrossRef]
3. Cichosz, M.; Wallenburg, C.M.; Knemeyer, A.M. Digital Transformation at Logistics Service Providers: Barriers, Success Factors and Leading Practices. *Int. J. Logist. Manag.* **2020**, *31*, 209–238. [CrossRef]
4. Gupta, S.; Kushwaha, P.S.; Badhera, U.; Chatterjee, P.; Gonzalez, E.D.R.S. Identification of Benefits, Challenges, and Pathways in E-Commerce Industries: An Integrated Two-Phase Decision-Making Model. *Sustain. Oper. Comput.* **2023**, *4*, 200–218. [CrossRef]
5. Reinartz, W.; Wiegand, N.; Imschloss, M. The Impact of Digital Transformation on the Retailing Value Chain. *Int. J. Res. Mark.* **2019**, *36*, 350–366. [CrossRef]
6. Surucu-Balci, E.; Iris, Ç.; Balci, G. Digital Information in Maritime Supply Chains with Blockchain and Cloud Platforms: Supply Chain Capabilities, Barriers, and Research Opportunities. *Technol. Forecast. Soc. Chang.* **2024**, *198*, 122978. [CrossRef]
7. Jain, A.; van der Heijden, R.; Marchau, V.; Bruckmann, D. Towards Rail-Road Online Exchange Platforms in EU-Freight Transportation Markets: An Analysis of Matching Supply and Demand in Multimodal Services. *Sustainability* **2020**, *12*, 10321. [CrossRef]
8. Chanpuypetch, W.; Supeekit, T.; Niemsakul, J. IOT-Based Business Process Management for Temperature-Controlled Logistics of Laboratory Specimens. In Proceedings of the 37th ECMS International Conference on Modelling and Simulation, ECMS 2023, Florence, Italy, 20–23 June 2023; Vicario, E., Bandinelli, R., Fani, V., Mastroianni, M., Eds.; European Council for Modeling and Simulation: Caserta, Italy, 2023; pp. 359–365.
9. Mishrif, A.; Khan, A. Technology Adoption as Survival Strategy for Small and Medium Enterprises during COVID-19. *J. Innov. Entrep.* **2023**, *12*, 53. [CrossRef]
10. Kraus, S.; Jones, P.; Kailer, N.; Weinmann, A.; Chaparro-Banegas, N.; Roig-Tierno, N. Digital Transformation: An Overview of the Current State of the Art of Research. *Sage Open* **2021**, *11*, 215824402110475. [CrossRef]
11. Ueasangkomsate, P. Adoption E-Commerce for Export Market of Small and Medium Enterprises in Thailand. *Procedia Soc. Behav. Sci.* **2015**, *207*, 111–120. [CrossRef]
12. Zamani, S.Z. Small and Medium Enterprises (SMEs) Facing an Evolving Technological Era: A Systematic Literature Review on the Adoption of Technologies in SMEs. *Eur. J. Innov. Manag.* **2022**, *25*, 735–757. [CrossRef]
13. Yadav, H.; Soni, U.; Gupta, S.; Kumar, G. Evaluation of Barriers in the Adoption of E-Commerce Technology in SMEs. *J. Electron. Commer. Organ.* **2021**, *20*, 1–18. [CrossRef]
14. Taherdoost, H. A Review of Technology Acceptance and Adoption Models and Theories. *Procedia Manuf.* **2018**, *22*, 960–967. [CrossRef]
15. Venkatesh, V.; Morris, M.G.; Davis, G.B.; Davis, F.D. User Acceptance of Information Technology: Toward a Unified View. *MIS Q.* **2003**, *27*, 425. [CrossRef]
16. He, Y.; Chen, Q.; Kitkuakul, S. Regulatory Focus and Technology Acceptance: Perceived Ease of Use and Usefulness as Efficacy. *Cogent Bus. Manag.* **2018**, *5*, 1459006. [CrossRef]
17. Chen, L.; Rashidin, M.S.; Song, F.; Wang, Y.; Javed, S.; Wang, J. Determinants of Consumer's Purchase Intention on Fresh E-Commerce Platform: Perspective of UTAUT Model. *Sage Open* **2021**, *11*, 215824402110278. [CrossRef]
18. Razif, M.; Miraja, B.A.; Persada, S.F.; Nadlifatin, R.; Belgiawan, P.F.; Redi, A.A.N.P.; Lin, S.-C. Investigating the Role of Environmental Concern and the Unified Theory of Acceptance and Use of Technology on Working from Home Technologies Adoption during COVID-19. *Entrep. Sustain. Issues* **2020**, *8*, 795–808. [CrossRef]
19. Sari, D.M.F.P.; Suprapti, N.W.S.; Sukaatmadja, I.P.G.; Sukawati, T.G.R. The Implementation of Purchasing Omnichannel Marketing Based through the Expansion of the UTAUT 2 Model. *Uncertain Supply Chain Manag.* **2023**, *11*, 1441–1450. [CrossRef]
20. Mensah, I.K.; Khan, M.K. Unified Theory of Acceptance and Use of Technology (UTAUT) Model: Factors Influencing Mobile Banking Services' Adoption in China. *Sage Open* **2024**, *14*, 1–18. [CrossRef]
21. Alduais, F.; Al-Smadi, M.O. Intention to Use E-Payments from the Perspective of the Unified Theory of Acceptance and Use of Technology (UTAUT): Evidence from Yemen. *Economies* **2022**, *10*, 259. [CrossRef]

22. Attuquayefio, S.; Addo, H. Using the UTAUT Model to Analyze Students' ICT Adoption. *Int. J. Educ. Dev. Using Inf. Commun. Technol.* **2014**, *10*, 75–86.
23. Cai, L.; Yuen, K.F.; Xie, D.; Fang, M.; Wang, X. Consumer's Usage of Logistics Technologies: Integration of Habit into the Unified Theory of Acceptance and Use of Technology. *Technol. Soc.* **2021**, *67*, 101789. [CrossRef]
24. Park, K.O. A Study on Sustainable Usage Intention of Blockchain in the Big Data Era: Logistics and Supply Chain Management Companies. *Sustainability* **2020**, *12*, 10670. [CrossRef]
25. Shahzad, K.; Zhang, Q.; Khan, M.K. Blockchain Technology Adoption in Supply Chain Management: An Investigation from UTAUT and Information System Success Model. *Int. J. Shipp. Transp. Logist.* **2024**, *18*, 165–190. [CrossRef]
26. Nguyen, L.-T.; Nguyen, D.-T.; Ngoc, K.N.-N.; Duc, D.T.V. Blockchain Adoption in Logistics Companies in Ho Chi Minh City, Vietnam. *Cogent Bus. Manag.* **2023**, *10*, 2216436. [CrossRef]
27. Gan, W.; Fu, C.; Chen, Z.; Zhong, R. Research on Logistics Platform Resource Integration Based on UTAUT Model. In *Proceeding of the 24th International Conference on Industrial Engineering and Engineering Management 2018*; Springer: Singapore, 2019; pp. 542–548.
28. HM, A.J.; KB, A.; VR, H. Technology Adoption in Material Procurement: An Empirical Study Applying the UTAUT Model Among Construction Companies in India. *Glob. Bus. Rev.* **2024**. [CrossRef]
29. Moryson, H.; Moeser, G. Consumer Adoption of Cloud Computing Services in Germany: Investigation of Moderating Effects by Applying an UTAUT Model. *Int. J. Mark. Stud.* **2016**, *8*, 14. [CrossRef]
30. Senk, C. Adoption of Security as a Service. *J. Internet Serv. Appl.* **2013**, *4*, 11. [CrossRef]
31. Queiroz, M.M.; Pereira, S.C.F. Intention to Adopt BIG DATA in Supply Chain Management: A Brazilian Perspective. *Rev. Adm. Empresas* **2019**, *59*, 389–401. [CrossRef]
32. Uddin, M.A.; Alam, M.S.; Mamun, A.A.; Khan, T.-U.-Z.; Akter, A. A Study of the Adoption and Implementation of Enterprise Resource Planning (ERP): Identification of Moderators and Mediator. *J. Open Innov. Technol. Mark. Complex.* **2020**, *6*, 2. [CrossRef]
33. Chayomchai, A.; Phonsiri, W.; Junjit, A.; Boongapim, R.; Suwannapusit, U. Factors Affecting Acceptance and Use of Online Technology in Thai People during COVID-19 Quarantine Time. *Manag. Sci. Lett.* **2020**, *10*, 3009–3016. [CrossRef]
34. Esawe, A.T. Exploring Retailers' Behavioural Intentions Towards Using M-Payment: Extending UTAUT with Perceived Risk and Trust. *Paradig. A Manag. Res. J.* **2022**, *26*, 8–28. [CrossRef]
35. Namahoot, K.S.; Jantasri, V. Integration of UTAUT Model in Thailand Cashless Payment System Adoption: The Mediating Role of Perceived Risk and Trust. *J. Sci. Technol. Policy Manag.* **2023**, *14*, 634–658. [CrossRef]
36. Bai, B.; Guo, Z. Understanding Users' Continuance Usage Behavior Towards Digital Health Information System Driven by the Digital Revolution Under COVID-19 Context: An Extended UTAUT Model. *Psychol. Res. Behav. Manag.* **2022**, *15*, 2831–2842. [CrossRef]
37. Misra, R.; Mahajan, R.; Singh, N.; Khorana, S.; Rana, N.P. Factors Impacting Behavioural Intentions to Adopt the Electronic Marketplace: Findings from Small Businesses in India. *Electron. Mark.* **2022**, *32*, 1639–1660. [CrossRef]
38. Odusanya, K.; Aluko, O.; Lal, B. Building Consumers' Trust in Electronic Retail Platforms in the Sub-Saharan Context: An Exploratory Study on Drivers and Impact on Continuance Intention. *Inf. Syst. Front.* **2022**, *24*, 377–391. [CrossRef]
39. Antwi-Boampong, A.; Boison, D.; Doumbia, M.; Boakye, A.; Osei-Fosua, L.; Owiredu Sarbeng, K. Factors Affecting Port Users' Behavioral Intentions to Adopt Financial Technology (Fintech) in Ports in Sub-Saharan Africa: A Case of Ports in Ghana. *FinTech* **2022**, *1*, 362–375. [CrossRef]
40. Persada, S.F.; Afandi, F.; Redi, A.A.N.P.; Nadlifatin, R.; Prasetyo, Y.T.; Kurniawan, A.C. Mix Method Analysis for Analyzing User Behavior on Logistic Company Mobile Pocket Software. *J. Sist. Dan Manaj. Ind.* **2023**, *7*, 69–81. [CrossRef]
41. Bajunaied, K.; Hussin, N.; Kamarudin, S. Behavioral Intention to Adopt FinTech Services: An Extension of Unified Theory of Acceptance and Use of Technology. *J. Open Innov. Technol. Mark. Complex.* **2023**, *9*, 100010. [CrossRef]
42. Sun, W.; Shin, H.Y.; Wu, H.; Chang, X. Extending UTAUT2 with Knowledge to Test Chinese Consumers' Adoption of Imported Spirits Flash Delivery Applications. *Heliyon* **2023**, *9*, e16346. [CrossRef]
43. Ali, M.B.; Tuhin, R.; Alim, M.A.; Rokonuzzaman, M.; Rahman, S.M.; Nuruzzaman, M. Acceptance and Use of ICT in Tourism: The Modified UTAUT Model. *J. Tour. Futures* **2022**, *10*, 334–349. [CrossRef]
44. Alalwan, A.A. Investigating the Impact of Social Media Advertising Features on Customer Purchase Intention. *Int. J. Inf. Manag.* **2018**, *42*, 65–77. [CrossRef]
45. Abbad, M.M.M. Using the UTAUT Model to Understand Students' Usage of e-Learning Systems in Developing Countries. *Educ. Inf. Technol.* **2021**, *26*, 7205–7224. [CrossRef]
46. Alwahaishi, S.; Snášel, V. Modeling the Determinants Affecting Consumers' Acceptance and Use of Information and Communications Technology. *Int. J. E-Adopt.* **2013**, *5*, 25–39. [CrossRef]
47. Alwahaishi, S.; Snásel, V. Consumers' Acceptance and Use of Information and Communications Technology: A UTAUT and Flow Based Theoretical Model. *J. Technol. Manag. Innov.* **2013**, *8*, 9–10. [CrossRef]
48. Sarfaraz, J. Unified Theory of Acceptance and Use of Technology (UTAUT) Model-Mobile Banking. *J. Internet Bank. Commer.* **2017**, *22*, 1–20.
49. Zhou, T. Understanding Online Community User Participation: A Social Influence Perspective. *Internet Res.* **2011**, *21*, 67–81. [CrossRef]
50. Kamal, S.A.; Shafiq, M.; Kakria, P. Investigating Acceptance of Telemedicine Services through an Extended Technology Acceptance Model (TAM). *Technol. Soc.* **2020**, *60*, 101212. [CrossRef]

51. Bagozzi, R.P.; Lee, K.-H. Multiple Routes for Social Influence: The Role of Compliance, Internalization, and Social Identity. *Soc. Psychol. Q.* **2002**, *65*, 226. [CrossRef]
52. Rathore, B.; Gupta, R.; Biswas, B.; Srivastava, A.; Gupta, S. Identification and Analysis of Adoption Barriers of Disruptive Technologies in the Logistics Industry. *Int. J. Logist. Manag.* **2022**, 33, 136–169. [CrossRef]
53. Zhang, X.Y.; Lee, S.Y. A Research on Users' Behavioral Intention to Adopt Internet of Things (IoT) Technology in the Logistics Industry: The Case of Cainiao Logistics Network. *J. Int. Logist. Trade* **2023**, *21*, 41–60. [CrossRef]
54. Zhang, X.; Yu, X. The Impact of Perceived Risk on Consumers' Cross-Platform Buying Behavior. *Front. Psychol.* **2020**, *11*, 592246. [CrossRef] [PubMed]
55. Lee, J.-H.; Song, C.H. Effects of Trust and Perceived Risk on User Acceptance of a New Technology Service. *Soc. Behav. Pres.* **2013**, *41*, 587–598. [CrossRef]
56. Qalati, S.A.; Vela, E.G.; Li, W.; Dakhan, S.A.; Hong Thuy, T.T.; Merani, S.H. Effects of Perceived Service Quality, Website Quality, and Reputation on Purchase Intention: The Mediating and Moderating Roles of Trust and Perceived Risk in Online Shopping. *Cogent Bus. Manag.* **2021**, *8*, 1869363. [CrossRef]
57. Al-Gahtani, S.S. Computer Technology Acceptance Success Factors in Saudi Arabia: An Exploratory Study. *J. Glob. Inf. Technol. Manag.* **2004**, *7*, 5–29. [CrossRef]
58. Islam, S.; Islam, M.F.; Zannat, N.-E. Behavioral Intention to Use Online for Shopping in Bangladesh: A Technology Acceptance Model Analysis. *Sage Open* **2023**, *13*, 1–19. [CrossRef]
59. Camilleri, M.A. The Online Users' Perceptions toward Electronic Government Services. *J. Inf. Commun. Ethics Soc.* **2020**, *18*, 221–235. [CrossRef]
60. Radicic, D.; Petković, S. Impact of Digitalization on Technological Innovations in Small and Medium-Sized Enterprises (SMEs). *Technol. Forecast. Soc. Chang.* **2023**, *191*, 122474. [CrossRef]
61. Lertwongsatien, C.; Wongpinunwatana, N. E-Commerce Adoption in Thailand: An Empirical Study of Small and Medium Enterprises (SMEs). *J. Glob. Inf. Technol. Manag.* **2003**, *6*, 67–83. [CrossRef]
62. Omoruyi, O. Influence of Information Technology on Logistics Integration and Delivery Reliability of Small and Medium Enterprises in Gauteng Province. *Int. J. Ebusiness Egovernment Stud.* **2018**, *10*, 34–50.
63. Setiawan, M.D.; Adhariani, D.; Harymawan, I.; Widodo, M. E-commerce and Micro and Small Industries Performance: The Role of Firm Size as a Moderator. *J. Open Innov.: Technol. Mark. Complex.* **2023**, *9*, 100142. [CrossRef]
64. Andrade, C. Sample Size and Its Importance in Research. *Indian. J. Psychol. Med.* **2020**, *42*, 102–103. [CrossRef] [PubMed]
65. Johnston, K.M.; Lakzadeh, P.; Donato, B.M.K.; Szabo, S.M. Methods of Sample Size Calculation in Descriptive Retrospective Burden of Illness Studies. *BMC Med. Res. Methodol.* **2019**, *19*, 9. [CrossRef] [PubMed]
66. Dash, G.; Paul, J. CB-SEM vs PLS-SEM Methods for Research in Social Sciences and Technology Forecasting. *Technol. Forecast. Soc. Chang.* **2021**, *173*, 121092. [CrossRef]
67. Taherdoost, H. Validity and Reliability of the Research Instrument; How to Test the Validation of a Questionnaire/Survey in a Research. *SSRN Electron. J.* **2016**, *5*, 28–36. [CrossRef]
68. Taber, K.S. The Use of Cronbach's Alpha When Developing and Reporting Research Instruments in Science Education. *Res. Sci. Educ.* **2018**, *48*, 1273–1296. [CrossRef]
69. Bhattacherjee, A. Understanding Information Systems Continuance: An Expectation-Confirmation Model. *MIS Q.* **2001**, *25*, 351. [CrossRef]
70. Haddad, S.S.; Nasib, N.F. *The Role of Online Platforms in Enhancing Logistics Activity Performance*; IGI Global: Hershey, PA, USA, 2023; pp. 186–203.
71. Hameed, M.A.; Arachchilage, N.A.G. *A Conceptual Model for the Organisational Adoption of Information System Security Innovations*; IGI Global: Hershey, PA, USA, 2017.
72. Anderson, J.C.; Gerbing, D.W. Structural Equation Modeling in Practice: A Review and Recommended Two-Step Approach. *Psychol. Bull.* **1988**, *103*, 411–423. [CrossRef]
73. Byrne, B.M. *Structural Equation Modeling With AMOS*; Routledge: London, UK, 2016; ISBN 9781315757421.
74. Usakli, A.; Rasoolimanesh, S.M. Which SEM to Use and What to Report? A Comparison of CB-SEM and PLS-SEM. In *Cutting Edge Research Methods in Hospitality and Tourism*; Emerald Publishing Limited: Bingley, UK, 2023; pp. 5–28.
75. Ghasemi, A.; Zahediasl, S. Normality Tests for Statistical Analysis: A Guide for Non-Statisticians. *Int. J. Endocrinol. Metab.* **2012**, *10*, 486–489. [CrossRef]
76. Kim, H.-Y. Statistical Notes for Clinical Researchers: Assessing Normal Distribution (2) Using Skewness and Kurtosis. *Restor. Dent. Endod.* **2013**, *38*, 52. [CrossRef]
77. Mishra, P.; Pandey, C.; Singh, U.; Gupta, A.; Sahu, C.; Keshri, A. Descriptive Statistics and Normality Tests for Statistical Data. *Ann. Card. Anaesth.* **2019**, *22*, 67. [CrossRef] [PubMed]
78. Fornell, C.; Larcker, D.F. Evaluating Structural Equation Models with Unobservable Variables and Measurement Error. *J. Mark. Res.* **1981**, *18*, 39–50. [CrossRef]
79. Marsh, H.W.; Hocevar, D. Application of Confirmatory Factor Analysis to the Study of Self-Concept: First- and Higher Order Factor Models and Their Invariance across Groups. *Psychol. Bull.* **1985**, *97*, 562–582. [CrossRef]
80. Kline, R.B. *Principles and Practice of Structural Equation Modeling*, 4th ed.; Methodology in the Social Sciences; Guilford Press: New York, NY, USA, 2016; ISBN 978-1-4625-2334-4 (Paperback); 978-1-4625-2335-1 (Hardcover); 978-1-4625-2300-9 (PDF).

81. MacCallum, R.C.; Browne, M.W.; Sugawara, H.M. Power Analysis and Determination of Sample Size for Covariance Structure Modeling. *Psychol. Methods* **1996**, *1*, 130–149. [CrossRef]
82. Moss, T.P.; Lawson, V.; White, P. Identification of the Underlying Factor Structure of the Derriford Appearance Scale 24. *PeerJ* **2015**, *3*, e1070. [CrossRef] [PubMed]
83. Hu, L.; Bentler, P.M. Fit Indices in Covariance Structure Modeling: Sensitivity to Underparameterized Model Misspecification. *Psychol. Methods* **1998**, *3*, 424–453. [CrossRef]
84. Browne, M.W.; Cudeck, R. Alternative Ways of Assessing Model Fit. *Sociol. Methods Res.* **1992**, *21*, 230–258. [CrossRef]
85. Tucker, L.R.; Lewis, C. A Reliability Coefficient for Maximum Likelihood Factor Analysis. *Psychometrika* **1973**, *38*, 1–10. [CrossRef]
86. Bentler, P.M. Comparative Fit Indexes in Structural Models. *Psychol. Bull.* **1990**, *107*, 238–246. [CrossRef]
87. Liu, Y.; Shang, M.; Jia, C.; Lim, X.-J.; Ye, Y. Understanding Consumers' Continuous-Use Intention of Crowdsourcing Logistics Services: Empirical Evidence from China. *Heliyon* **2024**, *10*, e29819. [CrossRef]
88. Boonsothonsatit, G.; Vongbunyong, S.; Chonsawat, N.; Chanpuypetch, W. Development of a Hybrid AHP-TOPSIS Decision-Making Framework for Technology Selection in Hospital Medication Dispensing Processes. *IEEE Access* **2024**, *12*, 2500–2516. [CrossRef]
89. Graf-Vlachy, L.; Buhtz, K.; König, A. Social Influence in Technology Adoption: Taking Stock and Moving Forward. *Manag. Rev. Q.* **2018**, *68*, 37–76. [CrossRef]
90. Kalia, P.; Zia, A.; Kaur, K. Social Influence in Online Retail: A Review and Research Agenda. *Eur. Manag. J.* **2023**, *41*, 1034–1046. [CrossRef]
91. Luo, Y. A General Framework of Digitization Risks in International Business. *J. Int. Bus. Stud.* **2022**, *53*, 344–361. [CrossRef] [PubMed]
92. Abrahão, R.d.S.; Moriguchi, S.N.; Andrade, D.F. Intention of Adoption of Mobile Payment: An Analysis in the Light of the Unified Theory of Acceptance and Use of Technology (UTAUT). *RAI Rev. Adm. Inovação* **2016**, *13*, 221–230. [CrossRef]
93. Kwarteng, M.A.; Ntsiful, A.; Diego, L.F.P.; Novák, P. Extending UTAUT with Competitive Pressure for SMEs Digitalization Adoption in Two European Nations: A Multi-Group Analysis. *Aslib J. Inf. Manag.* **2023**. [CrossRef]
94. Lee, J.; Suh, T.; Roy, D.; Baucus, M. Emerging Technology and Business Model Innovation: The Case of Artificial Intelligence. *J. Open Innov. Technol. Mark. Complex.* **2019**, *5*, 44. [CrossRef]
95. Saarikko, T.; Westergren, U.H.; Blomquist, T. Digital Transformation: Five Recommendations for the Digitally Conscious Firm. *Bus. Horiz.* **2020**, *63*, 825–839. [CrossRef]
96. Kazancoglu, I.; Ozbiltekin-Pala, M.; Mangla, S.K.; Kumar, A.; Kazancoglu, Y. Using Emerging Technologies to Improve the Sustainability and Resilience of Supply Chains in a Fuzzy Environment in the Context of COVID-19. *Ann. Oper. Res.* **2023**, *322*, 217–240. [CrossRef]
97. George, G.; Schillebeeckx, S.J.D. Digital Transformation, Sustainability, and Purpose in the Multinational Enterprise. *J. World Bus.* **2022**, *57*, 101326. [CrossRef]
98. Adouani, Y.; Khenissi, M.A. Investigating Computer Science Students' Intentions towards the Use of an Online Educational Platform Using an Extended Technology Acceptance Model (e-TAM): An Empirical Study at a Public University in Tunisia. *Educ. Inf. Technol.* **2024**. [CrossRef]
99. Saksono, A.S.; Untoro, W. Consumer Perceived Ease of Use and Consumer Perceived Usefulness in Using the Shopee Application in Surakarta with Discount as a Moderation Variable. *Eur. J. Bus. Manag. Res.* **2023**, *8*, 13–19. [CrossRef]
100. Kyriazos, T.A. Applied Psychometrics: Sample Size and Sample Power Considerations in Factor Analysis (EFA, CFA) and SEM in General. *Psychology* **2018**, *9*, 2207–2230. [CrossRef]

modelling

Article

Enhancing Highway Driving: High Automated Vehicle Decision Making in a Complex Multi-Body Simulation Environment

Ali Rizehvandi [1], Shahram Azadi [1] and Arno Eichberger [2,*]

[1] Faculty of Mechanical Engineering, K. N. Toosi University of Technology, Tehran 15418-49611, Iran; ali.rizehvandi75@gmail.com (A.R.)
[2] Institute of Automotive Engineering, Graz University of Technology, 8010 Graz, Austria
* Correspondence: arno.eichberger@tugraz.at

Abstract: Automated driving is a promising development in reducing driving accidents and improving the efficiency of driving. This study focuses on developing a decision-making strategy for autonomous vehicles, specifically addressing maneuvers such as lane change, double lane change, and lane keeping on highways, using deep reinforcement learning (DRL). To achieve this, a highway driving environment in the commercial multi-body simulation software IPG Carmaker 11 version is established, wherein the ego vehicle navigates through surrounding vehicles safely and efficiently. A hierarchical control framework is introduced to manage these vehicles, with upper-level control handling driving decisions. The DDPG (deep deterministic policy gradient) algorithm, a specific DRL method, is employed to formulate the highway decision-making strategy, simulated in MATLAB software. Also, the computational procedures of both DDPG and deep Q-network algorithms are outlined and compared. A set of simulation tests is carried out to evaluate the effectiveness of the suggested decision-making policy. The research underscores the advantages of the proposed framework concerning its convergence rate and control performance. The results demonstrate that the DDPG-based overtaking strategy enables efficient and safe completion of highway driving tasks.

Keywords: decision making; deep reinforcement learning; carmaker simulator; deep deterministic policy gradient algorithm; highway

Citation: Rizehvandi, A.; Azadi, S.; Eichberger, A. Enhancing Highway Driving: High Automated Vehicle Decision Making in a Complex Multi-Body Simulation Environment. *Modelling* **2024**, *5*, 951–968. https://doi.org/10.3390/modelling5030050

Academic Editors: Tomasz Nowakowski, Artur Kierzkowski, Agnieszka A. Tubis, Franciszek Restel, Tomasz Kisiel, Anna Jodejko- Pietruczuk and Mateusz Zając

Received: 18 June 2024
Revised: 29 July 2024
Accepted: 6 August 2024
Published: 15 August 2024

1. Introduction

Automated vehicles, SAE levels 3, 4, and 5, driven by artificial intelligence (AI), are becoming increasingly popular as they aim to reduce road accidents and improve traffic efficiency [1,2]. To achieve high automation, four key modules are required: perception, decision making, planning, and control [3]. However, further research and development are needed to accomplish high automation, especially in complex driving environments.

Automated driving involves a continuous series of maneuvers to accomplish specific navigation tasks. These maneuvers typically require adjustments to the accelerator pedal and steering angle. Researchers have made numerous efforts to develop suitable decision-making policies for automated driving (AD). For instance, Hole et al. employed a Monte Carlo tree search to develop decision-making strategies for AD [4]. They modeled the driving scenario as a partially observable Markov decision process (POMDP) and compared the outcomes with those obtained using a neural network (NN) policy. The authors also explored helpful lane-changing decisions to optimize the utilization of limited road properties and mitigate competition. In [5], the authors also discussed the decisions of high AD way exits for autonomous vehicles. They claimed that the proposed decision-making controller significantly increases the likelihood of successful highway exits based on 6000 stochastic simulations.

Reinforcement learning (RL), particularly deep reinforcement learning (DRL) methods, has shown excessive potential in tackling decision-making challenges in AD [6]. For example, in [7], deep Q-learning (DQL) was used to manage lane-changing decision making in uncertain highway environments. Similarly, for the lane-changing problem, Zhang et al. [8] introduced a model-based exploration policy based on surprise intrinsic rewards. Additionally, it provided a comprehensive overview of RL or DRL applications in automated vehicles, covering agent training, evaluation techniques, and robust estimation [9]. Despite their promise, DRL-based decision-making strategies face several limitations that hinder their real-world applicability, including issues with sample efficiency, slow learning rates, and operational safety.

In [10], the authors developed an advanced decision-making capability for urban road traffic scenarios. The decision-making policy presented incorporates multiple criteria, enabling city cars to make practical choices in different situations. Moreover, Nie et al. explored a lane-change decision-making strategy for connected automated vehicles [11]. This strategy incorporates cooperative car-following models and a candidate decision-making module. Additionally, in [12], the authors introduced the concept of a human-like driving system capable of adjusting driving decisions based on the demands of human drivers.

Deep reinforcement learning (DRL) techniques are becoming increasingly popular for solving complex problems involving sequential decision making. In the field of AD, several studies have explored the use of DRL-based approaches. For example, Duan et al. [13] developed a hierarchical construction for learning decision-making policies using reinforcement learning (RL) methods, which does not require historical labeled driving data. In [14,15], DRL methods were employed to tackle collision avoidance and path-following issues in automated vehicles, achieving better control performance than conventional RL methods. Additionally, refs. [16,17] extended considerations beyond path planning to include fuel consumption optimization for autonomous vehicles. These studies employed the deep Q-learning (DQL) algorithm, which proved to be effective in accomplishing driving missions. Furthermore, Han et al. [18] used the DQL algorithm to make lane-change or lane-keeping decisions for connected autonomous cars, utilizing feedback knowledge from nearby vehicles as input to the network. However, conventional DRL methods face challenges in addressing highway overtaking problems due to the continuous action space and large state space. Also, in [19], reinforcement learning has been widely used in the field of unmanned driving, but how to improve the stability of unmanned vehicles and meet the requirements of path tracking and vehicle obstacle avoidance under different working conditions is still a difficult problem. Aiming at the functional requirements of path tracking and obstacle avoidance of unmanned vehicles, an anti-collision control strategy of unmanned vehicles based on a deep deterministic policy gradient (DDPG) algorithm is proposed in this paper.

In [20], the authors propose a deep reinforcement learning (DRL)-based motion planning strategy for AD tasks in highway scenarios where an AV merges into two-lane road traffic flow and realizes the lane-changing (LC) maneuvers. They integrate the DRL model into the AD system relying on the end-to-end learning method. They used a DRL algorithm based on deep deterministic policy gradient (DDPG) with well-defined reward functions.

This study aims to create a decision-making policy that is both efficient and safe for highway AD. To achieve this, this study introduces a deep reinforcement learning (DRL) approach enhanced by deep deterministic policy gradient (DDPG). Also, this paper employs the DDPG algorithm for the first time to solve the highway navigation problem for long driving scenarios including lane change, double lane change and lane keeping for highway navigation. One of the primary advantages of using the DDPG algorithm in AD is its ability to operate effectively in continuous action spaces for long scenarios. Unlike traditional discrete action algorithms such as the DQN, which limit the agent to a predefined set of actions, DDPG allows for fine-grained control over the vehicle's steering, acceleration, and braking. This capability is particularly crucial in the context of highway navigation, where smooth and precise control is essential for maintaining safety and comfort. By leveraging an actor–critic framework, DDPG can learn a policy that outputs continuous actions in the complex highway environment, enabling the vehicle to make more natural and efficient driving decisions. This leads to improved performance in dynamic and complex driving environments, where discrete actions could result in less fluid and more abrupt maneuvers.

The approach is personalized for continuous action horizons in highway scenarios, as shown in Figure 1.

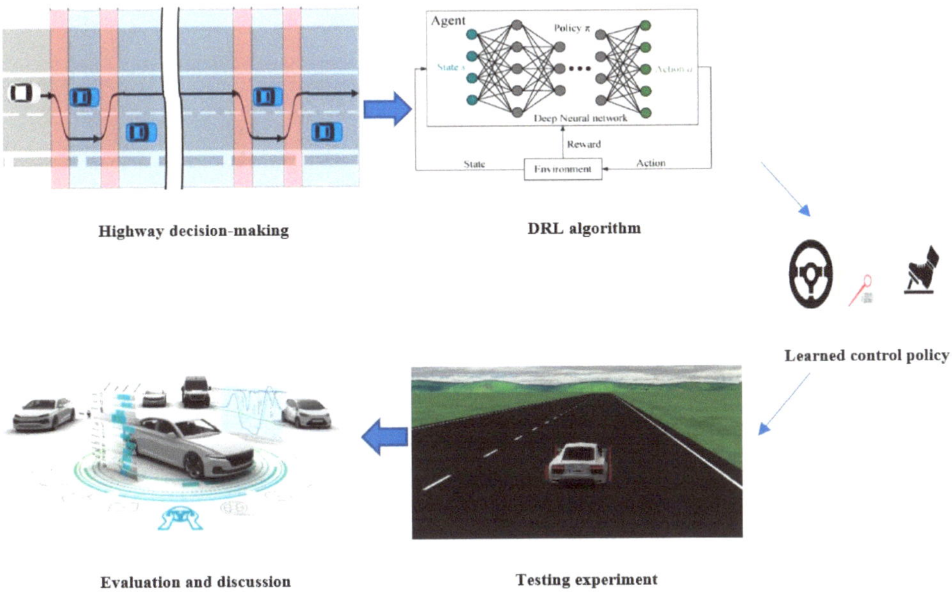

Figure 1. The developed highway driving policy for autonomous vehicles, enabled by deep reinforcement learning [evaluation and discussion: [21]].

Based on Figure 1, highway decision making is performed by the powerful DRL method, which means the DDPG algorithm. To accomplish this purpose, the driving environment will be simulated in the Carmaker simulator, while the DDPG algorithm learns the driving actions like steering angle and accelerating or decelerating.

The study begins by defining the real vehicle dynamic in [22] and driving scenarios by the Carmaker scenario [22] to ensure the automated vehicle operates safely and efficiently. The DDPG-enhanced DRL method leverages the actor–critic method to directly obtain control actions, establishing a trust region with clipped objectives. The DRL algorithm's implementation details are elaborated upon in the subsequent sections. Finally, the performance of the decision-making algorithm for a specific scenario including lane change, double lane change, and lane keeping is evaluated, and, also, for different scenarios, the algorithm adaptability is validated. This paper presents three key innovations that aim to improve the safety and efficiency of AD on highways for real maneuvering. These contributions are as follows:

1. Development of an advanced, safe, and efficient decision-making policy for AD on highways for a real maneuver using the DDPG algorithm;
2. Using real vehicle dynamics and real scenarios in the Carmaker simulator.

To elaborate on these contributions, this paper is structured as follows:

Section 2 provides a description of vehicle dynamics and driving scenarios on highways. Section 3 details the research on DDPG-enhanced DRL. Also, Section 4 evaluates the simulation results relevant to the presented decision-making strategy. Finally, concluding remarks are presented in Section 5.

2. Vehicle Dynamic and Driving Scenario

This section outlines the testing scenario for highway driving, which includes the autonomous ego vehicle (AEV) and the other vehicles on the road. It also describes the vehicle dynamics of the AEV involved. Additionally, reference models for driving maneuvers in the CarMmaker simulator are introduced.

2.1. *Vehicle Dynamic*

This study utilized the vehicle dynamics framework in the commercial multi-body simulation IPG CarMmaker as presented in [22].

2.2. *Driving Scenario*

This study involves creating a driving scenario in the CarMmaker simulator with three lanes per direction of travel. The objective is to simulate real-world driving conditions on the highway. An AEV will navigate through the scenario by ascertaining the control actions for vehicle speed and steering angle at each time step. The AEV aims to travel as swiftly as possible while ensuring safety on the road and avoiding collisions with surrounding vehicles. The ultimate goal of the AEV is to reach the highway exit (Figure 2).

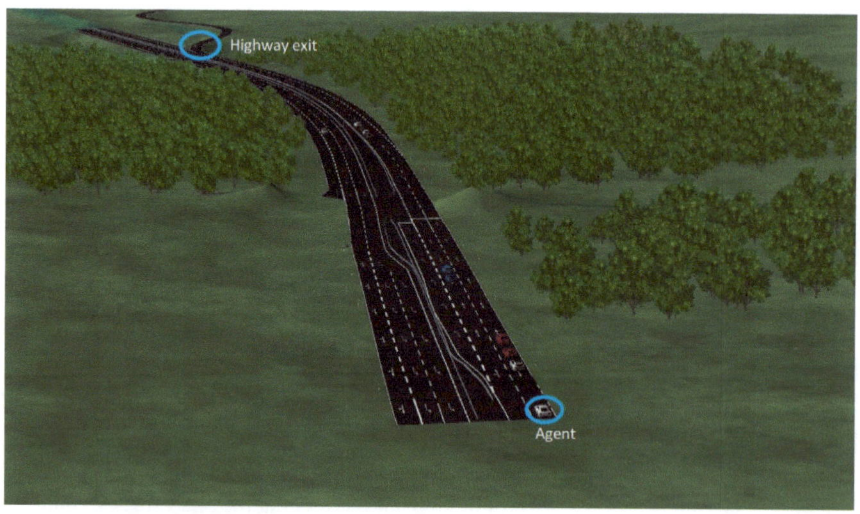

Figure 2. Driving scenario.

Typically, three criteria are commonly employed to assess the effectiveness of decision-making policies: safety, efficiency, and comfort. Safety entails ensuring that the AEV avoids collisions. Efficiency suggests that the AEV aims to enhance speed. Comfort entails regulating the lane-changing frequency and the extent of deceleration of the vehicle. In this study, the main priorities for the AEV are safety and efficiency. The vehicle typically positions itself in the fast lane, as depicted in Figure 3. The AEV is represented by the white vehicle, while the other vehicles represent the surrounding traffic. Each lane has different types of traffic. In this context, an episode refers to the AEV overtaking all adjacent vehicles to reach its destination on the highway. To ensure simplicity and versatility, we assume that the highway has three lanes (N = 3). Each lane contains different surrounding vehicles. The autonomous vehicle (AEV) is programmed to drive in the right lane. The simulation runs at a frequency of 20 Hz, with the AEV making decisions every second (sampling time of 1 s). Each episode lasts for 50 s. The driving behavior of the adjacent vehicles is governed by random models including autonomous vehicles and human driver vehicles.

Figure 3. RL approach [23].

3. RL Method

One of the sub-items of artificial intelligence (AI) is machine learning (ML), aiming to develop computational algorithms performance with data. Machine learning (ML) is classified into three main types: reinforcement learning (RL), supervised learning, and unsupervised learning. In the RL method, by trying to maximize a predefined reward function, an autonomous agent learns to fulfill a task in an environment. When the agent opts for a desirable action when interacting with its environment, it is rewarded. In contrast, a punishment or a negative reward is assigned to the autonomous agent, if the selected action is undesired.

Supervised learning concisely involves learning from labeled examples from experts. Since finding a label that exactly denotes an interaction is complicated, the approach is not appropriate for solving interactive problems [24].

As unsupervised learning, the method is learning to find a hidden configuration in unlabeled data. Though finding construction in collected data through learning is useful, the approach cannot optimize a reward, which is an objective of the RL [24].

Some RL problems have huge numbers of states and actions in the environment. In these RL problems, an artificial neural network (ANN) can be used as a function approximator. Utilizing an ANN as a function approximator in RL is called the deep reinforcement learning (DRL) method. The RL problems are typically depicted as Markov decision processes (MDPs). An MDP is a mathematical structure employed to represent decision-making challenges where results are influenced by both random factors and the decisions made by an agent. This framework finds extensive application in fields such as RL and operations research.

The key components of a Markov decision process are as follows:

States (S): The states of MDPs represent all possible situations or configurations of the environment. These states can be discrete or continuous, depending on the problem domain.

Actions (A): For each state in the MDP, there is a range of potential actions that the decision making can take. Actions represent the choices or decisions available to the decision making at each state.

Transition Probabilities (P): The transition probabilities ascertain the likelihood of transitioning from one state to another after executing a particular action. In other words, they specify the likelihood distribution across potential next states given the current state and action.

Rewards (R): At each state–action pair, there is an associated reward that represents the immediate benefit or cost of taking that action in that state. Rewards can be positive (rewards) or negative (penalties), and they may be deterministic or stochastic.

Policy (π): A policy is a mapping from states to actions, specifying the decision maker's strategy for selecting actions at each state. Reinforcement learning algorithms often strive to discover an optimal policy that maximizes the expected total reward over time.

Value Function (Q): The value function signifies the anticipated total reward achievable by adhering to a specific policy or executing a particular action within a given state. It helps in evaluating the quality of different policies or actions.

MDPs satisfy the Markov property, which implies that the subsequent state relies solely on the current state and action, not on the history of previous states and actions. This property simplifies the modeling and analysis of decision-making problems and allows for the application of various solution techniques, encompassing dynamic programming, Monte Carlo methods, and temporal difference learning.

Markov decision processes supply a formal framework for studying decision making under uncertainty and are widely used in areas such as robotics, autonomous systems, game theory, economics, and more.

The expected discounted reward R_t after the t time step can be considered as:

$$R_t = \sum_t^{\infty} \gamma^t \cdot r_t \quad (1)$$

where, when γ is in the range [0, 1], it is a discount factor. Parameter t, depending on the problem, is the finite value. Also, policy π $(a \mid s)$ maps from states to action probabilities. $V^{\pi}(s)$ can be recognized as an expected return regarding policy π from state s and is a value function and it is as follows:

$$V^{\pi}(s_t) = E_{\pi}[R_t | S_t, \pi] \quad (2)$$

$Q^{\pi}(s, a)$ is an action-value function as follows:

$$Q^{\pi}(s_t, a_t) = E_{\pi}[R_t | S_t, a_t, \pi] \quad (3)$$

The iterative Bellman equation is satisfied:

$$Q^{\pi}(s_t, a_t) = E_{\pi}[r_t + \gamma \max(Q^{\pi}(s_{t+1}, a_{t+1}))] \quad (4)$$

However, not all RL problems can be formulated as MDP. In specific situations, if state S can only be observed partly from the environment or cannot be observed directed from the defined environment, then our problems are able to be formulated as a partially observable Markov decision process (POMDP) for such events. A way to solve the problem is to produce observations that include past knowledge by containing previous observations or prior information together with a current observation and therefore solve the problem as an MDP. Learning a policy that maximizes the expected return is the main objective of the RL algorithm.

3.1. Deep Deterministic Policy Gradient (DDPG)

The deep deterministic policy gradient (DDPG) algorithm [24] is a reinforcement learning algorithm that combines elements of both value-based and policy-based methods. It is particularly well-suited for addressing problems with continuous action spaces in reinforcement learning.

DDPG utilizes an actor–critic architecture in which there are two main networks:

- Actor network: This network learns the policy function, which maps states to actions. It aims to maximize the expected return by directly selecting actions based on the current state;
- Critic network: This network learns the value function, which approximates the expected return (cumulative reward) of following a particular policy. It helps to evaluate the actions chosen by the actor network.

Also, DDPG is an off-policy algorithm, meaning it learns from data sampled from an experience replay buffer without explicitly following a specific policy. It is also model free, meaning it does not involve knowledge of the underlying dynamics of the environment. Unlike some other algorithms that are more suited to discrete action spaces, DDPG is designed to handle continuous action spaces, making it applicable to a wide range of problems, including robotics and control tasks.

A mean-squared Bellman error (MSBE) is defined as:

$$L(\varnothing, \theta) = E_D\left[\left(Q_\varnothing(s,a) - \left(r + \gamma(1-d)\max Q_\varnothing(\acute{s}, \acute{a})\right)\right)^2\right] \quad (5)$$

The DDPG algorithm combines elements of policy gradient methods and Q-learning. The actor network is trained using policy gradient methods to directly maximize the expected return, while the critic network is trained using Q-learning to approximate the value of state–action pairs.

In addition, DDPG introduces target networks to stabilize training. These are copies of the actor and critic networks that are updated less frequently than the main networks. Additionally, DDPG utilizes an experience replay buffer to store and sample experiences during training (based on Figure 4). This helps to decorate the data and improve sample efficiency.

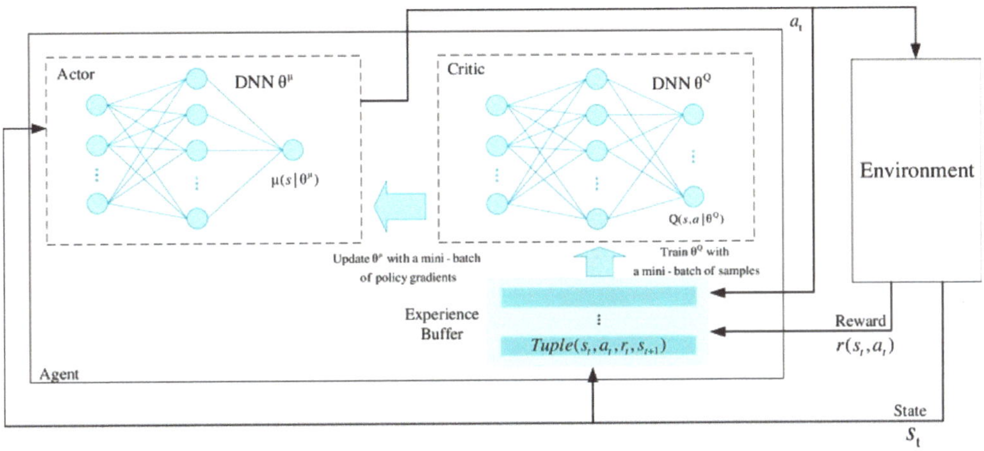

Figure 4. DDPG algorithm [25]. Copyright © 2022, Hu Z. et al. This open-access article is distributed under the Creative Commons Attribution License, which permits unrestricted use, distribution, and reproduction in any medium, provided the original work is properly cited.

Also, the reward function is as follow:

$$R_t = 0.5|\theta| + 0.1|a| - \text{collision} + M \quad (6)$$

where θ is the steering angle, a is longitudinal acceleration, and collision is 1, if $d_{relative} = 0$, and M is a constant.

Based on Equation (6), the agent is trained in the highway environment to cross the highway exit.

3.2. *Deep Q-Network (DQN)*

The Deep Q-network (DQN) algorithm is a deep reinforcement learning technique that integrates Q-learning with deep neural networks to approximate the optimal action-value function in a continuous state space. The DQN operates by learning to map states to actions directly from raw sensory inputs, typically images, enabling it to handle high-dimensional input spaces. It uses experience replay and a target network for stabilizing training and preventing over fitting by storing and randomly sampling past experiences from a replay buffer and periodically updating a target network with the weights of the trained Q-network. Through iterative updates using gradient descent, the DQN aims to minimize the temporal difference error between the forecasted Q-values and the target Q-values, ultimately learning a policy that maximizes cumulative rewards in the environment.

4. Discussion

This section assesses the performance of a proposed decision-making strategy for the AEV using the deep deterministic policy gradient (DDPG) method. The evaluation covers three main aspects. Firstly, it compares and verifies the efficacy of this decision-making approach against an alternative method using detailed simulation outcomes to demonstrate its superiority. Secondly, it validates the DDPG algorithm's learning capability by examining the accumulated rewards. Lastly, it assesses the derived decision-making strategy in two comparable driving scenarios on the highway to showcase its adaptability.

Figure 5 illustrates the specific scenario performed by the DDPG agent in the CarMmaker simulator.

1 2 3 4 5 6 7 8

Figure 5. Specific scenario including lane change, double lane change, and lane keeping (1: preparing for lane change, 2: executing lane change, 3: preparing for double lane change, 4: executing double lane change, 5: preparing for second lane change, 6: executing second lane change, 7: executing third lane change, 8: executing lane keeping).

4.1. Performance Evaluation

In this section, we are comparing the performance of DDPG and the DQN, both of which employ a hierarchical control framework. However, the upper levels differ. The default parameters for both DDPG and the DQN are identical. Figures 6 and 7 display the average rewards and episode rewards obtained by the DDPG and DQN agents, respectively. According to Figures 6 and 7, a higher reward signifies, when driving on the preferred lane, more efficiently maneuvering.

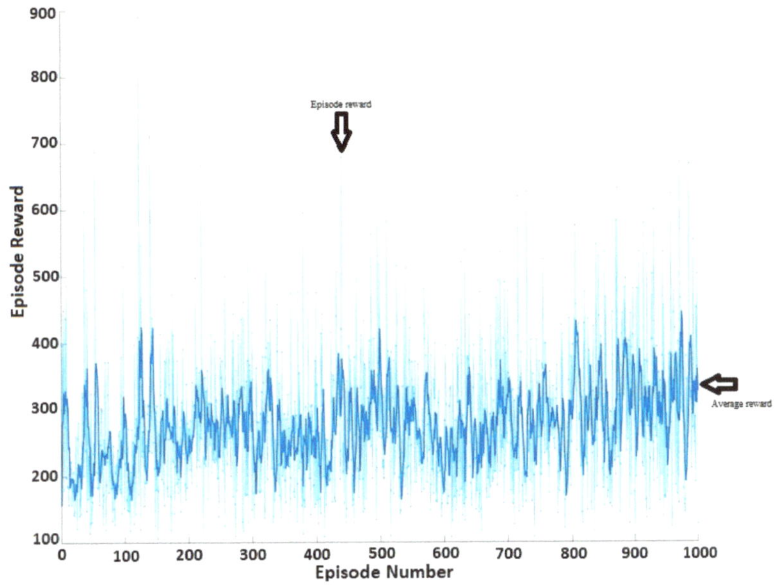

Figure 6. DDPG agent rewards.

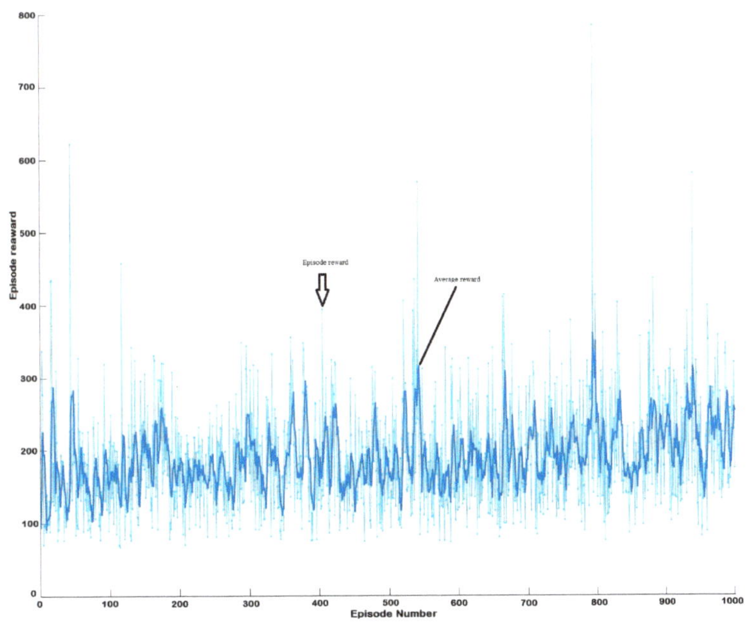

Figure 7. DQN agent rewards.

The DDPG approach exhibits superior training stability and learning speed than the DQN approach (based on Table 1), resulting in consistently higher rewards after approximately 1000 episodes. The main reason why DDPG is considered superior is due to the actor–critic network it uses. These networks can calculate the value of the chosen action at each step, allowing the ego vehicle to quickly identify a better decision-making policy.

Table 1. Comparative analysis of DDPG and DQN algorithms.

Parameter	DDPG	DQN
Max episode reward	892	791
Max average reward	443	355
Max speed (m/s)	16.9	14.7
Max distance (m)	961	907

To analyze the trajectories of the state variable in this study, Figure 8 displays the average vehicle speed for DDPG and DQN agents for 200 s.

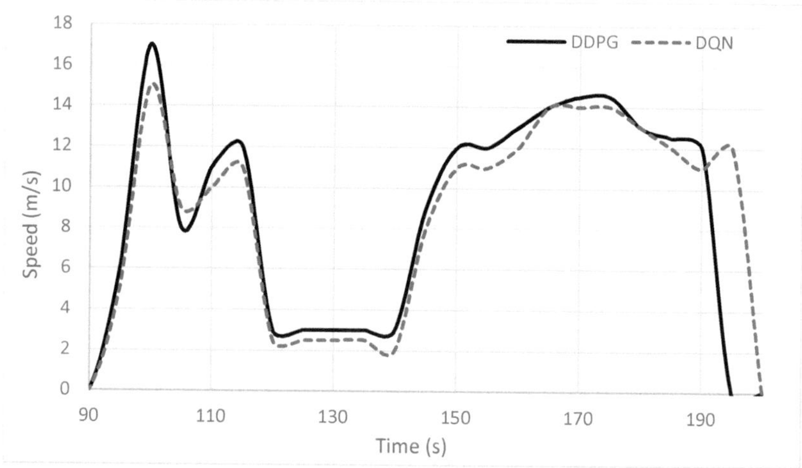

Figure 8. DDPG and DQN agent velocity.

A high speed means that the ego vehicle can maneuver through the driving environment more quickly, resulting in the acquisition of greater total rewards. Also, higher average vehicle speed indicates that the DDPG agent has learned a better policy for controlling the vehicles in the environment. This could suggest that the DDPG algorithm is more effective in this particular scenario for maximizing speed-related objectives. In other words, DDPG is an actor–critic algorithm that utilizes deterministic policy gradients. It tends to be more sample efficient and stable than the DQN.

The greater travel distance achieved by the DDPG agent suggests that it navigated the environment more efficiently than the DQN agent (based on Figure 9). This indicates that the DDPG agent made better use of its actions and state information to move through the environment, avoiding unnecessary delays or detours. Also, DDPG's continuous action space allows for more nuanced control over the agent's actions compared with the DQN's discrete actions. This finer-grained control has enabled the DDPG agent to select actions that result in smoother, more continuous movement, facilitating longer travel distances.

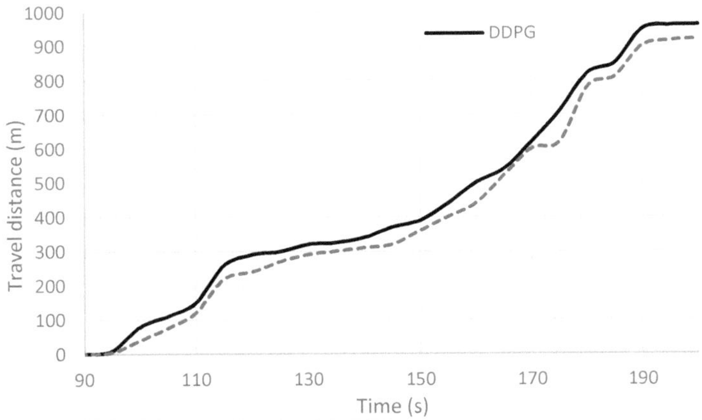

Figure 9. DDPG and DQN agent travel distance.

As mentioned earlier, DDPG often requires fewer samples to learn compared with the DQN. If DDPG achieved higher speeds and greater travel distances, it indicates that DDPG has learned a more effective policy more efficiently than the DQN in this particular scenario. In addition, the combination of higher average speeds and greater travel distances demonstrates that the DDPG agent outperformed the DQN agent regarding overall movement efficiency and effectiveness in the environment. This indicates that DDPG may be better suited for tasks where minimizing travel time is important.

The differences observed in the performance of the decision-making policies in this study were highlighted by the trajectories displayed. These results indicate that the policy proposed in this study outperforms the DQN benchmark methods. The superiority of the decision-making strategy enabled by DDPG is evident when considering all the results presented in this subsection, as it achieved successful episodes consistently.

4.2. Criticality Evaluation

According to Figure 5, part 7, the DDPG agent has a near condition for collision; then, we consider criticality metrics as follows:

To calculate the time to collision (TTC) in this scenario, we use the following formula:

$$TTC_{low} = \min\left(\frac{\Delta X(t)}{\Delta V(t)}\right) \quad (7)$$

where

$$\text{Agent velocity } (v_{\text{agent}}) = 2.8 \text{ m/s}$$

$$\text{Distance to the static vehicle } (d) = 4.4 \text{ m}$$

The relative velocity (v_{relative}) in this case would be the sum of the agent's velocity and the static vehicle's velocity (since it is not moving):

$$v_{\text{relative}} = v_{\text{agent}} - 0 = 2.8 \text{ m/s} - 0$$

$$v_{\text{relative}} = 2.8 \text{ m/s}$$

Now, we can calculate TTC based on (6):

$$TTC_{low} = \frac{4.4}{2.8} = 1.57 \text{ s} \quad (8)$$

So, the TTC_{low} in this scenario is approximately 1.57 s.

A time to collision (TTC) of 1.57 s indicates that there is still a reasonable (based on UN regulation No. 157 [26]) amount of time before a potential collision occurs between the agent and the static vehicle.

Figure 10 shows the behavior of the calculated TTC(t) according to Equation (6) in the presented scenario (based on Figure 5, parts 6–7), when following a target vehicle. We observe the lowest value for TTC at 1.57 s, a mean value of 2.89 s., and a standard deviation of 0.90 s when approximated with Gaussian behavior. Typically, autonomous emergency braking (AEB) systems include an intervention strategy that warns the driver if the TTC falls below approximately 2.5 s and initiates partial braking at approximately 1.5 s. Full braking would be applied at TTC_{low} values below 0.6 to 0.8 s. The vehicle manufacturer defines the thresholds of these intervention strategies and is also sometimes adjusted by the driver, changing the setup of the AEB system in the configuration menu of the car. Hence, the algorithm is usually in an acceptable range but sometimes requires braking interventions to stay in a safe area.

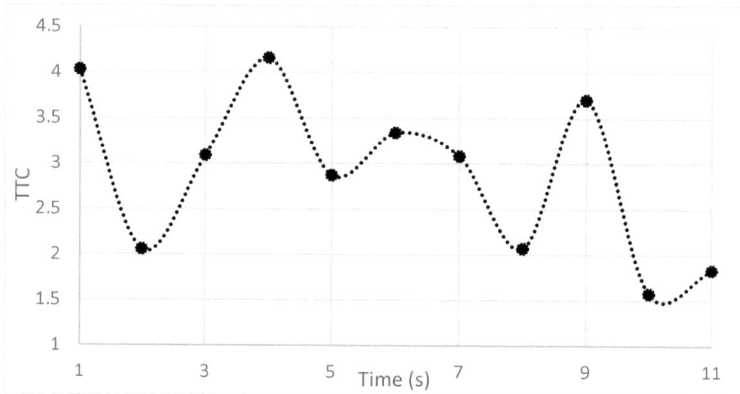

Figure 10. Time to collision (TTC) behavior.

In addition, the time gap between the ego vehicle and the target vehicle was calculated according to the following Equation (8),

$$t_{Gap} = \frac{\Delta X}{V_{Ego}} \quad (9)$$

where, again, ΔX is the relative distance between the target and ego vehicle and V_{Ego} is the ego vehicle speed.

Figure 11 shows the behavior of the time gap during the same scenario (based on Figure 5, parts 6–7). The lowest value reads 0.86 s, the average value is 2.60 s, and the standard deviation is 0.95 s. According to [26], most drivers prefer a time gap of 1.8 s compared with 1.0 and 1.3 s. Hence, we conclude that, in the current setting of the algorithm, the driving scenario is usually uncritical and acceptable for most drivers. A more detailed consideration of criticality is part of future investigations and will include more advanced criticality metrics as compared with the TTC. TTC is limited to the constant relative velocity between the target and ego vehicle and therefore not always suitable for criticality evaluation.

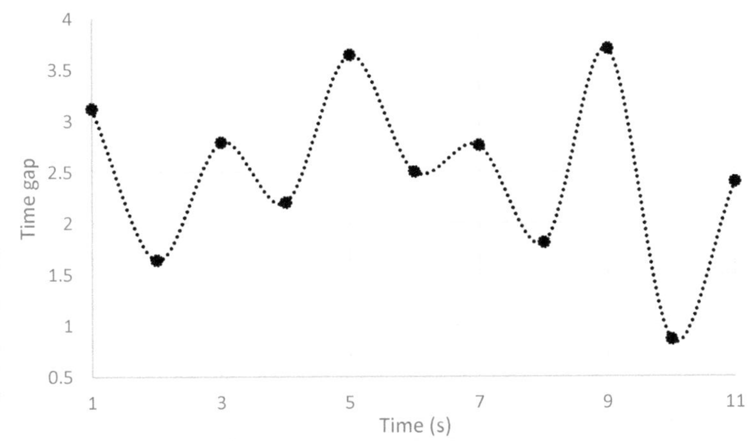

Figure 11. Time gap behavior.

4.3. Comparison between DDPG and DQN

This experiment evaluates the DDPG and DQN algorithms, both established techniques in deep reinforcement learning (DRL), to determine their effectiveness in learning and training methods.

To demonstrate how the dueling network can be applied to future decisions in AD, Figure 12 displays the trajectory of cumulative rewards for 100 episodes.

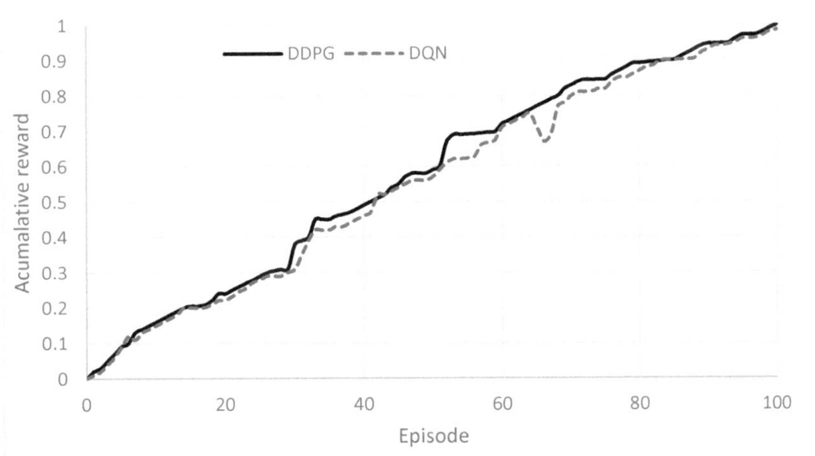

Figure 12. DDPG and DQN agent accumulative reward.

According to Figure 12, higher cumulative rewards indicate that the DDPG agent's learned policy was more effective at maximizing rewards in the environment compared with the DQN agent. This suggests that DDPG was able to make better decisions over time, leading to higher overall rewards. Also, DDPG's continuous action space and deterministic policy gradients may have allowed it to optimize its policy more efficiently than the DQN. This could have enabled the DDPG agent to explore and exploit the state–action space more effectively, leading to higher cumulative rewards.

In other words, DDPG's ability to learn a deterministic policy and its actor–critic architecture has enabled it to consider the long-term consequences of its actions better than the DQN. This has contributed to its ability to accumulate higher rewards over time. Moreover, DDPG's deterministic policy gradients have provided a more stable learning process compared with the epsilon-greedy exploration strategy used in the DQN. This has allowed DDPG to learn a more optimal policy with less variance, leading to higher cumulative rewards.

Figure 12 shows a decreasing trend, which indicates that both ego vehicles are becoming more acquainted with the driving environment as they interact with it. Moreover, the DDPG algorithm can gain a greater understanding of traffic situations in the same number of episodes, resulting in a faster learning process. As a result, the ego vehicle can navigate more efficiently and safely with the guidance of the DDPG algorithm.

4.4. Adaptability Estimation

After training automated vehicles for highway driving scenarios, a phase is implemented to evaluate their ability to adapt. The testing phase consists of 10 episodes with the initial configurations remaining consistent with the training phase. The neural networks' parameters acquired during training are preserved and can be directly applied in new circumstances. During testing, the average reward achieved and instances of collisions are analyzed to ensure that the vehicles can adapt to new situations. During the testing phase, Figure 13 shows the normalized average reward achieved by the DDPG and DQN methods. The reward is mainly determined by the vehicle's speed and collision occurrences.

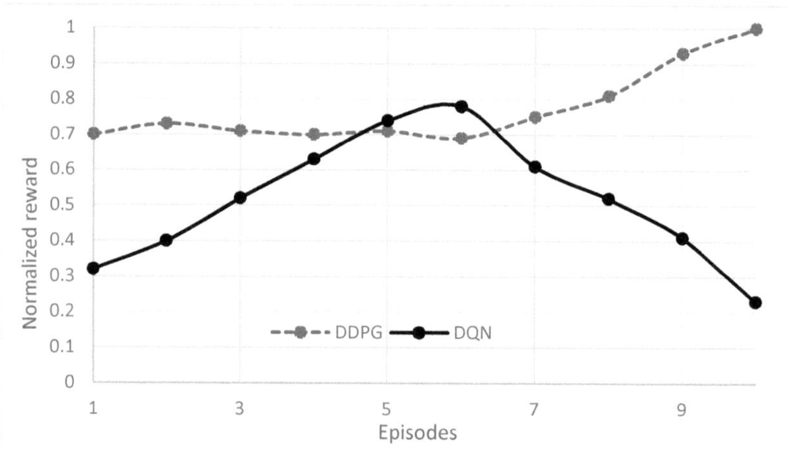

Figure 13. Normalized reward in the testing experiment of two compared methods.

The average reward might not reach its maximum score of 100 in this study due to instances where the ego vehicle must decelerate to prevent collisions. Additionally, the ego vehicle may need to switch lanes to facilitate overtaking maneuvers. To illustrate decision-making performance, three representative situations (A and B indicated in Figures 14 and 15) are selected for analysis without loss of generality.

Figure 14 demonstrates a driving scenario where the ego vehicle is surrounded by three vehicles in front of it. The ego vehicle needs to engage in extended car-following maneuvers until it finds an opportunity to overtake them. Consequently, the vehicle may not achieve its maximum speed, and it might not surpass all surrounding vehicles before reaching its destination.

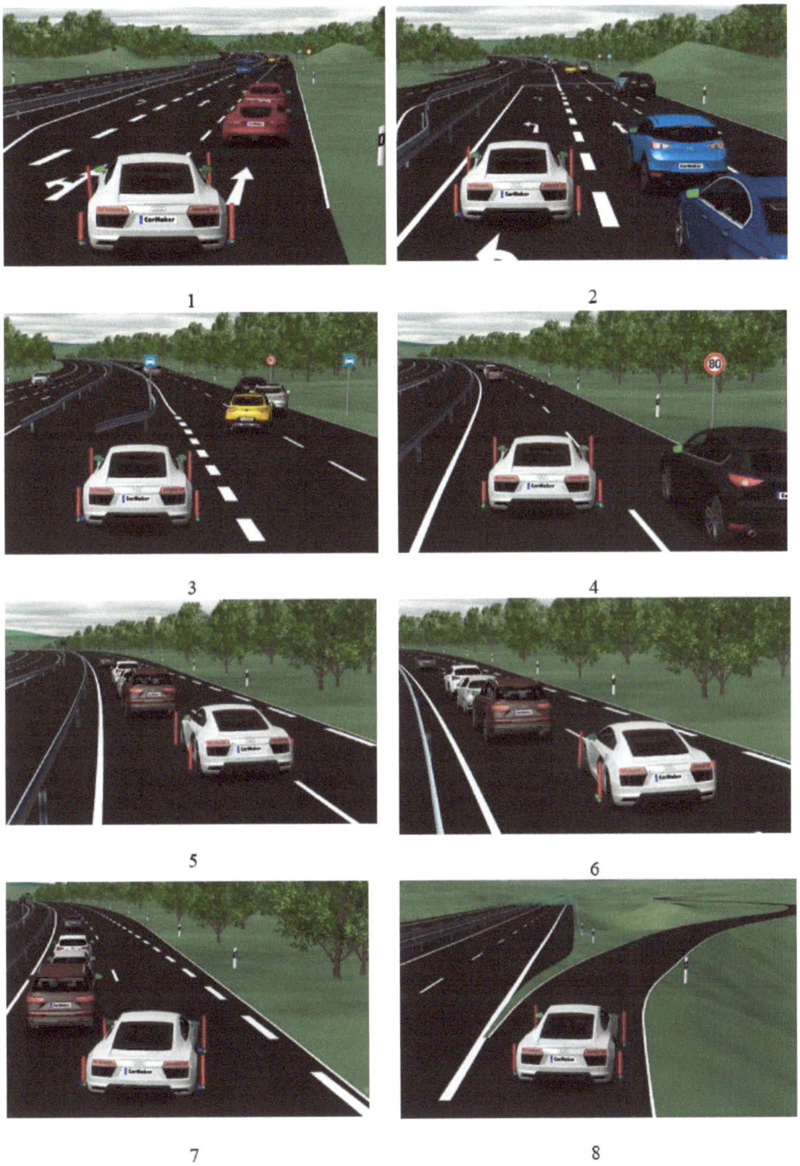

Figure 14. Testing scenario (A) (1: preparing for double lane change, 2: executing double lane change, 3: preparing for lane change, 4: executing lane change, 5: preparing for second lane change, 6: executing second lane change, 7: executing second lane change, 8: executing lane keeping).

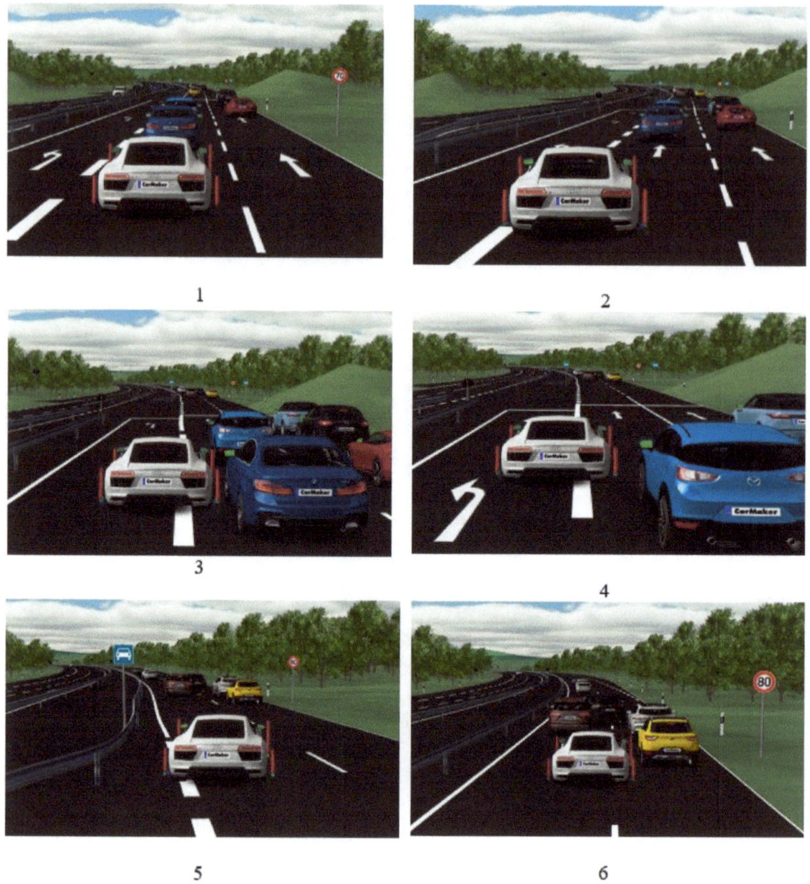

Figure 15. Testing scenario (B) (1: preparing for overtaking, 2: executing overtaking, 3: executing overtaking, 4: executing overtaking, 5: preparing for lane keeping, 6: executing lane keeping).

Moreover, Figure 15 depicts an uncommon driving scenario where the ego vehicle attempts a risky lane change to gain higher rewards. Then, due to insufficient operating space, it overtakes nearby vehicles while it has minimum speed. These kinds of occurrences are infrequent during training and can potentially lead the ego vehicle to cause collisions.

The comprehensive analysis presented in Figures 14 and 15 suggests the need for further refinement of the adaptable decision-making strategy through extended training. These findings also underscore the applicability of the corresponding control policy in real-world settings.

The DDPG algorithm succeeded in two different scenarios tested; it indicates a high level of adaptability and generalization ability.

The fact that DDPG performed well across multiple scenarios suggests that it was able to generalize its learned policy effectively. Generalization refers to the capability of an algorithm to apply knowledge gained from one scenario to perform well in new, unseen scenarios. DDPG's success in two scenarios indicates that it learned a policy that is robust and applicable across a range of environments. Moreover, DDPG's success in diverse scenarios highlights its flexibility as a reinforcement learning algorithm. It was able to adapt its policy to different environmental conditions and task requirements, demonstrating its ability to handle varying complexities and dynamics. DDPG's success across multiple scenarios also reflects its learning dynamics, including exploration strategies, update rules, and memory mechanisms. These learning dynamics have enabled DDPG to efficiently learn and adapt its policy to different environmental conditions.

Eventually, DDPG's consistent success across different scenarios also presents its robustness as an algorithm. Robustness refers to an algorithm's ability to maintain good performance despite variations in the environment or perturbations in the learning process. DDPG's ability to perform well in diverse scenarios suggests that it is robust to changes and uncertainties.

5. Conclusions

The study employs DRL techniques to explore the challenge of highway decision making. In this study, we introduce an innovative approach by incorporating a unique combination of driving scenarios, specifically lane change, double lane change, and lane keeping, which have not been collectively utilized in prior studies. This combination allows for a comprehensive evaluation of the decision-making DDPG algorithm under varied and complex driving conditions. By addressing this complex scenario, the present study offers a more realistic testing framework, enhancing the reliability adaptability of AD systems in real-world situations. This innovation not only fills a gap in the existing research but also contributes to the advancement of safer and more efficient AD technologies. A tailored control framework is established using the DDPG algorithm within the driving environments to ensure safety and effectiveness. The paper presents the proposed approach's performance, convergence rate, and adaptability through a sequence of simulation experiments. According to the results, the DDPG algorithm is more efficient and safer than the DQN technique. Furthermore, the testing results are thoroughly evaluated, showcasing the potential of the proposed approach to be successfully implemented in real-world driving scenarios. Future work involves implementing online highway decision making via hardware-in-loop research and using real-world highway databases to estimate relevant overtaking strategies.

Author Contributions: Conceptualization, A.R. and S.A.; methodology, A.R. and S.A.; software, A.R. and S.A.; validation, A.R. and A.E.; formal analysis, A.R.; investigation, A.R.; resources, A.R. and A.E.; data curation, A.E.; writing—original draft preparation, A.R.; writing—review and editing, A.E. All authors have read and agreed to the published version of the manuscript.

Funding: This research received no external funding.

Data Availability Statement: Data are contained within the article.

Acknowledgments: We acknowledge the use of online translation tools to enhance the quality of our text.

Conflicts of Interest: The authors declare no conflict of interest.

References

1. Liu, T.; Huang, B.; Deng, Z.; Wang, H.; Tang, X.; Wang, X.; Cao, D. Heuristics-oriented overtaking decision making for autonomous vehicles using reinforcement learning. *IET Electr. Syst. Transp.* **2020**, *10*, 417–424. [CrossRef]
2. Rasouli, A.; Tsotsos, J.K. Autonomous vehicles that interact with pedestrians: A survey of theory and practice. *IEEE Trans. Intell. Transp. Syst.* **2019**, *21*, 900–918. [CrossRef]
3. Liu, T.; Tian, B.; Ai, Y.; Chen, L.; Liu, F.; Cao, D. Dynamic States Prediction in Autonomous Vehicles: Comparison of Three Different Methods. In Proceedings of the IEEE Intelligent Transportation Systems Conference (ITSC 2019), Auckland, New Zealand, 27–30 October 2019.
4. Hoel, C.; Driggs-Campbell, K.; Wolff, K.; Laine, L.; Kochenderfer, M. Combining planning and deep reinforcement learning in tactical decision making for autonomous driving. *IEEE Trans. Intell. Veh.* **2019**, *5*, 294–305. [CrossRef]
5. Cao, Z.; Yang, D.; Xu, S.; Peng, H.; Li, B.; Feng, S.; Zhao, D. Highway Exiting Planner for Automated Vehicles Using Reinforcement Learning. *IEEE Trans. Intell. Transp. Syst.* **2020**, *22*, 990–1000. [CrossRef]
6. Sakib, N. Highway Lane Change under Uncertainty with Deep Reinforcement Learning Based Motion Planner. 2020. Available online: https://era.library.ualberta.ca/items/501e8502-0e1c-4ab9-adbe-aeb2da0e29fd (accessed on 17 June 2024).

7. Alizadeh, A.; Moghadam, M.; Bicer, Y.; Ure, N.; Yavas, U.; Kurtulus, C. Automated Lane Change Decision Making using Deep Reinforcement Learning in Dynamic and Uncertain Highway Environment. In Proceedings of the 2019 IEEE Intelligent Transportation Systems Conference (ITSC), Auckland, New Zealand, 27–30 October 2019; pp. 1399–1404.
8. Zhang, S.; Peng, H.; Nageshrao, S.; Tseng, E. Discretionary Lane Change Decision Making using Reinforcement Learning with Model-Based Exploration. In Proceedings of the 2019 18th IEEE International Conference on Machine Learning and Applications (ICMLA), Boca Raton, FL, USA, 16–19 December 2019; pp. 844–850.
9. Kiran, B.R.; Sobh, I.; Talpaert, V.; Mannion, P.; Sallab, A.; Yogamani, S.; Pérez, P. Deep reinforcement learning for autonomous driving: A survey. *arXiv* **2020**, arXiv:2002.00444. [CrossRef]
10. Furda, A.; Vlacic, L. Enabling safe autonomous driving in real-world city traffic using multiple criteria decisions making. *IEEE Intell. Transp. Syst. Mag.* **2011**, *3*, 4–17. [CrossRef]
11. Nie, J.; Zhang, J.; Ding, W.; Wan, X.; Chen, X.; Ran, B. Decentralized cooperative lane-changing decision-making for connected autonomous Vehicles. *IEEE Access* **2016**, *4*, 9413–9420. [CrossRef]
12. Li, L.; Ota, K.; Dong, M. Humanlike driving: Empirical decisionmaking system for autonomous vehicles. *IEEE Trans. Veh. Technol.* **2018**, *67*, 6814–6823. [CrossRef]
13. Duan, J.; Li, S.E.; Guan, Y.; Sun, Q.; Cheng, B. Hierarchical reinforcement learning for self-driving decision-making without reliance on labeled driving data. *IET Intell. Transp. Syst.* **2020**, *14*, 297–305. [CrossRef]
14. Li, G.; Li, S.; Li, S.; Qu, X. Continuous decision-making for autonomous driving at intersections using deep deterministic policy gradient. *IET Intel. Transp. Syst.* **2021**, *16*, 1669–1681. [CrossRef]
15. Zhang, Q.; Lin, J.; Sha, Q.; He, B.; Li, G. Deep interactive reinforcement learning for path following of autonomous underwater vehicle. *IEEE Access* **2020**, *8*, 24258–24268. [CrossRef]
16. Chen, C.; Jiang, J.; Lv, N.; Li, S. An intelligent path planning scheme of autonomous vehicles platoon using deep reinforcement learning on the network edge. *IEEE Access* **2020**, *8*, 99059–99069. [CrossRef]
17. Yang, C.; Zha, M.; Wang, W.; Liu, K.; Xiang, C. Efficient energy management strategy for hybrid electric vehicles/plug-in hybrid electric vehicles: Review and recent advances under intelligent transportation system. *IET Intell. Transp. Syst.* **2020**, *14*, 702–711. [CrossRef]
18. Han, S.; Miao, F. Behavior planning for connected autonomous vehicles using feedback deep reinforcement learning. *arXiv* **2020**, arXiv:2003.04371. Available online: http://arxiv.org/abs/2003.04371 (accessed on 4 September 2022).
19. Nageshrao, S.; Tseng, H.E.; Filev, D. Autonomous highway driving using deep reinforcement learning. In Proceedings of the IEEE International Conference on Systems, Man and Cybernetics (SMC), Bari, Italy, 6–9 October 2019; pp. 2326–2331.
20. Lv, K.; Pei, X.; Chen, C.; Xu, J. A Safe and Efficient Lane Change Decision-Making Strategy of Autonomous Driving Based on Deep Reinforcement Learning. *Mathematics* **2022**, *10*, 1551. [CrossRef]
21. Available online: https://www.avl.com/en/engineering/vehicle-engineering/vehicle-development/global-vehicle-benchmarking-and-technology (accessed on 17 November 2022).
22. Reichmann-Blaga, E. 'Validierung von Fahrzeugdynamischen Simulationsmodellen anhand von 546 Fahrzeugmessungen. Master's Thesis, Graz University of Technology, Graz, Austria, 2024. Available online: https://repository.tugraz.at/publications/3kttg-zmr02 (accessed on 17 June 2024).
23. The Reinforcement Learning Framework—Hugging Face Deep RL Course. Available online: https://huggingface.co/learn/deep-rl-course/unit1/rl-framework (accessed on 4 May 2022).
24. Song, W.; Xiong, G.; Chen, H. Intention-aware autonomous driving decision-making in an uncontrolled intersection. *Math. Probl. Eng.* **2016**, *2016*, 1025349. [CrossRef]
25. Hu, Z.; Gao, H.; Wang, T.; Han, D.; Lu, Y. Joint Optimization for Mobile Edge Computing-Enabled Blockchain Systems: A Deep Reinforcement Learning Approach. *Sensors* **2022**, *22*, 3217. [CrossRef]
26. Lai, J.-P.; Li, H.; Shi, Y.; Xu, L.-M.; Yan, H. Anti Collision Control Strategy of Unmanned Vehicle Based on DDPG Algorithm. *Wuhan Ligong Daxue Xuebao/J. Wuhan Univ. Technol.* **2021**, *43*, 68–76.

modelling

MDPI

Article

Integrating Null Controllability and Model-Based Safety Assessment for Enhanced Reliability in Drone Design

Zahra Motahari Rad and Jonathan Liscouët *

Mechanical, Industrial, and Aerospace Engineering Department, Gina Cody School of Engineering and Computer Science, Concordia University, Montreal, QC H3G 1M8, Canada
* Correspondence: jonathan.liscouet@concordia.ca

Abstract: The increasing use of drones for safety-critical applications, particularly beyond visual lines of sight and over densely populated areas, necessitates safer and more reliable designs. To address this need, this paper introduces a novel methodology integrating Null Controllability with the Model-Based Safety Assessment (MBSA) framework *AltaRica* 3.0 to optimize propulsor configurations and system architectures. The main advancement of this method lies in the automation of reliability modeling and the integration of controllability assessment, eliminating restrictions on the types of propulsor configurations and system architectures that can be evaluated and significantly reducing the effort required for each design iteration. Through a hexarotor drone case study, the proposed method enabled a high number of design iterations, efficiently exploring various aspects of the design problem simultaneously, such as configuration, system architecture, and controllability hypothesis, which is not possible with state-of-the-art techniques. This approach demonstrated significant reliability improvements by implementing and optimizing redundancies, reducing the probability of loss of control by up to 99%. The case study also highlighted the increasing difficulty of enhancing reliability with each iteration and confirmed that it is unnecessary to consider more than two simultaneous failures for design optimization. A comparison of reliability figures with previous studies highlights the crucial role of system architecture in effectively enhancing drone design reliability. This work advances the field by providing an effective multidisciplinary modeling framework for drone design, enhancing reliability in safety-critical applications.

Keywords: drone; reliability; Null Controllability; Model-Based Safety Assessment (MBSA); system architec-ture; actuator configuration; redundancy techniques; hexarotor; *AltaRica*; *System Analyst*

Citation: Motahari Rad, Z.; Liscouët, J. Integrating Null Controllability and Model-Based Safety Assessment for Enhanced Reliability in Drone Design. *Modelling* **2024**, *5*, 1009–1030. https://doi.org/10.3390/modelling5030053

Academic Editors: Tomasz Nowakowski, Artur Kierzkowski, Agnieszka A. Tubis, Franciszek Restel, Tomasz Kisiel, Anna Jodejko-Pietruczuk and Mateusz Zając

Received: 14 June 2024
Revised: 24 July 2024
Accepted: 14 August 2024
Published: 23 August 2024

1. Introduction

The increasing use of drones for critical applications, particularly for operations beyond visual lines of sight and over densely populated areas, highlights the need for safe and reliable drone designs. This requires incorporating safety and reliability considerations from the outset to achieve an optimal balance between performance, safety, and reliability. Consequently, this study focuses on integrating reliability considerations into the preliminary design of drones, encompassing both propulsor configuration and system architecture optimizations.

Various methodologies have been developed and applied over the years to enhance unmanned aerial vehicle (UAV) design reliability. The following literature review discusses the evolution of these methodologies, emphasizing key contributions and their limitations in integrating reliability, controllability, and redundancy optimization. The review focuses on methodologies that address changes to propulsor configuration and system architecture to achieve higher reliability, excluding works that explore other means, such as preventive maintenance as discussed in [1].

In 2007, Franco et al. [2] illustrated a conventional approach using Failure Modes and Effects Analysis (FMEA) and Fault Tree Analysis (FTA) to enhance the reliability of fixed-wing UAVs. This method focused on identifying potential failures and implementing preventive measures to minimize their impact on mission-critical systems. The analyses were conducted manually, requiring the expertise of control, system, and reliability specialists.

In 2015, Venkataraman et al. [3] introduced probabilistic models to assess fault impacts and analyzed the UAV's flight envelope to manage fault levels without catastrophic failure. Despite its insightful evaluation of the remaining flight envelope after failures, the scope was limited to propulsor configurations and did not address system architecture.

A significant advancement was achieved in 2016 when Shi et al. [4] investigated the reliability of multicopter configurations using the Null Controllability theory, assessing various designs based on the Available Control Authority Index (ACAI). This pioneering work integrated controllability and reliability for configuration optimization but required manual reliability modeling and did not consider system architecture.

In 2018, Wang et al. [5] applied fault tree modeling and Monte Carlo simulation to assess quadrotor UAV reliability, evaluating component importance for design improvement. This approach did not include controllability evaluation, as the loss of any rotor in a quadrotor is catastrophic, and it focused solely on propulsor configuration.

In 2019, building on Shi et al.'s work, Aslansefat et al. [6], and later in 2023, Thanaraj et al. [7], proposed frameworks integrating ACAI-based controllability analysis with Markov chain modeling for evaluating the reliability of multirotor UAVs. These frameworks integrated dynamic reliability models based on state transitions (i.e., Markov chain models) but required manual reliability modeling from the controllability assessment and did not include system architecture considerations. Also in 2019, Guo et al. [8] presented a framework that integrates controllability analysis based on operational balance with k-out-of-n requirements and state transition modeling. This work automated the reliability evaluation of propulsor configurations but was limited to multi-level balanced configurations and did not address system architecture.

Another significant advancement was achieved in 2022 when Nazarudeen et al. [9] and Liscouët et al. [10] introduced a method to integrate Null Controllability, reliability, and sizing of multirotor UAVs. They combined ACAI-based controllability analysis and Reliability Block Diagrams (RBD) for reliability analysis. This method enabled automatic reliability modeling by integrating propulsor configuration and system architecture, combining models of generic power/control systems with propulsor systems for overall UAV reliability. This integrated approach optimized and compared various configurations and architectures in terms of reliability, weight, and mission performance. However, it was limited to system architectures with generic power/control systems and propulsor configurations, whose reliability can be expressed independently. Sections 4.3.1 and 4.3.6 describe independent and dependent generic power/control systems and propulsor configurations, respectively. In the system outlined in Section 4.3.6, the intricate connection between the batteries and the ESCs prevents the overall system reliability from being expressed as a product of the power/control system and propulsor configuration reliabilities.

Building on previous work and addressing the limitations of system architectures, this article proposes a novel methodology that integrates ACAI-based controllability analysis with the Model-Based Safety Assessment (MBSA) framework *AltaRica* 3.0 [11]. The aim is not only to optimize propulsor configuration and system architecture jointly without any restrictions but also to facilitate extensive design iterations. This enables the designer to explore a higher number and more intricate design solutions, thereby broadening the scope and depth of the investigation. Through a case study of a hexarotor drone designed for a safety-critical mission, the paper illustrates the modeling advancements facilitated by the proposed methodology, showcasing its potential to enhance drone reliability through effective configuration and system redundancy optimization. Table 1 summarizes the previous literature review, comparing the proposed methodology with state-of-the-art techniques and highlighting its advancement: seamlessly integrating both control effector configuration and system architecture reliability considerations into a single design framework.

Table 1. Comparison of the proposed methodology and state-of-the-art techniques.

Reference	Approach	Address Control Effector Configuration	Address System Architecture
2007, Franco et al. [2]	Conventional FMEA and FTA	Yes, but not integrated [1]	Yes, but not integrated [1]
2015, Venkataraman et al. [3]	Flight envelope assessment and probabilistic models	Yes, but not integrated [1]	No
2016, Shi et al. [4]	ACAI and RBD	Yes, but not integrated [1]	No
2018, Wang et al. [5]	FTA and Monte Carlos Simulation	No controllability evaluation	No
2019, Aslansefat et al. [6] 2023, Thanaraj et al. [7]	ACAI and Markov Chain Models	Yes, but not integrated [1]	No
2019, Guo et al. [8]	Operational balance with k-out-of-n requirements and state transition models	Yes, but with a limited range of configurations [2]	No
2022, Nazarudeen et al. [9], Liscouët et al. [10]	ACAI and RBD	Yes	Yes, but with a limited range of architectures [3]
Proposed Methodology	ACAI and MBSA	Yes	Yes

[1] Reliability modeling must be manually performed by a reliability expert in collaboration with a control or system expert for each significant design change. [2] Limited to multi-level balanced propulsor configurations. [3] Limited to system architectures with generic power/control systems and propulsor configurations, whose reliability can be expressed independently.

The structure of this paper is organized as follows: Section 2 outlines the methodological approach used to evaluate the reliability and controllability of UAV flight control systems. Section 3 describes the custom-designed hexarotor case study. Section 4 demonstrates the effectiveness of the developed method and tool by analyzing and optimizing the reliability of the hexarotor case study. Finally, Section 5 concludes the study and suggests directions for future work.

2. Method

This section outlines the methodology used to evaluate the reliability and controllability of UAV flight control systems. This methodology is divided into four main steps, detailed as follows:

Step 1: Determine the propulsor configuration (detailed in Section 2.1) and establish the reliability hypothesis, including the failure rates of system components, while considering any limitations on the reconfigurability.

Step 2: Design the physical system architecture (detailed in Section 2.2). Define the components, their quantities, and their interconnections. Utilize a graphical interface for efficient task execution, resulting in a physical representation of the design. The selected graphical modeling environment is *System Analyst* 1.3 [12]. Convert this graphical representation into an *AltaRica* 3.0 language-based model using the embedded compiler.

Step 3: Assess the design's controllability for all possible failure combinations using the ACAI method (detailed in Section 2.3). This analysis and the next step are conducted in *MATLAB* R2022b.

Step 4: Integrate the controllability assessment into the *AltaRica* 3.0 model (detailed in Section 2.4). First, convert the controllability assessment into logical instructions for each combination of working propulsors that provide sufficient control. Importantly, these instructions must be written to enable their compilation into Boolean equations. Then, rewrite the *AltaRica* 3.0 model of the physical system to include these instructions. Finally, compile and execute the updated model in the *AltaRicaWizard* 1.2.0 modeling environment [13] to compute the UAV's probability of loss of control and perform a reliability sensitivity analysis.

The designer's main tasks are determining the propulsor configuration in Step 1 and designing the system's physical architecture in Step 2 to evaluate reliability. Steps 3 and 4 are primarily automated for efficiency. Figure 1 illustrates the methodology and its integration into the case study optimization process presented in Section 4.

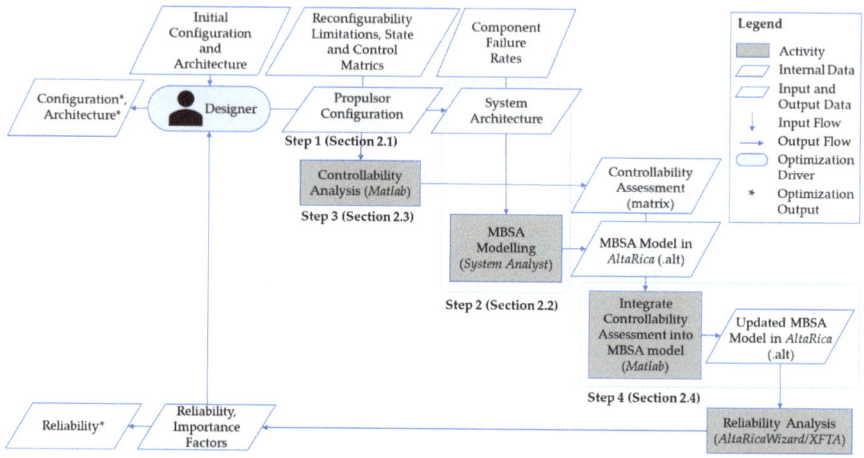

Figure 1. Overview of the methodology for multirotor UAV reliability and controllability evaluation in design optimization using extended design structure matrix (XDSM) representation [14,15].

2.1. Step 1—UAV Configuration Model

The configuration of a multirotor UAV refers to its geometric layout, encompassing the number of propulsors (motors and propellers), their spatial positions, their orientations relative to the vehicle's center of gravity, and the direction of rotation of each propulsor. This configuration determines how the propulsors' thrusts and torques contribute to the overall forces and moments acting on the UAV. For instance, in a hexarotor configuration, six propulsors are arranged in a hexagonal pattern around the center of gravity. These propulsors alternate rotational directions (clockwise and counterclockwise) to provide yaw controllability. Each propulsor's position, orientation, and rotation direction significantly influence the UAV's control dynamics and effectiveness.

To mathematically capture the influence of the propulsors on the UAV's dynamics, we utilize the control effectiveness matrix [16]. This matrix establishes the relationship between the individual propulsor thrusts and the resultant forces and moments acting on the vehicle, as follows:

$$u = B_f f, \tag{1}$$

where $u \in \mathbb{R}^n$ represents the forces and moments acting on the vehicle, $B_f \in \mathbb{R}^{n \times p}$ is the control effectiveness matrix, $f \in \mathbb{R}^p$ is the rotor thrust vector, and p is the number of control effectors, in effect rotors for multirotor UAVs. Typically, for rotorcraft design, $u = \begin{bmatrix} F_z & M_x & M_y & M_z \end{bmatrix}^T$ where F_z denotes the vertical lift, and M_x, M_y, and M_z represent the pitch, roll, and yaw moments, respectively. Hence, typically, $n = 4$.

Deriving B_f for a multirotor UAV involves mapping the thrusts generated by each propulsor to the resultant forces and moments acting on the vehicle. This matrix is constructed by first identifying the position vectors of the propulsors relative to the UAV's center of gravity and their thrust directions. The total thrust F_z is the sum of all propulsor thrusts, while the roll M_x and pitch M_y moments are calculated based on the propulsors' y and x-coordinates, respectively. The yaw moment M_z considers the reactive torque of the propellers, typically alternating in direction. Each row of B_f corresponds to a specific component of the virtual control input. Examples of control effectiveness matrices for hexarotor UAVS are provided in Section 3.1.

Finally, the controllability assessment, detailed in Section 2.3, requires defining the state matrix A_s and control matrix B from the dynamics model presented in Equation (3). An example is provided in Section 3.1.

2.2. Step 2—Physical System Architecture Model

This research aims to simplify the designer's task by deriving the reliability model from an intuitive representation of the physical system architecture, eliminating the need for expertise in reliability modeling. A physical system architecture model includes hardware components, their physical interfaces, system layout, component positioning, wiring, and critical design constraints [17,18]. This visualization helps designers understand and optimize component relationships, enhancing design efficiency and effectiveness. The recommended approach is to create, assess, and compare several candidate physical architecture models to select the most suitable one [17].

The proposed approach leverages Model-Based Safety Assessment (MBSA). MBSA is an approach for evaluating the safety and reliability of complex systems using formal models. It involves modeling component faults and their propagation, automatically computing safety analyses, and ensuring consistency between design and safety models [19]. It facilitates automated analysis, including reliability modeling derived from the physical system architecture. Utilizing a graphical modeling environment within an MBSA framework, the approach presented in this study allows designers to focus on defining the physical system architecture while providing a visual model representation. Implemented in *System Analyst* 1.3 [12], which uses the *AltaRica* 3.0 language for MBSA [11], it facilitates intuitive and effective design processes.

The *AltaRica* 3.0 language is an object-oriented modeling language dedicated to probabilistic risk and safety analyses [20] using guarded transition formalism [21]. In this study, the component models are designed following the hierarchical architectural pattern [22], mirroring the physical architecture and connections between components. Each component's simplified operational dynamics are modeled using a non-repairable component class [23], extended with a flow variable. The formal definition of each non-repairable component model is represented by the quintuple $\langle V, E, T, A, i \rangle$, where:

- V is a set of state and flow variables. A state variable indicates whether the component is working or failed. Flow variables connect the component model to other models.
- E is a set of events. For simplicity, each component model includes one event e, characterizing the component's inherent failure, associated with a probabilistic function.
- T is a set of transitions. Each component model includes one transition t_r, defined as a triple $\langle e, g, a \rangle$ and denoted by $g \xrightarrow{e} a$. Here, g is a Boolean condition on variables in V (the guard of the transition), and a is an instruction (the action of the transition) that changes the state variable from working to failed.
- A is a set of instructions, called the assertion, that defines the values of flow variables based on the state variable values.
- i is a function that gives the initial value of the state and flow variables.

The component models used in this study are illustrated by an ESC model created in *System Analyst* 1.3, as shown in Figure 2 The corresponding *AltaRica* 3.0 code is also provided in Figure 2. This model includes a state variable (*ESC1_29_working*, line 53), and eight flow variables: external inputs from the flight controller command (*In_cmd_32*, line 49) and battery DC power supply (*In_DC_34*, line 51), an external output to the motor (*Out_AC_33*, line 50), internal input (*ESC1_29_in*, line 54) and output (*ESC1_29_out*, line 55), and a logic operator variable (*Logic_310*, line 52).

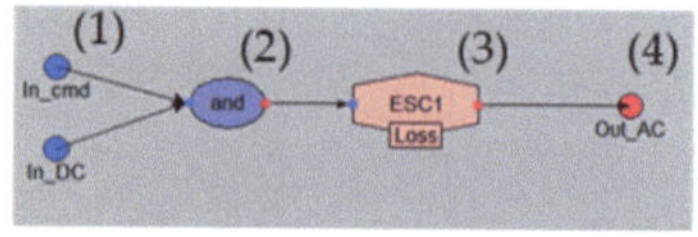


```
48      block ESC_1_28
49          Integer In_cmd_32 (reset = 0);
50          Integer Out_AC_33 (reset = 0);
51          Integer In_DC_34 (reset = 0);
52          Integer Logic_310 (reset = 0);
53          Integer ESC1_29_working (init = 1);
54          Integer ESC1_29_in (reset = 0);
55          Integer ESC1_29_out (reset = 0);
56          event ESC1_29_failure (delay=exponential(1e-05));
57          transition
58              ESC1_29_failure: ESC1_29_working==1 -> ESC1_29_working := 0;
59          assertion
60              ESC1_29_out := if ESC1_29_working==1 or (ESC1_29_in>6 and
    ESC1_29_in!=0) then ESC1_29_in else 6;
61              Logic_310 := if max(In_cmd_32, In_DC_34)<1 then min(In_cmd_32,
    In_DC_34) else max(In_cmd_32, In_DC_34);
62              ESC1_29_in := Logic_310;
63              Out_AC_33 := ESC1_29_out;
64      end
```

Figure 2. *AltaRica* 3.0 code generated from ESC block model in *System Analyst* 1.3 (top right) The model components are: (1) Model inputs, (2) Logic assertion, (3) State transition, and (4) Model output.

Event *ESC1_29_failure* (line 56) represents the inherent failure of the ESC, characterized by an exponential cumulative probability distribution with a constant failure rate of 10^{-5} per hour. The transition *ESC1_29_failure* (line 58) is as follows:

$$t_r : (ESC1\_29\_working = 1) \xrightarrow{ESC1\_29\_failure} (ESC1\_29\_working := 0), \tag{2}$$

where := denotes the Walrus operator assigning values to variables.

The assertion includes multiple instructions (lines 60 to 63) based on the dependency-oriented modeling principle used in *System Analyst* 1.3 [12]. This approach allows for the definition of operation modes as follows: working mode (value of 0), failure mode (value of 6), and alternate working, failure, and testing modes (values other than 0 and 6). This framework enables the future integration of various performance and failure modes, including cascading failures. In the ESC example, if both external inputs *In_DC_34* and *In_cmd_32* indicate working modes, and the state variable *ESC1_29_working* shows the component has not failed, the working mode is communicated to the motor via *Out_AC_33*. Otherwise, a failure mode is communicated to the motor.

The initialization sets all flow variables to the default value of zero and the component state to work. Manual coding of this model could optimize it further, for example, by removing internal flow variables and creating a class for each component instead of using a block model.

The integration of various system components, such as the ESC, to model the entire system replicates the conventional physical system architecture. Examples of these physical system architectures, modeled in *System Analyst* 1.3, are described in Sections 4.3.1 and 4.3.6. Notably, the proposed modeling method includes a generic controllability model that links the control effectors to the flight control function, represented by an observer block, for analysis purposes.

2.3. *Step 3—Controllability Assessment*

The controllability assessment is based on the ACAI method from [24]. The ACAI computation is based on a linear state-space model of multirotor drones in hovering conditions, such as:

$$\dot{x} = A_s x + B(u - u_0), \tag{3}$$

where $x \in \mathbb{R}^m$ is the state vector, m is the number of state variables, $A_s \in \mathbb{R}^{m \times m}$ is the state matrix, $B \in \mathbb{R}^{m \times n}$ represents the control matrix, and u_0 represents the forces and moments acting on the vehicle resulting in the reference state x_0. The main component of the control authority is the control vector u, which results from the rotor thrust vector f as described in Equation (1). Each rotor provides a purely positive and limited thrust, leading to the following constraint set for the rotor thrust vector $\mathcal{F}$ [24]:

$$\mathcal{F} = \left\{ f = [f_1 \cdots f_p]^T \middle| 0 \le f_i \le f_{max,i}, \text{ for } i = 1,2,\cdots,p \right\}, \tag{4}$$

where f_i is the thrust of i-th rotor and $f_{max,i}$ is the maximum thrust of the i-th rotor. The virtual control vector constraint set is expressed as follows [24]:

$$\Omega = \left\{ u \in \mathbb{R}^n \middle| u = B_f f, f \in \mathcal{F} \right\}. \tag{5}$$

The ACAI quantifies the maximum effort achievable in all control directions taking into account the impact of constrained rotor thrusts and the influence of a disturbance, such as the weight of the vehicle. More specifically, if u_0 is contained within Ω, then the ACAI is the radius of the largest enclosed four-dimensional sphere centered at u_0 in Ω. If u_0 is not contained within Ω, then the ACAI is the radius of the largest enclosed sphere centered at u_0 in the complementary set of Ω, denoted as Ω^c [24]:

$$\rho(u_0, \partial\Omega) \triangleq \begin{cases} \min\{\|u - u_0\| \,|\, u \in \partial\Omega\} & \text{if } u_0 \in \Omega \\ -\min\{\|u - u_0\| \,|\, u \in \partial\Omega\} & \text{if } u_0 \in \Omega^c \end{cases}, \tag{6}$$

where $\partial\Omega$ denotes the boundary of Ω, and $\|\cdot\|$ is the Euclidean norm. Detailed information on the computation of the ACAI is provided in [24]. Finally, as demonstrated in [24], a configuration is locally controllable in all directions of u only if

$$rank\ C(A_s, B) = m \text{ and } \rho(u_0, \partial\Omega) > 0, \tag{7}$$

where $C(A_s, B)$ is the system controllability matrix.

The proposed methodology involves assessing the controllability of each potential failure case. Initially, the full set of failure cases is defined by systematically combining one to k rotor failure(s). Limiting the number of failed rotors to k prevents overdesigning for highly improbable scenarios. This assumes the UAV flight control system will not adapt to more than k rotor failures, leading to loss of control for any combination exceeding k. The total number of failure cases to be assessed, j, is given by [10]:

$$j = \sum_{i=1}^{k} \frac{p!}{i!(p-i)!}, \tag{8}$$

where p denotes the total number of rotors (i.e., control effectors), and $k \in [1,\cdots,p-1]$ is the maximum number of simultaneous rotor failures to be considered in the design. The second step is to define a failure matrix for each possible failure case as follows [10]:

$$H_j = diag(\eta_1,\cdots,\eta_p), \tag{9}$$

where H_j represents the failure matrix of the j-th failure case, and $\eta_i \in \{0,1\}$ with $i = 1,\cdots,p$ represents available ($\eta_i = 1$) and failed ($\eta_i = 0$) rotors. The third step is to define the control effectiveness matrix from the failure matrix for each failure case [10]:

$$B'_{f,j} = B_f H_j, \tag{10}$$

where $B'_{f,j}$ is the control effectiveness matrix reflecting the effect of the j-th failure case.

The fourth step is to define the system total thrust/torque constraint set for each failure case as follows [10]:

$$\Omega'_j = \left\{ u'_{f,j} \middle| u'_{f,j} = B'_{f,j} f, f \in \mathcal{F} \right\}, \tag{11}$$

where Ω'_j is the remaining system total thrust/torque constraint in the j-th failure case.

Finally, the ACAI can be calculated for each failure case by revising Equation (7) as follows [10]:

$$\rho'_j(u_0, \partial\Omega') \triangleq \begin{cases} \min\left\{ \left\| u'_{f,j} - u_0 \right\| \middle| u'_{f,j} \in \partial\Omega'_j \right\} & \text{if } u_0 \in \Omega'_j \\ -\min\left\{ \left\| u'_{f,j} - u_0 \right\| \middle| u'_{f,j} \in \partial\Omega'_j \right\} & \text{if } u_0 \in \Omega'^{c}_j \end{cases}, \tag{12}$$

where $\rho'_j(u_0, \partial\Omega')$ is the ACAI for the j-th failure case, $\partial\Omega'_j$ is the boundary of Ω'_j, and Ω'^{c}_j is the complementary set of Ω'_j. The result of the controllability analysis represents the controllability assessment of all failure combinations, ranging from no failure to every combination of k failures.

The controllability assessment is conducted in *MATLAB* R2022b and outputs a matrix $M \in \mathbb{R}^{q \times p}$ that lists all the controllable cases, where q is the number of controllable cases. This matrix is exemplified here with the PPNNPN hexarotor case study, described in Section 3 with a system assumed to be capable of reconfiguring for up to two simultaneous failures:

$$M = \begin{bmatrix} 1 & 1 & 1 & 1 & 1 & 1 \\ 0 & 1 & 1 & 1 & 1 & 1 \\ 1 & 0 & 1 & 1 & 1 & 1 \\ 1 & 1 & 0 & 1 & 1 & 1 \\ 1 & 1 & 1 & 0 & 1 & 1 \\ 0 & 1 & 0 & 1 & 1 & 1 \\ 0 & 1 & 1 & 0 & 1 & 1 \\ 1 & 0 & 0 & 1 & 1 & 1 \\ 1 & 0 & 1 & 0 & 1 & 1 \end{bmatrix}. \tag{13}$$

Some applications may involve transporting a payload. The controllability assessment can accommodate any payload by adding this weight component to the UAV's empty weight u_0. Additionally, the controllability hypothesis can either aim to maintain control of all control directions or to sacrifice one of them, such as yaw as proposed in various fault tolerant control studies [25–28], before engaging in an emergency landing. This hypothesis can significantly impact UAV reliability. Sacrificing a control direction is achieved in the ACAI calculation by removing it from the corresponding row in u, u_0, and B_f before proceeding to the previous computations.

2.4. Step 4—Integrating Controllability Assessment into the Safety Assessment Model

The integration of the controllability assessment involves converting all controllable cases into Boolean expressions, which are then transformed into an assertion instruction for the controllability block model in *AltaRica* 3.0. Initially, the controllability block is designed in *System Analyst* 1.3 as an empty block model with an input flow variable for each propulsor and one output to observe the resulting controllability for reliability analysis. At this stage, the block contains no events, transitions, or assertions.

Once the *MATLAB* script completes the controllability analysis, as described in Section 2.3, it edits the *AltaRica* 3.0 code of the system model generated using the method in Section 2.2. Specifically, to populate the assertion of the controllability block, each row of the controllability assessment matrix M is converted into logic flow variables and logical instructions. These instructions evaluate the input from each propulsor to determine whether each controllable failure case is true or not using an AND operator, represented by the maximum function. If a controllable failure case is true, the associated logic flow variable is set to 0; otherwise, it is set to 6. The assessments are then combined using an OR operator, represented by the minimum function: if any controllable case is true, the UAV controllability variable is set to 0 (working); otherwise, it is set to 6 (failed) before being communicated to the observer. An example of an edited controllability block model in *AltaRica* 3.0 for a hexarotor PPNNPN configuration is shown in Figure 3.

```
49  block Controllability_35
50      Integer _1_45 (reset = 0);
51      Integer Out_46 (reset = 0);
52      Integer _2_47 (reset = 0);
53      Integer _3_48 (reset = 0);
54      Integer _4_49 (reset = 0);
55      Integer _5_50 (reset = 0);
56      Integer _6_51 (reset = 0);
57      Integer Logic_1 (reset = 0);
58      Integer Logic_2 (reset = 0);
59      Integer Logic_3 (reset = 0);
60      Integer Logic_4 (reset = 0);
61      Integer Logic_5 (reset = 0);
62      Integer Logic_6 (reset = 0);
63      Integer Logic_7 (reset = 0);
64      Integer Logic_8 (reset = 0);
65      Integer  Test1 ,  Test2 ,  Test3 (reset = 0);
66      assertion
67          Logic_1 := max ( _1_45  , _2_47  , _3_48  , _4_49  , _5_50  , _6_51  );
68          Logic_2 := max ( _2_47  , _3_48  , _4_49  , _5_50  , _6_51  );
69          Logic_3 := max ( _1_45  , _3_48  , _4_49  , _5_50  , _6_51  );
70          Logic_4 := max ( _1_45  , _2_47  , _4_49  , _5_50  , _6_51  );
71          Logic_5 := max ( _1_45  , _2_47  , _3_48  , _5_50  , _6_51  );
72          Logic_6 := max ( _2_47  , _4_49  , _5_50  , _6_51  );
73          Logic_7 := max ( _2_47  , _3_48  , _5_50  , _6_51  );
74          Logic_8 := max ( _1_45  , _3_48  , _5_50  , _6_51  );
75          Test1 := min ( Logic_1 , Logic_2 , Logic_3 );
76          Test2 := min ( Logic_4 , Logic_5 , Logic_6 );
77          Test3 := min ( Logic_7 , Logic_8 );
78          Out_46   := min (Test1,  Test2,  Test3 );
79  end
```

Figure 3. *AltaRica* 3.0 code of the controllability block model, automatically edited to include flow variables (lines 57 to 65) and instructions (lines 67 to 78) representing the controllability assessment logic for the hexarotor PPNNPN case study provided in Section 3.1.

An instruction such as the minimum and maximum function requires testing all combinations of the values of their inputs to compile them into Boolean equations. For example, the PPNNPN hexarotor configuration with two extra coaxial rotors discussed in Section 4.3.4 has 72 controllable cases when considering up to 4 failures simultaneously. Assuming that each logic integer representing one controllable case can have seven different values (from 0 to 6), if these 72 cases are combined using a minimum function in a final instruction, this instruction requires testing 7^{72} (~7×10^{60}) possible combinations of inputs. This combinatorial explosion of tests is not computationally viable.

To minimize this effect, the combination of logic integers is divided into multiple instructions associated with test integers. To optimize computational effort, we minimize the maximum number of inputs for these intermediate instructions. Although multiple layers of intermediate instructions could further reduce the maximum number of inputs, a single-layer approach is chosen for simplicity.

Thus, minimizing the maximum number of inputs is achieved when the number of logic integers allocated to each instruction, denoted n_{logic}, and the number of test integers that are input to the final instruction, denoted n_{test}, are equal. Since $n_{test} = q/n_{logic}$ the optimal number of allocated logic integers is given by $n_{logic} = \sqrt{q}$. In the previous example, the problematic instruction is divided into 9 intermediate instructions handling 8 logic integers, and one final instruction handling 9 test integers. This way, the maximum number of possible combinations to test for this final instruction is significantly reduced to a manageable number of 7^9 (~40,353,607). The example shown in Figure 3 has 9 controllable cases, and therefore the optimal number of test integers is 3.

The fundamental issue of combinatorial explosion arises because the *AltaRica* 3.0 model is automatically generated from another model in *System Analyst* 1.3, resulting in it not being optimized for the assessment tools *AltaRicaWizard* 1.2.0. While the implemented solution has proven efficient in its application to the case study, it could be limited to more complex reliability models. Addressing this limitation should be included in future work, which is expected to further increase the complexity of the reliability model. A possible solution could be to use Boolean variables instead of logic and test flow variables to optimize the generation of Boolean equations. However, the integer approach, inherent to the *System Analyst* 1.3 modeling environment, provides a solid basis for implementing various working and failure modes and their logical propagation throughout the system components. Future work will need to further address this risk of complexity issues while enabling the propagation of different working and failure modes.

In integrating *AltaRica* 3.0 and *MATLAB* R2022b, several coding challenges were encountered, including difficulty in distinguishing unordered inputs and outputs, ensuring proper syntax with commas, semicolons, and parentheses, and separating sections like Transition and Assertion into different lines within *AltaRica* 3.0 code. To address these issues, automated identification of inputs and outputs was implemented, and both syntax schemes and separate sections were stored in text variables to structure the code correctly.

3. Case Study

The hexarotor drone design was chosen for this case study due to its balance of simplicity and potential for varied fault tolerance capabilities, offering different reliability outcomes based on its configuration compared to quadrotors and octorotors. For instance, the conventional hexarotor PNPNPN configuration lacks fault tolerance under the requirement to maintain control of each axis, while the unconventional PPNNPN configuration provides some fault tolerance [24]. Additionally, the hexarotor was selected because of the availability of an in-house prototype, which provides detailed and accurate data for controllability and weight analysis. This makes the hexarotor design an ideal basis to illustrate the capability of the proposed methodology to effectively explore and evaluate both configuration and system design changes.

The case study involves the custom-designed hexarotor depicted in Figure 4. This design has a total weight of 3.48 kg, a rotor-to-rotor center diameter of 102 cm, a thrust-to-weight ratio of 2, and an autonomy of approximately 7 min in hover conditions. The maximum number of simultaneous rotor failures to be considered in the design is assumed to be two. This is because increasing the system's capability to reconfigure for more failures also increases its complexity, development work, and the number of tests required to verify its effectiveness. Two failures are assumed to be a good balance between reliability and these drawbacks.

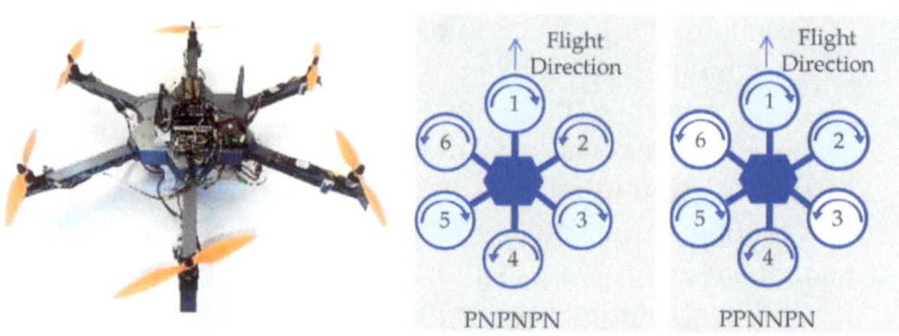

Figure 4. Hexarotor case study photograph (**left**) extracted from [29], PNPNPN-configuration (**center**), and PPNNPP-configuration (**right**). The "P" designation represents rotors rotating clockwise, while "N" designates rotors rotating counterclockwise.

3.1. *UAV Configuration, State, and Control Matrices*

For a hexarotor design with propulsors located at equal distances d (here, $d = 1.02$ m) from the center of gravity at angles of 0°, 60°, 120°, 180°, 240°, and 300°, as illustrated in Figure 4, the positions in Cartesian coordinates are: $r_1 = [d, 0, 0]^T$, $r_2 = \left[d/2, \sqrt{3}d/2, 0\right]^T$, $r_3 = \left[-d/2, \sqrt{3}d/2, 0\right]^T$, $r_4 = [-d, 0, 0]^T$, $r_5 = \left[-d/2, -\sqrt{3}d/2, 0\right]^T$, and $r_6 = \left[d/2, -\sqrt{3}d/2\right.$
For the hexarotor PNPNPN configuration, based on alternating propeller rotations for yaw control, the control effectiveness matrix $B_{f,\mathrm{PNPNPN}}$ is:

$$B_{f,\mathrm{PNPNPN}} = \begin{bmatrix} 1 & 1 & 1 & 1 & 1 & 1 \\ 0 & \sqrt{3}d/2 & \sqrt{3}d/2 & 0 & -\sqrt{3}d/2 & -\sqrt{3}d/2 \\ d & d/2 & -d/2 & -d & -d/2 & d/2 \\ -k_t & k_t & -k_t & k_t & -k_t & k_t \end{bmatrix}, \tag{14}$$

where k_t is the ratio of aerodynamic torque to propeller thrust. This coefficient captures the relationship between the thrust generated by a propeller and the resulting drag torque, which contributes to the yaw moment. Similarly, for the PPNNPN hexarotor configuration, the control effectiveness matrix $B_{f,\mathrm{PPNNPN}}$ is:

$$B_{f,\mathrm{PPNNPN}} = \begin{bmatrix} 1 & 1 & 1 & 1 & 1 & 1 \\ 0 & \sqrt{3}d/2 & \sqrt{3}d/2 & 0 & -\sqrt{3}d/2 & -\sqrt{3}d/2 \\ d & d/2 & -d/2 & -d & -d/2 & d/2 \\ k_t & k_t & -k_t & -k_t & k_t & -k_t \end{bmatrix}. \tag{15}$$

The state matrix A and control matrix B are identical for both the PNPNPN and PPNNPN hexarotor configurations. These matrices are derived from a linearized dynamics model for hover flight conditions, as described in Equation (3), with the assumption of negligible drag forces. The state vector is $x = \begin{bmatrix} h\ \varphi\ \theta\ \psi\ \dot{h}\ \dot{\varphi}\ \dot{\theta}\ \dot{\psi} \end{bmatrix}^T \in \mathbb{R}^8$, where h represents altitude, and φ, θ, and ψ are the roll, pitch, and yaw angles, respectively [30]:

$$A_s = \begin{bmatrix} 0_{4\times4} & I_{4\times4} \\ 0_{4\times4} & 0_{4\times4} \end{bmatrix} \epsilon \mathbb{R}^{8\times8},\ B = \begin{bmatrix} 0_{4\times4} \\ J^{-1} \end{bmatrix} \epsilon \mathbb{R}^{8\times4}, \tag{16}$$

where $J = \mathrm{diag}\left(-m_{kg},\ I_{xx},\ I_{yy},\ I_{zz}\right) \in \mathbb{R}^{4\times4}$, m_{kg} is the total mass of the UAV (in kg), and I_{xx}, I_{yy}, and I_{zz} are the moments of inertia around the x-, y-, and z-axes (in kg·m^2), respectively. The moments of inertia and total mass for the custom-designed hexarotor, obtained from a 3D computer-aided design model, are:

$$J = \begin{bmatrix} -3.877 & 0 & 0 & 0 \\ 0 & 7.998\ 10^{-2} & 0 & 0 \\ 0 & 0 & 7.884\ 10^{-2} & 0 \\ 0 & 0 & 0 & 0.148 \end{bmatrix}. \tag{17}$$

3.2. Physical System Architecture and Failure Rates

In this study, we consider a simplified power, propulsion, and control system for multirotor UAVs, illustrated in Figure 5, which employs a simplex architecture—an architecture without redundancy. The components of this simplified architecture include:

- Battery: Provides the necessary electrical direct current (DC) power to all components of the UAV. They are typically high-capacity lithium polymer (LiPo) batteries.
- Power Distribution Board (PDB): Distributes DC power from the battery to various components of the UAV, ensuring each component receives the appropriate power supply.
- Flight Sensors: Include inertial measurement units (IMUs), GPS modules, and other sensors that provide real-time feedback data on the UAV's orientation, position, altitude, and velocity to the flight controller.
- Flight Controller: The central processing unit of the UAV responsible for processing input from the flight sensors and executing control algorithms to stabilize and navigate the UAV. It sends motor speed commands to the electronic speed controllers (ESCs) to maintain desired flight characteristics.

- ESC: An electronic component that acts as an electric power converter between the battery and the electric motor. The ESC receives motor speed commands from the flight controller and adjusts the power supplied to the motors, thereby controlling their speed.
- Electric Motor: Converts electrical power from the ESC into mechanical power to drive the propeller. For weight efficiency, it is typically a three-phase alternate current (AC) motor with permanent magnets and an outrunner arrangement, meaning the casing is part of the rotor.
- Propeller: Attached to the electric motor, the propeller generates thrust by spinning at high speeds. For simplicity and cost efficiency, it typically has a fixed pitch and an aerodynamic profile optimized for rotation in one direction, generating upward thrust only.

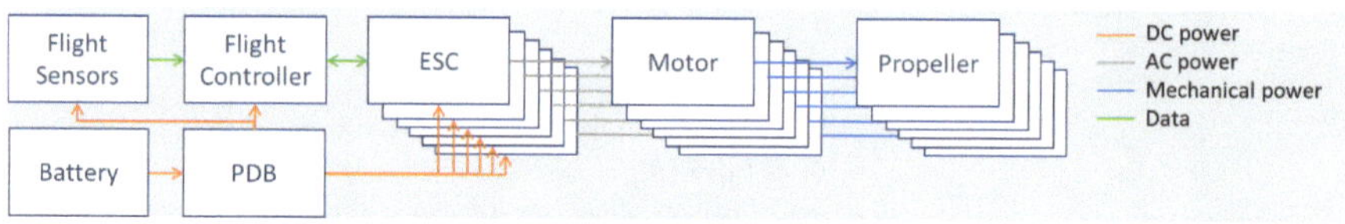

Figure 5. Simplified simplex power, propulsion, and control system architecture for multirotor UAVs.

We assume that the reliability of each component follows an exponential cumulative probability distribution with constant failure rates, as provided in Table 2. These failure rate values are derived from transport category airplane data in [10], with an arbitrarily applied derating factor of 100 to reflect the quality of commercial drones.

Table 2. Normalized weights and failure rate hypothesis.

Component	Normalized Weight (%)	Failure Rate (per Flight Hour)
Airframe	38.49	-
Battery	24.37	10^{-4}
Databus	0.39	10^{-5}
Electric Motor	2.07	10^{-4}
ESC	0.55	10^{-3}
Flight Controller	1.52	5×10^{-3}
Flight Sensors	1.20	3×10^{-4}
Majority Voter	0.21	10^{-5}
PDB	1.83	10^{-5}
Propeller	0.31	10^{-8}
Wiring	5.23	-

4. Case Study Analysis and Optimization

To demonstrate the effectiveness of the developed method and tool, we analyze and optimize the reliability of the hexarotor case study presented in Section 3 following the method detailed in Section 2. The design is optimized iteratively, with each iteration involving a reliability sensitivity analysis to guide the implementation of redundancy measures. Alongside the reliability evaluation, we qualitatively assess the weight impact of these redundancy measures to provide perspective on the trade-offs involved in increasing the UAV design reliability.

We investigate two design scenarios. In the first scenario, we assume a highly dynamic typical flight mission that requires maintaining control of all axes [9,10] to ensure the drone can continue safe flight and landing after a failure. The iterations of this scenario are detailed in Sections 4.3.1–4.3.6. In the second scenario, to further demonstrate the capability of the proposed method and tool, the design optimization process is repeated with a change in the controllability hypothesis. Here, it is assumed that the drone performs rather static typical flight missions and can execute an emergency landing immediately, sacrificing yaw control if necessary [25–28]. Thus, Scenario 2 requires revising the controllability analysis to exclude the yaw control axis requirement, as discussed in Section 2.3. The iterations of Scenario 2 are not detailed for conciseness. The results of both scenarios are summarized and compared in Section 4.3.7.

For simplicity, we assume components have a single mode of failure, such as an electric motor stopping without affecting the connected ESC and flight controller. Future work should revise this assumption to include more complex failure modes for more realistic design improvements.

4.1. Sensitivity Analysis: Fussell–Vesely Importance Factor

The sensitivity analysis aims to identify components with the greatest impact on system reliability that would benefit most from redundancy. This is achieved using the Fussell–Vesely Importance Factor [31,32], also referred to as the diagnostic importance factor. The Fussell–Vesely Importance Factor, denoted as I_i^{FV}, evaluates the probability that the UAV loss of control, $F_0(t)$, is due to the failure of the i-the component.

Specifically, it represents the probability that at least one minimal cut set containing component i has failed at time t, given that the system has failed at time t. It is expressed as [33]:

$$I_i^{FV}(t) = \frac{F_i(t) \cdot F_0(1_i, t)}{N}, \tag{18}$$

where $F_i(t)$ is the probability of failure of the i-th component, $F_0(1_i, t)$ is the conditional probability of UAV loss of control at time t given the i-th component failure, and N is the normalization factor. A higher I_i^{FV} value indicates that the component contributes more to the risk of UAV loss of control.

The normalization factor N is defined as the sum of all the I_i^{FV} scores, ensuring that the values are scaled to provide a relative measure of importance for each component:

$$N = \sum_{i=1}^{n_c} (F_i(t) \cdot F_0(1_i, t)), \tag{19}$$

where n_c is the total number of components.

The sensitivity is performed using the software tool *XFTA* 1.0 [33] embedded in the integrated modeling environment *AltaRicaWizard Editor* 1.2.0 [13], assuming a flight duration of t = 1 h.

4.2. Qualitative Weight Evaluation

For simplicity, component weights are qualitatively evaluated using actual weight data from the in-house drone design presented in Figure 4. The weight of each component is normalized with respect to the total UAV weight. These normalized component weights are listed in Table 2.

When adding a component for redundancy, its weight is simply added to the total UAV weight. An exception is made for the battery weight, which is adjusted based on the number of ESCs it supplies. The battery weight in Table 2 corresponds to a design where a single battery powers six ESCs, and this weight is linearly scaled according to the actual number of ESCs supplied.

To account for the wiring weight of additional components, the total wiring weight is adjusted based on the number of electrical components in the architecture. The wiring weight in Table 2 corresponds to a design with sixteen electrical components, and this weight is linearly scaled according to the actual number of electrical components.

4.3. Results

4.3.1. Scenario 1—Iteration 1

Iteration 1 compares the PNPNPN and PPNNPN configurations. The control effectiveness, state, and control matrices of the simplex PNPNPN and PPNNPN hexarotor configurations, as described in Section 3.1, are input into the *MATLAB* script for the controllability assessment detailed in Section 2.3. The physical system architecture modeled in *System Analyst* 1.3 is shown in Figure 6. The controllability and reliability assessments are executed automatically and evaluate a probability of loss of control of 1.199×10^{-2} for the PNPNPN configuration and a probability of loss of control of 7.6×10^{-3} for the PPNNPN configuration. The better results for the PPNNPN configuration align with its propulsor fault tolerance capability, while the PNPNPN has no fault tolerance as demonstrated in [24].

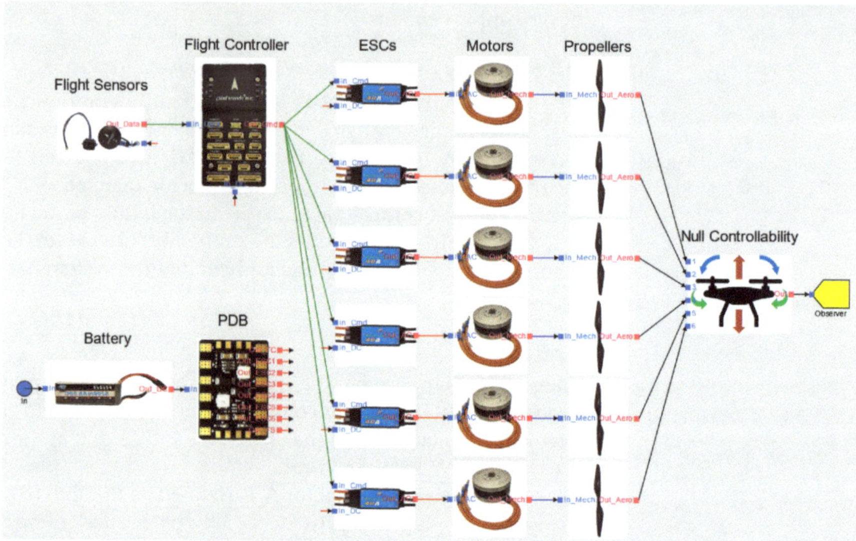

Figure 6. Iterations 1 and 2—MBSA model of the power, propulsion, and control system architecture for multirotor UAVs in *System Analyst* 1.3.

Decision: Based on these results, the PPNNPN configuration is selected for further analysis and development.

4.3.2. Scenario 1—Iteration 2

Iteration 2 initiates the design optimization of the PPNNPN configuration. The controllability assessment performed in iteration 1 does not need to be repeated in iteration 2, as there is no change to the PPNNPN configuration at this stage. Since the system architecture serves as the baseline for qualitative weight evaluation, this qualitative weight is set to one. The results of the sensitivity analysis are presented in Table 3, highlighting only the components with a first-order effect on UAV loss of control, indicating single points of failure. In other words, if any of these components fail, UAV control is lost. The data in Table 3 highlights that the flight controller has the highest criticality.

Decision: Based on these findings, redundancy will be applied to the flight controller. A critical failure mode of the flight controller involves providing corrupted commands to the ESCs, necessitating a strategy to ensure its integrity. Two viable strategies that limit the addition of flight controllers to two are pair-and-spare and triple majority redundancy (TMR) [34], also known as majority-voting-based redundancy. The majority-voting-based redundancy is selected to avoid the delays associated with starting the spare controller in the pair-and-spare approach, which could lead to a loss of control.

Table 3. Fussell–Vesely Importance Factors (first-order only) for Scenario 1 with full control axes requirement. The highest importance factors for each iteration are shown in bold.

Component	$I_i^{FV,N}$ (%)				
	Iteration 2	Iteration 3	Iteration 4	Iteration 5	Iteration 6
Flight Controller	**65.6**	2	2	2	2
ESC 5	13.2	**36.9**	2	2	2
ESC 6	13.2	**36.9**	2	2	2
Flight Sensor	3.9	11.1	**58.8**	2	2
Motor 5	1.3	3.7	2	2	2
Motor 6	1.3	3.7	2	2	2
Battery	1.3	3.7	19.6	**47.6**	2
PDB	0.1	0.4	2.0	4.8	8.9
Databus	1	0.4	2.0	1	1
Majority Voter	1	0.4	2.0	4.8	8.9
Propeller 5	1.3×10^{-4}	3.7×10^{-4}	2	2	2
Propeller 6	1.3×10^{-4}	3.7×10^{-4}	2	2	2

[1] Not used in this architecture. [2] Not a first-order importance factor.

4.3.3. Scenario 1—Iteration 3

The controllability assessment in iteration 1 remains unchanged due to the unaltered PPNNPN configuration. The system architecture now includes three flight controllers and a majority voter. The PDB has been updated to supply DC power from the battery to each flight controller. A databus distributes flight sensor data to each flight controller. The qualitative weight evaluation now accounts for the two additional flight controllers for majority-voting-based redundancy. The design's probability of loss of control and qualitative weight are now 2.707×10^{-3} and 1.055, respectively. The sensitivity analysis results in Table 3 show that ESCs 5 and 6 have the highest criticality, as propulsors 5 and 6 are the only non-fault-tolerant propulsors in the PPNNPN configuration.

Decision: Based on these findings, redundancy will be applied to propulsors 5 and 6 by adding two sets of ESCs, motors, and propellers. These new motors and propellers will be installed coaxially with propulsors 5 and 6 and will rotate in the same direction as their coaxial counterparts.

4.3.4. Scenario 1—Iteration 4

The controllability effectiveness and assessment are updated to account for the coaxial propulsors 5 and 6. The qualitative weight evaluation now includes the two additional propulsors, resulting in a total qualitative weight of 1.224. The probability of loss of control is now 5.107×10^{-4}. The results of the sensitivity analysis in Table 3 indicate that the flight sensors have the highest criticality.

Decision: Based on this result, redundancy will be applied to the flight sensors in the next iteration. To keep the design simple and avoid adding another majority voter, a set of flight sensors will be associated with each flight controller. This change will also allow for the removal of the databus used in design iteration 3.

4.3.5. Scenario 1—Iteration 5

The controllability assessment has been updated to account for the added flight sensors. The qualitative weight increases to 1.250 and the probability of loss of control has improved to 2.099×10^{-4}. The sensitivity analysis results in Table 3 now indicate that the battery has the highest criticality.

Decision: Based on this result, redundancy will be applied to the battery in the next and final iteration.

4.3.6. Scenario 1—Iteration 6

The controllability assessment performed in iteration 4 remains unchanged. Using the graphical interface of *System Analyst* 1.3, the physical system architecture model is modified, and reliability is assessed multiple times for architectures with two and three batteries, each with various connection schemes to the flight controllers and ESCs. This analysis reveals that adding a single redundant battery, resulting in a two-battery design, offers negligible improvement in UAV flight control reliability unless both batteries are connected to all eight propulsors, which incurs a significant weight penalty. In contrast, a three-battery design significantly improves reliability without requiring each battery to connect to all flight controllers and ESCs. The optimal three-battery system architecture is illustrated in Figure 7. The connection between the batteries and the ESCs is managed by the PDB block, with the following connections: the three batteries to the flight sensors and the three flight controllers; battery 1 to ESCs 1, 3, 5, and 6; battery 2 to ESCs 2, 4, 5, and 6; and battery 3 to ESCs 7 and 8. With this final design change, the qualitative weight is now 1.347, and the probability of loss of control lowered to 1.120×10^{-4}. The primary risk of failure is not due to a first-order issue but rather the simultaneous failure of any two flight controllers ($I^{FV,N} = 68.22\%$).

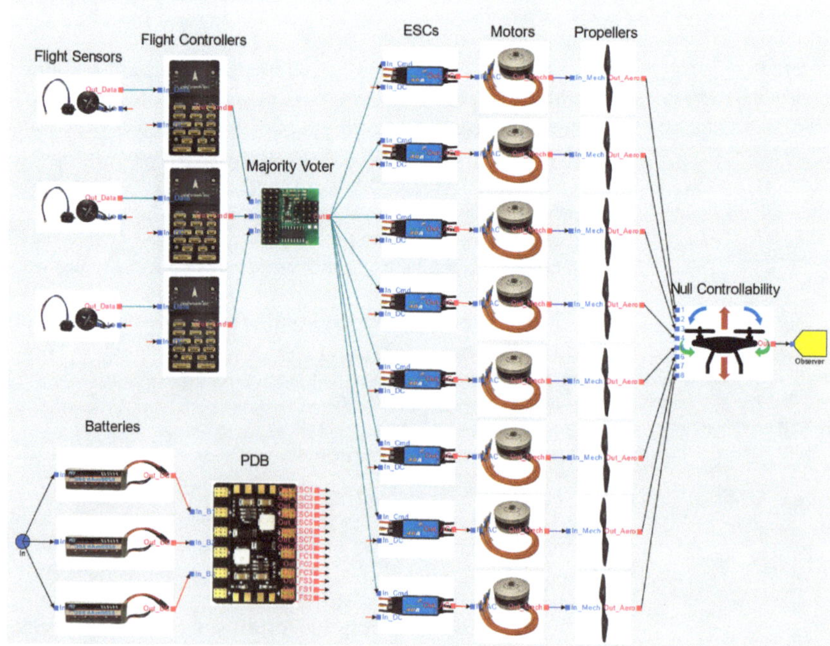

Figure 7. Iteration 6—MBSA model of the power, propulsion, and control system architecture for multirotor UAVs in *System Analyst* 1.3.

To evaluate the impact of the hypothesis limiting the number of simultaneous failures that the system can accommodate by reconfiguring itself, we repeated the controllability analysis and integration assuming a maximum of three simultaneous failures, and then four. The controllability analysis shows that the hexarotor PPNNPN configuration with two extra coaxial rotors is not controllable with any combination of five or more rotor failures. With a three-simultaneous-failures limit, the probability of failure decreases slightly to 1.1061×10^{-4}, and with a four-simultaneous-failures limit, it reduces insignificantly to 1.1060×10^{-4}. These results confirm that it is not worth significantly increasing the complexity of the system design and its development to accommodate more than two simultaneous failures.

4.3.7. Summary

The results of the six design optimization iterations of Scenario 1 are summarized in Table 3 and Figure 8. Figure 8 illustrates the evolution of the probability of loss of control and the qualitative weight throughout the iterations, highlighting the impact of each design decision. Notably, it demonstrates the well-known principle that improving reliability becomes increasingly difficult with each design iteration [34], reaching, here, a plateau by iteration 4. In contrast, the design complexity and weight impact continue to rise. Additionally, as discussed in Section 4.3.6, the evaluations of increasing the number of simultaneous failures the system can handle show that accommodating more than two simultaneous failures results in marginal reliability improvements while significantly increasing system complexity and development effort.

The results of Scenario 2 are summarized in Table 4 and Figure 8. Additionally, Table 5 provides a comparison of the design changes in Scenarios 1 and 2. Scenario 2 displays a lower probability of failure due to the alleviated controllability requirements. There is also no advantage for the PPNNPN configuration, and the selection of this configuration is motivated by the comparison with Scenario 1. The change in controllability requirement removes the criticality of propulsors 5 and 6, thus no coaxial rotors are added in this scenario, providing significant weight savings. These results demonstrate that a change in the controllability hypothesis can significantly impact the optimal design, reliability, and weight outcomes and that the proposed methodology can rapidly evaluate these impacts.

Table 4. Fussell–Vesely Importance Factors (first-order only) for Scenario 2 sacrificing yaw control requirement for emergency landing. The highest importance factors for each iteration are shown in bold.

Component	$I_i^{FV,N}$ (%)			
	Iteration 2	Iteration 3	Iteration 4	Iteration 5
Flight Controller	**92.2**	[2]	[2]	[2]
Flight Sensor	5.5	**58.6**	[2]	[2]
Battery	1.8	19.5	**47.6**	[2]
PDB	0.1	1.9	4.7	8.8
Databus	[1]	1.9	[1]	[1]
Majority Voter	[1]	1.9	4.7	4.8

[1] Not used in this architecture. [2] Not a first-order importance factor.

Table 5. Summary of the design changes.

Design Iteration	Scenario 1: Controllability Incl. All Control Axes	Scenario 2: Controllability Excl. Yaw Axis
1	$F = 1.199 \times 10^{-2}$, $W_q = 1$ Initial PNPNPN configuration Simplex architecture	$F = 5.405 \times 10^{-3}$, $W_q = 1$ Initial PNPNPN configuration Simplex architecture
2	$F = 7.6 \times 10^{-3}$, $W_q = 1$ Changed to PPNNPN configuration	$F = 5.405 \times 10^{-3}$, $W_q = 1$ Changed to PPNNPN configuration for comparison
3	$F = 2.707 \times 10^{-3}$, $W_q = 1.055$ Added majority voting redundancy with 3 flight controllers	$F = 5.118 \times 10^{-4}$, $W_q = 1.055$ Added majority voting redundancy with 3 flight controllers
4	$F = 5.107 \times 10^{-4}$, $W_q = 1.224$ Added 2 coaxial rotors	$F = 2.111 \times 10^{-4}$, $W_q = 1.081$ Added 2 sets of flight sensors
5	$F = 2.099 \times 10^{-4}$, $W_q = 1.25$ Added 2 sets of flight sensors	$F = 1.125 \times 10^{-4}$, $W_q = 1.088$ Added 2 batteries
6	$F = 1.12 \times 10^{-4}$, $W_q = 1.347$ Added 2 batteries	-

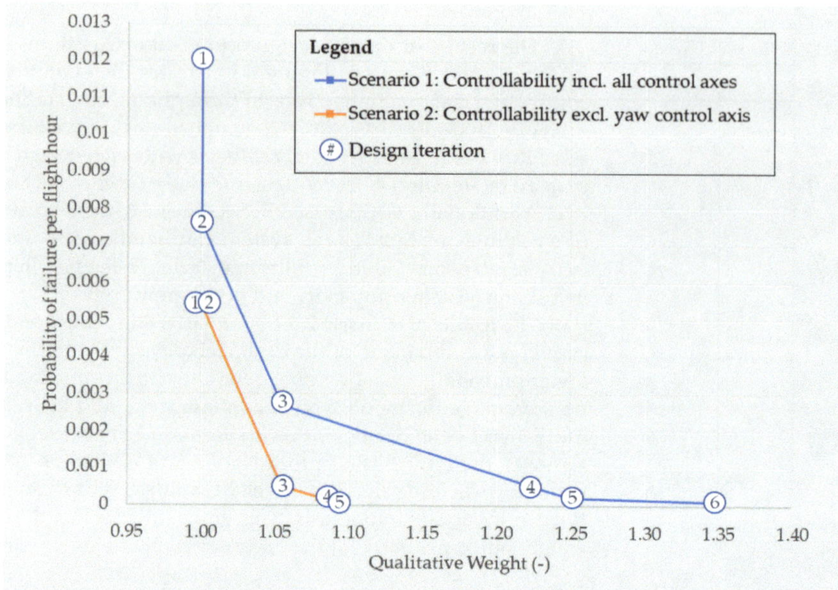

Figure 8. Evolution of probability of loss of control and qualitative weight throughout design optimization iterations. Refer to Table 5 for a description of the design iterations.

There are few publications providing reliability figures for multirotor drones. Petritoli et al. [1] report a reliability level of approximately 10^{-3} per flight hour for generic drones. For conventional PNPNPN hexarotor designs, assuming the requirement for control of all axes in case of failure, Shi et al.'s model [4] returns a probability of failure of 6.58×10^{-3} per flight hour, Nazarudeen et al. [9] extrapolated result is 1.75×10^{-2} assuming a failure rate derating factor of 100 for commercial applications, and Liscouët et al. [10] report 1.44×10^{-2}. These results are consistent with the present study's finding of 1.2×10^{-2} for PNPNPN hexarotor designs. For PPNNPN hexarotor designs, under similar assumptions and without redundancy optimization, Shi et al.'s model [4] provides a probability of failure of 2.2×10^{-3}, while the present study finds 7.6×10^{-3}. This indicates an increase in reliability with the PPNNPN configuration, though the present study shows only half the improvement compared to Shi's study, likely due to the non-optimized system architecture. Under the assumption that yaw control can be sacrificed for an immediate emergency landing, Shi et al.'s model [4] provides a probability of failure of 7.25×10^{-6} per flight hour for both PNPNPN and PPNNPN hexarotor designs, while the present study finds 5.405×10^{-3} with a conventional system architecture and 1.125×10^{-4} with an optimized architecture. This discrepancy suggests that Shi et al.'s model prediction may not be realistic and highlights the importance of considering system architecture to effectively improve reliability.

5. Conclusions

This study presents a novel methodology that integrates advanced controllability assessment techniques, specifically Null Controllability (ACAI), with the MBSA framework, *AltaRica* 3.0, using the *System Analyst* 1.3 and *AltaRicaWizard* 1.2.0 modeling environments. This approach optimizes propulsor configurations and system architectures for multirotor UAVs. The key advancement is the joint optimization of propulsor configuration and system architecture, allowing for the evaluation of any configuration without limitations and accommodating changes in controllability hypotheses, such as accepting the loss of yaw control for immediate emergency landing. Combining these capabilities is not possible with state-of-the-art techniques. This method significantly reduces the effort required for each design iteration and enables efficient exploration of numerous and diverse design options.

The methodology was demonstrated through a hexarotor drone case study. Integrating Null Controllability with the MBSA framework provided an efficient way to model a wide range of configurations and system architectures, enabling the computation of reliability metrics like Fussell–Vesely Importance Factors to guide the implementation and optimization of redundancies. This approach offered valuable design insights, reducing the probability of loss of control by up to 99% while balancing qualitative weight impacts. Notably, the iterative design optimization process highlighted the increasing difficulty of enhancing reliability with each iteration and confirmed the rationale for limiting system design complexity to accommodate a maximum of two simultaneous failures for reconfigurability. It also demonstrated that a change in the controllability hypotheses can significantly impact optimal design, reliability, and weight outcomes. For conventional non-redundant designs, assuming the need for control of all axes in case of failure, the present study's results align with previous studies, although those that disregard system architecture report slightly higher reliabilities. When assuming yaw control can be sacrificed for emergency landings, the discrepancy becomes significant, highlighting the crucial role of system architecture in effectively enhancing drone design reliability.

Regarding limitations, the study primarily considers a single failure mode for each component and a binary assessment of controllability, which may not encompass all real-world scenarios. A more granular controllability assessment, evaluating different levels of criticality such as hazardous, major, and minor [35], would allow for a more comprehensive safety and reliability analysis. Additionally, the qualitative sizing of components does not account for detailed weight impacts, which may affect the assessment of trade-offs between reliability improvements and overall system weight. Another limitation is the issue of combinatorial explosion, which arises because the *AltaRica* 3.0 model is automatically generated from another model, resulting in it not being optimized for the assessment tools. Although the implemented solution has proven efficient for the case study, it may be limited to more complex reliability models.

Therefore, future work should prioritize addressing various failure modes and incorporating wiring reliability for a more accurate reliability evaluation. Integrating advanced sizing tools such as *FAST-UAV* [10,36] would provide more precise and comprehensive weight and performance evaluations. Extending the controllability assessment to evaluate performance degradation across different levels of criticality will refine the safety and reliability analysis. Lastly, future work will need to tackle the complexity issues stemming from the conversion of the *AltaRica* 3.0 model, optimizing it for the assessment tools while enabling the propagation of different working and failure modes.

Author Contributions: Conceptualization, Z.M.R. and J.L.; methodology, Z.M.R. and J.L.; software, Z.M.R. and J.L.; validation, Z.M.R.; formal analysis, Z.M.R. and J.L.; investigation, Z.M.R. and J.L.; resources, J.L.; data curation, Z.M.R. writing—original draft preparation, Z.M.R.; writing—review and editing, J.L.; visualization, Z.M.R. and J.L.; supervision, J.L.; project administration, J.L.; funding acquisition, J.L. All authors have read and agreed to the published version of the manuscript.

Funding: This research was funded by the Natural Sciences & Engineering Research Council (NSERC) grant number [RGPIN-2022-05166].

Data Availability Statement: The primary codes and models used in this study are available in the GitHub repository: https://github.com/jliscouet/NullControllability_MBSA_DroneReliability (accessed 13 June 2024).

Acknowledgments: We sincerely thank Antoine B. Rauzy from the Norwegian University of Science and Technology and Tatiana Prosvirnova from The French Aerospace Lab ONERA for their invaluable support and prompt, helpful responses to our technical questions regarding *AltaRica* 3.0.

Conflicts of Interest: The authors declare no conflict of interest.

Nomenclature

Abbreviations	
AC	Alternate Current
ACAI	Available Control Authority Index
DC	Direct Current
ESC	Electronic Speed Controller
FMEA	Failure Mode and Effect Analysis
FTA	Fault Tree Analysis
GPS	Global Positioning System
IMU	Inertial Measurement Unit
MBSA	Model-Based Safety Assessment
PDB	Power Distribution Board
PPNNPN	Rotor configuration: two clockwise, two counterclockwise, two clockwise
PNPNPN	Rotor configuration: alternating clockwise and counterclockwise
RBD	Reliability Block Diagram
UAV	Unmanned Aerial Vehicle
Latin Variables	
A	Set of instructions
A_s	State matrix
a	Instruction
B	Control matrix
B_f	Control effectiveness matrix
C	Controllability matrix
d	Propulsor center's distance to center of gravity
E	Set of events
e	Event
F	Probability of failure per flight hour
$\mathcal{F}$	Constraint set for the rotor thrust vector
f	Rotor thrust vector
f_i	Thrust of i-th rotor
F_0	Conditional probability of UAV loss of control
F_i	Probability of failure of the i-th component
F_z	Vertical lift
g	Boolean condition on variables
H_j	Failure matrix of the j-th failure case
h	Altitude
i	Initialization function
I_i^{FV}	Fussell–Vesely Importance Factor of i-th component
I_{xx}, I_{yy}, I_{zz}	Moments of inertia
J	Inertia matrix
j	Total number of failure cases to be assessed
k	Maximum number of simultaneous rotor failures
k_t	Aerodynamic torque/thrust coefficient
M	Matrix of controllable cases
M_x, M_y, M_z	Pitch, roll, and yaw moments
m	Number of state variables
m_{kg}	Total mas of the UAV
n	Number of forces and moments acting on the vehicle
n_c	Total number of components
n_{logic}	Number of logic integers allocated to each instruction
n_{test}	Number of test integers that are inputs to the final instruction
N	Normalization factor
p	Number of control effectors (i.e., rotors)
q	Number of controllable cases
r_i	Cartesian coordinates of i-th rotor
T	Set of transitions
t	Time
t_r	Transition

u	Vector of forces and moments acting on the vehicle
V	Set of state and flow variables
v_h	Vertical velocity
W_q	Qualitative weight
x	State vector
Greek Variables	
η_i	Rotor availability
φ, θ, ψ	Roll, pitch, and yaw angles
Ω	Virtual control vector constraint set
Ω^c	Complementary set of Ω
$\partial\Omega$	Boundary of Ω
ρ	Available Control Authority Index (ACAI)
Superscripts and Subscripts	
j	Denotes the effect of the j-th failure case
max	Denotes the maximum value
0	Denotes the reference state

References

1. Petritoli, E.; Leccese, F.; Ciani, L. Reliability and Maintenance Analysis of Unmanned Aerial Vehicles. *Sensors* **2018**, *18*, 3171. [CrossRef] [PubMed]
2. Franco, B.; Góes, L.C.S. Failure Analysis Methods in Unmanned Aerial Vehicle (UAV) Applications. In Proceedings of the COBEM 2007, Brasilia, Brazil, 5–8 November 2007; ABCM: Brasília, Brazil, 2017; p. 11.
3. Venkataraman, R.; Lukátsi, M.; Vanek, B.; Seiler, P. Reliability Assessment of Actuator Architectures for Unmanned Aircraft. *IFAC-PapersOnLine* **2015**, *48*, 398–403. [CrossRef]
4. Shi, D.; Yang, B.; Quan, Q. Reliability Analysis of Multicopter Configurations Based on Controllability Theory. In Proceedings of the 2016 35th Chinese Control Conference (CCC), Chengdu, China, 27–29 July 2016; pp. 6740–6745.
5. Wang, Q.; Mao, J.; Wei, H. Reliability Analysis of Multi-Rotor UAV Based on Fault Tree and Monte Carlo Simulation. In *Advances in Mechanical Design, Proceedings of the 2017 International Conference on Mechanical Design (ICMD2017), Prague, Czech Republic, 6–8 September 2017*; Tan, J., Gao, F., Xiang, C., Eds.; Springer: Singapore, 2018; pp. 1525–1534.
6. Aslansefat, K.; Marques, F.; Mendonça, R.; Barata, J. A Markov Process-Based Approach for Reliability Evaluation of the Propulsion System in Multi-Rotor Drones. In *Technological Innovation for Industry and Service Systems, Proceedings of the 10th IFIP WG 5.5/SOCOLNET Advanced Doctoral Conference on Computing, Electrical and Industrial Systems, DoCEIS 2019, Costa de Caparica, Portugal, 8–10 May 2019*; Camarinha-Matos, L.M., Almeida, R., Oliveira, J., Eds.; Springer International Publishing: Cham, Switzerland, 2019; pp. 91–98.
7. T., T.; Govind, S.; Roy, A.; Ng, B.F.; Low, K.H. A Reliability Framework for Safe Octorotor UAV Flight Operations. In Proceedings of the 2023 International Conference on Unmanned Aircraft Systems (ICUAS), Warsaw, Poland, 6–9 June 2023; pp. 1013–1020.
8. Guo, J.; Elsayed, E.A. Reliability of Balanced Multi-Level Unmanned Aerial Vehicles. *Comput. Oper. Res.* **2019**, *106*, 1–13. [CrossRef]
9. Nazarudeen, S.B.; Liscouët, J. State-of-the-Art and Directions for the Conceptual Design of Safety-Critical Unmanned and Autonomous Aerial Vehicles. In Proceedings of the 2021 IEEE International Conference on Autonomous Systems (ICAS), Montreal, QC, Canada, 11–13 August 2021; IEEE: Piscataway, NJ, USA, 2021; pp. 1–5.
10. Liscouët, J.; Pollet, F.; Jézégou, J.; Budinger, M.; Delbecq, S.; Moschetta, J.-M. A Methodology to Integrate Reliability into the Conceptual Design of Safety-Critical Multirotor Unmanned Aerial Vehicles. *Aerosp. Sci. Technol.* **2022**, *127*, 107681. [CrossRef]
11. Batteux, M.; Prosvirnova, T.; Rauzy, A.; Kloul, L. The AltaRica 3.0 Project for Model-Based Safety Assessment. In Proceedings of the 11th IEEE International Conference on Industrial Informatics (INDIN), Bochum, Germany, 29–31 July 2013; pp. 741–746.
12. Breton, S.; Le-Com, P.; Thomas, T.; Clement, E. System-Analyst—Un Outil MBSA Pour L'analyse Des Risques, Libre De Diffusion Et Compatible Avec Arbre-Analyste Et Open-Altarica. In Proceedings of the Congrès Lambda Mu 21 «Maîtrise des Risques et Transformation Numérique: Opportunités et Menaces», Reims, France, 16–18 October 2018; p. hal-02063690.
13. Batteux, M.; Prosvirnova, T.; Rauzy, A. A Guided Tour of AltaRica Wizard, the AltaRica 3.0 Integrated Modeling Environment. In Proceedings of the 32nd European Safety and Reliability Conference (ESREL 2022), Dublin, Ireland, 28 August–1 September 2022; p. hal-03785784.
14. Lambe, A.B.; Martins, J.R.R.A. Extensions to the Design Structure Matrix for the Description of Multidisciplinary Design, Analysis, and Optimization Processes. *Struct. Multidisc. Optim.* **2012**, *46*, 273–284. [CrossRef]
15. Martins, J.; Ning, A. *Engineering Design Optimization*; Cambridge University Press: Cambridge, UK, 2021; ISBN 978-1-108-83341-7.
16. Quan, Q. *Introduction to Multicopter Design and Control*; Springer: Berlin/Heidelberg, Germany, 2017.
17. Faisandier, A. *Systems Architecture and Design*; Practical Guidelines; Sinergy'Com: Belberaud, France, 2013; Volume 3; ISBN 979-10-91699-03-7.
18. *ISO/IEC 26702:2007*; Systems Engineering—Application and Management of the Systems Engineering Process. ISO: Geneva, Switzerland, 2007.

19. Sun, M.; Gautham, S.; Ge, Q.; Elks, C.; Fleming, C. Defining and Characterizing Model-Based Safety Assessment: A Review. *Saf. Sci.* **2024**, *172*, 106425. [CrossRef]
20. Batteux, M.; Prosvirnova, T.; Rauzy, A. *AltaRica 3.0 Language Specification*; AltaRica Association: Les Essarts le Roi, France, 2015. Available online: https://www.openaltarica.fr/docs/AltaRica%203.0%20Language%20Specification%20-%20v1.1.pdf (accessed on 13 June 2024).
21. Rauzy, A.B. Guarded Transition Systems: A New States/Events Formalism for Reliability Studies. *Proc. Inst. Mech. Eng. Part O J. Risk Reliab.* **2008**, *222*, 495–505. [CrossRef]
22. Batteux, M.; Prosvirnova, T.; Rauzy, A.B. AltaRica 3.0 in Ten Modelling Patterns. *Int. J. Crit. Comput.-Based Syst.* **2019**, *9*, 133–165. [CrossRef]
23. Batteux, M.; Prosvirnova, T.; Rauzy, A.; Yang, L. Reliability Assessment of Phased-Mission Systems with AltaRica 3.0. In Proceedings of the 2018 3rd International Conference on System Reliability and Safety (ICSRS), Barcelona, Spain, 23–25 November 2018; pp. 400–407.
24. Du, G.-X.; Quan, Q.; Yang, B.; Cai, K.-Y. Controllability Analysis for Multirotor Helicopter Rotor Degradation and Failure. *J. Guid. Control Dyn.* **2015**, *38*, 978–985. [CrossRef]
25. Freddi, A.; Lanzon, A.; Longhi, S. A Feedback Linearization Approach to Fault Tolerance in Quadrotor Vehicles. *IFAC Proc. Vol.* **2011**, *44*, 5413–5418. [CrossRef]
26. Lippiello, V.; Ruggiero, F.; Serra, D. Emergency Landing for a Quadrotor in Case of a Propeller Failure: A Backstepping Approach. In Proceedings of the 2014 IEEE/RSJ International Conference on Intelligent Robots and Systems, Chicago, IL, USA, 14–18 September 2014; pp. 4782–4788.
27. Mueller, M.W.; D'Andrea, R. Stability and Control of a Quadrocopter despite the Complete Loss of One, Two, or Three Propellers. In Proceedings of the 2014 IEEE International Conference on Robotics and Automation (ICRA), Hong Kong, China, 31 May–7 June 2014; pp. 45–52.
28. Lanzon, A.; Freddi, A.; Longhi, S. Flight Control of a Quadrotor Vehicle Subsequent to a Rotor Failure. *J. Guid. Control Dyn.* **2014**, *37*, 580–591. [CrossRef]
29. Liscouët, J.; Uwantare, I.; Remoundos, A.; Desrosiers, J.; Heit, Z.; Senouci, A.; Chen, O. Validation of Reliability-Based Flight Control Optimization for UAVs. In Proceedings of the AIAA SCITECH 2024 Forum, Orlando, FL, USA, 8–12 January 2024; p. AIAA 2024-0341.
30. Du, G.-X.; Quan, Q.; Cai, K.-Y. Controllability Analysis and Degraded Control for a Class of Hexacopters Subject to Rotor Failures. *J. Intell. Robot. Syst.* **2015**, *78*, 143–157. [CrossRef]
31. Fussell, J.B. How to Hand-Calculate System Reliability and Safety Characteristics. *IEEE Trans. Reliab.* **1975**, *R-24*, 169–174. [CrossRef]
32. Vesely, W.E. A Time-Dependent Methodology for Fault Tree Evaluation. *Nucl. Eng. Des.* **1970**, *13*, 337–360. [CrossRef]
33. Rauzy, A.B. *Probabilistic Safety Analysis with XFTA*; AltaRica Association: Les Essarts le Roi, France, 2020; ISBN 978-82-692273-0-7.
34. Hecht, H. *Systems Reliability and Failure Prevention*; Artech House Inc.: Norwood, MA, USA, 2004; ISBN 1-58053-372-8.
35. SAE International. *Guidelines and Methods for Conducting the Safety Assessment on Civil Airborne Systems and Equipment*; SAE International: Warrendale, PA, USA, 1996.
36. Pollet, F.; Delbecq, S.; Budinger, M.; Moschetta, J.-M.; Liscouët, J. A Common Framework for the Design Optimization of Fixed-Wing, Multicopter and VTOL UAV Configurations. In Proceedings of the ICAS 2022 Congress, Stockholm, Sweden, 4–9 September 2022.

applied sciences

MDPI

Article

An Algorithm for Predicting Vehicle Behavior in High-Speed Scenes Using Visual and Dynamic Graphical Neural Network Inference

Menghao Li [1], Miao Liu [1,*], Weiwei Zhang [2], Wenfeng Guo [3], Enqing Chen [4], Chunguang Hu [5] and Maomao Zhang [6]

1 School of Mechanical and Automotive Engineering, Shanghai University of Engineering Science, Shanghai 201620, China; lmh000606@163.com
2 Shanghai Smart Vehicle Cooperating Innovation Center Co., Ltd., Shanghai 201805, China; weiweiz@smarvcte.com
3 School of the Vehicle and Mobility, Tsinghua University, Beijing 100084, China; gwf0330@163.com
4 School of Education and Foreign Languages, Wuhan Donghu University, Wuhan 430212, China; chenenqing0601@126.com
5 School of Urban Planning and Design, Peking University, Shenzhen 518055, China; chunguangh@foxmail.com
6 College of Public Administration, Huazhong University of Science and Technology, Wuhan 430079, China; star_mzhang@mails.ccnu.edu.cn
* Correspondence: liumiao@sues.edu.cn

Abstract: Accidents caused by vehicles changing lanes occur frequently on highways. Moreover, frequent lane changes can severely impact traffic flow during peak commuting hours and on busy roads. A novel framework based on a multi-relational graph convolutional network (MR-GCN) is herein proposed to address these challenges. First, a dynamic multilevel relational graph was designed to describe interactions between vehicles and road objects at different spatio-temporal granularities, with real-time updates to edge weights to enhance understanding of complex traffic scenarios. Second, an improved spatio-temporal interaction graph generation method was introduced, focusing on spatio-temporal variations and capturing complex interaction patterns to enhance prediction accuracy and adaptability. Finally, by integrating a dynamic multi-relational graph convolutional network (DMR-GCN) with dynamic scene sensing and interaction learning mechanisms, the framework enables real-time updates of complex vehicle relationships, thereby improving behavior prediction's accuracy and real-time performance. Experimental validation on multiple benchmark datasets, including KITTI, Apollo, and Indian, showed that our algorithmic framework achieves significant performance improvements in vehicle behavior prediction tasks, with Map, Recall, and F1 scores reaching 90%, 88%, and 89%, respectively, outperforming existing algorithms. Additionally, the model achieved a Map of 91%, a Recall of 89%, and an F1 score of 90% under congested road conditions in a self-collected high-speed traffic scenario dataset, further demonstrating its robustness and adaptability in high-speed traffic conditions. These results show that the proposed model is highly practical and stable in real-world applications such as traffic control systems and self-driving vehicles, providing strong support for efficient vehicle behavior prediction.

Keywords: graph neural network; vehicle intention; autonomous driving; forecasting; intelligent transportation system

Citation: Li, M.; Liu, M.; Zhang, W.; Guo, W.; Chen, E.; Hu, C.; Zhang, M. An Algorithm for Predicting Vehicle Behavior in High-Speed Scenes Using Visual and Dynamic Graphical Neural Network Inference. *Appl. Sci.* **2024**, *14*, 8873. https://doi.org/10.3390/app14198873

Academic Editors: Artur Kierzkowski, Tomasz Kisiel, Agnieszka A. Tubis, Franciszek Restel, Mateusz Zajac, Tomasz Nowakowski and Anna Jodejko-Pietruczuk

Received: 12 August 2024
Revised: 26 September 2024
Accepted: 28 September 2024
Published: 2 October 2024

1. Introduction

According to statistics, there are approximately 1.2 million high-speed traffic accidents yearly, resulting in numerous casualties. More than 30% of these accidents are caused by vehicle lane changing. The high incidence and severity of these accidents have prompted researchers to focus more on the study of vehicle behavior prediction. In recent years, with the rapid development of autonomous driving technology, the application of vehicle behavior prediction in high-speed scenarios has attracted considerable attention. Frequent vehicle lane changes lead to a decrease in overall traffic mobility and increase the likelihood of traffic accidents. Therefore, accurate vehicle behavior prediction can enhance the safety and efficiency of autonomous driving systems and lay a solid foundation for future intelligent transportation systems.

In autonomous driving, the acquisition of environmental information through advanced sensing technologies such as LiDAR, radar, and cameras is essential for predicting vehicle behavior [1,2]. Convolutional neural networks (CNNs) are widely employed to extract visual features for this purpose. However, reliance solely on visual data presents limitations, particularly in high-speed scenarios characterized by complex vehicle interactions [3,4], often resulting in diminished prediction accuracy in dynamic environments [5]. To enhance prediction accuracy, researchers have investigated dynamic graph neural networks (DGNNs), which effectively model spatio-temporal dependencies by capturing dynamic interactions between vehicles [6,7].

Approaches that combine graph convolutional networks (GCNs) with long short-term memory (LSTM) or temporal convolutional networks (TCNs) have shown promise in modeling these spatio-temporal dependencies [8,9]. Despite these advancements, existing methods often struggle to accurately represent non-Euclidean spatial relationships between vehicles and to account for critical factors such as lane positioning and traffic flow in high-speed contexts [10,11]. To address these challenges, researchers have proposed a method that integrates visual data with dynamic graph neural networks, employing CNNs to extract vehicle features and DMR-GCN to more effectively capture spatio-temporal dependencies [12,13]. This approach constructs a traffic scene graph to represent vehicle layout and motion, thereby improving predictions in complex, dynamic environments [14,15]. By effectively modeling non-Euclidean spatial relationships, the proposed method overcomes the limitations of existing techniques and enhances the accuracy of vehicle behavior predictions in high-speed scenarios [16,17].

The integration of advanced graph neural networks with visual features opens new possibilities for precise and reliable behavior prediction, even under challenging conditions [18,19]. Studies indicate that incorporating factors such as vehicle distance, lane position, and dynamic traffic flow significantly enhances prediction models [20,21]. Future research may further explore hybrid methods that combine visual data with advanced neural network architectures to improve the robustness of vehicle behavior prediction [22].

The aim of the current study was to enhance the accuracy of vehicle behavior prediction in high-speed scenarios, addressing traffic accidents caused by lane changes. By integrating visual perception with a dynamic graph neural network (DMR-GCN) to construct a multilevel graph model that captures vehicle interactions, an adaptive network tuning mechanism and a dynamic scene learning strategy are herein introduced to improve prediction accuracy and safety. This approach addresses the shortcomings of existing methods in modeling complex interactions. It critically supports advancing autonomous driving and intelligent transportation systems, demonstrating significant practical value and broad application potential.

The contributions made include the following:

(1) Visual and graph neural combination: integrating visual perception with dynamic graph neural networks (DMR-GCN) to enhance behavior prediction in high-speed scenes;
(2) Multilayer graph modeling: creating a graph model to capture vehicle interactions at different spatio-temporal levels. Nodes represent vehicles or road objects, while edges reflect real-time interactions, with DMR-GCN dynamically adjusting edge weights;
(3) Adaptive network tuning: introducing an adaptive mechanism in DMR-GCN for real-time interaction capture and feature processing to improve prediction accuracy;
(4) Dynamic scene learning: implementing real-time learning strategies based on vehicle position, speed, and road conditions to improve accuracy and safety.

2. Related Work

2.1. Vehicle and Traffic Element Identification

In autonomous driving and advanced driver assistance systems (ADAS), the recognition of vehicles and traffic elements is a pivotal task. Recently, numerous research endeavors have focused on crafting efficient object detection and recognition algorithms, prominently including YOLO (You Only Look Once) [23], SSD (Single-Shot MultiBox Detector) [24], and Faster R-CNN (Faster Region-based Convolutional Neural Network) [25]. YOLO, a deep learning-powered object detection approach, operates by segmenting the input image into $S \times S$ grids, where each grid cell forecasts B bounding boxes along with their confidence scores and C conditional class probabilities. The object detection is facilitated through end-to-end regression optimization.

YOLO distinguishes itself in terms of speed, accuracy, and computational efficiency. It achieves a remarkable detection speed of 45–60 frames per second (FPS) on standard GPUs, coupled with a mean average precision (Map) of 57.9%, surpassing Faster R-CNN's 42.7% Map. Furthermore, compared to SSD, YOLO offers faster processing rates (30–40 FPS for SSD) and a higher Map (46.5% for SSD). Notably, YOLO boasts a compact model size and minimal computational resource demands, rendering it an ideal candidate for deployment on embedded systems and mobile devices.

Given our reliance solely on monocular camera data and the imperative to drastically curtail computational overhead while maintaining high recognition accuracy, we opted for YOLO as the core object detection and recognition algorithm for our real-time, efficient vehicle and traffic element recognition framework in the context of autonomous driving and ADAS. Inspired by recent advances in vehicle tracking methods [26], we chose FairMOT-MCVT as our core algorithm for real-time efficient vehicle tracking for applications in autonomous driving and ADAS.

2.2. Vehicle Behavior Understanding

Understanding vehicle behavior and road scenarios is a crucial challenge in autonomous driving. Traditional approaches typically rely on data from multiple sensors to achieve this goal. Many studies have employed rule-based approaches and probabilistic modeling techniques to classify driving behavior. For example, Sivaraman et al. [27] used probabilistic modeling techniques to predict vehicle behavior. This approach effectively handles uncertainty but is hampered by model complexity and high computational resource demands. Numerous early studies primarily focused on predicting future vehicle trajectories rather than behavioral classification. For instance, Kitani et al. [28] utilized sensor data to predict vehicle trajectories, providing short-term driving path predictions but lacking a comprehensive understanding of driving behavior. Similarly, Lee et al. [29] proposed an approach based on trajectory history data, which, although adequate for short-term trajectory prediction, performs poorly in long-term behavior prediction.

This study takes a concise yet ambitious approach to predicting vehicle behavior by capturing and interpreting scenes from a single camera. While previous research has utilized single-camera data, most efforts have focused on predicting the behavior of the ego vehicle. In contrast, this research categorizes the behavior of surrounding vehicles from the perspective of the ego vehicle. Neumann and Vedaldi [30] demonstrated the potential of using monocular camera data for pedestrian and ego-vehicle trajectory prediction, illustrating how single-camera systems can offer valuable insights into dynamic road scenarios. Building on this, our approach enhances autonomous driving systems' behavioral planning and state estimation capabilities and has potential applications in traffic violation detection within video footage. By meticulously analyzing the behavioral patterns of neighboring vehicles, this technique improves traffic safety, optimizes decision-making in autonomous driving systems, and ultimately enhances both the driving experience and the overall reliability of the transportation system.

2.3. Graph Neural Networks

In recent years, graph data modeling has successfully integrated deep learning methodologies, leading to the development of numerous algorithms for unstructured data. Meyer et al. [31] introduced a deep learning framework using heterogeneous graph representations to model complex traffic scenarios in autonomous driving. Lee et al. [32] devised a multi-agent trajectory prediction model based on graph neural networks, enhancing prediction performance for multi-modal behaviors. Li et al. [33] presented the STS-DGNN, a dynamic graph neural network that extracts spatial and temporal features for vehicle trajectory prediction, significantly improving accuracy and real-time performance. Zhang et al. [34] integrated the Graph Attention Transformer (Gatformer) into trajectory prediction, effectively capturing the spatio-temporal interplay between traffic agents and road infrastructure. Despite these advancements, these methods still encounter computational efficiency and scalability challenges, mainly when dealing with real-time dynamic data and large-scale, intricate scenes.

A framework based on multi-relational graph convolutional networks (MR-GCN) was developed to capture and update real-time spatio-temporal relationships between vehicles. As shown in Figure 1, the framework's incorporation of MR-GCN capabilities and interactive learning mechanisms enhances real-time performance and prediction accuracy, addressing the complexities of modeling and predicting vehicle behavior in dynamic environments.

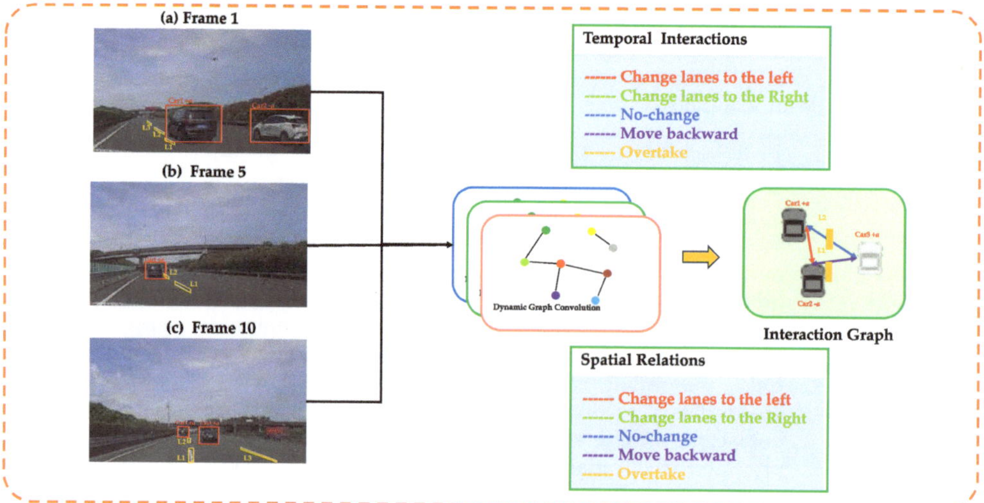

Figure 1. Schematic Framework for Predicting Vehicle Behavior in High-Speed Scenarios Using Vision and Dynamic Graph Neural Networks.

2.4. Current Challenges and Future Directions of Work

Despite significant advancements, predicting vehicle behavior and recognizing traffic elements in dynamic and complex environments remain challenging. These challenges include meeting real-time requirements, adapting to diverse traffic scenarios, and managing data noise and incompleteness. Additionally, reducing computational costs and energy consumption while maintaining model performance is a pressing concern.

To further improve model performance and practicality, future research could explore the following areas:

(1) Developing more robust graph neural network architectures that adapt to varying traffic complexities;

(2) Integrating diverse data streams from multiple sources, such as radar, LiDAR, and telematics, to enhance model comprehensiveness and accuracy;
(3) Investigating online learning and self-adaptive mechanisms to enable dynamic adjustment to shifting traffic environments;
(4) Optimizing computational resource usage to meet real-time application requirements.

Exploring these areas is expected to create a more intelligent and reliable system for predicting vehicle behavior and recognizing traffic elements.

3. Materials and Methods

The proposed methodological framework builds on and improves the data modeling methodology introduced in [35] (MR-GCN-LSTM), as shown in Figure 2. This multi-stage behavioral prediction process involves the following: first, modeling the video as a series of spatial scene graphs; second, generating a multi-relational interaction graph from this spatio-temporal information to represent the evolution of spatial relationships between entities over time; finally, designing a dynamic multi-relational graph convolutional network (DMR-GCN) to predict vehicle behavior using a graph-based learning model.

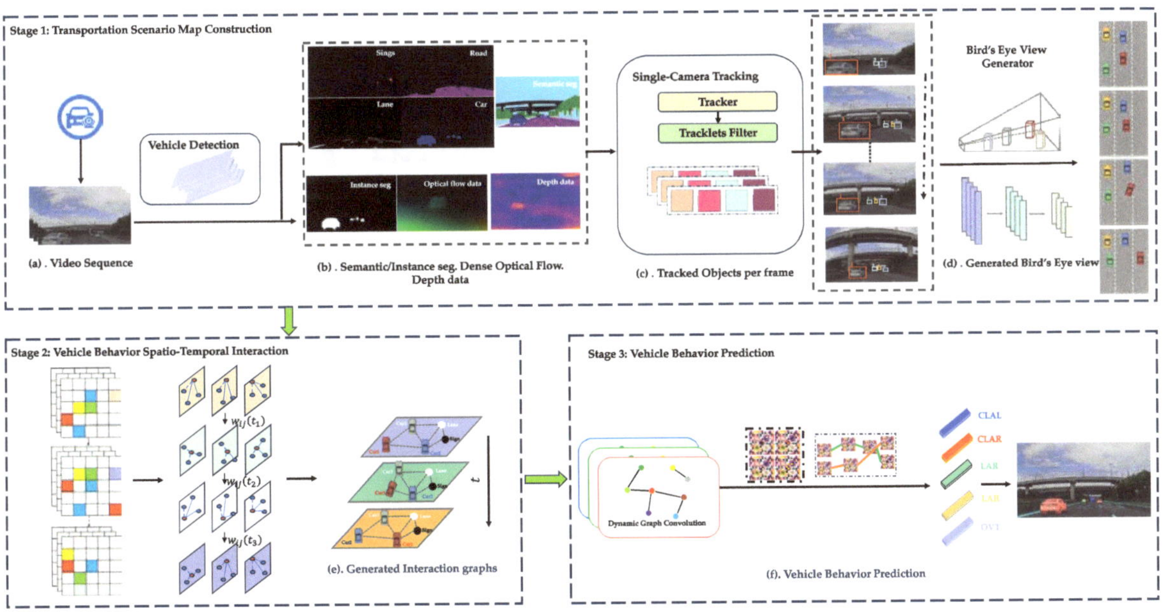

Figure 2. The pipeline of overall network architecture.

The proposed methodological framework's performance enhancement is due to three key innovations:

(1) The dynamic multilevel relational graph (DMRG) design accurately captures the complex dynamics of vehicle lane changes by representing different spatio-temporal granularities through multiple layers and updating edge weights in real time;
(2) The method for generating temporal interaction graphs focuses on capturing temporal changes and complex interaction patterns such as vehicle acceleration, sharp braking, and behaviors under challenging traffic conditions (e.g., rainy days and nights). This comprehensive approach enhances the model's understanding of dynamic vehicle interactions, improving prediction accuracy and adaptability to diverse traffic scenarios;
(3) The DMR-GCN structure includes a dynamic multi-relational graph convolutional network (DMR-GCN), dynamic scene perception, and an interactive learning mechanism.

DMR-GCN employs an adaptive adjacency matrix updating mechanism and a multi-relational feature fusion mechanism to achieve dynamic relational capture and comprehensive feature processing, enhancing adaptability to real-time data. The interactive learning mechanism allows the model to adjust to changing traffic environments, improving vehicle behavior prediction accuracy and real-time performance.

3.1. *Traffic Scenario Map Construction*

Traffic scene graph construction involves multiple steps. First, objects are identified and tracked across consecutive video frames. Next, these objects are oriented in a bird's-eye view to facilitate the identification of their relative positions. Finally, for each frame, the relative position information of the objects is encoded with a spatial map, as shown in Figure 3.

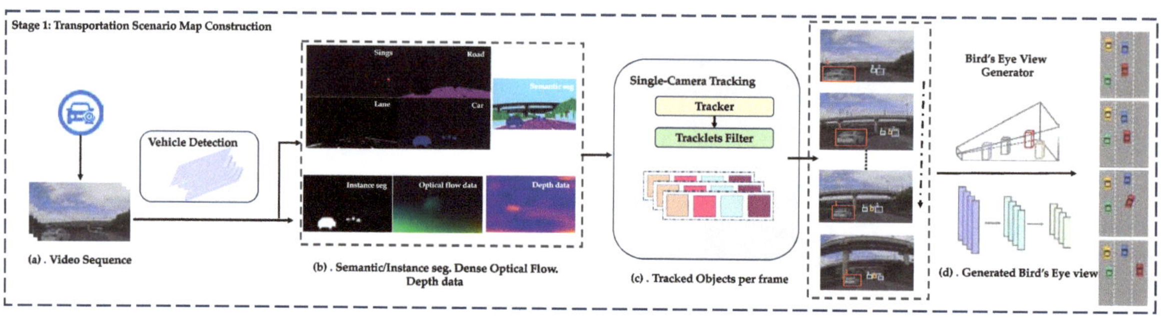

Figure 3. Traffic scenario map construction.

3.1.1. Object Tracking

To fully understand the dynamic scene, three main feature extraction pipelines are used as inputs to recognize different objects and determine the spatial relationships of vehicles in the video. The key components for dynamic feature extraction include instance segmentation, semantic segmentation, pixel-by-pixel optical flow, and depth data in video frames. First, Mask R-CNN [36] is used, for instance, as segmentation to detect and track moving objects over time. Second, Detectron2 [37] is employed to achieve a semantic understanding of the scene, particularly for static objects such as traffic signs and lane markings, which serve as important landmarks. Finally, RAFT (Recurrent All-Pairs Field Transforms) [38] is used to compute the optical flow between consecutive video frames, enabling the tracking of each object in the scene by combining optical flow information with instance features and semantic segmentation.

3.1.2. Monocular to Bird's-Eye View

After obtaining the stabilized trajectories in the image space, these trajectories are converted to a bird's-eye view to capture the relative positions of different objects better. The geometric relation formulation described in GitNet [39] is used for object localization to achieve this conversion. Each object is assigned a reference point, namely the centroid of the lane marker for lanes and the point adjacent to the road for vehicles and other landmarks, as shown in Figure 2d. Let $b = (x, y, 1)$ represent the reference point in homogeneous coordinates in the image. These homogeneous coordinates b are then converted to coordinates in the bird's-eye view as follows:

$$B = (Bx, By, Bz) = -\frac{hK^{-1}b}{\eta^T K^{-1}b} \tag{1}$$

where K is the camera internal reference matrix, η is the surface normal, h is the height of the camera to the ground plane, and B is the 3D coordinates of point b in the bird's eye view.

The specific process is as follows: A perspective transformation is performed using the intrinsic matrix K and the extrinsic parameters of the camera. Assume a point in the camera's coordinate system is (X, Y, Z). The ground space is represented as $(X, 0, Z)$ by setting the y -coordinate of the camera's coordinate system to 0. The camera's coordinate system is simplified by removing the y-dimension, resulting in (X, Z). Consequently, the coordinates in the bird's-eye view also remove the y-dimension and are represented as (X, Z). The one-to-one correspondence between image coordinates and ground coordinates can be expressed by the following equation:

$$\begin{pmatrix} u \\ v \\ 1 \end{pmatrix} = K \begin{pmatrix} X \\ Y \\ Z \end{pmatrix} \tag{2}$$

The inverse transformation from image coordinates to ground coordinates is given by the following:

$$\begin{pmatrix} X \\ Y \\ Z \end{pmatrix} = K^{-1} \begin{pmatrix} u \\ v \\ 1 \end{pmatrix} \tag{3}$$

After obtaining the trajectories of these objects in image space, they are relocated in the bird's-eye view (top view) by projecting the image coordinates to 3D coordinates. This repositioning helps determine the spatial relationships between different entities. Each object is assigned a reference point to account for height differences. For lane markers, the reference point is at the center, while for vehicles, it is a point close to the road. Using the geometric relations described above, the transformation from perspective space to a bird's-eye view can be achieved, allowing for the recovery of approximate ground coordinates from the image coordinates and the camera height [40]. However, in a real monocular perception system, the camera height is not directly available. This issue is addressed by learning the camera height parameters through the network. By combining optical flow information with instance features and semantic segmentation, each object in the scene can be accurately tracked in a bird's-eye view, providing a more comprehensive understanding of the environment.

3.1.3. Spatial Scene Maps

At each frame t of the video, spatial information is captured based on the relative positions of different objects in the bird's-eye view and represented as a scene graph $S_t^{i,j}$. At time t, the spatial relationship between subject i and object j $S_t^{i,j}$ is defined by the quadrant where the object is located relative to the subject, i.e., $S_t^{i,j} \in R_s$, where $R_s =$ {top left, top right, bottom left, bottom right}. An example of a spatial scene graph is illustrated in Figure 2e. These spatial relationships are determined by the 3D position information of the objects in the bird's-eye view. At each time step t, the spatial relationship $S_t^{i,j}$ between subject i and object j is represented as one of the four quadrants: upper left, upper right, lower left, and lower right. An example of a spatial scene graph is shown in Figure 2e.

3.2. *Vehicle Behavior Spatio-Temporal Interaction*

3.2.1. Dynamic Multi-Level Relationship Diagram Modeling

As shown in Figure 4, after generating the spatial scene graph, the dynamic multi-level relationship graph modeling method is introduced. This method is based on a multi-level structure that represents the relationships between vehicles at different spatial and temporal granularities, thereby capturing and understanding the dynamic interactions of vehicles in a more comprehensive manner. Compared with static spatial scene graphs, the dynamic multi-level relationship graph can update the interactions between vehicles in real time, improving the model's prediction accuracy and robustness for vehicle lane-changing behavior. By representing relationships at different spatio-temporal granularities, the model better adapts to complex traffic situations, including rapidly changing vehicle behaviors on highways.

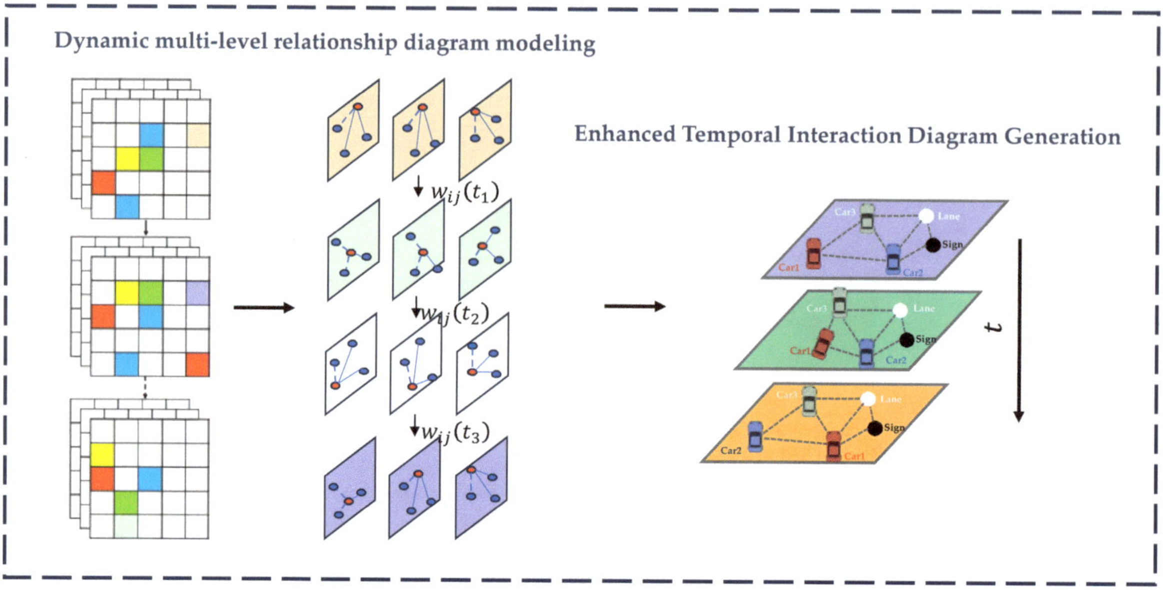

Figure 4. Modeling spatial and temporal interaction of vehicle behavior.

The interaction between vehicles and road objects such as traffic signs and obstacles can be modeled in complex traffic scenarios using the following equation:

$$G = (V, E) \tag{4}$$

where V represents the set of nodes, including vehicles and road objects, and E denotes the edges, reflecting the relationships between these nodes.

In the case of multi-level relationships, each level is characterized by a specific type of interaction. For instance, the first hierarchical level represents direct proximity, the second reflects extended proximity, and the third captures more distant or complex relationships. The overall relationship diagram of the transportation scenario can be represented as a combination of the relationship diagrams at each level:

$$G = \sum_{l=1}^{L} G^l = \sum_{l=1}^{L} \left(V, E^l\right) \tag{5}$$

where L represents the total number of relationship layers.

For dynamic edge weights, each edge e_{ij}^l, connecting nodes v_i and v_j on layer lll, is assigned a time-dependent weight $w_{ij}^l(t)$. This weight is dynamically computed using the following equation:

$$w_{ij}^l(t) = \frac{1}{1 + exp\left(-\left(\alpha d_{ij}(t) + \beta \Delta v_{ij}(t) + \gamma \Delta a_{ij}(t)\right)\right)} \tag{6}$$

where $d_{ij}(t)$ denotes the distance between vehicles v_i and v_j at time t; $\Delta v_{ij}(t)$ represents the relative speed between the two vehicles; $a_{ij}(t)$ indicates their relative acceleration; and α, β, and γ are coefficients that regulate the influence of these factors.

Ultimately, the complete mathematical expression for modeling dynamic multi-level relationship graphs is obtained as follows:

$$G(t) = (V, E(t)) = \left(V, \textstyle\sum_{l=1}^{L} \theta_l \cdot E^l(t)\right) = \left(V, \textstyle\sum_{l=1}^{L} \theta_l \cdot \left\{ \left(v_i, v_j, \frac{1}{1+exp\left(-\left(\alpha d_{ij}(t)+\beta \Delta v_{ij}(t)+\gamma \Delta a_{ij}(t)\right)\right)} \right) \right\}\right) \quad (7)$$

3.2.2. Enhanced Temporal Interaction Diagram Generation

In modern traffic prediction and vehicle interaction analysis, traditional temporal interaction map generation methods often consider only simple temporal evolution information. They cannot adequately capture complex vehicle dynamics and variable traffic scenarios. An enhanced temporal interaction graph generation method was designed to address this issue. This method focuses on temporal changes and captures complex interactions such as vehicle acceleration, hard braking, hard acceleration, and behavioral patterns under complex traffic conditions (e.g., rainy days and nighttime). The generated temporal interaction diagram includes information such as "Moving Forward", "Moving Backward", "Moving Left to Right", "Moving Right to Left", "No Change", and more complex interactions. This comprehensive approach enables the model to capture and understand dynamic vehicle interactions better, thereby improving prediction accuracy and adaptability to diverse traffic scenarios.

To represent the time series graph, the traffic scenario at each time step t is modeled using the following equation:

$$G_t = (V_t, E_t) \quad (8)$$

where V_t represents the set of nodes at time t, including all vehicles and road objects involved in the interaction at that moment, and E_t denotes the set of edges at time t, representing the interactions between these nodes.

To capture the dynamic interaction of vehicles over time, we consider the dynamic changes in edge weights $w_{ij}(t)$. The relationship between vehicles v_i and v_j is assumed to depend not only on their current state but also their behavior in previous frames. This relationship can be modeled using the following equation:

$$w_{ij}(t) = \textstyle\sum_{\tau=0}^{\Delta t} \lambda^{\tau} \cdot f(v_i(t-\tau), v_j(t-\tau)) \quad (9)$$

where Δt represents the time window indicating the number of past time steps considered, λ is a decay factor that controls the influence of past time steps on the current edge weights, and $f(v_i(t-\tau), v_j(t-\tau))$ is a function that calculates the interactions between vehicles v_i and v_j at time $t-\tau$, which may be based on differences in velocity, relative positions, accelerations, and other factors.

Furthermore, we introduce the concept of time series convolution to capture complex interaction patterns over time. This technique aggregates vehicle interaction information over time, generating a new edge weight. The convolution operation can be expressed as follows:

$$\widetilde{w}_{ij}(t) = \sigma\left(\textstyle\sum_{k=0}^{K-1} W_k \cdot w_{ij}(t-k) + b_k\right) \quad (10)$$

where $\widetilde{w}_{ij}(t)$ represents the edge weight after time convolution; K is the size of the convolution kernel, indicating the number of time steps considered; W_k is the convolution kernel weight for the k th time step; b_k is the bias term; and σ is the nonlinear activation function.

At each time step t, we obtain a temporally convolved graph $\widetilde{G}_t = (V_t, \widetilde{E}_t)$, where $\widetilde{E}$ represents the set of edges obtained after convolving the time series. To capture the dynamic interactions over the entire period, we aggregate the convolved graphs from all time steps to obtain the final temporal interaction graph G_{temp}:

$$G_{temp} = \textstyle\sum_{t=1}^{T} \widetilde{G}_t \quad (11)$$

where T denotes the total number of time steps. This temporal interaction graph synthesizes the complex behavioral patterns of vehicles over time, effectively reflecting dynamic changes throughout the entire time series. It is well suited for accurate behavioral prediction and analysis. Time series convolution enhances the model's ability to capture complex interactions in the temporal dimension, thereby improving its performance in time-dynamic scenarios .

3.3. Vehicle Behavior Prediction

The traditional MR-GCN framework fails to utilize dynamic knowledge to capture dynamic spatio-temporal correlations fully. A DMR-GCN structure was designed to address this issue, incorporating a multi-relational graph convolutional network (MR-GCN) and a dynamic scene perception and interactive learning mechanism. This structure models the spatio-temporal relationships between the vehicle, other vehicles, and landmarks in the scene over time to obtain a comprehensive spatio-temporal representation. This representation is achieved by stacking different neural network layers, effectively enhancing the capture of dynamic spatio-temporal correlations and improving the model's expressiveness and accuracy, as shown in Figure 5.

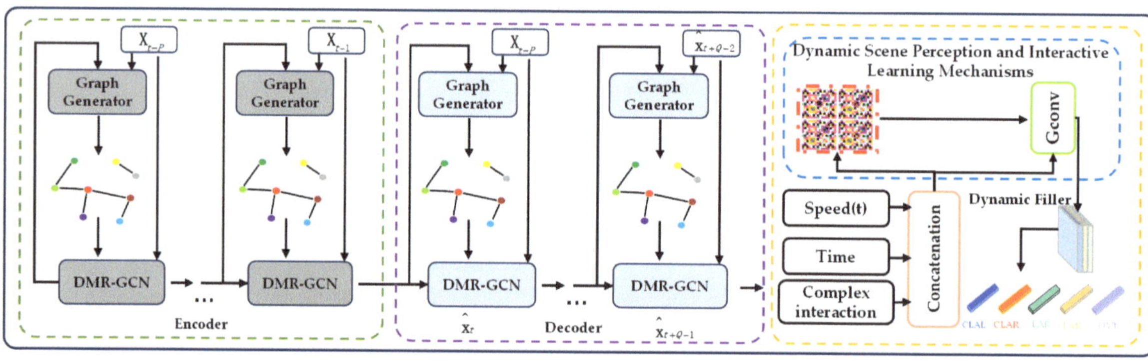

Figure 5. DMR-GCN structure.

3.3.1. DMR-GCN

Multi-relational graph convolutional networks (MR-GCNs) are a typical means of modeling interactions in knowledge graphs and spatio-temporal data. However, traditional MR-GCN frameworks use a fixed weight matrix for different relationship types, which limits adaptability to dynamic environments. A dynamic multi-relational graph convolutional network (DMR-GCN) was designed to address this. The core components of this network include an adaptive neighbor matrix updating mechanism and a multi-relationship feature fusion mechanism. These mechanisms allow for more flexible dynamic relationship capture and comprehensive multi-relationship feature processing, enhancing the network's adaptability to real-time data. The specific steps are as follows:

(1) Adaptive Neighborhood Matrix Update Mechanism:

(1) Node Feature Conversion

First, the node features h_G^{k-1} of the $(k-1)$ th layer are converted into a query vector Q and a key vector K using the query weight matrix W_q and the key weight matrix W_k, respectively:

$$Q = h_G^{k-1} W_q, K = h_G^{k-1} W_k \tag{12}$$

where $W_q, W_k \in R^{d \times d}$.

(2) Similarity calculation

Next, the similarity between the nodes is calculated using the dot product, which is then scaled to prevent tremendous values:

$$Similarity_{ij} = \frac{Q_i \cdot K_j^T}{\sqrt{d_k}} \tag{13}$$

where d_k is the dimension of the key vector.

(3) Weight normalization

Attention weights are obtained by normalizing the similarity using the following function:

$$\alpha_{ij} = softmax\left(\frac{Q_i \cdot K_j^T}{\sqrt{d_k}}\right) = \frac{exp\left(\frac{Q_i \cdot K_j^T}{\sqrt{d_k}}\right)}{\sum_{j=1}^{n} exp\left(\frac{Q_i \cdot K_j^T}{\sqrt{d_k}}\right)} \tag{14}$$

(4) Dynamic neighborhood matrix update

Using the attention weights α_{ij}, we thus update the adjacency matrix of the relationship r:

$$\hat{A}_r[i,j] = \alpha_{ij} \tag{15}$$

(2) Multi-Relational Feature Fusion Mechanisms

A multi-relational feature fusion mechanism was designed to synthesize features from different relationships. This mechanism enhances the model's ability to capture richer interaction information by introducing adjacency matrices for multiple relationship types. Specifically, adjacency matrices are separately constructed for each relationship type, and these matrices are used to perform weighted fusion among the different relationship features, generating a comprehensive feature representation. This approach more accurately reflects the multiple interactions between nodes and improves the model's expressiveness and predictive performance in complex scenarios.

(1) Multi-relational graph convolution

A multi-relationship graph convolution operation is performed on the updated adjacency matrix $\hat{A}_r$. For relationship r, the node feature representation h_G^k at the k th layer is computed using the following equation:

$$h_G^k = ReLU\left(\sum\nolimits_{r \in R} \hat{A}_r h_G^{k-1} W_r^k + h_G^{k-1} W_s^k\right) \tag{16}$$

where $\hat{A}_r$ is the adjacency matrix of relationship r dynamically updated through the attention mechanism, W_r^k is the weight matrix of the k th layer associated with relationship r, W_s^k is the weight matrix of the k th layer for calculating the node information (self-loop), and $ReLU$ is the activation function.

(3) Feature Fusion

The results of the convolution for all relationships are weighted and fused to combine the effects of the different relationships:

$$h_G^k = ReLU\left(\sum\nolimits_{r \in R} \beta_r \hat{A}_r h_G^{k-1} W_r^k + h_G^{k-1} W_s^k\right) \tag{17}$$

where β_r is the fusion weight of relationship r, which is learned through training dynamics.

(4) Representation of Interlayer Stacking and Multi-Hop Relationships

The multi-hop relational representation of a node can be obtained by stacking multiple layers of MR-GCNs. If K layers of MR-GCN are stacked, the K-hop relational representation h_G^k is provided as follows:

$$h_G^{k+1} = ReLU\left(\sum\nolimits_{r \in R} \beta_r \hat{A}_r h_G^k W_r^{k+1} + h_G^k W_s^{k+1}\right) \tag{18}$$

Overall, DMR-GCN achieves more flexible dynamic relationship capture and comprehensive multi-relational feature processing by introducing an adaptive adjacency matrix updating mechanism and a multi-relational feature fusion mechanism. These enhancements significantly improve the model's adaptability and predictive ability in complex scenes.

3.3.2. Dynamic Scene Perception and Interactive Learning Mechanism

After encoding the spatial relationships using DMR-GCN, a dynamic scene sensing and interactive learning mechanism is introduced. This mechanism dynamically adjusts the model's learning process by sensing real-time changes in the traffic scene and interaction patterns of vehicles. Through interactive learning, the model can adjust parameters and weights in real time during the prediction process, thereby improving predictions' real-time performance and accuracy based on the current traffic situation and vehicle interaction patterns.

4. Experimentation and Analysis

4.1. Datasets

To comprehensively evaluate the performance of our framework, three representative datasets were selected: ApolloScapes [41], KITTI [42], the Indian dataset, and a self-collected dataset. Following the criteria outlined in [35], the ApolloScapes, KITTI, and Indian datasets were partitioned into training, testing, and validation sets. This partitioning method was innovatively extended to the self-collected dataset, with specialized data annotation tailored to the task requirements.

In terms of datasets, the ApolloScapes dataset serves as the core. Its rich driving scenarios, especially overtaking and lane-changing behaviors, are fully demonstrated by the selected 4000 frames of images, providing a solid foundation for model training. We focused on highway and comprehensive road environments for the KITTI dataset, selecting 700 frames from tracking sequences 4, 5, and 10 to evaluate the model's generalization ability under complex road conditions. The Indian dataset, with its unique vehicle types (e.g., tricycles and trucks) and congested driving environments, presents a severe test of the model's transfer learning capability; 600 frames of images provide a small but challenging sample. The 65 km section of the S20 Outer Ring Highway and G2 Beijing-Shanghai Highway from Pudong to Jiading in Shanghai, characterized by high traffic flow and various vehicle types, frequently experiences traffic congestion due to lane-changing collisions, making it an ideal route for data acquisition, as shown in Figure 6. Video captured by the front monocular camera of the data collection vehicle on this road section constituted the experimental dataset.

Figure 6. Data collection equipment and collection routes.

Prior to training the network, extensive data preprocessing was performed, including normalizing the image pixel values to the [0, 1] range, adjusting the images to a consistent resolution, and applying data enhancement methods (e.g., rotation, flipping, and luminance adjustments) to improve the robustness of the model under varying lighting and environmental conditions. The network training process included dividing the dataset into a training set (70%), validation set (15%), and test set (15%), and training was performed using a cross-entropy loss function and an Adam optimizer with a learning rate of 0.001. The training lasted 300 epochs, and an early stopping strategy was implemented to prevent overfitting. Key hyperparameters include a batch size of 32, a learning rate of 0.001, a discard rate of 0.5 (to mitigate overfitting), and an architecture consisting of five convolutional layers and three fully connected layers.

At the behavior prediction level, five categories of crucial vehicle behaviors are defined: continuous left–right lane change (CLAL/R), left–right lane change (LCL/LCR), and overtaking (OVT). These categories ensure comprehensive coverage of vehicle dynamic behaviors. Through monocular video processing, the framework was rigorously evaluated using the KITTI, ApolloScapes, and Indian datasets. The selected driving scenarios span various urban and highway conditions and include both standard and non-standard vehicles. This comprehensive evaluation validated the predictive accuracy and adaptability of the model. By integrating multivariate datasets with detailed data annotation, our evaluation framework enhanced the model's predictive ability under complex highway driving conditions and improved its adaptability to various vehicles and environments. This integration lays a solid foundation for developing safer, more efficient autonomous driving systems.

4.2. *Qualitative Results*

4.2.1. Ablation Experiments

Ablation studies were conducted to verify the effectiveness of the DMR-GCN framework and its components. The experimental results are shown in Table 1 and Figure 9a. The base model (MR-GCN) performed mediocrely on all metrics, indicating its limitations in capturing vehicle lane-changing behavior and complex interaction patterns. After adding the dynamic multilevel relationship graph (DRG), the accuracy and recall of the model significantly improved from 0.85 and 0.84 to 0.87 and 0.86 for CLAL and CLAR, respectively; from 0.91 to 0.93 for Map; and from 0.90 to 0.92 for F1. These improvements suggest that the DRG is effective in capturing complex dynamics.

Table 1. DMR-GCN framework ablation experiments.

Architecture \ Classes	CLAL	CLAR	LCL	LCR	OVT	Map	Recall	F1
MR-GCN	0.85	0.84	0.83	0.82	0.80	0.91	0.89	0.90
MR-GCN + DRG	0.87	0.86	0.85	0.84	0.82	0.93	0.91	0.92
MR-GCN + ETIG	0.88	0.87	0.86	0.85	0.83	0.94	0.92	0.93
MR-GCN + GAMU	0.86	0.85	0.84	0.83	0.81	0.92	0.90	0.91
MR-GCN + DRG + ETIG	0.89	0.88	0.87	0.86	0.84	0.94	0.92	0.93
MR-GCN + DRG + GAMU + ETIG	0.89	0.88	0.87	0.86	0.84	0.94	0.92	0.93

Introducing the enhanced temporal interaction graph generation method (ETIG) improves the model's ability to capture complex interaction patterns significantly, with CLAL and CLAR reaching 0.88 and 0.87, respectively, and Map and F1 reaching 0.94 and 0.93. Additionally, graph attention's dynamic neighborhood matrix updating mechanism (GAMU) further improved dynamic relationship capturing and feature processing, achieving CLAL and CLAR scores of 0.86 and 0.85 and Map and F1 scores of 0.92 and 0.91, respectively.

The model's performance significantly improved when combined with the dynamic multi-level relational graph (DRG) and enhanced temporal interaction graph generation method (ETIG) components. Adding these components boosted the model's accuracies on continuous left-lane alteration (CLAL) and continuous right-lane alteration (CLAR) to 0.89 and 0.88, respectively. At the same time, the mean average precision (Map) and F1 scores increased to 0.94 and 0.93. These enhancements validate the effectiveness of the DRG and ETIG components in capturing and predicting complex traffic dynamics.

Following integrating all components, the model showed robust performance across all metrics, with CLAL and CLAR scores of 0.89 and 0.88 and Map and F1 scores of 0.94 and 0.93. This demonstrates the components' effectiveness in enhancing the model's understanding and prediction of complex traffic scenarios. In conclusion, integrating these components significantly improved the MR-GCN framework's capacity to handle intricate interaction patterns and various traffic scenarios.

4.2.2. Comparison Experiments

To validate the proposed method, it was trained and tested on several datasets, including the Apollo, KITTI, and Indian databases. Comparisons were made with the baseline model, the LSTM-based graph network Structural-RNN, and the LSTM + Multi-head Attention model. The experimental results are shown in Table 2 and Figure 9b. These comparison experiments indicate that the proposed method outperforms other models across the various datasets.

Table 2. Pairwise comparison results of different models for Apollo, KITTI, and Indian datasets.

Train and Test On	Apollo				KITTI				Indian			
Action \ Method	Baseline	St-RNN	LSTM+ Mutil-Head Attention	Ours	Baseline	St-RNN	LSTM+ Mutil-Head Attention	Ours	Baseline	St-RNN	LSTM+ Mutil-Head Attention	Ours
CLAL	0.80	0.82	0.85	0.88	0.81	0.83	0.86	0.89	0.82	0.84	0.87	0.90
CLAR	0.79	0.81	0.84	0.87	0.80	0.82	0.85	0.88	0.81	0.83	0.86	0.89
LCL	0.77	0.80	0.83	0.86	0.78	0.81	0.84	0.87	0.79	0.82	0.85	0.88
LCR	0.76	0.79	0.82	0.85	0.77	0.80	0.83	0.86	0.78	0.81	0.84	0.87
OVT	0.74	0.77	0.80	0.83	0.75	0.78	0.81	0.84	0.76	0.79	0.82	0.85
Map	0.87	0.90	0.92	0.94	0.87	0.90	0.92	0.94	0.87	0.90	0.92	0.94
Recall	0.85	0.88	0.90	0.92	0.86	0.88	0.90	0.92	0.85	0.88	0.90	0.92
F1	0.86	0.89	0.91	0.93	0.86	0.89	0.91	0.93	0.86	0.89	0.91	0.93
IT	50	45	40	35	52	46	41	36	51	44	39	34
FPS	20	22.2	25	28.6	19.2	21.7	24.4	27.8	19.6	22.7	25.6	29.4

Specifically, for the continuous left- and right-lane change (CLAL and CLAR) metrics, our method improved from 0.80 and 0.79 to 0.88 and 0.87 on the Apollo dataset. Similarly, it improved from 0.81 and 0.80 to 0.89 and 0.88 on the KITTI dataset and from 0.82 and 0.81 to 0.90 and 0.89 on the Indian dataset. These improvements highlight the advantages of dynamic multilayer relational maps in effectively capturing complex interactions between vehicles and road objects.

For the left- and right-lane change (LCL and LCR) metrics, the MR-GCN framework showed enhancements from 0.77 and 0.76 to 0.86 and 0.85 on the Apollo dataset. It also improved from 0.78 and 0.77 to 0.87 and 0.86 on the KITTI dataset and from 0.79 and 0.78 to 0.88 and 0.87 on the Indian dataset, demonstrating higher accuracy and robustness. This illustrates the efficacy of the MR-GCN framework in addressing diverse datasets and intricate interaction behaviors.

In terms of overtaking (OVT) metrics, improvements were observed on the Apollo dataset, increasing from 0.74 to 0.83. Similarly, the KITTI dataset increased from 0.75 to 0.84, while the Indian dataset improved from 0.76 to 0.85. These results underscore the method's effectiveness in complex traffic scenarios, confirming the MR-GCN framework's ability to handle real-time dynamic data.

Additionally, the MR-GCN framework exhibited excellent performance in average inference time and processing speed, achieving an average inference time of 35 milliseconds per frame and a processing speed of 28.6 frames per second, a significant improvement over other models. This short inference time and high frame rate enable it to respond to input data rapidly, making it well suited for real-time applications such as autonomous driving and traffic management systems. Reacting quickly and accurately is crucial for reducing accidents and enhancing road safety. Consequently, the MR-GCN framework's outstanding performance in dynamic traffic environments indicates its broad application potential in intelligent transportation systems. Furthermore, future optimizations of the algorithm and hardware will likely enhance its real-time processing capabilities, providing a stronger foundation for practical applications.

Finally, the MR-GCN framework demonstrated superior performance across all datasets for critical metrics such as mean average precision (Map), Recall, and F1. On the Indian dataset, the framework achieved scores of 0.94, 0.92, and 0.93, respectively. On the Apollo dataset, Map improved from 0.87 to 0.94, Recall from 0.85 to 0.92, and F1 from 0.86 to 0.93. On the KITTI dataset, Map increased from 0.87 to 0.94, Recall from 0.86 to 0.92, and F1 from 0.86 to 0.93. These results demonstrate the adaptability and prediction accuracy of the MR-GCN framework in diverse traffic scenarios, showcasing its significant advantages and reliability in practical applications.

4.2.3. Visualization of Results

The model's performance was further validated on the KITTI and our self-collected datasets. Figures 7 and 8 show the qualitative results of these datasets. A consistent color-coding convention describes vehicle behavior: green for vehicles moving away, red for vehicles driving usually, and blue for vehicles changing lanes to the right. In Figure 7a,b show examples of vehicles accelerating away from their lane. Figure 7b,c depict a vehicle ahead changing lanes to the right, with Figure 7c showing the initial lane change and Figure 7d showing the vehicle completing the maneuver.

(a) Frame 1 **(b) Frame 5** **(c) Frame 10** **(d) Frame 15**

Figure 7. Model validation visualization results on KITTI dataset.

(a) Frame 1 **(b) Frame 5** **(c) Frame 10** **(d) Frame 15**

Figure 8. Model validation visualization results on self-selected dataset.

In Figure 8a illustrates a car changing lanes, while Figure 8b shows a motorcycle accelerating. Figure 8c captures a car on the right side of the lane merging into the left lane, and Figure 8d depicts a vehicle on the right side of the road completing a lane change and preparing to change lanes again. These qualitative results further validate the classification and generalization capabilities of the proposed model on the self-collected dataset, even with fewer or no observed test vehicles.

4.3. Transfer Learning

To validate the transfer learning capability of the DMR-GCN model, experiments were conducted across multiple datasets. Specifically, the model was trained on the KITTI dataset and then tested on the Apollo, Indian, and a validation set of self-collected datasets. During testing, categories that did not exist in the corresponding dataset were excluded to ensure a fair evaluation. The experimental results are shown in Table 3 and Figure 9c.

Table 3. Results of the transfer learning experiment.

Train On		KITTI	
Test On	**Apollo**	**Indian**	**Ours**
CLAL	0.88	0.87	0.89
CLAR	0.87	0.86	0.88
LCL	0.86	0.85	0.87
LCR	0.85	0.84	0.86
OVT	0.83	0.82	0.84
Map	0.94	0.94	0.94
Recall	0.92	0.92	0.92
F1	0.93	0.93	0.93

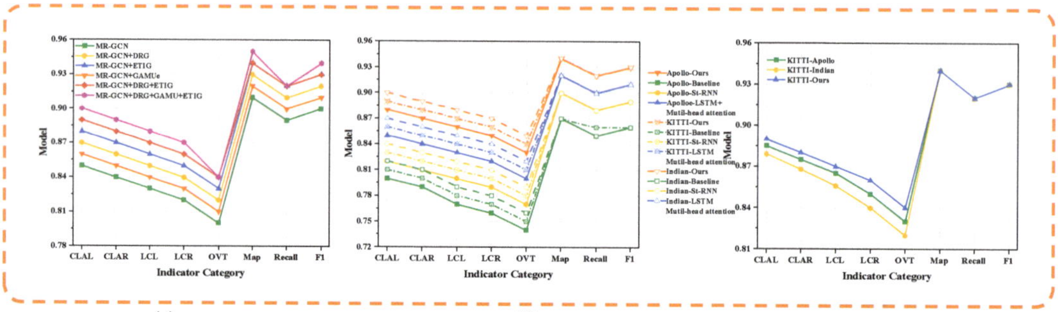

(a) (b) (c)

Figure 9. Comparative line graph of experimental results. (**a**) DMR-GCN framework ablation experiments, (**b**) Pairwise comparison results of different models for Apollo, KITTI and Indian datasets, (**c**) Results of the transfer learning experiment.

The experimental results demonstrated that the transfer learning model exhibits robust performance across all target datasets, underscoring the model's capacity for generalization. The transfer learning model achieved CLAL and CLAR scores of 0.88 and 0.87, LCL and LCR scores of 0.86 and 0.85, and accuracy and Recall rates of 0.94 and 0.92 on the Apollo dataset. These results outperform the model trained and tested solely on the KITTI dataset, demonstrating its resilience and adaptability to diverse scenarios.

On the Indian dataset, the transfer learning model showed high prediction accuracy in various traffic scenarios, with CLAL and CLAR scores of 0.87 and 0.86, LCL and LCR scores of 0.85 and 0.84, and accuracy and Recall rates of 0.94 and 0.92, respectively.

For the self-collected dataset, the transfer learning model achieved CLAL and CLAR scores of 0.89 and 0.88, LCL and LCR scores of 0.87 and 0.86, and accuracy and Recall rates of 0.94 and 0.92. These results further validate the model's high fidelity and adaptability across different traffic datasets, demonstrating its broad applicability and reliability in practical applications.

The results of the ablation experiment, comparison experiment, and transfer learning experiment are shown in Figure 9. It is clear from the figure that the DMR-GCN model outperformed the other models across all metrics. Specifically, the ablation experiments demonstrated the contribution of each component to the overall model performance; the comparison experiments showed that DMR-GCN significantly outperforms the baseline model and other existing methods in terms of accuracy, Recall, and F1 scores; and the transfer learning results validated the ability of DMR-GCN to generalize across different datasets. These findings fully demonstrate the superiority and robustness of the DMR-GCN model in handling complex traffic scenarios.

4.4. Discussion

The experimental results indicate that the DMR-GCN framework outperforms the baseline model and other existing methods across various traffic scenarios. Comparisons across the Apollo, KITTI, and Indian datasets revealed that the model achieved accuracy scores of 0.89 and 0.88 for continuous left-lane alteration (CLAL) and continuous right-lane alteration (CLAR), respectively, while the baseline model scored only 0.80 and 0.79. This improvement underscores the effectiveness of the dynamic multi-level relational graph (DRG) and the enhanced temporal interaction graph generation method (ETIG).

Specifically, the MR-GCN framework demonstrated increases in mean accuracy (Map) and F1 scores from 0.91 and 0.90 to 0.94 and 0.93, respectively, highlighting the contributions of these components in capturing complex traffic dynamics. In transfer learning experiments, the model achieved CLAL and CLAR scores of 0.87 and 0.86 on the unseen Indian dataset, in contrast to the lower performance of models trained solely on the KITTI dataset, further validating the model's generalization capabilities.

However, the limitations of this study include the necessity for improved adaptability to non-standard vehicles and extreme traffic conditions. Future research could address these issues by expanding the dataset to encompass various vehicle types and more complex traffic scenarios, enhancing the model's robustness. Additionally, incorporating richer external knowledge graphs may further improve the model's generalization under complex conditions.

In summary, the results of this study demonstrate that the DMR-GCN framework achieves high accuracy and adaptability in predicting complex traffic behaviors, providing a valuable foundation for the advancement of autonomous driving technologies.

5. Conclusions

An innovative MR-GCN-based framework is herein proposed to address frequent accidents and traffic flow issues caused by sudden lane changes in highway scenarios. The complex dynamics of vehicle lane changing in highway environments are accurately captured using continuous time-series images from a moving monocular camera, combined with a designed dynamic multi-level relational graph (DMLRG). The enhanced temporal interaction graph generation (ETIG) method further improved the model's accuracy in capturing dynamic vehicle interactions, especially under congested road conditions. The graph attention-based dynamic adjacency matrix update mechanism (GAMU) achieves dynamic adaptation to real-time traffic data by integrating an adaptive adjacency matrix update strategy with a multi-relational feature fusion technique.

Experimental validation on multiple benchmark datasets, including KITTI, Apollo, and Indian datasets, showed that the proposed DMR-GCN framework significantly improves vehicle behavior prediction performance, with Map, Recall, and F1 scores of 90%, 88%, and 89%, respectively, outperforming existing algorithms. Additionally, tests on a custom high-speed traffic scene dataset further demonstrated the model's robustness and adaptability, achieving a Map of 91%, a Recall of 89%, and an F1 score of 90%. In summary, the DMR-GCN framework not only theoretically enhances traffic behavior modeling methodology but also shows strong application potential and stability in practice, providing robust support for efficient vehicle behavior prediction.

Future work will optimize the model's real-time and generalization capabilities to adapt to more complex and dynamic traffic environments. Additionally, further research will explore integrating advanced machine learning techniques such as reinforcement learning to enhance decision-making processes in real-time traffic scenarios. Investigating the incorporation of multi-agent systems may also offer new insights into cooperative vehicle behaviors, ultimately contributing to safer and more efficient autonomous driving systems.

Author Contributions: Conceptualization, M.L. (Menghao Li) and W.Z.; methodology, M.L. (Menghao Li); software, M.L. (Menghao Li); validation, M.L. (Menghao Li), W.Z., M.L. (Miao Liu) and E.C.; formal analysis, M.L. (Menghao Li); investigation, M.L. (Miao Liu), E.C., C.H. and M.Z.; resources, M.L. (Miao Liu); data curation, M.L. (Miao Liu), C.H. and M.Z.; writing—original draft preparation, M.L. (Menghao Li); writing—review and editing, E.C.; supervision, W.G.; project administration, M.L. (Miao Liu); funding acquisition, W.Z. and W.G. All authors have read and agreed to the published version of the manuscript.

Funding: This research was funded by Shanghai Special Funds for Centralized Guided Local Science and Technology Development (YDZX20233100002002) and supported by the Postdoctoral Fellowship Program of CPSF under Grant Number GZB20240351.

Institutional Review Board Statement: Not applicable.

Informed Consent Statement: Not applicable.

Data Availability Statement: The data presented in the study are available on request from the corresponding author. The data are not publicly available due to privacy.

Conflicts of Interest: Author Weiwei Zhang was employed by the company Shanghai Smart Vehicle Cooperating Innovation Center Co., Ltd. The remaining authors declare that the re-search was conducted in the absence of any commercial or financial relationships that could be construed as a potential conflict of interest.

References

1. Feng, D.; Haase-Schütz, C.; Rosenbaum, L.; Hertlein, H.; Glaeser, C.; Timm, F.; Wiesbeck, W.; Dietmayer, K. Deep multi-modal object detection and semantic segmentation for autonomous driving: Datasets, methods, and challenges. *IEEE Trans. Intell. Transp. Syst.* **2020**, *22*, 1341–1360. [CrossRef]
2. Kuefler, A.; Morton, J.; Wheeler, T.; Kochenderfer, M. Imitating driver behavior with generative adversarial networks. In Proceedings of the 2017 IEEE Intelligent Vehicles Symposium (IV), Los Angeles, CA, USA, 11–14 June 2017; pp. 204–211.
3. Cui, Z.; Ke, R.; Pu, Z.; Wang, Y. Stacked bidirectional and unidirectional LSTM recurrent neural network for forecasting network-wide traffic state with missing values. *Transp. Res. Part C Emerg. Technol.* **2020**, *118*, 102674. [CrossRef]
4. Sharma, S.; Das, A.; Sistu, G.; Halton, M.; Eising, C. BEVSeg2TP: Surround View Camera Bird's-Eye-View Based Joint Vehicle Segmentation and Ego Vehicle Trajectory Prediction. *arXiv* **2023**, arXiv:2312.13081.
5. Messaoud, K.; Yahiaoui, I.; Verroust-Blondet, A.; Nashashibi, F. Attention based vehicle trajectory prediction. *IEEE Trans. Intell. Veh.* **2020**, *6*, 175–185. [CrossRef]
6. Chen, F.; Li, P.; Wu, C. Dgc: Training dynamic graphs with spatio-temporal non-uniformity using graph partitioning by chunks. *Proc. ACM Manag. Data* **2023**, *1*, 1–25. [CrossRef]
7. Zheng, Y.; Wei, Z.; Liu, J. Decoupled graph neural networks for large dynamic graphs. *arXiv* **2023**, arXiv:2305.08273. [CrossRef]
8. Mo, X.; Xing, Y.; Lv, C. Graph and recurrent neural network-based vehicle trajectory prediction for highway driving. In Proceedings of the 2021 IEEE International Intelligent Transportation Systems Conference (ITSC), Indianapolis, IN, USA, 19–22 September 2021; pp. 1934–1939.
9. Yang, M.; Zhu, H.; Wang, T.; Cai, J.; Weng, X.; Feng, H.; Fang, K. Vehicle Interactive Dynamic Graph Neural Network Based Trajectory Prediction for Internet of Vehicles. *IEEE Internet Things J.* **2024**. [CrossRef]
10. Xu, X.; Zhang, L.; Liu, B.; Liang, Z.; Zhang, X. Transport-Hub-Aware Spatial-Temporal Adaptive Graph Transformer for Traffic Flow Prediction. *arXiv* **2023**, arXiv:2310.08328.
11. Han, X.; Gong, S. LST-GCN: Long Short-Term Memory embedded graph convolution network for traffic flow forecasting. *Electronics* **2022**, *11*, 2230. [CrossRef]
12. Kumar, R.; Mendes Moreira, J.; Chandra, J. DyGCN-LSTM: A dynamic GCN-LSTM based encoder-decoder framework for multistep traffic prediction. *Appl. Intell.* **2023**, *53*, 25388–25411. [CrossRef]
13. Katayama, H.; Yasuda, S.; Fuse, T. Traffic density based travel-time prediction with GCN-LSTM. In Proceedings of the 2022 IEEE 25th International Conference on Intelligent Transportation Systems (ITSC), Macau, China, 8–12 October 2022; pp. 2908–2913.
14. Zhang, T.; Guo, G. Graph attention LSTM: A spatiotemporal approach for traffic flow forecasting. *IEEE Intell. Transp. Syst. Mag.* **2020**, *14*, 190–196. [CrossRef]
15. Kosaraju, V.; Sadeghian, A.; Martín-Martín, R.; Reid, I.; Rezatofighi, H.; Savarese, S. Social-bigat: Multimodal trajectory forecasting using bicycle-gan and graph attention networks. In Proceedings of the Advances in Neural Information Processing Systems 32, Vancouver, BC, Canada, 8–14 December 2019.
16. Sun, J.; Jiang, Q.; Lu, C. Recursive social behavior graph for trajectory prediction. In Proceedings of the IEEE/CVF Conference on Computer Vision and Pattern Recognition, Seattle, WA, USA, 13–19 June 2020; pp. 660–669.
17. Ivanovic, B.; Pavone, M. The trajectron: Probabilistic multi-agent trajectory modeling with dynamic spatiotemporal graphs. In Proceedings of the IEEE/CVF International Conference on Computer Vision, Seoul, Republic of Korea, 27 October–2 November 2019; pp. 2375–2384.
18. Li, X.; Ying, X.; Chuah, M.C. Grip: Graph-based interaction-aware trajectory prediction. In Proceedings of the 2019 IEEE Intelligent Transportation Systems Conference (ITSC), Auckland, New Zealand, 27–30 October 2019; pp. 3960–3966.
19. Chandra, R.; Bhattacharya, U.; Bera, A.; Manocha, D. Traphic: Trajectory prediction in dense and heterogeneous traffic using weighted interactions. In Proceedings of the IEEE/CVF Conference on Computer Vision and Pattern Recognition, Long Beach, CA, USA, 15–20 June 2019; pp. 8483–8492.
20. Lu, Y.; Chen, Y.; Zhao, D.; Liu, B.; Lai, Z.; Chen, J. CNN-G: Convolutional neural network combined with graph for image segmentation with theoretical analysis. *IEEE Trans. Cogn. Dev. Syst.* **2020**, *13*, 631–644. [CrossRef]

21. Gao, J.; Sun, C.; Zhao, H.; Shen, Y.; Anguelov, D.; Li, C.; Schmid, C. Vectornet: Encoding hd maps and agent dynamics from vectorized representation. In Proceedings of the IEEE/CVF Conference on Computer Vision and Pattern Recognition, Seattle, WA, USA, 13–19 June 2020; pp. 11525–11533.
22. Chaochen, Z.; Zhang, Q.; Li, D.; Li, H.; Pang, Z. Vehicle trajectory prediction based on graph attention network. In Proceedings of the Cognitive Systems and Information Processing: 6th International Conference, ICCSIP 2021, Suzhou, China, 20–21 November 2021; pp. 427–438.
23. Redmon, J.; Divvala, S.; Girshick, R.; Farhadi, A. You only look once: Unified, real-time object detection. In Proceedings of the IEEE Conference on Computer Vision and Pattern Recognition, Las Vegas, NV, USA, 27–30 June 2016; pp. 779–788.
24. Liu, W.; Anguelov, D.; Erhan, D.; Szegedy, C.; Reed, S.; Fu, C.-Y.; Berg, A.C. Ssd: Single shot multibox detector. In Proceedings of the Computer Vision–ECCV 2016: 14th European Conference, Amsterdam, The Netherlands, 11–14 October 2016; pp. 21–37.
25. Siradjuddin, I.A.; Muntasa, A. Faster region-based convolutional neural network for mask face detection. In Proceedings of the 2021 5th International Conference on Informatics and Computational Sciences (ICICoS), Semarang, Indonesia, 24–25 November 2021; pp. 282–286.
26. Li, M.; Liu, M.; Zhang, W.; Guo, W.; Chen, E.; Zhang, C. A Robust Multi-Camera Vehicle Tracking Algorithm in Highway Scenarios Using Deep Learning. *Appl. Sci.* **2024**, *14*, 7071. [CrossRef]
27. Sivaraman, S.; Trivedi, M.M. Looking at vehicles on the road: A survey of vision-based vehicle detection, tracking, and behavior analysis. *IEEE Trans. Intell. Transp. Syst.* **2013**, *14*, 1773–1795. [CrossRef]
28. Kitani, K.M.; Ziebart, B.D.; Bagnell, J.A.; Hebert, M. Activity forecasting. In Proceedings of the Computer Vision–ECCV 2012: 12th European Conference on Computer Vision, Florence, Italy, 7–13 October 2012; pp. 201–214.
29. Lee, J.; Balakrishnan, A.; Gaurav, A.; Czarnecki, K.; Sedwards, S. W ise m ove: A framework to investigate safe deep reinforcement learning for autonomous driving. In Proceedings of the Quantitative Evaluation of Systems: 16th International Conference, QEST 2019, Glasgow, UK, 10–12 September 2019; Proceedings 16, 2019, pp. 350–354.
30. Neumann, L.; Vedaldi, A. Pedestrian and ego-vehicle trajectory prediction from monocular camera. In Proceedings of the IEEE/CVF Conference on Computer Vision and Pattern Recognition, Nashville, TN, USA, 19–25 June 2021; pp. 10204–10212.
31. Meyer, E.; Brenner, M.; Zhang, B.; Schickert, M.; Musani, B.; Althoff, M. Geometric deep learning for autonomous driving: Unlocking the power of graph neural networks with CommonRoad-Geometric. In Proceedings of the 2023 IEEE Intelligent Vehicles Symposium (IV), Anchorage, AK, USA, 4–7 June 2023; pp. 1–8.
32. Lee, D.; Gu, Y.; Hoang, J.; Marchetti-Bowick, M. Joint interaction and trajectory prediction for autonomous driving using graph neural networks. *arXiv* **2019**, arXiv:1912.07882.
33. Li, F.-J.; Zhang, C.-Y.; Chen, C.P. STS-DGNN: Vehicle Trajectory Prediction Via Dynamic Graph Neural Network with Spatial-Temporal Synchronization. *IEEE Trans. Instrum. Meas.* **2023**, *72*, 1–13. [CrossRef]
34. Zhang, K.; Feng, X.; Wu, L.; He, Z. Trajectory prediction for autonomous driving using spatial-temporal graph attention transformer. *IEEE Trans. Intell. Transp. Syst.* **2022**, *23*, 22343–22353. [CrossRef]
35. Mylavarapu, S.; Sandhu, M.; Vijayan, P.; Krishna, K.M.; Ravindran, B.; Namboodiri, A. Towards accurate vehicle behaviour classification with multi-relational graph convolutional networks. In Proceedings of the 2020 IEEE Intelligent Vehicles Symposium (IV), Las Vegas, NV, USA, 19 October–13 November 2020; pp. 321–327.
36. He, K.; Gkioxari, G.; Dollár, P.; Girshick, R. Mask r-cnn. In Proceedings of the IEEE International Conference on Computer Vision, Venice, Italy, 22–29 October 2017; pp. 2961–2969.
37. Pham, V.; Pham, C.; Dang, T. Road damage detection and classification with detectron2 and faster r-cnn. In Proceedings of the 2020 IEEE International Conference on Big Data (Big Data), Atlanta, GA, USA, 10–13 December 2020; pp. 5592–5601.
38. Teed, Z.; Deng, J. Raft: Recurrent all-pairs field transforms for optical flow. In Proceedings of the Computer Vision–ECCV 2020: 16th European Conference, Glasgow, UK, 23–28 August 2020; Proceedings, Part II 16, 2020. pp. 402–419.
39. Gong, S.; Ye, X.; Tan, X.; Wang, J.; Ding, E.; Zhou, Y.; Bai, X. Gitnet: Geometric prior-based transformation for birds-eye-view segmentation. In Proceedings of the European Conference on Computer Vision, Tel Aviv, Israel, 23–27 October 2022; pp. 396–411.
40. Ammar Abbas, S.; Zisserman, A. A geometric approach to obtain a bird's eye view from an image. In Proceedings of the IEEE/CVF International Conference on Computer Vision Workshops, Seoul, Republic of Korea, 27 October–2 November 2019.
41. Huang, X.; Cheng, X.; Geng, Q.; Cao, B.; Zhou, D.; Wang, P.; Lin, Y.; Yang, R. The apolloscape dataset for autonomous driving. In Proceedings of the IEEE Conference on Computer Vision and Pattern Recognition Workshops, Salt Lake City, UT, USA, 18–22 June 2018; pp. 954–960.
42. Geiger, A.; Lenz, P.; Urtasun, R. Are we ready for autonomous driving? The kitti vision benchmark suite. In Proceedings of the 2012 IEEE Conference on Computer Vision and Pattern Recognition, Providence, RI, USA, 16–21 June 2012; pp. 3354–3361.

applied sciences

MDPI

Article

A Large Bridge Traffic Operation Status Impact Assessment Model Based on AHP–Delphi–SVD Method

Jianxing Guo [1], Yunrui Zhang [2], Guanhu Yuan [3], Yanbo Li [2,*], Longfei Wang [4,*] and Zhi Dong [5,*]

1 Infrastructure Department, Chang'an University, Xi'an 710064, China; gjxing@chd.edu.cn
2 School of Energy and Electrical Engineering, Chang'an University, Xi'an 710064, China; da0858213@163.com
3 Huawei Technologies Co., Ltd., Shenzhen 518116, China; tiger.2@163.com
4 School of Engineering, Dali University, Dali 671003, China
5 School of Transportation Engineering, Chang'an University, Xi'an 710064, China
* Correspondence: ybl@chd.edu.cn (Y.L.); wanglongfei@dali.edu.cn (L.W.); zdong@chd.edu.cn (Z.D.)

Abstract: As an important component of road traffic facilities, bridges play a crucial role in daily traffic operations, and changes in their status can have an impact on traffic operation. The existing research mainly focuses on monitoring the status of bridges themselves or analyzing the operation status of road traffic, and rarely considers the changes in traffic operation status caused by changes in bridge status. Therefore, in order to evaluate the impact relationship between the two, this article designs an algorithm that combines the Analytic Hierarchy Process (AHP), the Delphi method, and Singular Value Decomposition (SVD) based on the traditional evaluation of bridges and road traffic operation status, and establishes a bridge traffic operation status impact assessment model. Then, simulation analysis and actual data verification will be conducted based on the specific situation of Ma'anshi Bridge on the Chongqing Wuhan Expressway. The experimental results show that the evaluation model established in this paper conforms to the characteristics of traffic operation, can reflect the impact of bridge state changes on traffic operation status well, effectively promote the automation level of bridge traffic impact management, and has high reliability and accuracy.

Keywords: bridge traffic operation; evaluation model; analytic hierarchy process; Delphi method; singular value decomposition

Citation: Guo, J.; Zhang, Y.; Yuan, G.; Li, Y.; Wang, L.; Dong, Z. A Large Bridge Traffic Operation Status Impact Assessment Model Based on AHP–Delphi–SVD Method. *Appl. Sci.* **2024**, *14*, 9327. https://doi.org/10.3390/app14209327

Academic Editors: Giuseppe Lacidogna, Artur Kierzkowski, Tomasz Nowakowski, Agnieszka A. Tubis, Franciszek Restel, Tomasz Kisiel, Anna Jodejko-Pietruczuk and Mateusz Zając

Received: 12 July 2024
Revised: 26 September 2024
Accepted: 11 October 2024
Published: 13 October 2024

1. Introduction

China is a country with complex terrain and numerous mountains and rivers. In order to improve transportation efficiency and meet the needs of economic construction, China has built a large number of bridges in highway and railway construction projects. Bridges are indispensable infrastructure for road transportation, alleviating the difficulties caused by complex terrain in China's transportation development and ensuring the sustainable development of various industries and economies [1]. Significant achievements have been made in technology related to roads and bridges in our country, gradually moving towards maturity and promoting the continuous development of our social economy [2]. With the continuation of the operation of bridge facilities and factors such as environmental and climate changes, the status of these facilities will inevitably undergo certain changes, thereby affecting the operation status of transportation. Therefore, a real-time understanding of bridge status and an effective scientific evaluation of the impact of bridges on traffic during operation are of great significance for traffic safety and the economy.

In some developed countries, the construction of road traffic networks has been effectively completed, and a relatively mature and complete analysis and evaluation system has been established for various components of highways. They use existing infrastructure to analyze and evaluate the operation status of highways, improve the quality and service level of high-speed highway operations, which is the direction of future transportation development. However, research into an evaluation system for traffic operation status in

China started relatively late. If the assessment of bridge traffic impact is not strengthened, more bridge traffic accidents will occur, causing incalculable impacts on traffic operation. With the deepening of people's understanding of transportation issues and the increasing emphasis on the role of bridges in traffic operations, a large amount of related research has also been conducted.

In one study [3], a long-term oriented full bridge traffic load monitoring system was proposed for the long-term monitoring of the distribution of full bridge traffic load on large-span bridges. The authors of [4] proposed a stochastic traffic flow simulation method that considered the correlation between vehicle arrival order and axle load. The results show that the correlation of these traffic flow parameters has a significant impact on the response of large-span suspension bridges. The study carried out in [5] uses the generalized extreme value theory to predict and evaluate the extreme value effects of bridges under heavy load random traffic loads in Wuhan during different regression periods. The study carried out in [6] establishes a fitting relationship model for coordination of traffic operations on large-span bridge sections under crosswind action by analyzing the relationship between traffic accidents on bridges and the distribution pattern of crosswinds. In order to effectively determine and identify abnormal driving vehicles, the study carried out [7] has proposed a highway traffic safety estimation model based on vehicle operation status. In order to solve the problem of traffic congestion in the context of smart cities, the study carried out in [8] establishes a traffic ontology called "TrafCsOnto" using OWL language and the protégé tool. The study carried out in [9] proposed an innovative model that utilizes Convolutional Neural Networks (CNNs), Bidirectional Long Short Term Memory (BiLSTM), and an Attention Mechanism (AM) to predict highway traffic flow with high accuracy and feasibility. The study carried out in [10] constructs a bridge digital twin system connected by measured traffic flow loads to evaluate the health status of bridge structures during operation. The experimental results showed that the system integrates the response coordination relationship between various components within the overall structure in the finite element model and the real change law of the structure in the monitoring data, and is capable of carrying out the health status assessments of bridge structures throughout their entire life cycle. The study carried out in [11] proposed a probabilistic fatigue assessment method based on the simulation of random traffic flow and on-site measured strain influence lines, and verified the effectiveness of the proposed method through a case study of a large-span suspension bridge. The study carried out in [12] proposed a truck queue load model and a mixed traffic flow simulation method to study the load effect level of different truck queues and the impact of mixed traffic flow formed with ordinary vehicles on bridge structures, and evaluated the applicability of current regulations. The study carried out in [13] proposed a method for evaluating the dynamic influence factor (IM) from the dynamic response of bridges considering local surface damage in traffic flow. The method was validated by applying it to two existing bridges with different road surface damage and traffic conditions.

In summary, the above-mentioned research is still focused on the evaluation of bridges themselves and traffic operation status, or the impact of traffic flow on bridge structures, lacking an assessment of the impact of changes in bridge status on traffic operation status. However, a timely and accurate understanding of bridge information, effective evaluation of the impact of changes in bridge status on traffic conditions, and the timely adoption of effective measures play an important role in ensuring the safe and reliable operation of traffic and improving the operational efficiency of the highway network. Therefore, this article starts with the impact of bridges on traffic operation, and uses an algorithm model that combines Analytic Hierarchy Process (AHP), Delphi method, and Singular Value Decomposition (SVD) to establish a model for evaluating the impact of bridge traffic operation status, which is applied in practical engineering projects. This model can timely and accurately understand bridge information, effectively evaluate the impact of changes in bridge status on traffic conditions, and play an important role in ensuring the safe and reliable operation of traffic and improving the operational efficiency of the highway network.

2. Requirements for Detection Indicators of Bridge Facility Operation Status

In actual bridge monitoring, real-time monitoring is preferred for large bridges, and this monitoring method generates many data indicators that need to be screened for indicators related to traffic operation status. This article designs a Hierarchical Bridge Evaluation model, as shown in Figure 1.

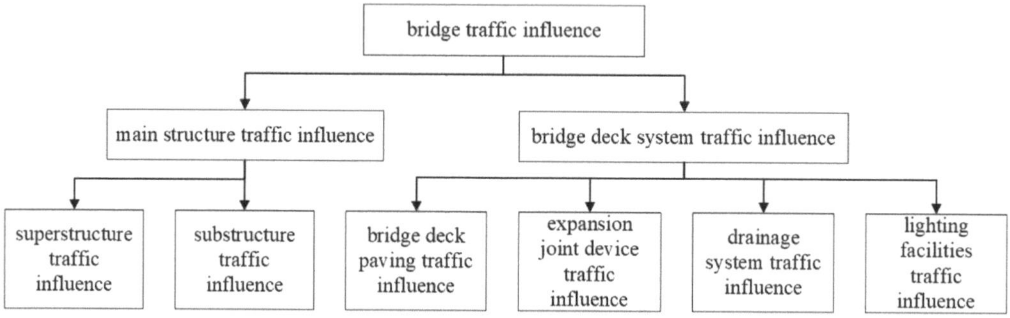

Figure 1. Hierarchy of bridge traffic impact assessment.

2.1. Determination of Existing Information and Analysis of Factors Affecting Traffic Operation Status

The existing information about large bridges is defined as the data for the current evaluation object held in various information systems included in the road network. From the perspective of information access, analysis and processing, and system evaluation, the existing information about large bridges can be divided into two categories: static information and dynamic information:

(1) Static information refers to the data of regarding the attributes of routes and bridges included in the road network, which are generally not changed after the completion of the bridge.
(2) Dynamic information refers to the real-time monitoring results of the dynamic monitoring system for large bridges in the highway network, the detection results of the maintenance, management, and monitoring of systems, and various results of daily inspections. This part of the data usually has a certain update frequency according to management needs.

The factors affecting bridge traffic are mainly divided into two categories: the impact of each component of the main structure and each component of the bridge deck system on traffic.

Among them, the impact of each component of the main structure on traffic includes:

(1) Superstructure traffic-influencing factors refer to the detection results of the bridge located above the support, including the bridge span structure and bridge deck structure.
(2) Substructure traffic-influencing factors refer to the detection results of the supporting structure of the bridge located below the support.

The impacts of various components of the bridge deck system on traffic include:

(1) Traffic-influencing factors of bridge deck paving refer to the detection results for bridge deck pavement.
(2) Traffic-influencing factors of expansion joint devices refer to the detection results for expansion joint devices in bridge deck systems.
(3) Traffic-influencing factors of drainage system refer to the results of an inspection of the drainage system in the bridge deck system.

(4) Traffic-influencing factors of lighting facilities refer to the results of an inspection of the lighting facilities on bridge decks.

2.2. Classification of Existing Factors Influencing Information Transportation

The existing information types and various traffic-influencing factors that have been identified can be basically divided into two categories based on their impact on traffic operation status. One category is considered to have direct influencing factors, and the other category is considered to have indirect influencing factors. When determining the weight of the impact of infrastructure on traffic, the weight determination methods can be divided into two categories: the expert scoring method and experimental determination method.

Determine the weights of the impacts of various traffic-influencing factors on existing information using expert scoring and experimental analysis methods. Among them, the expert scoring method mainly focuses on determining the weights of the impacts of traffic-influencing factors that are closely related to other factors, such as the long detection cycle of bridge pavement scales. The experimental analysis method mainly focuses on the traffic influencing factors that can be measured in the field and lead to direct changes in vehicle speed and driving. By measuring relevant parameters and designing corresponding experimental processes, the weights of the impacts of the above traffic-influencing factors can be measured.

This article analyzes various traffic influencing factors of existing information, and combines experimental data from the analysis method and expert scoring results to obtain the traffic influencing factors and degree of bridge infrastructure, as listed in Table 1. The degree of impact increases from 0 to 1. The classification and definition of each scale standard in this Table are consistent with the current "Technical Condition Evaluation Standards for Highway Bridges".

Table 1. Factors and degrees of impact on bridge infrastructure traffic.

Influencing Factors	Traffic Influence Degree (0–1)				
	Class 1	Class 2	Class 3	Class 4	Class 5
superstructure	0.00	0.15	0.25	0.40	1.00
substructure	0.00	0.15	0.25	0.40	1.00
bridge deck paving	0.00	0.10	0.20	0.25	-
expansion joint device	0.00	0.10	0.20	0.30	-
drainage system	0.00	0.05	0.15	0.30	-
lighting facilities	0.00	0.05	0.10	0.15	-

Note: A degree of impact of 0 indicates no impact on traffic, while a degree of impact of 1 indicates a significant impact on traffic and the need to interrupt it.

2.2.1. Traffic-Influencing Factors of the Superstructure

The results of the inspection of the superstructure of the bridge are classified into Class 1–5 based on their technical condition evaluation level, as shown in Table 2:

Table 2. Traffic-influencing factors of bridge superstructure inspection results.

Superstructure Inspection Results	Specific Meaning
Class 1	The technical condition of the main components of the superstructure is intact, and the evaluation results are all Class 1.
Class 2	The technical condition assessment results of the main components of the superstructure are Class 2 or below, with good technical condition of each component, no obvious deformation, only a small number of cracks, no defects or detachment, and no impact on normal driving.

Table 2. *Cont.*

Superstructure Inspection Results	Specific Meaning
Class 3	The technical condition assessment result of the main components of the superstructure is Class 3 or below, with obvious downward deflection, deflection less than the limit value, or individual components experiencing bending deformation, slight vibration or shaking when driving.
Class 4	The technical condition assessment result of the main components of the superstructure is Class 4 or below, and the components have obvious permanent deformation, defects or deformations less than or equal to the specification value.
Class 5	The technical condition assessment result of the main components of the superstructure is Class 5, with deflection or other deformations greater than the limit value, causing significant permanent deformation of the structure and significantly affecting driving safety.

2.2.2. Traffic-Influencing Factors of the Substructure

The results of the inspection of the substructure of the bridge are classified into Class 1–5 based on their technical condition evaluation level, as shown in Table 3:

Table 3. Traffic-influencing factors of bridge substructure inspection results.

Substructure Inspection Results	Specific Meaning
Class 1	The technical condition of the main components of the substructure is intact, and the evaluation results are all Class 1.
Class 2	The technical condition assessment results of the main components of the substructure are Class 2 or below, with good technical condition of each component, no obvious deformation, only a small number of cracks, no defects or detachment, and no impact on normal driving.
Class 3	The technical condition assessment result of the main components of the substructure is Class 3 or below, with obvious downward deflection, deflection less than the limit value, or individual components experiencing bending deformation, slight vibration or shaking when driving.
Class 4	The technical condition assessment result of the main components of the substructure is Class 4 or below, and the components have obvious permanent deformation, defects or deformations less than or equal to the specification value.
Class 5	The technical condition assessment result of the main components of the substructure is Class 5, with deflection or other deformations greater than the limit value, causing significant permanent deformation of the structure and significantly affecting driving safety.

2.2.3. Traffic-Influencing Factors of the Bridge Deck Paving

The results of the inspection of the bridge deck paving are classified into Class 1–4 based on their technical condition evaluation level, as shown in Table 4:

Table 4. Traffic-influencing factors of bridge deck paving inspection results.

Bridge Deck Paving	Specific Meaning
Class 1	The evaluation results of the technical condition of the bridge deck paving are all on Class 1 and in good condition.
Class 2	The evaluation of the technical condition of bridge deck paving has reached Class 2, with local cracks, damages, and deformations, but within a reasonable state.

Table 4. *Cont.*

Bridge Deck Paving	Specific Meaning
Class 1	The evaluation results of the technical condition of the bridge deck paving are all on Class 1 and in good condition.
Class 2	The evaluation of the technical condition of bridge deck paving has reached Class 2, with local cracks, damages, and deformations, but within a reasonable state.

2.2.4. Traffic-Influencing Factors of the Expansion Joint Device

The results of the inspection of the expansion joint device are classified into Class 1–4 based on their technical condition evaluation level, as shown in Table 5:

Table 5. Traffic-influencing factors of expansion joint device inspection results.

Expansion Joint Device	Specific Meaning
Class 1	The evaluation results of the technical condition of the expansion joint device are all on Class 1 and in good condition.
Class 2	The evaluation of the technical condition of expansion joint device has reached Class 2, with local cracks, damages, and deformations, but within a reasonable state.
Class 3	The evaluation of the technical condition of expansion joint device has reached Class 3, with some cracks, damages, and deformations, reaching or approaching the standard values.
Class 4	The evaluation result of the technical condition of the expansion joint device has reached Class 4, with significant cracks, damages, and deformations, the number of which is greater than the standard value.

2.2.5. Traffic Influencing Factors of the Drainage System

The results of the inspection of the drainage system are classified into Class 1–4 based on their technical condition evaluation level, as shown in Table 6:

Table 6. Traffic-influencing factors of drainage system inspection results.

Drainage System	Specific Meaning
Class 1	The evaluation results of the technical condition of the drainage system are all Class 1 and in good condition.
Class 2	The evaluation result of the technical condition of the drainage system has reached Class 2, and there are local drainage blockages, blockages, or damages in the drainage system, but within a reasonable state.
Class 3	The evaluation result of the technical condition of the drainage system has reached Class 3, with some local drainage problems such as obstruction, blockage or damage, reaching or approaching the standard values.
Class 4	The evaluation of the technical condition of the drainage system has reached Class 4, and there are obvious local drainage blockages, blockages, and damages in various components of the bridge deck, with a quantity greater than the standard value.

2.2.6. Traffic-Influencing Factors of the Lighting Facilities

The inspection results of the lighting facilities are classified into Class 1–4 based on their technical condition evaluation level, as shown in Table 7:

Table 7. Traffic-influencing factors of lighting facilities inspection results.

Lighting Facilities	Specific Meaning
Class 1	The evaluation results of the technical condition of the lighting facilities are all on Class 1 and in good condition.
Class 2	The evaluation result of the technical condition of the lighting facilities has reached Class 2, and there are a few deficiencies in the lighting facilities, but within a reasonable state.
Class 3	The evaluation result of the technical condition of lighting facilities has reached Class 3, and there are many deficiencies in the lighting facilities, reaching or approaching the standard value.
Class 4	The evaluation result of the technical condition of lighting facilities has reached Class 4, with a large number of lighting facilities missing and exceeding the standard value.

3. Impact Assessment Model for Traffic Operation Status of Large Bridges

Based on the established indicators of infrastructure, the degree of impact of each indicator item on traffic, and the weight values of each traffic impact indicator, this paper proposes a traffic state impact model that combines the Analytic Hierarchy Process (AHP), the Delphi method, and Singular Value Decomposition (SVD) to evaluate the impact of infrastructure state changes on traffic operation status. The overall process of this evaluation system is shown in Figure 2.

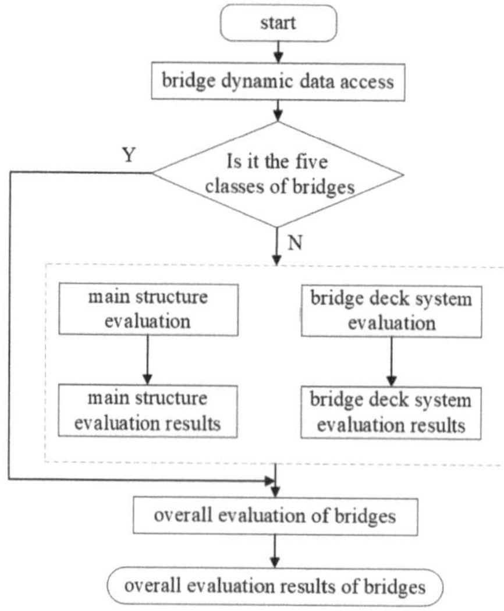

Figure 2. Flowchart of bridge traffic impact assessment system.

3.1. Calculation of Traffic Impact on the Bridge

The weights of the influence of different components of the bridge are ω_i, $i = 1, 2 \ldots$. The results of the inspection of different components of the bridge are a_i, $i = 1, 2 \ldots$. Then, the impact of the bridge on traffic is:

$$y(f) = 1 - \sum_{i=1}^{n} \omega_i a_i \tag{1}$$

The maximum traffic capacity of the road section is:

$$C_s = C \times (1 - y(f)) \tag{2}$$

Among them, C_s is the current maximum traffic capacity of the evaluated section, C is the traffic capacity designed for the evaluated section, and $y(f)$ is the traffic impact value from the evaluation section.

The degree of traffic impact on different influencing factors of bridges is f_{bi}, $i = 1, 2$. The weights of the influence of different influencing factors are β_i, $i = 1, 2$. The traffic impact result is recorded as f_2. The value of the impact f_2 of the bridge on traffic can be calculated using the following formula:

$$f_2 = \begin{cases} 1 - \sum\limits_{i=1}^{2} \beta_i (1 - f_{bi}) & f_{b1} < 1.0,\ f_{b2} < 1.0 \\ 1 & f_{b1} = 1.0 \end{cases} \tag{3}$$

In the formula, $f_{b1} = 1.0$ refers to the maximum impact of the main structure inspection results on traffic, and f_{b2} refers to the impact of the bridge deck system inspection results on traffic.

3.2. Calculation of Traffic Impact on the Main Structure

The different influencing factors of the main structure of the bridge and the degree of impact of the traffic are x_i, $i = 1, 2$, respectively. The weights of the influence of different influencing factors are β_{1i}, $i = 1, 2$. The traffic impact result is recorded as f_{b1}. The value of the impact f_{b1} of the main structure of the bridge on traffic can be calculated using the following formula:

$$f_{b1} = \begin{cases} 1 - \sum\limits_{i=1}^{2} \beta_{1i} (1 - x_i) & x_i < 1.0 \\ 1 & x_i = 1.0 \end{cases} \tag{4}$$

In the formula, $x_1 < 1.0$ represents the maximum impact of the superstructure detection results on traffic, and $x_2 < 1.0$ represents the maximum impact of the substructure detection results on traffic.

3.3. Calculation of Traffic Influencing Factors for Bridge Deck Systems

The degree of impact on transportation of different influencing factors in bridge deck systems is x_i, $i = 1, 2, 3, 4$. The influence weights of different influencing factors are β_{2i}, $i = 1, 2, 3, 4$. The traffic impact result is recorded as f_{b2}. The impact value f_{b2} of the bridge deck system on traffic can be calculated using the following formula:

$$f_{b2} = 1 - \sum_{i=1}^{4} \beta_{2i} (1 - x_i) \tag{5}$$

The traffic impact degree values of each component of the bridge deck system are expressed as a percentage, and the weighted sum of the impact weight and impact degree is used to obtain the overall impact degree value of the bridge deck system.

4. Theory and Method of Traffic Operation Status Evaluation

According to the analysis of the traffic-influencing factors mentioned earlier, the dynamic evaluation of bridge traffic operations mainly includes the real-time operation status and the impact of the bridge itself on the operation status. In the long run, the working status of various components of the bridge is in a dynamic and changing state, which will generate a series of dynamic parameter data about the bridge; however, in a relatively short period of time, the state of various components of the bridge can be seen as unchanged to a certain extent. If there is a sudden change under certain conditions, it will have a significant impact on the traffic operation status of the bridge in a short period of time.

The key to multi-objective system evaluation lies in how to scientifically and reasonably weight and comprehensively calculate various indicators, and integrate a multi-objective problem into a single-indicator form [14]. This article improves the Analytic Hierarchy Process (AHP) and Delphi combined with Singular Value Decomposition (SVD) to obtain the weight values of each evaluation-influencing factor in the evaluation model. A partial pseudocode of the algorithm can be found in Appendix A.

4.1. AHP

The Analytic Hierarchy Process (AHP) decomposes elements related to decision-making into levels such as goals, criteria, and plans, and makes decisions based on these levels. The method is concise and flexible. The Analytic Hierarchy Process (AHP) is a comprehensive evaluation method that combines qualitative and quantitative analysis. Its main idea is to decompose complex problems into several levels and compare the importance of each two indicators, create a judgment matrix, and calculate the weight values of different factors in the system by calculating the eigenvalues and eigenvectors of the judgment matrix [15].

In the impact assessment of bridge traffic operation status, according to the theory of AHP, the influencing factors of the bridge components to be evaluated can be divided into n different levels of factor sets, $U = \{u_1, u_2, \ldots, u_n\}$. The corresponding weight set is $W = \{w_1, w_2, \ldots, w_n\}$.

In addition, in order to effectively evaluate the overall impact of the bridge on traffic, based on the importance of the impact on traffic, the overall traffic impact of the bridge is divided into two categories: the results of the main structure inspection and the results of the bridge deck system inspection. As shown in Table 8:

Table 8. Traffic-influencing factors in lighting facilities inspection results.

Influencing Factors	Specific Meaning
Main structure inspection results	The results of daily, regular, special, and specialized inspections of the upper and lower structures of bridges are classified into Class 1–5.
Bridge deck system inspection results	The inspection or testing results of various components of the bridge deck system are divided into Class 1–4.

Establish a comment set $V = \{v_1, v_2, \ldots, v_n\}$ based on the above table. Based on the factor set and comment set, define the evaluation result set $R = \{r_1, r_2, \ldots, r_n\}$, which includes:

$$r_i = \sum_{j=1}^{n} \sum_{k=1}^{m} v_{ij} w_{jk} \tag{6}$$

In the formula, the selection of weight vector W is carried out using the Delphi method.

4.2. Delphi Method

The Delphi method, also known as the expert inquiry survey method, is a decision-making method that involves conducting several rounds of anonymous consultations to concentrate the opinions of expert groups and ultimately make conclusions that are in line with development trends [16]. The prediction and evaluation leadership group summarizes and organizes the opinions in each round, and sends them as reference materials to each expert for analysis and judgment for them to propose new opinions. After repeated attempts, the opinions tend to be consistent, resulting in a more reliable and unified conclusion. The essence of the Delphi method is to use the knowledge and experience of a collective of experts to seek multiple opinions from a group of relevant experts for complex problems that cannot be directly quantitatively analyzed, and to obtain measurement conclusions [17].

Due to the anonymity, feedback, and statistical characteristics of the Delphi method, the statistical analysis and feedback of expert opinions during the survey process fully reflect the role of information feedback and information control [18]. Therefore, this article uses the Delphi method to determine the weight of bridge traffic impact. The specific flowchart is shown in Figure 3:

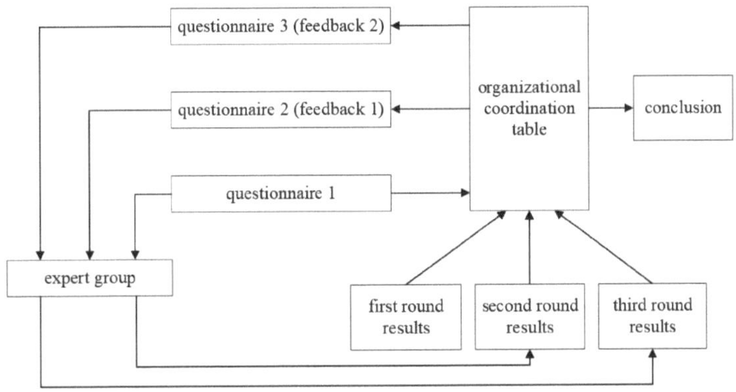

Figure 3. Application flowchart of Delphi method.

Based on the results of the Delphi method, the traffic impact weights of the influencing factors of the bridge, the main structure of the bridge, and the bridge deck system are finally determined. The specific steps are as follows:

(1) Define judgment matrix $B = (b_{ij})_{n \times n}$ and $b_{ij} = 0$; $i, j = 1, 2 \ldots n$, and eigenvectors $\beta = (\beta_1, \beta_2, \ldots \beta_n)^T$ and $\beta_i = 0$, $i = 1, 2 \ldots n$:

(2) Based on the expert experience database, compare different traffic influencing factors $A_1, A_2 \ldots A_n$ pairwise and rank them in a non-decreasing manner, denoted as $A_1 \geq A_2 \ldots \geq A_n$;

(3) Compare the sorted non-decreasing sequence $A_1 \geq A_2 \ldots \geq A_n$ with A_i and A_{i+1} based on $i = 1, 2 \ldots n-1$, and write the scaling value t_i;

(4) Construct judgment matrix B:
Assign initial values to i, j; let $i = 1, j = 1$, and $i, j = 1, 2 \ldots n$;
If $i = j$, then $b_{ij} = 1$;
If $i \neq j$, then $b_{ij} = t_i \times t_{i+1} \times \ldots \times t_{j-1}, (i \neq j)$, and $b_{ij} = \frac{1}{b_{ji}}$;

(5) Solve the matrix to obtain the maximum eigenvector $\lambda_{\max}$ and its corresponding eigenvector β;

(6) Normalize feature vectors, whose corresponding values are the weights in the non-decreasing sequence $A_1 \geq A_2 \ldots \geq A_n$.

Based on the results of the expert scoring method, establish a judgment matrix B. Using the above method, determine the traffic impact weights of each factor in the bridge deck system, as listed in Tables 9–11.

Table 9. Bridge-influencing factors and traffic impact weights.

Influencing Factors	Traffic Influence Weight
main structure	0.8
bridge deck system	0.2

Table 10. The influence weight of the main structure of the bridge.

Influencing Factors	Traffic Influence Weight
superstructure	0.5
substructure	0.5

Table 11. The influence weight of the bridge deck system.

Influencing Factors	Traffic Influence Weight
bridge deck paving	0.40
expansion joint device	0.25
drainage system	0.20
lighting facilities	0.15

4.3. SVD

The Singular Value Decomposition (SVD) is an important matrix factorization method in linear algebra, which has important applications in signal processing, statistics, and other fields [19].

In the establishment of the bridge traffic impact assessment model, the relative importance values of each evaluation factor are obtained through the Delphi method. In order to facilitate the calculation of the impact degree later, decoupling and normalization processing are needed, and the SVD method is used here [20].

Firstly, construct the initial matrix. The data obtained from the Delphi method expert scoring are a set of relative importance values, which are a set of one-dimensional data. In order to apply the SVD algorithm, a two-dimensional matrix needs to be constructed. If the original data are $b_i(i = 1, 2, \ldots, r)$ and r is the number of evaluation model factors, then the constructed matrix is $A \in R^{r \times r}$. The construction method is as follows:

(1) $a_i = 1(i = 1, 2, \ldots, r)$. The diagonal elements of the matrix have values of 1;
(2) If $i \neq j(i, j = 1, 2, \ldots, r)$, set $k = i : 1 : j - 1(i, j = 1, 2, \ldots, r)$, then $a_{ij} = a_{ij} * b_k$;
(3) If $a_{ji} \neq 0$, then $a_{ij} = 1/a_{ji}$.

This constructs an $r \times r$ matrix and also solves the problem of data coupling in the original one-dimensional array. By solving matrix $A \in R^{r \times r}$, a set of independent data are obtained.

Secondly, the singular value decomposition algorithm is applied to decompose matrix $A \in R^{r \times r}$. In order to maintain the sum of the weights of each factor as 1, after performing singular value decomposition on the matrix, the results also need to be normalized. The implementation steps are as follows:

(1) Solve the eigenvectors pv and the eigenvalues po of the matrix;
(2) Find the maximum eigenvalue and the eigenvector corresponding to the maximum eigenvalue;
(3) Normalize the eigenvector.

By following the steps above, the weight values of each evaluation factor in the evaluation model can be calculated. Since the model is implemented using the Analytic Hierarchy Process, this method has been used more than once in model evaluation, and will not be repeated here.

5. Simulation and Results of Impact of Parameters on Bridge Traffic Operation

5.1. Classification of Bridge Traffic Impact Levels

According to some regulations in the "Technical Condition Evaluation Standards for Highway Bridges", the impact of changes in the main structure and deck system of bridges on traffic is divided into four levels.

Level 1: Changes in the status of road and bridge infrastructure have a minor impact on traffic, with the bridge capacity changing from 90% to 100% of its original design capacity.

Level 2: Changes in the status of road and bridge infrastructure have a certain impact on traffic, with the bridge capacity changing from 80% to 89% of its original design capacity.

Level 3: Changes in the status of road and bridge infrastructure have a significant impact on traffic, with bridge capacity changing from 70% to 79% of its original design capacity.

Level 4: Changes in the status of road and bridge infrastructure have a significant impact on traffic, with the bridge capacity changing by less than 69% compared to the original design capacity.

5.2. Introduction to Simulation Environment

This article chooses to simulate the section of road from Yujiawan in Yubei District to Beibei, with a length of approximately 23 km, including the Ma'anshi Bridge. The simulation experiment is conducted using TransModeler software (version 6.0) [21,22], with a simulation time of 1 h (10:00–11:00). The occurrence of each obstacle started at 10:20 and ended at 10:40. The free flow speed of the Chongqing Wuhan Expressway is set to 120 km/h, with a free flow speed of 70 km/h for the bridge. The simulation time is from 10:00 to 11:00, and the entrance flow of the simulated section is 2500 vehicles/h and the exit flow is 2500 vehicles/h.

To simulate and analyze the impact of changes in road infrastructure on traffic capacity, changes in sensor data are used to simulate changes in bridge status on simulated road sections. Three traffic conditions are set for road capacity, excellent, good, and poor, and the impact of changes in traffic operation influencing factors on traffic operation status is verified. The specific meanings of the different road traffic capacities are as follows:

(1) Excellent: The road and bridge facilities have not undergone any changes, or have undergone minor changes, and these changes have almost no impact on the vehicles driving on the road. Vehicle traffic is basically smooth, and variables such as traffic flow, speed, and lane occupancy rate change steadily.
(2) Good: The road and bridge facilities have undergone certain changes. By changing the sensor data on the road, the state of the bridge facilities can be simulated to change. For example, obstacles can be set between sensors, and the maximum speed for passing through this obstacle is set to 60 km/h. However, the impact is not significant, and vehicle traffic is relatively smooth. The variables of traffic flow, speed, and occupancy rate change steadily, with occasional small changes, but the fluctuations are not significant.
(3) Poor: There have been significant changes in road and bridge facilities, such as a malfunction in a section of the road between sensors, resulting in a speed of 0 km/h and a significant impact on vehicle traffic, leading to poor traffic flow. The variables of traffic volume, speed, and occupancy rate have shown obvious abrupt changes.

5.3. Simulation Results and Analysis

We simulated three different conditions of road capacity, compared the conditions of good and bad traffic conditions with the conditions of excellent traffic conditions, and added a comparison of adjacent lane changes. Figures 4–7 show the comparison results of the number of vehicles and average speed collected during good traffic conditions with excellent traffic capacity:

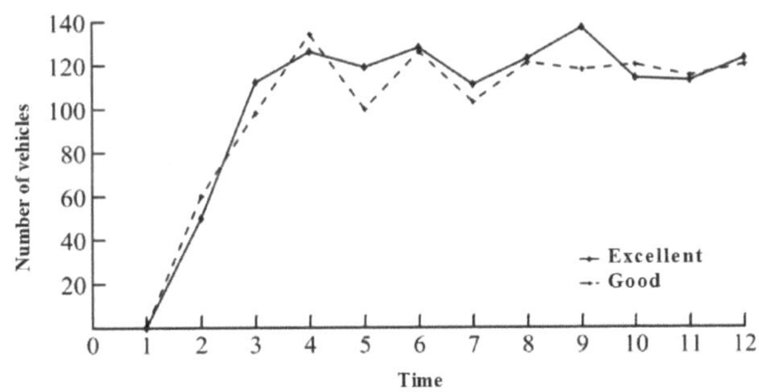

Figure 4. The number of vehicles recorded by sensor 378 when the traffic status is excellent or good.

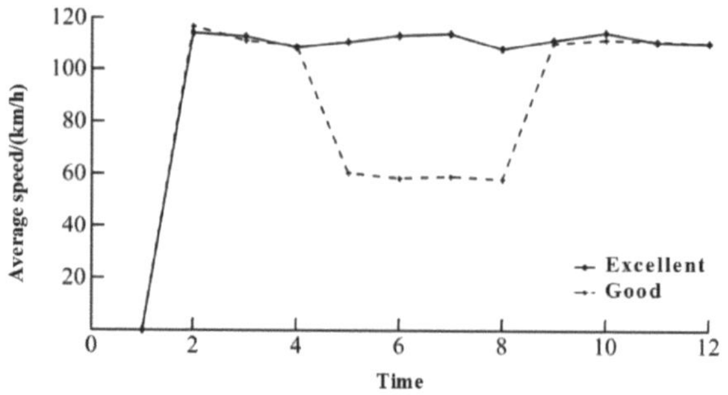

Figure 5. The average vehicle speed recorded by sensor 378 when the traffic status is excellent or good.

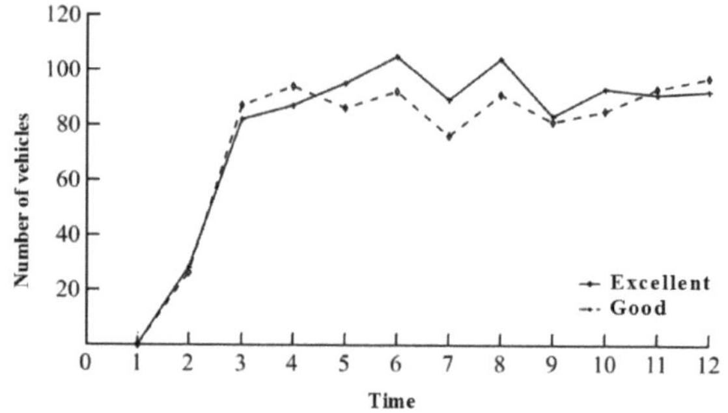

Figure 6. The number of vehicles recorded by sensor 379 when the traffic status is excellent or good.

It can be clearly seen from Figures 4 and 5 that, when there are slight changes in road and bridge facilities, the number of vehicles passing through the road does not change significantly and the changes are relatively stable. It has had a certain impact on the speed of the vehicle, with some changes occurring, but the changes are not significant, which can ensure the normal passage of the vehicle. From Figures 6 and 7, it can be seen that, when there are slight changes in the highway bridge facilities, they also have a certain impact on adjacent lanes, but the changes are not significant, and vehicles can pass through relatively smoothly.

In summary, when there are minor changes in bridge facilities, it will not have a significant impact on the driving conditions of vehicles, and they can pass smoothly as a whole. The number of vehicles passing through remains stable, and the speed is also higher than the minimum speed specified for highways (60 km/h).

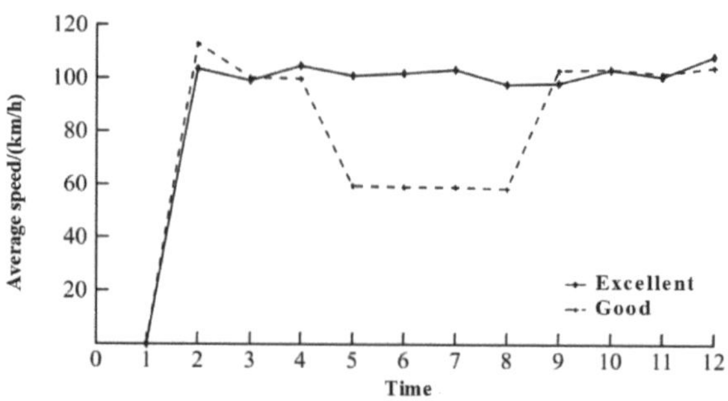

Figure 7. The average vehicle speed recorded by sensor 379 when the traffic status is excellent or good.

Figures 8–11 reflect the comparison of results of the number of vehicles and average speed collected when the traffic capacity is poor compared to when the traffic capacity is excellent:

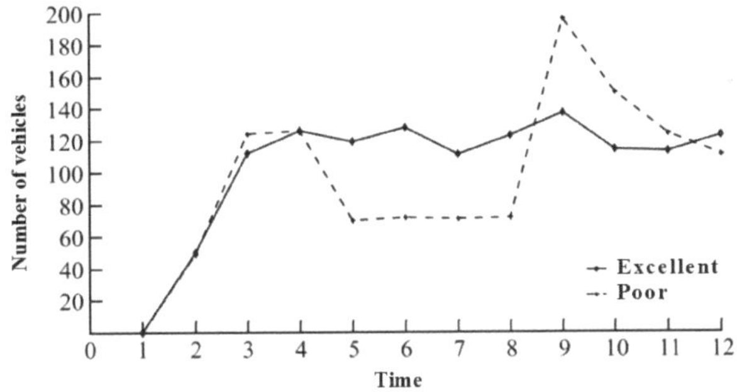

Figure 8. The number of vehicles recorded by sensor 378 when the traffic status is excellent or poor.

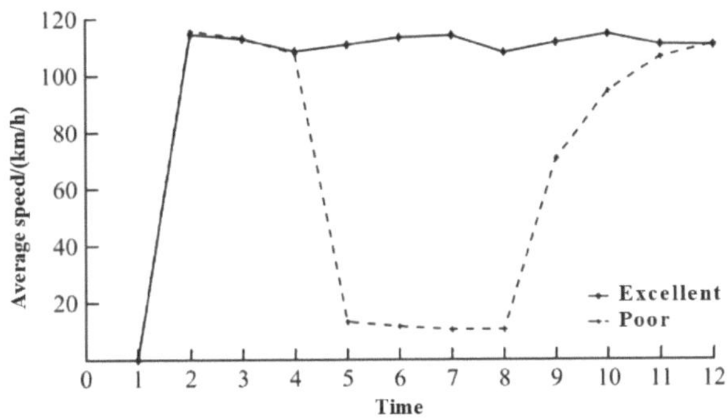

Figure 9. The average vehicle speed recorded by sensor 378 when the traffic status is excellent or poor.

From Figures 8 and 9, it can be seen that when there are significant changes in road and bridge facilities, the number of vehicles passing through the road undergoes a sudden change, and the number of vehicles rapidly increases, resulting in significant road congestion and seriously affecting the efficiency of vehicle traffic. From Figures 10 and 11, it can be seen that when the bridge infrastructure undergoes significant changes, it also has a significant impact on adjacent lanes, causing vehicle traffic to be obstructed and resulting in congestion.

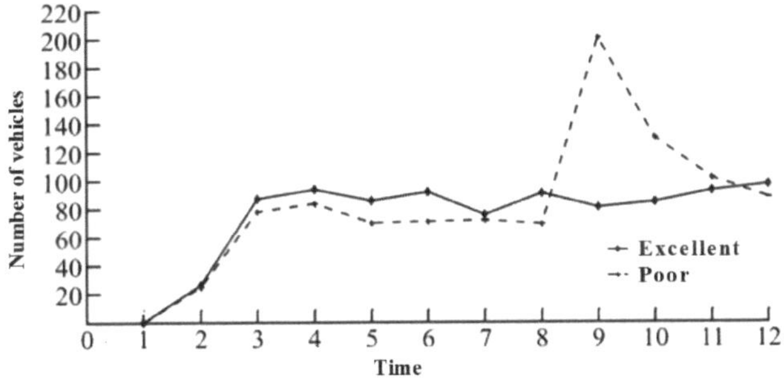

Figure 10. The number of vehicles recorded by sensor 379 when the traffic status is excellent or poor.

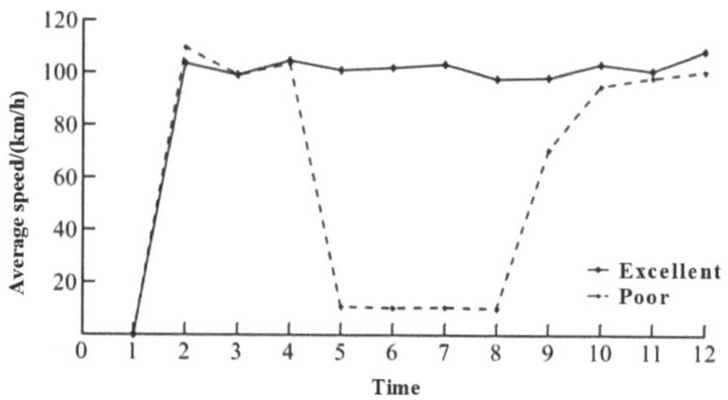

Figure 11. The average vehicle speed recorded by sensor 379 when the traffic status is excellent or poor.

In summary, when there are significant changes in bridge infrastructure, it has a significant impact on the driving conditions of vehicles. Overall, the traffic efficiency is low, the number of vehicles passing through is small, and it leads to serious congestion in adjacent lanes, resulting in a sharp decrease in vehicle passing speed.

Looking at the simulation results of the entire traffic model, it can be seen from the data collected by sensor 378 that when there are slight changes in the road and bridge infrastructure, the road capacity will be affected to a certain extent, but the overall impact is not significant and the recovery is relatively fast. When there are significant changes in the infrastructure of roads and bridges, the road capacity will undergo drastic changes, which will have a huge impact on vehicles and traffic, and the recovery of traffic will also be relatively slow.

According to the data recorded by sensor 379 arranged for adjacent lanes, when there are significant changes in bridge infrastructure, the road capacity of adjacent lanes will also be affected. This conclusion conforms to the general rules of transportation operation and has high reliability and accuracy.

6. Conclusions

This article aimed to promote the automation of bridge traffic impact management, effectively evaluate the impact of bridges on traffic operation status, maintain bridges in a timely manner, and avoid traffic accidents by modeling the bridge traffic impact assessment system, screening evaluation model factors, and processing parameters. On this basis, a bridge traffic operation status evaluation model is proposed based on existing bridge data, with the Analytic Hierarchy Process (AHP) as the main model structure, combined with the Delphi and Singular Value Decomposition (SVD) methods. However, the current research on the operation status of bridge traffic is still not sufficient, and improvements can be made to the following aspects in the future: Firstly, there is still significant room for optimization in terms of the accuracy and applicability of the model, and parameter refinement can be carried out according to different regions and types of roads and bridges. Secondly, reference can also be made to the actual road conditions and relevant standards of different countries and industries to further consider more traffic influencing factors and achieve more accurate assessments.

Author Contributions: Conceptualization, J.G. and Y.L.; methodology, Y.Z.; software, G.Y.; validation, L.W. and Z.D. All authors have read and agreed to the published version of the manuscript.

Funding: Shaanxi Natural Science Basic Research Program (No. 2022JQ-735).

Institutional Review Board Statement: Not applicable.

Informed Consent Statement: Not applicable.

Data Availability Statement: The original contributions presented in the study are included in the article, further inquiries can be directed to the corresponding authors.

Conflicts of Interest: Author Guanhu Yuan was employed by the company Huawei Technologies Co., Ltd. The remaining authors declare that the re-search was conducted in the absence of any commercial or financial relationships that could be construed as a potential conflict of interest.

Appendix A

Algorithm A1: Traffic Impact Assessment of Large Bridges

Input: X
Output: F

1. Initialization: β
2. **for** i = 1:n
3. **for** j = 1:m
4. $fbj = 1 - \sum_{1}^{j} \beta j(1 - xj)$
5. **End for**
6. $fi = 1 - \sum_{1}^{i} \beta i(1 - fbi)$
7. **End for**
8. **Return** F

Algorithm A2: Determination of Bridge Traffic Impact Weights

Input: A
Output: W

1. Initialization: A
2. **for** i = 1:n
3. Sort (A) --> A1 ≥ A2 ≥ A3 . . . ≥ An
4. **End for**
5. **For** i = 1:n
6. Compare (Ai, Ai + 1) --> T (t1, t2, t3 . . . tn)
7. **End for**
8. Compute (B):
9. **If** i = j, Bij = 1
10. **Else** Bij $= b_{ij} = t_i \times t_{i+1} \times \ldots \times t_{j-1}, (i \neq j)$
11. Eig (B) -- > β
12. W = Normalization (β)
13. **Return** W

References

1. Zhou, J.T.; Zheng, D. Reflections on Bridge Safety Guarantee. *Chin. Highw.* **2020**, *21*, 30–35. [CrossRef]
2. Liu, J. On the Current Situation and Development Trends of Road and Bridge Construction Technology. *Bulk Cement* **2024**, *2*, 95–97.
3. Ge, L.; Dan, D.; Koo, K.Y.; Chen, Y. An improved system for long-term monitoring of full-bridge traffic load distribution on long-span bridges. *Structures* **2023**, *54*, 1076–1089. [CrossRef]
4. Wang, D.; Zhao, Y.; Wang, J.F.; Wang, Q.; Liu, X. Establishment and effect analysis of traffic load for long-span bridge via fusion of parameter correlation. *Structures* **2023**, *55*, 1992–2002.
5. Li, C.; Zhong, J.W.; Wang, Y.F.; Yu, L. Prediction and evaluation of bridge extreme effects under heavy traffic loads. *Bridge Constr.* **2023**, *53*, 71–78. [CrossRef]
6. Shen, Q.J.; Wang, Z.J.; Xing, X.L.; Zhang, Y.; Gao, L.; Chen, L. Analysis of Traffic Operation Safety Impact of Long Span Bridges Based on Cross Wind Effect. *Chin. Foreign Highw.* **2024**, *44*, 115–122. [CrossRef]

7. Xiong, D.; Li, Y.Y.; Geng, Z.S. Research on Highway Traffic Safety Control Technology Based on Vehicle Operation Status. *Traffic Eng.* **2024**, *24*, 16–20. [CrossRef]
8. Saafi, N.; Dhouib, K. An Ontological Model to Enhance Traffic Conditions in Smart City Domain. *Spectr. Eng. Manag. Sci.* **2024**, *2*, 70–84. [CrossRef]
9. Kan, H.Y.; Li, C.; Wang, Z.Q. An integrated Convolutional Neural Network-Bidirectional Long Short-Term Memory-Attention Mechanism model for enhanced highway traffic flow prediction. *J. Urb. Dev. Manag.* **2024**, *3*, 18–33. [CrossRef]
10. Xu, Z.F.; Song, J.; Zhang, X.; Zhang, J. Application of a digital twin system considering measured traffic flow in the structural state evaluation of cable-stayed bridges. *Highw. Transp. Technol.* **2023**, *40*, 93–102.
11. Yang, D.H.; Guan, Z.X.; Yi, T.H.; Li, H.N.; Ni, Y.S. Fatigue evaluation of bridges based on strain influence line loaded by elaborate stochastic traffic flow. *J. Bridge Eng.* **2022**, *27*, 04022082. [CrossRef]
12. Wei, Y.; Zhang, T.K.; Ruan, X.; Jin, Z. Analysis of Vehicle Load Effects on Long span Bridges under Mixed Traffic Flow. *J. Tongji Univ.* **2023**, *51*, 1859–1867+1909.
13. Ho, H.; Nishio, M. Evaluation of dynamic impact factor of existing bridges with road surface damages based on dynamic response under traffic flow loading. *Struct. Infrastruct. Eng.* **2024**, *20*, 1390–1403. [CrossRef]
14. Chen, P.Y. A New Composite Method for Multi indicator Comprehensive Evaluation and Its Testing. *Stat. Decis. Mak.* **2021**, *37*, 33–37. [CrossRef]
15. Leal, J.E. AHP-express: A simplified version of the analytical hierarchy process method. *MethodsX* **2020**, *7*, 100748. [CrossRef]
16. Rahmani, A.; Vaziri Nezhad, R.; Ahmadi Nia, H.; Rezaeian, M. Methodological principles and applications of the Delphi method: A narrative review. *J. Rafsanjan Univ. Med. Sci.* **2020**, *19*, 515–538. [CrossRef]
17. Barrios, M.; Guilera, G.; Nuño, L.; Gómez-Benito, J. Consensus in the delphi method: What makes a decision change? *Technol. Forecast. Soc. Chang.* **2021**, *163*, 120484. [CrossRef]
18. Fu, J.J. Research on Key Detection Indicators of Rail Transit Bridges Based on Delphi Method. *Highw. Automob. Transp.* **2022**, *1*, 147–150.
19. Dongarra, J.; Gates, M.; Haidar, A.; Kurzak, J.; Luszczek, P.; Tomov, S.; Yamazaki, I. The singular value decomposition: Anatomy of optimizing an algorithm for extreme scale. *SIAM Rev.* **2018**, *60*, 808–865. [CrossRef]
20. Cao, Y.J. Comparative study on identifying moving loads using conjugate gradient least squares method and singular value decomposition method. *Henan Sci. Technol.* **2023**, *42*, 27–30. [CrossRef]
21. Feng, C.; Chen, H.; Dai, J.L.; Zou, X.L. Simulation Evaluation of Guangzhou Airport High Speed Flow Control Based on Transmodeler. *Highw. Automob. Transp.* **2020**, 32–34+45.
22. Chun, G.; Rouphail, N.; Samandar, M.S.; List, G.F.; Yang, G.; Akçelik, R. Analytical and microsimulation model calibration and validation: Application to roundabouts under sight-restricted conditions. *Transp. Res. Rec.* **2023**, *2677*, 274–288. [CrossRef]

logistics

MDPI

Article

Analysing the Influence of Augmented Reality on Organization Performance via Supply and Logistics Value Chain Functions: A Hybrid ANN-PLS Model Assessment in the Gulf Cooperation Council Region

Ahmad Aburayya

MBA Department, College of Business, City University Ajman, Ajman P.O. Box 18484, United Arab Emirates; a.aburrayya@cu.ac.ae; Tel.: +971-6-711-0000

Citation: Aburayya, A. Analysing the Influence of Augmented Reality on Organization Performance via Supply and Logistics Value Chain Functions: A Hybrid ANN-PLS Model Assessment in the Gulf Cooperation Council Region. *Logistics* **2024**, *8*, 110. https://doi.org/10.3390/logistics8040110

Academic Editors: Tomasz Nowakowski, Artur Kierzkowski, Agnieszka A. Tubis, Franciszek Restel, Tomasz Kisiel, Anna Jodejko-Pietruczuk and Mateusz Zając

Received: 3 October 2024
Revised: 24 October 2024
Accepted: 25 October 2024
Published: 5 November 2024

Abstract: *Background*: Despite the resurgence of interest in augmented reality (AR) due to Industry 4.0 and its ability to resolve several challenges faced by current business models, comprehensive research examining the capabilities of AR in supply chain management (SCM) and logistics remains limited. This article aims to investigate the potential effects of AR technology on organizational performance through the mediation role of SCM and logistics value chain functions to address the existing knowledge gap. *Methods*: This research employed a cross-sectional design and an explanatory survey as a deductive approach for hypothesis development. The primary data collection method involved the self-administration of a questionnaire to furniture suppliers located in the Gulf Cooperation Council (GCC), including six countries. Of the 656 questionnaires submitted to suppliers, 483 were considered usable, yielding a response rate of 73.6%. The research utilized partial least squares structural equation modelling (PLS-SEM) and artificial neural network (ANN) techniques to evaluate the gathered data. *Results*: The current paper's statistical evidence demonstrates that AR implementation has a positive impact on the supply and logistics value chain activities and organizational performance of furniture suppliers in the GCC region. Moreover, it illustrates that the design and planning variable of supply chain value dominates as the primary predictor of organization performance. The results indicated that the ANN strategy provided a more comprehensive explanation of internally generated constructs compared to the PLS-SEM technique. *Conclusions*: This study demonstrates its usefulness by advising furniture industry decision-makers on what to avoid and what aspects to consider when creating plans and regulations. The report also suggests operations managers apply machine learning (ANN) for prediction and decision-making in supply and operations value chains. This essay looks at how the AR and resource-based supply value chain view may affect company performance across countries, firm sizes, and ages.

Keywords: augmented reality; artificial intelligence; supply and logistics value chain; chain functions; organization performance; machine learning; artificial neural network analysis

1. Introduction

Augmented reality (AR) is a revolutionary technology that can profoundly influence various business domains, including operations management, supply chain management (SCM), and marketing management [1,2]. Indeed, it ranks among the most rapidly advancing technologies and will integrate into our daily existence. This technology is blurring the distinctions between the digital and physical realms [3,4]. It superimposes multiple layers of digital information onto the surrounding environment, rendering it rich, significant, and interactive [5,6]. Consumers are enthusiastic in utilizing AR technology for shopping [2]. Retailers are employing AR technology to deliver an exceptional and engaging shopping experience for their clientele [4]. The Amazon AR application enables customers to perceive a virtual representation of several real-world products and assess their appearance inside their home environment. Ikea Place, an application by Ikea, enables users to visualize Ikea furniture in their homes [3,6].

AR's ability to present data on three-dimensional screens allows for intricate and captivating data visualization techniques [4]. Utilizing virtualized platforms, individuals may see, evaluate, and collaborate on their data within their designated virtual environment [5]. AR can aid production personnel by delivering immediate information regarding equipment failures, maintenance concerns, and safety alerts. AR enables supply chain enterprises to improve digital experiences for both employees and customers [2,4,7]. These encompass improved repair and maintenance functionalities in manufacturing, storage, and logistics [2]. These immersive technologies will improve product visualization for consumers, as well as store layout and planning for businesses [8]. Businesses can utilize several technologies, like artificial intelligence (AI), virtual reality (VR), and AR, to acquire insights on their customers' interests and activities. Companies can utilize this data for supply chain demand predictions, targeted marketing, and customized product recommendations [8,9]. As a result of the swift progression of information technology, AR has gained significant strategic relevance and has emerged as a crucial asset for numerous organizations [4,5,8,9].

For supply chain systems to stay competitive in a constantly shifting environment and volatile markets, they must include developing technology and transform into sustainable operations. Businesses suffer severe consequences if they are unable to adapt to this fast-paced environment and fierce competition [10]. In summary, digital technology has changed how businesses operate and transformed many industries. This new trend opens up new opportunities for businesses and has a big impact on SCM [2,11]. Companies must restructure and modify their supply chain strategy in order to fully realise their digitalisation potential. SCM is essential across all enterprises, applications, and sectors because it facilitates the procurement of accurate product and service data. This, consequently, enhances the delivery of superior services and communication protocols for users [12,13]. SCM efficiently manages an increased number of data points, therefore reducing the likelihood of undesirable problems and failures in service delivery to users. The implementation of digital techniques such as AI, AR, and VR enhances the ability to store, access, process, and manage data in SCM, leading to improved efficiency for applications and organizations to bolster the firm's competitive advantage [1,4,7,14,15]. Numerous studies [5,9,16–19] indicate that the application of big data methodologies improves the ability to store, access, analyse, and manage data in supply SCM.

In SCM, the increase in energy costs and the continued use of outdated manufacturing methods have significantly raised operational costs. Consequently, corporations found themselves motivated to reduce production costs while upholding their quality standards within a specific range. AR is an emerging technology that provides cost-effective solutions to the rising operational expenses of enterprises. This technology assists various participants in supply chains, including "truck drivers, warehouse personnel, supervisors, and managers, by overlaying digital information onto the physical environment". This computer-generated information aids players in monitoring the movement of items from one location to another throughout a supply chain. The use of AR in SCM is progressively transforming traditional, sluggish, and paper-dependent logistics and supply chain operations into a rapid and technology-oriented sector [5,15].

Prior studies investigated the application of AR in SCM and logistics, concentrating on assessing the efficacy of AR solutions in particular and constrained process stages via a systematic literature review [5,15,20]. Still, there is not a full study of how it can be used and how it affects different supply chain management tasks, based on real-world data from the manufacturing industry around the world, especially in the Gulf Cooperation Council (GCC) region. Despite the rapid advancement and increasing popularity of AR across various human activities [6,21,22], a thorough literature review did not exist. We executed a comprehensive initiative in the GCC region to address this research deficit and ascertain new research domains and applications. We conducted a systematic literature review and survey-based research that encompassed not only definite procedure phases in SCM and logistics, such as order picking, but also all SCM functions and processes, including "warehousing, manufacturing, outdoor logistics, planning and design, and personnel training". The aim of this study was to evaluate the maturity level of AR technology utilized by suppliers in the furniture sector. Furthermore, it sought to assess the impact of this technology on suppliers' organizational performance, which is mediated by the prospective effects of SCM functions.

The main purpose of this preliminary section is to provide the reader with an overview of the research that is presented through the body of this work. It commences with a brief discussion of the background and the significance of the study, followed by an exploration of the literature review, research methodology, data analysis, and a discussion of the findings.

2. Literature Review

AR in SCM

Digitization refers to the conversion of operations, functions, models, processes, or activities through the utilization of digital technologies. Digitization facilitates new business models and serves as a significant catalyst for innovation, possessing the capacity to instigate the forthcoming wave of innovation [10,23]. The era of digital transformation has introduced numerous novel technologies that significantly affect supply chains [5]. People call technology that merges the real and virtual worlds 'AR' [24]. It describes a real-world environment augmented by visualization and specialized hardware and software [25]. AR is an emerging technology that offers cost-effective solutions to enterprises' rising operational expenses. This technology assists various participants in supply chains, including truck drivers, warehouse personnel, supervisors, and managers, by overlaying digital information onto the physical environment [15]. In fact, these computer-generated data aid players in monitoring the movement of commodities along a supply chain. AR applications in enterprises are progressively revolutionizing conventional, sluggish, and paper-dependent logistics and supply chain operations into a rapid and technology-oriented sector. Pick-and-pack services, collaborative logistics, maintenance services, procurement, and last-mile delivery are sectors presently employing AR technology [26–28].

Supply chain systems must integrate emerging technology and evolve into sustainable operations to remain competitive in a continuously changing environment and turbulent markets. The inability to adjust to this rapid climate and intense competition leads to dire repercussions for businesses. In essence, digital technology has revolutionized numerous industries and altered company operations. This emerging tendency generates new chances for enterprises, significantly affecting supply chain management. Businesses cannot completely achieve their digitization potential without reorganizing and adapting their supply chain strategy. This reimagined supply chain needs to be more interconnected, scalable, intelligent, and expedited compared to conventional supply chains [15,26]. Over 1 billion AR-enabled smartphones and tablets are in use, according to Koul [28]. Thus, supply and logistics organizations can benefit from AR technology without waiting for low-cost glasses [4]. AR technologies utilized in logistics and SCM vary. These include marker-based, marker-less, projection-based, and overlaid AR.

AR can be utilized in contemporary SCM, which contains numerous layers of sourcing, planning, multi-modality, and information exchanges at every step. The emergence of new business models involves mass personalization. Production and operational processes must evolve to enable specialization and customization. This evolution will involve the dynamic reconfiguration and adjustment of machine and replenishment stock locations and production line architecture to provide a flexible workplace and ecology [15]. AR can improve supply chain and logistics efficiency [29], adding value and improving user experiences. AR-based industrial systems should enable storage, maintenance, and assembly. This paper systematically reviews AR technologies' potential to digitize supply chains and enable smart industries [30].

A previous study by Stoltz et al. [27] indicated that the rapid rise in e-commerce transactions, the need to reduce inventory, and increased customer responsiveness have made storage strategy increasingly important for businesses. For instance, a company's warehouse operation strategy is vital to supply chain performance. Computer technologies and ICT have solved warehouse management problems for decades. New technology like AR will change warehouse management. AR can be used in warehouse operations for receiving, storage, order picking, and shipping. Koul [28] found that sensors in packing materials allow companies to assess their products' market success. Based on sensor data in packaging materials, a manufacturer can change commodity production in real time, according to the author. Data scientists and logistics specialists in the supply chain network can gain new information from things anywhere. If sales match estimates, producers can decide whether to reduce production and focus distribution on high-demand locations to reduce losses.

Attaran [5] reviewed the literature to determine how digital technologies might enable, improve, and streamline digital supply chain (DSC) performance. That study found that AR and VR can present data in 3D, enabling detailed and compelling data visualization. Virtualized platforms let users see, analyse, and collaborate on their data. Manufacturing workers can use AR to obtain real-time machinery malfunction, maintenance, and safety warnings. Supply chain companies may improve staff and consumer digital experiences with AR and VR. These include manufacturing, storage, and logistics repair and maintenance improvements. Immersive technologies will improve customer product visualization and retailer store layout and planning. Saunders [8] further supported this finding by demonstrating how AR/VR can enhance supply chains. Managers can monitor manufacturing in real time. They can constantly check facilities, distribution centres, and warehouses to maintain process efficiency; speed up and reduce errors in order picking; and provide the finest transportation routes to optimize delivery efficiency and on-time delivery.

Accordingly, the subsequent statement concisely outlines the principal contributions of the present work. First, this study examined the maturity level of AR technology used by furniture industry suppliers. Employing a creative and comprehensive study methodology will improve our understanding of the elements affecting AR tool application. We developed a theoretical model utilizing the technology readiness level model (TRLM) to underscore the significance and predictability of the results. This study employed all SCM functions as mediators to investigate the impact of AR adoption on organizational performance. To our knowledge, no current research has examined how various aspects facilitate AR deployment and how this implementation may affect SCM functions' ability to achieve long-lasting performance in the furniture sector. As a result, there is a compelling need to increase research on AR deployment and its impact on organizational performance. Second, based on the primary goal of the current work, this subject holds considerable importance for both furniture suppliers and customers. Within the furniture industry, both vendors and clients are increasingly motivated to enhance their benefits and embrace a wider range of recommendations. Both suppliers and customers recognize the impact of AR's acceptability and adoptability on their technology usage, which extends beyond their decision-making processes. Third, this study uniquely used the TRLM to examine the importance of AR in the furniture field, particularly in SCM. This study differentiates itself from prior research by Rejeb et al. [2], Attaran [5], Cao et al. [20], and Demir et al. [15], which focused on availability and mobility, by assessing its influence on suppliers' organization performance through the TRL model. This study fundamentally examined AR's effectiveness from a strictly supplier-centric perspective. The National Aeronautics and Space Administration (NASA) has developed a well-established model that this work applied, setting it apart from previous studies and encouraging the adoption of such technologies. In fact, this is the first study that applied such a model in the SCM context. Finally, this study utilized an innovative and versatile method of data analysis by employing a machine learning (ML) analytical tool. Almarzouqi et al. [31] widely acknowledge ML as the most efficient instrument for predicting different dependent variables in various business contexts.

3. Theoretical Framework Development

3.1. AR Applications in SCM

This study employed a conceptual framework to systematically categorize the potential of AR applications in SCM functions and then moved on to examine its effect on organization performance. AR technology's adaptability transcends a singular organizational role, facilitating many activities across the value chain. Consequently, we ground our approach in Porter's [32] value chain model. A value chain model serves as a fundamental tool, breaking down a business into strategically relevant and value-enhancing activities. "It links an organization's supply side (which includes raw materials, inbound logistics, and production processes) with its demand side (outbound logistics, marketing, and sales), which includes supporting activities (firm infrastructure, human resource management, technology development, and procurement)" [2]. We modified the value chain model in our research to show the progression of activities from planning and design to sales and external logistics. In essence, several scholars, such as Rejeb et al. [2], adopted the value chain model in the current AR-SCM literature and classified AR applications in SCM as warehousing, manufacturing, sales and outdoor logistics, planning and design, and human resource management.

Advanced production computer and communication tools are transforming organizations. These developments greatly affect product design, factory automation, cost and quality control, planning, processing, workforce skills, and output management [11,33,34]. According to Petruse et al. [35], AR-aided manufacturing involves more workers in production. AR can aid in product conceptualization and design by enabling staff to view computer-generated content in a factory setting; help companies move to sustainable manufacturing; improve product development efficiency; reduce faults, waste, and resource consumption; and speed up manufacturing and delivery. In addition, several authors stressed the importance of AR-assisted assembly instruction and supervision [36]. Real-time AR with tracking technologies and a display in the operator's field of view improves assembly design and planning, simulates product assembly and disassembly before manufacturing, and provides virtual instructions for monitoring and guidance [37]. AR can help maintenance personnel by enabling visual interactions and superimposing virtual production equipment instructions. AR technology can increase remote maintenance in severe environments and workflow efficiency compared to paper instructions [2,38,39].

The widespread existence of multi-product and multi-brand enterprises and manufacturing fragmentation drive product visualization and identification. Visualizing or digitizing a product, also called a digital twin, can reduce production errors and non-conformance costs [28,34,40]. AR solutions replicate reality and enable data-driven visualization. AR settings can also give operators 3D images of the target object (e.g., a warehouse shelving system) and relate it to reality, such as a warehouse facility. An AR solution can reduce errors and damage and allow visual management and monitoring of warehouse products moving to the assembly bay [11,28]. AR apps also provide mobility, position, pervasiveness, and context awareness data to help logistics track smart things. For example, an AR-based smart palletization solution improves warehouse visibility and navigation. The solution reduces palletization errors, increases productivity, and improves product identification and visibility [34,41]. In SCM, order picking is one of the most labour-intensive warehouse tasks because of the time required to handle each item and travel between sites. According to Chen et al. [42], order picking systems generally suffer from crowded temporary storage, long wait times, inefficient operations, and poor communication. AR enhances order picking efficiency by lowering item retrieval times and increasing output. Compared to pick-by-voice, AR's navigation capability reduces product location search times and mispacks [2].

Outside logistics includes all operations needed to deliver products to customers in the outdoors. Transportation relies on digital data and planning tools. AR can improve logistics, package identification, truck sequencing, loading, and routing, according to Ginters and Gutierrez [40]. Indeed, sensing devices and AR systems work together to improve outdoor logistics processes, ensuring security, product identification, and item presentation. These capabilities can boost international trade because merchandise shipments must meet industry standards and government laws and trade operations' paperwork can be more reliable. In addition, AR technology enables creative business models like value co-creation with customers, helping businesses stand out, customize their offerings, and gain a competitive edge. Ro et al. [43] showed that AR-enabled wearable gadgets boost brand value by facilitating consumer engagement. Therefore, AR technologies improve product presentation and enable unique and realistic situations to boost marketing and sales.

Facility layout is crucial to long-term operational efficiency. A practical layout can help smart production and digital manufacturing succeed [44], increasing flexibility, efficiency, and optimization. AR technologies are transforming companies into flexible, reconfigurable manufacturing systems [44,45] using AR technology's immersive nature. Using AR methods, AR helps companies swiftly examine their storage space. This lets them overlay the layout plan and superimpose virtual elements on the real-world layout facilities for an optimal arrangement. A library of robots, machines, and racks can be used to blend their functionality and location with real equipment and things. This method optimizes layout design and reduces costs from inefficient routings, machine installations, and manufacturing and storage space usage. Thus, AR technology can simplify layout planning and design by delivering a clear, precise image. This helps make decisions and trade-offs when multiple designs conflict [44,45]. Additionally, AR tools help clients visualize product design and adjustments before buying. Thus, AR can improve users' product design comprehension and engage them in a new value co-creation paradigm. Thus, AR tools attempt to involve clients in mass customization and integrate industrial design. They also help design and create items [2,44]. Rohacz and Strassburger [24] demonstrated a Daimler AG application that enables final assembly logistics planning using AR. Mobile intralogistics planning helps systems using AR technology like smartphones and tablets that are convenient and effective.

AR tools also help organizations consider modern technological innovations to maintain workforce training and expand employee knowledge, skills, and competency to contribute to value-generating activities. Recent research by Gangabissoon et al. [46] emphasized using AR to improve participation in training. They emphasized the use of technology in learning and development, indicating a growing trend of integrating creative methods into training approaches. AR's unique qualities can improve training, according to Gangabissoon et al. [46]. AR allows students to interact with real items under virtual instructions in mobile just-in-time learning. Using an AR system for training in real time can save time and money, improve training processes, and provide interactive feedback [47,48]. Businesses can streamline training with AR-based solutions, providing real-time, location-independent capabilities to inexperienced personnel during seminars and workshops [47]. To accelerate learning, add AR to supply chain and logistics training. Research by Hořejší [49] suggested that AR architecture can accelerate learning for organizations with significant worker turnover, resulting in faster learning for more employees.

The link between AR and SCM functions has been extensively studied in various organizations and sectors. The benefits of AR implementation will drive suppliers to maintain their technology, impacting their operations. This suggests the following hypothesis.

H1: *Implementing AR in SCM has a positive impact on supply value chain activities.*

3.2. SCM and Organization Performance

Organizational performance refers to a company's capacity to achieve its financial and strategic goals in relation to its competitors [50]. It refers to the manner in which a firm evolves in terms of the quality and quantity of its organizational activities, as well as how these advancements have favourable direct or indirect financial impacts on the firm. The literature presents numerous metrics to evaluate corporate success. Researchers have analysed it from multiple perspectives. Financial performance is a common term for evaluating corporate performance and is assessed by a variety of indicators, including return on investment and return on assets. Furthermore, scholars have examined non-financial success through many variables, including customer satisfaction, employee engagement, and social responsibility [44,50,51]. The alternative viewpoint, termed the resource-based view, asserts that a company's resources and capabilities influence its performance. Resources and capabilities encompass technology, corporate repute, and highly skilled personnel [52]. The final method is dynamic capabilities, which denote a firm's capacity to adjust to its business environment in order to attain greater performance [53]. Cao and Zhang [54] proposed assessing business performance by sales growth, profit margin on sales, return on investment, operating revenue, and market share.

Previous research examined a positive correlation between SCM functions and organizational achievement [16,55–58]. People recognize supply chain digital capabilities for their significant impact on enhancing corporate performance. By investing in and developing supply chain analytics skills, a company acquires the capacity to derive important insights from the extensive data produced throughout the supply chain. Currently, nearly every sector generates substantial amounts of data on a continuous basis. Companies can pinpoint their bottlenecks and modify their plans in response to market fluctuations using innovative technologies. This will lead to increased customer satisfaction and improved profitability. The adoption of digital supply chains entails more than just new technology. It goes further. Several scholars advise the corporation to link its digital initiatives with its supply chain goals. Every company recognizes the potential of the most recent digital supply chain technology. It can boost company performance and set the stage for outperforming competitors [56]. To achieve exceptional organizational performance, the corporation must implement digital supply chain techniques. Heizer et al. [50] and Alicke and Forsting [59] found that many organizations want to improve their supply chains but are only using a few digital technologies. Most enterprises believe that deploying digital SCM could increase earnings before interest and taxes and raise their yearly revenue. Thus, the subsequent hypotheses were proposed.

H2: *Digitalized value chain activities in SCM have a positive impact on organization performance.*

H3: *Digitalized value chain activities in SCM significantly moderate the relationship between AR and organization performance.*

3.3. Model Formulation

Figure 1 illustrates the development of a research theoretical framework. Based on the above three hypotheses, a theoretical model of supply chain value functions that mediates the relationship between AR and organizational performance was developed and is displayed in Figure 1. The links among AR, supply chain value functions, and organizational performance are incorporated in one single model. In these three hypotheses, AR is an independent variable and supply chain value functions are a mediating variable, while organizational performance is a dependent variable. In the three hypotheses, the relationships among the independent variable (AR), the mediating variable (supply chain value functions), and the dependent variable (organizational performance) are examined.

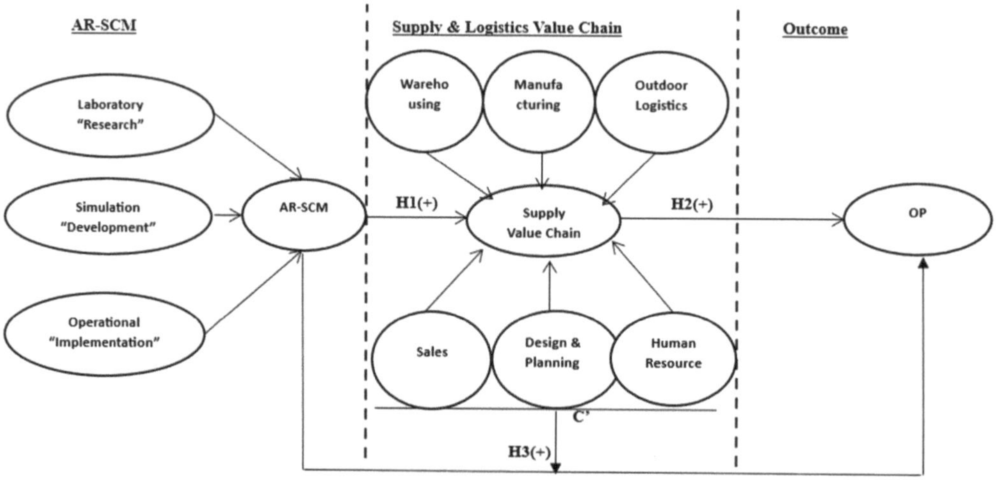

Figure 1. Study theoretical model. Note: AR-SCM: augmented reality-supply chain management; OP: organizational performance; C′: mediation effect.

4. Research Methodology

4.1. Data Collection

The need for furniture in the GCC is increasing significantly. Mordor Intelligence [60] predicts that the GCC furniture industry will grow by 7.54% from 2022 to 2029, reaching USD 20.88 billion. GCC countries have historically led in adopting new technology, especially in the furniture industry. Furniture products are in demand due to rising purchasing power and lifestyle trends. As more people use technologies like AI, AR, VR, data analysis, the internet of things (IoT), and more, the furniture industry is expanding. For example, furniture can be virtually designed using 3D modelling before it is physically built. This method enables the designers to preview the finished product in advance. When designing furniture, designers can test, experiment, and modify the design to meet the unique needs of the customers using a virtual prototype. It is possible to experience the texture, colour, location, load-bearing capacity, and aesthetics in advance. Overall, the GCC has 536 furniture and supply manufacturers (FSMs) [60,61].

The study employed a cross-sectional design with an explanatory survey to deductively formulate hypotheses, enhancing data variability and generalizability [62]. We employed quantitative research to evaluate models and hypotheses for conclusions and generalizations [63]. The primary data-gathering method for GCC FSMs was self-administered Google Form questionnaires. This research included 355 of the 536 registered FSMs, yielding a response rate of 66%, adequate for any SEM and ANN-ML investigation. In this investigation, we found that Hair et al. [64] employed a probability sampling technique utilizing a basic random systematic approach. The particular policies of the selected FSMs affected the choice of sampling methodology, as simple random sampling is an effective approach for time, cost, and resources for acquiring large samples [63]. The survey was disseminated to all SCM managers, directors, and senior management professionals who were formally engaged in SCM operations and possessed a comprehensive awareness of the financial and marketing dimensions of each participating organization. Consequently, we employed "the company" as the analytical unit. These experts represent a substantial repository of expertise on AR-SCM processes inside any supplier organization.

4.2. Questionnaire Design

The principal data-gathering tool utilized in this investigation was a structured questionnaire. This study's use of the NASA-developed TRL model underscores the necessity of AR in the furniture industry, particularly in SCM. NASA developed the TRL scale, a recognized metric that delineates the degree of technical progress [65,66]. This study employed the NASA TRL scale to integrate TRL criteria for AR systems intended for practical application. Additionally, the NASA TRL offers a spreadsheet application that allows the user to answer a series of questions pertaining to the AR technology program. The spreadsheet computes and displays the attained TRL. The calculator calculates the weighted geometric mean for all question categories possessing both a non-zero category TRL and a non-zero category weight as follows:

$$TRL_{wgtd} = \sum_{i=0}^{1} \sqrt[w_i]{\Pi_{L_i}^{w_i}}$$

where TRL_{wgtd} is the weighted geometric mean; w_i is the assigned category weights; with n being the number of categories included in the calculation; and L_i is the category TRLs.

The existing TRL scale comprises three primary components, namely, research, simulation, and operational implementation, which collectively encompass nine levels. Each level represents the degree of advancement in the development of a technology, starting from the first concept (level 1) and extending to its complete implementation in the market (level 9) [67]. We assessed the aspects described above using a 5-point Likert-type scale, where 1 signifies "strongly disagree" and 5 indicates "strongly agree". Practically speaking, this work introduces a new technique based on examples to classify and evaluate AR technology by mapping them onto the TRL, which indicates their level of maturity and availability.

Finally, research conducted by Blankemeyer et al. [7], Berkemeier et al. [14], Hammerschmid [39], Ustundag et al. [47], Schweigkofler et al. [68], Elbert and Sarnow [69], Hannola et al. [70], and Kokkas and Vosniakos [71] laid the foundation for the SCM and logistics value chain measure, which comprises 27 elements for six main factors: warehousing, manufacturing, outdoor logistics, sales, design and planning, and human resource activities. Akgün et al. [72] and Ellinger et al. [73] derived the organization performance (OP) scale, which consists of seven items. We assessed each of the SCM and logistics value chain and OP scores using a 5-point Likert-type scale, where 1 signifies "strongly disagree" and 5 indicates "strongly agree".

5. Data Analysis

5.1. FSMs' Demographic Information

The data shown in Table 1 indicate that the majority of the companies included in this study were situated in the United Arab Emirates (50.0%). Moreover, Table 1 shows that the FSMs exhibited a very uniform distribution of sizes and enjoyed a rather extensive use of AR systems. Approximately 75.8% of them are classified as medium in size; out of them, 65% have been utilizing an AR system for a duration exceeding 1 year. Regarding the respondents' job types, Table 1 shows that 50.0% of them are employed as SCM managers.

Table 1. FSMs' characteristics information.

Criterion	Factor	Frequency	Percentage (%)
Country	Saudi Arabia	135	38.0
	UAE	178	50.0
	Kuwait	13	3.7
	Oman	9	2.5
	Qatar	16	4.5
	Bahrain	4	1.3

Table 1. *Cont.*

Criterion	Factor	Frequency	Percentage (%)
Company size	Small	72	20.3
	Medium	269	75.8
	Large	14	3.9
Respondent's position	Logistics Manager	66	18.6
	Procurement Manager	43	12.0
	SCM Manager	178	50.0
	MRP/ERP Manager	9	2.5
	CEO	4	1.3
	Other	55	15.6
AR system utilization	Less Than 1 year	155	43.7
	1 To 5 Years	185	52.0
	More Than 5 Years	15	4.3
Total		483	100%

5.2. *Measurement Model Analysis*

When examining the measurement model, it was crucial to determine the construct validity, including convergent validity, and the construct reliability, inclusive of Cronbach's alpha. The following table (Table 2) displays Cronbach's alpha values ranging from 0.716 to 0.878, which, according to Nunnally and Bernstein [74], appear to exceed the established threshold of 0.7 for determining construct dependability. However, the results in the table confirmed the construct's robustness, establishing it as a construct devoid of errors. Based on the table, it is evident that the composite reality (CR) value exceeded the recommended threshold of 0.7 as proposed by Kline [75], as it fell within the range of 0.701 to 0.871. An essential aspect of assessing convergent validity is determining the average variance extracted (AVE) and component loading. The table shows that the factor loading values were higher than the usual 0.5 level and the AVR was between 0.722 and 0.788, which was also above the usual 0.7 level. This confirmed the construct validity and generally met the requirements for convergent validity.

Table 2. Internal consistency, FA, CR, and AVR tests.

Factor	Item	FA	CR	AVE	Cronbach's Alpha
AR-SCM	AR 1	0.734	0.871	0.764	0.878
	AR 2	0.687			
	AR 3	0.602			
	AR4	0.718			
	AR5	0.611			
	AR6	0.509			
	AR7	0.692			
	AR8	0.640			
	AR 9	0.799			

Table 2. *Cont.*

Factor	Item	FA	CR	AVE	Cronbach's Alpha
Warehousing	WR1	0.738	0.798	0.732	0.801
	WR2	0.682			
	WR3	0.705			
	WR4	0.599			
	WR5	0.831			
Manufacturing	MA1	0.764	0.846	0.722	0.811
	MA2	0.719			
	MA3	0.626			
	MA4	0.683			
	MA5	0.729			
	MA6	0.724			
Logistics	LOG1	0.637	0.728	0.751	0.716
	LOG2	0.633			
	LOG3	0.715			
Sales	SL1	0.599	0.701	0.772	0.724
	SL2	0.602			
	SL3	0.728			
Design and Planning	DP1	0.816	0.774	0.794	0.786
	DP2	0.727			
	DP3	0.826			
	DP4	0.773			
	DP5	0.783			
Human Resource	HR1	0.557	0.785	0.755	0.805
	HR2	0.637			
	HR3	0.828			
	HR4	0.789			
	HR5	0.783			
Organizational Performance	OP1	0.759	0.806	0.788	0.844
	OP2	0.738			
	OP3	0.824			
	OP4	0.790			
	OP5	0.834			
	OP6	0.748			
	OP7	0.689			

5.3. PLS-SEM Model Fit Indicators

We assessed the PLS-SEM goodness-of-fit for the study (Table 3). For the table below, the saturated model evaluated each construct's association with the others. The model structure was taken into consideration in the calculated model, which was based on a total impact scheme. The study's RMS theta value of 0.068 confirmed the validity of the PLS model. According to Hair et al. [64], the PLS-SEM model had a satisfactory fit when the RMS theta value fell between 0.00 and 0.12.

Table 3. Model fit indicators.

	Complete Model	
Indicator	Saturated Model	Estimated Model
SRMR	0.033	0.034
d_ULS	0.783	1.424
d_G	0.579	0.582
Chi-Square	441.337	446.793
NFI	0.848	0.852
RMS	0.068	

5.4. Structural Model and Hypothesis Testing

5.4.1. Hypotheses Testing Using PLS-SEM

This study offered an auxiliary way for hypothesis testing. Specifically, we used ANN techniques and PLS-SEM. This paper presents a methodology to assess the impact of AR-SCM on operational supply value chain (SVC), which in turn will affect OP. Based on well-known theory, Calcagno et al. [76] wrote that the PLS-SEM model should be used to test conceptual models and make predictions about dependent variables. The ANN algorithm efficiently identifies a dependent variable based on independent factors.

The route analysis, shown in Table 4, utilizing *p*-values, t-values, and path coefficients, empirically corroborated the study's hypotheses. Chin [77] categorized route coefficient values over 0.67 as strong, those ranging from 0.33 to 0.67 as moderate, and those between 0.19 and 0.33 as weak, while deeming any value below 0.19 as unsuitable for further investigation. According to the table results, AR in SCM strongly influenced SVC, which in turn positively affected OP. Consequently, we endorsed H1 and H2. Additionally, Table 5 corroborates the mediating function of the SVC between AR and OP. Consequently, we endorsed H3. Moreover, Table 6 below illustrates that the employed model possessed significantly high predictive capability for the variation in forecasting AR in SCM.

Table 4. Structural standardized path coefficients.

Model Relations	Std. Path Coefficient	t-Value	*p*-Value	Hs. Decision
AR → SVC	0.654	7.313 **	0.002	Accepted (+)
SVC → OP	0.472	4.262 **	0.006	Accepted (+)

** $p < 0.01$. Note: + (positive).

Table 5. Mediation analysis.

Mediation Relations	Test Statistic	St. Error	*p*-Value	Hs. Decision
AR > SVC > OP	3.280 **	0.094	0.001	Accepted (+)

** $p < 0.01$.

Table 6. R square of the endogenous latent variables.

Model Constructs	R Square	Predictive Power
AR on SVC	0.765	High
SVC on OP	0.734	High
AR and SVC on OP	0.682	High

5.4.2. Artificial Neural Network (ANN) Model Analysis

This study used the ANN model because it estimates accurately, making it preferable to PLS-SEM and regression modelling. ANN tools excel at modelling complicated connections with variable non-linear response values, leading to better predictions than traditional methods [78]. According to Lie'bana-Cabanillas et al. [79], ANN modelling finds latent non-linear practical ties in statistics and lets the modelling tool utilize the correlations on a new set of data. The ANN analysis was carried out operating SPSS. In the ANN analysis, only the relevant predictors derived from the PLS-SEM results were utilized. The ANN analysis considered the latent variables, as depicted in Figure 2 (network diagrams 1–3). Diagram 1 comprises one input neurone (AR) and one output neurone (SVC). Diagram 2 comprises one input neurone (SVC) and one output neurone (OP). Diagram 3 comprises two input neurones (AR and SVC "6 constructs") and one output neurone (OP). The results indicated that the predictive capability of the ANN analysis (R^2 = 83%) significantly surpassed that of PLS-SEM (R^2 = 68.2%). The results indicated that the ANN technique more effectively elucidated endogenous constructs compared to the PLS-SEM approach.

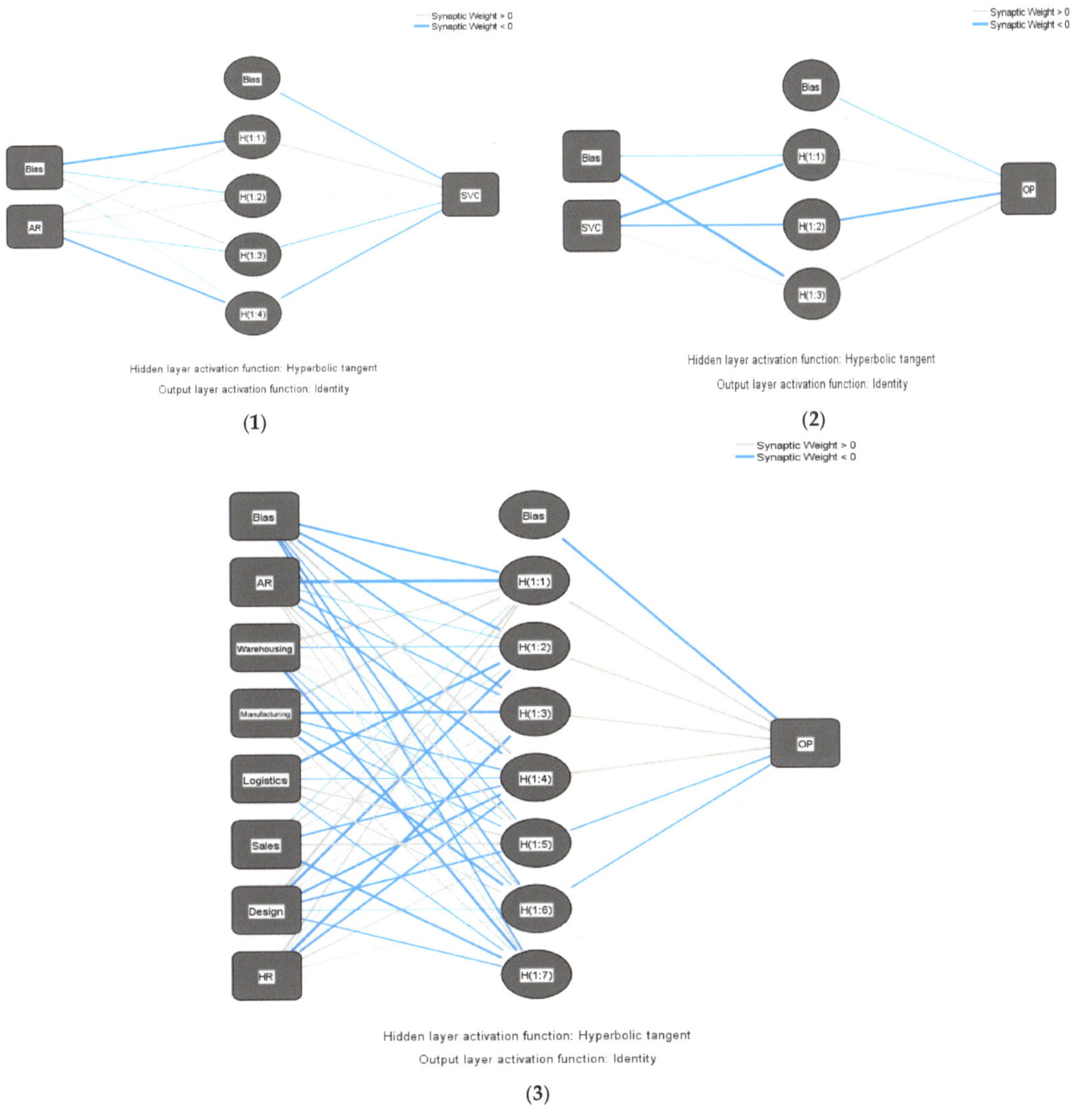

Figure 2. ANN model network diagrams (**1**–**3**).

To evaluate normalized significance, the average of each predictor was correlated with the highest mean value, represented as a fraction. Table 7 and Figure 3 display the average relevance and normalized significance of each predictor in the ANN models. Table 7's "sensitivity analysis" demonstrates that design and planning serve as the most significant predictors of OP.

Table 7. Independent variable importance of OP.

Construct	Importance	Normalized Importance
AR-SCM	0.079	17.6%
Warehousing	0.009	2.0%
Manufacturing	0.391	87.3%
Outdoor Logistics	0.043	9.6%
Sales	0.016	3.6%
Design and Planning	0.448	100.0%
Human Resource	0.014	3.2%

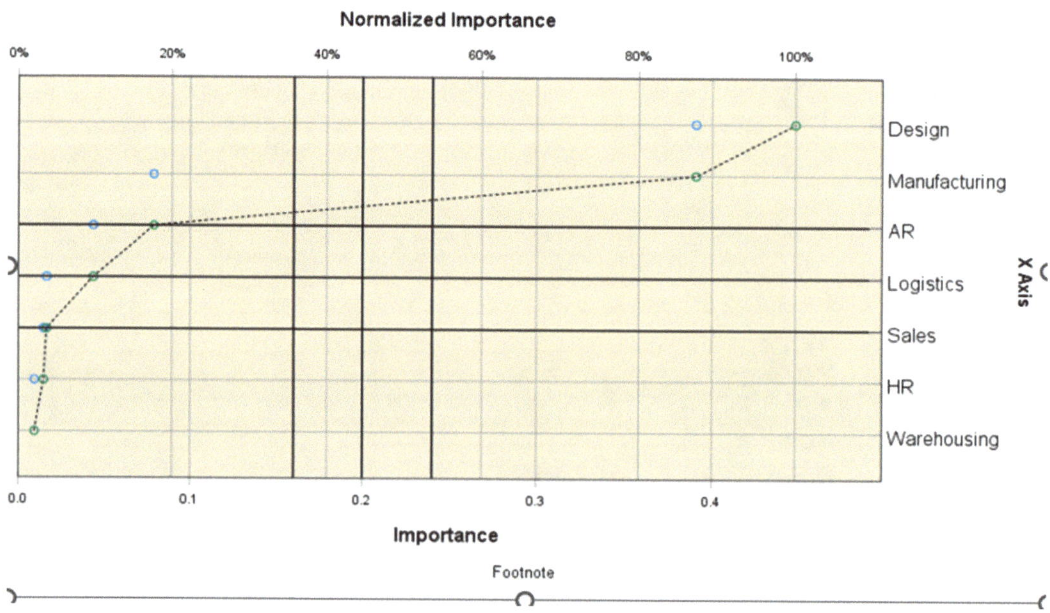

Figure 3. Independent variable importance.

6. Discussion

This study investigated the effects of AR on organizational performance in SCM furniture organizations as well as the role of supply and logistics value chain as a mediator. We used two well-known analysis models, PLS-SEM and ANN, as theoretical lenses. We found that the AR implementation process significantly improved SVC and OP among furniture suppliers in the GCC countries.

In fact, the study found that SVC was significantly affected by the adoption and the implementation of AR. Our findings are consistent with previous studies that showed that the integration of virtual elements into a real-world environment can help to enhance supply and logistics value chain activities [1,2,28,35,38,39].

Indeed, AR has several supply chain applications. AR enables novel value-added solutions for SCM and logistical processes. AR technologies can improve value chain efficiency and help firms overcome issues including poor planning and scheduling, process integration, and resource waste. The potentials we expect are that AR will enhance operations and reduce losses. The possibilities are based on primary and support tasks within the value chain, which include planning and designing the firm's infrastructure and supporting human resource management, inbound logistics, operations, marketing, and sales. These functions are projected to improve significantly with AR deployment, necessitating additional training. This study shows that AR's greater visualization, navigation, and immersive engagement can bridge the virtual and real worlds, simplify processes, and aid decision-making. AR pick-by-vision guiding often improves task performance and reduces workload. The organization may improve operating efficiency, cut expenses, and adapt to changes via AR-enabled visualization. AR helps logistics workers control industrial processes and gain situational awareness. This is possible with real-time data generation throughout the supply chain. We create a logical framework for AR implementation in a firm, bringing significant commercial value.

In detail, the results of the study specifically supported the assertion that organizations are changing as a result of sophisticated production computers and communication technologies. In essence, AR can aid in product conceptualization and design by enabling staff to view computer-generated content in a factory setting; help companies move to sustainable manufacturing; improve product development efficiency; reduce faults, waste, and resource consumption; and speed up manufacturing and delivery. In addition, several authors stressed the importance of AR-assisted assembly instruction and supervision [36]. Real-time AR with tracking technologies and a display in the operator's field of view improves assembly design and planning, simulates product assembly and disassembly before manufacturing, and provides virtual instructions for monitoring and guidance [37]. AR can help maintenance personnel by enabling visual interactions and superimposing virtual production equipment instructions. AR technology can increase remote maintenance in severe environments and workflow efficiency compared to paper instructions [1,38,39]. Furthermore, the study results proved that AR solutions replicate reality and enable data-driven visualization. AR settings can also give operators 3D images of the target object (e.g., a warehouse shelving system) and relate it to reality, such as a warehouse facility. An AR solution can reduce errors and damage and allow visual the management and monitoring of warehouse products moving to the assembly bay [11,28]. AR apps also provide mobility, position, pervasiveness, and context awareness data to help logistics track smart things. For example, an AR-based smart palletization solution improves warehouse visibility and navigation. The solution reduces palletization errors, increases productivity, and improves product identification and visibility [34,41].

The study's results further indicated that AR systems work together to improve outdoor logistics processes, ensuring security, product identification, and item presentation. These capabilities can boost international trade because merchandise shipments must meet industry standards and government laws and trade operations' paperwork can be more reliable. In addition, AR technology enables creative business models like value co-creation with customers, helping businesses stand out, customize their offerings, and gain a competitive edge. However, these results were consistent with Ro et al. [43]. They showed that AR-enabled wearable gadgets boost brand value by facilitating consumer engagement. Therefore, AR technologies improve product presentation and enable unique and realistic situations to boost marketing and sales. In addition, the study's findings confirmed that AR helps companies swiftly examine their storage space. This lets them overlay the layout plan and superimpose virtual elements on the real-world layout facilities for an optimal arrangement. A library of robots, machines, and racks can be used to blend their functionality and location with real equipment and things. This method optimizes layout design and reduces costs from inefficient routings, machine installations, and manufacturing and storage space usage. Thus, AR technology can simplify layout planning and design by delivering a clear, precise image. This helps make decisions and trade-offs when multiple designs conflict [44,45]. These findings are consistent with recent research. Rohacz and Strassburger [24] provide an example of a Daimler AG application that uses augmented reality to facilitate logistics planning for final assembly. AR-enabled mobile intralogistics planning apps for tablets and smartphones are practical and efficient. AR tools are also beneficial.

This study, moreover, indicated that organizations have had to consider modern technological innovations to maintain workforce training and expand employee knowledge, skills, and competency to contribute to value-generating activities. This argument mirrors recent research by Gangabissoon et al. [46], which emphasized using AR to improve participation in training. Furthermore, Al Harthy [80] emphasised the use of technology in learning and development, indicating a growing trend of integrating creative methods into training approaches. AR's unique qualities can improve training, according to Gangabissoon et al. [46]. AR allows employees to interact with real items under virtual instructions in mobile just-in-time learning. Using an AR system for training in real time can save time and money, improve training processes, and provide interactive feedback [47,48]. Therefore, businesses can streamline training with AR-based solutions, providing real-time, location-independent capabilities to inexperienced personnel during seminars and workshops [47]. To accelerate learning, add AR to supply chain and logistics training. Research by Hořejší [49] suggested that AR architecture can accelerate learning for organizations with significant worker turnover, resulting in faster learning for more employees.

This study also showed that AI-SVC integration improves OP. Previous research found a positive association between SCM functions and organizational success. People recognize the potential of supply chain digital capabilities to improve corporate performance. By investing in and developing supply chain analytics expertise, a company can gain valuable insights from significant supply chain data. Nearly every industry continuously creates large amounts of data. Innovative technology helps companies identify obstacles and adapt to market changes. This improves consumer satisfaction and revenue. Adopting digital supply chains requires more than new technology. It goes further. Several scholars recommend that the company align its digital and supply chain goals. All companies value innovative digital supply chain technology. Improved company performance can pave the way for outperforming the competition [56]. A company needs digital supply chain techniques to excel. Heizer et al. [50] and Alicke and Forsting [59] revealed that many companies aspire to strengthen their supply networks but use little digital technology. Most companies believe digital SCM can boost earnings before interest, taxes, and annual sales.

6.1. Theoretical and Practical Contributions

This work provides empirical proof to both academic and management knowledge in many ways. This study systematically assessed the impact of AR installation on operational performance for GCC furniture suppliers, rendering it distinctive. This research enhances theories on AR implementation within the GCC FSM provider industry. Consequently, the study broadens the scope of AR-SVC in SCM research. This research analysed AI-SVC dimensions and operational performance in the furniture sector in a developing economy, drawing upon prior discoveries.

This study is the first to examine AR elements and organizational performance measures in emerging nations' furniture systems. Several manufacturing industries, particularly in underdeveloped nations, have focused more on AR than furniture. Using the established model, this study validates the AR model in FSMs. In addition to the commonly held idea that AR-SVC can improve corporate performance, this study provides a crucial fresh perspective on SVC activities. As a result, future researchers will have a more complete and broad understanding of these factors, which may help them develop more effective and empirically validated AR-SVC models. SVC moderates the association between corporate performance and AR implementation, as this study shows. The literature under-represents this topic. This study is the first to examine mediation in the relationship between OP and SVC activities at GCC furniture suppliers. Additionally, the study's paradigm provides a foundation for future research in this sector. We could replicate this study in the fields of healthcare, finance, education, and hospitality.

To improve AR-SCM research methods, this study evaluated parameter data fluctuations using PLS-ANN. An ANN analytical instrument, an innovative and comprehensive data analysis tool, is best for furniture technology adoption forecasts. SCM leaders and managers should prioritize an AR-SVC work environment that builds strong relationships with key clients to optimize furniture industry supply and on-time delivery. AR can increase value chain efficiency and help organizations overcome poor planning, scheduling, and process integration and resource waste. The potential AR should improve operations and cut losses. The possibilities depend on the primary and support functions of the value chain, such as infrastructure planning and design, human resource management, inbound logistics, operations, marketing, and sales. Collaboration amongst FSMs should create structures and procedures. Finally, this study enhances furniture suppliers' managers' awareness of AR-SVC, which may help them comprehend its benefits and optimum implementation strategies. This study's model prepares GCC suppliers for AR. The developed model could highlight AR-SVC aspects that enhance OP.

6.2. Limtations and Future Resaerch

The fact that the study could only analyse the association in a cross-sectional timeframe is one of its limitations. Identification of the evolving business environment is necessary. In order to determine whether or not the association between the variables taken into consideration in the current study has altered, future research must use longitudinal design flow. Apart from temporal and financial constraints, the data used in this study were exclusively collected from the furniture industry, which reflects a remarkable service culture. Furthermore, the proposed model in this study is quite straightforward because it looks at how AR completely affects OP through the SVC functions. Thus, more advanced supplier orientation models should be developed in future research on this subject. Understanding how each SVC function affects OP, for example, will be fascinating. In addition, the survey questionnaire was the sole method used to collect the necessary data from the employees for this study. For supply managers and customers to have a comprehensive understanding of the AR applications in SCM, it is advised that various data-gathering methods or data-triangulation techniques, such as observations and interviews, be used.

7. Conclusions

This work employed the PLS-ANN hybrid model to illustrate the influence of AR in SCM on OP via SVC functions. The proposed model exhibited strong internal consistency, reliability, and predictive validity. The utilization of AR in business has risen owing to its capacity to discern the elements affecting the efficacy of AR-SVC in SCM firms, therefore bolstering the communication and resource-based perspective. This study effectively extracted AR results, highlighting the significance of supplier data assimilation for enhanced performance and competitive advantage. This study aimed to furnish senior management with insights for securing a competitive advantage via effective organizational performance.

Funding: The research work received funding from City University Ajman, Ajman, UAE. Number 39/2024.

Data Availability Statement: The data that support the findings of this study are available from the corresponding author upon reasonable request.

Conflicts of Interest: The author declares no conflicts of interest. The funder had no role in the design of the study; in the collection, analyses, or interpretation of data; in the writing of the manuscript; or in the decision to publish the results.

References

1. Abramovici, M.; Wolf, M.; Neges, M. Graph-based model for context-aware maintenance assistance with augmented reality and 3D visualization. In *Advances in Through-life Engineering Services*; Springer: Cham, Switzerland, 2017; pp. 367–386.
2. Rejeb, A.; Keogh, J.G.; Wamba, S.F.; Treiblmaier, H. The potentials of augmented reality in supply chain management: A state-of-the-art review. *Manag. Rev. Q.* **2020**, *71*, 819–856. [CrossRef]
3. Rüßmann, M.; Lorenz, M.; Gerbert, P.; Waldner, M.; Justus, J.; Engel, P.; Harnisch, M. Industry 4.0: The future of productivity and growth in manufacturing industries. *Boston Consult. Group* **2015**, *9*, 54–89.
4. Arnold, A. How AR and VR are revolutionizing the supply chain. Forbes. 29 January 2018. Available online: https://www.forbes.com/sites/andrewarnold/2018/01/29/how-ar-and-vr-arerevolutionizing-the-supplychain/#497b0fbf4cbf (accessed on 21 August 2019).
5. Attaran, M. Digital technology enablers and their implications for supply chain management. In *Supply Chain Forum: An International Journal*; Taylor & Francis: Oxfordshire, UK, 2020; Volume 21, pp. 158–172.
6. Knauer-Arnold, T. Augmented reality in training processes with Viscopic. In *Augmented Reality and Virtual Reality: Changing Realities in a Dynamic World*; Springer International Publishing: Cham, Switzerland, 2020; pp. 99–112.
7. Blankemeyer, S.; Wiemann, R.; Raatz, A. Intuitive assembly support system using augmented reality. In *Tagungsband des 3. Kongresses Montage Handhabung Industrieroboter*; Springer: Berlin/Heidelberg, Germany, 2018; pp. 195–203.
8. Saunders, K. How AR and VR Create Effective Supply Chains. 16 November 2018. Available online: http://morailogistics.com/augmented-reality-virtualreality-create-effective-supply-chains/ (accessed on 14 September 2019).
9. Implementing Blockchain in Your Enterprise. Available online: https://link.springer.com/chapter/10.1007/978-3-030-27798-7_14 (accessed on 21 March 2019).
10. Almaiah, M.A.; Alhumaid, K.; Aldhuhoori, A.; Alnazzawi, N.; Aburayya, A.; Alfaisal, R.; Salloum, S.A.; Lutfi, A.; Al Mulhem, A.; Alkhdour, T.; et al. Factors affecting the adoption of digital information technologies in higher education: An empirical study. *Electronics* **2022**, *11*, 3572. [CrossRef]
11. Aburayya, A. Predicting the supplier relationship management system's magnificence and investigating its effect on corporate shareholder value via the mediating role of marketing brand strength: A SEM-ANN hypered approach. *J. Infrastruct. Policy Dev.* **2024**, *8*, 5190. [CrossRef]
12. Jha, A.K.; Agi, M.A.; Ngai, E.W. A note on big data analytics capability development in supply chain. *Decis. Support Syst.* **2020**, *138*, 113382. [CrossRef]
13. Zeng, X.; Yi, J. Analysis of the Impact of Big Data and Artificial Intelligence Technology on Supply Chain Management. *Symmetry* **2023**, *15*, 1801. [CrossRef]
14. Berkemeier, L.; Zobel, B.; Werning, S.; Ickerott, I.; Thomas, O. Engineering of augmented reality-based information systems: Design and implementation for intralogistics services. *Bus. Inf. Syst. Eng.* **2019**, *61*, 67–89. [CrossRef]
15. Demir, S.; Yilmaz, I.; Paksoy, T. Augmented reality in supply chain management. In *Logistics 4.0*; CRC Press: Boca Raton, FL, USA, 2020; pp. 136–145.
16. Wamba, S.F.; Dubey, R.; Gunasekaran, A.; Akter, S. The performance effects of big data analytics and supply chain ambidexterity: The moderating effect of environmental dynamism. *Int. J. Prod. Econ.* **2020**, *222*, 107498. [CrossRef]
17. Wang, S.C.; Tsai, Y.T.; Ciou, Y.S. A hybrid big data analytical approach for analyzing customer patterns through an integrated supply chain network. *J. Ind. Inf. Integr.* **2020**, *20*, 100177. [CrossRef]
18. Yu, W.; Wong, C.Y.; Chavez, R.; Jacobs, M.A. Integrating big data analytics into supply chain finance: The roles of information processing and data-driven culture. *Int. J. Prod. Econ.* **2021**, *236*, 108135. [CrossRef]
19. Chatterjee, S.; Rana, N.P.; Tamilmani, K.; Sharma, A. The effect of AI-based CRM on organization performance and competitive advantage: An empirical analysis in the B2B context. *Ind. Mark. Manag.* **2021**, *97*, 205–219. [CrossRef]
20. Cao, M.; Zheng, L.; Jia, W.; Liu, X. Real-time video stabilization via camera path correction and its applications to augmented reality on edge devices. *Comput. Commun.* **2020**, *158*, 104–115. [CrossRef]
21. Moriuchi, E.; Landers, V.M.; Colton, D.; Hair, N. Engagement with chatbots versus augmented reality interactive technology in e-commerce. *J. Strateg. Mark.* **2021**, *29*, 375–389. [CrossRef]
22. Sáez-López, J.M.; Cózar-Gutiérrez, R.; González-Calero, J.A.; Gómez Carrasco, C.J. Augmented reality in higher education: An evaluation program in initial teacher training. *Educ. Sci.* **2020**, *10*, 26. [CrossRef]
23. Gürdür, D.; El-Khoury, J.; Törngren, M. Digitalizing Swedish industry: What is next?: Data analytics readiness assessment of Swedish industry, according to survey results. *Comput. Ind.* **2019**, *105*, 153–163. [CrossRef]
24. Rohacz, A.; Strassburger, S. Augmented reality in intralogistics planning of the automotive industry: State of the art and practical recommendations for applications. In Proceedings of the 2019 IEEE 6th International Conference on Industrial Engineering and Applications (ICIEA), Tokyo, Japan, 12–15 April 2019; IEEE: Piscataway, NJ, USA, 2019; pp. 203–208.
25. Cirulis, A.; Ginters, E. Augmented reality in logistics. *Procedia Comput. Sci.* **2013**, *26*, 14–20. [CrossRef]
26. Merlino, M.; Sproģe, I. The augmented supply chain. *Procedia Eng.* **2017**, *178*, 308–318. [CrossRef]
27. Stoltz, M.H.; Giannikas, V.; McFarlane, D.; Strachan, J.; Um, J.; Srinivasan, R. Augmented reality in warehouse operations: Opportunities and barriers. *IFAC-PapersOnLine* **2017**, *50*, 12979–12984. [CrossRef]
28. Koul, S. Augmented reality in supply chain management and logistics. *Int. J. Recent Sci. Res.* **2019**, *10*, 30732–30734.
29. Lupu, M.; List, J. Conferences 2017. *World Pat. Inf.* **2018**, *52*, 68–71. [CrossRef]

30. Illa, P.K.; Padhi, N. Practical guide to smart factory transition using IoT, big data and edge analytics. *IEEE Access* **2018**, *6*, 55162–55170. [CrossRef]
31. Almarzouqi, A.; Aburayya, A.; Salloum, S.A. Determinants predicting the electronic medical record adoption in healthcare: A SEM-Artificial Neural Network approach. *PLoS ONE* **2022**, *17*, e0272735. [CrossRef]
32. Porter, M.E. *Competitive Advantage: Creating and Sustaining Superior Performance*; Simon and Schuster: New York, NY, USA, 2008.
33. Ross, D.F. *Introduction to E-Supply Chain Management: Engaging Technology to Build Market-Winning Business Partnerships*; CRC Press: Boca Raton, FL, USA, 2016.
34. Song, Y.; Yu, F.R.; Zhou, L.; Yang, X.; He, Z. Applications of the Internet of Things (IoT) in smart logistics: A comprehensive survey. *IEEE Internet Things J.* **2020**, *8*, 4250–4274. [CrossRef]
35. Petruse, R.E.; Grecu, V.; Chiliban, B.M. Augmented reality applications in the transition towards the sustainable organization. In *Computational Science and Its Applications—ICCSA 2016: 16th International Conference, Beijing, China, 4–7 July 2016, Proceedings, Part III 16*; Springer International Publishing: Berlin/Heidelberg, Germany, 2016; pp. 428–442.
36. Zhang, J.; Ong, S.K.; Nee, A.Y.C. RFID-assisted assembly guidance system in an augmented reality environment. *Int. J. Prod. Res.* **2011**, *49*, 3919–3938. [CrossRef]
37. Wang, Z.B.; Ong, S.K.; Nee, A.Y.C. Augmented reality aided interactive manual assembly design. *Int. J. Adv. Manuf. Technol.* **2013**, *69*, 1311–1321. [CrossRef]
38. Elia, V.; Gnoni, M.G.; Lanzilotto, A. Evaluating the application of augmented reality devices in manufacturing from a process point of view: An AHP based model. *Expert Syst. Appl.* **2016**, *63*, 187–197. [CrossRef]
39. Hammerschmid, S. Chances for virtual and augmented reality along the value chain. In *Systems, Software and Services Process Improvement: 24th European Conference, EuroSPI 2017, Ostrava, Czech Republic, 6–8 September 2017*; Proceedings 24; Springer International Publishing: Berlin/Heidelberg, Germany, 2017; pp. 352–359.
40. Ginters, E.; Martin-Gutierrez, J. Low cost augmented reality and RFID application for logistics items visualization. *Procedia Comput. Sci.* **2013**, *26*, 3–13. [CrossRef]
41. Kretschmer, V.; Plewan, T.; Rinkenauer, G.; Maettig, B. Smart palletisation: Cognitive ergonomics in augmented reality based palletising. In *Intelligent Human Systems Integration: Proceedings of the 1st International Conference on Intelligent Human Systems Integration (IHSI 2018): Integrating People and Intelligent Systems, Dubai, United Arab Emirates, 7–9 January 2018*; Springer International Publishing: Berlin/Heidelberg, Germany, 2018; pp. 355–360.
42. Chen, Y.; Li, Y. Storage Location Assignment for Improving Human–Robot Collaborative Order-Picking Efficiency in Robotic Mobile Fulfillment Systems. *Sustainability* **2024**, *16*, 1742. [CrossRef]
43. Ro, Y.K.; Brem, A.; Rauschnabel, P.A. Augmented reality smart glasses: Definition, concepts and impact on firm value creation. In *Augmented Reality and Virtual Reality: Empowering Human, Place and Business*; Springer: Cham, Switzerland, 2018; pp. 169–181.
44. Chopra, S. Supply Chain Management. Pearson Higher Education. 16 April 2019. Available online: http://books.google.ie/books?id=g2WvvAEACAAJ&dq=chopra+supply+chain&hl=&cd=2&source=gbs_api (accessed on 23 August 2024).
45. Soatti, G.; Savazzi, S.; Nicoli, M.; Alvarez, M.A.; Kianoush, S.; Rampa, V.; Spagnolini, U. Distributed signal processing for dense 5G IoT platforms: Networking, synchronization, interference detection and radio sensing. *Ad Hoc Netw.* **2019**, *89*, 9–21. [CrossRef]
46. Gangabissoon, T.; Bekaroo, G.; Moedeen, W. Application of augmented reality in aviation: Improving engagement of cabin crew during emergency procedures training. In Proceedings of the 2nd International Conference on Intelligent and Innovative Computing Applications, Plaine Magnien, Mauritius, 24–25 September 2020; pp. 1–8.
47. Ustundag, A.; Cevikcan, E.; Esengün, M.; İnce, G. The role of augmented reality in the age of industry 4.0. In *Industry 4.0: Managing the Digital Transformation*; Springer: Cham, Switzerland, 2018; pp. 201–215.
48. Esengün, M.; Üstündağ, A.; İnce, G. Development of an augmented reality-based process management system: The case of a natural gas power plant. *IISE Trans.* **2022**, *55*, 201–216. [CrossRef]
49. Hořejší, P. Augmented reality system for virtual training of parts assembly. *Procedia Eng.* **2015**, *100*, 699–706. [CrossRef]
50. Heizer, J.; Render, B.; Munson, C. *Operations Management: Sustainability and Supply Chain Management*; Pearson: London, UK, 2020.
51. Ahmad, K.; Zabri, S.M. The application of non-financial performance measurement in Malaysian manufacturing firms. *Procedia Econ. Financ.* **2016**, *35*, 476–484. [CrossRef]
52. Wong, C.W.; Lai, K.H.; Shang, K.C.; Lu, C.S.; Leung, T.K.P. Green operations and the moderating role of environmental management capability of suppliers on manufacturing firm performance. *Int. J. Prod. Econ.* **2012**, *140*, 283–294. [CrossRef]
53. Protogerou, A.; Caloghirou, Y.; Lioukas, S. Dynamic capabilities and their indirect impact on firm performance. *Ind. Corp. Change* **2012**, *21*, 615–647. [CrossRef]
54. Cao, M.; Zheng, Q. Supply chain collaboration: Impact on collaborative advantage and firm performance. *J. Oper. Manag.* **2011**, *29*, 163–180. [CrossRef]
55. Kim, G.; Shin, B.; Kwon, O. Investigating the value of sociomaterialism in conceptualizing IT capability of a firm. *J. Manag. Inf. Syst.* **2012**, *29*, 327–362. [CrossRef]
56. Srivastava, A.K.; Sushil. Modeling strategic performance factors for effective strategy execution. *Int. J. Product. Perform. Manag.* **2013**, *62*, 554–582. [CrossRef]
57. Akter, S.; Wamba, S.F.; Gunasekaran, A.; Dubey, R.; Childe, S.J. How to improve firm performance using big data analytics capability and business strategy alignment? *Int. J. Prod. Econ.* **2016**, *182*, 113–131. [CrossRef]

58. Lee, C.; Ha, Y.K. Applying an extended investment model for enhancing supply chain performance. *J. Bus. Ind. Mark.* **2023**, *38*, 2533–2544. [CrossRef]
59. Alicke, K.; Forsting, M. *Supply chain segmentation scientific frameworks. Supply Chain Segmentation: Best-in-Class Cases, Practical Insights and Foundations*; Springer: Cham, Switzerland, 2017; pp. 5–13.
60. Mordor Intelligence. Furniture Market Analysis-Industry Growth, Size & Forecast Report (2024–2029). 2024. Available online: https://www.mordorintelligence.com/industry-reports/furniture-market?network=g&source_campaign=&utm_source=google&utm_medium=cpc&matchtype=p&device=c&gad_source=1&gclid=Cj0KCQiA_qG5BhDTARIsAA0UHSKvUtOqjAh6ml2nnjNVsPEz73wxtpfPszvS8sX-cWkxRpT131ovoHUaAv09EALw_wcB (accessed on 2 October 2024).
61. MarkNtel. UAE and Saudi Arabia Furniture Market Research Report: Forecast (2021–2026). 2024. Available online: https://www.marknteladvisors.com/research-library/uae-saudi-arabia-furniture-market.html (accessed on 2 October 2024).
62. Saunders, M.; Lewis, P. *and Thornhill, A. Research Methods for Business Students*, 7th ed.; Pearson: Harlow, UK, 2016.
63. Easterby-Smith, M.; Thorpe, R. *and Jackson, P. Management Research*, 4th ed.; SAGE Publication Ltd.: London, UK, 2012.
64. Hair Jr, J.F.; Matthews, L.M.; Matthews, R.L.; Sarstedt, M. PLS-SEM or CB-SEM: Updated guidelines on which method to use. *Int. J. Multivar. Data Anal.* **2017**, *1*, 107–123. [CrossRef]
65. Mankins, J.C. Technology readiness levels. *White Pap. April.* **1995**, *6*, 1995.
66. Lavin, A.; Renard, G. Technology readiness levels for AI & ML. *arXiv* **2020**, arXiv:2006.12497.
67. Martínez-Plumed, F.; Gómez, E.; Hernández-Orallo, J. Futures of artificial intelligence through technology readiness levels. *Telemat. Inform.* **2021**, *58*, 101525. [CrossRef]
68. Schweigkofler, A.; Monizza, G.P.; Domi, E.; Popescu, A.; Ratajczak, J.; Marcher, C.; Riedl, M.; Matt, D. Development of a digital platform based on the integration of augmented reality and BIM for the management of information in construction processes. In *Product Lifecycle Management to Support Industry 4.0: 15th IFIP WG 5.1 International Conference, PLM 2018, Turin, Italy, 2–4 July 2018, Proceedings 15*; Springer International Publishing: Berlin/Heidelberg, Germany, 2018; pp. 46–55.
69. Elbert, R.; Sarnow, T. Augmented reality in order picking—Boon and bane of information (over-) availability. In *Intelligent Human Systems Integration 2019: Proceedings of the 2nd International Conference on Intelligent Human Systems Integration (IHSI 2019): Integrating People and Intelligent Systems, San Diego, CA, USA, 7–10 February 2019*; Springer International Publishing: Berlin/Heidelberg, Germany, 2019; pp. 400–406.
70. Hannola, L.; Richter, A.; Richter, S.; Stocker, A. Empowering production workers with digitally facilitated knowledge processes–a conceptual framework. *Int. J. Prod. Res.* **2018**, *56*, 4729–4743. [CrossRef]
71. Kokkas, A.; Vosniakos, G.C. An Augmented Reality approach to factory layout design embedding operation simulation. *Int. J. Interact. Des. Manuf. (IJIDeM)* **2019**, *13*, 1061–1071. [CrossRef]
72. Akgün, A.E.; Keskin, H.; Byrne, J.C.; Aren, S. Emotional and learning capability and their impact on product innovativeness and firm performance. *Technovation* **2007**, *27*, 501–513. [CrossRef]
73. Ellinger, A.D.; Ellinger, A.E.; Yang, B.; Howton, S.W. The relationship between the learning organization concept and firms' financial performance: An empirical assessment. *Hum. Resour. Dev. Q.* **2002**, *13*, 5–22. [CrossRef]
74. Nunnally, J.C.; Bernstein, I.H. *Psychometric Theory*; McGraw-Hill: New York, NY, USA, 1994.
75. Kline, R.B. *Principles and Practice of Structural Equation Modeling*; Guilford Publications: New York, NY, USA, 2023.
76. Calcagno, M.V.; Barklund, P.J.; Zhao, L.; Azzam, S.; Knoll, S.S.; Chang, S. Semantic Analysis System for Interpreting Linguistic Structures Output by a Natural Language Linguistic Analysis System. U.S. Patent No. 7,177,799, 2007.
77. Chin, W.W. The partial least squares approach to structural equation modeling. In *Modern Methods for Business Research/Lawrence Erlbaum Associates*; University of Houston: Houston, TX, USA, 1998.
78. Taneja, A.; Arora, A. Modeling user preferences using neural networks and tensor factorization model. *Int. J. Inf. Manag.* **2019**, *45*, 132–148. [CrossRef]
79. Liébana-Cabanillas, F.; Marinkovic, V.; De Luna, I.R.; Kalinic, Z. Predicting the determinants of mobile payment acceptance: A hybrid SEM-neural network approach. *Technol. Forecast. Soc. Change* **2018**, *129*, 117–130. [CrossRef]
80. Al Harthy, M.A. Leveraging technology in learning & development. In Proceedings of the Abu Dhabi International Petroleum Exhibition and Conference, Abu Dhabi, United Arab Emirates, 31 October–3 November 2022; SPE: New Orleans, LA, USA, 2022; p. D031S100R002.

logistics

MDPI

Article

Formalizing Sustainable Urban Mobility Management: An Innovative Approach with Digital Twin and Integrated Modeling

Andrea Grotto [1,2,*], Pau Fonseca i Casas [1], Alyona Zubaryeva [2] and Wolfram Sparber [2]

[1] Department of Statistics and Operations Research, Universitat Politècnica de Catalunya (UPC), Campus Nord, C/Jordi Girona 1-3, 08034 Barcelona, Spain
[2] Institute of Renewable Energies, Eurac Research, Viale Druso 1, 39100 Bolzano, Italy; pau@fib.upc.edu (P.F.i.C.); alyona.zubaryeva@eurac.edu (A.Z.); wolfram.sparber@eurac.edu (W.S.)
* Correspondence: andrea.grotto@eurac.edu

Abstract: *Background*: Urban mobility management faces growing challenges that require the analysis and optimization of sustainable solutions. Digital twins (DTs) have emerged as innovative tools for this assessment, but their implementation requires standardized procedures and languages; *Methods*: As part of a broader methodology for continuous DT validation, this study focuses on the conceptual validation phase, presenting a conceptualization approach through formalization using Specification and Description Language (SDL), agnostic to simulation tools. The conceptual validation was achieved through stakeholder engagement in the Bolzano context, producing 41 SDL diagrams that define both elements common to different urban realities and specific local data collection procedures; *Results*: The feasibility of implementing this stakeholder-validated conceptualization was demonstrated using Simulation of Urban MObility (SUMO) for traffic simulation and optimization criteria calculation, and its framework SUMO Activity GenerAtion (SAGA) for generating an Activity-Based Modeling (ABM) mobility demand that can be improved through real sensor data; *Conclusions*: The SDL approach, through its graphical representation (SDL/GR), enables conceptual validation by enhancing stakeholder communication while defining a framework that, while adapting to the monitoring specificities of different urban realities, maintains a common and rigorous structure, independent of the chosen implementation tools and programming languages.

Keywords: urban mobility; digital twin; SDL; formalization; SUMO; SAGA

Citation: Grotto, A.; Casas, P.F.i.; Zubaryeva, A.; Sparber, W. Formalizing Sustainable Urban Mobility Management: An Innovative Approach with Digital Twin and Integrated Modeling. *Logistics* **2024**, *8*, 117. https://doi.org/10.3390/logistics8040117

Academic Editors: Tomasz Nowakowski, Artur Kierzkowski, Agnieszka A. Tubis, Franciszek Restel, Tomasz Kisiel, Anna Jodejko-Pietruczuk and Mateusz Zając

Received: 6 September 2024
Revised: 25 October 2024
Accepted: 6 November 2024
Published: 11 November 2024

1. Introduction

In cities worldwide, effective urban mobility management has become a crucial challenge in the context of growing urbanization [1]. With approximately 55% of the world's population residing in urban areas [2], covering nearly 11% [3] of the Earth's land area, understanding and planning for urban growth becomes increasingly critical. In many high-income countries, such as in Western Europe, the Americas, Australia, Japan, and the Middle East, over 80% of the population resides in urban areas [4].

Transportation systems play a vital role in modern cities, intricately linked with essential services such as healthcare, education, and utilities. Referred to as a 'subsystem' rather than an isolated entity, their efficiency holds paramount importance for the overall functioning of the urban ecosystem. Current challenges, including rising road traffic, highlight the need for sustainable urban transport solutions [5]. Issues such as traffic congestion [6], air pollution [7], and energy consumption [8] are just a few of the problems plaguing modern cities. Embracing technology and forecasting methods becomes essential for effective policy planning in the ever-evolving urban landscape [5].

In response to these challenges, digital twins have emerged as an innovative solution [9]. A DT is a virtual replica of a physical system that enables the simulation, monitoring, and optimization of real-world processes [10]. The primary objective of this research is to formalize the process of creating a DT for urban mobility management. Within the scope of this research, the focus is placed on creating a DT to address the challenges of urban mobility and assess the environmental and socio-economic sustainability of mobility scenarios, promoting the improvement in air quality, mitigating traffic congestion, reducing travel times and optimizing energy consumption. Specifically, we aim to develop a novel approach that expands the scope of the conceptual model to include not only the definition of the system but also the procedures for validating and executing the model.

The process of constructing a DT necessitates the validation of assumptions made by the various stakeholders who will be involved in the process. This validation is not a one-time event, but rather an ongoing process that must be carried out continuously before and once the digital twin is deployed in a production environment. A comprehensive validation and verification process for modeling can be found in [11]. One of the key aspects of this process is Solution Validation. This is a continuous process that must be carried out once the digital twin is used in production. The purpose of Solution Validation is to ensure that the digital twin continues to perform accurately and reliably under changing conditions and over time. It involves regularly checking the digital twin against its physical counterpart and updating the model as necessary to reflect any changes in the physical system. It is also essential to ensure that all stakeholders understand and accept the different assumptions that the models of the digital twin hold. This is because these assumptions can significantly impact the performance and accuracy of the digital twin. Therefore, it is crucial that these assumptions are clearly communicated to all stakeholders and that they are fully understood and accepted. In this regard, stakeholder engagement and communication are key. All stakeholders should be involved in the validation process and should have a clear understanding of the assumptions and limitations of the digital twin model. This will help to ensure that the digital twin is accepted and trusted by all stakeholders, and that it can effectively support decision-making and operational processes.

Moreover, a digital twin is a comprehensive system that typically encompasses several distinct components (see Figure 1). These components include the digital shadows, which encapsulate both synthetic data derived from the models and real-world data often obtained from sensors. The digital twin also includes conceptual models and their respective implementations, aligning with the principles of Industry 4.0 (I40 in Figure 1). The digital shadows serve as a bridge between the physical system and its digital counterpart, embodying a key aspect of I40's cyber–physical systems. They contain synthetic data generated by the models, which simulate the behavior of the physical system under various conditions. Additionally, they also contain real-world data collected from the physical system, often through sensors. This combination of synthetic and real-world data allows the digital twin to accurately represent and predict the behavior of the physical system, often through sensors, reflecting I40's emphasis on data-driven decision-making and interconnectivity. This combination of synthetic and real-world data allows the digital twin to accurately represent and predict the behavior of the physical system, enabling the smart factories and systems central to I40.

Conceptual models form the theoretical foundation of the digital twin, often named Digital Master. They represent abstract concepts and relationships that define the behavior and characteristics of the physical system, incorporating I40 principles such as interoperability and decentralized decision-making [12]. These conceptual models are then implemented in the digital twin, translating the abstract concepts into concrete algorithms and data structures that can simulate the behavior of the physical system.

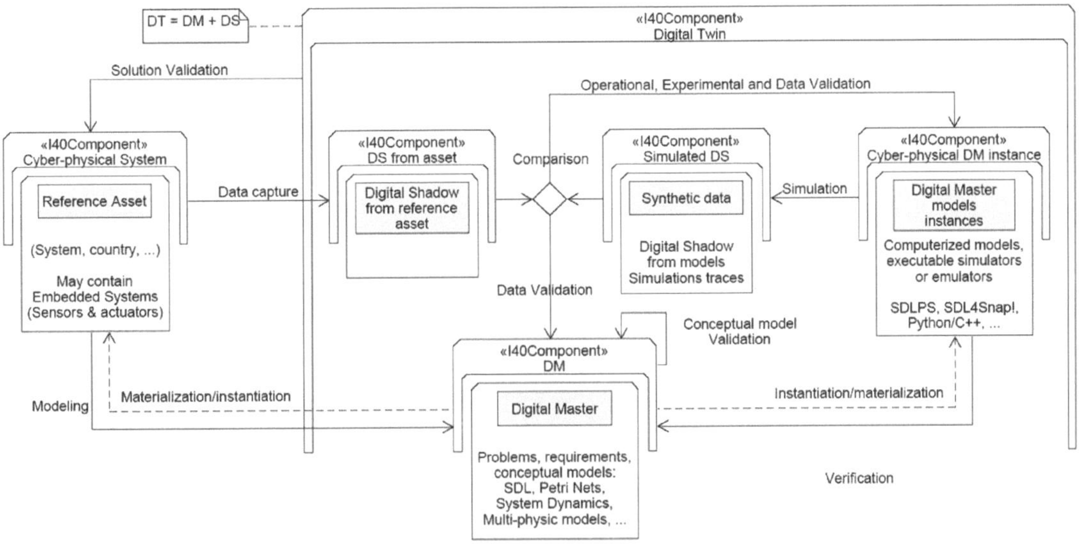

Figure 1. Digital twin architecture, from [13].

However, it is important to note that the validation and experimental aspects of the models are key to the success of the digital twin. Therefore, a primary objective of this research is to propose a novel approach that includes these aspects in the conceptual model. This is a novel approach that expands the scope of the conceptual model to include not only the definition of the system but also the procedures for validating and executing the model. This approach has been similarly proposed in [14], though not for the validation and execution procedures, but for the representation of the models.

With this novel approach, the conceptual model becomes a comprehensive framework that includes the definition of the system, the validation procedures to ensure the accuracy of the model, and the execution procedures to simulate the behavior of the system. The aim is to ensure that the digital twin is not only an accurate representation of the physical system but also a reliable tool for predicting its behavior and supporting decision-making processes.

Specifically for our area of research, we aim to develop a methodology that combines activity-based modeling (ABM), traffic microsimulation, and optimization techniques to achieve these goals. The innovation is based on the use of integration techniques through the SDL language: this research compares data from various sources, including real-time flows from IoT traffic sensors, historical data, and static information (such as schedules of local public transportation), with traffic simulation results generated by SUMO [15]. These simulations are based on an Activity-Based Model (ABM), which, in turn, is generated by SAGA [16], an internal framework within SUMO specialized in creating ABM models from OpenStreetMap files. This comparison leads to a continuous validation process, ensuring constant improvement and alignment of the model with the dynamics of the urban mobility system. In this way, a DT is generated that, in addition to monitoring real-time data, can perform predictive analyses and optimizations for future mobility scenarios.

Structure of This Research

The structure of this research is organized as follows: Section 2 reviews the relevant literature, with a focus on urban environmental modeling and innovative aspects of mobility modeling and formal languages. Section 3 describes the methodology, including the phased approach, data collection, and the sources of data used in this study. Section 4 presents the case study used for testing the proposed framework. Section 5 outlines the results, focusing on the conceptual validation and the implementation of the conceptual model. Section 6 provides a discussion of the findings, including an analysis of the results, limitations, and future research directions. Finally, Section 7 concludes this study by summarizing the key insights and suggesting areas for further investigation.

2. Literature Review

2.1. Urban Environmental Modeling

Urban environmental modeling is a process through which representations or simulations of cities or urban areas are created for the purposes of analysis, planning, design, and evaluation. These models are crucial for understanding the functioning of the city and predicting urban development, assessing the impact of planning and design decisions. They also serve as a testing ground for various urban strategies and policies and can be divided into various categories or subfields, each focusing on specific aspects of the urban ecosystem. Here, are some of the main subdivisions:

1. Land use models focus on representing the distribution and land use within an urban area. They can be used to predict how urban land will change over time, such as identifying areas for residential, commercial, or industrial development [17,18];
2. Transportation models concentrate on urban mobility, studying how people and goods move through a city. They are useful for optimizing public transportation systems, managing traffic congestion, and improving connectivity [19];
3. Environmental impact models assess the impact of human activities on air, water, and soil quality in urban areas. They can help evaluate environmental risks and design mitigation measures [20];
4. Urban resilience models are oriented toward preparing cities for extreme events such as natural disasters or economic crises. They help assess a city's capacity to cope with and recover from such events [21];
5. Urban growth models have been created and widely embraced for examining the growth of cities and its consequences on the surrounding environment. These models find application in urban policy formulation and the analysis of various developmental scenarios;
6. Socio-economic models analyze the economic and social aspects of a city, including factors like employment, income, access to services, and residents' quality of life [22];
7. Visualization models provide visual representations of cities or projects, allowing stakeholders to have a clear view of proposed changes [23].

Urban modeling analysis employs various combinations of the illustrated models. Taking SUMO as an example, it not only models' traffic within the examined city but also facilitates the visualization and calculation of emissions or energy consumption related to different mobility technologies.

2.2. Existing Mobility Modeling and Formal Languages

In this section, the analysis will explore both the mobility model and the specific languages employed to generate a digital twin through a formalism and standardization process. This understanding is critical for the next steps, such as methodology and data collection, as it lays the foundation for the research process. In the field of simulations, several tools have emerged to facilitate modeling and analysis of complex systems:

1. Demand models are used in the fields of transportation to predict and understand the demand for trips between different origins and destinations in a specific geographical area. These models provide a mathematical or statistical representation of the factors influencing travelers' choices, such as demographic characteristics, transportation accessibility, economic conditions, and individual preferences through a synthetic population [24]. The mobility demand can be calculated through two main approaches [25]: the traditional four-step method and Activity-Based Modeling (ABM). The first is a conventional approach and it is divided into four stages: trip generation, trip distribution, mode choice, and traffic assignment. The second is based on the daily activities of individuals and models travelers' behavior [26];

2. Microsimulations models [27] are traffic models that simulate the behavior of each individual vehicle (including the powertrain and the interactions with pedestrians, cyclists, public transport, etc.), which are evaluated at intervals of less than a second. Their journeys through the network are derived from the analysis of their interactions with the road network (lanes, speeds, signage, toll areas, etc.), traffic control systems (like traffic lights), and other vehicles on the network. These models are often used to analyze a small area in detail. Some tools that use microscopic simulations are SUMO [15], PTV Vissim [28], Citilab Cube [29], and TransModeler SE [30];
3. Macrosimulation models [31] deal with the overall traffic flow and consider aggregated variables such as traffic density, average speed, and vehicular flow. They are often used to assess the performance of road networks on a large scale. Traveler demand is analyzed at the flow level (sets of users moving between different origin–destination pairs throughout the environment) on a map-based network. The network has features such as capacity, directional speeds, turning restrictions, and generally, the models operate with cost equilibrium functions. Some tools that use macroscopic simulations are Transmodeler [32], PTV Visum [33], and EMME [34];
4. Traffic assignment models [35] allocate traffic flows to a road network based on traffic conditions, travel costs, and capacity constraints. They are useful for optimizing traffic flow and evaluating the impact of infrastructure changes. Based on the consideration of the temporal dimension, traffic assignment models are classified into three categories: dynamic traffic assignment (DTA), semi-dynamic traffic assignment, and static traffic assignment (STA) models [36]. A tool that uses traffic assignment models is DYNAMEQ [37];
5. Modal choice models [38] focus on individuals' decisions regarding the choice of transportation mode. They consider factors such as travel time, costs, convenience, and personal preferences to predict how people choose to commute. A tool that uses modal choice model is Biogeme [39]. In addition, there are tools such as Adopt [40] that allow for the evaluation of large-scale media choice without considering the road network;
6. Agent-based models [41] represent individual "agents" (vehicles, people) and simulate their behavior based on local rules and interactions. They are particularly useful for studying emerging behaviors and chain effects. Some tools that use agent-based simulations are Matsim [42], and SUMO [15].

Given the digital twin's objective to conduct an energy analysis for seamless integration, for example, with urban energy subsystems, the recent classification of mobility models based on a broader understanding of energy systems is introduced into the literature. In accordance with [43], simulation models of mobility can be categorized in the following manner:

1. Economy-wide energy scenario models are typically used to evaluate costs, energy use, and emissions impacts over broad geographic (national or global) and temporal (decades to centuries) scales under various assumptions about future technology evolution and policy;
2. Accounting and exploratory tools are often used to investigate the implications of "what if?" scenarios on energy use and emissions, typically without explicit representation of the effects of costs or other factors in consumer decision-making. Some tools in this category include representations of all end-use demand sectors, while others focus on specific sectors;
3. Vehicle choice models are typically used to project the evolution of the vehicle stock based on assumptions about consumer preferences and future technology cost and performance;
4. Transportation system simulators are primarily used for detailed analyses of the traffic and energy use implications of local infrastructure projects and policies;

5. Refueling infrastructure models are used to assess optimal network design (station location) and use for the alternative fuel production centers and refueling stations needed to meet specified levels of demand.

In the following section, a literature review of various formalization languages is presented. Formalism must be independent of simulation tools and models, allowing for the execution of analyses to determine relationships among different components comprising the urban mobility demand model. This approach facilitates clear and specific communication of how the integration has occurred, ensuring an understanding of the system.

1. SDL language: The Specification and Description Language (SDL) is an object-oriented formal language used in the field of computer science and engineering and standardized by ITU-T (Comité Consultatif International Télégraphique et Téléphonique) [CCITT]), established in 1968. SDL is primarily employed to model and describe the behavior of complex systems, including those related to communication protocols, distributed systems, and real-time systems. It provides a standardized way to define system components, their interactions, and their states;
2. The Discrete Event System Specification (DEVS) is a formal modeling and simulation language used to describe and analyze discrete-event systems, introduced by Bernard Zeigler in 1976 [44]. DEVS provides a structured framework for modeling complex systems where events occur at distinct points in time and where system states change in response to these events. Key features of DEVS include its ability to model systems with discrete events, concurrency, and hierarchical organization. DEVS models consist of atomic models (representing individual system components) and coupled models (describing the interactions between these components);
3. Petri nets (also known as place/transition nets: PTNs) are a formal mathematical modeling language used to represent and analyze systems with concurrent and distributed processes. PTNs were introduced by Carl Adam Petri in the 1960s [45] and have since found applications in various fields, including manufacturing, discrete event systems and networks, specification, simulation and validation of software. Petri nets consist of two main components: places and transitions. Places are represented as circles, while transitions are represented as rectangles. Tokens, usually depicted as dots, move between places and trigger transitions when certain conditions are met. The interactions between places, transitions, and tokens allow Petri nets to model dynamic and concurrent processes. Petri nets provide an effective approach in modeling dynamic systems, capturing relationships and interactions precisely. Their graphical representation simplifies the understanding of complex systems, offering an integrated environment for formal analysis. The ability to explicitly represent states and events, coupled with their practical applicability in scenarios like real-time control systems, underscores their utility in addressing complex challenges [46].

Innovative Aspects

While many studies have delved into exploring the use of digital twins in urban mobility [47,48], this research in Bolzano introduces an innovative approach focused on two key components: the formalization of the DT creation process and the optimization of mobility technologies through a multicriteria analysis (MCA).

The first innovative element in this research lies in the formalization process, facilitated by the Specification and Description Language (SDL). Formalization transforms intricate concepts into a clear structure, ensuring a comprehensive understanding for stakeholders in the urban mobility system. Additionally, formalization enables the standardization of digital twin creation, ensuring analysis repeatability over time and extending its applicability to other facets like urban utilities, energy systems, education, and health, thereby contributing to the development of a comprehensive urban digital twin.

The second key element of innovation involves the optimization of mobility technologies through a multicriteria analysis. This analysis considers crucial parameters such as air quality, traffic congestion, travel time reduction, and energy consumption. This approach allows for a comprehensive and integrated consideration of all aspects of urban sustainability, leading to tangible improvements in urban mobility management and the quality of life for citizens.

3. Materials and Methods

3.1. Continuous Validation Process

This research adopts a phased methodology for continuously developing and validating a digital twin for urban mobility. The methodology consists of several sequential validation phases, as illustrated in Figure 2, where each phase builds upon the results of the previous one, ensuring a systematic and comprehensive approach to digital twin development.

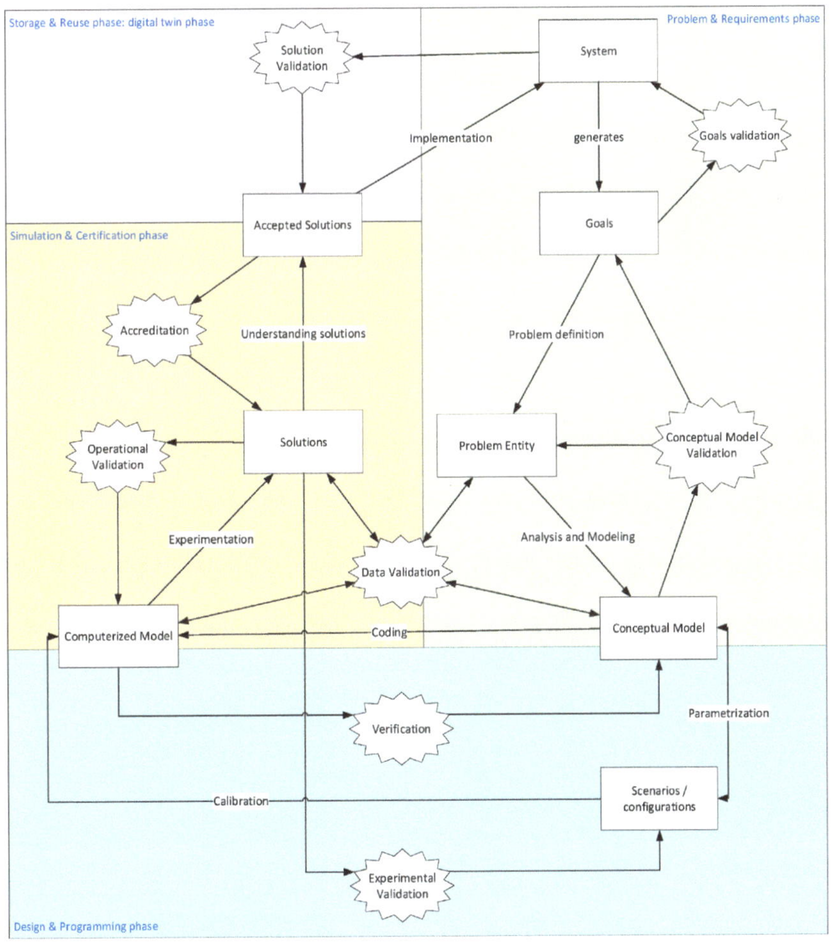

Figure 2. Phased methodology for continuous validation process [11].

The current work focuses on the Conceptual Model Validation, which is part of the Problem and Requirements phase (green part in Figure 2). This phase is crucial as it establishes the theoretical foundation upon which subsequent validations will build. The validation process integrates various components including the Problem Entity (System), Goals, Conceptual Model, Computerized Model, and Solutions, connected through different validation and verification activities in distinct phases (Problem and Requirements, Design and Programming, Simulation and Certification, and Storage and Reuse), as shown in Figure 2:

1. Goals' Validation (carried out): verified the system objectives and requirements, which are typically defined by stakeholders as city managers and mobility planners as part of their strategic planning process;
2. Conceptual Model Validation (current phase): validates the theoretical framework against the real system, ensuring that assumptions, theories, and mathematical representations adequately capture the Problem Entity's characteristics;
3. Computerized Model Verification (future phase): will verify that the computer programming and implementation accurately represent the Conceptual Model's specifications;
4. Operational Validation (future phase): will involve comparing the simulation results with the actual system behavior, ensuring the model's predictions align with real-world observations;
5. Experimental Validation (future phase): To validate the model's results, sound experimental procedures are necessary. This includes verifying the design of experiments, the number of replications, and the methods used to obtain them.
6. Data Validity (continuous process): ensures the quality and appropriateness of data used throughout all validation phases;
7. Solution Validation (future phase): will verify the effectiveness of implemented solutions in production environment, ensuring continuous alignment with real-world conditions;
8. Accreditation (future phase): will establish the formal acceptance of the model for its intended use, certifying its fitness for purpose.

3.2. Data Collection

The authors propose an innovative approach centered around data collection for real-time validation of urban mobility demand calculations. This phase involves a synergistic blend of data sourced from IoT traffic sensors and information integrated from online platforms. The process unfolds in two complementary modalities: utilizing static information such as public transportation schedules and the location of bus stops and leveraging dynamic data from IoT systems embedded in the streets to monitor real-time vehicle flows.

Data collection is facilitated using the Specification and Description Language for Graphical Representation (SDL/GR), detailing the integration process. As a formal language, SDL/GR provides a clear representation of system specifications, enabling efficient communication among different components. Integrating data from various sources becomes a structured and coherent process through SDL/GR, defining specifications and requirements related to data collection. This includes outlining information flows, synchronizing static and dynamic data, and configuring acquisition parameters.

Being a language commonly used in the telecommunications sector, SDL/GR can be successfully extended to the collection of data from signals generated by IoT sensors. In this way, the adoption of SDL/GR, as proposed by the authors, represents a significant step in the strategy to implement a digital twin of urban mobility, becoming a tool for analysis based on real information.

3.3. Source of the Data

As described in the previous paragraph, data are acquired from both static and dynamic perspectives, with each contributing crucial insight. Static data primarily encompass information extracted from online platforms, such as public transportation schedules, forming a fundamental element for assessing baseline mobility demand. On the other hand, dynamic data originate from real-time monitoring systems, including road sensors and embedded IoT devices. These sensors capture and transmit immediate information on vehicular movements, enhancing the understanding of current traffic patterns and mobility trends.

In the case of the Bolzano study, dynamic data are obtained from an open data platform directly connected to traffic sensors. This platform, named Analytics [49], collects traffic data through Bluetooth sensors embedded in the streets and cycling lanes. The sampling frequency is approximately every ten minutes for vehicular traffic and extends to about one hour for bicycles passing through the city's cycling lanes. The flows detected by the sensors do not directly represent actual traffic but instead signify the number of Bluetooth-enabled devices (smartphones) passing over the sensor.

The company managing these devices has validated that these flows account for approximately 15% of the total traffic, with this value continuously rising over time due to an increasing number of people passing over the sensors with Bluetooth enabled.

3.4. Methodology

The methodology is grounded in the process of formalization, facilitated by the structure of SDL/GR (Specification and Description Language/Graphical Representation). This approach concentrates on systematically formalizing various aspects, empowering the implementation of the conceptual model [50]. As specified by [51], the construction of the model sometimes reveals essential lacunae in the understanding of the model before its implementation.

The formalization approach leverages SDL/GR to create a comprehensive and unambiguous representation of the digital twin for urban mobility. This choice was made for these reasons:

1. SDL/GR provides a clear and non-ambiguous representation of system specifications, including both formal elements (such as states, transitions, events, variables, and processes) and a graphical representation that enhances comprehension through diagrams;
2. The visual aspect of SDL/GR promotes effective information exchange among diverse model components, facilitating communication among stakeholders with varying levels of technical expertise;
3. The graphical nature of SDL/GR enhances the clarity and accessibility of the specified system, making it an asset in the design and communication of complex systems in urban mobility applications;

The use of SDL/GR specifically enhances data communication and synchronization in the digital twin through its signal-based architecture. SDL treats different data streams as signals with varying priorities, enabling:

1. Dynamic triggering of specific model components or procedures when signals (data) arrive from different sources;
2. Prioritized handling of different data streams (sensors, historical data, static information);
3. Automated synchronization of model components based on signal reception and processing.

The methodology integrates various tools and techniques within the SDL/GR framework:

1. A traffic simulator for urban vehicular flow micro-simulation;
2. A trip generator for creating Activity-Based Models (ABMs) from OSM (Open Street Map);
3. IoT traffic sensors for real-time data collection;
4. Historical data and static information (such as public transportation schedules) for comprehensive model development.

This integration of diverse tools within the SDL/GR framework allows for a comprehensive and flexible representation of the urban mobility system. The specific details of the implementation, including the tools used, will be described in Section 5.2.

The formalization process in the methodology follows these steps:

1. Initial system analysis and requirement gathering;
2. Development of high-level SDL/GR diagrams representing the overall system structure;
3. Detailed modeling of individual components and their interactions;
4. Integration of various data sources and simulation tools within the SDL/GR framework;

5. Iterative refinement based on stakeholder feedback and validation results.

The agnostic approach in the formalization phase enables greater adaptability of the model to different urban and technological realities, ensuring that the framework remains valid regardless of the specific tools used in the implementation.

Formalization not only provides guidance for implementing the conceptual model but also creates a tangible outcome in the form of formalized documentation. This documentation can be considered a product itself because it comprehensively represents the structure and behavior of the simulations system [52]. In other words, the formalization itself, using a language like SDL/GR, becomes an output that can be shared, understood, and utilized by the stakeholders involved in the urban mobility project.

The approach goes beyond traditional modeling by incorporating validation procedures directly into the conceptual model. This novel aspect ensures that the digital twin remains accurate and reliable throughout its lifecycle, adapting to changing urban conditions.

As mentioned in Section 2.2, the history of SDL dates to the mid-1970s when it was developed under the ITU's guidance, a specialized agency of the United Nations [53] dealing with telecommunications and information technologies. SDL underwent a standardization process, resulting in its formalization as Z.100 in the 1980s. Since then, it has evolved to meet the changing needs many times the applications of telecommunications, distributed systems, and embedded systems. The language is currently described in ITU-T Recommendations Z.100–Z.109, playing a crucial role in specifying and describing complex systems in various sectors.

Before delving into the application of urban mobility demand, let us provide a sequential overview of the SDL language. SDL is composed of two distinct components: a static part and a dynamic part. The static aspect focuses on outlining the system's structure by defining stable elements such as states, transitions, and variables. In contrast, the dynamic facet models the system's behavior, encompassing events, processes, and dynamic interactions among static elements. This dual nature enables SDL to offer a comprehensive and well-defined perspective for specifying and describing intricate systems. Communication among components is facilitated through signals and communication channels. In summary, SDL can be categorized into three fundamental parts: structure, behavior, and communication.

In SDL, agents represent entities within the system that have a defined role or responsibility. These entities could be components, subsystems, or other functional units that play specific roles in the overall system architecture. The structure of SDL involves defining and organizing these agents, along with specifying their interactions and relationships. In SDL, designating the system as a top-level agent underscores its central role, representing the highest organizational entity that encompasses various components. It is crucial to note that the system, as a top-level agent, contains everything within its scope but cannot be contained by any other entity. Outside the system, there is the environment, which may or may not communicate with the system. In this sense, we can say that the environment is a bottom-level agent since it cannot contain other blocks or processes.

In Figure 3, there is a depiction of the urban mobility system and its communication channels with the external environment. The system is represented by a rectangle, while the input and output channels are depicted with arrows. Signal names are enclosed in square brackets. Specifically:

- [Technologies] indicates the types of mobility technology to be integrated into various scenarios;
- [Optimiz_tech] represents the quantification of technologies through optimization;
- [Real_Flow] shows traffic flows monitored by sensors;
- [FirstPop] contains information about the characteristics of individuals;
- [OSM] indicates an OpenStreetMap file for simulations;
- [GTFS] provides public transport schedules.

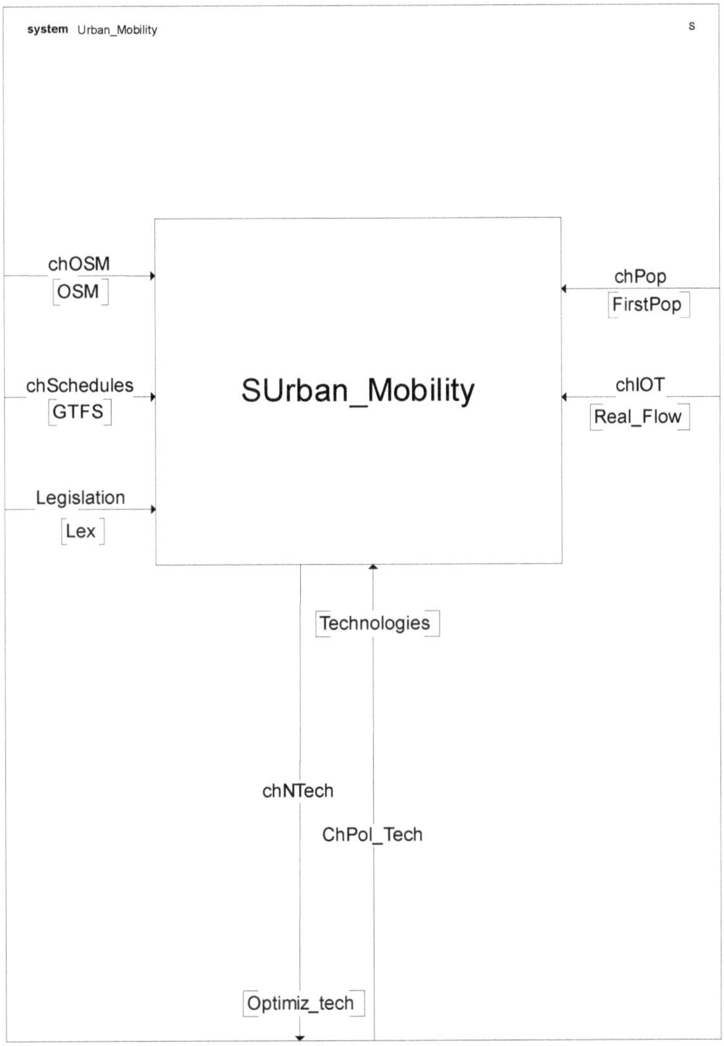

Figure 3. Urban mobility system.

Upon entering the system, a division into blocks is represented in Figure 4, representing different modules. These modules include the mobility demand model and the block dedicated to the optimization process. This optimization involves quantifying selected technologies for different scenarios based on a specific objective function. The objective function considers results obtained in the simulation block in terms of emissions, traffic, and energy consumption.

As evident from Figure 2, there are five blocks:

- Bmap_Net: This block involves representing the network on a discrete basis with nodes V (1, 2, 3, 4, ...) and arcs N (2-3, 3-2, ...), reflecting the physical structure of the network. The network encompasses various elements such as roads, bike lanes, railways, and traffic flow influencers like traffic lights, intersections, and road signs. Each node in this network is treated as an agent capable of generating (node 1-2) and/or regulating traffic flow. Additionally, each connection is assigned a weight corresponding to the imposed speed limit, a characteristic typical of discrete networks. The data are sourced from OpenStreetMap. Furthermore, this block is responsible for extracting points of interest on the map, subsequently passed to the activity block to evaluate the attractiveness of different areas of the city. This assessment aids in quantifying the flows between these areas;

- BPop_ABM: This block generates activities classified as primary, secondary, and home activities based on population characteristics such as population size and information received from the map block regarding the attractiveness of different areas of the city. Consequently, the city is divided into multiple areas, and an origin–destination matrix is generated. Using this matrix, trips are assigned to activities in space and time. These activity data are then provided to the simulator, which determines the most suitable routes to connect the various activities, thus generating the routes that each city inhabitant will take. Additionally, it creates a simplified synthetic population of street users who undertake various journeys within the city. Everyone in this population can be considered an agent, and the resulting model is referred to as Activity-Based Modeling (ABM);
- BTraff_Sim: In this block, routing, mode choice modeling, and traffic simulation operations are integrated. Routing involves selecting the most optimal path for each vehicle, while mode choice modeling models individuals' decision-making behaviors using a utility function to optimize their selections from multiple options. Subsequently, traffic simulation through microsimulation models simulates individual vehicles and their movements within a road system over a 24 h period. Each vehicle (or agent) is represented as an individual entity, considering specific behaviors such as acceleration, deceleration, lane changing, and interaction with other vehicles. Additionally, this simulation evaluates emissions, energy consumption, and traffic dynamics;
- BComp_Val: This block contributes to validating mobility demand before quantifying the different technologies associated with various scenarios. Routing involves selecting the most optimal path for each vehicle, while mode choice analysis models individuals' decision-making behaviors using a utility function to optimize their selections from multiple options. As depicted in Figure 4, a comparison is made between the actual monitored data of traffic flow and the results obtained from simulations. Real-time validation is essential, enhancing the overall accuracy and reliability of the digital twin. It offers several benefits, including the timely detection of problems and anomalies, ensuring the digital twin remains aligned with changes and updates in the mobility system, and supporting real-time decision-making;
- BSc: This block systematically adjusts the experimental factors for technologies within each scenario. Its role encompasses creating technology scenarios, with some of these technologies being quantified and optimized based on an objective function considering simulation results in terms of emissions, energy consumption, and traffic outcomes. The aim is to provide qualitative scenarios that can be quantified and optimized based on various criteria, such as emissions reduction, cost-effectiveness, energy efficiency, and traffic flow optimization, thereby facilitating informed decision-making.

In this article, not all graphs are presented. However, for illustrative purposes, the graph related to the modeling of transport infrastructures is included as an example. Entering the BPop_ABM block, the corresponding processes, channels and signals are shown (Figure 5).

Entering the processes PBuilding, P_Weight_TAZ, PDef_TAZ, POD_Matrix, PActivity_Gen SDL is utilized to explicitly outline the various phases involved in obtaining the activities for conducting simulations.

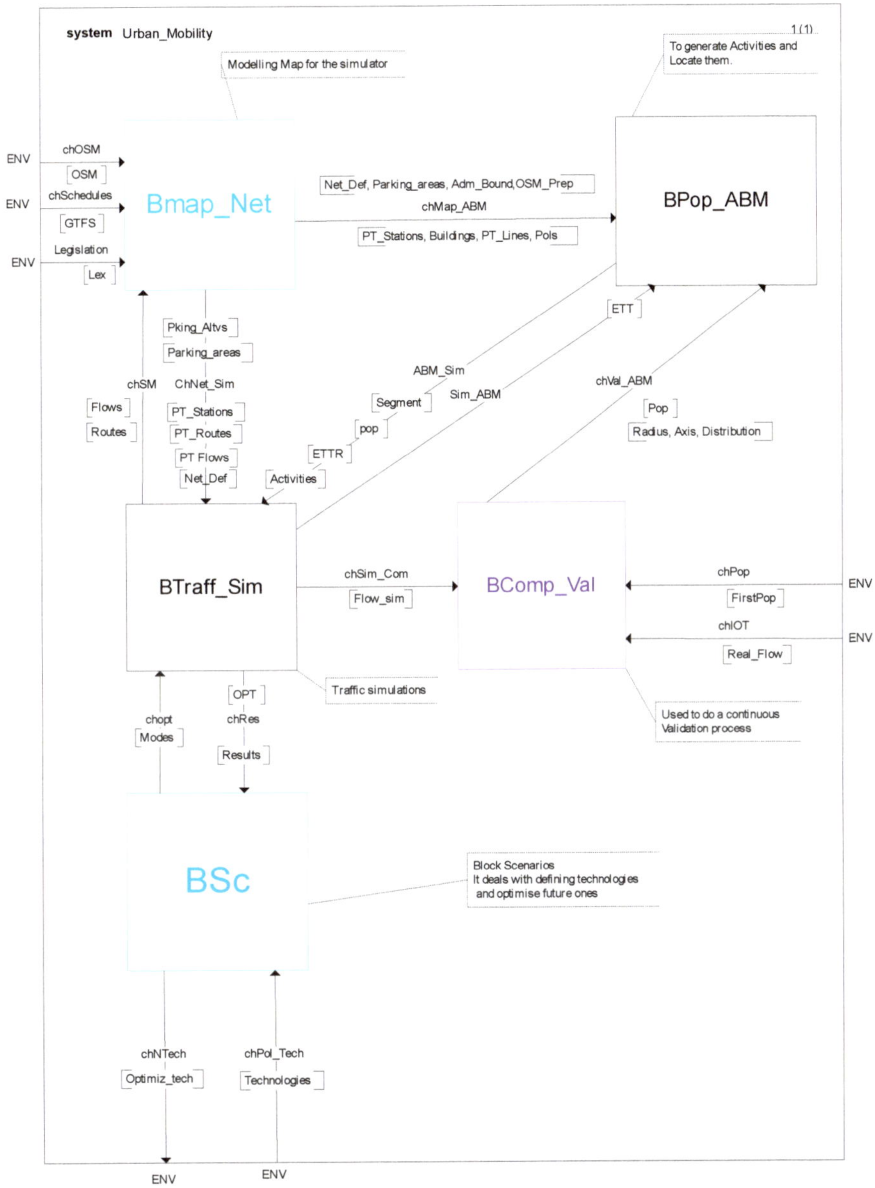

Figure 4. Inside an urban mobility system.

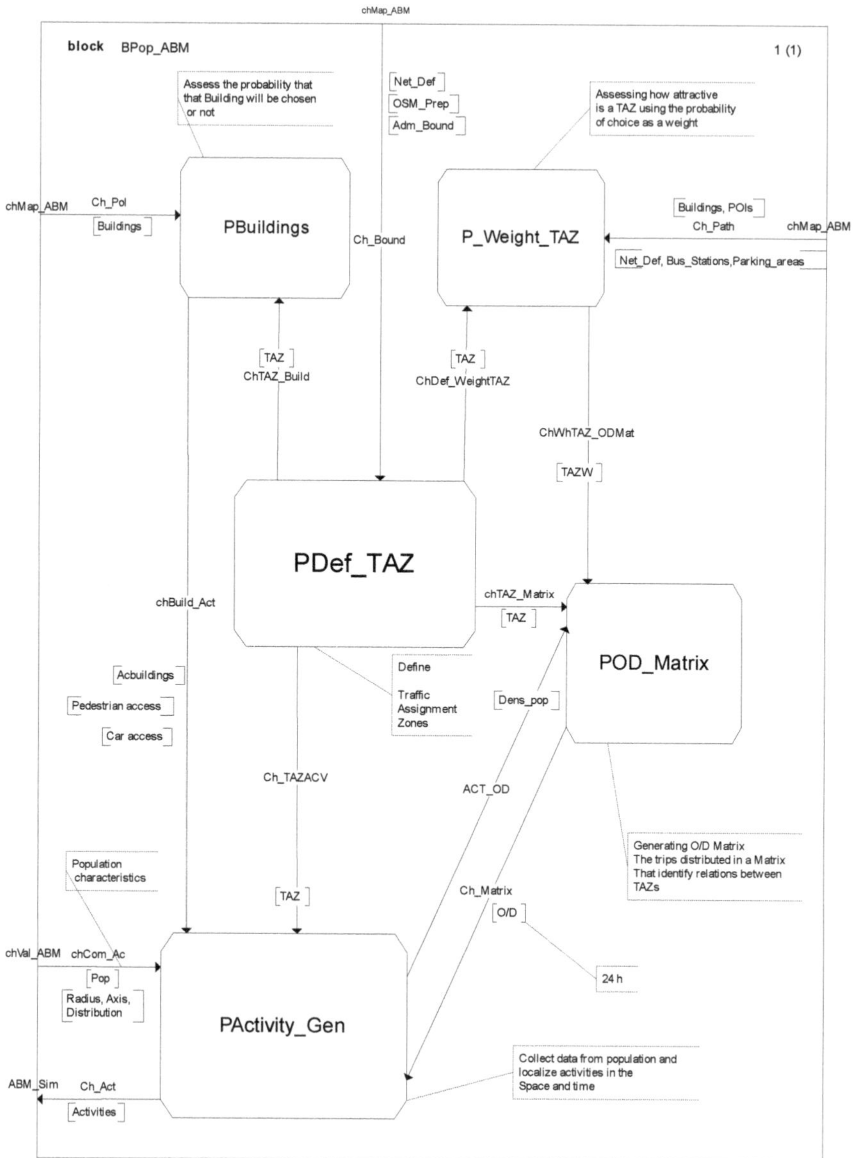

Figure 5. Block BPop_ABM.

4. Case Study

To demonstrate how the formalized process can be applied in a real urban context, this study focuses on the city of Bolzano. Located in northern Italy, Bolzano represents an interesting case study for several reasons. First, its size (106,107 inhabitants, 52.34 square kilometers) is representative of many medium-sized European cities. Second, the city demonstrates a strong commitment to sustainable mobility through its extensive transportation infrastructure, including 269.3 km of roads, 70.7 km of bike paths, and an innovative public transport system operating 115 buses (70 electric/hybrid and 12 hydrogen-powered).

Following the digital twin blocks structure presented in Section 3.3 of the methodology, the formalization process has been applied as follows:

1. Bmap_Net block: formalization of road networks structure, exemplified by elements such as public transport lines (14 lines, 348 stops) and limited traffic zones (351,715 square meters);
2. BPop_ABM block: formalization of mobility behavior patterns, characterized in Bolzano's case by specific modal splits (29% walking, 26% cycling, 10% public transport, 29% cars, 5% motorcycles/other);
3. BTraff_Sim block: formalization of traffic dynamics elements, including geographical features (valley location, confluence of Isarco, Sarentino, and Adige rivers), infrastructure (A22 motorway), and public transport usage (8,327,107 annual passengers);
4. BSc block: formalization of future scenarios focused on sustainable transport modes, exemplified through planned expansions of electric/hydrogen fleet and bike paths;
5. BComp_Val block: formalization of the validation process structure, implemented using data streams from the Analytics Open Data Hub and Bluetooth sensors on major roads and bicycle lanes.

The data above were provided by [54]. As described in Sections 3.2 and 3.3, the city monitors traffic flows on major roads and bicycle lanes through the open platform Analytics Open Data Hub. This platform can visualize the Bluetooth sensors present in the city (the simulation system) and those in the surrounding areas.

5. Results

The presented work focuses on the conceptual formalization of the urban mobility DT, using SDL as formal language and validating an ABM model, with a case study in Bolzano. This approach is extendable to other urban contexts where mobile phone data for activity tracking are limited and in situations where a DT based on road sensors is crucial for validating simulation results.

5.1. Conceptual Validation

Constructing a DT requires effective collaboration among stakeholders involved in the process. Validation, as specified in the introduction, is not an isolated event but a continuous process that must be carried out before and after the implementation of the DT in the operational environment. Unlike verification, which focuses on the correctness of the model and its software, ensuring they are built and function properly, validation aims to ensure that the model accurately reflects the observed reality. The combination of continuous validation and verification is crucial to ensure that the DT is accurate, reliable, and useful for making informed decisions and supporting operational processes. This article deals only with validation and not verification, which will be addressed separately in a subsequent analysis. Urban Resilience SL [55] (UR) has taken on the role of validating the assumptions of the DT conceptual model, collaborating closely with local authorities and other public and private stakeholders. Their involvement has ensured the validity and acceptance of the model assumptions, which are articulated through SDL diagrams. UR's specific interest lies in the deployment of SUMOSU charging stations. This infrastructure is designed not only as a charging point for electric vehicles but also as a comprehensive hub for urban mobility services, supporting the needs of electric buses, private electric vehicles, shared electric vehicles, and e-taxis.

For example, through the analysis of the SDL/GR model, UR was able to understand that the digital twin could perform energy-related calculations and emissions evaluations. This realization led UR to recognize that the model could be used to assess the impact of their SUMOSU charging stations on the wider urban environment. Specifically, UR could use the model to:

1. Evaluate the energy demand of the charging stations;
2. Estimate potential reductions in emissions due to increased use of electric vehicles;
3. Analyze the effects on traffic flow in the areas surrounding the charging stations.

This comprehensive understanding would not have been possible by simply reviewing code or running an isolated simulation. The SDL/GR model provided UR with a visual and interconnected representation of the urban mobility system, allowing them to see how their charging stations fit into the larger picture of urban transportation and energy use.

This integration aims to bridge the gap between public and private mobility sectors through inter-modality, making the hub a space for exchange. Given the diversity of users involved, it is crucial for UR to collaborate with various stakeholders, both public and private, to ensure the success and efficiency of these hub operations.

The use of the DT conceptualized model, as formalized in this article, can effectively assess the impact that an infrastructure can have on an urban level through the Scenario Block (BSc) and the Simulation Block (BTraff_Sim), as shown in Figure 4. However, to ensure that the model is accurate and meets the diverse needs and competencies of stakeholders, it is essential that the underlying assumptions of the conceptual model are validated by all involved parties, each with different interests and expertise.

The SDL diagrams, which provide a visual approach, facilitated explicit representation of the diverse logical relationships and assumptions within the model while also promoting communication among all stakeholders. For example, a public administration might be particularly interested in assessing the infrastructure's impact in terms of traffic and pollution. Conversely, a public transport company might focus more on reducing operational costs or optimizing fleet management. However, there are common assumptions regarding public transport management that can unite both stakeholders. A hypothesis could be, for example, that both stakeholders could validate that local citizens will adopt electric public transportation at a behavioral level in block BPop_ABM.

The formalization led to the creation of 41 diagrams, divided into 16 processes and 5 blocks, highlighting a combination of procedural and model-related aspects. During the development of SDL diagrams for urban mobility modeling, a significant distinction emerged between procedural and model-related components. In particular, the Bmap_Net and Bsc blocks, highlighted in light blue (as shown in Figure 4), exhibit a combination of both features. The highlighted blocks, Bmap_Net and Bsc, show this fusion of procedural and model-related aspects. Within these blocks, SDL diagrams are outlined with partial coloring: those in black refer to the pure model, while those in light blue indicate procedural aspects. In Figure 4, the Bcom_val diagram responsible for the validation of the DT is also colored in purple. In this case, it is intended to explicitly state that this block is aimed at improving the model but does not represent the actual model itself. In addition, once the SUMOSU station is implemented, it could provide real-time data from their monitoring devices to the DT necessary for the Solution Validation described in the introduction. Solution Validation is essential for verifying the effectiveness of proposed or implemented solutions in achieving their goals, meeting the needs of end users, and addressing identified challenges.

In contrast, Experimental Validation occurs before the implementation of new infrastructure, using, for example, existing data from road sensors to test and refine the DT conceptual model. This step ensures that the initial assumptions and predictions are accurate and reliable based on historical data. The BComp_Val block performs experimental validation by comparing real data from road sensors with simulated data and improving the ABM Block by modifying some parameters.

Since Solution Validation is the final step in the validation cycle, the real-time data from SUMOSU enable continuous validation, which is an essential ongoing process once the DT is operational. This validation process verifies that the DT remains accurate and reliable over time, adapting to changing conditions in the physical system. By regularly comparing the DT's simulations with real-world data from SUMOSU, stakeholders can assess performance metrics such as traffic flow, energy consumption, and environmental impact based on increasingly accurate forecasts owing to continuous interaction with the real world.

5.2. Implementation Tools: Identification and Initial Testing

SUMO and its SAGA framework have been identified as key tools for implementing the conceptualization developed with SDL, representing the component Digital Master models instances, as illustrated in Figure 1. These tools have been tested only to verify their basic capabilities in line with our conceptual model requirements. This testing phase represents an initial verification of their suitability, while the actual implementation of a digital twin, which requires continuous validation through real-time data integration and repeated simulations, has not yet been performed and remains for future development phases.

SUMO can assess emissions, energy consumption, and traffic effects in detail through microscopic simulation (as conceptualized in block BTraff_Sim), providing a criterion for the block BSc, where the different technologies identified qualitatively in each scenario are quantified. SAGA, on the other hand, can implement ABM represented by the block BPop_ABM using information extracted from the map generated as explicated in block Bmap_Net, thus enriching the understanding and predictive capacity of the DT. This integration combines SUMO's microscopic simulation capabilities, which calculate emissions and energy consumption using the Handbook of Emission Factors [56], with SAGA's ability to generate trips based on real traffic volume data. This complementary approach enhances the digital twin's predictive capabilities: SUMO provides accurate environmental impact calculations at the individual vehicle level, while SAGA ensures that the simulated mobility patterns reflect real-world traffic conditions.

Such data are synthetic, meaning activities and trips are created in the BPop_ABM and have not been collected, as described previously, from mobile phones.

Owing to its ability to simulate the behavior of individual vehicles and agents on the road based on social parameters, SUMO allows for a detailed representation of traffic flow in an urban context. This approach enables a more nuanced understanding of mobility patterns influenced by social factors.

As evident in Figure 6, the visualization output of the SUMO-SAGA simulation captures the road network of Bolzano, including major arteries and secondary roads as in the block Bmap_Net. The trips of individuals are evaluated based on the activities they perform, which have been calculated using data from Open Street Map (see block BPop_ABM. The distribution of the several vehicles (colored rectangles) across the city provides insight into traffic flow patterns, potential congestion areas, and the impact of the city's topography on mobility. This visualization not only demonstrates the practical implementation of the conceptual model but also serves as a tool for urban planners and policymakers to assess real-time traffic conditions, analyze road congestion, and identify critical points in the city's transportation network.

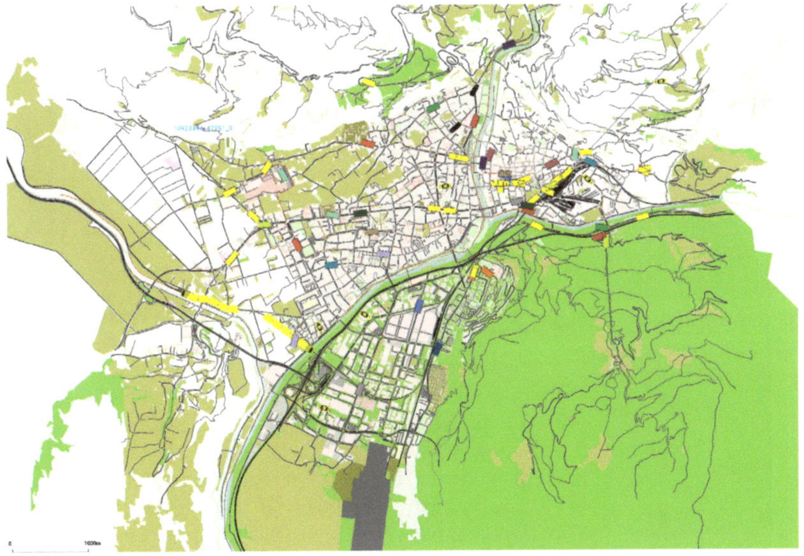

Figure 6. Bolzano's urban mobility visualization through SUMO-SAGA implementation.

To further test SUMO's capabilities in line with the conceptual model, the authors conducted additional tests on a smaller scale. Specifically, SUMO was applied to a particular scenario in a small neighborhood of Bolzano (Figure 7) to assess its ability to calculate energy consumption and emissions (block BSc).

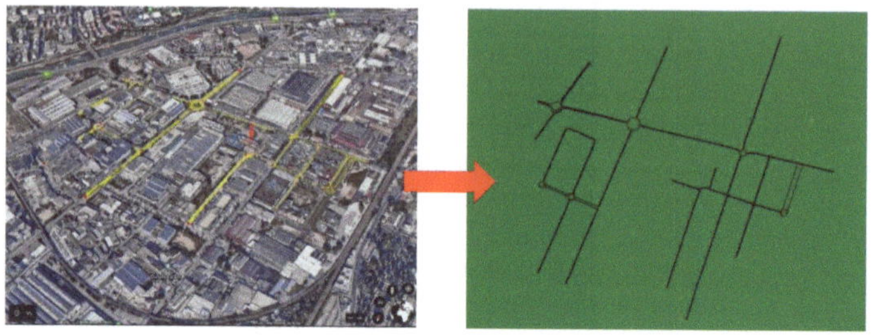

Figure 7. Road network discretization form OSM to test SUMO in a Bolzano district.

On this network, the authors simulated two basic scenarios to demonstrate SUMO's capabilities in analyzing different vehicle types. The first scenario involved a traditional internal combustion bus, while the second featured an electric vehicle. These simple, contrasting scenarios were chosen to illustrate SUMO's ability to model and compare different propulsion technologies.

As shown in Figures 8 and 9, the following charts present the results of these simulations, showcasing key performance metrics for each bus type:

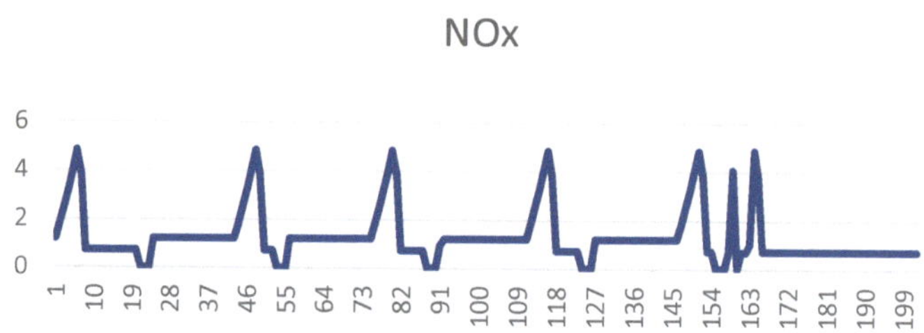

Figure 8. Bus NOx emissions (mg/s) over time (s).

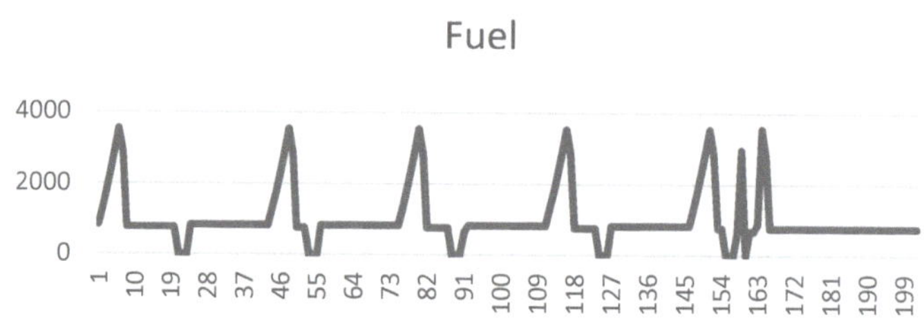

Figure 9. Bus fuel consumption (mg/s) over time (s).

It is important to note that these charts represent instantaneous values from microscopic simulations, capturing moment-to-moment variations in vehicle performance. To optimize different technologies (as conceptualized in BSc) and assess their overall impact, it is necessary to integrate these curves over time and aggregate the contributions from multiple vehicles. Figure 10 demonstrates SUMO's capability to model complex aspects of electric mobility, specifically showcasing an electric car's energy consumption pattern. The graph's positive values indicate power consumption during acceleration and cruising, while the negative values represent regenerative braking (the electric motor acts as a generator to recharge the battery during deceleration).

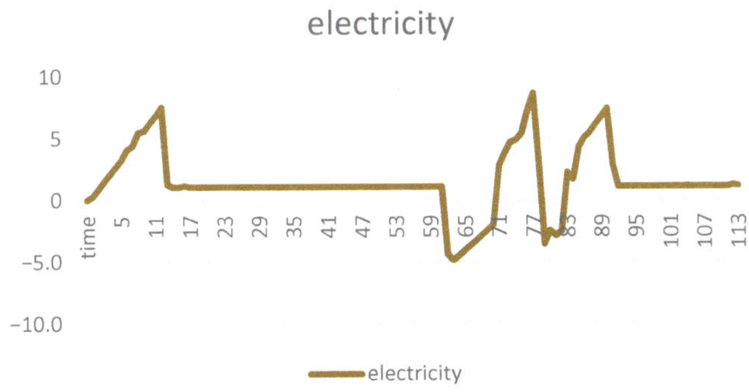

Figure 10. Electricity consumption (kW) of electric vehicle over time (s).

The comparison of real traffic data with simulated and synthetic data—crucial for refining the mobility demand generated in the BPop_ABM block—is handled within the BComp_Val block. Unlike SUMO and SAGA, which focus on specific aspects of simulation and modeling, BComp_Val is responsible for the integration and evaluation of real-time traffic conditions and simulated scenarios. This comprehensive assessment enables iterative refinement of the mobility demand model, a process beyond the capabilities of SUMO and SAGA alone.

6. Discussion

6.1. Analysis of Results and Strengths

The outcomes of this study derive from the formalization of the process of constructing a DT for urban mobility, with Bolzano as a case study. As demonstrated through the 41 SDL diagrams presented in Section 3.4, and particularly evident in the formalization of the validation process (Figure 3), the success of this approach is based on the use of the language SDL, which standardizes the DT through appropriate graphical symbols, making the process clear and independent of the simulation tools used. This standardization, following the principles established by [51], and demonstrated in our case study through stakeholder engagement (Section 5.1), facilitates understanding and communication among the involved parties, as described in the previous chapter.

Although SUMO and SAGA demonstrate capabilities in simulating urban mobility and calculating emissions and energy consumption, as shown in Section 5.2, it is important to note that they implement an Activity-Based Modeling approach but do not constitute a true DT. As defined in Section 3.1, a true DT requires real data to continuously improve simulations through an ongoing validation process. In line with this requirement, the proposed DT relies on real data from traffic sensors to continuously improve and update the activity model. This continuous improvement process, formalized in the model through the BComp_Val block (Section 3.4), constitutes a key difference from traditional simulation approaches. These relationships, which constitute a process of continuous improvement or ongoing validation, are explicitly detailed in the conceptual model but are not directly implemented by SUMO and SAGA. This requires the repeated use of the SUMO simulator and thus the ability to use it multiple times to compare real data with simulated data. Based on this distinction between traditional simulation tools and true DTs (Section 2), a crucial element in bridging this gap and enabling true DT functionality is TraCI (Traffic Control Interface for SUMO). TraCI uses a TCP-based client/server architecture to provide access to SUMO, allowing for dynamic interaction with the simulator. As demonstrated in our implementation (Section 5.2), this interface enables the integration of real data into the simulation process and continuous validation of the model, thus improving responsiveness while maintaining the standardization provided by SDL.

The approach presented in this study offers several key strengths that set it apart from traditional urban mobility modeling techniques:

1. Real-time integration: Unlike conventional simulation tools, as evidenced by Section 3.4, the method continuously incorporates live data from traffic sensors and other sources. This real-time integration enables a true digital twin representation of the urban environment, allowing the model to accurately reflect current conditions and adapt to changes as they occur;
2. Flexibility and adaptability: As shown by Section 3.4, the approach offers remarkable responsiveness to changing urban conditions through its SDL-based formalization. The modular structure of SDL diagrams enables the integration with other digital twins (such as weather monitoring systems) and allows for the expansion of the model through new blocks and processes to analyze specific phenomena. This adaptability makes it possible to study less frequent events, such as traffic congestion due to accidents or behavioral changes during adverse weather conditions, by capturing real data when these events occur. While SUMO and TraCI serve as implementation tools, it is the extensible nature of the SDL formalization that provides the true flexibility, allowing the model to evolve and incorporate new components as needed in the ever-changing landscape of urban mobility;
3. Balanced standardization and dynamism: Following established digital twin principles process [11], the approach strikes a careful balance between structure and flexibility. On the one hand, the use of SDL provides a standardized framework for representing the digital twin, ensuring consistency and clarity in the model's structure. On the other hand, the dynamic interactions enabled by the operational validation (Section 3.1) that reside in the block BComp_Val (Section 3.4) allow for real-time adjustments and updates. This combination creates a robust yet flexible approach to urban mobility modeling, capable of maintaining a standardized structure while adapting to the dynamic nature of urban traffic systems. The present study leverages TraCI to enable a dynamic interaction between real-world traffic data and the simulated environment. This feedback mechanism, as shown in Figure 5, demonstrates that as new data are collected from traffic sensors and other sources, these can be incorporated into the ongoing simulation, helping to keep the model current. As highlighted in recent DT research [57], this updating process aligns with the concept of a digital twin, as it allows the virtual representation to evolve alongside the physical system it represents.

6.2. Technical Limitations and Challenges

Despite its strengths, the approach presented in this study faces limitations and challenges that warrant consideration:

1. Computational intensity: As evidenced in the block BSc (see Supplementary Materials), the method's reliance on continuous validation through repeated use of SUMO simulations could present a significant computational challenge. While this aspect will be fully assessed during the implementation phase, considering that this computational intensity would apply even to a medium-sized city like Bolzano, this process may require substantial computing resources, potentially limiting the model's scalability or real-time performance in larger urban settings or when dealing with more complex scenarios;
2. Data dependency: Consistent with findings from other urban digital twin studies [58], while the integration of real-time sensor data is a key strength, it also introduces a notable dependency. The approach's effectiveness is closely tied to the availability and quality of real-time data from traffic sensors and other urban monitoring systems. This dependency may limit the model's applicability in areas with less developed data infrastructure or in cities where comprehensive sensor networks are not yet in place;
3. Potential for expanded use of synthetic data: As explained in Section 3.4, the digital twin uses only synthetic traffic volume data to compare with real data. In the future, this setup will allow for operational validation during implementation. Additionally, given the presence of air quality sensors in Bolzano, these data could also be used

to achieve a more comprehensive operational validation. Both synthetic and real traffic volume data are generated by the simulator, though other synthetic data from different simulators could also be integrated to assess extreme traffic conditions, such as catastrophic natural events [59].

6.3. Study Limitations

In addition to these technical challenges, this study presents broader limitations that warrant acknowledgment:

1. Focus on conceptual validation: This study focuses specifically on the conceptual validation of the digital twin formalization process. Other forms of validation, such as operational validation and technical implementation validation, will be addressed in subsequent studies;
2. Single urban context: The formalization process has been developed considering Bolzano's characteristics. Its applicability to cities with different scales, mobility patterns, and infrastructure will be investigated in future validations;
3. Stakeholder scope: The current validation process involved a specific set of stakeholders (Section 5.2) in the Bolzano context. Future validation phases will need to consider different stakeholder groups and institutional frameworks;
4. Validation process scope: while this study establishes the foundation for Digital Twin formalization, additional validation aspects including economic feasibility and organizational integration will be explored in future research;
5. Mobility patterns influenced by social factors (Block BComp_Val): The synthetic population's activities in the model are limited to primary, secondary, and domestic categories, lacking differentiation for specific types like shopping or business. This reduces the model's precision in capturing true mobility patterns. Furthermore, journeys rely on polylines from Open Street Map (OSM), which connect to building areas. While this avoids using mobile phone data, the assumption that travel volume correlates with building area may lead to inaccuracies, especially in urban environments with many skyscrapers.

6.4. Future Research Directions

Building on the validation framework presented in Section 3.1, future research will address the subsequent phases of the validation process. While this study has focused on the Conceptual Model Validation phase, the next steps will involve the Computerized Model Verification and Operational Validation phases. These phases will include the development and verification of computer implementation to ensure it accurately represents the formalized conceptual model, followed by validation of the digital twin in a real urban environment. This progression will also concentrate on enhancing the continuous data validation processes, focusing on real-time data integration and quality assurance. Moreover, the measurements of environmental impacts, such as NOx emissions and fuel consumption, will be extended to consider seasonal variations (summer/winter), providing a more comprehensive understanding of how temporal factors affect urban mobility patterns and environmental impacts. Through these validation phases, we aim to address the technical and methodological challenges identified in this study, working towards the development of a fully operational DT for urban mobility management.

7. Conclusions

This research has focused on the conceptual validation phase of developing a DT for urban mobility. Through SDL formalization, we have established a theoretical framework that has been validated with stakeholders, particularly Urban Resilience, ensuring that our assumptions and theoretical representations adequately capture the characteristics of the urban mobility system.

While identifying SUMO and SAGA as potential implementation tools, this study has primarily focused on validating the conceptual model structure and relationships.

Initial tests with these tools in Bolzano have demonstrated their basic capabilities, but their actual implementation as part of a Digital Twin—which requires continuous validation and real-time data integration—remains for future phases.

The SDL formalization has also identified TraCI as a key component for future implementation, as its architecture aligns with the conceptual model's requirements for enabling real-time data integration and continuous validation. However, this integration, like the actual implementation of continuous simulation and validation processes, is a part of future development phases.

While the conceptual validation has proven effective with a moderately sized city like Bolzano, the integration of digital twins into broader urban systems can enhance the overall efficiency and sustainability of urban management. The technologies considered in this study primarily focus on replacing traditional technologies, such as cars, with alternative technologies. However, technologies like connectivity and intramodality have not been considered. Future research should aim to include these technologies to obtain a more comprehensive and integrated view of urban mobility.

Future research should also aim to extend the SDL-based formalization process to encompass a more extensive range of urban services, ensuring that the digital twin framework remains scalable and adaptable to different city sizes and configurations. By focusing on a holistic approach to urban planning and management, the potential benefits of digital twins can be fully realized, contributing to smarter, more resilient, and sustainable urban environments.

Supplementary Materials: The following supporting information can be downloaded at: https://www.mdpi.com/article/10.3390/logistics8040117/s1, This supplementary material contains SDL (Specification and Description Language) diagrams formalizing the concept of the digital twin. These diagrams provide a graphical representation of the structure and interactions of the digital twin, offering a detailed overview of its functioning and features. Figures 3–5 in the main text are derived from it.

Author Contributions: Conceptualization, A.G. and P.F.i.C.; methodology, P.F.i.C. and A.G.; software, A.G.; validation, A.G. and P.F.i.C.; formal analysis, A.G. and P.F.i.C.; investigation, A.G. and P.F.i.C.; resources, W.S. and A.Z.; data curation, A.G., A.Z. and W.S.; writing—original draft preparation, A.G., P.F.i.C., A.Z. and W.S.; writing—review and editing, A.G. and P.F.i.C.; visualization, A.G.; supervision, W.S. and P.F.i.C.; project administration, A.Z. and W.S.; funding acquisition, W.S. and A.Z. All authors have read and agreed to the published version of the manuscript.

Funding: This study was funded by the European Union—NextGenerationEU, in the framework of the consortium iNEST—Interconnected Nord-Est Innovation Ecosystem (PNRR, Missione 4 Componente 2, Investimento 1.5 D.D. 1058 23/06/2022, ECS_00000043—Spoke1, RT3, CUP I43C22000250006). The authors thank the Department of Innovation, Research University and Museums of the Autonomous Province of Bozen/Bolzano for covering the Open Access publication costs.

Data Availability Statement: No new data were created in this study.

Conflicts of Interest: The authors declare no conflict of interest.

References

1. Hannah, F. *Mobility Management—Insights and Examples to Successful Implementation Eltis*; European Commission: Brussels, Belgium, 2021.
2. Solé-Ribalta, A.; Gómez, S.; Arenas, A. Decongestion of Urban Areas with Hotspot Pricing. *Netw. Spat. Econ.* **2018**, *18*, 33–50. [CrossRef]
3. Ritchie, H.; Roser, M. *Land Use*; Our World in Data: Wales, UK, 2013; Available online: https://ourworldindata.org/land-use (accessed on 22 March 2023).
4. Urbanization; Our World in Data: Wales, UK. 2023. Available online: https://ourworldindata.org/urbanization#:~:text=Across%20most%20high-income%20countries%20%E2%80%93%20across%20Western%20Europe,,80%25%20of%20the%20population%20lives%20in%20urban%20areas (accessed on 22 March 2023).
5. Salcedo-Sanz, S.; Cuadra, L.; Alexandre-Cortizo, E.; Jiménez-Fernández, S.; Portilla-Figueras, A. Soft-Computing: An Innovative Technological Solution for Urban Traffic-Related Problems in Modern Cities. *Technol. Forecast. Soc. Chang.* **2014**, *89*, 236–244. [CrossRef]

6. Ilahi, A.; Belgiawan, P.F.; Axhausen, K.W. Influence of Pricing on Mode Choice Decision Integrated with Latent Variable. In *Mapping the Travel Behavior Genome*; Elsevier: Amsterdam, The Netherlands, 2020; pp. 125–143. [CrossRef]
7. Dong, S.; Ren, G.; Xue, Y.; Liu, K. How Does Green Innovation Affect Air Pollution? An Analysis of 282 Chinese Cities. *Atmos. Pollut. Res.* **2023**, *14*, 101863. [CrossRef]
8. Eicker, U. *Urban Energy System for Low-Carbon Cities*; Elsevier: Amsterdam, The Netherlands, 2019; pp. 1–15. [CrossRef]
9. Modular Metacognitive Digital Twin Technologies for Greener Cities & Cleaner Mobility. *Adv. Appl. Energy* **2022**, *5*, 100081. [CrossRef]
10. Sharma, A.; Kosasih, E.; Zhang, J.; Brintrup, A.; Calinescu, A. Digital Twins: State of the Art Theory and Practice, Challenges, and Open Research Questions. *J. Ind. Inf. Integr.* **2022**, *30*, 100383. [CrossRef]
11. Fonseca i Casas, P. A Continuous Process for Validation, Verification, and Accreditation of Simulation Models. *Mathematics* **2023**, *11*, 845. [CrossRef]
12. Schmidt, R.; Möhring, M.; Härting, R.C.; Reichstein, C.; Neumaier, P.; Jozinović, P. Industry 4.0—Potentials for Creating Smart Products: Empirical Research Results. *Lect. Notes Bus. Inf. Process.* **2015**, *208*, 16–27. [CrossRef]
13. Fonseca i Casas, P.; Garcia i Subirana, J.; Garcia i Carrasco, V. Modeling SARS-CoV-2 True Infections in Catalonia through a Digital Twin. *Adv. Theory Simul.* **2023**, *6*, 2200917. [CrossRef]
14. Fonseca i Casas, P.; Pi Palomés, X.; Casanovas Garcia, J.; Jové, J. Definition of Virtual Reality Simulation Models Using Specification and Description Language Diagrams. In Proceedings of the SDL 2013: Model. Driven Dependability Engineering, Montreal, QC, Canada, 26–28 June 2013; pp. 258–274. [CrossRef]
15. SUMO. Available online: https://eclipse.dev/sumo/ (accessed on 6 November 2023).
16. Codecà, L.; Erdmann, J.; Cahill, V.; Härri, J. SAGA: An Activity-Based Multi-Modal Mobility Scenario Generator for SUMO. In Proceedings of the Vol. 1 (2020): SUMO User Conference 2020, Virtual Event, 26–28 October 2020.
17. Oryani, K.; Harris, B. Review of Land Use Models: Theory and Application. 1997. Available online: https://rosap.ntl.bts.gov/view/dot/33825 (accessed on 22 March 2023).
18. Ustaoglu, E.; Aydinoglu, A. Theory, Data, and Methods: A Review of Models of Land-Use Change. 2019. Available online: https://www.igi-global.com/chapter/theory-data-and-methods/230528 (accessed on 22 November 2023).
19. van Nes, R.; de Jong, G. Transport Models. *Adv. Transp. Policy Plan.* **2020**, *6*, 101–128. [CrossRef]
20. Apreda, C.; D'Ambrosio, V.; Di Martino, F. A Climate Vulnerability and Impact Assessment Model for Complex Urban Systems. *Environ. Sci. Policy* **2019**, *93*, 11–26. [CrossRef]
21. Tang, X.; Chen, N. A Multi-Scenario Urban Resilience Simulation Study Based on the System Dynamics Model. *Phys. Chem. Earth Parts A/B/C* **2023**, *132*, 103467. [CrossRef]
22. Zhang, Y.; Zhang, F.; Fang, L.; Chen, N. Inferring Socioeconomic Environment from Built Environment Characteristics Based Street View Images: An Approach of Seq2Seq Method. *Int. J. Appl. Earth Obs. Geoinf.* **2023**, *123*, 103458. [CrossRef]
23. Deng, Z.; Weng, D.; Liu, S.; Tian, Y.; Xu, M.; Wu, Y. A Survey of Urban Visual Analytics: Advances and Future Directions. *Comput. Vis. Media* **2022**, *9*, 3–39. [CrossRef]
24. Mueller, K.; Axhausen, K.W. *Hierarchical IPF: Generating a Synthetic Population for Switzerland*; ERSA Conference Papers; ETH Library: Zürich, Switzerland, 2011; Available online: https://www.research-collection.ethz.ch/bitstream/handle/20.500.11850/39167/eth-3088-01.pdf (accessed on 28 November 2023)ERSA Conference Papers.
25. Calvert, S.; Minderhoud, M.; Taale, H.; Wilmink, I.; Knoop, V. Traffic Assignment and Simulation Models. 2016. Available online: https://www.researchgate.net/publication/287958614_Traffic_Assignment_and_Simulation_Models (accessed on 29 November 2023).
26. Shan, R.; Zhong, M.; Lu, C. Comparison Between Traditional Four-Step & Activity-Based Travel Demand Modeling—A Case Study of Tampa, Florida. In Proceedings of the Second International Conference on Transportation Information and Safety, Wuhan, China, 29 June–2 July 2013; pp. 627–633. [CrossRef]
27. Khan, N.A.; Habib, M.A. Microsimulation of Mobility Assignment within an Activity-Based Travel Demand Forecasting Model. *Transp. A Transp. Sci.* **2023**, *19*, 1983664. [CrossRef]
28. PTV Vissim. Available online: https://www.ptvgroup.com/en/products/ptv-vissim (accessed on 28 November 2023).
29. Citilab. Available online: https://www.bentley.com/software/cube/ (accessed on 28 November 2023).
30. TransModeler SE. Available online: https://www.caliper.com/TransModeler/transmodeler-se-analysis-software.htm (accessed on 28 November 2023).
31. Drabicki, A.; Szarata, A.; Kucharski, R. Suppressing the Effects of Induced Traffic in Urban Road Systems: Impact Assessment with Macrosimulation Tools—Results from the City of Krakow (Poland). *Transp. Res. Procedia* **2020**, *47*, 131–138. [CrossRef]
32. TransModeler. Available online: https://www.caliper.com/TransModeler/default.htm (accessed on 28 November 2023).
33. PTV Visum. Available online: https://www.ptvgroup.com/en/products/ptv-visum (accessed on 28 November 2023).
34. EMME. Available online: https://www.bentley.com/software/emme/ (accessed on 28 November 2023).
35. Wang, Y.; Szeto, W.Y.; Han, K.; Friesz, T.L. Dynamic Traffic Assignment: A Review of the Methodological Advances for Environmentally Sustainable Road Transportation Applications. *Transp. Res. Part B Methodol.* **2018**, *111*, 370–394. [CrossRef]
36. Bliemer, M.C.J.; Raadsen, M.P.H.; Brederode, L.J.N.; Bell, M.G.H.; Wismans, L.J.J.; Smith, M.J. Genetics of Traffic Assignment Models for Strategic Transport Planning. *Transp. Rev.* **2017**, *37*, 56–78. [CrossRef]
37. DYNAMEQ. Available online: https://www.bentley.com/software/dynameq/ (accessed on 28 November 2023).

38. Ben-Akiva, M.; Bierlaire, M. Discrete Choice Methods and Their Applications to Short Term Travel Decisions. In *Handbook of Transportation Science*; Hall, R.W., Ed.; International Series in Operations Research & Management Science; Springer: Boston, MA, USA, 1999; pp. 5–33. ISBN 978-1-4615-5203-1.
39. Bierlaire, M. A Short Introduction to Biogeme. 2023. Available online: https://transp-or.epfl.ch/documents/technicalReports/Bier23.pdf (accessed on 29 November 2023).
40. Brooker, A.; Gonder, J.; Lopp, S.; Ward, J. ADOPT: A Historically Validated Light Duty Vehicle Consumer Choice Model. 2015. Available online: https://www.sae.org/publications/technical-papers/content/2015-01-0974/ (accessed on 29 November 2023).
41. Mehdizadeh, M.; Nordfjaern, T.; Klöckner, C.A. A Systematic Review of the Agent-Based Modelling/Simulation Paradigm in Mobility Transition. *Technol. Forecast. Soc. Chang.* **2022**, *184*, 122011. [CrossRef]
42. Axhausen, K.W.; Horni, A.; Nagel, K. (Eds.) *The Multi-Agent Transport Simulation MATSim*; Ubiquity Press: London, UK, 2016.
43. Muratori, M.; Jadun, P.; Bush, B.; Bielen, D.; Vimmerstedt, L.; Gonder, J.; Gearhart, C.; Arent, D. Future Integrated Mobility-Energy Systems: A Modeling Perspective. *Renew. Sustain. Energy Rev.* **2020**, *119*, 109541. [CrossRef]
44. Preface to the Third Edition. In *Theory of Modeling and Simulation*, 3rd ed.; Zeigler, B.P., Muzy, A., Kofman, E., Eds.; Academic Press: Cambridge, MA, USA, 2019; pp. xxi–xxii. ISBN 978-0-12-813370-5.
45. Petri, C.A. Kommunikation mit Automaten. 1962. Available online: http://edoc.sub.uni-hamburg.de/informatik/volltexte/2011/160/pdf/diss_petri.pdf (accessed on 29 November 2023).
46. Casas, P.F.I. *Formal Languages for Computer Simulation: Transdisciplinary Models and Applications*, 1st ed.; IGI Global: Hershey, PA, USA, 2013; ISBN 978-1-4666-4369-7.
47. Yeon, H.; Eom, T.; Jang, K.; Yeo, J. DTUMOS, Digital Twin for Large-Scale Urban Mobility Operating System. *Sci. Rep.* **2023**, *13*, 5154. [CrossRef]
48. Semanjski, I.C. Smart Urban Mobility: Transport Planning in the Age of Big Data and Digital Twins. In *Smart Urban Mobility: Transport Planning in the Age of Big Data and Digital Twins*; Elsevier: Amsterdam, The Netherlands, 2023; pp. 1–257. [CrossRef]
49. Open Data Hub Analytics. Available online: https://analytics.opendatahub.com/ (accessed on 19 May 2024).
50. Sargent, R.G. Verification and Validation of Simulation Models. *J. Simul.* **2013**, *7*, 12–24. [CrossRef]
51. Fonseca Casas, P. Using Specification and Description Language to Define and Implement Discrete Simulation Models. 2010. Available online: https://upcommons.upc.edu/handle/2117/8341 (accessed on 29 November 2023).
52. Brade, D. Enhancing Modeling and Simulation Accreditation by Structuring Verification and Validation Results. In Proceedings of the 2000 Winter Simulation Conference Proceedings (Cat. No.00CH37165), Orlando, FL, USA, 10–13 December 2000; Volume 1, pp. 840–848.
53. ITU. Available online: https://www.itu.int/en/about/Pages/default.aspx (accessed on 28 January 2024).
54. Bolzano Andamento Demografico e "Bolzano 2023—La Città in Cifre"/Comunicati Stampa/Novità/Città Di Bolzano—Città Di Bolzano. Available online: https://opencity.comune.bolzano.it/Novita/Comunicati-stampa/Andamento-demografico-e-Bolzano-2023-La-citta-in-cifre (accessed on 16 May 2024).
55. Urban Resilience—A Global Infrastructure Initiative. Available online: https://urbanresilience.eu/es/ (accessed on 4 June 2024).
56. HBEFA—Handbook Emission Factors for Road Transport. Available online: https://www.hbefa.net/ (accessed on 25 October 2024).
57. Lydia Van Aert Digital Twinning Is SimCity with Real Cities. Available online: https://www.cursor.tue.nl/en/news/2021/maart/week-4/digital-twinning-is-simcity-with-real-cities/ (accessed on 7 November 2023).
58. Bachechi, C. Digital Twins for Urban Mobility. *Commun. Comput. Inf. Sci.* **2022**, *1652*, 657–665. [CrossRef]
59. Attaran, M.; Celik, B.G. Digital Twin: Benefits, Use Cases, Challenges, and Opportunities. *Decis. Anal. J.* **2023**, *6*, 100165. [CrossRef]

energies

Article

An Interdisciplinary Double-Diamond Design Thinking Model for Urban Transport Product Innovation: A Design Framework for Innovation Combining Mixed Methods for Developing the Electric Microvehicle "Leonardo Project"

Sara Viviani [1,*], Michelangelo-Santo Gulino [2], Alessandra Rinaldi [1] and Dario Vangi [2,*]

[1] Innovation in Design & Engineering (IDEE) Laboratory, Department of Architecture (DIDA), University of Florence, 50121 Florence, Italy; alessandra.rinaldi@unifi.it
[2] Department of Industrial Engineering (DIEF), University of Florence, 50139 Firenze, Italy; michelangelosanto.gulino@unifi.it
* Correspondence: sara.viviani@unifi.it (S.V.); dario.vangi@unifi.it (D.V.)

Abstract: The increase in greenhouse gas emissions prompts the transport sector towards new technological perspectives on personal mobility. Addressing sustainable mobility through electric micromobility requires interdisciplinary design research methods and approaches. In the context of the LEONARDO project, funded under the Horizon 2020 framework, this paper addresses a critical literature review on the design thinking, design research models, tools, and mixed methods to be undertaken for driving product mobility innovation in a cross-disciplinary context. Following the "research through design" research strategy, the authors applied the Double-Diamond design thinking model to frame the design research process in four phases, aligning with three overarching objectives, four specific research objectives, and 24 research tasks, supported by a total of 71 mixed methods and tools. As a result, the transdisciplinary process provides a co-designed energy-efficient stand-alone microvehicle and a scalable interdisciplinary design model for urban transport product innovation. In conclusion, this case study suggests the value of the Double-Diamond design thinking model as a design research instrument capable of addressing sustainable mobility and guiding interdisciplinary design research, design practice, and education in the industrial engineering and design disciplinary sectors.

Keywords: interdisciplinary design model for urban transport product innovation; double-diamond design thinking model; interdisciplinary research methods; mixed-methods; electric microvehicle; micromobility; sustainable mobility

Citation: Viviani, S.; Gulino, M.-S.; Rinaldi, A.; Vangi, D. An Interdisciplinary Double-Diamond Design Thinking Model for Urban Transport Product Innovation: A Design Framework for Innovation Combining Mixed Methods for Developing the Electric Microvehicle "Leonardo Project". *Energies* **2024**, *17*, 5918. https://doi.org/10.3390/en17235918

Academic Editors: Artur Kierzkowski, Tomasz Kisiel, Agnieszka A. Tubis, Franciszek Restel, Mateusz Zając, Tomasz Nowakowski and Anna Jodejko-Pietruczuk

Received: 30 August 2024
Revised: 5 November 2024
Accepted: 20 November 2024
Published: 25 November 2024

1. Introduction

The increase in greenhouse gas emissions prompts the transport sector towards new technological perspectives on personal mobility [1]. In the last few years, the major strategy adopted to achieve a decrease in pollutants and the adoption of more sustainable and active mobility lifestyles is electric micromobility. This term frames light and active vehicles, human-powered and electrically assisted, that weigh up to 350 kg and can reach a maximum speed between 25 and 45 km/h [2], according to the country's regulations.

The emerging interest in these convenient and efficient alternative means of transport has led to the exploration of the role of design and engineering researchers and practitioners in cross-disciplinary contexts. This paper, therefore, shows the design research results of the research project "MicrovehicLE fOr staNd-Alone and shaReD mObility—LEONARDO", funded by the European Commission under the Horizon 2020 framework.

MicrovehicLE fOr staNd-Alone and shaReD mObility—LEONARDO

The main programme H2020 on "Societal Challenges—Smart, Green, And Integrated Transport" sets the basis for understanding the urgency of achieving citizens' mobility needs, a climate-resilient economy, and an energy-efficient low-carbon society [3]. Focusing on the call for a disruptive change in public transport, the LEONARDO project addresses the topic "Next Generation electrified vehicles for urban and suburban use", proposing a work program tackling the subtopic "Urban light personal mobility" (see [4]). In this research context, the overarching objectives of the LEONARDO project are (i) to define a new microvehicle enhancing existing micromobility technologies to foster the actual paradigm shift in urban mobility [5] through designing a climate- and environmentally friendly and safe mobility solution; (ii) to define the compatibility with sharing mobility schemes; and (iii) to perform an extensive demonstration test in a real environment in Europe for a revision and re-design process to reach the Technology Readiness Level (TRL) 8–9 [6]. Moreover, this research project aims to develop the mass diffusion of microvehicles by implementing and prototyping a fully electric urban transport product with increased autonomy. Indeed, the research team addresses the issues related to the high cost and large need for charging infrastructure, rethinking the existing sharing systems. Notably, the researchers propose a solution where the battery is shared, instead of the microvehicle, reducing the daily time-wasting charging and the city's necessity to provide spaces for recharging parking stations. Therefore, the project foresees the development of an auxiliary battery to adapt to the "Battery pack sharing, monitoring, and management system" (Priority number 102018000001810, see [7]), previously patented by the University of Florence.

This project's consortium has an efficient industrial, administrative, road safety, and consultancy background, involving (i) manufacturing companies, in charge of the prototypes' production and of the design of solutions that consider cost constraints, and (ii) research centres for identifying and collecting technical and bureaucratic product requirements. Notably, WP1 aims to critically review the literature on existing microvehicles and urban vehicles, involving research centres to gather legislative European aspects, data from accidents, and requirements on safety. WP2 has several objectives, including (i) defining and designing the final vehicle concept through the generation of detailed vehicle specifications, involving the manufacturing companies; (ii) identifying and defining mechanical design aspects, such as the exterior and functional features, materials, and structural sizing, involving the manufacturing companies; and (iii) defining electrical design and purchasing or developing elements that have to be integrated into the microvehicle, involving the manufacturing companies. Lastly, WP3 aims to develop a running microvehicle prototype, all the mechanical parts, and components, and design the assembly for field tests, involving the manufacturing companies.

The work plan of this initial research stage let the researchers identify the need for further interdisciplinary collaboration, in addition to the existing partnership, between industrial engineering and design disciplinary sectors. The engineering research team embraces the potential, opportunities, and sources of value creation this cooperation could offer. Therefore, the research team was enlarged to support the design of this attractive, affordable, resource-efficient, safe, seamless, and new micromeans of transport.

In consequence, the current contribution is strictly related to the first three work packages, aiming at developing and providing the first version of the microvehicle for on-field pilots. This renovated systemic research vision also set the basis for broadening the research horizon through the development of an "interdisciplinary design model for urban transport product innovation", enabling research scalability. Starting from this cross-disciplinary perspective, the researchers identify further specific research objectives to address in parallel and support the project's objectives, which are as follows: (i) to define a scalable and replicable design model for urban transport product innovation; (ii) to correlate mixed methods that can be used from different scientific disciplinary sectors; (iii) to highlight the hierarchical structure of the design research phases and the corresponding research tasks to be undertaken; and (iv) to evaluate the design model, developing and prototyping the "Leonardo" microvehicle. Therefore, the research questions are as follows: (i) What are the main design frameworks for innovation capable of supporting the development of new urban mobility products? (ii) Which design research tools and methods enable the urban mobility product development? (iii) Which tools and methods, specific to the industrial engineering disciplinary sector, enable product development?

This paper is organized as follows. Section 2 critically reviews the current literature on the existing frameworks for innovation. The authors are oriented by the previously highlighted research questions and address the second and third specific research objectives. The results are discussed and summed up through the definition of a graphic visualization. Then, Section 3 is about the methodological approach, the design research strategy adopted, how the research tasks have been structured into work packages (WPs), and which tools and methods have been employed. Following this, Section 4 is about the "Leonardo Project" and its developing and prototyping results. In Sections 3 and 4, the research team tackles the third and fourth specific research objectives. Lastly, Section 5 discusses the scalability and replicability of the project through the definition of the design model addressing the first specific research objectives.

2. Design Framework for Innovation

The design research team frames the design thinking approach as a user-centred and iterative driver of "innovation and creative problem-solving" [8,9], capable of defining micromobility solutions valued by testing end-users. This section provides an overview of the present "Double-diamond" design thinking framework for innovation, highlighting its hierarchy, attributes, and related tools. Therefore, the research team sets the basis for delineating the design journey and accurately structuring WPs, research tasks, and activities, starting from dividing the process into a problem and solution space.

The authors consider "design thinking" as a visionary, inspiring, disruptive, creative, and organizational model that bridges the two disciplinary sectors and fosters the expected outcomes across working groups. In recent years, many researchers extended the design domain to the engineering field, discussing the potential of "design thinking" and how it can broaden engineering skills [10]. This approach is envisioned as a problem-oriented, socially positive, and high-performance approach to foster empathy, creativity, and innovation in engineering design [11–13]. The literature also highlights how design thinking could be supportive in a communicative way, facilitating interactions within interdisciplinary and informal social team environments, and capable of achieving the complexity of the research challenges [12,14,15]. Panke [16] discussed how fruitful and effective it is to apply "design thinking" in a varied scientific field in higher education and listed the following positive traits and effects: (i) encouraging tacit experiences; (ii) increasing empathy; (iii) reducing cognitive bias; (iv) promoting playful learning; (v) creating flow/verve; (vi) fostering inter/meta-disciplinary collaboration; (vii) inducing productive failure/increasing resilience; (viii) producing surprising and delightful solutions; and (ix) nurturing creative confidence. These multifaceted features highlight the reasons for integrating design thinking in higher education among different disciplines and demonstrate the possible emerging outputs. Similarly, Micheli et al. [8] provided further insights into "design thinking" by discussing the general attributes to better understand what constitutes it: (i) creativity and innovation; (ii) user-centeredness and involvement; (iii) problem-solving; (iv) iteration and experimentation; (v) interdisciplinary collaboration; (vi) ability to visualize; (vii) gestalt view; (viii) abductive reasoning; (ix) tolerance of ambiguity and failure; (x) blending rationality and intuition; and (xi) design tools and methods. The last feature refers to strategies that can enable the other characteristics. The development of personalized, human-centred, co-creative, and innovative novel ideas is, therefore, determined by tools and methods through balancing analytical and intuitive thinking [17]; understanding relationships, environmental factors, trends, backgrounds, and user needs [18]; accepting and embracing uncertainty [19]; integrating diverse research perspectives [20]; adopting a nonlinear, iterative, and alternative approach; and developing or prototyping visual insights.

Researchers frame tools and methods within design models and phases to better solve "wicked problems" (see [21]). The following subsection delves into the existing design thinking model and, notably, the authors explore the "Double-diamond design thinking model" in more detail.

Double-Diamond Design Thinking Model

Design philosophers, design companies, and councils have published many design thinking models since the design thinking perspective was theorized [22]. The literature presents about 15 similar models that slightly differentiate one another by addressing different challenges in varying research contexts and areas [22,23], among which the most known are as follows: (i) collective action toolkit [24]; (ii) design thinking [25]; (iii) design thinking bootleg [26]; (iv) human-centred design toolkit [27,28]; (v) the double diamond [29]; and (vi) service design [30].

The correlation between the design thinking models presented shows five clear correspondences and Kueh and Thom [22] framed them into the following phases: (i) the context or problem framing phase; (ii) the idea generation phase; (iii) the prototyping phase; (iv) the implementation phase; (v) and the reframing phase. These similarities enable every model to approach the research challenges. Notably, the authors explore the Double Diamond as the model visually capable of highlighting divergent and convergent thinking in the design process.

The straightforward Double-Diamond design thinking process, originally developed by the British Design Council [29], foresees four consequential, explorative (divergent), and exploitative (convergent) key phases: (i) discover; (ii) define; (iii) develop; and (iv) deliver [31].

The "Discover" and "Define" phases refer to the "Context or problem framing phase". The "Develop" phase refers to the "idea generation" phase. The "Deliver" phase refers to the "Prototyping" and "Implementation" phases.

In the first diamond, the "Discover" divergent phase is about "broadening our horizons" through research. This is the inspiring fuzzy starting point on the topic, the products or services constraints, and the related complex problems users are facing. The aim is to understand the context of the problem, identify the macro-trends and the related technological advancements, empathize, and engage whoever is affected by the issues of the research to perceive the end-user perspective, motivations, and latent needs and uncover disruptive ideas. Moreover, the literature provides much research on which tools and methods it could entail [8,16,32–37]: (i) systematic literature review; (ii) secondary research; (iii) case study research; (iv) ethnographic design methods; (v) identify stakeholder; (vi) interview; (vii) survey; (viii) questionnaires; (ix) card sorting; (x) field study; (xi) user observation; (xii) user shadowing; (xiii) empathy maps; (ixv) focus group; (xv) hierarchical task analysis; (xvi) benchmarking (see Figure 1).

The "Define" convergent phase is about "heading towards the core problem" and actionable tasks. This is a reflective phase regarding the previously collected knowledge resource and design insights. The aim is to analyze, synthesize, co-design insights, and organize the input that emerged from the "Discover" phase to define the strategic brief, a technical document reporting systematically the findings into opportunities and objectives [38]. Going on, as the literature stated [8,16,32–37], it could entail the following tools and methods: (i) thematic analysis (see [39]); (ii) content analysis (see [40]); (iii) personas; (iv) stakeholder analysis; (v) stakeholder map; (vi) task mapping; (vii) experience map; (viii) scenarios of use; (ix) user journey map; (x) service blueprint map; (xi) affinity mapping; (xii) mind maps; (xiii) miro.com (see [41]); (ixv) SWOT analysis; (xv) Kano model; (xvi) brainstorming; (xvii) "How might we" question; (xviii) focus group; (xix) co-design workshop; (xx) design brief (see Figure 1).

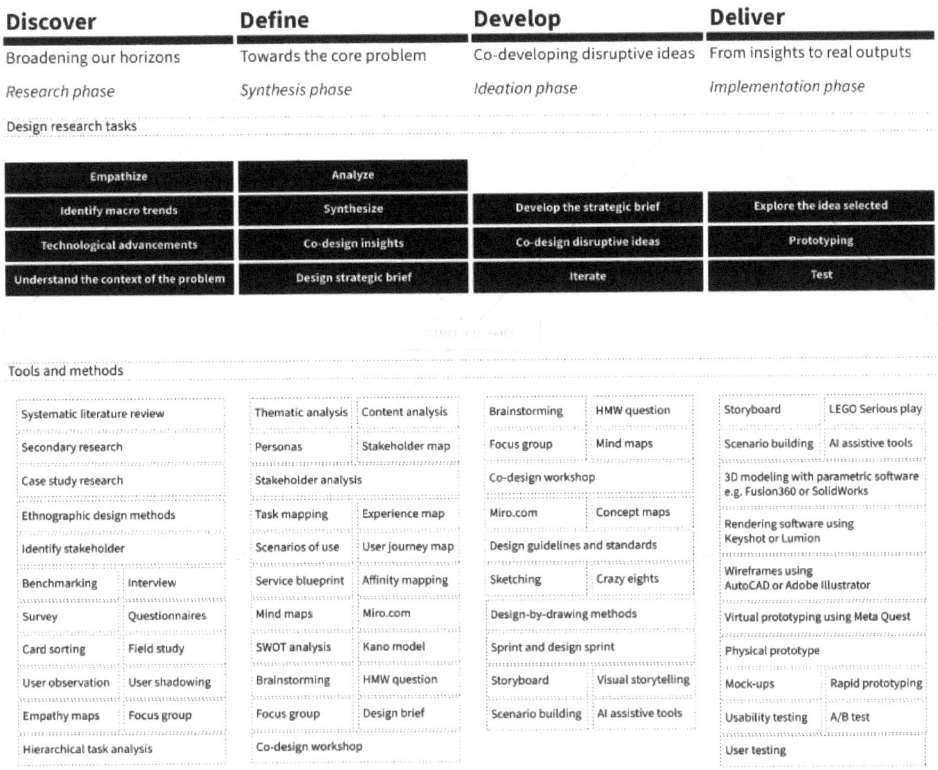

Figure 1. Design Framework for Innovation. The figure adopts the "Double-diamond design thinking model" [29] to summarize the results from the critical literature review.

In the second diamond, the "Develop" divergent phase is about "co-expressing and developing disruptive ideas in a free-thinking environment". This is a collaborative design phase of the strategic brief. The aim is to generate an iterative co-dialogue with involved stakeholders, researchers, designers, and experts on the data gathered to design the outcome's desirability, feasibility, and viability. Moving forward, as the literature stated [8,16,32–37], it entails the following tools and methods: (i) brainstorming; (ii) "How might we" question; (iii) focus group; (iv) co-design workshops; (v) mind maps; (vi) miro.com; (vii) concept maps; (viii) design guidelines and standards; (ix) sketching; (x) crazy eights; (xi) design-by-drawing method; (xii) sprint and design sprint (see [42–44]); (xiii) storyboard; (ixv) visual storytelling; (xv) scenario building; (xvi) AI assistive tools (see Figure 1).

The "Deliver" convergent phase is about "transforming insights into real outputs". This is about prototyping and testing promising emerging insights. The aim is to enable the designers to explore the idea selected, whether in low or high fidelity, select valuable ones, and receive feedback about the users' perception, usability, affordance, or accessibility of the outcome. Lastly, as the literature stated [8,16,32–37], it entails the following tools and methods: (i) storyboard; (ii) LEGO Serious Play (see [45]); (iii) scenario building; (iv) 3D modelling with parametric software (e.g., Fusion [46] or SolidWorks [47]); (v) rendering software using Keyshot [48] or Lumion [49]; (vi) wireframes using AutoCAD or Adobe Illustrator; (vii) virtual prototyping (e.g., Meta Quest); (viii) AI assistive tools; (ix) physical prototype; (x) mock-ups; (xi) rapid prototyping; (xii) 3D printing; (xiii) field experiments; (ixv) usability testing; (xv) user testing; (xvi) A/B test (see Figure 1).

In conclusion, the present section sets the basis for framing the methodological approach of the research project "Leonardo" through a critical literature review of the "Design thinking" attributes, process, models, phases, and tools. The authors, therefore, identified the "Design thinking" process as a framework for innovation and organized the research through the "Double diamond design thinking model".

3. Materials and Methods

The research strategy adopted is "research through design" (see [50]) and aims to produce knowledge to guide interdisciplinary design research, practice, and education in defining urban transport product innovation.

The Design Framework for Innovation, framed in Figure 1, is used as an interdisciplinary design research model for organizing data gathered from the initial brief, structuring the research phases, selecting specific mixed methods, developing research materials, and designing research activities. This model enables the research team to understand the context, opportunities, and insights to explore formal, technological, and functional microvehicle innovation ideas and define a scalable interdisciplinary design model for urban transport product innovation. Notably, the model was used for organizing the research design process in 4 WPs, aligning with 3 specific research objectives for each research direction and encompassing 14 research macro-tasks and 24 corresponding micro-tasks. The research tasks were addressed by employing quantitative and qualitative research methods from different disciplinary sectors.

3.1. WP1: Discover Phase

During the initial "Discover" phase, the team organized the research phase into 4 macro-tasks, presented in Figure 1: (i) identify macro-trends; (ii) understand the context of the problem; (iii) empathize; and (iv) conduct in-depth analysis of the technological advancements.

The first macro-task was addressed by the design research team and aimed at deepening analysis of the topic proposed, which is the research project "LEONARDO". The engineering research team presented their initial brief, the challenges, the research objectives, the expected outcomes, and the deadlines. Thus, the design research team employed the critical literature review and case study research methods to delve into the mobility field and electric micromobility strategy to quickly achieve sustainable goals.

The second macro-task is about understanding the context of the research project and the research team organized the work into the following micro-research tasks: (i) an in-depth analysis of the brief; (ii) an in-depth analysis of the project partners; (iii) an in-depth analysis of the disruptive micromobility product to define; (iv) the mapping of the research areas; and (v) the framing of the research methods. Firstly, the research team re-analyzed the project priorities and constraints, included in the Grant Agreement. Going on, they designed the first participatory activities, with the principal investigator, project experts, research partners, partner company, and design and engineering researchers, through the use of focus groups and semi-structured interviews to identify the corresponding stakeholders and relevant implicit factors. The research team also made comparable benchmarks of the kick scooters and monowheels as product-related market segments. The aim was to analyze the competitors, similar to the blended microvehicle proposed by the research project, and identify formal emerging trends, technological advancements, manufacturing processes adopted, materials employed, user needs, and user evaluations. Finally, brainstorming and workshops with the principal investigator, project experts, and design and engineering researchers were organized to map the research areas and directions to follow for defining the expected outcomes. The research team also framed the research methods into a graphic visualization that was highly iterated during the process and reported in Section 5.

Moving forward, the third macro-research task is about empathizing with the users and performing (i) an in-depth analysis of the stakeholders and (ii) the user interactions. The research team adopted ethnographic design methods from the design research literature review and used semi-structured interviews, hierarchical task analysis, user shadowing, and user observation tools to identify how users interact with kick scooters and monowheels. These qualitative methods set the basis for identifying new design insights regarding the positioning of the feet and the need to delve into anthropometric data. Particularly, the research team sized the components according to the anthropometric parameters for the European population (see [51]).

Lastly, the research team analyzed technological advancements through secondary research. Some previous tasks were iterated, involving creating benchmarks and performing critical academic literature reviews and case study research on materials and manufacturing processes adopted by companies in the mobility sector. The researchers also delved into patent research and technical literature reviews through comfort, mechanical strength, safety, and rigidity regulations to examine which materials could fulfil the project requirements fully. The team deepened the analysis of safety regulations, requirements, and framework conditions to let these vehicles circulate on public roads through accident databases and international literature. Going on, they address the safety analysis through (i) investigation of kick scooter-related regulations in European countries; (ii) mapping of kick scooter accidents in Germany; and (iii) a survey on the use of kick scooters. They addressed legal and technical frameworks and municipal regulations of the microvehicles through accident analysis. The latter investigation was carried out through the analysis and mapping of the (i) conflict and collision situations and constellations; (ii) data information on personal injuries; and (iii) related accident parameters regarding the location and the use of helmets and protective clothing.

3.2. WP2: Define Phase

The "Define" phase was about synthesizing the data collected in the previous stage of development. The team structured the work package into 4 macro-tasks, as presented in Figure 1: (i) analyze the data gathered; (ii) synthesize data; (iii) co-design insights; and (iv) design a strategic brief into a technical document. The first macro-task aimed to deepen the data collected through the use of thematic analysis and content analysis in a participative environment, with the principal investigator, project experts, and design and engineering researchers. Going on, the research team synthesized the knowledge resources through the use of SWOT analysis, brainstorming, and participatory affinity mapping. Coherently, they defined an overview of the product components framing the entire assembly through a task mapping tool to identify the main elements and the research directions that need to be focused on during the "Develop" and "Deliver" work packages. After defining the strengths, opportunities, weaknesses, and threats, the third macro-task fostered co-design activities through brainstorming, focus groups, and co-design workshops to define insights. Lastly, the team organized the work through (i) opportunity areas mapping and (ii) defining the correct research question to orient the second diamond. These two micro-tasks were supported by affinity mapping, HMW questions, and brainstorming tools in a co-design context to design the strategic brief.

3.3. WP3: Develop Phase

The third diverging phase aimed to develop disruptive and innovative ideas in a participatory context. This stage of development is organized as follows: (i) developing the strategic brief; (ii) co-designing disruptive ideas; and (iii) iterating the results that emerged from the previous macro-tasks. Initially, the research team structured the latter two work packages framing the research methods to be employed. HMW questions and a focus group with the principal investigator, the project experts, and design and engineering researchers oriented the following research phases, organizing and structuring the subsequent research tasks. Moving forward, the next research activity aimed to explore new formal solutions. The technological advancements, materials, and manufacturing processes identified in the strategic brief guided this formal exploration, and the researchers employed both design-by-drawing and design sprint methods, and the crazy eight, sketching, and visual storytelling tools. The research team organized and designed a "1-day" design sprint (see [44]), incorporating all the tools cited in the previous paragraph. The design sprint adapted foresees the following phases, tools, and methods: (i) the idea generation through the definition of disruptive and individual HMW questions in 10 min; (ii) the sharing of the HMW questions generated and the categorization in theme through the use of affinity diagram tool in 20 min; (iii) the voting of the most coherent, compelling and important HMW question, that could orient the next research phases in 5 min; (iv) the individual fast sketching exercise through crazy eight tools in 8 min; (v) the sharing and voting of the emerged ideas and sketches in 20 min; (vi) the re-elaboration of the selected sketch in 20 min; (vii) the definition of storyboard and storytelling about how the user should interact with the final product through the use of rendering, 3D modelling, and wireframes software in 40 min; and (viii) a brief presentation of the output in 30 min. In the end, the team iterated the previous tasks as long as they identified a coherent idea.

3.4. WP4: Deliver Phase

The fourth converging phase involved the exploration and successive implementation of the selected idea. The team organized this work package into 3 macro-tasks, as presented in Figure 1: (i) exploring the selected idea; (ii) product prototyping; and (iii) product testing. As already carried out in the previous stage, the researchers iterated the framing of the research methods through HMW questions and focus groups. The second macro-task aimed at prototyping the idea selected. The team divided the tasks into three different approaches: (i) prototyping in a virtual environment; (ii) rapid prototyping; and (iii) physical prototyping. Firstly, the team explored the product output through wireframes and 3D modelling with AutoCAD and the parametric software Solidworks. Subsequently, they explained how the users have to interact with the product through renderings, storyboards, and visual storytelling, defined with Lumion [49], Adobe Illustrator [52], and Photoshop [53]. The last macro-task is about testing and was structured in parallel with the prototyping one as follows: (i) remote quantitative testing and (ii) testing following regulations. Before possibly testing the product in person, the research team performed a usability test with Google Forms. The objective was to understand how attractive, perspicuous, efficient, dependable, stimulative, and innovative the product appears to be. Therefore, the team grouped 27 people to which they presented the Leonardo microvehicle and submitted a questionnaire, asking the following questions: (i) "An overall impression of the product. Do users like or dislike it?"; (ii) "Is it easy to get familiar with the product and learn how to use it?"; (iii) "Can users solve their tasks without unnecessary effort? Does it react fast?"; (iv) "Does the user feel in control of the interaction? Is it secure and predictable?"; (v) "Is it exciting and motivating to use the product? Is it fun to use?"; and (vi) "Is the design of the product creative? Does it catch the interest of users?". The researchers identified the Likert scale as a coherent question type to express the level of agreement or disagreement in the samples and quantitatively analyze the responses.

The first tangible prototypes were singular components of the assembly and were created through additive manufacturing for performing preliminary testing. Later, the research team developed different types of component prototypes for each research direction to be re-assembled into one product prototype, in collaboration with the partner company. Following regulations, the aim was to test safety, comfort, mechanical strengths, and rigidity of the materials identified in the research phase [54].

4. Results

As discussed in the Section 1, the overarching research objectives are as follows: (i) developing a new silent, clean, energy-efficient, and safe microvehicle; (ii) developing a microvehicle compatible with sharing mobility schemes; and (iii) performing an extensive demonstration (see [55]). Notably, the "LEONARDO" research project aims to enhance the features of monowheel and scooter concepts for defining a blended stand-alone microvehicle that is attractive and affordable to a wider market segment. Going on, the project foresees the design of an auxiliary battery as a touchpoint of the "battery-sharing mode" sharing system, patented and developed by the University of Florence, enhancing the project's aim to make the novel microvehicle compatible with sharing mobility schemes. Lastly, the project's expected outcomes strive to test a fleet of microvehicles in a real environment in five European cities, including Rome, Palermo, and Bruxelles.

4.1. *"Discover Phase": Towards Urban Transport Product Innovation*

The first "Discover" phase delves into the main objectives and research challenges. The first three macro-tasks—identifying macrotrends, understanding the context of the problem, and empathizing—set the basis for identifying that the novel microvehicle solution fosters the last mile concept and intermodal issues. In order to address these macrotrends, the design research team clusters the following product scopes: (i) increasing driving stability, ease of driving, ride comfort, and braking ability designing a highly specialized vehicle; (ii) guaranteeing flexibility through the design of a stand-alone and sharing mode; (iii) designing lightness and compactness through materials; (iv) supporting inclusivity in use, suitability, and safety; (v) reducing the dimension-weight and setting a target at 10 kg; (vi) increasing the ease of transport through the design of foldable mechanisms, trolley mode, or self-balancing; (vii) allowing the compatibility with shared mobility scheme, or rather a battery sharing, through the design of an auxiliary battery; and (vii) lowering the cost of sale and the related costs of maintenance. Starting from these product design criteria, the design research team gathers the data collected and designs a participatory workshop, focus groups, and semi-structured interviews involving the principal investigator, project experts, research partners, partner company, and design and engineering researchers. Therefore, the researchers empathize with the corresponding stakeholders, identifying their aim to achieve the initial overarching objectives, working on a hybrid vehicle conceptualized within the engineering research team (see Figure 2). According to this concept, the specific purposes are to design (i) a hub motor with a 14–16″ diameter front wheel; (ii) an easy handling, tilting, flexible, and high resistance platform; (iii) a steering system on the hub motor, lowering the weight; (iv) a high-efficient, easy-balancing handlebar; (v) a clean, minimal, and attractive geometry; (vi) an on-board battery; and (vii) an auxiliary battery, as an assembly component that makes the microvehicle compatible with shared mobility schemes.

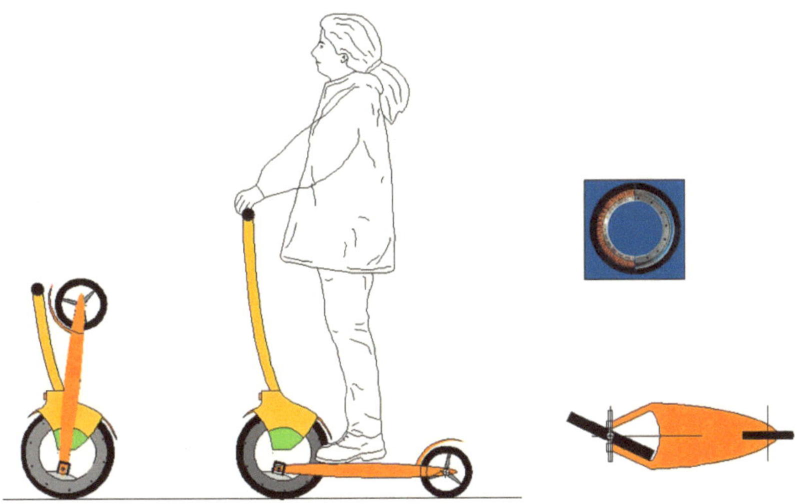

Figure 2. Microvehicle concept conceptualized by the engineering research team for the LEONARDO project (GA 101006687, see [56]).

This initial in-depth analysis of the macrotrend, brief, topic, stakeholders, and user interactions set the basis for dividing the design research process into four main research directions to simultaneously address the following factors: (i) platform designing; (ii) steering column designing; (iii) auxiliary and fixed battery case designing; and (iv) assembly designing. According to the specific project purposes, as previously outlined, the research team has to address the following factors and priorities for every research direction: (i) costs; (ii) lightness; (iii) compactness; and (iv) ease of transport.

During the development of the fourth microtask, the design research team mainly adopted the "Micromobility Landscape" (see [57]) to collect the technological advancements and implement the technical requirements for the microvehicle. In the following subsections, the research team presents in detail every aspect of the final work performed, dividing each one into three parts.

The first one outlines the strategic design brief of the corresponding research direction, including the collection of all information relating to technical/functional requirements and project constraints useful for defining the needs and problems that the product intends to solve. This is the result of the second "Define" phase, which synthesized the data gathered and analyzed in a participatory context.

The second one shows the results that emerged from the "Develop" phase. Notably, this paper presents the embodiment and detailed design, where the set of elaborated ideas is structured. The layouts and the preliminary drawings of the various components are visually depicted and the considerations made on the production processes are listed in technical-economic terms.

Lastly, the authors present the "Deliver" phase results, merging every research direction's prototype design and various tangible outputs in Section 4.4. The authors show the overall result of the "implementation" phase to provide an overview of the assembly.

4.2. Research Direction #1: Platform Designing

4.2.1. "Define Phase" Results: Platform Strategic Design Brief

The benchmark lets us compare similar products available on the market, such as kick scooters and monowheels. This method lays the foreground for defining four specific platform objectives: (i) providing flexibility to the platform, replacing the suspensions of traditional microvehicles for increased lightness; (ii) minimizing the weight of the entire vehicle; (iii) improving the user comfort; and (iv) supporting the "trolley mode", or rather, the ease of transport when the microvehicle is folded so that the platform does not collide with any other part in the folding process.

The first specific research objective refers to the assessment of the materials to be employed, among which the research team identifies (i) aluminum sheets; (ii) fibre-reinforced composites; and (iii) bamboo.

The second research objective refers to the ergonomic design of the platform. The researchers, therefore, identify that the minimum width of the platform in correspondence with the hinge is 190 mm. This measure is based on the front wheel encumbering and kinematics, consisting, respectively, of a 14-inch diameter and a total steering angle of 50 degrees. Starting from these dimensions, the research team defines the minimum width of the overall platform profile, which is 250 mm (Figures 3 and 4). Notably, the research team focuses their output on one relevant platform constraint: to be jointed to the front wheel and hinged a few centimetres below the centre of the wheel. The highlighted characteristic showed the platform should be wider than those commercially available and the ideal rotation angle for adequate maneuverability and driving comfort has to be set to a value of 25 degrees in both directions.

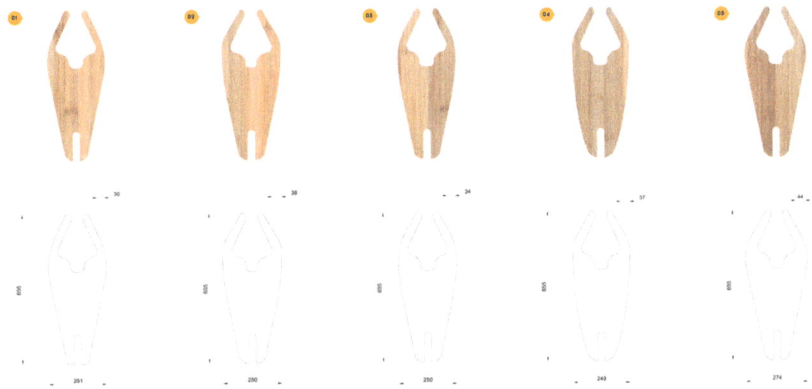

Figure 3. Five different concepts for the platform in terms of encumbering and shape. The research team chose the second concept (see Figure 4).

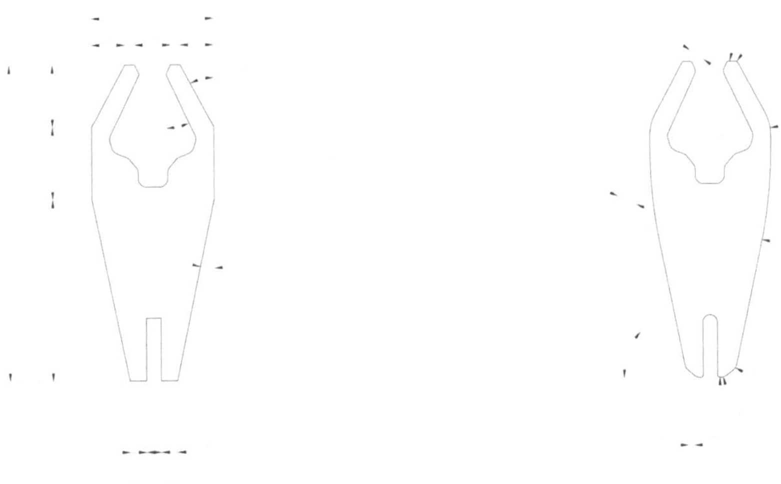

Figure 4. Final selected geometry and encumbering of the LEONARDO platform.

4.2.2. "Develop Phase" Results: Platform Embodiment and Detailed Design

The discussion on the "Design strategic brief" of the platform, presented in Section 4.2.1, is addressed and brainstormed by the research team through a focus group with all the experts on the project.

Starting from the anthropometric data collected, the authors set the basis for identifying a new design idea: users can stand on the platform by placing their feet next to each other. This foot position enhances comfort and allows the assumption of an upright, stable, and natural position compared to typical kick scooters, in which the platform size obliges the placement of the feet one in front of the other.

Therefore, the research team focuses on a renovated formal exploration through the design sprint method, brainstorming, HMW question, crazy eight, design-by-drawing method through AutoCAD [58] and Adobe Illustrator [52], etc. They drew inspiration from the lines of the automotive sector and developed the first fast sketch aiming at conveying dynamism, innovativeness, and propensity to the future to capture the attention of the public and define the image of the product. The selected details are inserted in each acute and protruding intersection point, respecting the symmetry and balancing the proportions.

Regarding the selection of materials and processes, the research team examines three light and structurally valid solution possibilities, including (i) a sandwich of fibreglass with a PVC core with inserts of wood at the height of the hinges; (ii) layers of bamboo, with diverse thicknesses, glued together with epoxy, with lower aluminum support; and (iii) aluminum sheets of minimum thickness with aluminum ribs. These three possible alternatives are conceptualized, physically prototyped, and tested in order to be inserted inside the overall LEONARDO vehicle model. The best solution to adopt is highly iterated and identified following the mechanical evaluations (see [48]), considering the partner company's manufacturing availability, as well as through focus groups, brainstorming, and co-design workshops within the interdisciplinary research team.

4.3. Research Direction #2: Steering Column Designing

4.3.1. "Define Phase" Results: Steering Column Strategic Design Brief

The main objectives are the development of unmovable and auxiliary battery packs, to increase the microvehicle's autonomy. According to these, the research team defines four specific steering column objectives strictly related to the battery packs: (i) enhancing the ease of use; (ii) widening the volume and size of the steering column to insert the auxiliary battery in its profile; (iii) minimizing the weight through the assessment of materials and formal exploration; and (iv) supporting the "trolley mode".

The first research objective is to improve the users' comfort. Therefore, the research team decides to position the auxiliary battery inside the steering column higher to facilitate assembly and disassembly.

The second objective is about dimensioning. Starting from the collection of standard height measures, commercially agreed, the research team dimensions the height of the steering column, extending by 1000 mm, concerning the platform level.

Going on, the weight objective regards the assessment of materials and manufacturing solutions for large-scale production, among which the research team identifies (i) polyurethane and aluminum core co-moulding; (ii) plastic injection moulding; (iii) aluminum calendaring; or (iv) laser cutting and welding.

Lastly, the research phase highlights that the hub is the only stable point on which the column can be fixed, capable of supporting the "trolley mode" development. Therefore, the research team highlights the following requirements: designing (i) the folding mechanisms, coupling the platform and the column; (ii) the space, or rather a hole, along the column for positioning appropriately the rear wheel; and (iii) compactness, making sure that the final output does not impact the overall aesthetics.

4.3.2. "Develop Phase" Results: Steering Column Embodiment and Detailed Design

The "Design strategic brief" of the steering column, presented in Section 4.3.1, is brainstormed by the researchers through a focus group and HMW questions to determine which research methods are employed.

These measures set the basis to start the "co-design disruptive ideas" research task. The research team explores the form through design sprint methods, fast sketching, design-by-drawing methods, and iteration, setting the basis for the following design insights: (i) design a steering column that provides enough space for fitting the auxiliary and interchangeable battery case; (ii) design a slot at half the height of the column to accommodate the rear wheel inside the steering column profile and compact the folded position of the system while minimizing the overall volume; and (iii) shape and conform the aesthetic details, explored in the first research direction, into every designed component.

Therefore, an aesthetic recognizable visual element is created between the platform and the steering column through the use of the parametric software Solidworks and the fillet and chamfer function on the edges. This recognizable and harmonious geometry is adapted to every specific characteristic of the product.

The formal exploration research task leads to the development of an innovative, minimal, dynamic concept, whose curvature invites movement and maintains a homogeneous profile, which aligns with the shape of the platform and the other components. These concepts are displayed and illustrated in Figure 5.

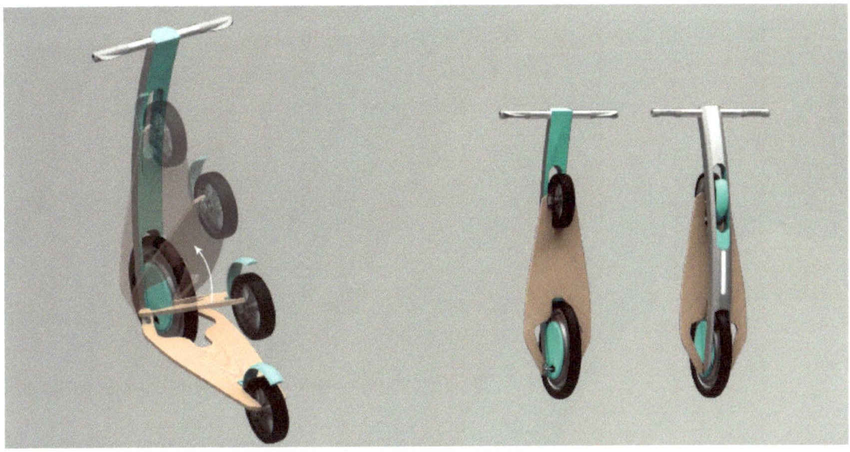

Figure 5. Possible movement of the platform concerning the steering column and front/rear view of the vehicle in "trolley mode".

Going on, the aesthetics of the column profile section is designed to be easily reproducible with low-cost technological processes, established within the panorama of manufacturers operating in Europe. Starting from these assumptions, three possible implementation solutions are formally and materially analyzed. The first involves the use of polyurethane, inside which an aluminum core is co-moulded, and consists of eight components welded together (Figure 6). The second involves the use of reinforced nylon through the injection moulding process (Figures 7 and 8). The last involves the creation of a customized extruded aluminum profile, the calendaring of the piece according to the radius of curvature considered, and laser cutting to form the central slot, the lower holes for bolts, and the housing slot for the auxiliary battery (Figure 9). In conclusion, the research team identifies the aluminum solution as the cheapest and the polyurethane alternative as the most appealing choice from an aesthetic standpoint.

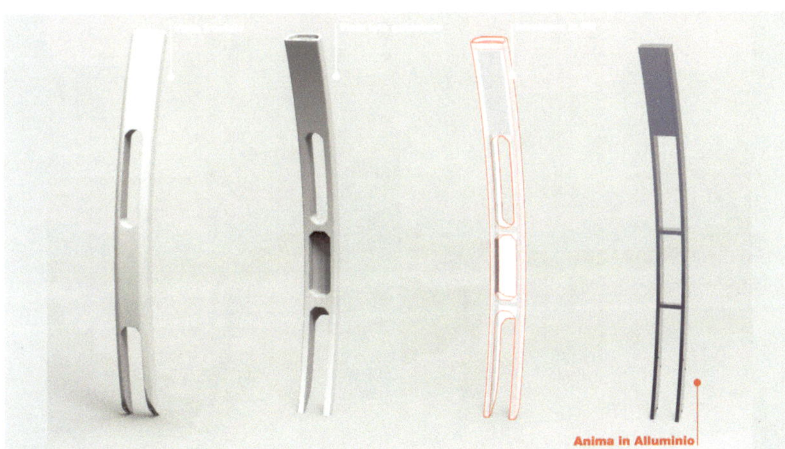

Figure 6. The concept for polyurethane and aluminum core co-moulding of the steering column.

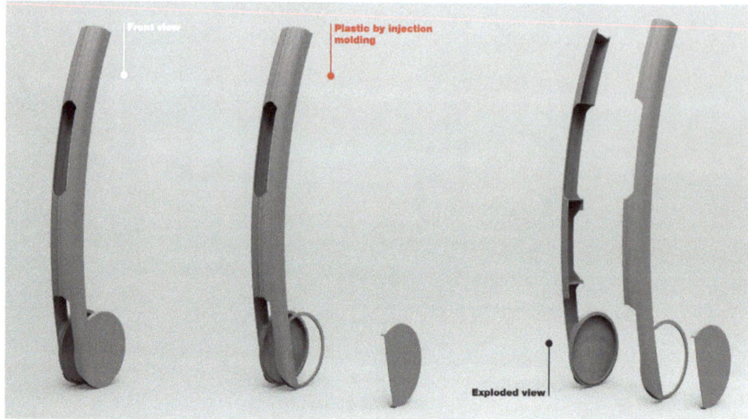

Figure 7. The concept for plastic injection moulding of the steering column.

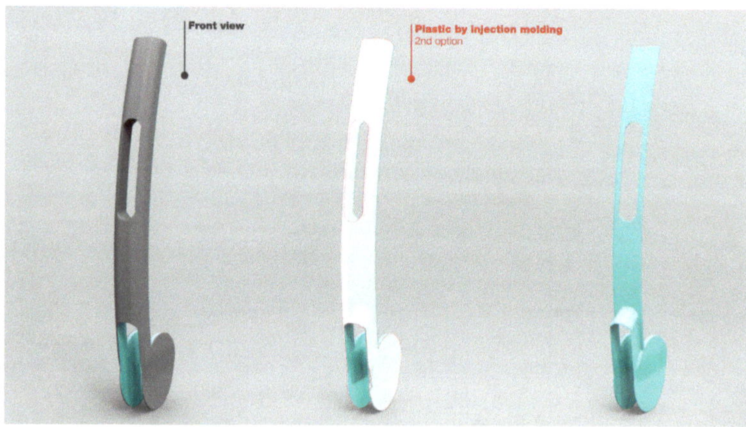

Figure 8. The concept for plastic injection moulding of the steering column (2nd option).

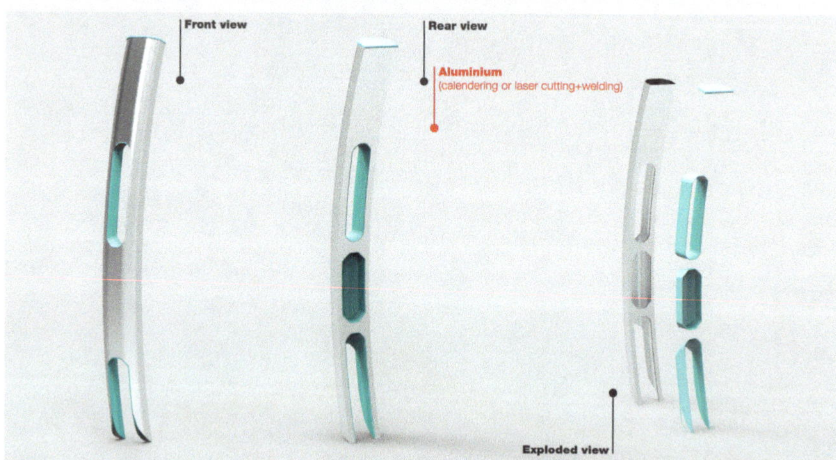

Figure 9. The concept for aluminum calendaring or laser cutting and welding of the steering column.

4.4. Research Direction #3: Auxiliary and Fixed Battery Case Designing

4.4.1. "Define Phase" Results: Auxiliary and Fixed Battery Case Strategic Design Brief

Starting from the map of the research opportunities areas, the research team frames the need to (i) dimension the auxiliary battery case and design the slot on the steering column, and (ii) dimension the fixed battery and measure the storage on the hub. The researchers define four specific auxiliary and fixed battery case objectives related to the following factors: (i) lowering the costs of the materials employed; (ii) increasing the user comfort in inserting the case in the storage; (ii) improving the compactness of the case profile to support the "trolley mode"; and (iv) increasing the ease of transport.

The design of the battery case depends on (i) the total volume occupied by the cells; (ii) the BMS; (iii) the connectors; and (iv) the adoption of the formal, recognizable, aesthetic characteristics determined in the previous subsections.

Lastly, the research team identifies the following constraints: (i) the auxiliary battery needs at least 10 batteries that measure 18.5 × 69 mm each; (ii) the fixed one needs at least 20 batteries that measure 18.5 × 69 mm each; (ii) the BMS measurements are 69 × 56 × 15 mm; and (iii) the connector is 23.85 × 8.6 × 7 mm.

4.4.2. "Develop Phase" Results: Auxiliary and Fixed Battery Case Embodiment and Detailed Design

The "Design strategic brief" of the auxiliary and fixed battery case, presented in Section 4.4.1, is brainstormed by the researchers through a focus group and HMW questions to determine which research methods are employed.

In preliminary dimensioning, the team explores the possible alternative configurations of the main auxiliary and fixed battery components, as shown in Figures 10 and 11. This sets the basis for pointing out that the steering column does not have sufficient space to insert a case containing more than 10 batteries. Therefore, the research team starts exploring the initial concepts by using graphical two/three-dimensional tools to better represent the inscribed auxiliary battery case. The peculiarity of the idea selected is the upper part of the case side profile, which describes the arc of a circle. The team designs this detail to foster ease of use in performing the downward movement for inserting the case and improving the user–product interaction. As depicted in Figures 12 and 13, its centre coincides with the lower end of the case housing inside the steering column. This point is identified as the case's pivot for insertion. The 3D models, of both the auxiliary and fixed battery case, are designed to prototype a preliminary 3D printing model. Additionally, the team selects the reinforced nylon and the injection moulding process as a manufacturing process. Therefore, the model is designed to be extracted from the mould during production.

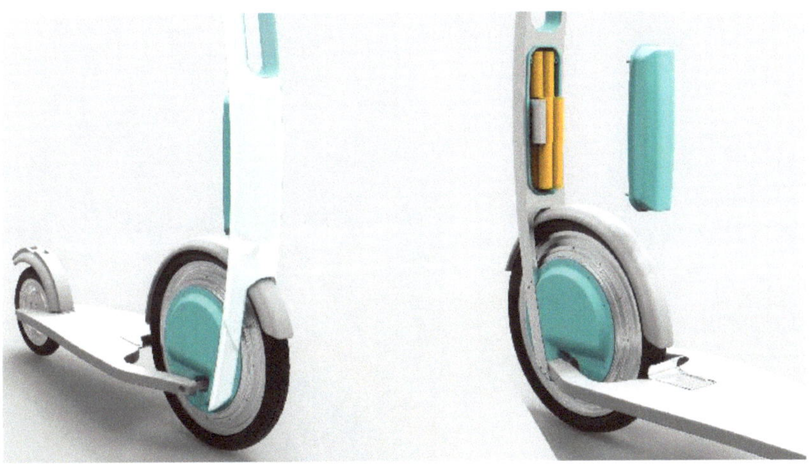

Figure 10. Overview of the auxiliary and fixed battery case configuration.

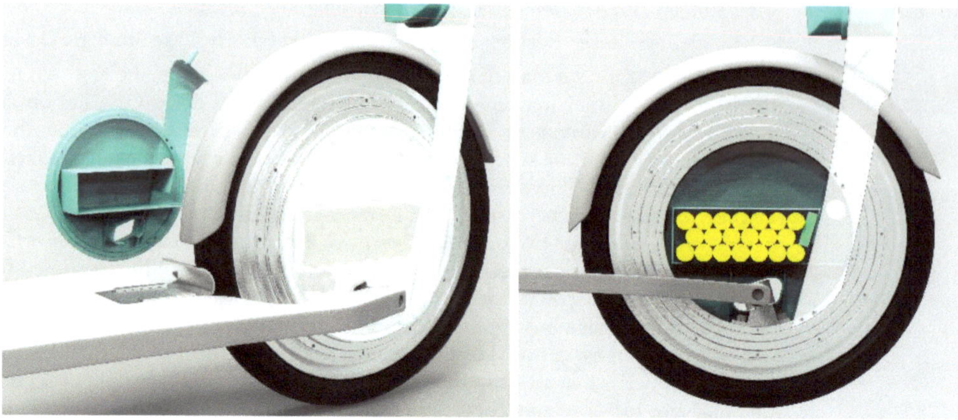

Figure 11. Overview of the fixed battery case configuration.

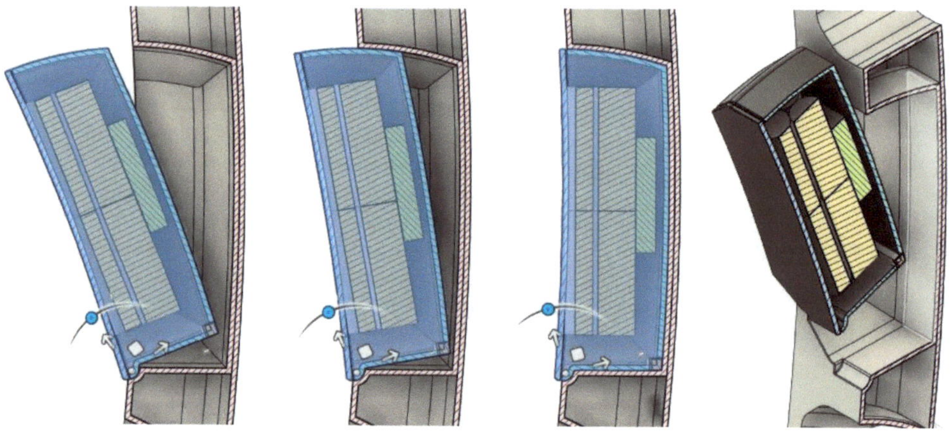

Figure 12. View of the case battery model profile and two/three-dimensional representation of how the user should fit it in the steering column.

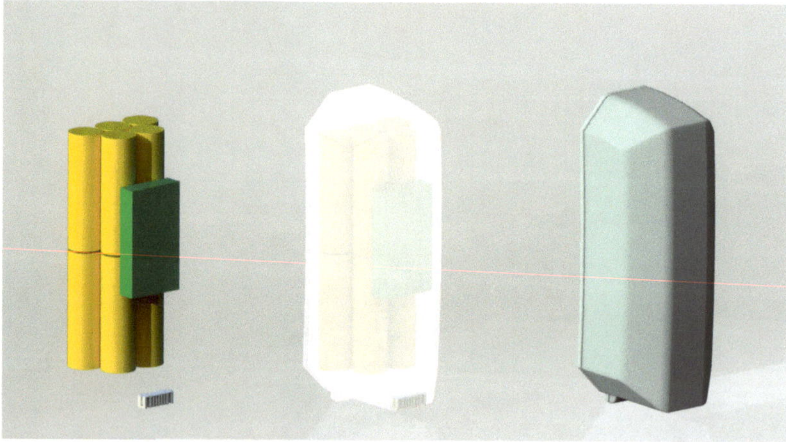

Figure 13. Overview of the 3 categories of components' configuration.

4.5. Research Direction #4: Assembly Designing

"Deliver Phase" Results: Assembly Prototyping and Testing

The closing of the research process, in correspondence to the end of the second diamond, combines the research directions findings in a single product assembly: the newly silent, clean, energy-efficient, and safe microvehicle. This blended version of the existing personal light electric vehicle employs a control mechanism, where, similar to the monowheel mechanisms, the users push the steering column to accelerate and pull it for decelerating.

This highly improves the perceived user experience. Indeed, the remote quantitative usability testing highlights a positive response to the question "Perspicuity. Is it easy to get familiar with the product and to learn how to use it?", with the following results: 14.8% voted 4/7; 33.3% voted 5/7; 37% voted 6/7; and 14.8% voted 7/7.

Following the manufacturing process conducted by the partner companies, the research team develops the first generation of the scooter physical prototype (presented in Figure 14), which consists of (i) the location of the main axle at the lower part of the front wheel; (ii) the steering column; (iii) the presence of the fixed battery compartment and the corresponding 3D printed case; and (iv) the platform, which is made of a sandwich structure.

Figure 14. First generation of the kick scooter prototype.

As shown in Figure 14, some components are rapidly prototyped through additive manufacturing, e.g., the fixed battery storage; others are only prototyped in the virtual environment through Solidworks and visual renderings, e.g., the auxiliary battery; and others are purchased to enable the assembly prototyping, e.g., the headlight, handlebar and rear wheel.

5. Discussion

This paper's introduction highlights how interdisciplinary collaboration in the design of urban mobility products could be a source of value creation. The researchers, therefore, structured a critical literature review on design thinking and highlighted the coherent Design Framework for Innovation and the Double-Diamond design thinking model as important methods to overcome research challenges and define disruptive and novel outputs in a transdisciplinary way. Hence, the cross-disciplinary research team evaluated the findings in the context of the H2020 LEONARDO project, which addresses the limitations of current micromobility solutions.

As discussed in the last two sections, the results of the present research context confirmed the assumption and set the basis for re-elaborating the design model, as presented in Figure 1. Notably, the application of the methods and tools relating to the first diamond widened the opportunities to conceptualize the overall model. The analysis and synthesis of the data gathered in the research phase indeed provided the researchers with different inquiry directions to be addressed, supporting the development of the electric microvehicle. These findings set the basis for splitting the current strategic design brief into four research directions, relating four different product components and the corresponding assembly prototyping, which fall under the mutual objective of designing an innovative micromobility vehicle based on the smart fusion of the monowheel and kick scooter concepts. Moreover, regarding how many inquiry directions were identified by the research team, the research tasks of the second diamond were addressed four times; they were addressed one additional time to merge the final results in the product assembly. These results were fundamental in clarifying the design research process for the innovative development of urban transport products.

Figure 15 presents a graphical summary of the discussions. In other words, the research team framed "an interdisciplinary Double-Diamond design thinking model for urban transport product innovation" characterized by (i) guidelines for organizing and structuring the research process, as shown in Figure 16; (ii) macro- and micro-design research tasks to be addressed, as shown in Figure 16; (iii) interdisciplinary methods and tools to be undertaken, as shown in Figure 16; and (iv) a novel interpretation of the second diamond, as shown in Figure 15. Notably, the macrotasks and microtasks related to the following steps: (i) identifying the macrotrends through in-depth analysis of the topic; (ii) understanding the context of the problem through in-depth analysis of the brief, in-depth analysis of the project partners, in-depth analysis of the product, mapping of the research areas, and framing of the research methods; (iii) empathizing through in-depth analysis of the stakeholders and in-depth analysis of the user interactions; (iv) technological advancements through secondary research; (v) analyzing through in-depth analysis of the data collected; (vi) synthesizing through mapping of the knowledge resources and framing "product assembly components"; (vii) co-designing insights through co-designing insights; (viii) designing the strategic brief through mapping of the research opportunities areas and defining research questions; (ix) developing the strategic brief through the framing of the research methods; (x) co-designing disruptive ideas through formal exploration; (xi) iterating through iteration tasks; (xii) exploring the idea selected through the framing of the research methods; (xiii) prototyping through prototyping in a virtual environment, rapid prototyping, and physical prototyping; and (xiv) testing through remote quantitative usability testing and testing in accordance with regulations.

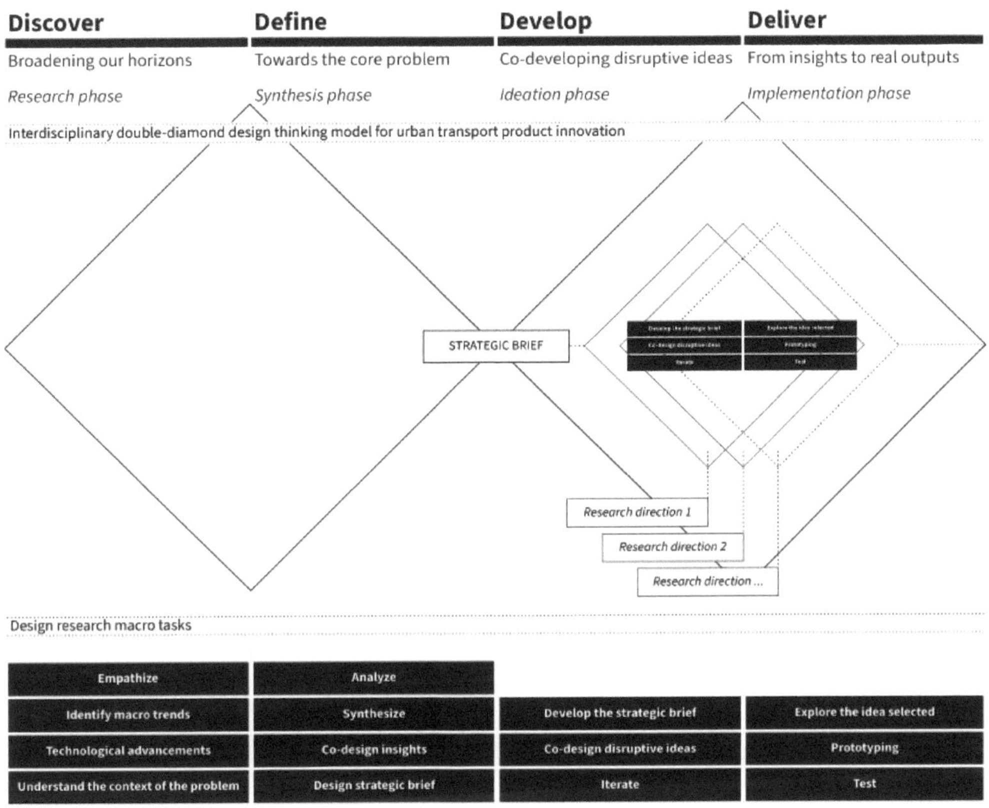

Figure 15. Interdisciplinary Double-Diamond Design Thinking model for urban transport product innovation. The figure presents the novelty of the design model and the corresponding research macro-tasks. Re-elaboration of the "Double-diamond design thinking model" [29].

The fourth feature of the design model highlights the actual diversification compared to the Double-Diamond presented in Figure 1. Notably, it emphasizes the parallel development process of the product components, the recurring adoption of the research activities for each research direction, and how these research directions are inscribed in the second diamond, between the "Develop" and "Deliver" phase, or rather in the stages where the research team should develop and implement the product assembly.

Therefore, the research team widens the vision of the second diamond, showing it foresees a constant and repetitive adoption of the research tasks for every research direction, identified by the design practitioners at the end of the initial phase of the process. The researchers highlight and support the need to identify and map the components of the mobility product assembly to delineate a wider variety of design potentials and give design experts more opportunities to visualize insights for interventions.

Figure 16. Interdisciplinary Double-Diamond Design Thinking model for urban transport product innovation. The figure presents the design research macro- and micro-tasks and tools and methods to employ for mobility product innovation. Re-elaboration of the "Double-diamond design thinking model" [29].

6. Conclusions

In conclusion, the research team adopted the Double-Diamond Design Thinking model in a transdisciplinary context to develop an innovative micromobility vehicle. Notably, they designed the hierarchical research process, the four stages of development, and the macro- and micro-tasks to undertake an innovative micromobility solution, involving tools and methods that are employed in the four stages of development. The project research context of "Leonardo" is an exploratory case study in which the team sought to answer the research questions through the application of the Double-Diamond model. During the process, the team correlated and employed mixed methods in both the design and engineering disciplinary sectors, involving stakeholders, project experts, design and engineering researchers, and practitioners through the design of participative research activities. The researchers obtained both qualitative and quantitative data. Consequently, the emerging findings set the basis for the iteration of the graphic representation of the design model and led to the framing of the "Interdisciplinary Double-Diamond Design Thinking model for urban transport product innovation". This design model aims to let design and engineering researchers and practitioners adopt, replicate, and scale the process the team developed. Finally, the research team argues this novel design model has the potential, as a design research instrument, to be capable of addressing sustainable mobility and guiding interdisciplinary design research, design practice, and education in the industrial engineering and design disciplinary sectors.

Author Contributions: Conceptualization, S.V., A.R. and D.V.; methodology, S.V.; investigation, M.-S.G. and D.V.; data curation, S.V. and M.-S.G.; writing—original draft preparation, S.V. and A.R.; writing—review and editing, M.-S.G. and D.V.; visualization, S.V. and M.-S.G.; supervision, A.R. and D.V.; project administration, D.V.; funding acquisition, D.V. All authors have read and agreed to the published version of the manuscript.

Funding: This research was funded by the European Commission, grant number 101006687. The APC was funded by the European Commission, grant number 101074307.

Data Availability Statement: Data are contained within this article.

Acknowledgments: The authors would like to thank the other partners of the LEONARDO research team for their valuable work and contribution to the development of the material included in this work.

Conflicts of Interest: The authors declare no conflicts of interest.

References

1. Rohilla, K.; Desai, A.; Patel, C.R. A Cutting-Edge Examination of the Dichotomy of Electric Vehicles as a Symbol of "Sustainable Mobility" and "Clean Energy". *J. Inst. Eng. India Ser. A* **2024**, *105*, 209–227. [CrossRef]
2. Santacre, A.; Yannis, G.; de Saint Leon, O.; Crist, P. *Safe Micromobility*; International Transport Forum: Paris, France, 2020. Available online: https://www.itf-oecd.org/sites/default/files/docs/safe-micromobility_1.pdf (accessed on 8 August 2024).
3. Cordis. SOCIETAL CHALLENGES—Smart, Green and Integrated Transport. Available online: https://cordis.europa.eu/programme/id/H2020-EU.3.4./en (accessed on 28 October 2024).
4. Cordis. Next Generation Electrified Vehicles for Urban and Suburban Use. Available online: https://cordis.europa.eu/programme/id/H2020_LC-GV-08-2020/en (accessed on 28 October 2024).
5. Khamis, A. *Smart Mobility: Exploring Foundational Technologies and Wider Impacts*; Apress: Berkeley, CA, USA, 2021.
6. Cordis. MicrovehicLE fOr staNd-Alone and shaReD mObility—LEONARDO. Available online: https://cordis.europa.eu/project/id/101006687 (accessed on 8 August 2024).
7. De Lucia, M.; Vangi, D.; Cialdai, C.; Messeri, M. *Sistema di Condivisione, Monitoraggio e Gestione di Pacchi Batterie Ricaricabili*; University of Florence: Florence, Italy, 2020. Available online: https://www.regione.toscana.it/documents/10180/163864099/Brevetto_UNIFI_Mobilit%C3%A0_Condivisione_batterie.pdf/878f961f-0c29-70b8-b74f-5699ab67d464?t=1691575332566 (accessed on 28 October 2024).
8. Micheli, P.; Wilner, S.J.S.; Bhatti, S.H.; Mura, M.; Beverland, M.B. Doing design thinking: Conceptual review, synthesis, and research agenda. *J. Prod. Innov. Manag.* **2019**, *36*, 124–148. [CrossRef]
9. Verganti, R.; Dell'Era, C.; Swan, K.S. Design thinking: Critical analysis and future evolution. *J. Prod. Innov. Manag.* **2021**, *38*, 603–622. [CrossRef]
10. Deng, Y.; Liu, W. How to develop engineering students as design thinkers: A systematic review of design thinking implementations in engineering education. In Proceedings of the ASEE Annual Conference & Exposition, Baltimore, MD, USA, 25–28 June 2023.

11. Alok, G.; Pothupogu, S.; Reddy, M.S.; Priya, P.S.; Devi, V.R.D. Trenchant Pathway to bring Innovation through Foundations to Product Design in Engineering Education. In Proceedings of the IEEE 6th International Conference on MOOCs, Innovation and Technology in Education (MITE), Hyderabad, India, 29–30 November 2018; pp. 43–47.
12. Kaygan, P. From forming to performing: Team development for enhancing interdisciplinary collaboration between design and engineering students using design thinking. *Int. J. Technol. Des. Educ.* **2023**, *33*, 457–478. [CrossRef]
13. Turcios-Esquivel, A.M.L.; Avilés-Rabanales, E.G.; Hernández-Rodriguez, F. Enhancing Empathy and Innovation in Engineering Education Through Design Thinking and Design of Experiments. In Proceedings of the IEEE Global Engineering Education Conference (EDUCON), Kos Island, Greece, 8–11 May 2024; pp. 1–5.
14. Leem, S.; Lee, S.W. Fostering collaboration and interactions: Unveiling the design thinking process in interdisciplinary education. *Think. Ski. Creat.* **2024**, *52*, 101520. [CrossRef]
15. Wang, C. Using design thinking for interdisciplinary curriculum design and teaching: A case study in higher education. *Humanit. Soc. Sci. Commun.* **2024**, *11*, 307. [CrossRef]
16. Panke, S. Design Thinking in Education: Perspectives, Opportunities and Challenges. *Open Educ. Stud.* **2019**, *1*, 281–306. [CrossRef]
17. Martin, R. Design thinking: Achieving insights via the "knowledge funnel". *Strategy Leadersh.* **2010**, *38*, 37–41. [CrossRef]
18. Rösch, N.; Tiberius, V.; Kraus, S. Design thinking for innovation: Context factors, process, and outcomes. *Eur. J. Innov. Manag.* **2023**, *26*, 160–176. [CrossRef]
19. Luchs, M.G. A brief introduction to design thinking. In *Design Thinking: New Product Development Essentials from the PDMA*; Luchs, M.G., Swan, S., Griffin, A., Eds.; Wiley: Hoboken, NJ, USA, 2016; pp. 1–12.
20. Carlgren, L.; Rauth, I.; Elmquist, M. Framing design thinking: The concept in idea and enactment. *Creat. Innov. Manag.* **2016**, *25*, 38–57. [CrossRef]
21. Buchanan, R. Wicked Problems in Design Thinking. *Des. Issues* **1992**, *8*, 5–21. [CrossRef]
22. Kueh, C.; Thom, R. Visualising Empathy: A Framework to Teach User-Based Innovation in Design. In *Visual Tools for Developing Cross-Disciplinary Collaboration, Innovation and Entrepreneurship Capacity*; Griffith, S., Carruthers, K., Bliemel, M., Eds.; Common Ground Research Networks: Champaign, IL, USA, 2018; pp. 177–196.
23. Peng, F. Design Thinking: From Empathy to Evaluation. In *Foundations of Robotics*; Herath, D., St-Onge, D., Eds.; Springer: Singapore, 2022; pp. 63–81. [CrossRef]
24. Frog. Empowering Communities with Design-Thinking. Available online: https://www.frog.co/designmind/collective-action-toolkit-empowering-communities (accessed on 8 August 2024).
25. Cross, N. *Design Thinking: Understanding How Designers Think and Work*, 2nd ed.; Bloomsbury Publishing: London, UK, 2023.
26. d.School. Design Thinking Bootleg. Available online: https://dschool.stanford.edu/resources/design-thinking-bootleg (accessed on 8 August 2024).
27. IDEO. Design KIT: The Human-Centered Design Toolkit. Available online: https://www.ideo.com/journal/design-kit-the-human-centered-design-toolkit (accessed on 8 August 2024).
28. IDEO. DESIGN KIT. *The Field Guide to Human-Centered Design*. Available online: https://www.designkit.org/resources/1.html (accessed on 8 August 2024).
29. Design Council. The Double Diamond. Available online: https://www.designcouncil.org.uk/our-resources/the-double-diamond/ (accessed on 8 August 2024).
30. Stickdorn, M.; Schneider, J. *This Is Service Design Thinking: Basics, Tools, Cases*; Wiley: Hoboken, NJ, USA, 2012.
31. Kochanowska, M.; Gagliardi, W.R.; Ball, J. The Double Diamond Model: In Pursuit of Simplicity and Flexibility. In *Perspectives on Design II*; Raposo, D., Neves, J., Silva, J., Eds.; Springer Series in Design and Innovation; Springer: Cham, Switzerland, 2021; Volume 16. [CrossRef]
32. Design Council. *Design Methods for Developing Services*; Design Council: London, UK, 2015. Available online: https://www.designcouncil.org.uk/fileadmin/uploads/dc/Documents/DesignCouncil_Design%2520methods%2520for%2520developing%2520services.pdf (accessed on 8 August 2024).
33. Tosi, F. *Design for Ergonomics*; Springer: Cham, Switzerland, 2020; pp. 111–128.
34. Xu, J.; Chao, C.; Fu, Z. Research on Intelligent Design Tools to Stimulate Creative Thinking. In *Cross-Cultural Design. User Experience of Products, Services, and Intelligent Environments, Proceedings of the 12th International Conference, CCD 2020, Held as Part of the 22nd HCI International Conference, HCII 2020, Copenhagen, Denmark, 19–24 July 2020*; Springer International Publishing: Cham, Switzerland, 2020; Volume 12192.
35. Cross, N. *Engineering Design Methods: Strategies for Product Design*, 5th ed.; Wiley: Hoboken, NJ, USA, 2021.
36. Guaman-Quintanilla, S.; Everaert, P.; Chiluiza, K.; Valcke, M. Fostering Teamwork through Design Thinking: Evidence from a Multi-Actor Perspective. *Educ. Sci.* **2022**, *12*, 279. [CrossRef]
37. Bordegoni, M.; Carulli, M.; Spadoni, E. User experience and user experience design. In *Prototyping User Experience in Extended Reality*; Bordegoni, M., Carulli, M., Spadoni, E., Eds.; Springer: Cham, Switzerland, 2023; pp. 11–28.
38. Rinaldi, A. *Innovare Attraverso il Design e la Tecnologia*; FrancoAngeli: Milan, Italy, 2020; pp. 49–69.
39. Braun, V.; Clarke, V. Using thematic analysis in psychology. *Qual. Res. Psychol.* **2006**, *3*, 77–101. [CrossRef]
40. Heidi, J. Content analysis. In *The Sage Encyclopedia of Qualitative Research Methods*; Given, L.M., Ed.; Sage Publications: Thousand Oaks, CA, USA, 2008; pp. 120–122.

41. Miro. Available online: https://miro.com/it/ (accessed on 8 August 2024).
42. Knapp, J.; Zeratsky, J.; Kowitz, B. *Sprint: How to Solve Big Problems and Test New Ideas in Just Five Days*; Simon and Schuster: New York, NY, USA, 2016.
43. Magistretti, S.; Dell'Era, C.; Doppio, N. Design sprint for SMEs: An organizational taxonomy based on configuration theory. *Manag. Decis.* **2020**, *58*, 1803–1817. [CrossRef]
44. Rinaldi, A.; Busciantella-Ricci, D.; Viviani, S. Design for Movability: A New Design Research Challenge for Sustainable Design Scenarios in Urban Mobility. In *For Nature/With Nature: New Sustainable Design Scenarios*; Gambardella, C., Ed.; Springer: Cham, Switzerland, 2024; Volume 38, pp. 929–949.
45. Primus, D.J.; Sonnenburg, S. Flow Experience in Design Thinking and Practical Synergies with Lego Serious Play. *Creat. Res. J.* **2018**, *30*, 104–112. [CrossRef]
46. Autodesk Fusion. Autodesk Fusion: More than CAD, It's the Future of Design and Manufacturing. Available online: https://www.autodesk.com/eu/products/fusion-360/overview?term=1-YEAR&tab=subscription (accessed on 19 November 2024).
47. Solidworks. SOLIDWORKS Is the Proven Solution for 3D Design and Product Development. Available online: https://www.solidworks.com/ (accessed on 19 November 2024).
48. Keyshot. Keyshot—Reinventing the Way Products Are Brought to Market. Available online: https://www.keyshot.com/ (accessed on 19 November 2024).
49. Lumion. See the Future—Design Faster and Visualize in Real-Time with Lumion 3D Rendering Software. Available online: https://lumion.com/ (accessed on 19 November 2024).
50. Godin, D.; Zahedi, M. Aspects of Research through Design: A Literature Review. In Proceedings of the Design's Big Debates—DRS International Conference 2014, Umea, Sweden, 16–19 June 2014.
51. Pheasant, S.; Haslegrave, C.M. *Bodyspace: Anthropometry, Ergonomics and the Design of Work*, 3rd ed.; CRC Press: Boca Raton, FL, USA, 2006.
52. Adobe Illustrator. AI Illustrator—More artwork. *Less Busywork.* Available online: https://www.adobe.com/uk/products/illustrator.html (accessed on 19 November 2024).
53. Adobe Photoshop. PS Photoshop—Create at the Speed of Imagination. Available online: https://www.adobe.com/uk/products/photoshop.html (accessed on 19 November 2024).
54. Gulino, S.M.; Vichi, G.; Zonfrillo, G.; Vangi, D. Comfort assessment for electric kick scooter decks. *IOP Conf. Ser. Mater. Sci. Eng.* **2022**, *1214*, 012043. [CrossRef]
55. Gulino, M.; Laschi, M.; Zonfrillo, G.; Fabbri, M.; Garcia, E.; Miklis, P.; Unger, T.; Damaziak, K.; Mazurkiewicz, L.; Sybilski, K.; et al. Development of an innovative microvehicle for urban mobility. *Open Res. Eur.* **2024**, *4*, 190. [CrossRef]
56. Trimis. MicrovehicLE fOr staNd-Alone and shaReD mObility—LEONARDO. Available online: https://trimis.ec.europa.eu/project/microvehicle-stand-alone-and-shared-mobility-leonardo (accessed on 28 October 2024).
57. Micromobility America. Micromobility Landscape. Available online: https://micromobility.io/landscape (accessed on 28 October 2024).
58. Autodesk AutoCAD. Autodesk AutoCAD: Design and Drafting Software Trusted by Millions. 2D and 3D CAD Tools Built to Accelerate your Creativity. Available online: https://www.autodesk.com/eu/products/autocad/overview?term=1-YEAR&tab=subscription (accessed on 19 November 2024).

MDPI AG
Grosspeteranlage 5
4052 Basel
Switzerland
Tel.: +41 61 683 77 34

MDPI Books Editorial Office
E-mail: books@mdpi.com
www.mdpi.com/books

www.ingramcontent.com/pod-product-compliance
Lightning Source LLC
Chambersburg PA
CBHW040854230926
51850CB00014BA/3430